GENETICS
Principles and Analysis
Fourth Edition

D A N I E L   L .   H A R T L
Harvard University

E L I Z A B E T H   W .   J O N E S
Carnegie Mellon University

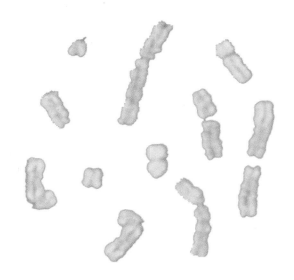

Jones and Bartlett Publishers
Sudbury, Massachusetts
BOSTON   LONDON   SINGAPORE

TO THE BEST TEACHERS WE EVER HAD—OUR PARENTS AND OUR STUDENTS

ABOUT THE AUTHORS

Daniel L. Hartl is a Professor of Biology at Harvard University. He received his B.S. degree and Ph.D. from the University of Wisconsin. His research interests include molecular genetics, molecular evolution, and population genetics. Elizabeth W. Jones is a Professor of Biological Sciences at Carnegie Mellon University. She received her B.S. degree and Ph.D. from the University of Washington in Seattle. Her research interests include gene regulation and the genetic control of cellular form. Currently she is studying the function and assembly of organelles in the yeast *Saccharomyces*.

ABOUT THE COVER

A human chromosome.

ABOUT THE PUBLISHER

Editorial, Sales, and Customer Service Offices

Jones and Bartlett Publishers
40 Tall Pine Drive
Sudbury, MA 01776
978-443-5000
info@jbpub.com
http://www.jbpub.com

Jones and Bartlett Publishers International
Barb House, Barb Mews
London W6 7PA
UK

ABOUT THE BOOK

Senior Managing Editor: Judith H. Hauck
Executive Editor: Brian L. McKean
Marketing Manager: Rich Pirozzi
Project Editor: Kathryn Twombly
Senior Production Editor: Mary Hill
Manufacturing Buyer: Jane Bromback
Web Site Design: Andrea Wasik

Development Editor: Richard Morel
Book and Cover Design: J/B Woolsey Associates
Art Development and Rendering: J/B Woolsey Associates
Composition and Book Layout: Thompson Steele, Inc.
Prepress: Westwords, Inc.
Cover Manufacture: Coral Graphic Services, Inc.
Book Manufacture: World Color Book Services

Library of Congress Cataloging-in-Publication Data

Hartl, Daniel L.
    Genetics : principles and analysis / Daniel L. Hartl, Elizabeth W. Jones. — 4th ed.
      p. cm.
    Includes bibliographical references (p.).
     ISBN 0-7637-0489-X
     1. Genetics.   I. Jones, Elizabeth W.   II. Title.
  QH430.H3733   1998
  576.5—dc21   97-40566
                     CIP

Printed in the United States

03 02 01 00 99 98 97   9 8 7 6 5 4 3 2 1

BRIEF CONTENTS

CONTENTS

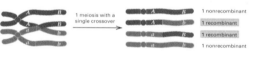

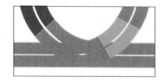

CHAPTER 11

Regulation of Gene Activity 458

U U U U U C U U C G C A U U C U U U U U A C C U U C

Phe Phe Phe Ala Phe Phe Phe Thr Phe

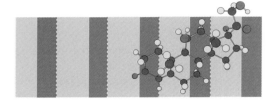

PAPERS EXCERPTED IN CONNECTIONS IN CHRONOLOGICAL ORDER

Gregor Mendel 1866
Monastery of St. Thomas,
Brno, Czech Republic
Experiments on Plant Hybrids
Verhandlungen des naturforschenden
den Vereines in Brünn 4: 3–47

Godfrey H. Hardy 1908
Trinity College,
Cambridge, England
*Mendelian Proportions in a Mixed
Population*
Science 28: 49–50

Thomas Hunt Morgan 1910
Columbia University, New York,
New York
Sex Limited Inheritance in Drosophila
Science 32: 120–122

E. Eleanor Carothers 1913
University of Kansas, Lawrence,
Kansas
*The Mendelian Ratio in Relation
to Certain Orthopteran Chromosomes*
Journal of Morphology
24: 487–511

Thomas Hunt Morgan 1913
Columbia University,
New York, New York
Heredity and Sex
Columbia University Press,
New York

Alfred H. Sturtevant 1913
Columbia University,
New York, New York
*The Linear Arrangement of Six Sex-Linked
Factors in* Drosophila, *as Shown by Their
Mode of Association*
Journal of Experimental Zoology
14: 43–59

Lilian V. Morgan 1922
Columbia University,
New York, New York
Non-crisscross Inheritance in Drosophila
Melanogaster
Biological Bulletin 42: 267–274

Hermann J. Muller 1927
University of Texas,
Austin, Texas
Artificial Transmutation of the Gene
Science 66: 84–87

Ronald Aylmer Fisher 1936
University College,
London, England
Has Mendel's Work Been Rediscovered?
Annals of Science 1: 115–137

**George W. Beadle and Edward L.
Tatum 1941**
Stanford University,
Stanford, California
*Genetic Control of Biochemical Reactions
in* Neurospora
Proceedings of the National
Academy of Sciences USA
27: 499–506

**Oswald T. Avery, Colin M.
MacLeod, and Maclyn McCarty
1944**
The Rockefeller University,
New York, New York
*Studies on the Chemical Nature of the
Substance Inducing Transformation
of Pneumococcal Types*
Journal of Experimental Medicine
79: 137–158

**Joshua Lederberg and Edward L.
Tatum 1946**
Yale University,
New Haven, Connecticut
Gene Recombination in Escherichia
coli
Nature 158: 558

**Alfred D. Hershey and Raquel
Rotman 1948**
Washington University,
St. Louis, Missouri
*Genetic Recombination Between Host-
Range and Plaque-Type Mutants of
Bacteriophage in Single Bacterial Cells*
Genetics 34: 44–71

Barbara McClintock 1950
Cold Spring Harbor Laboratory,
Cold Spring Harbor, New York
*The Origin and Behavior of Mutable
Loci in Maize*
Proceedings of the National Academy
of Sciences USA 36: 344–355

**Alfred D. Hershey and Martha
Chase 1952**
Cold Spring Harbor Laboratories,
Cold Spring Harbor, New York
*Independent Functions of Viral Protein
and Nucleic Acid in Growth of
Bacteriophage*
Journal of General Physiology
36: 39–56

**James D. Watson and
Francis H. C. Crick 1953**
Cavendish Laboratory,
Cambridge, England
A Structure for Deoxyribose Nucleic Acid
Nature 171: 737–738

Seymour Benzer 1955
Purdue University,
West Lafayette, Indiana
*Fine Structure of a Genetic Region in
Bacteriophage*
Proceedings of the National
Academy of Sciences USA
41: 344–354

**Matthew Meselson and Franklin
W. Stahl 1958**
California Institute of Technology,
Pasadena, California
The Replication of DNA in
Escherichia coli
Proceedings of the National
Academy of Sciences of the USA
44: 671–682

**Jerome Lejeune, Marthe Gautier,
and Raymond Turpin 1959**
National Center for Scientific
Research, Paris, France
*Study of the Somatic Chromosomes of
Nine Down Syndrome Children
(original in French)*
Comptes rendus des séances de
l'Académie des Sciences
248: 1721–1722

**François Jacob, David Perrin,
Carmen Sanchez and Jaques
Monod 1960**
Institute Pasteur, Paris, France
*The Operon: A Group of Genes Whose
Expression Is Coordinated by an Operator
(original in French)*
Comptes Rendus des Séances de
l'Academie des Sciences
250: 1727–1729

**Sydney Brenner[1], François
Jacob[2] and Matthew Meselson[3]
1961**
[1] Cavendish Laboratory,
   Cambridge, England;
[2] Institute Pasteur, Paris, France;
[3] California Institute of Technology,
   Pasadena, California
*An Unstable Intermediate Carrying
Information from Genes to Ribosomes
for Protein Synthesis*
Nature 190: 576–581

**Francis H. C. Crick, Leslie
Barnett, Sydney Brenner, and
R. J. Watts-Tobin, 1961**
Cavendish Laboratory,
Cambridge, England
*General Nature of the Genetic Code
for Proteins*
Nature 192: 1227–1232

Mary F. Lyon 1961
Medical Research Council, Harwell,
England
*Gene Action in the X Chromosome
of the Mouse* (Mus musculus L.)
Nature 190: 372

**Ruth Sager and Zenta Ramanis
1965**
Columbia University,
New York, New York
*Recombination of Nonchromosomal Genes
in* Chlamydomonas
Proceedings of the National Academy
of Sciences USA 53: 1053–1061

**Lynn Margulis (formerly Lynn
Sagan) 1967**
Boston University,
Boston, Massachusetts
The Origin of Mitosing Cells
Journal of Theoretical Biology
14: 225–274

**James B. Hicks, Jeffrey N.
Strathern, and Ira Herskowitz
1977**
University of Oregon,
Eugene, Oregon
*The Cassette Model of Mating-type
Interconversion*
Pages 457–462 in Ahmad I. Bukhari,
James A. Shapiro, and Sankar L.
Adhya (editors). DNA Insertion
Elements, Plasmids, and Episomes,
Cold Spring Harbor Laboratory,
Cold Spring Harbor, New York.

Joan Fisher Box 1978
Madison, Wisconsin
R. A. Fisher: The Life of a Scientist
John Wiley & Sons, New York

**Christiane Nüsslein-Volhard
and Eric Wieschaus 1980**
European Molecular Biology
Laboratory, Heidelberg, Germany
*Mutations Affecting Segment Number
and Polarity in* Drosophila
Nature 287: 795–801

**John E. Sulston[1], E. Schierenberg[2],
J. G. White[1], and J. N. Thomson[1]
1983**
[1] Medical Research Council Laboratory
for Molecular Biology, Cambridge,
England;
[2] Max-Planck Institute for
Experimental Medicine, Gottingen,
Germany
*The Embryonic Cell Lineage of the
Nematode* Caenorhabditis Elegans
Developmental Biology 100: 64–119

**Alec J. Jeffreys, John F. Y.
Brookfield, and Robert
Semeonoff 1985**
University of Leicester,
Leicester, England
*Positive Identification of an Immigration
Test-Case Using Human DNA Fingerprints*
Nature 317: 818–819

**David T. Burke, Georges F. Carle,
and Maynard V. Olson 1987**
Washington University,
St. Louis, Missouri
*Cloning of Large Segments of Exogenous
DNA into Yeast by Means of Artificial
Chromosome Vectors*
Science 236: 806–812

**Carol W. Greider and Elizabeth H.
Blackburn 1987**
University of California,
Berkeley, California
*The Telomere Terminal Transferase
of* Tetrahymena *Is a Ribonucleoprotein
with Two Kinds of Primer Specificity*
Cell 51: 887–898

Francis Galton 1989
42 Rutland Gate, South Kensington,
London, England
Natural Inheritance
Macmillan Publishers, London

**Micheline Strand[1], Tomas A.
Prolla[2], R. Michael Liskay[2],
and Thomas D. Petes[1] 1993**
[1] University of North Carolina,
Chapel Hill, North Carolina
[2] Yale University,
New Haven, Connecticut
*Destabilization of Tracts of Simple
Repetitive DNA in Yeast by Mutations
Affecting DNA Mismatch Repair*
Nature 365: 274–276

**The Huntington's Disease
Collaborative Research Group 1993**
Comprising 58 authors among
9 institutions
*A Novel Gene Containing a Trinucleotide
Repeat That is Expanded and Unstable on
Huntington's Disease Chromosomes*
Cell 72: 971–983

**Jeffrey C. Murry and
26 other investigators 1994**
University of Iowa and 9 other
research institutions
*A Comprehensive Human Linkage Map
with Centimorgan Density*
Science 265: 2049–2054

**Ian Wilmut, Anagelika E.
Schnieke, Jim McWhir, Alex J.
Kind, and Keith H. S. Campbell
1997**
Roslin Institute, Roslin, Midlothian,
Scotland
*Viable Offspring Derived from Fetal and
Adult Mammalian Cells*
Nature 385: 810–813

PREFACE

This book is titled *Genetics: Principles and Analysis*, Fourth Edition, because it embodies our belief that a good course in genetics should maintain the right balance between two important aspects of the science. The first aspect is that genetics is a body of knowledge pertaining to genetic transmission, function, and mutation. This constitutes the *Principles*. The second aspect is that genetics is an experimental approach, or a kit of "tools," for the study of biological processes such as development or behavior. This is *Analysis*.

The overall aim of *Genetics: Principles and Analysis*, Fourth Edition, is to provide a clear, comprehensive, rigorous, and balanced introduction to genetics at the college level. It is a guide to learning a critically important and sometimes difficult subject. The rationale of the book is that any student claiming a knowledge of genetics must:

- Understand the basic processes of gene transmission, mutation, expression, and regulation;
- Be able to think like a geneticist at the elementary level of being able to formulate genetic hypotheses, work out their consequences, and test the results against observed data;
- Be able to solve problems of several types, including problems that ask the student to verbalize genetic principles in his or her own words, single-concept exercises that require application of definitions or the basic principles of genetics, genetic analysis in which several concepts must be applied in logical order, and quantitative problems that call for some numerical calculation;
- Gain some sense of the social and historical context in which genetics has developed and is continuing to develop; and
- Have some familiarity with the genetic resources and information that are available through the Internet.

Genetics: Principles and Analysis, Fourth Edition, incorporates many special features to help students achieve these learning goals. The text is clearly and concisely written in a somewhat relaxed prose style without being chummy or excessively familiar. Each chapter is headed by a list of **Principles** that are related at numerous points to the larger whole. Each chapter contains two or three **Connections** in which the text material is connected to excerpts of classic papers that report key experiments in genetics or that raise important social, ethical, or legal issues in genetics. Each Connection has a brief introduction of its own, explaining the importance of the experiment and the historical context in which it was carried out. At the end of each chapter is a complete **Summary,**

Key Terms, GeNETtics on the web exercises that guide students in the use of Internet resources in genetics, and several different types and levels of **Problems.** These features are discussed individually below.

In recent decades, both the amount of genetic knowledge and its rate of growth have exploded. Many of the new discoveries have personal and social relevance through applications of genetics to human affairs in prenatal diagnosis, testing for carriers, and identification of genetic risk factors for complex traits, such as breast cancer and heart disease. There are also ethical controversies: Should genetic manipulation be used on patients for the treatment of disease? Should human fetuses be used in research? Should human beings be cloned? There are also social controversies—for example, when insurance companies exclude coverage of people because of their inherited risks of certain diseases.

Inspired in part by the controversies and the publicity, many of today's students come to a course in genetics with great enthusiasm. The challenges for the teacher are:

- To sustain this enthusiasm;
- To help motivate a desire to understand the principles of genetics in a comprehensive and rigorous way;
- To guide students in gaining an understanding that genetics is not only a set of principles but also an experimental approach to solving a wide range of biological problems; and
- To help students learn to think about genetic problems and about the wider social and ethical issues arising from genetics.

While addressing these challenges in *Principles and Analysis*, we have also tried to show the beauty, logical clarity, and unity of the subject. Endlessly fascinating, genetics is the material basis of the continuity of life.

CHAPTER ORGANIZATION

In order to help the student keep track of the main issues and avoid being distracted by details, each chapter begins with a list of the Principles that provide the main focus of the chapter. There is also an **Outline,** showing step by step the path along the way. An opening paragraph gives an overview of the chapter, illustrates the subject with some specific examples, and shows how the material is connected to genetics as a whole. The text makes liberal use of numbered lists and "bullets" in order to help students organize their learning, as well as summary statements set off in special type in order to emphasize important principles. Each chapter ends with a **Summary** and list of **Key Terms** as well as the **Problems.** There is a

Concise Dictionary of Genetics at the end of the book for students to check their understanding of the Key Terms or look up any technical terms they may have forgotten. The Dictionary includes not only the Key Terms but also genetic terms that students are likely to encounter in exploring the Internet or in their further reading. The Dictionary also includes page references for terms defined in the text.

CONTENTS

The organization of the chapters is that favored by the majority of instructors who teach genetics. It is the organization we use in our own courses. An important feature is the presence of an introductory chapter providing a broad overview of genes—what they are, how they function, how they change by mutation, and how they evolve through time. Today, most students learn about DNA in grade school or high school; in our teaching, we have found it rather strange to pretend that DNA does not exist until the middle of the term. The introductory chapter serves to connect the more advanced concepts that students are about to learn with what they already know. It also serves to provide each student with a solid framework for integrating the material that comes later.

Throughout each chapter, there is a balance between observation and theory, between principle and concrete example, and between challenge and motivation. Molecular, classical, and evolutionary genetics are integrated throughout. A number of points related to organization and coverage should be noted:

- **Chapter 1** is an overview of genetics designed to bring students with disparate backgrounds to a common level of understanding. This chapter enables classical, molecular, and evolutionary genetics to be integrated in the rest of the book. Included in Chapter 1 are the basic concepts of genetics: trait, gene, genotype, phenotype, gene interaction, and so forth. Chapter 1 also includes a discussion of the experimental evidence that DNA is the genetic material, as well as a description of DNA structure and how DNA codes for proteins.
- **Chapters 2 through 4** are the core of Mendelian genetics, including segregation and independent assortment, the chromosome theory of heredity, mitosis and meiosis, linkage and chromosome mapping, and tetrad analysis in fungi. Also included is the basic probability framework of Mendelian genetics and the testing of genetic models by means of the chi-square test.

    An important principle of genetics, too often ignored or given inadequate treatment, is that of the complementation test and how complementation differs from segregation or other genetic principles.

Chapter 2 includes a clear and concise description of complementation, with examples, showing how complementation is used in genetic analysis to group mutations into categories corresponding to genes. This chapter also introduces the use of molecular markers, especially with reference to human genetic analysis, because these are the principal types of genetic markers often used in modern genetics.

- **Chapters 5 and 6** deal with the molecular structure and replication of DNA and with the molecular organization of chromosomes. A novel feature is a description of how basic research that revealed the molecular mechanisms of DNA replication ultimately led to such important practical applications as DNA sequencing and the polymerase chain reaction. This example illustrates the value of basic research in leading, often quite unpredictably, to practical applications. The chapter on chromosome structure also includes a discussion of repetitive DNA sequences in eukaryotic genomes, including transposable elements.
- **Chapter 7** covers the principles of cytogenetics, including variation in chromosome number and the chromosome mechanics of deletions, duplications, inversions, and translocations. Also included is the subject of the human genome with special reference to human chromosome number and structure and the types of aberrations that are found in human chromosomes.
- **Chapter 8** deals with the principles of genetics in prokaryotes with special emphasis on *E. coli* and temperate and virulent bacteriophages. There is an extensive discussion of mechanisms of genetic recombination in microbes, including transformation, conjugation, transduction, and the horizontal transfer of genes present in plasmids, such as F' plasmids.
- **Chapter 9** focuses on recombinant DNA and genome analysis. Included are the use of restriction enzymes and vectors in recombinant DNA, cloning strategies, site-directed mutagenesis, "reverse genetics" (the production of genetically defined, transgenic animals and plants), and applications of genetic engineering. Also discussed are methods used in the analysis of complex genomes, such as the human genome, in which a gene that has been localized by genetic mapping to a region of tens of millions of base pairs must be isolated in cloned form and identified.
- **Chapters 10 through 12** deal with molecular genetics in the strict sense. These chapters include the principles of gene expression, gene regulation, and the genetic control of development. The chapter on development focuses especially on genetic analysis of development in nematodes (*Caenorhabditis elegans*) and *Drosophila*, and there is a thorough examination of the exciting new work on the genetic basis of floral development in *Arabidopsis thaliana*.

- **Chapter 13** covers the molecular details of mutation and the effects of mutagens, including new information on the genetic effects of the Chernobyl nuclear accident. It also covers the rapidly growing field of DNA repair mechanisms, as well as the molecular mechanisms of recombination.
- **Chapter 14** covers organelle genetics.
- **Chapters 15 and 16** deal with population and evolutionary genetics. The discussion of population genetics includes DNA typing in criminal investigations and paternity testing. The material on quantitative genetics includes a discussion of methods by which particular genes influencing quantitative traits (QTLs, or quantitative-trait loci) may be identified and mapped by linkage analysis. QTL mapping is presently one of the most important approaches for identifying the genetic basis of human disease.
- **Chapter 17,** entitled the *Genetics of Biorhythms and Behavior,* illustrates the genetic analysis of behavior with experimental models, including chemotaxis in bacteria, mating behavior in *Drosophila,* and learning in laboratory rodents. This chapter also includes a section on mad cow disease and its relation to the molecular basis of biological rhythms. There is also a section on the genetic determinants of human behavior with examples of the approach using "candidate" genes that led to the identification of the "natural Prozac" polymorphism in the human serotonin transporter gene.

Integrated throughout the book are frequent references to **human genetics,** including sections on the fragile-X syndrome, imprinting, the genetic basis of cancer, expansion of unstable repeats in diseases such as Huntington disease, the relationship of DNA repair enzymes to hereditary colon cancer, the controversial mitochondrial "Eve," genetic diseases associated with defects in biorhythms, and many other special topics, including the human genome project.

CONNECTIONS

A unique special feature of this book is found in boxes called **Connections.** Each chapter has two or three of these boxes. They are our way of connecting genetics to the world outside the classroom. All of the Connections include short excerpts from the original literature of genetics, usually papers, each introduced with a short explanatory passage. Many of the Connections are excerpts from classic papers, such as Mendel's paper, but by no means all of the "classic" papers are old papers. More than a quarter were published more recently than 1980, including the paper in which the cloning of the sheep *Dolly* was reported.

The pieces are called Connections because each connects the material in the text to something that broadens or enriches its implications. Some of the Connections raise issues of ethics in the application of genetic knowledge, social issues that need to be addressed, or issues related to the proper care of laboratory animals. They illustrate other things as well. Because each Connection names the place where the research was carried out, the student will learn that great science is done in many universities and research institutions throughout the world. Some of the pieces were published originally in French, others in German. These appear in English translation. In papers that use outmoded or unfamiliar terminology, or that use archaic gene symbols, we have substituted the modern equivalent because the use of a consistent terminology in the text and in the Connections makes the material more accessible to the student.

GENETICS ON THE INTERNET

More than in most fields of biology, genetic resources and genetic information are abundant on the Internet. The most useful sites are not always easy to find. A recent search of Internet sites using the Alta Vista search engine and the keyword *genetics* yielded about 500,000 hits. Most of these are of limited usefulness, but quite a few are invaluable to the student and to the practicing geneticist. The problem is how to find the really useful ones among the 500,000 sites.

To make the genetic information explosion on the Internet available to the student, we have developed **Internet Exercises,** called **GeNETics on the web,** which make use of Internet resources. One reason for developing these exercises is that genetics is a dynamic science, and most of the key Internet resources are kept up to date. Continually updated, the Internet exercises introduce the newest discoveries as soon as they appear, and this keeps the textbook up to date as well.

The addresses of the relevant genetic sites are not printed in the book. Instead, the sites are accessed through the use of key words that are highlighted in each exercise. The key words are maintained as hot links at the publisher's web site (http://www.jbpub.com/genetics) and are kept constantly up to date, tracking the address of each site if it should change. The use of key words also allows an innovation: one exercise in each chapter makes use of a **mutable site** that changes frequently in both the site accessed and the exercise. Students should look at the Internet Exercises. The instructor may wish to make short assignments from some of them, or use them for extra credit or as short term papers. We have included a suggested assignment for each of the exercises, but many instructors may wish to develop their own. We would be pleased to receive suggestions for new web exercises at the Jones and Bartlett home page: www.jbpub.com.

PROBLEMS

Each chapter provides numerous problems for solution, graded in difficulty, for the students to test their understanding. The problems are of three different types:

Review the Basics problems ask for genetic principles to be restated in the student's own words; some are matters of definition or call for the application of elementary principles.

Analysis and Applications problems are more traditional types of genetic problems in which several concepts must be applied in logical order and often require some numerical calculation. The level of mathematics is that of arithmetic and elementary probability as it pertains to genetics. None of the problems uses mathematics beyond elementary algebra.

Challenge Problems are similar to those in Analysis and Applications, but they are a degree more challenging, often because they require a more extensive analysis of data before the question can be answered.

Supplementary Problems, in a special section at the end of the book, consist of over 300 additional problems. These include representatives of all three types of problems found at the ends of the chapters, and they are graded in difficulty. The Supplementary Problems may be used for additional assignments, more practice, or even as examination questions. The problems were generously contributed by geneticist Elena R. Lozovskaya of Harvard University, and they were selected and edited by the authors. Unlike the other problems, the solutions to the Supplementary Problems are not included in the answer section at the end of the book. Solutions are available for the instructor in the *Test Bank and Solutions Manual.*

GUIDE TO PROBLEM SOLVING

Each chapter contains a **Guide to Problem Solving** that demonstrates problems worked in full. The concepts needed to solve the problem, and the reasoning behind the answer, are explained in detail. The Guide to Problem Solving serves as another level of review of the important concepts used in working problems. It also highlights some of the most common mistakes made by beginning students and gives pointers on how the student can avoid falling into these conceptual traps.

SOLUTIONS

All Analysis and Applications Problems and all Challenge Problems are answered in full, with complete methods and explanations, in the answer section at the end of the book. The rationale for giving all the answers is that problems are valuable opportunities to learn. Problems that the student cannot solve are usually more important than the ones that can be solved, because the sticklers usually identify trouble spots, areas of confusion, or gaps in understanding. As often as not, the conceptual difficulties are resolved when the problem is worked in full and the correct approach explained, and the student seldom stumbles over the same type of problem again.

FURTHER READING

Each chapter also includes recommendations for **Further Reading** for the student who either wants more information or who needs an alternative explanation for the material presented in the book. Some additional "classic" papers and historical perspectives are included. Complete author lists are also given for a few Connections that had too many authors to cite individually in the text.

ILLUSTRATIONS

The art program is spectacular, thanks to the creative efforts of J/B Woolsey Associates, with special thanks to John Woolsey and Patrick Lane. Every chapter is richly illustrated with beautiful graphics in which color is used functionally to enhance the value of each illustration as a learning aid. The illustrations are also heavily annotated with "process labels" explaining step-by-step what is happening at each level of the illustration. These labels make the art inviting as well as informative. They also allow the illustrations to stand relatively independently of the text, enabling the student to review material without rereading the whole chapter.

The art program is used not only for its visual appeal but also to increase the pedagogical value of the book:

- Characteristic colors and shapes have been used consistently throughout the book to indicate different types of molecules—DNA, mRNA, tRNA, and so forth. For example, DNA is illustrated in any one of a number of ways, depending on the level of resolution necessary for the illustration, and each time a particular level of resolution is depicted, the DNA is shown in the same way. It avoids a great deal of potential confusion that DNA, RNA, and proteins are represented in the same manner in Chapter 17 as they are in Chapter 1.
- There are numerous full-color photographs of molecular models in three dimensions; these give a strong visual reinforcement of the concept of macromolecules as physical entities with defined three-dimensional shapes and charge distributions that serve as the basis of interaction with other macromolecules.
- The page design is clean, crisp, and uncluttered. As a result, the book is pleasant to look at and easy to read.

FLEXIBILITY

There is no necessary reason to start at the beginning and proceed straight to the end. Each chapter is a self-contained unit that stands on its own. This feature gives the book the flexibility to be used in a variety of course formats. Throughout the book, we have integrated classical and molecular principles, so you can begin a course with almost any of the chapters. Most teachers will prefer starting with the overview in Chapter 1, possibly as suggested reading, because it brings every student to the same basic level of understanding. Teachers preferring the Mendel-early format should continue with Chapter 2; those preferring to teach the details of DNA early should continue with Chapter 5. Some teachers are partial to a chromosomes-early format, which would suggest continuing with Chapter 3, followed by Chapters 2 and 4. A novel approach would be a genomes-first format, which could be implemented by continuing with Chapter 9. Some teachers like to discuss mechanisms of mutation early in the course, and Chapter 13 can easily be assigned early. The writing and illustration program was designed to accommodate a variety of formats, and we encourage teachers to take advantage of this flexibility in order to meet their own special needs.

SUPPLEMENTS

An unprecedented offering of traditional and interactive multimedia supplements is available to assist instructors and aid students in mastering genetics. Additional information and review copies of any of the following items are available through your Jones and Bartlett Sales Representative.

For the Instructor

- *Test Bank and Solutions Manual.* This evaluation tool, authored by Michael Draper of Tufts University with contributions from Patrick McDermot of Tufts University, contains 850 test items, with 50 questions per chapter. There is a mix of factual, descriptive, and quantitative question types. A typical chapter file contains 20 multiple-choice objective questions, 15 fill-ins, and 15 quantitative problems. A Solutions Manual containing worked solutions of all the supplemental problems in the main text is bound together with the Test Bank.
- *Electronic Test Bank.* An electronic version of the test bank for preparing customized tests is available for Macintosh or Windows operating systems.
- *Genetics Lecture Success CD-ROM.* This easy-to-use multimedia tool contains over 300 figures from the text specially enhanced for classroom presentation. You select the images you need by chapter, topic, and figure

number. This lecture aid readily interfaces with other presentation tools. It also contains key simulated web sites that allow you to bring the Internet into the classroom without the need for a live Internet connection.
- *Visual Genetics Plus: Tutorial and Laboratory Simulations. Faculty Version.* This Mac/IBM CD-ROM, created by Alan W. Day and Robert L. Dean of the University of Western Ontario, is already in use at over 200 institutions worldwide. Visual Genetics 3.0 continues to provide a unique, dynamic presentation tool for viewing key genetic and molecular processes in the classroom. With this new, greatly expanded version of the Virtual Genetics Lab 2.0, instructors can now assign 17 comprehensive lab simulations. You can also bring the lab into the classroom, as the program allows you to perform on-screen tasks such as the selection of mutant colonies, using a pipette to make a dilution series, inoculating mutants to petri dishes to test for response to growth factors, and then to analyze and interpret the data. Through the testing feature and presentation capabilities, you can offer a complete lab environment. Site Licenses and Instructor Copies are available.
- *Video Resource Library.* A full complement of quality videos is available to qualified adopters. Genetics-related topics include: Origin and Evolution of Life, Human Gene Therapy, Biotechnology, the Human Genome Project, Oncogenes, and Science and Ethics.

For the Student

- *The Gist of Genetics: Guide to Learning and Review.* Written by Rowland H. Davis and Stephen G. Weller of the University of California, Irvine, this study aid uses illustrations, tables, and text outlines to review all of the fundamental elements of genetics. It includes extensive practice problems and review questions with solutions for self-check. The Gist helps students formulate appropriate questions and generate hypotheses that can be tested with classical principles and modern genetic techniques.
- *GeNETics on the web.* Corresponding to the end-of-chapter GeNETics on the web exercises, this World Wide Web site offers genetics-related links, articles and monthly updates to other genetics sites on the Web. Material for this site is carefully selected and updated by the authors. Jones and Bartlett Publishers ensures that links for the site are regularly maintained. Visit the *GeNETics on the web* site at http://www.jbpub.com/genetics.
- *Cogito: Electronic Companion to Genetics.* This Mac/IBM CD-ROM, by Philip Anderson and Barry Ganetzky of the University of Wisconsin, Madison, reviews important genetics concepts covered in class using state-of-the-art interactive multimedia. It

consists of hundreds of animations, diagrams, and videos that dynamically explain difficult concepts to students. In addition, it contains over 400 interactive multiple-choice, "drag and drop," true/false, and fill-in problems. These resources will prove invaluable to students in a self-study environment and to instructors as a lecture-enhancement tool. This CD-ROM is available for packaging exclusively with Jones and Bartlett Publishers texts.

- ***Visual Genetics Plus: Tutorial and Laboratory Simulations. Student Version.*** This Mac/IBM CD-ROM, created by Alan W. Day and Robert L. Dean of the University of Western Ontario, is already in use at over 200 institutions worldwide. Visual Genetics 3.0 affords a dynamic multimedia review of key genetic and molecular processes, including a greatly expanded version of the Virtual Genetics Lab 2.0, with which students can work on 17 comprehensive lab simulations. The lab allows students to perform tasks on-screen—such as selecting mutant colonies, making a dilution series, inoculating mutants into petri dishes to test for response to growth factors—and then guides them in analyzing and interpreting the data. The Student Version is available for purchase and can be packaged with our text.

ACKNOWLEDGMENTS

We are indebted to the many colleagues whose advice and thoughts were immensely helpful throughout the preparation of this book. These colleagues range from specialists in various aspects of genetics who checked for accuracy or suggested improvement to instructors who evaluated the material for suitability in teaching or sent us comments on the text as they used it in their courses.

Jeremy C. Ahouse, Brandeis University

John C. Bauer, Stratagene, Inc., La Jolla, CA

Mary K. B. Berlyn, Yale University

Pierre Carol, Université Joseph Fourier, Grenoble, France

John W. Drake, National Institute of Environmental Health Sciences, Research Triangle Park, NC

Jeffrey C. Hall, Brandeis University

Steven Henikoff, Fred Hutchinson Cancer Research Center, Seattle, WA

Joyce Katich, Monsanto, Inc., St. Louis, MO

Jeane M. Kennedy, Monsanto, Inc., St. Louis, MO

Jeffrey King, University of Berne, Switzerland

K. Brooks Low, Yale University

Gustavo Maroni, University of North Carolina

Jeffrey Mitton, University of Colorado, Boulder

Gisela Mosig, Vanderbilt University

Robert K. Mortimer, University of California, Berkeley

Ronald L. Phillips, University of Minnesota

Robert Pruitt, Harvard University

Pamela Reinagel, California Institute of Technology, Pasadena

Kenneth E. Rudd, National Library of Medicine

Leslie Smith, National Institute of Environmental Health Sciences, Research Triangle Park, NC

Johan H. Stuy, Florida State University

Irwin Tessman, Purdue University

Kenneth E. Weber, University of Southern Maine

We would also like to thank the reviewers, listed below, who reviewed one or more chapters and who, in several cases, reviewed the complete fourth edition manuscript. Their comments and recommendations helped improve the content, organization, and presentation of the material. We offer special thanks to Dick Morel, who carefully reviewed and commented on all of the illustrations as well as the text.

Laura Adamkewicz, George Mason University

Peter D. Ayling, University of Hull (UK)

Anna W. Berkovitz, Purdue University

John Celenza, Boston University

Stephen J. D'Surney, University of Mississippi

Kathleen Dunn, Boston College

David W. Francis, University of Delaware

Mark L. Hammond, Campbell University

Richard Imberski, University of Maryland

Sally A. MacKenzie, Purdue University

Kevin O'Hare, Imperial College (UK)

Peggy Redshaw, Austin College

Thomas F. Savage, Oregon State University

David Shepard, University of Delaware

Charles Staben, University of Kentucky

David T. Sullivan, Syracuse University

James H. Thomas, University of Washington

We also wish to acknowledge the superb art, production, and editorial staff who helped make this book possible: Mary Hill, Patrick Lane, Andrea Fincke, Judy Hauck, Bonnie Van Slyke, Sally Steele, John Woolsey, Brian McKean, Kathryn Twombly, Rich Pirozzi, Mike Campbell, and Tom Walker. Much of the credit for the attractiveness and readability of the book should go to them. Thanks also to Jones and Bartlett, the publishers, for the high quality of the book production. We are also grateful to the many people, acknowledged in the legends of the illustrations, who contributed photographs, drawings, and micrographs from their own research and publications, especially those who provided color photographs for this edition. Every effort has been made to obtain permission to use copyrighted material and to make full disclosure of its source. We are grateful to the authors, journal editors, and publishers for their cooperation. Any errors or omissions are wholly inadvertant and will be corrected at the first opportunity.

INTRODUCTION: FOR THE STUDENT

In signing up for a genetics course, our students often wonder how much work is going to be required, how much time it will take to do the reading and written assignments, how hard the examinations will be, and what is their likelihood of getting a good grade. These are perfectly legitimate issues, and you should not feel guilty if they are foremost in your mind.

You may also be wondering what you are going to learn by taking a course in genetics. Will the material be interesting? Is there any reason to study genetics other than to satisfy an academic requirement? At the end of the course, will you be glad that you took it? Will there be any practical value to what you will learn? This introduction is designed to reassure you that the answer to each question is yes. The study of genetics is relevant not only to biologists but to all members of our modern, complex, technological society. Understanding the principles of genetics will help you to make informed decisions about numerous matters of political, scientific, and personal concern.

At least 4000 years ago in the Caucasus, the Middle East, Egypt, South America, and other parts of the world, farmers recognized that they could improve their crops and their animals by selective breeding. Their knowledge was based on experience and was very incomplete, but they did recognize that many features of plants and animals were passed from generation to generation. They discovered that desirable traits—such as size, speed, and weight of animals—could sometimes be combined by controlled mating and that, in plants, crop yield and resistance to arid conditions could be combined by cross-pollination. The ancient breeding programs were not based on much solid information because nothing was known about genes or any of the principles of heredity. In a few instances, the pattern of hereditary transmission of a human trait came to be recognized. One example is *hemophilia,* or failure of the blood to clot, which results in life-threatening bleeding from small cuts and bruises. By the second century of the present era, rules governing exemptions from circumcision had been incorporated into the Talmud, indicating that several key features of the mode of inheritance of hemophilia were understood. The Talmud's exemptions apply in the case of a mother who lost two sons from excessive bleeding following circumcision: Subsequent boys born to the same mother, and all boys born to her sisters, were exempt. However, the paternal half brothers of a boy who had died from excessive bleeding were not exempt. (Paternal half brothers have the same father but a different mother.) These rules of exemption from circumcision make very good sense when judged in light of our modern understanding of the inheritance of hemophilia, as you will learn in Chapter 3.

The scientific study of heredity is called **genetics.** The modern approach to genetics can be traced to the mid-nineteenth century with Gregor Mendel's careful analyses of inheritance in peas. Mendel's experiments were simple and direct and brought forth the most significant principles that determine how traits are passed from one generation to the next. In Chapter 2, you will learn the rules followed by genes and chromosomes as they pass from generation to generation, and you will be able to calculate in many instances the probabilities by which organisms with particular traits will be produced. Mendel's kind of experiments, which occupied most of genetic research until the middle of the twentieth century, is called **transmission genetics.** Some people have called it formal genetics, because the subject can be understood and the rules clearly seen without any reference to the biochemical nature of genes or gene products.

Beginning about 1900, geneticists began to wonder about a subject we now call **molecular genetics.** Is the gene a known kind of molecule? How can genetic information be encoded in a molecule? How is the genetic information transmitted from one generation to the next? In what way is the genetic information changed in a mutant organism? At that time, there was no logical starting point for such an investigation, no experimental "handle." In the 1940s, critical observations were made that implicated the molecule deoxyribonucleic acid (DNA), first discovered in 1869. You will learn about these experiments in Chapter 1. With the discovery of the structure of DNA in 1953 by Watson and Crick, genetics entered the DNA age. Within a decade, there came an understanding of the chemical nature of genes and how genetic information is stored, released to a cell, and transmitted from one generation to the next. During the first three decades after the discovery of DNA structure, the body of genetic knowledge grew with a two-year doubling time. These were exciting times, and you will be presented with a distillation of these findings in the chapters of this book that deal with **molecular genetics.**

Since the early 1970s, genetics has undergone yet another revolution: the development of recombinant DNA technology. This technology is a collection of methods that enable genes to be transferred, at the will of the molecular geneticist, from one organism to another. This branch of genetics is known as **genetic engineering.** Genetic engineering has had an enormous impact in genetic research, particularly in our ability to understand

Balancing Act

Thomas Hunt Morgan 1913
Columbia University, New York,
New York

Genetics and cell biology have both advanced with surprising rapidity in recent years. Hardly a week goes by without a new discovery of notable importance being reported in the pages of Science *or* Nature *or some other major research journal. Nontechnical accounts of new discoveries are regularly reported in the popular press and on television. We are in the midst of a knowledge explosion—doubtless you remember being told this before. We are so often reminded that we live in a fast-paced world and should be proud to be speeding along. But hit the brakes, and pause for a moment, to reread the first sentence. It is an almost direct quotation of the words that Thomas Hunt Morgan wrote to introduce his first book in genetics. This was in 1913. Morgan was one of the pioneers of modern genetics, and genetics in 1913 was poised for truly spectacular advances. He could scarcely have imagined what modern genetics would be like—how much we would know about*

some things, how little we would know about others; how powerful the methods would be in some ways, how limited they would be in others. Morgan did see one thing clearly. It was that the key to understanding biology is to maintain the right balance among different ways of studying organisms—through genetics, cell biology, molecular biology, biochemistry, biophysics, developmental biology, neurobiology, evolutionary biology, and ecology. Maintaining the right balance for today's students has been our primary goal in writing this book.

Two lines of research have developed with surprising rapidity in recent years. Their development has been independent, but at many stages in their progress they have looked to each other for help. The study of the cell has furnished some fundamental facts connected with problems of heredity. The modern study of heredity has proven itself to be an instrument even more subtle in the analysis of the materials of the germ cells than actual observations on the germ cells themselves. The time has come, we think, when a failure to recognize the close

bond between these two modern lines of advance can no longer be interpreted as a wise or cautious skepticism. An anarchistic spirit in science does not always mean greater profundity, nor is our attitude toward science more correct because we are unduly skeptical toward every advance. To maintain the right balance is the hardest task we have to meet. What we most fear is that in attempting to formulate some of the difficult problems of present-day interest we may appear to make at times unqualified statements in a dogmatic spirit. All conclusions in science are relative and subject to change, for change in science does not mean so much that what has gone before is wrong, as the discovery of a better strategic position than the one last held.

Source: Heredity and Sex. NY: Columbia University Press.

gene expression and its regulation in plants and animals. Topics previously unapproachable suddenly became amenable to experimental investigation. Currently, genetic engineering is providing us with new tools of great economic importance and of value in medical practice. Current projects of great interest include the genetic modification of plants and domesticated animals and the production of clinically active substances.

Beginning in the 1980s came the new emphasis on **genomics,** the application of recombinant DNA strategies to the study of whole **genomes** (the totality of genetic information in an organism) rather than single genes. The complete set of DNA instructions has been determined by direct DNA sequencing in a number of viruses, cellular organelles such as mitochondria, several bacteria, and the yeast *Saccharomyces cerevisiae.* Programs are also underway to determine the complete DNA sequence of other model organisms. (In genetics, a **model organism** is a species that is studied as an example to learn basic principles that we hope will be applicable to other organisms.) Just on

the horizon is the capability of determining the complete DNA sequence in the human genome. The availability of genomic sequences opens up new approaches for genetics because it turns the subject on its head. Instead of starting with a mutant organism that has some physical abnormality, attempting to identify the gene responsible, and determining the DNA sequence, one can now start with a DNA sequence that has already been determined and try to learn what the gene does.

By far the greatest practical influence of genetics has been in the fields of medicine and agriculture. There have been many important contributions to modern clinical practice, and progress is accelerating because of the increased emphasis on genomic analysis. Genetic experiments have revealed thousands of new genetic markers in the human genome and have given us new methods for the detection of mutant genes—not only in affected individuals but also in their relatives and in members of the population at large. These methods have given genetic counseling new meaning. Human beings are at risk for any of several thousand different inherited diseases. Married couples can be informed of the possibility of their producing an affected offspring and can now make choices between childbearing and adoption. Consider the relief of a woman and man who learn that they do not carry a particular defective gene and can produce a child without worry. Even when an offspring might be affected with a genetic disorder, techniques are available to determine if a fetus does, in fact, carry a mutant gene.

In agriculture, studies of the genetic composition of economically important plants have enabled plant breeders to institute rational programs for developing new varieties. Among the more important plants that have been developed are high-yielding strains of corn and dwarf wheat, disease-resistant rice, corn with an altered and more nutritious amino acid composition (high-lysine corn), and wheat that grows faster, allowing crops to be grown in short-season regions such as Canada and Sweden. You will be introduced to the techniques for developing some of these strains in this book. Often new plant varieties have shortcomings, such as a requirement for increased amounts of fertilizer or a decreased resistance to certain pests. How to overcome these shortcomings is a problem for the modern geneticist, who has the job of manipulating the inherited traits. Genetic engineering is also providing new procedures for such manipulations, and quite recently there have been dramatic successes.

A few words about the book. Each chapter contains two or three **Connections** set off in special boxes. Each connects the material in the text to the real world of genetics outside the classroom. Some of the Connections are excerpts from classic papers, including Mendel's paper. Others are very recent, such as the paper that reports the cloning of an adult sheep. Some of the Connections raise issues of ethics in the application of genetic knowledge, social issues that need to be addressed, issues related to the proper care of laboratory animals, or other matters. We have included a Connection in this Introduction to give you a taste. For an appreciation of genetics in a broad historical context related at many points to contemporary research and social and ethical issues, we urge you to connect with the Connections. There is a complete listing, chapter by chapter, of all the Connections in the Table of Contents. Following the Table of Contents is a complete list of all the material excerpted, shown in chronological order.

Each chapter comes with a set of **Internet Exercises,** called **GeNETics on the web,** which will introduce you to the genetic resources and information that can be accessed through the Internet. These are important because genetics is more richly represented on the Internet than any other field of biology. Each exercise uses a key word in describing an issue or a problem. The key words are maintained as hot links at the publisher's web site (http://www.jbpub.com/genetics) and are kept constantly up to date. Each exercise comes with a short written component that your instructor may wish to assign. We urge you to go through the web exercises even if they are not assigned, as they will help you to become familiar with some of the extraordinary resources that are out there. We should mention two special types of exercises. One is the **mutable site** in which the site and the exercise are changed frequently. You can check back on a mutable site that you have explored before, and there will be a good chance that it will have changed in the meantime. The other special site is the **PIC site,** which connects you to a genetics site chosen for its visual appeal.

As a pedagogical aid, important terms are printed in **boldface** in the text. These terms are collected at the end of each chapter in a section entitled **Key Terms.** You should know their meanings because they form the basic vocabulary of genetics. If necessary, you can look them up in the **Concise Dictionary of Genetics** at the back of the book. Each chapter also includes a **Summary** at the end of the text. Sample problems are worked in the section titled **Guide to Problem Solving.** Each chapter ends with a fairly large collection of problems. These are of three types:

Review the Basics problems ask you to restate genetic principles or definitions in your own words or to apply elementary principles.

Analysis and Applications problems require you to apply several concepts in logical order and usually to do some numerical calculation. (The calculations use only simple arithmetic, so there is no reason to be intimidated even if higher mathematics is not a comfortable part of your repertoire.)

Challenge Problems are similar in nature but a little more difficult because you may need to analyze some data to solve the problem.

It is essential that you work as many of the problems as you can, because experience has shown that practice with problems is a good way to learn genetics and to identify particular points or concepts that have been misunderstood. Sometimes it is not even necessary to solve a problem completely but only to read the problem and decide whether you could solve it if asked to do so. The **Answers** to all of the problems, and full explanations, are given at the back of the book. A problem will be more useful to you if you take a fair shot at it before turning to the answer. The back of the book also includes a large set of **Supplementary Problems,** without answers, for still more practice. There is nothing better than solving problems not only to test your knowledge but to make it part of your long-term memory.

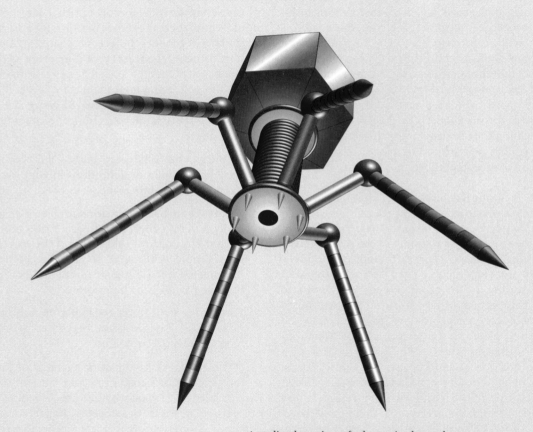

A stylized version of a bacteriophage that very much resembles the phage T2 used in the Hershey-Chase experiments. [Courtesy of Paul Dowrick, © Phage *et al* Ltd.]

CHAPTER **1**

The Molecular Basis of Heredity and Variation

CHAPTER OUTLINE

PRINCIPLES

- Genes control biologically inherited traits; a trait that is genetically determined can also be influenced by environmental factors.
- Genes are composed of the chemical deoxyribonucleic acid (DNA).
- DNA replicates to form (usually identical) copies of itself.
- DNA contains a code specifying what types of enzymes and other proteins are made in cells.
- DNA occasionally mutates, and the mutant forms specify altered proteins.
- Genes interact with one another in sometimes complex ways.
- Organisms change genetically through generations in the process of biological evolution.

CONNECTIONS

CONNECTION: It's the DNA!
Oswald T. Avery, Colin M. MacLeod,
and Maclyn McCarty 1944
Studies on the chemical nature of the substance inducing transformation of pneumococcal types

CONNECTION: Shear Madness
Alfred D. Hershey and Martha Chase 1952
Independent functions of viral protein and nucleic acid in growth of bacteriophage

The members of any biological species are similar in some characteristics but different in others. For example, all human beings share a set of observable characteristics, or **traits,** that define us as a species. We have a backbone and a spinal cord; these traits are among those that define us as a type of vertebrate. We are warm blooded and feed our young with milk from mammary glands; these traits are among those that define us as a type of mammal. We are, in finer detail, a type of primate that habitually stands upright and has long legs, relatively little body hair, a large brain, a flat face with a prominent nose, jutting chin, distinct lips, and small teeth. These traits set us apart from other primates, such as chimpanzees and gorillas. The biological characteristics that define us as a species are inherited, but they do not differ from one person to the next.

Within the human species, however, there is also much variation. Traits such as hair color, eye color, skin color, height, weight, and personality characteristics are tremendously variable from one person to the next. There is also variation in health-related traits, such as predisposition to high blood pressure, diabetes, chemical dependence, mental depression, and the Alzheimer disease. Some of these traits are inherited biologically, others are inherited culturally. Eye color results from biological inheritance; the native language we speak results from cultural inheritance. Many traits are influenced jointly by biological inheritance and environmental factors. For example, weight is determined in part by inheritance but also in part by eating habits and level of physical activity.

The study of biologically inherited traits is **genetics.** Among the traits studied in genetics are those that are influenced in part by the environment. The fundamental concept of genetics is

> Inherited traits are determined by elements of heredity, called **genes,** that are transmitted from parents to offspring in reproduction.

The elements of heredity and some basic rules governing their transmission from generation to generation were discovered by Gregor Mendel in experiments with garden peas. His results were pub-

lished in 1866. Mendel's experiments are among the most beautifully designed, carefully executed, and elegantly interpreted in the history of experimental science. Mendel interpreted his data in terms of a few abstract rules by which hereditary elements are transmitted from parents to offspring. Three years later, in 1869, Friedrich Miescher discovered a new type of weakly acid substance, abundant in the nuclei of salmon sperm and white blood cells. At the time he had no way of knowing that it would turn out to be the chemical substance of which genes are made. Miescher's weak acid, the chemical substance of the gene, is now called **deoxyribonucleic acid (DNA).** However, the connection between DNA and heredity was not demonstrated until about the middle of the twentieth century. How was this connection established?

1.1 DNA: The Genetic Material

The importance of the cell nucleus in inheritance became apparent in the 1870s with the observation that the nuclei of male and female reproductive cells fuse in the process of fertilization. This observation suggested that there was something inside the sperm and egg nucleus that was responsible for inherited characteristics. The next major advance was the discovery of thread-like objects inside the nucleus that become visible in the light microscope when stained with certain dyes; these threads were called **chromosomes.** As we shall see in Chapter 3, chromosomes have a characteristic "splitting" behavior in cell division, which ensures that each daughter cell receives an identical complement of chromosomes. By 1900 it had become clear that the number of chromosomes is constant within each species but differs among species. The characteristics of chromosomes made it seem likely that they were the carriers of the genes.

By the 1920s, more and more evidence suggested a close relationship between DNA and the genetic material. Studies using special stains showed that DNA, in addition to certain proteins, is present in chromosomes. Furthermore, investigations disclosed that almost all cells of a given species contain a constant amount of DNA, whereas the amount and kinds of proteins and other molecules differ greatly in different cell types. The indirect evidence that genes are DNA was rejected because crude chemical analyses of DNA had suggested (incorrectly) that it lacks the chemical diversity needed for a genetic substance. In contrast, proteins were known to be an exceedingly diverse collection of molecules. And so, on the basis of incorrect data, it became widely accepted that proteins were the genetic material and that DNA merely provided the structural framework of chromosomes. Against the prevailing opinion that genes are proteins, experiments purporting to demonstrate that DNA is the genetic material had also to demonstrate that proteins are *not* the genetic material. Two of the experiments regarded as decisive are described in this section.

Experimental Proof of the Genetic Function of DNA

The first evidence that genes are DNA came from studies of bacteria that cause pneumonia. Bacterial pneumonia in mammals is caused by strains of *Streptococcus pneumoniae* that are able to synthesize a slimy "capsule" around each cell. Strains that lack a capsule do not cause pneumonia. The capsule is composed of a complex carbohydrate (polysaccharide) that protects the bacterium from the immune response of the infected animal and enables the bacterium to cause the disease. When a bacterial cell is grown on solid medium, it undergoes repeated cell divisions to form a visible clump of cells called a **colony.** The enveloping capsule gives the colony a glistening or smooth (S) appearance. Some strains of *S. pneumoniae* are not able to synthesize a capsule. As a result, they form colonies that have a rough (R) surface (Figure 1.1). The R strains do not cause pneumonia, because without their capsules, the bacteria are attacked by the immune system of the host. Both types of bacteria "breed true" in the sense that the progeny formed by cell division have the capsular type of the parent, either S or R.

When mice are injected either with living R cells or with dead S cells killed by heat, they remain healthy. However, it was discovered in 1928 that mice often died of pneumonia when injected with a mixture containing a small number of living R cells and a large number of dead S cells. Bacteria isolated from blood samples of the mice infected with the mixture produced S cultures with a capsule typical of the injected S cells, even though the injected S cells had been killed by heat. Therefore, the material containing the dead S cells that was injected must have included a substance that could convert, or transform, otherwise harmless cells of the R bacterial strain into S strain cells with the ability to resist the immunological system of the mouse, multiply, and cause pneumonia. In other words, there was a genetic **transformation** of an R cell into an S cell. Furthermore, the new genetic characteristics were inherited by descendants of the transformed bacteria.

Figure 1.1

Colonies of rough (R, the small colonies) and smooth (S, the large colonies) strains of *Streptococcus pneumoniae*. The S colonies are larger because of the capsule on the S cells. [Photograph from O. T. Avery, C. M. MacLeod, and M. McCarty. 1944. *J. Exp. Med.* 79: 137.]

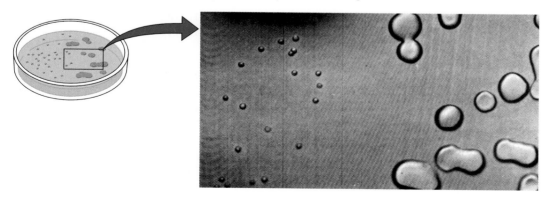

It's the DNA!

Oswald T. Avery, Colin M. MacLeod,
and Maclyn McCarty 1944

The Rockefeller University,
New York, New York
*Studies on the Chemical Nature of the
Substance Inducing Transformation of
Pneumococcal Types*

This paper is one of the milestones of molecular biology. Genetics and biochemistry were at last united through the finding that DNA was the chemical substance of the genetic material. There is very little biology in the paper beyond the use of Streptococcus (then called Pneumococcus) to ascertain whether a particular batch of extract, or an extract treated in some manner, contained the active substance able to transform type R into type S cells. The thrust of the paper is biochemistry: purifying the substance, showing that no known macromolecules other than DNA could be found in the extract, and demonstrating that the transforming activity could be destroyed by enzymes that attack DNA but not by protease or RNase enzymes.

Biologists have long attempted by chemical means to induce in higher organisms predictable and specific changes which thereafter could be transmitted as hereditary characters. Among microorganisms the most striking example of inheritable and specific alterations in cell structure and function that can be experimentally induced is the transformation of specific types of *Pneumococcus*. This phenomenon was first described by Griffith, who succeeded in transforming an attenuated [nonvirulent] and nonencapsulated (R) variant into fully encapsulated and virulent (S) cells. . . . The present paper is concerned

> *Within the limit of the analytical methods, the active fraction contains no demonstrable protein, lipid, or polysaccharide and consists principally, if not solely, of a highly polymerized form of deoxyribonucleic acid.*

with a more detailed analysis of the phenomenon of transformation of specific types of *Pneumococcus*. The major interest has centered in attempts to isolate the active principle from crude extracts and to identify its chemical nature, or at least to characterize it sufficiently to place it in a general group of known chemical substances. . . . A biologically active fraction has been isolated in highly purified form which in exceedingly minute amounts is capable under appropriate cultural conditions of inducing the transformation of unencapsulated R variants into fully encapsulated forms of the same specific type as that of the heat-killed microorganisms from which the inducing material was recovered. . . . Within the limit of the analytical methods, the active fraction contains no demonstrable protein, lipid, or polysaccharide and consists principally, if not solely, of a highly polymerized form of deoxyribonucleic acid. . . . Various enzymes have been tested for their capacity to destroy the transforming activity. Extracts to which were added crystalline trypsin and chymotrypsin [proteases], or combinations of both, suffered no loss in activity. . . . Prolonged treatment with crystalline ribonuclease under optimal conditions caused no demonstrable decrease in transforming activity. . . . The blood serum of several mammalian species contains an enzyme which causes the depolymerization of deoxyribonucleic acid; fresh dog and rabbit serum are capable of completely destroying transforming activity. . . . The evidence presented supports the belief that a nucleic acid of the deoxyribose type is the fundamental unit of the transforming principle.

Source: Journal of Experimental Medicine
79: 137–158

What substance was present in the dead S cells that made transformation possible? In the early 1940s, components of dead S cells were extracted and added to R cell cultures. The key experiment was one in which DNA was extracted from dead S cells and added to growing cultures of R cells and the resulting mixture spread onto an agar surface (Figure 1.2A). Among the R colonies, a few of type S appeared! Although the DNA preparations may still have contained traces of protein and RNA, the addition of an enzyme that destroys proteins (a *protease* enzyme) or one that destroys RNA (an *RNase* enzyme) did not eliminate the transforming activity (Figure 1.2B). On the other hand, the addition of an enzyme that destroys DNA completely eliminated the transforming activity (Figure 1.2C). These experiments were carried out by Oswald Avery, Colin MacLeod, and Maclyn McCarty at the Rockefeller University. They concluded their landmark report by noting that "the evidence presented supports the belief that a nucleic acid of the deoxyribose type is fundamental

(A) The transforming activity in S cells is not destroyed by heat

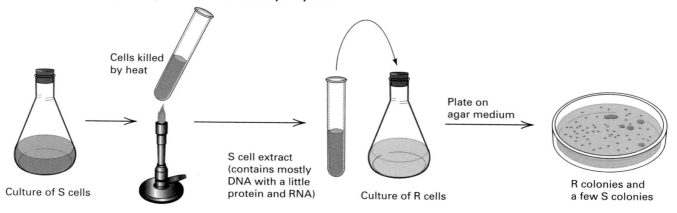

Culture of S cells

Cells killed by heat

S cell extract (contains mostly DNA with a little protein and RNA)

Culture of R cells

Plate on agar medium

R colonies and a few S colonies

(B) The transforming activity is not destroyed by either protease or RNAse

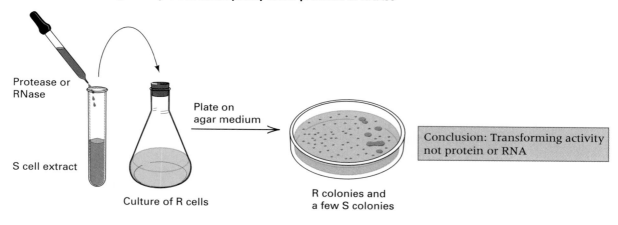

Protease or RNase

S cell extract

Culture of R cells

Plate on agar medium

R colonies and a few S colonies

Conclusion: Transforming activity not protein or RNA

(C) The transforming activity is destroyed by DNAse

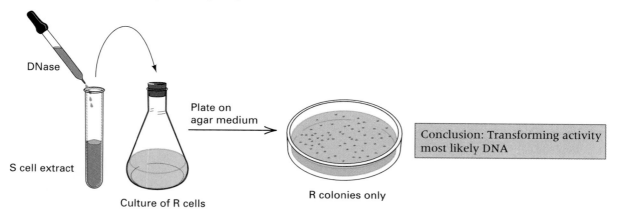

DNase

S cell extract

Culture of R cells

Plate on agar medium

R colonies only

Conclusion: Transforming activity most likely DNA

Figure 1.2

A diagram of the experiment that demonstrated that DNA is the active material in bacterial transformation. (A) Purified DNA extracted from heat-killed S cells can convert some living R cells into S cells, but the material may still contain undetectable traces of protein and/or RNA. (B) The transforming activity is not destroyed by either protease or RNase. (C) The transforming activity is destroyed by DNase and so probably consists of DNA.

unit of the transforming principle." In other words, DNA seems to be the genetic material.

Genetic Role of DNA in Bacteriophage

A second important finding concerned a type of virus that infects bacterial cells. The virus, T2 by name, is known as a **bacteriophage,** or **phage** for short, because it infects bacterial cells. Bacteriophage means "bacteria-eater." T2 infects cells of the intestinal bacterium *Escherichia coli*. A T2 particle is illustrated in Figure 1.3. It is exceedingly small, yet it has a complex structure composed of head (which contains the phage DNA), collar, tail, and tail fibers. (For comparison, consider that the head of a human sperm is about 30 to 50 times larger in both length and width than the T2 head.) T2 infection begins with attachment of a phage particle by the tip of its tail to the bacterial cell wall, entry of phage material into the cell, multiplication of this material to form a hundred or more progeny phage, and release of progeny by disruption of the bacterial host cell.

Because DNA contains phosphorus but no sulfur, and proteins usually contain some sulfur but no phosphorus, the DNA and proteins in a phage particle can be labeled differentially by the use of radioac-

tive isotopes of the two elements. This difference was put to use by Alfred Hershey and Martha Chase in 1952, working at the Cold Spring Harbor Laboratories. By that time it was already known that T2 particles are composed of DNA and protein in approximately equal amounts. Hershey and Chase produced particles with radioactive DNA by infecting *E. coli* cells that had been grown for several generations in a medium containing ^{32}P (a radioactive isotope of phosphorus) and then collecting the phage progeny. Other particles with labeled proteins were obtained in the same way, using a medium that contained ^{35}S (a radioactive isotope of sulfur).

The experiments are summarized in Figure 1.4. Nonradioactive *E. coli* cells were infected with phage labeled with *either* ^{32}P (Figure 1.4A) *or* ^{35}S (Figure 1.4B) in order to follow the proteins and DNA separately. Infected cells were concentrated by centrifugation, resuspended in fresh medium, and then agitated in a kitchen blender to shear attached phage material from the cell surfaces. The blending was found to have no effect on the subsequent course of the infection, which implies that the genetic material must enter the infected cells very soon after phage attachment. When intact bacteria were separated from the material removed by blending, most of the radio

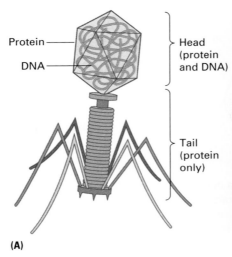

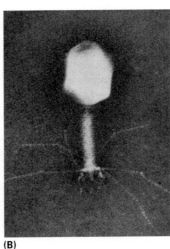

Figure 1.3
(A) Drawing of *E. coli* phage T2, showing various components. The DNA is confined to the interior of the head. (B) An electron micrograph of phage T4, a closely related phage. [Electron micrograph courtesy of Robley Williams.]

Protein —
DNA —

Head (protein and DNA)

Tail (protein only)

(A)

(B)

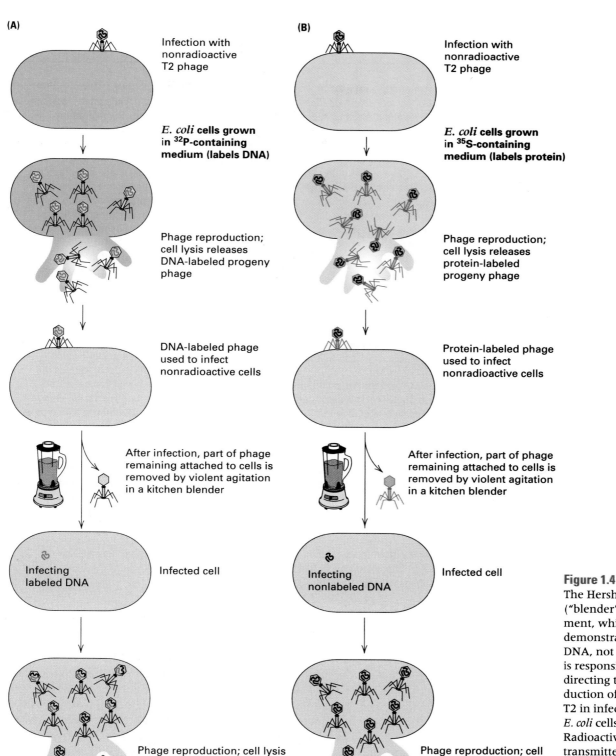

(A)

Infection with nonradioactive T2 phage

E. coli cells grown in ³²P-containing medium (labels DNA)

Phage reproduction; cell lysis releases DNA-labeled progeny phage

DNA-labeled phage used to infect nonradioactive cells

After infection, part of phage remaining attached to cells is removed by violent agitation in a kitchen blender

Infecting labeled DNA

Infected cell

Phage reproduction; cell lysis releases progeny phage that contain some ³²P-labeled DNA from the parental phage DNA

(B)

Infection with nonradioactive T2 phage

E. coli cells grown in ³⁵S-containing medium (labels protein)

Phage reproduction; cell lysis releases protein-labeled progeny phage

Protein-labeled phage used to infect nonradioactive cells

After infection, part of phage remaining attached to cells is removed by violent agitation in a kitchen blender

Infecting nonlabeled DNA

Infected cell

Phage reproduction; cell lysis releases progeny phage that contain almost no ³⁵S-labeled protein

Conclusion: DNA from an infecting parental phage is inherited in the progeny phage

Figure 1.4
The Hershey-Chase ("blender") experiment, which demonstrated that DNA, not protein, is responsible for directing the reproduction of phage T2 in infected *E. coli* cells. (A) Radioactive DNA is transmitted to progeny phage in substantial amounts. (B) Radioactive protein is transmitted to progeny phage in negligible amounts.

Alfred D. Hershey and Martha Chase 1952
Cold Spring Harbor Laboratories,
Cold Spring Harbor, New York
*Independent Functions of Viral Protein and
Nucleic Acid in Growth of Bacteriophage*

*Published a full eight years after the paper
of Avery, MacLeod and McCarty, the
experiments of Hershey and Chase get
equal billing. Why? Some historians of science suggest that the Avery et al. experiments were "ahead of their time." Others
suggest that Hershey had special standing
because he was a member of the "in
group" of phage molecular geneticists.
Max Delbrück was the acknowledged
leader of this group, with Salvador Luria
close behind. (Delbrück, Luria and Hershey
shared a 1969 Nobel Prize.) Another possible reason is that whereas the experiments
of Avery et al. were feats of strength in biochemistry, those of Hershey and Chase
were quintessentially genetic. Which
macromolecule gets into the hereditary
action, and which does not? Buried in the
middle of this paper, and retained in the
excerpt, is a sentence admitting that an
earlier publication by the researchers was
a misinterpretation of their preliminary
results. This shows that even first-rate scientists, then and now, are sometimes mis-
led by their preliminary data. Hershey later
explained, "We tried various grinding
arrangements, with results that weren´t
very encouraging. When Margaret*
*McDonald loaned us her kitchen blender
the experiment promptly succeeded."*

The work [of others] has shown that bacteriophages T2, T3, and T4 multiply in the bacterial cell in a non-infective [immature]
form. Little else is known about the vegetative [growth] phase of these viruses. The
experiments reported in this paper show
that one of the first steps in the growth of
T2 is the release from its protein coat of the

> **Our experiments show
> clearly that a physical
> separation of the phage T2
> into genetic and nongenetic
> parts is possible.**

nucleic acid of the virus particle, after
which the bulk of the sulfur-containing protein has no further function. . . . Anderson
has obtained electron micrographs indicating that phage T2 attaches to bacteria by its
tail. . . . It ought to be a simple matter to
break the empty phage coats off the
infected bacteria, leaving the phage DNA
inside the cells. . . . When a suspension of
cells with ^{35}S- or ^{32}P-labeled phage was
spun in a blender at 10,000 revolutions per
minute, . . . 75 to 80 percent of the phage
sulfur can be stripped from the infected
cells. . . . These facts show that the bulk of
the phage sulfur remains at the cell surface

during infection. . . . Little or no ^{35}S is contained in the mature phage progeny. . . .
Identical experiments starting with phage
labeled with ^{32}P show that phosphorus is
transferred from parental to progeny phage
at yields of about 30 phage per infected
bacterium. . . . [Incomplete separation of
phage heads] explains a mistaken preliminary report of the transfer of ^{35}S from
parental to progeny phage. . . . The following questions remain unanswered. (1) Does
any sulfur-free phage material other than
DNA enter the cell? (2) If so, is it transferred to the phage progeny? (3) Is the
transfer of phosphorus to progeny direct or
indirect? . . . Our experiments show clearly
that a physical separation of the phage T2
into genetic and nongenetic parts is possible. The chemical identification of the
genetic part must wait until some of the
questions above have been answered. . . .
The sulfur-containing protein of resting
phage particles is confined to a protective
coat that is responsible for the adsorption
to bacteria, and functions as an instrument
for the injection of the phage DNA into the
cell. This protein probably has no function
in the growth of the intracellular phage.
The DNA has some function. Further chemical inferences should not be drawn from
the experiments presented.

*Source: Journal of General Physiology
36: 39–56*

activity from ^{32}P-labeled phage was found
to be associated with the bacteria; however,
when the infecting phage was labeled with
^{35}S, only about 20 percent of the radioactivity was associated with the bacterial
cells. From these results, it was apparent
that a T2 phage transfers most of its DNA,
but not much of its protein, to the cell it
infects. The critical finding (Figure 1.4) was
that about 50 percent of the transferred
^{32}P-labeled DNA, but less than 1 percent of
the transferred ^{35}S-labeled protein, was
inherited by the *progeny* phage particles.
Because some protein was transferred to
infected cells and transmitted to the progeny phage, the Hershey-Chase experiment
was not nearly so rigorous as the transformation experiments in implicating DNA as
the genetic material. Nevertheless, owing
to its consistency with the DNA hypothesis,
the experiment was very influential.

The transformation experiment and the
Hershey-Chase experiment are regarded as
classics in the demonstration that genes
consist of DNA. At the present time, many
research laboratories throughout the world
carry out the equivalent of the transformation experiment on a daily basis, generally

using bacteria, yeast, or animal or plant cells grown in culture. These experiments indicate that DNA is the genetic material in these organisms as well as phage T2.

There are no known exceptions to the generalization that DNA is the genetic material in all cellular organisms.

It is worth noting, however, that in a few types of viruses, the genetic material consists of another type of nucleic acid called RNA.

1.2 DNA Structure: The Double Helix

Even with the knowledge that genes are DNA, many questions still remained. How does the DNA in a gene duplicate when a cell divides? How does the DNA in a gene control a hereditary trait? What happens to the DNA when a mutation (a change in the DNA) takes place in a gene? In the early 1950s, a number of researchers began to try to understand the detailed molecular structure of DNA in hopes that the structure alone would suggest answers to these questions. The first essentially correct three-dimensional structure of the DNA molecule was proposed in 1953 by James Watson and Francis Crick at Cambridge University. The structure was dazzling in its elegance and revolutionary in suggesting how DNA duplicates itself, controls hereditary traits, and undergoes mutation. Even while the tin sheet and wire model of the DNA molecule was still incomplete, Crick could be heard boasting in his favorite pub that "we have discovered the secret of life."

In the Watson-Crick structure, DNA consists of two long chains of subunits twisted around one another to form a double-stranded helix. The double helix is right-handed, which means that as one looks along the barrel, each chain follows a clockwise path as it progresses. You can see the right-handed coiling in Figure 1.5A if you imagine yourself looking up into the structure from the bottom: The "backbone" of each individual strand coils in a clockwise direction. The subunits of each strand are **nucleotides,** each of which contains

any one of four chemical constituents called **bases.** The four bases in DNA are

- **Adenine (A)**
- **Thymine (T)**
- **Guanine (G)**
- **Cytosine (C)**

The chemical structures of the nucleotides and bases are included in Chapter 5. A key point for present purposes is that the bases in the double helix are paired as shown in Figure 1.5B.

At any position on the paired strands of a DNA molecule, if one strand has an A, the partner strand has a T; and if one strand has a G, the partner strand has a C.

The pairing between A—T and G—C is said to be **complementary:** The complement

Figure 1.5
Molecular structure of a DNA double helix. (A) A "space-filling" model, in which each atom is depicted as a sphere. (B) A diagram highlighting the helical strands around the outside of the molecule and the A—T and G—C base pairs inside.

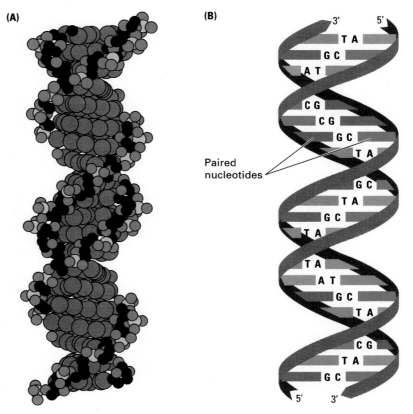

(A)

(B)

Paired nucleotides

of A is T, and the complement of G is C. The complementary pairing in the duplex molecule means that each base along one strand of the DNA is matched with a base in the opposite position on the other strand. Furthermore,

> Nothing restricts the sequence of bases in a single strand, so any sequence could be present along one strand.

This principle explains how only four bases in DNA can code for the huge amount of information needed to make an organism. It is the *sequence* of bases along the DNA that encodes the genetic information, and the sequence is completely unrestricted.

The complementary pairing is also called **Watson-Crick pairing.** In the three-dimensional structure (Figure 1.5A), the base pairs are represented by the spheres filling the interior of the double helix. The base pairs lie almost flat, stacked on top of one another perpendicular to the long axis of the double helix, like pennies in a roll. When discussing a DNA molecule, biologists frequently refer to the individual strands as **single-stranded DNA** and to the double helix as **double-stranded DNA** or **duplex DNA.**

Each DNA strand has a *polarity,* or directionality, like a chain of circus elephants linked trunk to tail. In this analogy, each elephant corresponds to one nucleotide along the DNA strand. The polarity is determined by the direction in which the nucleotides are pointing. The "trunk" end of the strand is called the *5' end* of the strand, and the "tail" end is called the *3' end.* In double-stranded DNA, the paired strands are oriented in opposite directions, the 5' end of one strand aligned with the 3' end of the other. The molecular basis of the polarity, and the reason for the opposite orientation of the strands in duplex DNA, is explained in Chapter 5.

Beyond the most optimistic hopes, knowledge of the structure of DNA immediately gave clues to its function:

1. The sequence of bases in DNA could be copied by using each of the separate "partner" strands as a pattern for the creation of a new partner strand with a complementary sequence of bases.

2. The DNA could contain genetic information in coded form in the sequence of bases, analogous to letters printed on a strip of paper.

3. Changes in genetic information (mutations) could result from errors in copying in which the base sequence of the DNA became altered.

In the remainder of this chapter, some of the implications of these clues are discussed.

1.3 An Overview of DNA Replication

In their first paper on the structure of DNA, Watson and Crick remarked that "it has not escaped our notice that the specific base pairing we have postulated immediately suggests a copying mechanism for the genetic material." The copying mechanism they had in mind is illustrated in Figure 1.6; the process is now called **replication.** In replication, the strands of the original (parent) duplex separate, and each individual strand serves as a pattern, or **template,** for the synthesis of a new strand (replica). The replica strands are synthesized by the addition of successive nucleotides in such a way that each base in the replica is complementary (in the Watson-Crick pairing sense) to the base across the way in the template strand. Although the model in Figure 1.6 is simple in principle, it is a complex process with chemical and geometrical problems that require a large number of enzymes and other proteins to resolve. The details are discussed in Chapter 5. For purposes of this overview, the important point is that the replication of a duplex molecule results in two duplex daughter molecules, each with a sequence of nucleotides identical to the parental strand.

In Figure 1.6A, the backbones in the parental DNA strands and those in the newly synthesized strand are shown in contrasting colors. In the process on the left, the top strand is the template present in the parental molecule, and the bottom strand is the newly synthesized partner. In

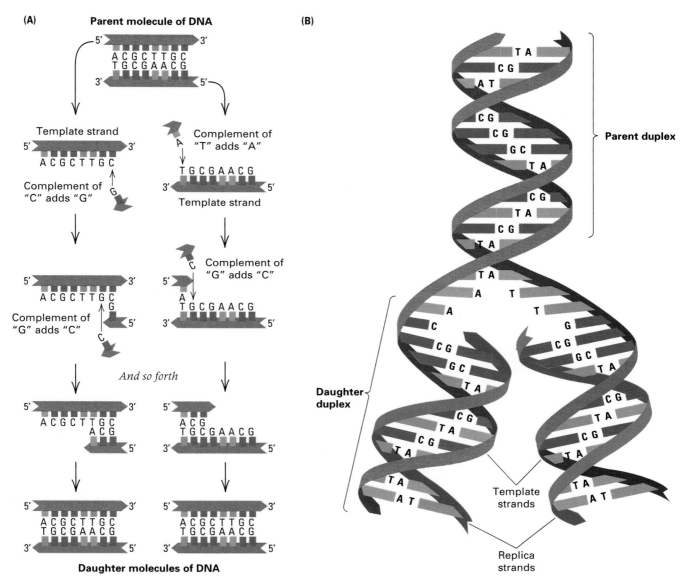

(A)

Parent molecule of DNA

5′   3′
A C G C T T G C
T G C G A A C G
3′   5′

Template strand

5′   3′
A C G C T T G C

Complement of
"C" adds "G"

5′   3′
A C G C T T G C

Complement of
"G" adds "C"

And so forth

5′   3′
A C G C T T G C
       A C G
5′

5′   3′
A C G C T T G C
T G C G A A C G
3′   5′

Complement of
"T" adds "A"

T G C G A A C G
3′   5′
Template strand

Complement of
"G" adds "C"

5′
T G C G A A C G
3′   5′

5′
A C G
T G C G A A C G
3′   5′

5′   3′
A C G C T T G C
T G C G A A C G
3′   5′

Daughter molecules of DNA

(B)

Parent duplex

Daughter
duplex

Template
strands

Replica
strands

Figure 1.6

Replication of DNA. (A) Each of the parental strands serves as a template for the production of a complementary daughter strand, which grows in length by the successive addition of single nucleotides. (B) Replication in a long DNA duplex as originally proposed by Watson and Crick. As the parental strands separate, each parental strand serves as a template for the formation of a new daughter strand by means of A−T and G−C base pairing.

the process on the right, the bottom strand is the template from the parental molecule, and the top strand is the newly synthesized partner. How the process of replication occurs in a long duplex molecule is shown in Figure 1.6B. The separation of the parental strands and the synthesis of the daughter strands take place simultaneously in different parts of the molecule. In each successive region along the parental duplex, as the parental strands come apart, each of the separated parental strand serves as a template for the synthesis of a new daughter strand.

1.4
Genes and Proteins

By the beginning of the twentieth century, it had already become clear that proteins were responsible for most of the metabolic activities of cells. Proteins were known to be essential for the breakdown of organic molecules to generate the chemical energy needed for cellular activities. They were also known to be required for the assembly of small molecules into more complex molecules and cellular structures. In 1878, the term **enzyme** was introduced to refer to the biological catalysts that accelerate biochemical reactions in cells. By 1900, owing largely to the genius of the German biochemist Emil Fischer, enzymes had been shown to be proteins. Other proteins are key components of cells; for example, structural proteins give the cell form and mobility, other proteins form pores in the cell membrane and control the traffic of small molecules into and out of the cell, and still other proteins regulate cellular activities in response to molecular signals from the external environment or from other cells.

In 1908, the British physician Archibald Garrod had an important insight into the relationship between enzymes and disease:

> Any hereditary disease in which cellular metabolism is abnormal results from an inherited defect in an enzyme.

Such hereditary diseases became known as **inborn errors of metabolism,** a term still in use today. Although the full implications of Garrod's suggestion could not be explored experimentally until many years afterward, the coupling of the concepts of inheritance (gene) with enzyme (protein) was a brilliant simplification of the problem of biochemical genetics because it put the emphasis on the question "How do genes control the structure of proteins?" How biologists pursued this question is summarized in the following sections.

Transcription of DNA Makes RNA

Watson and Crick were quite right in suggesting that the genetic information in DNA is contained in the sequence of bases; it is encoded in a manner analogous to letters (the bases) printed on a strip of paper. However, learning the details of the genetic code and the manner in which it is deciphered took about 20 years of additional work. The long series of investigations showed that in a region of DNA that directs the synthesis of a protein, the genetic code for the protein is contained in a DNA strand. The coded genetic information in this strand is decoded in a linear order in which each successive "word" in the DNA strand specifies the next chemical subunit to be added to the protein as it is being made. The protein subunits are called **amino acids.** Each "word" in the genetic code consists of three adjacent bases.

For example, the base sequence ATG in a DNA strand specifies the amino acid methionine (Met), TTT specifies phenylalanine (Phe), GGA specifies glycine (Gly), and GTG specifies valine (Val). How the genetic information is transferred from the base sequence of a DNA strand into the amino acid sequence of the corresponding protein is shown in Figure 1.7. This scheme, in which DNA codes for RNA and RNA codes for proteins, is known as the **central dogma** of molecular genetics. (The term *dogma* means a set of beliefs. The term dates from the time when the idea was first advanced as a theory; since then, the "dogma" has been confirmed experimentally, but the term persists.)

The main concept in the central dogma is that DNA does not code for protein directly but acts through an intermediary molecule called **ribonucleic acid (RNA).** The structure of RNA is similar, but not identical, to that of DNA. The sugar is ribose rather than deoxyribose. RNA is usually single-stranded (not a duplex), and RNA contains a base, **uracil (U),** that takes the place of thymine (T) in DNA (Chapter 5). In the synthesis of proteins, there are actually three types of RNA that participate and that play different roles:

- A **messenger RNA (mRNA),** which carries the genetic information from DNA and is used as a template for protein synthesis.

- The **ribosomal RNA (rRNA),** which is a major constituent of the cellular parti-

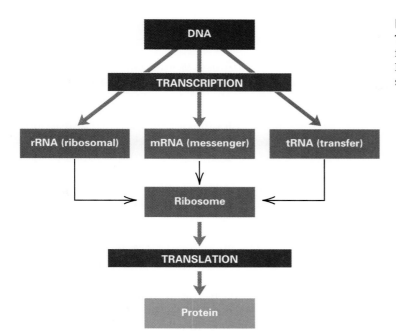

Figure 1.7
The "central dogma" of molecular genetics: DNA codes for RNA, and RNA codes for protein. The DNA → RNA step is transcription, and the RNA → protein step is translation.

cles called **ribosomes** on which protein synthesis actually takes place.

- A set of **transfer RNA (tRNA)** molecules, each of which incorporates a particular amino acid subunit into the growing protein when it recognizes a specific group of three adjacent bases in the mRNA.

Why on Earth should a process as functionally simple as DNA coding for protein have the additional complexity of RNA intermediaries? Certain biochemical features of RNA suggest a hypothesis: that RNA played a central role in the earliest forms of life and that it became locked into the processes of information transfer and protein synthesis. So it remains today: The participation of RNA in protein synthesis is a relic of the earliest stages of evolution—a "molecular fossil." The hypothesis that the first forms of life used RNA both for carrying information (in the base sequence) and as catalysts (accelerating chemical reactions) is supported by a variety of observations. Two examples: (1) DNA replication requires an RNA molecule in order to get started (Chapter 5), and (2) some RNA molecules act to catalyze biochemical reactions important in protein synthesis (Chapter 10). In the later evolution of the early life forms, additional complexity could have been added. The function of information storage and replication could have been transferred from RNA to DNA, and the function of RNA catalysis in metabolism could have been transferred from RNA to protein by the evolution of RNA-directed protein synthesis.

The manner in which genetic information is transferred from DNA to RNA is straightforward (Figure 1.8). The DNA opens up, and one of the strands is used as a template for the synthesis of a complementary strand of RNA. (How the template strand is chosen is discussed in Chapter 10.) The process of making an RNA strand from a DNA template is **transcription,** and the RNA molecule that is made is the **transcript.** The base sequence in the RNA is complementary (in the Watson-Crick pairing sense) to that in the DNA template, except that U (which pairs with A) is present in the RNA in place of T. The base-pairing rules between DNA and RNA are summarized in Figure 1.9. Like DNA, an RNA strand also has a polarity, exhibiting a 5' end and a 3' end determined by the orientation of the nucleotides. The 5' end of the RNA transcript is synthesized first and, in the RNA−DNA duplex formed in transcription, the polarity of the RNA strand is opposite to that of the DNA strand.

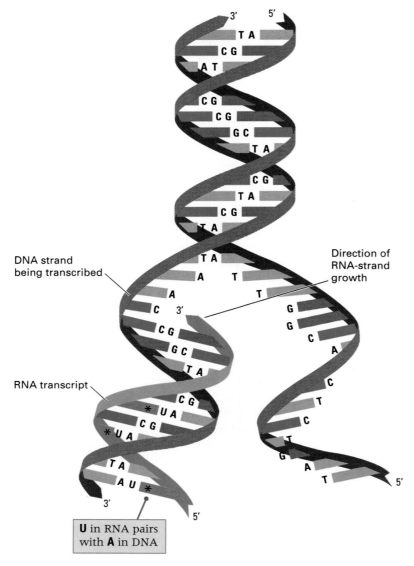

DNA strand being transcribed

Direction of RNA-strand growth

RNA transcript

U in RNA pairs with **A** in DNA

Figure 1.8

Transcription is the production of an RNA strand that is complementary in base sequence to a DNA strand. In this example, the DNA strand at the bottom left is being transcribed into a strand of RNA. Note that in an RNA molecule, the base U (uracil) plays the role of T (thymine) in that it pairs with A (adenine). Each A−U pair is marked.

Figure 1.9

Pairing between bases in DNA and in RNA. The DNA bases A, T, G, and C pair with the RNA bases U, A, C, and G, respectively.

	Adenine	Thymine	Guanine	Cytosine
Base in DNA template	A	T	G	C
Base in RNA transcript	U	A	C	G
	Uracil	Adenine	Cytosine	Guanine

Translation of RNA Makes Protein

The synthesis of a protein under the direction of an mRNA molecule is **translation.** Although the sequence of mRNA bases codes for the sequence of amino acids, the molecules that actually do the "translation" are the tRNA molecules. The mRNA molecule is translated in groups of three bases called **codons.** For each codon in the mRNA, there is a tRNA molecule that contains a complementary group of three adjacent bases that can pair with those in the codon. At each step in protein synthesis, when the proper tRNA with an attached amino acid comes into line along the mRNA, the incomplete protein chain is attached to the amino acid on the tRNA, increasing the length of the protein chain by one amino acid. When the next tRNA comes into line, the protein chain is detached from the previous tRNA and attached to the amino acid of the next in line, again increasing the length of the protein chain by one amino acid. A protein is therefore synthesized in a stepwise manner, one amino acid at a time. By way of analogy, the process of protein synthesis by the addition of consecutive amino acids is like the construction of a chain of pop-together plastic beads by the addition of consecutive beads.

The role of tRNA in translation is illustrated in Figure 1.10 and can be described as follows:

> The mRNA is read codon by codon. Each codon specifying an amino acid matches with a complementary group of three adjacent bases in a single tRNA molecule, which brings the correct amino acid into line.

The tRNA molecules used in translation do not line up along the mRNA simultaneously as shown in Figure 1.10. The process of translation takes place on a ribosome, which combines with a single mRNA molecule and moves along it from one end (the 5' end) to the other (the 3' end) in steps of three adjacent nucleotides (codon by codon). As each new codon comes into place, the correct tRNA attaches to the ribosome, and the growing chain of amino

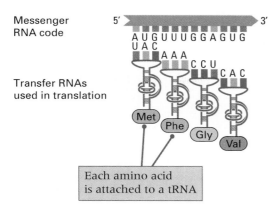

Messenger RNA code

Transfer RNAs used in translation

Each amino acid is attached to a tRNA

Figure 1.10
The role of transfer RNA in the synthesis of proteins. The sequence of bases in the messenger RNA determines the order in which transfer RNA molecules are lined up. Each group of three adjacent bases in the messenger RNA attracts a transfer RNA containing a complementary sequence of three bases. Each transfer RNA molecule carries a particular amino acid, and the amino acids in the protein join together in the same order in which the transfer RNA molecules line up along the messenger RNA. Because the transfer RNA molecules are aligned in this manner, the sequence of bases in the messenger RNA determines the sequence of amino acids in the protein. Polypeptide chains are synthesized by the sequential addition of amino acids, one at a time. As each transfer RNA molecule is brought into line, the incomplete polypeptide chain grows one amino acid longer by becoming attached to the amino acid linked to the transfer RNA.

acids becomes attached to the amino acid on the tRNA. As the ribosome moves along the mRNA, successive amino acids are added to the growing chain until any one of three particular codons specifying "stop" is encountered. At this point, synthesis of the chain of amino acids is finished, and the protein is released from the ribosome. (Chapter 10 gives a more detailed treatment of translation.)

Technically speaking, the chain of amino acids produced in translation is a **polypeptide.** The distinction between a polypeptide and a protein is that a protein can consist of several polypeptide chains that come together after translation. Some proteins are composed of two or more identical polypeptide chains (encoded in the same gene); others are composed of two or more different polypeptide chains (encoded in different genes). For example, the protein **hemoglobin,** which is the oxygen-carrying protein in red blood cells, is composed of four polypeptide chains encoded in two different genes: Two of the chains are β polypeptide chains translated from the β-globin gene, and the other two are α polypeptide chains translated from the α-globin gene.

1.5 Mutation

Mutation means any heritable change in a gene. The Watson-Crick structure of DNA also suggested that, chemically speaking, a mutation is a change in the sequence of bases along the DNA. The change may be simple, such as the substitution of one pair of bases in a duplex molecule for a different pair of bases. For example, an A—T pair in a duplex molecule may mutate to either T—A, C—G, or G—C. The change in base sequence may also be more complex, such as the deletion or addition of base pairs. These and other types of mutations are discussed in Chapter 13.

One possible consequence of a mutation is illustrated in Figure 1.11. Part A shows a region of duplex DNA and the mRNA transcribed from the bottom strand. The tRNA molecules used in translation result in the amino acid sequence

Met—Phe—Gly—Val

What happens if the T—A base pair marked with the "sunburst" mutates to become a C—G base pair? The result is shown in part B. The second codon in the mRNA is now CUU, which codes for leucine (Leu), instead of the codon UUU, which codes for phenylalanine (Phe). In translation, the CUU codon in the mRNA combines with the leucine-bearing tRNA, and the result is the mutant amino acid sequence

Met—Leu—Gly—Val

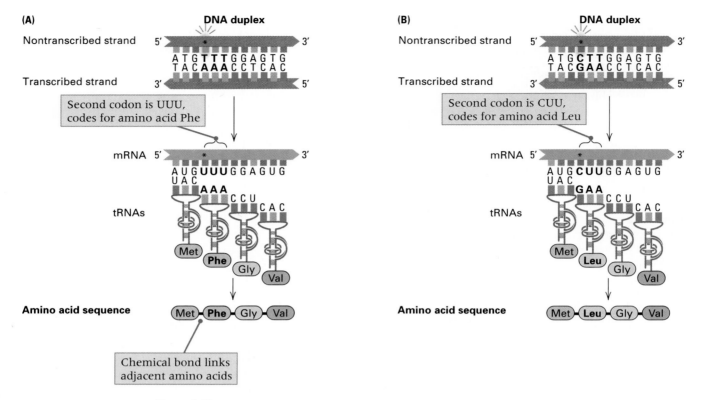

(A)

Nontranscribed strand

Transcribed strand

DNA duplex

ATG**TTT**GGAGTG
TAC**AAA**CCTCAC

Second codon is UUU, codes for amino acid Phe

mRNA 5′ ... 3′

AUG**UUU**GGAGUG
UAC
AAA CCU CAC

tRNAs

Met Phe Gly Val

Amino acid sequence

Met Phe Gly Val

Chemical bond links adjacent amino acids

(B)

Nontranscribed strand

Transcribed strand

DNA duplex

ATG**CTT**GGAGTG
TAC**GAA**CCTCAC

Second codon is CUU, codes for amino acid Leu

mRNA 5′ ... 3′

AUG**CUU**GGAGUG
UAC
GAA CCU CAC

tRNAs

Met Leu Gly Val

Amino acid sequence

Met Leu Gly Val

Figure 1.11

A mutation is a change in base sequence in the DNA. Any mutation that causes the insertion of an incorrect amino acid in a protein can impair the function of the protein. (A) The DNA molecule is transcribed into a messenger RNA that codes for the sequence of amino acids Met−Phe−Gly−Val. In the DNA molecule, the marked T−A base pair results in the initial U in the messenger RNA codon UUU for Phe (phenylalanine). (B) Substitution of a C−G base pair for the normal T−A base pair results in a messenger RNA containing the codon CUU instead of UUU. The CUU codon codes for Leu (leucine), which therefore replaces Phe in the mutant polypeptide chain.

1.6
How Genes Determine Traits

We have seen that the key principle of molecular genetics is the central dogma:

> The sequence of nucleotides in a gene specifies the sequence of amino acids in a protein using messenger RNA as the intermediary molecule in the coding process.

It is one of the ironies of genetics (and a consequence of the biochemical complexity of organisms) that whereas the connection between genes and proteins is conceptually simple, the connection between genes and traits is definitely not

simple. Most visible traits of organisms are the net result of many genes acting together in combination with environmental factors. Therefore, the relationship between genes and traits is often complex for one or more of the following reasons:

1. One gene can affect more than one trait.

2. One trait can be affected by more than one gene.

3. Many traits are affected by environmental factors as well as by genes.

Now let us examine each of these principles, with examples.

Pleiotropy: One Gene Can Affect More Than One Trait

A mutant gene may affect a number of seemingly unrelated traits. The mutation is then said to show **pleiotropy,** and the various manifestations of the mutation are known as **pleiotropic effects.** An example of a mutation in human beings with manifold pleiotropic effects is **sickle-cell anemia,** which affects the major oxygen-carrying protein of the red blood cells. The major organs and organ systems affected by the pleiotropic effects of the mutation are shown in Figure 1.12.

The underlying mutation in sickle-cell anemia is in the gene for β-globin, which codes for the β polypeptide chains present in the oxygen-carrying protein of red blood cells. The molecular basis of the disease is shown in Figure 1.13. Figure 1.13A shows the region of the β-globin gene that codes for amino acids 5 through 8 (the complete polypeptide chain is 146 amino acids in length). The sickle-cell mutation changes the base pair marked with the sunburst. As shown in part B, the mutant form of the gene contains a T—A base pair instead of the normal A—T base pair. As a result of the mutation, the mRNA contains a GUG codon instead of the normal GAG codon. Because GUG codes for valine (Val), this amino acid is incorporated into the polypeptide in place of the normal glutamic acid (Glu) at position number 6.

The defective β polypeptide chain gives the hemoglobin protein a tendency to form long, needle-like polymers. Red blood cells in which polymerization happens become deformed into crescent, sickle-like shapes. Some of the deformed red blood cells are destroyed immediately (reducing the oxygen-carrying capacity of the blood and causing the anemia), whereas others may clump together and clog the blood circulation in the capillaries.

The consequences of the Glu → Val replacement are a profound set of pleiotropic effects. All of these effects are related to the breakdown of red blood cells, to the decreased oxygen-carrying capacity of the blood, or to physiological adjustments the body makes to try to compensate for the disease (such as enlargement of the spleen). Patients with sickle-cell anemia

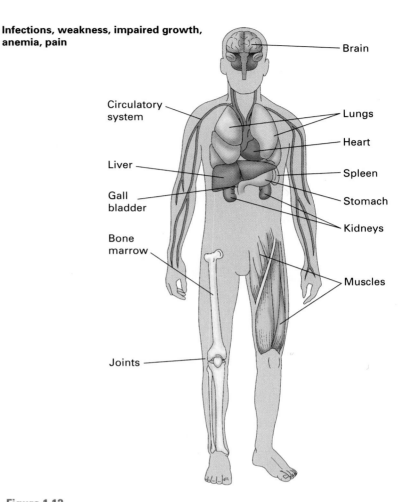

Figure 1.12

Sickle-cell anemia has multiple, seemingly unrelated symptoms known as pleiotropic effects. The primary defect is a mutant form of hemoglobin in the blood. The resulting destruction of red blood cells and the impaired ability of the blood to carry oxygen affect the circulatory system, bone marrow, muscles, brain, and virtually all major internal organs. The symptoms of the disease are anemia, recurrent pain, weakness, susceptibility to infections, and slowed growth.

suffer bouts of severe pain. The anemia causes impaired growth, weakness, and jaundice. Affected people are so generally weakened that they are susceptible to bacterial infections, which are the most common cause of death in children with the disease.

Although sickle-cell anemia is a severe genetic disease that often results in premature death, it is relatively frequent in areas of Africa and the Middle East in which a type of malaria caused by the protozoan parasite *Plasmodium falciparum* is widespread. The association between sickle-cell anemia and malaria is not coincidental. The

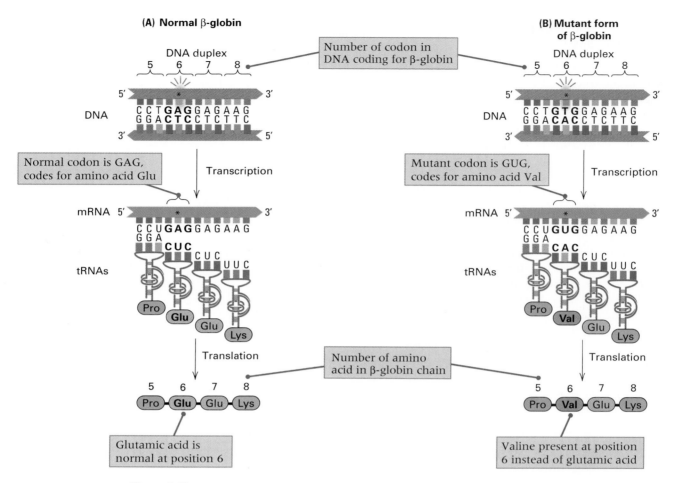

(A) Normal β-globin

Number of codon in DNA coding for β-globin

(B) Mutant form of β-globin

DNA duplex

5 6 7 8

DNA

5' C C T **GAG** G A G A A G 3'
3' G G A **CTC** C T C T T C 5'

Normal codon is GAG, codes for amino acid Glu

Transcription

mRNA 5' C C U **GAG** G A G A A G 3'
           G G A **CUC** C U C U U C

tRNAs

Pro Glu Glu Lys

Translation

5   6   7   8
Pro Glu Glu Lys

Glutamic acid is normal at position 6

DNA duplex

5 6 7 8

DNA

5' C C T **GTG** G A G A A G 3'
3' G G A **CAC** C T C T T C 5'

Mutant codon is GUG, codes for amino acid Val

Transcription

mRNA 5' C C U **GUG** G A G A A G 3'
           G G A **CAC** C U C U U C

tRNAs

Pro Val Glu Lys

Translation

5   6   7   8
Pro Val Glu Lys

Valine present at position 6 instead of glutamic acid

Number of amino acid in β-globin chain

Figure 1.13

Genetic basis of sickle-cell anemia: (A) Part of the DNA in the normal β-globin gene is transcribed into a messenger RNA coding for the amino acid sequence Pro−Glu−Glu−Lys. The T in the marked A−T base pair is transcribed as the A in the GAG codon for Glu (glutamic acid). (B) Mutation of the normal A−T base pair to a T−A base pair results in the codon GUG instead of GAG. The codon GUG codes for Val (valine), so the polypeptide sequence in this part of the molecule is Pro−Val−Glu−Lys. The resulting hemoglobin is defective and tends to polymerize at low oxygen concentration.

association results from the ability of the mutant β-hemoglobin to afford some protection against malarial infection. In the life cycle of the parasite, it passes from a mosquito to a human through the mosquito's bite. The initial stages of infection take place in cells in the liver, where specialized forms of the parasite are produced that are able to infect and multiply in red blood cells. Widespread infection of red blood cells impairs the ability of the blood to carry oxygen, causing the weakness, anemia, and jaundice characteristic of malaria. In people with the mutant β-hemoglobin, however, it is thought that the infected blood cells undergo sickling and are rapidly removed from circulation. The prolifera-

tion of the parasite among the red blood cells is thereby checked, and the severity of the malarial infection is reduced. There is consequently a genetic balancing act between the prevalence of the genetic disease sickle-cell anemia and that of the parasitic disease malaria. If the mutant β-hemoglobin becomes too frequent, more lives are lost from sickle-cell anemia than are gained by the protection against malaria; on the other hand, if the mutant β-hemoglobin becomes too rare, fewer lives are lost from sickle-cell anemia but the gain is offset by more deaths from malaria. The end result of this kind of genetic balancing act is discussed in quantitative terms in Chapter 15.

Epistasis: One Trait Can Be Affected By More Than One Gene

Every trait requires numerous genes for its proper development, metabolism, and physiology. Consequently, one trait can be affected by more than one gene. An example of this principle is illustrated in Figure 1.14, which shows the effects of two genes that function in eye pigmentation in *Drosophila*. The genes are *vermilion (v)* and *cinnabar (cn)*. These genes encode enzymes, denoted V and Cn, respectively, that are used in the **biochemical pathway** that converts the amino acid tryptophan into the brown eye pigment *xanthommatin* through a series of intermediate substances I1, I2, and so forth (Figure 1.14A). Each step in the pathway is catalyzed by a different enzyme encoded by a different gene. The nonmutant, or **wildtype,** eye color of *Drosophila* is a brick-like red because the

Figure 1.14

The *Drosophila* mutants *vermilion* and *cinnabar* exemplify epistasis between mutant genes affecting eye color. (A) Metabolic pathway for the production of the brown pigment xanthommatin. The intermediate substances are denoted I1, I2, and so forth, and each single arrow represents one step in the pathway. (The multiple arrows at the end represent an unspecified number of steps.) (B) The *cn* gene codes for an enzyme, Cn, that converts I2 to I3. In flies mutant for *cn*, the pathway is blocked at this step. (C) The *v* gene codes for a different enzyme, V, that catalyzes the conversion of tryptophan into intermediate I1. In flies mutant for *v*, the pathway is blocked at this step. (D) In *v cn* double mutants, the pathway is blocked at the earlier step, in this case the conversion of tryptophan to I1.

(A) Wildtype: pathway for synthesis of xanthommatin

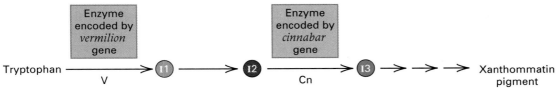

Different pathway produces *drosopterin* pigment.

Wildtype eye color

(B) Mutation in *cinnabar (cn):* biochemical pathway blocked at *cinnabar* step

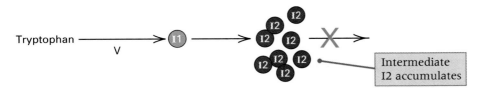

Intermediate I2 accumulates

Bright red eye

(C) Mutation in *vermilion (v):* biochemical pathway blocked at *vermilion* step

Bright red eye

(D) Mutations in both *vermilion (v)* and *cinnabar (cn):* pathway blocked at *vermilion* step

Bright red eye

pigment cells contain not only xanthommatin but also a bright red pigment called *drosopterin* synthesized by a different biochemical pathway.

As indicated in Figure 1.14B, flies that are mutant for *cn* lack xanthommatin. They have bright red eyes because of the drosopterin. Flies mutant for *cn* have a nonfunctional Cn enzyme, so the pathway is blocked at the step at which Cn should function. Because there is no functional Cn enzyme to convert intermediate I2 into the next intermediate along the way, I2 accumulates in *cn* flies. Mutant *v* flies also lack xanthommatin but for a different reason (Figure 1.14C). In these flies the pathway is blocked because there is no functional V enzyme. It does not matter whether the Cn enzyme is present, because without the V enzyme, there is no I2 for Cn to work on.

The pathway in flies with a mutation in both *v* and *cn* is illustrated in Figure 1.14D. The situation is identical to that in flies with a *v* mutation only because, lacking functional V enzyme, the pathway is blocked at this step. The general term for gene interaction is **epistasis.** Freely translated from the Greek, epistasis means "standing over." Epistasis means that the presence of one mutation "stands over,"or conceals, the effects of a different mutation. In the example in Figure 1.14, we would say that *v* is *epistatic* to *cn*, because in flies with a *v* mutation, it is impossible to determine from the status of the xanthommatin pathway whether the *cn* gene is mutant or wildtype. The converse is not true: In flies with a *cn* mutation, the presence or absence of intermediate I2 shows whether the *v* gene is mutant or wildtype. If I2 accumulates, the V enzyme must be present (and the *v* gene wildtype); whereas if I2 is absent, the V enzyme must be nonfunctional (and the *v* gene mutant.)

The example in Figure 1.14 also illustrates an important feature of genetic terminology. Although both *vermilion* and *cinnabar* are needed for the synthesis of the *brown* pigment, the names of the genes are shades of *bright red*. At first this seems illogical, but *mutant genes are named for their effects on the organism.* Because mutations in either *vermilion* or *cinnabar* result in bright red eyes, the gene names make sense even though the products of both genes function in the brown-pigment pathway.

Effects of the Environment

Genes and environment also interact. To appreciate the interaction between genes and environment, consider the trait "anemia," which refers to a generalized weakness resulting from an insufficient number of red blood cells or from an inadequate volume of blood. There are many different types of anemia. Some forms of anemia are genetically determined, such as sickle-cell anemia (Figure 1.13). Other forms of anemia are caused by the environment; an example is anemia resulting from chronic deficiency of dietary iron or from infection with malaria. Still other forms of anemia are caused by genetic and environmental factors acting together. For example, people with a mutant form of the enzyme **glucose-6-phosphate dehydrogenase (G6PD),** an enzyme important in maintaining the integrity of the membrane of red blood cells, become severely anemic when they eat fava beans, because a substance in the beans triggers destruction of red cells. Because of its association with fava beans, the disease is called **favism,** but a more common name is **G6PD deficiency.** Red-cell destruction in people with G6PD deficiency can also be triggered by various chemicals such as naphthalene (used in mothballs) as well as by certain antibiotics and other drugs. G6PD deficiency, which affects primarily males, has a relatively high frequency in populations in coastal regions around the Mediterranean Sea. It is thought that the defect in the red blood cells may increase resistance to malaria.

With these examples as background, consider this question: Is anemia caused by heredity or environment? There is no simple answer. As we have seen, a complex trait such as anemia has many possible causes. Some types are genetically determined, some environmental in origin, and some require both genes and environment for their expression. The genes-versus-environment issue is exceptionally clear in the example of anemia only because various forms of the disorder have already been sorted out and assigned causes, whether they be genetic or environmental or both. However, before the various forms were distinguished, anemia was regarded as a tremendously complex condition, and all varieties were lumped together. Without separating the disorder into categories, all that one could conclude was that family history seemed to be important in some cases, but not all, and that the environment certainly played a role as well.

Most complex traits are analogous to anemia in consisting of different conditions lumped together because of their overall similarity. A familiar example is heart disease. It is well known that inherited risk factors in heart disease are related to the metabolism of saturated fats and cholesterol. Some rare forms of the disease with a strong genetic component have already been identified. There are also environmental risk factors in heart disease—cigarette smoking, being overweight, lack of exercise, high dietary intake of saturated fats and cholesterol, and so forth. In the population as a whole, the overall risk of heart disease is determined by both genetic and environmental factors, and some of the factors act synergistically, which means that the risk from two factors together is greater than would be predicted from the risk of each factor considered by itself.

The example of heart disease also illustrates that genetic and environmental effects can be offsetting. For example, a person with a family history of heart disease can considerably mitigate the risk by careful diet, exercise, abstention from smoking, and other behaviors. Taking drugs to control high blood pressure is also an example of an environmental intervention that reduces the overall risk of heart disease.

Heart disease is a typical example of a complex trait influenced by multiple genes as well as by many environmental factors. Most of the variation found in human beings falls into this category, including personality and other behavioral characteristics. Some traits are more strongly influenced by genetic factors than others, and it is extremely difficult to sort out the forms of a trait that might share a single cause. An illustration of complex genetic and environmental causation is shown in Figure 1.15. The boxes labeled mild, moderate, and so forth represent various different severities in which a trait can be expressed; these are analogous to the different forms of anemia. Across the top are three genes and three environmental factors that influence the trait. The heavy lines represent major effects, the thin lines minor effects. If the four types of expression of the trait

Figure 1.15

Most complex traits are affected by multiple genetic and environmental factors, not all of them equal in influence. In this example, the severity of expression of a complex disease is affected by three genes (1, 2, 3) and three environmental factors (X, Y, Z) that, in various combinations, determine the particular manner in which the disease will be expressed. Heavy arrows depict major influences, light arrows minor influences. For example, the mild expression of the disease is determined primarily by gene 1 with a minor influence of environmental factor X. The moderate expression of the disease is determined by two genes (2 and 3) and two environmental factors (X having a major effect and Y a minor effect). In a complex trait, therefore, some forms of expression of the disease (mild in this example) may have a relatively simple form of genetic causation, whereas other forms of the same disease (moderate in this example) may have a more complex causation that even includes different genes. The genetic basis of such diseases is difficult to determine unless the different forms of the disease can be distinguished.

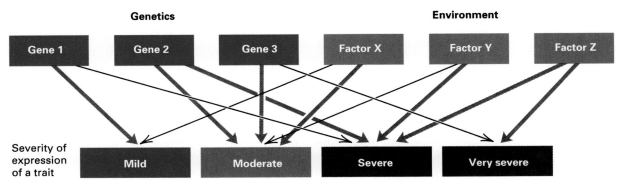

were regarded as a single entity without being distinguished, then the genetic and environmental causation could be characterized only as "three genes and three environmental factors, each with major effects." However, when the different levels of severity of the trait are considered separately, the situation can be clarified. For example, the mild form is determined by one major genetic factor and one minor environmental factor, and the very severe form is determined by one minor genetic factor and one major environmental factor. Real complexity remains in the moderate and severe forms, however: The moderate form is determined by two major genes together with one major and one minor environmental factor, and the severe form is determined by two major environmental factors together with one major and one minor genetic factor. Figure 1.15 also illustrates the more general point that traits do not present themselves already classified in the most informative manner. Progress in genetics has often resulted from the proper subdivision of a complex trait into distinct types that differ in their genetic or environmental causation.

1.7 Evolution

One of the remarkable discoveries of molecular genetics is that organisms that seem very different (for example, plants and animals) share many common features in their genetics and biochemistry. These similarities indicate a fundamental "unity of life":

> All creatures on Earth share many features of the genetic apparatus, including genetic information encoded in the sequence of bases in DNA, transcription into RNA, and translation into protein on ribosomes via transfer RNAs. All creatures also share certain characteristics in their biochemistry, including many enzymes and other proteins that are similar in amino acid sequence.

The Molecular Continuity of Life

The molecular unity of life comes about because all creatures share a common origin through **evolution,** the process by which populations of organisms that are descended from a common ancestor gradually become more adapted to their environment and sometimes split into separate species. In the evolutionary perspective, the unity of fundamental molecular processes is derived by inheritance from a distant common ancestor in which many mechanisms were already in place.

Not only the unity of life but also many other features of living organisms become comprehensible from an evolutionary perspective. The importance of the evolutionary perspective in understanding aspects of biology that seem pointless or needlessly complex is summed up in a famous aphorism of the evolutionary biologist Theodosius Dobzhansky: "Nothing in biology makes sense except in the light of evolution."

One indication of the common ancestry among Earth's creatures is illustrated in Figure 1.16. The tree of relationships was inferred from similarities in nucleotide sequence in a type of ribosomal RNA molecule common to all these organisms. Three major kingdoms of organisms are distinguished:

1. *Bacteria* This group includes most bacteria and cyanobacteria (formerly called blue-green algae). Cells of these organisms lack a membrane-bounded nucleus and mitochondria, are surrounded by a cell wall, and divide by binary fission.

2. *Archaea* This group was initially discovered among microorganisms that produce methane gas or that live in extreme environments, such as hot springs or high salt concentrations; they are widely distributed in more normal environments as well. Like Bacteria, the cells of Archaea lack internal membranes. DNA sequence analysis indicates that the machinery for DNA replication and transcription resembles that of Eukarya whereas metabolism in Archaea strongly resembles that of Bacteria. About half of their genes are unique to Archaea, however.

3. *Eukarya* This group includes all organisms whose cells contain an elaborate network of internal membranes, a membrane-bounded nucleus, and mito-

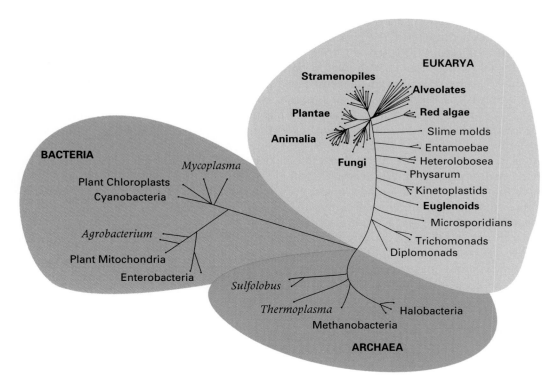

Figure 1.16
Evolutionary relationships among the major life forms as inferred from similarities in nucleotide sequence in an RNA molecule found in the small subunit of the ribosome. The three major kingdoms of Bacteria, Archaea, and Eukarya are apparent. Plants, animals, and fungi are more closely related to each other than to members of either of the other kingdoms. Note the diverse groups of undifferentiated, relatively simple organisms that diverged very early in the eukaryote lineage. [Courtesy of Mitchell L. Sogin.]

chondria. Their DNA is organized into true chromosomes, and cell division takes place by means of mitosis (discussed in Chapter 3). The eukaryotes include plants and animals as well as fungi and many single-celled organisms, such as amoebae and ciliated protozoa.

The Bacteria and Archaea are often grouped together into a larger assemblage called **prokaryotes,** which literally means "before [the evolution of] the nucleus." This terminology is convenient for designating prokaryotes as a group in contrast with **eukaryotes,** which literally means "good [well-formed] nucleus."

Adaptation and Diversity

Figure 1.16 illustrates the unity of life, but it also illustrates the diversity. Frogs are different from fungi, and beetles are different

from bacteria. As a human being, it is sobering to consider that complex, multicellular organisms came relatively late onto the evolutionary scene of life on Earth. Animals came later still and primates very late indeed. What about human evolution? In the time scale of Earth history, human evolution is a matter of a few million years—barely a snap of the fingers.

If common ancestry is the source of the unity of life, what is the source of diversity? Because differences among species are inherited, the original source of the differences must be mutation. However, mutations alone are not sufficient to explain why organisms are adapted to living in their environments—why ocean mammals have special adaptations that make swimming and diving possible, or why desert mammals have special adaptations that enable them to survive on minimal amounts of water. Mutations are chance events not directed toward any particular adaptive goal, like longer fur among mammals living in the Arctic. The process that accounts for adaptation was outlined by Charles Darwin in his 1859 book *On the Origin of Species*. Darwin proposed that adaptation is the result of **natural selection,** the process in which individual organisms that carry particular mutations or combinations of mutations that equip them to survive or reproduce more effectively in the prevailing environment will leave more offspring than other organisms and so contribute their favorable genes disproportionately to future generations. If this process is repeated throughout the course of many generations, the entire species becomes genetically transformed because a gradually increasing proportion of the population inherits the favorable mutations. The genetic basis of natural selection is discussed in Chapter 15.

The Role of Chance in Evolution

Natural selection is undoubtedly the key process in bringing about the genetic adaptation of organisms to their environments. Hence there is a great deal of appeal in being able to explain why particular traits are adaptive. Unfortunately, the ingenuity of the human imagination makes it all too easy to make up an adaptive story for any trait whatsoever. One example is the adaptive argument that the reason why blood is red is that seeing it scares one's enemies when one is injured. This explanation sounds almost plausible, but the truth is that blood is red for the same reason as rust; it contains oxidized iron. Each hemoglobin chain carries an atom of iron, and oxidized iron is red as a matter of physics, not biological evolution.

Made-up adaptive stories not supported by hard evidence are called **just-so stories** after the title (*Just So Stories*) of a 1902 book by Rudyard Kipling. The stories tell how animal traits came to be: how the elephant got its trunk (a crocodile caught a baby elephant by his nose and "pulled and pulled and pulled it out into a really truly trunk same as all elephants have today"); how the whale got his throat (because he swallowed a sailor who wedged a grate at the back of his throat that prevented him from eating anything except "very, very small fish—and this is the reason why whales nowadays never eat men or boys or little girls"); how the camel got his hump (a wizard cursed it on him for not doing his work); and so forth.

The rationale for inventing evolutionary just-so stories is the assumption that all traits are adaptive by necessity and one needs only to find a reason why. But this is not necessarily so. For example, some traits exist not because they are selectively advantageous in themselves but because they are pleiotropic effects of genes selected for other reasons. Chance may also play a large role in some major events in the history of life. Many evolutionary biologists now believe that a mass extinction was precipitated 65.3 million years ago when an asteroid smashed into the Pacific Ocean off the Yucatan Peninsula and spewed so much debris into the air that Earth went dark for years. The mass extinction triggered by this event was not by any means the largest in Earth history, but it led to the extinction of all dinosaur species and about 90 percent of other species. Until then, dinosaurs were a wonderfully diverse and well-adapted group of organisms. The demise of the dinosaurs made way for the evolutionary diversification and success of mammals, so one could argue that a chance asteroid impact explains in part why human beings are here.

Chapter Summary

Organisms of the same species have some traits (characteristics) in common, but they may differ from one another in innumerable other traits. Many of the differences between individual organisms result from genetic differences, the effects of the environment, or both. Genetics is the study of inherited traits, including those influenced in part by the environment. The elements of heredity consist of genes, which are transmitted from parents to offspring in reproduction. Although the sorting of genes in successive generations was first put into numerical form by Mendel, the chemical basis of genes was discovered by Miescher in the form of a weak acid now called deoxyribonucleic acid (DNA). However, experimental proof that DNA is the genetic material did not come until about the middle of the twentieth century.

The first convincing evidence of the role of DNA in heredity came from experiments of Avery, MacLeod, and McCarty, who showed that genetic characteristics in bacteria could be altered from one type to another by treatment with purified DNA. In studies of *Streptococcus pneumoniae,* they transformed mutant cells unable to cause pneumonia into cells that could by treatment with pure DNA from disease-causing forms. A second important line of evidence was the Hershey-Chase experiment, which showed that the T2 bacterial virus injects primarily DNA into the host bacterium (*Escherichia coli*) and that a much higher proportion of parental DNA, as compared with parental protein, is found among the progeny phage.

The three-dimensional structure of DNA, proposed in 1953 by Watson and Crick, gave many clues about the manner in which DNA functions as the genetic material. A molecule of DNA consists of two long chains of nucleotide subunits twisted around one another to form a right-handed helix. Each nucleotide subunit contains any one of four bases: A (adenine), T (thymine), G (guanine), or C (cytosine). The bases are paired in the two strands of a DNA molecule. Wherever one strand has an A, the partner strand has a T, and wherever one strand has a G, the partner strand has a C. The base pairing means that the two paired strands in a DNA duplex molecule have complementary base sequences along their lengths. The structure of the DNA molecule suggested that genetic information could be coded in DNA in the sequence of bases. Mutations (changes in the genetic material) could result from changes in the sequence of bases, such as by the substitution of one nucleotide for another or by the insertion or deletion of one or more nucleotides. The structure of DNA also suggested a mode of replication in which the two strands of the parental DNA molecule separate and each individual strand serves as a template for the synthesis of a new complementary strand.

Most genes code for proteins. More precisely stated, most genes specify the sequence of amino acids in a polypeptide chain. The transfer of genetic information from DNA into protein is a multistep process that includes several types of RNA (ribonucleic acid). Structurally, an RNA strand is similar to a DNA strand except that the "backbone" contains a different sugar (ribose instead of deoxyribose) and RNA contains the base uracil (U) instead of thymine (T). Also, RNA is usually present in cells in the form of single, unpaired strands. The initial step in gene expression is transcription, in which a molecule of RNA is synthesized that is complementary in base sequence to whichever DNA strand is being transcribed. In polypeptide synthesis, which takes place on a ribosome, the base sequence in the RNA transcript is translated in groups of three adjacent bases (codons). The codons are rec-

ognized by different types of transfer RNA (tRNA) through base pairing. Each type of tRNA is attached to a particular amino acid, and when a tRNA base-pairs with the proper codon on the ribosome, the growing end of the polypeptide chain is attached to the amino acid on the tRNA. There are special codons that specify the "start" and "stop" of polypeptide synthesis. The most probable reason why various types of RNA are an intimate part of transcription and translation is that the earliest forms of life used RNA for both genetic information and enzyme catalysis.

A mutation that alters one or more codons in a gene may change the amino acid sequence of the resulting protein synthesized in the cell. Often the altered protein is functionally defective, so an inborn error of metabolism results. The particular manner in which an inborn error of metabolism is expressed can be very complex, because metabolism consists of an intricate branching network of biochemical pathways. Most visible traits of organisms result from many genes acting together in combination with environmental factors. The relationship between genes and traits is often complex because (1) every gene potentially affects many traits (hence a gene may show pleiotropy), (2) every trait is potentially affected by many genes (hence two different genes may interact, or show epistasis), and (3) many traits are significantly affected by environmental factors as well as by genes. Many complex traits include unrecognized subtypes that differ in their genetic or environmental causation. Progress in genetics has often resulted from finding ways to distinguish the subtypes.

All living creatures are united by sharing many features of the genetic apparatus (for example, transcription and translation) and many metabolic features. The unity of life results from common ancestry and is one of the evidences for evolution. There is also great diversity among living creatures. The three major kingdoms of organisms are the bacteria (which lack a membrane-bounded nucleus), the archaea (which share features with both eukarya and bacteria but form a distinct group), and eukarya (all "higher" organisms, whose cells have a membrane-bounded nucleus containing DNA organized into discrete chromosomes). The bacteria and archaea collectively are often called prokaryotes.

The ultimate source of diversity among organisms is mutation. However, natural selection is the process by which mutations that are favorable for survival and reproduction are retained and mutations that are harmful are eliminated. Natural selection, first proposed by Charles Darwin, is therefore the primary mechanism by which organisms become progressively more adapted to their environments. It is not always easy to determine how (or whether) a particular trait is adaptive.

Key Terms

adenine (A)	base	colony
amino acid	biochemical pathway	complementary
Archaea	central dogma	cytosine (C)
Bacteria	chromosome	deoxyribonucleic acid (DNA)
bacteriophage	codon	double-stranded DNA

duplex DNA
enzyme
epistasis
Eukarya
eukaryote
evolution
favism
genes
genetics
glucose-6-phosphate dehydrogenase
G6PD
G6PD deficiency
guanine (G)
hemoglobin
inborn error of metabolism
just-so story

messenger RNA
mRNA
mutation
natural selection
nucleotide
phage
pleiotropic effect
pleiotropy
polypeptide
prokaryote
replication
ribonucleic acid
ribosomal RNA
ribosome
RNA
rRNA

sickle-cell anemia
single-stranded DNA
template
thymine (T)
trait
transcript
transcription
transfer RNA
transformation
translation
tRNA
uracil (U)
Watson-Crick pairing
wildtype

Review the Basics

- What is a trait? Give five examples of human traits. How could you determine whether each of these traits was genetically transmitted?

- How is it possible for a trait to be determined by *both* heredity and environment? Give an example of such a trait.

- How did understanding the molecular structure of DNA give clues to its ability to replicate, to code for proteins, and to undergo mutations?

- Why is pairing of complementary bases a key feature of DNA replication? What is the process of transcription and in what ways does it differ from DNA replication?

- How is the sequence of amino acids in a protein encoded in the sequence of nucleotides in a messenger RNA?

- What is "an inborn error of metabolism"? How did this concept serve as a bridge between genetics and biochemistry?

- What does it mean to say that any gene potentially affects more than one trait? What does it mean to say that one trait is potentially affected by more than one gene?

- If a species A is more closely related evolutionarily to species B than it is to species C, would you expect the DNA sequence of B to be closer to that of A or that of C? Why?

Guide to Problem Solving

Problem 1: A double-stranded DNA molecule has the sequence

```
5'-ATGCTTCATTTCAGCTCGAATTTTGCC-3'
3'-TACGAAGTAAAGTCGAGCTTAAAACGG-5'
```

When this molecule is replicated, what is the base sequence of the new partner strand that is synthesized to pair with the upper strand? What is the base sequence of the new partner strand that pairs with the lower strand?

Answer: Both new strands have a nucleotide sequence that follows the Watson-Crick base-pairing rules of A with T and G with C. Therefore, the newly synthesized partner of the upper strand has a base sequence identical to that of the old lower strand, including the same 3' → 5' polarity. Similarly, the newly synthesized partner of the lower strand has a base sequence identical to that of the old upper strand.

Problem 2: Certain enzymes isolated from bacteria can recognize specific, short DNA sequences in duplex DNA and cleave both strands. The enzyme *Alu*I is an example. It rec-

ognizes the sequence 5'-AGCT-3' in double-stranded DNA and cleaves both strands at the chemical bond connecting the G and C nucleotides. If the DNA duplex in Worked Problem 1 were cleaved with *Alu*I, what DNA fragments would result?

Answer: There is only one 5'-AGCT-3' site in the duplex, so both strands would be cleaved once at the position between the G and the C in this sequence. Each cleavage generates a new 5' end and a new 3' end, which must maintain the polarity of the strand cleaved. Therefore, the resulting double-stranded DNA fragments are

```
5'-ATGCTTCATTTCAG-3'      5'-CTCGAATTTTGCC-3'
3'-TACGAAGTAAAGTC-5'      3'-GAGCTTAAAACGG-5'
```

Problem 3: Suppose that one strand of the DNA duplex in Worked Problem 1 is transcribed from left to right as the molecule is illustrated. Which strand is the one transcribed? What is the sequence of the resulting transcript? (Hint: The 5' end of the RNA transcript is synthesized first.)

GeNETics on the web will introduce you to some of the most important sites for finding genetic information on the Internet. To complete the exercises below, visit the Jones and Bartlett home page at

http://www.jbpub.com/genetics

Select the link to *Genetics: Principles and Analysis* and then choose the link to *GeNETics on the web*. You will be presented with a chapter-by-chapter list of highlighted keywords.

GeNETics EXERCISES

Select the highlighted keyword in any of the exercises below, and you will be linked to a web site containing the genetic information necessary to complete the exercise. Each exercise suggests a specific, written report that makes use of the information available at the site. This report, or an alternative, may be assigned by your instructor.

1. James D. Watson once said that he and Francis Crick had no doubt that their proposed DNA structure was essentially correct, because the structure was so beautiful it had to be true. At an internet site accessed by the keyword **DNA,** you can view a large collection of different types of models of DNA structure. Some models highlight the sugar-phosphate backbones, others the A—T and G—C base pairs, still others the helical structure of double-stranded DNA. If assigned to do so, pick one of the models that appeals to you. Make a sketch of the model (or, alternatively, print the model), label the major components, and write a paragraph explaining why you find this representation appealing.

2. One of the first inborn errors of metabolism studied by Archibald Garrod (1902) was a condition called **alkaptonuria**. Use this keyword to learn about the symptoms of this condition and its molecular basis. What enzyme is defective in alkaptonuria? What substance is present in the urine of patients that causes it to turn dark upon standing? If assigned to do so, write a 200-word summary of what you have learned.

3. Perhaps surprisingly, the history of the bacteriophage T2 that figures so prominently in the experiments of Hershey and Chase is clouded in mystery. Use the keyword **T2** to learn what is known about its origin and the sleuthing required to find it out. If assigned, prepare a timeline (chronology) of T2 phage from the time of its first isolation (under a different name) and its passage from researcher to researcher until it received its "final" name, phage T2.

MUTABLE SITE EXERCISES

The Mutable Site Exercise changes frequently. Each new update includes a different exercise that makes use of genetics resources available on the World Wide Web. Select the **Mutable Site** for Chapter 1, and you will be linked to the current exercise that relates to the material presented in this chapter.

PIC SITE

The Pic Site showcases some of the most visually appealing genetics sites on the World Wide Web. To visit the showcase genetics site, select the **Pic Site** for Chapter 1.

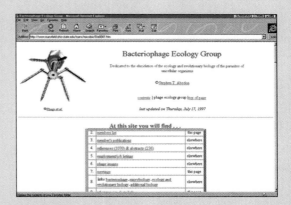

Answer: Because the 5' end of the RNA transcript is synthesized first, and transcription proceeds from left to right as the template molecule is drawn, the transcribed strand must be the lower strand. This is necessary so that the template DNA strand and the RNA transcript will have opposite polarities. In addition, the base uracil (U) in RNA replaces the base thymine (T) in DNA. Therefore, the RNA transcript (shown paired with the DNA template to illustrate the polarity relation) is

RNA 5'-AUGCUUCAUUUCAGCUCGAAUUUUGCC-3'
DNA 3'-TACGAAGTAAAGTCGAGCTTAAAACGG-5'

Problem 4: RNA transcripts can be translated *in vitro* using ribosomes, transfer RNAs, and other necessary constituents extracted from cells, but the first codon can be any sequence of three nucleotides (instead of AUG, which is used *in vivo*). A synthetic mRNA consists of the repeating tetranucleotide 5'-AUGC-3' and hence has the sequence

5'-AUGCAUGCAUGCAUGCAUGCAUGC · · · -3'.

When this molecule is translated *in vitro*, the resulting polypeptide has the repeating sequence

Met—His—Ala—Cys—Met—His—Ala—Cys · · ·

What does this result tell you about the number of nucleotides in a codon? Using the fact that the only methionine codon is 5'-AUG-3', deduce a codon for histidine (His), alanine (Ala), and cysteine (Cys). Would the result differ if the mRNA were translated from the 3' end to the 5' end instead of in the actual direction from the 5' and to the 3' end?

Answer: The result means that each codon consists of three nucleotides and that they are translated in nonoverlapping groups of three. A repeating sequence of four nucleotides repeats four codons; in this case, the sequence

5'-AUGCAUGCAUGCAUGCAUGCAUGCAUGC-3'

is translated by grouping into the codons

5'-AUG CAU GCA UGC AUG CAU GCA UGC AUG CAU GC-3'

(You should verify that it does not matter at which nucleotide in the mRNA the translation begins, because all three possible reading frames yield the same set of repeating codons.) Because 5'-AUG-3' codes for Met, it follows that 5'-CAU-3' codes for His (the next amino acid after Met), 5'-GCA-3' codes for Ala (the next in line), and 5'-UGC-3' codes for Cys. Translation of the RNA in the 3' → 5' direction is precluded, because in this direction, there is no AUG codon, so the resulting polypeptide could not contain methionine.

Analysis and Applications

1.1 Considering that favism is brought on by eating broad beans, would you consider this a "genetic" trait or a trait caused by the environment? Why?

1.2 What is the end result of replication of a duplex DNA molecule?

1.3 What is the role of the messenger RNA in translation? What is the role of the ribosome? What is the role of transfer RNA? Is there more than one type of ribosome? Is there more than one type of transfer RNA?

1.4 What important observation about S and R strains of *Streptococcus pneumoniae* prompted Avery, MacLeod, and McCarty to study this organism?

1.5 In the transformation experiments of Avery, MacLeod, and McCarty, what was the strongest evidence that the substance responsible for the transformation was DNA rather than protein?

1.6 What feature of the physical organization of bacteriophage T2 made it suitable for use in the Hershey-Chase experiments?

1.7 Although the Hershey-Chase experiments were widely accepted as proof that DNA is the genetic material, the results were not completely conclusive. Why not?

1.8 The DNA extracted from a bacteriophage contains 28 percent A, 28 percent T, 22 percent G, and 22 percent C. What can you conclude about the structure of this DNA molecule?

1.9 The DNA extracted from a bacteriophage consists of 24 percent A, 30 percent T, 20 percent G, and 26 percent C. What is unusual about this DNA? What can you conclude about its structure?

1.10 A double-stranded DNA molecule is separated into its constituent strands, and the strands are separated in an ultracentrifuge. In one of the strands the base composition is 24 percent A, 28 percent T, 22 percent G, and 26 percent C. What is the base composition of the other strand?

1.11 While studying sewage, you discover a new type of bacteriophage that infects *E. coli*. Chemical analysis reveals protein and RNA but no DNA. Is this possible?

1.12 One strand of a DNA duplex has the base sequence 5'-ATCGTATGCACTTTACCCGG-3'. What is the base sequence of the complementary strand?

1.13 A region along one strand of a double-stranded DNA molecule consists of tandem repeats of the trinucleotide 5'-TCG-3', so the sequence in this strand is

5'-TCGTCGTCGTCGTCG · · · -3'

What is the sequence in the other strand?

1.14 A duplex DNA molecule contains a random sequence of the four nucleotides with equal proportions of each. What is the average spacing between consecutive occurrences of the sequence 5'-GGCC-3'? Between consecutive occurrences of the sequence 5'-GAATTC-3'?

1.15 A region along a DNA strand that is transcribed contains no A. What base will be missing in the corresponding region of the RNA?

1.16 The duplex nucleic acid molecule shown here consists of a strand of DNA paired with a complementary strand of RNA. Is the RNA the top or the bottom strand? One of the base pairs is mismatched. Which pair is it?

5'- AUCGGUUACAUUCCGACUGA-3'
3'- TAGCCAATGTAAGGGTGACT-5'

1.17 The sequence of an RNA transcript that is initially synthesized is 5'-UAGCUAC-3', and successive nucleotides are added to the 3' end. This transcript is produced from a DNA strand with the sequence

3'-AAGTCGCATATCGATGCTAGCGCAACCT-5'

What is the sequence of the RNA transcript when synthesis is complete?

1.18 An RNA molecule folds back upon itself to form a "hairpin"structure held together by a region of base pairing. One segment of the molecule in the paired region has the base sequence 5'-AUACGAUA-3'. What is the base sequence with which this segment is paired?

1.19 A synthetic mRNA molecule consists of the repeating base sequence

5'-UUUUUUUUUUUU · · · ·-3'

When this molecule is translated *in vitro* using ribosomes, transfer RNAs, and other necessary constituents from *E. coli*, the result is a polypeptide chain consisting of the repeating amino acid Phe−Phe−Phe−Phe · · · ·. If you assume that the genetic code is a triplet code, what does this result imply about the codon for phenylalanine (Phe)?

1.20 A synthetic mRNA molecule consisting of the repeating base sequence

5'-UUUUUUUUUUUU · · · -3'

is terminated by the addition, to the right-hand end, of a single nucleotide bearing A. When translated *in vitro*, the resulting polypeptide consists of a repeating sequence of phenylalanines terminated by a single leucine. What does this result imply about the codon for leucine?

1.21 With *in vitro* translation of an RNA into a polypeptide chain, the translation can begin anywhere along the RNA molecule. A synthetic RNA molecule has the sequence

5'-CGCUUACCACAUGUCGCGAACUCG-3'

How many reading frames are possible if this molecule is translated *in vitro*? How many reading frames are possible if this molecule is translated *in vivo*, in which translation starts with the codon AUG?

1.22 You have sequenced both strands of a double-stranded DNA molecule. To inspect the potential amino acid coding content of this molecule, you conceptually transcribe it into RNA and then conceptually translate the RNA into a polypeptide chain. How many reading frames will you have to examine?

1.23 A synthetic mRNA molecule consists of the repeating base sequence 5'-UCUCUCUCUCUCUCUC · · · -3'. When this molecule is translated *in vitro*, the result is a polypeptide chain consisting of the alternating amino acids Ser−Leu−Ser−Leu−Ser−Leu · · · . Why do the amino acids alternate? What does this result imply about the codons for serine (Ser) and leucine (Leu)?

1.24 A synthetic mRNA molecule consists of the repeating base sequence 5'-AUCAUCAUCAUCAUC · · · -3'. When this molecule is translated *in vitro*, the result is a mixture of three different polypeptide chains. One consists of repeating isoleucines (Ile−Ile−Ile−Ile · · ·), another of repeating serines (Ser−Ser−Ser−Ser · · ·), and the third of repeating histidines (His−His−His−His · · ·). What does this result imply about the manner in which an mRNA is translated?

1.25 How is it possible for a gene with a mutation in the coding region to encode a polypeptide with the same amino acid sequence as the nonmutant gene?

Further Reading

Bearn, A. G. 1994. Archibald Edward Garrod, the reluctant geneticist. *Genetics* 137: 1.

Birge, R. R. 1995. Protein-based computers. *Scientific American*, March.

Calladine, C. R. 1997. *Understanding DNA: The Molecule and How It Works*. New York: Academic Press.

Erwin, D. H. 1996. The mother of mass extinctions. *Scientific American*, July.

Gehrig, A., S. R. Schmidt, C. R. Muller, S. Srsen, K. Srsnova, and W. Kress. 1997. Molecular defects in alkaptonuria. *Cytogenetics & Cell Genetics* 76: 14.

Gould, S. J. 1994. The evolution of life on the earth. *Scientific American*, October.

Horgan, J. 1993. Eugenics revisited. *Scientific American*, June.

Horowitz, N. H. 1996. The sixtieth anniversary of biochemical genetics. *Genetics* 143: 1.

Judson, H. F. 1996. *The Eighth Day of Creation: The Makers of the Revolution in Biology*. Cold Spring Harbor, NY: Cold Spring Harbor Laboratory Press.

Mirsky, A. 1968. The discovery of DNA. *Scientific American*, June.

Olson, G. J., and C. R. Woese. 1997. Archaeal genomics: An overview. *Cell* 89: 991.

Radman, M., and R. Wagner. 1988. The high fidelity of DNA duplication. *Scientific American*, August.

Rennie, J. 1993. DNA's new twists. *Scientific American*, March.

Scazzocchio, C. 1997. Alkaptonuria: From humans to moulds and back. *Trends in Genetics* 13: 125.

Smithies, O. 1995. Early days of electrophoresis. *Genetics* 139: 1.

Stadler, D. 1997. Ultraviolet-induced mutation and the chemical nature of the gene. *Genetics* 145: 863.

Susman, M. 1995. The Cold Spring Harbor phage course (1945–1970): A 50th anniversary remembrance. *Genetics* 139: 1101.

Vulliamy, T., P. Mason, and L. Luzzatto. 1992. The molecular basis of glucose-6-phosphate dehydrogenase deficiency. *Trends in Genetics* 8: 138.

Watson, J. D. 1968. *The Double Helix*. New York: Atheneum.

In this small garden plot adjacent to the monastery of St. Thomas, Gregor Mendel grew more than 33,500 pea plants in the years 1856–1863, including more than 6,400 plants in one year alone. He received some help from two fellow monks who assisted in the experiments. Inside the monastery wall on the right is the Mendel museum (called the Mendelianum). The flowers are maintained as a memorial to Mendel's experiments.

Principles of Genetic Transmission

PRINCIPLES

- Inherited traits are determined by the genes present in the reproductive cells united in fertilization.
- Genes are usually inherited in pairs—one from the mother and one from the father.
- The genes in a pair may differ in DNA sequence and in their effect on the expression of a particular inherited trait.
- The maternally and paternally inherited genes are not changed by being together in the same organism.
- In the formation of reproductive cells, the paired genes separate again into different cells.
- Random combinations of reproductive cells containing different genes result in Mendel's ratios of traits appearing among the progeny.
- The ratios actually observed for any traits are determined by the types of dominance and gene interaction.
- In genetic analysis, the complementation test is used to determine whether two recessive mutations that cause a similar phenotype are alleles of the same gene. The mutant parents are crossed, and the phenotype of the progeny is examined. If the progeny phenotype is nonmutant (complementation occurs), the mutations are in different genes; if the progeny phenotype is mutant (lack of complementation), the mutations are in the same gene.

CONNECTIONS

CONNECTION: What Did Gregor Mendel Think He Discovered?
Gregor Mendel 1866
Experiments on plant hybrids

CONNECTION: This Land Is Your Land, This Land Is My Land
The Huntington's Disease Collaborative Research Group
1993
*A novel gene containing a trinucleotide repeat that is expanded
and unstable on Huntington's disease chromosomes*

Gregor Mendel's story is one of the inspiring legends in the history of modern science. Living as a monk at the distinguished monastery of St. Thomas in the town of Brno (Brünn), in what is now the Czech Republic, Mendel taught science at a local trade school and also carried out biological experiments. The most important experiments were crosses of sweet peas carried out from 1856 to 1863 in a small garden plot nestled in a corner of the monastery grounds. He reported his experiments to a local natural history society, published the results and his interpretation in its scientific journal in 1866, and began exchanging letters with Carl Nägeli in Munich, one of the leading botanists of the time. However, no one understood the significance of Mendel's work. By 1868, Mendel had been elected abbot of the monastery, and his scientific work effectively came to an end. Shortly before his death in 1884, Mendel is said to have remarked to one of the younger monks, "My scientific work has brought me a great deal of satisfaction, and I am convinced that it will be appreciated before long by the whole world." The prophecy was fulfilled 16 years later when Hugo de Vries, Carl Correns, and Erich von Tschermak, each working independently and in a different European country, published results of experiments similar to Mendel's, drew attention to Mendel's paper, and attributed priority of discovery to him.

Although modern historians of science disagree over Mendel's intentions in carrying out his work, everyone concedes that Mendel was a first-rate experimenter who performed careful and exceptionally well-documented experiments. His paper contains the first clear exposition of *transmission genetics,* or the statistical rules governing the transmission of hereditary elements from generation to generation. The elegance of Mendel's experiments explains why they were embraced as the foundation of genetics, and the rules of hereditary transmission inferred from his results are often referred to as **Mendelian genetics.** Mendel's breakthrough experiments and concepts are the subject of this chapter.

2.1 The Monohybrid Crosses

The principal difference between Mendel's approach and that of other plant hybridizers of his era is that Mendel thought in quantitative terms about traits that could be classified into two contrasting categories, such as round seeds versus wrinkled seeds. He proceeded by carrying out quite simple crossing experiments and then looked for statistical regularities that might suggest general rules. In his own words, he wanted to "determine the number of different forms in which hybrid progeny appear" and to "ascertain their numerical interrelationships."

Mendel selected peas for his experiments for two reasons. First, he had access to varieties that differed in observable alternative characteristics, such as round versus wrinkled seeds and yellow versus green seeds. Second, his earlier studies had indicated that peas usually reproduce by self-pollination, in which pollen produced in a flower is used to fertilize the eggs in the same flower (Figure 2.1). To produce hybrids by cross-pollination, he needed only to open the keel petal (enclosing the reproductive structures), remove the immature anthers (the pollen-producing structures) before they shed pollen, and dust the stigma (the female structure) with pollen taken from a flower on another plant.

Mendel recognized the need to study inherited characteristics that were uniform within any given variety of peas but different between varieties (for example, round seeds always observed in one variety and wrinkled seeds always observed in another). For this reason, at the beginning of his experiments, he established **true-breeding** varieties in which the plants produced only progeny like themselves when allowed to self-pollinate normally. These different varieties, which bred true for seed shape, seed color, flower color, pod shape, or any of the other well-defined characters that Mendel had selected for investigation (Figure 2.2), provided the parents for subsequent hybridization. A **hybrid** is the off-

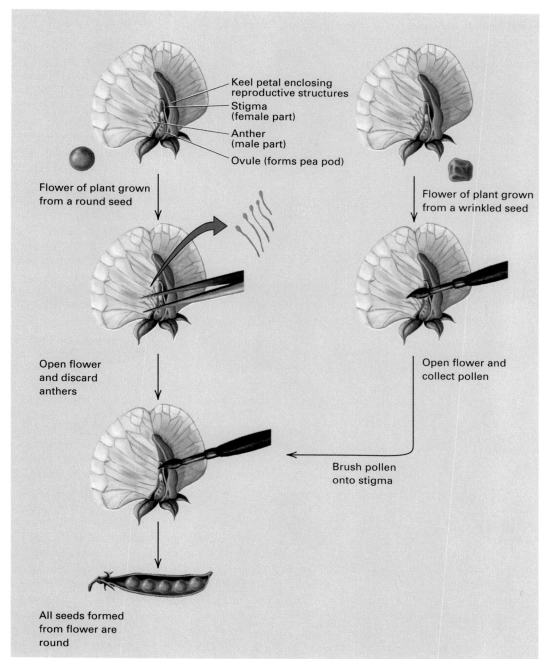

Keel petal enclosing
reproductive structures
Stigma
(female part)
Anther
(male part)
Ovule (forms pea pod)

Flower of plant grown
from a round seed

Flower of plant grown
from a wrinkled seed

Open flower
and discard
anthers

Open flower and
collect pollen

Brush pollen
onto stigma

All seeds formed
from flower are
round

Figure 2.1
Crossing pea plants
requires some minor
surgery in which the
anthers of a flower are
removed before they
produce pollen. The
stigma, or female part
of the flower, is not
removed. It is fertil-
ized by brushing with
mature pollen grains
taken from another
plant.

spring of a cross between parents that differ in one or more traits. A **monohybrid** is a hybrid in which the parents differ in only one trait of interest. (They may differ in other traits as well, but the other differences are ignored for the purposes of the experiment.)

It is worthwhile to examine a few of Mendel's original experiments to learn what his methods were and how he interpreted his results. One pair of characters that he studied was round versus wrinkled seeds. When pollen from a variety of plants with wrinkled seeds was used to cross-pollinate plants from a variety with round seeds, all of the resulting hybrid seeds were round. Geneticists call the hybrid seeds or plants the **F₁ generation** to distinguish

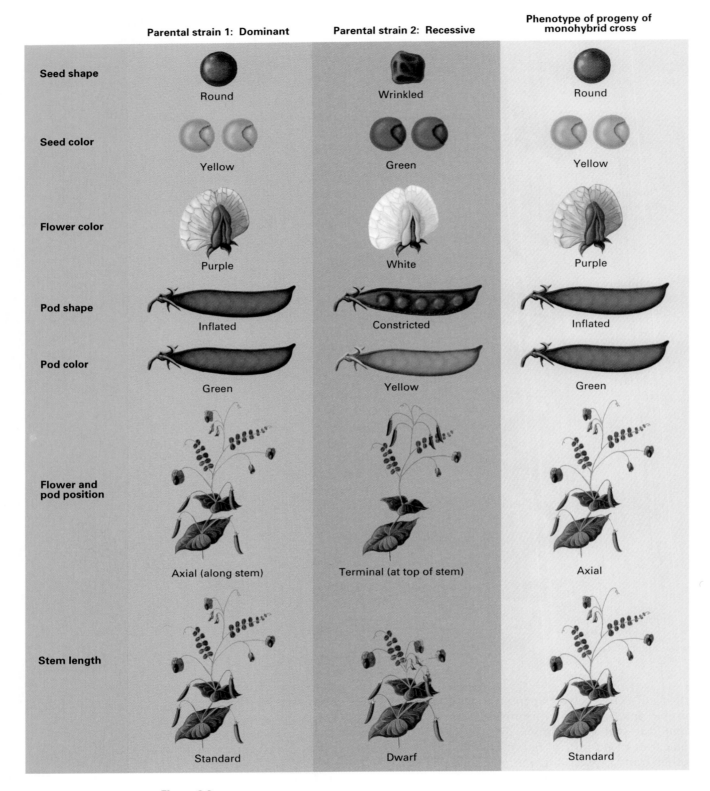

	Parental strain 1: Dominant	Parental strain 2: Recessive	Phenotype of progeny of monohybrid cross
Seed shape	Round	Wrinkled	Round
Seed color	Yellow	Green	Yellow
Flower color	Purple	White	Purple
Pod shape	Inflated	Constricted	Inflated
Pod color	Green	Yellow	Green
Flower and pod position	Axial (along stem)	Terminal (at top of stem)	Axial
Stem length	Standard	Dwarf	Standard

Figure 2.2

The seven character differences in peas studied by Mendel. The characteristic shown at the far right is the dominant trait that appears in the hybrid produced by crossing.

them from the pure-breeding parents, the **P₁ generation.** Mendel also performed the **reciprocal cross,** in which plants from the variety with round seeds were used as the pollen parents and those from the variety with wrinkled seeds as the female parents. As before, all of the F_1 seeds were round (Figure 2.3). The principle illustrated by the equal result of reciprocal crosses is that, with a few important exceptions that will be discussed in later chapters,

> The outcome of a genetic cross does not depend on which trait is present in the male and which is present in the female; reciprocal crosses yield the same result.

Similar results were obtained when Mendel made crosses between plants that differed in any of the pairs of alternative characteristics. In each case, all of the F_1 progeny exhibited only one of the parental traits, and the other trait was absent. The trait expressed in the F_1 generation in each of the monohybrid crosses is shown at the right in Figure 2.2. The trait expressed in the hybrids Mendel called the **dominant** trait; the trait not expressed in the hybrids he called **recessive.**

Traits Present in the Progeny of the Hybrids

Although the recessive trait is not expressed in the hybrid progeny of a monohybrid cross, it reappears in the next generation when the hybrid progeny are allowed to undergo self-fertilization. For example, when the round hybrid seeds from the round × wrinkled cross were grown into plants and allowed to undergo self-fertilization, some of the resulting seeds were round and others wrinkled. The two types were observed in definite numerical

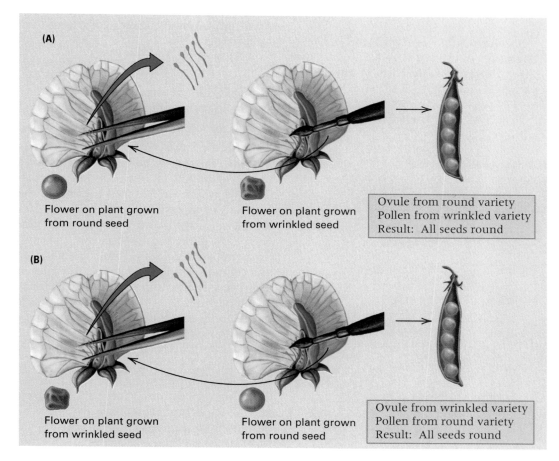

(A)

Flower on plant grown from round seed

Flower on plant grown from wrinkled seed

Ovule from round variety
Pollen from wrinkled variety
Result: All seeds round

(B)

Flower on plant grown from wrinkled seed

Flower on plant grown from round seed

Ovule from wrinkled variety
Pollen from round variety
Result: All seeds round

Figure 2.3
Mendel was the first to show that the characteristics of the progeny produced by a cross do not depend on which parent is the male and which the female. In this example, the seeds of the hybrid offspring are round whether the egg came from the round variety and the pollen from the wrinkled variety (A) or the other way around (B).

proportions. Mendel counted 5474 seeds that were round and 1850 that were wrinkled. He noted that this ratio was approximately 3 : 1.

The progeny seeds produced by self-fertilization of the F_1 generation constitute the **F_2 generation.** Mendel found that the dominant and recessive traits appear in the F_2 progeny in the proportions 3 round : 1 wrinkled. The results of crossing the round and wrinkled varieties are summarized in the following diagram.

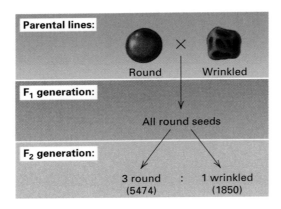

Similar results were obtained in the F_2 generation of crosses between plants that differed in any of the pairs of alternative characteristics (Table 2.1). Note that the

first two traits (round versus wrinkled seeds and yellow versus green seeds) have many more observations than any of the others; the reason is that these traits can be classified directly in the seeds, whereas the others can be classified only in the mature plants. The principal observations from the data in Table 2.1 were

- The F_1 hybrids express only the dominant trait.

- In the F_2 generation, plants with either the dominant or the recessive trait are present.

- In the F_2 generation, there are approximately three times as many plants with the dominant trait as plants with the recessive trait. In other words, the F_2 ratio of dominant : recessive equals approximately 3 : 1.

In the remainder of this section, we will see how Mendel followed up these basic observations and performed experiments that led to his concept of discrete genetic units and to the principles governing their inheritance.

Mendel's Genetic Hypothesis and its Experimental Tests

In Mendel's monohybrid crosses, the recessive trait that was not expressed in the F_1 hybrids reappeared in unchanged form in the F_2 generation, differing in no discernible way from the trait present in the original P_1 recessive plants. In a letter describing this finding, Mendel noted that in the F_2 generation, "the two parental traits appear, separated and unchanged, and there is nothing to indicate that one of them has either inherited or taken over anything from the other." From this finding, Mendel concluded that the hereditary determinants for the traits in the parental lines were transmitted as two different elements that retain their purity in the hybrids. In other words, the hereditary determinants do not "mix" or "contaminate" each other. Hence, a plant with the dominant trait might carry, in unchanged form, the hereditary determinant for the recessive trait.

Table 2.1 Results of Mendel's nonhybrid experiments

Parental traits	F_1 trait	Number of F_2 progeny	F_2 ratio
round × wrinkled (seeds)	round	5474 round, 1850 wrinkled	2.96 : 1
yellow × green (seeds)	yellow	6022 yellow, 2001 green	3.01 : 1
purple × white (flowers)	purple	705 purple, 224 white	3.15 : 1
inflated × constricted (pods)	inflated	882 inflated, 299 constricted	2.95 : 1
green × yellow (unripe pods)	green	428 green, 152 yellow	2.82 : 1
axial × terminal (flower position)	axial	651 axial, 207 terminal	3.14 : 1
long × short (stems)	long	787 long, 277 short	2.84 : 1

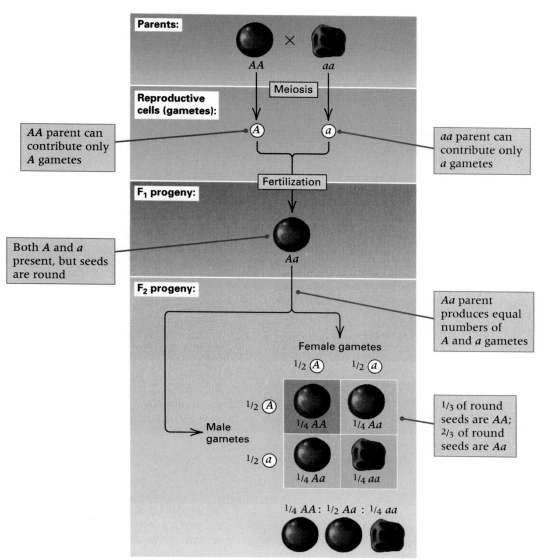

Figure 2.4

A diagrammatic explanation of Mendel's genetic hypothesis to explain the 3 : 1 ratio of dominant : recessive phenotypes observed in the F$_2$ generation of a monohybrid cross. Note that the ratio of *AA* : *Aa* : *aa* genetic types in the F$_2$ generation is 1 : 2 : 1.

To explain his results, Mendel developed a genetic hypothesis that can be understood with reference to Figure 2.4. He assumed that each reproductive cell, or *gamete*, contains one representative of each kind of hereditary determinant in the plant. The hereditary determinant for round seeds he called *A*; that for wrinkled seeds he called *a*. Mendel proposed that in the true-breeding variety with round seeds, all of the reproductive cells would contain *A*; in the true-breeding variety with wrinkled seeds, all of the reproductive cells would contain *a*. When the varieties are crossed, the F$_1$ hybrid should receive one of each of *A* and *a* and so should have the genetic constitution *Aa* (Figure 2.4). If *A* is dominant to *a*, the F$_1$ seeds should be round.

When an F$_1$ plant is self-fertilized, the *A* and *a* determinants would separate from one another and be included in the gametes in equal numbers. Hence, as shown in Figure 2.4, random combinations of the gametes should result in an F$_2$ generation with the genetic composition 1/4 *AA*, 1/2 *Aa*, and 1/4 *aa*. The *AA* and *Aa* types should have round seeds, and the *aa* types should have wrinkled seeds, and so the predicted ratio of round : wrinkled seeds would be 3 : 1. (The genetic types *AA*, *Aa*, and *aa* can also be written with slashes as *A/A*, *A/a* and *a/a*, respectively; the two types of symbolism are equivalent.)

The genetic hypothesis in Figure 2.4 also illustrates another of Mendel's important

What Did Gregor Mendel Think He Discovered?

Gregor Mendel 1866
Monastery of St. Thomas, Brno
[then Brünn], Czech Republic
Experiments on Plant Hybrids
(original in German)

Mendel's paper is remarkable for its precision and clarity. It is worth reading in its entirety for this reason alone. Although the most important discovery attributed to Mendel is segregation, he never uses this term. His description of segregation is found in the first passage in italics in the excerpt. (All of the italics are reproduced from the original.) In his description of the process, he takes us carefully through the separation of A and a in gametes and their coming together again at random in fertilization. One flaw in the description is Mendel's occasional confusion between genotype and phenotype, which is illustrated by his writing A instead of AA and a instead of aa in the display toward the end of the passage. Most early geneticists made no consistent distinction between genotype and phenotype until 1909, when the terms themselves were first coined.

Artificial fertilization undertaken on ornamental plants to obtain new color variants initiated the experiments reported here. The striking regularity with which the same hybrid forms always reappeared whenever fertilization between like species took place suggested further experiments whose task it was to follow that development of hybrids in their progeny. . . . This paper discusses the attempt at such a detailed experiment. . . . Whether the plan by which the individual experiments were set up and carried out was adequate to the assigned task should be decided by a benevolent judgment. . . . [Here the experimental results are described in detail.] Thus experimentation also justifies the assumption *that pea hybrids form germinal and pollen cells that in their composition correspond in equal numbers to all the constant forms resulting from the combination of traits united through fertilization.* The difference of forms among the progeny of hybrids, as well as the ratios in which they

are observed, find an adequate explanation in the principle [of segregation] just deduced. The simplest case is given by the series for *one pair of differing traits.* It is shown that this series is described by the expression: $A + 2Aa + a$, in which A and a signify the forms with constant differing traits, and Aa the form hybrid for both. The series contains four individuals in three different terms. In their production, pollen and germinal cells of form A and a participate, on the average, equally in fertilization; therefore each form manifests itself twice, since four individuals are produced. Participating in fertilization are thus:

Pollen cells	$A + A + a + a$
Germinal cells	$A + A + a + a$

Whether the plan by which the individual experiments were set up and carried out was adequate to the assigned task should be decided by a benevolent judgment

---

It is entirely a matter of chance which of the two kinds of pollen combines with each single germinal cell. However, according to the laws of probability, in an average of many cases it will always happen that every pollen form A and a will unite equally often with every germinal-cell form A and a; therefore, in fertilization, one of the two pollen cells A will meet a germinal cell A, the other a germinal cell a, and equally, one pollen cell a will become associated with a germinal cell A, and the other a.

Pollen cells	A	A	a	a
Germinal cells	A	A	a	a

The result of fertilization can be visualized by writing the designations for associated germinal and pollen cells in the form of fractions, pollen cells above the line, germi-

nal cells below. In the case under discussion one obtains

$$\frac{A}{A} + \frac{A}{a} + \frac{a}{A} + \frac{a}{a}$$

In the first and fourth terms germinal and pollen cells are alike; therefore the products of their association must be constant, namely A and a; in the second and third, however, a union of the two differing parental traits takes place again, therefore the forms arising from such fertilizations are absolutely identical with the hybrid from which they derive. *Thus, repeated hybridization takes place.* The striking phenomenon, that hybrids are able to produce, in addition to the two parental types, progeny that resemble themselves is thus explained: Aa and aA both give the same association, Aa, since, as mentioned earlier, it makes no difference to the consequence of fertilization which of the two traits belongs to the pollen and which to the germinal cell. Therefore

$$\frac{A}{A} + \frac{A}{a} + \frac{a}{A} + \frac{a}{a} = A + 2Aa + a$$

This represents the *average* course of self-fertilization of hybrids when two differing traits are associated in them. In individual flowers and individual plants, however, the ratio in which the members of the series are formed may be subject to not insignificant deviations. . . . Thus it was proven experimentally that, in *Pisum,* hybrids form *different kinds* of germinal and pollen cells and that this is the reason for the variability of their offspring.

Source: Verhandlungen des naturforschenden den Vereines in Brünn 4: 3–47

deductions: Two plants with the same outward appearance, such as round seeds, might nevertheless differ in their hereditary makeup as revealed by the types of progeny observed when they are crossed. For example, in the true-breeding round variety, the genetic composition of the seeds is AA, whereas in the F_1 hybrid seeds of the round × wrinkled cross, the genetic composition of the seeds is Aa.

But how could the genetic hypothesis be tested? Mendel realized that a key prediction of his hypothesis concerned the genetic composition of the round seeds in the F_2 generation. If the hypothesis is correct, then one-third of the round seeds should have the genetic composition AA and two-thirds of the round seeds should have the genetic composition Aa. This principle is shown in Figure 2.5. The ratio of $AA : Aa : aa$ in the F_2 generation is $1 : 2 : 1$, but if we disregard the recessives, then the ratio of $AA : Aa$ is $1 : 2$; in other words, $1/3$ of the round seeds are AA and $2/3$ are Aa. Upon self-fertilization, plants grown from the AA types should be true breeding for round seeds, whereas those from the Aa types should yield round and wrinkled seeds in the ratio $3 : 1$. Furthermore, among the wrinkled seeds in the F_2 generation, all should have the genetic composition aa, and so, upon self-fertilization, they should be true breeding for wrinkled seeds.

For several of his traits, Mendel carried out self-fertilization of the F_2 plants in order to test these predictions. His results for round versus wrinkled seeds are summarized in the diagram below:

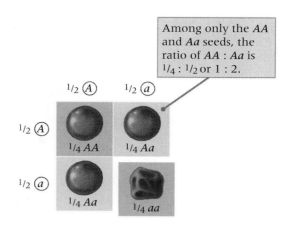

Among only the AA and Aa seeds, the ratio of $AA : Aa$ is $1/4 : 1/2$ or $1 : 2$.

Figure 2.5

In the F_2 generation, the ratio of $AA : Aa$ is $1 : 2$. Therefore, *among those seeds that are round*, $1/3$ should be AA and $2/3$ should be Aa.

As predicted from Mendel's genetic hypothesis, the plants grown from F_2 wrinkled seeds were true breeding for wrinkled seeds. They produced only wrinkled seeds in the F_3 generation. Moreover, among 565 plants grown from F_2 round seeds, 193 were true breeding, producing only round seeds in the F_3 generation, whereas the other 372 plants produced both round and wrinkled seeds in a proportion very close to $3 : 1$. The ratio $193 : 372$ equals $1 : 1.93$, which is very close to the ratio $1 : 2$ of $AA : Aa$ types predicted theoretically from the genetic hypothesis in Figure 2.4. Overall, taking all of the F_2 plants into account, the ratio of genetic types observed was very close to the predicted $1 : 2 : 1$ of $AA : Aa : aa$ expected from Figure 2.4.

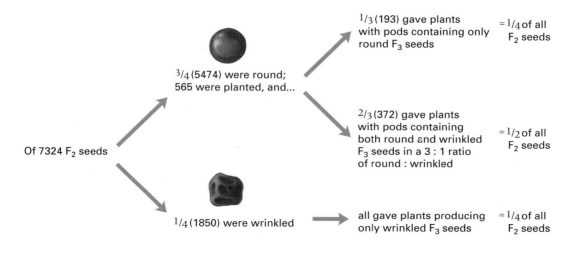

Of 7324 F_2 seeds

$3/4$ (5474) were round; 565 were planted, and...

$1/4$ (1850) were wrinkled

$1/3$ (193) gave plants with pods containing only round F_3 seeds — $= 1/4$ of all F_2 seeds

$2/3$ (372) gave plants with pods containing both round and wrinkled F_3 seeds in a 3 : 1 ratio of round : wrinkled — $= 1/2$ of all F_2 seeds

all gave plants producing only wrinkled F_3 seeds — $= 1/4$ of all F_2 seeds

The Principle of Segregation

The diagram in Figure 2.4 is the heart of Mendelian genetics. You should master it and be able to use it to deduce the progeny types produced in crosses. Be sure you thoroughly understand the meaning, and the biological basis, of the ratios 3 : 1 and 1 : 2 and 1 : 2 : 1. The following list highlights Mendel's key assumptions in formulating his model of inheritance.

1. For each of the traits that Mendel studied, a pea plant contains two hereditary determinants.

2. For each pair of hereditary determinants present in a plant, the members may be identical (for example, *AA*) or different (for example, *Aa*).

3. Each reproductive cell (gamete) produced by a plant contains only one of each pair of hereditary determinants (that is, either *A* or *a*).

4. In the formation of gametes, any particular gamete is equally likely to include either hereditary determinant (hence, from an *Aa* plant, half the gametes contain *A* and the other half contain *a*).

5. The union of male and female reproductive cells is a random process that reunites the hereditary determinants in pairs.

The essential feature of Mendelian genetics is the separation, technically called **segregation,** in unaltered form, of the two hereditary determinants in a hybrid plant in the formation of its reproductive cells (points 3 and 4 in the foregoing list). The principle of segregation is sometimes called *Mendel's first law,* although Mendel never used this term.

> **The Principle of Segregation:** In the formation of gametes, the paired hereditary determinants separate (segregate) in such a way that each gamete is equally likely to contain either member of the pair.

Apart from the principle of segregation, the other key assumption, implicit in points 1 and 5 in the list, is that the hereditary determinants are present as pairs in both the parental organisms and the progeny organisms but as single copies in the reproductive cells.

Important Genetic Terminology

One of the handicaps under which Mendel wrote was the absence of an established vocabulary of terms suitable for describing his concepts. Hence he made a number of seemingly elementary mistakes, such as occasionally confusing the outward appearance of an organism with its hereditary constitution. The necessary vocabulary was developed only after Mendel's work was rediscovered, and it includes the following essential terms.

1. A hereditary determinant of a trait is called a **gene.**

2. The different forms of a particular gene are called **alleles.** In Figure 2.4, the alleles of the gene for seed shape are *A* for round seeds and *a* for wrinkled seeds. *A* and *a* are alleles because they are alternative forms of the gene for seed shape. Alternative alleles are typically represented by the same letter or combination of letters, distinguished either by uppercase and lowercase or by means of superscripts and subscripts or some other typographic identifier.

3. The **genotype** is the genetic constitution of an organism or cell. With respect to seed shape in peas, *AA*, *Aa*, and *aa* are examples of the possible genotypes for the *A* and *a* alleles. Because gametes contain only one allele of each gene, *A* and *a* are examples of genotypes of gametes.

4. A genotype in which the members of a pair of alleles are different, as in the *Aa* hybrids in Figure 2.4, is said to be **heterozygous;** a genotype in which the two alleles are alike is said to be **homozygous.** A homozygous organism may be homozygous dominant (*AA*) or homozygous recessive (*aa*). The terms *homozygous* and *heterozygous* can-

not apply to gametes, which contain only one allele of each gene.

5. The *observable* properties of an organism constitute its **phenotype.** Round seeds, wrinkled seeds, yellow seeds, and green seeds are all phenotypes. The phenotype of an organism does not necessarily tell you anything about its genotype. For example, a seed with the phenotype "round" could have the genotype *AA* or *Aa*.

Verification of Mendelian Segregation by the Testcross

A second way in which Mendel tested the genetic hypothesis in Figure 2.4 was by crossing the F₁ heterozygous genotypes with plants that were homozygous recessive. Such a cross, between an organism that is heterozygous for one or more genes (for example, *Aa*), and an organism that is homozygous for the recessive alleles (for example, *aa*), is called a **testcross.** The result of such a testcross is shown in Figure 2.6. Because the heterozygous parent is expected to produce *A* and *a* gametes in equal numbers, whereas the homozygous recessive produces only *a* gametes, the expected progeny are 1/2 with the genotype *Aa* and 1/2 with the genotype *aa*. The former have the dominant phenotype (because *A* is dominant to *a*) and the latter have the recessive phenotype. A testcross is often extremely useful in genetic analysis because

> In a testcross, the relative frequencies of the different gametes produced by the heterozygous parent can be observed directly in the phenotypes of the progeny, because the recessive parent contributes only recessive alleles.

Mendel carried out a series of testcrosses with the genes for round versus wrinkled seeds, yellow versus green seeds, purple versus white flowers, and long versus short stems. The results are shown Table 2.2. In all cases, the ratio of phenotypes among the progeny is very close to the 1 : 1 ratio expected from segregation of the alleles in the heterozygous parent.

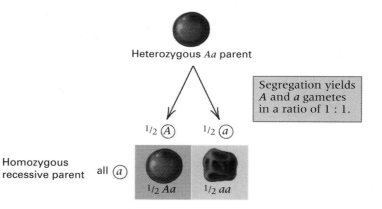

Figure 2.6

In a testcross of an *Aa* heterozygous parent with an *aa* homozygous recessive, the progeny are *Aa* and *aa* in the ratio of 1 : 1. A testcross shows the result of segregation.

Another valuable type of cross is a **backcross,** in which hybrid organisms are crossed with one of the parental genotypes. Backcrosses are commonly used by geneticists and by plant and animal breeders, as we will see in later chapters. Note that the testcrosses in Table 2.2 are also backcrosses, because in each case, the F₁ heterozygous parent came from a cross between the homozygous dominant and the homozygous recessive.

Table 2.2 Mendel's testcross results

Testcross (F₁ heterozygote × homozygous recessive)	Progeny from testcross	Ratio
Round × wrinkled seeds	193 round, 192 wrinkled	1.01 : 1
Yellow × green seeds	196 yellow, 189 green	1.04 : 1
Purple × white flowers	85 purple, 81 white	1.05 : 1
Long × short stems	87 long 79 short	1.10 : 1

2.2 Segregation of Two or More Genes

Mendel also carried out experiments in which he examined the inheritance of two or more traits simultaneously to determine whether the same pattern of inheritance applied to each pair of alleles separately when more than one allelic pair was segregating in the hybrids. For example, plants from a true-breeding variety with round and yellow seeds were crossed with plants from a variety with wrinkled and green seeds. The F_1 progeny were hybrid for both characteristics, or **dihybrid,** and the phenotype of the seeds was round and yellow. The F_1 phenotype was round and yellow because round is dominant to wrinkled and yellow is dominant to green (Figure 2.2).

Then Mendel self-fertilized the F_1 progeny to obtain seeds in the F_2 generation. He observed four types of seed phenotypes in the progeny and, in counting the seeds, obtained the following numbers:

round, yellow	315
round, green	108
wrinkled, yellow	101
wrinkled, green	32
Total	556

In these data, Mendel noted the presence of the expected monohybrid 3 : 1 ratio for each trait separately. With respect to each trait, the progeny were

round : wrinkled
$$= (315 + 108) : (101 + 32)$$
$$= 423 : 133$$
$$= 3.18 : 1$$

yellow : green
$$= (315 + 101) : (108 + 32)$$
$$= 416 : 140$$
$$= 2.97 : 1$$

Furthermore, in the F_2 progeny of the dihybrid cross, the separate 3 : 1 ratios for the two traits were combined at random, as shown in Figure 2.7. When the phenotypes of two traits are combined at random, then, among the 3/4 of the progeny that are round, 3/4 will be yellow and 1/4 green; similarly, among the 1/4 of the progeny that are wrinkled, 3/4 will be yellow and 1/4 green. The overall proportions of round yellow to round green to wrinkled yellow to wrinkled green are therefore expected to be $3/4 \times 3/4$ to $3/4 \times 1/4$ to $1/4 \times 3/4$ to $1/4 \times 1/4$ or

$$9/16 : 3/16 : 3/16 : 1/16$$

The observed ratio of 315 : 108 : 101 : 32 equals 9.84 : 3.38 : 3.16 : 1, which is reasonably close to the 9 : 3 : 3 : 1 ratio expected from the cross-multiplication of the separate 3 : 1 ratios in Figure 2.7.

Figure 2.7

The 3 : 1 ratio of round : wrinkled, when combined at random with the 3 : 1 ratio of yellow : green, yields the 9 : 3 : 3 : 1 ratio that Mendel observed in the F_2 progeny of the dihybrid cross.

Seed color phenotypes

3/4 Yellow   1/4 Green

Seed shape phenotypes

3/4 Round

9/16 Round, yellow   3/16 Round, green

1/4 Wrinkled

3/16 Wrinkled, yellow   1/16 Wrinkled, green

Ratio of phenotypes in the F_2 progeny of a dihybrid cross is 9 : 3 : 3 : 1.

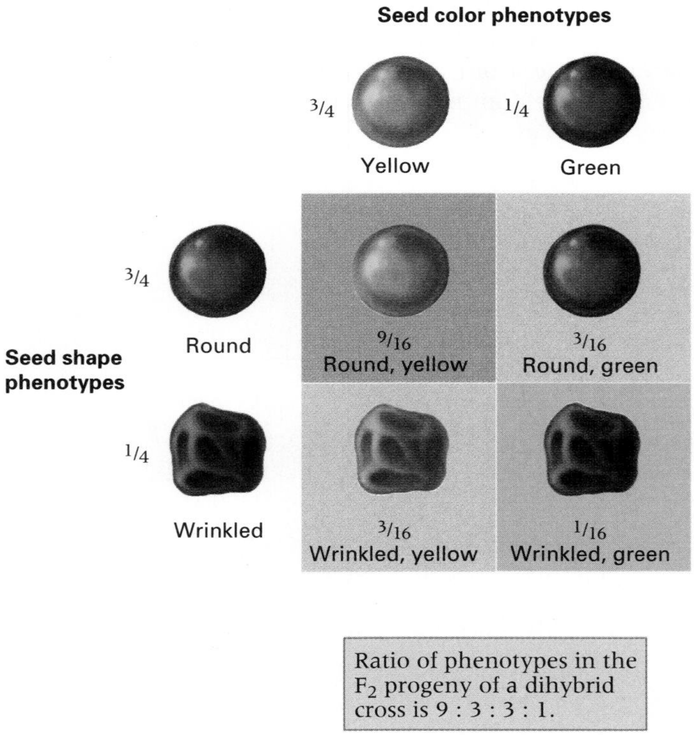

The Principle of Independent Assortment

Mendel carried out similar experiments with other combinations of traits and, for each pair of traits he examined, consistently observed the 9 : 3 : 3 : 1 ratio. He also deduced the biological reason for this observation. To illustrate his explanation using the dihybrid round × wrinkled cross, we can represent the dominant and recessive alleles of the pair that affect seed shape as *W* and *w*, respectively, and the allelic pair that affect seed color as *G* and *g*. Mendel proposed that the underlying reason for the 9 : 3 : 3 : 1 ratio in the F$_2$ generation is that the segregation of the alleles *W* and *w* for round or wrinkled seeds has no effect on the segregation of the alleles *G* and *g* for yellow or green seeds. Each pair of alleles undergoes segregation into the gametes independently of the segregation of the other pair of alleles. In the P$_1$ generation, the parental genotypes are *WW GG* (round, yellow seeds) and *ww gg* (wrinkled, green seeds). Then, the genotype of the F$_1$ is the double heterozygote *Ww Gg*. Note that this genotype can also be designated using the symbolism

$$W/w; G/g \quad \text{or}$$

$$\frac{W}{w} \frac{G}{g}$$

in which the slash (also called a virgule) is replaced with a short horizontal line.

The result of independent assortment in the F$_1$ plants is that the *W* allele is just as likely to be included in a gamete with *G* as with *g*, and the *w* allele is just as likely to be included in a gamete with *G* as with *g*. The independent segregation is illustrated in Figure 2.8. When two pairs of alleles undergo independent assortment, the gametes produced by the double heterozygote are

$$1/4 \; WG \quad 1/4 \; Wg \quad 1/4 \; wG \quad 1/4 \; wg$$

When the four types of gametes combine at random to form the zygotes of the next generation, the result of independent assortment is shown in Figure 2.9. The cross-multiplication-like format, which is used to show how the F$_1$ female and male gametes may combine to produce the F$_2$ genotypes, is called a **Punnett square.** In the Punnett square, the possible phenotypes of the F$_2$ progeny are indicated. Note that the ratio of phenotypes is 9 : 3 : 3 : 1.

The Punnett square in Figure 2.9 also shows that the ratio of *genotypes* in the F$_2$ generation is not 9 : 3 : 3 : 1. With independent assortment, the ratio of genotypes in the F$_2$ generation is

$$1 : 2 : 1 : 2 : 4 : 2 : 1 : 2 : 1$$

The reason for this ratio is shown in Figure 2.10. Among seeds with the *WW* genotype,

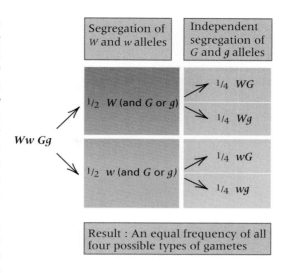

Figure 2.8

Independent segregation of the *Ww* and *Gg* allele pairs means that, among each of the *W* and *w* classes, the ratio of *G* : *g* is 1 : 1. Likewise, among each of the *G* and *g* classes, the ratio of *W* : *w* is 1 : 1.

the ratio of *GG* : *Gg* : *gg* equals 1 : 2 : 1. Among seeds with the *Ww* genotype, the ratio is 2 : 4 : 2 (the 1 : 2 : 1 is multiplied by 2 because there are twice as many *Ww* genotypes as either *WW* or *ww*). And among seeds with the *ww* genotype, the ratio of *GG* : *Gg* : *gg* equals 1 : 2 : 1. The phenotypes of the seeds are shown beneath the genotypes. The combined ratio of phenotypes is 9 : 3 : 3 : 1. From Figure 2.10, one can also see that among seeds that are *GG*, the ratio of *WW* : *Ww* : *ww* equals 1 : 2 : 1; among seeds that are *Gg*, it is 2 : 4 : 2; and among seeds that are *gg*, it is 1 : 2 : 1. Therefore, the independent segregation means that, among each of the possible genotypes formed by one allele pair, the ratio of homozygous dominant to heterozygous to homozygous recessive for the other allele pair is 1 : 2 : 1.

Mendel tested the hypothesis of independent segregation by ascertaining whether the predicted genotypes were actually present in the expected proportions. He did the tests by growing plants from the F$_2$ seeds and obtaining F$_3$ progeny by self-pollination. To illustrate the tests, consider one series of crosses in which he grew plants from F$_2$ seeds that were round, green. Note in Figures 2.9 and 2.10 that round, green F$_2$ seeds are expected to have the genotypes *Ww gg* and *WW gg* in the ratio 2 : 1. Mendel grew 102 plants from such seeds and found that 67 of them produced both round, green and wrinkled, green seeds (indicating that the parental plants must have been *Ww gg*) and 35 of them produced only round, green seeds (indicating that the parental genotype was

Figure 2.9

Diagram showing the basis for the 9 : 3 : 3 : 1 ratio of F₂ phenotypes resulting from a cross in which the parents differ in two traits determined by genes that undergo independent assortment.

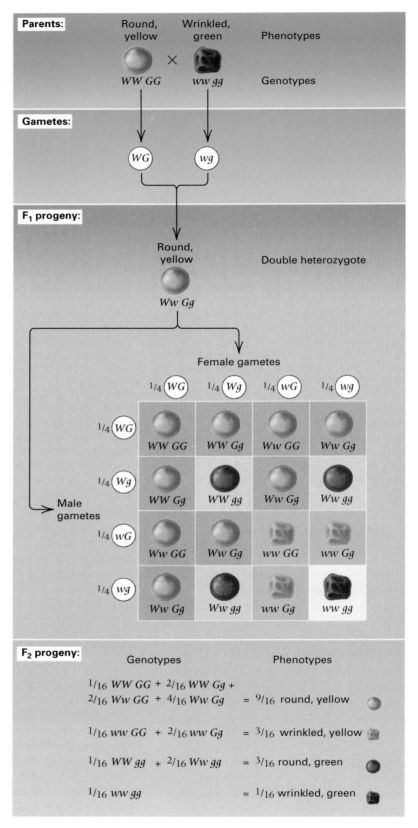

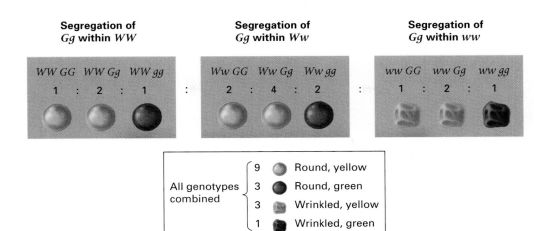

Segregation of *Gg* within *WW*				Segregation of *Gg* within *Ww*				Segregation of *Gg* within *ww*		
WW GG	*WW Gg*	*WW gg*		*Ww GG*	*Ww Gg*	*Ww gg*		*ww GG*	*ww Gg*	*ww gg*
1	2	1	:	2	4	2	:	1	2	1

All genotypes combined	9	Round, yellow
	3	Round, green
	3	Wrinkled, yellow
	1	Wrinkled, green

Figure 2.10
The F$_2$ progeny of the dihybrid cross for seed shape and seed color. In each of the genotypes for one of the allelic pairs, the ratio of homozygous dominant, heterozygous, and homozygous recessive genotypes for the other allelic pair is 1 : 2 : 1.

WW gg). The ratio 67 : 35 is in good agreement with the expected 2 : 1 ratio of genotypes. Similar good agreement with the predicted relative frequencies of the different genotypes was found when plants were grown from round, yellow or from wrinkled, yellow F$_2$ seeds. (As expected, plants grown from the wrinkled, green seeds, which have the predicted homozygous recessive genotype *ww gg*, produced only wrinkled, green seeds.)

Mendel's observation of independent segregation of two pairs of alleles has come to be known as the principle of independent assortment, or sometimes as *Mendel's second law:*

> **The Principle of Independent Assortment:** Segregation of the members of any pair of alleles is independent of the segregation of other pairs in the formation of reproductive cells.

Although the principle of independent assortment is of fundamental importance in Mendelian genetics, in later chapters we will see that there are important exceptions.

Dihybrid Testcrosses

A second way in which Mendel tested the hypothesis of independent assortment was by carrying out a testcross with the F$_1$ genotypes that were heterozygous for both genes (*Ww Gg*). In a testcross, one parental genotype is always multiple homozygous recessive, in this case *ww gg*. As shown in

Figure 2.11, the double heterozygotes produce four types of gametes—*WG, Wg, wG,* and *wg*—in equal frequencies, whereas the *ww gg* plants produce only *wg* gametes. Thus the progeny phenotypes are expected to consist of round yellow, round green, wrinkled yellow, and wrinkled green in a ratio of 1 : 1 : 1 : 1; the ratio of phenotypes is a direct demonstration of the ratio of gametes produced by the double heterozygote because no dominant alleles are contributed by the *ww gg* parent to obscure the results. In the actual cross, Mendel obtained 55 round yellow, 51 round green, 49 wrinkled yellow, and 53 wrinkled green, which is in good agreement with the predicted 1 : 1 : 1 : 1 ratio. The results were the same in the reciprocal cross—that is, with the double heterozygote as the female parent and the homozygous recessive as the male parent. This observation confirmed Mendel's assumption that the gametes of both sexes included each possible genotype in approximately equal proportions.

The Big Experiment

Taking his hypothesis a step further, Mendel also carried out crosses between varieties that differed in three traits: seed shape (round or wrinkled, alleles *W* and *w*), seed color (yellow or green, alleles *G* and *g*), and flower color (purple or white, alleles *P* and *p*). The phenotype of the trihybrid F$_1$ seeds was round and yellow, and the plants grown from these seeds had purple flowers. By analogy with the dihybrid

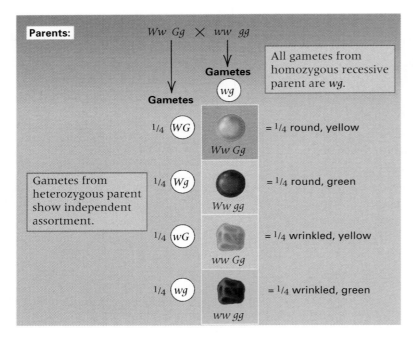

Parents: *Ww Gg* ✕ *ww gg*

Gametes

Gametes
wg

All gametes from homozygous recessive parent are *wg*.

Gametes from heterozygous parent show independent assortment.

1/4 *WG* — *Ww Gg* = 1/4 round, yellow

1/4 *Wg* — *Ww gg* = 1/4 round, green

1/4 *wG* — *ww Gg* = 1/4 wrinkled, yellow

1/4 *wg* — *ww gg* = 1/4 wrinkled, green

Figure 2.11
Genotypes and phenotypes resulting from a testcross of a *Ww Gg* double heterozygote.

cross, if the alleles of all three genes undergo independent assortment, then self-fertilization of the F_1 flowers should result in combinations of phenotypes given by successive terms in the multiplication of

$[(3/4)D + (1/4)R]^3$, which yields the ratio 27 : 9 : 9 : 9 : 3 : 3 : 3 : 1. For Mendel's cross, the multiplication is carried out in Figure 2.12. The most frequent phenotype (27/64) has the dominant form of all three traits, the next most frequent (9/64) has the dominant form of two of the traits, the next most frequent (3/64) has the dominant form of only one trait, and the least frequent (1/64) is the triple recessive. Observe that if you consider any one of the traits and ignore the other two, then the ratio of phenotypes is 3 : 1; and if you consider any two of the traits, then the ratio of phenotypes is 9 : 3 : 3 : 1. This means that all of the possible one- and two-gene independent segregations are present in the overall three-gene segregation.

The observed and expected numbers in Figure 2.12 indicate that agreement with the hypothesis of independent assortment is very good. This, however, did not satisfy Mendel. He realized that there should be 27 different genotypes present in the F_2 progeny, so he self-fertilized each of the 639 plants to determine its genotype for each of the three traits. Mendel alludes to the amount of work this experiment entailed by noting that "of all the experiments, it required the most time and effort."

The result of the experiment is shown in Figure 2.13. From top to bottom, the

Figure 2.12
With independent assortment, the expected ratio of phenotypes in a trihybrid cross is obtained by multiplying the three independent 3 : 1 ratios of the dominant and recessive phenotypes. A dash used in a genotype symbol indicates that either the dominant or the recessive allele is present; for example, *W*— refers collectively to the genotypes *WW* and *Ww*. (The expected numbers total 640 rather than 639 because of round-off error.)

(3/4 *W*— + 1/4 *ww*) ✕ (3/4 *G*— + 1/4 *gg*) ✕ (3/4 *P*— + 1/4 *pp*)			Observed number	Expected number
27/64	*W*— *G*— *P*—	Round, yellow, purple	269	270
9/64	*W*— *G*— *pp*	Round, yellow, white	98	90
9/64	*W*— *gg* *P*—	Round, green, purple	86	90
9/64	*ww* *G*— *P*—	Wrinkled, yellow, purple	88	90
3/64	*W*— *gg* *pp*	Round, green, white	27	30
3/64	*ww* *G*— *pp*	Wrinkled, yellow, white	34	30
3/64	*ww* *gg* *P*—	Wrinkled, green, purple	30	30
1/64	*ww* *gg* *pp*	Wrinkled, green, white	7	10

● For any one gene, the ratio of phenotypes is 48 : 16 = 3 : 1

● For any pair of genes, the ratio of phenotypes is 36 : 12 : 12 : 4 = 9 : 3 : 3 : 1

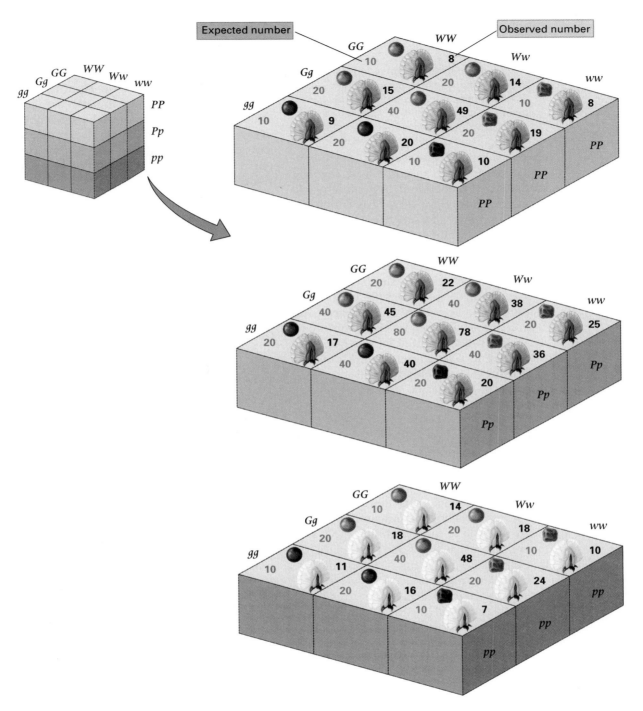

Figure 2.13
Results of Mendel's analysis of the genotypes formed in the F₂ generation of a trihybrid cross with the allelic pairs *W, w* and *G, g* and *P, p*. In each pair of numbers, the red entry is the expected number and the black entry is the observed. Note that each gene, by itself, yields a 1 : 2 : 1 ratio of genotypes and that each pair of genes yields a 1 : 2 : 1 : 2 : 4 : 2 : 1 : 2 : 1 ratio of genotypes.

three Punnett squares show the segregation of *W* and *w* from *G* and *g* in the genotypes *PP, Pp,* and *pp.* In each cell, the number in red is the expected number of plants of each genotype, assuming independent assortment, and the number in black is the observed number of each genotype of plant. The excellent agreement confirmed Mendel in what he regarded as the main conclusion of his experiments:

> "Pea hybrids form germinal and pollen cells that in their composition correspond in equal numbers to all the constant forms resulting from the combination of traits united through fertilization."

In this admittedly somewhat turgid sentence, Mendel incorporated both segregation and independent assortment. In modern terms, what he means is that the gametes produced by any hybrid plant consist of equal numbers of all possible combinations of the alleles present in the original true-breeding parents whose cross produced the hybrids. For example, the cross $WW\,gg \times ww\,GG$ produces F_1 progeny of genotype $Ww\,Gg$, which yields the gametes WG, Wg, wG, and wg in equal numbers. Segregation is illustrated by the $1:1$ ratio of $W:w$ and $G:g$ gametes, and independent assortment is illustrated by the equal numbers of WG, Wg, wG, and wg gametes.

2.3
Mendelian Inheritance and Probability

A working knowledge of the rules of probability for predicting the outcome of chance events is basic to understanding the transmission of hereditary characteristics. In the first place, the proportions of the different types of offspring obtained from a cross are the cumulative result of numerous independent events of fertilization. Furthermore, in each fertilization, the particular combination of dominant and recessive alleles that come together is random and subject to chance variation.

In the analysis of genetic crosses, the probability of an event may be considered as equivalent to the proportion of times that the event is expected to be realized in numerous repeated trials. Likewise, the proportion of times that an event is expected to be realized in numerous repeated trials is equivalent to the probability that it is realized in a single trial. For example, in the F_2 generation of the hybrid between pea varieties with round seeds and those with wrinkled seeds, Mendel observed 5474 round seeds and 1850 wrinkled seeds (Table 2.1). In this case, the proportion of wrinkled seeds was $1850/(1850 + 5474) = 1/3.96$, or very nearly $1/4$. We may therefore regard $1/4$ as the approximate proportion of wrinkled seeds to be expected among a large number of progeny from this cross. Equivalently, we can regard $1/4$ as the probability that any particular seed chosen at random will be wrinkled.

Evaluating the probability of a genetic event usually requires an understanding of the mechanism of inheritance and knowledge of the particular cross. For example, in evaluating the probability of obtaining a round seed from a particular cross, you need to know that there are two alleles, W and w, with W dominant over w; you also need to know the particular cross, because the probability of round seeds is determined by whether the cross is

$WW \times ww$, in which all seeds are expected to be round,

$Ww \times Ww$, in which $3/4$ are expected to be round, or

$Ww \times ww$, in which $1/2$ are expected to be round.

In many genetic crosses, the possible outcomes of fertilization are equally likely. Suppose that there are n possible outcomes, each as likely as any other, and that in m of these, a particular outcome of interest is realized; then the probability of the outcome of interest is m/n. In the language of probability, an outcome of interest is typically called an *event*. As an example, consider the progeny produced by self-pollination of an Aa plant; four equally likely progeny genotypes (*outcomes*) are possible: namely AA, Aa, aA, and aa. Two of the four possible outcomes are heterozygous, so the probability of a heterozygote is $2/4$, or $1/2$.

Mutually Exclusive Events: The Addition Rule

Sometimes an outcome of interest can be expressed in terms of two or more possibilities. For example, a seed with the phenotype of "round" may have either of two genotypes, WW and Ww. A seed that is round cannot have both genotypes at the same time. With events such as the formation of the WW or Ww genotypes, only one event can be realized in any one organism, and the realization of one event in an organism precludes the realization of others in the same organism. In this example, realization of the genotype WW in a plant precludes realization of the genotype Ww in the same plant, and the other way around. Events that exclude each other in this manner are said to be *mutually exclusive*. When events are mutually exclusive, their probabilities are combined according to the addition rule.

> **Addition Rule:** The probability of the realization of one or the other of two mutually exclusive events, A or B, is the sum of their separate probabilities.

In symbols, where *Prob* is used to mean *probability,* the addition rule is written

$$Prob \{A \text{ or } B\} = Prob \{A\} + Prob \{B\}$$

The addition rule can be applied to determine the proportion of round seeds expected from the cross $Ww \times Ww$. The round-seed phenotype results from the expression of either of two genotypes, WW and Ww, which are mutually exclusive. In any particular progeny organism, the probability of genotype WW is 1/4 and that of Ww is 1/2. Hence the overall probability of either WW or Ww is

$$
\begin{aligned}
Prob \{WW \text{ or } Ww\} \\
= Prob \{WW\} + Prob \{Ww\} \\
= 1/4 + 1/2 = 3/4
\end{aligned}
$$

Because 3/4 is the probability of an individual seed being round, it is also the expected proportion of round seeds among a large number of progeny.

Independent Events: The Multiplication Rule

Events that are not mutually exclusive may be *independent*, which means that the realization of one event has no influence on the possible realization of any others. For example, in Mendel's crosses for seed shape and color, the two traits are independent, and the ratio of phenotypes in the F_2 generation is expected to be 9/16 round yellow, 3/16 round green, 3/16 wrinkled yellow, and 1/16 wrinkled green. These proportions can be obtained by considering the traits separately, because they are independent. Considering only seed shape, we can expect the F_2 generation to consist of 3/4 round and 1/4 wrinkled seeds. Considering only seed color, we can expect the F_2 generation to consist of 3/4 yellow and 1/4 green. Because the traits are inherited independently, among the 3/4 of the seeds that are round, there should be 3/4 that are yellow, so the overall proportion of round yellow seeds is expected to be $3/4 \times 3/4 = 9/16$. Likewise, among the 3/4 of the seeds that are round, there should be 1/4 green, yielding $3/4 \times 1/4 = 3/16$ as the expected proportion of round green seeds. The proportions of the other phenotypic classes can be deduced in a similar way. The principle is that when events are independent, the probability that they are realized together is obtained by multiplication.

Successive offspring from a cross are also independent events, which means that the genotypes of early progeny have no influence on the relative proportions of genotypes in later progeny. The independence of successive offspring contradicts the widespread belief that in each human family, the ratio of girls to boys must "even out" at approximately 1 : 1, and so, if a family already has, say, four girls, they are somehow more likely to have a boy the next time around. But this belief is not supported by theory, and it is also contradicted by actual data on the sex ratios in human sibships. (The term **sibship** refers to a group of offspring from the same parents.) The data indicate that a human family is no more likely to have a girl on the next birth if it already has five boys than if it already has five girls. The statistical reason is that, though the sex ratios tend to balance out when they are averaged across a large number of sibships, they do not need to balance within individual sibships. Thus, among families in which there are five children, the sibships consisting of five boys balance those consisting of five girls, for an overall sex ratio of 1 : 1. However, both of these sibships are unusual in their sex distribution.

When events are independent (such as independent traits or successive offspring from a cross), the probabilities are combined by means of the multiplication rule.

Multiplication Rule: The probability of two independent events, A and B, being realized simultaneously is given by the product of their separate probabilities.

In symbols, the multiplication rule is

$$Prob\{A \text{ and } B\} = Prob\{A\} \cdot Prob\{B\}$$

The multiplication rule can be used to answer questions like the following one: Of two offspring from the mating $Aa \times Aa$, what is the probability that both have the dominant phenotype? Because the mating is $Aa \times Aa$, the probability that any particular offspring has the dominant phenotype equals 3/4. The multiplication rule says that the probability that both of two offspring have the dominant phenotype is $3/4 \times 3/4 = 9/16$.

Here is a typical genetic question that can be answered by using the addition and multiplication rules together: Of two offspring from the mating $Aa \times Aa$, what is the probability of one dominant phenotype and one recessive? Sibships of one dominant phenotype and one recessive can come about in two different ways—with the dominant born first or with the dominant born second—and these outcomes are mutually exclusive. The probability of the first case is $3/4 \times 1/4$ and that of the second is $1/4 \times 3/4$; because the events are mutually exclusive, the probabilities are added. The answer is therefore

$$(3/4 \times 1/4) + (1/4 \times 3/4) = 2(3/4)(1/4)$$

The addition and multiplication rules are very powerful tools for calculating the probabilities of genetic events. Figure 2.14 shows how the rules are applied to determine the expected proportions of the nine different genotypes possible among the F_2 progeny produced by self-pollination of a $Ww\ Gg$ dihybrid.

In genetics, independence applies not only to the successive offspring formed by a mating, but also to genes that segregate according to the principle of independent assortment (Figure 2.15). The independence means that the multiplication rule can be used to determine the probability of

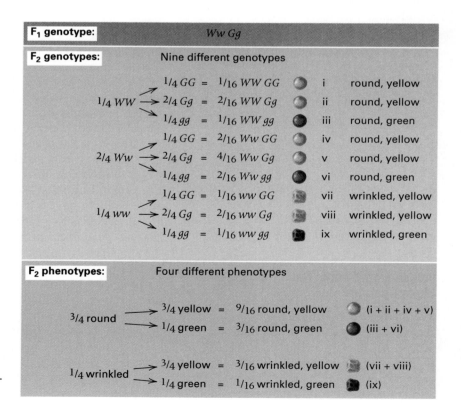

Figure 2.14
Example of the use of the addition and multiplication rules to determine the probabilities of the nine genotypes and four phenotypes in the F_2 progeny obtained from self-pollination of a dihybrid F_1. The Roman numerals are arbitrary labels identifying the F_2 genotypes.

(A)

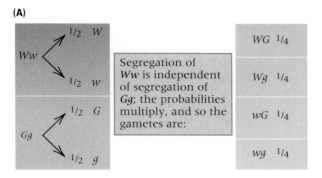

Ww ⟨ $1/2$ W / $1/2$ w ⟩	Segregation of *Ww* is independent of segregation of *Gg*; the probabilities multiply, and so the gametes are:
Gg ⟨ $1/2$ G / $1/2$ g ⟩	

WG $1/4$

Wg $1/4$

wG $1/4$

wg $1/4$

(B)

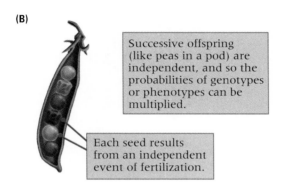

Successive offspring (like peas in a pod) are independent, and so the probabilities of genotypes or phenotypes can be multiplied.

Each seed results from an independent event of fertilization.

Figure 2.15
In genetics, two important types of independence are (A) independent segregation of alleles that show independent assortment and (B) independent fertilizations resulting in successive offspring. In these cases, the probabilities of the individual outcomes of segregation or fertilization are multiplied to obtain the overall probability.

the various types of progeny from a cross in which there is independent assortment among numerous pairs of alleles. This principle is the theoretical basis for the expected progeny types from a trihybrid cross, shown in Figure 2.12. One can also use the multiplication rule to calculate the probability of a specific genotype among the progeny of a cross. For example, if a quadruple heterozygote of genotype *Aa Bb Cc Dd* is self-fertilized, the probability of a quadruple heterozygote *Aa Bb Cc Dd* offspring is $(1/2)(1/2)(1/2)(1/2) = (1/2)^4$, or 1/16, assuming independent assortment of all four pairs of alleles.

2.4
Segregation in Human Pedigrees

Determining the genetic basis of a trait from the kinds of crosses that we have considered requires that we control matings between organisms and obtain large numbers of offspring to classify with regard to phenotype. The analysis of segregation by this method is not possible in human

beings, and it is not usually feasible for traits in large domestic animals. However, the mode of inheritance of a trait can sometimes be determined by examining the appearance of the phenotypes that reflect the segregation of alleles in several generations of related individuals. This is typically done with a family tree that shows the phenotype of each individual; such a diagram is called a **pedigree.** An important application of probability in genetics is its use in pedigree analysis.

Figure 2.16 depicts most of the standard symbols used in drawing human pedigree. Females are represented by circles and males by squares. (A diamond is used if the sex of an individual is unknown.) Persons with the phenotype of interest are indicated by colored or shaded symbols. For recessive alleles, heterozygous carriers are depicted with half-filled symbols. A mating between a female and a male is indicated by joining their symbols with a horizontal line, which is connected vertically to a second horizontal line below that connects the symbols for their offspring. The offspring within a sibship, called **siblings** or **sibs** regardless of sex, are represented from left to right in order of their birth.

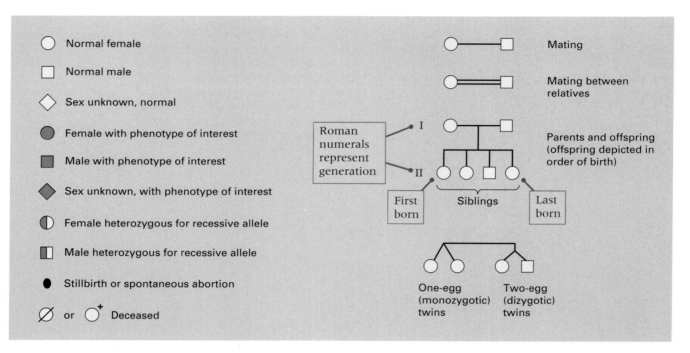

Figure 2.16
Conventional symbols used in depicting human pedigrees.

A pedigree for the trait **Huntington disease,** which is due to a dominant allele, is shown in Figure 2.17. The numbers in the pedigree are for convenience in referring to particular persons. The successive generations are designated by Roman numerals. Within any generation, all of the persons are numbered consecutively from left to right. The pedigree starts with the woman I-1 and the man I-2. He has Huntington disease, which is a progressive nerve degeneration that usually begins about middle age. It results in severe physical and mental disability and then death. The pedigree shows that the trait affects both sexes, that it is transmitted from affected parent to affected offspring, and that about half of all the offspring of an

Figure 2.17
Pedigree of a human family showing the inheritance of the dominant gene for Huntington disease. Females and males are represented by circles and squares, respectively. Red symbols indicate persons affected with the disease.

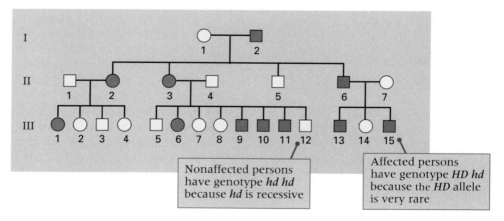

In Memoriam: This Land Is Your Land, This Land Is My Land

The Huntington's Disease Collaborative Research Group 1993
Comprising 58 authors
among 9 institutions
A Novel Gene Containing a Trinucleotide Repeat That is Expanded and Unstable on Huntington's Disease Chromosomes

Modern genetic research is sometimes carried out by large collaborative groups in a number of research institutions scattered across several countries. This approach is exemplified by the search for the gene responsible for Huntington disease. The search was highly publicized because of the severity of the disease, the late age of onset, and the dominant inheritance. Famed folk singer Woody Guthrie, who wrote "This Land Is Your Land" and other well-known tunes, died of the disease in 1967. When the gene was identified, it turned out to encode a protein (now called huntingtin) of unknown function that is expressed in many cell types throughout the body and not, as expected, exclusively in nervous tissue. Within the coding sequence of this gene is a trinucleotide repeat ($5'$-CAG-$3'$) that is repeated in tandem a number of times according to the general formula ($5'$-CAG-$3'$)$_n$. Among normal alleles, the number n of repeats ranges from 11 to 34 with an average of 18; among mutant alleles, the number of repeats ranges from 40 to 86. This tandem repeat is genetically unstable in that it can, by some unknown mechanism, increase in copy number ("expand"). In two cases in which a new mutant allele was analyzed, one had increased in repeat number from 36 to 44

and the other from 33 to 49. This is a mutational mechanism that is quite common in some human genetic diseases. The excerpt cites several other examples. The authors also emphasize that their discovery raises important ethical issues, including genetic testing, confidentiality, and informed consent.

Huntington's disease (HD) is a progressive neurodegenerative disorder characterized by motor disturbance, cognitive loss, and

---

We consider it of the utmost importance that the current internationally accepted guidelines and counseling protocols for testing people at risk continue to be observed, and that samples from unaffected relatives should not be tested inadvertently or without full consent.

---

psychiatric manifestations. It is inherited in an autosomal dominant fashion and affects approximately 1 in 10,000 individuals in most populations of European origin. The hallmark of HD is a distinctive choreic [jerky] movement disorder that typically has a subtle, insidious onset in the fourth to fifth decade of life and gradually worsens over a course of 10 to 20 years until death. . . . The genetic defect causing HD was assigned to chromosome 4 in one of the first successful linkage analyses using

DNA markers in humans. Since that time, we have pursued an approach to isolating and characterizing the HD gene based on progressively refining its localization. . . . [We have found that a] 500 kb segment is the most likely site of the genetic defect. [The abbreviation kb stands for kilobase pairs; 1 kb equals 1000 base pairs.] Within this region, we have identified a large gene, spanning approximately 210 kb, that encodes a previously undescribed protein. The reading frame contains a polymorphic (CAG)$_n$ trinucleotide repeat with at least 17 alleles in the normal population, varying from 11 to 34 CAG copies. On HD chromosomes, the length of the trinucleotide repeat is substantially increased. . . . Elongation of a trinucleotide repeat sequence has been implicated previously as the cause of three quite different human disorders, the fragile-X syndrome, myotonic dystrophy, and spino-bulbar muscular atrophy. . . . It can be expected that the capacity to monitor directly the size of the trinucleotide repeat in individuals "at risk" for HD will revolutionize testing for the disorder. . . . We consider it of the utmost importance that the current internationally accepted guidelines and counseling protocols for testing people at risk continue to be observed, and that samples from unaffected relatives should not be tested inadvertently or without full consent. . . . With the mystery of the genetic basis of HD apparently solved, [it opens] the next challenges in the effort to understand and to treat this devastating disorder.

Source: Cell 72: 971–983

---

affected parent are affected. These are characteristic features of simple Mendelian dominance. The dominant allele, *HD*, that causes Huntington disease is very rare. All affected persons in the pedigree have the heterozygous genotype *HD hd*, whereas nonaffected persons have the homozygous normal genotype *hd hd*.

A pedigree pattern for a trait due to a homozygous recessive allele is shown in Figure 2.18. The trait is **albinism**, absence of pigment in the skin, hair, and iris of the eyes. Both sexes can be affected, but the affected individuals need not have affected parents. The nonaffected parents are called **carriers** because they are heterozygous for the recessive allele; in a mating between carriers (*Aa* × *Aa*), each offspring has a 1/4 chance of being affected. The pedigree also illustrates another feature found with

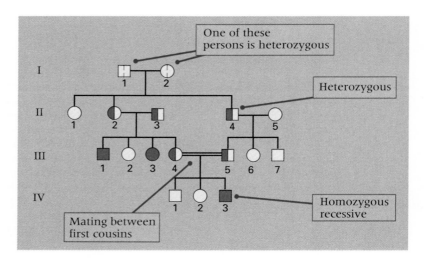

Figure 2.18
Pedigree of albinism. With recessive inheritance, affected persons (filled symbols) often have unaffected parents. The double horizontal line indicates a mating between relatives—in this case, first cousins.

recessive traits, particularly rare traits, which is that the parents of affected individuals are often related. A mating between relatives, in this case first cousins, is indicated with a double line connecting the partners.

Matings between relatives are important for observing rare recessive alleles, because when a recessive allele is rare, it is more likely to become homozygous through inheritance from a common ancestor than from parents who are completely unrelated. The reason is that the carrier of a rare allele may have many descendants who are carriers. If two of these carriers should mate (for example, in a first-cousin mating) the recessive allele can become homozygous with a probability of 1/4. Mating between relatives constitutes *inbreeding,* and the consequences of inbreeding are discussed further in Chapter 15.

2.5
Genetic Analysis

When a geneticist makes the statement that a single gene with alleles *P* and *p* determines whether the color of the flowers on a pea plant will be purple or white, the statement does not imply that this gene is the only one responsible for flower color. The statement means only that this particular gene affecting flower color has been identified owing to the discovery of the recessive *p* mutation that, when homozygous, changes the color from purple to white. Many genes beside *P* are also necessary for purple flower coloration. Among these are genes that encode enzymes in the biochemical pathway for the synthesis of the purple pigment, anthocyanin. A geneticist interested in understanding the genetic basis of flower color would rarely be satisfied in having identified only the *P* gene necessary for purple coloration. The ultimate goal of a genetic analysis of flower color would be to isolate at least one mutation in every gene necessary for purple coloration and then, through further study of the mutant phenotypes, determine the normal function of each of the genes that affect the trait.

The Complementation Test in Gene Identification

In a genetic analysis of flower color, a geneticist would begin by isolating many new mutants with white flowers. Although mutations are usually very rare, their frequency can be increased by treatment with radiation or certain chemicals. The isolation of a set of mutants, all of which show the same type of defect in phenotype, is called a **mutant screen.** Among the mutants that are isolated, some will contain mutations in genes already identified. For example, a genetic analysis of flower color in peas might yield one or more new mutations that changed the wildtype P allele into a defective p allele that prevents the formation of the purple pigment. Each of the $P \rightarrow p$ mutations might differ in DNA sequence, but all of the newly isolated p alleles would be defective forms of P that prevent formation of the purple pigment. On the other hand, a mutant screen should also yield mutations in genes not previously identified. Each of the new genes might be also represented by several recurrences of mutation, analogous to the multiple $P \rightarrow p$ mutations.

In a mutant screen for flower color, all of the new mutations are identified in plants with white flowers. Most of the new mutant alleles will encode an inactive protein needed for the formation of the purple pigment. The mutant alleles will be recessive, because in the homozygous recessive genotype, neither of the mutant alleles can produce the wildtype protein needed for pigmentation, and so the flowers will be white. In the heterozygous genotype, which carries one copy of the mutant allele along with one copy of the wildtype allele, the flowers will be purple because the wildtype allele codes for a functional protein that compensates for the defective protein encoded by the mutant.

Because white flowers may be caused by mutations in any of several genes, any two genotypes with white flowers may be homozygous recessive for alleles of the same gene or for alleles of different genes. After a mutant screen, how can the geneticist determine which pairs of mutations are alleles and which pairs of mutations are not alleles? The issue is illustrated for three particular white-flower mutations in Figure 2.19. Each of the varieties is homozygous for a recessive mutation that causes the flowers to be white. On the one hand, the varieties might carry separate occurrences of a mutation in the same gene; the mutations would be alleles. On

the other hand, each mutation might be in a different gene; they would not be alleles.

The issue of possible allelism of the mutations is resolved by observing the phenotype of the progeny produced from a cross between the varieties. As indicated in Figure 2.19, there are two possible outcomes of the cross. The F_1 progeny have either the wildtype phenotype (Figure 2.19A, purple flowers) or the mutant phenotype (Figure 2.19B, white flowers). If the progeny have purple flowers, it means that the mutations in the parental plants are in different genes; this result is called **complementation.** When complementation is observed, it implicates two different genes needed for purple flowers. In this example, mutant strain 1 is homozygous pp for the recessive p allele. The other parent is homozygous for a mutant allele in a different gene, designated cc. The complete genotype of the parental strains should therefore be written $pp\ CC$ for mutant strain 1 and $PP\ cc$ for mutant strain 2. The cross yields the F_1 genotype $Pp\ Cc$, which is heterozygous Pp and heterozygous Cc. Because p and c are both recessive, the phenotype of the F_1 progeny is purple flowers.

The other possible outcome of the cross is shown in Figure 2.19B. In this case, the F_1 progeny have white flowers, which is the mutant phenotype. This outcome is called **noncomplementation.** The lack of complementation indicates that both parental strains have a mutation in the same gene, because neither mutant strain can provide the genetic function missing in the other. In this example, mutant strain 1 is known to have the genotype pp. Mutant strain 3 is homozygous for a different mutation (possibly a recurrence of p). Because the F_1 has white flowers, the genotype of the F_1 must be pp. But the only source of the second p allele must be mutant strain 3, which means that mutant strain 3 is also homozygous pp. In the figure, the particular allele in mutant strain 3 is designated $p*$ to indicate that this mutation arose independently of the original p allele.

The kind of cross illustrated in Figure 2.19 is a **complementation test.** As we have seen, it is used to determine whether recessive mutations in each of two different strains are alleles of the same gene.

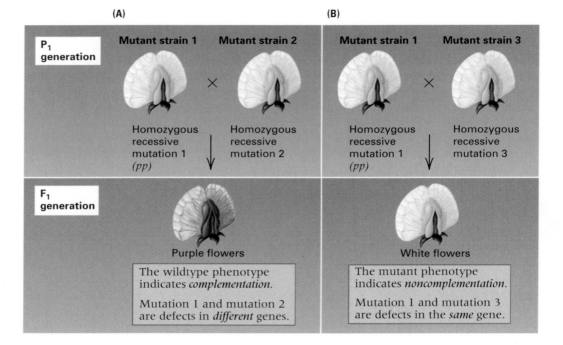

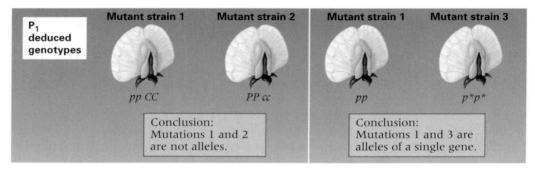

Figure 2.19

The complementation test reveals whether two recessive mutations are alleles of the same gene. In the complementation test, homozygous recessive genotypes are crossed. If the phenotype of the F_1 progeny is nonmutant (A), it means that the mutations in the parental strains are alleles of different genes. If the phenotype of the F_1 progeny is mutant (B), it means that the mutations in the parental strains are alleles of the same gene.

Because the result indicates the presence or absence of allelism, the complementation test is one of the key experimental operations in genetics. To illustrate the application of the test in practice, suppose a mutant screen were carried out to isolate new mutations for white flowers in peas. Starting with a true-breeding strain with purple flowers, we treat pollen with x rays and use the irradiated pollen to fertilize ovules to obtain seeds. The F_1 seeds are grown and the resulting plants allowed to self-fertilize, after which the F_2 plants are grown. A few of the F_1 seeds may contain a new mutation for white flowers, but because the white phenotype is recessive, the flower will be purple. However, the resulting F_1 plant will be heterozygous for the new white mutation, so self-fertilization will result in the formation of F_2 plants with a 3 : 1 ratio of purple : white flowers. Because mutations resulting in a particular phenotype are quite rare, even when induced by radiation, only a few among many thousands of self-fertilized plants, will be found to have a new white-flower mutation. Let us suppose that we were lucky enough to obtain four new mutations, in addition to the *p* and *c* mutations identified by the complementation test in Figure 2.19.

How are we going to name these four new mutations? We can make no assumptions about the number of genes represented. All four could be recurrences of either *p* or *c*. On the other hand, each of the four could be a new mutation in a different gene needed for flower color. For the moment, let us call the new mutations *x1*, *x2*, *x3*, and *x4*, where the "*x*" does not imply a gene but rather that the mutation was obtained with *x* irradiation. Each mutation is recessive and was identified through the white flowers of the homozygous recessive F_2 seeds (for example, *x1 x1*).

Now the complementation test is used to classify the "*x*" mutations into groups.

Figure 2.20 shows that the results of a complementation test are typically reported in a triangular array of + and − signs. The crosses that yield F_1 progeny with the wild-type phenotype (in this case, purple flowers) are denoted with a + in the box where imaginary lines from the male parent and the female parent intersect. The crosses that yield F_1 progeny with the mutant phenotype (white flowers) are denoted with a − sign. The + signs indicate complementation between the mutant alleles in the parents; the − signs indicate lack of complementation. The bottom half of the triangle is unnecessary because the reciprocal of each cross produces F_1 progeny with the same genotype and phenotype as the cross that is shown. The diagonal elements are also unnecessary, because a cross between any two organisms carrying the identical mutation, for

Figure 2.20

Results of complementation tests among six mutant strains of peas, each homozygous for a recessive allele resulting in white flowers. Each box gives the phenotype of the F_1 progeny of a cross between the male parent whose genotype is indicated in the far left column and the female parent whose genotype is indicated in the top row.

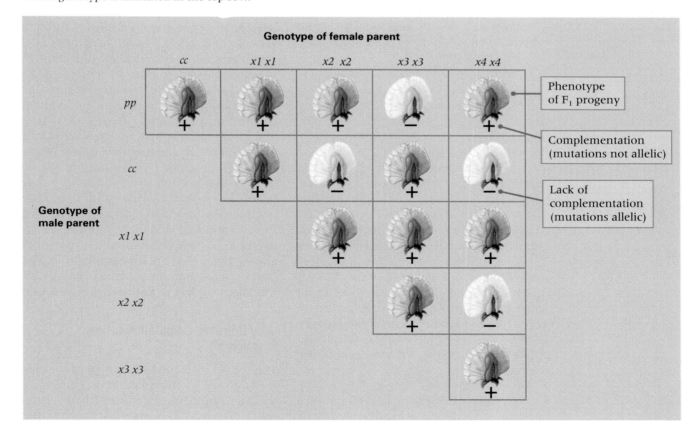

example, *x1 x1* × *x1 x1*, must yield homozygous recessive *x1 x1* progeny, which will be mutant. As we have seen in Figure 2.19, complementation in a cross means that the parental strains have their mutations in different genes. Lack of complementation means that the parental mutations are in the same gene. The principle underlying the complementation test is

> **The Principle of Complementation:** If two recessive mutations are alleles of the same gene, then the phenotype of an organism containing both mutations is mutant; if they are alleles of different genes, then the phenotype of an organism containing both mutations is wildtype (nonmutant).

In interpreting complementation data such as those in Figure 2.20, we actually apply the principle the other way around. Examination of the phenotype of the F_1 progeny of each possible cross reveals which of the mutations are alleles of the same gene:

> In a complementation test, if the combination of two recessive mutations results in a mutant phenotype, then the mutations are regarded as alleles of the same gene; if the combination results in a wildtype phenotype, then the mutations are regarded as alleles of different genes.

A convenient way to analyze the data in Figure 2.20 is to arrange the alleles in a circle as shown in Figure 2.21A. Then, for each possible pair of mutations, connect the pair by a straight line if the mutations *fail* to complement (Figure 2.21B). According to the principle of complementation, the lines must connect mutations that are alleles of each other because, in a complementation test, lack of complementation means that the mutations are alleles. In this example, mutation *x3* is an allele of *p*, so *x3* and *p* are different mutant alleles of the gene *P*. Similarly, the mutations *x2*, *x4*, and *c* are different mutant alleles of the gene *C*. The mutation *x1* does complement all of the others. It represents a third gene, different from *P* and *C*, that affects flower coloration.

In an analysis like that in Figure 2.21, each of the groups of noncomplementing mutations is called a **complementation group.** As we have seen, each complementation group defines a gene.

> A gene is defined experimentally as a set of mutations that make up one complementation group. Any pair of mutations in such a group fail to complement one another and result in an organism with an observable mutant phenotype.

The mutations in Figure 2.21 therefore represent three genes, a mutation in any one of which results in white flowers. The gene *P* is represented by the alleles *p* and *x3*; the gene *C* is represented by the alleles *c, x2,* and *x4*; and the allele *x1* represents a third gene different from either *P* or *C*. Each gene coincides with one of the complementation groups.

At this point in a genetic analysis, it is possible to rename the mutations to indicate which ones are true alleles. Because the *p* allele already had its name before the mutation screen was carried out to obtain more flower-color mutations, the new allele of *p*, *x3*, should be renamed to reflect its allelism with *p*. We might rename the *x3* mutation p_3, for example, using the subscript to indicate that p_3 arose independently of *p*. For similar reasons, we might rename the *x2* and *x4* mutations c_2 and c_4 to reflect their allelism with the original *c* mutation and to convey their independent origins. The *x1* mutation represents an allele of a new gene to which we can assign a name arbitrarily. For example, we might call the mutation *albus* (Latin for white) and assign the *x1* allele the new name *alb*. The wildtype dominant allele of *alb*, which is necessary for purple coloration, would then be symbolized as *Alb* or as alb^+. The procedure of sorting new mutations into complementation groups and renaming them according to their allelism is an example of how geneticists identify genes and name alleles. Such renaming of alleles is the typical manner in which genetic terminology evolves as knowledge advances.

Why Does the Complementation Test Work?

There is an old Chinese saying that the correct naming of things is the beginning of wisdom, and this is certainly true in the case of genes. The proper renaming of the *p, c,* and *alb* mutations to indicate which mutations are alleles of the three genes is a wise way to create a terminology that indicates, for each possible genotype, what the phenotype will be with regard to flower color. A purple flower requires the presence of at least one copy of each of the wildtype *P, C,* and *Alb* alleles. Any genotype that contains two mutant alleles of *P* will have white flowers. These genotypes are pp, p_3p_3, and pp_3. Likewise, any genotype that contains two mutant alleles of *C* will have white flowers. These genotypes are cc, c_2c_2, c_4c_4, cc_2, cc_4, and c_2c_4. Finally, any genotype that contains two mutant alleles of *Alb* will have white flowers. In this case there is only one such genotype, *alb alb.*

The biological reason why the screen for flower-color mutants yielded mutations in each of three genes is based on the biochemical pathway by which the purple pigment is synthesized in the flowers. Examination of the biochemical pathway also explains why the complementation test works. The pathway is illustrated in Figure 2.22. The purple pigment anthocyanin is produced from a colorless precursor by way of two colorless intermediate compounds denoted X and Y. Each arrow represents a "step" in the pathway, a biochemical conversion from one substance to the next along the way. Each step requires an enzyme encoded by the wildtype allele of the gene indicated at the top. The allele *P,* for example, codes for the enzyme required in the last step in the pathway, which converts intermediate compound Y into anthocyanin. If this enzyme is missing (or is present in an inactive form), the intermediate substance Y cannot be converted into anthocyanin. The pathway is said to be "blocked" at this step. Although the block causes an increase in the concentration of Y inside the cell (because the precursor is still converted into X, and X into Y), no purple pigment is produced and the flower remains white.

Each of the mutant alleles listed across the bottom of the pathway codes for an inactive form of the corresponding enzyme. Any genotype that is homozygous for any of the mutations fails to produce an active form of the enzyme. For example, because the mutant *alb* allele codes for an inactive form of the enzyme for the first

This connecting line means that *p* and *x3* fail to complement one another when the parents are crossed; they are alleles of the *P* gene.

These connecting lines mean that *c, x2,* and *x4* fail to complement one another in all combinations in which the parents are crossed; they are alleles of the *C* gene.

x1 complements all the other alleles; it represents a third gene, different from *P* and *C,* that affects flower coloration.

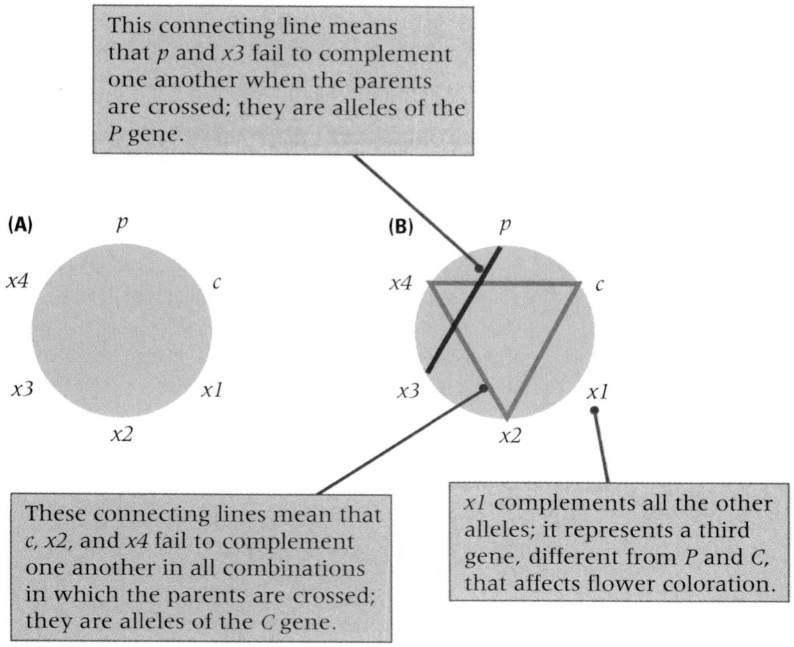

Figure 2.21
A method for interpreting the results of complementation tests. (A) Arrange the mutations in a circle. (B) Connect by a straight line any pair of mutations that fails to complement (that yields a mutant phenotype); any pair of mutations so connected are alleles of the same gene. In this example, there are three complementation groups, each of which represents a single gene needed for purple flower coloration.

step in the pathway, the genotype *alb alb* has no active enzyme for this step. In the homozygous mutant, the pathway is blocked at the first step, so the precursor is not converted to intermediate X. Because no X is produced, there can be no Y, and without Y there can be no anthocyanin, and so homozygous *alb alb* results in white flowers. The *alb* allele is recessive, because in the heterozygous genotype *Alb alb,* the wildtype *Alb* allele codes for a functional enzyme for the first step in the pathway, and so the pathway is not blocked.

Mutant alleles of the *C* gene block the second step in the pathway. In this case, an inactive enzyme is produced not only in the homozygous genotypes cc, c_2c_2, and c_4c_4, but also in the genotypes cc_2, cc_4, and c_2c_4. In the last three genotypes, each mutant allele encodes a *different* (but still inactive) form of the enzyme, so the pathway is blocked at step 2, and the color of the flowers is white. The c, c_2, and c_4 alleles are all in the same complementation group (they fail to complement one another) because they all encode inactive forms of the same enzyme.

A similar situation holds for mutations in the *P* gene. The wildtype *P* allele encodes the enzyme for the final step in the pathway to anthocyanin. Any of the genotypes pp, pp_3, and p_3p_3 lacks a functional form of the enzyme, which blocks the pathway at the last step and results in white flowers. The alleles p and p_3 are in the same complementation group because they are both mutations in the *P* gene.

Multiple Alleles

The *C* and *P* genes in Figure 2.22 also illustrate the phenomenon of **multiple alleles,** in which there are more than two allelic forms of a given gene. Because the wildtype form of each gene also counts as an allele, there are two alleles of the *Alb* gene (*Alb* and *alb*), four alleles of the *C* gene (C, c, c_2, and c_4), and three alleles of the *P* gene (P, p, and p_3).

When a complementation test reveals that two independent mutations are alleles of the same gene, one does not know whether the mutant alleles have identical nucleotide sequences in the DNA. Recall from Chapter 1 that, at the level of DNA, a gene is a sequence of nucleotides that specifies the sequence of amino acids in a protein. Each nucleotide contains a base,

either A (adenine), T (thymine), G (guanine), or C (cytosine), so a gene of *n* nucleotides can theoretically mutate at any of the positions to any of the three other nucleotides. The number of possible single-nucleotide differences in a gene of length *n* is therefore $3 \times n$; each of these DNA sequences, if it exists in the population, is an allele. When $n = 5000$, for example, there are potentially 15,000 alleles (not counting any of the possibilities with more than one nucleotide substitution). Most of the potential alleles may not actually exist at any one time, but many of them may be present in any population. The following rules govern the number of alleles.

- A gamete may contain only one allele of each gene.

- Any particular organism or cell may contain up to two different alleles.

- A population of organisms may contain any number of alleles

Many genes have multiple alleles. For example, the human blood groups designated A, B, O, or AB are determined by three types of alleles denoted I^A, I^B, and I^O, and the blood group of any person is determined by the particular pair of alleles present in his or her genotype. (Actually, there are two slightly different variants of the I^A

Figure 2.22
Biochemical pathway for the synthesis of the purple pigment anthocyanin from a colorless precursor and colorless intermediates X and Y. Each step (arrow) in the pathway is a biochemical conversion that requires an enzyme encoded in the wildtype allele of the gene indicated.

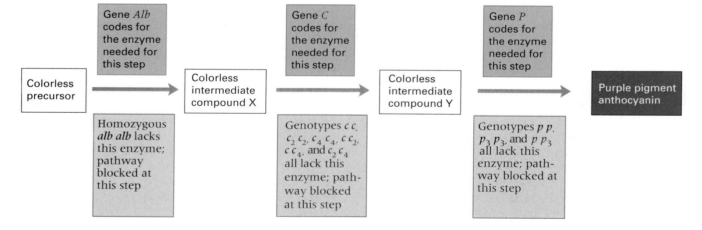

allele, so four alleles can be distinguished in this case.)

In modern genetics, multiple alleles are encountered in two major settings. One is in genetic analysis when a mutant screen potentially yields two or more mutant alleles of each of a large number of genes. For example, in the early 1980s, mutant screens were carried out in *Drosophila* to obtain new recessive mutations that blocked embryonic development and so led to the death of homozygous recessive embryos. The screens resulted in the identification of approximately 18,000 such mutations, the study of which ultimately earned a 1995 Nobel Prize for Christiane Nüsslein-Volhard and Eric Wieschaus (they shared the prize with Edward B. Lewis, who had already done pioneering work in the genetics of *Drosophila* development).

Geneticists also encounter multiple alleles in studies of natural populations of organisms. In most populations, including the human population, each gene may have many alleles that differ slightly in nucleotide sequence. Most of these alleles, even though they differ in one or more nucleotides in the DNA sequence, are able to carry out the normal function of the gene and produce no observable difference in phenotype.

In human populations, it is not unusual for a gene to have many alleles. Genes used in DNA typing, such as those employed in criminal investigations, usually have multiple alleles in the population. For each of these genes, any person can have no more than two alleles, but often there are 20 or more alleles in the population as a whole. Hence, any two unrelated people are not likely to have the same genotype, especially if several different genes, each with multiple alleles, are examined. Similarly, in the inherited recessive diseases *cystic fibrosis* and *phenylketonuria,* more than 200 different defective alleles of each gene have been identified in studies of affected children throughout the world. The "normal" form of each gene also exists in many alternative forms. Indeed, for most genes in most populations, the "normal" or "wildtype" allele is not a single nucleotide sequence but rather a set of different nucleotide sequences, each capable of carrying out the normal function of the gene.

In some cases, the multiple alleles of a gene exist merely by chance and reflect the history of mutations that have taken place in the population and the dissemination of these mutations among population subgroups by migration and interbreeding. In other cases, there are biological mechanisms that favor the maintenance of a large number of alleles. For example, genes that control self-sterility in certain flowering plants can have large numbers of allelic types. This type of self-sterility is found in species of red clover that grow wild in many pastures. The self-sterility genes prevent self-fertilization because a pollen grain can undergo pollen tube growth and fertilization only if it contains a self-sterility allele different from either of the alleles present in the flower on which it lands. In other words, a pollen grain containing an allele already present in a flower will not function on that flower. Because all pollen grains produced by a plant must contain one of the self-sterility alleles present in the plant, pollen cannot function on the same plant that produced it, and self-fertilization cannot take place. Under these conditions, any plant with a new allele has a selective advantage, because pollen that contains the new allele can fertilize all flowers except those on the same plant. Through evolution, populations of red clover have accumulated hundreds of alleles of the self-sterility gene, many of which have been isolated and their DNA sequences determined. Many of the alleles differ at multiple nucleotide sites, which implies that the alleles in the population are very old.

2.6
Modified Dihybrid Ratios Caused by Epistasis

In Figure 2.22 we saw how the products of several genes may be necessary to carry out all the steps in a biochemical pathway. In genetic crosses in which two mutations that affect different steps in a single pathway are both segregating, the typical F_2 dihybrid ratio of 9 : 3 : 3 : 1 is not observed. One way in which the ratio may be modified is illustrated by the interaction of the *C,*

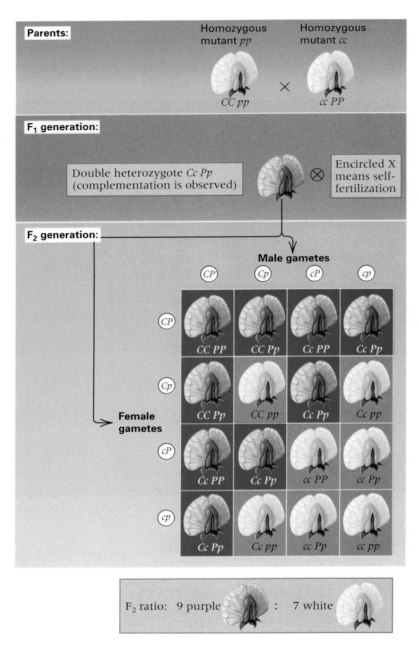

Figure 2.23

A cross showing epistasis in the determination of flower color in sweet peas. Formation of the purple pigment requires the dominant allele of both the *C* and *P* genes. With this type of epistasis, the dihybrid F₂ ratio is modified to 9 purple : 7 white.

in different genes. Self-fertilization of the F₁ plants (indicated by the encircled cross sign) results in the F₂ progeny genotypes shown in the Punnett square. Because only the progeny with at least one *C* allele and at least one *P* allele have purple flowers and all the rest have white flowers, the ratio of purple flowers to white flowers in the F₂ generation is 9 : 7.

Any type of gene interaction that results in the F₂ dihybrid ratio of 9 : 3 : 3 : 1 being modified into some other ratio is called *epistasis*. For a trait determined by the interaction of two genes, each with a dominant allele, there are only a limited number of ways in which the 9 : 3 : 3 : 1 dihybrid ratio can be modified. The possibilities are illustrated in Figure 2.24. In part A are the genotypes produced in the F₂ generation by independent assortment and the ratios in which the genotypes occur. In the absence of epistasis, the F₂ ratio of phenotypes is 9 : 3 : 3 : 1. The possible modified ratios are shown in part B of the figure. In each row, the color coding indicates phenotypes that are indistinguishable because of epistasis, and the resulting modified ratio is given. For example, in the modified ratio at the bottom, the phenotypes of the "3 : 3 : 1" classes are indistinguishable, resulting in a 9 : 7 ratio. This is the ratio observed in the segregation of the *C, c* and *P, p* alleles in Figure 2.23, and the 9 : 7 ratio is the ratio of purple flowers to white flowers. Taking all the possible modified ratios in Figure 2.24B together, there are nine possible dihybrid ratios when both genes show complete dominance. Examples are known of each of the modified ratios. However, the most frequently encountered modified ratios are 9 : 7, 12 : 3 : 1, 13 : 3 , 9 : 4 : 3, and 9 : 6 : 1. The types of epistasis that result in these modified ratios are illustrated in the following examples, taken from a variety of organisms. Other examples can be found in the problems at the end of the chapter.

9 : 7 This is the ratio observed when a homozygous recessive mutation in either or both of two different genes results in the same mutant phenotype. It is exemplified by the segregation of purple and white flowers in Figure 2.23. Genotypes that are *C—* for the *C* gene

c and *P, p* allele pairs affecting flower coloration. Figure 2.23 shows a cross between the homozygous mutants *pp* and *cc*. The phenotype of the plants in the F₁ generation is purple flowers; complementation is observed because the *p* and *c* mutations are

F₂ ratio: 9 purple : 7 white

and $P-$ for the P gene have purple flowers; all other genotypes have white flowers. In this notation, the dash in $C-$ means that the unspecified allele could be either C or c, and so $C-$ refers collectively to CC and Cc. Similarly, the dash in $P-$ means that the unspecified allele could be either P or p.

12 : 3 : 1 A modified dihybrid ratio of the 12 : 3 : 1 variety results when the presence of a dominant allele of one gene masks the genotype of a different gene. For example, if the $A-$ genotype renders the $B-$ and bb genotypes indistinguishable, then the dihybrid ratio is 12 : 3 : 1 because the $A- B-$ and $A- bb$ genotypes are expressed as the same phenotype.

In a genetic study of the color of the hull in oat seeds, a variety having white hulls was crossed with a variety having black hulls. The F_1 hybrid seeds had black hulls. Among 560 progeny in the F_2 generation produced by self-fertilization of the F_1, the following seed phenotypes were observed in the indicated numbers:

418 black hulls
106 gray hulls
 36 white hulls

Note that the observed ratio of phenotypes is 11.6 : 2.9 : 1, or very nearly 12 : 3 : 1. These results can be explained by a genetic hypothesis in which the black-hull phenotype results from the

Figure 2.24
Modified F_2 dihybrid ratios. (A) The F_2 genotypes of two independently assorting genes with complete dominance result in a 9 : 3 : 3 : 1 ratio of phenotypes if there is no interaction between the genes (epistasis). (B) If there is epistasis that renders two or more of the phenotypes indistinguishable, indicated by the colors, then the F_2 ratio is modified. The most frequently encountered modified ratios are 9 : 7, 12 : 3 : 1, 13 : 3, 9 : 4 : 3, and 9 : 6 : 1.

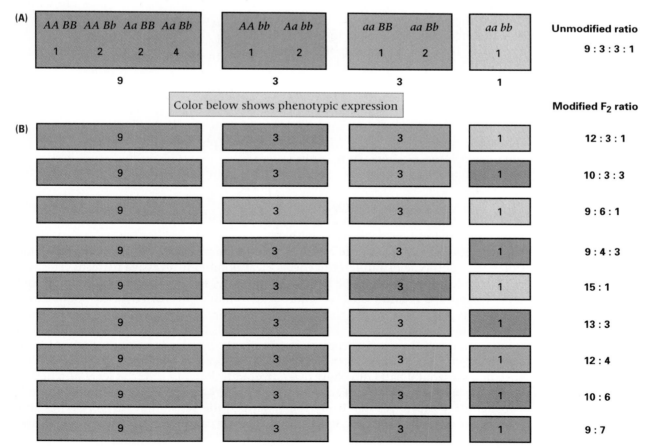

presence of a dominant allele (say, A) and the gray-hull phenotype results from another dominant allele (say, B) whose effect is apparent only in the aa homozygotes. On the basis of this hypothesis, the original true-breeding varieties must have had genotypes $aa\ bb$ (white) and $AA\ BB$ (black). The F_1 has genotype $Aa\ Bb$ (black). If the A,a allele pair and the B,b allele pair undergo independent assortment, then the F_2 generation is expected to have the following composition of genotypes:

9/16	$A-$	$B-$	(black hull)
3/16	$A-$	bb	(black hull)
3/16	aa	$B-$	(gray hull)
1/16	aa	bb	(white hull)

This type of epistasis accounts for the 12 : 3 : 1 ratio.

13 : 3 This type of epistasis is illustrated by the difference between White Leghorn chickens (genotype $CC\ II$) and White Wyandotte chickens (genotype $cc\ ii$. Both breeds have white feathers because the C allele is necessary for colored feathers, but the I allele in White Leghorns is a dominant inhibitor of feather coloration. The F_1 generation of a dihybrid cross between these breeds has the genotype $Cc\ Ii$, which is expressed as white feathers because of the inhibitory effects of the I allele. In the F_2 generation, only the $C-\ ii$ genotype has colored feathers, so there is a 13 : 3 ratio of white : colored.

9 : 4 : 3 This dihybrid ratio (often stated as 9 : 3 : 4) is observed when homozygosity for a recessive allele with respect to one gene masks the expression of the genotype of a different gene. For example, if the aa genotype has the same phenotype regardless of whether the genotype is $B-$ or bb, then the 9 : 4 : 3 ratio results.

In the mouse, the grayish coat color called "agouti" is produced by the presence of a horizontal band of yellow pigment just beneath the tip of each hair. The agouti pattern results from the presence of a dominant allele A, and in

aa animals the coat color is black. A second dominant allele, C, is necessary for the formation of hair pigments of any kind, and cc animals are albino (white fur). In a cross of $AA\ CC$ (agouti) $\times aa\ cc$ (albino), the F_1 progeny are $Aa\ Cc$ and agouti. Crosses between F_1 males and females produce F_2 progeny in the following proportions:

9/16	$A-$	$C-$	(agouti)
3/16	$A-$	cc	(albino)
3/16	aa	$C-$	(black)
1/16	aa	cc	(albino)

The dihybrid ratio is therefore 9 agouti : 4 albino : 3 black.

9 : 6 : 1 This dihybrid ratio is observed when homozygosity for a recessive allele of either of two genes results in the same phenotype, but the phenotype of the double homozygote is distinct. For example, red coat color in Duroc-Jersey pigs requires the presence of two dominant alleles R and S. Pigs of genotype $R-\ ss$ and $rr\ S-$ have sandy-colored coats, and $rr\ ss$ pigs are white. The F_2 dihybrid ratio is therefore

9/16	$R-$	$S-$	(red)
3/16	$R-$	ss	(sandy)
3/16	rr	$S-$	(sandy)
1/16	rr	ss	(white)

The 9 : 6 : 1 ratio results from the fact that both single recessives have the same phenotype.

2.7
Complications in the Concept of Dominance

In Mendel's experiments, all traits had clear dominant-recessive patterns. This was fortunate, because otherwise he might not have made his discoveries. Departures from strict dominance are also frequently observed. In fact, even for such a classical trait as round versus wrinkled seeds in peas, it is an oversimplification to say that round is dominant. At the level of whether a seed is round or wrinkled, round is dominant in the sense that the genotypes WW and Ww cannot be distinguished by the outward appearance of the seeds. However, as emphasized in Chapter 1, every gene potentially affects many traits. It often happens that the same pair of alleles shows complete dominance for one trait but not complete dominance for another trait. For example, in the case of round versus wrinkled seeds, the genetic defect in wrinkled seeds is the absence of an active form of an enzyme called starch-branching enzyme I (SBEI), which is needed for the synthesis of a branched-chain form of starch known as amylopectin. Compared with homozygous *WW*, seeds that are heterozygous *Ww* have only half as much SBEI, and seeds that are homozygous *ww* have virtually none (Figure 2.25A). Homozygous *WW* peas contain large, well-rounded starch grains, with the result that the seeds retain water and shrink uniformly as they ripen, so they do not become wrinkled. In homozygous *ww* seeds, the starch grains lack amylopectin; they are irregular in shape, and when these seeds ripen, they lose water too rapidly and shrink unevenly, resulting in the wrinkled phenotype observed (Figure 2.25B and C).

The *w* allele also affects the shape of the starch grains in *Ww* heterozygotes. In heterozygous seeds, the starch grains are intermediate in shape (Figure 2.25B). Nevertheless, their amylopectin content is high enough to result in uniform shrinking of the seeds and no wrinkling (Figure 2.25C). Thus there is an apparent paradox of dominance. If we consider only the overall shape of the seeds, round is dominant over wrinkled. There are only two phenotypes. If we examine the shape of the starch grains with a microscope, all three genotypes can be distinguished from each other: large, rounded starch grains in *WW*; large, irregular grains in *Ww*; and small, irregular grains in *ww*. If we consider the amount of the SBEI enzyme, the *Ww* genotype has an amount about halfway between the amounts in *WW* and *ww*.

The round, wrinkled pea example in Figure 2.25 makes it clear that "dominance" is not simply a property of a particular pair of alleles no matter how the resulting phenotypes are observed. When a gene affects multiple traits (as most genes do), then a particular pair of alleles might

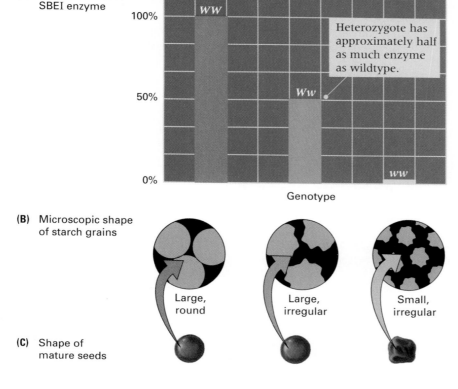

(A) Amount of active SBEI enzyme

100%

WW

Heterozygote has approximately half as much enzyme as wildtype.

50%

Ww

0%

ww

Genotype

(B) Microscopic shape of starch grains

Large, round

Large, irregular

Small, irregular

(C) Shape of mature seeds

Figure 2.25
Phenotypic expression of three traits affected by Mendel's alleles *W* and *w* determining round versus wrinkled seeds. (A) Relative amounts of starch-branching enzyme I (SBEI); the enzyme level in the heterozygous genotype is about halfway between the levels in the homozygous genotypes. (B) Size and shape of the microscopic starch grains; the heterozygote is intermediate. (C) Effect on shape of mature seeds; for seed shape, *W* is dominant over *w*.

show simple dominance for some traits but not others. The general principle illustrated in Figure 2.25 is:

> The phenotype consists of many different physical and biochemical attributes, and dominance may be observed for some of these attributes and not for others. Thus dominance is a property of a pair of alleles in relation to a particular attribute of phenotype.

Amorphs, Hypomorphs, and Other Types of Mutations

We have seen that a wildtype allele can potentially undergo mutation at any of a large number of nucleotide sites in the DNA, resulting in multiple alleles of a gene. In a series of multiple alleles, some alleles may have a more drastic effect on the phenotype than others. For example, one mutant allele may render the corresponding enzyme completely inactive, whereas another mutant allele may impair the enzyme in such a way as to cause only a partial loss of enzyme activity. Geneticists sometimes classify mutations according to the severity of their effects.

A mutation such as Mendel's *wrinkled* mutation, which encodes an inactive form of the SBEI enzyme, is often called an **amorph.** At the molecular level, an amorphic mutation may result from an amino acid replacement that inactivates the enzyme or even from a deletion of the gene so that no enzyme is produced. A mutation that reduces the enzyme level, but does not eliminate it, is called a **hypomorph.** Hypomorphic mutations typically result from amino acid replacements that impair enzyme activity or that prevent the enzyme from being produced at the normal level. As the prefix *hyper* implies, a **hypermorph** produces a greater-than-normal enzyme level, typically because the mutation changes the regulation of the gene in such a way that the gene product is overproduced.

Relative to their effects on the protein product of the gene they affect, most mutations can be classified as amorphs, hypomorphs, or hypermorphs. They result in none, less, or more of the enzyme activity produced by the wildtype, nonmutant allele. But other types of mutations also arise. A **neomorph** is a type of mutation that qualitatively alters the action of a gene. For example, a neomorph may cause a gene to become active in a type of cell or tissue in which the gene is not normally active. Or a neomorph can result in the expression of a gene in development at a time during which the wildtype gene is not normally expressed. Neomorphic mutations in a *Drosophila* gene called *eyeless*, which cause the wildtype gene product to be expressed in non–eye-forming tissues, can result in the development of parts of compound

(A)

(B)

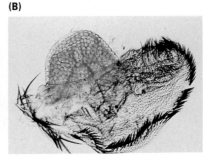

Figure 2.26

Ecoptic expression of the wildtype allele of the *eyeless* gene in *Drosophila* results in misplaced eye tissue. (A) An adult head in which both antennae form eye structures. (B) A wing with eye tissue growing out from it. (C) A single antenna in which most of the third segment consists of eye tissue. (D) Middle leg with an eye outgrowth at the base of the tibia. [Courtesy of G. Halder and W. J. Gehring. From G. Halder, P. Callaerts, and W. J. Gehring, *Science* 1995. 267: 1788.]

(C)

(D)

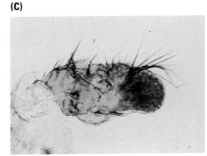 

eyes, complete with eye pigments, in abnormal locations. The locations can be anywhere the wildtype *eyeless* gene is expressed, including on the legs or mouthparts, in the abdomen, or on the wings (Figure 2.26). Expression of a wildtype gene in an abnormal location is called **ectopic expression.**

Another type of mutation is an **antimorph,** whose mutant product antagonizes the normal product of the gene. In some cases this occurs through an amino acid replacement that causes the mutant protein to combine with the wildtype protein into an inactive complex. These various terms for mutations were coined by Herman J. Muller in 1931. Muller also discovered that mutations can be caused by x rays (Chapter 13). Many x-ray–induced mutations are associated with major disruptions or rearrangements of the DNA sequence, which result in unusual types of patterns of expression of the affected genes. Muller's terms were useful for describing such mutations, and they have come into widespread use for discussing other types of mutations as well.

Incomplete Dominance

When the phenotype of the heterozygous genotype lies in the range between the phenotypes of the homozygous genotypes, there is said to be **incomplete dominance.** Most genes code for enzymes, and each allele in a genotype often makes its own contribution to the total level of the enzyme in the cell or organism. In such cases, the phenotype of the heterozygote falls in the range between the phenotypes of the corresponding homozygotes, as illustrated in Figure 2.27. There is no settled terminology for the situation: the terms *incomplete dominance, partial dominance,* and *semidominance* are all in use.

A classical example of incomplete dominance concerns flower color in the snapdragon *Antirrhinum* (Figure 2.28). In wildtype flowers, a red type of anthocyanin pigment is formed by a sequence of enzymatic reactions. A wildtype enzyme, encoded by the *I* allele, is limiting to the rate of the overall reaction, so the amount of red pigment is determined by the amount of enzyme that the *I* allele pro-

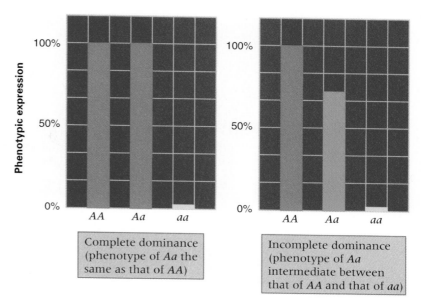

Figure 2.27

Levels of phenotypic expression in heterozygotes with complete dominance and with incomplete dominance.

duces. The alternative *i* allele codes for an inactive enzyme, and *ii* flowers are ivory in color. Because the amount of the critical enzyme is reduced in *Ii* heterozygotes, the amount of red pigment in the flowers is reduced also, and the effect of the dilution is to make the flowers pink.

The result of Mendelian segregation is observed directly when snapdragons that differ in flower color are crossed. For example, a cross between plants from a true-breeding red-flowered variety and a true-breeding ivory-flowered variety results in F_1 plants with pink flowers. In the F_2 progeny obtained by self-pollination of the F_1 hybrids, one experiment resulted in 22 plants with red flowers, 52 with pink flowers, and 23 with ivory flowers. The numbers agree fairly well with the Mendelian ratio of 1 dominant homozygote : 2 heterozygotes : 1 recessive homozygote. In agreement with the predictions from simple Mendelian inheritance, the red-flowered F_2 plants produced only red-flowered progeny, the ivory-flowered plants produced only ivory-flowered progeny, and the pink-flowered plants produced red, pink, and ivory progeny in the proportions 1/4 red : 1/2 pink : 1/4 ivory.

Incomplete dominance is often observed when the phenotype is quantitative

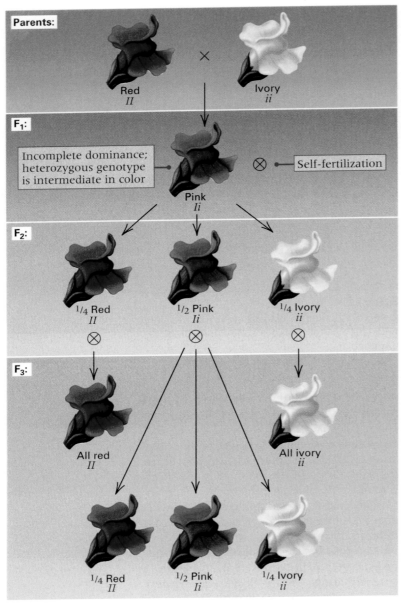

Figure 2.28
Absence of dominance in the inheritance of flower color in snapdragons.

Parents:
Red
II
×
Ivory
ii

F₁:
Incomplete dominance; heterozygous genotype is intermediate in color
Self-fertilization
Pink
Ii

F₂:
¼ Red
II
½ Pink
Ii
¼ Ivory
ii

F₃:
All red
II
All ivory
ii
¼ Red
II
½ Pink
Ii
¼ Ivory
ii

homozygotes, and therefore there is incomplete dominance.

Codominance and the Human ABO Blood Groups

A special term, **codominance,** refers to a situation in which the phenotype of a heterozygous genotype is a mixture of the phenotypes of both of the corresponding homozygous genotypes. In such cases, the heterozygous phenotype is not intermediate between the homozygous genotypes (like pink snapdragons) but rather has the characteristics of both homozygous genotypes.

What we mean by "has the characteristics of both homozygous genotypes" is illustrated by one of the classical examples of codominance. These are the alleles that determine the A, B, AB, and O human blood groups, which were discussed earlier in the context of multiple alleles. Blood type is determined by the types of polysaccharides (polymers of sugars) present on the surface of red blood cells. Two different polysaccharides, A and B, can be formed. Both are formed from a precursor substance that is modified by the enzyme product of either the I^A or the I^B allele. The gene products are transferase enzymes that attach either of two types of sugar units to the precursor (Figure 2.29). People of genotype $I^A I^A$ produce red blood cells having only the A polysaccharide and are said to have blood type A. Those of genotype $I^B I^B$ have red blood cells with only the B polysaccharide and have blood type B. Heterozygous $I^A I^B$ people have red cells with both the A and B polysaccharides and have blood type AB. The $I^A I^B$ genotype illustrates codominance, because the heterozygous genotype has the characteristics of both homozygous genotypes—in this case the presence of both the A and the B carbohydrate on the red blood cells.

The third allele, I^O, does not show codominance. It encodes a defective enzyme that leaves the precursor unchanged; neither the A nor the B type of polysaccharide is produced. Homozygous $I^O I^O$ persons therefore lack both the A and the B polysaccharides; they are said to have blood type O. In $I^A I^O$ heterozygotes, pres-

rather than discrete. A trait that is *quantitative* can be measured on a continuous scale; examples include height, weight, number of eggs laid by a hen, time of flowering of a plant, and amount of enzyme in a cell or organism. A trait that is *discrete* is all or nothing; examples include round versus wrinkled seeds, and yellow versus green seeds. With a phenotype that is quantitative, the measured value of a heterozygote usually falls in the range between the

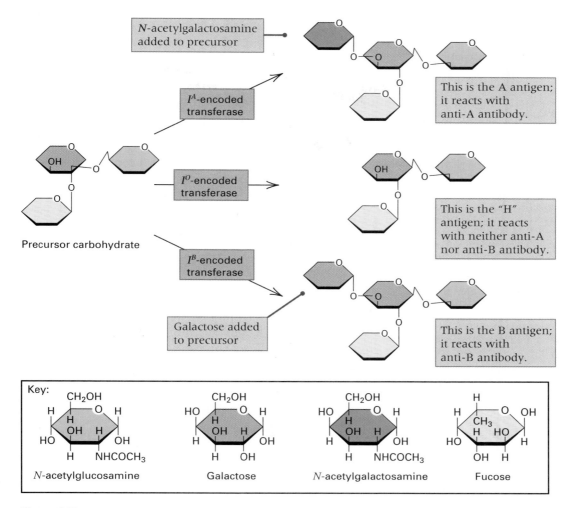

N-acetylgalactosamine
added to precursor

I^A-encoded
transferase

This is the A antigen;
it reacts with
anti-A antibody.

Precursor carbohydrate

I^O-encoded
transferase

This is the "H"
antigen; it reacts
with neither anti-A
nor anti-B antibody.

I^B-encoded
transferase

Galactose added
to precursor

This is the B antigen;
it reacts with
anti-B antibody.

Key:

CH_2OH

N-acetylglucosamine

CH_2OH

Galactose

CH_2OH

N-acetylgalactosamine

Fucose

Figure 2.29

The ABO antigens on the surface of human red blood cells are carbohydrates. They are formed from a precursor carbohydrate by the action of transferase enzymes encoded by alleles of the I gene. Allele I^O codes for an inactive enzyme and leaves the precursor unmodified. The unmodified form is called the H substance. The I^A allele encodes an enzyme that adds N-acetylgalactosamine (purple) to the precursor. The I^B allele encodes an enzyme that adds galactose (green) to the precursor. The other colored sugar units are N-acetylglucosamine (orange), and fucose (yellow). The sugar rings also have side groups attached to one or more of their carbon atoms; these are shown in the detailed structures inside the box.

ence of the I^A allele results in production of the A polysaccharide; and in $I^B I^O$ heterozygotes, presence of the I^B allele results in production of the B polysaccharide. The result is that $I^A I^O$ persons have blood type A and $I^B I^O$ persons have blood type B, so I^O is recessive to both I^A and I^B. The genotypes and phenotypes of the ABO blood group system are summarized in the first three columns of Table 2.3.

The ABO blood groups are important in medicine because of the frequent need for blood transfusions. A crucial feature of the ABO system is that most human blood contains antibodies to either the A or the B polysaccharide. An **antibody** is a protein that is made by the immune system in response to a stimulating molecule called an **antigen** and is capable of binding to the antigen. An antibody is usually specific in that it recognizes only one antigen. Some antibodies combine with antigen and form large molecular aggregates that may precipitate.

Antibodies act in the body's defense against invading viruses and bacteria, as well as other cells, and help in removing such invaders from the body. Although antibodies do not normally form without prior stimulation by the antigen, people capable of producing anti-A and anti-B antibodies do produce them. Production of these antibodies may be stimulated by antigens that are similar to polysaccharides A and B and that are present on the surfaces of many common bacteria. However, a mechanism called *tolerance* prevents an organism from producing antibodies against its own antigens. This mechanism ensures that A antigen or B antigen elicits antibody production only in people whose own red blood cells do not contain A or B, respectively. The end result:

> People of blood type O make both anti-A and anti-B antibodies; those of blood type A make anti-B antibodies; those of blood type B make anti-A antibodies; and those of blood type AB make neither type of antibody.

The antibodies found in the blood fluid of people with each of the ABO blood types are shown in the fourth column in Table 2.3. The clinical significance of the ABO blood groups is that transfusion of blood containing A or B red-cell antigens into persons who make antibodies against them results in an agglutination reaction in which the donor red blood cells are clumped. In this reaction, the anti-A antibody will agglutinate red blood cells of either blood type A or blood type AB, because both carry the A antigen (Figure 2.30). Similarly, anti-B antibody will agglutinate red blood cells of either blood type B or blood type AB. When the blood cells agglutinate, many blood vessels are blocked, and the recipient of the transfusion goes into shock and may die. Incompatibility in the other direction, in which the donor blood contains antibodies against the recipient's red blood cells, is usually acceptable because the donor's antibodies are diluted so rapidly that clumping is avoided. The types of compatible blood transfusions are shown in the last two columns of Table 2.3. Note that a person of blood type AB can receive blood from a person of any other ABO type; type AB is called a *universal recipient.* Conversely, a person of blood type O can donate blood to a person of any ABO type; type O is called a *universal donor.*

Incomplete Penetrance and Variable Expressivity

Monohybrid Mendelian ratios, such as 3 : 1 (or 1 : 2 : 1 when the heterozygote is intermediate), are not always observed even when a trait is determined by the action of a single recessive allele. Regular ratios such as these indicate that organisms with the same genotype also exhibit the same phenotype. Although the phenotypes of organisms with a particular genotype are often very similar, this is not always the case—particularly in natural populations in which

Table 2.3 Genetic control of the human ABO blood groups

Genotype	Antigens present on red blood cells	ABO blood goup phenotype	Antibodies present in blood fluid	Blood types that can be tolerated in transfusion	Blood types that can accept blood for transfusion
$I^A I^A$	A	Type A	Anti-B	A & O	A & AB
$I^A I^O$	A	Type A	Anti-B	A & O	A & AB
$I^B I^B$	B	Type B	Anti-A	B & O	B & AB
$I^B I^O$	B	Type B	Anti-A	B & O	B & AB
$I^A I^B$	A & B	Type AB	Neither anti-A nor anti-B	A, B, AB & O	AB only
$I^O I^O$	Neither A nor B	Type O	Anti-A & anti-B	O only	A, B, AB & O

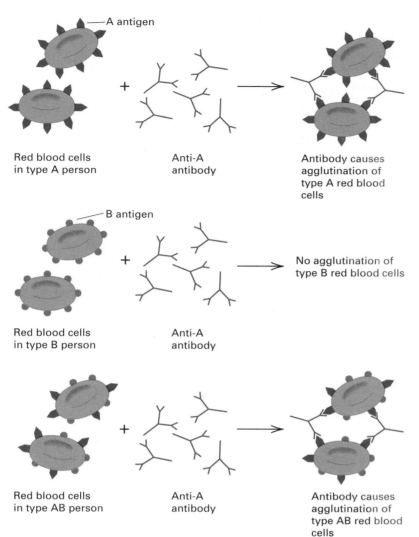

Figure 2.30
Antibody against type-A antigen will agglutinate red blood cells that carry the type-A antigen, whether or not they also carry the type-B antigen. Blood fluid containing anti-A antibody will agglutinate red blood cells of type A and type AB, but not red blood cells of type B or type O.

A antigen

Red blood cells in type A person — Anti-A antibody — Antibody causes agglutination of type A red blood cells

B antigen

Red blood cells in type B person — Anti-A antibody — No agglutination of type B red blood cells

Red blood cells in type AB person — Anti-A antibody — Antibody causes agglutination of type AB red blood cells

neither the matings nor the environmental conditions are under an experimenter's control. Variation in the phenotypic expression of a particular genotype may happen because other genes modify the phenotype or because the biological processes that produce the phenotype are sensitive to environmental conditions.

The types of variable gene expression are usually grouped into two categories:

• **Variable expressivity** refers to genes that are expressed to different degrees in different organisms. For example, inherited genetic diseases in human beings are often variable in expression from one person to the next. One patient may be very sick, whereas another with the same disease may be less severely affected. Variable expressivity means that the same mutant gene can result in a severe form of the disease in one person but a mild form in another. The different degrees of expression often form a continuous series from full expression to almost no expression of the expected phenotypic characteristics.

• **Incomplete penetrance** means that the phenotype expected from a particular genotype is not always expressed. For example, a person with a genetic predisposition to lung cancer may not get the disease if he or she does not smoke tobacco. A lack of gene expression may result from environmental conditions, such as in the example of not smoking, or from the effects of other genes. Incomplete penetrance is but an extreme of variable expressivity, in which the expressed phenotype is so mild as to be undetectable. The proportion of organisms whose phenotype matches their genotype for a given character is called the **penetrance** of the genotype. A genotype that is always expressed has a penetrance of 100 percent.

Chapter Summary

Inherited traits are determined by particulate elements called genes. In a higher plant or animal, the genes are present in pairs. One member of each gene pair is inherited from the maternal parent and the other member from the paternal parent. A gene can have different forms owing to differences in DNA sequence. The different forms of a gene are called alleles. The particular combination of alleles present in an organism constitutes its genotype. The observable characteristics of an organism constitute its phenotype. In an organism, if the two alleles of a gene pair are the same (for example, AA or aa), then the genotype is homozygous for the A or a allele; if the alleles are different (Aa), then the genotype is heterozygous. When the phenotype of a heterozygote is the same as that of one of the homozygous genotypes, the allele that is expressed is called dominant and the hidden allele is called recessive.

In genetic studies, the organisms produced by a mating constitute the F_1 generation. Matings between members of the F_1 generation produce the F_2 generation. In a cross such as $AA \times aa$, in which only one gene is considered (a monohybrid cross), the ratio of genotypes in the F_2 generation is 1 dominant homozygote (AA) : 2 heterozygotes (Aa) : 1 recessive homozygote (aa). The phenotypes in the F_2 generation appear in the ratio 3 dominant : 1 recessive. The Mendelian ratios of genotypes and phenotypes result from segregation in gamete formation, when the members of each allelic pair segregate into different gametes, and random union of gametes in fertilization.

The processes of segregation, independent assortment, and random union of gametes follow the rules of probability, which provide the basis for predicting outcomes of genetic crosses. Two basic rules for combining probabilities are the addition rule and the multiplication rule. The addition rule applies to mutually exclusive events; it states that the probability of the realization of either one or the other of two events equals the sum of the respective probabilities. The multiplication rule applies to independent events; it states that the probability of the simultaneous realization of both of two events is equal to the product of the respective probabilities. In some organisms—for example, human beings—it is not possible to perform controlled crosses, and genetic analysis is accomplished through the study of several generations of a family tree, called a pedigree. Pedigree analysis is the determination of the possible genotypes of the family members in a pedigree and of the probability that an individual member has a particular genotype.

The complementation test is the functional definition of a gene. Two recessive mutations are considered alleles of different genes if a cross between the homozygous recessives results in nonmutant progeny. Such alleles are said to complement one another. On the other hand, two recessive mutations are considered alleles of the *same* gene if a cross between the homozygous recessives results in mutant progeny. Such alleles are said to fail to complement. For any group of recessive mutations, a complete complementation test entails crossing the homozygous recessives in all pairwise combinations.

Multiple alleles are often encountered in natural populations or as a result of mutant screens. Multiple alleles means that more than two alternative forms of a gene exist. Examples of large numbers of alleles include the genes used in DNA typing and the self-sterility alleles in some flowering plants. Although there may be multiple alleles in a population, each gamete can carry only one allele of each gene, and each organism can carry at most two different alleles of each gene.

Dihybrid crosses differ in two genes—for example, AA $BB \times aa$ bb. The phenotypic ratios in the dihybrid F_2 are $9 : 3 : 3 : 1$, provided that both the A and the B alleles are dominant and that the genes undergo independent assortment. The $9 : 3 : 3 : 1$ ratio can be modified in various ways by interaction between the genes (epistasis). Different types of epistasis may result in dihybrid ratios such as $9 : 7$, $12 : 3 : 1$, $13 : 3$, $9 : 4 : 3$, and $9 : 6 : 1$.

In heterozygous genotypes, complete dominance of one allele over the other is not always observed. In most cases, a heterozygote for a wildtype allele and a mutant allele encoding a defective gene product will produce less gene product than in the wildtype homozygote. If the phenotype is determined by the amount of wildtype gene product rather than by its mere presence, the heterozygote will have an intermediate phenotype. This situation is called incomplete dominance. Codominance means that both alleles in a heterozygote are expressed, so the heterozygous genotype exhibits the phenotypic characteristics of both homozygous genotypes. Codominance is exemplified by the I^A and I^B alleles in persons with blood group AB. Codominance is often observed for proteins when each alternative allele codes for a different amino acid replacement, because it may be possible to distinguish the alternative forms of the protein by chemical or physical means. Genes are not always expressed to the same extent in different organisms; this phenomenon is called variable expressivity. A genotype that is not expressed at all in some organisms is said to have incomplete penetrance.

Key Terms

addition rule
albinism
allele
amorph
antibody
antigen
antimorph
backcross
carrier
codominance
complementation group
complementation test
dihybrid
dominant
ectopic expression
F_1 generation
F_2 generation
gamete
gene
genotype

heterozygous
homozygous
Huntington disease
hybrid
hypermorph
hypomorph
incomplete dominance
incomplete penetrance
independent assortment
Mendelian genetics
monohybrid
multiple alleles
multiplication rule
mutant screen
neomorph
P_1 generation
partial dominance
pedigree
penetrance
phenotype

Punnett square
recessive
reciprocal cross
segregation
sib
sibling
sibship
testcross
true breeding
variable expressivity
zygote

Review the Basics

- What is segregation? How would the segregation of a pair of alleles be exhibited in the progeny of a testcross?

- Explain the following statement: "Among the F_2 progeny of a dihybrid cross, the ratio of genotypes is $1 : 2 : 1$, but among the progeny that express the dominant phenotype, the ratio of genotypes is $1 : 2$."

- What is a mutant screen and how is it used in genetic analysis?

- What is a complementation test? How does this test enable a geneticist to determine whether two different mutations are or are not mutations in the same gene?

- What do we mean by a "modified dihybrid F_2 ratio"? Give two examples of a modified dihybrid F_2 ratio and explain the gene interactions that result in the modified ratio.

- What is the distinction between incomplete dominance and codominance? Give an example of each.

Guide to Problem Solving

Problem 1: In tomatoes, the shape of the fruit is inherited, and both round fruit and elongate fruit are true breeding. The cross round × elongate produces F_1 progeny with round fruit, and the cross $F_1 \times F_1$ produces 3/4 progeny with round fruit and 1/4 progeny with elongate fruit. What kind of genetic hypothesis can explain these data?

Answer: In this kind of problem, a good strategy is to look for some indication of Mendelian segregation. The 3 : 1 ratio in the F_2 generation is characteristic of Mendelian segregation when there is dominance. This observation suggests the genetic hypothesis of a dominant gene *R* for round fruit and a recessive allele *r* for elongate fruit. If the hypothesis were correct, then the true-breeding round and elongate genotypes would be *RR* and *rr,* respectively. The F_1 progeny of the cross *RR* (round) × *rr* (elongate) would be *Rr,* which has

round fruit, as observed. The $F_1 \times F_1$ cross (*Rr* × *Rr*) yields 1/4 *RR*, 1/2 *Rr*, and 1/4 *rr*. Because both *RR* and *Rr* have round fruit, the expected F_2 ratio of round : elongate phenotypes is 3 : 1, as expected from the single-gene hypothesis.

Problem 2: In Shorthorn cattle, both red coat color and white coat color are true breeding. Crosses of red × white produce progeny that are uniformly reddish brown but thickly sprinkled with white hairs; this type of coat color is called *roan.* Crosses of roan × roan produce 1/4 red : 1/2 roan : 1/4 white. What kind of genetic hypothesis can explain these data?

Answer: In this case, the 1 : 2 : 1 ratio of phenotypes in the cross roan × roan suggests Mendelian segregation, because this is the ratio expected from a mating between

GeNETics EXERCISES

Select the highlighted keyword in any of the exercises below, and you will be linked to a web site containing the genetic information necessary to complete the exercise. Each exercise suggests a specific, written report that makes use of the information available at the site. This report, or an alternative, may be assigned by your instructor.

1. Mendel's paper is one of the few nineteenth-century scientific papers that reads almost as clearly as though it had been written today. It is important reading for every aspiring geneticist. You can access a conveniently annotated text by using the keyword **Mendel**. Although modern geneticists make a clear distinction between genotype and phenotype, Mendel made no clear distinction between these concepts. If assigned to do so, make a list of three specific instances in Mendel's paper, each supported by a quotation, in which the concepts of genotype and phenotype are not clearly separated; rewrite each quotation in a way that makes the distinction clear.

2. Although the incidence of **Huntingon disease** is only 30 to 70 per million people in most Western countries, it has received great attention in genetics because of its late age of onset and autosomal dominant inheritance. Use the keyword to learn more about this condition. Under the heading *History* there is an *Editor's Note* quoting the blind seer Tiresias confronting Oedipus with the following paradox: "It is sorrow to be wise when wisdom profits not." If assigned to do so, write a 250-word essay explaining what this means in reference to Huntington disease and why DNA-based diagnosis is regarded as an ethical dilemma.

3. The red and purple colors of flowers, as well as of autumn leaves, result from members of a class of pigments called anthocyanins. The biochemical pathway for **anthocyanin** synthesis

heterozygotes when dominance is incomplete. Supposing that roan is heterozygous (say, *Rr*). Then the cross roan × roan (*Rr* × *Rr*) is expected to produce 1/4 *RR*, 1/2 *Rr*, and 1/4 *rr* genotypes. The observed result, that 1/2 of the progeny are roan (*Rr*), fits this hypothesis, which implies that the *RR* and *rr* genotypes correspond to red and white. The problem states that red and white are true breeding, which is consistent with their being homozygous genotypes. Additional confirmation comes from the cross *RR* × *rr*, which yields *Rr* (roan) progeny, as expected. Note that the gene symbols *R* and *r* are assigned to red and white arbitrarily, so it does not matter whether *RR* stands for red and *rr* for white, or the other way around.

Problem 3: The tailless trait in the mouse results from an allele of a gene in chromosome 17. The cross tailless × tailless produces tailless and wildtype progeny in a ratio of 2 tailless : 1 wildtype. All tailless progeny from this cross, when mated with wildtype, produce a 1 : 1 ratio of tailless to wildtype progeny.

(a) Is the allele for the tailless trait dominant or recessive?

(b) What genetic hypothesis can account for the 2 : 1 ratio of tailless : wildtype and the results of the crosses between the tailless animals?

Answer: **(a)** If the tailless phenotype were homozygous recessive, then the cross tailless × tailless should produce only tailless progeny. This is not the case, so the tailless phenotype must result from a dominant allele, say *T*. **(b)** Because the cross tailless × tailless produces both tailless and wildtype progeny, both parents must be heterozygous *Tt*. The expected ratio of genotypes among the zygotes is 1/4 *TT*, 1/2 *Tt*, and 1/4 *tt*. Because *T* is dominant, the *Tt* animals are tailless and the *tt* animals are wildtype. The 2 : 1 ratio can be explained if the *TT* zygotes do not survive (that is, the *TT* genotype is lethal). Because all surviving tailless animals must be *Tt*, this genetic hypothesis would also explain why all of the tailless animals from the cross, when mated with *tt*, give a 1 : 1 ratio of tailless (*Tt*) to wildtype (*tt*). (Developmental studies confirm that about 25 percent of the embryos do not survive.)

Problem 4: The accompanying illustration shows four alternative types of combs in chickens; they are called rose, pea, single, and walnut. The following data summarize the results of crosses. The rose and pea strains used in crosses 1, 2, and 5 are true breeding.

1. rose × single → rose

2. pea × single → pea

3. (rose × single) F_1 × (rose × single) F_1 → 3 rose : 1 single

in the snapdragon, *Antirrhinum majus,* can be found at this keyword site. The enzyme responsible for the first step in the pathway limits the amount of pigment formed, which explains why red and white flowers in *Antirrhinum* show incomplete dominance. If assigned to do so, identify the enzyme responsible for the first step in the pathway, and give the molecular structures of the substrate (or substrates) and product. Also, examine all of the intermediates in the anthocyanin pathway, and identify which atom that is so prominent in the purine and pyrimidine bases is not found in anthocyanin.

MUTABLE SITE EXERCISES

The Mutable Site Exercise changes frequently. Each new update includes a different exercise that makes use of genetics resources available on the World Wide Web. Select the **Mutable Site** for Chapter 2, and you will be linked to the current exercise that relates to the material presented in this chapter.

PIC SITE

The Pic Site showcases some of the most visually appealing genetics sites on the World Wide Web. To visit the showcase genetics site, select the **Pic Site** for Chapter 2.

4. (pea × single) F_1 × (pea × single) F_1 → 3 pea : 1 single

5. rose × pea → walnut

6. (rose × pea) F_1 × (rose × pea) F_1 → 9 walnut : 3 rose : 3 pea : 1 single

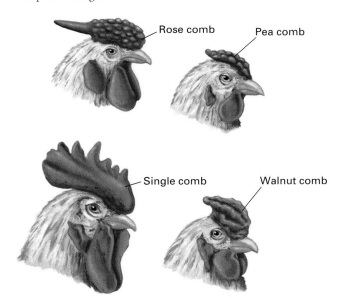

Rose comb

Pea comb

Single comb

Walnut comb

(a) What genetic hypothesis can explain these results?

(b) What are the genotypes of parents and progeny in each of the crosses?

(c) What are the genotypes of true-breeding strains of rose, pea, single, and walnut?

Answer:

(a) Cross 6 gives the Mendelian ratios expected when two genes are segregating, so a genetic hypothesis with two genes is necessary. Crosses 1 and 3 give the results expected if rose comb were due to a dominant allele (say, *R*). Crosses 2 and 4 give the results expected if pea comb were due to a dominant allele (say, *P*). Cross 5 indicates that walnut comb results from the interaction of *R* and *P*. The segregation in cross 6 means that *R* and *P* are not alleles of the same gene.

(b) 1. *RR pp* × *rr pp* → *Rr pp*.
2. *rr PP* × *rr pp* → *rr Pp*.
3. *Rr pp* × *Rr pp* → 3/4 *R− pp* : 1/4 *rr pp*.
4. *rr Pp* × *rr Pp* → 3/4 *rr P−* : 1/4 *rr pp*.
5. *RR pp* × *rr PP* → *Rr Pp*.
6. *Rr Pp* × *Rr Pp* →
     9/16 *R− P−* : 3/16 *R− pp* : 3/16 *rr P−* : 1/16 *rr pp*.

(c) The true-breeding genotypes are *RR pp* (rose), *rr PP* (pea), *rr pp* (single), and *RR PP* (walnut).

Problem 5: The pedigree in the accompanying illustration shows the inheritance of coat color in a group of cocker spaniels. The coat colors and genotypes are as follows:

Black   *A−*   *B−*   (black symbols)
Liver   *aa*   *B−*   (pink symbols)
Red   *A−*   *bb*   (red symbols)
Lemon   *aa*   *bb*   (yellow symbols)

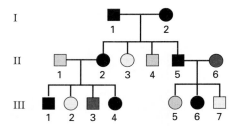

(a) Specify in as much detail as possible the genotype of each dog in the pedigree.

(b) What are the possible genotypes of the animal III-4, and what is the probability of each genotype?

(c) If a single pup is produced from the mating of III-4 × III-7, what is the probability that the pup will be red?

Answer:

(a) All three matings (I-1 × I-2, II-1 × II-2, and II-5 × II-6) produce lemon-colored offspring *aa bb*, so each parent must carry at least one *a* allele and at least one *b* allele. Therefore, in consideration of the phenotypes, the genotypes must be as follows: I-1 *Aa Bb*, I-2 *Aa Bb*, II-1 *aa Bb*, II-2 *Aa Bb*, II-5 *Aa Bb*, and II-6 *Aa bb*. The genotypes of the offspring can be deduced from their own phenotypes and the genotypes of the parents. These are as follows: II-3 *aa bb*, II-4 *aa B−*, III-1 *Aa B−*, III-2 *aa bb*, III-3 *Aa bb*, III-4 *Aa B−*, III-5 *aa Bb*, III-6 *A− Bb*, III-7 *aa bb*.

(b) Animal III-4 is either *Aa BB* or *Aa Bb*, and the probabilities of these genotypes are 1/3 and 2/3, respectively.

(c) If the animal III-4 is *Aa BB*, then the probability of a red pup is 0; and if the animal III-4 is *Aa Bb*, then the probability of a red pup is $1/2 \times 1/2 = 1/4$ (that is, the probability of an *A b* gamete from III-4). Overall, the probability of a red pup from the mating is $1/3 \times 0 + 2/3 \times 1/4 = 1/6$.

Problem 6: From the F_2 generation of a cross between mouse genotypes *AA* × *aa*, one male progeny of genotype *A−* was chosen and mated with an *aa* female. All of the progeny in the resulting litter were *A−*. How large a litter is required for you to be able to assert, with 95 percent confidence, that the father's genotype is *AA*? How large a litter is required for 99 percent confidence?

Answer: The a priori ratio of the probabilities that the father is *AA* versus *Aa* is 1/3 : 2/3, because the father was chosen at random from among the *A−* progeny in the F_2 generation. With one *A−* progeny in a testcross, the ratio of probabilities drops to 1/3 : (2/3) × (1/2), because 1/2 of the *A−* fathers in such a testcross will yield an *aa* progeny and so identify themselves as *Aa*. Similarly, with *n* progeny, the ratio of *AA* : *Aa* probabilities is 1/3 : (2/3) × $(1/2)^n$, because the probability that an *Aa* father has *n* consecutive *A−* offspring in a testcross is $(1/2)^n$. For 95 percent confidence we need

$$1/3 : (2/3) \times (1/2)^n \geq 0.95 : 0.05$$

or

$$n \geq \log\left[(0.95 \times 2)/0.05\right]/\log(2) = 5.25$$

hence, $n = 6\,A−$ progeny are necessary for 95 percent confidence that the father is *AA*. For 99 percent confidence, the corresponding formula is

$$n \geq \log\left[(0.99 \times 2)/0.01\right]/\log(2) = 7.6$$

so in this case, $n = 8\,A−$ progeny are required.

Analysis and Applications

2.1 With respect to homozygosity and heterozygosity, what can be said about the genotype of a strain or variety that breeds true for a particular trait?

2.2 What gametes can be formed by an individual organism of genotype *Aa*? Of genotype *Bb*? Of genotype *Aa Bb*?

2.3 How many different gametes can be formed by an organism with genotype *AA Bb Cc Dd Ee* and, in general, by an organism that is heterozygous for *m* genes and homozygous for *n* genes?

2.4 Mendel summarized his conclusions about heredity by describing the gametes produced by the F_1 generation in the following manner: "Pea hybrids form germinal and pollen cells that in their composition correspond in equal numbers to all the constant forms resulting from the combination of traits united through fertilization." Explain this statement in terms of the principles of segregation and independent assortment.

2.5 Round pea seeds are planted that were obtained from the F_2 generation of a cross between a true-breeding strain with round seeds and a true-breeding strain with wrinkled seeds. The pollen was collected and used *en masse* to fertilize plants from the true-breeding wrinkled strain. What fraction of the progeny is expected to have wrinkled seeds?

2.6 If an allele *R* is dominant over *r*, how many different phenotypes are present in the progeny of a cross between *Rr* and *Rr,* and in what ratio? How many phenotypes are there, and in what ratio, if there is no dominance between *R* and *r*?

2.7 In genetically self-sterile plants like red clover, why are all plants heterozygous for the self-sterility alleles?

2.8 Assuming equal numbers of boys and girls, if a mating has already produced a girl, what is the probability that the next child will be a boy? If a mating has already produced two girls, what is the probability that the next child will be a boy? On what type of probability argument do you base your answers?

2.9 Assuming equal numbers of boys and girls, what is the probability that a family that has two children has two girls? One girl and one boy?

2.10 In the following questions, you are asked to deduce the genotype of certain parents in a pedigree. The phenotypes are determined by dominant and recessive alleles of a single gene.

(a) A homozygous recessive results from the mating of a heterozygote and a parent with the dominant phenotype. What does this tell you about the genotype of the parent with the dominant phenotype?

(b) Two parents with the dominant phenotype produce nine offspring. Two have the recessive phenotype. What does this tell you about the genotype of the parents?

(c) One parent has a dominant phenotype and the other has a recessive phenotype. Two offspring result, and both have the dominant phenotype. What genotypes are possible for the parent with the dominant phenotype?

2.11 Pedigree analysis tells you that a particular parent may have the genotype *AA BB* or *AA Bb*, each with the same probability. Assuming independent assortment, what is the probability of this parent's producing an *Ab* gamete? What is the probability of the parent's producing an *AB* gamete?

2.12 Assume that the trihybrid cross *AA BB rr × aa bb RR* is made in a plant species in which *A* and *B* are dominant but there is no dominance between *R* and *r*. Consider the F_2 progeny from this cross, and assume independent assortment.

(a) How many phenotypic classes are expected?

(b) What is the probability of the parental *aa bb RR* genotype?

(c) What proportion would be expected to be homozygous for all three genes?

2.13 In the cross *Aa Bb Cc Dd × Aa Bb Cc Dd*, in which all genes undergo independent assortment, what proportion of offspring are expected to be heterozygous for all four genes?

2.14 The pattern of coat coloration in dogs is determined by the alleles of a single gene, with *S* (solid) being dominant over *s* (spotted). Black coat color is determined by the domi-

nant allele *A* of a second gene, tan by homozygosity for the recessive allele *a*. A female having a solid tan coat is mated with a male having a solid black coat and produces a litter of six pups. The phenotypes of the pups are 2 solid tan, 2 solid black, 1 spotted tan, and 1 spotted black. What are the genotypes of the parents?

2.15 In the human pedigree shown here, the daughter indicated by the red circle(II-1) has a form of deafness determined by a recessive allele. What is the probability that the phenotypically normal son (II-3) is heterozygous for the gene?

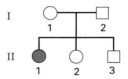

2.16 Huntington disease is a rare neurodegenerative human disease determined by a dominant allele, *HD*. The disorder is usually manifested after the age of forty-five. A young man has learned that his father has developed the disease.

(a) What is the probability that the young man will later develop the disorder?

(b) What is the probability that a child of the young man carries the *HD* allele?

2.17 The Hopi, Zuni, and some other Southwest American Indians have a relatively high frequency of albinism (absence of skin pigment) resulting from homozygosity for a recessive allele, *a*. A normally pigmented man and woman, each of whom has an albino parent, have two children. What is the probability that both children are albino? What is the probability that at least one of the children is albino?

2.18 Which combinations of donor and recipient ABO blood groups are compatible for transfusion? (Consider a combination to be compatible for transfusion if all the antigens in the donor red blood cells are also present in the recipient.)

2.19 Red kernel color in wheat results from the presence of at least one dominant allele of each of two independently segregating genes (in other words, *R– B–* genotypes have red kernels). Kernels on *rr bb* plants are white, and the genotypes *R– bb* and *rr B–* result in brown kernel color. Suppose that plants of a variety that is true breeding for red kernels are crossed with plants true breeding for white kernels.

(a) What is the expected phenotype of the F_1 plants?

(b) What are the expected phenotypic classes in the F_2 progeny and their relative proportions?

2.20 Heterozygous *Cp cp* chickens express a condition called creeper, in which the leg and wing bones are shorter than normal (*cp cp*). The dominant *Cp* allele is lethal when homozygous. Two alleles of an independently segregating gene determine white (*W–*) versus yellow (*ww*) skin color. From matings between chickens heterozygous for both of these genes, what phenotypic classes will be represented among the viable progeny, and what are their expected relative frequencies?

2.21 White Leghorn chickens are homozygous for a dominant allele, C, of a gene responsible for colored feathers, and also for a dominant allele, I, of an independently segregating gene that prevents the expression of C. The White Wyandotte breed is homozygous recessive for both genes $cc\ ii$. What proportion of the F_2 progeny obtained from mating White Leghorn × White Wyandotte F_1 hybrids would be expected to have colored feathers?

2.22 The F_2 progeny from a particular cross exhibit a modified dihybrid ratio of $9:7$ (instead of $9:3:3:1$). What phenotypic ratio would be expected from a testcross of the F_1?

2.23 Phenylketonuria is a recessive inborn error of metabolism of the amino acid phenylalanine that results in severe mental retardation of affected children. The female II-3 (red circle) in the pedigree shown here is affected. If persons III-1 and III-2 (they are first cousins) mate, what is the probability that their offspring will be affected? (Assume that persons II-1 and II-5 are homozygous for the normal allele.)

2.24 Black hair in rabbits is determined by a dominant allele, B, and white hair by homozygosity for a recessive allele, b. Two heterozygotes mate and produce a litter of three offspring.

(a) What is the probability that the offspring are born in the order white-black-white? What is the probability that the offspring are born in either the order white-black-white or the order black-white-black?

(b) What is the probability that exactly two of the three offspring will be white?

2.25 Assuming equal sex ratios, what is the probability that a sibship of four children consists entirely of boys? Of all boys or all girls? Of equal numbers of boys and girls?

2.26 Andalusian fowls are colored black, splashed white (resulting from an uneven sprinkling of black pigment through the feathers), or slate blue. Black and splashed white are true breeding, and slate blue is a hybrid that segregates in the ratio 1 black : 2 slate blue : 1 splashed white. If a pair of blue Andalusians is mated and the hen lays three eggs, what is the probability that the chicks hatched from these eggs will be one black, one blue, and one splashed white?

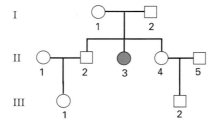

Challenge Problems

2.27 In the mating $Aa \times Aa$, what is the smallest number of offspring, n, for which the probability of at least one aa offspring exceeds 95 percent?

2.28 From the F_2 generation of a cross between mouse genotypes $AA \times aa$, one male progeny of genotype $A-$ was chosen and mated with an aa female. All of the progeny in the resulting litter were $A-$. From this result you would like to conclude that the sire's genotype is AA. How much confidence could you have in this conclusion for each litter size from 1 to 15? (In other words what is the probability that the sire's genotype is AA, given that the a priori probability is $1/3$ and that a litter of n pups resulted in all $A-$ progeny?)

2.29 Meiotic drive is an unusual phenomenon in which two alleles do not show Mendelian segregation from the heterozygous genotype. Examples are known from mammals, insects, fungi, and other organisms. The usual mechanism is one in which both types of gametes are formed, but one of them fails to function normally. The excess of the driving

allele over the other can range from a small amount to nearly 100 percent. Suppose that D is an allele showing meiotic drive against its alternative allele d, and suppose that Dd heterozygotes produce functional D-bearing and d-bearing gametes in the proportions $3/4 : 1/4$. In the mating $Dd \times Dd$,

(a) What are the expected proportions of DD, Dd, and dd genotypes?

(b) If D is dominant, what are the expected proportions of $D-$ and dd phenotypes?

(c) Among the $D-$ phenotypes, what is the ratio of $DD : Dd$?

(d) Answer parts (a) through (c), assuming that the meiotic drive takes place in only one sex.

Ashley, C. T., and S. T. Warren. 1995. Trinucleotide repeat expansion and human disease. *Annual Review of Genetics* 29: 703.

Bowler, P. J. 1989. *The Mendelian Revolution.* Baltimore, MD: Johns Hopkins University Press.

Carlson, E. A. 1987 . *The Gene: A Critical History.* 2d ed. Philadelphia: Saunders.

Dunn, L. C. 1965. *A Short History of Genetics.* New York: McGraw-Hill.

Hartl, D. L., and V. Orel. 1992. What did Gregor Mendel think he discovered? *Genetics* 131: 245.

Huntington's Disease Collaborative Research Group: M. E. MacDonald, C. M. Ambrose, M. P. Duyao, R. H. Myers, C. Lin, L. Srinidhi, G. Barnes, S. A. Taylor, M. James, N. Groot, H. MacFarlane, B. Jenkins, M. A. Anderson, N. S. Wexler, J. F. Gusella; G. P. Bates, S. Baxendate, H. Hummerich, S. Kirby, M. North, S. Youngman, R. Mott, G. Zehetner, Z. Sedlacek, A. Poustka, A.-M. Frischauf, H. Lehrach; A. J. Buckler, D. Church, L. Doucette-Stamm, M. C. O'Donovan, L. Ribe-Ramirez, M. Shah, V. P. Stanton, S. A. Strobel, K. M. Draths, J. L. Wales, P. Dervan, D. E. Housman; M. Altherr, R. Shiang, L. Thompson, T. Fielder, J. J. Wasmuth; D. Tagle, J. Valdes, L. Elmer, M. Allard, L. Castilla, M. Swaroop, K. Blanchard, F. S. Collins; R. Snell, T. Holloway, K. Gillespie, N. Datson, D. Shaw, P S. Harper. 1993. A novel gene containing a trinucleotide repeat that is expanded and unstable on Huntington's disease chromosomes. *Cell* 72: 971.

Judson, H. F. 1996. *The Eighth Day of Creation: The Makers of the Revolution in Biology.* Cold Spring Harbor, NY: Cold Spring Harbor Laboratory Press.

Mendel, G. 1866. Experiments in plant hybridization. (Translation.) In *The Origins of Genetics: A Mendel Source Book,* ed. C. Stern and E. Sherwood. 1966. New York: Freeman.

Olby, R. C. 1966. *Origins of Mendelism.* London: Constable.

Orel, V. 1996. *Gregor Mendel: The First Geneticist.* Oxford, England: Oxford University Press.

Orel, V., and D. L. Hartl. 1994. Controversies in the interpretation of Mendel's discovery. *History and Philosophy of the Life Sciences* 16: 423.

Stern, C., and E. Sherwood. 1966. *The Origins of Genetics: A Mendel Source Book.* New York: Freeman.

Sturtevant, A. H. 1965. *A Short History of Genetics.* New York: Harper & Row.

The calico cat illustrates the phenomenon of codominance. This female is heterozygous for an allele for black fur and for an allele for orange (also called "yellow") fur. Some patches of fur are black, whereas other patches are yellow. Because both alleles express their characteristic phenotype when heterozygous, they are considered codominant. Why the black and orange alleles are expressed in alternate patches of cells, rather than in overlapping patches, is explained in Chapter 7. The white spots are caused by an allele of a different gene that prevents any color formation.

Genes and Chromosomes

PRINCIPLES

- Chromosomes in eukaryotic cells are usually present in pairs.
- The chromosomes of each pair separate in meiosis, one going to each gamete.
- In meiosis, the chromosomes of different pairs undergo independent assortment because nonhomologous chromosomes move independently.
- In many animals, sex is determined by a special pair of chromosomes—the X and Y.
- The "criss-cross" pattern of inheritance of X-linked genes is determined by the fact that a male receives his X chromosome only from his mother and transmits it only to his daughters.
- Irregularities in the inheritance of an X-linked gene in *Drosophila* gave experimental proof of the chromosomal theory of heredity.
- The progeny of genetic crosses follow the binomial probability formula.
- The chi-square statistical test is used to determine how well observed genetic data agree with expectations derived from a hypothesis.

CONNECTIONS

CONNECTION: Grasshopper, Grasshopper
E. Eleanor Carothers 1913
The Mendelian ratio in relation to certain Orthopteran chromosomes

CONNECTION: The White-Eyed Male
Thomas Hunt Morgan 1910
Sex limited inheritance in Drosophila

CONNECTION: The Case Against Mendel's Gardener
Ronald Aylmer Fisher 1936
Has Mendel's work been rediscovered?

Mendel's experiments made it clear that in heterozygous genotypes, neither allele is altered by the presence of the other. The hereditary units remain stable and unchanged in passing from one generation to the next. However, at the time, the biological basis of the transmission of genes from one generation to the next was quite mysterious. Neither the role of the nucleus in reproduction nor the details of cell division had been discovered. Once these phenomena were understood, and when microscopy had improved enough that the chromosomes could be observed and were finally recognized as the carriers of the genes, new understanding came at a rapid pace. This chapter examines both the relationship between chromosomes and genes and the mechanism of chromosome segregation in cell division.

3.1
The Stability of Chromosome Complements

The importance of the cell nucleus and its contents was suggested as early as the 1840s when Carl Nägeli observed that in dividing cells, the nucleus divided first. This was the same Nägeli who would later fail to understand Mendel's discoveries. Nägeli also failed to see the importance of nuclear division when he discovered it. He regarded the cells in which he saw nuclear division as aberrant. Nevertheless, by the 1870s it was realized that nuclear division is a universal attribute of cell division. The impor-

tance of the nucleus in inheritance was reinforced by the nearly simultaneous discovery that the nuclei of two gametes fuse in the process of fertilization. The next major advance came a decade later with the discovery of **chromosomes,** which had been made visible by light microscopy when stained with basic dyes. A few years later, chromosomes were found to segregate by an orderly process into the daughter cells formed by cell division as well as into the gametes formed by the division of reproductive cells. Finally, three important regularities were observed about the **chromosome complement** (the complete set of chromosomes) of plants and animals.

1. The nucleus of each **somatic cell** (a cell of the body, in contrast with a **germ cell,** or gamete) contains a fixed number of chromosomes typical of the particular species. However, the numbers vary tremendously among species and bear little relation to the complexity of the organism (Table 3.1).

2. The chromosomes in the nuclei of somatic cells are usually present in pairs. For example, the 46 chromosomes of human beings consist of 23 pairs (Figure 3.1). Similarly, the 14 chromosomes of peas consist of 7 pairs. Cells with nuclei of this sort, containing two similar sets of chromosomes, are called **diploid.** The chromosomes are present in pairs because one chromosome of each pair derives from the maternal parent and the other from the paternal parent of the organism.

Table 3.1 Somatic chromosome numbers of some plant and animal species

Organism	Chromosome number	Organism	Chromosome number
Field hosetail	216	Yeast (*Saccharomyces cerevisiae*)	32
Bracken fern	116	Fruit fly (*Drosophilia melanogaster*)	8
Giant sequoia	22	Nematode (*Caenorhabditis elegans*)	11 ♂, 12 ♀
Macaroni wheat	28	House fly	12
Bread wheat	42	Scorpion	4
Fava bean	12	Geometrid moth	224
Garden pea	14	Common toad	22
Wall cress (*Arabidopsis thaliana*)	10	Chicken	78
Corn (*Zea mays*)	20	Mouse	40
Lily	12	Gibbon	44
Snapdragon	16	Human being	46

3. The germ cells, or gametes, that unite in fertilization to produce the diploid state of somatic cells have nuclei that contain only one set of chromosomes, consisting of one member of each of the pairs. The gamete nuclei are **haploid.**

In multicellular organisms that develop from single cells, the presence of the diploid chromosome number in somatic cells and the haploid chromosome number in germ cells indicates that there are *two* different processes of nuclear division. One of these, mitosis, maintains the chromosome number; the other, meiosis, halves the number. These two processes are examined in the following sections.

3.2 Mitosis

Mitosis is a precise process of nuclear division that ensures that each of two daughter cells receives a diploid complement of chromosomes identical with the diploid complement of the parent cell. Mitosis is usually accompanied by **cytokinesis,** the process in which the cell itself divides to yield two daughter cells. The essential details of mitosis are the same in all organisms, and the basic process is remarkably uniform:

1. Each chromosome is already present as a duplicated structure at the beginning of nuclear division. (The duplication of each chromosome coincides with the replication of the DNA molecule contained within it.)

2. Each chromosome divides longitudinally into identical halves that become separated from each other.

3. The separated chromosome halves move in opposite directions, and each becomes included in one of the two daughter nuclei that are formed.

In a cell not undergoing mitosis, the chromosomes are not visible with a light microscope. This stage of the cell cycle is called **interphase.** In preparation for mitosis, the genetic material (DNA) in the chromosomes is replicated during a period of interphase called **S** (Figure 3.2). (The S stands for *synthesis* of DNA.) DNA replication is accompanied by chromosome dupli-

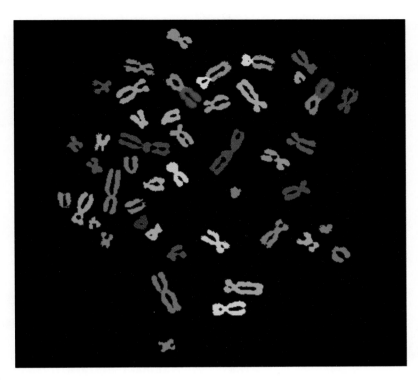

Figure 3.1
Chromosome complement of a human male. There are 46 chromosomes, present in 23 pairs. At the stage of the division cycle in which these chromosomes were observed, each chromosome consists of two identical halves lying side by side longitudinally. Except for the members of one chromosome pair (the pair that determines sex), the members of each of the other chromosome pairs are the same color because they contain DNA molecules that were labeled with the same mixture of fluorescent dyes. The colors differ from one pair to the next because the dye mixtures for each chromosome differ in color. In some cases, the long and the short arms have been labeled with different colors. [Courtesy of David C. Ward and Michael R. Speicher.]

cation. Before and after S, there are periods, called G_1 and G_2, respectively, in which DNA replication does not take place. The **cell cycle,** or the life cycle of a cell, is commonly described in terms of these three interphase periods followed by mitosis, **M.** The order of events is therefore $G_1 \rightarrow S \rightarrow G_2 \rightarrow M$, as shown in Figure 3.2. In this representation, cytokinesis, the division of the cytoplasm into two approximately equal parts containing the daughter nuclei, is included in the M period. The length of time required for a complete life cycle varies with cell type. In higher eukaryotes, the majority of cells require from 18 to 24 hours. The relative duration of the different periods in the cycle also varies considerably with cell type. Mitosis, requiring from 1/2 hour to 2 hours, is usually the shortest period.

Figure 3.2

The cell cycle of a typical mammalian cell growing in tissue culture with a generation time of 24 hours. The critical control points for the G₁S and G₂M transitions are governed by a p34 kinase that is activated by stage-specific cyclins and that regulates the activity of its target proteins through phosphorylation.

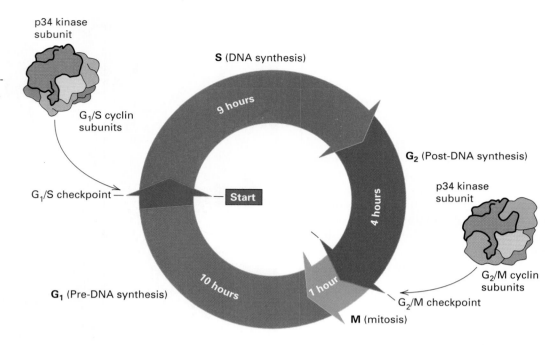

p34 kinase subunit

G₁/S cyclin subunits

G₁/S checkpoint

Start

S (DNA synthesis)

9 hours

G₂ (Post-DNA synthesis)

p34 kinase subunit

4 hours

G₂/M cyclin subunits

G₂/M checkpoint

M (mitosis)

1 hour

10 hours

G₁ (Pre-DNA synthesis)

The cell cycle itself is under genetic control. The mechanisms of control appear to be essentially identical in all eukaryotes. There are two critical transitions—from G₁ into S and from G₂ into M (Figure 3.2). The G₁/S and G₂/M transitions are called "checkpoints" because the transitions are delayed unless key processes have been completed. For example, at the G₁/S checkpoint, either sufficient time must have elapsed since the preceding mitosis (in some cell types) or the cell must have attained sufficient size (in other cell types) for DNA replication to be initiated. Similarly, the G₂/M checkpoint requires that DNA replication and repair of any DNA damage be completed for the M phase to commence. Both major control points are regulated in a similar manner and make use of a specialized protein kinase (called the p34 kinase subunit in Figure 3.2) that regulates the activity of target proteins by phosphorylation (transfer of phosphate groups). The p34 kinase is one of numerous types of protein kinases that are used to regulate cellular processes. To become activated, the p34 polypeptide subunit must combine with several other polypeptide chains that are known as **cyclins** because their abundance cycles in phase with the cell cycle. At the G₁/S control point, one set of cyclins combines with the p34 subunit to yield the active kinase that triggers DNA replication and other events of the S period. Similarly, at the G₂/M control point, a second set of cyclins combines with the p34 subunit to yield the active kinase that initiates condensation of the chromosomes, breakdown of the nuclear envelope, and reorganization of the cytoskeleton in preparation for cytokinesis.

Illustrated in Figure 3.3 are the essential features of chromosome behavior in

Figure 3.3   *(facing page)*

Diagram of mitosis in an organism with two pairs of chromosomes (red/rose versus green/blue). At each stage, the smaller inner diagram represents the entire cell, and the larger diagram is an exploded view showing the chromosomes at that stage. Interphase is usually not considered part of mitosis proper; it is typically much longer than the rest of the cell cycle, and the chromosomes are not yet visible. In early prophase, the chromosomes first become visible as fine strands, and the nuclear envelope and one or more nucleoli are intact. As prophase progresses, the chromosomes condense and each can be seen to consist of two sister chromatids; the nuclear envelope and nucleoli disappear. In metaphase, the chromosomes are highly condensed and aligned on the central plane of the spindle, which forms at the end of prophase. In anaphase, the centromeres split longitudinally, and the sister chromatids of each chromosome move to opposite poles of the spindle. In telophase, the separation of sister chromatids is complete, the spindle breaks down, new nuclear envelopes are formed around each group of chromosomes, the condensation process of prophase is reversed, and the cell cycles back into interphase.

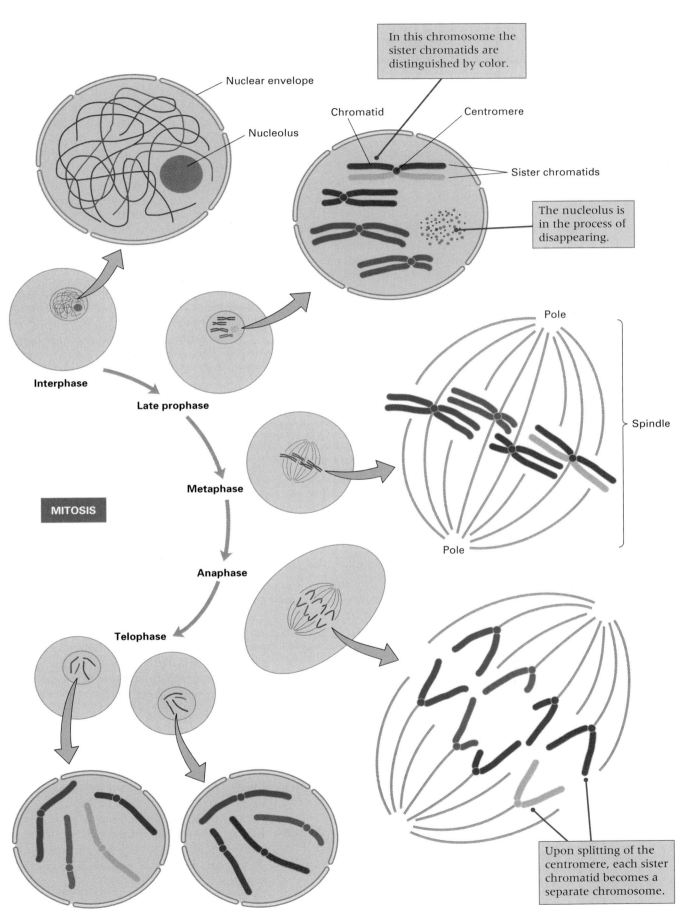

Nuclear envelope

Nucleolus

In this chromosome the sister chromatids are distinguished by color.

Chromatid

Centromere

Sister chromatids

The nucleolus is in the process of disappearing.

Interphase

Late prophase

Pole

Spindle

Metaphase

MITOSIS

Anaphase

Pole

Telophase

Upon splitting of the centromere, each sister chromatid becomes a separate chromosome.

mitosis. Mitosis is conventionally divided into four stages: **prophase, metaphase, anaphase,** and **telophase.** (If you have trouble remembering the order, you can jog your memory with *p*eas *m*ake *a*wful *t*arts.) The stages have the following characteristics:

1. *Prophase* In interphase, the chromosomes have the form of extended filaments and cannot be seen with a light microscope as discrete bodies. Except for the presence of one or more conspicuous dark bodies (**nucleoli**), the nucleus has a diffuse, granular appearance. The beginning of prophase is marked by the condensation of chromosomes to form visibly distinct, thin threads within the nucleus. Each chromosome is already longitudinally double, consisting of two closely associated subunits called **chromatids.** The longitudinally bipartite nature of each chromosome is readily seen later in prophase. Each pair of chromatids is the product of the duplication of one chromosome in the S period of interphase. The chromatids in a pair are held together at a specific region of the chromosome called the **centromere.** As prophase progresses, the chromosomes become shorter and thicker as a result of intricate coiling. At the end of prophase, the nucleoli disappear and the nuclear envelope, a membrane surrounding the nucleus, abruptly disintegrates.

2. *Metaphase* At the beginning of metaphase, the **mitotic spindle** forms. The spindle is a bipolar structure consisting of fiber-like bundles of microtubules that extend through the cell between the poles of the spindle. Each chromosome becomes attached to several spindle fibers in the region of the centromere. The structure associated with the centromere to which the spindle fibers attach is technically known as the **kinetochore.** After the chromosomes are attached to spindle fibers, they move toward the center of the cell until all the kinetochores lie on an imaginary plane equidistant from the spindle poles. This imaginary plane is called the **metaphase plate**. Aligned on the metaphase plate, the chromosomes reach their maximum contraction and are easiest to count and examine for differences in morphology.

Proper chromosome alignment is an important cell cycle control checkpoint at metaphase in both mitosis and meiosis. In a cell in which a chromosome is attached to only one pole of the spindle, the completion of metaphase is delayed. By grasping such a chromosome with a micromanipulation needle and pulling, one can mimic the tension that the chromosome would experience were it attached on both sides; the mechanical tension allows the metaphase checkpoint to be passed, and the cell enters the next stage of division. The signal for chromosome alignment comes from the kinetochore, and the chemical nature of the signal seems to be the dephosphorylation of certain kinetochore-associated proteins. The role of the kinetochore is demonstrated by the finding that metaphase is not delayed by an unattached chromosome whose kinetochore has been destroyed by a focused laser beam. The role of dephosphorylation is demonstrated through the use of an antibody that reacts specifically with some kinetochore proteins only when they are phosphorylated. Unattached kinetochores combine strongly with the antibody, but attachment to the spindle weakens the reaction. In chromosomes that have been surgically detached from the spindle, the antibody reaction with the kinetochore reappears. Through the signaling mechanism, when all of the kinetochores are under tension and aligned on the metaphase plate, the metaphase checkpoint is passed and the cell continues the process of division.

3. *Anaphase* In anaphase, the centromeres divide longitudinally, and the two **sister chromatids** of each chromosome move toward opposite poles of the spindle. Once the centromeres divide, each sister chromatid is regarded as a separate chromosome in its own right. Chromosome movement results in part from progressive shortening of the spindle fibers attached to the centromeres, which pulls the chromosomes in opposite directions toward the poles. At the completion of anaphase, the chromosomes lie in two groups near opposite poles of the spindle. Each group contains the same number of chromosomes that was present in the original interphase nucleus.

4. *Telophase* In telophase, a nuclear envelope forms around each compact group of chromosomes, nucleoli are formed, and the spindle disappears. The chromosomes undergo a reversal of condensation until they are no longer visible as discrete entities. The two daughter nuclei slowly assume a typical interphase appearance as the cytoplasm of the cell divides into two by means of a gradually deepening furrow around the periphery. (In plants, a new cell wall is synthesized between the daughter cells and separates them.)

3.3
Meiosis

Meiosis is a mode of cell division in which cells are created that contain only one member of each pair of chromosomes present in the premeiotic cell. When a diploid cell with two sets of chromosomes undergoes meiosis, the result is four daughter cells, each genetically different and each containing one haploid set of chromosomes.

Meiosis consists of two successive nuclear divisions. The essentials of chromosome behavior during meiosis are outlined in Figure 3.4. This outline affords an overview of meiosis as well as an introduction to

Figure 3.4
Overview of the behavior of a single pair of homologous chromosomes in meiosis. (A) The homologous chromosomes form a pair by coming together; each chromosome consists of two chromatids joined at a single centromere. (B) The members of each homologous pair separate. (C) At the end of the first meiotic division, each daughter nucleus carries one or the other of the homologous chromosomes. (D) In the second meiotic division, in each of the daughter nuclei formed in meiosis I, the sister chromatids separate. (E) The end result is four products of meiosis, each containing one of each pair of homologous chromosomes. For clarity, this diagram does not incorporate crossing-over, an interchange of chromosome segments that takes place at the stage depicted in part A. If crossing-over were included, each chromatid would consist of one or more segments of red and one or more segments of blue. (Crossing-over is depicted in Figure 3.7.)

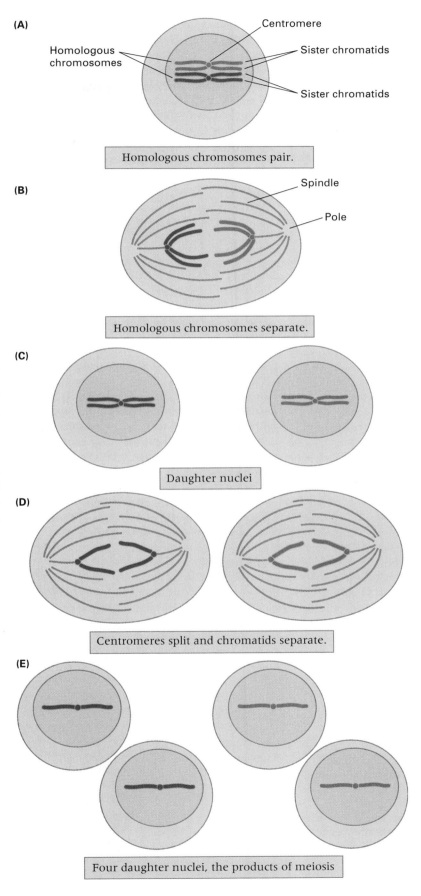

(A) Homologous chromosomes pair.

(B) Homologous chromosomes separate.

(C) Daughter nuclei

(D) Centromeres split and chromatids separate.

(E) Four daughter nuclei, the products of meiosis

the process as it takes place in a cellular context.

1. Prior to the first nuclear division, the members of each pair of chromosomes become closely associated along their length (Figure 3.4). The chromosomes that pair with each other are said to be **homologous** chromosomes. Because each member of a pair of homologous chromosomes is already replicated, each member consists of two sister chromatids joined at the centromere. The pairing of the homologous chromosomes therefore produces a four-stranded structure.

2. In the first nuclear division, the homologous chromosomes are separated from each another, one member of each pair going to opposite poles of the spindle (Figure 3.4B). Two nuclei are formed, each containing a haploid set of duplex chromosomes (Figure 3.4C) with two chromatids.

3. The second nuclear division loosely resembles a mitotic division, *but there is no chromosome replication.* At metaphase, the chromosomes align on the metaphase plate; and at anaphase, the chromatids of each chromosome are separated into opposite daughter nuclei (Figure 3.4D). The net effect of the two divisions in meiosis is the creation of four haploid daughter nuclei, each containing the equivalent of a single sister chromatid from each pair of homologous chromosomes (Figure 3.4E).

Figure 3.4 does not show that at the time of chromosome pairing, the homologous chromosomes can exchange genes. The exchanges result in the formation of chromosomes that consist of segments from one homologous chromosome intermixed with segments from the other. In Figure 3.4, the exchanged chromosomes would be depicted as segments of alternating color. The exchange process is one of the critical features of meiosis, and it will be examined in the next section.

In animals, meiosis takes place in specific cells called **meiocytes,** a general term for the primary oocytes and spermatocytes in the gamete-forming tissues (Figure 3.5). The oocytes form egg cells, and the spermatocytes form sperm cells. Although the process of meiosis is similar in all sexually reproducing organisms, in the female of both animals and plants, only one of the four products develops into a functional cell (the other three disintegrate). In animals, the products of meiosis form gametes (sperm or eggs).

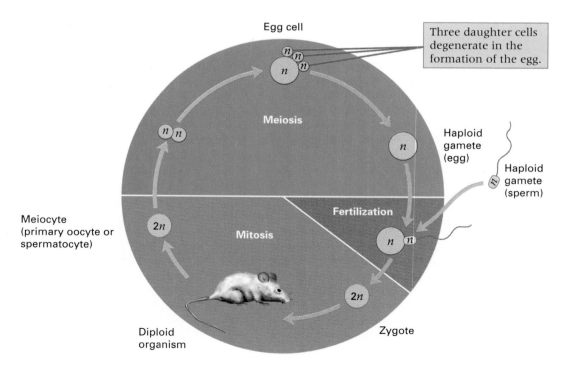

Figure 3.5
The life cycle of a typical animal. The number *n* is the number of chromosomes in the haploid chromosome complement. In males, the four products of meiosis develop into functional sperm; in females, only one of the four products develops into an egg.

In plants, the situation is slightly more complicated:

1. The products of meiosis typically form **spores,** which undergo one or more mitotic divisions to produce a haploid **gametophyte** organism. The gametophyte produces gametes by mitotic division of a haploid nucleus (Figure 3.6).

2. Fusion of haploid gametes creates a diploid zygote that develops into the **sporophyte** plant, which undergoes meiosis to produce spores and so restarts the cycle.

Meiosis is a more complex and considerably longer process than mitosis and usually requires days or even weeks. The entire process of meiosis is illustrated in its cellular context in Figure 3.7. The essence is that *meiosis consists of two divisions of the nucleus but only one duplication of the chromosomes.* The nuclear divisions—called the **first meiotic division** and the **second meiotic division**—can be separated into a sequence of stages similar to those used to describe mitosis. The distinctive events of this important process occur during the first division of the nucleus; these events are described in the following section.

The First Meiotic Division: Reduction

The first meiotic division (meiosis I) is sometimes called the **reductional division** because it divides the chromosome number in half. By analogy with mitosis, the first meiotic division can be split into the four stages of **prophase I, metaphase I, anaphase I,** and **telophase I.** These stages are generally more complex than their counterparts in mitosis. The stages

Figure 3.6

The life cycle of corn, *Zea mays.* As is typical in higher plants, the diploid spore-producing (sporophyte) generation is conspicuous, whereas the gamete-producing (gametophyte) generation is microscopic. The egg-producing spore is the *megaspore,* and the sperm-producing spore is the *microspore.* Nuclei participating in meiosis and fertilization are shown in yellow and green.

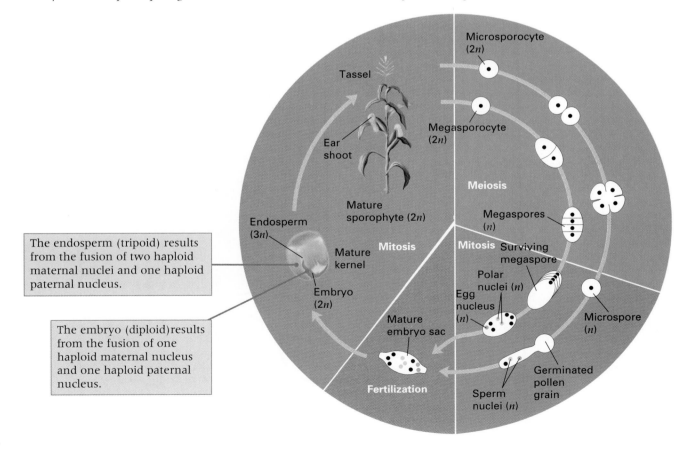

The endosperm (tripoid) results from the fusion of two haploid maternal nuclei and one haploid paternal nucleus.

The embryo (diploid)results from the fusion of one haploid maternal nucleus and one haploid paternal nucleus.

Figure 3.7

Diagram illustrating the major features of meiosis in an organism with two pairs of homologous chromosomes. At each stage, the small diagram represents the entire cell and the larger diagram is an expanded view of the chromosomes at that stage.

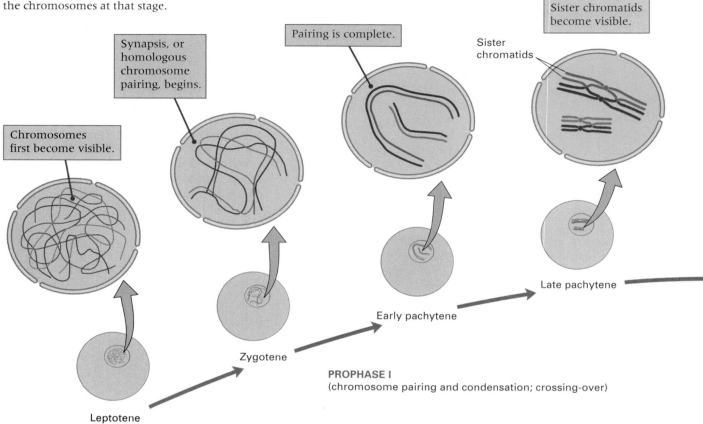

Chromosomes first become visible.

Synapsis, or homologous chromosome pairing, begins.

Pairing is complete.

Sister chromatids become visible.

Sister chromatids

Late pachytene

Early pachytene

Zygotene

PROPHASE I
(chromosome pairing and condensation; crossing-over)

Leptotene

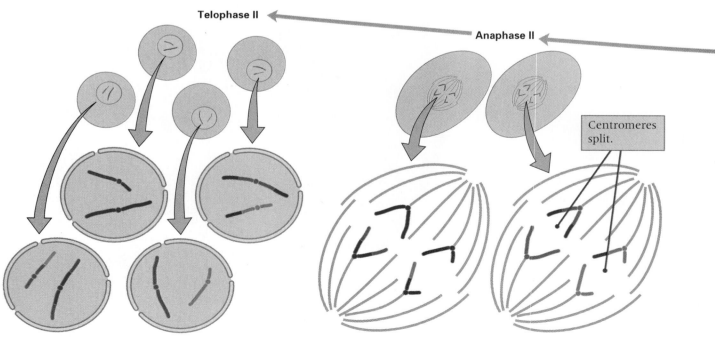

Telophase II

Anaphase II

Centromeres split.

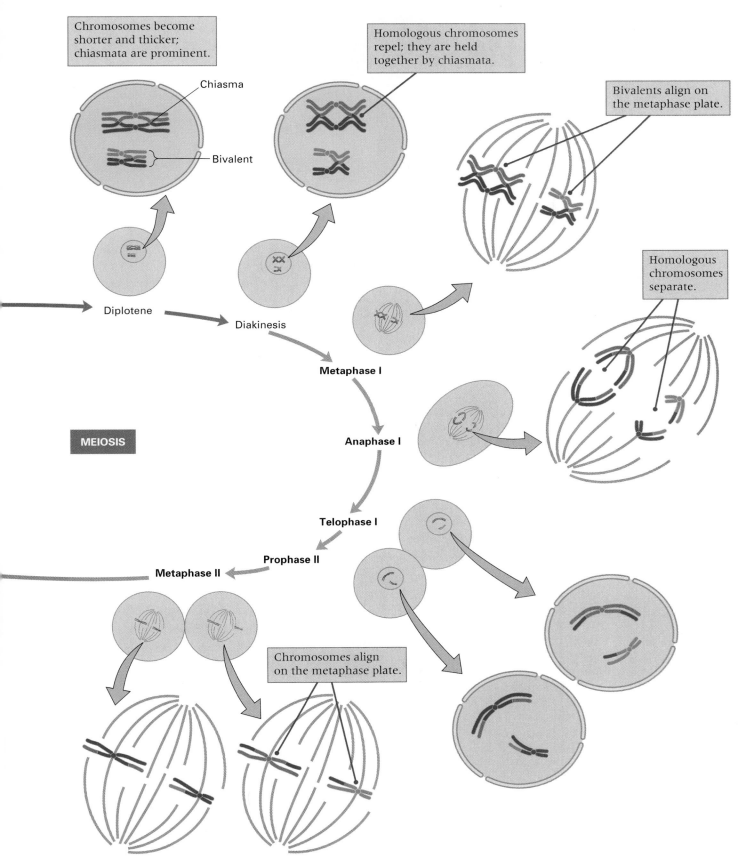

Chromosomes become shorter and thicker; chiasmata are prominent.

Chiasma

Bivalent

Homologous chromosomes repel; they are held together by chiasmata.

Bivalents align on the metaphase plate.

Diplotene

Diakinesis

Metaphase I

Homologous chromosomes separate.

MEIOSIS

Anaphase I

Telophase I

Prophase II

Metaphase II

Chromosomes align on the metaphase plate.

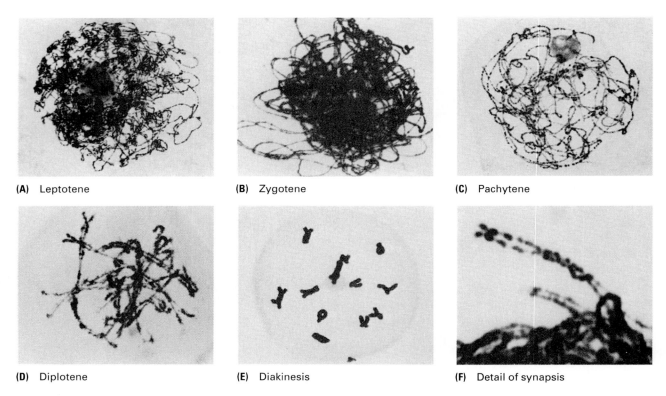

(A) Leptotene   **(B)** Zygotene   **(C)** Pachytene

(D) Diplotene   **(E)** Diakinesis   **(F)** Detail of synapsis

Figure 3.8

Substages of prophase of the first meiotic division in microsporocytes of a lily (*Lilium longiflorum*): (A) leptotene, in which condensation of the chromosomes is initiated and bead-like chromomeres are visible along the length of the chromosomes; (B) zygotene, in which pairing (synapsis) of homologous chromosomes occurs (paired and unpaired regions can be seen particularly at the lower left in this photograph); (C) pachytene, in which crossing-over between homologous chromosomes occurs; (D) diplotene, characterized by mutual repulsion of the paired homologous chromosomes, which remain held together at one or more cross points (chiasmata) along their length; (E) diakinesis, in which the chromosomes reach their maximum contraction; (F) zygotene (at higher magnification in another cell) showing paired homologs and matching of chromomeres during synapsis. [Courtesy of Marta Walters (parts A, B, C, E, and F) and Herbert Stern (part D).]

and substages can be visualized with reference to Figures 3.7 and 3.8.

1. *Prophase I* This long stage lasts several days in most higher organisms and is commonly divided into five substages: **leptotene, zygotene, pachytene, diplotene,** and **diakinesis.** These terms describe the appearance of the chromosomes at each substage.

In **leptotene,** which literally means "thin thread," the chromosomes first become visible as long, thread-like structures. The pairs of sister chromatids can be distinguished by electron microscopy. In this initial phase of condensation of the chromosomes, numerous dense granules

appear at irregular intervals along their length. These localized contractions, called **chromomeres,** have a characteristic number, size, and position in a given chromosome (Figure 3.8A).

The **zygotene** period is marked by the lateral pairing, or **synapsis,** of homologous chromosomes, beginning at the chromosome tips. (The term *zygotene* means "paired threads.") As the pairing process proceeds along the length of the chromosomes, it results in a precise chromomere-by-chromomere association (Figure 3.8B and F). Each pair of synapsed homologous chromosomes is referred to as a **bivalent.**

During **pachytene** (Figure 3.8C), condensation of the chromosomes continues.

Pachytene literally means "thick thread" and, throughout this period, the chromosomes continue to shorten and thicken (Figure 3.7). By late pachytene, it can sometimes be seen that each bivalent (that is, each set of paired chromosomes) actually consists of a **tetrad** of four chromatids, but the two sister chromatids of each chromosome are usually juxtaposed very tightly. The important event of genetic exchange, which is called **crossing-over,** takes place during pachytene, but crossing-over does not become apparent until the transition to diplotene. In Figure 3.7, the sites of exchange are indicated by the points where chromatids of different colors cross over each other.

At the onset of **diplotene,** the synapsed chromosomes begin to separate. Diplotene means "double thread," and the diplotene chromosomes are clearly double (Figure 3.8D and F). However, the homologous chromosomes remain held together at intervals along their length by cross-connections resulting from crossing-over. Each cross-connection, called a **chiasma** (plural, chiasmata), is formed by a breakage and rejoining between nonsister chromatids. As shown in the chromosome and diagram in Figure 3.9, *a chiasma results from physical exchange between chromatids of homologous chromosomes.* In normal meiosis, each bivalent usually has at least one chiasma, and bivalents of long chromosomes often have three or more.

The final period of prophase I is **diakinesis,** in which the homologous chromosomes seem to repel each other and the segments not connected by chiasmata move apart. Diakinesis means "moving apart." It is at this substage that the chromosomes attain their maximum condensation (Figure 3.8E). The homologous chromosomes in a bivalent remain connected by at least one chiasma, which persists until the first meiotic anaphase. Near the end of diakinesis, the formation of a spindle is initiated, and the nuclear envelope breaks down.

2. *Metaphase I* The bivalents become positioned with the centromeres of the two homologous chromosomes on opposite sides of the metaphase plate (Figure 3.10A). As each bivalent moves onto the metaphase plate, its centromeres are oriented at random with respect to the poles of the spindle. As shown in Figure 3.11, the bivalents formed from nonhomologous pairs of chromosomes can be oriented on the metaphase plate in either of two ways. The orientation of the centromeres determines which member of each bivalent will subsequently move to each pole. If each of the nonhomologous chromosomes is heterozygous for a pair of alleles, then one type of alignment results in *AB* and *ab* gametes and the other type results in *Ab* and *aB* gametes (Figure 3.11). Because the metaphase alignment takes place at random, the two types of alignment—and

Figure 3.9

Light micrograph (A) and interpretative drawing (B) of a bivalent consisting of a pair of homologous chromosomes. This bivalent was photographed at late diplotene in a spermatocyte of the salamander *Oedipina poelzi.* It shows two chiasmata where the chromatids of the homologous chromosomes appear to exchange pairing partners. [From F. W. Stahl. 1964. *The Mechanics of Inheritance.* Prentice-Hall, Inc.; courtesy of James Kezer.]

(A)

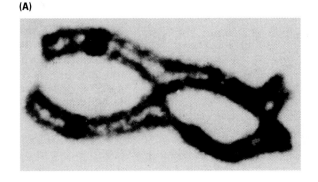

(B)

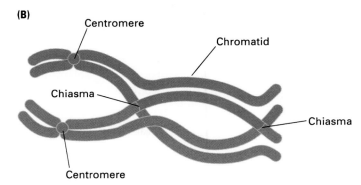

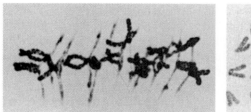
(A) Metaphase I

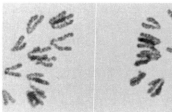
(B) Anaphase I

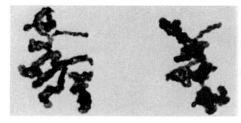
(C) Metaphase II (telophase I
and prophase II not shown)

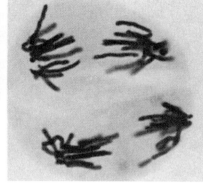
(D) Anaphase II

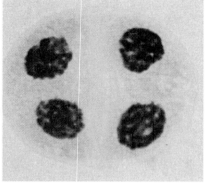
(E) Telophase II

Figure 3.10
Later meiotic stages in microsporocytes of the lily *Lilium longiflorum:* (A) metaphase I;
(B) anaphase I; (C) metaphase II; (D) anaphase II; (E) telophase II. Cell walls have begun to form
in telophase, which will lead to the formation of four pollen grains. [Courtesy of Herbert Stern.]

therefore the four types of gametes—are equally frequent. The ratio of the four types of gametes is 1 : 1 : 1 : 1, which means that the *A, a* and *B, b* pairs of alleles undergo independent assortment. In other words,

> Genes on different chromosomes undergo independent assortment because non-homologous chromosomes align at random on the metaphase plate in meiosis I.

3. *Anaphase I* In this stage, homologous chromosomes, each composed of two chromatids joined at an undivided centromere, separate from one another and move to opposite poles of the spindle (Figure 3.10B). Chromosome separation at anaphase is the cellular basis of the segregation of alleles:

> The physical separation of homologous chromosomes in anaphase is the physical basis of Mendel's principle of segregation.

4. *Telophase I* At the completion of anaphase I, a haploid set of chromosomes consisting of one homolog from each bivalent is located near each pole of the spindle (Figure 3.6). In telophase, the spindle breaks down and, depending on the species, either a nuclear envelope briefly forms around each group of chromosomes or the chromosomes enter the second meiotic division after only a limited uncoiling.

The Second Meiotic Division: Equation

The second meiotic division (meiosis II) is sometimes called the **equational division** because the chromosome number remains the same in each cell before and after the second division. In some species, the chromosomes pass directly from telophase I to **prophase II** without loss of condensation; in others, there is a brief pause between the two meiotic divisions and the chromo-

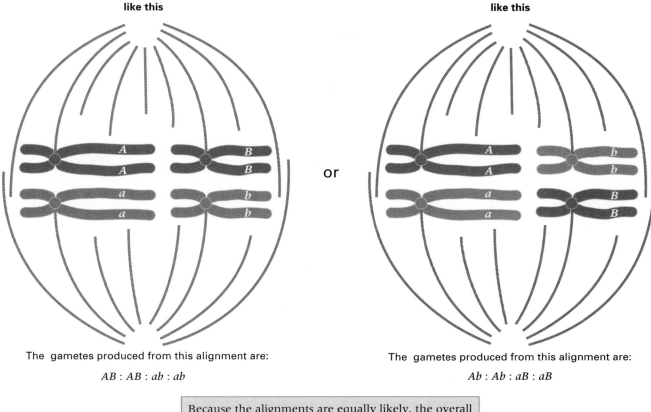

Metaphase alignment of genes on different chromosomes may be:

like this   like this

or

The gametes produced from this alignment are: The gametes produced from this alignment are:

AB : AB : ab : ab   *Ab : Ab : aB : aB*

> Because the alignments are equally likely, the overall ratio of gametes is:
>
> *AB : Ab : aB : ab* = 1 : 1 : 1 : 1
>
> This ratio is characteristic of independent assortment.

Figure 3.11
Random alignment of nonhomologous chromosomes at metaphase I results in the independent assortment of genes on nonhomologous chromosomes.

somes may "decondense" (uncoil) somewhat. *Chromosome replication never takes place between the two divisions;* the chromosomes present at the beginning of the second division are identical to those present at the end of the first division.

After a short prophase (prophase II) and the formation of second-division spindles, the centromeres of the chromosomes in each nucleus become aligned on the central plane of the spindle at **metaphase II** (Figure 3.10C). In **anaphase II,** the centromeres divide longitudinally and the chromatids of each chromosome move to opposite poles of the spindle (Figure 3.10D). Once the centromere has split at anaphase II, each chromatid is considered to be a separate chromosome.

Telophase II (Figure 3.10E) is marked by a transition to the interphase condition of the chromosomes in the four haploid nuclei, accompanied by division of the cytoplasm. Thus the second meiotic division superficially resembles a mitotic division. However, there is an important difference: *The chromatids of a chromosome are usually not genetically identical sisters along their entire length because of crossing-over associated with the formation of chiasmata during prophase of the first division.*

Grasshopper, Grasshopper

E. Eleanor Carothers 1913
University of Kansas,
Lawrence, Kansas
The Mendelian Ratio in Relation to Certain Orthopteran Chromosomes

As an undergraduate researcher, Carothers showed that nonhomologous chromosomes undergo independent assortment in meiosis. For this purpose she studied a grasshopper in which one pair of homologous chromosomes had members of unequal length. At the first anaphase of meiosis in males, she could determine by observation whether the longer or the shorter chromosome went in the same direction as the X chromosome. As detailed in this paper, she found 154 of the former and 146 of the latter, a result in very close agreement with the 1 : 1 ratio expected from independent assortment. There is no mention of the Y chromosome because in the grasshopper she studied, the females have the sex chromosome constitution XX, whereas the males have the sex chromosome constitution X. In the males she examined, therefore, the X chromosome did not have a pairing partner. The instrument referred to as a camera lucida was at that time in widespread use for studying chromosomes and other microscopic objects. It is an optical instrument containing a prism or an arrangement of mirrors that, when mounted on a microscope, reflects an image of the microscopic object onto a piece of paper where it may be traced.

The aim of this paper is to describe the behavior of an unequal bivalent in the primary spermatocytes of certain grasshoppers. The distribution of the chromosomes of this bivalent, in relation to the X chromosome, follows the laws of chance; and, therefore, affords direct cytological support of Mendel's laws. This distribution is easily traced on account of a very distinct difference in size of the homologous chromosomes. Thus another link is added to the already long chain of evidence that the chromosomes are distinct morphological individuals continuous from generation to generation, and, as

> *Another link is added to the already long chain of evidence that the chomosomes are distinct morphological individuals continuous from generation to generation, and, as such, are the bearers of the hereditary qualities.*

such, are the bearers of the hereditary qualities. . . . This work is based chiefly on *Brachystola magna* [a short-horned grasshopper]. . . . The entire complex of chromosomes can be separated into two groups, one containing six small chromosomes and the other seventeen larger ones. [One of the larger ones is the X chromosome.] Examination shows that this group of six small chromosomes is composed of five of about equal size and one decidedly larger. [One of the small ones is the homolog of the decidedly larger one, making this pair of chromosomes unequal in size.] . . . In early metaphases the chromosomes appear as twelve separate individuals [the bivalents]. Side views show the X chromosome in its characteristic position near one pole. . . . Three hundred cells were drawn under the camera lucida to determine the distribution of the chromosomes in the asymmetrical bivalent in relation to the X chromosome. . . . In 228 cells the bivalent and the X chromosome were in the same section [the cells had been embedded in wax and thinly sliced]. In 107 cells the smaller chromosome was going to the same pole as the X chromosome, and in the remaining 121 the larger chromosome occupied this position. In the other 72 cells the X chromosome and the bivalent were in different sections, but great care was used to make sure that there was no mistake in identifying the cell or in labeling the drawings. The smaller chromosome is accompanying the X chromosome in 39 of the cells, and the larger in 33. As a net result, then, in the 300 cells drawn, the smaller chromosome would have gone to the same nucleus as the X chromosome 146 times, or in 48.7 percent of the cases; and the larger one, 154 times, or in 51.3 percent of the cases. . . . A consideration of the limited number of chromosomes and the large number of characters in any animal or plant will make it evident that each chromosome must control numerous different characters. . . . Since the rediscovery of Mendel's laws, increased knowledge has been constantly bringing into line facts that at first seemed utterly incompatible with them. There is no cytological explanation of any other form of inheritance. . . . It seems to me probable that all inheritance is, in reality, Mendelian.

Source: Journal of Morphology 24: 487–511

3.4 Chromosomes and Heredity

Shortly after the rediscovery of Mendel's paper, it became widely assumed that genes were physically located in the chromosomes. The strongest evidence was that Mendel's principles of segregation and independent assortment paralleled the behavior of chromosomes in meiosis. But the first undisputable proof that genes are parts of chromosomes was obtained in experiments concerned with the pattern of transmission of the **sex chromosomes,** the chromosomes responsible for the determination of the separate sexes in some plants and in almost all animals. We will examine these results in this section.

Chromosomal Determination of Sex

The sex chromosomes are an exception to the rule that all chromosomes of diploid organisms are present in pairs of morphologically similar homologs. As early as 1891, microscopic analysis had shown that one of the chromosomes in males of some insect species does not have a homolog. This unpaired chromosome was called the **X chromosome,** and it was present in all somatic cells of the males but in only half the sperm cells. The biological significance of these observations became clear when females of the same species were shown to have two X chromosomes.

In other species in which the females have two X chromosomes, the male has one X chromosome along with a morphologically different chromosome. This different chromosome is referred to as the **Y chromosome,** and it pairs with the X chromosome during meiosis in males because the X and Y share a small region of homology. The difference in chromosomal constitution between males and females is a chromosomal mechanism for determining sex at the time of fertilization. Whereas every egg cell contains an X chromosome, half the sperm cells contain an X chromosome and the rest contain a Y chromosome. Fertilization of an X-bearing egg by an X-bearing sperm results in an XX zygote, which normally develops into a female; and fertilization by a Y-bearing sperm results in an XY zygote, which normally develops into a male (Figure 3.12). The result is a criss-cross pattern of inheritance of the X chromosome in which a male receives his X chromosome from his mother and transmits it only to his daughters.

The XX-XY type of chromosomal sex determination is found in mammals, including human beings, many insects, and other animals, as well as in some flowering plants. The female is called the **homogametic** sex because only one type of gamete (X-bearing) is produced, and the male is called the **heterogametic** sex because two different types of gametes (X-bearing and Y-bearing) are produced. When the union of gametes in fertilization is random, a sex ratio at fertilization of 1 : 1 is expected because males produce equal numbers of X-bearing and Y-bearing sperm.

The X and Y chromosomes together constitute the sex chromosomes; this term distinguishes them from other pairs of chromosomes, which are called **autosomes.** Although the sex chromosomes control the developmental switch that determines the earliest stages of female or male development, the developmental process itself requires many genes scattered throughout the chromosome complement, including genes on the autosomes. The X chromosome also contains many genes with functions unrelated to sexual differentiation, as will be seen in the next section. In most organisms, including human beings, the Y chromosome carries few genes other than those related to male determination.

X-linked Inheritance

The compelling evidence that genes are in chromosomes came from the study of a *Drosophila* gene for white eyes, which proved to be present in the X chromosome. Recall that in Mendel's crosses, it did not matter which trait was present in the male parent and which in the female parent. Reciprocal crosses gave the same result. One of the earliest exceptions to this rule was found by Thomas Hunt Morgan in 1910, in an early study of a mutant in the fruit fly *Drosophila melanogaster* that had white eyes. The wildtype eye color is a brick-red combination of red and brown pigments (Figure 3.13). Although white eyes can result from certain combinations of autosomal genes that eliminate the pigments individually, the white-eye mutation that Morgan studied results in a metabolic block that knocks out both pigments simultaneously.

Morgan's study started with a single male with white eyes that appeared in a wildtype laboratory population that had been maintained for many generations. In a mating of this male with wildtype females, all of the F_1 progeny of both sexes had red eyes, which showed that the allele for white eyes is recessive. In the F_2 progeny from the mating of F_1 males and females, Morgan observed 2459 red-eyed females, 1011 red-eyed males, and 782 white-eyed males. The white-eyed phenotype was somehow connected with sex because all of the white-eyed flies were males.

(A)

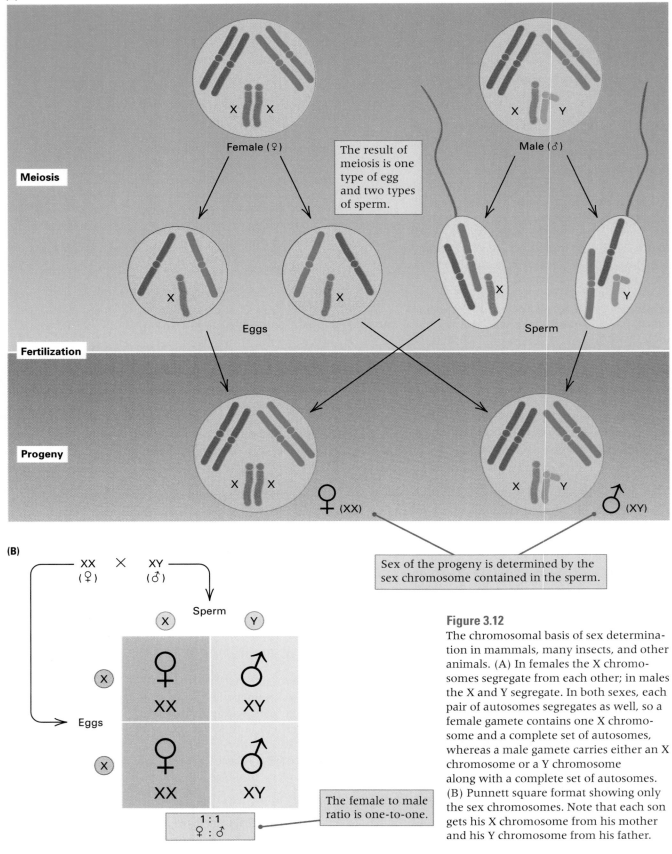

Figure 3.12

The chromosomal basis of sex determination in mammals, many insects, and other animals. (A) In females the X chromosomes segregate from each other; in males the X and Y segregate. In both sexes, each pair of autosomes segregates as well, so a female gamete contains one X chromosome and a complete set of autosomes, whereas a male gamete carries either an X chromosome or a Y chromosome along with a complete set of autosomes. (B) Punnett square format showing only the sex chromosomes. Note that each son gets his X chromosome from his mother and his Y chromosome from his father.

(A)   **(B)**

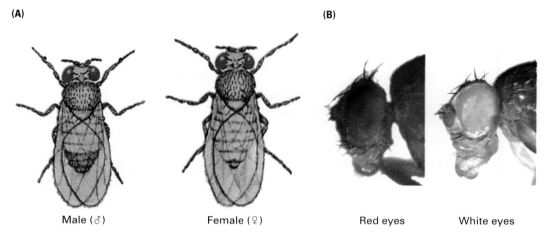

Male (♂)   Female (♀)   Red eyes   White eyes

Figure 3.13
Drawings of a male and a female fruit fly, *Drosophila melanogaster*. The photographs show the eyes of a wildtype red-eyed male and a mutant white-eyed male. [Drawings courtesy of Carolina Biological Supply Company; photographs courtesy of E. R. Lozovskaya.]

On the other hand, white eyes were not restricted to males. For example, when red-eyed F_1 females from the cross of wildtype ♀♀ × white ♂♂ were backcrossed with their white-eyed fathers, the progeny consisted of both red-eyed and white-eyed females and red-eyed and white-eyed males in approximately equal numbers.

A key observation came from the mating of white-eyed females with wildtype males. All the female progeny had wildtype eyes, but all the male progeny had white eyes. This is the reciprocal of the original cross of wildtype ♀♀ × white ♂♂, which had given only wildtype females and wildtype males, so the reciprocal crosses gave different results.

Morgan realized that reciprocal crosses would yield different results if the allele for white eyes were present in the X chromosome. The reason is that the X chromosome is transmitted in a different pattern by males and females. Figure 3.12B shows that a male transmits his X chromosome only to his daughters, whereas a female transmits one of her X chromosomes to the offspring of both sexes. Figure 3.14 shows the normal chromosome complement of *Drosophila melanogaster*. Females have an XX chromosome complement; the males are XY, and the Y chromosome does not contain a counterpart of the *white* gene. A gene on the X chromosome is said to be **X-linked**.

Figure 3.15 illustrates the chromosomal interpretation of the reciprocal crosses wildtype ♀ × white ♂ (Cross A) and white ♀ × wildtype ♂ (Cross B). The symbols w and w^+ denote the mutant and wildtype forms of the *white* gene present in the X chromosome. The genotype of a white-eyed male is wY, and that of a wildtype male is w^+Y. Because the w allele is recessive, white-eyed females are of genotype ww and wildtype females are either

Male

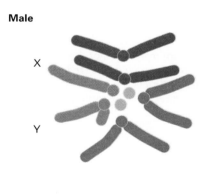

X

Y

Female

X

X

Figure 3.14
The diploid chromosome complements of a male and a female *Drosophila melanogaster*. The centromere of the X chromosome is nearly terminal, but that of the Y chromosome divides the chromosome into two unequal arms. The large autosomes (chromosomes 2 and 3, shown in blue and green) are not easily distinguishable in these types of cells. The tiny autosome (chromosome 4, shown in yellow) appears as a dot.

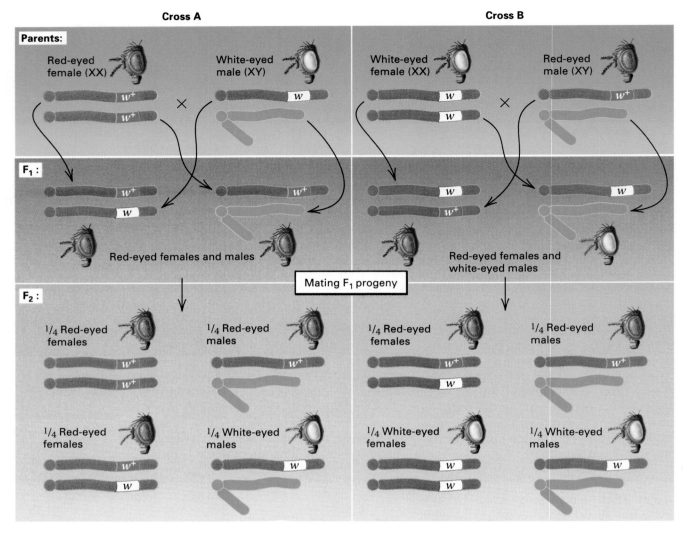

Figure 3.15

A chromosomal interpretation of the results obtained in F_1 and F_2 progenies in crosses of *Drosophila*. Cross A is a mating of a wildtype (red-eyed) female with a white-eyed male. Cross B is the reciprocal mating of a white-eyed female with a red-eyed male. In the X chromosome, the wildtype w^+ allele is shown in red and the mutant w allele in white. The Y chromosome does not carry either allele of the w gene.

heterozygous w^+w or homozygous w^+w^+. The diagrams in Figure 3.15 account for the different phenotypic ratios observed in the F_1 and F_2 progeny from the crosses. Many other genes were later found in *Drosophila* that also follow the X-linked pattern of inheritance.

The characteristics of X-linked inheritance can be summarized as follows:

1. Reciprocal crosses resulting in different phenotypic ratios in the sexes often indicate X linkage; in the case of white eyes in *Drosophila*, the cross of a red-eyed female with a white-eyed male

yields all red-eyed progeny (Figure 3.15, Cross A), whereas the cross of a white-eyed female with a red-eyed male yields red-eyed female progeny and white-eyed male progeny (Figure 3.15, Cross B).

2. Heterozygous females transmit each X-linked allele to approximately half their daughters and half their sons; this is illustrated in the F_2 generation of Cross B in Figure 3.15.

3. Males that inherit an X-linked recessive allele exhibit the recessive trait because the Y chromosome does not contain a

The White-Eyed Male

Thomas Hunt Morgan 1910
Columbia University,
New York, New York
Sex Limited Inheritance in Drosophila

Morgan's genetic analysis of the white-eye mutation marks the beginning of Drosophila genetics. It is in the nature of science that as knowledge increases, the terms used to describe things change also. This paper affords an example, because the term sex limited inheritance is used today to mean something completely different from Morgan's usage. What Morgan was referring to is now called X-linked inheritance or sex-linked inheritance. To avoid confusion, we have taken the liberty of substituting the modern equivalent wherever appropriate. Morgan was also unaware that Drosophila males had a Y chromosome. He thought that females were XX and males X, as in grasshoppers (see the Carothers paper). We have also supplied the missing Y chromosome. On the other hand, Morgan's gene symbols have been retained as in the original. He uses R for the wildtype allele for red eyes and W for the recessive allele for white eyes. This is a curious departure from the convention, already introduced by Mendel, that dominant and recessive alleles should be represented by the same symbol. Today we use w for the recessive allele and w^+ for the dominant allele.

In a pedigree culture of *Drosophila* which had been running for nearly a year through a considerable number of generations, a male appeared with white eyes. The normal flies have brilliant red eyes. The white-eyed male, bred to his red-eyed sisters, produced 1,237 red-eyed offspring. . . . The F_1 hybrids, inbred, produced

2,459	red-eyed females
1,011	red-eyed males
782	white-eyed males

No white-eyed females appeared.

No white-eyed females appeared. The new character showed itself to be sex-linked in the sense that it was transmitted only to the grandsons. But that the character is not incompatible with femaleness is shown by the following experiment. The white-eyed male (mutant) was later crossed with some of his daughters (F_1), and produced

129	red-eyed females
132	red-eyed males
88	white-eyed females
86	white-eyed males

The results show that the new character, white eyes, can be carried over to the females by a suitable cross, and is in consequence in this sense not limited to one sex. It will be noted that the four classes of individuals occur in approximately equal numbers (25 percent). . . . The results just described can be accounted for by the following hypothesis. Assume that all of the spermatozoa of the white-eyed male carry the "factor" for white eyes "W"; that half of the spermatozoa carry a sex factor "X," the other half lack it, *i. e.,* the male is heterozygous for sex. [The male is actually XY.] Thus, the symbol for the male is "WXY", and for his two kinds of spermatozoa WX—Y. Assume that all of the eggs of the red-eyed female carry the red-eyed "factor" R; and that all of the eggs (after meiosis) carry one X each, the symbol for the red-eyed female will be therefore RRXX and that for her eggs will be RX. . . . The hypothesis just utilized to explain these results first obtained can be tested in several ways. [There follow four types of crosses, each yielding the expected result.] . . . In order to obtain these results it is necessary to assume that, when the two classes of spermatozoa are formed in the RXY male, R and X go together. . . . The fact is that this R and X are combined and have never existed apart.

Source: Science 32: 120–122

wildtype counterpart of the gene. Affected males transmit the recessive allele to all of their daughters but none of their sons; this principle is illustrated in the F_1 generation of Cross A in Figure 3.15. Any male that is not affected carries the wildtype allele in his X chromosome.

An example of a human trait with an X-linked pattern of inheritance is **hemophilia A,** a severe disorder of blood clotting determined by a recessive allele. Affected persons lack a blood-clotting protein called factor VIII needed for normal clotting, and they suffer excessive, often life-threatening bleeding after injury. A famous pedigree of hemophilia starts with Queen Victoria of England (Figure 3.16). One of her sons, Leopold, was hemophilic, and two of her daughters were heterozygous carriers of the gene. Two of Victoria's granddaughters were also carriers, and by marriage they introduced the gene into the royal families of Russia and Spain. The heir to the Russian throne of the Romanoffs, Tsarevich Alexis, was afflicted with the condition. He inherited the gene from his mother, the Tsarina Alexandra, one of Victoria's granddaughters. The Tsar, the Tsarina, Alexis, and his four sisters were all executed by the Bolsheviks in the 1918 Russian revolution. Ironically, the present royal family of England is descended from a normal son of Victoria and is free of the disease.

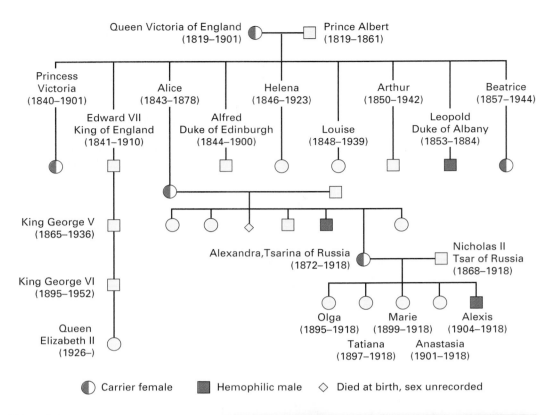

Queen Victoria of England
(1819–1901)   Prince Albert
(1819–1861)

Princess
Victoria
(1840–1901)   Alice
(1843–1878)   Helena
(1846–1923)   Arthur
(1850–1942)   Beatrice
(1857–1944)

Edward VII
King of England
(1841–1910)   Alfred
Duke of Edinburgh
(1844–1900)   Louise
(1848–1939)   Leopold
Duke of Albany
(1853–1884)

King George V
(1865–1936)

Alexandra, Tsarina of Russia
(1872–1918)   Nicholas II
Tsar of Russia
(1868–1918)

King George VI
(1895–1952)

Olga
(1895–1918)   Marie
(1899–1918)   Alexis
(1904–1918)

Queen
Elizabeth II
(1926–)   Tatiana
(1897–1918)   Anastasia
(1901–1918)

◐ Carrier female   ■ Hemophilic male   ◇ Died at birth, sex unrecorded

Figure 3.16
Genetic transmission of hemophilia A among the descendants of Queen Victoria of England, including her granddaughter, Tsarina Alexandra of Russia, and Alexandra's five children. The photograph is that of Tsar Nicholas II, Tsarina Alexandra, and the Tsarevich Alexis, who was afflicted with hemophilia. [*Source:* Culver Pictures.]

In some organisms, the homogametic and heterogametic sexes are reversed; that is, the males are XX and the females are XY. This type of sex determination is found in birds, in some reptiles and fish, and in moths and butterflies. The reversal of XX and XY in the sexes results in an opposite pattern of nonreciprocal inheritance of X-linked genes. For example, some breeds of chickens have feathers with alternating transverse bands of light and dark color, resulting in a phenotype referred to as barred. The feathers are uniformly colored in the nonbarred phenotypes of other breeds. Reciprocal crosses between true-breeding barred and nonbarred types give the following outcomes:

Nonbarred ♀ × Barred ♂
(WZ)   (WW)

Barred ♀♀ and Barred ♂♂
(WZ)   (WW)

Barred ♀ × Nonbarred ♂
(WZ)   (WW)

Nonbarred ♀♀ and Barred ♂♂
(WZ)   (WW)

These results indicate that the gene that determines barring is on the chicken X chromosome and is dominant. To distinguish sex determination in birds, butterflies, and moths from the usual XX-XY mechanism, in these organisms the sex chromosome constitution in the homogametic sex is sometimes designated WW and that in the heterogametic sex as WZ. Hence in birds, butterflies, and moths, males are chromosomally WW and females are chromosomally WZ.

Nondisjunction as Proof of the Chromosome Theory of Heredity

The parallelism between the inheritance of the *Drosophila white* mutation and the genetic transmission of the X chromosome supported the chromosome theory of heredity that genes are parts of chromosomes. Other experiments with *Drosophila* provided the definitive proof.

One of Morgan's students, Calvin Bridges, discovered rare exceptions to the expected pattern of inheritance in crosses with several X-linked genes. For example, when white-eyed *Drosophila* females were mated with red-eyed males, most of the progeny consisted of the expected red-eyed females and white-eyed males. However, about one in every 2000 F_1 flies was an exception, either a white-eyed female or a red-eyed male. Bridges showed that these rare exceptional offspring resulted from occasional failure of the two X chromosomes in the mother to separate from each other during meiosis—a phenomenon called **nondisjunction.** The consequence of nondisjunction of the X chromosome is the formation of some eggs with two X chromosomes and others with none. Four classes of zygotes are expected from the fertilization of these abnormal eggs (Figure 3.17). Animals with no X chromosome are not detected because embryos that lack an X are not viable; likewise, most progeny with three X chromosomes die early in development. Microscopic examination of the chromosomes of the exceptional progeny from the cross white ♀♀ × wildtype ♂♂ showed that the exceptional white-eyed females had two X chromosomes *plus* a Y chromosome, and the

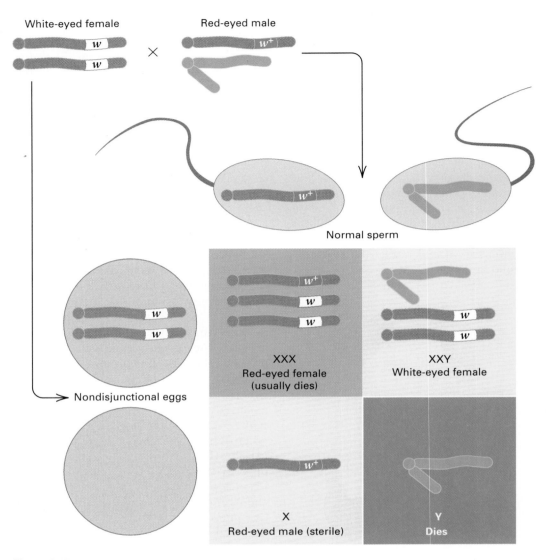

White-eyed female   Red-eyed male

×

Normal sperm

Nondisjunctional eggs

XXX
Red-eyed female
(usually dies)

XXY
White-eyed female

X
Red-eyed male (sterile)

Y
Dies

Figure 3.17
The results of meiotic nondisjunction of the X chromosomes in a female *Drosophila.*

exceptional red-eyed males had a single X but were *lacking* a Y. The latter, with a sex-chromosome constitution denoted XO, were sterile males.

These and related experiments demonstrated conclusively the validity of the chromosome theory of heredity.

> **Chromosome theory of heredity:** Genes are contained in the chromosomes.

Bridges's evidence for the chromosome theory was that exceptional behavior on the part of chromosomes is precisely paralleled by exceptional inheritance of their genes. This proof of the chromosome theory ranks among the most important and elegant experiments in genetics.

Sex Determination in *Drosophila*

In the XX-XY mechanism of sex determination, the Y chromosome is associated with the male. In some organisms, including human beings, this association occurs because the presence of the Y chromosome triggers events in embryonic development that result in the male sexual characteristics. *Drosophila* is unusual among organisms with an XX-XY type of sex determination because the Y chromosome, although associated with maleness, is not male-determining. This is demonstrated by the finding, shown in Figure 3.17, that in *Drosophila*, XXY embryos develop into morphologically normal, fertile females, whereas XO embryos develop into morphologically normal, but sterile, males. (The "O" is written in the formula XO to emphasize that a sex chromosome is missing.) The sterility of XO males shows that the Y chromosome, though not necessary for male development, is essential for male fertility; in fact, the *Drosophila* Y chromosome contains six genes required for the formation of normal sperm.

The genetic determination of sex in *Drosophila* depends on the number of X chromosomes present in an individual fly compared with the number of sets of autosomes. In *Drosophila*, a haploid set of autosomes consists of one copy each of chromosomes 2, 3, and 4 (the autosomes). Normal diploid flies have two haploid sets of autosomes (a homologous pair each of chromosomes 2, 3, and 4) plus either two X chromosomes (in a female) or one X and one Y chromosome (in a male). We will use A to represent a complete haploid complement of autosomes; hence

$$A = \text{Chromosome } 2 \\ + \text{ Chromosome } 3 \\ + \text{ Chromosome } 4$$

In these terms, a normal male has the chromosomal complement XYAA, and the ratio of X chromosomes to sets of autosomes (the X/A ratio) equals 1 X : 2 A, or 1 : 2. Normal females have the chromosomal complement XXAA, and in this sex the X/A ratio is 2 X : 2 A, or 1 : 1. Flies with X/A ratios smaller than 1 : 2 (for example, XAAA—one X chromosome and three sets of autosomes) are male; those with X/A ratios greater than 1 : 1 (for example, XXXAA—three X chromosomes and two sets of autosomes) are female. Intermediate X/A ratios such as 2 : 3 (for example, XXAAA—two X chromosomes and three sets of autosomes) develop as intersexes with some characteristics of each sex.

Sexual differentiation in *Drosophila* is controlled by a gene called *Sex-lethal (Sxl)*. The *Sxl* gene codes for two somewhat different proteins, depending on whether a male-specific coding region is included in the messenger RNA. Furthermore, the amount of Sxl protein present in the early embryo regulates the expression of the *Sxl* gene by a feedback mechanism. At low levels of Sxl protein, the male-specific form of the protein is made and shuts off further expression of the gene. At higher levels of the Sxl protein, the female-specific form of the protein is made and the gene continues to be expressed. In some unknown manner, the products of certain genes are sensitive to the X/A ratio and determine the amount of Sxl protein available to regulate the *Sxl* gene. The genes for sensing the number of X chromosomes are called *numerator genes* because they determine the "numerator" of the ratio X/A, and the genes for sensing the number of sets of autosomes are known as *denominator genes* because they determine the "denominator" in the ratio X/A. In normal males (X/A = 1 : 2), there is too little Sxl protein and the *Sxl* gene shuts down; in the absence of *Sxl* expression, sexual differentiation follows the male pathway, which is the "default" pathway. In normal females (X/A = 1 : 1), there is enough Sxl protein that the *Sxl* gene continues to be expressed. Continued expression of the *Sxl* gene initiates a cascade of genetic events, each gene in the cascade controlling one or more other genes downstream, and results in the expression of female-specific gene products and the repression of male-specific gene products. In intermediate situations when the X/A ratio is between 1 : 2 and 1 : 1, some genes specific to each sex are expressed, and the resulting sexual phenotype is ambiguous—an intersex. The Sxl protein is an RNA-binding protein that determines the type of mRNA produced by some of the sex-determining genes. An outline of the genetic control of sex determination in *Drosophila* is shown in Figure 3.18.

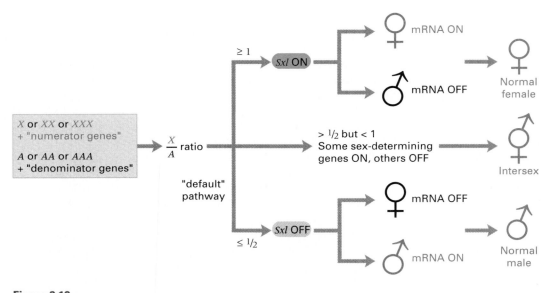

Figure 3.18
Early steps in the genetic control of sex determination in *Drosophila* through the activity of the Sex-lethal protein and ultimately through the "numerator" and "denominator" genes that signal the ratio of X chromosomes to sets of autosomes (the X/A ratio).

3.5 Probability in Prediction and Analysis of Genetic Data

Genetic transmission includes a large component of chance. A particular gamete from an *Aa* organism might or might not include the *A* allele, depending on chance. A particular gamete from an *Aa Bb* organism might or might not include both the *A* and *B* alleles, depending on the chance orientation of the chromosomes on the metaphase I plate. Genetic ratios result not only from the chance assortment of genes into gametes, but also from the chance combination of gametes into zygotes. Although exact predictions are not possible for any particular event, it is possible to determine the probability that a particular event might be realized, as we have seen in Chapter 2. In this section, we consider some of the probability methods used in interpreting genetic data.

Using the Binomial Distribution in Genetics

The addition rule of probability deals with outcomes of a genetic cross that are mutually exclusive. Outcomes are "mutually exclusive" if they are incompatible in the sense that they cannot occur at the same time. For example, there are four mutually exclusive outcomes of the sex distribution of sibships with three children—namely, the inclusion of 0, 1, 2, or 3 girls. These have probability 1/8, 3/8, 3/8, and 1/8, respectively. The addition rule states that the overall probability of any combination of mutually exclusive events is equal to the sum of the probabilities of the events taken separately. For example, the probability that a sibship of size 3 contains *at least one girl* includes the outcomes 1, 2, and 3 girls, so the overall probability of at least one girl equals 3/8 + 3/8 + 1/8 = 7/8.

The multiplication rule of probability deals with outcomes of a genetic cross that are independent. Any two outcomes are independent if the knowledge that one outcome is actually realized provides no information about whether the other is realized also. For example, in a sequence of births, the sex of any one child is not affected by the sex distribution of any children born earlier and has no influence whatsoever on the sex distribution of any siblings born later. Each successive birth is independent of all the others. When possible outcomes are independent, the multiplication rule states that the probability of any combination of outcomes being realized equals the product of the probabilities of all of the individual outcomes taken separately. For example, the probability that a sibship of three children will consist of three girls equals $1/2 \times 1/2 \times 1/2$, because the probability of each birth resulting in a girl is 1/2, and the successive births are independent.

Probability calculations in genetics frequently use the addition and multiplication rules together. For example, the probability that all three children in a family will be of the same sex uses both the addition and the multiplication rules. The probability that all three will be girls is $(1/2)(1/2)(1/2) = 1/8$, and the probability that all three will be boys is also 1/8. Because these outcomes are mutually exclusive (a sibship of size three cannot include three boys *and* three girls), the probability of either three girls or three boys is the sum of the two probabilities, or $1/8 + 1/8 = 1/4$. The other possible outcomes for sibships of size three are that two of the children will be girls and the other a boy, and that two will be boys and the other a girl. For each of these outcomes, three different orders of birth are possible—for example, GGB, GBG, and BGG—each having a probability of $1/2 \times 1/2 \times 1/2 = 1/8$. The probability of two girls and a boy, disregarding birth order, is the sum of the probabilities for the three possible orders, or 3/8; likewise, the probability of two boys and a girl is also 3/8. Therefore, the distribution of probabilities for the sex ratio in families with three children is

GGG	GGB	GBB	BBB
	GBG	BGB	
	BGG	BBG	

$(1/2)^3 + 3(1/2)^2(1/2) + 3(1/2)(1/2)^2 + (1/2)^3 =$

1/8 $\quad + \quad$ 3/8 $\quad + \quad$ 3/8 $\quad + $ 1/8 $\;= 1$

The sex ratio information in this display can be obtained more directly by expanding the binomial expression $(p + q)^n$, in which p is the probability of the birth of a girl (1/2), q is the probability of the birth of a boy (1/2), and n is the number of children. In the present example,

$$(p + q)^3 = 1p^3 + 3p^2q + 3pq^2 + 1q^3$$

in which the red numerals are the possible number of birth orders for each sex distribution. Similarly, the binomial distribution of probabilities for the sex ratios in families of five children is

$$(p + q)^5 = 1p^5 + 5p^4q + 10p^3q^2 + $$
$$10p^2q^3 + 5pq^4 + 1q^5$$

Each term tells us the probability of a particular combination. For example, the third term is the probability of three girls (p^3) and two boys (q^2) in a family that has five children—namely,

$$10(1/2)^3(1/2)^2 = 10/32 = 5/16$$

There are $n + 1$ terms in a binomial expansion. The exponents of p decrease by one from n in the first term to 0 in the last term, and the exponents of q increase by one from 0 in the first term to n in the last term. The coefficients generated by successive values of n can be arranged in a regular triangle known as **Pascal's triangle** (Figure 3.19). Note that the horizontal rows of the triangle are symmetrical, and that each number is the sum of the two numbers on either side of it in the row above.

In general, if the probability of event A is p and that of event B is q, and the two events are independent and mutually exclusive (see Chapter 2), the probability that A will be realized four times and B two times—in a specific order—is p^4q^2, by the multiplication rule. However, suppose that we were interested in the combination of events "four of A and two of B," regardless of order. In that case, we multiply the

Figure 3.19
Pascal's triangle. The numbers are the coefficients of each term in the expansion of the polynomial $(p + q)^n$ for successive values of n from 0 through 6.

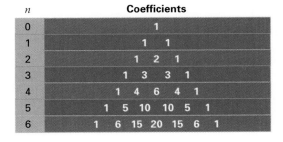

n	Coefficients
0	1
1	1 1
2	1 2 1
3	1 3 3 1
4	1 4 6 4 1
5	1 5 10 10 5 1
6	1 6 15 20 15 6 1

probability that the combination 4A : 2B will be realized in any one specific order by the number of possible orders. The number of different combinations of six events, four of one kind and two of another, is

$$\frac{6!}{4!\,2!} = \frac{1 \times 2 \times 3 \times 4 \times 5 \times 6}{(1 \times 2 \times 3 \times 4) \times (1 \times 2)} = 15$$

The symbol ! stands for *factorial,* or the product of all positive integers from 1 through a given number. Except for $n = 0$, the formula for factorial is

$$n! = 1 \times 2 \times 3 \times 4 \times \cdots \times (n - 1) \times n$$

The case $n = 0$ is an exception because 0! is defined as equal to 1. The first few factorials are

$$0! = 1$$
$$1! = 1$$
$$2! = 1 \times 2 = 2$$
$$3! = 1 \times 2 \times 3 = 6$$
$$4! = 1 \times 2 \times 3 \times 4 = 24$$
$$5! = 1 \times 2 \times 3 \times 4 \times 5 = 120$$
$$6! = 1 \times 2 \times 3 \times 4 \times 5 \times 6 = 720$$

The factorial formula

$$6!/(4! \times 2!) = 720/(24 \times 2) = 15$$

is the coefficient of the term $p^4 q^2$ in the expansion of the binomial $(p + q)^6$. Therefore, the probability that event A will be realized four times and event B two times is $15p^4 q^2$.

The general rule for repeated trials of events with constant probabilities is as follows:

> If the probability of event A is p and the probability of the alternative event B is q, the probability that, in n trials, event A is realized s times and event B is realized t times is
>
> $$\frac{n!}{s!\,t!} p^s q^t \qquad (1)$$

in which $s + t = n$ and $p + q = 1$. Equation (1) applies even when either s or t equals 0 because 0! is defined to equal 1. (Remember also that any number raised to the zero power equals 1; for example, $(1/2)^0 = 1$.) Any individual term in the expansion of the binomial $(p + q)^n$ is given by Equation (1) for the appropriate values of s and t.

It is worth taking a few minutes to consider the meaning of the factorial part of the binomial expansion in Equation (1), which equals $n!/(s!t!)$. This ratio enumerates all possible ways in which s elements of one kind and t elements of another kind can be arranged in order, provided that the s elements and the t elements are not distinguished among themselves. A specific example might include s yellow peas and t green peas. Although the yellow peas and the green peas can be distinguished from each other because they have different colors, the yellow peas are not distinguishable from one another (because they are all yellow) and the green peas are not distinguishable from one another (because they are all green).

The reasoning behind the factorial formula begins with the observation that the total number of elements is $s + t = n$. Given n elements, each distinct from the next, the number of different ways in which they can be arranged is

$$n \times (n - 1) \times (n - 2) \times \cdots \times 3 \times 2 \times 1$$

Why? Because the first element can be chosen in n ways, and once this is chosen, the next can be chosen in $n - 1$ ways (because only $n - 1$ are left to choose from), and once the first two are chosen, the third can be chosen in $n - 2$ ways, and so forth. Finally, once $n - 1$ elements have been chosen, there is only 1 way to choose the last element. The $s + t$ elements can be arranged in $n!$ ways, provided that the elements are all distinguished among themselves. However, applying again the argument we just used, each of the $n!$ particular arrangements must include $s!$ different arrangements of the s elements and $t!$ different arrangements of the t elements, or $s! \times t!$ altogether. Dividing $n!$ by $s! \times t!$ therefore yields the exact number of ways in which the s elements and the t elements can be arranged when the elements of each type are not distinguished among themselves.

Let us consider a specific application of Equation (1), in which we calculate the probability that a mating between two heterozygous parents yields exactly the expected 3 : 1 ratio of the dominant and recessive traits among sibships of a particular size. The probability p of a child showing the dominant trait is 3/4, and the probability q of a child showing the recessive trait is 1/4. Suppose we wanted to know how often families with eight children would contain exactly six children with the dominant phenotype and two with the recessive phenotype. This is the "expected" Mendelian ratio. In this case, $n = 8$, $s = 6$, $t = 2$, and the probability of this combination of events is

$$\frac{8!}{6!\,2!}\, p^6 q^2 = \frac{6! \times 7 \times 8}{6! \times 2!}(3/4)^6(1/4)^2 = 0.31$$

That is, in only 31 percent of the families with eight children would the offspring exhibit the expected 3 : 1 phenotypic ratio; the other sibships would deviate in one direction or the other because of chance variation. The importance of this example is in demonstrating that, although a 3 : 1 ratio is the "expected" outcome (and is also the single most probable outcome), the majority of the families (69 percent) actually have a distribution of offspring different from 3 : 1.

Evaluating the Fit of Observed Results to Theoretical Expectations

Geneticists often need to decide whether an observed ratio is in satisfactory agreement with a theoretical prediction. Mere inspection of the data is unsatisfactory because different investigators may disagree. Suppose, for example, that we crossed a plant having purple flowers with a plant having white flowers and, among the progeny, observed 14 plants with purple flowers and 6 with white flowers. Is this result close enough to be accepted as a 1 : 1 ratio? What if we observed 15 plants with purple flowers and 5 with white flowers? Is this result consistent with a 1 : 1 ratio? There is bound to be statistical variation in the observed results from one experiment to the next. Who is to say what results are consistent with a particular genetic hypothesis? In this section, we describe a test of whether observed results deviate too far from a theoretical expectation. The test is

called a test for **goodness of fit,** where the word *fit* means how closely the observed results "fit," or agree with, the expected results.

The Chi-square Method

A conventional measure of goodness of fit is a value called **chi-square** (symbol, χ^2), which is calculated from the number of progeny observed in each of various classes, compared with the number expected in each of the classes on the basis of some genetic hypothesis. For example, in a cross between plants with purple flowers and those with white flowers, we may be interested in testing the hypothesis that the parent with purple flowers is heterozygous for a pair of alleles determining flower color and that the parent with white flowers is homozygous recessive. Suppose further that we examine 20 progeny plants from the mating and find that 14 are purple and 6 are white. The procedure for testing this genetic hypothesis (or any other genetic hypothesis) by means of the chi-square method is as follows:

1. *State the genetic hypothesis in detail, specifying the genotypes and phenotypes of the parents and the possible progeny.* In the example using flower color, the genetic hypothesis implies that the genotypes in the cross *purple* × *white* could be symbolized as $Pp \times pp$. The possible progeny genotypes are either Pp or pp.

2. *Use the rules of probability to make explicit predictions of the types and proportions of progeny that should be observed if the genetic hypothesis is true. Convert the proportions to numbers of progeny (percentages are not allowed in a χ^2 test).* If the hypothesis about the flower-color cross is true, then we should expect the progeny genotypes Pp and pp to occur in a ratio of 1 : 1. Because the hypothesis is that Pp flowers are purple and pp flowers are white, we expect the phenotypes of the progeny to be purple or white in the ratio 1 : 1. Among 20 progeny, the expected numbers are 10 purple and 10 white.

3. *For each class of progeny in turn, subtract the expected number from the observed number. Square this difference and divide the result by the expected number.* In our example, the calculation for the purple progeny is $(14 - 10)^2/10 = 1.6$, and that for the white progeny is $(6 - 10)^2/10 = 1.6$.

4. *Sum the result of the numbers calculated in step 3 for all classes of progeny. The summation is the value of χ^2 for these data.* The sum for the purple and white classes of progeny is $1.6 + 1.6 = 3.2$, and this is the value of χ^2 for the experiment, calculated on the assumption that our genetic hypothesis is correct.

In symbols, the calculation of χ^2 can be represented by the expression

$$\chi^2 = \sum \frac{(\text{Observed} - \text{Expected})^2}{\text{Expected}}$$

in which Σ means the summation over all the classes of progeny. Note that χ^2 is calculated using the observed and expected *numbers,* not the proportions, ratios, or percentages. Using something other than the actual numbers is the most common beginner's mistake in applying the χ^2 method. The χ^2 value is reasonable as a measure of goodness of fit, because the closer the observed numbers are to the expected numbers, the smaller the value of χ^2. A value of $\chi^2 = 0$ means that the observed numbers fit the expected numbers perfectly.

As another example of the calculation of χ^2, suppose that the progeny of an $F_1 \times F_1$ cross includes two contrasting phenotypes observed in the numbers 99 and 45. In this case the genetic hypothesis might be that the trait is determined by a pair of alleles of a single gene, in which case the expected ratio of dominant : recessive phenotypes among the F_2 progeny is 3 : 1. Considering the data, the question is whether the observed ratio of 99 : 45 is in satisfactory agreement with the expected 3 : 1. Calculation of the value of χ^2 is illustrated in Table 3.2. The total number of progeny is $99 + 45 = 144$. The *expected* numbers in the two classes, on the basis of the genetic hypothesis that the true ratio is 3 : 1, are calculated as $(3/4) \times 144 = 108$ and $(1/4) \times 144 = 36$. Because there are two classes of data, there are two terms in the χ^2 calculation:

$$\chi^2 = \frac{(99 - 108)^2}{108} + \frac{(45 - 36)^2}{36}$$
$$= 0.75 + 2.25$$
$$= 3.00$$

Once the χ^2 value has been calculated, the next step is to interpret whether this value represents a good fit or a bad fit to the expected numbers. This assessment is done with the aid of the graphs in Figure 3.20. The x-axis gives the χ^2 values that reflect goodness of fit, and the y-axis gives the probability P that a worse fit (or one equally bad) would be obtained by chance, assuming that the genetic hypothesis is true. If the genetic hypothesis is true, then the observed numbers should be reasonably close to the expected numbers. Suppose that the observed χ^2 is so large that the probability of a fit as bad or worse is very small. Then the observed results do *not* fit the theoretical expectations. This means that the genetic hypothesis used to calculate the expected numbers of progeny must be rejected, because the observed numbers of progeny deviate too much from the expected numbers.

In practice, the critical values of P are conventionally chosen as 0.05 (the 5 percent level) and 0.01 (the 1 percent level). For P values ranging from 0.01 to 0.05, the probability that chance alone would lead to a fit as bad or worse is between 1 in 20 experiments and between 1 in 100, respectively. This is the purple region in Figure 3.20; if the P value falls in this range, the

Table 3.2 Calculation of χ^2 for a monohybrid ratio

Phenotype (class)	Observed number	Expected number	Deviation from expected	$\frac{(\text{Deviation})^2}{\text{expected number}}$
Wildtype	99	108	−9	0.75
Mutant	45	36	+9	2.25
Total	144	144		$\chi^2 = 3.00$

correctness of the genetic hypothesis is considered very doubtful. The result is said to be **significant** at the 5 percent level. For P values smaller than 0.01, the probability that chance alone would lead to a fit as bad or worse is less than 1 in 100 experiments. This is the green region in Figure 3.20; in this case, the result is said to be **highly significant** at the 1 percent level, and the genetic hypothesis is rejected outright. If

the terminology of statistical significance seems backward, it is because the term "significant" refers to the magnitude of the deviation between the observed and the expected numbers; in a result that is statistically significant, there is a large ("significant") difference between what is observed and what is expected.

To use Figure 3.20 to determine the P value corresponding to a calculated χ^2, we

Figure 3.20

Graphs for interpreting goodness of fit to genetic predictions using the chi-square test. For any calculated value of χ^2 along the x-axis, the y-axis gives the probability P that chance alone would produce a fit as bad as or worse than that actually observed, when the genetic predictions are correct. Tests with P in the purple region (less than 5 percent) or in the green region (less than 1 percent) are regarded as statistically significant and normally require rejection of the genetic hypothesis leading to the prediction. Each χ^2 test has a number of degrees of freedom associated with it. In the tests illustrated in this chapter, the number of degrees of freedom equals the number of classes in the data minus 1.

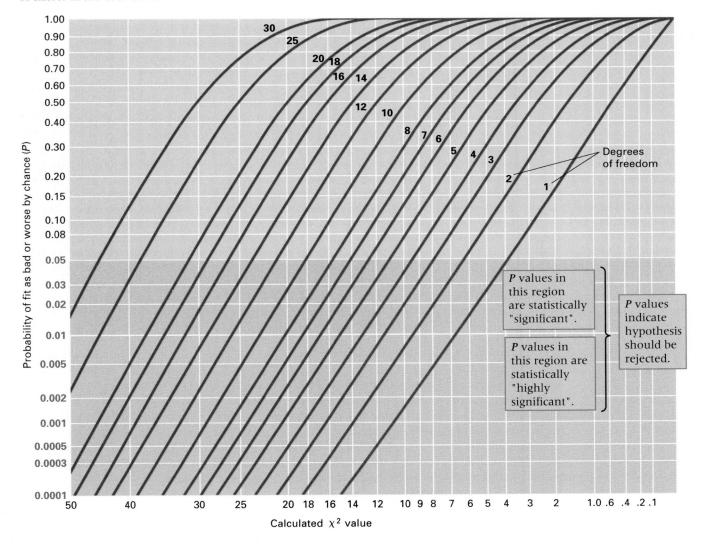

The Case Against Mendel's Gardener

Ronald Aylmer Fisher 1936
University College, London, England
Has Mendel's Work Been Rediscovered?

R. A. Fisher, one of the founders of modern statistics, was also interested in genetics. He gave Mendel's data a thorough going over and made an "abominable discovery." Fisher's unpleasant discovery was that some of Mendel's experiments yielded a better fit to the wrong expected values than they did to the right expected values. At issue are two series of experiments consisting of progeny tests in which F_2 plants with the dominant phenotype were self-fertilized and their progeny examined for segregation to ascertain whether each parent was heterozygous or homozygous. In the first series of experiments, Mendel explicitly states that he cultivated 10 seeds from each plant. What Mendel did not realize, apparently, is that inferring the genotype of the parent on the basis of the phenotypes of 10 progeny introduces a slight bias. The reason is shown in the accompanying illustration. Because a fraction $(3/4)^{10}$ of all progenies from a heterozygous parent will not exhibit segregation, purely as a result of chance, this proportion of Aa parents gets misclassified as AA. The expected proportion of "apparent" AA plants is $(1/3) + (2/3)(3/4)^{10}$ and that of Aa plants is $(2/3)[1 - (3/4)^{10}]$, for a ratio of 0.37 : 0.63. In the first series of experiments, among 600 plants tested,

Mendel reports a ratio of 0.335 : 0.665, which is in better agreement with the incorrect expectation of 0.33 : 0.67 than with 0.37 : 0.63. In the second series of experiments, among 473 progeny, Mendel reports a ratio of 0.32 : 0.68, which is again in better agreement with 0.33 : 0.67 than

---

The reconstruction [of Mendel's experiments] gives no doubt whatever that his report is to be taken entirely literally, and that his experiments were carried out in just the way and much in the order that they are recounted.

---

with 0.37 : 0.63. This is the "abominable discovery." The reported data differ highly significantly from the true expectation. How could this be? Fisher suggested that Mendel may have been deceived by an overzealous assistant. Mendel did have a gardener who tended the fruit orchards, a man described as untrustworthy and excessively fond of alcohol, and Mendel was also assisted in his pea experiments by two fellow monks. Another possibility, also suggested by Fisher, is that in the second series of experiments, Mendel cultivated more than 10 seeds from each plant.

(Mendel does not specify how many seeds were tested from each plant in the second series.) If he cultivated 15 seeds per plant, rather than 10, then the data are no longer statistically significant and the insinuation of data tampering evaporates.

In connection with these tests of homozygosity by examining ten offspring formed by self-fertilization, it is disconcerting to find that the proportion of plants misclassified by this test is not inappreciable. Between 5 and 6 percent of the heterozygous plants will be classified as homozygous. . . . Now among 600 plants tested by Mendel 201 were classified as homozygous and 399 as heterozygous. . . . The deviation [from the true expected values of 222 and 378] is one to be taken seriously. . . . A deviation as fortunate as Mendel's is to be expected once in twenty-nine trials. . . . [In the second series of experiments], a total deviation of the magnitude observed, and in the right direction, is only to be expected once in 444 trials; there is therefore a serious discrepancy. . . . If we could suppose that larger progenies, say fifteen plants, were grown on this occasion, the greater part of the discrepancy would be removed. . . . Such an explanation, however, could not explain the discrepancy observed in the first group of experiments, in which the procedure is specified, without the occurrence of a coincidence of considerable

---

need the number of **degrees of freedom** of the particular χ^2 test. For the type of χ^2 test illustrated in Table 3.2, the number of degrees of freedom equals the number of classes of data minus 1. Table 3.2 contains two classes of data (wildtype and mutant), so the number of degrees of freedom is $2 - 1 = 1$. The reason for subtracting 1 is that, in calculating the expected numbers of progeny, we make sure that the total number of progeny is the same as that actually observed. For this reason, one of

the classes of data is not really "free" to contain any number we might specify; because the expected number in one class must be adjusted to make the total come out correctly, one "degree of freedom" is lost. Analogous χ^2 tests with three classes of data have 2 degrees of freedom, and those with four classes of data have 3 degrees of freedom.

Once we have decided the appropriate number of degrees of freedom, we can interpret the χ^2 value in Table 3.2. Refer to

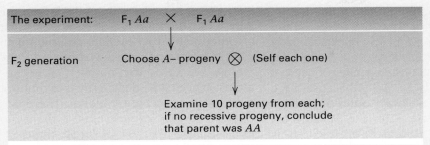

The experiment:	$F_1 \, Aa$	$\times$	$F_1 \, Aa$

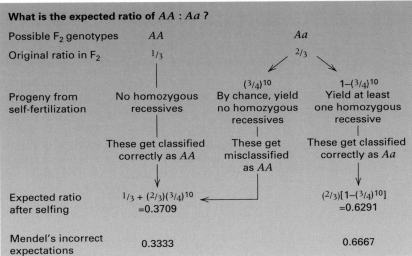

F_2 generation	Choose $A-$ progeny $\otimes$ (Self each one)

Examine 10 progeny from each; if no recessive progeny, conclude that parent was AA

What is the expected ratio of $AA : Aa$?

Possible F_2 genotypes	AA			Aa
Original ratio in F_2	$1/3$			$2/3$
Progeny from self-fertilization	No homozygous recessives	$(3/4)^{10}$ By chance, yield no homozygous recessives	$1-(3/4)^{10}$ Yield at least one homozygous recessive	
	These get classified correctly as AA	These get misclassified as AA	These get classified correctly as Aa	
Expected ratio after selfing	$1/3 + (2/3)(3/4)^{10}$ =0.3709		$(2/3)[1-(3/4)^{10}]$ =0.6291	
Mendel's incorrect expectations	0.3333		0.6667	

appeared, in that in two series of results the numbers observed agree excellently with the two to one ratio, which Mendel himself expected, but differ significantly from what should have been expected had his theory been corrected to allow for the small size of his test progenies. . . . Although no explanation can be expected to be satisfactory, it remains a possibility among others that Mendel was deceived by some assistant who knew too well what was expected.

Source: Annals of Science 1: 115–137

improbability. . . . The reconstruction [of Mendel's experiments] gives no doubt whatever that his report is to be taken entirely literally, and that his experiments were carried out in just the way and much in the order that they are recounted. The detailed reconstruction of his programme on this assumption leads to no discrepancy whatsoever. A serious and almost inexplicable discrepancy has, however,

Figure 3.20, and observe that each curve is labeled with its degrees of freedom. To determine the P value for the data in Table 3.2, in which the χ^2 value is 3 (3.00), first find the location of $\chi^2 = 3$ along the x-axis in Figure 3.20. Trace vertically from 3 until you intersect the curve with 1 degree of freedom. Then trace horizontally to the left until you intersect the y-axis, and read the P value; in this case, $P = 0.08$. This means that chance alone would produce a χ^2 value as great as or greater than 3 in about 8 percent of experiments of the type in Table 3.2; and, because the P value is within the blue region, the goodness of fit to the hypothesis of a 3 : 1 ratio of wild-type: mutant is judged to be satisfactory.

As a second illustration of the χ^2 test, we will determine the goodness of fit of Mendel's round versus wrinkled data to the expected 3 : 1 ratio. Among the 7324 seeds that he observed, 5474 were round and 1850 were wrinkled. The expected numbers are $(3/4) \times 7324 = 5493$ round and

$(1/4) \times 7324 = 1831$ wrinkled. The χ^2 value is calculated as

$$\chi^2 = \frac{(5474 - 5493)^2}{5493} + \frac{(1850 - 1831)^2}{1831}$$
$$= 0.26$$

The fact that the χ^2 is less than 1 already implies that the fit is very good. To find out how good, note that the number of degrees of freedom equals $2 - 1 = 1$ because there are two classes of data (round and wrinkled). From Figure 3.20, the P value for $\chi^2 = 0.26$ with 1 degree of freedom is approximately 0.65. This means that in about 65 percent of all experiments of this type, a fit as bad or worse would be expected simply because of chance; only about 35 percent of all experiments would yield a better fit.

3.6
Are Mendel's Data Too Good to Be True?

Many of Mendel's experimental results are very close to the expected values. For the ratios listed in Table 2.1 in Chapter 2, the χ^2 values are 0.26 (round versus wrinkled seeds), 0.01 (yellow versus green seeds), 0.39 (purple versus white flowers), 0.06 (inflated versus constricted pods), 0.45 (green versus yellow pods), 0.35 (axial versus terminal flowers), and 0.61 (long versus short stems). (As an exercise in χ^2, you should confirm these calculations for yourself.) All of the χ^2 tests have P values of 0.45 or greater (Figure 3.20), which means that the reported results are in excellent agreement with the theoretical expectations.

The statistician Ronald Fisher pointed out in 1936 that Mendel's results are *suspiciously* close to the theoretical expectations. In a large number of experiments, some experiments can be expected to yield fits that appear doubtful simply because of chance variation from one experiment to the next. In Mendel's data, the doubtful values that are to be expected appear to be missing. Figure 3.21 shows the observed deviations in Mendel's experiments compared with the deviations expected by

chance. (The measure of deviation is the square root of the χ^2 value, assigned either a plus or a minus sign according to whether the dominant or the recessive phenotypic class was in excess of the expected number.) For each magnitude of deviation, the height of the yellow bar gives the number of experiments that Mendel observed with such a magnitude of deviation, and the orange bar gives the number of experiments expected to deviate by this amount as a result of chance alone. There are clearly too few experiments with deviations smaller than -1 or larger than $+1$. This type of discrepancy could be explained if Mendel discarded or repeated a few experiments with large deviations that made him suspect that the results were not to be trusted.

Did Mendel cheat? Did he deliberately falsify his data to make them appear better? Mendel's paper reports extremely deviant ratios from individual plants, as well as experiments repeated a second time when the first results were doubtful. These are not the kinds of things that a dishonest person would admit. Only a small bias is necessary to explain the excessive goodness of fit in Figure 3.21. In a count of seeds or individual plants, only about 2 phenotypes per 1000 would need to be assigned to the wrong category to account for the bias in the 91 percent of the data generated by the testing of monohybrid ratios. The excessive fit could also be explained if three or four entire experiments were discarded or repeated because deviant results were attributed to pollen contamination or other accident. After careful reexamination of Mendel's data in 1966, the evolutionary geneticist Sewall Wright concluded,

Mendel was the first to count segregants at all. It is rather too much to expect that he would be aware of the precautions now known to be necessary for completely objective data. . . . Checking of counts that one does not like, but not of others, can lead to systematic bias toward agreement. I doubt whether there are many geneticists even now whose data, if extensive, would stand up wholly satisfactorily under the χ^2 text. . . . Taking everything into account, I am confident that there was no deliberate effort at falsification.

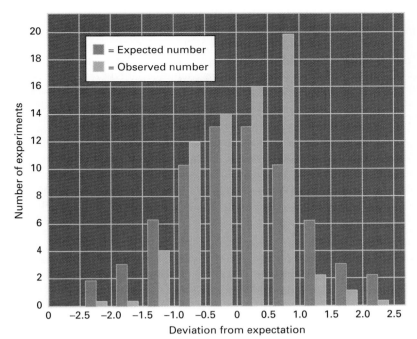

Figure 3.21

Distribution of deviations observed in 69 of Mendel's experiments (yellow bars) compared with expected values (orange bars). There is no suggestion that the data in the middle have been adjusted to improve the fit. However, several experiments with large deviations may have been discarded or repeated, because there are not so many experiments with large deviations as might be expected.

Mendel's data are some of the most extensive and complete "raw data" ever published in genetics. Additional examinations of the data will surely be carried out as new statistical approaches are developed. However, the principal point to be emphasized is that up to the present time, no reputable statistician has alleged that Mendel knowingly and deliberately adjusted his data in favor of the theoretical expectation.

Chapter Summary

The chromosomes in somatic cells of higher plants and animals are present in pairs. The members of each pair are homologous chromosomes, and each member is a homolog. Pairs of homologs are usually identical in appearance, whereas nonhomologous chromosomes often show differences in size and structural detail that make them visibly distinct from each other. A cell whose nucleus contains two sets of homologous chromosomes is diploid. One set of chromosomes comes from the maternal parent and the other from the paternal parent. Gametes are haploid. A gamete contains only one set of chromosomes, consisting of one member of each pair of homologs.

Mitosis is the process of nuclear division that maintains the chromosome number when a somatic cell divides. Before mitosis, each chromosome replicates, forming a two-part structure consisting of two sister chromatids joined at the centromere (kinetochore). At the onset of mitosis, the chromosomes become visible and, at metaphase, become aligned on the metaphase plate perpendicular to the spindle. At anaphase, the centromere of each chromosome divides, and the sister chromatids are pulled by spindle fibers to opposite poles of the cell. The separated sets of chromosomes present in telophase nuclei are genetically identical.

Meiosis is the type of nuclear division that takes place in germ cells, and it reduces the diploid number of chromosomes to the haploid number. The genetic material is replicated before the onset of meiosis, so each chromosome consists of two sister chromatids. The first meiotic division is the reduction division, which reduces the chromosome number by half. The homologous chromosomes first pair (synapsis) and then, at anaphase I, separate. The resulting products contain chromosomes that consist of two chromatids attached to a common centromere. However, as a result of crossing-over, which takes place in prophase I, the chromatids may not be genetically identical along their entire length. In the second meiotic division, the centromeres divide and the homologous chromatids separate. The end result of meiosis is the formation of four genetically different haploid nuclei.

A distinctive feature of meiosis is the synapsis, or side-by-side pairing, of homologous chromosomes in the zygotene substage of prophase I. During the pachytene substage, the paired chromosomes become connected by chiasmata (the physical manifestations of crossing-over) and do not separate until anaphase I. This separation is called disjunction (unjoining), and failure of chromosomes to separate is called nondisjunction. Nondisjunction results in a gamete that contains either two copies or no copies of a particular chromosome. Meiosis is the physical basis of the segregation and independent assortment of genes. In *Drosophila*, an unexpected pattern of inheritance of the X-linked *white* gene was shown to be accompanied by nondisjunction of the X chromosome; these observations gave experimental proof of the chromosome theory of heredity.

Unlike other chromosome pairs, the X and Y sex chromosomes are visibly different and contain different genes. In mammals and in many insects and other animals, as well as in some flowering plants, the female contains two X chromosomes (XX) and hence is homogametic, and the male contains one X chromosome and one Y chromosome (XY) and hence is heterogametic. In birds, moths, butterflies, and some

reptiles, the situation is the reverse: Females are the heterogametic sex (WZ) and males the homogametic sex (WW). The Y chromosome in many species contains only a few genes. In human beings and other mammals, the Y chromosome includes a male-determining factor. In *Drosophila*, sex is determined by a male-specific or female-specific pattern of gene expression that is regulated by the ratio of the number of X chromosomes to the number of sets of autosomes. In most organisms, the X chromosome contains many genes unrelated to sexual differentiation. These X-linked genes show a characteristic pattern of inheritance that is due to their location in the X chromosome.

The progeny of genetic crosses often conform to the theoretical predictions of the binomial probability formula. The degree to which the observed numbers of different genetic classes of progeny fit theoretically expected numbers is usually found with a chi-square (χ^2) test. On the basis of the criterion of the χ^2 test, Mendel's data fit the expectations somewhat more closely than chance would dictate. However, the bias in the data is relatively small and is unlikely to be due to anything more than recounting or repeating certain experiments whose results were regarded as unsatisfactory.

Key Terms

anaphase	heterogametic	spore
anaphase I	highly significant	sporophyte
anaphase II	homogametic	statistical significance
autosomes	homologous	synapsis
bivalent	interphase	telophase
cell cycle	kinetochore	telophase I
centromere	leptotene	telophase II
chiasma	M period	tetrad
chi-square	meiocyte	X chromosome
chromatid	meiosis	X-linked
chromomere	metaphase	Y chromosome
chromosome	metaphase plate	zygotene
chromosome complement	metaphase I	
crossing-over	metaphase II	
cyclins	mitosis	
cytokinesis	mitotic spindle	
degrees of freedom	nondisjunction	
diakinesis	nucleolus	
diploid	pachytene	
diplotene	Pascal's triangle	
equational division	prophase	
first meiotic division	prophase I	
G_1 period	prophase II	
G_2 period	reductional division	
gametophyte	S phase	
germ cell	second meiotic division	
goodness of fit	sister chromatids	
haploid	sex chromosome	
hemophilia A	somatic cell	

- Explain the following statement: "Independent alignment of nonhomologous chromosomes at metaphase I of meiosis is the physical basis of independent assortment of genes on different chromosomes."

- Draw a diagram of a bivalent, and label the following parts: centromere, sister chromatids, nonsister chromatids, homologous chromosomes, chiasma.

- What is the genetic consequence of the formation of a chiasma in a bivalent?

- T. H. Morgan discovered X-linkage by following up his observation that reciprocal crosses in which one parent was wildtype for eye color and the other had white eyes yielded different types of progeny. Diagram the reciprocal crosses, indicating the X and Y chromosomal genotypes of each parent and each class of offspring.

- What does it mean to say that two outcomes of a cross are mutually exclusive? What does it mean to say that two outcomes of a cross are independent?

- In what way does the chi-square value indicate "goodness of fit"?

- What are the conventional P values for "significant" and "highly significant" and what do these numbers mean?

- If Mendel did discard the results of some experiments because he considered them excessively deviant as a result of to pollen contamination or some other factor, do you consider this a form of "cheating"? Why or why not?

Guide to Problem Solving

Problem 1: The black and yellow pigments in the fur of cats are determined by an X-linked pair of alleles, c^b (black) and c^y (yellow). Males are black (c^b) or yellow (c^y), and females are either homozygous black ($c^b c^b$), homozygous yellow ($c^y c^y$), or heterozygous ($c^b c^y$). The phenotype of the heterozygous female has patches of black and patches of yellow, a pattern knows as *calico*. (The white spotting usually also present in domestic short-hair cats is caused by a separate gene.)

(a) What genotypes and phenotypes would be expected among the offspring of a cross between a black female and a yellow male?

(b) In a litter of eight kittens, there are two calico females, one yellow female, two black males, and three yellow males. What are the genotypes and phenotypes of the parents?

(c) Rare calico males are the result of nondisjunction. What are their sex-chromosome constitution and their genotype?

Answer: **(a)** The black female has genotype $c^b c^b$ and the yellow male has genotype c^y. The female offspring receive an X chromosome from each parent, so their genotype is $c^b c^y$ and their phenotype is calico. The male offspring receive an X chromosome from their mother, so their genotype is c^b and their phenotype is black. **(b)** The male offspring provide information about the X chromosomes in the mother. Because some males are black (c^b) and some yellow (c^y), the mother must have the genotype $c^b c^y$ and have a calico coat. The fact that one of the offspring is a yellow female means that both parents carry an X chromosome with the c^y allele, so the father must have the genotype c^y and have a yellow coat. The

occurrence of the calico female offspring is also consistent with these parental genotypes. **(c)** Nondisjunction in a female of genotype $c^b c^y$ can produce an XX egg with the alleles c^b and c^y. When the XX egg is fertilized by a Y-bearing sperm, the result is an XXY male of genotype $c^b c^y$, which is the genotype of a male calico cat. (The XXY males are sterile but otherwise are similar to normal males.)

Problem 2: Certain breeds of chickens have reddish gold feathers because of a recessive allele, *g*, on the W chromosome (the avian equivalent of the X chromosome); presence of the dominant allele results in silver plumage. An autosomal recessive gene, *s*, results in feathers called silkie that remain soft like chick down. What genotypes and phenotypes of each sex would be expected from a cross of a red rooster heterozygous for silkie and a silkie hen with silver plumage? (Remember that, in birds, females are the heterogametic sex, WZ, and males the homogametic sex, WW.)

Answer: In this problem, you need to keep track of both the W-linked and the autosomal inheritance, and the situation has the additional complication that males are WW and females WZ for the sex chromosomes. The parental red rooster that is heterozygous for silkie has the genotype *gg* for the W chromosome and *Ss* for the relevant autosome, and the silver, silkie hen has the genotype *G* for the W chromosome and *ss* for the autosome. The female (WZ) offspring from the cross receive their W chromosome from their father (the reverse of the situation in most animals), so they have genotype *g* (reddish gold feathers) and half of them are *Ss* (wildtype) and half *ss* (silkie). The male (WW) offspring have genotype *Gg* (silver feathers), and again, half are *Ss* (wildtype) and half *ss* (silkie).

Chapter 3

GeNETics on the web will introduce you to some of the most important sites for finding genetic information on the Internet. To complete the exercises below, visit the Jones and Bartlett home page at

http://www.jbpub.com/genetics

Select the link to *Genetics: Principles and Analysis* and then choose the link to *GeNETics on the web.* You will be presented with a chapter-by-chapter list of highlighted keywords.

GeNETics EXERCISES

Select the highlighted keyword in any of the exercises below, and you will be linked to a web site containing the genetic information necessary to complete the exercise. Each exercise suggests a specific, written report that makes use of the information available at the site. This report, or an alternative, may be assigned by your instructor.

1. See meiosis in action by using this keyword. Then locate the still photographs of chromosomes in various stages of prophase I. How many bivalents are formed in this organism? What is the diploid chromosome number of the organism? If assigned to do so, identify each of the prophase I stages depicted as leptotene, zygotene, pachytene, diplotene, or diakinesis.

2. Some of the main characteristics of X-linked inheritance can be examined at this site. Mate wildtype females with white males and observe the results. Then cross the F_1 females with their white-eye fathers and observe the results. Make a diagram of the crosses, giving the genotypes and phenotypes of all the flies and the numbers observed in each class of offspring. If assigned to do so, prepare a similar report of an initial mating of white-eyed females with wildtype males, followed by mating of the F_1 female progeny with their wildtype fathers.

3. *FlyBase* is the main internet repository of information about *Drosophila* genetics. Using the keyword white, you can learn about the metabolic defect in the mutation that Thomas Hunt Morgan originally discovered. Enter the keyword into the search engine, and

Problem 3: Ranch mink with the dark gray coat color known as aleutian are homozygous recessive, *aa*. Genotypes *AA* and *Aa* have the standard deep brown color. A mating of *Aa* × *Aa* produces eight pups.

(a) What is the probability that none of them has the aleutian coat color?

(b) What is the probability of a perfect 3 : 1 distribution of standard to aleutian?

(c) What is the probability of a 1 : 1 distribution in this particular litter?

Answer: This kind of problem demonstrates the effect of chance variation in segregation ratios in small sibships. In the mating *Aa* × *Aa*, the probability that any particular pup has the standard coat color is 3/4, and that of aleutian is 1/4. Therefore, in a litter of size 8, the probability of 0, 1, 2, 3, . . . pups with the standard coat color is given by successive terms in the binomial expansion of $[(3/4) + (1/4)]^8$. The specific probability of *r* standard pups and $8 - r$ aleutian pups is given by $8!/[r!(8 - r)!] \times (3/4)^r \times (1/4)^{(8 - r)}$.

(a) The probability that none is aleutian means that $r = 8$, so the probability is

$$(3/4)^8 = 6561/65{,}536 = 0.10$$

(b) A perfect 3 : 1 ratio in the litter means that $r = 6$, and the probability is

$$[8!/(6!\,2!)](3/4)^6(1/4)^2 = 28(729/4096)(1/16) = 0.31$$

(That is, a little less than a third of the litters would have the "expected" Mendelian ratio.)

(c) A 1 : 1 distribution in the litter means that $r = 4$, and the probability is

$$[8!/(4!\,4!)](3/4)^4(1/4)^4 = 70(81/256)(1/256)$$
$$= 0.09$$

Problem 4: Certain varieties of maize are true-breeding either for colored aleurone (the outer layer of the seed) or for colorless aleurone. A cross of a colored variety with a colorless variety gave an F_1 with colored seeds. Among 1000 seeds produced by the cross F_1 × colored, all were colored; and, among 1000 seeds produced by the cross F_1 × colorless, 525 were colored, and 475 were colorless.

(a) What genetic hypothesis can explain these data?

(b) Using the criterion of a χ^2 test, evaluate whether the data are in satisfactory agreement with the model.

Answer: **(a)** Mendelian segregation in a heterozygote is suggested by the 525 : 475 ratio, and dominance is implied by

then select *w*. Near the bottom of the report, select *Full*, and then read the section on phenotypic information. If assigned to do so, select the allele number 1 (not +1) and write a 100-word report on the molecular basis of the w^1 mutation. In preparing this report, you will find it helpful to return to the search engine and enter the keyword *Doc*.

4. Using the keyword **controversies**, you can learn about at least nine controversial issues concerning Mendel's motivation for doing his work or about aspects of the work itself. If assigned to do so, pick any three of these controversial issues and write a paragraph about each, describing the issue and summarizing the opinions of those on opposite sides of the matter.

MUTABLE SITE EXERCISES

The Mutable Site Exercise changes frequently. Each new update includes a different exercise that makes use of genetics resources available on the World Wide Web. Select the **Mutable Site** for Chapter 3, and you will be linked to the current exercise that relates to the material presented in this chapter.

PIC SITE

The Pic Site showcases some of the most visually appealing genetics sites on the World Wide Web. To visit the showcase genetics site, select the **Pic Site** for Chapter 3.

the crosses of colored × colorless and F$_1$ × colored. Therefore, a hypothesis that seems to fit the data is that there is a dominant allele for colored aleurone (say, *C*) and the true-breeding colored and colorless varieties are *CC* and *cc*, respectively. The F$_1$ has the genotype *Cc*. The cross F$_1$ × colored (*CC*) yields 1 *CC* : 1 *Cc* progeny (all colored), and the cross F$_1$ × colorless (*cc*) is expected to produce a 1 : 1 ratio of colored (*Cc*) to colorless (*cc*). **(b)** The expected numbers are 500 colored and 500 colorless, and therefore the χ^2 equals $(525 - 500)^2/500 + (475 - 500)^2/500 = 2.5$. Because there are two classes of data, there is 1 degree of freedom, and the *P* value is approximately 0.12 (see Figure 3.20). The *P* value is greater than 0.05, so the goodness of fit is regarded as satisfactory.

Analysis and Applications

3.1 At what stage in mitosis and meiosis are the chromosomes replicated? When do the chromosomes first become visible in the light microscope?

3.2 If a cell contains 23 pairs of chromosomes immediately after completion of mitotic telophase, how many chromatids were present in metaphase?

3.3 The Greek roots of the terms *leptotene, zygotene, pachytene, diplotene,* and *diakinesis* literally mean *thin thread, paired thread, thick thread, doubled thread* and *moving apart,* respectively. How are these terms appropriate in describing

the appearance and behavior of the chromosomes during prophase I? (Incidentally, the ending *-tene* denotes the adjective; the nouns are formed by dropping *-tene* and adding *-nema,* yielding the alternative terms *leptonema, zygonema, pachynema, diplonema,* and *diakinesis,* which are preferred by some authors.)

3.4 The first meiotic division is often called the *reductional* division and the second meiotic division the *equational* division. Which feature of the chromosome complement is reduced in the first meiotic division and which is kept equal in the second?

3.5 Maize is a diploid organism with 10 pairs of chromosomes. How many chromatids and chromosomes are present in the following stages of cell division:

(a) Metaphase of mitosis?

(b) Metaphase I of meiosis?

(c) Metaphase II of meiosis?

3.6 Sweet peas have a somatic chromosome number of 14. If the centromeres of the 7 homologous pairs are designated as Aa, Bb, Cc, Dd, Ee, Ff, and Gg,

(a) how many different combinations of centromeres can be produced during meiosis?

(b) what is the probability that a gamete will contain only centromeres designated by capital letters?

3.7 Emmer wheat (*Triticum dicoccum*) has a somatic chromosome number of 28, and rye (*Secale cereale*) has a somatic chromosome number of 14. Hybrids produced by crossing these cereal grasses are highly sterile and have many characteristics intermediate between the parental species. How many chromosomes do the hybrids possess?

3.8 X-linked inheritance is occasionally called *crisscross* inheritance. In what sense does the X chromosome move back and forth between the sexes every generation? In what sense is the expression misleading?

3.9 The most common form of color blindness in human beings results from an X-linked recessive gene. A phenotypically normal couple have a normal daughter and a son who is color-blind. What is the probability that the daughter is heterozygous?

3.10 The mutation for Bar-shaped eyes in *Drosophila* has the following characteristics of inheritance:

(a) Bar males × wildtype females produce wildtype sons and Bar daughters,

(b) The Bar females from the mating in part **a,** when mated with wildtype males, yield a 1 : 1 ratio of Bar : wildtype sons and Bar : wildtype daughters.

What mode of inheritance do these characteristics suggest?

3.11 Vermilion eye color in *Drosophila* is determined by the recessive allele, *v*, of an X-linked gene, and the wildtype eye color determined by the v^+ allele is brick red. What genotype and phenotype ratios would be expected from the following crosses:

(a) vermilion male × wildtype female

(b) vermilion female × wildtype male

(c) daughter from mating in part **a** × wildtype male

(d) daughter from mating in part **a** × vermilion male

3.12 The autosomal recessive allele, *bw* for brown eyes in *Drosophila*, interacts with the X-linked recessive allele, *v*, for vermilion, to produce white eyes. What eye-color phenotypes, and in what proportions, would be expected from a cross of a white-eyed female (genotype *v v;bw bw*) with a brown-eyed male (genotype v^+ Y;*bw bw*)?

3.13 It is often advantageous to be able to determine the sex of newborn chickens from their plumage. How could this be done by using the W-linked dominant allele *S* for silver plumage and the recessive allele *s* for gold plumage? (Remember that, in chickens, the homogametic and heterogametic sexes are the reverse of those in mammals.)

3.14 A recessive mutation of an X-linked gene in human beings results in hemophilia, marked by a prolonged increase in the time needed for blood clotting. Suppose that phenotypically normal parents produce two normal daughters and a son affected with hemophilia.

(a) What is the probability that both of the daughters are heterozygous carriers?

(b) If one of the daughters mates with a normal man and produces a son, what is the probability that the son will be affected?

3.15 In the pedigree illustrated, the shaded symbols represent persons affected with an X-linked recessive form of mental retardation. What are the genotypes of all the persons in this pedigree?

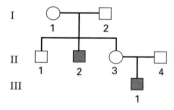

3.16 For an autosomal gene with five alleles, there are fifteen possible genotypes (five homozygotes and ten heterozygotes). How many genotypes are possible with five alleles of an X-linked gene?

3.17 Attached-X chromosomes in *Drosophila* are formed from two X chromosomes attached to a common centromere. Females of genotype C(1)RM/Y, in which C(1)RM denotes the attached-X chromosomes, produce C(1)RM-bearing and Y-bearing gametes in equal proportions. What progeny are expected to result from the mating between a male carrying the X-linked allele, *w*, for white eyes and an attached-X female with wildtype eyes? How does this result differ from the typical pattern of X-linked inheritance? (Note: *Drosophila* zygotes containing three X chromosomes or no X chromosomes do not survive.)

3.18 People who have the sex chromosome constitution XXY are phenotypically male. A woman heterozygous for an X-linked mutation for color blindness mates with a normal man and produces an XXY son, who is color-blind. What kind of nondisjunction can explain this result?

3.19 Mice with a single X chromosome and no Y chromosome (an XO sex-chromosome constitution) are fertile females. Assuming that at least one X chromosome is required for viability, what sex ratio is expected among surviving progeny from the mating XO ♀ × XY ♂ ?

3.20 In the accompanying pedigree, the shaded symbols represent persons affected with X-linked hemophilia, a blood-clotting disorder.

(a) If the woman identified as II-2 has two more children, what is the probability that neither will be affected?

(b) What is the probability that the first child of the mating II-4 × II-5 will be affected?

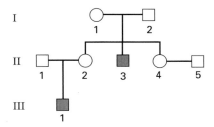

3.21 Assume a sex ratio at birth of 1 : 1 and consider two sibships, A and B, each with three children.

(a) What is the probability that A consists only of girls and B only of boys?

(b) What is the probability that one sibship consists only of girls and the other only of boys?

Challenge Problems

3.22 A hybrid corn plant with green leaves was testcrossed with a plant having yellow-striped leaves. When 250 seedlings were grown, 140 of the seedlings had green leaves and 110 had yellow-striped leaves. Using the chi-squared method, test this result for agreement with the expected 1 : 1 ratio.

3.23 A cross was made to produce *D. melanogaster* flies heterozygous for two pairs of alleles: dp^+ and dp, which determine long versus short wings, and e^+ and e, which determine gray versus ebony body color. The following F_2 data were obtained:

Long wing, gray body	462
Long wing, ebony body	167
Short wing, gray body	127
Short wing, ebony body	44

Test these data for agreement with the 9 : 3 : 3 : 1 ratio expected if the two pairs of alleles segregate independently.

Further Reading

Allshire, R. C. 1997. Centromeres, checkpoints and chromatid cohesion. *Current Opinion in Genetics & Development* 7: 264.

Chandley, A. C. 1988. Meiosis in man. *Trends in Genetics* 4: 79.

Cohen, J. S., and M. E. Hogan. 1994. The new genetic medicine. *Scientific American,* December.

McIntosh, J. R., and K. L. McDonald. 1989. The mitotic spindle. *Scientific American,* October.

McKusick, V. A. 1965. The royal hemophilia. *Scientific American,* August.

Miller, O. J. 1995. The fifties and the renaissance of human and mammalian genetics. *Genetics* 139: 484.

Page, A. W. and T. L. Orr-Weaver. 1997. Stopping and starting the meiotic cell cycle. *Current Opinion in Genetics & Development* 7: 23.

Sokal, R. R., and F. J. Rohlf. 1969. *Biometry.* New York: Freeman.

Sturtevant, A. H. 1965. *A Short History of Genetics.* Harper & Row.

Voeller, B. R., ed. 1968. *The Chromosome Theory of Inheritance: Classical Papers in Development and Heredity.* New York: Appleton-Century-Crofts.

Welsh, M. J., and A. E. Smith. 1995. Cystic fibrosis. *Scientific American,* December.

Zielenski, J., and L. C. Tsui. 1995. Cystic fibrosis: Genotypic and phenotypic variations. *Annual Review of Genetics* 29: 777.

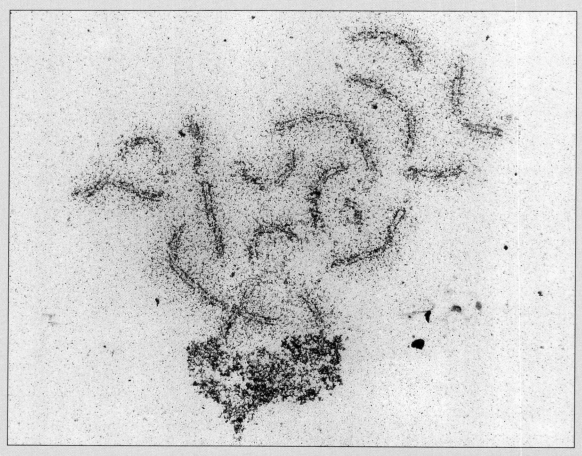

Electron micrograph of the chromosomes in a haploid yeast cell (*Saccharomyces cerevisiae*) in prophase of mitosis, showing the full complement of 16 chromosomes. The darkly stained mass at the bottom is the nucleus, which is associated with chromosome XII. [Courtesy of Kuei-Shu Tung and Shirleen Roeder.]

Genetic Linkage and Chromosome Mapping

CHAPTER OUTLINE

PRINCIPLES

- Genes that are located in the same chromosome and that do not show independent assortment are said to be linked.
- The alleles of linked genes that are present together in the same chromosome tend to be inherited as a group.
- Crossing-over between homologous chromosomes results in recombination that breaks up combinations of linked alleles.
- A genetic map depicts the relative positions of genes along a chromosome.
- The map distance between genes in a genetic map is related to the rate of recombination between the genes.
- Physical distance along a chromosome is often—but not always—correlated with map distance.
- Tetrads are sensitive indicators of linkage because each contains all the products of a single meiosis.
- Recombination can also take place between nucleotides within a gene.
- The complementation test is the experimental determination of whether two mutations are, or are not, alleles of the same gene.

CONNECTIONS

CONNECTION: Genes All in a Row
Alfred H. Sturtevant 1913
The linear arrangement of six sex-linked factors in Drosophila, *as shown by their mode of association*

CONNECTION: Dos XX
Lilian V. Morgan 1922
Non-crisscross inheritance in Drosophila melanogaster

In meiosis, homologous chromosomes form pairs, and the individual members of each pair separate from one another. The observation that homologous chromosomes behave as complete units when they separate led to the expectation that genes located in the same chromosome would not undergo independent assortment but rather would be transmitted together with complete **linkage.** As we shall see, Thomas Hunt Morgan examined this issue using two genes that he knew were both present in the X chromosome of *Drosophila*. One was a mutation for white eyes, the other a mutation for miniature wings. Morgan did observe linkage, but it was incomplete. Morgan found that the *white* and *miniature* alleles present in each X chromosome of a female tended to remain together in inheritance, but he also observed that some X chromosomes were produced that had new combinations of the *white* and *miniature* alleles.

In this chapter, we will see that Morgan's observation of incomplete linkage is the rule for genes present in the same chromosome. The reason why linkage is incomplete is that the homologous chromosomes, when they are paired, can undergo an exchange of segments. An exchange event between homologous chromosomes, crossing-over, results in the **recombination** of genes in the homologous chromosomes. The probability of crossing-over between any two genes serves as a measure of genetic distance between the genes and makes possible the construction of a **genetic map,** a diagram of a chromosome showing the relative positions of the genes. The genetic mapping of linked genes is an important research tool in genetics because it enables a new gene to be assigned to a chromosome and often to a precise position relative to other genes within the same chromosome. Genetic mapping is usually a first step in the identification and isolation of a new gene and the determination of its DNA sequence. Genetic mapping is essential in human genetics for the identification of genes associated with hereditary diseases, such as the genes whose presence predisposes women carriers to the development of breast cancer.

4.1
Linkage and Recombination of Genes in a Chromosome

As we saw in Chapter 3, a direct test of independent assortment is to carry out a testcross between an F_1 double heterozygote (*Aa Bb*) and the double recessive homozygote (*aa bb*). When the genes are on different chromosomes, the expected gametes from the *Aa Bb* parent are as shown in Figure 4.1. Because the pairs of homologous chromosomes segregate independently of each other in meiosis, the double heterozygote produces all four possible types of gametes—*AB, Ab, aB,* and *ab*—in equal proportions.

Independent assortment takes place in the *Aa Bb* genotype whether the *parents* were genotypically *AA BB* and *aa bb* or genotypically *AA bb* and *aa BB*. The four products of meiosis are still expected in equal proportions. An expected 50 percent of the testcross progeny result from gametes with the same combination of alleles present in the parents of the double heterozygote (**parental combinations**), and 50 percent result from gametes with new combinations of the alleles (**recombinants**). For example, if the double heterozygote came from the mating *AA BB* × *aa bb*, then the *AB* and *ab* gametes would be parental and the *Ab* and *aB* gametes recombinant. On the other hand, if the double heterozygote came from the mating *AA bb* × *aa BB*, then the *Ab* and *aB* gametes would be parental and the *ab* and *AB* gametes recombinant. In either case, with independent assortment, the genotypes of the testcross progeny are expected in the ratio of 1 : 1 : 1 : 1. In this chapter, we will examine phenomena that cause deviations from this expected ratio.

In his early experiments with *Drosophila*, Morgan found mutations in each of several X-linked genes that provided ideal materials for studying the inheritance of genes in the same chromosome. One of these genes, with alleles w^+ and w, determined normal red eye color versus white eyes, as discussed in Chapter 3; another such gene, with the alleles m^+ and m, determined whether the size of the wings was normal or miniature. The initial cross was between females with white eyes and normal wings and males with red eyes and miniature wings. We will use the slash in this instance to help us follow these X-linked traits:

$$w\,m^+/w\,m^+ \times w^+\,m/Y$$

The resulting F_1 progeny consisted of wild-type females and white-eyed, nonminiature males. When these were crossed,

$$w\,m^+/w^+\,m \times w\,m^+/Y$$

the female progeny consisted of a 1 : 1 ratio of red : white eyes (all were nonminiature), and the male progeny were as follows:

white eye, normal wing ($w\,m^+/Y$)	226	66.5 percent are parental phenotypes
red eye, miniature wing ($w^+\,m/Y$)	202	
red eye, normal wing ($w^+\,m^+/Y$)	114	33.5 percent are recombinant phenotypes
white eye, miniature wing ($w\,m/Y$)	102	
	644	

Because each male receives his X chromosome from his mother, the phenotype reveals the genotype of the X chromosome that he inherited. The results of the experiment show a great departure from the 1 : 1 : 1 : 1 ratio of the four male phenotypes expected with independent assortment. If genes in the same chromosome tended to remain together in inheritance but were not completely linked, this pattern of deviation might be observed. In this case, the combinations of phenotypic traits in the parents of the original cross (parental phenotypes) were present in 428/644 (66.5 percent) of the F_2 males, and nonparental combinations (recombinant phenotypes) of the traits were present in 216/644 (33.5 percent). The 33.5 percent recombinant X chromosomes is called the **frequency of recombination,** and it should be contrasted with the 50 percent recombination expected with independent assortment.

The recombinant X chromosomes $w^+\,m^+$ and $w\,m$ result from crossing-over in meiosis in F_1 females. In this example, the frequency of recombination between the linked w and m genes was 33.5 percent, but

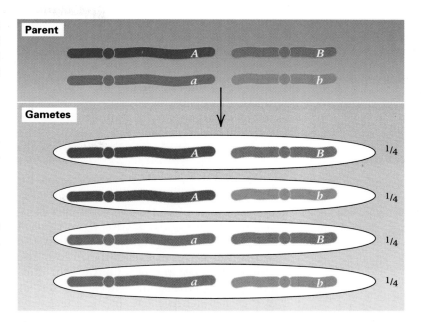

Figure 4.1
Alleles of genes in different chromosomes undergo independent assortment. The pairs of homologous chromosomes segregate at random with respect to one another, so an A-bearing chromosome is as likely to go to the same anaphase pole with a B-bearing chromosome as with a b-bearing chromosome. The result is that each possible combination of chromatids is equally likely among the gametes: 1/4 each for $A\,B$, $A\,b$, $a\,B$, and $a\,b$.

with other pairs of linked genes it ranges from near 0 to 50 percent. Even genes in the same chromosome can undergo independent assortment (frequency of recombination equal to 50 percent) if they are sufficiently far apart. This implies the following principle:

> Genes with recombination frequencies smaller than 50 percent are present in the same chromosome (linked). Two genes that undergo independent assortment, indicated by a recombination frequency equal to 50 percent, either are in nonhomologous chromosomes or are located far apart in a single chromosome.

Geneticists use a notation for linked genes that has the general form $w^+\,m/w\,m^+$. This notation is a simplification of a more descriptive but cumbersome form:

$$\frac{w^+ \qquad m}{w \qquad m^+}$$

In this form of notation, the horizontal line separates the two homologous chromosomes in which the alleles of the genes are located. The linked genes in a chromosome are always written in the same order for consistency. In the system of gene notation used for *Drosophila*, and in a similar system used for other organisms, this convention makes it possible to indicate the wildtype allele of a gene with a plus sign in the appropriate position. For example, the genotype $w\,m^+/w^+\,m$ can be written without ambiguity as $w\,+/+\,m$.

A genotype that is heterozygous for each of two linked genes can have the alleles in either of two possible configurations, as shown in Figure 4.2 for the w and m genes. In one configuration, called the **trans,** or **repulsion,** configuration, the mutant alleles are in opposite chromosomes, and the genotype is written as $w\,+/+\,m$ (Figure 4.2A). In the alternative configuration, called the **cis,** or **coupling,** configuration, the mutant alleles are present in the same chromosomes, and the genotype is written as $w\,m/+\,+$.

Morgan's study of linkage between the *white* and *miniature* alleles began with the *trans* configuration. He also studied progeny from the *cis* configuration of the w and m alleles, which results from the cross of white, miniature females with red, non-miniature males:

$$\frac{w\,m}{w\,m} \times \frac{+\,+}{\mathrm{Y}}$$

In this case, the F_1 females were phenotypically wildtype double heterozygotes, and the males had white eyes and miniature wings. When these F_1 progeny were crossed,

$$\frac{w\,m}{+\,+} \times \frac{w\,m}{\mathrm{Y}}$$

they produced the following progeny:

red eye, normal wing ($++/w\,m$ ♀♀ and $++/$Y ♂♂)	395	62.3 percent are parental phenotypes
white eye, miniature wing ($w\,m/w\,m$ ♀♀ and $w\,m/$Y ♂♂)	382	
white eye, normal wing ($w\,+/w\,m$ ♀♀ and $w\,+/$Y ♂♂)	223	37.7 percent are recombinant phenotypes
red eye, miniature wing ($+\,m/w\,m$ ♀♀ and $+\,m/$Y ♂♂)	247	
	1247	

Compared to the preceding experiment with w and m, the frequency of recombination between the genes is approximately the same: 37.7 percent versus 33.5 percent. The difference is within the range expected from random variation from experiment to experiment. However, in this case, the phenotypes constituting the parental and recombinant classes of offspring are reversed. They are reversed because the original parents of the F_1 female were different. In the first cross, the F_1 female was the *trans* double heterozygote ($w\,+/+\,m$); in the second cross, the F_1 female had the *cis* configuration ($w\,m/+\,+$). The repeated finding of equal recombination frequencies in experiments of this kind leads to the following conclusion:

> Recombination between linked genes takes place with the same frequency whether the alleles of the genes are in the *trans* configuration or in the *cis* configuration; it is the same no matter how the alleles are arranged.

The recessive allele y of another X-linked gene in *Drosophila* results in yellow body color instead of the usual gray color determined by the y^+ allele. When white-eyed females were mated with males having yellow bodies, and the wildtype F_1 females were testcrossed with yellow-bodied, white-eyed males,

$$+\,w/+\,w \;\times\; y\,+/\mathrm{Y}$$
$$\downarrow$$
$$+\,w/y\,+ \;\times\; y\,w/\mathrm{Y}$$

Figure 4.2
There are two possible configurations of the mutant alleles in a genotype that is heterozygous for both mutations. (A) The *trans,* or repulsion, configuration has the mutant alleles on opposite chromosomes. (B) The *cis,* or coupling, configuration has the mutant alleles on the same chromosome.

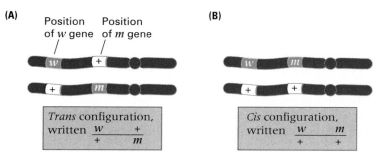

the progeny were

wildtype body, white eye (maternal gamete, + w)	4292	} 98.6 percent are parental phenotypes
yellow body, red eye (maternal gamete, y +)	4605	
wildtype body, red eye (maternal gamete, + +)	86	} 1.4 percent are recombinant phenotypes
yellow body, white eye (maternal gamete, y w)	44	
	9027	

In a second experiment, yellow-bodied, white-eyed females were crossed with wildtype males, and the F_1 wildtype females and F_1 yellow-bodied, white-eyed males were intercrossed:

$$y w/y w \ \times \ + +/Y$$
$$\downarrow$$
$$+ +/y w \ \times \ y w/Y$$

In this case, 98.6 percent of the F_2 progeny had parental phenotypes and 1.3 percent had recombinant phenotypes. The parental and recombinant phenotypes were reversed in the reciprocal crosses, but the recombination frequency was virtually the same. Females with the *trans* genotype $y +/+ w$ produced about 1.4 percent recombinant progeny, carrying either of the recombinant chromosomes $y w$ or $+ +$; similarly, females with the *cis* genotype $y w/+ +$ produced about 1.4 percent recombinant progeny, carrying either of the recombinant chromosomes $y +$ or $+ w$. However, the recombination frequency was much lower between the genes for yellow body and white eyes than between the genes for white eyes and miniature wings (1.4 percent versus about 35 percent). These and other experiments have led to the following conclusions:

- The recombination frequency is a characteristic of a particular pair of genes.

- Recombination frequencies are the same in *cis* (coupling) and *trans* (repulsion) heterozygotes.

In experiments with other genes, Morgan also discovered that *Drosophila* is unusual in that recombination does not take place in males. Although it is not known how (or why) crossing-over is prevented in males, the result of the absence of recombination in *Drosophila* males is that all alleles located in a particular chromosome show complete linkage in the male. For example, the genes *cn* (cinnabar eyes) and *bw* (brown eyes) are both in chromosome 2 but are so far apart that, in females, there is 50 percent recombination. Thus the cross

$$\frac{cn \ bw}{+ +} \female \ \times \ \frac{cn \ bw}{cn \ bw} \male$$

yields progeny of genotype $+ +/cn \ bw$ and $cn \ bw/cn \ bw$ (the nonrecombinant types) as well as $cn +/cn \ bw$ and $+ bw/cn \ bw$ (the recombinant types) in the proportions $1 : 1 : 1 : 1$. However, because there is no crossing-over in males, the reciprocal cross

$$\frac{cn \ bw}{cn \ bw} \female \ \times \ \frac{cn \ bw}{+ +} \male$$

yields progeny only of the nonrecombinant genotypes $+ +/cn \ bw$ and $cn \ bw/cn \ bw$ in equal proportions. The absence of recombination in *Drosophila* males is a convenience often made use of in experimental design; as shown in the case of *cn* and *bw*, all the alleles present in any chromosome in a male must be transmitted as a group, without being recombined with alleles present in the homologous chromosome. The absence of crossing-over in *Drosophila* males is atypical; in most other animals and plants, recombination takes place in both sexes.

4.2
Genetic Mapping

The linkage of the genes in a chromosome can be represented in the form of a **genetic map,** which shows the linear order of the genes along the chromosome with the distances between adjacent genes proportional to the frequency of recombination between them. A genetic map is also called a **linkage map** or a **chromosome map.** The concept of genetic mapping was first developed by Morgan's student, Alfred H. Sturtevant, in 1913. The early geneticists understood that recombination between genes takes place by an exchange of segments between homologous chromosomes in the process now called crossing-over. Each crossing-over is manifested physically as a chiasma, or cross-shaped configuration, between homologous chromosomes; chiasmata are observed in prophase I of meiosis (Chapter 3). Each chiasma results from the breaking and rejoining of chromatids during synapsis, with the result that there is an exchange of corresponding segments between them. The theory of crossing-over is that each chiasma results in a new association of genetic markers. This process is illustrated in Figure 4.3. When there is no crossing-over (Figure 4.3A), the alleles present in each homologous chromosome remain in the same combination. When crossing-over does take place (Figure 4.3B), the outermost alleles in two of the chromatids are interchanged (recombined).

The unit of distance in a genetic map is called a **map unit;** 1 map unit is equal to 1 percent recombination. For example, two genes that recombine with a frequency of 3.5 percent are said to be located 3.5 map units apart. One map unit is also called a **centimorgan,** abbreviated cM, in honor of T. H. Morgan. A distance of 3.5 map units therefore equals 3.5 centimorgans and indicates 3.5 percent recombination between the genes. For ease of reference, we list the four completely equivalent ways in which a genetic distance between two genes may be represented.

- As a *frequency of recombination* (in the foregoing example, 0.035)

- As a *percent recombination* (here 3.5 percent)

- As a *map distance* in map units (in this case, 3.5 map units)

- As a map distance in *centimorgans* (here 3.5 centimorgans, abbreviated 3.5 cM)

Physically, 1 map unit corresponds to a length of the chromosome in which, on the

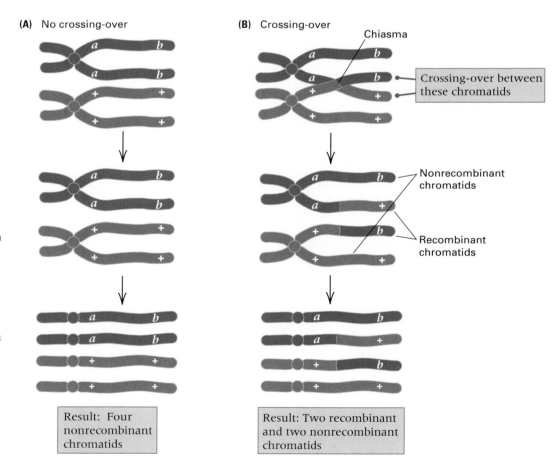

Figure 4.3
Diagram illustrating crossing-over between two genes. (A) When there is no crossing-over between two genes, the alleles are not recombined. (B) When there is crossing-over between them, the result of the crossover is two recombinant and two nonrecombinant products, because the exchange is between only two of the four chromatids.

(A) No crossing-over

Result: Four nonrecombinant chromatids

(B) Crossing-over

Chiasma

Crossing-over between these chromatids

Nonrecombinant chromatids

Recombinant chromatids

Result: Two recombinant and two nonrecombinant chromatids

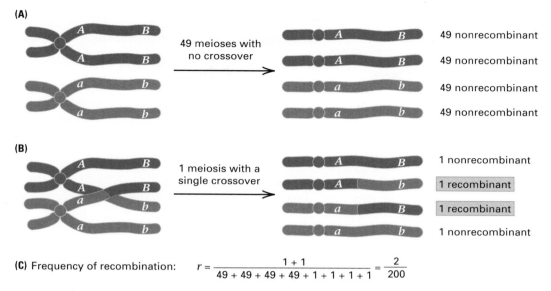

(A)

49 meioses with
no crossover

49 nonrecombinant

49 nonrecombinant

49 nonrecombinant

49 nonrecombinant

(B)

1 meiosis with a
single crossover

1 nonrecombinant

1 recombinant

1 recombinant

1 nonrecombinant

(C) Frequency of recombination: $r = \dfrac{1 + 1}{49 + 49 + 49 + 49 + 1 + 1 + 1 + 1} = \dfrac{2}{200}$

Figure 4.4

Diagram of chromosomal configurations in 50 meiotic cells, in which one has a crossover between two genes. (A) The 49 cells without a crossover result in 98 *A B* and 98 *a b* chromosomes; these are all nonrecombinant. (B) The cell with a crossover yields chromosomes that are *A B*, *A b*, *a B*, and *a b*, of which the middle two types are recombinant chromosomes. (C) The recombination frequency equals 2/200, or 1 percent, which is also called 1 map unit or 1 cM. Hence 1 percent recombination means that 1 meiotic cell in 50 has a crossover in the region between the genes.

average, one crossover is formed in every 50 cells undergoing meiosis. This principle is illustrated in Figure 4.4. If one meiotic cell in 50 has a crossing-over, the frequency of *crossing-over* equals 1/50, or 2 percent. Yet the frequency of *recombination* between the genes is 1 percent. The correspondence of 1 percent recombination with 2 percent crossing-over is a little confusing until you consider that a crossover results in two recombinant chromatids and two nonre-combinant chromatids (Figure 4.4). A frequency of crossing-over of 2 percent means that of the 200 chromosomes that result from meiosis in 50 cells, exactly 2 chromo-somes (the two involved in the exchange) are recombinant for genetic markers span-ning the particular chromosome segment. To put the matter in another way, 2 percent crossing-over corresponds to 1 percent recombination because only half the chro-matids in each cell with an exchange are actually recombinant.

In situations in which there are **genetic markers** along the chromosome, such as the *A, a* and *B, b* pairs of alleles in Figure 4.4, recombination between the marker genes takes place only when crossing-over

occurs *between* the genes. Figure 4.5 illus-trates a case in which crossing-over takes place between the gene A and the cen-tromere, rather than between the genes *A* and *B*. The crossing-over does result in the physical exchange of segments between the innermost chromatids. However, because it is located outside the region between *A* and *B*, all of the resulting gametes must carry either the *A B* or *a b* allele combinations. These are nonrecombinant chromosomes. The presence of the crossing-over is unde-tected because it is not in the region between the genetic markers.

In some cases, the region between genetic markers is large enough that two (or even more) crossovers can be formed in a single meiotic cell. One possible configu-ration for two crossovers is shown in Figure 4.6. In this example, both crossovers are between the same pair of chromatids. The result is that there is a physical exchange of a segment of chromosome between the marker genes, but the double crossover remains undetected because the markers themselves are not recombined. The absence of recombination results from the fact that the second crossover reverses the

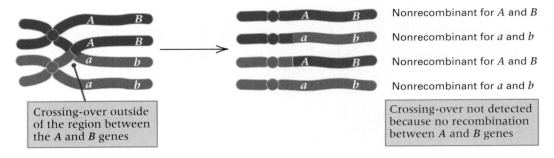

Nonrecombinant for *A* and *B*

Nonrecombinant for *a* and *b*

Nonrecombinant for *A* and *B*

Nonrecombinant for *a* and *b*

Crossing-over outside of the region between the *A* and *B* genes

Crossing-over not detected because no recombination between *A* and *B* genes

Figure 4.5
Crossing-over outside the region between two genes is not detectable through recombination. Although a segment of chromosome is exchanged, the genetic markers stay in the nonrecombinant configurations, in this case *A B* and *a b*.

effect of the first, insofar as recombination between *A* and *B* is concerned. The resulting chromosomes are either *A B* or *a b*, both of which are nonrecombinant.

Given that double crossing-over in a region between two genes can remain undetected because it does not result in recombinant chromosomes, there is an important distinction between the distance between two genes as measured by the recombination frequency and as measured in map units. Map units measure how much crossing-over takes place between the genes. For any two genes, the map distance between them equals one-half times the average number of crossover events that take place in the region per meiotic cell. The recombination frequency, on the other hand, reflects how much recombination is actually *observed* in a particular

experiment. Double crossovers that do not yield recombinant gametes, such as the one in Figure 4.6, *do* contribute to the map distance but *do not* contribute to the recombination frequency. The distinction is important only when the region in question is large enough that double crossing-over can occur. If the region between the genes is so short that no more than one crossover can be formed in the region in any one meiosis, then map units and recombination frequencies are the same (because there are no multiple crossovers that can undo each other). This is the basis for defining a map unit as being equal to 1 percent recombination. Over an interval so short as to yield 1 percent observed recombination, multiple crossovers are usually precluded, so the map distance equals the recombination frequency in this case.

Figure 4.6
If two crossovers take place between marker genes, and both involve the same pair of chromatids, then neither crossover is detected because all of the resulting chromosomes are nonrecombinant *A B* or *a b*.

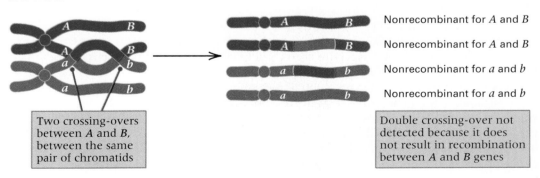

Nonrecombinant for *A* and *B*

Nonrecombinant for *A* and *B*

Nonrecombinant for *a* and *b*

Nonrecombinant for *a* and *b*

Two crossing-overs between *A* and *B*, between the same pair of chromatids

Double crossing-over not detected because it does not result in recombination between *A* and *B* genes

Genes All in a Row

Alfred H. Sturtevant 1913
Columbia University,
New York, New York
The Linear Arrangement of Six Sex-Linked Factors in Drosophila, *As Shown by Their Mode of Association*

Genetic mapping remains the cornerstone of genetic analysis. It is the principal technique used in modern human genetics to identify the chromosomal location of mutant genes associated with inherited diseases, as we saw with Huntington disease in Chapter 2. The genetic markers used in human genetics are homologous DNA fragments that differ in length from one person to the next, but the basic principles of genetic mapping are the same as those originally enunciated by Sturtevant. In this excerpt, we have substituted the symbols presently in use for the genes, y *(yellow body),* w *(white eyes),* v *(vermilion eyes),* m *(miniature wings), and* r *(rudimentary wings). (The sixth gene mentioned is another mutant allele of white, now called white-eosin.) In this paper, Sturtevant uses the term* crossing-over *instead of* recombination *and* crossovers *instead of* recombinant chromosomes. We have retained his original terms but, in a few cases, have put the modern equivalent in brackets.*

Morgan, by crossing white eyed, long winged flies to those with red eyes and rudimentary wings (the new sex-linked character), obtained, in F₂, white eyed rudimentary winged flies. This could happen only if "crossing-over" [recombination] is possible; which means, on the assumption that both of these factors are in the X chromosome, that an interchange of materials

between homologous chromosomes occurs (in the female only, since the male has only one X chromosome). A point not noticed at this time came out later in connection with other sex-linked factors in *Drosophila*. It became evident that some of the sex-linked factors are associated, i. e., that crossing-over does not occur freely between some factors, as shown by the fact that the combinations present in the F₁ flies are much more frequent in the F₂ than

---

These results form a new argument in favor of the chromosome view of inheritance, since they strongly indicate that the factors investigated are arranged in a linear series.

---

are new combinations of the same characters. This means, on the chromosome view, that the chromosomes, or at least certain segments of them, are much more likely to remain intact during meiosis than they are to interchange materials. . . . It would seem, if this hypothesis be correct, that the proportion of "crossovers" [recombinant chromosomes] could be used as an index of the distance between any two factors. Then by determining the distances (in the above sense) between A and B and between B and C, one should be able to predict AC. . . . Just how far our theory stands the test is shown by the data below, giving observed percent of crossovers [recombinant chromosomes] and the distances calculated [from the summation of shorter intervals].

Factors	Calculated distance	Observed percentage of crossovers
$y-v$	30.7	32.2
$y-m$	33.7	35.5
$y-r$	57.6	37.6
$w-m$	32.7	33.7
$w-r$	56.6	45.2

It will be noticed at once that the longer distances, $y-r$ and $w-r$, give smaller per cent of crossovers, than the calculation calls for. This is a point which was to be expected and is probably due to the occurrence of two breaks in the same chromosome, or "double crossing-over." But in the case of the shorter distances the correspondence with expectation is perhaps as close as was to be expected with the small numbers that are available. . . . It has been found possible to arrange six sex-linked factors in *Drosophila* in a linear series, using the number of crossovers per 100 cases [the frequency of recombination] as an index of the distance between any two factors. A source of error in predicting the strength of association between untried factors is found in double crossing-over. The occurrence of this phenomenon is demonstrated. . . . These results form a new argument in favor of the chromosome view of inheritance, since they strongly indicate that the factors investigated are arranged in a linear series.

Source: Journal of Experimental Zoology 14: 43–59

When adjacent chromosome regions separating linked genes are sufficiently short that multiple crossovers are not formed, the recombination frequencies (and hence the map distances) between the genes are additive. This important feature of recombination, and also the logic used in genetic mapping, is illustrated by the example in Figure 4.7. The genes are all in the X chromosome of *Drosophila*: *y* (yellow body), *rb* (ruby eye color), and *cv* (shortened wing crossvein). The recombination frequency between genes *y* and *rb* is 7.5 percent, and that between *rb* and *cv* is 6.2 percent. The genetic map might be any one of three possibilities, depending on which gene is in the middle (*y, cv,* or *rb*). Map A, which has *y* in the middle, can be excluded

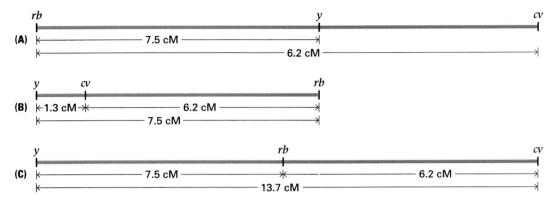

Figure 4.7

In *Drosophila*, the genes *y* (yellow body) and *rb* (ruby eyes) have a recombination frequency of 7.5 percent, and *rb* and *cv* (shortened wing crossvein) have a recombination frequency of 6.2 percent. There are three possible genetic maps, depending on whether *y* is in the middle (A), *cv* is in the middle (B), or *rb* is in the middle (C). Map A can be excluded because it implies that *rb* and *y* are closer then *rb* and *cv*, whereas the observed recombination frequency between *rb* and *y* is larger than that between *rb* and *cv*. Maps B and C are compatible with the data given.

because it implies that the recombination frequency between *rb* and *cv* should be greater than that between *rb* and *y*, and this contradicts the observed data.

Maps B and C are both consistent with the recombination frequencies. They differ in their predictions regarding the recombination frequency between *y* and *cv*. In map B the predicted distance is 1.3 map units, whereas in map C the predicted distance is 13.3 map units. In reality, the observed recombination frequency between *y* and *cv* is 13.3 percent. Map C is therefore correct.

There are actually two genetic maps corresponding to map C. They differ only in whether *y* is placed at the left or the right. One map is

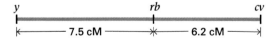

The other map is

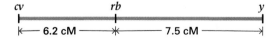

These two ways of depicting the genetic map are completely equivalent.

A genetic map can be expanded by this type of reasoning to include all the known genes in a chromosome; these genes consti-

tute a **linkage group.** The number of linkage groups is the same as the haploid number of chromosomes of the species. For example, cultivated corn *(Zea mays)* has ten pairs of chromosomes and ten linkage groups. A partial genetic map of chromosome 10 is shown in Figure 4.8, along with the dramatic phenotypes shown by some of the mutants. The ears of corn in the two photographs (Figure 4.8C and 4.8F) demonstrate the result of Mendelian segregation. The photograph in Figure 4.8C shows a 3 : 1 segregation of yellow : orange kernels produced by the recessive *orange pericarp-2 (orp-2)* allele in a cross between two heterozygous genotypes. The ear in the photograph in Figure 4.8F shows a 1 : 1 segregation of marbled : white kernels produced by the dominant allele *R1-mb* in a cross between a heterozygous genotype and a homozygous normal.

Crossing-over

The orderly arrangement of genes represented by a genetic map is consistent with the conclusion that each gene occupies a well-defined site, or **locus,** in the chromosome, with the alleles of a gene in a heterozygote occupying corresponding locations in the pair of homologous chromosomes. Crossing-over, which is brought about by a physical exchange of segments

Figure 4.8

Genetic map of chromosome 10 of corn, *Zea mays*. The map distance to each gene is given in standard map units (centimorgans) relative to a position 0 for the telomere of the short arm (lower left). Mutations in the gene *lesion-6 (les6)*, result in many small to medium-sized, irregularly spaced, discolored spots on the leaf blade and sheath; (A) shows the phenotype of a heterozygote for *Les6*, a dominant allele. Mutations in the gene *oil yellow-1 (oy1)* result in a yellow-green plant. In (B), the plant in front is heterozygous for the dominant allele *Oy1*; behind is a normal plant. The *orp2* allele is a recessive expressed as orange pericarp, a maternal tissue that surrounds the kernels; (C) shows the segregation of *orp2* in a cross between two heterozygous genotypes, yielding a 3 : 1 ratio of yellow : orange seeds. The gene *zn1* is *zebra necrotic-1*, in which dying tissue appears in transverse leaf bands; in (D), the left leaf is homozygous *zn1*, the right leaf wildtype. Mutations in the gene *teopod-2 (tp2)* result in many small, partially podded ears and a simple tassle; one of the ears in a plant heterozygous for the dominant allele *Tp2* is shown (E). The mutation *R1-mb* is an allele of the *r1* gene resulting in red or purple color in the aleurone layer of the seed; (F) shows the marbled color in kernels of an ear segregating for *R1-mb*. [Photographs courtesy of M. G. Neuffer; genetic map courtesy of E. H. Coe.]

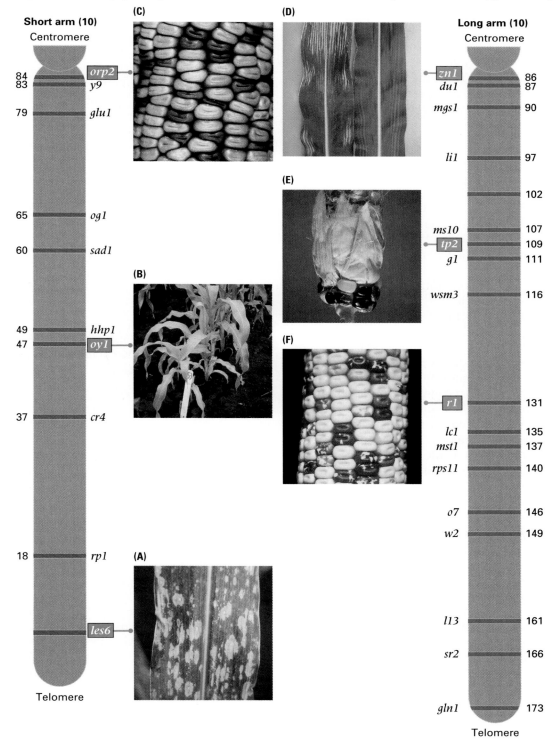

Short arm (10)
Centromere

84	*orp2*
83	*y9*
79	*glu1*
65	*og1*
60	*sad1*
49	*hhp1*
47	*oy1*
37	*cr4*
18	*rp1*
	les6

Telomere

Long arm (10)
Centromere

zn1	86
du1	87
mgs1	90
li1	97
	102
ms10	107
tp2	109
g1	111
wsm3	116
r1	131
lc1	135
mst1	137
rps11	140
o7	146
w2	149
l13	161
sr2	166
gln1	173

Telomere

(A) (B) (C) (D) (E) (F)

that results in a new association of genes in the same chromosome, has the following features:

1. The exchange of segments between parental chromatids takes place in the first meiotic prophase, *after the chromosomes have duplicated.* The four chromatids (strands) of a pair of homologous chromosomes are closely synapsed at this stage. Crossing-over is a physical exchange between chromatids in a pair of homologous chromosomes.

2. The exchange process consists of the breaking and rejoining of the two chromatids, resulting in the *reciprocal* exchange of equal and corresponding segments between them (see Figure 4.3).

3. The sites of crossing-over are more or less random along the length of a chromosome pair. Hence the probability of crossing-over between two genes increases as the physical distance between the genes along the chromosome becomes larger. This principle is the basis of genetic mapping.

The demonstration that crossing-over (as detected by the recombination of two heterozygous markers) is associated with a physical exchange of segments between homologous chromosomes was made possible by the discovery of two structurally altered chromosomes that permitted the microscopic recognition of parental and recombinant chromosomes. In 1931, Curt Stern discovered two X chromosomes of *Drosophila* that had undergone structural changes that made them distinguishable from each other and from a normal X chromosome. He used these structurally altered X chromosomes in an experiment that provided one of the classical proofs of the physical basis of crossing-over (Figure 4.9). One of the altered X chromosomes was missing a segment that had become attached to chromosome 4. This altered X chromosome could be identified by its missing terminal segment. The second aberrant X chromosome had a small piece of a Y chromosome attached as a second arm. The mutant alleles *car* (abbreviated *c*, a recessive allele resulting in carnation eye color instead of wildtype red) and *B* (a

dominant allele resulting in bar-shaped eyes instead of round) were present in the first altered X chromosome, and the wild-type alleles of these genes were in the second altered X. Females with the two structurally and genetically marked X chromosomes were mated with males having a normal X that carried the recessive alleles of the genes (Figure 4.9). In the progeny from this cross, flies with parental or recombinant combinations of the phenotypic traits were recognized by their eye color and shape, and their chromosomal makeup could be determined by microscopic examination of the offspring they produced in a testcross. In the genetically recombinant progeny from the cross, the X chromosome had the morphology that would be expected if recombination of the genes were accompanied by an exchange that recombined the chromosome markers; that is, the progeny with wildtype (red), bar-shaped eyes had an X chromosome with a missing terminal segment and the attached Y arm; similarly, progeny with carnation-colored, round eyes had a structurally normal X chromosome with no missing terminus and no Y arm. As expected, the nonrecombinant progeny were found to have an X chromosome morphologically identical with one in their mothers.

Crossing-over Takes Place at the Four-Strand Stage of Meiosis

So far we have asserted, without citing experimental evidence, that crossing-over takes place in meiosis after the chromosomes have duplicated, at the stage when each bivalent has four chromatid strands. One experimental proof that crossing-over takes place after the chromosomes have duplicated came from a study of laboratory stocks of *D. melanogaster* in which the two X chromosomes in a female are joined to a common centromere to form an aberrant chromosome called an **attached-X,** or **compound-X,** chromosome. The normal X chromosome in *Drosophila* has a centromere almost at the end of the chromosome, and the attachment of two of these chromosomes to a single centromere results in a chromosome with two equal arms, each consisting of a virtually complete X. Females with a compound-X chro-

(A)

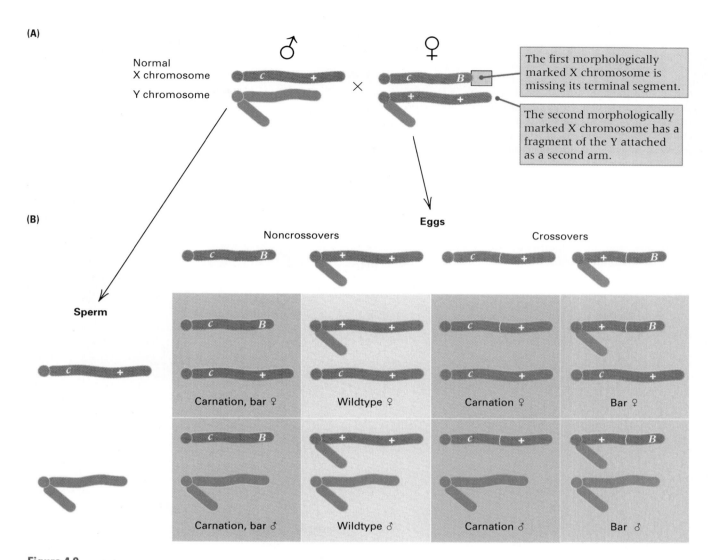

The first morphologically marked X chromosome is missing its terminal segment.

The second morphologically marked X chromosome has a fragment of the Y attached as a second arm.

(B)

Figure 4.9
(A) Diagram of a cross in which the two X chromosomes in a *Drosophila* female are morphologically distinguishable from each other and from a normal X chromosome. One X chromosome has a missing terminal segment, and the other has a second arm consisting of a fragment of the Y chromosome. (B) Result of the cross. The carnation offspring contain a structurally normal X chromosome, and the bar offspring contain an X chromosome with both morphological markers. The result demonstrates that genetic recombination between marker genes is associated with physical exchange between homologous chromosomes. Segregation of the missing terminal segment of the X chromosome, which is attached to chromosome 4, is not shown.

mosome usually contain a Y chromosome as well, and they produce two classes of viable offspring: females who have the maternal compound-X chromosome along with a paternal Y chromosome, and males with the maternal Y chromosome along with a paternal X chromosome (Figure 4.10). Attached-X chromosomes are frequently used to study X-linked genes in *Drosophila* because a male carrying any X-linked mutation, when crossed with an attached-X female, produces sons who also carry the mutation and daughters who carry the attached-X chromosome. In matings with attached-X females, therefore, the inheritance of an X-linked gene in the male passes from father to son to grandson, and so forth, which is the opposite of usual X-linked inheritance.

In an attached-X chromosome in which one X carries a recessive allele and the other carries the wildtype nonmutant allele,

Figure 4.10

Attached-X (compound-X) chromosomes in *Drosophila*. (A) A structurally normal X chromosome in a female. (B) An attached-X chromosome, with the long arms of two normal X chromosomes attached to a common centromere. (C) Typical attached-X females also contain a Y chromosome. (D) Outcome of a cross between an attached-X female and a normal male. The eggs contain either the attached-X or the Y chromosome, which combine at random with X-bearing or Y-bearing sperm. Genotypes with either three X chromosomes or no X chromosomes are lethal. Note that a male fly receives its X chromosome from its father and its Y chromosome from its mother—the opposite of the usual situation in *Drosophila*.

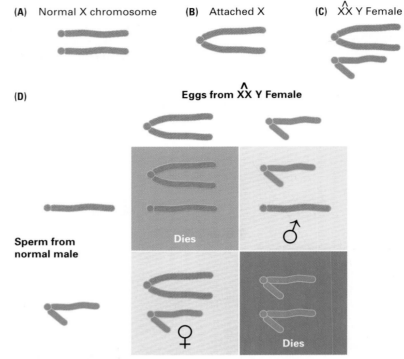

(A) Normal X chromosome (B) Attached X (C) X̂X̂ Y Female

(D) Eggs from X̂X̂ Y Female

Sperm from normal male

Dies

♂

♀

Dies

Figure 4.11

Diagram showing that crossing-over must take place at the four-strand stage in meiosis to produce a homozygous attached-X chromosome from one that is heterozygous for an allele. To yield homozygosity, the exchange must take place between the centromere and the gene.

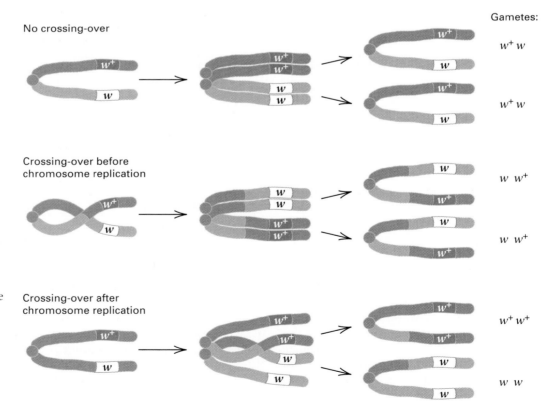

No crossing-over

Gametes:

$w^+\ w$

$w^+\ w$

Crossing-over before chromosome replication

$w\ w^+$

$w\ w^+$

Crossing-over after chromosome replication

$w^+\ w^+$

$w\ w$

Dos XX

Lilian V. Morgan 1922
Columbia University,
New York, New York
Non-crisscross Inheritance in Drosophila
melanogaster

Lilian V. Morgan was a first-rate geneticist long associated with T. H. Morgan as his collaborator and wife. She discovered the first attached-X chromosome as a single exceptional female in a routine mapping cross. She realized instinctively that this female was extremely important. There is an old story, of uncertain validity, that the female temporarily escaped, causing consternation and a mad search by everyone in the laboratory, until finally it was found resting on a window pane. The attached-X chromosome is still one of the most important genetic tools available to Drosophila *geneticists.*

A complete reversal of the ordinary criss-cross inheritance of recessive X-linked characters occurs in a line of *Drosophila* recently obtained. In ordinary X-linked inheritance, the recessive X-linked characters of the mother are transmitted to the sons, while the daughters show the domi-nant allele of the father. In the present case, the daughters show a recessive X-linked character of the mother and the sons show the dominant allele of the father. The reversal is explicable on the assumption that the two X-chromosomes of the mother are united and behave in meiosis as a single body. The cytological evidence verifies the genetic deduction. The eggs of these

---

The reversal is explicable on the assumption that the two X-chromosomes of the mother are united and behave in meiosis as a single body.

---

females do have two united X-chromosomes. . . . [A single female fly with a yellow abdomen was found in a cross between a homozygous nonyellow (gray) female and a yellow male.] She was mated to a gray male and produced 43 daughters and 59 sons. The daughters were, without exception, all yellow and the sons were all gray. The conclusion was at once evident that the [mother] had received from its father two yellow-bearing chromosomes, inseparable from one another, and that these inseparable chromosomes were transmitted together to the next generation producing (wherever they occurred) females, because there were always two of them. No male offspring could be yellow, because no single yellow chromosome was transmitted. . . . The F_1 females were fertile. . . . The daughters were all yellow, but differed from their yellow mother in having, besides the "yellow-bearing" double chromosome [the attached-X], a Y-bearing chromosome from their father. . . . The genetic behavior of the line of flies having the two inseparable X chromosomes is in entire accord with the condition of the chromosomes as seen in cytological preparations. . . . The origin of the [attached-X] can be explained if at some division in spermatogenesis of the father (perhaps at the equational division) the two halves of the X chromosome failed to become completely detached, but remained fastened together at one of their ends, producing the V-shaped chromosome found in the germ cells of the female descendants.

Source: Biological Bulletin 42: 267–274

---

crossing-over between the X-chromosome arms can yield attached-X products in which the recessive allele is present in both arms of the attached-X chromosome (Figure 4.11). Hence, attached-X females that are heterozygous can produce some female progeny that are homozygous for the recessive allele. The frequency with which homozygosity is observed increases with increasing map distance of the gene from the centromere. From the diagrams in Figure 4.11, it is clear that homozygosity can result only if the crossover between the gene and the centromere takes place after the chromosome has duplicated. The implication of finding homozygous attached-X female progeny is therefore that crossing-over takes place at the *four-strand* stage of meiosis. If this were not the case, and crossing-over happened before duplication of the chromosome (at the *two-strand* stage), it would result only in a swap of the alleles between the chromosome arms and would never yield the homozygous products that are actually observed.

The Molecular Basis of Crossing-over

As we will see in Chapter 6, each chromosome in a eukaryote contains a single, long molecule of duplex DNA complexed with proteins that undergoes a process of **condensation,** forming a hierarchy of coils upon coils that becomes progressively tighter as the chromosome progresses through nuclear division and reaches a state of maximum condensation at metaphase. Crossing-over along a chromosome must therefore correspond to some type of exchange of genetic information between DNA molecules.

The first widely accepted model of recombination between DNA molecules was proposed by Robin Holliday in 1964. Although it is overly simplistic in some of the details, the model has formed the basis of more realistic models favored today that account for most observations related to recombination. These models, and the evidence on which they are based, are discussed in detail in Chapter 13 in the context of DNA breakage and repair. It is, however, appropriate to introduce the Holliday model at this point to connect crossing-over observed in chromosomes to exchange between DNA molecules as envisaged in the Holliday model.

An outline of the Holliday model is illustrated in Figure 4.12. The DNA molecules depicted are those present in the chromatids that participate in the recombination event. The DNA duplexes in the other two chromatids, which are also present at the time of recombination, are not shown. The exchange is initiated by a single-stranded break in each molecule (Figure 4.12A), the ends of which are joined crosswise (Figure 4.12B). DNA is a dynamic molecule that "breathes" in the sense that local regions of paired bases frequently come apart and form again. Such "breathing" in the region of the exchange allows the molecules to exchange pairing partners along a region near the point of exchange (Figure 4.12C); the exchange of pairing partners is called *branch migration*. At any time, breaks at the positions of the arrows in part C, followed by crosswise rejoining, result in separate DNA molecules (Figure 4.12D) that are recombinant for the outside genetic markers—namely, *Ab* and *aB*. In part E, the second pair of breaks rejoin to resolve the interconnected Holliday structure in part C.

We need to make one additional comment relative to scale. The molecular events in Figure 4.12 are submicroscopic, and the Holliday structure can be observed only under favorable conditions through an electron microscope. Therefore, the cross-shaped exchange between the DNA strands indicated in part C is invisible through the light microscope. What, then, is a chiasma, the cross-shaped structure that connects nonsister chromatids in a bivalent? In pachytene, at the time of crossing-over, the chromatids are already condensed enough to be visible through the light microscope. In Figure 4.12, the DNA is shown in an elongated form rather than in the highly convoluted form actually present in condensed chromatin. The events in Figure 4.12 take place in a local region of DNA where the molecules are able to undergo the exchange. The events themselves are invisible. However, the resulting connection between the chromatids forms a visible chiasma between nonsister chromatids. Like a loose knot sliding along a rope, a chiasma can also slide along a chromosome, so the physical position of a chiasma may not necessarily represent the physical location of the DNA exchange that led to its formation.

Multiple Crossing-over

When two genes are located far apart along a chromosome, more than one crossover can be formed between them in a single meiosis, and this complicates the interpretation of recombination data. The probability of multiple crossovers increases with the distance between the genes. Multiple crossing-over complicates genetic mapping because map distance is based on the number of physical exchanges that are formed, and some of the multiple exchanges between two genes do not result in recombination of the genes and hence are not detected. As we saw in Figure 4.6, the effect of one crossover can be canceled by another crossover farther along the way. If two exchanges between the same two chromatids take place between the genes *A* and *B*, then their net effect will be that all chromosomes are nonrecombinant, either *A B* or *a b*. Two of the products of this meiosis have an interchange of their middle segments, but the chromosomes are not recombinant for the genetic markers, and so are genetically indistinguishable from noncrossover chromosomes. The possibility of such canceling events means that the observed recombination value is an *underestimate* of the true exchange frequency and the map distance between the genes. In higher organisms, double crossing-over is effectively precluded in chromosome segments that are sufficiently short. Therefore,

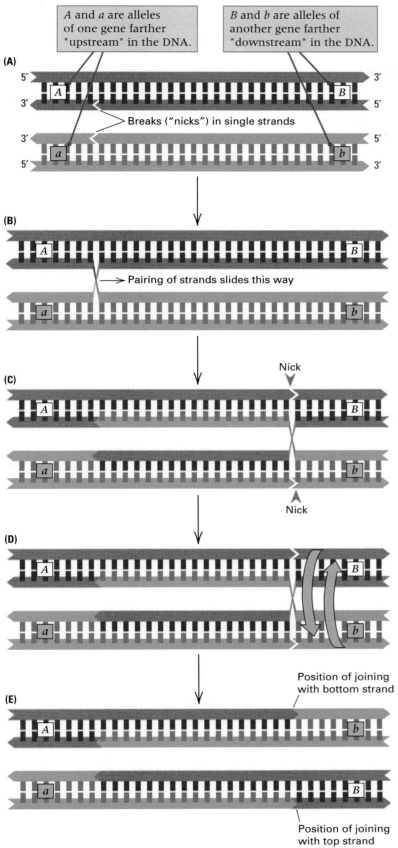

(A)

A and a are alleles of one gene farther "upstream" in the DNA.

B and b are alleles of another gene farther "downstream" in the DNA.

Breaks ("nicks") in single strands

(B)

Pairing of strands slides this way

(C)

Nick

Nick

(D)

(E)

Position of joining with bottom strand

Position of joining with top strand

Figure 4.12

The Holliday model of recombination. (A) In the participating DNA molecules, the exchange process is initiated by a single-stranded break in one strand of each duplex. (B) The ends of the broken strands are joined crosswise, resulting in a connection between the molecules. (C) The newly joined strands "unzip" a little and exchange pairing partners (the exchange of pairing partners is called *branch migration*). The exchange can be resolved by the breaking and rejoining of the outer strands. (D) In resolving the structure, the nicked outer strands exchange places. (E) Sealing of the gaps results in molecules that are recombinant for the outside genetic markers (*A b* and *a B*).

by using recombination data for closely linked genes to build up genetic linkage maps, we can avoid multiple crossovers that cancel each other's effects.

The *minimum* recombination frequency between two genes is 0. The recombination frequency also has a maximum:

No matter how far apart two genes may be, the maximum frequency of recombination between any two genes is 50 percent.

Fifty percent recombination is the same value that would be observed if the genes were on nonhomologous chromosomes and assorted independently. The maximum frequency of recombination is observed when the genes are so far apart in the chromosome that at least one crossover is almost always formed between them.

Figure 4.3B, shows that a single exchange in every meiosis would result in half of the products having parental combinations and the other half having recombinant combinations of the genes. Two exchanges between two genes have the same effect, as shown in Figure 4.13. Figure 4.13A shows a two-strand double crossover, in which the same chromatids participate in both exchanges; no recombination of the marker genes is detectable. When the two exchanges have one chromatid in common (three-strand double crossover, Figure 4.13B and C), the result is indistinguishable from that of a single exchange; two products with parental combinations and two with recombinant combinations are produced. Note that there are two types of three-strand doubles, depending on which three chromatids participate. The final pos-

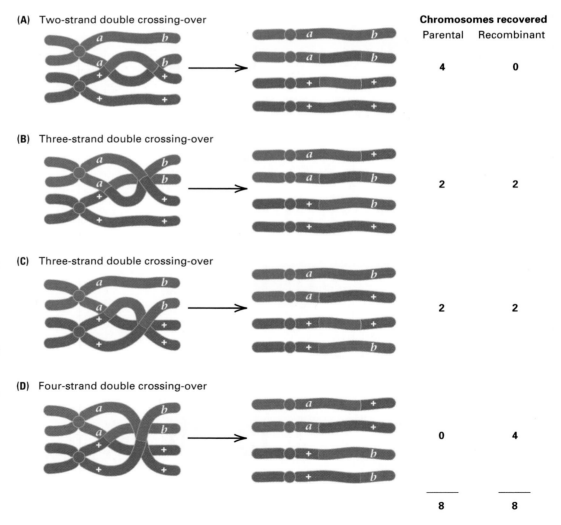

Figure 4.13
Diagram showing that the result of two exchanges in the interval between two genes is indistinguishable from independent assortment of the genes, provided that the chromatids participate at random in the exchanges. (A) A two-strand double crossing-over. (B and C) The two types of three-strand double crossing-overs. (D) A four-strand double crossing-over.

sibility is that the second exchange connects the chromatids that did not participate in the first exchange (four-strand double crossover, Figure 4.13D), in which case all four products are recombinant.

In most organisms, when double crossovers are formed, the chromatids that take part in the two exchange events are selected at random. In this case, the expected proportions of the three types of double exchanges are 1/4 four-strand doubles, 1/2 three-strand doubles, and 1/4 two-strand doubles. This means that, on the average,

$$(1/4)(0) + (1/2)(2) + (1/4)(4) = 2$$

recombinant chromatids will be found among the 4 chromatids produced from meioses with two exchanges between a pair of genes. This is the same proportion obtained with a single exchange between the genes. Moreover, a maximum of 50 percent recombination is obtained for any number of exchanges.

In the discussion of Figure 4.13, we emphasized that, in most organisms, the chromatids taking part in double-exchange events are selected at random. Then the maximum frequency of recombination is 50 percent. When there is a *nonrandom* choice of chromatids in successive crossovers, the phenomenon is called **chromatid interference.** It can be seen in Figure 4.13 that, relative to a random choice of chromatids, an excess of four-strand double crossing-over (*positive chromatid interference*) results in a maximum frequency of recombination greater than 50 percent; likewise, an excess of two-strand double crossing-over (*negative chromatid interference*) results in a maximum frequency of recombination smaller than 50 percent. Therefore, the finding that the maximum frequency of recombination between two genes in the same chromosome is not 50 percent can be regarded as evidence for chromatid interference. Positive chromatid interference has not yet been observed in any organism; negative chromatid interference has been reported in some fungi.

Double crossing-over is detectable in recombination experiments that employ **three-point crosses,** which include three pairs of alleles. If a third pair of alleles, c^+ and c, is located between the two with

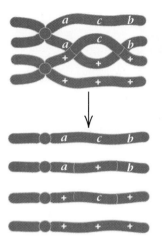

Figure 4.14
Diagram showing that two exchanges between the same chromatids and spanning the middle pair of alleles in a triple heterozygote will result in a reciprocal exchange of that pair of alleles between the two chromatids.

which we have been concerned (the outermost genetic markers), then double exchanges in the region can be detected when the crossovers flank the c gene (Figure 4.14). The two crossovers, which in this example take place between the same pair of chromatids, would result in a reciprocal exchange of the c^+ and c alleles between the chromatids. A three-point cross is an efficient way to obtain recombination data; it is also a simple method for determining the order of the three genes, as we will see in the next section.

4.3 Gene Mapping from Three-Point Testcrosses

The data in Table 4.1, which result from a testcross in corn with three genes in a single chromosome, illustrates the analysis of a three-point cross. The recessive alleles of the genes in this cross are *lz* (for lazy or prostrate growth habit), *gl* (for glossy leaf), and *su* (for sugary endosperm), and the multiply heterozygous parent in the cross has the genotype

$$\frac{Lz \; Gl \; Su}{lz \; \; gl \; \; su}$$

Therefore, the two classes of progeny that inherit noncrossover (parental-type) gametes are the normal plants and those with the lazy-glossy-sugary phenotype. These classes are far larger than any of the crossover classes. If the combination of dominant and recessive alleles in the chromosomes of the heterozygous parent were

Table 4.1 Progeny from a three-point testcross in corn

Phenotype of testcross progeny	Genotype of gamete from hybrid parent	Number
Normal (wildtype)	*Lz Gl Su*	286
Lazy	*lz Gl Su*	33
Glossy	*Lz gl Su*	59
Sugary	*Lz Gl su*	4
Lazy, glossy	*lz gl Su*	2
Lazy, sugary	*lz Gl su*	44
Glossy, sugary	*Lz gl su*	40
Lazy, glossy, sugary	*lz gl su*	272

unknown, then we could deduce from their relative frequency in the progeny that the noncrossover gametes were *Lz Gl Su* and *lz gl su*. This is a point important enough to state as a general principle:

> In any genetic cross involving linked genes, no matter how complex, the two most frequent types of gametes with respect to any pair of genes are *nonrecombinant*; these provide the linkage phase (*cis* versus *trans*) of the alleles of the genes in the multiply heterozygous parent.

In mapping experiments, the gene sequence is usually not known. In this example, the order in which the three genes are shown is entirely arbitrary. However, there is an easy way to determine the correct order from three-point data. Simply identify the genotypes of the double-crossover gametes produced by the heterozygous parent and compare them with the nonrecombinant gametes. Because the probability of two simultaneous exchanges is considerably smaller than that of either single exchange, the double-crossover gametes will be the least frequent types. Table 4.1 shows that the classes composed of four plants with the sugary phenotype and two plants with the lazy-glossy phenotype (products of the *Lz Gl su* and *lz gl Su* gametes, respectively) are the least frequent and therefore constitute the double-crossover progeny.

The effect of double crossing-over, as Figure 4.14 shows, is to interchange the members of the *middle* pair of alleles between the chromosomes.

This means that if the parental chromosomes are

Lz Gl Su   and   *lz gl su*

and the double-crossover chromosomes are

Lz Gl su   and   *lz gl Su*

then *Su* and *su* are interchanged by the double crossing-over and must be the middle pair of alleles. Therefore, the genotype of the heterozygous parent in the cross should be written as

$$\frac{Lz \quad\quad Su \quad\quad Gl}{lz \quad\quad su \quad\quad gl}$$

which is now diagrammed correctly with respect to both the order of the genes and the array of alleles in the homologous chromosomes. A two-strand double crossover between chromatids of these parental types is diagrammed below, and the products can be seen to correspond to the two types of gametes identified in the data as the double crossovers.

From this diagram, it can also be seen that the reciprocal products of a single crossover between *lz* and *su* would be *Lz su gl* and *lz Su Gl* and that the products of a single exchange between *su* and *gl* would be *Lz Su gl* and *lz su Gl*.

We can now summarize the data in a more informative way, writing the genes in correct order and identifying the numbers of the different chromosome types produced by the heterozygous parent that are present in the progeny.

Lz	*Su*	*Gl*	286	Parental types
lz	*su*	*gl*	272	
Lz	*su*	*gl*	40	Single crossovers
lz	*Su*	*Gl*	33	between *lz* and *su*
Lz	*Su*	*gl*	59	Single crossovers
lz	*su*	*Gl*	44	between *su* and *gl*
Lz	*su*	*Gl*	4	Double-crossover
lz	*Su*	*gl*	2	types
			740	

Note that each class of single recombinants consists of two reciprocal products and that these are found in approximately equal frequencies (40 versus 33 and 59 versus 44). This observation illustrates an important principle:

> The two reciprocal products that result from any crossover, or any combination of crossovers, are expected to appear in approximately equal frequencies among the progeny.

In calculating the frequency of recombination from the data, remember that the double-recombinant chromosomes result from *two* exchanges, one in each of the chromosome regions defined by the three genes. Therefore, chromosomes that are recombinant between *lz* and *su* are represented by the following chromosome types:

Lz	su	gl	40
lz	Su	Gl	33
Lz	su	Gl	4
lz	Su	gl	2
			79

That is, 79/740, or 10.7 percent, of the chromosomes recovered in the progeny are recombinant between the *lz* and *su* genes, so the map distance between these genes is 10.7 map units or 10.7 centimorgans. Similarly, the chromosomes that are recombinant between *su* and *gl* are represented by

Lz	Su	gl	59
lz	su	Gl	44
Lz	su	Gl	4
lz	Su	gl	2
			109

The recombination frequency between this second pair of genes is 109/740, or 14.8 percent, so the map distance between them indicated by these data is 14.8 map units or 14.8 centimorgans. The genetic map of the chromosome segment in which the three genes are located is therefore

lz   su   gl

|←10.7 map units ─→|←──── 14.8 map units ────→|

The error that students most commonly make as they are learning how to interpret three-point crosses is to forget to include the double recombinants when calculating the recombination frequency between adjacent genes. You can keep from falling into this trap by remembering that the double recombinant chromosomes have single recombination in *both* regions.

Chromosome Interference in Double Crossing-over

The detection of double crossing-over makes it possible to determine whether exchanges in two different regions of a pair of chromosomes are formed independently of each other. Using the information from the example with corn, we know from the recombination frequencies that the probability of recombination is 0.107 between *lz* and *su* and 0.148 between *su* and *gl*. If crossing-over is independent in the two regions (which means that the formation of one exchange does not alter the probability of the second exchange), then the probability of an exchange in both regions is the product of these separate probabilities, or $0.107 \times 0.148 = 0.0158$ (1.58 percent). This implies that in a sample of 740 gametes, the expected number of double crossovers would be 740×0.0158, or 12, whereas the number actually observed was only 6. Such deficiencies in the observed number of double crossovers are common and identify a phenomenon called **chromosome interference,** in which crossing-over in one region of a chromosome reduces the probability of a second crossover in a nearby region. Because chromosome interference is nearly universal, and chromatid interference is virtually unknown, the term *interference,* when used without qualification, almost always refers to chromosome interference.

The **coefficient of coincidence** is the observed number of double recombinant chromosomes divided by the expected number. Its value provides a quantitative measure of the degree of interference, defined as

i = interference

    = 1 − coefficient of coincidence

From the data in our example, the coefficient of coincidence is $6/12 = 0.50$, which means that the observed number of double crossovers was only 50 percent of the number we would expect to observe if crossing-over in the two regions were independent. The value of the interference depends on the distance between the genetic markers and on the species. In some species, the interference increases as the distance between the two outside markers becomes smaller, until a point is reached at which double crossing-over is eliminated; that is, no double crossovers are found, and the coefficient of coincidence equals 0 (or, to say the same thing, the interference equals 1). In *Drosophila*, this distance is about 10 map units. In yeast, by contrast, interference is incomplete even over short distances. For markers separated by 3 map units, the interference is in the range 0.3 to 0.6; for those separated by 7 map units, it is in the range 0.1 to 0.3. In most organisms, when the total distance between the genetic markers is greater than about 30 map units, interference essentially disappears and the coefficient of coincidence approaches 1.

Genetic Mapping Functions

The effect of interference on the relationship between genetic map distance and the frequency of recombination is illustrated in Figure 4.15. Each curve in Figure 4.15 is an example of a **mapping function,** which is the mathematical relation between the genetic distance across an interval in map units (centimorgans) and the observed frequency of recombination across the interval. In other words, a mapping function tells you how to convert a *map distance* between genetic markers into a *recombination frequency* between the markers. As we have seen, when the map distance between the markers is small, the recombination frequency equals the map distance. This principle is reflected in the curves in Figure 4.15 in the region in which the map distance is smaller than about 10 cM. At less than this distance, all of the curves are nearly straight lines, which means that map distance and recombination frequency are equal; 1 map unit equals 1 percent recombination, and 10 map units equal 10 percent recombination.

For distances greater than 10 map units, the recombination frequency becomes smaller than the map distance. How much smaller it is, for any given map distance, depends on the pattern of interference along the chromosome. Each pattern of interference yields a different mapping function. In Figure 4.15, three types of mapping functions are shown. The upper curve is based on the assumption of com-

Figure 4.15

A mapping function is the relation between genetic map distance across an interval and the observed frequency of recombination across the interval. Map distance is defined as one-half the average number of crossovers converted into a percentage. The three mapping functions correspond to different assumptions about interference, i. In the top curve, $i = 1$ (complete interference); in the bottom curve, $i = 0$ (no interference). The mapping function in the middle is based on the assumption that i decreases as a linear function of distance.

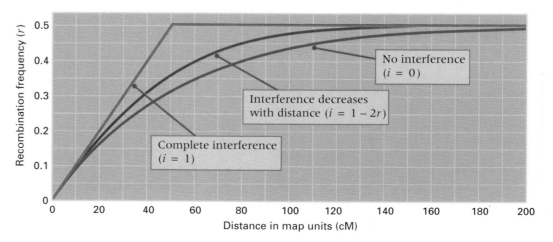

Chapter 4 Genetic Linkage and Chromosome Mapping

plete interference i, so that $i = 1$. With this mapping function, the linear relation holds all the way to a map distance of 50 cM, for which the recombination frequency is 50 percent; for map distances larger than 50 cM, the recombination frequency remains constant at 50 percent.

The bottom curve in Figure 4.15 is usually called **Haldane's mapping function** after its inventor. It assumes no interference ($i = 0$), and the mathematical form of the function is $r = (1/2)(1 - e^{-d/50})$, where d is the map distance in centimorgans. Any mapping function for which i is between 0 and 1 must lie in the interval between the top and bottom curves. The example shown is **Kosambi's mapping function,** in which the interference is assumed to decrease as a linear function of distance according to $i = 1 - 2r$. Although simple in its underlying assumptions, the formula for Kosambi's function is not simple. (The formula is in one of the problems at the end of the chapter.)

Haldane developed his mapping function in 1919, Kosambi his in 1943. Between these years and long afterward, geneticists had little interest in different mapping functions other than as curiosities, because there were few sets of data large enough to distinguish one reasonable function from the next. In recent years, with the explosion in the number of genetic markers available in virtually all organisms, and with the resurgence of interest in genetic mapping because of its role in identifying the position of mutations as precisely as possible prior to cloning (isolating the DNA), mapping functions have again become moderately fashionable. Checked against large data sets, none of the simple mapping functions in Figure 4.15 fits perfectly, but alternatives that fit better are much more complex even than Kosambi's mapping function.

Most mapping functions are almost linear near the origin, as are those in Figure 4.15. This near linearity implies that for map distances smaller than about 10 cM, whatever the pattern of chromosome interference, there are so few double recombinants that the recombination frequency in percent essentially equals the map distance. Hence the map distance between two widely separated genetic markers can be estimated with some confidence by summing the map distances across smaller segments between the markers, provided that each of the smaller segments is less than about 10 map units in length.

Genetic Distance and Physical Distance

Generally speaking, the greater the physical separation between genes along a chromosome, the greater the map distance between them. Physical distance and genetic map distance are usually correlated because a greater distance between genetic markers affords a greater chance for a crossover to take place; crossing-over is a physical exchange between the chromatids of paired homologous chromosomes.

On the other hand, the general correlation between physical distance and genetic map distance is by no means absolute. We have already noted that the frequency of recombination between genes may differ in males and females. An unequal frequency of recombination means that the sexes can have different map distances in their genetic maps, although the physical chromosomes of the two sexes are the same and the genes must have the same linear order. An extreme example of a sex difference in recombination is in *Drosophila*, in which there is no recombination in males (as we noted earlier). Hence, in *Drosophila* males, the map distance between any pair of genes located in the same chromosome is 0. (Genes on different chromosomes do undergo independent assortment in males.)

The general correlation between physical distance and genetic map distance can even break down in a single chromosome. For example, crossing-over is much less frequent in certain regions of the chromosome than in other regions. The term **heterochromatin** refers to certain regions of the chromosome that have a dense, compact structure in interphase; these regions take up many of the standard dyes used to make chromosomes visible. The rest of the chromatin, which becomes visible only after chromosome condensation in mitosis or meiosis, is called **euchromatin.** In most organisms, the major heterochromatic regions are adjacent to the centromere; smaller blocks are present at the ends of the chromosome arms (the telomeres) and interspersed with the euchromatin. In general, crossing-over is much less frequent in regions of heterochromatin than in regions of euchromatin.

Because there is less crossing-over in heterochromatin, a given length of heterochromatin will appear much shorter in the genetic map than an equal length of euchromatin. In heterochromatic regions, therefore, the genetic map gives a distorted picture of the physical map. An example of such distortion appears in Figure 4.16, which compares the physical map and the genetic map of chromosome 2 in *Drosophila*. The physical map is depicted as the chromosome appears in metaphase of mitosis. Two genes near the tips and two near the euchromatin-heterochromatin junction are indicated in the genetic map. The map distances across the euchromatic arms are 54.5 and 49.5 map units, respectively, for a total euchromatic map distance of 104.0 map units. However, the heterochromatin, which constitutes approximately 25 percent of the entire chromosome, has a genetic length in map units of only 3.0 percent. The distorted length of the heterochromatin in the genetic map results from the reduced frequency of crossing-over in the heterochromatin. In spite of the distortion of the genetic map across the heterochromatin, in the regions of euchromatin there is a good correlation between the physical distance between genes and their distance in map units in the genetic map.

4.4
Genetic Mapping in Human Pedigrees

Before the advent of recombinant DNA, mapping genes in human beings was very tedious and slow. There were numerous practical obstacles to genetic mapping in human pedigrees:

1. Most genes that cause genetic diseases are rare, so they are observed in only a small number of families.

2. Many genes of interest in human genetics are recessive, so they are not detected in heterozygous genotypes.

3. The number of offspring per human family is relatively small, so segregation cannot usually be detected in single sibships.

4. The human geneticist cannot perform testcrosses or backcrosses, because human matings are not manipulated by an experimenter.

In recent years, because recombinant-DNA techniques allow direct access to the DNA, genetic mapping in human pedigrees has been carried out primarily by using genetic markers present in the DNA itself, rather than through the phenotypes produced by mutant genes. There are many minor differences in DNA sequence from one person to the next. On the average, the DNA sequences at corresponding positions in any two chromosomes, taken from any two people, differ at approximately one in every thousand base pairs. Most of the differences in DNA sequence are not associated with any inherited disease or disability. Indeed, many of the differences

Figure 4.16
Chromosome 2 in *Drosophila* as it appears in metaphase of mitosis (physical map, top) and in the genetic map (bottom). Heterochromatin and euchromatin are in contrasting colors. The genes indicated on the map are *net* (net wing veins), *pr* (purple eye color), *cn* (cinnabar eye color), and *sp* (speck of wing pigment). The genes *pr* and *cn* are actually in euchromatin but are located near the junction with heterochromatin. The total map length is 54.5 + 49.5 + 3.0 = 107.0 map units. The heterochromatin accounts for 3.0/107.0 = 2.8 percent of the total map length but constitutes approximately 25 percent of the physical length of the metaphase chromosome.

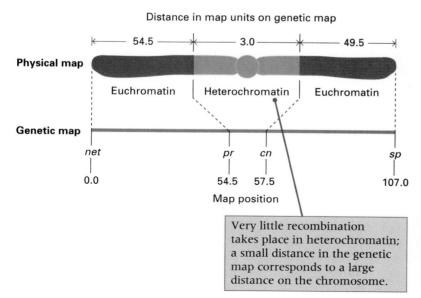

Distance in map units on genetic map

| 54.5 | 3.0 | 49.5 |

Physical map

Euchromatin | Heterochromatin | Euchromatin

Genetic map

net   pr   cn   sp
0.0   54.5 57.5   107.0
Map position

Very little recombination takes place in heterochromatin; a small distance in the genetic map corresponds to a large distance on the chromosome.

are found in DNA sequences that do not code for proteins. Nevertheless, all of these differences can serve as convenient genetic markers, and differences that are genetically linked to genes causing hereditary diseases are particularly important. Some differences in nucleotide sequence are detected by means of a type of enzyme called a **restriction endonuclease,** which cleaves double-stranded DNA molecules wherever a particular, short sequence of bases is present. For example, the restriction enzyme *Eco*RI cleaves DNA wherever the sequence GAATTC appears in either strand, as illustrated in Figure 4.17. Restriction enzymes will be considered in detail in Chapter 5. For now, we simply note that their significance is related to the fact that a difference in DNA sequence that eliminates a cleavage site can be detected because the region lacking the cleavage site will be cleaved into one larger fragment instead of two smaller ones (Figure 4.18). More rarely, a mutation in DNA sequence will create a new site rather than destroy one already present. Organisms, including human beings, frequently have minor differences in DNA sequence that are present in homologous, and otherwise completely identical, regions of DNA; any difference that alters a cleavage site will also change the length of the DNA fragments produced by cleavage with the corresponding restriction enzyme. The different DNA fragments can be separated by size by an electric field in a supporting gel and detected by various means. Differences in DNA fragment length produced by the presence or absence of the cleavage sites in DNA molecules are known as **restriction fragment length polymorphisms (RFLPs).**

RFLPs are typically formed in one of two ways. A mutation that changes a base sequence may result in loss or gain of a cleavage site that is recognized by the restriction endonuclease in use. Figure 4.19A gives an example. On the left is shown the relevant region in the homologous DNA molecules in a person who is heterozygous for such a sequence polymorphism. The homologous chromosomes in the person are distinguished by the letters *a* and *b*. In the region of interest, chromosome *a* contains two cleavage sites and chromosome *b* contains three. On the right is shown the position of the DNA fragments produced by cleavage after separation in an electric field. Each fragment appears as a horizontal band in the gel. The fragment from chromosome *a* migrates more slowly than those from chromosome *b* because it is larger, and larger fragments move more slowly through the gel. In this example, DNA from a person heterozygous for the *a* and *b* types of chromosomes would yield three bands in a gel. Similarly,

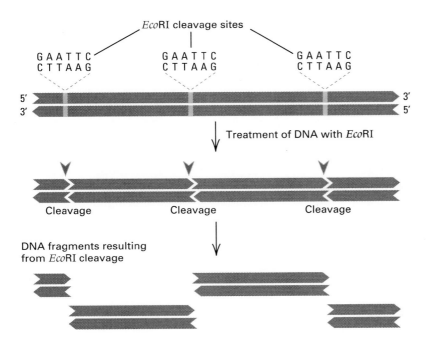

Figure 4.17

The restriction enzyme *Eco*RI cleaves double-stranded DNA wherever the sequence 5'-GAATTC-3' is present. In the example shown here, the DNA molecule contains three *Eco*RI cleavage sites, and it is cleaved at each site, producing a number of fragments.

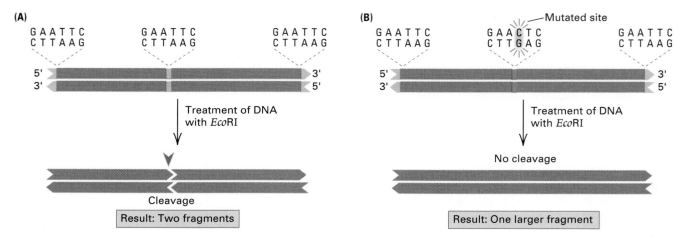

Figure 4.18

A minor difference in the DNA sequence of two molecules can be detected if the difference eliminates a restriction site. (A) This molecule contains three restriction sites for *Eco*RI, including one at each end. It is cleaved into two fragments by the enzyme. (B) This molecule has a mutated base sequence in the *Eco*RI site in the middle. It changes 5'-GAATTC-3' into 5'-GAACTC-3', which is no longer cleaved by *Eco*RI. Treatment of this molecule with *Eco*RI results in one larger fragment.

DNA from homozygous *aa* would yield one band, and that from homozygous *bb* would yield two bands.

A second type of RFLP results from differences in the number of copies of a short DNA sequence that may be repeated many times in tandem at a particular site in a chromosome (Figure 4.19B). In a particular chromosome, the tandem repeats may contain any number of copies, typically ranging from ten to a few hundred. When a DNA molecule is cleaved with a restriction endonuclease that cleaves at sites flanking the tandem repeat, the size of the DNA fragment produced is determined by the number of repeats present in the molecule. Figure 4.19B illustrates homologous DNA sequences in a heterozygous person containing one chromosome *a* with two copies of the repeat and another chromosome *b* with five copies of the repeat. When cleaved and separated in a gel, chromosome *a* yields a shorter fragment than that from chromosome *b*, because *a* contains fewer copies of the repeat. An RFLP resulting from a **variable number of tandem repeats** is called a **VNTR.** The utility of VNTRs in human genetic mapping derives from the very large number of alleles that may be present in the human population. The large number of alleles also implies that most people will be heterozygous, so

their DNA will yield two bands upon cleavage with the appropriate restriction endonuclease. Because of their high degree of variation among people, VNTRs are also widely used in DNA typing in criminal investigations (Chapter 15).

In genetic mapping, the phenotype of a person with respect to an RFLP is a pattern of bands in a gel. As with any other type of gene, the genotype of a person with respect to RFLP alleles is inferred, insofar as it is possible, from the phenotype. Linkage between different RFLP loci is detected through lack of independent assortment of the alleles in pedigrees, and recombination and genetic mapping are carried out using the same principles that apply in other organisms except that, in human beings, because of the small family size, different pedigrees are pooled together for analysis. Primarily through the use of RFLP and VNTR polymorphisms, genetic mapping in humans has progressed rapidly.

A three-generation pedigree of a family segregating for several alleles at a VNTR locus is illustrated in Figure 4.20. In this example, each of the parents is heterozygous, as are all of the children. Yet every person can be assigned his or her genotype because the VNTR alleles are codominant. At present, DNA polymorphisms are the principal types of genetic markers used in

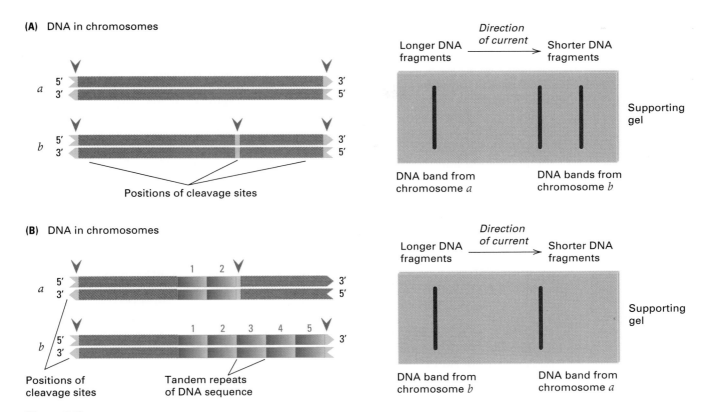

(A) DNA in chromosomes

Direction of current

Longer DNA fragments → Shorter DNA fragments

Positions of cleavage sites

DNA band from chromosome *a*

DNA bands from chromosome *b*

Supporting gel

(B) DNA in chromosomes

Direction of current

Longer DNA fragments → Shorter DNA fragments

Positions of cleavage sites

Tandem repeats of DNA sequence

DNA band from chromosome *b*

DNA band from chromosome *a*

Supporting gel

Figure 4.19
Two types of genetic variation that are widespread in most natural populations of animals and plants. (A) RFLP (restriction fragment length polymorphism), in which alleles differ in the presence or absence of a cleavage site in the DNA. The different alleles yield different fragment lengths (shown in the gel pattern at the right) when the molecules are cleaved with a restriction enzyme. (B) VNTR (variable number of tandem repeats), in which alleles differ in the number of repeating units present between two cleavage sites

genetic mapping in human pedigrees. Such polymorphisms are prevalent, are located in virtually all regions of the chromosome set, and have multiple alleles and so yield a high proportion of heterozygous genotypes. Furthermore, only a small amount of biological material is needed to perform the necessary tests. Many of the polymorphisms that are most useful in genetic mapping result from variation in the number of

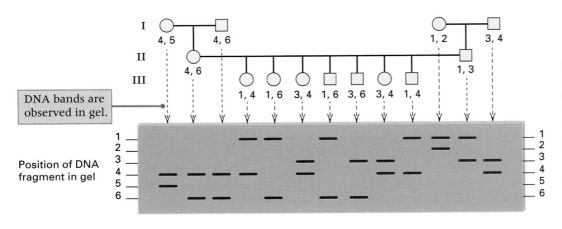

DNA bands are observed in gel.

Position of DNA fragment in gel

Figure 4.20
Human pedigree showing segregation of VNTR alleles. Six alleles (1–6) are present in the pedigree, but any one person can have only one allele (if homozygous) or two alleles (if heterozygous).

tandem copies of a simple repeating sequence present in the DNA, such as

5'- · · · TGTGTGTGTG · · · -3'

A simple-sequence polymorphism of this type is called a **simple tandem repeat polymorphism,** or **STRP.** The utility of STRPs in genetic mapping derives from the large number of alleles present in the population and the high proportion of genotypes that are heterozygous for two different alleles.

The present human genetic map is based on more than 5000 genetic markers, primarily STRPs, each heterozygous in an average of 70 percent of people tested. Because there is more recombination in females than in males, the female and male genetic maps differ in length. The female map is about 4400 cM, the male map about 2700 cM. Averaged over both sexes, the length of the human genetic map for all 23 pairs of chromosomes is about 3500 cM. Because the total DNA content per haploid set of chromosomes is 3154 million base pairs, there is, very roughly, 1 cM per million base pairs in the human genome.

4.5
Mapping by Tetrad Analysis

In some species of fungi, each meiotic tetrad is contained in a sac-like structure called an **ascus** and can be recovered as an intact group. Each product of meiosis is included in a reproductive cell called an **ascospore,** and all of the ascospores formed from one meiotic cell remain together in the ascus (Figure 4.21). The advantage of using these organisms to study recombination is the potential for analyzing all of the products from each meiotic division. Two other features of the organisms are especially useful for genetic analysis: (1) They are haploid, so dominance is not a complicating factor because the genotype is expressed directly in the phenotype; and (2) they produce very large numbers of progeny, making it possible to detect rare events and to estimate their frequencies accurately.

The life cycles of these organisms tend to be short. The only diploid stage is the zygote, which undergoes meiosis soon after it is formed; the resulting haploid meiotic products (which form the ascospores) germinate to regenerate the vegetative stage (Figure 4.22). In some species, each of the four products of meiosis subsequently undergoes a mitotic division, with the result that each member of the tetrad yields a *pair* of genetically identical ascospores. In most of the organisms, the meiotic products, or their derivatives, are not arranged in any particular order in the ascus. However, bread molds of the genus *Neurospora* and related organisms have the useful characteristic that the meiotic products are arranged in a definite order directly related to the planes of the meiotic divisions. We will examine the ordered system after first looking at unordered tetrads.

The Analysis of Unordered Tetrads

In the tetrads, when two pairs of alleles are segregating, three patterns of segregation are possible. For example, in the cross $AB \times ab$, the three types of tetrads are

AB AB ab ab   referred to as **parental ditype,** or PD. Only two genotypes are represented, and their alleles have the same combinations found in the parents.

Ab Ab aB aB   referred to as **nonparental ditype,** or NPD. Only two genotypes are represented, but their alleles have nonparental combinations.

AB Ab aB ab   referred to as **tetratype,** or TT. All four of the possible genotypes are present.

It is because of the following principle that tetrad analysis is an effective way to determine whether two genes are linked.

When genes are *unlinked,* the parental ditype tetrads and nonparental ditype tetrads are expected in equal frequencies (PD = NPD).

The reason for the equality PD = NPD for unlinked genes is shown in Figure 4.23A for two pairs of alleles, *A, a* and *B, b,* located

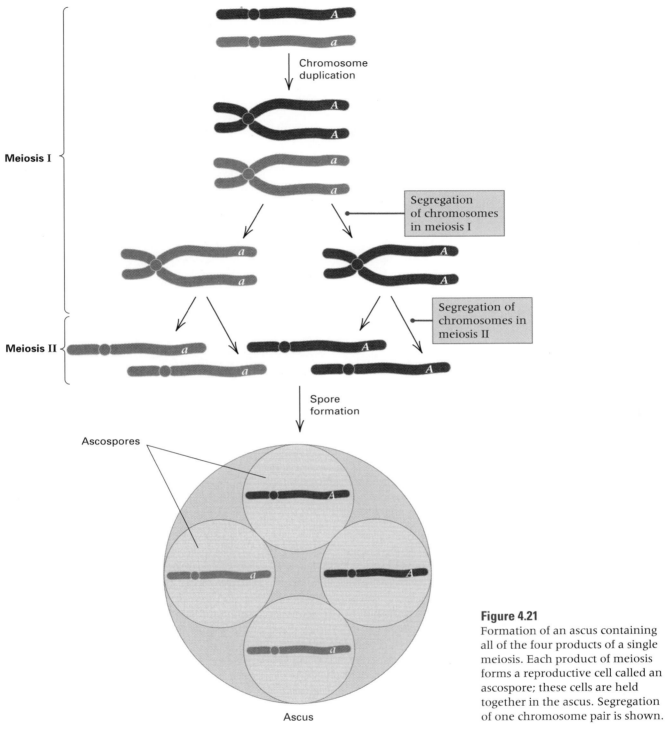

Meiosis I

Chromosome duplication

Segregation of chromosomes in meiosis I

Meiosis II

Segregation of chromosomes in meiosis II

Spore formation

Ascospores

Ascus

Figure 4.21
Formation of an ascus containing all of the four products of a single meiosis. Each product of meiosis forms a reproductive cell called an ascospore; these cells are held together in the ascus. Segregation of one chromosome pair is shown.

in different chromosomes. In the absence of crossing-over between either gene and its centromere, the two chromosomal configurations are equally likely at metaphase I, so PD = NPD. When there is crossing-over between either gene and its centromere (Figure 4.23B), a tetratype tetrad

results, but this does not change the fact that PD = NPD.

In contrast, when genes are linked, parental ditypes are far more frequent than nonparental ditypes. To see why, assume that the genes are linked and consider the events required for the production of the

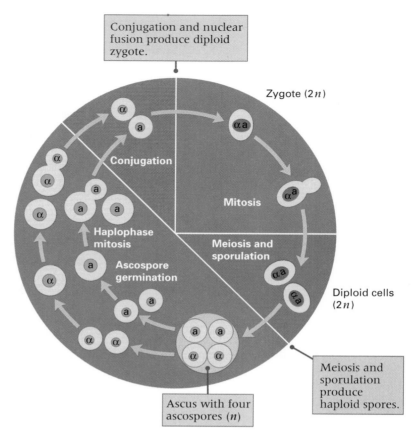

Figure 4.22

Life cycle of the yeast *Saccharomyces cerevisiae*. Mating type is determined by the alleles **a** and α. Both haploid and diploid cells normally multiply by mitosis (budding). Depletion of nutrients in the growth medium induces meiosis and sporulation of cells in the diploid state. Diploid nuclei are red; haploid nuclei are yellow.

three types of tetrads. Figure 4.24 shows that when no crossing-over takes place between the genes, a PD tetrad is formed. Single crossing-over between the genes results in a TT tetrad. The formation of a two-strand, three-strand, or four-strand double crossover results in a PD, TT, or NPD tetrad, respectively. With linked genes, meiotic cells with no crossovers will always outnumber those with four-strand double crossovers. Therefore,

> Linkage is indicated when nonparental ditype tetrads appear with a much lower frequency than parental ditype tetrads (NPD << PD).

The relative frequencies of the different types of tetrads can be used to determine the map distance between two linked genes, assuming that the genes are sufficiently close that triple crossovers and higher levels of crossing-over can be neglected. To determine the map distance, the number of NPD tetrads is first used to estimate the frequency of double crossovers. The four types of double crossovers in Figure 4.24 are expected in equal frequency. Thus the number of tetrads resulting from double crossovers (parts C, D, E, and F of Figure 4.24) should equal four times the number of NPD tetrads (part F), or 4 × [NPD], in which the square brackets denote the observed number of NPD tetrads. At the same time, the number of TT tetrads resulting from three-strand double crossovers (parts D and E) should equal twice the number of NPD tetrads, or 2 × [NPD] (part F). Subtracting 2 × [NPD] from the total number of tetratypes yields

(A) No crossing-over

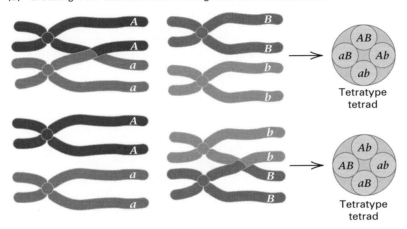

(B) Crossing-over between one of the genes and its centromere

Figure 4.23
Types of unordered asci produced with two genes in different chromosomes. (A) In the absence of crossing-over, random arrangement of chromosome pairs at metaphase I yields two different combinations of chromatids, one yielding PD tetrads and the other NPD tetrads. (B) When crossing-over takes place between one gene and its centromere, the two chromosome arrangements yield TT tetrads. If both genes are closely linked to their centromeres (so that crossing-over is rare), then few TT tetrads are produced.

the number of tetratype tetrads that originate from single crossovers, or [TT] − 2[NPD]. By definition, the map distance between two genes equals one-half the frequency of single-crossover tetrads ([TT] − 2[NPD]) plus the frequency of double-crossover tetrads (4[NPD]). Thus

Map distance

$$= \frac{(1/2)([TT] - 2[NPD]) + 4[NPD]}{\text{Total number of tetrads}} \times 100$$

$$= \frac{(1/2)[TT] + 3[NPD]}{\text{Total number of tetrads}} \times 100$$

$$= \frac{(1/2)([TT] + 6[NPD])}{\text{Total number of tetrads}} \times 100 \qquad (1)$$

The 1/2 in this equation corrects for the fact that only half of the meiotic products in a TT tetrad produced by single crossing-over are recombinant. In the special case where the linked genes are close enough that [NPD] = 0 (no double crossovers), the map distance equals the percentage of TT tetrads divided by 2. (The division by 2 is necessary because only half the ascospores in TT tetrads are recombinant for the alleles.)

A systematic format for analyzing linkage relationships for allelic pairs in unordered tetrads is presented in the form of a tree in Figure 4.25. All of the equations are based on the assumption that there is no chromatid interference. The right-hand side summarizes the equations for linked genes that we derived on the basis of Figure 4.25. The left-hand side pertains to unlinked genes, and it shows how the frequency of tetratypes can be used to make

Figure 4.24

Types of tetrads produced with two linked genes. In the absence of crossing-over (A), a PD tetrad is produced. With a single crossover between the genes (B), a TT tetrad is produced. Among the four possible types of double crossovers between the genes (C, D, E, and F), only the four-strand double crossover in part F yields an NPD tetrad.

(A) No crossing-over

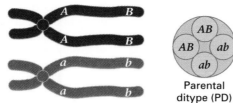

Parental
ditype (PD)

(B) Single crossing-over

Tetratype (TT)

(C) 2-strand double crossing-over

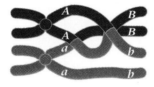

Parental
ditype (PD)

(D) 3-strand double crossing-over

Tetratype (TT)

(E) 3-strand double crossing-over

Tetratype (TT)

(F) 4-strand double crossing-over

Nonparental
ditype (NPD)

inferences about the distance of each gene from its own centromere.

As an example of the calculations for linked genes, consider a two-factor cross that yields 112 PD, 4 NPD, and 24 TT tetrads. The fact that NPD << PD (that is, 4 << 112) indicates that the two genes are linked and leads us down the right branch of the tree. Because both TT and NPD tetrads are recovered, we take the first left fork, and because there are NPD tetrads, we also take the next left fork. The appropriate equation is therefore Equation (1). Substitution of the values yields a map distance of

Map distance

$$= \frac{[(1/2) \times 24] + (3 \times 4)}{112 + 4 + 24} \times 100$$

$$= 17.1 \text{ cM}$$

Note that this mapping procedure differs from that presented earlier in the chapter in that recombination frequencies are not calculated directly from the number of recombinant and nonrecombinant chromatids, but rather from the inferred types of crossovers.

As Figure 4.25 indicates, tetrad analysis yields a great deal of information about the linkage relationship between genetic markers and the distance of unlinked markers to their respective centromeres. However, it is not necessary to carry out a full tetrad analysis for estimating linkage. The alternative is to examine spores chosen at random after allowing the tetrads to break open and disseminate their spores. This procedure is called **random-spore analysis,** and the linkage relationships are determined exactly as described earlier for *Drosophila* and *Zea mays.* In particular, the frequency of recombination equals the number of spores that are recombinant for the genetic markers divided by the total number of spores.

The Analysis of Ordered Tetrads

In *Neurospora crassa,* a species used extensively in genetic investigations, the products of meiosis are contained in an *ordered* array of ascospores (Figure 4.26). A zygote nucleus contained in a sac-like ascus undergoes meiosis almost immediately

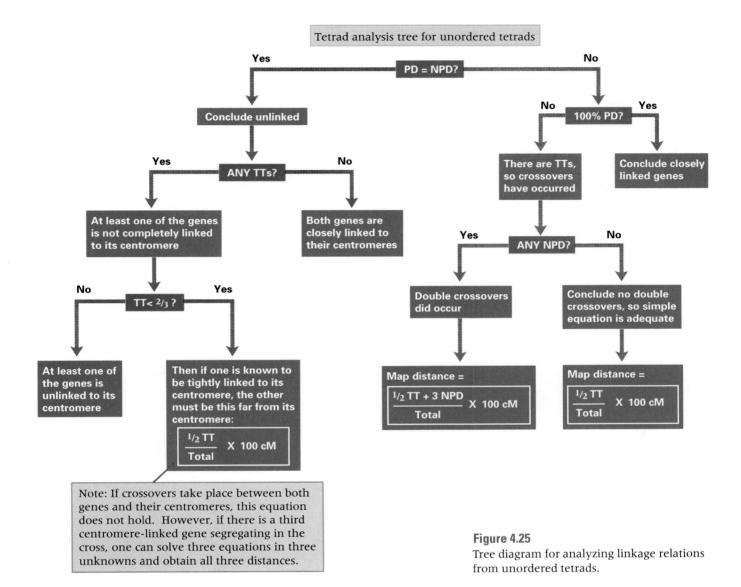

Figure 4.25
Tree diagram for analyzing linkage relations
from unordered tetrads.

after it is formed. The four nuclei produced by meiosis are in a linear, ordered sequence in the ascus, and each of them undergoes a mitotic division to form two genetically identical and adjacent ascospores. Each mature ascus contains eight ascospores arranged in four pairs, each pair derived from one of the products of meiosis. The ascospores can be removed one by one from an ascus and each germinated in a culture tube to determine its genotypes.

Ordered asci also can be classified as PD, NPD, or TT with respect to two pairs of alleles; hence the tree diagram in Figure 4.25 can be used to analyze the linkage data. In addition, the ordered arrangement of meiotic products makes it possible to determine

the recombination frequency between any particular gene and its centromere. The logic of the mapping technique is based on the feature of meiosis shown in Figure 4.27:

Homologous centromeres of parental chromosomes separate at the first meiotic division; the centromeres of sister chromatids separate at the second meiotic division.

Thus, in the absence of crossing-over between a gene and its centromere, the alleles of the gene (for example, *A* and *a*) must separate in the first meiotic division; this separation is called **first-division segregation.** If, instead, a crossover is formed between the gene and its centromere, the *A*

Figure 4.26
The life cycle of *Neurospora crassa*. The vegetative body consists of partly segmented filaments called hyphae. Conidia are asexual spores that function in the fertilization of organisms of the opposite mating type. A protoperithecium develops into a structure in which numerous cells undergo meiosis.

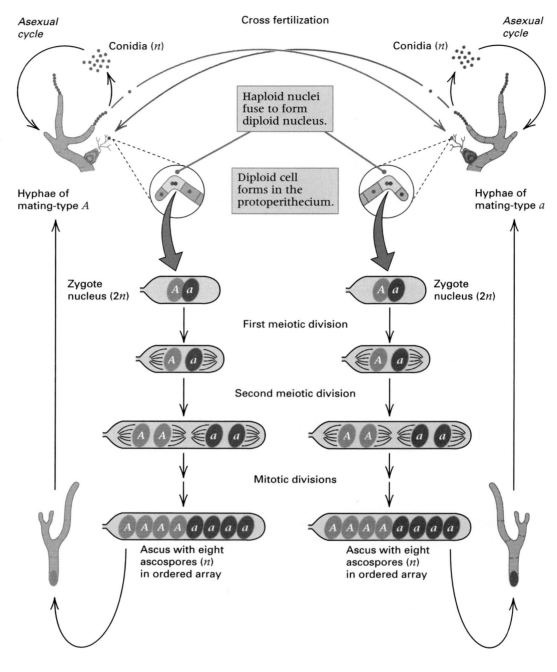

and *a* alleles do not become separated until the second meiotic division; this separation is called **second-division segregation.** The distinction between first-division and second-division segregation is shown in Figure 4.27. As shown in Figure 4.27A, only two possible arrangements of the products of meiosis can yield first-division segregation: *A A a a* and *a a A A*. However, four patterns of second-division segregation are possible because of the random arrangement of homologous chromosomes at metaphase I and of the chromatids at metaphase II. These four arrangements, shown in Figure 4.27B, are *A a A a, a A a A, A a a A,* and *a A A a*.

The percentage of asci with second-division segregation patterns for a gene can be used to map the gene with respect to its centromere. For example, let us assume that 30 percent of a sample of asci from a cross have a second-division segregation pattern for the *A* and *a* alleles. This means that 30 percent of the cells undergoing

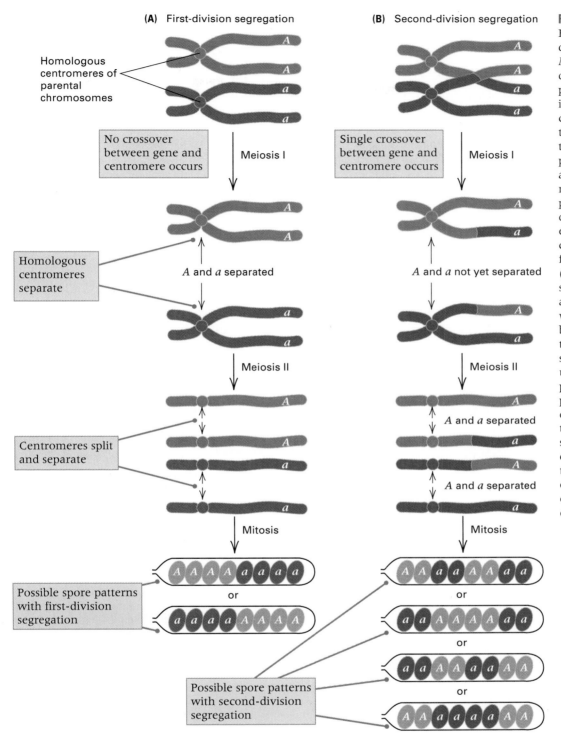

(A) First-division segregation

Homologous centromeres of parental chromosomes

No crossover between gene and centromere occurs → Meiosis I

A and *a* separated

Homologous centromeres separate

Meiosis II

Centromeres split and separate

Mitosis

Possible spore patterns with first-division segregation

or

(B) Second-division segregation

Single crossover between gene and centromere occurs → Meiosis I

A and *a* not yet separated

Meiosis II

A and *a* separated

A and *a* separated

Mitosis

or

or

or

Possible spore patterns with second-division segregation

Figure 4.27
First- and second-division segregation in *Neurospora*. (A) First-division segregation patterns are found in the ascus when crossing-over between the gene and centromere does not take place. The alleles separate (segregate) in meiosis I. Two spore patterns are possible, depending on the orientation of the pair of chromosomes on the first-division spindle. (B) Second-division segregation patterns are found in the ascus when crossing-over between the gene and the centromere delays separation of *A* from *a* until meiosis II. Four patterns of spores are possible, depending on the orientation of the pair of chromosomes on the first-division spindle and that of the chromatids of each chromosome on the second-division spindle.

meiosis had a crossover between the *A* gene and its centromere. Furthermore, in each cell in which crossing-over takes place, two of the chromatids are recombinant and two are nonrecombinant. In other words, a frequency of crossing-over of 30 percent corresponds to a recombination frequency of 15 percent. By convention, map distance refers to the frequency of recombinant meiotic products rather than to the frequency of cells with crossovers. Therefore, the map distance between a gene and its centromere is given by the equation

$$\frac{(1/2)(\text{Asci with second-division segregation patterns})}{\text{Total number of asci}} \times 100 \qquad (2)$$

This equation is valid as long as the gene is close enough to the centromere that multiple crossing-over can be neglected. Reliable linkage values are best determined for genes that are near the centromere. The location of more distant genes is then accomplished by the mapping of these genes relative to genes nearer the centromere.

If a gene is far from its centromere, crossing-over between the gene and its centromere will be so frequent that the *A* and *a* alleles become randomized with respect to the four chromatids. The result is that the six possible spore arrangements shown in Figure 4.27 are all equally frequent. This equal likelihood reflects the patterns that can result from choosing randomly among 2 *A* and 2 *a* spore pairs in the ascus, as shown by the branching diagram in Figure 4.28. Therefore, in the absence of chromatid interference,

The maximum frequency of second-division segregation asci is 2/3.

4.6 Mitotic Recombination

Genetic exchange can also take place in mitosis, although at a frequency about 1000-fold lower than in meiosis and probably by a somewhat different mechanism than meiotic recombination. The first evidence for mitotic recombination was obtained by Curt Stern in 1936 in experiments with *Drosophila*, but the phenomenon has been studied most carefully in fungi such as yeast and *Aspergillus,* in which the frequencies are higher than in most other organisms. Genetic maps can be constructed from mitotic recombination frequencies. In some organisms in which a sexual cycle is unknown, mitotic recombination is the only method of obtaining linkage data. In organisms in which both meiotic recombination and mitotic recom-

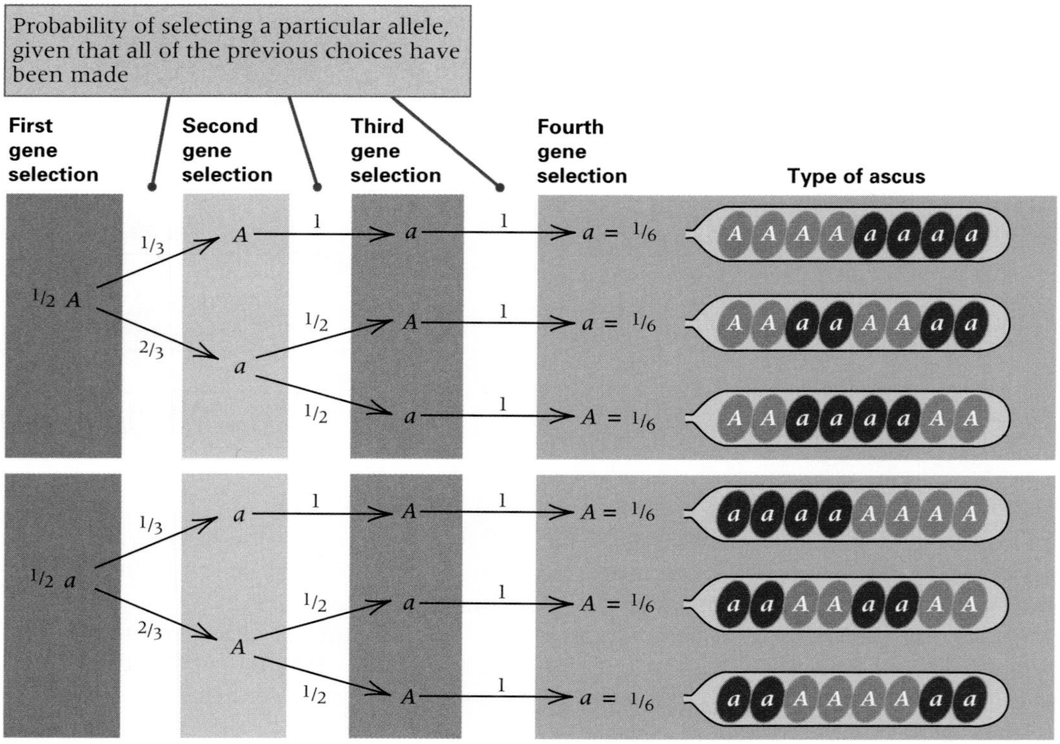

Figure 4.28
Diagram showing the result of free recombination between an allelic pair, *A* and *a,* and the centromere. The frequency of second-division segregation equals 2/3. This is the maximum frequency of second-division segregation, provided that there is no chromatid interference.

bination are found, mitotic recombination is always at a much lower frequency. In mitotic recombination, the relative map distances between particular genes sometimes correspond to those based on meiotic recombination frequencies, but for unknown reasons, the distances are often markedly different. The discrepancies may reflect different mechanisms of exchange and perhaps a nonrandom distribution of potential sites of exchange.

The genetic implications of mitotic recombination are illustrated in Figure 4.29. Each of the homologous chromosomes has two genetic markers distal to the site of the breakage and reunion. At the following anaphase, each centromere splits and the daughter cells receive one centromere of each color along with the chromatid attached to it. The separation can happen in two ways, as shown on the right. If the two nonexchanged and the two exchanged chromatids go together (top), the result of the exchange is not detectable genetically. However, if each nonexchanged chromatid goes with one of the exchanged chromatids, then the result is that the region of the chromosome distal to the site of the exchange becomes homozygous. With an appropriate configuration of genetic markers, the exchange is detectable as a **twin spot** in which cells of one homozygous genotype (in this example, *aB aB*) are adjacent to cells of a different homozygous genotype (in this example, *Ab Ab*). In the original experiments with *Drosophila*, Stern used the X-linked markers $A = y^+$ and $a = y$ (*y* is *yellow* body color) along with $B = sn^+$ and $b = sn$ (*sn* is *singed* bristles). In females, the result of mitotic recombination was observed as a twin spot in which a patch of cuticle with a yellow color and normal bristles was adjacent to a patch of cuticle with normal color and singed bristles.

The rate of mitotic recombination can be increased substantially through x-ray treatment and the use of certain mutations.

Figure 4.29

Mitotic recombination. Homologous chromosomes in prophase of mitosis are shown in light and dark blue. At anaphase, each centromere splits, and each daughter cell receives one centromere of each color and its attached chromatid. When a rare reciprocal exchange takes place between non-sister chromatids of homologous chromosomes, the daughter cells can be either of two types. If the nonexchanged chromatids and the exchanged chromatids go together (top right), both cells are genetically like the parent; in this example, *A b/a B*. If the nonexchanged and exchanged chromatids go together (bottom right), the result is that alleles distal to the point of exchange become homozygous. In this example, one daughter cell is *a B/a B* and the other *A b/A b*. Adjacent cell lineages are therefore homozygous for either *aa* or *bb* and can be detected phenotypically as a "twin spot" of *aa* somatic tissue adjacent to *bb* somatic tissue.

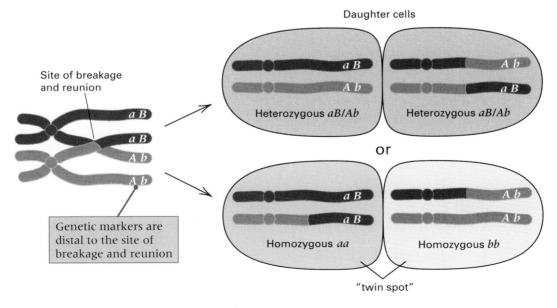

Mitotic recombination is a useful tool in genetics because it results in the production of **somatic mosaics,** organisms that contain two or more genetically different types of tissue. For example, the mosaic *Drosophila* females discussed in the previous paragraph consist of predominantly wildtype tissue but include some patches of $y^+ sn/y^+ sn$ and some patches of $y sn^+/y sn^+$. Each twin spot of mutant tissue derives from repeated division of a single cell that became homozygous through a mitotic exchange. Hence the mutant tissue resulting from mitotic recombination can be used to trace the movement and differentiation of particular cell lineages in development.

4.7
Recombination Within Genes

Genetic analyses and microscopic observations made before 1940 led to the view that a chromosome was a linear array of particulate units (the genes) joined in some way resembling a string of beads. The gene was believed to be the smallest unit of genetic material capable of alteration by mutation and the smallest unit of inheritance. Recombination had been seen in all regions of the chromosome but not within genes, and the idea of the indivisibility of the gene developed. This classical concept of a gene began to change when geneticists realized that mutant alleles of a gene might be the consequence of alterations at different mutable sites within the gene. The evidence that led to this idea was the finding of rare recombination events within several genes in *Drosophila*. The initial observation came in 1940 from investigations by C. P. Oliver of a gene in the X chromosome of *D. melanogaster* known as *lozenge* (*lz*). Mutant alleles of *lozenge* are recessive, and their phenotypic effects include a disturbed arrangement of the facets of the compound eye and a reduction in the eye pigments. Numerous *lz* alleles with distinguishable phenotypes are known.

Females heterozygous for two different *lz* alleles, one in each homologous chromosome, have lozenge eyes. When such heterozygous females were crossed with males that had one of the *lz* mutant alleles in the X chromosome, and large numbers of progeny were examined, flies with normal eyes were occasionally found. For example, one cross carried out was

$$\text{X} \frac{+ \quad lz^{BS}+ \quad +}{ct \quad + \quad lz^g \quad v} \text{X} \times \frac{+ \quad + lz^g \quad v}{\qquad} \begin{array}{l} \text{X} \\ \text{Y} \end{array}$$

in which lz^{BS} and lz^g are mutant alleles of *lozenge*, and *ct* (cut wing) and *v* (vermilion eye color) are genetic markers that map 7.7 units to the left and 5.3 units to the right, respectively, of the *lozenge* locus. From this cross, 134 males and females with wildtype eyes were found among more than 16,000 progeny, for a frequency of 8×10^{-3}. These exceptional progeny might have resulted from a reverse mutation of one or of the other of the mutant *lozenge* alleles to lz^+, but the observed frequency, though small, was much greater than the known frequencies of reverse mutations. The genetic constitution of the maternally derived X chromosomes in the normal-eyed offspring was also inconsistent with such an explanation: The male offspring had cut wings and the females were $ct/+$ heterozygotes. That is, all of the rare nonlozenge progeny had an X chromosome with the constitution

$$\underline{ct \qquad ++ \qquad +}$$

which could be accounted for by recombination between the two *lozenge* alleles. Proof of this conclusion came from the detection of the reciprocal recombinant chromosome

$$\underline{+ \quad lz^{BS}lz^g \quad v}$$

in five male progeny that had a lozenge phenotype distinctly different from the phenotype resulting from the presence of either a lz^{BS} or lz^g allele alone. The exceptional males also had vermilion eyes, as expected from reciprocal recombination. The observation of intragenic recombination indicated that genes have a fine structure and that the multiplicity of allelic forms of some genes might be due to mutations at different sites in the gene. As we have seen in Chapter 1, these sites are the nucleotide constituents that make up DNA.

4.8
A Closer Look at Complementation

The discovery of intragenic recombination emphasized the question "What is a gene?" A gene cannot be defined solely on the basis of a mutant phenotype. On the one hand, mutations in different genes can yield similar phenotypes. On the other hand, multiple alleles of the same gene can have quite different phenotypes, depending on the residual function of the mutation and the time- and tissue-specificity of its expression.

An alternative definition of a gene is based on recombination: Two mutations with similar phenotypes are alleles of the same gene if they fail to recombine. The discovery of recombination between alleles of *lozenge* and between alleles of other genes in *Drosophila* showed that this definition of a gene is unacceptable. Allelic mutations can recombine. The theoretical problem posed by recombination between alleles is indicated by the term *pseudoalleles* used to designate alleles that recombined. However, we now know that there is nothing "pseudo" about pseudoalleles. They are perfectly legitimate alleles that are far enough apart along the DNA to allow the detection of recombination.

The experimental resolution of the problem of the definition of a gene was the complementation test discussed in Chapter 2. The test is the same sort of test one would use to find out whether holes in each member of a pair of socks were in the same place. Put both socks on the same foot. If the holes are in the same place, bare skin shows through where the holes overlap. If the holes are in different places, each hole is covered by the fabric present in the other sock. In the complementation test, two recessive alleles are regarded as alleles of the same gene if, when they are present together in the same organism, they fail to complement each other and so yield a mutant phenotype, the "bare skin."

When a set of recessive mutations is tested in all pairwise combinations, the complementation test enables them to be separated into groups called complementation groups (Chapter 2). For example, in a genetic analysis of flower color in peas, a geneticist may isolate many independent recessive mutations, each of which causes the flowers to be white instead of purple. To group the mutations into complementation groups, strains that are homozygous for different mutations are crossed in all possible pairs. If the progeny of a cross has white flowers, then the mutations in the parents are said to fail to complement. These mutations are classified as belonging to the same complementation group, which means that they are alleles of the same gene. On the other hand, if the progeny of a cross between two mutant strains has purple flowers, then the mutations in the parents are said to complement. The mutations are assigned to different complementation groups, which means that they are alleles of different genes.

The molecular basis of complementation is illustrated in Figure 4.30. Figure 4.30A depicts the situation when two mutations, m_1 and m_2, each resulting in white flowers, are different mutations of the same gene. (The site of each mutation is indicated by a cross.) In a genotype in which m_1 and m_2 are in the *trans*, or repulsion, configuration, m_1 codes for a protein with one type of defect and m_2 codes for a protein with a different type of defect, but both types of protein are nonfunctional. Hence alleles in the same gene yield a mutant phenotype (white flowers), because neither mutation is able to support the production of a wildtype form of the protein.

When the mutations are alleles of different genes, the situation is as depicted in Figure 4.30B. Because the mutations are in different genes, the homozygous m_1 genotype is also homozygous for the wildtype allele of the gene mutated in m_2; likewise, the homozygous m_2 genotype is also homozygous for the wildtype allele of the gene mutated in m_1. Hence the same cross that yields the genotype m_1/m_2 in the case of allelic mutations (Figure 4.30A), in the case of different genes yields the genotype $m_1 + / + m_2$, where the plus signs represent the wildtype alleles of the genes not mutated in m_1 and m_2. In this case, the mutations do complement each other and yield an organism with a wildtype phenotype (purple flowers). With respect to the protein rendered defective by m_1, there is a functional form encoded by the wildtype allele brought in from the $m_2 m_2$ parent.

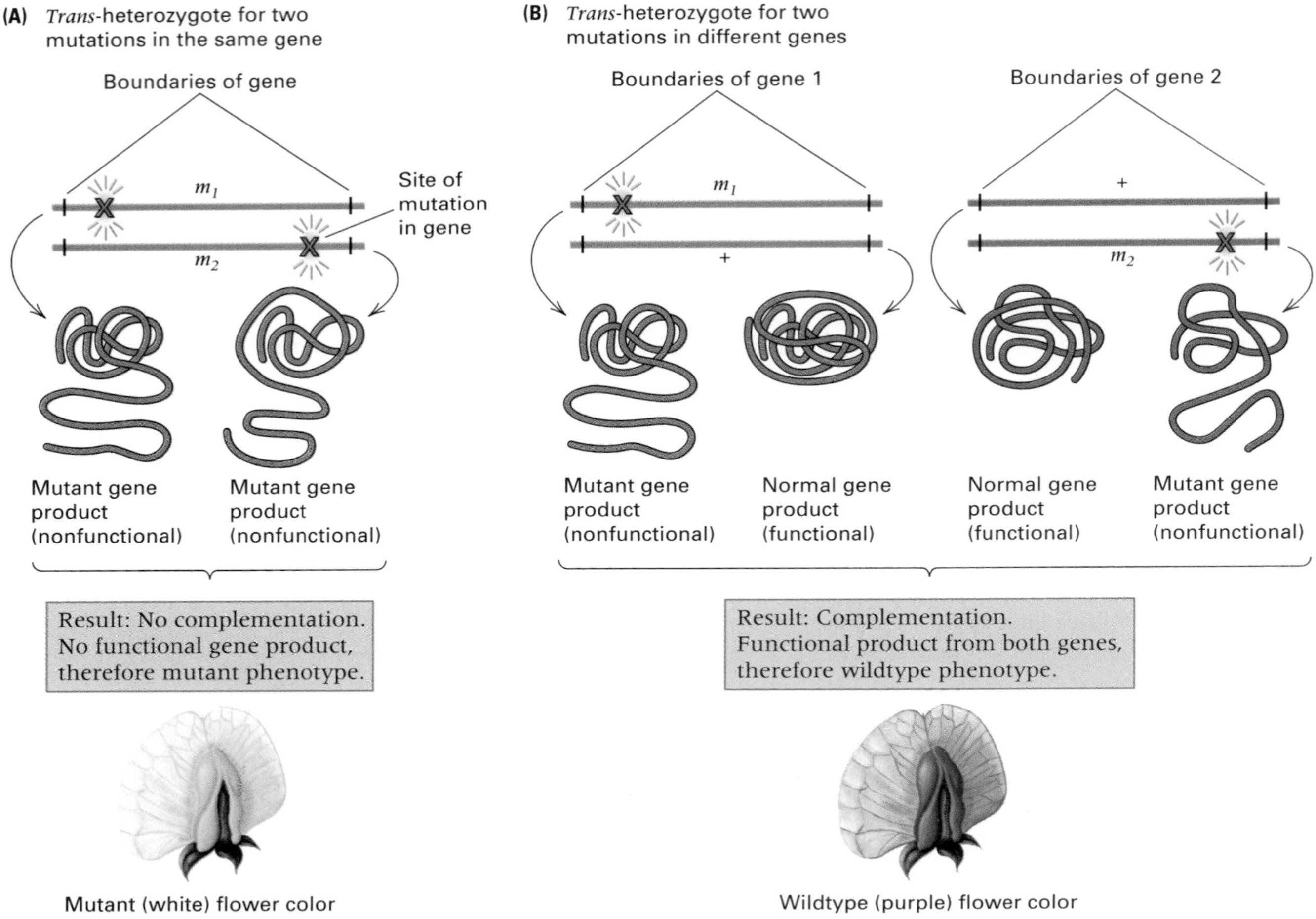

(A) *Trans*-heterozygote for two
mutations in the same gene

Boundaries of gene

m_1

Site of
mutation
in gene

m_2

Mutant gene
product
(nonfunctional)

Mutant gene
product
(nonfunctional)

Result: No complementation.
No functional gene product,
therefore mutant phenotype.

Mutant (white) flower color

(B) *Trans*-heterozygote for two
mutations in different genes

Boundaries of gene 1

m_1

+

Boundaries of gene 2

+

m_2

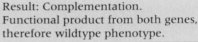

Mutant gene
product
(nonfunctional)

Normal gene
product
(functional)

Normal gene
product
(functional)

Mutant gene
product
(nonfunctional)

Result: Complementation.
Functional product from both genes,
therefore wildtype phenotype.

Wildtype (purple) flower color

Figure 4.30
The basis for interpre-
tation of a comple-
mentation test used to
determine whether
two mutations are
alleles of the same
gene (A) or alleles of
different genes (B).

With respect to the protein rendered defective by m_2, there is again a functional form
encoded by the wildtype allele brought in
from the m_1m_1 parent. Because a functional form of both proteins is produced,
the result is a normal phenotype, or complementation.

Complementation and recombination
must not be confused. Complementation is
inferred from the phenotype of the *trans*-
heterozygote having a mutation in each of
two different genes (Figure 4.30B). Recombination is inferred from genetic analysis of
the progeny of heterozygous genotypes.

Chapter Summary

Nonallelic genes located in the same chromosome tend to
remain together in meiosis rather than to undergo independent assortment. This phenomenon is called linkage. The
indication of linkage is a significant deviation from the
1 : 1 : 1 : 1 ratio of phenotypes in the progeny of a cross of
the form *Aa Bb* × *aa bb*. When alleles of two linked genes
segregate, more than 50 percent of the gametes produced
have parental combinations of the segregating alleles, and
fewer than 50 percent have nonparental (recombinant) combinations of the alleles. The recombination of linked genes
results from crossing-over, a process in which nonsister chromatids of the homologous chromosomes exchange corresponding segments in the first meiotic prophase. At the

molecular level, the Holliday model explains recombination
as a result of a single-strand break in each of two homologous DNA molecules, interchange of pairing partners
between the broken strands, and repair of the gaps, followed
by strand migration and a second breakage and reunion to
connect the originally unbroken DNA strands. This model,
although not correct in all its details, is nevertheless the basis
on which more accurate and complex models of recombination have been developed.

The frequency of recombination between different genes
can be used to determine the relative order and locations of
the genes in chromosomes. This type of analysis is called
genetic mapping. Distance between adjacent genes in such a

map (a genetic or linkage map) is defined to be proportional to the frequency of recombination between them; the unit of map distance (the map unit or centimorgan) is defined as 1 percent recombination. One map unit corresponds to a physical length of the chromosome in which a crossover event takes place, on the average, once in every 50 meiotic divisions. For short distances, map units are additive. (For example, for three genes with order *a b c*, if the map distances *a* to *b* and *b* to *c* are 2 and 3 map units, respectively, then the map distance *a* to *c* is 2 + 3 = 5 map units.) The recombination frequency underestimates actual genetic distance if the region between the genes being considered is too great. This discrepancy results from multiple crossover events, which yield either no recombinants or the same number produced by a single event. For example, two crossovers in the region between two genes may yield no recombinants, and three crossover events may yield recombinants of the same type as that from a single crossover.

When many genes are mapped in a particular species, they form linkage groups equal in number to the haploid chromosome number of the species. The maximum frequency of recombination between any two genes in a cross is 50 percent; this happens when the genes are in nonhomologous chromosomes and assort independently or when the genes are sufficiently far apart in the same chromosome that at least one crossover is formed between them in every meiosis. The map distance between two genes may be considerably greater than 50 centimorgans because the map distance is equal to half of the average number of crossovers per chromosome times 100. A mapping function is the mathematical relation between the genetic map distance across an interval and the observed percent recombination in the interval.

In many organisms, including model experimental organisms, agricultural animals and plants, and human beings, the genetic map includes hundreds or thousands of genetic markers distributed more or less uniformly throughout the euchromatin. Some of the most useful genetic markers are changes in base sequence present in wildtype organisms that are not associated with any phenotypic abnormalities. Prominent among these are nucleotide substitutions that create or destroy a particular cleavage site recognized by a restriction endonuclease. Such mutations can be detected because different chromosomes yield restriction fragments that differ in size according to the positions of the cleavage sites. Genetic variation of this type is called restriction fragment length polymorphism (RFLP). Most species also have considerable genetic variation in which one allele differs from the next according to the number of copies of a tandemly repeated DNA sequence it contains (variable number of tandem repeats, or VNTR). Many of the genetic markers in the genetic map of human beings consist of RFLPs and VNTRs. Because of their high degree of variability among people, VNTRs are also employed in DNA typing.

The four haploid products of each of a series of meiotic divisions can be used to analyze linkage and recombination in some species of fungi and unicellular algae. The method is called tetrad analysis. In *Neurospora* and related fungi, the meiotic tetrads are contained in a tubular sac, or ascus, in a linear order, which makes it possible to determine whether a pair of alleles segregated in the first or the second meiotic division. With such asci, it is possible to use the centromere as a genetic marker; in fact, the centromere serves as a reference point to which all genes in the same chromosome can be mapped. Linkage analysis in unordered tetrads is based on the frequencies of parental ditype (PD), nonparental ditype (NPD), and tetratype (TT) tetrads. The observation of NPD << PD is a sensitive indicator of linkage.

Recombination between alleles of the same gene demonstrated that genes are composed of a linear array of subunits, now known to be the nucleotides in the DNA. The functional definition of a gene is therefore not based on mutation or on recombination but on complementation (by means of the complementation test). If two recessive mutations, *a* and *b*, are present in a cell or organism in the *trans* configuration and the phenotype is wildtype (complementation), the mutations are in different genes. If the phenotype is mutant (noncomplementation), the mutations are in the same gene. At the molecular level, lack of complementation implies that allelic mutations impair the function of the same protein molecule.

Key Terms

ascospore
ascus
attached-X chromosome
centimorgan
chromosome interference
chromosome map
chromatid interference
coefficient of coincidence
compound-X chromosome
condensation
coupling
euchromatin
first-division segregation
frequency of recombination
genetic map
genetic marker

Haldane's mapping function
heterochromatin
Kosambi's mapping function
linkage
linkage group
linkage map
locus
map unit
mapping function
negative chromatid interference
nonparental ditype (NPD)
parental combinations
parental ditype (PD)
positive chromatid interference
random-spore analysis
recombinant

recombination
repulsion
restriction endonuclease
restriction fragment length polymorphism (RFLP)
second-division segregation
simple tandem repeat polymorphism (STRP)
somatic mosaics
tetratype (TT)
three-point cross
twin spot
variable number of tandem repeats (VNTR)

GeNETics on the web will introduce you to some of the most important sites for finding genetic information on the Internet. To complete the exercises below, visit the Jones and Bartlett home page at

http://www.jbpub.com/genetics

Select the link to *Genetics: Principles and Analysis* and then choose the link to *GeNETics on the web.* You will be presented with a chapter-by-chapter list of highlighted keywords.

GeNETics EXERCISES

Select the highlighted keyword in any of the exercises below, and you will be linked to a web site containing the genetic information necessary to complete the exercise. Each exercise suggests a specific, written report that makes use of the information available at the site. This report, or an alternative, may be assigned by your instructor

1. The absence of crossing-over in males of *Drosophila* is a great convenience for detecting linkage. You can see this for yourself at the keyword site. Mate females having a black body and dumpy wings with wildtype males. Then cross the F_1 male progeny with the black, dumpy females. What are the phenotypes of the progeny and in what numbers do they occur? Does this result tell you anything about the linkage of black and dumpy? If assigned to do so, mate wildtype females with black, dumpy males, and then cross the F_1 female progeny with the black, dumpy males. Make a diagram of the matings, indicating all genotypes and phenotypes and the numbers observed in each generation of progeny. Calculate the frequency of recombination between black and dumpy.

2. The keyword Saccharomyces will connect you with the genome database of the yeast *Saccharomyces cerevisiae*. Select *Maps* for graphical views of yeast chromosomes and *Genetic Map* to access the linkage map of each chromosome individually. If assigned to do so, use the site to find the relation between map length (in centimorgans) and physical length (in base pairs) of each of the yeast chromsosomes. Make a bar graph of this information.

3. Detecting genetic linkage in human pedigrees requires special methods because human sibships are typically quite small. For each

Review the Basics

- What does it mean to say that two genes have a recombination frequency of 12 percent? How is the recombination frequency estimated?

- For a region between two genes in which no more than one crossing over can take place in each cell undergoing meiosis, the frequency of recombination is equal to one-half the frequency of meiotic cells in which a chiasma occurs in the region. Explain why the factor of one-half occurs when we compare recombination frequency to chiasma frequency.

- What is a double crossover? How many different kinds of double crossovers are possible? Draw a diagram of each

kind of double crossover in a bivalent in a region between two genes, each heterozygous.

- If each cell undergoing meiosis had two crossovers between the genes, and the types of double crossovers were equally frequent, what would be the frequency of recombination between the genes?

- What is interference? In a region of high interference, would you observe more or fewer double crossovers?

- Explain why the recovery of ordered tetrads allows any gene to be mapped relative to the centromere of the chromosome on which the gene is located.

Guide to Problem Solving

Problem 1: In *Drosophila*, the genes *cn* (cinnabar eyes) and *bw* (brown eyes) are located in the same chromosome but so far apart that they appear unlinked. Compare the possible offspring produced by the cross

$$\frac{cn\ bw}{+\ +}\ ♀ \times \frac{cn\ bw}{cn\ bw}\ ♂$$

and the cross

$$\frac{cn\ bw}{cn\ bw}\ ♀ \times \frac{cn\ bw}{cn\ bw}\ ♂$$

The + symbols denote the wildtype alleles. Both *cn* and *bw* are recessive. Flies with the genotype *cn cn* have cinnabar

genetic marker in a pedigree in which a trait, for example, a genetic disease, is present, the likelihood of linkage is assessed by a "lod score." The principal advantage of lod scores is that data from individual pedigrees can be summed. Use the keyword to learn more about this approach and examine the sample lod score calculation for a simple pedigree. If assigned to do so, write a 250-word summary of the method and the logic on which it is based.

MUTABLE SITE EXERCISES

The Mutable Site Exercise changes frequently. Each new update includes a different exercise that makes use of genetics resources available on the World Wide Web. Select the Mutable Site for Chapter 4, and you will be linked to the current exercise that relates to the material presented in this chapter.

PIC SITE

The Pic Site showcases some of the most visually appealing genetics sites on the World Wide Web. To visit the showcase genetics site, select the Pic Site for Chapter 4.

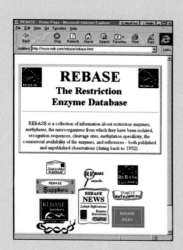

(bright red) eyes, *bw bw* flies have brown eyes, and the double mutant *cn bw/cn bw* has white eyes.

Answer: The key to this problem is to remember that crossing-over does not take place in *Drosophila* males. When the male is *cn bw/+ +*, his gametes are 1/2 *cn bw* and 1/2 + + , so the progeny are *cn bw/cn bw* (white eyes) and + +/*cn bw* (wildtype eyes) in equal proportions. Crossing-over does happen in female *Drosophila*, but *cn* and *bw* are so far apart that they undergo independent assortment. When the female is *cn bw/+ +*, her gametes are 1/4 *cn bw*, 1/4 *cn* +, 1/4 + *bw*, and 1/4 + + , so the progeny are *cn bw/cn bw* (white eyes), *cn* +/*cn bw* (bright red), + *bw/cn bw* (brown), and + +/*cn bw* (wildtype) in equal proportions.

Problem 2: One of the earliest reported cases of linkage was in peas in the year 1905 between the gene for purple versus red flowers and the gene for elongate versus round pollen. These genes are separated by about 12 map units. Plants of genotype *B−* have purple flowers, and plants of the genotype *bb* have red flowers. Plants of genotype *E−* have elongate pollen, and those of genotype *ee* have round pollen. Determine what genotypes and phenotypes among the progeny would be expected from crosses of F_1 hybrids × red, round when the F_1 hybrids are obtained as follows

(a) purple, elongate × red, round → purple, elongate F_1

(b) purple, round × red, elongate → purple, elongate F_1

Answer: The red, round parents have genotype *b e/b e*. You must deduce the genotype of the F_1 parent. Although both F_1 hybrids have the dominant purple, elongate phenotype, they are different kinds of double heterozygotes. **(a)** In this mating, both the *b* and *e* alleles come from the red, round parent, so the F_1 genotype is the *cis*, or coupling, heterozygote *B E/b e*. **(b)** In this mating, the *b* allele comes from one parent and the *e* allele from the other, so the F_1 genotype is the *trans*, or repulsion, heterozygote *B e/b E*. To calculate the types of progeny expected, use the fact that 1 map unit corresponds to 1 percent recombination: Because the genes are 12 map units apart, 12 percent of the gametes will be recombinant (6 percent of each of the reciprocal classes) and 88 percent will be nonrecombinant (44 percent of each of the reciprocal classes). Therefore, the two types of heterozygotes produce gametes with the following expected percentages:

Gametes	*BE/be*	*Be/bE*
BE	44	6
Be	6	44
bE	6	44
be	44	6

The answers to the specific questions are as follows.

(a) *BE/be* (purple, elongate) 44 percent; *Be/be* (purple, round) 6 percent; *bE/be* (red, elongate) 6 percent; *be/be* (red, round) 44 percent.

(b) BE/be (purple, elongate) 6 percent; Be/be (purple, round) 44 percent; bE/be (red, elongate) 44 percent; be/be (red, round) 6 percent.

Problem 3: In *Drosophila*, the genes *ct* (cut wing margin), *y* (yellow body), and *v* (vermilion eye color) are X-linked. Females heterozygous for all three markers were mated with wildtype males, and the following male progeny were obtained. As is conventional in *Drosophila* genetics, the wild-type allele of each gene is designated by a + sign in the appropriate column.

ct	*y*	*v*	4
ct	*y*	+	93
ct	+	*v*	54
ct	+	+	349
+	*y*	*v*	331
+	*y*	+	66
+	+	*v*	97
+	+	+	6
		Total	1000

(a) What was the genotype of the female parents?

(b) What is the order of the genes?

(c) Calculate the recombination frequencies and the interference between the genes.

(d) Draw a genetic map of the genes.

Answer: This is a conventional kind of three-point linkage problem. The way to solve it is to start by deducing the genotype of the triply heterozygous parent. This is done by noting that *the most common classes of offspring are the nonrecombinants*, in this case *ct* + + and +*y v*. Then deduce the order of the genes. This is done by noting that *the least common classes of offspring are the double recombinants*, in this case *ct y v* and + + +. Moreover, *each class of double recombinants will be identical with one of the nonrecombinants except for the marker that is in the middle*. In this case, comparing *ct y v* with + *y v* and + + + with *ct* + + shows that *ct* is in the middle. **(a)** Therefore, the answer to part (a) is that the mothers' genotype was + *ct* +/*y* + *v*. **(b)** The answer to part (b) is that *ct* is in the middle. (It is not possible to determine from these data whether the gene order is *y ct v* or *v ct y*.)

The next step is to rearrange the data with the genes in the correct order, with reciprocal chromosomes adjacent and with nonrecombinants at the top, single recombinants for region 1 and 2 in the middle, and nonrecombinants at the bottom.

+	*ct*	+	349	Nonrecombinant
y	+	*v*	331	Nonrecombinant
y	*ct*	+	93	Region 1 crossing-over
+	+	*v*	97	Region 1 crossing-over
+	*ct*	*v*	54	Region 2 crossing-over
y	+	+	66	Region 2 crossing-over
y	*ct*	*v*	4	Region 1 + Region 2 crossing-over
+	+	+	6	Region 1 + Region 2 crossing-over

(c) Now, to answer part (c), calculate the recombination frequencies. The most common mistake is to ignore the double recombinants. In fact, the double recombinants are recombinant in *both* regions and so must be counted twice—once in calculating the recombination frequency in region 1 and again in calculating the recombination frequency in region 2. The frequency of recombination in region 1 (*y−ct*) is $(93 + 97 + 4 + 6)/1000 = 0.20$, or 20 percent. The frequency of recombination in region 2 (*ct−v*) is $(54 + 66 + 4 + 6)/1000 = 0.13$, or 13 percent. To calculate the interference, note that the expected number of double recombinants is $0.20 \times 0.13 \times 1000 = 26$, whereas the actual number is 10. The coincidence is the ratio $10/26 = 0.385$, and therefore the interference across the region is $1 - 0.385 = 0.615$. **(d)** For part (d), the genetic map based on these data is shown below.

Problem 4: In the yeast *Saccharomyces cerevisiae*, the gene *MET14* codes for an enzyme used in synthesis of the amino acid methionine and is very close to the centromere of chromosome XI. For simplicity, denote *MET14* with the symbol *a*, and suppose that *a* is used in mapping a new mutation—say, *b*—whose location in the genome is unknown. Diploids are made by mating *a B* × *A b*, in which the uppercase symbols denote the wildtype alleles, and asci are examined. In this mating, what kinds of asci correspond to parental ditype (PD), nonparental ditype (NPD) and tetratype (TT)? What relative proportions of PD, NPD and TT asci would be expected in each of the following cases:

(a) *b* closely linked with *a*?

(b) *b* on a different chromosome from the one *a* is on but close to the centromere?

(c) *b* on a different chromosome from the one *a* is on and far from the centromere?

Answer: The genotype of the diploid is the *trans*, or repulsion, heterozygote *a B/A b*, so the parental ditype (PD) tetrads contain two *a B* spores and two *A b* spores; the nonparental ditype (NPD) tetrads contain two *a b* spores and two *A B* spores; and the tetratype (TT) tetrads contain one spore each of *a B*, *A b*, *a b*, and *A B*. In this problem, the specific questions illustrate the principle that *any marker tightly linked to its centromere in a fungus with unordered tetrads serves to identify first-division and second-division segregation of every other gene in the organism.* This principle makes possible the mapping of other genes with respect to their centromeres. Reasoning through problems of tetrad analysis is best approached by considering the possible chromosome configurations during the meiotic divisions. The various possibilities are illustrated in the adjoining figure, in which parts A through C correspond to questions (a) through (c).

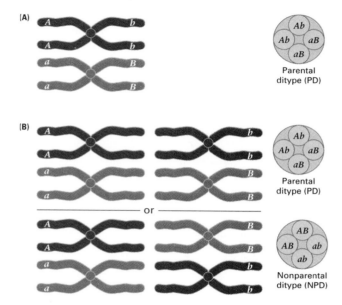

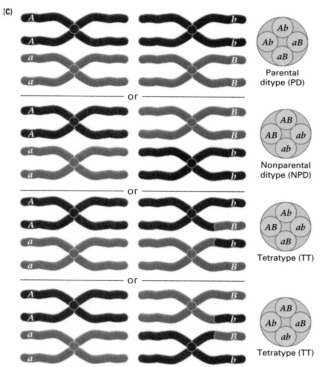

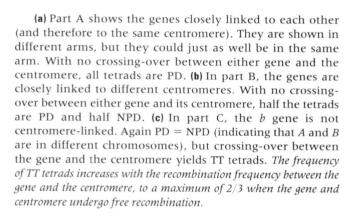

(a) Part A shows the genes closely linked to each other (and therefore to the same centromere). They are shown in different arms, but they could just as well be in the same arm. With no crossing-over between either gene and the centromere, all tetrads are PD. (b) In part B, the genes are closely linked to different centromeres. With no crossing-over between either gene and its centromere, half the tetrads are PD and half NPD. (c) In part C, the *b* gene is not centromere-linked. Again PD = NPD (indicating that *A* and *B* are in different chromosomes), but crossing-over between the gene and the centromere yields TT tetrads. *The frequency of TT tetrads increases with the recombination frequency between the gene and the centromere, to a maximum of 2/3 when the gene and centromere undergo free recombination.*

Analysis and Applications

4.1 Each of two recessive mutations, m_1 and m_2, can produce albino guinea pigs. The genes are close together in one chromosome and may be allelic. The cross $m_1m_1 \times m_2m_2$ produces progeny with wildtype pigmentation. Are the mutations in the same gene?

4.2 What gametes are produced by an individual of genotype $A\ b/a\ B$ if the genes are

(a) in different chromosomes?

(b) in the same chromosome with no recombination between them?

4.3 What gametes are produced by an individual of genotype $A\ b/a\ B$ if recombination can occur between the genes? Which gametes are the most frequent?

4.4 In the absence of recombination, what genotypes of progeny are expected from the cross $A\ B/a\ b \times A\ b/a\ B$? If *A* and *B* are dominant, what ratio of phenotypes is expected?

4.5 Which of the following arrays represents the *cis* configuration of the alleles *A* and *B*: $a\ b/A\ B$ or $a\ B/A\ b$?

4.6 What gametes, and in what ratios, are produced by male and female *Drosophila* of genotype $A\ B/a\ b$ when the genes are present in the same chromosome and the frequency of recombination between them is 5 percent?

4.7 If a large number of genes were mapped, how many linkage groups would be found in

(a) a haploid organism with 17 chromosomes per somatic cell?

(b) a bacterial cell with a single, circular DNA molecule?

(c) the rat, with 42 chromosomes per diploid somatic cell?

4.8 If the recombination frequency between genes A and B is 6.2 percent, what is the distance between the genes in map units in the linkage map?

4.9 Consider an organism heterozygous for two genes located in the same chromosome, $A\,B/a\,b$. If a single crossing-over occurs between the genes in every cell undergoing meiosis, and no multiple crossing-overs occur between the genes, what is the recombination frequency between the genes? What effect would multiple crossing-over have on the recombination frequency?

4.10 Normal plant height in wheat requires the presence of either or both of two dominant alleles, A and B. Plants homozygous for both recessive alleles ($a\,b/a\,b$) are dwarfed but otherwise normal. The two genes are in the same chromosome and recombine with a frequency of 16 percent. From the cross $A\,b/a\,B \times A\,B/a\,b$, what is the expected frequency of dwarfed plants among the progeny?

4.11 In the yellow-fever mosquito, *Aedes aegypti*, sex is determined by a single pair of alleles, M and m. Heterozygous Mm mosquitos are male, homozygous mm are female. The recessive gene bz for bronze body color (the dominant phenotype is black) is linked to the sex-determining gene and recombines with it at a frequency of 6 percent. In the cross of a heterozygous $bz^+\,M/bz\,m$ male with a bronze female, what genotypes and phenotypes of progeny are expected?

4.12 Two genes in chromosome 7 of corn are identified by the recessive alleles gl (glossy), determining glossy leaves, and ra (ramosa), determining branching of ears. When a plant heterozygous for each of these alleles was crossed with a homozygous recessive plant, the progeny consisted of the following genotypes with the numbers of each indicated:

$Gl\,ra/gl\,ra$	88	$gl\,Ra/gl\,ra$	103
$Gl\,Ra/gl\,ra$	6	$gl\,ra/gl\,ra$	3

Calculate the frequency of recombination between these genes.

4.13 The recessive alleles b and cn of two genes in chromosome 2 of *Drosophila* determine black body color and cinnabar eye color, respectively. If the genes are 8 map units apart, what are the expected genotypes and phenotypes and their relative frequencies among the progeny of the cross $++/b\,cn \times ++/b\,cn$? (Remember that crossing-over does not occur in *Drosophila* males.)

4.14 In a testcross of an individual heterozygous for each of three linked genes, the most frequent classes of progeny were $A\,B\,c/a\,b\,c$ and $a\,b\,C/a\,b\,c$, and the least frequent classes were $A\,B\,C/a\,b\,c$ and $a\,b\,c/a\,b\,c$. What was the genotype of the triple heterozygote parent, and what is the order of the genes?

4.15 Construct a map of a chromosome from the following recombination frequencies between individual pairs of genes: $r–c$, 10; $c–p$, 12; $p–r$, 3; $s–c$, 16; $s–r$, 8. You will discover that the distances are not strictly additive. Why aren't they?

4.16 Yellow versus gray body color in *D. melanogaster* is determined by the alleles y and y^+, vermilion versus wildtype eyes by the alleles v and v^+, and singed versus straight bristles by the alleles sn and sn^+. When females heterozygous for each of these X-linked genes were testcrossed with yellow, vermilion, singed males, the following classes and numbers of progeny were obtained:

yellow, vermilion, singed	53
yellow, vermilion	108
yellow, singed	331
yellow	5
vermilion, singed	3
vermilion	342
singed	95
wildtype	63

(a) What is the order of the three genes? Construct a linkage map with the genes in their correct order, and indicate the map distances between the genes.

(b) How does the frequency of double crossovers observed in this experiment compare with the frequency expected if crossing-over occurs independently in the two chromosome regions? Determine the coefficient of coincidence and the interference.

4.17 In corn, the alleles C and c result in colored versus colorless seeds, Wx and wx in nonwaxy versus waxy endosperm, and Sh and sh in plump versus shrunken endosperm. When plants grown from seeds heterozygous for each of these pairs of alleles were testcrossed with plants from colorless, waxy, shrunken seeds, the progeny seeds were as follows:

colorless, nonwaxy, shrunken	84
colorless, nonwaxy, plump	974
colorless, waxy, shrunken	20
colorless, waxy, plump	2349
colored, waxy, shrunken	951
colored, waxy, plump	99
colored, nonwaxy, shrunken	2216
colored, nonwaxy, plump	15
Total	6708

Determine the order of the three genes, and construct a linkage map showing the genetic distances between adjacent genes.

4.18 The accompanying diagram summarizes the recombination frequencies observed in a large experiment to study three linked genes.

What was the observed frequency of double crossing-over in this experiment? Calculate the interference.

4.19 The recessive mutations *b* (black body color), *st* (scarlet eye color), and *hk* (hooked bristles) identify three autosomal genes in *D. melanogaster*. The following progeny were obtained from a testcross of females heterozygous for all three genes.

black, scarlet	243
black	241
black, hooked	15
black, hooked, scarlet	10
hooked	235
hooked, scarlet	226
scarlet	12
wildtype	18

What conclusions are possible concerning the linkage relations of these three genes? Calculate any appropriate map distances.

4.20 In the nematode *Caenorhabditis elegans,* the mutations *dpy*-21 (dumpy) and *unc*-34 (uncoordinated) identify linked genes that affect body conformation and coordination of movement. The frequency of recombination between the genes is 24 percent. If the heterozygote *dpy*-21 +/+ *unc*-34 undergoes self-fertilization (the normal mode of reproduction in this organism), what fraction of the progeny is expected to be both dumpy and uncoordinated?

4.21 Dark eye color in rats requires the presence of a dominant allele of each of two genes, *R* and *P*. Animals homozygous for either or both of the recessive alleles have light-colored eyes. In one experiment, homozygous dark-eyed rats were crossed with doubly recessive light-eyed rats, and the resulting F_1 animals were testcrossed with rats of the homozygous light-eyed strain. The progeny from the testcross consisted of 628 dark-eyed and 889 light-eyed rats. In another experiment, *R p/R p* animals were crossed with *r P/r P* animals, and the F_1 progeny were crossed with animals from an *r p/r p* strain. The progeny consisted of 86 dark-eyed and 771 light-eyed rats. Determine whether the genes are linked, and if they are, estimate the frequency of recombination between the genes from each of the experiments.

4.22 A *Drosophila* geneticist exposes flies to a mutagenic chemical and obtains nine mutations in the X chromosome that are lethal when homozygous. The mutations are tested in pairs in complementation tests, with the results shown in the accompanying table. A + indicates complementation (that is, flies carrying both mutations survive), and a − indicates noncomplementation (flies carrying both mutations die). How many genes (complementation groups) are represented by the mutations, and which mutations belong to each complementation group?

	1	2	3	4	5	6	7	8	9
1	−	+	−	+	+	+	−	+	+
2		−	+	+	+	+	+	+	+
3			−	+	+	+	−	+	+
4				−	+	−	+	+	+
5					−	+	+	+	−
6						−	+	+	+
7							−	+	+
8								−	+
9									−

4.23 The following classes and frequencies of ordered tetrads were obtained from the cross $a^+ b^+ \times a\ b$ in *Neurospora*. (Only one member of each pair of spores is shown.)

Spore pair				Number of asci
1–2	3–4	5–6	7–8	
a^+b^+	a^+b^+	$a\ b$	$a\ b$	1766
a^+b^+	$a\ b$	a^+b^+	$a\ b$	220
a^+b^+	$a\ b^+$	a^+b	$a\ b$	14

What is the order of the genes in relation to the centromere?

4.24 Three genes in chromosome 9 of corn determine shrunken *sh* versus plump *Sh* kernels, waxy *wx* versus nonwaxy *Wx* endosperm, and glossy *gl* versus nonglossy *Gl* leaves. The genetic map of this chromosomal region is *sh*−30−*wx*−10−*gl*. From a plant of genotype *Sh wx Gl/sh Wx gl*, what is the expected frequency of *sh wx gl* gametes:

(a) in the absence of interference?

(b) assuming 60 percent interference?

4.25 The following spore arrangements were obtained in the indicated frequencies from ordered tetrads in a cross between a *Neurospora* strain *com val* (*c v*), which exhibits a compact growth form and is unable to synthesize the amino acid valine, and a wildtype strain, ++. (Only one member of each pair of spores is shown.)

Spore pair	Ascus composition				
1–2	c v	c +	c v	+ v	c v
3–4	c v	c +	c +	c +	+ v
5–6	+ +	+ v	+ v	c v	c +
7–8	+ +	+ v	+ +	+ +	+ +
Number:	34	36	20	1	9

What can you conclude about the linkage and location of the genes with respect to each other and to the centromere?

Challenge Problems

4.26 In *Neurospora* several genes are needed for production of black pigment in spores. Recessive alleles at any one of these can result in tan spores. Six possible arrangements of spore colors are possible in crosses segregating for one of these allelic pairs, as shown in the accompanying diagram. Three different allelic pairs were analyzed in crosses 1, 2, and 3, respectively. The results are tabulated in the illustration.

Asci observed	Cross 1	Cross 2	Cross 3
①	6	0	20
②	6	0	20
③	48	60	20
④	6	0	20
⑤	6	0	20
⑥	48	60	20

For each cross, indicate how the different arrangements arise and what you can deduce about the location of the segregating gene. In crosses in which it is relevant, calculate a map distance.

4.27 In baker's yeast (*Saccharomyces cerevisiae*), which produces unordered asci, the following cross was made:

$$leu2 \quad trp1 \quad + \quad \times \quad + \quad + \quad met14$$

The resulting diploid was induced to undergo meiosis, and the asci were dissected and spore clones analyzed. The following tetrad types were found, with the observed number of each shown below. For convenience, *leu2, trp1,* and *met14* are abbreviated *l, t,* and *m,* respectively.

l t m	l t +	+ t +	+ t m	+ t m	+ t +
l t m	l t +	+ t +	+ t m	l t m	l t +
+ + +	+ + m	l + m	l + +	+ + +	+ + m
+ + +	+ + m	l + m	l + +	l + +	l + m
230	235	215	220	54	46

Analyze these data as fully as you can with respect to the linkage of each gene with respect to every other gene and with respect to its centromere. Where a genetic linkage is indicated, calculate the map distance.

4.28 In baker's yeast (*Saccharomyces cerevisiae*), which produces unordered asci, the following cross was made:

$$rad6 \quad trp5 \quad leu1 \quad + \quad \times \quad + \quad + \quad + \quad met14$$

The resulting diploid was induced to undergo meiosis, the asci were dissected, and the spore clones analyzed. The following tetrad types were found in the numbers shown below each tetrad. For convenience, *rad6, trp5, leu1,* and *met14* are abbreviated as *r, t, l,* and *m,* respectively.

r t l +	r t l m	r t l +	r t l m	r t l +
r t l +	r t l m	+ t l +	+ t l m	+ + l +
+ + + m	+ + + +	r + + m	r + + +	r t + m
+ + + m	+ + + +	+ + + m	+ + + +	+ + + m
188	206	105	92	140

r t l m	r t l +	r t l m	r + l +	+ t l m
+ + l m	+ + + +	+ + l m	+ t l +	+ + l m
r t + +	r t l m	+ t + +	r t + m	r + + +
+ + + +	+ + + m	r + + +	+ + + m	r t + +
154	109	3	2	1

Use what you learned about *met14* in Challenge Problem 4.27 to analyze these data as fully as you can for linkage of the genes to one another and of each to its respective centromere. When genetic linkage is indicated, calculate the map distance.

Botstein, D., R. L. White, M. Skolnick, and R. W. Davis. 1980. Construction of a genetic linkage map in man using restriction fragment length polymorphisms. *American Journal of Human Genetics* 32: 314.

Carlson, E. A. 1987. *The Gene: A Critical History.* 2d ed. Philadelphia: Saunders.

Creighton, H. S., and B. McClintock. 1931. A correlation of cytological and genetical crossing over in *Zea mays.* *Proceedings of the National Academy of Sciences, USA* 17: 492.

Fincham, J. R. S., P. R. Day, and A. Radford. 1979. *Fungal Genetics.* Oxford, England: Blackwell.

Green, M. M. 1996. The "Genesis of the White-Eyed Mutant" in *Drosophila melanogaster*: A reappraisal. *Genetics* 142: 329.

Kohler, R. E. 1994. *Lords of the Fly.* University of Chicago Press.

Levine, L. 1971. *Papers on Genetics.* St. Louis, MO: Mosby.

Lewis, E. B. 1995. Remembering Sturtevant. *Genetics* 141: 1227.

Morton, N. E. 1995. LODs past and present. *Genetics* 140: 7.

Stewart, G. D., T. J. Hassold, and D. M. Kurnit. 1988. Trisomy 21: Molecular and cytogenetic studies of nondisjunction. *Advances in Human Genetics* 17: 99.

Sturtevant, A. H. 1965. *A History of Genetics.* New York: Harper & Row.

Sturtevant, A. H., and G. W. Beadle. 1962. *An Introduction to Genetics.* New York: Dover.

Voeller, B. R., ed. 1968. *The Chromosome Theory of Inheritance: Classical Papers in Development and Heredity.* New York: Appleton-Century-Crofts.

White, R., and J.-M. Lalouel. 1988. Chromosome mapping with DNA markers. *Scientific American,* February.

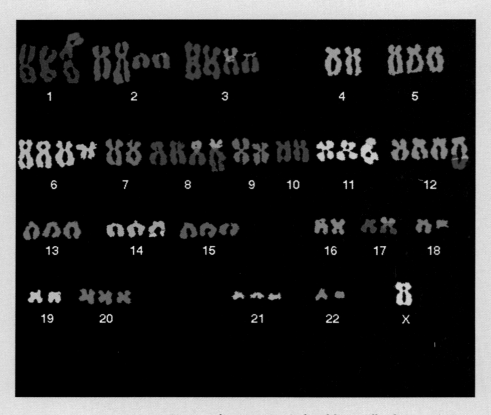

Human chromosomes isolated from cells of an ovarian cancer. Each chromosome is labeled with a different color. Note the many abnormal chromosomes, containing two or more colors, that have been formed by breakage of other chromosomes and reunion of their broken parts in abnormal combinations. Note also that this cell has 63 chromosomes (instead of the normal 46) and that one of the X chromosomes is missing. [Courtesy of David C. Ward and Michael R. Speicher.]

The Molecular Structure and Replication of the Genetic Material

PRINCIPLES

- A DNA strand is a polymer of A, T, G and C deoxyribonucleotides joined 3' to 5' by phosphodiester bonds.
- The two DNA strands in a duplex are held together by hydrogen bonding between the A−T and G−C base pairs.
- DNA replication is semiconservative and takes place only in the 5' to 3' direction; successive nucleotides are added only at the 3' end.
- Each type of restriction endonuclease enzyme cleaves double-stranded DNA at a particular sequence of bases usually four or six nucleotides in length.
- Separated strands of DNA or RNA that are complementary in nucleotide sequence can come together (hybridize) spontaneously to form duplexes.
- In the polymerase chain reaction, short oligonucleotide primers are used in successive cycles of DNA replication to amplify selectively a particular region of a DNA duplex.
- The DNA fragments produced by a restriction enzyme can be separated by electrophoresis, isolated, sequenced, and manipulated in other ways.

CONNECTIONS

CONNECTION: The Double Helix
James D. Watson and Francis H. C. Crick 1953
A structure for deoxyribose nucleic acid

CONNECTION: Replication by Halves
Matthew Meselson and Franklin W. Stahl 1958
The replication of DNA in Escherichia coli

Analysis of the patterns of inheritance and even the phenotypic expression of genes reveals nothing about gene structure at the molecular level, how genes are copied to yield exact replicas of themselves, or how they determine cellular characteristics. Understanding these basic features of heredity requires identification of the chemical nature of the genetic material and the processes through which it is replicated. In Chapter 1, we reviewed the experimental evidence demonstrating that the genetic material is DNA. The structure of DNA was described as a helix of two paired, complementary strands, each composed of an ordered string of nucleotides bearing A (adenine), T (thymine), G (guanine), or cytosine (C). Watson-Crick base pairing between A and T and between G and C in the complementary strands holds the strands together. The complementarity also holds the key to replication, because each strand can serve as a template for the synthesis of a new complementary strand. In this chapter, we take a closer look at DNA structure and its replication. We also consider how our knowledge of DNA structure and replication has been used in the development of laboratory techniques for isolating fragments that contain genes or parts of genes of particular interest and for determining the sequence of bases in DNA fragments.

5.1
The Chemical Composition of DNA

DNA is a polymer—a large molecule that contains repeating units—composed of 2'-deoxyribose (a five-carbon sugar), phosphoric acid, and the four nitrogen-containing bases denoted A, T, G, and C. The chemical structures of the bases are shown in Figure 5.1. Note that two of the bases have a double-ring structure; these are called **purines.** The other two bases have a single-ring structure; these are called **pyrimidines.**

- The purine bases are adenine (A) and guanine (G).

- The pyrimidine bases are thymine (T) and cytosine (C).

In DNA, each base is chemically linked to one molecule of the sugar deoxyribose, forming a compound called a **nucleoside.** When a phosphate group is also attached to the sugar, the nucleoside becomes a **nucleotide** (Figure 5.2). Thus a nucleotide is a nucleoside plus a phosphate. In the conventional numbering of the carbon atoms in the sugar in Figure 5.2, the carbon atom to which the base is attached is the 1' carbon. (The atoms in the sugar are given

Figure 5.1
Chemical structures of adenine, thymine, guanine, and cytosine, the four nitrogen-containing bases in DNA. In each base, the nitrogen atom linked to the deoxyribose sugar is indicated. The atoms shown in red participate in hydrogen bonding between the DNA base pairs, as explained in Section 5.2.

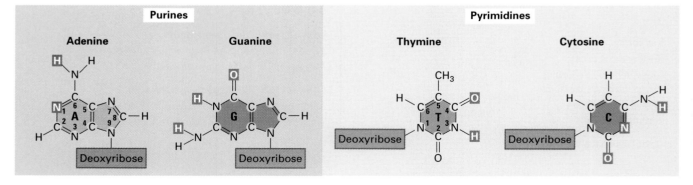

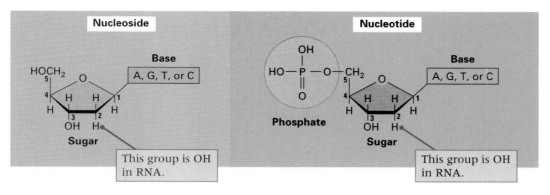

Figure 5.2
A typical nucleotide showing the three major components (phosphate, sugar, and base), the difference between DNA and RNA, and the distinction between a nucleoside (no phosphate group) and a nucleotide (with phosphate). Nucleotides may contain one phosphate unit (monophosphate), two such units (diphosphate) or three (triphosphate).

primed numbers to distinguish them from atoms in the bases.) The nomenclature of the nucleoside and nucleotide derivatives of the DNA bases is somewhat complicated and is summarized in Table 5.1. Most of these terms are not needed in this book; they are included because they are likely to be encountered in further reading.

In nucleic acids, such as DNA and RNA, the nucleotides are joined to form a **polynucleotide chain,** in which the phosphate attached to the 5' carbon of one sugar is linked to the hydroxyl group attached to the 3' carbon of the next sugar in line (Figure 5.3). The chemical bonds by which the sugar components of adjacent nucleotides are linked through the phosphate groups are called **phosphodiester bonds.** The 5'–3'–5'–3' orientation of these linkages continues throughout the chain, which typically consists of millions of nucleotides. Note that the terminal groups of each polynucleotide chain are a 5'-phosphate (**5'-P**) group at one end and a 3'-hydroxyl (**3'-OH**) group at the other. The asymmetry of the ends of a DNA strand implies that each strand has a **polarity** determined by which end bears the 5'-phosphate and which end bears the 3' hydroxyl.

Three years before Watson and Crick proposed their essentially correct three-dimensional structure of DNA as a double helix, Erwin Chargaff developed a chemical technique to measure the amount of each base present in DNA. As we describe his technique, we will let the molar concentration of any base be represented by the symbol for the base in square brackets; for example, [A] denotes the molar concentration of adenine. Chargaff used his technique to measure the [A], [T], [G], and [C] content of the DNA from a variety of sources. He found that the **base composition** of the DNA, d efined as the **percent G + C,** differs among species but is

Table 5.1 DNA nomenclature

Base	Nucleoside	Nucleotide
Adenine (A)	Deoxyadenosine	Deoxyadenosine-5' monophosphate (dAMP) diphosphate (dADP) triphosphate (dATP)
Guanine (G)	Deoxyguanosine	Deoxyguanosine-5' monophosphate (dGMP) diphosphate (dGDP) triphosphate (dGTP)
Thymine (T)	Deoxythymidine	Deoxythymidine-5' monophosphate (dTMP) diphosphate (dTDP) triphosphate (dTTP)
Cytosine (C)	Deoxycytidine	Deoxycytidine-5' monophosphate (dCMP) diphosphate (dCDP) triphosphate (dCTP)

(A)

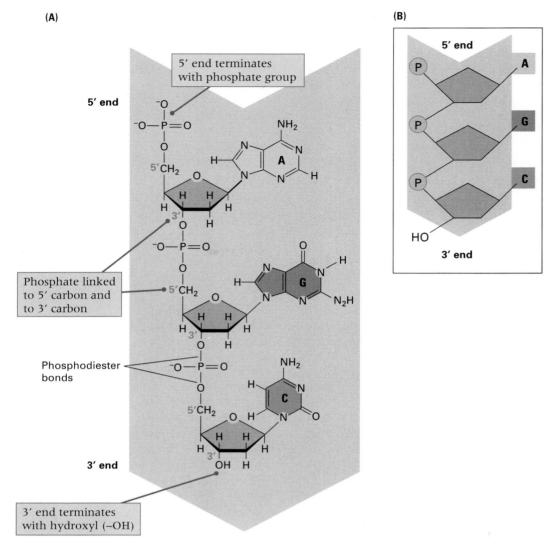

5' end terminates
with phosphate group

5' end

Phosphate linked
to 5' carbon and
to 3' carbon

Phosphodiester
bonds

3' end

3' end terminates
with hydroxyl (–OH)

(B)

5' end

P — A

P — G

P — C

HO

3' end

Figure 5.3

Three nucleotides at the 5' end of a single polynucleotide strand. (A) The chemical structure of the sugar-phosphate linkages, showing the 5'-to-3' orientation of the strand (the red numbers are those assigned to the carbon atoms). (B) A common schematic way to depict a polynucleotide strand.

constant in all cells of an organism and within a species. Data on the base composition of DNA from a variety of organisms are given in Table 5.2.

Chargaff also observed certain regular relationships among the molar concentrations of the different bases. These relationships are now called **Chargaff's rules**:

• The amount of adenine equals that of thymine: [A] = [T].

• The amount of guanine equals that of cytosine: [G] = [C].

• The amount of purine base equals that of pyrimidine bases:

$$[A] + [G] = [T] + [C].$$

Although the chemical basis of these observations was not known at the time, one of the appealing features of the Watson-Crick structure of paired comple-

Table 5.2 Base composition of DNA from different organisms

Organism	Base (and percentage of total bases)				Base composition (percent G + C)
	Adenine	Thymine	Guanine	Cytosine	
Bacteriophage T7	26.0	26.0	24.0	24.0	48.0
Bacteria					
Clostridium perfringens	36.9	36.3	14.0	12.8	26.8
Streptococcus pneumoniae	30.2	29.5	21.6	18.7	40.3
Escherichia coli	24.7	23.6	26.0	25.7	51.7
Sarcina lutea	13.4	12.4	37.1	37.1	74.2
Fungi					
Saccharomyces cerevisiae	31.7	32.6	18.3	17.4	35.7
Neurospora crassa	23.0	22.3	27.1	27.6	54.7
Higher plants					
Wheat	27.3	27.2	22.7	22.8*	45.5
Maize	26.8	27.2	22.8	23.2*	46.0
Animals					
Drosophila melanagaster	30.8	29.4	19.6	20.2	39.8
Pig	29.4	29.6	20.5	20.5	41.0
Salmon	29.7	29.1	20.8	20.4	41.2
Human being	29.8	31.8	20.2	18.2	38.4

*Includes one-fourth 5-methylcytosine, a modified form of cytosine found in most plants more complex than algae and in many animals

mentary strands was that it explained Chargaff's rules. Because A is always paired with T in double-stranded DNA, it must follow that [A] = [T]. Similarly, because G is paired with C, [G] = [C]. The third rule follows by addition of the other two: [A] + [G] = [T] + [C]. In the next section, we examine the molecular basis of base pairing in more detail.

5.2
The Physical Structure of the Double Helix

In the three-dimensional structure of the DNA molecule proposed in 1953 by Watson and Crick, the molecule consists of two polynucleotide chains twisted around one another to form a double-stranded helix in which adenine and thymine, and guanine and cytosine, are paired in opposite strands (Figure 5.4). In the standard structure, which is called the **B form** of DNA, each chain makes one complete turn every 34 Å. The helix is right-handed, which means that as you look down the barrel, each chain follows a clockwise path as it progresses. The bases are spaced at 3.4 Å, so there are ten bases per helical turn in each strand and ten base pairs per turn of the double helix. Each base is paired to a complementary base in the other strand by hydrogen bonds, which provide the main force holding the strands together. (A

(A)

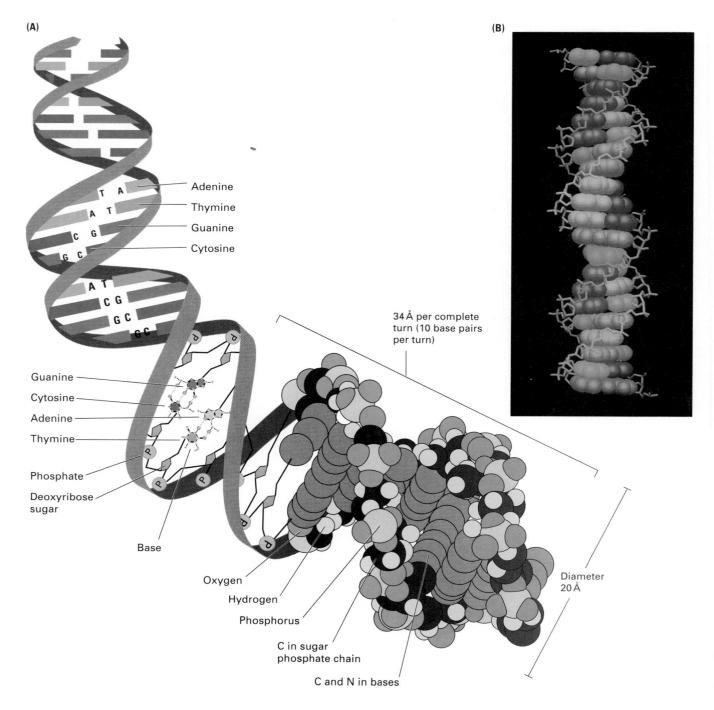

Adenine

Thymine

Guanine

Cytosine

Guanine

Cytosine

Adenine

Thymine

Phosphate

Deoxyribose
sugar

Base

Oxygen

Hydrogen

Phosphorus

C in sugar
phosphate chain

C and N in bases

(B)

34 Å per complete
turn (10 base pairs
per turn)

Diameter
20 Å

Figure 5.4

Two representations of DNA illustrating the three-dimensional structure of the double helix. (A) In a "ribbon diagram" the sugar-phosphate backbones are depicted as bands, with horizontal lines used to represent the base pairs. (B) A computer model of the B form of a DNA molecule. The stick figures are the sugar-phosphate chains winding around outside the stacked base pairs, forming a major groove and a minor groove. The color coding for the base pairs is A, red or pink; T, dark green or light green; G, dark brown or beige; C, dark blue or light blue. The bases depicted in dark colors are those attached to the blue sugar-phosphate backbone; bases depicted in light colors are attached to the beige backbone. [B, courtesy of Antony M. Dean.]

hydrogen bond is a weak bond in which two negatively charged atoms share a hydrogen atom.) The paired bases are planar, parallel to one another, and perpendicular to the long axis of the double helix. When discussing a DNA molecule, molecular biologists frequently refer to the individual strands as single strands or as single-stranded DNA; they refer to the double helix as double-stranded DNA or as a *duplex* molecule. The two grooves spiraling along outside of the double helix are not symmetrical; one groove, called the **major groove,** is larger than the other, which is called the **minor groove.**

Proteins that interact with double-stranded DNA often have regions that make contact with the base pairs by fitting into the major groove, into the minor groove, or into both grooves.

The central feature of DNA structure is the pairing of complementary bases, A with T and G with C. The hydrogen bonds that form in the adenine-thymine base pair and in the guanine-cytosine pair are illustrated in Figure 5.5. Note that an A−T pair (Figure 5.5A and B) has two hydrogen bonds and that a G−C pair (Figure 5.5C and D) has three hydrogen bonds. This means that the hydrogen bonding between G and C is

Figure 5.5

Normal base pairs in DNA. On the left, the hydrogen bonds (dotted lines) and the joined atoms are shown in red. (A, B) An A−T base pair. (C, D) A G−C base pair. In the space-filling models (B and D), the colors are C, gray; N, blue; O, red; and H (shown in the bases only), white. Each hydrogen bond is depicted as a white disk squeezed between the atoms sharing the hydrogen. The stick figures on the outside represent the backbones winding around the stacked base pairs. [Space-filling models courtesy of Antony M. Dean.]

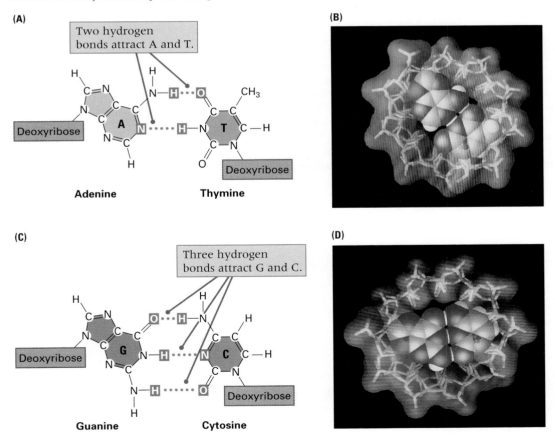

The Double Helix

James D. Watson and
Francis H. C. Crick 1953
Cavendish Laboratory,
Cambridge, England
A Structure for Deoxyribose Nucleic Acid

This is one of the watershed papers of twentieth-century biology. After its publication, nothing in genetics was the same. Everything that was known, and everything still to be discovered, would now need to be interpreted in terms of the structure and function of DNA. The importance of the paper was recognized immediately, in no small part because of its lucid and concise description of the structure. This excerpt is unusual in that it includes the acknowledgment. Watson and Crick benefited tremendously in knowing that their structure was consistent with the unpublished structural studies of Maurice Wilkins and Rosalind Franklin. The same issue of Nature *that included the Watson and Crick paper also included, back to back, a paper from the Wilkins group and one from the Franklin group detailing their data and the consistency of their data with the proposed structure. It has been said that Franklin was poised a mere two half-steps from making the discovery herself, alone. In any event, Watson and Crick and Wilkins were awarded the 1962 Nobel Prize for their discovery of DNA structure. Rosalind Franklin, tragically, died of cancer in 1958 at the age of 38.*

We wish to suggest a structure for the salt of deoxyribose nucleic acid (DNA). . . . The structure has two helical chains each coiled round the same axis. . . . Both chains follow right-handed helices, but the two chains run in opposite directions. . . . The bases are on the inside of the helix and the phosphates on the outside. . . . There is a residue on each chain every 3.4 A and the structure repeats after 10 residues. . . . The

---

If only specific pairs of bases can be formed, it follows that if the sequences of bases on one chain is given, then the sequence on the other chain is automatically determined.

---

novel feature of the structure is the manner in which the two chains are held together by the purine and pyrimidine bases. The planes of the bases are perpendicular to the fiber axis. They are joined together in pairs, a single base from one chain being hydrogen-bonded to a single base from the other chain, so that the two lie side by side. One of the pair must be a purine and the other a pyrimidine for bonding to occur. . . . Only specific pairs of bases can bond together. These pairs are: adenine (purine) with thymine (pyrimidine), and guanine

(purine) with cytosine (pyrimidine). In other words, if an adenine forms one member of a pair, on either chain, then on these assumptions the other member must be thymine; similarly for guanine and cytosine. The sequence of bases on a single chain does not appear to be restricted in any way. However, if only specific pairs of bases can be formed, it follows that if the sequence of bases on one chain is given, then the sequence on the other chain is automatically determined. . . . It has not escaped our notice that the specific pairing we have postulated immediately suggests a plausible copying mechanism for the genetic material. . . . We are much indebted to Dr. Jerry Donohue for constant advice and criticism, especially on interatomic distances. We have also been stimulated by a knowledge of the general nature of the unpublished experimental results and ideas of Dr. Maurice H. F. Wilkins, Dr. Rosalind Franklin and their co-workers at King's College, London.

Source: Nature 171: 737–738

---

stronger in the sense that it requires more energy to break; for example, the amount of heat required to separate the paired strands in a DNA duplex increases with the percent of G + C. Because nothing restricts the sequence of bases in a single strand, any sequence could be present along one strand. This explains Chargaff's observation that DNA from different organisms may have different base compositions. However, because the strands in duplex DNA are complementary, Chargaff's rules of

[A] = [T] and [G] = [C] are true whatever the base composition.

Each backbone in a double helix consists of deoxyribose sugars alternating with phosphate groups that link the 3' carbon atom of one sugar to the 5' carbon of the next in line (Figure 5.3). The two polynucleotide strands of the double helix are oriented in opposite directions in the sense that the bases that are paired are attached to sugars lying above and below the plane of pairing, respectively. The sugars are offset because the phosphate linkages in the backbones run in opposite directions (Figure 5.6), and the strands are said to be **antiparallel.** This means that each terminus of the double helix possesses one 5'-P group (on one strand) and one 3'-OH group (on the other strand), as shown in Figure 5.6.

The diagrams of the DNA duplexes in Figures 5.4 and 5.6 are static and so some-what misleading. DNA is a dynamic molecule, constantly in motion. In some regions, the strands can separate briefly and then come together again in the same conformation or in a different one. Although the right-handed double helix in Figure 5.4 is the standard form, DNA can form more than 20 slightly different variants of right-handed helices, and some regions can even form helices in which the strands twist to the left (called the **Z form** of DNA). If there are complementary stretches of nucleotides in the same strand, then a single strand, separated from its partner, can fold back upon itself like a hairpin. Even triple helices consisting of three strands can form in regions of DNA that contain suitable base sequences.

Figure 5.6

A segment of a DNA molecule showing the antiparallel orientation of the complementary strands. The overlying blue arrows indicate the 5'-to-3' direction of each strand. The phosphates (P) join the 3' carbon atom of one deoxyribose (horizontal line) to the 5' carbon atom of the adjacent deoxyribose.

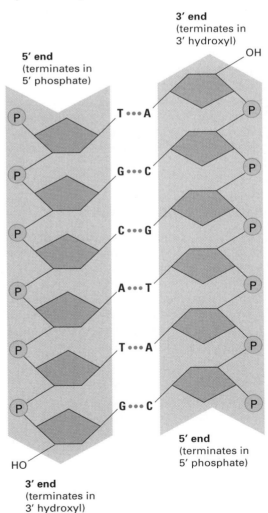

3' end
(terminates in 3' hydroxyl)

5' end
(terminates in 5' phosphate)

OH

T •••A

G •••C

C •••G

A •••T

T •••A

G •••C

HO

3' end
(terminates in 3' hydroxyl)

5' end
(terminates in 5' phosphate)

5.3
What a Genetic Material Needs That DNA Supplies

Not every polymer would be useful as genetic material. However, DNA is admirably suited to a genetic function because it satisfies the three essential requirements of a genetic material. First, any genetic material must be able to be replicated accurately, so that the information it contains is precisely replicated and inherited by daughter cells. The basis for exact duplication of a DNA molecule is the complementarity of the A−T and G−C pairs in the two polynucleotide chains. Unwinding and separation of the chains, with each free chain being copied, results in the formation of two identical double helices (Figure 5.7).

A genetic material must also have the capacity to carry all of the information needed to direct the organization and metabolic activities of the cell. As seen in Chapter 1, the product of most genes is a protein molecule—a polymer composed of molecular units called amino acids. The sequence of amino acids in the protein determines its chemical and physical properties. A gene is expressed when its protein product is synthesized, and one requirement of the genetic material is that it direct the order in which amino acid units are

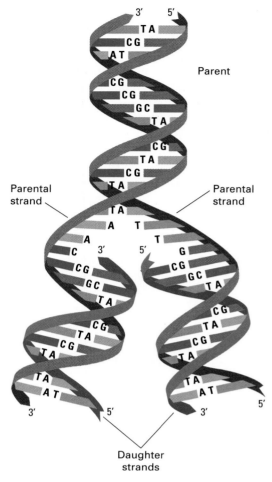

A genetic material must also be capable of undergoing occasional mutations in which the information it carries is altered. Furthermore, the mutant molecules must be capable of being replicated as faithfully as the parental molecule, so that mutations are heritable. Watson and Crick suggested that heritable mutations might be possible in DNA by rare mispairing of the bases, with the result that an incorrect nucleotide becomes incorporated into a replicating DNA strand.

5.4
The Replication of DNA

The process of replication, in which each strand of the double helix serves as a template for the synthesis of a new strand (Figure 5.7), is simple in principle. It requires only that the hydrogen bonds joining the bases break to allow separation of the chains and that appropriate free nucleotides of the four types pair with the newly accessible bases in each strand. However, it is a complex process with geometric problems requiring a variety of enzymes and other proteins. These processes are examined in this section.

The Basic Rule for the Replication of Nucleic Acids

The primary function of any mode of DNA replication is to reproduce the base sequence of the parent molecule. The specificity of base pairing—adenine with thymine (replaced by uracil in RNA) and guanine with cytosine—provides the mechanism used by all genetic replication systems. Furthermore,

- Nucleotide monomers are added one by one to the end of a growing strand by an enzyme called a DNA polymerase.

- The sequence of bases in each newly replicated strand, or **daughter strand,** is complementary to the base sequence in the old strand, or **parental strand,** being replicated. For example, wherever

Figure 5.7
Watson-Crick model of DNA replication. The newly synthesized strands are in red. We will see in Section 5-6 that the details of DNA replication are considerably more complex, because any new DNA strand can be synthesized only in the 5'-to-3' direction.

added to the end of a growing protein molecule. In DNA, this is done by means of a genetic code in which groups of three bases specify amino acids. Because the four bases in a DNA molecule can be arranged in any sequence, and because the sequence can vary from one part of the molecule to another and from organism to organism, DNA can contain a great many unique regions, each of which can be a distinct gene. A long DNA chain can direct the synthesis of a variety of different protein molecules.

an adenine nucleotide is present in the parental strand, a thymine nucleotide will be added to the growing end of the daughter strand.

The following section explains how the two strands of a daughter molecule are physically related to the two strands of the parental molecule.

The Geometry of DNA Replication

The production of daughter DNA molecules from a single parental molecule gives rise to several problems that result from the helical structure and great length of typical DNA molecules and the circularity of many DNA molecules. These problems and their solutions are described here.

Semiconservative Replication of Double-Stranded DNA In the **semiconservative mode** of replication, each parental DNA strand serves as a template for one new strand, and as each new strand is formed, it is hydrogen-bonded to its parental template (Figure 5.7). As replication proceeds, the parental double helix unwinds and then rewinds again into two new double helices, each of which contains one originally parental strand and one newly formed daughter strand.

In theory, DNA could be replicated by a number of mechanisms other than the semiconservative mode. However, the reality of semiconservative replication was demonstrated experimentally by Matthew Meselson and Franklin Stahl in 1958. The experiment made use of a newly developed high-speed centrifuge (an **ultracentrifuge**) that could spin a solution so fast that molecules differing only slightly in density could be separated. In their experiment, the heavy ^{15}N isotope of nitrogen was used for physical separation of parental and daughter DNA molecules. DNA isolated from the bacterium *E. coli* grown in a medium containing ^{15}N as the only available source of nitrogen is denser than DNA from bacteria grown in media with the normal ^{14}N isotope. These DNA molecules can be separated in an ultracentrifuge, because

they have about the same density as a very concentrated solution of cesium chloride (CsCl).

When a CsCl solution containing DNA is centrifuged at high speed, the Cs^+ ions gradually sediment toward the bottom of the centrifuge tube. This movement is counteracted by diffusion (the random movement of molecules), which prevents complete sedimentation. At equilibrium, a linear gradient of increasing CsCl concentration—and of density—is present from the top to the bottom of the centrifuge tube. The DNA also moves upward or downward in the tube to a position in the gradient at which the density of the solution is equal to its own density. At equilibrium, a mixture of ^{14}N-containing ("light") and ^{15}N-containing ("heavy") *E. coli* DNA will separate into two distinct zones in a density gradient even though they differ only slightly in density. DNA from *E. coli* containing ^{14}N in the purine and pyrimidine rings has a density of 1.708 g/cm^3, whereas DNA with ^{15}N in the purine and pyrimidine rings has a density of 1.722 g/cm^3. These molecules can be separated because a solution of 5.6 molar CsCl has a density of 1.700 g/cm^3. When spun in a centrifuge, the CsCl solution forms a gradient of density that brackets the densities of the light and heavy DNA molecules. It is for this reason that the separation technique is called **equilibrium density-gradient centrifugation.**

The result of the Meselson-Stahl experiment is shown in Figure 5.8. Prior to the experiment, bacteria were grown for many generations in a ^{15}N-containing medium. Therefore, at the beginning of the experiment, essentially all the DNA was uniformly labeled with ^{15}N and had a heavy density. The cells were then transferred to a ^{14}N-containing medium, and DNA was isolated from samples of cells taken from the culture at intervals and subjected to equilibrium density-gradient centrifugation. Each photograph in Figure 5.8 shows the image of a solution within a centrifuge tube taken in ultraviolet light of wavelength 260 nm (nanometers), which is absorbed by DNA. The positions of the DNA molecules in the density gradient are therefore indicated by the dark bands that absorb the

Replication by Halves

**Matthew Meselson and
Franklin W. Stahl 1958**
California Institute of Technology,
Pasadena, California
The Replication of DNA in
Escherichia coli

*Replication of DNA by separation of its
strands, followed by the synthesis of a
new partner strand for each separated
parental strand, is so fundamental a biological mechanism that it is easy to forget
that it had to be proved experimentally
somewhere along the way. Although semiconservative replication is a simple and
logical way to proceed, biological systems
might have found a different way to replicate their DNA. The Meselson-Stahl
experiment that demonstrated semiconservative replication made use of a novel
technique, density-gradient centrifugation,
which was invented by the authors specifically for the purpose of separating DNA
molecules differing slightly in density. The
use of cesium chloride was mandated,
because in concentrated solution, it has a
density similar to that of DNA. The method
of cesium chloride density centrifugation is
still important in modern molecular biology; it is used in the isolation of DNA
molecules.*

Studies of bacterial transformation and
bacteriophage infection strongly indicate
that deoxyribonucleic acid (DNA) can
carry and transmit hereditary information
and can direct its own replication.
Hypotheses for the mechanism of DNA
replication differ in the predictions they
make concerning the distribution among

progeny molecules of atoms derived from
parental molecules. . . . We anticipated
that a label which imparts to the DNA
molecule an increased density might
permit an analysis of this distribution by
sedimentation techniques. To this end, a
method was developed for the detection of
small density differences among
macromolecules. . . . A small amount of
DNA in a concentrated solution of cesium

**The results of the present
experiment are in exact
accord with the expectation of
the Watson-Crick model for
DNA duplication.**

---

chloride is centrifuged until equilibrium is
closely approached. The opposing
processes of sedimentation and diffusion
have then produced a stable concentration
gradient of the cesium chloride, with a
continuous increase in density along the
direction of centrifugal force. The
macromolecules of DNA present in this
density gradient are driven by the centrifugal field into the region where the solution
density is equal to their own buoyant
density. . . . Bacteria uniformly labeled with
N^{15} were abruptly changed into N^{14}
medium. . . . Samples were withdrawn from
the culture immediately and afterward at
intervals for several generations. . . . Until
one generation time has elapsed, half-
labeled molecules accumulate. One
generation after the switch to N^{14}, these

half-labeled or "hybrid" molecules alone
are observed. Subsequently, only half-
labeled DNA and completely unlabeled
DNA are found. When two generation
times have elapsed, half-labeled and
unlabeled DNA are present in equal
amounts. These results permit the
following conclusions to be drawn: 1.
*The nitrogen of a DNA molecule is
divided equally between two subunits
which remain intact through many generations. . . . 2. Following replication, each
daughter molecule has received one
parental subunit. . . . 3. The replicative act
results in a molecular doubling. . . .* A
molecular structure for DNA has been
proposed by Watson and Crick that . . .
suggested to them a definite and
structurally plausible hypothesis for the
duplication of the DNA molecule. . . .
The results of the present experiment
are in exact accord with the expectation of
the Watson-Crick model for DNA duplication. . . . The results presented here direct
our attention to [other important problems].
What are the molecular structures of the
subunits of DNA which are passed on
intact to each daughter molecule? What is
the relationship of these subunits to each
other in a DNA molecule? What is the
mechanism of the synthesis and dissociation of the subunits in vivo?

*Source: Proceedings of the National Academy
of Sciences of the USA* 44: 671–682

light. Each photograph is oriented such that
the bottom of the tube is at the right and
the top is at the left. To the right of each
photograph is a graph showing the absorption of the ultraviolet light from the top of
the centrifuge tube to the bottom. In each
trace, the peaks correspond to the positions
of the bands in the photographs, but the

height and width of each peak make it possible to quantify the DNA in each band.

At the start of the experiment (time 0),
all of the DNA was heavy (^{15}N). After the
transfer to ^{14}N, a band of lighter density
began to appear, and it gradually became
more prominent as the cells replicated their
DNA and divided. After one generation of

growth (one round of replication of the DNA molecules and a doubling of the number of cells), all of the DNA had a "hybrid" density exactly intermediate between the densities of ^{15}N-DNA and ^{14}N-DNA. The finding of molecules with a hybrid density indicates that the replicated molecules contain equal amounts of the two nitrogen isotopes. After a second generation of replication in the ^{14}N medium (1.9 generations in the original experiment), half of the DNA had the density of DNA with ^{14}N in both strands ("light" DNA) and the other half had the hybrid density. After three generations, the ratio of light to hybrid DNA was approximately 3 : 1, and after four generations (4.1 in the original experiments), it was approximately 7 : 1 (Figure 5.8). This distribution of ^{15}N atoms is precisely the result predicted from semiconservative replication of the Watson-Crick structure, as illustrated in Figure 5.9. Subsequent experiments with replicating DNA from numerous viruses, bacteria, and higher organisms have indicated that semiconservative replication is universal.

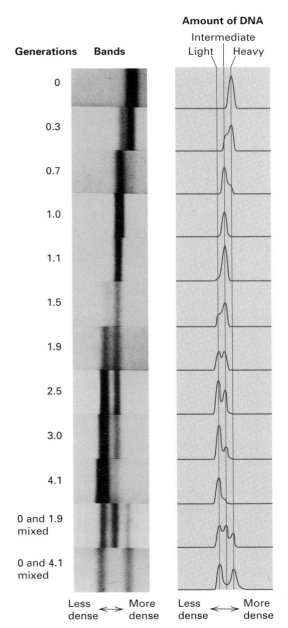

Figure 5.8
The DNA replication experiment of Meselson and Stahl. Cells whose DNA contained heavy nitrogen were transferred into growth medium containing only light nitrogen. Samples were taken at intervals, and the DNA was subjected to equilibrium density-gradient centrifugation in a solution of CsCl. During centrifugation, each type of DNA molecule moves until it comes to rest at a position in the centrifuge tube at which its density equals the density of the CsCl solution at that position. Photographs of the centrifuge tubes taken with ultraviolet light are shown at the left. The top of the tube is at the left, the bottom at the right. The smooth curves show quantitatively the amount of absorption of the ultraviolet light across the tube. [Photograph courtesy of Matthew Meselson. From M. Meselson and F. Stahl, *Proc. Natl. Acad. Sci. USA* 1958. 44:671.]

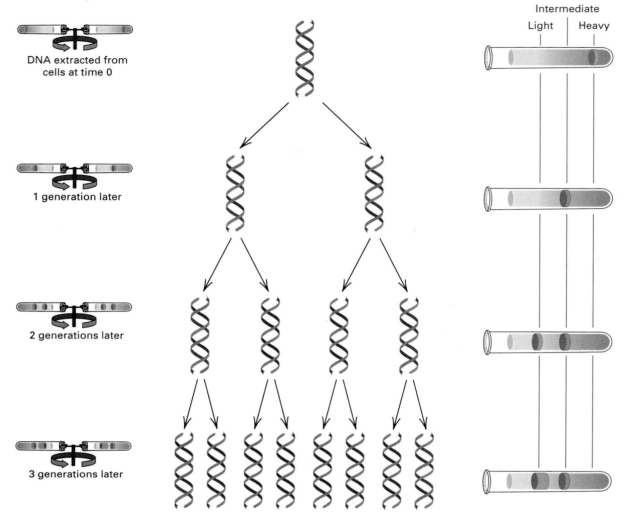

Figure 5.9
Interpretation of the data in Figure 5.8 on the basis of semiconservative replication. The diagrams show the composition of the DNA duplexes after 0, 1, 2, and 3 rounds of replication. DNA strands labeled with ^{15}N are shown in red, those labeled with ^{14}N in blue. The diagrams at the right show the positions at which bands are expected in the CsCl gradient.

Geometry of the Replication of Circular DNA Molecules

In the Meselson-Stahl experiment, *E. coli* DNA was extensively fragmented when isolated, so the form of the molecule was unknown. Later, the isolation of unbroken molecules and their examination by two techniques—autoradiography and electron microscopy—showed that the DNA in *E. coli* cells is circular.

The first proof that *E. coli* DNA replicates as a circle came from an autoradiographic experiment. (Genetic mapping experiments to be described in Chapter 8 had already suggested that the bacterial chromosome is circular.) Cells were grown in a medium containing radioactive thymine (^{3}H-thymine) so that all DNA synthesized would be radioactive. The DNA was isolated without fragmentation and placed on photographic film. Each radioactive decay caused a tiny black spot to appear in the film, and after several months there were enough spots to visualize the DNA with a microscope; the pattern of black spots on the film located the molecule. One of the now-famous autoradiograms from this experiment is shown in Figure 5.10. The actual length of the *E. coli* chromosome is 1.6 mm (4.7 million base pairs). On a smaller scale, Figure 5.11 shows a replicating circular molecule of a plasmid found inside certain bacterial cells. The total length of the molecule is only 0.01 mm (3000 base pairs). The replicating circle is schematically like the Greek letter θ (theta), and so this mode of replication is usually called θ **replication.**

The circularity of the replicating molecules in Figures 5.10 and 5.11 brings out an important geometric feature of semiconservative replication. There are about 400,000 turns in an *E. coli* double helix, and because the two chains of a replicating molecule must make a full rotation to unwind each of these gyres, some kind of swivel must exist to avoid tangling the entire structure (Figure 5.12). The axis of rotation for unwinding is provided by nicks made in the backbone of *one* strand of the double helix during replication, nicks that are rapidly repaired after unwinding. Enzymes capable of making such nicks and then rapidly repairing them have been isolated from both bacterial and mammalian cells; they are called **topoisomerases.** How they work is suggested in Figure 5.13, which shows the structure of part of the topoisomerase I enzyme from *E. coli* when it is in contact with duplex DNA. Note that the molecule is wrapped completely around the DNA molecule, which passes through the enzyme like a train through a tunnel.

The position along a molecule at which DNA replication begins is called a **replication origin,** and the region in which parental strands are separating and new strands are being synthesized is called a **replication fork.** The process of generating a new replication fork is **initiation.** In some organisms, replication is initiated at many positions; in others, the origin of replication may be a unique site. In most bacteria, bacteriophage, and viruses, *DNA replication is initiated at a unique origin of replication.* Furthermore, with only a few exceptions, two replication forks move in opposite directions from the origin (Figure

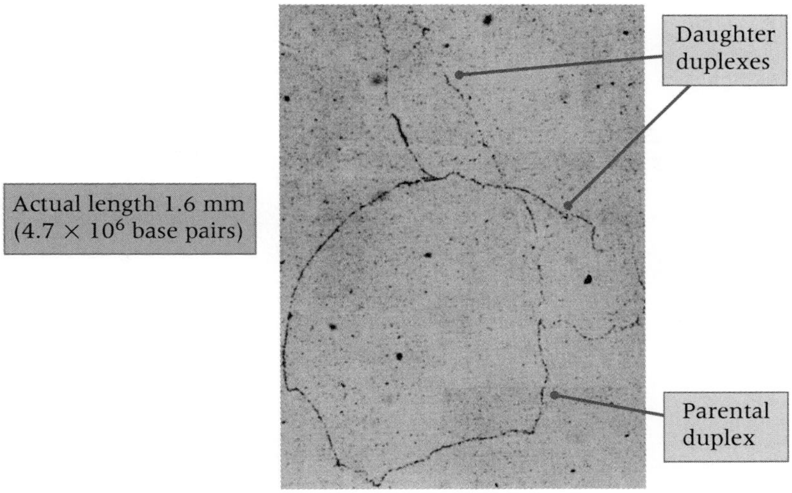

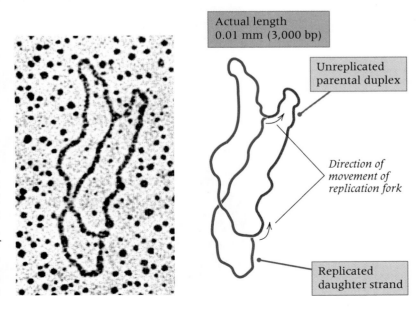

Actual length 1.6 mm (4.7×10^6 base pairs)

Daughter duplexes

Parental duplex

Figure 5.10
Autoradiogram of the intact replicating chromosome of an *E. coli* cell that has grown in a medium containing ³H-thymine for slightly less than two generations. The continuous lines of dark grains were produced by electrons emitted by decaying ³H atoms in the DNA molecule. The pattern is seen by light microscopy. [From J. Cairns, *Cold Spring Harbor Symp. Quant. Biol.* 1963. 28:44].

Figure 5.11
Electron micrograph of a small circular DNA molecule replicating by the θ mode. The parental and daughter segments are shown in the drawing. [Electron micrograph courtesy of Donald Helinski.]

Actual length 0.01 mm (3,000 bp)

Unreplicated parental duplex

Direction of movement of replication fork

Replicated daughter strand

(A)

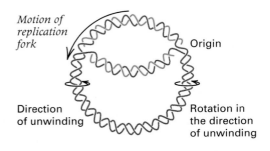

Motion of replication fork

Origin

Direction of unwinding

Rotation in the direction of unwinding

(B)

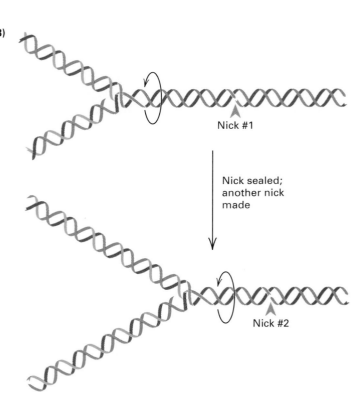

Nick #1

Nick sealed; another nick made

Nick #2

Figure 5.12

Replication of a circular DNA molecule. (A) The unwinding motion of the branches of a replicating circle, without positions at which free rotation can occur, causes overwinding of the unreplicated part. (B) Mechanism by which a single-strand break (a nick) ahead of a replication fork allows rotation.

Figure 5.13

Proposed structure of an association between duplex DNA and topoisomerase I in which the DNA passes through a hole formed in the enzyme. The two views are perpendicular to each other. The structure of the enzyme includes a pair of jaw-like projections that can open and close. The intermediate shown here is created when the enzyme closes its jaws around a DNA molecule, completely enclosing it. What happens then is that one DNA strand is cleaved, swiveled, and reconnected, whereupon the jaws open and the DNA molecule is released. [Courtesy of C. D. Lima, J. C. Wang, and A. Móndragon. From C. D. Lima, J. C. Wang, and A. Móndragon, *Nature* 1994. 367:138.]

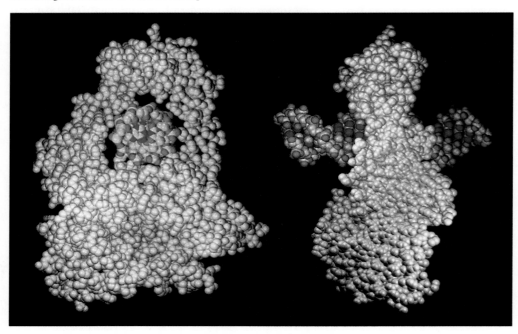

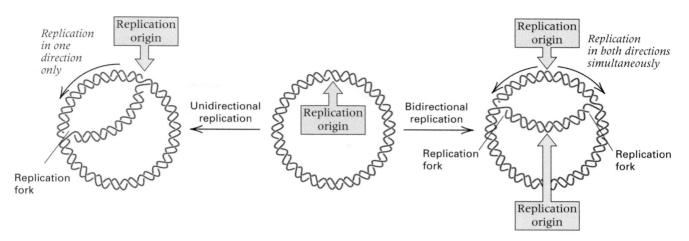

Figure 5.14
The distinction between unidirectional and bidirectional DNA replication. In unidirectional replication, there is only one replication fork; bidirectional replication requires two replication forks. The curved arrows indicate the direction of movement of the forks. Most DNA replicates bidirectionally.

5.14); that is, *DNA almost always replicates bidirectionally in both prokaryotes and eukaryotes.*

Multiple Origins of Replication in Eukaryotes
The DNA duplex in a eukaryotic chromosome is linear and replicates **bidirectionally.** Furthermore, replication is initiated at many sites in the DNA. The structures that result from the numerous origins are seen in electron micrographs as multiple loops along a DNA molecule (Figure 5.15). Multiple initiation is a means of reducing the total replication time of a large molecule. In eukaryotic cells, movement of each replication fork proceeds at a rate of approximately 10 to 100 nucleotide pairs per second. For example, in *D. melanogaster,* the rate of replication is about 50 nucleotide pairs per second at 25°C. Because the DNA molecule in the largest chromosome in *Drosophila* contains about 7×10^7 nucleotide pairs, replication from a single bidirectional origin of replication would take about 8 days. Developing *Drosophila* embryos actually use about 8500 replication origins per chromosome, which reduces the replication time to a few minutes. In a typical eukaryotic cell, origins are spaced about 40,000 nucleotide pairs apart, which allows each chromosome to replicate in 15 to 30 minutes. Because chromosomes do not all replicate

simultaneously, complete replication of all chromosomes in eukaryotes usually takes from 5 to 10 hours.

DNA replication is much faster in prokaryotes than in eukaryotes. In *E. coli,* for example, replication of a DNA molecule takes place at a rate of approximately 1500 nucleotide pairs per second; similar rates are found in the replication of all phage DNA molecules studied. Complete replication of the *E. coli* DNA molecule from its single bidirectional origin of replication requires approximately 30 minutes because the molecule contains about 4.7×10^6 nucleotide pairs. Yet under optimal culture conditions, an *E. coli* cell can divide every 20 minutes. How is this possible? It is possible because a new round of replication is started at the origin every 20 minutes, even though the previous round of replication has not yet been completed. In a rapidly dividing culture of *E. coli* cells, the DNA molecule passed to each daughter cell in division is already in the process of replicating in preparation for the next division.

Rolling-Circle Replication Some circular DNA molecules, including those of a number of bacterial and eukaryotic viruses, replicate by a process that does not include a θ-shaped intermediate. This replication mode is called **rolling-circle replication.**

Figure 5.15

Replicating DNA of *Drosophila melanogaster*. (A) An electron micrograph of a 30,000-nucleotide-pair segment showing seven replication loops. (B) An interpretive drawing showing how loops merge. Two replication origins are shown in the drawing. The small arrows indicate the direction of movement of the replication forks. [Electron micrograph courtesy of David Hogness.]

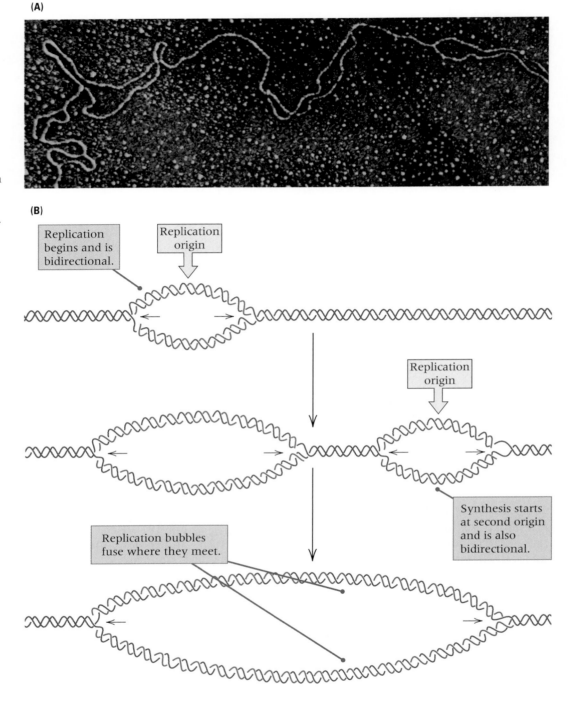

(A)

(B)

Replication begins and is bidirectional.

Replication origin

Replication origin

Synthesis starts at second origin and is also bidirectional.

Replication bubbles fuse where they meet.

In this process, replication starts with a cut at a specific sugar-phosphate bond in a double-stranded circle (Figure 5.16). This cut produces two chemically distinct ends: a 3' end (at which the nucleotide has a free 3'-OH group) and a 5' end (at which the nucleotide has a free 5'-P group). The DNA is synthesized by the addition of deoxynu-cleotides to the 3' end with simultaneous displacement of the 5' end from the circle. As replication proceeds around the circle, the 5' end rolls out as a tail of increasing length.

In most cases, as the tail is extended, a complementary chain is synthesized, which results in a double-stranded DNA

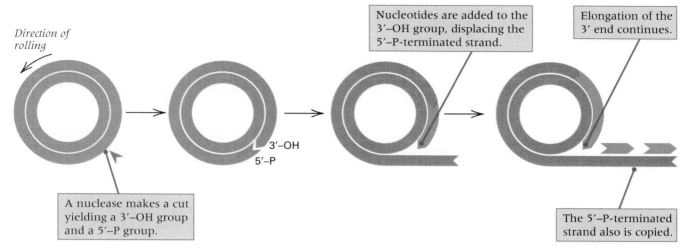

Direction of rolling

A nuclease makes a cut yielding a 3'–OH group and a 5'–P group.

3'–OH
5'–P

Nucleotides are added to the 3'–OH group, displacing the 5'–P-terminated strand.

Elongation of the 3' end continues.

The 5'–P-terminated strand also is copied.

Figure 5.16
Rolling-circle replication. Newly synthesized DNA is in red. The displaced strand is replicated in short fragments, as explained in Section 5.6.

tail. Because the displaced strand is chemically linked to the newly synthesized DNA in the circle, replication does not terminate, and extension proceeds without interruption, forming a tail that may be many times longer than the circumference of the circle. Rolling-circle replication is a common feature in late stages of replication of double-stranded DNA phages that have circular intermediates. An important example of rolling-circle replication will also be seen in Chapter 8, where matings between donor and recipient *E. coli* cells are described.

So far, we have considered only certain geometrical features of DNA replication. In the next section, the enzymes and protein factors used in DNA replication ar described.

5.5 DNA Synthesis

Nucleic acids are synthesized in chemical reactions controlled by enzymes, as is the case with most metabolic reactions in living cells. An enzyme that forms the sugar-phosphate bond (the phosphodiester bond) between adjacent nucleotides in a nucleic acid chain is called a **DNA polymerase.** A variety of DNA polymerases have been purified, and DNA synthesis has been carried out *in vitro* in a cell-free system pre-

pared by disrupting cells and combining purified components in a test tube under precisely defined conditions.

Three principal requirements must be met for DNA polymerases to catalyze synthesis of DNA.

- *The 5'-triphosphates of the four deoxynucleosides must be present.* These are the compounds denoted in Table 5.1 as dATP, dGTP, dTTP, and dCTP, which contain the bases adenine, guanine, thymine, and cytosine, respectively. Details of the structures of dCTP and dGTP are shown in Figure 5.17, in which the phosphate groups cleaved off during DNA synthesis are indicated. DNA synthesis requires all four nucleoside 5'-triphosphates and does not take place if any of them is omitted.

- *A preexisting single strand of DNA to be replicated must be present.* Such a strand is called a **template** strand.

- *A nucleic acid segment, which may be very short, must be present and must be hydrogen-bonded to the template strand.* This segment is called a **primer.** *No known DNA polymerase is able to initiate chains,* so the presence of a primer chain with a free 3'-OH group is absolutely essential for the initiation of replication. In living cells, the primer is

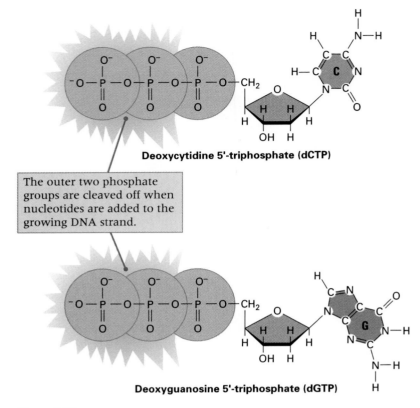

Deoxycytidine 5'-triphosphate (dCTP)

The outer two phosphate groups are cleaved off when nucleotides are added to the growing DNA strand.

Deoxyguanosine 5'-triphosphate (dGTP)

Figure 5.17
Two deoxynucleoside triphosphates used in DNA synthesis. The outer two phosphate groups are removed during synthesis.

a short segment of RNA; in cell-free replication *in vitro,* the primer may be either RNA or DNA.

The reaction catalyzed by the DNA polymerases is the formation of a phosphodiester bond between the free 3'-OH group of the chain being extended and the innermost phosphorus atom of the nucleoside triphosphate being incorporated at the 3' end (Figure 5.18). The result is as follows:

DNA synthesis proceeds by the elongation of primer chains, *always in the 5' to 3' direction.*

Recognition of the appropriate incoming nucleoside triphosphate in replication depends on base pairing with the opposite nucleotide in the template chain. DNA polymerase will usually catalyze the polymerization reaction that incorporates the new nucleotide at the primer terminus only when the correct base pair is present. The

same DNA polymerase is used to add each of the four deoxynucleoside phosphates to the 3'-OH terminus of the growing strand.

Two DNA polymerases are needed for DNA replication in *E. coli*: **DNA polymerase I,** which is often written **Pol I,** and **DNA polymerase III (Pol III).** Polymerase III is the major replication enzyme. In the cell, it exists as a large complex (molecular mass of about 1×10^6 daltons) composed of two Pol III protein subunits plus at least seven other proteins. Although each cell has only 10–20 copies of the Pol III complex, it is responsible not only for the elongation of DNA chains but also for the initiation of the replication fork at origins of replication and the addition of deoxynucleotides to the RNA primers. Polymerase I plays an essential, but secondary, role in replication that will be described in a later section. Eukaryotic cells also contain several DNA polymerases. The enzyme responsible for the replication of chromosomal DNA is called **polymerase α.** Mitochondria have their own DNA polymerase, **polymerase γ,** which replicates the mitochondrial DNA.

In addition to their ability to polymerize nucleotides, most DNA polymerases are capable of **nuclease** activities that break phosphodiester bonds in the sugarphosphate backbones of nucleic acid chains. The many other enzymes that have nuclease activity, are of two types: (1) **exonucleases** can remove a nucleotide only from the end of a chain, and (2) **endonucleases** break bonds within the chains. DNA polymerases I and III of *E. coli* have an exonuclease activity that acts only at the 3' terminus (a 3'-to-5' exonuclease activity). This exonuclease activity provides a built-in mechanism for correcting rare errors in polymerization. Occasionally, a polymerase adds to the end of the growing chain an incorrect nucleotide, which cannot form a proper base pair with the base in the template strand. The presence of an unpaired nucleotide activates the 3'-to-5' exonuclease activity, which cleaves the unpaired nucleotide from the 3'-OH end of the growing chain (Figure 5.19). Because it cleaves off an incorrect nucleotide and gives the polymerase another chance to get it right, the 3'-to-5' exonuclease activity of DNA polymerase is also called the **proof-**

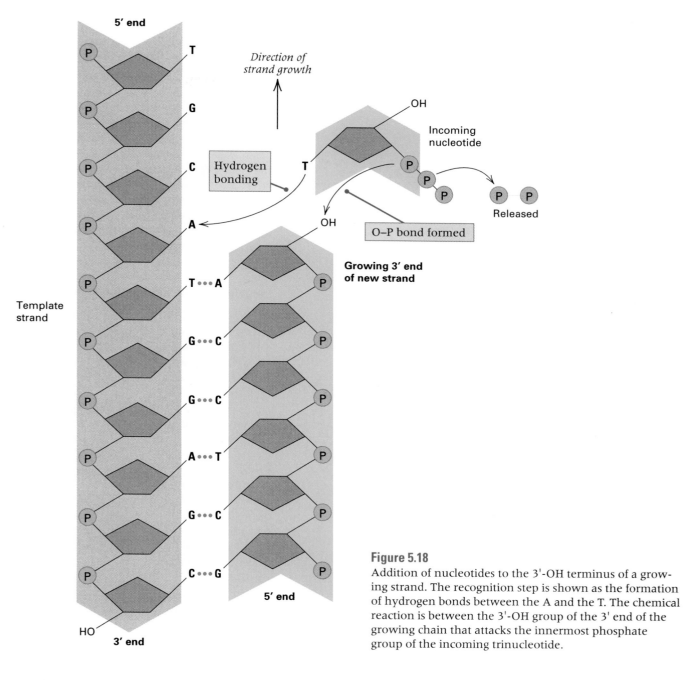

Figure 5.18
Addition of nucleotides to the 3'-OH terminus of a growing strand. The recognition step is shown as the formation of hydrogen bonds between the A and the T. The chemical reaction is between the 3'-OH group of the 3' end of the growing chain that attacks the innermost phosphate group of the incoming trinucleotide.

reading or **editing function.** The proofreading function can "look back" only one base (the one added last). Nevertheless,

> The genetic significance of the proofreading function is that it is an error-correcting mechanism that serves to reduce the frequency of mutation resulting from the incorporation of incorrect nucleotides in DNA replication.

Two unexpected features of DNA replication result from functional constraints that are present in all known DNA polymerases. One constraint is that a polymerase can elongate a newly synthesized DNA strand only at its 3' end (Figure 5.20). Hence the polymerase can move along the template strand only in the 3'-to-5' direction. The second constraint is that DNA polymerase is unable to initiate new chains but rather requires a preexisting primer. How the process of DNA replication deals with these constraints is described in the next section.

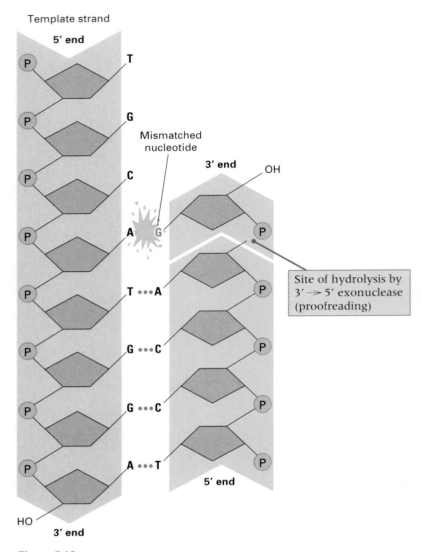

Template strand

5′ end

Mismatched nucleotide

3′ end

OH

A ✦ G

Site of hydrolysis by 3′ → 5′ exonuclease (proofreading)

T ••• A

G ••• C

G ••• C

A ••• T

5′ end

HO

3′ end

Figure 5.19
The 3′-to-5′ exonuclease activity of the proofreading function. The growing strand is cleaved to release a nucleotide containing the base G, which does not pair with the base A in the template strand.

Figure 5.20
The geometry of DNA replication. The new strand (red) is elongated by the addition of successive nucleotides to the 3′ end as the polymerase moves along the template strand in the 3′-to-5′ direction.

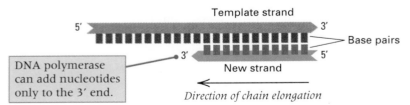

Template strand

5′   3′

Base pairs

3′   5′

DNA polymerase can add nucleotides only to the 3′ end.

New strand

Direction of chain elongation

5.6 Discontinuous Replication

In the model of replication suggested by Watson and Crick, which is illustrated in Figure 5.7, both daughter strands were supposed to be replicated as continuous units. However, no known DNA molecule replicates in this way. Because DNA polymerase can elongate a newly synthesized DNA strand only at its 3′ end, *one of the daughter strands is made in short fragments, which are then joined together.* The reason for this mechanism and the properties of these fragments are described next.

Fragments in the Replication Fork

As we have seen, all known DNA polymerases can add nucleotides only to a 3′-OH group. Thus if both daughter strands grew in the same overall direction, then each growing strand would need a 3′-OH terminus. However, the two strands of DNA are antiparallel, so only one of the growing strands can terminate in a free 3′-OH group; the other must terminate in a free 5′ end. The solution to this topological problem is that within a single replication fork, both strands grow in the 5′-to-3′ orientation, which requires that they grow in opposite directions along the parental strands. One strand of the newly made DNA is synthesized continuously (in the lower fork in Figure 5.21). The other strand (in the upper fork in Figure 5.21) is made in small **precursor fragments.** The precursor fragments are also known as **Okazaki fragments,** after their discoverer. The size of the precursor fragments is from 1000 to 2000 base pairs in prokaryotic cells and from 100 to 200 base pairs in eukaryotic cells. Because synthesis of the discontinuous strand is initiated only at intervals, at least one single-stranded region of the parental strand is always present on one side of the replication fork (the upper side in Figure 5.21). Single-stranded regions have been seen in high-resolution electron micrographs of replicating DNA molecules, as indicated by the arrows in Figure 5.22. Another implication of discontinuous replication of one strand is that the

3'-OH terminus of the continuously replicating strand is always ahead of the 5'-P terminus of the discontinuous strand; this is the physical basis of the terms **leading strand** and **lagging strand** that are used for the continuously and discontinuously replicating strands, respectively.

Next, we examine how synthesis of a precursor fragment is initiated.

Initiation by an RNA Primer

As emphasized earlier, DNA polymerases cannot initiate the synthesis of a new strand, so a free 3'-OH is needed. In most organisms, initiation is accomplished by a special type of RNA polymerase. RNA is usually a single-stranded nucleic acid consisting of four types of nucleotides joined together by 3'-to-5' phosphodiester bonds (the same chemical bonds as those in DNA). Two chemical differences distinguish RNA from DNA (Figure 5.23). The first difference is in the sugar component. RNA contains **ribose,** which is identical to the deoxyribose of DNA except for the

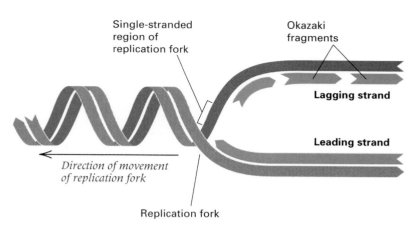

Single-stranded region of replication fork

Okazaki fragments

Lagging strand

Leading strand

Direction of movement of replication fork

Replication fork

Figure 5.21
Short fragments in the replication fork. For each tract of base pairs, the lagging strand is synthesized later than the leading strand.

presence of an -OH group on the 2' carbon atom. The second difference is in one of the four bases: The thymine found in DNA is replaced by the closely related pyrimidine *uracil* (*U*) in RNA. In RNA synthesis, a DNA strand is used as a template to form a complementary strand in which the bases

Figure 5.22
Electron micrograph (A) and an interpretive drawing (B) of a replicating θ molecule of phage λ DNA. In the electron micrograph, each arrow points to a short, single-stranded region of unreplicated DNA near the replication fork; the single-stranded regions are somewhat difficult to see, but they are apparent on close inspection. (Electron micrograph courtesy of Manuel Valenzuela.)

(A)

(B)

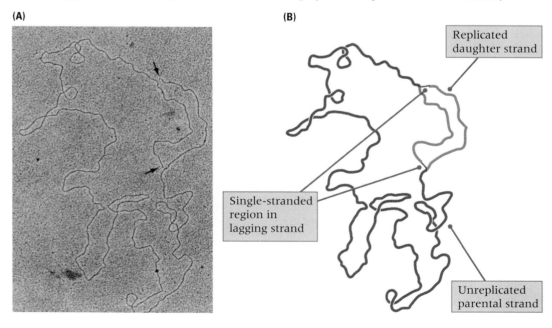

Replicated daughter strand

Single-stranded region in lagging strand

Unreplicated parental strand

in the DNA are paired with those in the RNA. Synthesis is catalyzed by an enzyme called an **RNA polymerase.** RNA polymerases differ from DNA polymerases in that they can initiate the synthesis of RNA chains without a primer.

DNA synthesis is initiated by using a short stretch of RNA that is base-paired with its DNA template. The size of the primer differs according to the initiation event. In *E. coli*, the length is typically from 2 to 5 nucleotides; in eukaryotic cells, it is usually from 5 to 8 nucleotides. This short stretch of RNA provides a primer onto which a DNA polymerase can add deoxynucleotides (Figure 5.24). The RNA polymerase that produces the primer for DNA synthesis is called **primase.** The primase is usually found in a multienzyme complex composed of 15 to 20 polypeptide chains and called a **primosome.** While it is being synthesized, each precursor fragment in the lagging strand has the structure shown in Figure 5.25.

The Joining of Precursor Fragments

The precursor fragments are ultimately joined to yield a continuous strand of DNA. This strand contains no RNA sequences, so the final stitching together of the lagging strand requires

- Removal of the RNA primer

- Replacement with a DNA sequence

- Joining where adjacent DNA fragments come into contact

In *E. coli*, the first two processes are accomplished by DNA polymerase I, and joining is catalyzed by the enzyme **DNA ligase,** which can link adjacent 3'-OH and 5'-P groups at a nick. How this is done is shown in Figure 5.26. Pol III extends the growing strand until the RNA of the primer of the previously synthesized precursor fragment is reached. Where the DNA and RNA segments meet, there is a single-strand interruption, or **nick.** The *E. coli* DNA ligase cannot seal the nick because a triphosphate is present (it can link only a 3'-OH and a 5'-*mono*phosphate). Here, however, DNA polymerase I takes over. This enzyme has an exonuclease activity that can remove nucleotides from the 5' end of a base-paired fragment. It is effective with both DNA and RNA. The activity is a **5' → 3' exonuclease** activity. Pol I acts at the nick and displaces it in the 5' → 3' direction by removing RNA nucleotides one by one and adding DNA nucleotides to the 3' end of the DNA strand. When all of the RNA nucleotides have been removed, DNA ligase joins the 3'-OH group to the terminal 5'-P of the precursor fragment. By this sequence of events, the precursor fragment is assimilated into the lagging strand. When the next precursor fragment reaches the RNA primer of the fragment just joined, the sequence begins again. Polymerase I is essential for DNA replication, because the 5'-to-3' exonuclease activity required for removing the RNA primer and joining the precursor fragments is not present in Pol III.

Other Proteins Needed For DNA Replication

The roles of some of the most important components of DNA replication are depicted in Figure 5.27. The replication process includes, in addition to the

Figure 5.23

Differences between DNA and RNA. The chemical groups that are highlighted are the distinguishing features of deoxyribose and ribose and of thymine and uracil.

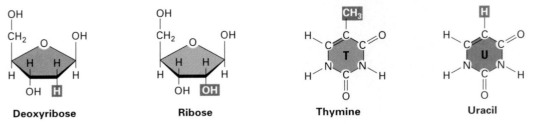

Deoxyribose   Ribose   Thymine   Uracil

Chapter 5 The Molecular Structure and Replication of the Genetic Material

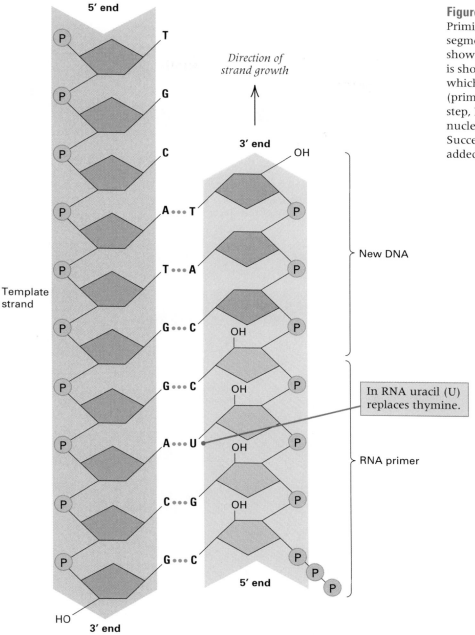

5′ end

T

G

C

A ••• T   OH

T ••• A

G ••• C   OH

G ••• C   OH

A ••• U   OH

C ••• G   OH

G ••• C

3′ end

Direction of strand growth

3′ end

New DNA

In RNA uracil (U) replaces thymine.

RNA primer

Template strand

HO

3′ end

5′ end

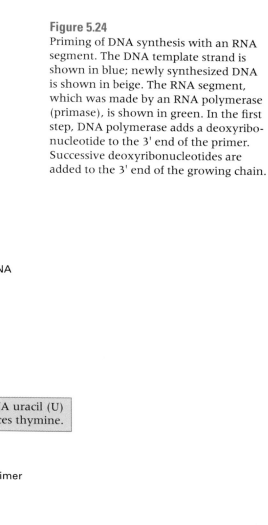

Figure 5.24
Priming of DNA synthesis with an RNA segment. The DNA template strand is shown in blue; newly synthesized DNA is shown in beige. The RNA segment, which was made by an RNA polymerase (primase), is shown in green. In the first step, DNA polymerase adds a deoxyribonucleotide to the 3′ end of the primer. Successive deoxyribonucleotides are added to the 3′ end of the growing chain.

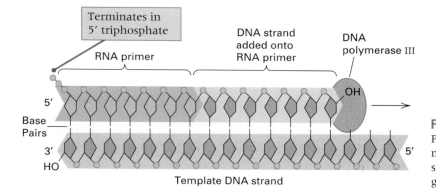

Terminates in 5′ triphosphate

RNA primer

DNA strand added onto RNA primer

DNA polymerase III

OH

5′

Base Pairs

3′

HO

Template DNA strand

5′

Figure 5.25
Prior to their being joined, each precursor fragment in the lagging strand has the structure shown here. The short RNA primer is shown in green.

Figure 5.26
Sequence of events in the joining of adjacent precursor fragments. (A) The DNA polymerase of the upstream fragment meets the RNA primer from the next precursor fragment downstream. (B) The RNA nucleotide immediately adjacent is excised and replaced with a DNA nucleotide. (C) Each RNA nucleotide is cleaved and replaced in turn. (D) When all the RNA nucleotides have been replaced, the adjacent ends of the DNA strand are joined.

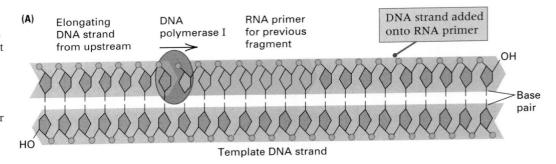

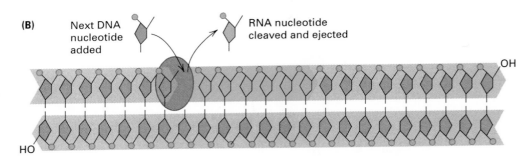

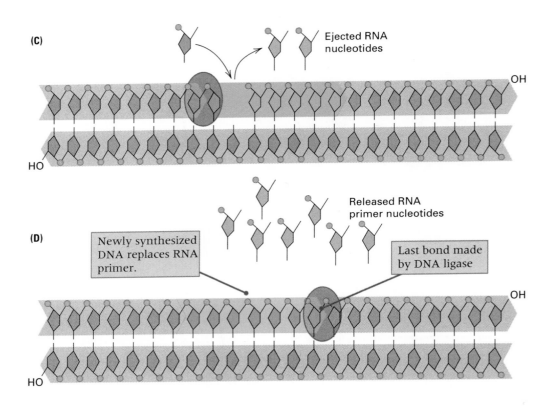

polymerase III complex and the RNA primase complex, at least one type of topoisomerase, **helicase** proteins that bind at the replication fork and unwind the double helix, **single-strand binding proteins** that bind with and stabilize the single-

stranded DNA at the replication fork, the DNA ligase that joins DNA fragments, and the polymerase I complex (not shown in Figure 5.27) that eliminates the RNA primers from precursor fragments before joining can take place.

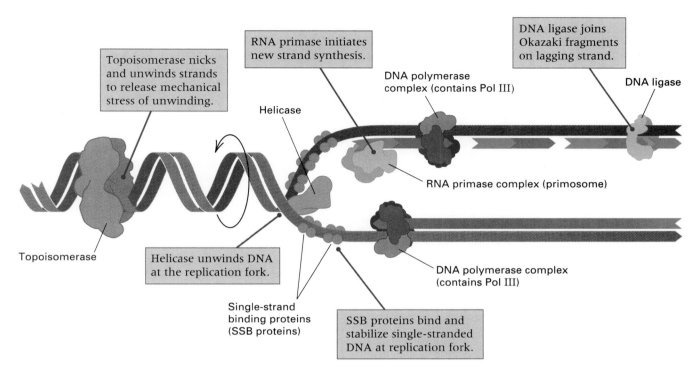

Figure 5.27
Role of some of the key proteins in DNA replication. The DNA polymerase III complex and the primase complex are both composed of multiple different polypeptide subunits. DNA polymerase I, which joins precursor fragments where they meet, is not shown.

5.7
The Isolation and Characterization of Particular DNA Fragments

This section and the following sections show how our knowledge of DNA structure and replication has been put to practical use in the development of procedures for the isolation and manipulation of DNA. One set of procedures is based on the ability of duplex DNA to be separated into single strands that, under the proper conditions, can form duplexes with other single strands having complementary sequences. This phenomenon is described next.

Denaturation and Renaturation

The double-stranded helical structure of DNA is maintained by forces that include hydrogen bonds between the bases of complementary pairs. When solutions of DNA are exposed to temperatures considerably higher than those normally encountered in most living cells or to excessively high pH, the hydrogen bonds break and the paired strands separate. Unwinding of the helix happens in less than a few minutes, the time depending on the length of the molecule. When the helical structure of DNA is disrupted and the strands are separated, the molecule is said to be **denatured.** A common way to detect denaturation is by measuring the capacity of DNA in solution to absorb ultraviolet light of wavelength 260 nm. The absorption at 260 nm (A_{260}) of a solution of single-stranded molecules is 37 percent higher than the absorption of the double-stranded molecules at the same concentration. When a DNA solution is slowly heated and the value of A_{260} is recorded at various temperatures, a curve called a **melting curve** is obtained. An example is shown in Figure 5.28. The melting transition is usually described in terms of the temperature at which the increase in the value of A_{260} is half complete. This temperature is called the **melting temperature** and is denoted by T_m.

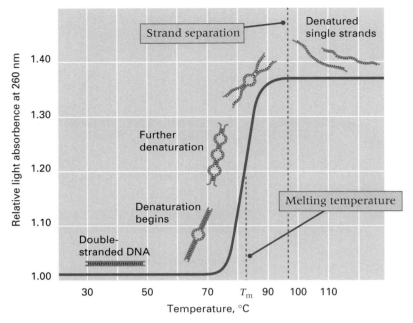

Figure 5.28
A melting curve of DNA showing the T_m and possible shapes of a DNA molecule at various degrees of denaturation.

The value of T_m increases with G + C content, because G−C pairs, joined by three hydrogen bonds, are stronger than A−T pairs, which are joined by two hydrogen bonds.

The single strands in a solution of denatured DNA can, under certain conditions, re-form double-stranded DNA. This process is called **renaturation** or **annealing.** For renaturation to happen, two requirements must be met: (1) The salt concentration must be high (>0.25 M) to neutralize the negative charges of the phosphate groups, which would otherwise cause the complementary strands to repel one another; and (2) the temperature must be high enough to disrupt hydrogen bonds that form at random between short sequences of bases within the same strand, but not so high that stable base pairs between the complementary strands would be disrupted. A temperature about 20°C below T_m is usually optimal. The initial phase of renaturation is a slow process, because its rate is limited by the random chance that a region of two complementary strands will come together to form a short sequence of cor-

rect base pairs. This initial pairing step is followed by a rapid pairing of the remaining complementary bases and rewinding of the helix. Rewinding is accomplished in a matter of seconds, and its rate is independent of DNA concentration (because the complementary strands have already found each other). Correct initial base pairing of all molecules in a sample is concentration-dependent and may require several minutes to many hours when standard conditions are used.

An example utilizing a hypothetical DNA molecule will enable us to understand some of the molecular details of renaturation. Consider the double-stranded DNA molecule in Figure 5.29A containing 30,000 base pairs, in which a specific sequence of six base pairs appears only twice. (In general, a six-base sequence would be expected to be found an average of about seven times in a molecule of this length.) This molecule is heat-denatured (Figure 5.29B) and then the temperature is lowered to promote renaturation. In the solution of these denatured molecules, a random collision between noncomplementary base sequences cannot initiate renaturation, but a collision that brings together sequences I and II' or I' and II can result in base pairing (Figure 5.29C). This pairing will be transient at the elevated temperatures used for renaturation, because the paired region is short and the adjacent bases in the two strands are out of register and unable to pair, as is required to form a double-stranded molecule. On the other hand, if a collision results in the pairing of sequence I with I' or II with II'—or any other short complementary sequences—then pairing of the adjacent bases (and all other bases in the strands) will proceed in a zipperlike action (Figure 5.29D). The main point is that only base pairing that brings the complementary sequences into register will cause renaturation.

Nucleic Acid Hybridization

Another important point is that because a solution of denatured DNA usually contains a large number of identical DNA molecules, a double-stranded molecule formed by renaturation is rarely composed of the same two strands that were paired before

(A)

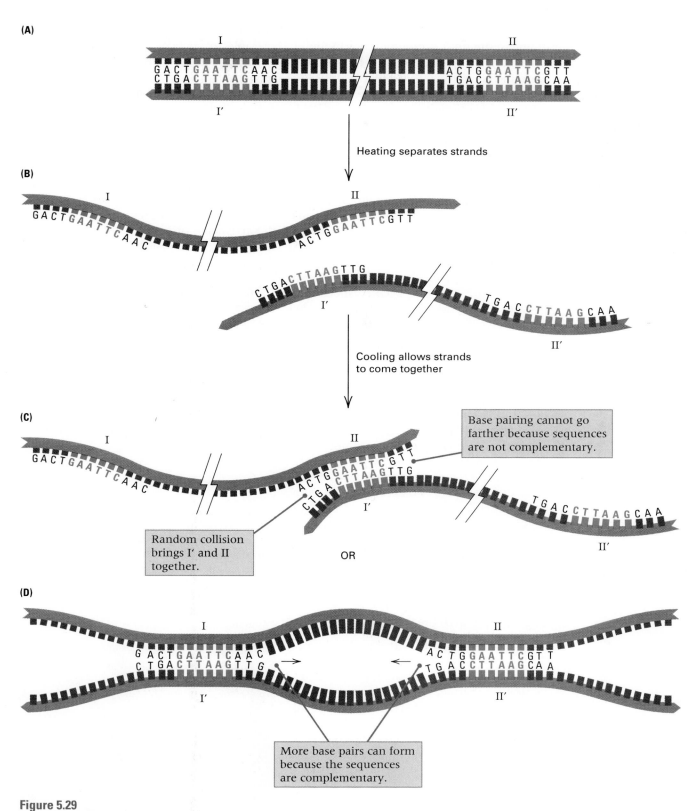

I
II

G A C T G A A T T C A A C
C T G A C T T A A G T T G

A C T G G A A T T C G T T
T G A C C T T A A G C A A

I′
II′

Heating separates strands

(B)

I
II

G A C T G A A T T C A A C
A C T G G A A T T C G T T

C T G A C T T A A G T T G
T G A C C T T A A G C A A

I′
II′

Cooling allows strands
to come together

(C)

I
II

G A C T G A A T T C A A C

A C T G G A A T T C G T T
C T G A C T T A A G T T G

Base pairing cannot go
farther because sequences
are not complementary.

Random collision
brings I′ and II
together.

I′

T G A C C T T A A G C A A

II′

OR

(D)

I
II

G A C T G A A T T C A A C
C T G A C T T A A G T T G

A C T G G A A T T C G T T
T G A C C T T A A G C A A

I′
II′

More base pairs can form
because the sequences
are complementary.

Figure 5.29

Denaturation and renaturation of a DNA duplex. (A) Original duplex. (B) High temperature causes the strands to come apart. Lowering the temperature allows short complementary stretches to come together. (C) If the sequences flanking the paired region are not complementary, then the pairing is unstable and the strands come apart again. (D) If the sequences flanking the paired region are complementary, then further base pairing stabilizes the renatured duplex.

renaturation. This phenomenon is the basis of one of the most useful methods in molecular genetics—**nucleic acid hybridization.** In this context, molecular hybrids are molecules formed when two single nucleic acid strands obtained by denaturing DNA from *different* sources have sufficient sequence complementarity to form a duplex. An example of the use of DNA-DNA hybridization is determination of the fraction of DNA in two different species that have common base sequences. To illustrate one procedure that is used, let us assume that two bacterial species have been grown in media that cause the DNA of one species to contain only the nonradioactive isotope of phosphorus, ^{31}P, and cause the DNA of the second species to contain the radioactive isotope, ^{32}P. The DNA molecules of the two species are isolated, broken into many small fragments, and denatured. The nonradioactive molecules are immobilized on a nitrocellulose filter, and the radioactive DNA is added. The temperature and salt concentration are raised to promote renaturation, and then, after a suitable period of time, the filter is washed. The existence of common base sequences will be detected by the presence of renatured fragments of radioactive ^{32}P-DNA on the filter. The extent of common base sequences determined in this manner in the DNA of two species generally agrees with the evolutionary relatedness of the species as indicated by direct DNA sequencing of a sample of their genes.

Restriction Enzymes and Site-Specific DNA Cleavage

One of the problems with breaking large DNA molecules into smaller fragments by random shearing is that the fragments containing a particular gene, or part of a gene, will all be of different sizes. With random shearing, owing to the random length of each fragment, it is not possible to isolate and identify a *particular* DNA fragment. However, there is an important enzymatic technique, described in this section, that can be used for cleaving DNA molecules at specific sites.

As we saw in Chapter 4, members of a class of enzymes known as **restriction endonucleases** or **restriction enzymes** are able to cleave DNA molecules at the positions at which particular, short sequences of bases are present. For example, the enzyme *Bam*HI recognizes the double-stranded sequence

$$5'-GGATCC-3'$$
$$3'-CCTAGG-5'$$

and cleaves each strand between the G-bearing nucleotides shown in red. Figure 5.30 shows how the regions that make up the active site of *Bam*HI contact the recognition site (blue) just prior to cleavage, and the cleavage reaction is indicated in Figure 5.31.

Table 5.3 lists six of the several hundred restriction enzymes that are known. Most restriction enzymes are isolated from bacteria, and they are named after the species in which they were found. *Bam*HI, for example, was isolated from *Bacillus amyloliquefaciens* strain H, and it is the first (I) restriction enzyme isolated from this organism. Most restriction enzymes recognize only one short base sequence, usually

Figure 5.30

Part of the restriction enzyme *Bam*HI in contact with its recognition site in the DNA (blue). The pink and green cylinders represent regions of the enzyme in which the amino acid chain is twisted in the form of a right-handed helix. [Courtesy of A. A. Aggarwal. From M. Newman, T. Strzelecka, L. F. Dorner, I. Schildkraut, and A. A. Aggarwal, 1995. *Science* 269:656.]

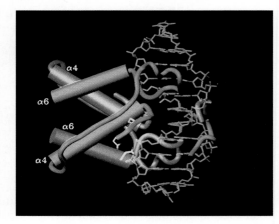

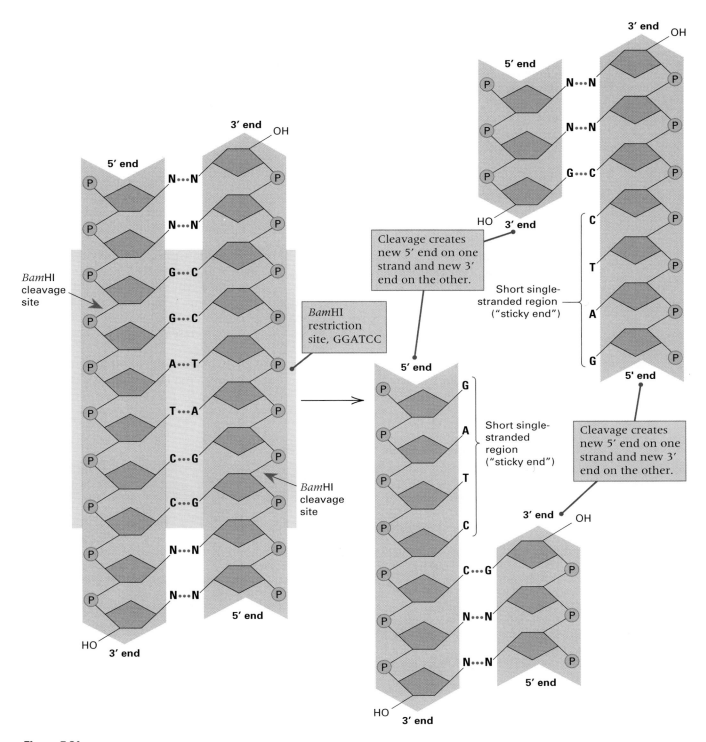

Figure 5.31

Mechanism of DNA cleavage by the restriction enzyme *Bam*HI. The enzyme makes a single cut in the backbone of each DNA strand wherever the duplex contains a *Bam*HI restriction site. Each cut creates a new 3' end and a new 5' end, separating the upper and lower parts of the duplex. In the case of *Bam*HI the cuts are staggered cuts, so the resulting ends terminate in single-stranded regions, each four base pairs in length.

four or six nucleotide pairs. The enzyme binds with the DNA at these sites and makes a break in each strand of the DNA molecule, producing 3'-OH and 5'-P groups at each position. The nucleotide sequence recognized for cleavage by a restriction enzyme is called the **restriction site** of the enzyme. The restriction enzymes in Table 5.3 all cleave their restriction site asymmetrically (at different sites in the two DNA strands), but some restriction enzymes cleave symmetrically (at the same site in both strands). The former leave **sticky ends** because each end of the cleaved site has a small, single-stranded overhang that is complementary in base sequence to the other end (Figure 5.31). In contrast, enzymes that have symmetrical cleavage sites yield DNA fragments that have **blunt ends.** In virtually all cases, the restriction site of a restriction enzyme reads the same on both strands, provided that the opposite polarity of the strands is taken into account; for example, each strand in the restriction site of *Bam*HI reads 5'-GGATCC-3' (Figure 5.31). A DNA sequence with this type of symmetry is called a **palindrome.** (In ordinary English, a palindrome is a word or phrase that reads the same forwards and backwards, such as "madam.")

Restriction enzymes have the following important characteristics:

- Most restriction enzymes recognize a single restriction site.

- The restriction site is recognized without regard to the source of the DNA.

- Because most restriction enzymes recognize a unique restriction site sequence, the number of cuts in the DNA from a particular organism is determined by the number of restriction sites present.

The DNA fragment produced by a pair of adjacent cuts in a DNA molecule is called a

Table 5.3 Some restriction endonucleases, their sources, and their cleavage sites

Enzyme	Microorganism	Target sequence and cleavage sites	Enzyme	Microorganism	Target sequence and cleavage sites
EcoRI	Escherichia coli	GAATTC / CTTAAG	HindIII	Haemophilus influenzae	AAGCTT / TTCGAA
BamHI	Bacillus amyloliquefaciens H	GGATCC / CCTAGG	PstI	Providencia stuartii	CTGCAG / GACGTC
HaeII	Haemophilus aegyptus	PuGCGCPy / PyCGCGPu	TaqI	Thermus aquaticus	TCGA / AGCT

Note: The vertical dashed line indicates the axis of symmetry in each sequence. Red arrows indicate the sites of cutting. The enzyme *Taq*I yields cohesive ends consisting of two nucleotides, whereas the cohesive ends produced by the other enzymes contain four nucleotides. Pu and Py refer to any purine and pyrimidine, respectively.

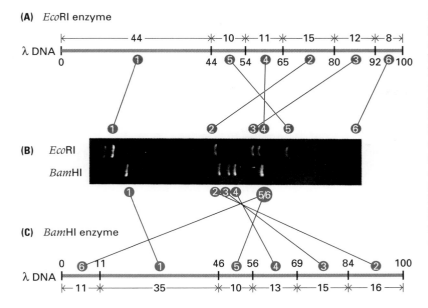

(A) *Eco*RI enzyme

(B) *Eco*RI
    *Bam*HI

(C) *Bam*HI enzyme

Figure 5.32
Restriction maps of λ DNA for the restriction enzymes. (A) *Eco*RI and (C) *Bam*HI. The vertical bars indicate the sites of cutting. The black numbers indicate the approximate percentage of the total length of λ DNA measured from the end of the molecule that is arbitrarily designated the left end. The numbers within the arrows are the lengths of the fragments, each expressed as percentage of the total length. (B) An electrophoresis gel of *Bam*HI and *Eco*RI enzyme digests of λ DNA. Numbers indicate fragments in order from largest (1) to smallest (6); the circled numbers on the maps correspond to the numbers beside the gel. The DNA has not undergone electrophoresis long enough to separate bands 5 and 6 of the *Bam*HI digest.

restriction fragment. A large DNA molecule will typically be cut into many restriction fragments of different sizes. For example, an *E. coli* DNA molecule, which contains 4.7×10^6 base pairs, is cut into several hundred to several thousand fragments, and mammalian nuclear DNA is cut into more than a million fragments. Although these numbers are large, they are actually quite small relative to the number of sugar-phosphate bonds in the DNA of an organism. Restriction fragments are usually short enough that they can be separated by electrophoresis and manipulated in various ways—for example, using DNA ligase to insert them into self-replicating molecules such as bacteriophage, plasmids, or even small artificial chromosomes. These procedures constitute **DNA cloning** and are the basis of one form of *genetic engineering,* discussed further in Chapter 9.

Because of the sequence specificity, *a particular restriction enzyme produces a unique set of fragments for a particular DNA molecule.* Another enzyme will produce a different set of fragments from the same DNA molecule. Figure 5.32A and 5.32C show the sites of cutting of *E. coli* phage λ DNA by the enzymes *Eco*RI and *Bam*HI. A map showing the unique sites of cutting of the DNA of a particular organism by a single enzyme is called a **restriction map.** The family of fragments produced by a single enzyme can be detected easily by gel electrophoresis of enzyme-treated DNA (Figure 5.32B), and particular DNA fragments can be isolated by cutting out the small region of the gel that contains the fragment and removing the DNA from the gel. Gel electrophoresis for the separation of DNA fragments is described next.

Gel Electrophoresis

The physical basis for the separation of the DNA fragments in Figure 5.32 is that DNA molecules are negatively charged and can move in an electric field. If the terminals of an electrical power source are connected to the opposite ends of a horizontal tube containing a DNA solution, then the molecules will move toward the positive end of the tube, at a rate that depends on the electric field strength and on the shape and size of the molecules. The movement of charged molecules in an electric field is called **electrophoresis.**

The type of electrophoresis most commonly used in genetics is **gel electrophoresis.** An experimental arrangement for gel electrophoresis of DNA is

Figure 5.33

Apparatus for gel electrophoresis capable of handling seven samples simultaneously. Liquid gel is allowed to harden in place, with an appropriately shaped mold placed on top of the gel during hardening in order to make "wells" for the samples (purple). After electrophoresis, the samples, located at various positions in the gel, are made visible by removing the plastic frame and immersing the gel in a solution containing a reagent that binds to or reacts with the separated molecules. The separated components of a sample appear as bands, which may be either visibly colored or fluorescent when illuminated with fluorescent light, depending on the particular reagent used. The region of a gel in which the components of one sample can move is called a *lane*. Thus, this gel has seven lanes.

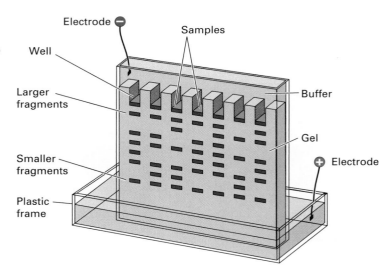

shown in Figure 5.33. A thin slab of a gel, usually agarose or acylamide, is prepared containing small slots (called wells) into which samples are placed. An electric field is applied, and the negatively charged DNA molecules penetrate and move through the gel. A gel is a complex molecular network that contains narrow, tortuous passages, so smaller DNA molecules pass through more easily; hence the rate of movement increases as the molecular weight decreases. For each molecule, the rate of movement depends primarily on the molecular size, provided the molecule is linear and not too large. Figure 5.34 shows the result of electrophoresis of a collection of double-stranded DNA molecules in an agarose gel. Each discrete region containing DNA is called a **band.**

The Southern Blot

Several techniques enable a researcher to locate a particular DNA fragment in a gel. One of the most generally applicable procedures is the **Southern blot.** In this procedure, a gel in which DNA molecules have been separated by electrophoresis is treated with alkali to denature the DNA and render it single-stranded. Then the DNA is transferred to a sheet of nitrocellulose in such a way that the relative positions of the DNA bands are maintained (Figure 5.35). The nitrocellulose, to which the single-stranded DNA binds tightly, is then exposed to denatured radioactive complementary RNA or

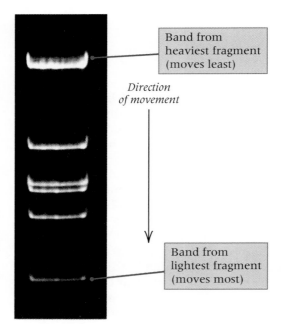

Figure 5.34

Gel electrophoresis of DNA. Molecules of different sizes were mixed and placed in a well. Electrophoresis was in the vertical direction. The DNA has been made visible by the addition of a dye (ethidium bromide) that binds only to DNA and that fluoresces when the gel is illuminated with short-wavelength ultraviolet light.

DNA (the **probe**) in a way that leads the complementary strands to anneal to form duplex molecules. Radioactivity becomes stably bound (resistant to removal by washing) to the DNA only at positions at

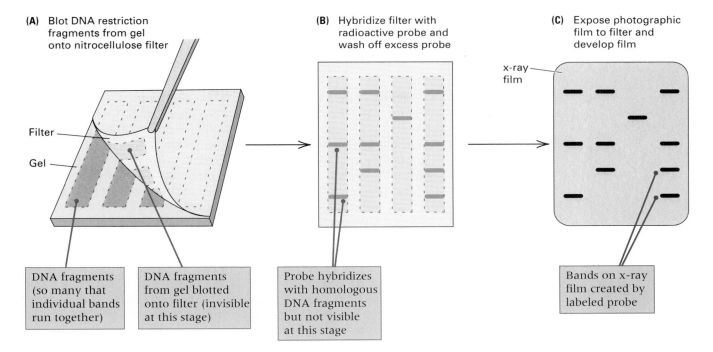

(A) Blot DNA restriction fragments from gel onto nitrocellulose filter

Filter

Gel

DNA fragments (so many that individual bands run together)

DNA fragments from gel blotted onto filter (invisible at this stage)

(B) Hybridize filter with radioactive probe and wash off excess probe

Probe hybridizes with homologous DNA fragments but not visible at this stage

(C) Expose photographic film to filter and develop film

x-ray film

Bands on x-ray film created by labeled probe

Figure 5.35

Southern blot. (A) DNA restriction fragments are separated by electrophoresis, blotted from the gel onto a nitrocellulose or nylon filter, and chemically attached by the use of ultraviolet light. (B) The strands are denatured and mixed with radioactive probe DNA, which binds with complementary sequences present on the filter. The bound probe remains, whereas unbound probe washes off. (C) Bound probe is revealed by darkening of photographic film placed over the filter. The positions of the bands indicate which restriction fragments contain DNA sequences homologous with those in the probe.

which base sequences complementary to the radioactive molecules are present, so that duplex molecules can form. The radioactivity is located by placing the paper in contact with x-ray film; after development of the film, blackened regions indicate positions of radioactivity. For example, a cloned DNA fragment from one species may be used as probe DNA in a Southern blot with DNA from another species; the probe will hybridize only with restriction fragments containing DNA sequences that are complementary enough to allow stable duplexes to form.

5.8
The Polymerase Chain Reaction

It is also possible to obtain large quantities of a particular DNA sequence merely by selective replication. The method for selec-

tive replication is called the **polymerase chain reaction (PCR),** and it uses DNA polymerase and a pair of short, synthetic **oligonucleotide primers,** usually about 20 nucleotides in length, that are complementary in sequence to the ends of any DNA sequence of interest. Starting with a mixture containing as little as one molecule of the fragment of interest, repeated rounds of DNA replication increase the number of molecules exponentially. For example, starting with a single molecule, 25 rounds of DNA replication will result in $2^{25} = 3.4 \times 10^7$ molecules. This number of molecules of the amplified fragment is so much greater than that of the other unamplified molecules in the original mixture that the amplified DNA can often be used without further purification. For example, a single fragment of 3000 base pairs in *E. coli* accounts for only 0.06 percent of the total DNA in this organism. However, if this single fragment were replicated through 25 rounds of replication, then 99.995 percent of the resulting mixture would consist of the amplified sequence.

An outline of the polymerase chain reaction is shown in Figure 5.36. The DNA sequence to be amplified and the oligonucleotide sequences are shown in contrasting colors. The oligonucleotides act as primers for DNA replication, because they anneal to the ends of the sequence to be amplified and become the substrates for chain elongation by DNA polymerase. In the first cycle of PCR amplification, the DNA is denatured to separate the strands. The denaturation temperature is usually around 95°C. Then the temperature is decreased to allow annealing in the presence of a vast excess of the primer oligonucleotides. The annealing temperature is typically in the range from 50°C to 60°C, depending largely on the G + C content of the oligonucleotide primers. The temperature is raised slightly, to about 70°C, for the elongation of each primer. The first cycle in PCR produces two copies of each molecule containing sequences complementary to the primers. The second cycle of PCR is similar to the first. The DNA is denatured, then renatured in the presence of an excess of primer oligonucleotides, and the primers are elongated by DNA polymerase; after this cycle, there are four copies of each molecule present in the original mixture. The steps of denaturation, renaturation, and replication are repeated from 20–30 times, and in each cycle, the number of molecules of the amplified sequence is doubled. The theoretical result of 25 rounds of amplification is 2^{25} copies of each template molecule present in the original mixture.

Implementation of PCR with conventional DNA polymerases is not practical, because at the high temperature necessary for denaturation, the polymerase is itself irreversibly unfolded and becomes inactive. However, DNA polymerase isolated from certain archaebacteria is heat stable because the organisms normally live in hot springs at temperatures well above 90°C, such as are found in Yellowstone National Park. Such organisms are said to be **thermophiles**. The most widely used heat-stable DNA polymerase is called *Taq* polymerase, because it was originally isolated from the thermophilic archaebacterium *Thermus aquaticus.*

Figure 5.37 presents a closer look at the primer oligonucleotide sequences and their relationship to the DNA sequence to be amplified. The primers are designed to anneal to opposite DNA strands at the extreme ends of the region to be amplified. The primers are oriented with their 3' ends facing the sequence to be amplified so that, in DNA synthesis, the new strands grow toward each other but along complementary template strands. In this way, each newly synthesized strand terminates in a sequence that can anneal with the complementary primer and so can be used for further amplification.

PCR amplification is very useful for generating large quantities of a specific DNA sequence. The principal limitation of the technique is that the DNA sequences at the ends of the region to be amplified must be known so that primer oligonucleotides can be synthesized. In addition, sequences longer than about 5000 base pairs cannot be replicated efficiently by conventional PCR procedures. On the other hand, there are many applications in which PCR amplification is useful. PCR can be employed to study many different mutant alleles of a gene whose wildtype sequence is known in order to identify the molecular basis of the mutations. Similarly, DNA sequence variation among alleles present in natural populations can easily be determined by using PCR. The PCR procedure has also come into widespread use in clinical laboratories for diagnosis. To take just one very important example, the presence of the human immunodeficiency virus (HIV), which causes acquired immune deficiency syndrome (AIDS), can be detected in trace quantities in blood banks via PCR by using primers complementary to sequences in the viral genetic material. These and other applications of PCR are facilitated by the fact that the procedure lends itself to automation—for example, by the use of mechanical robots to set up the reactions.

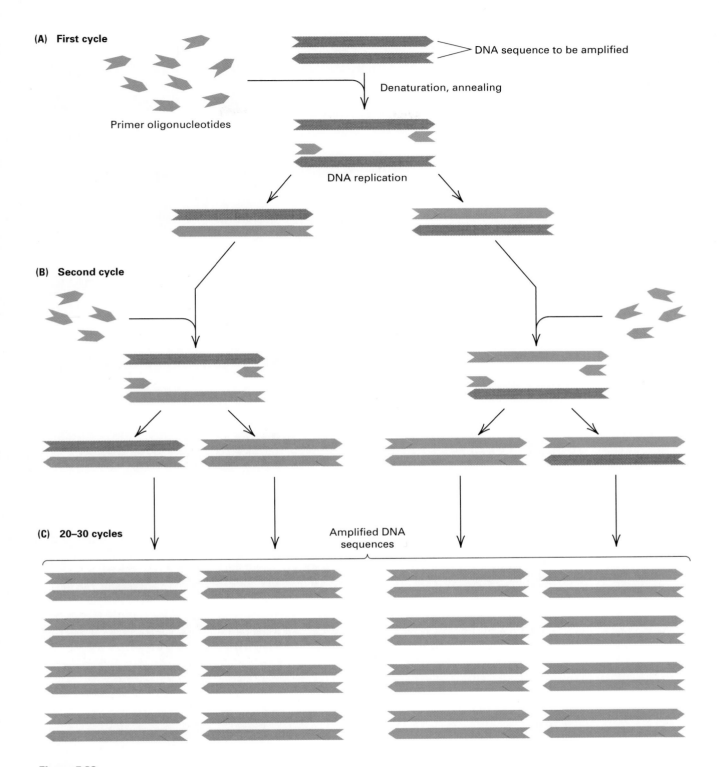

Figure 5.36
Polymerase chain reaction (PCR) for amplification of particular DNA sequences. Oligonucleotide primers (green) that are complementary to the ends of the sequence of interest (blue) are used in repeated rounds of denaturation, annealing, and DNA replication. Copies of the target sequence are shown in pink. The number of copies of the target sequence doubles in each round of replication, eventually overwhelming any other sequences that may be present.

Figure 5.37
Example of primer sequences used in PCR amplification. Newly synthesized DNA is shown in pink. (A) Original DNA duplex (blue) and primers (green). (B) The original DNA duplex is denatured and the primers annealed. (C) After one round of binding of the primers and elongation. Each primer has been extended at the 3' end. (D) After another round of replication. Each primer is again extended at the 3' end, but the elongation terminates at the 5' end of the primer oligonucleotide on the other strand. After the first round of replication, each newly synthesized DNA molecule forms a new template, so the number of copies increases exponentially.

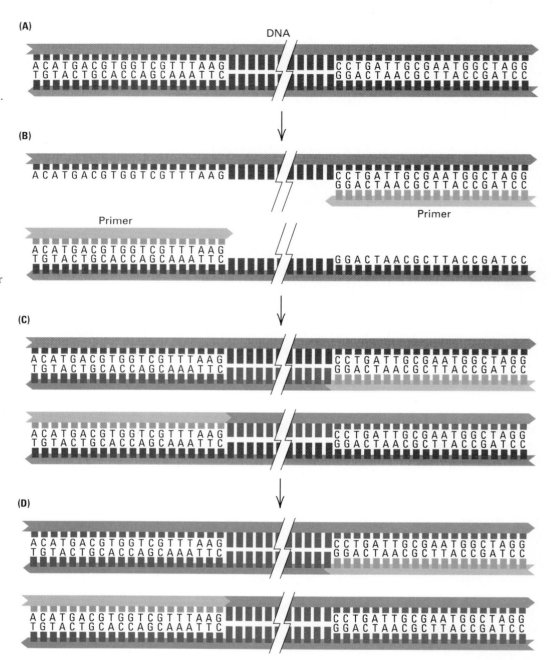

5.9
Determination of the Sequence of Bases in DNA

A great deal of information about gene structure and gene expression can be obtained by direct determination of the sequence of bases in a DNA molecule. Several techniques are available for base sequencing; the most widely used method is described in this section. No technique can determine the sequence of bases in an entire chromosome in a single experiment, so chromosomes are first cut into fragments a few hundred base pairs long, a size that can be sequenced easily. To obtain the sequence of a long stretch of DNA, a set of overlapping fragments is prepared, the sequence of each is determined, and all sequences are then combined. The procedures are straightforward, and DNA sequences have accumulated at such a rapid rate that large computer databases are necessary to manage the hundreds of millions of nucleotides of DNA sequence already determined for a variety of genes from many organisms. A still larger database would be required to handle the DNA in human sperm or eggs, which consists of 3×10^9 nucleotide pairs. Although the sequencing technique described here is the manual method still in use in many research laboratories, in Chapter 9 we will see how DNA sequencing can be automated and how DNA-sequencing machines work.

The **dideoxy sequencing method** employs DNA synthesis in the presence of small amounts of nucleotides that contain the sugar **dideoxyribose** instead of deoxyribose (Figure 5.38). Dideoxyribose lacks the 3'-OH group, which is essential for attachment of the next nucleotide in a growing DNA strand, so incorporation of a dideoxynucleotide instead of a deoxynucleotide immediately terminates further synthesis of the strand. To sequence a DNA strand, four DNA synthesis reactions are carried out. Each reaction contains the single-stranded DNA template to be sequenced, a single oligonucleotide primer complementary to a stretch of the template strand, all four deoxyribonucleoside triphosphates, and a small amount of *one* of the nucleoside triphosphates in the dideoxy form. Each reaction produces a set of fragments that terminates at the point at which a dideoxynucleotide was randomly incorporated in place of the normal deoxynucleotide. Therefore, in each of the four reactions, the lengths of the fragments are determined by the positions in the daughter strand at which the particular dideoxynucleotide present in that reaction was incorporated. The sizes of the fragments produced by chain termination are determined by gel electrophoresis, and the base sequence is then determined by the following rule:

> If a fragment containing *n* nucleotides is generated in the reaction containing a particular dideoxynucleotide, then position *n* in the *daughter strand* is occupied by the base present in the dideoxynucleotide. The numbering is from the 5' nucleotide of the primer.

For example, if a 93-base fragment is present in the reaction containing the dideoxy

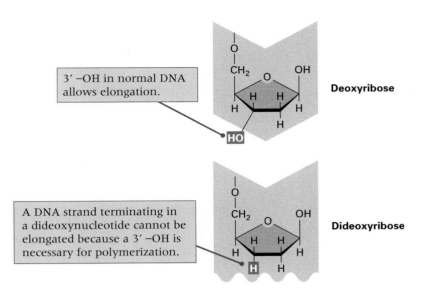

Figure 5.38
Structures of normal deoxyribose and the dideoxyribose sugar used in DNA sequencing. The dideoxyribose has a hydrogen atom (red) attached to the 3' carbon, in contrast with the hydroxyl group (red) at this position in deoxyribose. Because the 3' hydroxyl group is essential for the attachment of the next nucleotide in line in a growing DNA strand, the incorporation of a dideoxynucleotide immediately terminates synthesis.

form of dATP, then the 93rd base in the daughter strand produced by DNA synthesis must be an adenine (A). Because most native duplex DNA molecules consist of complementary strands, it does not matter whether the sequence of the template strand or that of the daughter strand is determined. The sequence of the template strand can be deduced from the daughter strand because their nucleotide sequences are complementary. However, in practice, both strands of a molecule are usually sequenced independently and compared in order to eliminate most of the errors that can be made by misreading the gels.

With short, single-stranded fragments, molecules that differ in size by a single base can be separated by electrophoresis in an acrylamide gel. For example, if a reaction mixture contains a set of DNA fragments consisting of 20, 21, 22, . . . , 100 nucleotides, then electrophoresis will yield separate bands—one containing the fragment of size 20, the next the fragment of size 21, and so forth. This extraordinary sensitivity to size is the basis of the DNA-sequencing procedure.

The Sequencing Procedure

The procedure for sequencing a DNA fragment is diagrammed in Figure 5.39. In this example, the primer length is 20 nucleotides. Termination at G produces fragments of 21, 24, 27, and 29 nucleotides; termination at A produces fragments of 22 and 30 nucleotides; termination at T produces a fragment of 26 nucleotides; and termination at C produces fragments of 23, 25, and 28 nucleotides. Typically, either the primer or one of the deoxynucleotides used contains a radioactive atom, so that after electrophoresis to separate the fragments by size, the fragments can be visualized as bands on photographic film placed over the gel. After electrophoresis and autoradiography, the positions of the bases in the daughter strand can be read directly from the gel, starting with position 21, as GACGCT-GCGA. This corresponds to the template strand sequence CTGCGACGCT.

Figure 5.40 is an autoradiogram of an actual sequencing gel. The shortest fragments are those that move the fastest and farthest. Each fragment contains the primer fragment at the 5' end of the daughter strand. The sequence can be read

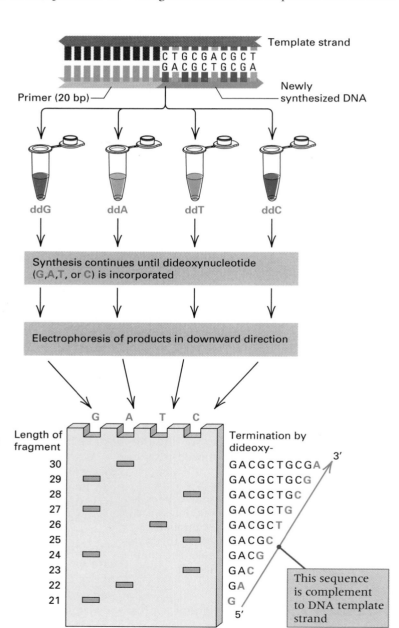

Figure 5.39

Dideoxy method of DNA sequencing. Four DNA synthesis reactions are carried out in the presence of all normal nucleotides plus a small amount of *one* of the dideoxynucleotides containing G, A, T, or C. Synthesis continues along the template strand until a dideoxynucleotide is incorporated. The products that result from termination at each dideoxynucleotide are indicated at the right. The fragments are separated by size by electrophoresis, and the positions of the nucleotides are determined directly from the gel. In this example, the length of the primer needed to initiate DNA synthesis is 20 nucleotides. Lengths of the DNA fragments are shown at the left of the gel. The sequence of the daughter strand is read from the bottom of the gel as 5'-GACGCTGCGA-3'.

directly from the bottom to the top of the gel. The sequence of the first 16 bases in the segment in this gel, read from bottom to top, is

5'-CACTGCCTGCGCCCAG-3'

and so forth.

Clinical Use of Dideoxynucleoside Analogs

Our knowledge of DNA structure and replication has applications not only in procedures for the manipulation of DNA but also in the development of drugs for clinical use. An example is in one approach to the treatment of AIDS. A number of dideoxynucleoside analogs are effective in inhibiting replication of the viral genetic material. A few of these are illustrated in Figure 5.41. Recall that a *nucleoside* is a base attached to a sugar without a phosphate. A nucleoside **analog** is a molecule similar, but not identical, in structure to a nucleoside. In Figure 5.41, ddC is the normal dideoxyribocytidine nucleoside. It is effective against AIDS, as are the dideoxynucleoside analogs AZT, D4T, and ddI (and other such analogs). The nucleoside, rather than the nucleotide, is used in therapy because the nucleotide, having a highly charged phosphate group, cannot cross the cell membrane as easily. The emergence of these drugs from our basic knowledge of DNA structure and replication is one of the prime examples of the fact that "pure" science may have many unforeseen practical applications. The basic experiments on DNA were carried out long before the recognition of AIDS as a distinct infectious disease and the discovery that HIV is the causative agent.

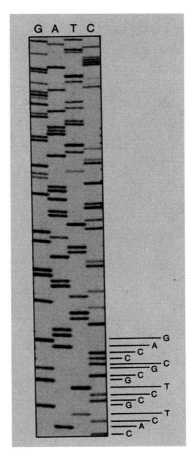

Figure 5.40
A section of a dideoxy sequencing gel. The sequence is read from the bottom to the top. Each horizontal row represents a single nucleotide position in the DNA strand synthesized from the template. The vertical columns result from termination by the dideoxy forms of G, A, T, or C. The sequence from the lower part of the gel is indicated.

Figure 5.41
A few of the drugs that have been found to be effective in the treatment of AIDS by interfering with the replication of HIV virus. The technical names of the substances are as follows: ddI is 2',3'-dideoxyinosine; ddC is 2',3'-dideoxycytidine; AZT is 3'-azido-2',3'-dideoxythymidine; and D4T is 2',3'-didehydro-2',3'-dideoxythymidine.

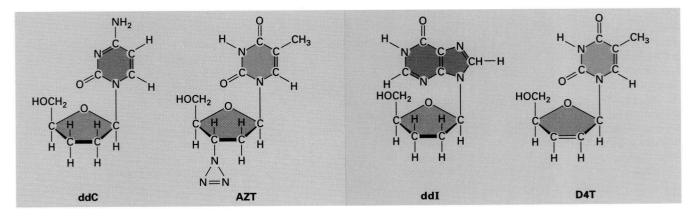

Chapter Summary

DNA is a double-stranded polymer consisting of deoxyribonucleotides. A nucleotide has three components: a base, a sugar (deoxyribose in DNA, ribose in RNA), and a phosphate. Sugars and phosphates alternate in forming a single polynucleotide chain with one terminal 3'-OH group and one terminal 5'-P group. In double-stranded (duplex) DNA, the two strands are antiparallel: Each end of the double helix carries a terminal 3'-OH group in one strand and a terminal 5'-P group in the other strand. Four bases are found in DNA: adenine (A) and guanine (G), which are purines; and cytosine (C) and thymine (T), which are pyrimidines. Equal numbers of purines and pyrimidines are found in double-stranded DNA, because the bases are paired as A−T pairs and G−C pairs (Chargaff's rules). This pairing holds the two polynucleotide strands together in a double helix. The base composition of DNA varies from one organism to the next. The information content of a DNA molecule resides in the sequence of bases along the chain, and each gene consists of a unique sequence.

The double helix replicates by using enzymes called DNA polymerases, but many other proteins also are needed. Replication is semiconservative in that each parental single strand, called a template strand, is found in one of the double-stranded progeny molecules. Semiconservative replication was first demonstrated in the Meselson-Stahl experiment, which used equilibrium density-gradient centrifugation to separate DNA molecules containing two ^{15}N-labeled strands, two ^{14}N-labeled strands, or one of each. Replication proceeds by a DNA polymerase (1) bringing in a nucleotide triphosphate with a base capable of hydrogen-bonding with the corresponding base in the template strand and (2) joining the 5'-P group of the nucleotide to the free 3'-OH group of the growing strand. (The terminal P−P from the nucleotide triphosphate is cleaved off and released.) Because double-stranded DNA is antiparallel, only one strand (the leading strand) grows in the direction of movement of the replication fork. The other strand (the lagging strand) is synthesized in the opposite direction as short fragments (Okazaki fragments) that are subsequently joined together. DNA polymerases cannot initiate synthesis, so a primer is always needed. The primer is an RNA fragment made by an RNA-polymerizing enzyme called primase; the RNA primer is removed at later stages of replication. DNA molecules of prokaryotes usually have a single replication origin; eukaryotic DNA molecules usually have many origins.

Restriction enzymes cleave DNA molecules at the positions of specific sequences (restriction sites) of usually four or six nucleotides. Each restriction enzyme produces a unique set of fragments for any particular DNA molecule. These fragments can be separated by electrophoresis and used for purposes such as DNA sequencing. The positions of particular restriction fragments in a gel can be visualized by means of a Southern blot, in which radioactive probe DNA is mixed with denatured DNA made up of single-stranded restriction fragments that have been transferred to a filter membrane after electrophoresis. The probe DNA will form stable duplexes (anneal or renature) with whatever fragments contain sufficiently complementary base sequences, and the positions of these duplexes can be determined by autoradiography of the filter. Particular DNA sequences can also be amplified without cloning by means of the polymerase chain reaction (PCR), in which short, synthetic oligonucleotides are used as primers to replicate repeatedly and amplify the sequence between them.

The base sequence of a DNA molecule can be determined by dideoxynucleotide sequencing. In this method, the DNA is isolated in discrete fragments containing several hundred nucleotide pairs. Complementary strands of each fragment are sequenced, and the sequences of overlapping fragments are combined to yield the complete sequence. The dideoxy sequencing method uses dideoxynucleotides to terminate daughter strand synthesis and reveal the identity of the base present in the daughter strand at the site of termination.

Key Terms

analog
annealing
antiparallel
B form
band
base composition
bidirectional replication
blunt ends
Chargaff's rules
daughter strand
denaturation
dideoxyribose
dideoxy sequencing method
DNA cloning
DNA ligase

DNA polymerase
DNA polymerase I (Pol I)
DNA polymerase III (Pol III)
editing function
endonuclease
equilibrium density-gradient
    centrifugation
exonuclease
5'-P group
5' → 3' exonuclease
gel electrophoresis
helicase
initiation
lagging strand
leading strand

major groove
melting curve
melting temperature
minor groove
nick
nuclease
nucleic acid hybridization
nucleoside
nucleotide
Okazaki fragments
oligonucleotide primers
palindrome
parent strand
percent G + C
phosphodiester bond

polarity
polymerase α
polymerase chain reaction (PCR)
polymerase γ
polynucleotide chain
precursor fragment
primase
primer
primosome
probe DNA
proofreading function
purine
pyrimidine
renaturation

replication fork
replication origin
restriction endonuclease
restriction enzyme
restriction fragment
restriction map
restriction site
ribose
RNA polymerase
rolling-circle replication
semiconservative replication
single-strand binding proteins
Southern blot
sticky ends

template
thermophiles
θ replication
3'-OH group
topoisomerase
ultracentrifuge
Z-form DNA

Review the Basics

- What are the four bases commonly found in DNA? Which form base pairs? What five-carbon sugar is found in DNA? What is the difference between a nucleoside and a nucleotide?

- How many phosphate groups are there per base in DNA, and how many phosphates are there in each precursor for DNA synthesis?

- Which chemical groups are at the ends of a single polynucleotide strand?

- What is the relationship between the amount of DNA in a somatic cell and the amount in a gamete?

- Name four requirements for initiation of DNA synthesis. To what chemical group in a DNA chain is an incoming nucleotide added, and what group in the nucleotide reacts with the DNA terminus?

- In what sense are the two strands of DNA antiparallel?

- How does the polymerase chain reaction work? What is it used for? What information about the target sequence must be known in advance?

Guide to Problem Solving

Problem 1: A technique is used for determining the base composition of double-stranded DNA. Rather than giving the relative amounts of each of the four bases—[A], [T], [G], and [C]—it yields the value of the ratio [A]/[C]. If this ratio is 1/3, what are the relative amounts of each of the four bases?

Answer: In double-stranded DNA, [A] = [T] and [G] = [C] because of the base pairing between complementary strands. Therefore, if [A]/[C] = 1/3, then [C] = [G] = 3 × [A]. Because [A] + [T] + [G] + [C] = 1, everything can be put in terms of [A] as

$$[A] + [A] + 3[A] + 3[A] = 1$$

or

$$[A] = [T] = 0.125$$

This makes

$$[C] = [G] = (1 - 0.25)/2 = 0.375$$

In other words, the DNA is 12.5 percent A, 12.5 percent T, 37.5 percent G, and 37.5 percent C.

Problem 2: The restriction enzyme *Eco*RI cleaves double-stranded DNA at the sequence 5'-GAATTC-3', and the restriction enzyme *Hind*III cleaves at 5'-AAGCTT-3'. A 20-kilobase (kb) circular plasmid is digested with each enzyme individually and then in combination, and the resulting fragment sizes are determined by means of electrophoresis. The results are as follows:

*Eco*RI alone: fragments of 6 kb and 14 kb

*Hind*III alone: fragments of 7 kb and 13 kb

*Eco*RI and *Hind*III: fragments of 2 kb, 4 kb, 5 kb and 9 kb

How many possible restriction maps are compatible with these data? For each possible restriction map, make a diagram of the circular molecule and indicate the relative positions of the *Eco*RI and *Hind*III restriction sites.

Answer: Because the single-enzyme digests give two bands each, there must be two restriction sites for each enzyme in the molecule. Furthermore, because digestion with *Hind*III makes both the 6-kb and the 14-kb restriction fragments disappear, each of these fragments must contain one *Hind*III site. Considering the sizes of the fragments in the double digest, the 6-kb *Eco*RI fragment must be cleaved into 2-kb and 4-kb fragments, and the 14-kb *Eco*RI fragment must be cleaved into 5-kb and 9-kb fragments. Two restriction maps are compatible with the data, depending on which end of the 6-kb *Eco*RI fragment the *Hind*III site is nearest. The position of the remaining *Hind*III site is determined by the fact that the 2-kb and 5-kb fragments in the double digest must be adjacent in the intact molecule in order for a 13-kb fragment to be produced by *Hind*III digestion alone.

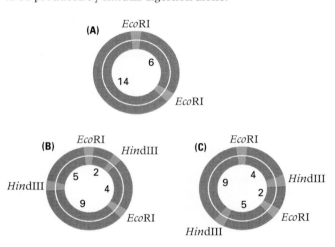

The accompanying figure shows the relative positions of the *Eco*RI sites (part A). Parts B and C are the two possible restriction maps, which differ according to whether the *Eco*RI site at the top generates the 2-kb or the 4-kb fragment in the double digest.

Problem 3: The DNA sequencing gels diagrammed in parts A and B of the accompanying illustration were obtained from human DNA by the dideoxy sequencing method. Part A comes from a person homozygous for the normal allele of the cystic fibrosis gene, part B from an affected person homozygous for a mutant allele. Use the bands in the gel to answer the following questions.

(a) Why are there discrete bands in the gel?

(b) Which end of the gel (top or bottom) corresponds to the 5' end of the DNA sequence?

(c) What nucleotide sequence is indicated by the bands in the gels?

(d) What is the sequence of the complementary strand in duplex DNA?

(e) Compare the sequences in parts A and B. What is special about the three nucleotides that give the bands indicated in red?

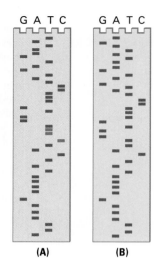

Answer:

(a) Each band results from a fragment of DNA whose replication was terminated by the incorporation of a dideoxy nucleotide. The fragments differ by one nucleotide in length, and they are arranged by size with the smallest fragments at the bottom, the largest at the top.

(b) Because DNA replication elongates at the 3' end, each fragment of increasing length has an additional 3' nucleotide. Therefore, the DNA sequence is oriented with the 5' end at the bottom of the gel and the 3' end at the top.

(c) The sequences, read directly from the gel from bottom to top, are (part A)

5'-ATTAAAGAAAATATCATCTTTGGTGTTTCCTATGATGAATAT-3'

and (part B),

5'-ATTAAAGAAAATATCATTGGTGTTTCCTATGATGAATATAGA-3'

(d) The complementary strands in duplex DNA are antiparallel and have A paired with T and G paired with C. For part A, the complementary strand is

3'-TAATTTCTTTTATAGTAGAAACCACAAAGGATACTACTTATA-5'

and, for part B, the complementary strand is

3'-TAATTTCTTTTATAGTAACCACAAAGGATACTACTTATATCT-5'

(e) The three nucleotides in red in the wildtype gene are missing in the mutant gene. This three-base deletion is found in about 70 percent of the mutations in cystic fibrosis patients, and it results in a missing amino acid at position 508 in the corresponding protein.

GeNETics on the web will introduce you to some of the most important sites for finding genetic information on the Internet. To complete the exercises below, visit the Jones and Bartlett home page at

http://www.jbpub.com/genetics

Select the link to *Genetics: Principles and Analysis* and then choose the link to *GeNETics on the web.* You will be presented with a chapter-by-chapter list of highlighted keywords.

GeNETics EXERCISES

Select the highlighted keyword in any of the exercises below, and you will be linked to a web site containing the genetic information necessary to complete the exercise. Each exercise suggests a specific, written report that makes use of the information available at the site. This report, or an alternative, may be assigned by your instructor.

1. The keyword **DNA** will connect you with a site that has an excellent collection of DNA structural models, including some animations. Browse through some of these and you will get a better intuitive feel for the structure and the manner in which the two strands are paired. If assigned to do so, take a close look at DNA model Number 24. There is no legend, but you should be able to deduce the color coding of the spheres. What type of atom corresponds to red? To green? To blue? What is the color of the phosphorous atoms?

2. One of the principal tools of the molecular geneticist is the **polymerase chain reaction**, which is unique in its ability to bring about exponential amplification of a particular DNA sequence present in a large background of other sequences. The keyword site summarizes the development of this key method. If assigned to do so, write a paragraph describing two innovations that were instrumental in the maturation of the method, and explain why each was important.

3. In DNA sequencing, there is sometimes uncertainty in the identification of a particular base because the gel pattern for that base is not completely unambiguous. An **ambiguous base** is denoted in a sequence by any one of a standard set of abbreviations. For example, the letter M stands for the possibilities "either A or C," V stands for "either A or C or G," and N (or X) stands for "unknown base." The keyword will access the complete set of abbreviations. If assigned to do so, organize the abbreviations in a systematic way of your choosing.

MUTABLE SITE EXERCISES

The Mutable Site Exercise changes frequently. Each new update includes a different exercise that makes use of genetics resources available on the World Wide Web. Select the **Mutable Site** for Chapter 5, and you will be linked to the current exercise that relates to the material presented in this chapter.

PIC SITE

The Pic Site showcases some of the most visually appealing genetics sites on the World Wide Web. To visit the showcase genetics site, select the **Pic Site** for Chapter 5.

Analysis and Applications

5.1 What is the base sequence of a DNA strand that is complementary to the hexanucleotide 3'-AGGCTC-5'? Label the termini as to 5' or 3'.

5.2 What is a nuclease enzyme, and how do endonucleases and exonucleases differ?

5.3 In what direction (5' → 3' or 3' → 5') does a DNA polymerase move along the template strand? How do organisms solve the problem that all DNA polymerases move in the same direction along a template strand, yet double-stranded DNA is antiparallel?

5.4 What is the chemical difference between the groups joined by DNA polymerase and DNA ligase?

5.5 What are three enzymatic activities of DNA polymerase I?

5.6 Why can RNA polymerases and primases initiate DNA replication, whereas DNA polymerases cannot?

5.7 In gel electrophoresis, do smaller double-stranded molecules move more slowly or more rapidly than larger molecules?

5.8 Consider a hypothetical phage whose DNA replicates exclusively by rolling-circle replication. A phage with radioactive DNA in both strands infects a bacterium and is allowed to replicate in a nonradioactive medium. Assume that only daughter DNA from the elongating branch ever gets packaged into progeny phage particles.

(a) What fraction of the parental radioactivity will appear in progeny phage?

(b) How many progeny phage will contain radioactive DNA, and how will the number be affected by recombination?

5.9 What is the fundamental difference between the initiation of θ replication and that of rolling-circle replication?

5.10 When the base composition of DNA from the bacterium *Mycobacterium tuberculosis* was determined, 18 percent of the bases were found to be adenine.

(a) What is the percentage of cytosine?

(b) What is the entire base composition of the DNA and the [G] + [C] content?

5.11 The double-stranded DNA molecule of a newly discovered virus was found by electron microscopy to have a length of 34 μm.

(a) How many nucleotide pairs are present in one of these molecules?

(b) How many complete turns of the two polynucleotide chains are present in such a double helix?

5.12 An elegant combined chemical and enzymatic technique enables one to identify "nearest neighbors" of bases (adjacent bases in a DNA strand). For example, if the single-stranded tetranucleotide 5'-AGTC-3' were treated in this way, the nearest neighbors would be AG, GT, and TC (they are always written with the 5' terminus at the left). Before

techniques were available for determining the complete base sequence of DNA, nearest-neighbor analysis was used to determine sequence relations. Nearest-neighbor analysis also indicated that complementary DNA strands are antiparallel, a phenomenon that you are asked to examine in this problem by predicting some nearest-neighbor frequencies. Assume that you have determined the frequencies of the following nearest neighbors: AG, 0.15; GT, 0.03; GA, 0.08; TT, 0.10.

(a) What are the nearest-neighbor frequencies of CT, AC, TC, and AA?

(b) If DNA had a parallel (rather than an antiparallel) structure, what nearest-neighbor frequencies could you deduce from the observed values?

5.13 What is the chemical group (3'-P, 5'-P, 3'-OH, or 5'-OH) at the sites indicated at the positions labeled a, b, and c in the accompanying figure?

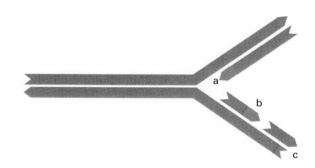

5.14 Identify the chemical group that is at the indicated terminus of the daughter strand of the extended branch of the rolling circle in the accompanying diagram.

Challenge Problems

5.15 The following sequence of bases is present along one chain of a DNA duplex that has opened up at a replication fork, and synthesis of an RNA primer on this template begins by copying the base in red.

3'- · · · TCTGATATCAGTACG · · · -5'

(a) If the RNA primer consists of eight nucleotides, what is its base sequence?

(b) In the intact RNA primer, which nucleotide has a free hydroxyl (-OH) terminus and what is the chemical group on the nucleotide at the other end of the primer?

(c) If replication of the other strand of the original DNA duplex proceeds continuously (with few or no intervening RNA primers), is the replication fork more likely to move in a left-to-right or a right-to-left direction?

5.16 A DNA fragment produced by the restriction enzyme *Sal*I is inserted into a unique *Sal*I cloning site in a vector molecule. Digestion with restriction enzymes produces the following fragment sizes that originate from the inserted DNA:

(a) *Sal*I: 20 kb

(b) *Sal*I + *Eco*RI: 7 kb, 13 kb

(c) *Sal*I + *Hind*III: 4 kb, 5 kb, 11 kb

(d) *Sal*I + *Eco*RI + *Hind*III: 3 kb, 4kb, 5 kb, 8 kb

What restriction map of the insert is consistent with these fragment sizes?

5.17 The illustrated DNA sequencing gel was obtained by the dideoxy sequencing method. What is the nucleotide sequence of the strand synthesized in the sequencing reactions? What is the nucleotide sequence of the complementary template strand? Label each end as 3' or 5'.

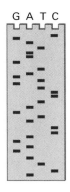

5.18 For the following restriction endonucleases, calculate the average distance between restriction sites in an organism whose DNA has a random sequence and equal proportions of all four nucleotides. The symbol R means any purine (A or G), and Y means any pyrimidine (T or C), but an R−Y pair must be either A−T or G−C.

*Taq*I   5'-TCGA-3'
          3'-AGCT-5'

*Bam*HI   5'-GGATCC-3'
          3'-CCTAGG-5'

*Hae*II   5'-RGCGCY-3'
          3'-YCGCGR-5'

Further Reading

Bauer, W. R., F. H. C. Crick, and J. H. White. 1980. Supercoiled DNA. *Scientific American,* July.

Cairns, J. 1966. The bacterial chromosome. *Scientific American,* January.

Danna, K., and D. Nathans. 1971. Specific cleavage of Simian Virus 40 DNA by restriction endonuclease of *Hemophilus influenzae. Proceedings of the National Academy of Sciences,* USA 68: 2913.

Davies, J. 1995. Vicious circles: Looking back on resistance plasmids. *Genetics* 139: 1465.

DePamphilis, M. L., ed. 1996. *DNA Replication in Eukaryotic Cells.* Cold Spring Harbor, NY: Cold Spring Harbor Press.

Donovan, S., and J. F. X. Diffley. 1996. Replication origins in eukaroytes. *Current Opinion in Genetics & Development* 6: 203.

Grimaldi, D. A. 1996. Captured in amber. *Scientific American,* April.

Grunstein, M. 1992. Histones as regulators of genes. *Scientific American,* October.

Hubscher, U., and J. M. Sogo. 1997. The eukaryotic DNA replication fork. *News in Physiological Sciences* 12: 125.

Kelley, T., ed. 1988. *Eukaryotic DNA Replication.* Cold Spring Harbor Laboratory.

Kornberg, A. 1995. *DNA Replication.* 2d ed. New York: Freeman.

Kornberg, R. D., and A. Klug. 1981. The nucleosome. *Scientific American,* February.

Mullis, K. B. 1990. The unusual origin of the polymerase chain reaction. *Scientific American,* April.

Neidhardt, F. C., R. Curtiss III, J. L. Ingraham, E. C. C. Lin, K. B. Low, B. Magasanik, W. S. Reznikoff, M. Riley, M. Schaechter, and H. E. Umbarger, eds. 1996. Escherichia coli *and* Salmonella typhimurium: *Cellular and Molecular Biology* (2 volumes). 2d ed. Washington, DC: American Society for Microbiology.

Singer, M., and P. Berg. 1991. *Genes & Genomes.* Mill Valley, CA: University Science Books.

Variegated kernel color in corn due to interaction between the transposable elements *Ac* and *Ds* discovered by Barbara McClintock. These kernels contain two copies of *Ds* located in chromosome 9 proximal (toward the centromere) to the locus of *C1*, which is reponsible for the purple anthocyanin pigment. The homologous chromosome carries an inactive mutant allele *c1*. The element *Ac* is present elsewhere in the genome. When *Ac* breaks chromosome 9 at the position of either *Ds* element, the tip of chromosome 9 containing the dominant *C1* allele is lost, and the portion of the kernel that develops from such a cell is colorless. The colorless patches are large or small depending on whether the breakage occurred early or late in development. [Courtesy of Clifford Weil and Susan Wessler. From C. F. Weil and S. R. Wessler. 1993. *The Plant Cell* 5: 515.]

The Molecular Organization of Chromosomes

PRINCIPLES

- Prokaryotes and lower eukaryotes have smaller genomes (less DNA) than higher eukaryotes.
- Chromosomes consist largely of DNA combined with histone proteins; each chromosome contains a single, usually very long, DNA molecule.
- Most eukaryotic genomes, in addition to the "unique" DNA sequences that make up the majority of genes, also contain DNA sequences that are highly repetitive or moderately repetitive.
- Transposable elements are DNA sequences able to change their location within a chromosome or to move between chromosomes.
- The centromere is a specialized DNA structure that functions as the center of chromosome movement in cell division.
- The telomere is another specialized DNA structure that serves to stabilize the chromosome tips from shortening through progressive loss of DNA; the telomere is elongated by a special enzyme, telomerase.

CONNECTIONS

CONNECTION: Her Feeling for the Organism
Barbara McClintock 1950
The origin and behavior of mutable loci in maize

CONNECTION: Telomeres—The Beginning of the End
Carol W. Greider and Elizabeth H. Blackburn 1987
The telomere terminal transferase of Tetrahymena *is a ribonucleoprotein enzyme with two kinds of primer specificity*

To understand genetic processes requires a knowledge of the organization of the genetic material at the level of chromosomes. In this chapter, we will see that chromosomes are diverse in size and structural properties and that their DNA differs in the composition and arrangement of nucleotide sequences. The most pronounced differences in structure and genetic organization are between the chromosomes of eukaryotes and those of prokaryotes. Some viral chromosomes are especially noteworthy in that they consist of one single-stranded (rather than double-stranded) DNA molecule and, in a small number of viruses, of one or more molecules of RNA instead of DNA. Also, eukaryotic cells contain several chromosomes, each of which contains one intricately coiled DNA molecule, whereas prokaryotes contain a single major chromosome (and, occasionally, several copies of one or more small, usually circular, DNA molecules called plasmids).

The genetic complement of a cell or virus constitutes its **genome.** In eukaryotes, this term is commonly used to refer to one complete haploid set of chromosomes, such as that found in a sperm or egg.

6.1
Genome Size and Evolutionary Complexity

Measurement of the nucleic acid content of the genomes of viruses, bacteria, and lower and higher eukaryotes has led to the following generalization:

Genome size increases roughly with evolutionary complexity.

This generalization is based on the observations that the single nucleic acid molecule of a typical virus is smaller than the DNA molecule in a bacterial chromosome; that unicellular eukaryotes, such as the yeasts, contain more DNA than a typical bacterium; and that multicellular eukaryotes have the greatest amount of DNA per genome. However, among the multicellular eukaryotes, no correlation exists between evolutionary complexity and amount of DNA. In higher eukaryotes, DNA content is not directly proportional to number of genes.

A summary of genome size in a sample of organisms is shown in Table 6.1. Bacteriophage MS2 is one of the smallest viruses; it has only four genes in a single-stranded RNA molecule containing 3569 nucleotides. SV40 virus, which infects monkey and human cells, has a genetic complement of five genes in a circular double-stranded DNA molecule consisting of about 5000 nucleotide pairs. Large DNA molecules are measured in **kilobase pairs (kb),** or thousands of base pairs. The genome of SV40 is about 5 kb. The more complex phages and animal viruses have as many as 250 genes and DNA molecules ranging from 50 to 300 kb. Bacterial genomes are substantially larger. For example, the chromosome of *E. coli* contains about 4000 genes in a DNA molecule composed of about 4700 kb.

Although the genomes of prokaryotes are composed of DNA, their DNA is not packaged into chromosomes. True chromosomes are found only in eukaryotes. The number of chromosomes is characteristic of the particular species, as we saw in Chapter 3. In moving up the evolutionary scale of animals or plants, the DNA content per haploid genome generally tends to increase, but there are many individual exceptions. The number of chromosomes shows no pattern. One of the smallest genomes in a multicellular animal is that of the nematode worm *Caenorhabditis elegans*, with a DNA content about 20 times that of the *E. coli* genome. The *D. melanogaster* and the human genomes have about 40 and 700 times as much DNA, respectively, as the *E. coli* genome. The genomes of some amphibians and fish are very large—many times the size of mammalian genomes. Such large genomes are measured in **megabase pairs (Mb),** or millions of base pairs. The human genome is large, but it is by no means the largest among animals or higher plants. At 3000 Mb, the human genome is only 67 percent the size of that in corn and only 4 percent the size of that in the salamander *Amphiuma* (Table 6.1).

Among higher animals and plants, a large genome size does not imply a large number of genes. For example, the 3000-Mb human genome is large enough to con-

Table 6.1 Genome size of some representative viral, bacterial, and eukaryotic genomes

Genome	Approximate length in thousands of nucleotides	Form
Virus		
MS2	4	Single-stranded RNA
SV40	5	Circular double-stranded DNA
ϕX174	5 ⎤	Circular single-stranded DNA;
M13	6 ⎦	double-stranded replicative form
λ	50 ⎤	
Herpes simplex	152 ⎥	
T2,T4,T6	165 ⎥	Linear double-stranded DNA
Smallpox	267 ⎦	
Bacteria		
Mycoplasma hominis	760	Circular double-stranded DNA
Escherichia coli	4700	
Eukaryotes		Haploid chromosome number
Saccharomyces cerevisiae (yeast)	13,000	16
Caenorhabditis elegans (nematode)	100,000	6
Arabidopsis thaliana (wall cress)	100,000	5
Drosophila melanogaster (fruit fly)	165,000	4
Homo sapiens (human being)	3,000,000	23
Zea Mays (maize)	4,500,000	10
Amphiuma sp. (salamander)	76,500,000	14

tain perhaps 10^6 genes; however, various lines of evidence suggest that the number of genes is no greater than approximately 10^5. Considering the number and size of proteins produced in human cells, it appears that no more than about 4 percent of the human genome actually codes for proteins. Similarly, although closely related species of salamanders are thought to have about the same number of genes, their genome size can differ by 30-fold in the total amount of DNA. Therefore, in higher animals and plants, the actual number of genes is much less than the theoretical maximum. The reason for the discrepancy is that in higher organisms, most of the DNA has functions other than coding for the amino acid sequence of proteins. (This issue is discussed further in Chapter 11).

A remarkable feature of the genetic apparatus of eukaryotes is that the enormous amount of genetic material contained in the nucleus of each cell is precisely divided in each cell division. A haploid human genome contained in a gamete has a DNA content equivalent to a linear DNA molecule 1 meter (10^6 μm) in length. The largest of the 23 chromosomes in the human genome contains a DNA molecule that is 82 mm (8.2×10^4 μm) long. However, at metaphase of a mitotic division, the DNA molecule is condensed into a compact structure about 10 μm long and less than 1 μm in diameter. An analogy may be helpful in appreciating the prodigious feat of packaging that such chromosome condensation represents. If the DNA molecule in human chromosome 1 (the longest chromosome) were a cooked spaghetti noodle 1 mm in diameter, it would stretch for 25 miles; in chromosome condensation, this noodle is gathered together, coil upon coil, until at metaphase it is a canoe-sized tangle of spaghetti 16 feet long and 2 feet wide. After cell division, the noodle is unwound again. Although the genomes of prokaryotes and viruses are much smaller than those of eukaryotes, they also are very compact. For example, an *E. coli* chromosome, which contains a DNA molecule about 1500 μm long, is contained in a cell about 2 μm long and 1 μm in diameter.

6.2
The Supercoiling of DNA

The DNA of prokaryotic and eukaryotic chromosomes is **supercoiled,** which means that segments of double-stranded DNA are twisted around one another, analogous to the manner in which a telephone cord can be twisted around itself. The

geometry of supercoiling can be illustrated by a simple example. Consider first a linear duplex DNA molecule whose ends are joined in such a way that each strand forms a continuous circle. Such a DNA molecule is called a **covalent circle,** and it is said to be **relaxed** if no twisting is present other than the helical twisting (Figure 6.1A). The individual polynucleotide strands of a relaxed circle form the usual right-handed (positive) helical structure with ten nucleotide pairs per turn of the helix. Suppose you were to cut one strand in a relaxed circle and unwind it one complete rotation of 360° so as to undo one complete turn of the double helix. When the ends were rejoined again, the result would be a circular helix that is "underwound." Because a DNA molecule has a strong tendency to maintain its standard helical form with ten nucleotide pairs per turn, the circular molecule would respond to the underwinding in one of two ways: (1) by forming regions with "bubbles" in which the bases are unpaired (Figure 6.1B) or (2) by twisting the circular molecule in the opposite sense from the direction of under-

Figure 6.1

Different states of a covalent circle. (A) A nonsupercoiled (relaxed) covalent circle with 36 helical turns. (B) An underwound covalent circle with only 32 helical turns. (C) The molecule in part B, but with four twists to eliminate the underwinding. (D) Electron micrograph showing nicked circular and supercoiled DNA of phage PM2. Note that no bases are unpaired in part C. In solution, parts B and C would be in equilibrium. [Electron micrograph courtesy of K. G. Murti.]

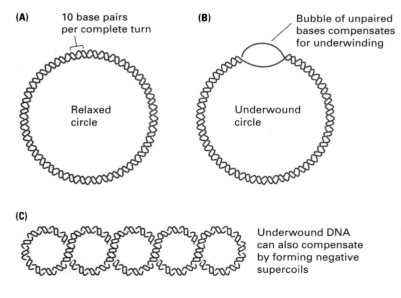

(A) 10 base pairs per complete turn

Relaxed circle

(B) Bubble of unpaired bases compensates for underwinding

Underwound circle

(C) Underwound DNA can also compensate by forming negative supercoils

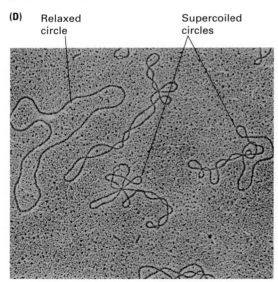

(D) Relaxed circle

Supercoiled circles

winding (Figure 6.1C). This twisting is called **supercoiling,** and a molecule with this sense of twisting is **negatively super-coiled.** Examples of supercoiled molecules are shown in Figure 6.1C and D. The two responses to underwinding are not independent, and underwinding is usually accommodated by a combination of the two processes: An underwound molecule contains some bubbles of unpaired bases and some supercoiling, with the supercoiling predominating. Although supercoiling occasionally plays a role in the expression of some genes, the overall biological function of supercoiling is unknown.

Topoisomerase Enzymes

The negative supercoiling of natural DNA molecules is produced by DNA topoisomerase enzymes (Chapter 5). The class of enzymes denoted **topoisomerase I** cause a single-stranded nick by breaking a phosphodiester bond in the backbone of one of the DNA strands; the nick produces a gap through which the intact DNA strand is passed, and then the nick is resealed. Depending on conditions, topoisomerase I enzymes can either increase or decrease the amount of supercoiling. As we saw in Section 5.4, topoisomerase I enzymes are essential to relieve the coiling produced by separation of the DNA strands in DNA replication.

A second class of topoisomerase enzymes is called **topoisomerase II.** These enzymes work by producing a double-stranded gap in one molecule through which another double-stranded molecule is passed. Therefore, topoisomerase II enzymes are able to pass one DNA duplex entirely through another or to separate two circular DNA molecules that are interlocked. The mechanism of action of topoisomerase II from the yeast *Saccharomyces cerevisiae* is illustrated in Figure 6.2. The molecular structure of the enzyme includes two sets of "jaws" set approximately at right angles (Figure 6.2A). The inner set clamps one of the duplexes, the outer set the other (Figure 6.2B and C). To allow the outer DNA molecule to pass completely through the inner one, the inner duplex is first cleaved, and then the outer duplex is passed through the gap (Figure 6.2D and E). After passage, the gap is repaired and both molecules are released (Figure 6.2F and G).

In a supercoiled DNA molecule free of proteins that maintain the supercoiling, any nick eliminates all supercoiling because the strain of underwinding is relaxed by free rotation of the intact strand about the sugar-phosphate bond opposite the break. Therefore, any treatment that nicks DNA relaxes the supercoiling. Single-stranded nicks can be produced by any of a variety of enzymes, such as **deoxyribonuclease (DNase),** that cleaves sugar-phosphate bonds.

6.3
The Structure of the Bacterial Chromosome

The chromosome of *E. coli* is a condensed unit called a **nucleoid** or **folded chromosome,** which is composed of a single circular DNA molecule and associated proteins. The term chromosome is a misnomer for this structure, because it is not a true "chromosome" in the sense of a eukaryotic chromosome. The most striking feature of the bacterial nucleoid is that the DNA is organized into a set of looped domains (Figure 6.3). As isolated from bacterial cells, the nucleoid contains, in addition to DNA, small amounts of several proteins, which are thought to be responsible in some way for the multiply looped arrangement of the DNA. The degree of condensation of the isolated nucleoid (that is, its physical dimensions) is affected by a variety of factors, and some controversy exists about the state of the nucleoid within the cell.

Figure 6.3 also shows that loops of the DNA of the *E. coli* chromosome are supercoiled. Note that some loops are not supercoiled; this is a result of the action of DNases during isolation, and it indicates that the loops are in some way independent of one another. In the preceding section, we stated that supercoiling is generally eliminated in a DNA molecule by one single-strand break. However, such a break in the *E. coli* chromosome does not

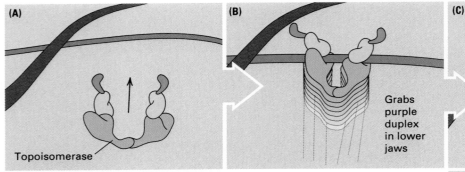

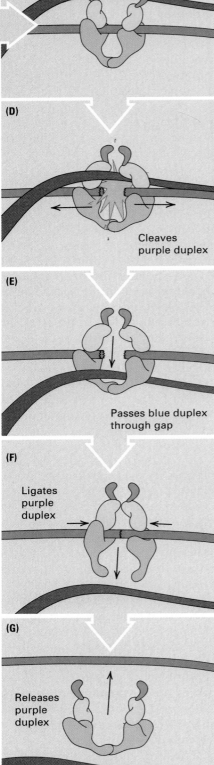

Figure 6.2
Topoisomerase II untangles a pair of DNA molecules by cleaving one DNA duplex and passing the other duplex through the gap. The enzyme illustrated here, from the yeast *Saccharomyces cerevisiae*, has two sets of "jaws." (A through C) The inner jaws (green) trap one duplex. The outer jaws (red) trap the other duplex. (D and E) The duplex in the inner jaws is cleaved, and the second, uncleaved duplex is passed through. (F) The uncleaved duplex is expelled through an opening that forms in the basal part of the enzyme. (G) Then the gap in the cleaved duplex is repaired, and it is released. [After J. M. Berger, S. J. Gamblin, S. C. Harrison and J. C. Wang. 1996. *Nature* 379:225.]

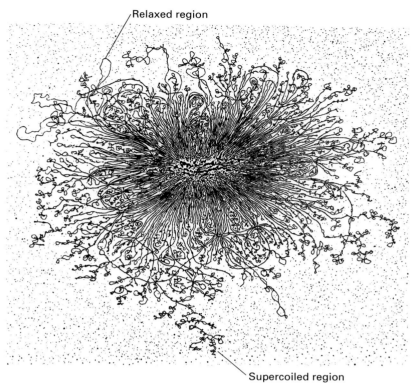

Relaxed region

Supercoiled region

Figure 6.3

An electron micrograph of an *E. coli* chromosome showing the multiple loops emerging from a central region. [Courtesy of Ruth Kavenoff. Copyright 1983 by Designergenes Posters Ltd.]

eliminate all supercoiling. If nucleoids, all of whose loops are supercoiled, are treated with a DNase and examined at various times after treatment, it is observed that supercoiling is relaxed in one loop at a time, not in all loops at once (Figure 6.4).

The loops must be isolated from one another in such a way that rotation in one loop is not transmitted to other loops. The independence is probably the result of proteins that bind to the DNA in a way that prevents rotation of the helix.

Figure 6.4

A schematic drawing of the folded supercoiled *E. coli* chromosome, showing 11 of the 40 to 50 loops attached to a protein core (blue shaded area) and the opening of loops by nicks.

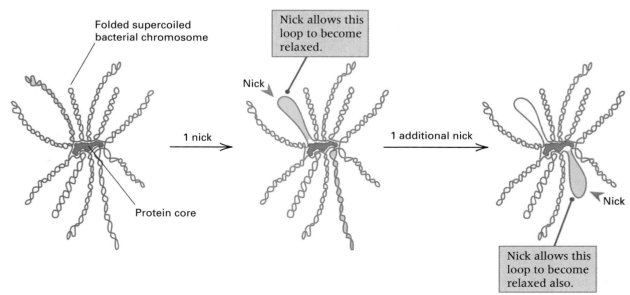

Folded supercoiled bacterial chromosome

Protein core

Nick allows this loop to become relaxed.

Nick

1 nick

1 additional nick

Nick

Nick allows this loop to become relaxed also.

6.4
The Structure of Eukaryotic Chromosomes

A eukaryotic chromosome contains a single DNA molecule of enormous length. For example, the largest chromosome in the *D. melanogaster* genome has a DNA content of about 65,000 kb (6.5×10^7 nucleotide pairs), which is equivalent to a continuous linear duplex about 22 mm long. These long molecules usually fracture during isolation, but some fragments that are recovered are still very long. Figure 6.5 is an autoradiograph of radioactively labeled *Drosophila* DNA more than 36,000 kb in length.

In organisms such as baker's yeast, *Saccharomyces cerevisiae*, which have small genomes, the DNA molecules present in the chromosomes can be separated by special types of electrophoretic methods. In conventional electrophoresis, which we examined in Section 5.7, the electric field is maintained in a constant state, usually at constant voltage. The fragments move in response to the field according to their size, smaller fragments moving faster. However, conventional electrophoresis can separate only molecules smaller than about 20 kb. All molecules larger than about 20 kb have the same electrophoretic mobility under these conditions and so form a single band in the gel. Simple modifications of the electrophoresis result in separation among much larger DNA fragments. The most common modifications alter the geometry of the electric field at periodic intervals during the course of the electrophoretic separation. Some electrophoretic apparatuses alternate the electric field between two sets of electrodes oriented at right angles or among three sets of electrodes forming a hexagon. In one type of large-fragment electrophoresis, called **pulsed-field gel electrophoresis (PFGE),** the apparatus is similar to that for conventional electrophoresis, but the orientation of the electric field is changed periodically; the improved separation of large DNA molecules apparently results from the additional time it takes for large molecules to reorient themselves when the orientation of the electric field is changed. Figure 6.6 illustrates the separation of the 16 chromosomes from yeast by means of PFGE. The chromosomes range in size from approximately 200 kb to 2.2 Mb. Electrophoretic separation yields a visual demonstration that each of the chromosomes contains a single DNA molecule that runs continuously throughout its length. Using the methods discussed in Section 5.7, we can determine the identity of each chromosomal band in the gel by hybridization of a Southern blot with probes for genes known to map to a particular chromosome.

The Nucleosome Is the Basic Structural Unit of Chromatin

The DNA of all eukaryotic chromosomes is associated with numerous protein molecules in a stable, ordered aggregate called **chromatin.** Some of the proteins present in chromatin determine chromosome structure and the changes in structure that occur during the division cycle of the cell. Other chromatin proteins appear to play important roles in regulating chromosome functions.

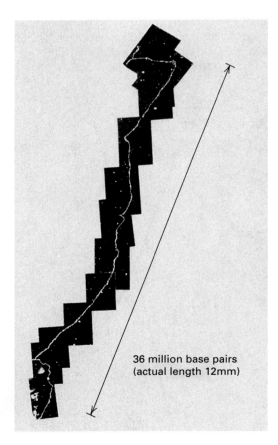

Figure 6.5
Autoradiogram of a DNA molecule from *D. melanogaster*. The molecule is 12 mm long (approximately 36,000 kb). [From R. Kavenoff, L. C. Klotz, and B. H. Zimm. 1974. *Cold Spring Harbor Symp. Quant. Biol.*, 38: 4.]

36 million base pairs (actual length 12mm)

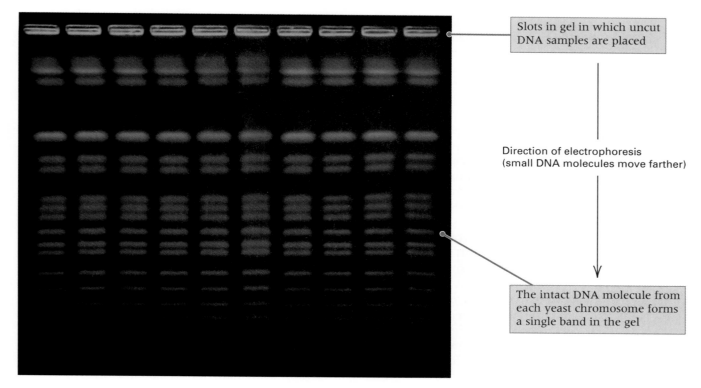

Slots in gel in which uncut DNA samples are placed

Direction of electrophoresis (small DNA molecules move farther)

The intact DNA molecule from each yeast chromosome forms a single band in the gel

Figure 6.6
Separation of the 16 chromosomes of yeast by pulsed-field gel electrophoresis, in which there is regular change in the orientation of the electric field. Bands for each of the chromosomes are clearly visible. They range in size from 200 kb to 2.2 Mb. [Courtesy of BioRad Laboratories, Hercules, California.]

Nucleosome Core Particles

The simplest form of chromatin is present in nondividing eukaryotic cells, when chromosomes are not sufficiently condensed to be visible by light microscopy. Chromatin isolated from such cells is a complex aggregate of DNA and proteins. The major class of proteins comprises the **histone** proteins. Histones are largely responsible for the structure of chromatin. Five major types—**H1, H2A, H2B, H3, and H4**—are present in the chromatin of nearly all eukaryotes in amounts about equal in mass to that of the DNA. Histones are small proteins (100–200 amino acids) that differ from most other proteins in that from 20 to 30 percent of the amino acids are lysine and arginine, both of which have a positive charge. (Only a few percent of the amino acids of a typical protein are lysine and arginine.) The positive charges enable histone molecules to bind to DNA, primarily by electrostatic attraction to the negatively charged phosphate groups in the sugar-phosphate backbone of DNA. Placing chromatin in a solution with a high salt concentration (for example, 2 molar NaCl) to eliminate the electrostatic attraction causes the histones to dissociate from the DNA. Histones also bind tightly to each other; both DNA-histone and histone-histone binding are important for chromatin structure.

The histone molecules from different organisms are remarkably similar, with the exception of H1. In fact, the amino acid sequences of H3 molecules from widely different species are almost identical. For example, the sequences of H3 of cow chromatin and pea chromatin differ by only 4 of 135 amino acids. The H4 proteins of all organisms also are quite similar; cow and pea H4 differ by only 2 of 102 amino acids. There are few other proteins whose amino acid sequences vary so little from one species to the next. When the variation between organisms is very small, we say that the sequence is highly **conserved.** The extraordinary conservation in histone composition through hundreds of millions of years of evolutionary divergence is consistent with the important role of these proteins in the structural organization of eukaryotic chromosomes.

In the electron microscope, chromatin resembles a regularly beaded thread (Figure 6.7). The bead-like units in chromatin are called **nucleosomes.** The organization of the nucleosomes in chromatin is illustrated in Figure 6.8A. Each unit has a definite composition, consisting of two molecules each of H2A, H2B, H3, and H4, a segment of DNA containing about 200 nucleotide pairs, and one molecule of histone H1. The complex of two subunits each of H2A, H2B, H3, and H4, as well as part of the DNA, forms each "bead," and the remaining DNA bridges between the beads. Histone H1 also appears to play a role in bridging between the beads, but it is not shown in Figure 6.8A.

Brief treatment of chromatin with certain DNases (for example, **micrococcal nuclease** from the bacterium *Staphylococcus aureus*) yields a collection of small particles of quite uniform size consisting only of histones and DNA (Figure 6.9). The DNA fragments in these particles are of lengths equal to about 200 nucleotide pairs or small multiples of that unit size (the precise size varies with species and tissue). These particles result from cleavage of the linker DNA segments between the beads (Figure 6.8B). More extensive treatment with DNase results in loss of the H1 histone and digestion of all the DNA except that protected by the histones in the bead. The resulting structure, called a **core particle,** consists of an octamer of pairs of H2A, H2B, H3, and H4, around which the remaining DNA, approximately 145 base pairs, is wound in about one and three-fourths turns (Figure 6.8B). Each nucleosome is composed of a core particle, additional DNA called **linker**

DNA that links adjacent core particles (the linker DNA is removed by extensive nuclease digestion), and one molecule of H1; the H1 binds to the histone octamer and to the linker DNA, causing the linkers extending from both sides of the core particle to cross and draw nearer to the octamer, though some of the linker DNA does not come into contact with any histones. The size of the linker ranges from 20 to 100 nucleotide pairs for different species and even in different cell types in the same organism (200 − 145 = 55 nucleotide pairs is usually considered an average size). Little is known about the structure of the linker DNA or about whether it has a special genetic function, and the cause of the variation in its length is also unknown.

The Arrangement of Chromatin Fibers in a Chromosome

The DNA molecule of a chromosome is folded and folded again in such a way that it is convenient to think of chromosomes as having several hierarchical levels of organization, each responsible for a particular degree of shortening of the enormously long strand (Figure 6.10). Assembly of DNA and histones can be considered the first level—namely, a sevenfold reduction in length of the DNA and the formation of a beaded flexible fiber 110 Å (11 nm) wide (Figure 6.10B), roughly five times the width of free DNA (Figure 6.10A). The structure of chromatin varies with the concentration of salts, and the 110 Å fiber is present only when the salt concentration is quite low. If the salt concentration is increased slightly, then the fiber becomes shortened somewhat by forming a zigzag arrangement of closely spaced beads between which the linking DNA is no longer visible in electron micrographs. If

Figure 6.7
Dark-field electron micrograph of chromatin showing the beaded structure at low salt concentration. The beads have diameters of about 100 Å. [Courtesy of Ada Olins.]

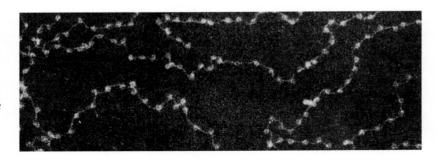

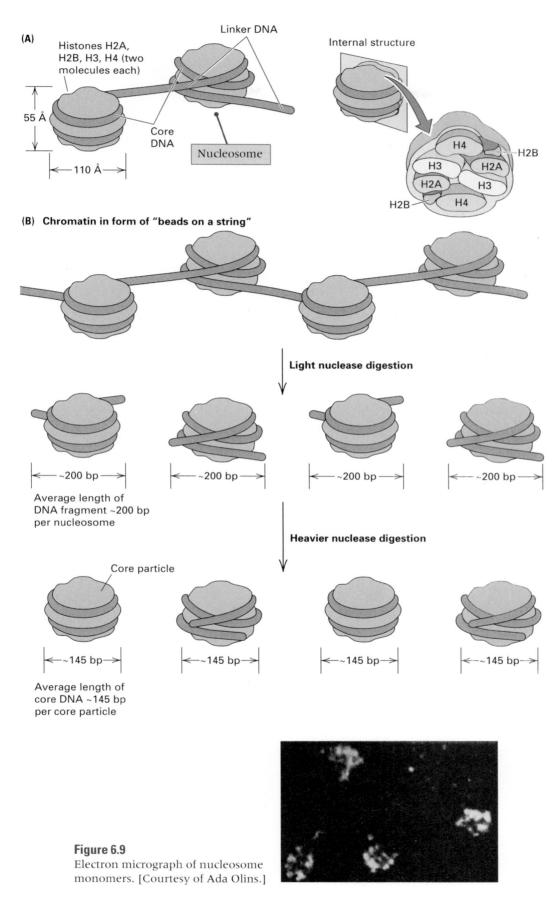

(A)

Histones H2A, H2B, H3, H4 (two molecules each)

Linker DNA

Core DNA

Nucleosome

55 Å

110 Å

Internal structure

H4

H2B

H3

H2A

H2A

H3

H2B

H4

(B) Chromatin in form of "beads on a string"

Light nuclease digestion

~200 bp ~200 bp ~200 bp ~200 bp

Average length of DNA fragment ~200 bp per nucleosome

Heavier nuclease digestion

Core particle

~145 bp ~145 bp ~145 bp ~145 bp

Average length of core DNA ~145 bp per core particle

Figure 6.8

(A) Organization of nucleosomes. The DNA molecule is wound one and three-fourths turns around a histone octamer called the core particle. If H1 were present, it would bind to the octamer surface and to the linkers, causing the linkers to cross. (B) Effect of treatment with micrococcal nuclease. Brief treatment cleaves the DNA between the nucleosomes and results in core particles associated with histone H1 and approximately 200 base pairs of DNA. More extensive treatment results in loss of H1 and digestion of all but 145 base pairs of DNA in intimate contact with each core particle.

Figure 6.9

Electron micrograph of nucleosome monomers. [Courtesy of Ada Olins.]

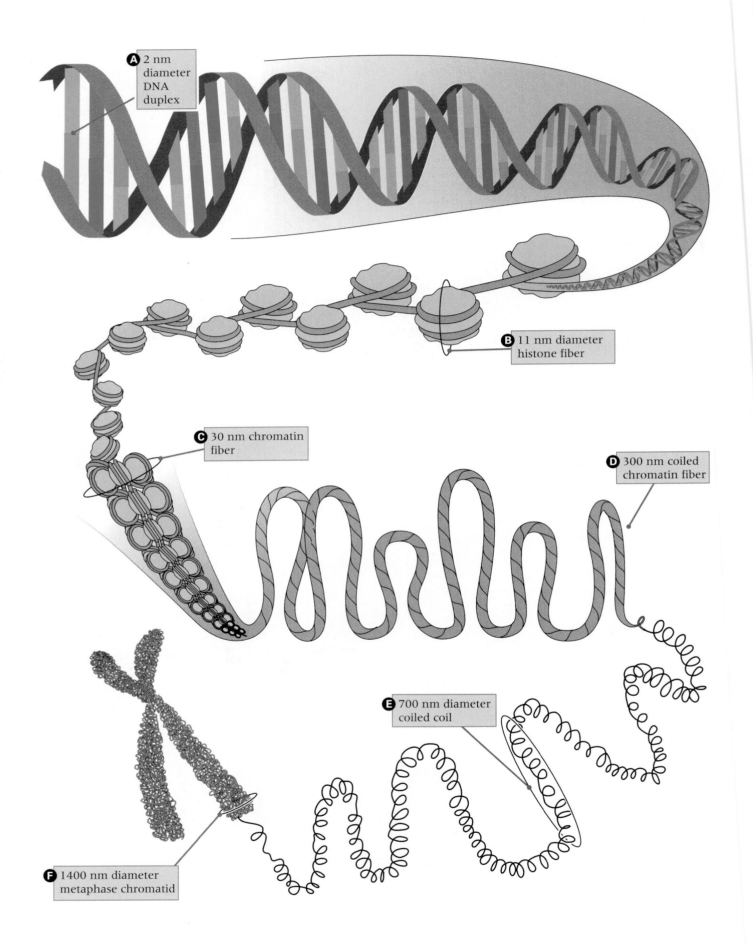

the salt concentration is further increased to that present in living cells, then a second level of compaction occurs: the organization of the 110 Å nucleosome fiber into a shorter, thicker fiber with an average diameter ranging from 300 to 350 Å, called the **30 nm fiber** (Figure 6.10C). In forming this structure, the 110 Å fiber apparently coils in a somewhat irregular left-handed superhelix or solenoidal supercoil with six nucleosomes per turn (Figure 6.11). It is believed that most intracellular chromatin has the solenoidal supercoiled configuration.

The final level of organization is that in which the 30 nm fiber condenses into a chromatid of the compact metaphase chromosome (Figure 6.10D through F). Little is known about this process other than that it seems to proceed in stages. In electron micrographs of isolated metaphase chromosomes from which histones have been removed, the partly unfolded DNA has the form of an enormous number of loops that seem to extend from a central core, or **scaffold,** composed of nonhistone chromosomal proteins (Figure 6.12). Electron microscopic studies of chromosome condensation in mitosis and meiosis suggest that the scaffold extends along the chromatid and that the 30 nm fiber becomes arranged into a helix of loops radiating from the scaffold. Details are not known about the additional folding that is required of the fiber in each loop to produce the fully condensed metaphase chromosome.

The genetic significance of the compaction of DNA and protein into chromatin and ultimately into the chromosome is that it greatly facilitates the movement of the genetic material during nuclear division. Relative to a fully extended DNA molecule, the length of a metaphase chromosome is reduced by a factor of approximately 10^4 as a result of chromosome condensation. Without chromosome condensation, the chromosomes would become so entangled

Figure 6.10 (*facing page*)
Various stages in the condensation of DNA (A) and chromatin (B through E) in forming a metaphase chromosome (F). The dimensions indicate known sizes of intermediates, but the detailed structures are hypothetical.

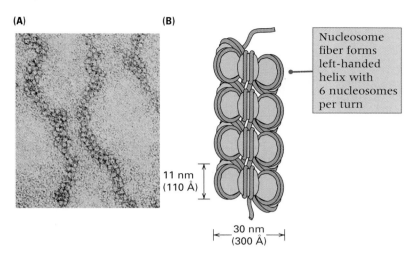

Figure 6.11
(A) Electron micrograph of the 30 nm component of mouse metaphase chromosomes. (B) A proposed solenoidal model of chromatin. The DNA (blue-gray) is wound around each nucleosome. It is unlikely that the real structure is so regular. [Electron micrograph courtesy of Barbara Hamkalo; drawing after J. T. Finch and A. Klug. 1976. *Proc. Nat. Acad. Sci. USA,* 73: 1900.]

Figure 6.12
Electron micrograph of a partially disrupted anaphase chromosome of the milkweed bug *Oncopeltus fasciatus,* showing multiple loops of 30-nm chromatin at the periphery. [From V. Foe, H. Forrest, L. Wilkinson, and C. Laird. 1982. *Insect Ultrastructure,* 1: 222.]

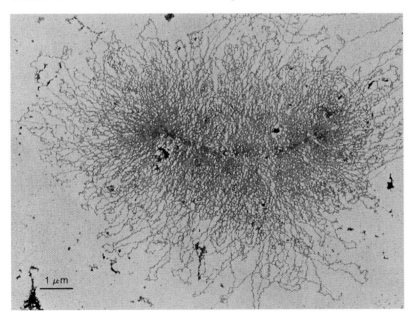

that there would be many more abnormalities in the distribution of genetic material into daughter cells.

6.5 Polytene Chromosomes

A typical eukaryotic chromosome contains only a single DNA molecule. However, in the nuclei of cells of the salivary glands and certain other tissues of the larvae of *Drosophila* and other two-winged (dipteran) flies, there are giant chromosomes, called **polytene chromosomes,** that contain about 1000 DNA molecules laterally aligned. Each of these chromosomes has a length and cross-sectional diameter many times greater than those of the corresponding chromosome at mitotic metaphase in ordinary somatic cells, as well as a constant and distinctive pattern of transverse banding (Figure 6.13). The polytene structures are formed by repeated replication of the DNA in a closely synapsed pair of homologous chromosomes without separation of the replicated chromatin strands or of the two chromosomes. Polytene chromosomes are atypical chromosomes and are formed in "terminal" cells; that is, the larval cells containing them do not divide and are eliminated in the formation of the pupa. Although they do not contribute to the tissues in the adult fly, the polytene tissues of larvae have been especially valuable in the genetics of *Drosophila*, as will become apparent in Chapter 7.

In polytene nuclei of *D. melanogaster* and other dipteran species, large blocks of heterochromatin adjacent to the centromeres are aggregated into a single compact mass called the **chromocenter.** Because the two largest chromosomes in *Drosophila* (chromosomes numbered 2 and 3) have centrally located centromeres, the chromosomes appear in the configuration shown in Figure 6.14: The paired X chromosomes (in a female), the left and right arms of chromosomes 2 and 3, and a short chromosome (chromosome 4) project from the chromocenter. In a male, the Y chromosome, which consists almost entirely of heterochromatin, is incorporated in the chromocenter.

The darkly staining transverse bands in polytene chromosomes have about a tenfold range in width. These bands result from the side-by-side alignment of tightly folded regions of the individual chromatin strands that are often visible in mitotic and meiotic prophase chromosomes as chromomeres. More DNA is present within the bands than in the interband (lightly stained) regions. About 5000 bands have been identified in the *D. melanogaster* polytene chromosomes. This linear array of bands, which has a pattern that is constant and characteristic for each species, provides a finely detailed **cytological map** of the chromosomes. The banding pattern is such that observers with sufficient training and experience can identify short regions in any of the chromosomes (Figure 6.14).

Because of their large size and finely detailed morphology, polytene chromosomes are exceedingly useful for a process called *in situ* nucleic acid hybridization. In the procedure of *in situ hybridization,* nuclei

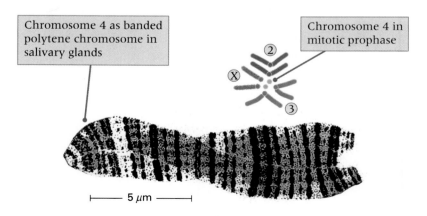

Chromosome 4 as banded polytene chromosome in salivary glands

Chromosome 4 in mitotic prophase

⊢—— 5 µm ——⊣

Figure 6.13
The polytene fourth chromosome of *Drosophila melanogaster* adhering to the chromocenter (positioned to the left, not shown). The somatic chromosomes of *Drosophila*, drawn to scale with respect to the polytene fourth chromosome, are shown at the upper right as they appear in mitotic prophase. [From C. Bridges. 1935. *J. Heredity,* 26: 60.]

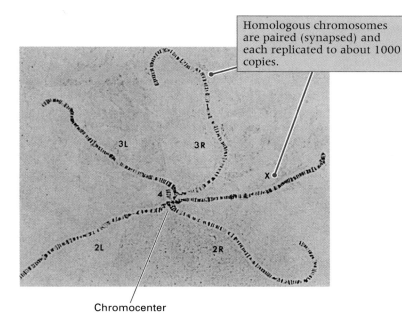

Homologous chromosomes are paired (synapsed) and each replicated to about 1000 copies.

Chromocenter

Figure 6.14
Polytene chromosomes from a larval salivary gland cell of *Drosophila melanogaster.* The chromocenter is the central region in which the centromeric regions of all chromosomes are united. [Courtesy of George Lefevre.]

containing polytene chromosomes are squashed and the chromosomal DNA denatured, after which labeled probe DNA or RNA is added under conditions that favor renaturation. After washing, the only probe that remains in the chromosomes has formed hybrid duplexes with chromosomal DNA, and its position can be identified cytologically (Figure 6.15).

Figure 6.15
Autoradiogram of *Drosophila melanogaster* polytene chromosomes hybridized *in situ* with radioactively labeled RNA copied from the histone genes, showing hybridization to a particular region (arrow). This region identifies the position of the histone genes in the chromosomes. [Courtesy of Mary Lou Pardue.]

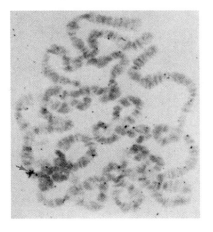

6.6
Repetitive Nucleotide Sequences in Eukaryotic Genomes

In bacteria, the variation in average base composition from one part of the genome to another is quite small. However, in eukaryotes, some components of the genome can be detected because their base composition is quite different from the average of the rest of the genome. For example, one component of crab DNA is only 3 percent G + C, compared to an average of approximately 50 percent G + C, in the rest of the genome. The components with unusually low or unusually high G + C contents are called **satellite DNA.** In the mouse, satellite DNA accounts for about 10 percent of the genome. A striking feature of satellite DNA is that it consists of fairly short nucleotide sequences that may be repeated *as many as a million times in a haploid genome.* Other **repetitive sequences** also are present in eukaryotic DNA. Because repetitive DNA consists of many highly similar or identical sequences, fragments of repetitive DNA renature more readily than fragments of nonrepetitive DNA. Information about the size of repeated sequences and the number of copies of a particular sequence can be obtained through studies of the rate of renaturation. The quantitative analysis of DNA renaturation is considered next.

Kinetics of DNA Renaturation

As we saw in Chapter 5, the rate-limiting step in the renaturation of separated DNA strands is the initial collision between two complementary single strands. Because the chance of initial collision is concentration-dependent, the rate of reassociation increases with DNA concentration (Figure 6.16A). Any increase in DNA concentration results in a corresponding increase in the number of potential pairing partners for a given strand. The study of DNA renaturation has contributed greatly to our understanding of the number and types of DNA sequences present in various genomes.

To illustrate the analysis of reassociation, we begin by comparing the renaturation rates of solutions of DNA molecules from the bacteriophages T7 and T4, which have no common base sequences and different molecular weights, and the T7 genome is smaller than that of T4. In solutions in which the number of grams of DNA per milliliter is the same, the molar concentration of T7 is greater than that of T4. Thus if each solution is separately denatured and renatured, the molecules of T7 will renature more rapidly than those of T4. If the two solutions are instead mixed, the T7 and T4 will renature independently of one another (because they are not homologous), and a curve such as that in Figure 6.16B will be obtained. Note that the curve consists of two steps, one for the more rapidly renaturing T7 molecules and the other for the T4 molecules. Each step in the curve accounts for half of the change in the absorption of the solution at 260 nanometers (A_{260}) because the initial concentration (in μg of DNA/ml, which is proportional to A_{260}) of each type of molecule was the same.

As is apparent in Figure 6.16A, the renaturation curve of a molecule that contains no repeating base sequences consists of only a single step, and the rate of renaturation is a function of the size of the molecule. If such molecules are fragmented into many components of equal size, then the molar concentration of each component will be the same as that of the unbroken

Figure 6.16

(A) Dependence of renaturation time on the concentration of phage T7 DNA. After a period at 90°C to separate the strands, the DNA was cooled to 60°C. Renaturation is complete when the relative absorption reaches 1. (B) Renaturation of a mixture of T4 and T7 DNA, each at the same temperature. Extrapolation (black dashed line) yields the early portion of the T4 curve; the ratio of the absorptions at points x and y yields the fraction of the total DNA that is T4 DNA. The times required for half-completion of renaturation, $t_{1/2}$, are obtained by drawing the red horizontal lines, which divide each curve equally in the vertical direction, and then extending the red vertical lines to the time axis.

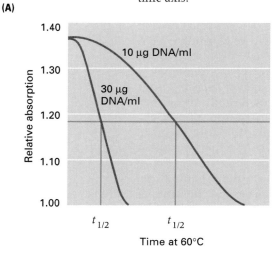

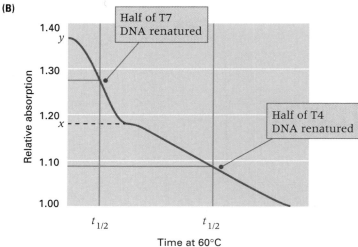

molecule, so the renaturation rate should be unchanged by fragmentation. The rate actually decreases somewhat, but the important point is that *if a DNA molecule contains only a single nonrepetitive sequence of base pairs, then breakage of the molecule does not yield a renaturation curve consisting of steps.* In contrast, if a molecule with an overall sequence that includes both a unique component and several copies of different repeated sequences is fragmented, *fragments containing the more numerous repeated sequences will renature more rapidly than the fragments containing portions of the unique sequence.* For example, consider a molecule containing 50,000 base pairs and consisting of 100 copies of a tandemly repeated sequence of 500 base pairs. If the molecules are broken into about 100 fragments of roughly equal size, each fragment will be about 500 base pairs. Although the renaturation curve for the fragments will have a single step, the renaturation rate will be characteristic of molecules 500 nucleotides in length. If a genome contains multiple families of repeated sequences whose abundances differ, the renaturation curve will have steps—one step for each repeating sequence. This principle is the basis of the analysis of renaturation kinetics.

Renaturation kinetics can be described in a simple mathematical form, because the reaction is one in which the rate-limiting step is the initial collision of two molecules. In such a case, the fraction of single strands remaining denatured at a time t after the start of renaturation is given by the expression

$$\frac{C}{C_0} = \frac{1}{1 + kC_0t}$$

in which C is the concentration of single-stranded DNA in moles of nucleotide per liter, C_0 is the initial concentration, and k is a constant. The expression C_0t is commonly called **Cot,** and a plot of C/C_0 versus C_0t is called a **Cot curve.** When renaturation is half completed, $C/C_0 = 1/2$ and

$$C_0t_{1/2} = 1/k$$

The value of $1/k$ depends on experimental conditions, but for a particular set of conditions, the value is proportional to the number of bases in the renaturing sequences. The longer the sequence, the greater will

be the time to achieve half-complete renaturation for a particular starting concentration (because the number of molecules will be smaller). The equation just stated applies to a single molecular species, a point to which we will return. If a molecule consists of several subsequences, then one needs to know C_0 for each subsequence, and a set of values of $1/k$ will be obtained (one for each step in the renaturation curve), each value depending on the length of the subsequence. What is meant by the length of the sequence that determines the rate is best described by example. A DNA molecule containing only adenine in one strand (and thymine in the other) has a repeating length of 1. The repeating tetranucleotide . . . GACTGACT . . . has a repeating length of 4. A nonrepeating DNA molecule containing n nucleotide pairs has a unique length of n.

Experimentally, the number of bases per repeating unit is not determined directly. Generally, renaturation curves for a series of molecules of known molecular weight *with no repeating elements in their sequences* (and hence yielding one-step renaturation curves) serve as standards. Molecules composed of short repeating sequences are also occasionally used. A set of curves of this kind is shown in Figure 6.17. Note that two of these simple curves represent the entire genomes of *E. coli* and phage T4. With the standard conditions for Cot analysis used to obtain this set of curves, the sequence length N (in base pairs) that yields a particular value of $C_0t_{1/2}$ is

$$N = (5 \times 10^5)C_0t_{1/2}$$

in which t is in seconds, C_0 is in nucleotides per liter, and 5×10^5 is a constant dependent on the conditions of renaturation. Through this formula, the experimentally determined value of Cot ($C_0t_{1/2}$) yields an estimate of the repeat length, N, of a repetitive sequence. Note again that C_0 is not the overall DNA concentration but the concentration of the individual sequence producing a particular step in a curve. How one obtains the necessary value of C_0 will become clear when we analyze a Cot curve. Such an analysis begins by first noting the number of steps in the curve (each step of which represents a sequence or class of sequences of a particular length) and the fraction of the material represented by each step. The observed value of $C_0t_{1/2}$ for each step must be corrected by first inferring the value of C_0 for each sequence class. The lengths of the sequences are then determined from these corrected values by comparison to standards, and the sequence lengths and sequence abundances, as a proportion of the total, are compared to obtain the number of copies of each sequence. This analysis is best understood by looking at an example, which we will do next.

Analysis of Genome Size and Repetitive Sequences by Renaturation

Figure 6.18 shows a Cot curve typical of those obtained in analyses of the renaturation kinetics of eukaryotic genomes. Three discrete steps are evident: 50 percent of the DNA has $C_0t_{1/2} = 10^3$, 30 percent has $C_0t_{1/2} = 10^0 = 1$, and 20 percent has $C_0t_{1/2} = 10^{-2}$. The scale at the top of the figure was obtained from Cot analysis of molecules that have unique sequences of known lengths, as in Figure 6.17. The

sequence sizes cannot be determined directly from the observed $C_0t_{1/2}$ values, because each value of C_0 used in plotting the horizontal axis in the body of the figure is the total DNA concentration in the renaturation mixture. Multiplying each $C_0t_{1/2}$ value by the fraction of the total DNA that it represents yields the necessary corrected $C_0t_{1/2}$ values (that is, 0.50×10^3, 0.30×1, and 0.20×10^{-2}. From the size scale at the top of Figure 6.18, the corresponding sequence sizes (repeat lengths) are approximately 3.0×10^8, 2.2×10^5, and 1×10^3 base pairs, respectively.

To determine the number of copies of each sequence, we make use of the fact that the number of copies of a sequence having a particular renaturation rate is inversely proportional to $t_{1/2}$ and hence to the observed (uncorrected) $C_0t_{1/2}$ values for each class. Thus if the haploid genome contains only one copy of the longest sequence (3.0×10^8 base pairs), it contains 10^3 copies of the sequence (or sequences) of length 2.2×10^5 base pairs and 10^5 copies of the sequence (or sequences) of length 1×10^3 base pairs. An estimate of

Figure 6.17

A set of Cot curves for various DNA samples. The black arrows pointing up to the red scale indicate the number of nucleotide pairs for each sample; they align with the intersection of each curve with the horizontal red line (the point of half-renaturation, or $C_0t_{1/2}$). The y axis on this graph can be related to that on Figure 6.16 by noting that maximum absorption represents totally single-stranded DNA and minimum absorption represents totally double-stranded DNA.

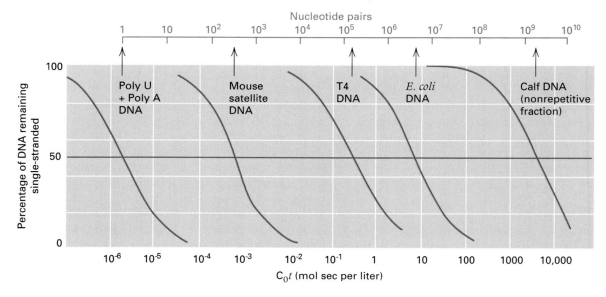

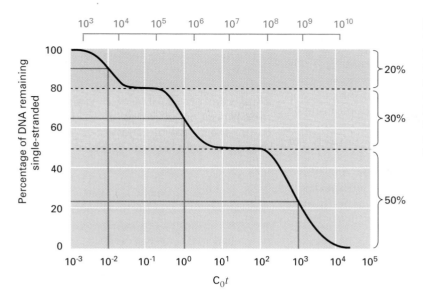

Figure 6.18
The Cot curve analyzed in the text. The scale at the top is the same as that in Figure 6.17. The black dashed lines indicate the fractional contribution of each class of molecules to the total DNA.

the total number of base pairs per genome would in this case be

$$(3.0 \times 10^{8)} + (10^3 \times 2.2 \times 10^{5)} +$$
$$(10^5 \times 1 \times 10^{3)} = 6.2 \times 10^8 \text{ base pairs}$$

The different sequence components of a eukaryotic DNA molecule can be isolated by procedures that recover the double-stranded molecules formed at different times during a renaturation reaction. The method used is to allow renaturation to proceed only to a particular $C_0 t_{1/2}$ value. The reassociated molecules present at that point are then separated from the remaining single stranded molecules, usually by passing the solution through a tube filled with a form of calcium phosphate crystal (hydroxylapatite) that preferentially binds double-stranded DNA.

6.7
Nucleotide Sequence Composition of Eukaryotic Genomes

The nucleotide sequence composition of many eukaryotic genomes has been examined via analysis of the renaturation kinetics of DNA. The principal finding is that eukaryotic organisms differ widely in the proportion of the genome that consists of repetitive DNA sequences and in the types of repetitive sequences that are present. In most eukaryotic genomes, the DNA consists of three major components:

- *Unique, or single-copy, sequences* This is usually the major component and typically comprises from 30 to 75 percent of the chromosomal DNA in most organisms.
- *Highly repetitive sequences* This component constitutes from 5 to 45 percent of the genome, depending on the species. Some of these sequences are the satellite DNA referred to earlier. The sequences in this class are typically from 5 to 300 base pairs per repeat and are duplicated as many as 10^5 times per genome.
- *Middle-repetitive sequences* This component constitutes from 1 to 30 percent of a eukaryotic genome and includes sequences that are repeated from a few times to 10^5 times per genome.

These different components can be identified not only by the kinetics of DNA reassociation but also by the number of bands that appear in Southern blots with the use of appropriate probes and by other methods. It should be clear from the preceding discussion of DNA reassociation that the dividing line between many middle-repetitive sequences and highly repetitive sequences is arbitrary.

Unique Sequences

Most gene sequences and the adjacent nucleotide sequences required for their expression are contained in the unique-sequence component. With minor exceptions (for example, the repetition of one or a few genes), the genomes of viruses and prokaryotes are composed entirely of single-copy sequences; in contrast, such sequences constitute only 38 percent of the total genome in some sea urchin species, a little more than 50 percent of the human genome, and about 70 percent of the *D. melanogaster* genome.

Highly Repetitive Sequences

Many highly repetitive sequences are localized in blocks of tandem repeats, whereas others are dispersed throughout the genome. An example of the dispersed type is a family of related sequences in the human genome called the **Alu** family because the sequences contain a characteristic restriction site for the enzyme *Alu*I (Section 5.7). The *Alu* sequences are about 300 base pairs in length and are present in approximately 500,000 copies in the human genome; this repetitive DNA family alone accounts for about 5 percent of human DNA.

Among the localized highly repetitive sequences, most are fairly short. Sequences of this type make up about 6 percent of the human genome and 18 percent of the *D. melanogaster* genome, but they account for 45 percent of the DNA of *D. virilis*. One of the simplest possible repetitive sequences is composed of an alternating . . . ATAT . . . sequence with about 3 percent G + C interspersed, which makes up 25 percent of the genomes of certain species of land crabs. In the *D. virilis* genome, the major components of the highly repetitive class are three different but related sequences of seven base pairs rich in A−T:

$$5'\text{-ACAAACT-}3'$$
$$5'\text{-ATAAACT-}3'$$
$$5'\text{-ACAAATT-}3'$$

Blocks of satellite (highly repetitive) sequences in the genomes of several organisms have been located by *in situ* hybridization with metaphase chromosomes (Figure 6.19). The satellite sequences located by this method have been found to be in the regions of the chromosomes called **heterochromatin.** These are regions that condense earlier in prophase than the rest of the chromosome and are darkly stainable by many standard dyes used to make chromosomes visible (Figure 6.20); sometimes

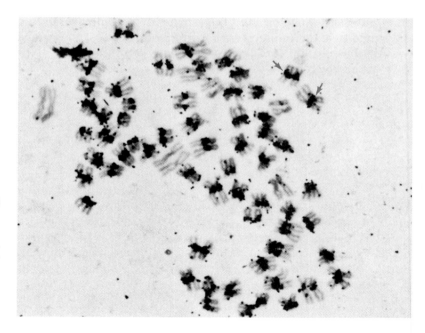

Figure 6.19
Autoradiogram of metaphase chromosomes of the kangaroo rat *Dipodomys ordii*; radioactive RNA copied from purified satellite DNA sequences has been hybridized to the chromosomes to show the localization of the satellite DNA. Hybridization is principally in the regions adjacent to the centromeres (arrows). Note that some chromosomes are apparently free of this satellite DNA. They contain a different satellite DNA not examined in this experiment. This cell is unusual in that each chromosome has undergone two rounds of chromosome duplication without an intervening mitosis; this is why each chromosome in the normal diploid set is present as a pair. [Courtesy of David Prescott.]

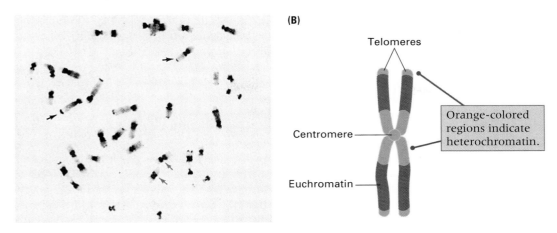

(B)

Telomeres

Centromere

Euchromatin

Orange-colored regions indicate heterochromatin.

Figure 6.20

(A) Metaphase chromosomes of the ground squirrel *Ammospermophilus harrissi,* stained to show the heterochromatic regions near the centromere of most chromosomes (red arrows) and the telomeres of some chromosomes (black arrows). (B) An interpretive drawing. [Micrograph courtesy of T. C. Hsu.]

the heterochromatin remains highly condensed throughout the cell cycle. The **euchromatin,** which makes up most of the genome, is visible only in the mitotic cycle. The major heterochromatic regions are adjacent to the centromere; smaller blocks are present at the ends of the chromosome arms (the telomeres) and interspersed with the euchromatin. In many species, an entire chromosome, such as the Y chromosome in *D. melanogaster* and in mammals, is almost completely heterochromatic. Different highly repetitive sequences have been purified from *D. melanogaster,* and in situ hybridization has shown that each sequence has its own distinctive distribution among the chromosomes.

The genetic content of heterochromatin is summarized in the following generalization:

> The number of genes located in heterochromatin is small relative to the number in euchromatin.

The relatively small number of genes means that many large blocks of heterochromatin are genetically almost inert, or devoid of function. Indeed, heterochromatic blocks can often be rearranged in the genome, duplicated, or even deleted without major phenotypic consequences.

Middle-Repetitive Sequences

Middle-repetitive sequences constitute about 12 percent of the *D. melanogaster* genome and 40 percent or more of the human and other eukaryotic genomes. These sequences differ greatly in the number of copies and their distribution within a genome. They comprise many families of related sequences and include several groups of genes. For example, the genes for the RNA components of the ribosomes (the particles on which proteins are synthesized; Chapter 10) and the genes for tRNA molecules (which also participate in protein synthesis) are repeated in the genomes of all organisms. The two major ribosomal RNA molecules come from a tandem pair of genes that is repeated several hundred times in most eukaryotic genomes. The genomes of all eukaryotes also contain multiple copies of the histone genes. Each histone gene is repeated about 10 times per genome in chickens, 20 times in mammals, about 100 times in *Drosophila,* and as many as 600 times in certain sea urchin species.

The dispersed middle-repetitive DNA of the *D. melanogaster* genome consists of about 50 families of related sequences, and from 20 to 60 copies of each family are widely scattered throughout the chromosomes. The positions of these sequences differ from one fly to the next except in completely homozygous laboratory strains. There is variability in position, because many of these sequences are able to move from one location to another in a chromosome and between chromosomes; they are said to be **transposable elements.** Analogous types of sequences are also found in the genomes of yeast, maize, and bacteria (Chapter 8) and are probably present in all organisms. An important dimension has been added to our understanding of the genome as a structural and functional unit by the discovery of these mobile genetic elements, because they can, in some cases, cause chromosome breakage, chromosome rearrangements, modification of the expression of genes, and novel types of mutations.

6.8
Transposable Elements

In the 1940s, in a study of the genetics of kernel mottling in maize (Figure 6.21), Barbara McClintock discovered an element that not only regulated the mottling but also caused breakage of the chromosome carrying the genes for color and consistency of the kernels. The element was called Dissociation (*Ds*). Mapping data showed that the chromosome breakage always occurs at or very near the location of *Ds*. McClintock's critical observation was

Figure 6.21

Sectors of purple and yellow tissue in the endosperm of maize kernels resulting from the presence of the transposable elements *Ds* and *Ac*. The heavier sectoring in some ears results from dosage effects of *Ac*. The least speckled ear has one copy of *Ac*; that in the middle has two (*Ac Ac*); and the most speckled ear has three (*Ac Ac Ac*). [Courtesy of Jerry L. Kermicle.]

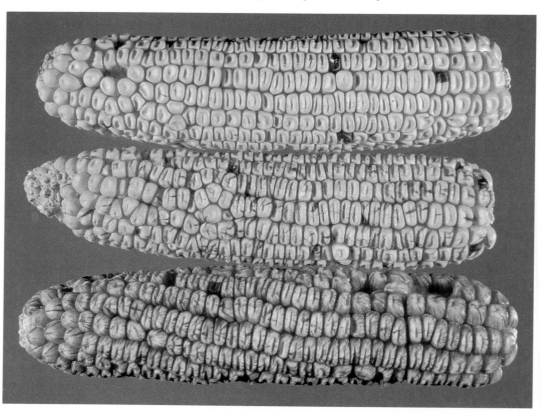

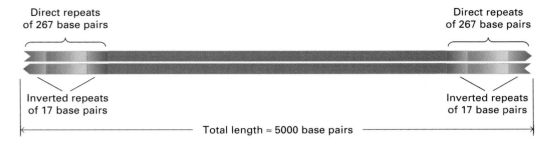

Figure 6.22

Sequence organization of a *copia* transposable element of *Drosophila melanogaster.*

that *Ds* does not have a constant location but occasionally moves to a new position (**transposition**), causing chromosome breakage at the new site. Furthermore, *Ds* moves only if a second element, called Activator (*Ac*), is also present. In addition, *Ac* itself moves within the genome and can cause, in the expression of genes at or near its insertion site, alterations similar to the modifications resulting from the presence of *Ds*.

Other **transposable elements** with characteristics and genetic effects similar to those of *Ac* and *Ds* are known in maize. Much of the color variegation seen in the kernels of varieties used for decorative purposes are attributable to the presence of one or more of these elements.

Since McClintock's discovery, transposable nucleotide sequences have been observed to be widespread in eukaryotes and prokaryotes. In *D. melanogaster,* they constitute from 5 to 10 percent of the genome and comprise about 50 distinct families of sequences. One well-studied family of closely related, but not identical, sequences is called *copia.* This element is present in about 30 copies per genome. The copia element (Figure 6.22) contains about 5000 base pairs with two identical sequences of 267 base pairs located terminally and in the same orientation. Repeated DNA sequences with the same orientation are called **direct repeats** (Figure 6.23A). The ends of each of these copia direct repeats contain two segments of 17 base pairs, whose sequences are also nearly identical; these shorter segments have opposite orientations. Repeated DNA sequences with opposite orientations are called **inverted repeats** (Figure 6.23B).

(A) Direct repeat

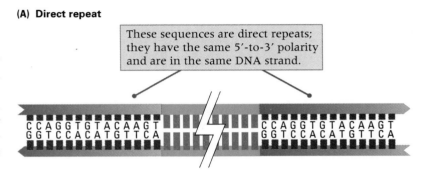

(B) Inverted repeat

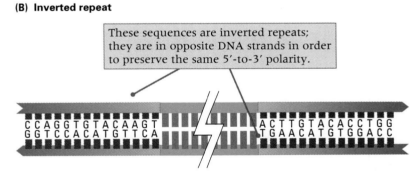

Figure 6.23

(A) In a direct repeat, a DNA sequence is repeated in the same left-to-right orientation. (B) In an inverted repeat, the sequence is repeated in the reverse left-to-right orientation *in the opposite strand.* The opposite strand is stipulated in order to maintain the correct 5'-to-3' polarity.

Other transposable elements have a similar organization with direct or inverted terminal repeats, as do many such elements in other organisms, such as the transposable elements in *E. coli* described in Chapter 8.

The molecular processes responsible for the movement of transposable elements are not well understood (some information will be presented in Chapter 8). A common feature is that transposition of the element is usually accompanied by the duplication of a small number of base pairs originally present at the insertion site, with the result that a copy of this short chromosomal sequence is found immediately adjacent to both ends of the inserted element (Figure 6.24). The length of the duplicated segment

Figure 6.24
The sequence arrangement of one type of transposable element (in this case, *Ds* of maize) and the changes that take place when it inserts into the genome. *Ds* is inserted into the maize *sh* gene at the position indicated. In the insertion process, a sequence of eight base pairs next to the site of insertion is duplicated and flanks the *Ds* element.

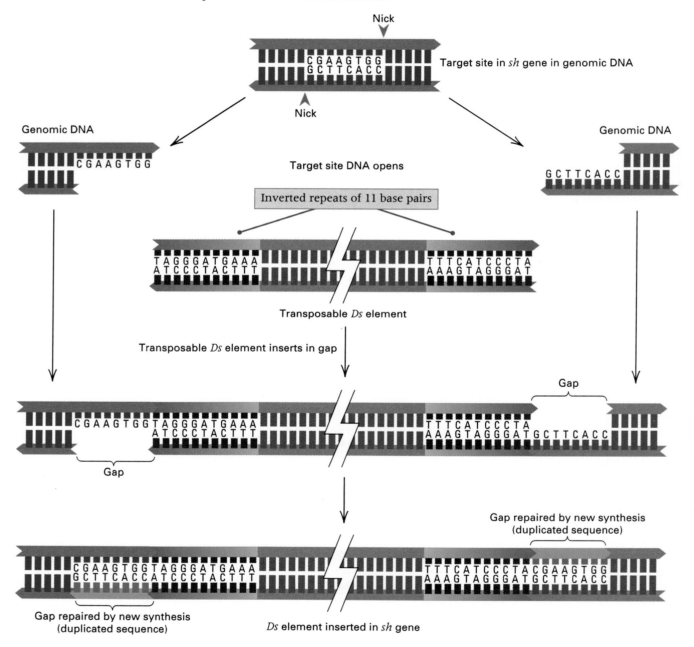

Her Feeling for the Organism

Barbara McClintock 1950
Cold Spring Harbor Laboratory,
Cold Spring Harbor, New York
The Origin and Behavior of Mutable Loci in Maize

Many geneticists regard McClintock's papers on transposable elements as difficult. Her discoveries were completely novel. Genes that could move from one place to another in the genome were unheard of. She had no terminology with which to discuss such things, so she had to adapt the conventional terminology to describe a unique situation. What we now call a transposable element, and believe to be universal among organisms, McClintock calls a "chromatin element." McClintock was a superb geneticist and cytogeneticist, perhaps the best of her generation. In tracking down transposable elements by genetic means, McClintock had to use considerable ingenuity in designing crosses that were sometimes quite complex. Her writing style is also uniquely McClintock. The excerpt that follows is a good example. It is a relentless marshaling of observation and hypothesis, experiment and result, interpretation and deduction. McClintock was awarded a Nobel Prize in 1983. The following passage deals with the discovery of Ds *(Dissociation) and* Ac *(Activator). In modern terminology, we call* Ac *an autonomous transposable element because it codes for all the proteins it needs for its own transposition;* Ds *is a nonautonomous transposable element because it requires* Ac *to provide the proteins necessary for it to move.*

In the course of an experiment designed to reveal the genic composition of the short arm of chromosome 9, a phenomenon of rare occurrence (or recognition) in maize began to appear with remarkably high frequencies in the cultures. The terms mutable genes, unstable genes, variegation, mosaicism, or mutable loci have been applied to this phenomenon. Its occurrence in a wide variety of organisms has been recognized. . . . A fortunate discovery was made early in the study of the mutable loci which proved to be of singular importance in showing the kinds of events that are associated with their origin and behavior. A locus was found in the short arm of chromosome 9 at which breaks were occurring in somatic cells. The time and frequency of the breakage events occurring at this *Ds* (Dissociation) locus appeared to be the same as the time and frequency of the mutation-producing mutable loci. An extensive study of the *Ds* locus

---

Transposition of Ac *takes place from one position in the chromosomal complement to another — very often from one chromosome to another.*

---

has indicated the reason for this relationship and has produced the information required to interpret the events occurring at mutable loci. It has been concluded that the changed phenotypic expression of such loci is related to changes in a chromatin element other than that composing the genes themselves, and that mutable loci arise when such chromatin is inserted adjacent to the genes that are affected. . . . The *Ds* locus is composed of this kind of material. Various types of alterations are observed as a consequence of events occurring at the *Ds* locus. . . . They involve chromosome breakage and fusion. The breaks are related, however, to events occurring at this one specific locus in the chromosome — the *Ds* locus. . . . [Among the known events is] transposition of *Ds* activity from one position to another in the chromosome complement with or without an associated gross chromosomal rearrangement. . . . It is from the transpositions of *Ds* that some of the new mutable loci may arise. . . . In one case the transposed *Ds* locus appeared in a single gamete of a plant carrying chromosome 9 with a dominant *C* (colored aleurone) allele. . . . The new position of *Ds* corresponded to the known location of *C*. . . . Significantly, the appearance of *Ds* activity at this new position was correlated with the disappearance of the normal action of the *C* locus. The resulting phenotype was the same as that produced by the known recessive allele *c*. . . . That the *c* phenotype in this case was associated with the appearance of *Ds* at the *C* locus was made evident because mutations from *c* to *C* occurred [in which] *Ds* action concomitantly disappeared. . . . The origin and behavior of this mutation at the *c* locus have been interpreted as follows: Insertion of the chromatin composed of *Ds* adjacent to the *C* locus is responsible for complete inhibition of the action of *C*. Removal of this foreign chromatin can occur. In many cases, the mechanism associated with the removal results in restoration of former genic organization and action. . . . The mutation-producing mechanisms involve only *Ds*. No gene mutations occur at the *C* locus; the restoration of its action is due to the removal of the inhibiting *Ds* chromatin. . . . [The movement] of *Ds* requires an activator. This activator has been designated *Ac*. . . . *Ac* shows a very important characteristic not exhibited in studies of the usual genetic factors. This characteristic is the same as that shown by *Ds*. Transposition of *Ac* takes place from one position in the chromosomal complement to another — very often from one chromosome to another. Again, as in *Ds*, changes in state may occur at the *Ac* locus. . . . It should be emphasized that when no *Ac* is present in a nucleus, no mutation-producing events occur at *Ds*.

Source: Proceedings of the National Academy of Sciences of the USA 36: 344–355

ranges from 2 to 12 base pairs, depending on the particular transposable element. Insertion of a transposable element is not a sequence-specific process in that the element is flanked by a different duplicated sequence at each location in the genome; however, the *number* of duplicated base pairs is usually the same at each location in the genome and is characteristic of a particular transposable element. Experimental deletion or mutation of part of the terminal base sequences of several different elements has shown that the short terminal inverted repeats are essential for transposition, probably because they are necessary for binding an enzyme called a **transposase** that is required for transposition. Many transposable elements code for their own transposase by means of a gene located in the central region between the terminal repeats, so these elements are able to promote their own transposition. Elements in which the transposase gene has been lost or inactivated by mutation are transposable only if a related element is present in the genome to provide this activity. The inability of the maize *Ds* element to transpose without *Ac* results from the absence of a functional transposase gene in *Ds*. Transposable elements are sometimes referred to as **selfish DNA,** because each type of element maintains itself in the genome as a result of its ability to replicate and transpose.

Transposable elements are responsible for many visible mutations. A transposable element is even responsible for the wrinkled-seed mutation in peas studied by Gregor Mendel. The wildtype allele of the gene codes for starch-branching enzyme I (SBEI), which is used in the synthesis of amylopectin (starch with branched chains). In the wrinkled mutation, a transposable element inserted into the gene renders the enzyme nonfunctional. The pea transposable element has terminal inverted repeats that are very similar to those in the maize *Ac* element, and the insertion site in the *wrinkled* allele is flanked by a duplication of eight base pairs of the SBEI coding sequence. This particular insertion appears to be genetically quite stable: The transposable element does not seem to have been excised in the long history of wrinkled peas.

6.9 Centromere and Telomere Structure

Eukaryotic chromosomes contain regions specialized for maneuvering the chromosomes in cell division and for capping the ends. These regions are discussed next.

Molecular Structure of the Centromere

The centromere is a specific region of the eukaryotic chromosome that becomes visible as a distinct morphological entity along the chromosome during condensation. It serves as a central component of the kinetochore, the complex of DNA and proteins to which the spindle fibers attach as they move the chromosomes in both mitosis and meiosis. The kinetochore is also the site at which the spindle fibers shorten, causing the chromosomes to move toward the poles. Electron microscopic analysis has shown that in some organisms—for example, the yeast *Saccharomyces cerevisiae*—a single spindle-protein fiber is attached to centromeric chromatin. Most other organisms have multiple spindle fibers attached to each centromeric region.

Centromeres exhibit considerable structural variation among species. At one extreme are **holocentric chromosomes,** which appear to have centromeric sequences spread throughout their length (constituting what is called a **diffuse centromere**). The nematode *Caenorhabiditis elegans,* which is widely used in genetic research, has holocentric chromosomes. In a holocentric chromosome, microtubules attach along the entire length of each chromatid. If a holocentric chromosome is broken into fragments by x rays, each fragment behaves as a separate, smaller holocentric chromosome that moves normally in cell division. Diffuse centromeres are very poorly understood and will not be discussed further.

The conventional type of centromere is the **localized centromere,** in which microtubules attach to a single region of the chromosome (the kinetochore). Localized centromeres fall into two types, *point centromeres* and *regional centromeres.*

Point centromeres are found in a number of different yeasts, including *Saccharomyces cerevisiae*, and they are relatively small in terms of their DNA content. Other eukaryotes, including higher eukaryotes, have regional centromeres that may contain hundreds of kilobases of DNA.

The chromatin segment of the centromeres of *Saccharomyces cerevisiae* has a unique structure that is exceedingly resistant to the action of various DNases; it has been isolated as a protein-DNA complex containing from 220 to 250 base pairs. The nucleosomal constitution and DNA base sequences of all of the yeast centromeres have been determined. Several common features of the base sequences are shown in Figure 6.25A. There are four regions, labeled CDE1, CDE2, CDE3, and CDE4. All yeast centromeres have sequences highly similar to those indicated for regions 1, 2, and 3, but the sequence of region CDE4 varies from one centromere to another. Region 2 is noteworthy in that approximately 90 percent of the base pairs are A−T pairs. The centromeric DNA is contained in a structure (the centromeric core particle) that contains more DNA than a typical yeast nucleosome core particle (which contains 160 base pairs) and is larger. This structure is responsible for the resistance of centromeric DNA to DNase. The spindle fiber is believed to be attached directly to this particle (Figure 6.25B).

The base-sequence arrangement of the yeast centromeres is not typical of other eukaryotic centromeres. In higher eukaryotes, the chromosomes are about 100 times as large as yeast chromosomes, and several spindle fibers are usually attached to each region of the centromere. Furthermore, the centromeric regions of the chromosomes of many higher eukaryotes contain large amounts of heterochromatin, consisting of repetitive satellite DNA, as described in Section 6.7. For example, the centromeric regions of human chromosomes contain a tandemly repeated DNA sequence of about 170 base pairs called the **alpha satellite** (Figure 6.26). The number of alpha-satellite copies in the centromeric region ranges from 5000 to 15,000, depending on the chromosome. The DNA sequences needed for spindle fiber attachment may be interspersed among the alpha-satellite sequences, but whether the alpha-satellite sequences themselves contribute to centromere activity is unknown.

Molecular Structure of the Telomere

Each end of a linear chromosome is composed of a special DNA-protein structure called a **telomere** that is essential for chromosome stability. Genetic and microscopic

Figure 6.25

A yeast centromere. (A) Diagram of centromeric DNA showing the major regions (CDE1 through CDE4) common to all yeast centromeres. The letter R stands for any purine (A or G), and the letter N indicates any nucleotide. Inverted-repeat segments in region 3 are indicated by arrows. The sequence of region CDE4 varies from one centromere to the next. (B) Positions of the centromere core and the nucleosomes on the DNA. The DNA is wrapped around histones in the nucleosomes, but the detailed organization and composition of the centromere core are unknown. [After K. S. Bloom, M. Fitzgerald-Hayes, and J. Carbon. 1982. *Cold Spring Harbor Symp. Quant. Biol.*, 47: 1175.]

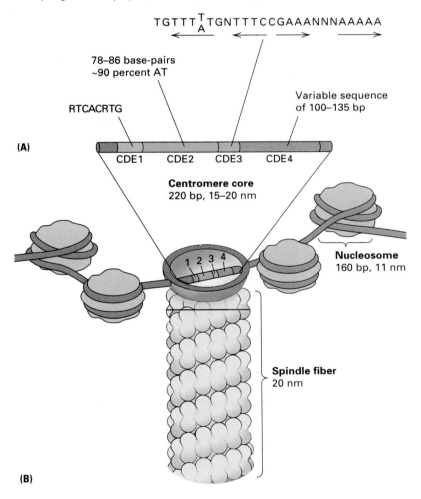

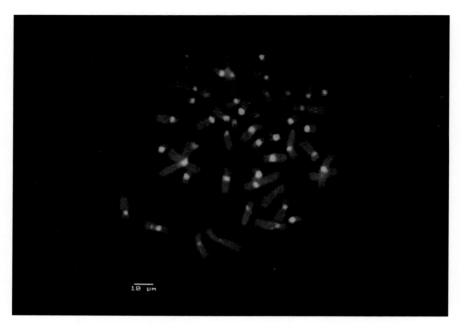

Figure 6.26
Hybridization of human metaphase chromosomes (red) with alpha-satellite DNA. The yellow areas result from hybridization with the labeled DNA. The sites of hybridization of the alpha satellite coincide with the centromeric regions of all 46 chromosomes. [Courtesy of Paula Coelho and Claudio E. Sunkel.]

observations first indicated that telomeres are special structures. In *Drosophila,* Hermann J. Muller found that chromosomes without ends could not be recovered after chromosomes were broken by treatment with x rays. In maize, Barbara McClintock observed that broken chromosomes frequently fuse with one another and form new chromosomes with abnormal structures (often having two centromeres). As we saw in Chapter 5, DNA polymerases cannot initiate DNA synthesis but instead require an RNA primase, so at least one end of each chromosome must have a short (8 to 12 nucleotides) stretch of single-stranded DNA that remains after the RNA primer at the tip has been removed. It turns out that both ends of linear chromosomes terminate in a stretch of single-stranded DNA.

Telomeric DNA in most organisms consists of tandem repeats of simple sequences such as 5'-TTAGGG-3'. These special telomere DNA sequences are *added* to the ends of eukaryotic chromosomes by an enzyme called **telomerase.** The substrate

for the telomerase is a telomere addition sequence consisting of a short repetitive sequence. In mammals, the telomere addition sequence consists of many tandem repeats of 5'-TTAGGG-3'. In the ciliate protozoan *Tetrahymena,* it consists of repeats of the similar sequence 5'-TTGGGG-3'. Relatively few copies of the repeat are necessary to prime the telomerase to add additional copies and form a telomere. Remarkably, the telomerase enzyme incorporates an essential RNA molecule, called a **guide RNA,** that contains sequences complementary to the telomere repeat and that serves as a template for telomere synthesis and elongation. For example, the *Tetrahymena* guide RNA contains the sequence 3'-AACCCCAAC-5'. The guide RNA undergoes base pairing with the telomere repeat and serves as a template for telomere elongation by the addition of more repeating units (Figure 6.27). The complementary DNA strand of the telomere is analogous to the lagging strand at a replication fork and is synthesized by DNA polymerase in the usual manner with the

Telomeres: The Beginning of the End

Carol W. Greider and Elizabeth H. Blackburn 1987
University of California, Berkeley, California

The Telomere Terminal Transferase of Tetrahymena *Is a Ribonucleoprotein enzyme with Two Kinds of Primer Specificity*

What a wonderful surprise that an RNA is a key ingredient in the formation of telomeres! Two limitations of DNA polymerase are that it requires a primer oligonucleotide and that it can elongate a DNA strand only at the 3' end. The limitations imply that the 3' end of a DNA strand should become progressively shorter with each round of replication, owing to the need for a primer at that extremity. (Chromosomes without proper telomeres do, in fact, become progressively shorter.) The organism used in this study, Tetrahymena, *is a ciliated protozoan. Each cell has a specialized type of nucleus called a macronucleus that contains many hundreds of small chromosomes. The level of telomerase activity is high because each of these tiny chromosomes needs a pair of telomeres. The convenience of using* Tetrahymena *for the study of telomere function illustrates a principle that runs through the history of genetics: Breakthroughs often come from choosing just the right organism to study. The authors speculate about the possible presence of a "guide" RNA in the telomerase. They were exactly right.*

Precise recognition of nucleic acids is often carried out by enzymes that contain both RNA and protein components. For some of these ribonucleoproteins (RNPs), the RNA components provide specificity to the reaction by base pairing with the substrate. The recognition that RNA can act catalytically has led to the increase in the number of known RNP-catalyzed reactions. We report here that an RNP is involved in synthesizing the telomeric sequences

---

. . . it is tempting to speculate that the RNA component of the telomerase might be involved in determining the sequence of the telomeric repeats that are synthesized . . .

---

found at the ends of *Tetrahymena* macronuclear chromosomes. . . . We have previously reported the identification of an activity in *Tetrahymena* cell extracts that adds telomeric repeats onto appropriate telomeric sequence primers in a nontemplated manner. . . . Repeats of the *Tetrahymena* telomeric sequence TTGGGG are added, 1 nucleotide at a time, onto the 3' end of the input primer. . . . We have begun characterizing and purifying the telomerase enzyme in order to investigate the mechanisms controlling the specificity of the reaction. . . . We propose that the

RNA component(s) of telomerase may play a role in specifying the sequence of the added telomeric TTGGGG repeats, in recognizing the structure of the G-rich telomeric sequence primers, or both. . . . In the course of purifying the telomerase from crude extracts, we noted a marked sensitivity to salts [in which high] concentrations inactivated the telomeric elongation activity. The salt sensitivity and the large size of the enzyme suggested that the telomerase may be a complex containing a nucleic acid component. To test whether the telomerase contained an essential nucleic acid, we treated active fractions with either micrococcal nuclease or RNAase A. The nuclease activity of each of these enzymes abolished the telomeric elongation activity. . . . These experiments suggest that the telomerase contains an essential RNA component. . . . The RNA of telomerase may simply provide a scaffold for the assembly of proteins in the active enzyme complex; however, . . . it is tempting to speculate that the RNA component of the telomerase might be involved in determining the sequence of the telomeric repeats that are synthesized and/or the specific primer recognition. If the RNA of telomerase contains the sequence CCCCAA, this sequence could act as an internal guide sequence.

Source: Cell 51: 887–898

---

help of an RNA primer. In the telomeric regions in most eukaryotes, there are also longer, moderately repetitive DNA sequences just preceding the terminal repeats. These sequences differ among organisms and even among different chromosomes in the same organism.

What limits the length of a telomere? In most organisms the answer is unknown. In yeast, however, a protein called Rap1p has been identified that appears to be important in regulating telomere length. The Rap1p protein binds to the yeast telomere sequence. Molecules of Rap1p bind to the

telomere sequence as it is being elongated until about 17 Rap1p molecules have been bound. At this point, telomere elongation stops, probably because the accumulation of Rap1p inhibits telomerase activity. Because each Rap1p molecule binds to approximately 18 base pairs of the telomere, the predicted length of a yeast telomere is $17 \times 18 = 306$ base pairs, which is very close to the value observed. Additional evidence for the role of Rap1p comes from mutations in the *RAP 1* gene producing a protein that cannot bind to telomere sequences; in these mutant

Figure 6.27
Telomere formation in *Tetrahymena*. The telomerase enzyme contains an internal RNA with a sequence complementary to the telomere repeat. The RNA undergoes base pairing with the telomere repeat and serves as a template for telomere elongation. The newly forming DNA strand is produced by DNA polymerase.

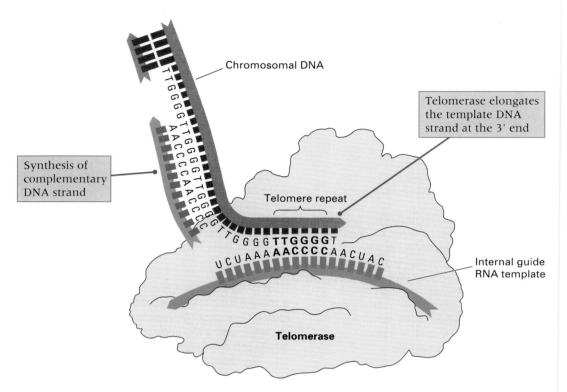

Chromosomal DNA

Synthesis of complementary DNA strand

Telomerase elongates the template DNA strand at the 3' end

Telomere repeat

Internal guide RNA template

GGGG**TTGGGG**T
UCUAAA**AACCCC**AACUAC

Telomerase

strains, massive telomere elongation is observed.

Telomeric sequences found in a variety of organisms are shown in Table 6.2. The table does not include *Drosophila* because its telomeres, rather than consisting of simple repeats as in most other organisms, are composed of specialized transposable elements. Each telomere sequence in Table 6.2 in written as if it were at the left-hand end of a chromosome. Both strands are written in the 5'-to-3' direction. The top strand is the 5' end of the telomere, and the bottom strand is the 3' end of the telomere. It is the 3' end of the bottom strand (in each entry, the 3' end is at the far right) that is elongated by the telomerase and that is single-stranded. Runs of nucleotides are indicated by subscripts; for example, the notation C_3TA_2/T_2AG_3 for vertebrate telomeres means the sequence

$$\xleftarrow{\text{telomerase}} \begin{array}{l} 5'\text{-CCCTAA-}3' \\ 3'\text{-CCCATT-}5' \end{array}$$

In many cases, there is variation in the length of a run; for example, C_{1-8} means variation ranging from as few as one C to as many as eight C's from one repeat to the next. There is also occasional variation in which nucleotide occupies a given position, which is indicated by small stacked letters.

There is a strong tendency for the elongated strand of the telomere repeats in Table 6.2 to be rich in guanine nucleotides. The guanine bases are special in that they have the capacity to hydrogen-bond to one another in a variety of ways. When the elongated 3' strand of the telomere folds back upon itself, the guanine bases can even pair to give the G-quartet structure illustrated in Figure 6.28. The hydrogen bonding in the G-quartet is called **Hoogstein base pairing** to distinguish it from ordinary Watson-Crick base pairing. A protein has been isolated from the ciliated protozoan *Oxytricha* that binds specifically to the telomeric DNA of linear chromosomes. The β subunit of this telomere-binding protein promotes

Table 6.2 Sequences of telomeric DNAs.

Organism	Sequence
Protozoa	
Tetrahymena	C_4A_2/T_2G_4
Paramecium	$C_3{}^C_AA_2/T_2{}^G_TG_3$
Oxytricha	C_4A_4/T_4G_4
Plasmodium	$C_3T{}^A_GA_2/T_2{}^T_CAG_3$
Trypanosoma	C_3TA_2/T_2AG_3
Giardia	C_3TA/TAG_3
Slime molds	
Physarum	C_3TA_2/T_2AG_3
Didymium	C_3TA_2/T_2AG_3
Dictyostelium	$C_{1-8}T/AG_{1-8}$
Fungi	
Saccharomyces	$C_{2-3}ACA_{1-6}/T_{1-6}GTG_{2-3}$
Kluyveromyces	$ACAC_2ACATAC_2TA_2TCA_3TC_2GA/TCG_2AT_3GAT_2AG_2TATGTG_2TGT$
Candida	$ACAC_2A_2GA_2GT_2AGACATC_2GT/ACG_2ATGTCTA_2CT_2CT_2G_2TGT$
Schizosaccharomyces	$C_{1-6}G_{0-1}T_{0-1}GTA_{1-2}/T_{1-2}ACA_{0-1}C_{0-1}G_{1-6}$
Neurospora	C_3TA_2/T_2AG_3
Podospora	C_3TA_2/T_2AG_3
Cryptococcus	$A_2C_{3-5}T/AG_{3-5}T_2$
Cladosporium	C_3TA_2/T_2AG_3
Invertebrates	
Caenorhabditis	GC_2TA_2/T_2AG_2C
Ascaris	GC_2TA_2/T_2AG_2C
Parascaris	$TGCA_2/T_2GCA$
Bombyx, other insects	C_2TA_2/T_2AG_2
Vertebrates	
	C_3TA_2/T_2AG_3
Plants	
Chlamydomonas	C_3TA_4/T_4AG_3
Chlorella	C_3TA_3/T_3AG_3
Arabidopsis	C_3TA_3/T_3AG_3
Tomato	$C_3A{}^T_AT_2/A_2{}^A_TTG_3$

Source: From V.A. Zakian. 1995, *Science* 270:1602

G-quartet formation by the telomeric DNA of *Oxytricha:*

    . . . 5'-TTTTGGGGTTTTGGGGT-3' . . .

or *Tetrahymena:*

    . . . 5'-TTGGGGTTGGGGT-3' . . .

It is hypothesized that within telomeres, telomeric DNA may be organized in special three-dimensional conformations that include such G-quartets. Models for *Oxytricha* and *Tetrahymena* telomeric DNA, showing the possible positions of G-quartets, are presented in Figure 6.29.

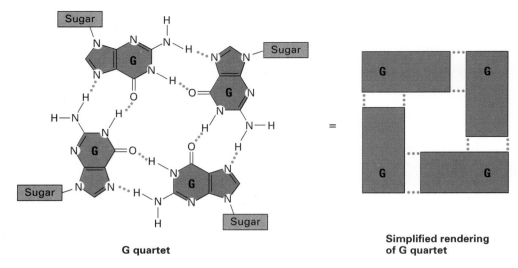

G quartet

**Simplified rendering
of G quartet**

Figure 6.28
G-quartet structure formed by hydrogen bonding between four guanine bases present in a single DNA strand folded back upon itself. The rectangles on the right show how the guanines are oriented, with the deoxyribose sugars attached at the outside corners. On the left, the sugars are indicated by the small blue boxes. There is also a monovalent cation in the center of the quartet (not shown). [From J. R. Williamson, M. K. Raghuraman and T. R. Cech. 1982. *Cell* 59: 871.]

Figure 6.29
Models of telomere structure in *Oxytricha* (A) and *Tetrahymena* (B) that incorporate the G-quartet structure. The arrowhead indicates the 3' end of the DNA strand. The indicated G bases participate in the Hoogstein base pairing, and the G-quartets are drawn as green squares, as in Figure 6.28. There are two G-quartets postulated in the *Oxytricha* telomere, three in *Tetrahymena*. [From J. R. Williamson, M. K. Raghuraman, and T. R. Cech. 1982. *Cell* 59: 871.]

(A) *Oxytricha*

(B) *Tetrahymena*

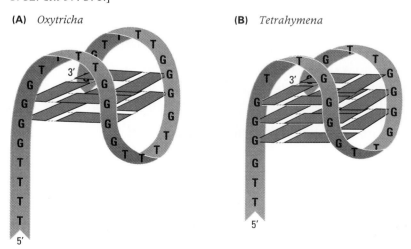

Chapter Summary

The DNA content of organisms varies widely. Small viruses exist whose DNA contains only a few thousand nucleotides, and among the higher animals and plants, the DNA content can be as large as 1.5×10^{11} nucleotides. Generally speaking, DNA content increases with the complexity of the organism, but within particular groups of organisms, the DNA content can vary as much as tenfold.

DNA molecules come in a variety of forms. Except for a few of the smallest viruses, whose DNA is single-stranded, and some viruses in which RNA is the genetic material, all organisms contain double-stranded DNA. The chromosomal DNA of higher organisms is almost always linear. Bacterial DNA is circular, as is the DNA of many animal viruses and of some bacteriophages. Circular DNA molecules are invariably supercoiled. The bacterial chromosome consists of independently supercoiled domains.

The DNA of both prokaryotic and eukaryotic cells and of viruses is never in a fully extended state but rather is folded in an intricate way, which reduces its effective volume. In viruses, the DNA is tightly folded but without bound protein molecules. In bacteria, the DNA is folded to form a multiply looped structure called a nucleoid, which includes several proteins that are essential for folding. In eukaryotes, the DNA is compacted into chromosomes, which contain several proteins and which are thick enough to be visible by light microscopy during the mitotic phase of the cell cycle. The DNA-protein complex of eukaryotic chromosomes is called chromatin. The protein component of chromatin consists primarily of five distinct proteins: histones H1, H2A, H2B, H3, and H4. The last four histones aggregate to form an octameric protein that contains two molecules of each. DNA is wrapped around the histone octamer, forming a particle called a nucleosome. This wrapping is the first level of compaction of the DNA in chromosomes. Each nucleosome unit contains about 200 nucleotide pairs, of which about 145 are in contact with the protein. The remaining 55 nucleotide pairs link adjacent nucleosomes. Histone H1 binds to the linker segment and draws the nucleosomes nearer to one another. The DNA in its nucleosome form is further compacted into a helical fiber, the 30-nm fiber. In forming a visible chromosome, this unit undergoes several additional levels of folding, producing a highly compact visible chromosome. The result is that a eukaryotic DNA molecule, whose length and width are about 50,000 and 0.002 μm, respectively, is folded to form a chromosome with a length of about 5 μm and a width of about 0.5 μm.

Polytene chromosomes are found in certain organs in insects. These gigantic chromosomes consist of about 1000 molecules of partly folded chromatin aligned side by side. Seen by microscopy, they have about 5000 transverse bands. Polytene chromosomes do not replicate further, and cells that contain them do not divide. They are useful to geneticists primarily as morphological markers for particular genes and chromosome segments.

The number of copies of individual base sequences in a DNA molecule can vary tremendously. In prokaryotic DNA, most sequences are unique. However, in eukaryotic DNA, only a fraction of the DNA consists of unique sequences present once per haploid genome. The sequence composition of complex genomes can be studied by DNA renaturation kinetics, or Cot analysis. In eukaryotes, many sequences are present in hundreds to millions of copies. Some highly repetitive sequences are primarily located in the centromeric regions of the chromosomes, whereas others are dispersed. A significant fraction of the DNA, the middle-repetitive DNA, consists of sequences of which from 10 to 1000 copies are present per cell. Much of middle-repetitive DNA in higher eukaryotes consists of transposable elements, which are sequences able to move from one part of the genome to another. A typical transposable element is a sequence ranging in length from one to several thousand nucleotide pairs terminating in short sequences that are repeated, either in the same orientation (direct repeats) or in reverse orientation (inverted repeats). The terminal repeats and a transposase enzyme are necessary for the movement of these elements, a process known as transposition. Many transposable elements contain a gene coding for their own transposase. Insertion of most transposable elements causes duplication of a short chromosomal nucleotide sequence flanking the point of insertion. Insertions of transposable elements are the cause of many visible mutations.

Centromeres and telomeres are regions of eukaryotic chromosomes specialized for spindle fiber attachment and stabilization of the tips, respectively. The centromeres of most higher eukaryotes are associated with localized, highly repeated, satellite DNA sequences. Telomeres are formed by a telomerase enzyme that contains a guide RNA that serves as a template for the addition of nucleotides to the 3' end of a telomerase addition site. The complementary strand is synthesized in a manner analogous to the replication of the lagging strand at an ordinary replication fork. In mammals and other vertebrates, the 3' strand of the telomere terminates in tandem repeats of the simple sequence 5'-TTAGGG-3'. Relatively few copies of this sequence are needed to prime the telomerase. The special properties of the telomere may be related to the ability of guanine nucleotides to form complex hydrogen-bonded structures—in particular, the G-quartet.

Key Terms

alpha satellites	Cot	diffuse centromere
chromatin	Cot curve	direct repeat
chromocenter	covalent circle	DNase
conserved sequence	cytological map	euchromatin
core particle	deoxyribonuclease	folded chromosome

genome
guide RNA
H1 histone
H2A histone
H2B histone
H3 histone
H4 histone
heterochromatin
highly repetitive sequences
histone
holocentric chromosome
Hoogstein base pairing
inverted repeat

in situ hybridization
kilobase pairs (kb)
localized centromere
megabase pairs (Mb)
micrococcal nuclease
middle-repetitive sequences
negative supercoiling
nucleoid
nucleosome
polytene chromosome
pulsed-field gel electrophoresis
relaxed DNA
repetitive sequence

satellite DNA
scaffold
selfish DNA
single-copy sequence
supercoiled DNA
telomerase
30-nm fiber
topoisomerase I
topoisomerase II
transposable element
transposase
transposition
unique sequence

Review the Basics

- What are the principal differences between prokaryotes and eukaryotes in the organization of the genetic material?

- In a eukaryotic chromosome, where are the telomeres located? Is it possible to have a chromosome with the centromere exactly at the tip? What shape of chromosome has no telomeres?

- What is a transposable element?

- Are all repetitive DNA sequences transposable elements? If a transposable element were present somewhere in a long stretch of DNA, what features in the nucleotide sequence would help to identify it?

- Histone proteins are positively charged (basic) molecules, and they interact with negatively charged (acidic) mole-

cules. What is the principal type of acid with which histones interact?

- How many polypeptide chains are present in each nucleosome? How many *different* polypeptide chains are present in each nucleosome? Which type of histone is not part of the nucleosome core particle?

- Are the ratios of the different histone types identical in the nucleosomes in all cells of a eukaryotic organism? In all eukaryotic organisms?

- Is it correct to say that no genes are present in heterochromatin? Explain your answer.

Guide to Problem Solving

Problem 1: *Dp(1;f)1187* is a minichromosome induced by x rays in *Drosophila*. It is about 1.3 megabases in length. What is the approximate physical length of this chromosome in millimeters? (*Note:* One megabase (Mb) equals 1×10^6 base pairs, and $1 \text{ Å} = 10^{-7}$ mm.)

Answer: The standard form of duplex DNA contains 10 nucleotide pairs per 34 Å, or 3.4 Å per nucleotide pair. Therefore, a 1.3-Mb molecule has a length of $1.3 \times 10^6 \times 3.4$ Å $= 4.4 \times 10^6$ Å $= 0.44$ mm; that is, the 1.3-megabase molecule of DNA in *Dp(1;f)1187* has a physical length just a little smaller than half a millimeter.

Problem 2: The genome size of *Drosophila melanogaster* is 165,000 kilobases (kb); approximately 2/3 of the genome is euchromatic and 1/3 is heterochromatic. In salivary gland chromosomes, the euchromatic part of the genome becomes polytene, and the giant chromosomes exhibit approximately 5000 transverse bands used as landmarks for the locations of

genes, chromosome breakpoints, and other cytogenetic features. What is the approximate DNA content of an average band in the salivary gland chromosomes?

Answer: The euchromatic part of the *Drosophila* genome consists of approximately 165,000 kb × 2/3 = 110,000 kb of DNA. This amount of DNA is distributed over 5000 bands, for an average of 110,000 kb/5000 = 22 kb per band.

Problem 3: Suppose that the DNA sequences repeated at the ends of a transposable element undergo pairing and recombination. What are the genetic consequences if:

(a) the sequences are direct repeats?

(b) the sequences are inverted repeats?

Draw diagrams to support your answers.

Answer: The diagrams show the consequences.

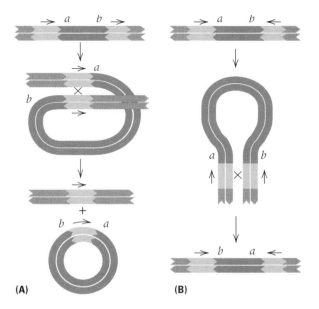

(a) When direct repeats pair and undergo crossing-over, the result is the deletion of one of the repeats and the DNA segment between them (part A).

(b) When the repeats are inverted repeats, the result is that the transposable element is inverted in the chromosome (part B).

(A)   (B)

Analysis and Applications _____

6.1 The w^{pch} mutant allele of the *white* gene in *Drosophila* contains an insertion of 1.3 kb within the controlling region of the gene. The w^{pch} allele undergoes further mutation at a high rate, including reverse mutations to w^+. The w^+ derivatives are found to lack the 1.3-kb insert. What kind of DNA sequence is the 1.3-kb insert likely to be?

6.2 Some transposable elements in animals are active only in the germ line and not in somatic cells. How might this restriction of activity be advantageous to the persistence of the transposable element?

6.3 Most transposable elements create a short duplication of host sequence when they insert. The duplication is always in a direct, rather than an inverted, orientation. What does this tell you about the process of insertion?

6.4 The *E. coli* chromosome is 4700 kb in length. What is its length in millimeters? The haploid human genome contains 3×10^9 base pairs of DNA. What is its length in millimeters?

6.5 If you had never seen a chromosome in a microscope but had seen nuclei, what feature of DNA structure would tell you that DNA must exist in a highly coiled state within cells?

6.6 What causes the bands in a polytene chromosome? How do polytene chromosomes undergo division?

6.7 Is there an analog of a telomere in the *E. coli* chromosome?

6.8 Many transposable elements contain a pair of repeated sequences. What is their position in the element? Are they in direct or inverted orientation?

6.9 Consider a long linear DNA molecule, one end of which is rotated four times with respect to the other end in the unwinding direction.

(a) If the two ends are joined to keep the molecule in the underwound state, how many base pairs will be broken?

(b) If the underwound molecule is allowed to form a super-coil, how many twists will be present?

6.10 Endonuclease S1 can break single stranded DNA but does not break double-stranded linear DNA. However, S1 can cleave supercoiled DNA, usually making a single break. Why does this occur?

6.11 Circular DNA molecules of the same size migrate at different rates in electrophoretic gels, depending on whether they are supercoiled or relaxed circles. Why?

6.12 Denaturation of DNA refers to breakdown of the double helix and ultimate separation of the individual strands, and it can be induced by heat and other treatments. Because G−C base pairs have three hydrogen bonds and A−T base pairs have only two, the temperature required for denaturation increases with the G−C content and also with the length of continuous G−C tracts in the molecule. Renaturation of DNA refers to the formation of double-stranded DNA from complementary single strands.

(a) Which process—denaturation or renaturation—is dependent on the concentration of DNA?

Chapter 6

GeNETics on the web will introduce you to some of the most important sites for finding genetic information on the Internet. To complete the exercises below, visit the Jones and Bartlett home page at

http://www.jbpub.com/genetics

Select the link to *Genetics: Principles and Analysis* and then choose the link to *GeNETics on the web.* You will be presented with a chapter-by-chapter list of highlighted keywords.

GeNETics EXERCISES

Select the highlighted keyword in any of the exercises below, and you will be linked to a web site containing the genetic information necessary to complete the exercise. Each exercise suggests a specific, written report that makes use of the information available at the site. This report, or an alternative, may be assigned by your instructor

1. The **centromere** consensus sequences for CDE1, CDE2, and CDE3 can be compared among all yeast chromosomes because the complete sequence of the entire genome is available. This exercise concerns only the centromeres of chromosomes IV, IX, XIII, and XVI. Find the CDE1, CDE2, and CDE3 sequences in each of these chromosomes, and compare them with each other and with the consensus sequences given in the chapter. If assigned to do so, copy the CDE1, CDE2, and CDE3 sequences and align them.

2. Another view of the **telomerase** at work can be found at this keyword site. Although the 5'-to-3' polarities of the DNA and RNA strands are not indicated, you should be able to deduce them. If assigned to do so, make a sketch of the telomerase, and indicate the polarity of each of the DNA and RNA strands.

3. You will find up-to-date descriptions of the **transposable element** that Barbara McClintock first described by searching the keyword site for the terms *Activator* and *Dissociation.* Use the hot links to locate the nucleotide sequence of *Ac,* and identify the

(b) Which of the two illustrated DNA molecules (1 or 2) would have the lower temperature for strand separation? Why?

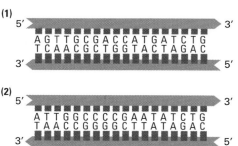

6.13 DNA from species A, labeled with ^{14}N and randomly fragmented, is renatured with an equal concentration of DNA from species B, labeled with ^{15}N and randomly fragmented, and then centrifuged to equilibrium in CsCl. Five percent of the total renatured DNA has a hybrid density. What fraction of the base sequences is common to the two species?

6.14 What is meant by the terms "direct repeat" and "inverted repeat"? Using the base sequence illustrated, diagram a duplex DNA molecule, showing the sequence as a direct repeat and as an inverted repeat.

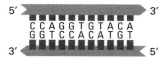

6.15 Mutations have not been observed that result in nonfunctional histones. Why should this be expected?

6.16 A sequence of middle-repetitive DNA from *Drosophila* is isolated and purified. It is used as a template in a polymerization reaction with DNA polymerase and radioactive substrates, and highly radioactive probe DNA is prepared. The probe DNA is then used in an *in situ* hybridization experiment with cells containing polytene chromosomes obtained from ten different flies of the same species. Autoradiography indicates that the radioactive material is localized to about 20 sites in the genome, but they are in different sites in each fly examined. What does this observation suggest about the DNA sequence being studied?

Challenge Problem

6.17 A sample of identical DNA molecules, each containing about 3000 base pairs per molecule, is mixed with histone octamers under conditions that allow formation of chromatin. The reconstituted chromatin is then treated with a nuclease, and enzymatic digestion is allowed to occur. The histones are removed, and the positions of the cuts in the DNA are identified by sequencing the fragments. It is found that the breaks have been made at random positions and, as

inverted repeats at its ends. If assigned to do so, write a 150-word report on these transposable elements, paying particular attention to the DNA sequence relations between the *Ac* and *Ds* elements and specifying the element that encodes the active transposase.

MUTABLE SITE EXERCISES

The Mutable Site Exercise changes frequently. Each new update includes a different exercise that makes use of genetics resources available on the World Wide Web. Select the **Mutable Site** for Chapter 6, and you will be linked to the current exercise that relates to the material presented in this chapter.

PIC SITE

The Pic Site showcases some of the most visually appealing genetics sites on the World Wide Web. To visit the showcase genetics site, select the **Pic Site** for Chapter 6.

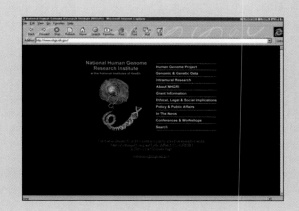

expected, at about 200-base-pair intervals. The experiment is then repeated with a single variation. A protein known to bind to DNA is added to the DNA sample before addition of the histone octamers. Again, reconstituted chromatin is formed and digested with nuclease. In this experiment, it is found that the breaks are again at intervals of about 200 base pairs, but they are localized at particular positions in the base

sequence. At each position, the site of breakage can vary over only a 2- to-3-base range. However, examination of the base sequences in which the breaks have occurred does not indicate that breakage occurs in a particular sequence; in other words, each 2- to-3-base region in which cutting occurs has a different sequence. Explain the difference between the two experiments.

Further Reading

Berg, D. E., and M. M. Howe. 1989. *Mobile DNA.* American Association for Microbiology. Washington, DC.

Blackburn, E. H. 1990. Telomeres and their synthesis. *Science* 249: 489.

Comfort, N. C. 1995. Two genes, no enzyme: A second look at Barbara McClintock and the 1951 Cold Spring Harbor Symposium. *Genetics* 140: 1161.

Curtis, B. C., and D. R. Johnson. 1969. Hybrid wheat. *Scientific American*, May.

Elgin, S. C. R. 1996. Heterochromatin and gene regulation in *Drosophila. Current Opinion in Genetics & Development* 6: 193.

Engels, W. R. 1997. Invasions of *P* elements. *Genetics* 145: 11.

Federoff, N. 1984. Transposable genetic elements in maize. *Scientific American,* June.

Green, M. M. 1980. Transposable elements in *Drosophila* and other Diptera. *Annual Review of Genetics* 14: 109.

Greider, C. W., and E. H. Blackburn. 1996. Telomeres, telomerase and cancer. *Scientific American,* February.

Hartl, D. L., E. R. Lozovskaya, D. I. Nurminsky, and A. R. Lohe. 1997. What restricts the activity of *mariner*-like transposable elements? *Trends in Genetics* 13: 197.

Haseltine, W. A., and F. Wong-Stahl. 1988. The molecular biology of the AIDS virus. *Scientific American,* October.

Hsu, T. H. 1979. *Human and Mammalian Cytogenetics.* New York: Springer-Verlag.

Jacks, T. 1996. Tumor suppressor gene mutations in mice. *Annual Review of Genetics* 30: 603.

Manning, C. H., and H. O. Goodman. 1981. Parental origin of chromosomes in Down's syndrome. *Human Genetics* 59: 101.

McClintock, B. 1965. Control of gene action in maize. *Brookhaven Symposium on Quantitative Biology* 18: 162.

Pardue, M. L., O. N. Danilevskaya, K. Lowenhaupt, F. Slot, and K. L. Traverse. 1996. *Drosophila* telomeres: New views on chromosome evolution. *Trends in Genetics* 12: 48.

Wagner, R. P., M. P. Maguire, and R. L. Stallings. 1993. *Chromosomes.* New York: Wiley-Liss.

White, M. J. D. 1977. *Animal Cytology and Evolution.* London: Cambridge University Press.

Zakian, V. A. 1996. Structure, function, and replication of *Saccharomyces cerevisiae* telomeres. *Annual Review of Genetics* 30: 141.

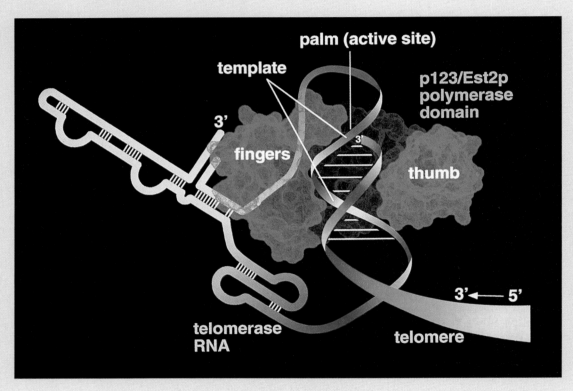

A model of the telomerase enzyme showing protein components in green and the RNA component in blue. The "palm" of the protein is the active site of telomere elongation, whereas the "fingers" and "thumb" are protruding. The DNA strand, whose telomere is being elongated, is shown in red. [Courtesy of Joachim Lingner and Thomas Cech, from J. Lingner, T. R. Hughes, A. Shevchenko, M. Mann, V. Lundblad and T. R. Cech. 1997. *Science* 276: 561.]

Variation in Chromosome Number and Structure

PRINCIPLES

- Duplication of the entire chromosome complement present in a species—or in a hybrid between species—is a major process in the evolution of higher plants.
- The genetic unbalance caused by a single chromosome that is extra or missing may have a more serious phenotypic effect than an entire extra set of chromosomes.
- Chromosome abnormalities are an important cause of human genetic disease and are a major factor in spontaneous abortions.
- Aneuploid (unbalanced) chromosome rearrangements usually have greater phenotypic effects than euploid (balanced) chromosome rearrangements.
- By a process of mispairing and unequal crossing-over, genes that are duplicated in tandem along the chromosome can give rise to chromosomes with even more copies.
- Gene-for-gene pairing between a wildtype chromosome and one that contains an inversion of a segment of genes results in the formation of a loop in one of the chromosomes; crossing-over within the "inversion loop" leads to chromosomal abnormalities.
- Reciprocal translocations result in abnormal gametes because they upset segregation.
- Some types of cancer are associated with particular chromosome rearrangements.

CONNECTIONS

CONNECTION: **The First Human Chromosomal Disorder**
Jerome Lejeune, Marthe Gautier, and Raymond Turpin 1959
Study of the somatic chromosomes of nine Down syndrome children

CONNECTION: **Lyonization of an X Chromosome**
Mary F. Lyon 1961
Gene action in the X chromosome of the mouse (Mus musculus L.)

I n all species, an occasional organism is found that has extra chromosomes or that lacks a particular chromosome. Such a deviation from the norm is an abnormality in chromosome *number*. Other organisms, usually rare, are found to have alterations in the arrangement of genes in the genome, such as by having a chromosome with a particular segment missing, reversed in orientation, or attached to a different chromosome. These variations are abnormalities in chromosome *structure*. This chapter will deal with the genetic effects of both numerical and structural chromosome abnormalities. We will see that animals are much less tolerant of such changes than are plants. Furthermore, in animals, numerical alterations often produce greater effects on phenotype than do structural alterations.

7.1
Centromeres and the Genetic Stability of Chromosomes

Chromosomes that have a single centromere are usually the only ones that are transmitted reliably from parental cells to daughter cells and from parental organisms to their progeny. When a cell divides, spindle fibers attach to the centromere of each chromosome and pull the sister chromatids to opposite poles. Occasionally, a chromosome arises that has an abnormal number of centromeres, as diagrammed in Figure 7.1A. The chromosome on the left has two centromeres and is said to be **dicentric.** A dicentric chromosome is genetically unstable, which means it is not transmitted in a predictable fashion. The dicentric chromosome is frequently lost from a cell when the two centromeres proceed to opposite poles in the course of cell division; in this case, the chromosome is stretched and forms a *bridge* between the daughter cells. This bridge may not be included in either daughter nucleus, or it may break, with the result that each daughter nucleus receives a broken chromosome. The chromosome on the right in Figure 7.1A is an **acentric** chromosome, which lacks a centromere. Acentric chromosomes also are genetically unstable because they cannot be maneuvered properly during cell division and tend to be lost.

In eukaryotic organisms, virtually all chromosomes have a single centromere and are rod-shaped. (Rarely, a **ring** chromosome is found, which results from breakage and loss of the telomere at each end of a rod chromosome and subsequent fusion of the broken ends.) Monocentric rod chromosomes are often classified according to the relative position of their centromeres. A chromosome with its centromere about in the middle is a **metacentric chromosome;** the arms are of approximately equal length and form a V shape at anaphase (Figure 7.1B). When the centromere is somewhat off center, the

Figure 7.1
(A) Diagram of a chromosome that is dicentric (two centromeres) and one that is acentric (no centromere). Dicentric and acentric chromosomes are frequently lost in cell division, the former because the two centromeres may bridge between the daughter cells, and the latter because the chromosome cannot attach to the spindle fibers. (B) Three possible shapes of monocentric chromosomes in anaphase as determined by the position of the centromere. The centromeres are shown in dark blue.

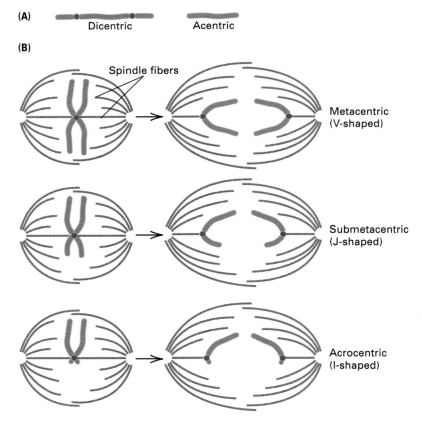

(A)

Dicentric   Acentric

(B)

Spindle fibers

Metacentric (V-shaped)

Submetacentric (J-shaped)

Acrocentric (I-shaped)

chromosome is a **submetacentric chromosome,** and the arms form a J shape at anaphase. A chromosome with the centromere very close to one end appears I-shaped at anaphase because the arms are grossly unequal in length; such a chromosome is **acrocentric.**

The distinction among metacentric, submetacentric, and acrocentric chromosomes is useful because it draws attention to the chromosome arms. In the evolution of chromosomes, often the number of chromosome *arms* is conserved without conservation of the individual *chromosomes*. For example, *Drosophila melanogaster* has two large metacentric autosomes, but many other *Drosophila* species have four acrocentric autosomes instead of the two metacentrics. Detailed comparison of the genetic maps of these species reveals that the acrocentric chromosomes in the other species correspond, arm for arm, with the large metacentrics in *Drosophila melanogaster* (Figure 7.2). Among higher primates, chimpanzees and human beings have 22 pairs of chromosomes that are morphologically similar, but chimpanzees have two pairs of acrocentrics not found in human beings, and human beings have one pair of metacentrics not found in chimpanzees. In this case, the human metacentric chromosome was formed by fusion of the telomeres between the short arms of the chromosomes that, in chimpanzees, remain acrocentrics. The metaphase chromosome resulting from the fusion is human chromosome 2.

7.2 Polyploidy

The genus *Chrysanthemum* illustrates **polyploidy,** an important phenomenon found frequently in higher plants in which a species has a genome composed of multiple complete sets of chromosomes. One *Chrysanthemum* species, a diploid species, has 18 chromosomes. A closely related species has 36 chromosomes. However, comparison of chromosome morphology indicates that the 36-chromosome species has two complete sets of the chromosomes found in the 18-chromosome species (Figure 7.3). The basic chromosome set in the group, from which all the other genomes are formed, is called the **monoploid** chromosome set. In *Chrysanthemum,* the monoploid chromosome number is 9. The diploid species has two complete copies of the monoploid set, or 18 chromosomes altogether. The 36-chromosome species has four copies of the monoploid set (4 × 9 = 36) and is a **tetraploid.** Other species of *Chrysanthemum* have 54 chromosomes (6 × 9, constituting the **hexaploid**), 72 chromosomes (8 × 9, constituting the **octoploid**), and 90 chromosomes (10 × 9, constituting the **decaploid**).

In meiosis, the chromosomes of all *Chrysanthemum* species synapse normally in pairs to form bivalents (Section 3.3). The 18-chromosome species forms 9 bivalents, the 36-chromosome species forms 18 bivalents, the 54-chromosome species forms 27

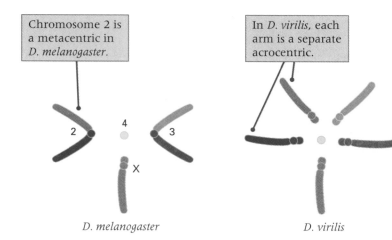

Chromosome 2 is a metacentric in *D. melanogaster.*

2   4   3

X

D. melanogaster

In *D. virilis*, each arm is a separate acrocentric.

D. virilis

Figure 7.2
The haploid chromosome complement of two species of *Drosophila.* Color indicates homology of chromosome arms. The large metacentric chromosomes of *Drosophila melanogaster* (chromosomes 2 and 3) correspond arm for arm with the four large acrocentric autosomes of *Drosophila virilis.*

Figure 7.3

Chromosome numbers in diploid and polyploid species of *Chrysanthemum*. Each set of homologous chromosomes is depicted in a different color.

Monoploid chromosome set

Diploid (18)

Tetraploid (36)

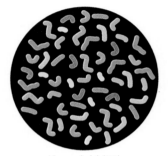

Hexaploid (54)

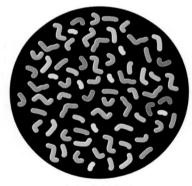

Octaploid (72)

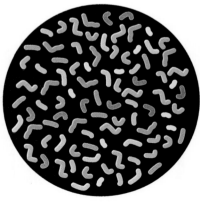

Decaploid (90)

bivalents, so forth. Gametes receive one chromosome from each bivalent, so the number of chromosomes in the gametes of any species is exactly half the number of chromosomes in its somatic cells. The chromosomes present in the gametes of a species constitute the **haploid** set of chromosomes. In the species of *Chrysanthemum* with 90 chromosomes, for example, the haploid chromosome number is 45; in meiosis, 45 bivalents are formed, so each gamete contains 45 chromosomes. When two such gametes come together in fertilization, the complete set of 90 chromosomes in the species is restored. Thus the gametes of a polyploid organism are not always monoploid, as they are in a diploid organism; for example, a tetraploid organism has diploid gametes.

The distinction between the terms *monoploid* and *haploid* is subtle:

- The *monoploid* chromosome set is the basic set of chromosomes that is multiplied in a polyploid series of species, such as *Chrysanthemum*.

- The *haploid* chromosome set is the set of chromosomes present in a gamete, irrespective of the chromosome number in the species.

The potential confusion arises because of diploid organisms, in which the monoploid chromosome set and the haploid chromosome set are the same. Considering the tetraploid helps to clarify the difference: It contains four monoploid chromosome sets, and the haploid gametes are diploid.

Polyploidy is widespread in certain plant groups. Among flowering plants, from 30 to 35 percent of existing species are thought to have originated as some form of polyploid. Valuable agricultural crops that are polyploid include wheat, oats, cotton, potatoes, bananas, coffee, and sugar cane. Polyploidy often leads to an increase in the size of individual cells, and polyploid plants are often larger and more vigorous than their diploid ancestors; however, there are many exceptions to these generalizations. Polyploidy is rare in vertebrate animals, but it is found in a few groups of invertebrates. One reason why polyploidy is rare in animals is the difficulty in regular segregation of the sex chromosomes. For example, a tetraploid animal with XXXX females and XXYY males would produce XX eggs and XY sperm (if all chromosomes paired to form bivalents), so the progeny would be exclusively XXXY and thus unlike either of the parents.

Polyploid plants found in nature almost always have an even number of sets of chromosomes, because organisms that have an odd number have low fertility. Organisms with three monoploid sets of chromosomes are known as **triploids.** As far as growth is concerned, a triploid is quite normal because the triploid condition does not interfere with mitosis; in mitosis in triploids (or any other type of polyploid), each chromosome replicates and divides just as in a diploid. However, because each chromosome has more than one pairing partner, chromosome segregation is severely upset in meiosis, and most gametes are defective. Unless the organism can perpetuate itself by means of asexual reproduction, it will eventually become extinct.

The infertility of triploids is sometimes of commercial benefit. For example, the seeds in commercial bananas are small and edible because the plant is triploid and most of the seeds fail to develop to full size. In oysters, triploids are produced by treating fertilized diploid eggs with a chemical that causes the second polar body of the egg to be retained. The triploid oysters are sterile and do not spawn, so they remain edible through the hot summer months of June, July, and August (the months that lack the letter *r*), when normal oysters are spawning. In Florida and in certain other states, weed control in waterways is aided by the release of weed-eating fish (the grass carp), which do not become overpopulated because the released fish are sterile triploids.

Tetraploid organisms can be produced in several ways. The simplest mechanism is a failure of chromosome separation in either mitosis or meiosis, which instantly doubles the chromosome number. Chromosome doubling through an abortive cell division is called **endoreduplication.** In a plant species that can undergo self-fertilization, endoreduplication creates a new, genetically stable species, because the chromosomes in the tetraploid can pair two by two in meiosis and therefore segregate regularly, each gamete receiving a full diploid

set of chromosomes. Self-fertilization of the tetraploid restores the chromosome number, so the tetraploid condition can be perpetuated. The genetics of tetraploid species, and that of other polyploids, is more complex than that of diploid species because the organism carries more than two alleles of any gene. With two alleles in a diploid, only three genotypes are possible: *AA, Aa,* and *aa.* In a tetraploid, by contrast, five genotypes are possible: *AAAA, AAAa, AAaa, Aaaa,* and *aaaa.* Among these genotypes, the middle three represent different types of tetraploid heterozygotes.

An octoploid species (eight sets of chromosomes) can be generated by failure of chromosome separation in mitosis in a tetraploid. If only bivalents form in meiosis, then an octoploid organism can be perpetuated sexually by self-fertilization or through crosses with other octoploids. Furthermore, cross-fertilization between an octoploid and a tetraploid results in a hexaploid (six sets of chromosomes). Repeated episodes of polyploidization and cross-fertilization may ultimately produce an entire polyploid series of closely related organisms that differ in chromosome number, as exemplified in *Chrysanthemum.*

Chrysanthemum represents a type of polyploidy, known as **autopolyploidy,** in which all chromosomes in the polyploid species derive from a single diploid ancestral species. In many cases of polyploidy, the polyploid species have complete sets of chromosomes from two or more *different* ancestral species. Such polyploids are known as **allopolyploids.** They derive from occasional hybridization between different diploid species when pollen from one species germinates on the stigma of another species and sexually fertilizes the ovule, followed by endoreduplication in the zygote to yield a hybrid plant in which each chromosome has a pairing partner in meiosis. The pollen may be carried to the wrong flower by wind, insects, or other pollinators. Figure 7.4 illustrates hybridization between species A and B in which endoreduplication leads to the formation of an allopolyploid (in this case, an *allotetraploid*), which carries a complete diploid genome from each of its two ancestral species. The formation of allopolyploids through hybridization and endoreduplication is an extremely important process in

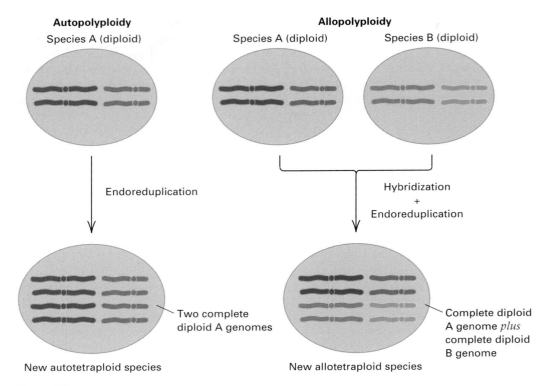

Autopolyploidy

Species A (diploid)

Allopolyploidy

Species A (diploid)   Species B (diploid)

Endoreduplication

Hybridization
+
Endoreduplication

Two complete
diploid A genomes

Complete diploid
A genome *plus*
complete diploid
B genome

New autotetraploid species   New allotetraploid species

Figure 7.4

Endoreduplication (the doubling of the chromosome complement) plays a key role in the formation of polyploid species. When it takes place in a diploid species, endoreduplication results in the formation of an autotetraploid species. When it takes place in the hybrid formed by cross-fertilization between distinct species, endoreduplication results in the formation of an allotetraploid species that has a complete diploid set of chromosomes from each of the parental species.

plant evolution and plant breeding. At least half of all naturally occurring polyploids are allopolyploids. Cultivated wheat provides an excellent example of allopolyploidy. Cultivated bread wheat is a hexaploid with 42 chromosomes constituting a complete diploid genome of 14 chromosomes from each of three ancestral species. The 42-chromosome allopolyploid is thought to have originated by the series of hybridizations and endoreduplications outlined in Figure 7.5.

The ancestral origin of the chromosome sets in an allopolyploid can often be revealed by the technique of **chromosome painting,** in which chromosomes are "painted" different colors by hybridization with DNA strands labeled with fluorescent dyes. DNA from each of the putative ancestral species is isolated, denatured, and labeled with a different fluorescent dye. Then the labeled single strands are spread on a microscope slide and allowed to renature with homologous strands present in the chromosomes of the allopolyploid species.

An example of chromosome painting is shown in Figure 7.6. The flower is from a variety of crocus called Golden Yellow. Its genome contains seven pairs of chromosomes, which are shown painted in yellow and green. Golden Yellow was thought to be an allopolyploid formed by hybridization of two closely related species followed by endoreduplication of the chromosomes in the hybrid. The putative ancestral species are *Crocus flavus,* which has four pairs of chromosomes, and *Crocus angustifolius,* which has three pairs of chromosomes. To paint the chromosomes of Golden Yellow, DNA from *C. flavus* was isolated and labeled with a fluorescent green dye, and that from *C. angustifolius* was isolated and labeled with a fluorescent yellow

(A)

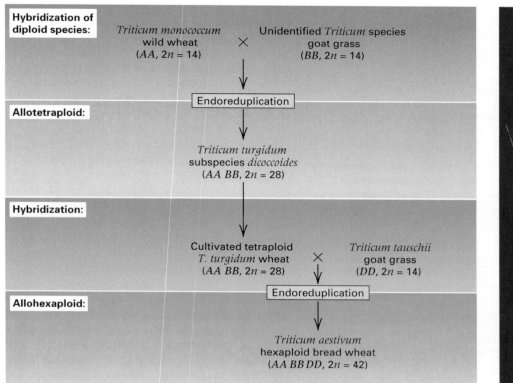

| Hybridization of diploid species: | *Triticum monococcum* wild wheat (*AA, 2n = 14*) | × | Unidentified *Triticum* species goat grass (*BB, 2n = 14*) |

Endoreduplication

Allotetraploid:

Triticum turgidum subspecies *dicoccoides* (*AA BB, 2n = 28*)

Hybridization:

Cultivated tetraploid *T. turgidum* wheat (*AA BB, 2n = 28*) × *Triticum tauschii* goat grass (*DD, 2n = 14*)

Endoreduplication

Allohexaploid:

Triticum aestivum hexaploid bread wheat (*AA BB DD, 2n = 42*)

(B)

Figure 7.5

(A) Repeated hybridization and endoreduplication in the ancestry of cultivated bread wheat (*Triticum aestivum*), which is an allohexaploid containing complete diploid genomes (*AA, BB, DD*) from three ancestral species. (B) The spike of *T. turgidum* in the photograph is an allotetraploid. The large grains of this species made it attractive to early hunter-gatherer societies in the Middle East. One of the earliest cultivated wheats, *T. turgidum*, is the progenitor of commercial macaroni wheat. [Photograph courtesy of Gordon Kimber.]

Figure 7.6

Flower of the crocus variety Golden Yellow and chromosome painting that reveals its origin as an allopolyploid. Its seven pairs of chromosomes are shown at the right. The chromosomes in green hybridized with DNA from *C. angustifolius*, which has three pairs of chromosomes, and those in yellow hybridized with DNA from *C. flavus*, which has four pairs of chromosomes. [Courtesy of J. S. Heslop-Harrison, John Innes Centre, Norwich, UK. With permission of the *Annals of Botany*.]

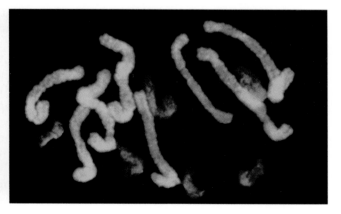

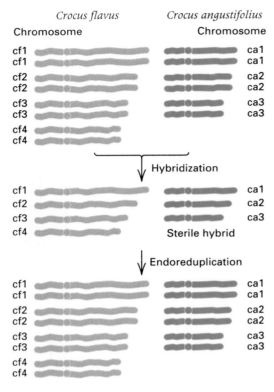

Crocus flavus Chromosome

cf1
cf1
cf2
cf2
cf3
cf3
cf4
cf4

Crocus angustifolius Chromosome

ca1
ca1
ca2
ca2
ca3
ca3

Hybridization

cf1
cf2
cf3
cf4

ca1
ca2
ca3

Sterile hybrid

Endoreduplication

cf1
cf1
cf2
cf2
cf3
cf3
cf4
cf4

ca1
ca1
ca2
ca2
ca3
ca3

Figure 7.7

Allopolyploidy in the evolutionary origin of the Golden Yellow crocus. The original hybrid between *C. flavus* and *C. angustifolius* produced a sterile monoploid with one copy of each chromosome from each species. Chromosome endoreduplication in the monoploid results in a fertile allotetraploid with a complete diploid genome from each species.

dye. The result of the chromosome painting is very clear: three pairs of chromosomes hybridize with the green-labeled DNA from *C. flavus,* and four pairs of chromosomes hybridize with the yellow-labeled DNA from *C. angustifolius.* This pattern of hybridization strongly supports the hypothesis of the autopolyploid origin of Golden Yellow diagrammed in Figure 7.7.

7.3
Monoploid Organisms

As noted, the monoploid chromosome set is the set of chromosomes multiplied in polyploid species. An organism is mono-

ploid if it develops from a monoploid cell. Meiosis cannot take place normally in the germ cells of a monoploid, because each chromosome lacks a pairing partner, and hence monoploids are usually sterile. Monoploid organisms are quite rare, but they occur naturally in certain insect species (ants, bees) in which males are derived from unfertilized eggs. These monoploid males are fertile because the gametes are produced by a modified meiosis in which chromosomes do not separate in meiosis I.

Monoploids are important in plant breeding because, in the selection of diploid organisms with desired properties, favorable recessive alleles may be masked by heterozygosity. This problem can be avoided by studying monoploids, provided that their sterility can be overcome. In many plants, the production of monoploids capable of reproducing can be stimulated by conditions that yield aberrant cell divisions. Two techniques make this possible.

With some diploid plants, monoploids can be derived from cells in the anthers (the pollen-bearing structures). Extreme chilling of the anthers causes some of the haploid cells destined to become pollen grains to begin to divide. These cells are monoploid as well as haploid. If the cold-shocked cells are placed on an agar surface containing suitable nutrients and certain plant hormones, then a small dividing mass of cells called an **embryoid** forms. A subsequent change of plant hormones in the growth medium causes the embryoid to form a small plant with roots and leaves that can be potted in soil and allowed to grow normally. Because monoploid cells have only a single set of chromosomes, their genotypes can be identified without regard to the dominance or recessiveness of individual alleles. A plant breeder can then select a monoploid plant with the desired traits. In some cases, the desired genes are present in the original diploid plant and are merely sorted out and selected in the monoploids. In other cases, the anthers are treated with mutagenic agents in the hope of producing the desired traits.

When a desired mutation is isolated in a monoploid, it is necessary to convert the monoploid into a homozygous diploid because the monoploid plant is sterile and

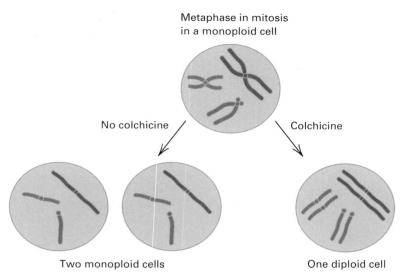

Metaphase in mitosis
in a monoploid cell

No colchicine

Colchicine

Two monoploid cells

One diploid cell

Figure 7.8
Production of a diploid from a monoploid by treatment with colchicine. The colchicine disrupts the spindle and thereby prevents separation of the chromatids after the centromeres (circles) divide.

does not produce seeds. Converting the monoploid into a diploid is possible by treatment of the meristematic tissue (the growing point of a stem or branch) with the substance **colchicine.** This chemical is an inhibitor of the formation of the mitotic spindle. When the treated cells in the monoploid meristem begin mitosis, the chromosomes replicate normally; however, the colchicine blocks metaphase and anaphase, so the result is endoreduplication (doubling of each chromosome in a given cell). Many of the cells are killed by colchicine, but, fortunately for the plant breeder, a few of the monoploid cells are converted into the diploid state (Figure 7.8). The colchicine is removed to allow continued cell multiplication, and many of the now-diploid cells multiply to form a small sector of tissue that can be recognized microscopically. If placed on a nutrient-agar surface, this tissue will develop into a complete plant. Such plants, which are completely homozygous, are fertile and produce normal seeds.

7.4
Extra or Missing Chromosomes

Occasionally, organisms are formed that have extra copies of individual chromosomes, rather than extra entire sets of chromosomes. This situation is called **polysomy.** In contrast to polyploids, which in plants are often healthy and in some cases are more vigorous than the diploid, polysomics are usually less vigorous than the diploid and have abnormal phenotypes. In most plants, a single extra chromosome (or a missing chromosome) has a more severe effect on phenotype than the presence of a complete extra set of chromosomes. Each chromosome that is extra or missing results in a characteristic phenotype. For example, Figure 7.9 shows the seed capsule of the Jimson weed *Datura stramonium,* beneath which is a series of capsules of strains, each having an extra copy of a different chromosome. The seed capsule of each of the strains is distinctive.

An otherwise diploid organism that has an extra copy of an individual chromosome is called a **trisomic.** In a trisomic organism, the segregation of chromosomes in meiosis is upset because the trisomic chromosome has two pairing partners instead of one. The behavior of the chromosomes in meiosis depends on the manner in which the homologous chromosome arms pair and on the chiasmata formed between them. In some cells, the three chromosomes form a **trivalent** in which distinct parts of one chromosome are paired with homologous parts of each of the others (Figure 7.10A). In metaphase, the trivalent is usually oriented with two centromeres pointing toward one pole and the other centromere pointing toward the other. The result is

Figure 7.9
Seed capsules of the normal diploid *Datura stramonium* (Jimson weed), which has a haploid number of 12 chromosomes, and 4 of the 12 possible trisomics. The phenotype of the seed capsule in trisomics differs according to the chromosome that is trisomic.

Diploid

Trisomics

Figure 7.10
Meiotic pairing in a trisomic. (A) Formation of a trivalent. (B) Formation of a univalent and a bivalent. Both types of pairing result in one pair of gametes containing two copies of the trisomic chromosome and the other pair of gametes containing one copy of the trisomic chromosome.

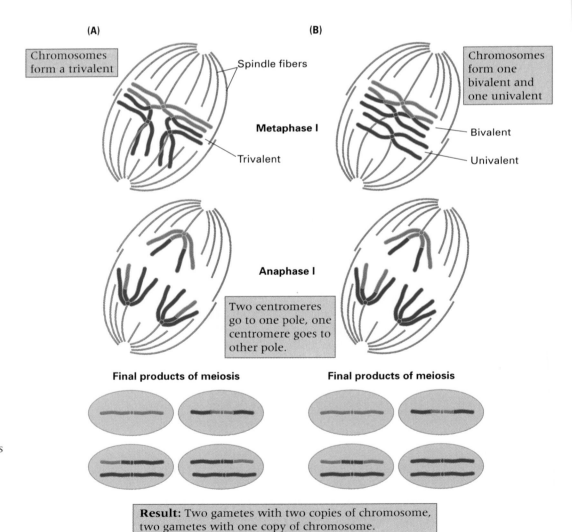

(A)

Chromosomes form a trivalent

Spindle fibers

Metaphase I

Trivalent

(B)

Chromosomes form one bivalent and one univalent

Bivalent

Univalent

Anaphase I

Two centromeres go to one pole, one centromere goes to other pole.

Final products of meiosis

Final products of meiosis

Result: Two gametes with two copies of chromosome, two gametes with one copy of chromosome.

that, at the end of both meiotic divisions, one pair of gametes contains two copies of the trisomic chromosome, and the other pair of gametes contains only a single copy. Alternatively, the trisomic chromosome can form one normal bivalent and one **univalent,** or unpaired chromosome, as shown in Figure 7.10B. In anaphase I, the bivalent disjoins normally and the univalent usually proceeds randomly to one pole or the other. Again, the end result is the formation of two products of meiosis that contain two copies of the trisomic chromosome and two products of meiosis that contain one copy. To state the matter in another way, a trisomic organism with three copies of a chromosome (say, $C\ C\ C$) will produce gametes among which half contain $C\ C$ and half contain C. The unequal segregation of trisomic chromosomes is also the cause of the infertility in triploids, because in a triploid, each chromosome behaves independently as though it were trisomic. For each of the homologous chromosomes in a triploid, a gamete can receive either one copy or two copies.

Polysomy generally results in more severe phenotypic effects than does polyploidy. The usual explanation is that the greater harmful phenotypic effects in trisomics are related to the imbalance in the number of copies of different genes. A polyploid organism has a "balanced" genome in the sense that the ratio of the numbers of copies of any pair of genes is the same as in the diploid. For example, in a tetraploid, each gene is present in twice as many copies as in a diploid, so no gene or group of genes is out of balance with the others. Balanced chromosome abnormalities, which retain equality in the number of copies of each gene, are said to be **euploid.** In contrast, gene equality is upset in a trisomic because three copies of the genes located in the trisomic chromosome are present, whereas two copies of the genes in the other chromosomes are present. Such unbalanced chromosome complements are said to be **aneuploid.** Aneuploid abnormalities are usually more severe than euploid abnormalities. For example, in *Drosophila*, triploid females are viable, fertile, and nearly normal in morphology, whereas trisomy for either of the two large autosomes is invariably lethal (the larvae die at an early stage).

Just as an occasional organism may have an extra chromosome, a chromosome may be missing. Such an individual is said to be **monosomic.** A missing copy of a chromosome results in more harmful effects than an extra copy of the same chromosome, and monosomy is often lethal. If the monosomic organism can survive to sexual maturity, then chromosomal segregation in meiosis is unequal because the monosomic chromosome forms a univalent. Half of the gametes will carry one copy of the monosomic chromosome, and the other half will contain no copies.

7.5 Human Chromosomes

The chromosome complement of a normal human male is illustrated in Figure 7.11. The chromosomes have been treated with a staining reagent called Giemsa, which causes the chromosomes to exhibit transverse bands that are specific for each pair of homologs. These bands permit the chromosome pairs to be identified individually. By convention, the autosome pairs are arranged and numbered from longest to shortest and, on the basis of size and centromere position, are separated into seven groups designated by the letters A through G. This conventional representation of chromosomes is called a **karyotype;** it is obtained by cutting each chromosome out of a photograph taken at metaphase and pasting it into place. In a karyotype of a normal human female, the autosomes would not differ from those of a male and hence would be identical to those in Figure 7.11 but there would be two X chromosomes instead of an X and a Y. Abnormalities in chromosome number and morphology are made evident by a karyotype.

The nomenclature of the banding patterns in human chromosomes is shown in Figure 7.12. For each chromosome, the short arm is designated with the letter p, which stands for "petite," and the long arm by the letter q, which stands for "not-p." Within each arm the regions are numbered. The first digit is the major region, numbered consecutively proceeding from the centromere toward the telomere;

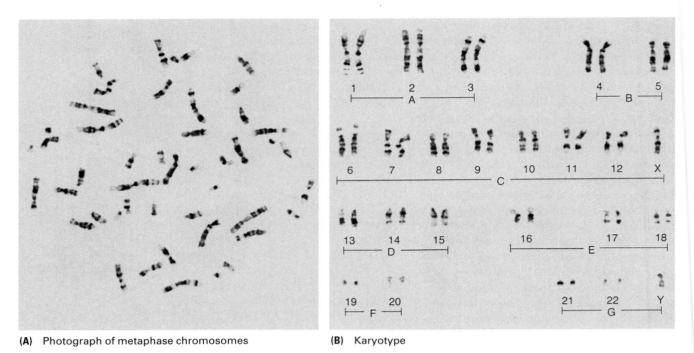

(A) Photograph of metaphase chromosomes
(B) Karyotype

Figure 7.11
A karyotype of a normal human male. Blood cells arrested in metaphase were stained with Giemsa and photographed with a microscope. (A) The chromosomes as seen in the cell by microscopy. (B) The chromosomes have been cut out of the photograph and paired with their homologs. [Courtesy of Patricia Jacobs.]

within each region, the second number indicates the next-smaller division, again numbered outward from the centromere. For example, the designation 1p34 indicates chromosome 1, short arm, division 34. Some divisions can be subdivided still further by the bands and interbands within them, which are numbered consecutively and indicated by a digit placed after a decimal point following the main division; for example, 1p36.2 means the second band in 1p36. Some familiar genetic landmarks in the human genome are the Rh (Rhesus) blood group locus, which is somewhere in the region between 1p34 and 1p36.2; the ABO blood group locus at 9q34; the red-green color-blindness genes at Xq28, and the male-determining gene on the Y chromosome, called *SRY* (*sex-determining region, Y*), at Yp11.3. At the base of each chromosome in Figure 7.12, the red number indicates the number of genes presently assigned a position on the chromosome. These numbers do not include the approximately 5000 DNA markers used in devel-

oping the human genetic map. Many of the genes enumerated in Figure 7.12 are associated with inherited diseases. There are a total of 3399 genes assigned a position on one of the autosomes, 251 assigned a position on the X chromosome, and 19 assigned a position on the Y chromosome.

The technique of chromosome painting seen earlier in Figure 7.6 has been applied to human chromosomes to achieve spectacular effects. An example is shown in the metaphase spread and karyotype in Figure 7.13. To produce this effect, chromosomes were hybridized simultaneously with 27 different fluorescent DNA probes, each specific for hybridization with a single chromosome or chromosome arm. The probes were obtained by microdissection of metaphase nuclei to isolate each individual chromosome or chromosome arm, and the minuscule amount of DNA so obtained was amplified by the polymerase chain reaction (Section 5.8). After hybridization, each chromosome was scanned along its length, and at each point, the fluorescent signal

The First Human Chromosomal Disorder

Jerome Lejeune, Marthe Gautier, and Raymond Turpin 1959
National Center for Scientific Research, Paris, France
Study of the Somatic Chromosomes of Nine Down Syndrome Children (original in French)

Down syndrome had been one of the greatest mysteries in human genetics. One of the most common forms of mental retardation, the syndrome did not follow any pattern of Mendelian inheritance. Yet some families had two or more children with Down syndrome. (Many of these cases are now known to be due to a translocation involving chromosome 21.) This paper marked a turning point in human genetics by demonstrating that Down syndrome actually results from the presence of an extra chromosome. It was the first identified chromosomal disorder. The excerpt uses the term telocentric, *which means a chromosome that has its centromere very near one end. In the human genome, the smallest chromosomes are three very small telocentric chromosomes. These are chromosomes 21, 22, and the Y. A normal male has five small telocentrics (21, 21, 22, 22, and Y); a normal female has four (21, 21, 22, and 22). (The X is a medium-sized chromosome with its centromere somewhat off center.) In the table that follows, note the variation in chromosome counts in the "doubtful" cells. The methods for counting chromosomes were then very difficult, and many errors were made either by counting two nearby chromosomes as one or by including in the count of one nucleus a chromosome that actually belonged to a nearby nucleus. Lejeune and collaborators wisely chose to ignore these doubtful counts and based their conclusion only on the "perfect" cells. Sometimes good science is a matter of knowing which data to ignore.*

The culture of fibroblast cells from nine Down syndrome children reveals the presence of 47 chromosomes, the supernumerary chromosome being a small telocentric one. The hypothesis of the chromosomal determination of Down syndrome is considered. . . . The observations made in these nine cases (five boys and four girls) are recorded in the table to the right.

The number of cells counted in each case may seem relatively small. This is due to the fact that only the pictures [of the spread chromosomes] that claim a minimum of interpretation have been retained in this table. The apparent variation in the chromosome number in the "doubtful" cells, that is to say, cells in

| | | Number of chromosomes | | | | | |
| | | "Doubtful" cells | | | "Perfect" cells | | |
		46	47	48	46	47	48
Boys	1	6	10	2	—	11	—
	2	—	2	1	—	9	—
	3	—	1	1	—	7	—
	4	—	3	—	—	1	—
	5	—	—	—	—	8	—
Girls	1	1	6	1	—	5	—
	2	1	2	—	—	8	—
	3	1	2	1	—	4	—
	4	1	1	2	—	4	—

Analysis of the chromosome set of the "perfect" cells reveals the presence in Down syndrome boys of 6 small telocentric chromosomes (instead of 5 in the normal man) and 5 small telocentric ones in Down syndrome girls (instead of 4 in the normal woman).

---

which each chromosome cannot be noted individually with certainty, has been pointed out by several authors. It does not seem to us that this phenomenon represents a cytological reality, but merely reflects the difficulties of a delicate technique. It therefore seems logical to prefer a small number of absolutely certain counts ("perfect" cells in the table) to a mass of doubtful observations, the statistical variance of which rests solely on the lack of precision of the observations. Analysis of the chromosome set of the "perfect" cells reveals the presence in Down syndrome boys of 6 small telocentric chromosomes (instead of 5 in the normal man) and 5 small telocentric ones in Down syndrome girls (instead of 4 in the normal woman). . . . It therefore seems legitimate to conclude that there exists in Down syndrome children a small supernumerary telocentric chromosome,

accounting for the abnormal figure of 47. To explain these observations, the hypothesis of nondisjunction of a pair of small telocentric chromosomes at the time of meiosis can be considered. . . . It is, however, not possible to say that the supernumerary small telocentric chromosome is indeed a normal chromosome and at the present time the possibility cannot be discarded that a fragment resulting from another type of aberration is involved.

Source: Comptes rendus des séances de l'Académie des Sciences 248: 1721–1722 Translation in S. H. Boyer. 1963. (Papers on Human Genetics). Englewood Cliffs, NJ: Prentice Hall.

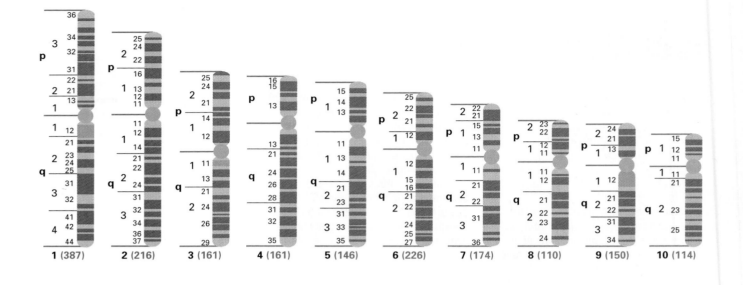

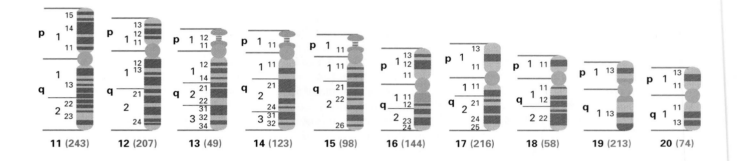

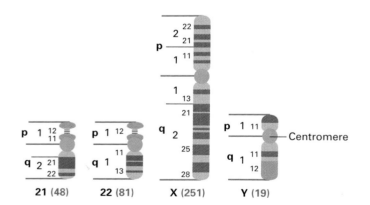

Figure 7.12

Designations of the bands and interbands in the human karyotype. Below each chromosome is the number of genes presently assigned a location on the chromosome. [Data from V. A. McKusick. 1988. *Mendelian Inheritance in Man,* 8th ed. Baltimore: Johns Hopkins University Press.]

was converted to a specific color in the visible spectrum. The chromosomes in Figure 7.13 are therefore "painted" in 27 different colors. This technique is of considerable utility in human cytogenetics because even complex chromosome rearrangements can be detected rapidly and easily. The painting technique makes it possible to decipher some chromosome rearrangements, particularly those involving small pieces of chromosome, that are not amenable to analysis by conventional banding procedures.

Figure 7.13
(A) Metaphase spread and (B) karyotype in which human chromosomes have been "painted" in 27 different colors according to their hybridization with fluorescent probes specific to individual chromosomes or chromosome arms. [Courtesy of M. R. Speicher and D. C. Ward. See M. R. Speicher, S. G. Ballard, and D. C. Ward. 1996. *Nature Genetics* 12: 368.]

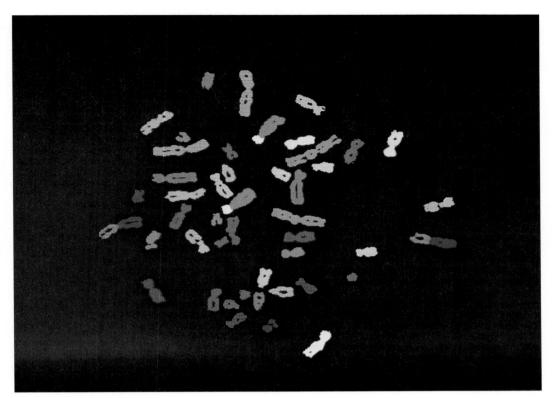

(A) Metaphase spread

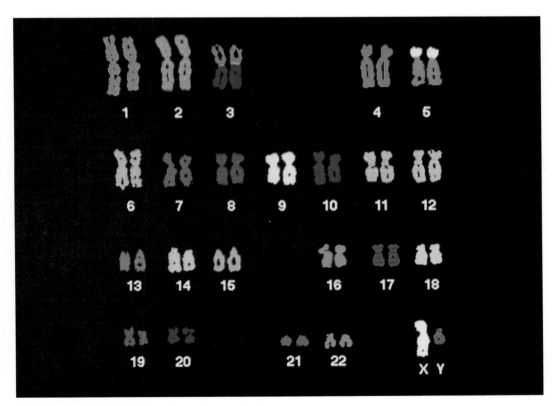

(B) Karyotype

Trisomy in Human Beings

Monosomy or trisomy of most human autosomes is usually incompatible with life. Most zygotes with missing chromosomes or extra chromosomes either fail to begin embryonic development or undergo spontaneous abortion at an early stage. There are a few exceptions. One exception is **Down syndrome,** which is caused by trisomy of chromosome 21. Down syndrome affects about 1 in 750 live-born children. Its major symptom is mental retardation, but there can be multiple physical abnormalities as well, such as major heart defects.

Most cases of Down syndrome are caused by nondisjunction, which means the failure of homologous chromosomes to separate in meiosis, as explained in Chapter 3. The result of chromosome-21 nondisjunction is one gamete that contains two copies of chromosome 21 and one that contains none. If the gamete with two copies participates in fertilization, then a zygote with trisomy 21 is produced. The gamete with one copy may also participate in fertilization, but zygotes with monosomy 21 do not survive even through the first few days or weeks of pregnancy. For unknown reasons, nondisjunction of chromosome 21 is more likely to happen in oogenesis than in spermatogenesis, and so the abnormal gamete in Down syndrome is usually the egg.

Chromosome 21 is a small chromosome and therefore is somewhat less likely to undergo meiotic crossing-over than a longer one. Noncrossover bivalents sometimes have difficulty aligning at the metaphase plate because they lack a chiasma to hold them together, so there is an increased risk of nondisjunction. Among the events of nondisjunction that result in Down syndrome, about 40 percent are derived from such nonexchange bivalents. (Students already familiar with trisomy 21 may know it as *Down's* syndrome, with an apostrophe *s*. This text follows current practice in human genetics in avoiding the possessive form of proper names used to designate syndromes.)

The risk of nondisjunction of chromosome 21 increases dramatically with the age of the mother, and the risk of Down syndrome reaches 6 percent in mothers of age 45 and older (Figure 7.14). Thus many physicians recommend that older women who are pregnant have cells from the fetus tested in order to detect Down syndrome prenatally. This can be done from 15 to 16 weeks after fertilization by **amniocentesis,** in which cells of a developing fetus are obtained by insertion of a fine needle through the wall of the uterus and into the sac of fluid (the *amnion*) that contains the fetus, or even earlier in pregnancy by sampling cells from another of the embryonic membranes (the *chorion*). In about 3 percent of families with a Down syndrome child, the risk of another affected child is very high—up to 20 percent of births. This high risk is caused by a chromosome abnormality called a *translocation* in one of the parents, which will be considered in Section 7.6.

Dosage Compensation

Abnormal numbers of sex chromosomes usually produce less severe phenotypic effects than do abnormal numbers of autosomes. For example, the effects of extra Y chromosomes are relatively mild. This is in part because the Y chromosome in mammals is largely heterochromatic. In human beings, there is a region at the tip of the short arm of the Y that is homologous with a corresponding region at the tip of the short arm of the X chromosome. It is in this region of homology that the X and Y chromosomes synapse in spermatogenesis, and an obligatory crossover in the region holds the chromosomes together and ensures proper separation during anaphase I. The crossover is said to be obligatory because it takes place somewhere in this region in every meiotic division. The shared X-Y homology defines the **pseudoautosomal** region. Genes within the pseudoautosomal region show a pattern of inheritance very similar to ordinary autosomal inheritance, because they are not completely linked to either the X chromosome or the Y chromosome but can exchange between the sex chromosomes by crossing-over. Near the pseudoautosomal region, but not within it, the Y chromosome contains the master sex-controller gene *SRY* at position Yp11.3. *SRY* codes for a protein transcription factor, the **testis-determining factor (TDF),** which triggers male embryonic development by inducing the undifferentiated embryonic genital ridge, the precursor of the gonad, to develop as a testis. A transcription factor, as the name implies, is a component that, together with other factors, stimulates transcription of its target genes. Apart from the pseudoautosomal pairing region and *SRY*, the Y chromosome contains very few genes, so extra Y chromosomes are milder in their effects on phenotype than are extra autosomes.

Extra X chromosomes have milder effects than extra autosomes because, in mammals, all X chromosomes except one are genetically inactivated very early in embryonic development. The inactivation tends to minimize the phenotypic effects of extra X chromosomes, but there are still some effects due to a block of genes near the tip of the short arm that are *not* inactivated.

In female mammals, X-chromosome inactivation is a normal process in embryonic development. In human beings, at an early stage of embryonic development, one of the two X chromosomes is inactivated in each somatic cell; different tissues undergo X inactivation at different times. The X chromosome that is inactivated in a particular somatic cell is selected at random, but once the decision is made, the same X chromosome remains inactive in all of the descendants of the cell.

X-chromosome inactivation has two consequences. First, it equalizes the number of active copies of X-linked genes in females and males. Although a female has two X chromosomes and a male has only one, because of inactivation of one X chromosome in each of the somatic cells of the female, the number of active X chromosomes in both sexes is one. In effect, gene dosage is equalized except for the block of genes in the short arm of the X that escapes inactivation. This equalization in dosage of active genes is called **dosage compensation.** The mammalian method of dosage compensation by means of X inactivation was originally proposed by Mary Lyon and is called the **single-active-X principle.**

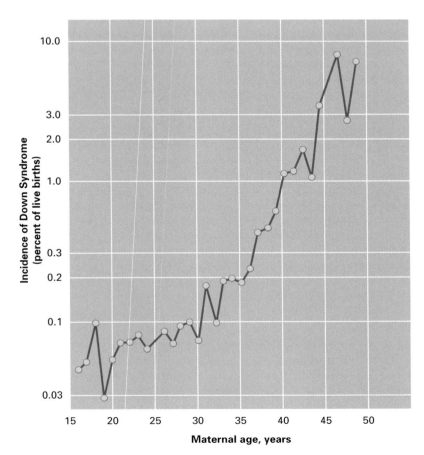

Figure 7.14
Frequency of Down syndrome (number of cases per 100 live births) related to age of mother. The graph is based on 438 Down syndrome births (among 330,859 total births) in Sweden in the period 1968 to 1970. [Data from E. B. Hook and A. Lindsjö. 1978. *Am. J. Human Genet.* 30:19.]

Lyonization of an X Chromosome

Mary F. Lyon 1961
Medical Research Council,
Harwell, England
*Gene Action in the X Chromosome
of the Mouse* (Mus musculus L.)

How do organisms solve the problem that females have two X chromosomes whereas males have only one? Unless there were some type of correction (called dosage compensation), the unequal number would mean that for all the genes in the X chromosome, cells in females would have twice as much gene product as cells in males. It would be difficult for the developing organism to cope with such a large difference in dosage for so many genes. The problem of dosage compensation has been solved by different organisms in different ways. The hypothesis put forward in this paper is that in the mouse (and, by inference, in other mammals), the mechanism is very simple: One of the X chromosomes, chosen at random in each cell lineage early in development, becomes inactivated and remains inactivated in all descendant cells in the lineage. In certain cells, the inactive X chromosome becomes visible in interphase as a deeply staining "sex-chromatin body." We now know that there are a few genes in the short arm of the X chromosome that are not inactivated. There is also good evidence that the inactivation of the rest of the X chromosome takes place sequentially from an "X-inactivation center." We also know that in marsupial mammals, such as the kangaroo, it is always the paternal X chromosome that is inactivated.

It has been suggested that the so-called sex chromatin body is composed of one heteropyknotic [that is, deeply staining during interphase] X chromosome. . . . The

---

The coat of the tortoiseshell cat, being a mosaic of the black and yellow colours of the two homozygous genotypes, fulfills this expectation.

---

present communication suggests that evidence of mouse genetics indicates: (1) that the heteropyknotic X chromosome can be either paternal or maternal in origin, in different cells of the same animal; (2) that it is genetically inactivated. The evidence has two main parts. First, the normal phenotype of XO females in the mouse shows that only one active X chromosome is necessary for normal development, including sexual development. The second piece of evidence concerns the mosaic phenotype of female mice heterozygous for some sex-linked [X-linked] mutants. All sex-linked mutants so far known affecting coat colour cause a "mottled" or "dappled" phenotype, with patches of normal and mutant colour. . . . It is here suggested that this mosaic phenotype is due to the inactivation of one or other X chromosome early in embryonic development. If this is true, pigment cells descended from the cells in which the chromosome carrying the mutant gene was inactivated will give rise to a normal-coloured patch and those in which the chromosome carrying the normal gene was inactivated will give rise to a mutant-coloured patch. . . . Thus this hypothesis predicts that for all sex-linked genes of the mouse in which the phenotype is due to localized gene action the heterozygote will have a mosaic appearance. . . . The genetic evidence does not indicate at what early stage of embryonic development the inactivation of the one X chromosome occurs. . . . The sex-chromatin body is thought to be formed from one X chromosome in the rat and in the opossum. If this should prove to be the case in all mammals, then all female mammals heterozygous for sex-linked mutant genes would be expected to show the same phenomena as those in the mouse. The coat of the tortoiseshell cat, being a mosaic of the black and yellow colours of the two homozygous genotypes, fulfills this expectation.

Source: Nature 190: 372

The second consequence of X-chromosome inactivation is that a normal female is a **mosaic** for X-linked genes (Figure 7.15A). That is, each somatic cell expresses the genes in only one X chromosome, but the X chromosome that is active genetically differs from one cell to the next. This mosaicism has been observed directly in females that are heterozygous for X-linked alleles that determine different forms of an enzyme, A and B; when cells from the heterozygous female are individually cultured in the laboratory, half of the clones are found to produce only the A form of the enzyme and the other half to produce only the B form. Mosaicism can be observed directly in women who are heterozygous for an X-linked recessive mutation that results in the absence of sweat glands; these women exhibit patches of skin in which sweat glands are present (these patches are derived from embryonic cells in which the normal X chromosome remained active and the mutant X was inactivated) and other patches of skin in which sweat glands are absent (these patches are derived from embryonic cells in which the normal X chromosome was inactivated and the mutant X remained active.)

(A)

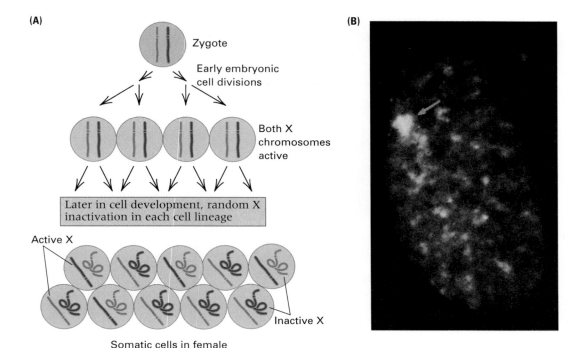

Zygote

Early embryonic cell divisions

Both X chromosomes active

Later in cell development, random X inactivation in each cell lineage

Active X

Inactive X

Somatic cells in female

(B)

Figure 7.15

(A) Schematic diagram of somatic cells of a normal female showing that the female is a mosaic for X-linked genes. The two X chromosomes are shown in red and blue. An active X is depicted as a straight chromosome, an inactive X as a tangle. Each cell has just one active X, but the particular X that remains active is a matter of chance. In human beings, the inactivation includes all but a few genes in the tip of the short arm. (B) Fluorescence micrograph of a human cell showing a Barr body (bright spot at the upper left; see arrow.) This cell is from a normal human female, and it has one Barr body. [Micrograph courtesy of A. J. R. de Jonge.]

In certain cell types, the inactive X chromosome in females can be observed microscopically as a densely staining body in the nucleus of interphase cells. This is called a **Barr body** (upper left in Figure 7.15B). Although cells of normal females have one Barr body, cells of normal males have none. Persons with two or more X chromosomes have all but one X chromosome per cell inactivated, and the number of Barr bodies equals the number of inactivated X chromosomes.

Although all mammals use X chromosomal inactivation for dosage compensation in females, the choice of which chromosome to inactivate is not always random. In marsupial mammals, which include the kangaroo, the koala, and the wombat, the X chromosome that is inactivated is always the one contributed by the father. The result is that female marsupials are not genetic mosaics of paternal and maternal X-linked genes. A female marsupial expresses the X-linked genes that she inherited from her mother.

The Calico Cat as Evidence for X-Chromosome Inactivation

In nonmarsupial mammals, the result of random X inactivation in females can sometimes be observed in the external phenotype. One example is the "calico" pattern of coat coloration in female cats. Two alleles affecting coat color are present in the X chromosome in cats. One allele results in an orange coat color (sometimes referred to as "yellow"), the other in a black coat color. Because he has only one X chromosome, a normal male has either the orange or the black allele. A female can be heterozygous for orange and black, and in this case the coat color is "calico"—a mosaic of orange and black patches mixed with patches of white. Figure 7.16 is a photograph of a female cat with the classic calico pattern.

Figure 7.16
A female cat that is heterozygous for the orange and black coat-color alleles and shows the classic "calico" pattern of patches of orange, black, and white fur.

The orange and black patches result from X chromosome inactivation. In cell lineages in which the X chromosome bearing the orange allele is inactivated, the X chromosome with the black allele is active and so the fur is black. In cell lineages in which the X chromosome with the black allele is inactivated, the orange allele in the active X chromosome results in orange fur.

The white patches have a completely different explanation. The white patches are due to an autosomal gene *S* for white spotting, which prevents pigment formation in the cell lineages in which it is expressed. Why the *S* gene is expressed in some cell lineages and not others is not known. Homozygous *SS* cats have more white than heterozygous *Ss* cats. The female in the photograph is homozygous.

Sex-Chromosome Abnormalities

Many types of sex-chromosome abnormalities have been observed. As noted, they are usually less severe in their phenotypic effects than are abnormal numbers of autosomes. The four most-common types are

- *47,XXX* This condition is often called the **trisomy-X syndrome.** The num-

ber 47 in the chromosome designation refers to the total number of chromosomes, and XXX indicates that the person has three X chromosomes. People with the karyotype 47,XXX are female. Many are phenotypically normal or nearly normal, though the frequency of mild mental retardation is somewhat greater than among 46,XX females.

- *47,XYY* This condition is often called the **double-Y syndrome.** These people are male and tend to be tall, but they are otherwise phenotypically normal. At one time it was thought that 47,XYY males developed severe personality disorders and were at a high risk of committing crimes of violence, a belief based on an elevated incidence of 47,XYY among violent criminals. Further study indicated that most 47,XYY males have slightly impaired mental function and that, although their rate of criminality is higher than that of normal males, the crimes are mainly nonviolent petty crimes such as theft. The majority of 47,XYY males are phenotypically and psychologically normal, have mental capabilities in the normal range, and have no criminal convictions.

- *XXY* This condition is called **Klinefelter syndrome.** Affected persons are male. They tend to be tall, do not undergo normal sexual maturation, are sterile, and in some cases have enlargement of the breasts. Mild mental impairment is common.

- *45,X* Monosomy of the X chromosome in females is called **Turner syndrome.** Affected persons are phenotypically female but short in stature and without sexual maturation. Mental abilities are typically within the normal range.

The Fragile-X Syndrome

An important form of inherited mental retardation is associated with a class of X chromosomes containing a site at Xq27−Xq28 (toward the end of the long arm) that tends to break in cultured cells that are starved for DNA precursors, such as the nucleotides. The X chromosomes containing this site are called **fragile-X** chromosomes, and the associated form of mental retardation is the **fragile-X syndrome.** The fragile-X syndrome affects about 1 in 2500 children. It accounts for about half of all cases of X-linked mental retardation and is second only to Down syndrome as a cause of inherited mental impairment. The fragile-X syndrome has an unusual pattern of inheritance in which approximately 1 in 5 males with the fragile-X chromosome are phenotypically normal and also have phenotypically normal children. However, the heterozygous daughters of such a "transmitting male"

often have affected sons, and about one-third of their heterozygous daughters are also affected. This pattern is illustrated in Figure 7.17. The transmitting male denoted I-2 is not affected, but the X chromosome that he transmits to his daughters (II-2 and II-5) somehow becomes altered in the female germ line in such a way that sons and daughters in the next generation (III) are affected. Affected and normal granddaughters of the transmitting male sometimes have affected progeny (generation IV). Males who are affected are severely retarded and do not reproduce. Among females, there is substantial variation in severity of expression (variable expressivity). In general, females are less severely affected than males, and, as noted, some heterozygous females are not affected at all (incomplete penetrance).

The molecular basis of the fragile-X chromosome has been traced to a **trinucleotide repeat** of the form $(CCG)_n$ present in the DNA at the site where the breakage takes place. Normal X chromosomes have from 6 to 54 tandem copies of the repeating unit (the average is about 30), and affected persons have more than 230 copies of the repeat. Transmitting males have an intermediate number, between 52 and 230 copies, which is called the fragile-X "premutation." The unprecedented feature of the premutation is that, in females, it invariably increases in copy number to reach a level of 230 copies or greater, at which stage the chromosome causes mental retardation. The amplification in the number of copies present in the germ line of daughters of transmitting males is related

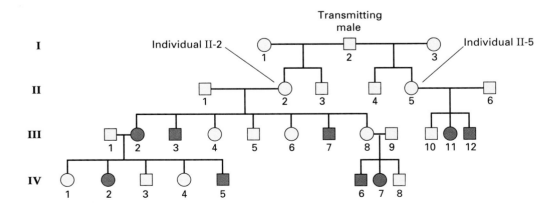

Figure 7.17
Pedigree showing transmission of the fragile-X syndrome. Male I-2 is not affected, but his daughters (II-2 and II-5) have affected children and grandchildren. [After C. D. Laird. 1987. *Genetics,* 117: 587.]

to the cycle of X-chromosome inactivation and reactivation. Chromosomes with the premutation that go through the X-inactivation cycle become permanently altered. In these chromosomes, the inactivation of a gene designated *FMR1* (fragile-site mental retardation-1) is irreversible, and the gene is unable to be reactivated in oogenesis. The X chromosome is said to have undergone a process of **imprinting.** Current evidence suggests that imprinting is associated with the addition of methyl groups ($-CH_3$) to certain cytosine bases in the DNA. Failure to remove the methyl groups irreversibly inactivates *FMR1* and also triggers amplification of the CCG repeats. The variable expressivity of the fragile-X syndrome in affected females also is related to X-chromosome inactivation: More severely affected females have a higher proportion of cells in which the normal X chromosome is inactivated.

Chromosome Abnormalities in Spontaneous Abortion

Approximately 15 percent of all recognized pregnancies in human beings terminate in spontaneous abortion, and in about half of

Table 7.1 Chromosome abnormalities per 100,000 recognized human pregnancies

	15,000 abort spontaneously 7500 chromosomally abnormal	85,000 live births 550 chromosomally abnormal
Trisomy		
A: 1	0	0
A: 2	159	0
A: 3	53	0
B: 4	95	0
B: 5	0	0
C: 6–12	561	0
D: 13	128	17
D: 14	275	0
D: 15	318	0
E: 16	1229	0
E: 17	10	0
E: 18	223	13
F: 19–20	52	0
G: 21	350	113
G: 22	424	0
Sex chromosomes		
XYY	4	46
XXY	4	44
XO	1350	8
XXX	21	44
Translocations		
Balanced	14	164
Unbalanced	225	52
Polyploid		
Triploid	1275	0
Tetraploid	450	0
Other (mosaics. etc.)	280	49
Total	7500	550

them, the fetus has a major chromosome abnormality. Table 7.1 summarizes the average rates of chromosome abnormality found per 100,000 recognized pregnancies in several studies. Many of the spontaneously aborted fetuses have trisomy of one of the autosomes. Triploids and tetraploids also are common in spontaneous abortions. Note that the majority of trisomy-21 fetuses are spontaneously aborted, as are the vast majority of 45,X fetuses. If all trisomy-21 fetuses survived to birth, the incidence of Down syndrome would rise to 1 in 250, approximately a threefold increase from the incidence observed.

Although many autosomal trisomies are found in spontaneous abortions, autosomal monosomies are not found. Monosomic embryos do exist. They are probably created in greater numbers than the trisomic fetuses because chromosome loss, leading to monosomy, is usually much more frequent than chromosome gain, leading to trisomy. The absence of autosomal monosomies in spontaneously aborted fetuses is undoubtedly due to these embryos being aborted so early in development that the pregnancy goes unrecognized. The spontaneous abortions summarized in Table 7.1, although they represent a huge fetal wastage, serve the important biological function of eliminating many fetuses that are grossly abnormal in their development because of major chromosome abnormalities.

7.6 Abnormalities in Chromosome Structure

Thus far, abnormalities in chromosome *number* have been described. The remainder of this chapter deals with abnormalities in chromosome *structure*. There are several principal types of structural aberrations, each of which has characteristic genetic effects. Chromosome aberrations were initially discovered through their genetic effects, which, though confusing at first, eventually came to be understood as resulting from abnormal chromosome structure. This interpretation was later confirmed directly by microscopic observations.

Deletions

A chromosome sometimes arises in which a segment is missing. Such a chromosome is said to have a **deletion** or a **deficiency.** Deletions are generally harmful to the organism, and the usual rule is the larger the deletion, the greater the harm. Very large deletions are usually lethal, even when heterozygous with a normal chromosome. Small deletions are often viable when they are heterozygous with a structurally normal homolog, because the normal homolog supplies gene products that are necessary for survival. However, even small deletions are usually homozygous lethal (lethal when both members of a pair of homologous chromosomes carry the deletion).

Deletions can be detected genetically by making use of the fact that a chromosome with a deletion no longer carries the wild-type alleles of the genes that have been eliminated. For example, in *Drosophila*, many *Notch* deletions are large enough to remove the nearby wildtype allele of *white*, also. When these deleted chromosomes are heterozygous with a structurally normal chromosome carrying the recessive *w* allele, the fly has white eyes because the wildtype w^+ allele is no longer present in the deleted *Notch* chromosome. This **uncovering** of the recessive allele implies that the corresponding wildtype allele of *white* has also been deleted. Once a deletion has been identified, its size can be assessed genetically by determining which recessive mutations in the region are uncovered by the deletion. This method is illustrated in Figure 7.18.

With the banded polytene chromosomes in *Drosophila* salivary glands, it is possible to study deletions and other chromosome aberrations physically. For example, all the *Notch* deletions cause particular bands to be missing in the salivary chromosomes. Physical mapping of deletions also allows individual genes, otherwise known only from genetic studies, to be assigned to specific bands or regions in the salivary chromosomes.

Physical mapping of genes in part of the *Drosophila* X chromosome is illustrated in Figure 7.19. The banded chromosome is shown near the top, along with the numbering system used to refer to specific bands. Each chromosome is divided into numbered sections (the X chromosome comprises sections 1 through 20), and each of the sections is divided into subdivisions designated A through F. Within each lettered subdivision, the bands are numbered in order, and so, for example, 3A6 is the sixth band in subdivision A of section 3 (Figure 7.19). On the average, each band contains about 20 kb of DNA, but there is considerable variation in DNA content from band to band.

In Figure 7.19, the mutant X chromosomes labeled I through VI have deletions. The deleted part of each chromosome is shown in red. These deletions define regions along the chromosome, some of which correspond to specific bands. For example, the deleted region in both chromosome I and II that is present in all the

other chromosomes consists of band 3A3. In crosses, only deletions I and II uncover the mutation *zeste (z)*, so the *z* gene must be in band 3A3, as indicated at the top. Similarly, the recessive-lethal mutation *zw2* is uncovered by all deletions except VI; therefore, the *zw2* gene must be in band 3A9. As a final example, the *w* mutation is uncovered only by deletions II, III, and IV; thus the *w* gene must be in band 3C2. The *rst* (rough eye texture) and *N* (notched wing margin) genes are not uncovered by any of the deletions. These genes were localized by a similar analysis of overlapping deletions in regions 3C5 to 3C10.

Duplications

Some abnormal chromosomes have a region that is present twice. These chromosomes are said to have a **duplication.**

Figure 7.18

Mapping of a deletion by testcrosses. The F_1 heterozygotes with the deletion express the recessive phenotype of all deleted genes. The expressed recessive alleles are said to be uncovered by the deletion.

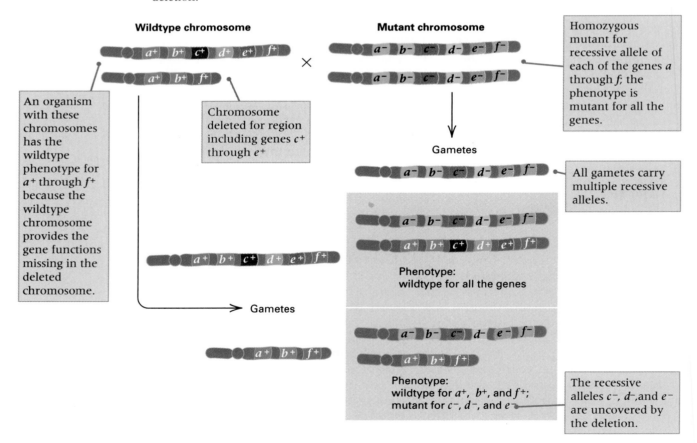

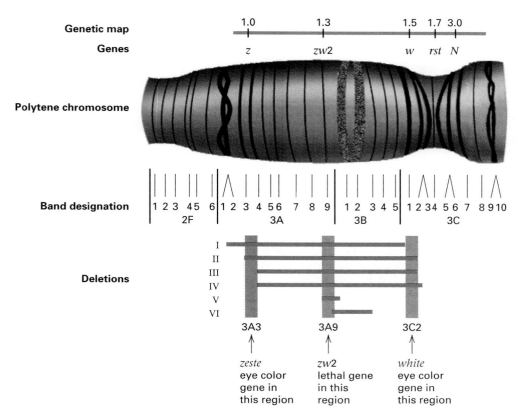

Figure 7.19

Part of the X chromosome in polytene salivary gland nuclei of *Drosophila melanogaster* and the extent of six deletions (I–VI) in a set of chromosomes. Any recessive allele that is uncovered by a deletion must be located inside the boundaries of the deletion. This principle can be used to assign genes to specific bands in the chromosome.

Certain duplications have phenotypic effects of their own. An example is the *Bar* duplication in *Drosophila*, which is a tandem duplication of bands 16A1 through 16A7 in the X chromosome. A **tandem duplication** is one in which the duplicated segment is present in the same orientation immediately adjacent to the normal region in the chromosome. In the case of *Bar*, the tandem duplication produces a dominant phenotype of bar-shaped eyes.

Tandem duplications are able to produce even more copies of the duplicated region by means of a process called **unequal crossing-over.** Figure 7.20A illustrates the chromosomes in meiosis of an organism that is homozygous for a tandem duplication (brown region). When they undergo synapsis, these chromosomes can mispair with each other, as illustrated in Figure 7.20B. A crossover within the mispaired part of the duplication (Figure

7.20C) will thereby produce a chromatid carrying a triplication, and a reciprocal product (labeled "single copy" in Figure 7.20D) that has lost the duplication. For the *Bar* region, the triplication can be recognized because it produces an even greater reduction in eye size than the duplication.

The most frequent effect of a duplication is a reduction in viability (probability of survival); in general, survival decreases with increasing size of the duplication. However, deletions are usually more harmful than duplications of comparable size.

Unequal Crossing-Over in Human Red-Green Color Blindness

Human color vision is mediated by three light-sensitive protein pigments present in the cone cells of the retina. Each of the pigments is related to **rhodopsin,** the pigment

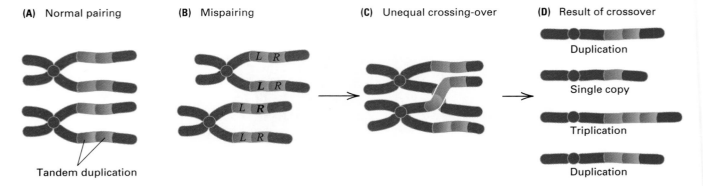

(A) Normal pairing **(B)** Mispairing **(C)** Unequal crossing-over **(D)** Result of crossover

Duplication

Single copy

Triplication

Duplication

Tandem duplication

Figure 7.20
An increase in the number of copies of a chromosome segment resulting from unequal crossing-over of tandem duplications (brown). (A) Normal synapsis of chromosomes with a tandem duplication. (B) Mispairing. The right-hand element of the lower chromosome is paired with the left-hand element of the upper chromosome. (C) Crossing-over within the mispaired duplication, which is called unequal crossing-over. (D) The outcome of unequal crossing-over. One product contains a single copy of the duplicated region, another chromosome contains a triplication, and the two strands not participating in the crossover retain the duplication.

that is found in the rod cells and mediates vision in dim light. The light sensitivities of the cone pigments are toward blue, red, and green. These are our primary colors. We perceive all other colors as mixtures of these primaries. The gene for the blue-sensitive pigment resides somewhere in 7q22-qter (which means 7q22 to the terminus), and the genes for the red and green pigments are on the X chromosome at Xq28 separated by less than 5 cM (roughly 5 Mb of DNA). Because the red and green pigments arose from the duplication of a single ancestral pigment gene and are still 96 percent identical in amino acid sequence, the genes are similar enough that they can pair and undergo unequal crossing-over. The process of unequal crossing-over is the genetic basis of red-green color blindness.

Almost everyone is familiar with **red-green color blindness;** it is one of the most common inherited conditions in human beings. Approximately 5 percent of males have some form of red-green color blindness. The preponderance of affected males immediately suggests X-linked inheritance, which is confirmed by pedigree studies (Section 3.4). Affected males have normal sons and carrier daughters, and the carrier daughters have 50 percent affected sons and 50 percent carrier daughters.

Actually, there are several distinct varieties of red-green color blindness. Defects in red vision go by the names of **protanopia,** an inability to perceive red, and **protanomaly,** an impaired ability to perceive red. The comparable defects in green perception are called **deuteranopia** and **deuteranomaly,** respectively. Isolation of the red and green pigment genes and study of their organization in people with normal and defective color vision have indicated quite clearly how the "-opias" and "-omalies" differ and have also explained why the frequency of color blindness is so relatively high.

The organization of the red and green pigment genes in men with normal vision is illustrated in Figure 7.21A. Unexpectedly, a significant proportion of normal X chromosomes contain two or three green-pigment genes. How these arise by unequal crossing-over is shown in Figure 7.21B. The red-pigment and green-pigment genes pair and the crossover takes place in the region of homology between the genes. The result is a duplication of the green-pigment gene in one chromosome and a deletion of the green-pigment gene in the other.

The recombinational origins of the defects in color vision are illustrated in Figure 7.22. The top chromosome in Figure 7.22A is the result of deletion of the green-

Figure 7.21

(A) Organization of red-pigment and green-pigment genes in normal X chromosomes. Some chromosomes contain one copy of the green-pigment gene, others two, still others three. (B) Origin of multiple green-pigment genes by unequal crossing-over in the region of DNA homology between the genes. Note that one product of unequal crossing-over is a chromosome containing a red-pigment gene but no green-pigment gene.

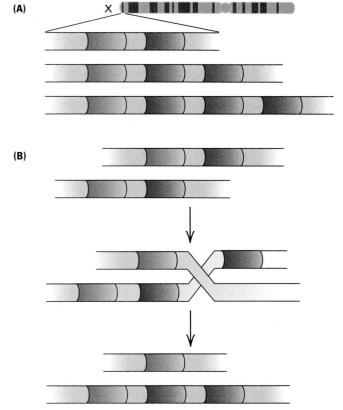

pigment gene shown earlier in Figure 7.21B. Males with such an X chromosome have deuteranopia, or "green-blindness." Other types of abnormal pigments result when crossing-over takes place within mispaired red-pigment and green-pigment genes. Crossing-over between the genes yields a **chimeric gene,** which is a composite gene: part of one joined with part of the other. The chimeric gene in Figure 7.22A joins the 5' end of the green-pigment gene with the 3' end of the red-pigment gene. If the crossover point is toward the 5' end (toward the left in the figure), then the resulting chimeric gene is mostly "red" in sequence, so the chromosome causes deuteranopia, or "green-blindness." However, if the crossover point is near the 3' end (toward the right in the figure), then most of the green-pigment gene remains intact, and the chromosome causes deuteranomaly.

Chromosomes associated with defects in red vision are illustrated in Figure 7.22B. The chimeric genes are the reciprocal prod-

ucts of the unequal crossovers that yield defects in green vision. In this case, the chimeric gene consists of the red-pigment gene at the 5' end and the green-pigment gene at the 3' end. If the crossover point is near the 5' end, most of the red-pigment gene is replaced with the green-pigment gene. The result is protanopia, or "red-

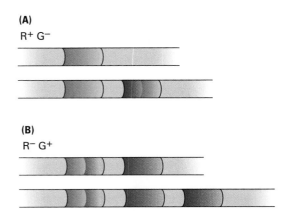

(A)
R+ G−

(B)
R− G+

Figure 7.22

Genetic basis of absent or impaired red-green color vision. (A) Defects in green vision result from unequal crossing-over between mispaired red-pigment and green-pigment genes, yielding a green-red chimeric gene. If the green-pigment gene is missing, or if the chimeric gene is largely "red" in its sequence, then deuteranopia is the result. If the chimeric gene is largely "green" in its sequence, then deuteranomaly is the result. (B) Defects in red vision result from unequal crossing-over between mispaired red-pigment and green-pigment genes, again yielding a green-red chimeric gene. If the chimeric gene is largely "green," then protanopia results; if it is largely "red," then protanomaly results. Note that the red gene cannot be eliminated altogether (as the green gene can), because the red gene is at the end of the region of homology between the chromosomes.

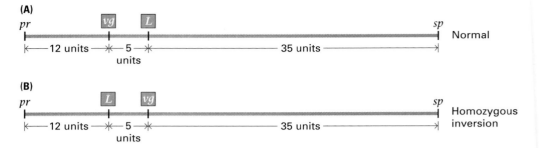

(A)

pr   vg   L   *sp*   Normal

|←—12 units—→|←5→| units |←————— 35 units —————→|

(B)

pr   L   vg   *sp*   Homozygous inversion

|←—12 units—→|←5→| units |←————— 35 units —————→|

Figure 7.23

Partial genetic map of chromosome 2 of *Drosophila melanogaster*. (A) Normal order. (B) Genetic map of the same region in a chromosome that has undergone an inversion in which the order of *vg* and *L* is reversed.

blindness." The same is true of the other chromosome indicated in Figure 7.22B. However, if the crossover point is near the 3′ end, then most of the red-pigment gene remains intact, and the result is protanomaly.

Inversions

Another important type of chromosome abnormality is an **inversion,** a segment of a chromosome in which the order of the genes is the reverse of the normal order. An example is shown in Figure 7.23.

In an organism that is heterozygous for an inversion, one chromosome is struc-

turally normal (wildtype), and the other carries an inversion. These chromosomes pass through mitosis without difficulty because each chromosome duplicates and its chromatids are separated into the daughter cells without regard to the other chromosome. There is a problem in meiosis, however. The problem is that the chromosomes are attracted gene for gene in the process of synapsis, as shown in Figure 7.24. In an inversion heterozygote, in order for gene-for-gene pairing to take place everywhere along the length of the chromosome, one or the other of the chromosomes must twist into a loop in the region in which the gene order is inverted. In

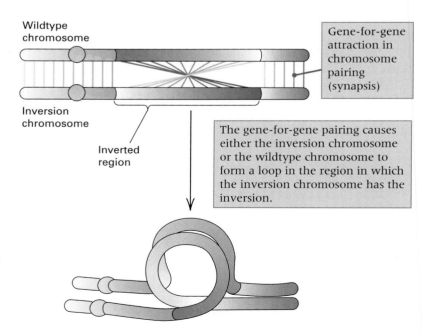

Wildtype chromosome

Inversion chromosome

Inverted region

Gene-for-gene attraction in chromosome pairing (synapsis)

The gene-for-gene pairing causes either the inversion chromosome or the wildtype chromosome to form a loop in the region in which the inversion chromosome has the inversion.

Figure 7.24

In an organism that carries a chromosome that is structurally normal along with a homologous chromosome with an inversion, the gene-for-gene attraction between the chromosomes during synapsis causes one of the chromosomes to form into a loop in the region in which the gene order is inverted. In this example, the structurally normal chromosome forms the loop.

Figure 7.24, it is the structurally normal chromosome that is shown as looped, but in other cells it may be the inverted chromosome that is looped. In either case, the loop is called an **inversion loop.**

The loop itself does not create a problem. The looping apparently takes place without difficulty and can be observed through the microscope. As long as there is no crossing-over within the inversion, the homologous chromosomes can separate normally at anaphase I, as illustrated in Figure 7.25. On the other hand, when there is crossing-over within the inversion loop, the chromatids involved in the crossing-over become physically joined, and the result is the formation of chromosomes containing large duplications and deletions. The products of the crossing-over can be deduced from Figure 7.26A by tracing along the chromatids. The outer chromatids are the ones not participating in the crossover. One of these contains the inverted sequence and the other the normal sequence, as shown in Figure 7.26B. Because of the crossover, the inner chromatids, which did participate in the crossover, are connected. If the centromere is not included in the inversion loop, as is the case here, then the result is a dicentric chromosome. The reciprocal product of the crossover produces an acentric chromosome. Neither the dicentric chromosome nor the acentric chromosome can be included in a normal gamete. The acentric chromosome is usually lost because it lacks a centromere and, in any case, has a deletion of the *a* region and a duplication of the *d* region. The dicentric chromosome is also often lost because it is held on the meiotic spindle by the chromatid bridging between the centromeres; in any case, this chromosome is deleted for the *d* region and duplicated for the *a* region. Hence, when there is a crossover in the inversion loop, the only chromatids that can be recovered in the gametes are the chromatids that did not participate on the crossover. One of these carries the inversion, and the other does not. It is for this reason that inversions prevent the recovery of crossover products. In the early years of genetics, before their identity as inversions was discovered, inversion-bearing chromosomes were known as "crossover suppressors."

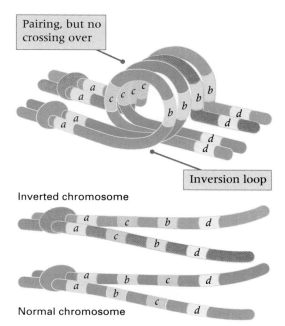

Figure 7.25
In an inversion heterozygote, if there is no crossing over within the inversion loop, then the homologous chromosomes disjoin without problems. Two of the resulting gametes carry the inverted chromosome, and two carry the structurally normal chromosome.

The inversion shown in Figure 7.26, in which the centromere is not included in the inverted region, is known as a **paracentric inversion,** which means inverted "beside" (*para-*) the centromere. As seen in the figure, the products of crossing-over include a dicentric and an acentric chromosome.

When the inversion does include the centromere, it is called a **pericentric inversion,** which means "around" (*peri-*) the centromere. Chromatids with duplications and deficiencies are also created by crossing-over within the inversion loop of a pericentric inversion, but in this case the crossover products are monocentric. The situation is illustrated in Figure 7.27A. The diagram is identical to that in Figure 7.26 except for the position of the centromere. The products of crossing-over can again be deduced by tracing the chromatids. In this case, both products of the crossover are monocentric, but one chromatid carries a duplication of *a* and a deletion of *d*, and the other carries a duplication of *d* and a deletion of *a* (Figure 7.27B). Although

(A) Paracentric inversion

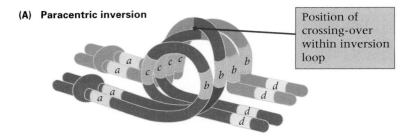

Position of crossing-over within inversion loop

(B) Anaphase I chromosome separation

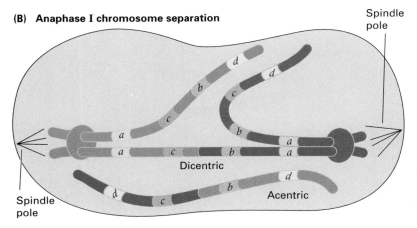

Spindle pole

Dicentric

Acentric

Spindle pole

Figure 7.26

(A) Synapsis between homologous chromosomes, one of which contains an inversion. There is a crossing-over within the inversion loop. (B) Anaphase I configuration resulting from the crossover. Because the centromere is not included in the inverted region, one of the crossover products is a dicentric chromosome, and the reciprocal product is an acentric chromosome. Among the two chromatids not involved in the crossover, one carries the inversion and the other the normal gene sequence.

either of these chromosomes could be included in a gamete, the duplication and deficiency usually result in inviability. Thus, as with the paracentric inversion, the products of recombination are not recovered, but for a different reason. Among the chromatids not participating in the crossing-over in Figure 7.27A, one carries the pericentric inversion, and the other has the normal sequence.

The looping of one chromosome in an inversion heterozygote creates a rather complex structure, but it can be depicted in simplified form as in Figure 7.28. In this diagram, only two of the four chromatids

are shown (the two that participate in the crossing-over), and the location of the centromere is not indicated. This kind of diagram is convenient because it shows that the essential result of the crossing-over is the production of the duplication and deletion products. These will also be dicentric or acentric if the centromere is located outside of the inverted region (a paracentric inversion).

Reciprocal Translocations

A chromosomal aberration resulting from the interchange of parts between nonhomologous chromosomes is called a **translocation.** In Figure 7.29, organism A is homozygous for two pairs of structurally normal chromosomes. Organism B contains one structurally normal pair of chromosomes and another pair of chromosomes that have undergone an interchange of terminal parts. The organism is said to be *heterozygous* for the translocation. The translocation is properly called a **reciprocal translocation** because it consists of two reciprocally interchanged parts. As indicated in Figure 7.29C, an organism can also be homozygous for a translocation if both pairs of homologous chromosomes undergo an interchange of parts.

An organism that is heterozygous for a reciprocal translocation usually produces only about half as many offspring as normal—a condition that is called **semisterility.** The reason for the semisterility is difficulty in chromosome segregation in meiosis. When meiosis takes place in a translocation heterozygote, the normal and translocated chromosomes must undergo synapsis, as shown in Figure 7.30. Ordinarily, there would also be chiasmata between nonsister chromatids in the arms of the homologous chromosomes, but these are not shown, as though the translocation were present in an organism with no crossing-over, such as a male *Drosophila.* Segregation from this configuration can take place in any of three ways. In the list that follows, the symbol $1 + 2 \longleftrightarrow 3 + 4$ means that at the first meiotic anaphase, the chromosomes in Figure 7.30 labeled 1 and 2 go to one pole and those labeled 3 and 4 go to the opposite pole. The red

(A) Pericentric inversion

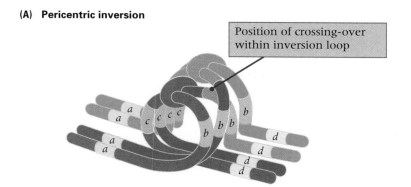

Position of crossing-over within inversion loop

(B) Anaphase I chromosome separation

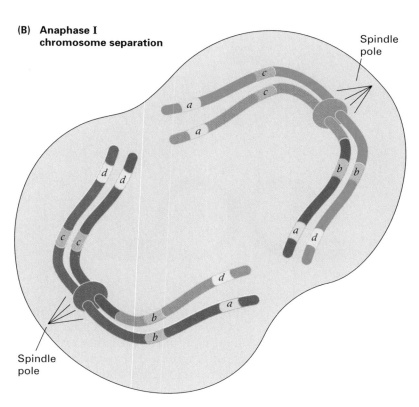

Spindle pole

Spindle pole

Figure 7.27
(A) Synapsis between homologous chromosomes, one of which carries a pericentric inversion. A crossing-over within the inversion loop is shown. (B) Anaphase I configuration resulting from the crossover. One of the crossover products is duplicated for *a* and deficient for *d;* the other is duplicated for *d* and deficient for *a*. Among the two chromatids not involved in the crossover, one carries the inversion and the other is normal.

numbers indicate the two parts of the reciprocal translocation. The three types of segregation are

- *1 + 2 ⟷ 3 + 4* This mode is called **adjacent-1** segregation. Homologous centromeres go to opposite poles, but each normal chromosome goes with one part of the reciprocal translocation. All gametes formed from adjacent-1 segregation have a large duplication

and deficiency for the distal part of the translocated chromosomes. (The *distal* part of a chromosome is the part farthest from the centromere.) The pair of gametes originating from the 1 + 2 pole are duplicated for the distal part of the blue chromosome and deficient for the distal part of the red chromosome; the pair of gametes from the 3 + 4 pole have the reciprocal deficiency and duplication.

Figure 7.28
Simplified diagram of an inversion loop showing the consequence of crossing-over within the loop. Only the two chromatids participating in the crossover are shown, and the location of the centromeres is not indicated.

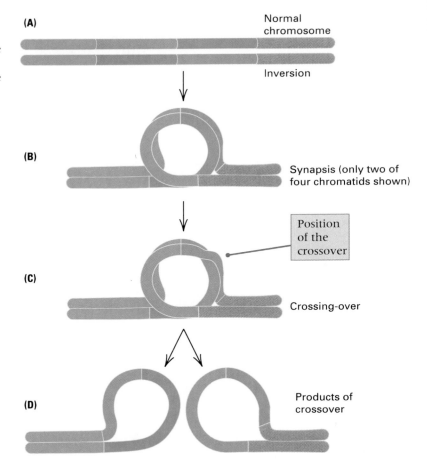

(A) Normal chromosome
Inversion

(B) Synapsis (only two of four chromatids shown)

Position of the crossover

(C) Crossing-over

(D) Products of crossover

- *1 + 3* ⟷ *2 + 4* This mode is **adjacent-2** segregation, in which homologous centromeres go to the same pole at anaphase I. In this case, all gametes have a large duplication and deficiency of the proximal part of the translocated chromosome. (The *proximal* part of a chromosome is the part closest to the centromere.) The pair of gametes from the 1 + 3 pole have a duplication of the proximal part of the red chromosome and a deficiency of the proximal part of the blue chromosome; the pair of gametes from the 2 + 4 pole have the reciprocal deficiency and duplication.

Figure 7.29
(A) Two pairs of nonhomologous chromosomes in a diploid organism. (B) Heterozygous reciprocal translocation, in which two nonhomologous chromosomes (the two at the top) have interchanged terminal segments. (C) Homozygous reciprocal translocation.

(A) Homozygous normal (both pairs normal)

(B) Heterozygous translocation (one pair interchanged, one pair normal)

(C) Homozygous translocation (both pairs interchanged)

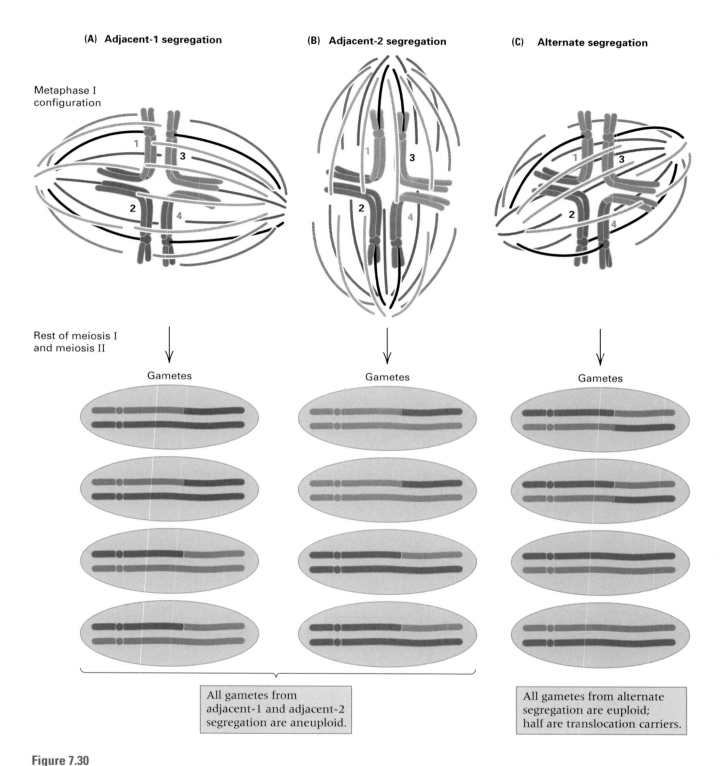

(A) Adjacent-1 segregation

(B) Adjacent-2 segregation

(C) Alternate segregation

Metaphase I configuration

Rest of meiosis I and meiosis II

Gametes

Gametes

Gametes

All gametes from adjacent-1 and adjacent-2 segregation are aneuploid.

All gametes from alternate segregation are euploid; half are translocation carriers.

Figure 7.30
Quadrivalent formed in the synapsis of a heterozygous reciprocal translocation. The translocated chromosomes are numbered in red, their normal homologs in black. No chiasmata are shown. (A) Adjacent-1 segregation, in which homologous centromeres separate at anaphase I; all of the resulting gametes have a duplication of one terminal segment and a deficiency of the other. (B) Adjacent-2 segregation, in which homologous centromeres go together at anaphase I; all of the resulting gametes have a duplication of one proximal segment and a deficiency of the other. (C) Alternate segregation, in which half of the gametes receive both parts of the reciprocal translocation and the other half receive both normal chromosomes.

- $1 + 4 \longleftrightarrow 2 + 3$ In this type of segregation, which is called **alternate** segregation, the gametes are all balanced (euploid), which means that none has a duplication or deficiency. The gametes from the $1 + 4$ pole have both parts of the reciprocal translocation; those from the $2 + 3$ pole have both normal chromosomes.

The semisterility of genotypes that are heterozygous for a reciprocal translocation results from lethality due to the duplication and deficiency gametes produced by adjacent-1 and adjacent-2 segregation. The frequency with which these types of segregation take place is strongly influenced by the position of the translocation breakpoints, by the number and distribution of chiasmata in the interstitial region between the centromere and each breakpoint, and by whether the quadrivalent tends to open out into a ring-shaped structure on the metaphase plate. A ring often forms if the breakpoints are in the middle region of the arms and there is no crossing-over in the interstitial regions; in this case, the frequencies of adjacent-1 : adjacent-2 : alternate segregation are approximately 1 : 1 : 2. If there is crossing-over in the interstitial regions, or if one breakpoint is near a telomere, then the orientation of the quadrivalent at metaphase tends to discourage adjacent-2 segregation. In some cases, adjacent-2 segregation is eliminated, and the ratio of adjacent-1 : alternate segregation is approximately 1 : 1. Adjacent-1 segregation is quite common in any event, which means that semisterility is to be expected from virtually all translocation heterozygotes.

Translocation semisterility is manifested in different life-history stages in plants and animals. Plants have an elaborate gametophyte phase of the life cycle, a haploid phase in which complex metabolic and developmental processes are necessary. In plants, large duplications and deficiencies are usually lethal in the gametophyte stage. Because the gametophyte produces the gametes, in higher plants the semisterility is manifested as pollen or seed lethality. In animals, by contrast, only minimal gene activity is necessary in the gametes, which function in spite of very large duplications

and deficiencies. In animals, therefore, the semisterility is usually manifested as zygotic lethality.

Certain groups of plants have reciprocal translocations present in natural populations without the semisterility usually expected. Among these are the evening primroses in the genus *Oenothera*, of which there are about 100 wild species native to North America and many cultivated varieties. *Oenothera* escapes the translocation semisterility because segregation is always in the alternate mode. In terms of Figure 7.30, the mode of segregation is always $1 + 4 \longleftrightarrow 2 + 3$. In some species of *Oenothera*, entire sets of chromosomes are interconnected through a chain of reciprocal translocations of the type illustrated in Figure 7.31A. The chromosomes at the top left are all normal; those at the top right are translocated in such a way that each chromosome has exchanged an arm with the next in line. When the chromosomes in the complex translocation heterozygote undergo synapsis, the result is a ring of chromosomes (Figure 7.31B). The astonishing feature of meiosis in such *Oenothera* heterozygotes is that the segregation is exclusively of the alternate type, so the only gametes formed contain either the entire set of normal chromosomes or the entire set of translocated chromosomes. This is not the end of the surprises. In *Oenothera*, one of the gametic types is inviable in the pollen and the other is inviable in the ovule, so fertilization restores the karyotype of the complex translocation heterozygote!

In species in which translocation heterozygotes exhibit semisterility, the semisterility can be used as the phenotype to map the breakpoint of the translocation just as though it were a normal gene. The mapping procedure can be made clear by means of an example. Translocation *TB-10L1* is a translocation with one breakpoint in the long arm of chromosome 10 in maize, and it results in semisterility when heterozygous. A cross is made between a translocation heterozygote and a genotype homozygous for both *zn1* (*zebra necrotic 1*) and *tp2* (*teopod 2*), and semisterile progeny are testcrossed with *zn1 tp2* homozygotes. (The phenotypes of *zn1* and *tp2* are shown in Figure 4.8.) The parental genotype is

(A)

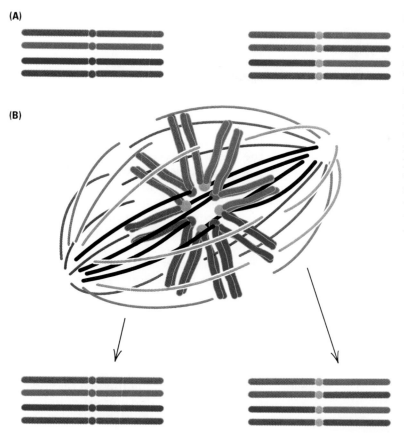

(B)

Figure 7.31

(A) Complex translocation heterozygote of the type found in some species of *Oenothera*. The chromosomes at the top left are not rearranged. Those at the top right are connected by a chain of translocations, each chromosome having exchanged an arm with the next chromosome in line. (B) At metaphase I in meiosis, the pairing configuration of the translocation heterozygote is a ring of chromosomes in which each arm is paired with its proper partner; note that each chromosome consists of two chromatids. Alternate segregation from the metaphase ring yields, after the second meiotic division, two types of gametes: those containing all normal chromosomes and those containing all translocated chromosomes.

therefore *Zn1 Tp2 TB-10L1/zn1 tp2 +*, where the + denotes the position of the translocation breakpoint in the homologous chromosome. The progeny phenotypes are as follows:

nonzebrastripe	nonteopod	semisterile	392
nonzebrastripe	nonteopod	fertile	3
nonzebrastripe	teopod	semisterile	42
nonzebrastripe	teopod	fertile	73
zebrastripe	nonteopod	semisterile	83
zebrastripe	nonteopod	fertile	34
zebrastripe	teopod	semisterile	1
zebrastripe	teopod	fertile	372

These data are analyzed exactly like the three-point crosses in Section 4.3. The double recombinants are present in the rarest classes of progeny and differ from the parental genotypes in exchanging the gene that is in the middle of the three. The double recombinants are *Zn1 Tp2 +* and

zn1 tp2 TB-10L1, which means that the translocation breakpoint lies between *zn1* and *tp2*. Hence, the map distance between *zn1* and the breakpoint is

$$(3 + 73 + 83 + 1)/1000 = 16 \text{ cM}$$

and that between the translocation breakpoint and *tp2* is

$$(3 + 42 + 34 + 1)/1000 = 8 \text{ cM}$$

The genetic map of the region containing the *TB-10L1* translocation breakpoint is

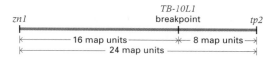

Robertsonian Translocations

A special type of *non*reciprocal translocation is a **Robertsonian translocation,** in which the centromeric regions of two

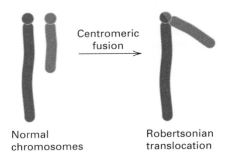

Centromeric
fusion →

Normal
chromosomes

Robertsonian
translocation

Figure 7.32
Formation of a Robertsonian translocation by
fusion of two acrocentric chromosomes in the
centromeric region.

Figure 7.33
A karyotype of a child with Down syndrome,
carrying a Robertsonian translocation of chromosomes 14 and 21 (arrow). Chromosomes 19
and 22 are faint in this photo; this has no significance. [Courtesy of Irene Uchida.]

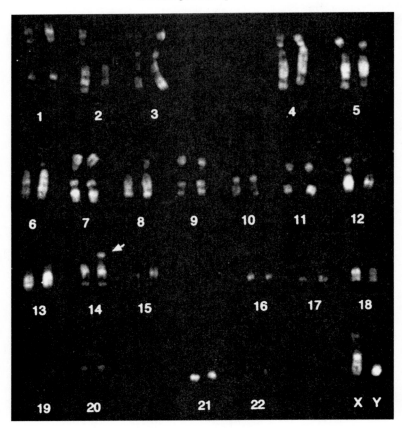

nonhomologous acrocentric chromosomes
become fused to form a single centromere
(Figure 7.32). Robertsonian translocations
are important in human genetics, especially as a risk factor to be considered in
Down syndrome. When chromosome 21 is
one of the acrocentrics in a Robertsonian
translocation, the rearrangement leads to a
familial type of Down syndrome in which
the risk of recurrence is very high.
Approximately 3 percent of children with
Down syndrome are found to have one
parent with such a translocation.

A Robertsonian translocation that joins
chromosome 21 with chromosome 14 is
shown in Figure 7.33 (arrow). The heterozygous carrier is phenotypically normal,
but a high risk of Down syndrome results
from aberrant segregation in meiosis. The
possible modes of segregation are shown in
Figure 7.34. The symbol *rob* refers to the
translocation, and the +, or −, preceding a
chromosome number designates an extra
copy, or a missing copy, of the entire
chromosome. Among the several possible
types of gametes that can arise, one contains a normal chromosome 21 along with
the 14/21 Robertsonian translocation
(Figure 7.34A). If this aberrant gamete is
used in fertilization, then the fetus will contain two copies of the normal chromosome
21 plus the 14/21 translocation. In effect,
the fetus contains three copies of chromosome 21 and hence has Down syndrome.
The other abnormal gametes that result
from adjacent-1 or adjacent-2 segregation
either are missing chromosome 21 or chromosome 14 or contain effectively two
copies of chromosome 14 (Figure 7.34A
and B). If these gametes participate in
ertilization, the result is monosomy 21,
monosomy 14, or trisomy 14, respectively.
The monosomic embryos undergo very
early spontaneous abortion; the trisomy 14-
fetus undergoes spontaneous abortion later
in pregnancy. Hence families with a high
risk of translocation Down syndrome also
have a high risk of spontaneous abortion
due to other chromosome abnormalities.
Alternate segregation of a Robertsonian
translocation yields gametes carrying either
the translocation or both normal chromosomes (Figure 7.34C). Because these

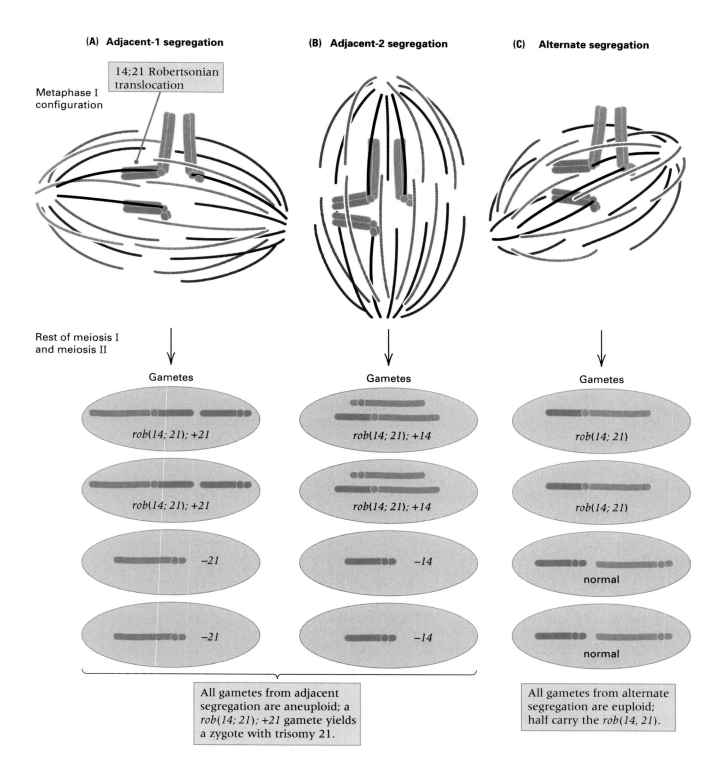

Figure 7.34

Segregation of Robertsonian translocation between chromosomes 14 and 21. (A) Adjacent-1 segregation, in which the gametes formed from the pole at the top have, effectively, an extra copy of chromosome 21. (B) Adjacent-2 segregation. The gametes are either duplicated or deficient for chromosome 14. (C) Alternate segregation. Half of the gametes give rise to phenotypically normal children who are carriers of the Robertsonian translocation.

gametes derive from reciprocal products of meiosis, the nonaffected children have a risk of 1/2 of carrying the translocation.

7.7 Position Effects on Gene Expression

Genes near the breakpoints of chromosomal rearrangement become repositioned in the genome and flanked by new neighboring genes. In many cases, the repositioning of a gene affects its level of expression or, in some cases, its ability to function; this is called **position effect.** These effects have been studied extensively in *Drosophila* and also in yeast. In *Drosophila*, the most common type of position effect results in a mottled (mosaic) phenotype that is observed as interspersed patches of wildtype cells, in which the wildtype gene is expressed, and mutant cells, in which the wildtype gene is inactivated. The phenotype is said to show **variegation,** and the phenomenon is called **position-effect variegation (PEV).** In the older literature, PEV is often referred to as variegated or V-type position effect.

PEV usually results from a chromosome aberration that moves a wildtype gene from a position in euchromatin to a new position in or near heterochromatin (Figure 7.35). (Euchromatin and heterochromatin are discussed in Chapter 6.) Figure 7.36 illustrates some of the patterns of wildtype (red) and mutant (white) facets that are observed in male flies that carry a rearranged X chromosome in which an inversion repositions the wildtype w^+ allele into heterochromatin. The same types of patterns are found in females heterozygous for the rearranged X chromosome and an X chromosome carrying the w allele. The patterns of w^+ expression coincide with the clonal lineages in the eye; that is, all of the red cells in a particular patch derive from a single ancestral cell in the embryo in which the w^+ allele was activated. In contrast with the pattern shown in Figure 7.36, other chromosome rearrangements with PEV yield a salt-and-pepper pattern of mosaicism, which consists of numerous very small patches of wildtype tissue, and still others yield a combination of many small and a few large patches. These patterns imply that gene activation can be very late in development as well as very early.

Although the mechanism of PEV is not understood in detail, it is thought to result from the unusual chromatin structure of heterochromatin interfering with gene activation. The determination of gene expression or nonexpression is thought to take place when the boundary between condensed heterochromatin and euchromatin is established. Where heterochromatin is juxtaposed with euchromatin, the chromatin condensation characteristic of heterochromatin may spread into the adjacent euchromatin, inactivating euchromatic genes in the cell and all of its descendants. A similar inactivation phenomenon takes place in cells of female mammals when euchromatic genes translocated to the X chromosome become heterochromatic and inactive. The length of the euchromatic region that is inactivated ranges from 1 to 50 bands in the *Drosophila* polytene chromosomes, depending on the particular chromosome abnormality. At the molecular level, this range is approximately 20 to 1000 kb. The term *spreading,* with its implication of smooth

Figure 7.35
Position-effect variegation (PEV) is often observed when an inversion or other chromosome rearrangement repositions a gene normally in euchromatin to a new location in or near heterochromatin. In this example, an inversion in the X chromosome of *Drosophila melanogaster* repositions the wildtype allele of the *white* gene near heterochromatin. PEV of the w^+ allele is observed as mottled red and white eyes.

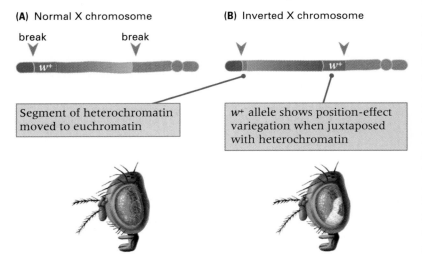

(A) Normal X chromosome

break   break

w^+

Segment of heterochromatin moved to euchromatin

(B) Inverted X chromosome

w^+

w^+ allele shows position-effect variegation when juxtaposed with heterochromatin

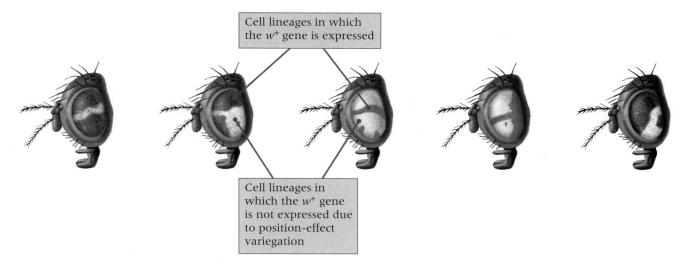

Cell lineages in which the w^+ gene is expressed

Cell lineages in which the w^+ gene is not expressed due to position-effect variegation

Figure 7.36
Patterns of red and white sectors in the eye of *Drosophila melanogaster* resulting from position-effect variegation. Each group of contiguous facets of the same color derives from a single cell in development. Such large patches of red are atypical. More often, one observes numerous very small patches of red or a mixture of many small and a few large patches.

continuity, should not be taken literally, however, because in both *Drosophila* and mammals, chromosomal regions affected by PEV are known to contain active genes interspersed with inactive ones. It is likely that heterochromatic regions associate to form a relatively compact "compartment" in the nucleus, and the presence of heterochromatin near a euchromatic gene may silence the gene simply by attracting it into the heterochromatic compartment.

The heterochromatin of different chromosomes differs in its ability to induce variegation of specific genes and in its response to environmental or genetic modifiers. Environmental modifiers include temperature; genetic modifiers include the presence of extra blocks of heterochromatin (such as a Y chromosome). Some mutations that affect chromatin structure also suppress PEV, and certain euchromatic DNA sequences, when repositioned into heterochromatin, act as "insulators" protecting nearby genes by binding with chromatin-associated proteins that prevent inactivation. Other than position-effect variegation, genes are rarely shut down completely because of their position in the genome. Most genes can be expressed irrespective of neighboring genes, but the level of expression may vary substantially according to position.

7.8 Chromosome Abnormalities and Cancer

Cancer is an unrestrained proliferation and migration of cells. In all known cases, cancer cells derive from the repeated division of a single mutant cell whose growth has become unregulated, and so cancer cells initially constitute a clone. With continued proliferation, many cells within such clones develop chromosomal abnormalities, such as extra chromosomes, missing chromosomes, deletions, duplications, or translocations. The chromosomal abnormalities found in cancer cells are diverse, and they may differ among cancer cells in the same person or among people with the same type of cancer. The accumulation of chromosome abnormalities is evidently one accompaniment of unregulated cell division.

Amid the large number of apparently random chromosome abnormalities found in cancer cells, a small number of aberrations are found consistently in certain types of cancer, particularly in blood diseases such as the leukemias. For example, chronic myelogenous leukemia is frequently associated with an apparent deletion of part of the long arm of chromosome 22. The abnormal chromosome 22 in this disease is called the **Philadelphia chromosome;** this chromosome is actually one part of a reciprocal translocation in which the missing segment of chromosome 22 is attached to either chromosome 8 or chromosome 9. Similarly, a deletion of part of the short arm of chromosome 11 is frequently associated with a kidney tumor called Wilms tumor, which is usually found in children.

For many of these characteristic chromosome abnormalities, the breakpoint in the chromosome is near the chromosomal location of a **cellular oncogene.** An **oncogene** is a gene associated with cancer. Cellular oncogenes, also called *protooncogenes,* are the cellular homologs of **viral oncogenes** contained in certain cancer-causing viruses. The distinction is one of location: Cellular oncogenes are part of the normal genome; viral oncogenes are derived from cellular oncogenes through some rare mechanism in which they become incorporated into virus particles. More than 50 different cellular oncogenes are known. They are apparently normal developmental genes that predispose cells to unregulated division when mutated or abnormally expressed. Many of the genes function in normal cells as growth factors that promote and regulate cell division. When a chromosome rearrangement happens near a cellular oncogene (or when the gene is incorporated into a virus), the gene may become expressed abnormally and result in unrestrained proliferation of the cell that contains it. However, abnormal expression by itself is usually not sufficient to produce cancerous growth. One or more additional mutations in the same cell are also required.

A sample of characteristic chromosome abnormalities found in certain cancers is given in Figure 7.37, along with the locations of known cellular oncogenes on the same chromosomes. The symbols are again *p* for the short arm and *q* for the long arm. The + and − signs are new designations:

- A + or − preceding a chromosome number indicates an extra (or missing) copy of the entire chromosome.

- A + or − following a chromosome designation means extra material (or missing material) corresponding to part of the designated chromosome or arm; for example, 11p− refers to a deletion of part of the short arm of chromosome 11.

The symbol *t* means reciprocal translocation; hence *t(9;22)* refers to a reciprocal translocation between chromosome 9 and chromosome 22.

In Figure 7.37, the location of the cellular oncogene, when known, is indicated by a red arrow at the left, and the chromosomal breakpoint is indicated by a black arrow at the right. In most cases for which sufficient information is available, the correspondence between the breakpoint and the cellular oncogene location is very close, and in some cases, the breakpoint is within the cellular oncogene itself. The significance of this correspondence is that the chromosomal rearrangement disturbs normal cellular oncogene regulation and ultimately leads to the onset of cancer.

Retinoblastoma and Tumor-Suppressor Genes

A second class of genes associated with inherited cancers consists of the **tumor-suppressor genes.** These are genes whose presence is necessary to suppress tumor formation. Absence of both normal alleles, through either mutational inactivation or deletion, results in tumor formation. An example of a tumor-suppressor gene is the human gene *Rb-1,* located in chromosome 13 in band 13q14. When the normal *Rb-1* gene product is absent, malignant tumors form in the retinas, and surgical removal of the eyes becomes necessary. The disease is known as **retinoblastoma.**

Retinoblastoma is unusual in that the predisposition to retinal tumors is dominant in pedigrees but the *Rb-1* mutation is recessive at the cell level; that is, a person who inherits one copy of the *Rb-1* mutation through the germ line is heterozygous, and the penetrance of retinoblastoma in this person is 100 percent. However, the retinal cells that become malignant have the genotype *Rb-1 Rb-1.* The explanation for this apparent paradox is illustrated in Figure 7.38. Part A of Figure 7.38 shows the genotype of an *Rb-1* heterozygote, along with a few other genes in the same chromosome. Parts B through E show four possible ways in which a second genetic event can result

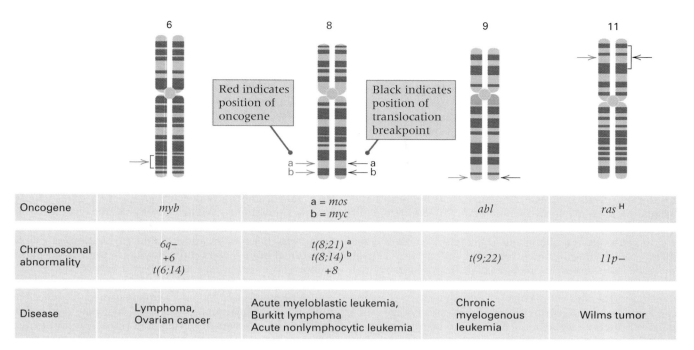

	6	8	9	11
Oncogene	*myb*	a = *mos* b = *myc*	*abl*	*ras* H
Chromosomal abnormality	*6q–* *+6* *t(6;14)*	*t(8;21)* a *t(8;14)* b *+8*	*t(9;22)*	*11p–*
Disease	Lymphoma, Ovarian cancer	Acute myeloblastic leukemia, Burkitt lymphoma Acute nonlymphocytic leukemia	Chronic myelogenous leukemia	Wilms tumor

Red indicates position of oncogene

Black indicates position of translocation breakpoint

Figure 7.37
Correlation between oncogene positions (red arrows) and chromosome breaks (black arrows) in aberrant human chromosomes frequently found in cancer cells. A number of breakpoints near *myb* result in a cancer-associated *t(6;14)*.

in *Rb-1Rb-1* cells in the retina. The simplest (part B) is mutation of the wildtype allele in the homolog. The wildtype allele could also be deleted (part C). Part D illustrates a situation in which the normal homolog of the *Rb-1* chromosome is lost and replaced by nondisjunction of the *Rb-1*-bearing chromosome. Mitotic recombination (part E) is yet another possibility for making *Rb-1* homozygous. Although each of these events takes place at a very low rate per cell division, there are so many cells in the retina (approximately 10^8) that the *Rb-1* allele usually becomes homozygous in at least one cell. (In fact, the average number of tumors per retina is three.)

Figure 7.38
Mechanisms by which (A) a single copy of *Rb-1* inherited through the germ line can become homozygous in cells of the retina: (B) new mutation; (C) deletion; (D) loss of the normal homologous chromosome and replacement by nondisjunction of the *Rb-1*-bearing chromosome; (E) mitotic recombination. Each of these events is rare, but there are so many cells in the retina that on average, there are three such events per eye.

(A) *Rb-1* heterozygous genotype (zygote)

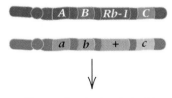

Types of retinal cells lacking *Rb-1* function

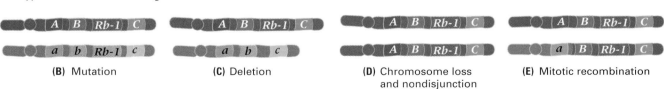

(B) Mutation **(C)** Deletion **(D)** Chromosome loss and nondisjunction **(E)** Mitotic recombination

A typical chromosome contains a single centromere, the position of which determines the shape of the chromosome as it is pulled to the poles of the cell during anaphase. Rare chromosomes with no centromere, and those with two or more centromeres, are usually lost within a few cell generations because of aberrant separation during anaphase. Ring chromosomes are genetically relatively stable but are very rare.

Polyploid organisms contain more than two complete sets of chromosomes. Polyploidy is widespread among higher plants and uncommon otherwise. Between 30 and 35 percent of all species of flowering plants are thought to have originated as some form of polyploid. An autopolyploid organism contains multiple sets of chromosomes from a single ancestral species; allopolyploid organisms contain complete sets of chromosomes from two or more ancestral species. Organisms occasionally arise in which an individual chromosome either is missing or is present in excess; in either case, the number of copies of genes in such a chromosome is incorrect. Departures from normal gene dosage (aneuploidy) often result in reduced viability of the zygote in animals or of the gametophyte in plants. In general, too many copies of genes or chromosomes have less severe effects than too few copies.

The normal human chromosome complement consists of 22 pairs of autosomes, which are assigned numbers 1 through 22 from longest to shortest, and one pair of sex chromosomes (XX in females and XY in males). Fetuses that contain an abnormal number of autosomes usually fail to complete normal embryonic development or die shortly after birth, though people with Down syndrome (trisomy 21) sometimes survive for several decades. Persons with excess sex chromosomes survive, because the Y chromosome contains relatively few genes other than the master sex-controller *SRY,* and because only one X chromosome is genetically active in the cells of females (dosage compensation through the single-active-X principle). The mosaic orange and black pattern of the female "calico" cat results from X inactivation, because these alternative coat color alleles are X-linked in cats.

Most structural abnormalities in chromosomes are duplications, deletions, inversions, or translocations. In a duplication, there are two copies of a chromosomal segment. In a deletion, a chromosomal segment is missing. An organism can often tolerate an imbalance of gene dosage resulting from small duplications or deletions, but large duplications or deletions are almost always harmful. Chromosome rearrangements may affect gene expression through position effects. Although most genes can be expressed whatever their location in the genome, the level of expression may vary substantially according to position. A major effect of gene location on expression is observed in position-effect variegation (PEV), in which a wildtype allele repositioned in or near heterochromatin is unable to be expressed in a fraction of the cell lineages.

A chromosome that contains an inversion has a group of adjacent genes in reverse of the normal order. Expression of the genes is usually unaltered, so inversions rarely affect viability. However, crossing-over between an inverted chromosome and its noninverted homolog in meiosis yields abnormal chromatids. Crossing-over within a heterozygous paracentric inversion yields an acentric chromosome and a dicentric chromosome, both of which also have a duplication and a deficiency. Crossing-over within a heterozygous pericentric inversion yields monocentric chromosomes, but both have a duplication and a deficiency.

Two nonhomologous chromosomes that have undergone an exchange of parts constitute a reciprocal translocation. Organisms that contain a reciprocal translocation, as well as the normal homologous chromosomes of the translocation, produce fewer offspring (this is called semisterility) because of abnormal segregation of the chromosomes in meiosis. The semisterility is caused by the aneuploid gametes produced in adjacent-1 and adjacent-2 segregation. Alternate segregation yields equal numbers of normal and translocation-bearing gametes. In genetic crosses, the semisterility of a heterozygous translocation behaves like a dominant genetic marker that can be mapped like any other gene; however, what is actually mapped is the breakpoint of the translocation.

A translocation may also be nonreciprocal. A Robertsonian translocation is a type of nonreciprocal translocation in which the long arms of two acrocentric chromosomes are attached to a common centromere. In human beings, Robertsonian translocations that include chromosome 21 account for about 3 percent of all cases of Down syndrome, and the parents have a high risk of recurrence of Down syndrome in a subsequent child.

Malignant cells in many types of cancer contain specific types of chromosome abnormalities. Frequently, the breakpoints in the chromosomes coincide with the chromosomal location of one of a group of cellular oncogenes coding for cell growth factors. Abnormal oncogene expression is implicated in cancer. Viral oncogenes, found in certain cancer-causing viruses, are derived from cellular oncogenes. Cells also contain tumor-suppressor genes, the absence of which predisposes to cancer. The gene for retinoblastoma codes for a tumor-suppressor in the retina of the eyes. The gene is dominant in predisposing to retinal malignancy but recessive at the cellular level. People who inherit one copy of the gene through the germ line develop retinal tumors when the gene becomes homozygous in cells in the retina. Homozygosity in the retina can result from any number of genetic events, including new mutation, deletion, chromosome loss and nondisjunction, and mitotic recombination.

Key Terms

acentric chromosome
acrocentric chromosome
adjacent-1 segregation
adjacent-2 segregation
allopolyploid
alternate segregation
amniocentesis
aneuploid
autopolyploidy
Barr body
cellular oncogene
chimeric gene
chromosome painting
colchicine
decaploid
deficiency
deletion
deuteranopia
deuteranomaly
dicentric chromosome
dosage compensation
double-Y syndrome
Down syndrome
duplication
embryoid
endoreduplication
euploid

fragile-X chromosome
fragile-X syndrome
haploid
hexaploid
imprinting
inversion
inversion loop
karyotype
Klinefelter syndrome
metacentric chromosome
monoploid
monosomic
mosaic
octoploid
oncogene
paracentric inversion
pericentric inversion
Philadelphia chromosome
position effect
position-effect variegation (PEV)
protanopia
protanomaly
polyploidy
polysomy
pseudoautosomal
reciprocal translocation
red-green color blindness

retinoblastoma
rhodopsin
ring chromosome
Robertsonian translocation
semisterility
single-active-X principle
submetacentric chromosome
tandem duplication
testis-determining factor (TDF)
tetraploid
translocation
trinucleotide repeat
triplication
triploid
trisomic
trisomy-X syndrome
trivalent
tumor-suppressor gene
Turner syndrome
uncovering
unequal crossing-over
univalent
variegation
viral oncogene
X inactivation

Review the Basics

- What is a metacentric chromosome? A submetacentric chromosome? An acrocentric chromosome?

- Why does an acentric chromosome fail to align at the metaphase plate and fail to move to one of the poles in anaphase? What type of abnormal chromosome forms a chromosomal "bridge" between the daughter cells at anaphase?

- How can a normal gamete have two sets of chromosomes? What possible gametes could be produced by a tetraploid plant with the genotype $AAaa\ BBbb$?

- Define each of the following terms: allopolyploid, aneuploid, deletion, duplication, paracentric inversion, pericentric inversion, reciprocal translocation, Robertsonian translocation.

- What does it mean to say that a deletion "uncovers" a recessive mutation?

- Which type of chromosome rearrangement can change a metacentric chromosome into a submetacentric chromosome? Which type of chromosome rearrangement can fuse two acrocentric chromosomes to make one metacentric chromosome?

- Inversions are often called "suppressors" of crossing-over. Is this term literally true? If not, what is meant by the term?

- What does *dosage compensation* mean with reference to X-linked genes?

- What are two broad classes of genes often associated with inherited cancers?

Guide to Problem Solving

Problem 1: The first artificial allotetraploid was created by the Russian agronomist G. D. Karpechenko in the 1930s by crossing the radish *Raphanus sativus* with the cabbage *Brassica oleracea*. Both species have a diploid chromosome number of 18. The initial F_1 hybrid was virtually sterile, but among the offspring was a rare, fully fertile allotetraploid that he called *Raphanobrassica*. What is the chromosome number of the F_1 of the cross *R. sativus* × *B. oleracea*? What is the chromosome number of *Raphanobrassica?* How many bivalents are formed in meiosis in *Raphanobrassica?*

Answer: Both *R. sativus* and *B. oleracea* produce gametes with 9 chromosomes, so the F_1 has 18 chromosomes. The allotetraploid results from a doubling of the F_1 chromosome complement, so *Raphanobrassica* has 36 chromosomes. Because each chromosome has a homolog, *Raphanobrassica* forms 18 bivalents.

Problem 2: Genes *a, b, c, d, e,* and *f* are closely linked in a chromosome, but their order is unknown. Three deletions in the region are found to uncover recessive alleles of the genes as follows:

> Deletion 1 uncovers *a, b,* and *d*.
>
> Deletion 2 uncovers *a, d, c,* and *e*.
>
> Deletion 3 uncovers *e* and *f*.

What is the order of the genes? In this problem, you will see that there is enough information to order most, but not all, of the genes. Suggest what experiments you might carry out to complete the ordering.

Answer: Problems of this sort are worked by noting that genes uncovered by a single deletion must be contiguous. The gene order is deduced from the overlaps between the deletions. The overlaps are *a* and *d* between the first and second deletions and *e* between the second and third. The gene order (as far as can be determined from these data) is diagrammed in part A of the accompanying illustration. The deletions are shown in red. Gene *b* is at the far left, then *a* and *d* (in unknown order), then *c, e,* and *f*. (Gene *c* must be to the left of *e,* because otherwise *c* would be uncovered by deletion 3.) Part B is a completely equivalent map with gene *b* at the right. The ordering can be completed with a three-point cross between *b, a,* and *d* or between *a, d,* and *c* or by examining additional deletions. Any deletion that uncovers either *a* or *d* (but not both) plus at least one other marker on either side would provide the information to complete the ordering.

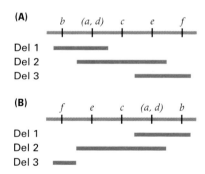

Problem 3: In each of the following cases, determine the consequences of a single crossover within the inverted region of a pair of homologous chromosomes with the gene order *A B C D* in one chromosome and *a c b d* in the other.
(a) The centromere is not included within the inversion.
(b) The centromere is included within the inversion.

Answer: These kinds of problems are most easily solved by drawing a diagram. The accompanying illustration shows how pairing within a heterozygous inversion results in a looped configuration. (Crossing-over occurs at the four-strand stage of meiosis, but for simplicity only the chromatids participating in the crossover are diagrammed.) Part A illustrates the situation when the centromere is not included within the inversion. The crossover chromatids consist of a dicentric (two centromeres) and an acentric (no centromere), and the products are duplicated for the terminal region containing *A* and deficient for the terminal region containing *D,* or the other way around. The noncrossover chromatids (not shown) are the parental *A B C D* and *a c b d* monocentric chromosomes.

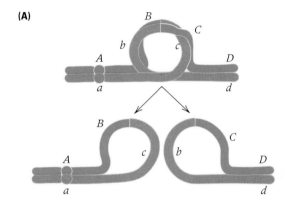

Part B illustrates the situation when the centromere is included in the inversion. The duplications and deficiencies are the same as in part A, but in this case both products are monocentrics. As before, the noncrossover chromatids are the parental *A B C D* and *a c b d* configurations.

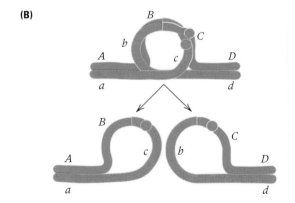

Analysis and Applications

7.1 How many Barr bodies would be present in each of the following human conditions?

(a) Klinefelter syndrome

(b) Turner syndrome

(c) Down syndrome

(d) XYY

(e) XXX

Chapter 7

GeNETics on the web will introduce you to some of the most important sites for finding genetic information on the Internet. To complete the exercises below, visit the Jones and Bartlett home page at

http://www.jbpub.com/genetics

Select the link to *Genetics: Principles and Analysis* and then choose the link to *GeNETics on the web*. You will be presented with a chapter-by-chapter list of highlighted keywords.

GeNETics EXERCISES

Select the highlighted keyword in any of the exercises below, and you will be linked to a web site containing the genetic information necessary to complete the exercise. Each exercise suggests a specific, written report that makes use of the information available at the site. This report, or an alternative, may be assigned by your instructor.

1. The original **Bar duplication** may itself have arisen from unequal crossing-over. You can access the details by searching for the keyword at this site. If assigned to do so, write a brief description of the proposed origin of the original duplication, and draw a diagram illustrating the events in the process.

2. Molecular studies of the red and green **opsin** genes indicate abnormalities in 15.7 percent of Caucasian men. This frequency is substantially smaller than the incidence of color blindness as shown by color-vision testing, which suggests that some chimeric genes created by unequal crossing-over may be compatible with normal color vision. More information about the molecular genetics and physiology of color vision can be found by searching the keyword site for "protanopia" and "deuteranopia." You can also find a description of the "Nagel anomaloscope" used in testing color vision. If assigned to do so, write a one-paragraph description of how this instrument works, and discuss some of its limitations.

3. There are many myths about **Down syndrome.** One is that people with Down syndrome are severely retarded. You can find more myths, and the truth about them, at the keyword site. If assigned to do so, prepare a double-column list giving a brief description of each myth and the correct story.

MUTABLE SITE EXERCISES

The Mutable Site Exercise changes frequently. Each new update includes a different exercise that makes use of genetics resources available on the World Wide Web. Select the **Mutable Site** for Chapter 7, and you will be linked to the current exercise that relates to the material presented in this chapter.

PIC SITE

The Pic Site showcases some of the most visually appealing genetics sites on the World Wide Web. To visit the showcase genetics site, select the **Pic Site** for Chapter 7.

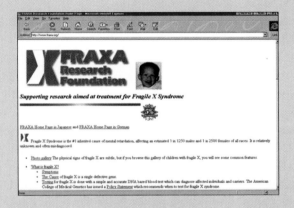

7.2 An autopolyploid series similar to *Chrysanthemum* consists of five species. The basic monoploid chromosome number in the group is 5. What chromosome numbers would be expected among the species?

7.3 A plant species S coexists with two related species A and B. All species are interfertile. In meiosis, S forms 26 bivalents, and A and B form 14 and 12 bivalents, respectively. Hybrids between S and A form 14 bivalents and 12 univalents, and hybrids between S and B form 12 bivalents and 14 univalents. Suggest a probable evolutionary origin of species S.

7.4 A spontaneously aborted human fetus was found to have the karyotype 92,XXYY. What might have happened to the chromosomes in the zygote to result in this karyotype?

7.5 A spontaneously aborted human fetus is found to have 45 chromosomes. What is the most probable karyotype? Had the fetus survived, what genetic disorder would it have had?

7.6 Color blindness in human beings is an X-linked trait. A man who is color-blind has a 45,X (Turner syndrome) daughter who is also color-blind. Did the nondisjunction that led to the 45,X child occur in the mother or the father? How can you tell?

7.7 A phenotypically normal woman has a child with Down syndrome. The woman is found to have 45 chromosomes. What kind of chromosome abnormality can account for these observations? How many chromosomes does the affected child have? How does this differ from the usual chromosome number and karyotype of a child with Down syndrome?

7.8 A chromosome has the gene sequence *A B C D E F G*. What is the sequence following a *C*-through-*E* inversion? Following a *C*-through-*E* deletion? Two chromosomes with the sequences *A B C D E F G* and *M N O P Q R S T U V* undergo a reciprocal translocation after breaks in *E−F* and *S−T*. What are the possible products? Which products are genetically stable?

7.9 Recessive genes *a, b, c, d, e,* and *f* are closely linked in a chromosome, but their order is unknown. Three deletions in the region are examined. One deletion uncovers *a, d,* and *e;* another uncovers *c, d,* and *f;* and the third uncovers *b* and *c.* What is the order of the genes?

7.10 Six bands in a salivary gland chromosome of *Drosophila* are shown in the accompanying figure, along with the extent of five deletions (Del1−Del5).

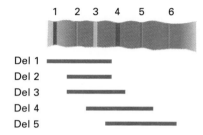

Recessive alleles *a, b, c, d, e,* and *f* are known to be in the region, but their order is unknown. When the deletions are heterozygous with each allele, the following results are obtained:

	a	*b*	*c*	*d*	*e*	*f*
Del 1	−	−	−	+	+	+
Del 2	−	+	−	+	+	+
Del 3	−	+	−	+	−	+
Del 4	+	+	−	−	−	+
Del 5	+	+	+	−	−	−

In this table, the − means that the deletion is missing the corresponding wildtype allele (the deletion uncovers the recessive allele) and + means that the corresponding wildtype allele is still present. Use these data to infer the position of each gene relative to the salivary gland chromosome bands.

7.11 A strain of corn that has been maintained by self-fertilization for many generations, when crossed with a normal strain, produces an F_1 in which many meiotic cells have dicentric chromatids and acentrics, and the amount of recombination in chromosome 6 is greatly reduced. However, in the original strain, no dicentrics or acentrics are found. What kind of abnormality in chromosome structure can explain these results?

7.12 Four strains of *Drosophila melanogaster* are isolated from different localities. The banding patterns of a particular region of salivary chromosome 2 have the following configurations (each letter denotes a band).

(a) a b f e d c g h i j
(b) a b c d e f g h i j

(c) a b f e h g i d c j
(d) a b f e h g c d i j

Assuming that part (c) is the ancestral sequence, deduce the evolutionary ancestry of the other chromosomes.

7.13 Two species of Australian grasshoppers coexist side by side. In meiosis, each has 8 pairs of chromosomes. When the species are crossed, the chromosomes in the hybrid form 6 pairs of chromosomes and one group of 4. What alteration in chromosome structure could account for these results?

7.14 Why are translocation heterozygotes semisterile? Why are translocation homozygotes fully fertile? If a translocation homozygote is crossed with an individual that has normal chromosomes, what fraction of the F_1 is expected to be semi-sterile?

7.15 Curly wings (*Cy*) is a dominant mutation in the second chromosome of *Drosophila*. A *Cy*/+ male was irradiated with x rays and crossed with +/+ females, and the *Cy*/+ sons were mated individually with +/+ females. From one cross, the progeny were

curly males	146
wildtype males	0
curly females	0
wildtype females	163

What abnormality in chromosome structure is the most likely explanation for these results? (Remember that crossing-over does not take place in male *Drosophila*.)

7.16 Yellow body (*y*) is a recessive mutation near the tip of the X chromosome of *Drosophila*. A wildtype male was irradiated with x rays and crossed with *yy* females, and one y^+ son was observed. This male was mated with *yy* females, and the offspring were

yellow females	256
yellow males	0
wildtype females	0
wildtype males	231

The yellow females were found to be chromosomally normal, and the y^+ males were found to breed in the same manner as their father. What type of chromosome abnormality could account for these results?

7.17 In the homologous chromosomes shown here, the red region represents an inverted segment of chromosome.

An individual of this genotype was crossed with an *a b c d e* homozygote. Most of the offspring were either *A B C D E* or *a b c d e*, but a few rare offspring were obtained that were *A B c D E*. What events occurring in meiosis in the inversion heterozygote can explain these rare progeny? Is the gene sequence in the rare progeny normal or inverted?

7.18 A wildtype strain of yeast, thought to be a normal haploid, was crossed with a different haploid strain carrying the mutation *his7*. This mutation is located in chromosome 2 and is an allele of a gene normally required for synthesis of the amino acid histidine. Among 15 tetrads analyzed from this cross, the following types of segregation were observed:

4 wildtype : 0 *his7*   4 tetrads
3 wildtype : 1 *his7*   4 tetrads
2 wildtype : 2 *his7*   1 tetrad

However, when the same wildtype strain was crossed with haploid strains with recessive markers on other chromosomes, segregation in the tetrads was always 2 : 2. What type of chromosome abnormality in the wildtype strain might account for the unusual segregation when the strain is mated with *his7*?

7.19 A strain of semisterile maize heterozygous for a reciprocal translocation between chromosomes 1 and 2 was crossed with chromosomally normal plants homozygous for the recessive mutations *brachytic* and *fine-stripe* on chromosome 1. When semisterile F_1 plants were crossed with plants of the *brachytic, fine-stripe* parental strain, the following phenotypes were found in a total of 682 F_2 progeny.

	Semisterile	Fertile
wildtype	333	19
brachytic	17	6
fine-stripe	1	8
brachytic, fine-stripe	25	273

What are the recombinant frequencies between *brachytic* and the translocation breakpoint and between *fine-stripe* and the translocation breakpoint?

Further Reading

Bickmore, W. A., and A. T. Sumner. 1989. Mammalian chromosome banding: An expression of genome organization. *Trends in Genetics* 5: 144.

Carson, H. L. 1970. Chromosome tracers of the origin of the species. *Science* 168: 1414.

Cavenee, W. K., and R. L. White. 1995. The genetic basis of cancer. *Scientific American,* March.

Curtis, B. C., and D. R. Johnson. 1969. Hybrid wheat. *Scientific American,* May.

Epstein, C. J. 1988. Mechanisms of the effects of aneuploidy in mammals. *Annual Review of Genetics* 22: 51.

Guerrero, I. 1987. Proto-oncogenes in pattern formation. *Trends in Genetics* 3: 269.

Hamel, P. A., B. L. Gallie, and R. A. Phillips. 1992. The retinoblastoma protein and cell cycle regulation. *Trends in Genetics* 8: 180.

Hsu, T. H. 1979. *Human and Mammalian Cytogenetics.* New York: Springer-Verlag.

Kamb, A. 1995. Cell-cycle regulators and cancer. *Trends in Genetics* 11: 136.

Kimber, G., and M. Feldman. 1987. *Wild Wheat: An Introduction.* Columbia: University of Missouri Press.

Manning, C. H., and H. O. Goodman. 1981. Parental origin of chromosomes in Down's syndrome. *Human Genetics* 59: 101.

Nathans, J.. 1989. The genes for color vision. *Scientific American,* February.

Richards, R. I., and G. R. Sutherland. 1992. Fragile X syndrome: The molecular picture comes into focus. *Trends in Genetics* 8: 249.

Stebbins, G. L. 1971. *Chromosome Evolution in Higher Plants.* Reading, MA: Addison-Wesley.

Stewart, G. D., T. J. Hassold, and D. M. Kurnit. 1988. Trisomy 21: Molecular and cytogenetic studies of nondisjunction. *Advances in Human Genetics* 17: 99.

Trichopoulos, D., F. P. Li, and D. J. Hunter. 1996. What causes cancer? *Scientific American,* September.

Vogt, P. K., ed. 1997. *Chromosomal Translocations and Oncogenic Transcription Factors.* New York: Springer-Verlag.

Wagner, R. P., M. P. Maguire, and R. L. Stallings. 1993. *Chromosomes.* New York: Wiley-Liss.

Wang, J. Y. J. 1997. Retinoblastoma protein in growth suppression and death protection. *Current Opinion in Genetics & Development* 7: 39.

Weinberg, R. A. 1996. How cancer arises. *Scientific American,* September.

White, M. J. D. 1977. *Animal Cytology and Evolution.* London: Cambridge University Press.

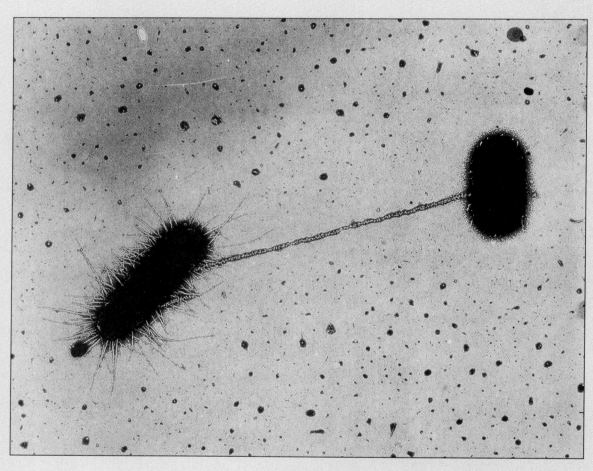

Two cells of *Escherichia coli* caught in the act of mating (conjugation). The male cell is at the lower left, and the female cell, at the upper right. The connecting tube is an F pilus, encoded by genes in the F plasmid, through which DNA from the male cell enters the female cell. (The pilus shortens somewhat prior to DNA transfer.) In this case, for ease of visualization, the F pilus has been coated with particles of a male-specific bacteriophage. The male cell has numerous other appendages, not F pili, that are used in colonizing the intestine. [Courtesy of C. C. Brinton, Jr., and J. Carnahan.]

CHAPTER 8

The Genetics of Bacteria and Viruses

PRINCIPLES

- Some bacteria are capable of DNA transfer and genetic recombination.
- In transformation, bacterial cells take up DNA from the surrounding medium and incorporate it into their genome by homologous recombination, replacing some host genes in the process.
- In *E. coli*, the F (fertility) plasmid can mobilize the chromosome for transfer to another cell in the process of conjugation.
- Some types of bacteriophages can incorporate bacterial genes and transfer them into new host cells in the process of transduction.
- DNA molecules from related bacteriophages that are present in the same host cell can undergo genetic recombination.
- Transposable elements and plasmids are widely used for genetic analysis and DNA manipulation in bacteria.

CONNECTIONS

CONNECTION: The Sex Life of Bacteria
Joshua Lederberg and Edward L. Tatum 1946
Gene recombination in Escherichia coli

CONNECTION: Is a Bacteriophage an "Organism"?
Alfred D. Hershey and Raquel Rotman 1948
Genetic recombination between host-range and plaque-type mutants of bacteriophage in single bacterial cells

CONNECTION: Artoo
Seymour Benzer 1955
Fine structure of a genetic region in bacteriophage

In earlier chapters, we examined the genetic properties of representative eukaryotic organisms. The genomes of these organisms consist of multiple chromosomes. In the course of sexual reproduction, genotypic variation among the progeny is achieved both by random assortment of the chromosomes and by crossing-over, processes that take place in meiosis. Two important features of crossing-over in eukaryotes are that (1) it results in a reciprocal exchange of material between two homologous chromosomes and (2) both products of a single exchange can often be recovered in different progeny. The situation is quite different in prokaryotes, as we will see in this chapter.

Bacteria and viruses bring to traditional types of genetic experiments four important advantages over multicellular plants and animals. First, they are haploid, so dominance or recessiveness of alleles is not a complication in identifying genotype. Second, a new generation is produced in minutes rather than weeks or months, which vastly increases the rate of accumulation of data. Third, they are easy to grow in enormous numbers under controlled laboratory conditions, which facilitates molecular studies and the analysis of rare genetic events. Fourth, the individual members of these large populations are genetically identical; that is, each laboratory population is a **clone** of genetically identical cells.

8.1
The Genetic Organization of Bacteria and Viruses

The organization of a typical prokaryotic cell is exemplified by the bacterial cell in Figure 8.1A. The genetic material is located in a region that lacks clear boundaries and is called the **nucleoid.** Contrast this relatively unstructured organization with the eukaryotic cell in Figure 8.1B. In eukaryotes, the genetic material is enclosed in the nucleus by a membrane envelope connected in special ways to the cytoplasm; eukaryotic cells also have other membrane systems that subdivide the cytoplasm into regions of specialized function.

A bacterial cell contains a major DNA molecule that almost never encounters another complete DNA molecule. Instead, genetic exchange is usually between a chromosomal fragment from one cell and an intact chromosome from another cell. Furthermore, a clear donor-recipient relationship exists: The donor cell is the source of a DNA fragment, which is transferred to the recipient cell by one of several mechanisms, and exchange of genetic material takes place in the recipient by means of reciprocal recombination between homologous DNA sequences. Incorporation of a part of the donor DNA into the chromosome requires at least two exchange events, one at each end. Because the recipient molecule is circular, only an even number of exchanges results in a viable product. The usual outcome of these events is the recovery of *only one* of the crossover products. However, in some situations, the transferred DNA is also circular, and a single exchange results in total incorporation of the circular donor DNA into the chromosome of the recipient.

Three major types of genetic transfer are found in bacteria: *transformation,* in which a DNA molecule is taken up from the external environment and incorporated into the genome; *transduction,* in which DNA is transferred from one bacterial cell to another by a bacterial virus; and *conjugation,* in which donor DNA is transferred from one bacterial cell to another by direct contact. Recombination frequencies that result from these processes are used to produce genetic maps of bacteria. Although the maps are exceedingly useful, they differ in major respects from the types of maps obtained from crosses in eukaryotes because genetic maps in eukaryotes are based on frequencies of crossing-over in meiosis.

A virus is a small particle, considerably smaller than a cell, that is able to infect a susceptible cell and multiply within it to form a large number of progeny virus particles. Few, if any, organisms are not subject to viral infection. Many human diseases are caused by viruses, including influenza, measles, AIDS, and the common cold. Viruses that infect bacterial cells are called **bacteriophages** or simply **phages.** Most viruses consist of a single molecule of genetic material enclosed in a protective coat composed of one or more kinds of protein molecules; however, their size, molecular constituents, and structural complexity vary greatly (Figure 8.2).

Isolated viruses possess no metabolic systems, so a virus can multiply only within a cell. It is "living" only in the sense that its genetic material directs its own multiplication; outside its host cell, a virus is an inert particle. Nevertheless, within an infected cell, many viruses and bacteriophages pos-sess mechanisms by which their DNA mol-ecules exchange genetic material resulting in genetic recombination. In the life cycle of a bacteriophage, a phage particle attaches to a host bacterium and injects its nucleic acid into the cell; the nucleic acid replicates many times; finally, newly synthesized nucleic acid molecules are packaged into protein shells (forming progeny phage), and then the particles are released from the cell (Figure 8.3). Except for new mutations, phage progeny from a bacterium infected

(A)

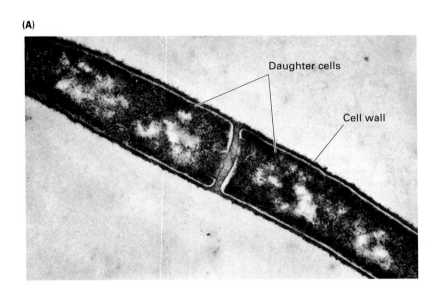

(B)

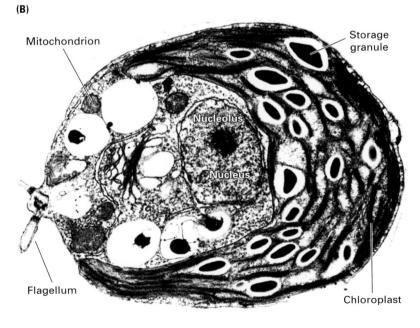

Figure 8.1
The organization of a prokaryotic cell and a eukaryotic cell. (A) An electron micrograph of a dividing bacterial cell, showing the dispersed genetic material (light areas). (B) An electron micrograph of a section through a cell of the eukaryotic alga *Tetraspora,* showing the membrane-bounded nucleus, the nucleolus (dark central body), and other membrane systems (chloroplasts and mitochondria) that subdivide the cytoplasm into regions of special-ized function. The dark sharply bounded regions in part B are starch storage granules, which store carbohydrate. Note the complexity of this structure compared with the simpler organization of the bacterial cell shown in part A. [Part A, courtesy of A. Benichou-Ryter; part B, courtesy of Jeremy Pickett-Heaps.]

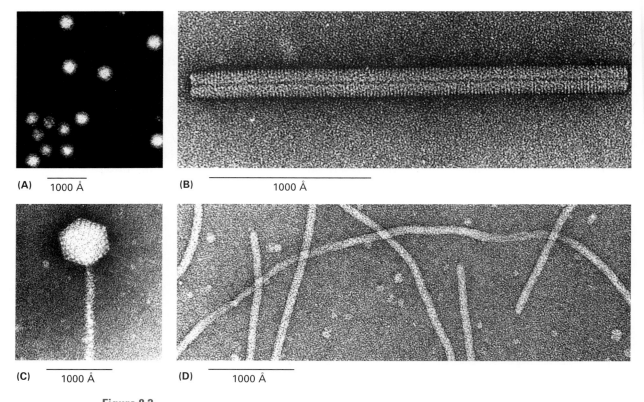

(A) 1000 Å **(B)** 1000 Å

(C) 1000 Å **(D)** 1000 Å

Figure 8.2
Electron micrographs of four different viruses: (A) poliovirus; (B) tobacco mosaic virus; (C) *E. coli* phage λ; (D) *E. coli* phage M13. In each case, the length of the bar is 10^{-5} cm, or 1000 Å units. [Courtesy of Robley Williams.]

Figure 8.3
A schematic diagram of the life cycle of a typical bacteriophage.

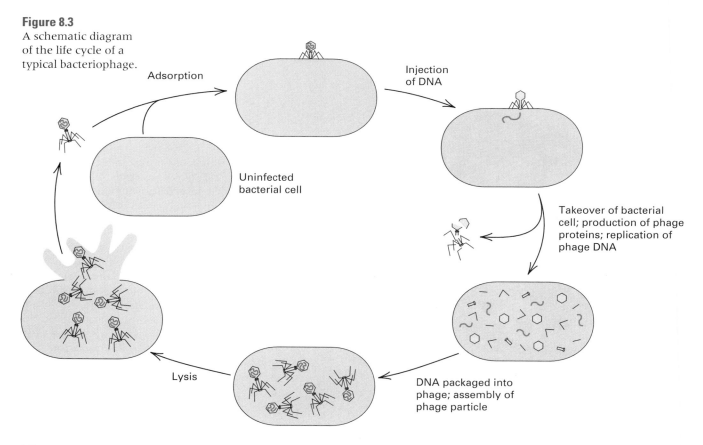

Adsorption

Injection of DNA

Uninfected bacterial cell

Takeover of bacterial cell; production of phage proteins; replication of phage DNA

DNA packaged into phage; assembly of phage particle

Lysis

by a single phage have the same genotype as the parental phage. However, if two phage particles with *different* genotypes infect a single bacterial cell, new genotypes can arise by genetic recombination. This process differs significantly from genetic recombination in eukaryotes in two ways: (1) the number of participating DNA molecules differs from one cell to the next, and (2) reciprocal recombinants are not always recovered in equal frequencies from a single infected cell. Some phages also possess systems that enable phage DNA to recombine with bacterial DNA. Phage-phage and phage-bacterium recombination are the topic of the second part of this chapter. The best-understood bacterial and phage systems are those of *E. coli*, and we will concentrate on these systems.

8.2
Bacterial Mutants

Bacteria can be grown in liquid medium or on the surface of a semisolid growth medium gelled with agar. Bacteria used in genetic analysis are usually grown on agar. A single bacterial cell placed on an agar medium will grow and divide many times, forming a visible cluster of cells called a *colony* (Figure 8.4). The number of bacterial cells present in a liquid culture can be determined by spreading a known volume of the culture on a solid medium and counting the number of colonies that form. Typical *E. coli* cultures contain up to 10^9 cells/ml. The appearance of colonies, or the ability or inability to form colonies on particular media can, in some cases, be used to identify the genotypes of bacterial cells.

As we have seen in earlier chapters, genetic analysis requires mutants; with bacteria, three types are particularly useful:

- *Antibiotic-resistant mutants* These mutants are able to grow in the presence of an antibiotic, such as streptomycin (Str) or tetracycline (Tet). For example, streptomycin-sensitive (Str-s) cells have the wildtype phenotype and fail to form colonies on medium that contains streptomycin, but streptomycin-resistant (Str-r) mutants can form colonies on such medium.

- *Nutritional mutants* Wildtype bacteria can synthesize most of the complex nutrients they need from simple

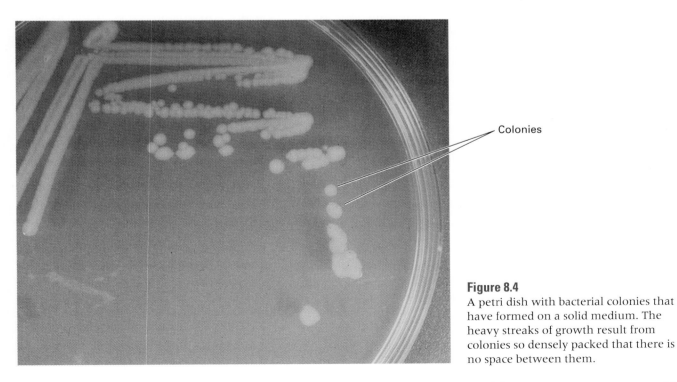

Colonies

Figure 8.4
A petri dish with bacterial colonies that have formed on a solid medium. The heavy streaks of growth result from colonies so densely packed that there is no space between them.

molecules present in the growth medium. The wildtype cells are said to be **prototrophs**. The ability to grow in simple medium can be lost by mutations that disable the enzymes used in synthesizing the complex nutrients. Mutant cells are unable to synthesize an essential nutrient and cannot grow unless the required nutrient is supplied in the medium. Such a mutant bacterium is said to be an **auxotroph** for the particular nutrient. For example, a methionine auxotroph cannot grow on a **minimal medium** that contains only inorganic salts and a source of energy and carbon atoms (such as glucose), but the methionine auxotroph *can* grow if the minimal medium is supplemented with methionine.

- *Carbon-source mutants* Such mutant cells cannot utilize particular substances as sources of carbon atoms or of energy. For example, Lac⁻ mutants cannot utilize the sugar lactose for growth and are unable to form colonies on minimal medium that contains only lactose as the carbon source.

A medium on which all wildtype cells form colonies is called a **nonselective medium.** Mutants and wildtype cells may or may not be distinguishable by growth on a nonselective medium. If the medium allows growth of only one type of cell (either wildtype or mutant), then it is said to be **selective.** For example, a medium containing streptomycin is selective for the Str-r phenotype and selective against the Str-s phenotype; similarly, minimal medium containing lactose as the sole carbon source is selective for Lac⁺ cells and against Lac⁻ cells.

In bacterial genetics, phenotype and genotype are designated in the following way. A phenotype is designated by three letters, the first of which is capitalized, with a superscript + or − to denote presence or absence of the designated phenotype, and with *s* or *r* for sensitivity or resistance. A genotype is designated by lowercase italicized letters. Thus a cell unable to grow without a supplement of leucine (a leucine auxotroph) has a Leu⁻ phenotype, and this

would usually result from a *leu⁻* mutation. Often the − superscript is omitted, but using it prevents ambiguity.

8.3
Bacterial Transformation

Bacterial transformation is a process in which recipient cells acquire genes from free DNA molecules in the surrounding medium. Transformation with purified DNA was the first experimental proof that DNA is the genetic material (Chapter 1). In these experiments, a rough-colony phenotype of *Streptococcus pneumoniae* was changed to a smooth-colony phenotype by exposure of the cells to DNA from a smooth-colony strain. In the laboratory, donor DNA is usually isolated from donor cells and then added to a suspension of recipient cells. In natural settings, such as soil, free DNA can become available by spontaneous breakage (lysis) of donor cells.

Transformation begins with uptake of a DNA fragment from the surrounding medium by a recipient cell and terminates with *one strand* of donor DNA replacing the homologous segment in the recipient DNA. Most bacterial species are probably capable of the recombination step, but many species have only a very limited ability to take up free DNA efficiently. Even in a species capable of transformation, DNA is able to penetrate only some of the cells in a growing population. However, many bacterial species can be made competent to take up DNA, provided that the cells are subjected to an appropriate chemical treatment (for example, treatment with $CaCl_2$).

Transformation is a convenient technique for gene mapping in some species. When DNA is isolated from a donor bacterium (Figure 8.5), it is invariably broken into small fragments. In most species, with suitable recipient cells and excess external DNA, transformation takes place at a frequency of about 1 transformed cell per 10^3 cells. If two genes, *a* and *b*, used as genetic markers, are so widely separated in the donor chromosome that they are always contained in two different DNA fragments, then the probability of simultaneous trans-

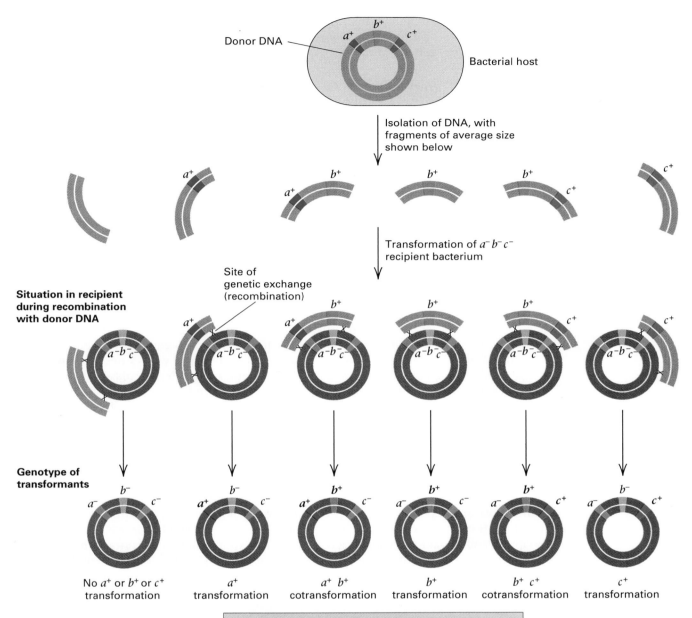

Figure 8.5

Cotransformation of linked markers. Markers *a* and *b* are near enough to each other that they are often present on the same donor fragment, as are markers *b* and *c*. Markers a and *c* are not near enough to undergo cotransformation. The gene order must therefore be *a b c*. The size of the transforming DNA, relative to that of the bacterial chromosome, is greatly exaggerated.

formation (**cotransformation**) of an $a^- \ b^-$ recipient into wildtype is the product of the probabilities of transformation of each genetic marker, or roughly $10^{-3} \times 10^{-3}$, which equals one $a^+ \ b^+$ transformant per 10^6 recipient cells. However, if the two genes are so near one another that they are often present in a single donor fragment, then the frequency of cotransformation is nearly the same as the frequency of single-gene transformation, or one wildtype transformant per 10^3 recipients. The general principle is as follows:

> Cotransformation of two genes at a frequency substantially greater than the product of the single-gene transformations implies that the two genes are close together in the bacterial chromosome.

Studies of the ability of various pairs of genes to be cotransformed also yield gene order. For example, if genes *a* and *b* can be cotransformed, and genes *b* and c can be cotransformed, but genes *a* and *c* cannot, the gene order must be *a b c* (Figure 8.5). Note that cotransformation frequencies are not equivalent to the recombination frequencies used in mapping eukaryotes, because they are determined by the size distribution of donor fragments and the likelihood of recombination between bacterial DNA molecules rather than by the occurrence of chiasmata in synapsed homologous chromosomes (Chapter 4).

8.4 Conjugation

Conjugation is a process in which DNA is transferred from a bacterial donor cell to a recipient cell by cell-to-cell contact. It has been observed in many bacterial species and is best understood in *E. coli*, in which it was discovered by Joshua Lederberg in 1951.

When bacteria conjugate, DNA is transferred to a recipient cell from a donor cell under the control of a set of genes that give the donor cell its transfer capability. These genes are often present in a nonchromosomal, circular DNA molecule called a **plasmid.** Plasmid-mediated recombination in *E. coli* usually results from the presence of a plasmid called the **F factor** or the **fertility factor.** In some instances, the F plasmid and similar types of plasmids can become incorporated into the bacterial chromosome so that the plasmid genes allowing DNA transfer are themselves part of the bacterial chromosome. In this case, the

plasmid is said to have been **integrated** into the chromosome. Because it can exist either separate from the chromosome or incorporated into it, the F factor is an example of an **episome,** a term that refers to any genetic element that can exist free in the cell or as a segment of DNA integrated into the chromosome.

Conjugation begins with physical contact between a donor cell and a recipient cell. A tubular projection from the donor cell forms a passageway between the donor and recipient cells. Through this passageway, a copy of the donor DNA moves from the donor to the recipient. In the final stage, which requires recombination if the donor contains an integrated plasmid, a segment of the transferred donor DNA becomes part of the genetic complement of the recipient. If the donor contains a free plasmid, then only the plasmid DNA is transferred and takes up residence in the recipient.

Let us begin with a description of the genetic properties of plasmids and their transfer and then examine plasmid-mediated chromosomal transfer.

Plasmids

Plasmids are circular DNA molecules that are capable of replicating independently of the chromosome and range in size from a few kilobases to a few hundred kilobases (Figure 8.6). The F factor is approximately 100 kb in length and contains many genes for its maintenance in the cell and its transmission between cells. Plasmids have been observed in many bacterial species and are usually not essential for growth of the cells. For studying plasmids in the laboratory, a culture of cells derived from a single plasmid-containing cell is used; because plasmids replicate and are inherited, all cells of the culture contain the plasmid of interest. Plasmids not only contain a diversity of genes for their own maintenance, which are not present in the bacterial chromosome, but plasmids can also acquire chromosomal genes from their host by several mechanisms. The presence of certain plasmids in bacterial cells is made evident by phenotypic characteristics of the host cell conferred by genes in the plasmid. For example, a plasmid containing the *tet-r* gene will make the recipient bacterial cell resistant to tetracycline.

Plasmids rely on the DNA-replication enzymes of the host cell for their reproduction, but *initiation* of replication is controlled by plasmid genes. The number of copies of a particular plasmid in a cell varies from one plasmid to the next, depending on its particular mode of regulation of initiation. High-copy-number plasmids are found in as many as 50 copies per host cell, whereas low-copy-number plasmids are present to the extent of 1 or 2 copies per cell. Plasmid DNA can be taken up by cells and become permanently established in the bacteria. The ability of plasmid DNA to transform cells genetically has made plasmids important in genetic engineering (Chapter 9).

From the point of view of bacterial genetics, the F plasmid is of greatest interest because of its role in mediating conjugation between cells of *E. coli*. Cells that contain F are donors and are designated F^+ ("F plus"); those lacking F are recipients and are designated F^- ("F minus"). The F plasmid is a low-copy-number plasmid. A typical F^+ cell contains one or two copies of F. These F plasmids replicate once per cell cycle and segregate to both daughter cells in cell division.

The F plasmid contains a set of genes for establishing conjugation between cells and for transferring DNA from donor to recipient. A copy of the F plasmid can be transferred in conjugation from an F^+ cell to an F^- cell. Transfer is always accompanied by replication of the plasmid. Contact between an F^+ and an F^- cell initiates rolling-circle replication of F (Section 5.4), which results in the transfer of a single-stranded linear branch of the rolling circle to the recipient cell. During transfer, DNA is synthesized in both donor and recipient (Figure 8.7). Synthesis in the donor replaces the transferred single strand, and synthesis in the recipient converts the transferred single strand into double-stranded DNA. When transfer is complete, the linear F strand becomes circular again in the recipient cell. Note that because one replica remains in the donor while the other is transferred to the recipient, after transfer *both cells contain F and can function as donors*. The transfer of F

Figure 8.6
Electron micrograph of a ruptured *E. coli* cell showing released chromosomal DNA and several plasmid molecules. [Courtesy of David Dressler and Huntington Potter.]

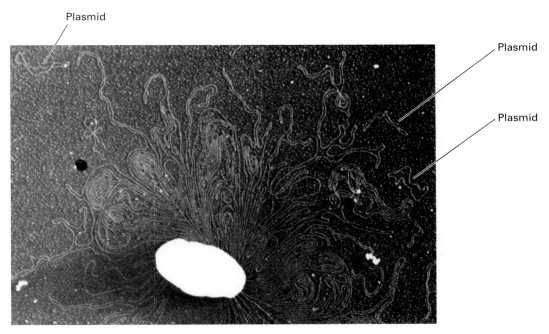

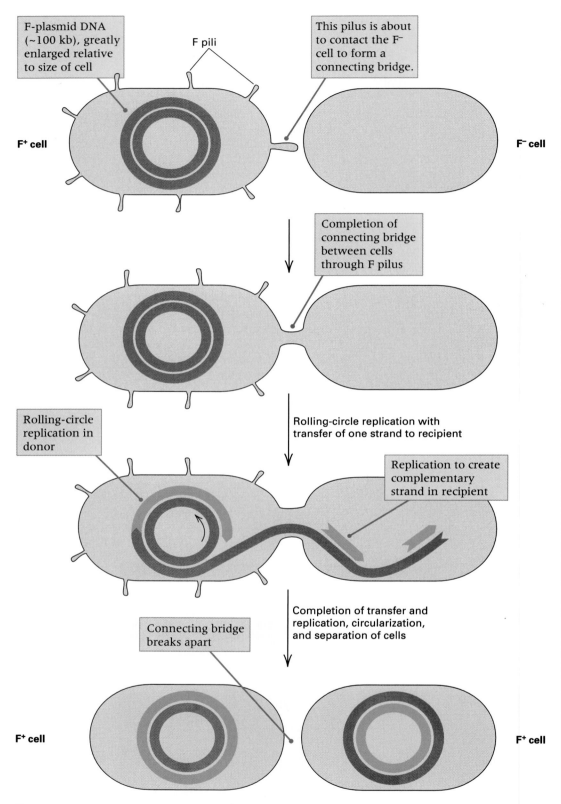

F-plasmid DNA (~100 kb), greatly enlarged relative to size of cell

F pili

This pilus is about to contact the F⁻ cell to form a connecting bridge.

F⁺ cell

F⁻ cell

Completion of connecting bridge between cells through F pilus

Rolling-circle replication with transfer of one strand to recipient

Rolling-circle replication in donor

Replication to create complementary strand in recipient

Completion of transfer and replication, circularization, and separation of cells

Connecting bridge breaks apart

F⁺ cell

F⁺ cell

Figure 8.7
Transfer of F from an F⁺ to an F⁻ cell. Pairing of the cells triggers rolling-circle replication. Pink represents DNA synthesized during pairing. For clarity, the bacterial chromosome is not shown, and the plasmid is drawn overly large; the plasmid is in fact much smaller than a bacterial chromosome.

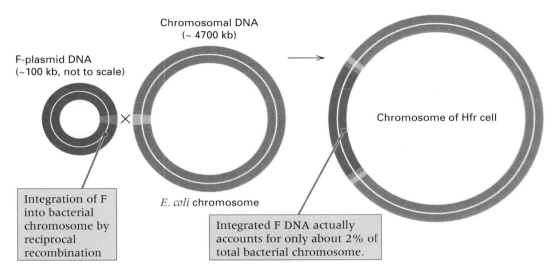

F-plasmid DNA
(~100 kb, not to scale)

Chromosomal DNA
(~ 4700 kb)

Chromosome of Hfr cell

Integration of F into bacterial chromosome by reciprocal recombination

E. coli chromosome

Integrated F DNA actually accounts for only about 2% of total bacterial chromosome.

Figure 8.8

Integration of F (blue circle) by recombination between a nucleotide sequence in F and a homologous sequence in the bacterial chromosome. The F plasmid DNA is shown greatly enlarged relative to the size of the bacterial chromosome. In reality, the length of the DNA molecule in F is about 2 percent of the length of the bacterial chromosome.

requires only a few minutes. In laboratory cultures, if a small number of donor cells is mixed with an excess of recipient cells, F spreads throughout the population in a few hours, and all cells ultimately become F^+. Transfer is not so efficient under natural conditions, and only about 10 percent of naturally occurring *E. coli* cells contain the F factor.

Hfr Cells

The F plasmid occasionally becomes integrated into the *E. coli* chromosome by an exchange between a sequence in F and a sequence in the chromosome (Figure 8.8). The bacterial chromosome remains circular, though enlarged about 2 percent by the F DNA. Integration of F is an infrequent event, but single cells containing integrated F can be isolated and cultured. The cells in such a strain are called **Hfr cells.** Hfr stands for **high frequency of recombination,** which refers to the relatively high frequency with which donor genes are transferred to the recipient. Integrated F mediates the transfer of DNA from the bacterial chromosome in an Hfr cell; thus a replica of part of the bacterial chromosome,

as well as part of the plasmid, is transferred to the F^- cell.

The Hfr $\times$ F^- conjugation process is illustrated in Figure 8.9. The stages of transfer are much like those by which F is transferred to F^- cells: coming together of donor and recipient cells, rolling-circle replication in the donor cell, and conversion of the transferred single-stranded DNA into double-stranded DNA by lagging-strand synthesis in the recipient. However, in the case of Hfr matings, the transferred DNA does not become circular and is not capable of further replication in the recipient because the transferred F factor is not complete. The replication and associated transfer of the chromosomal DNA are controlled by the integrated F and are initiated in the Hfr chromosome at the same point in F at which replication and transfer begin within an unintegrated F plasmid. A part of F is the first DNA transferred, chromosomal genes are transferred next, and the remaining part of F is the last DNA to enter the recipient. Because the conjugating cells usually break apart long before the entire bacterial chromosome is transferred, the final segment of F is almost never transferred into the recipient.

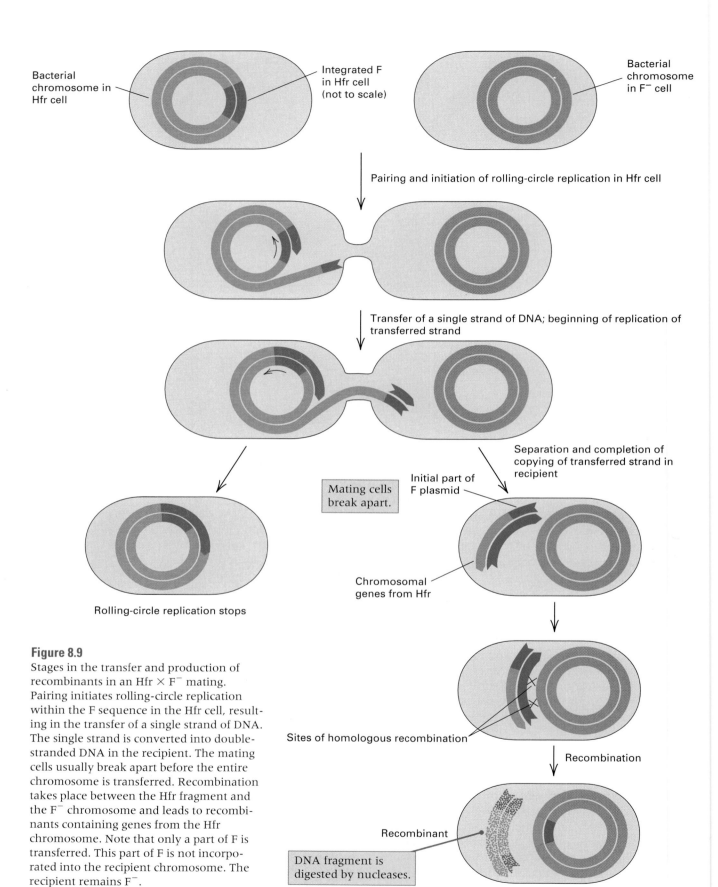

Bacterial chromosome in Hfr cell

Integrated F in Hfr cell (not to scale)

Bacterial chromosome in F⁻ cell

Pairing and initiation of rolling-circle replication in Hfr cell

Transfer of a single strand of DNA; beginning of replication of transferred strand

Mating cells break apart.

Separation and completion of copying of transferred strand in recipient

Initial part of F plasmid

Chromosomal genes from Hfr

Rolling-circle replication stops

Sites of homologous recombination

Recombination

Recombinant

DNA fragment is digested by nucleases.

Figure 8.9
Stages in the transfer and production of recombinants in an Hfr × F⁻ mating. Pairing initiates rolling-circle replication within the F sequence in the Hfr cell, resulting in the transfer of a single strand of DNA. The single strand is converted into double-stranded DNA in the recipient. The mating cells usually break apart before the entire chromosome is transferred. Recombination takes place between the Hfr fragment and the F⁻ chromosome and leads to recombinants containing genes from the Hfr chromosome. Note that only a part of F is transferred. This part of F is not incorporated into the recipient chromosome. The recipient remains F⁻.

Several differences between F transfer and Hfr transfer are notable.

- It takes 100 minutes under the usual conditions for an entire bacterial chromosome to be transferred, in contrast with about 2 minutes for the transfer of F. The difference in time is a result of the relative sizes of F and the chromosome (100 kb versus 4700 kb).

- During transfer of Hfr DNA into a recipient cell, the mating pair usually breaks apart before the entire chromosome is transferred. Under usual conditions, several hundred genes are transferred before the cells separate.

- In a mating between Hfr and F$^-$ cells, the F$^-$ recipient remains F$^-$ because cell separation usually takes place before the final segment of F is transferred.

- In Hfr transfer, some regions in the transferred DNA fragment become incorporated into the recipient chromosome. The incorporated regions replace homologous regions in the recipient chromosome. The result is that some F$^-$ cells become recombinants containing one or more genes from the Hfr donor cell. For example, in a mating between Hfr leu^+ and F$^-$ leu^-, some F$^-$ leu^+ cells arise. However, *the genotype of the donor Hfr cell remains unchanged.*

Genetic analysis requires that recombinant recipients be identified. Because the recombinants derive from recipient cells, a method is needed to eliminate the donor cells. The usual procedure is to employ an F$^-$ recipient containing an allele that can be selected. Genes that confer antibiotic resistance are especially useful for this purpose. For instance, after a mating between Hfr leu^+ str-s and F$^-$ leu^- str-r cells, the Hfr Str-s cells can be selectively killed by plating the mating mixture on medium containing streptomycin. A selective medium that lacks leucine can then be used to distinguish between the nonrecombinant and the recombinant recipients. The F$^-$ leu^- parent cannot grow in medium that lacks leucine, but recombinant F$^-$ leu^+ cells can grow because they possess a leu^+ gene. Only recombinant recipients—that is, cells having the genotype leu^+ str-r—form colonies on a selective medium containing streptomycin and lacking leucine. The selected allele should be located at such a place in the chromosome that most mating cells will have broken apart before the selected gene is transferred, and the selected allele must not be present in the Hfr cell. The selective agent can then be used to select the F$^-$ cells and eliminate the Hfr donors.

When a mating is done in this way, the transferred marker that is selected by the growth conditions (leu^+ in this case) is called a **selected marker,** and the marker used to prevent growth of the donor (str-s in this case) is called the **counterselected marker.** Selection and counterselection are necessary in bacterial matings because recombinants constitute only a small proportion of the entire population of cells (in spite of the name "high frequency of recombination").

Time-of-Entry Mapping

Genes can be mapped by Hfr × F$^-$ matings. However, the genetic map is quite different from all maps that we have seen so far in that it is not a linkage map but a transfer-order map. It is obtained by deliberate interruption of DNA transfer in the course of mating—for example, by violent agitation of the suspension of mating cells in a kitchen blender. The time at which a particular gene is transferred can be determined by breaking the mating cells apart at various times and noting the earliest time at which breakage no longer prevents recombinants from appearing. This procedure is called the **interrupted-mating technique.** When this is done with Hfr × F$^-$ matings, the number of recombinants of any particular allele increases with the time during which the cells are in contact. This phenomenon is illustrated in Table 8.1. The reason for the increase is that different Hfr × F$^-$ pairs initiate conjugation and chromosome transfer at slightly different times.

A greater understanding of the transfer process can be obtained by observing the

Table 8.1 Data showing the production of Leu⁺ Str-r recombinants in a cross between Hfr *leu⁺ str-s* and F⁻ *leu⁻ str-r* cells when mating is interrupted at various times

Minutes after mating	Number of Leu⁺ Str-r recombinants per 100 Hfr cells
0	0
3	0
6	6
9	15
12	24
15	33
18	42
21	43
24	43
27	43

Note: Minutes after mating means minutes after the Hfr and F⁻ cell suspensions are mixed. Extrapolation of the recombination data to a value of zero recombinants indicates that the earliest time of entry of the *leu⁺* marker is 4 minutes.

results of a mating with several genetic markers. For example, consider the mating

Hfr $a^+ b^+ c^+ d^+ e^+$ *str-s* ×

F⁻ $a^- b^- c^- d^- e^-$ *str-r*

in which a^- cells require nutrient A, b^- cells require nutrient B, and so forth. At various times after mixing of the cells, samples are agitated violently and then plated on a series of media that contain streptomycin and different combinations of the five substances A through E (in each medium, one of the five is left out). Colonies that form on the medium lacking A are a^+ *str-r*, those growing without B are b^+ *str-r*, and so forth. All of these data can be plotted on a single graph to give a set of curves, as shown in Figure 8.10A. Four features of this set of curves are notable.

1. The number of recombinants in each curve increases with length of time of mating.

2. For each marker, there is a time (the

time of entry) before which no recombinants are detected.

3. Each curve has a linear region that can be extrapolated back to the time axis, defining the time of entry of each gene $a^+, b^+, \ldots, e^+$.

4. The number of recombinants of each type reaches a maximum, the value of which decreases with successive times of entry.

The explanation for the time-of-entry phenomenon is the following: All donor cells do not start transferring DNA at the same time, so the number of recombinants increases with time. Transfer begins at a particular point in the Hfr chromosome (the replication origin of F). Genes are transferred in linear order to the recipient, and the time of entry of a gene is the time at which that gene first enters a recipient in the population. Separation of a mating pair prevents further transfer and limits the number of recombinants seen at a particular time.

The times of entry of the genes used in the mating just described can be placed on a map, as shown in Figure 8.10B. The numbers on this map and the others are genetic distances between the markers, *measured as minutes between their times of entry*. Mating with another F⁻ with genotype $b^- e^- f^- g^- h^-$ *str-r* could be used to locate the three genes *f, g,* and *h*. Data for the second recipient would yield a map such as that shown in Figure 8.10C. Because genes *b* and *e* are common to both maps, the two maps can be combined to form a more comprehensive map, as shown in Figure 8.10D.

Studies with different Hfr strains (Figure 8.10E) also are informative. It is usually found that different Hfr strains are distinguishable by their origins and directions of transfer, which indicates that F can integrate at numerous sites in the chromosome and in two different orientations. Combining the maps obtained with different Hfr strains yields a composite map that is *circular*, as illustrated in Figure 8.10F. The circularity of the map is a result of the circularity of the *E. coli* chromosome in F⁻ cells and the multiple points of integration

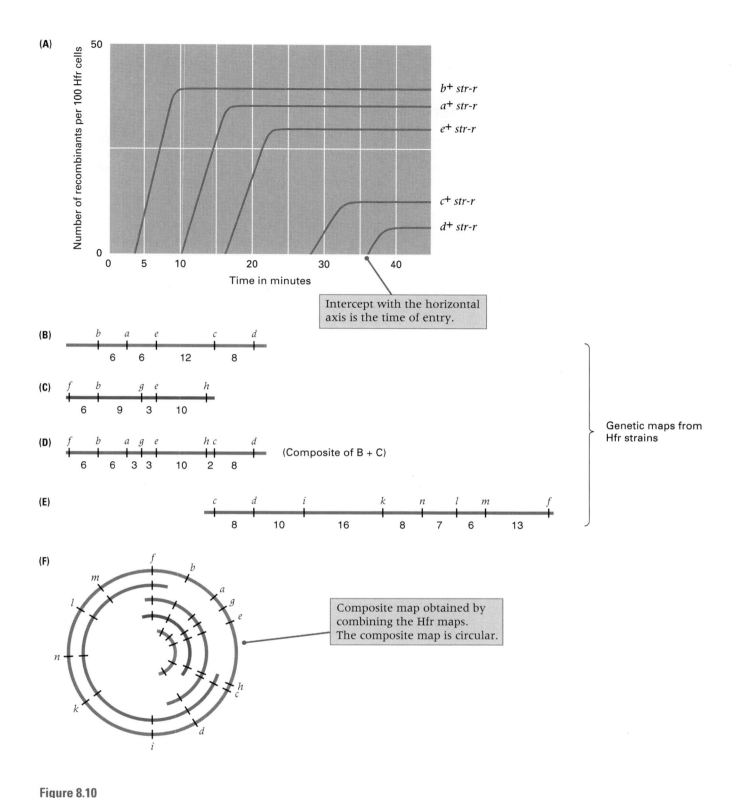

Figure 8.10

Time-of-entry mapping. (A) Time-of-entry curves for one Hfr strain. (B) The linear map derived from the data in part A. (C) A linear map obtained with the same Hfr but with a different F⁻ strain containing the alleles $b^-\ e^-\ f^-\ g^-\ h^-$. (D) A composite map formed from the maps in parts B and C. (E) A linear map from another Hfr strain. (F) The circular map (gold) obtained by combining the two maps (green and blue) of parts D and E.

The Sex Life of Bacteria

Joshua Lederberg and Edward L. Tatum
1946

Yale University,
New Haven, Connecticut
Gene Recombination in Escherichia coli

Since their discovery in the nineteenth century, bacteria were considered "things apart"—unlike other organisms in fundamental ways. Lederberg and Tatum's discovery of what at first appeared to be a conventional sexual cycle was a sensation, completely unexpected. It meant that bacteria could be considered "genetic organisms" along with yeast, Neurospora, Drosophila, and other genetic favorites. For this and related discoveries, Lederberg and Tatum were awarded the 1958 Nobel Prize along with George W. Beadle. In this excerpt, you will note that the authors discuss bacterial recombination as requiring a cell fusion that would bring both parental genomes together. This interpretation shows that it is possible to make exactly the right observation, and realize its significance, but not quite grasp what is really going on. The conclusion that bacterial recombination involved unidirectional transfer was reached much later, after the discovery of Hfr strains and the development of the interrupted-mating technique.

Analysis of mixed cultures of nutritional mutants has revealed the presence of new types which strongly suggest the occurrence of a sexual process in the bacterium *Escherichia coli*. The mutants consist of strains which differ from their parent wildtype strain K-12, in lacking the ability to synthesize growth factors. As a result of these deficiencies they will only grow in media supplemented with their specific nutritional requirements. In these mutants single nutritional requirements are established as single mutational steps under the influence of x-rays or ultraviolet light. By successive treatments, strains

---

> *These types can most reasonably be interpreted as instances of the assortment of genes in new combinations*

---

with several requirements have been obtained. In the recombination studies here reported, two triple mutants have been used, one requiring threonine, leucine and thiamin, the other requiring biotin, phenylalanine and cystine. The strains were grown in mixed culture in complete medium. The cells were washed with sterile water and inoculated heavily into synthetic agar medium, to which various supplements had been added to allow the growth of colonies of various nutritional types. This procedure readily allows the detection of very small numbers of cell types different from the parental forms. The only new types found in "pure" cultures of the individual mutants were occasional forms which had reverted for a single factor, giving strains which required only two of the original three substances. In mixed cultures, however, a variety of types has been found. These include wildtype strains with no growth-factor deficiencies and single mutant types requiring only thiamin or phenylalanine. . . . These types can most reasonably be interpreted as instances of the assortment of genes in new combinations. In order that various genes may have the opportunity to recombine, a cell fusion would be required. . . . The fusion presumably occurs only rarely, since in the cultures investigated only one cell in a million can be classified as a recombinant type. . . . These experiments imply the occurrence of a sexual process in the bacterium *Escherichia coli*.

Source: Nature 158: 558

of the F plasmid; if F could integrate at only one site and in one orientation, the map would be linear.

A great many such mapping experiments have been carried out, and the data have been combined to provide an accurate map of approximately 2000 genes throughout the *E. coli* chromosome. Figure 8.11 is a map of the chromosome of *E. coli* containing a sample of the mapped genes. Both the DNA molecule and the genetic map are circular. The entire chromosome requires 100 minutes to be transferred (it usually breaks first), so the total map length is 100 minutes. In the outer circle, the arrows indicate the direction of transcription and the coding region included in each transcript. The purple arrowheads show the origin and direction of transfer of a number of Hfr strains. Transfer from HfrC, for example, goes counterclockwise starting with *purE acrA lac.*

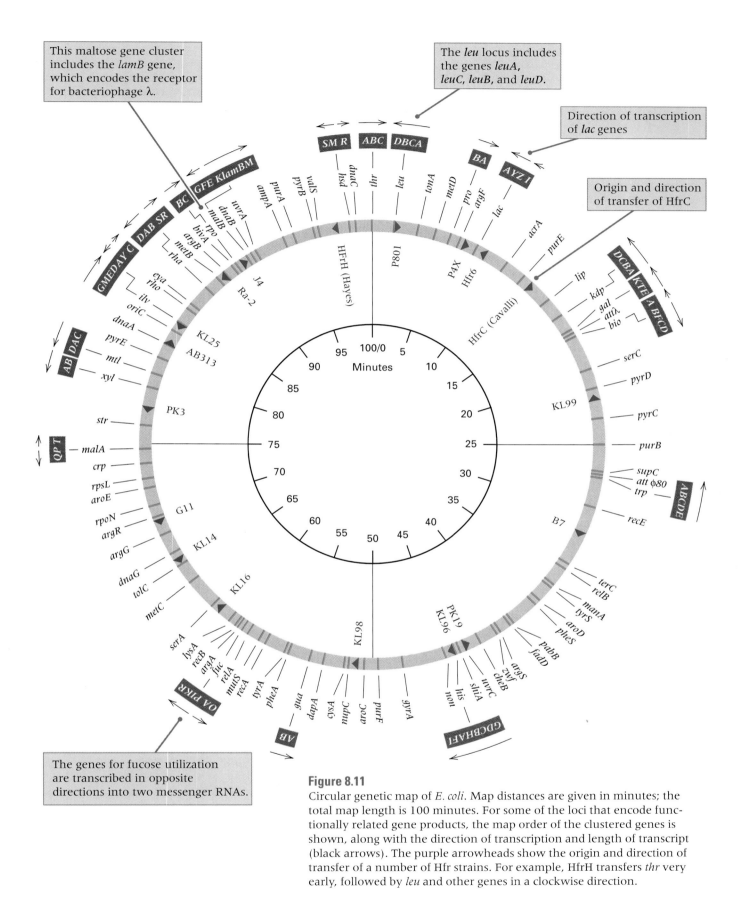

This maltose gene cluster includes the *lamB* gene, which encodes the receptor for bacteriophage λ.

The *leu* locus includes the genes *leuA*, *leuC*, *leuB*, and *leuD*.

Direction of transcription of *lac* genes

Origin and direction of transfer of HfrC

The genes for fucose utilization are transcribed in opposite directions into two messenger RNAs.

Figure 8.11

Circular genetic map of *E. coli*. Map distances are given in minutes; the total map length is 100 minutes. For some of the loci that encode functionally related gene products, the map order of the clustered genes is shown, along with the direction of transcription and length of transcript (black arrows). The purple arrowheads show the origin and direction of transfer of a number of Hfr strains. For example, HfrH transfers *thr* very early, followed by *leu* and other genes in a clockwise direction.

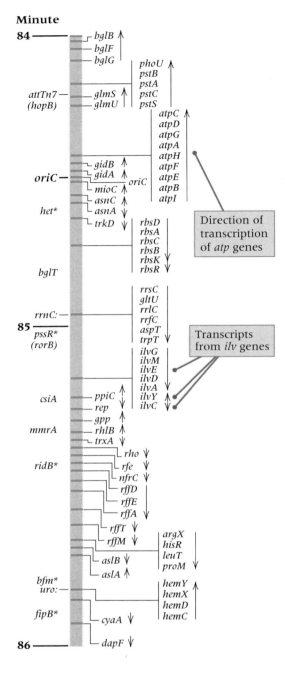

Minute

Figure 8.12
A more dense genetic map showing genes in the region between 84 and 86 minutes of the *E. coli* chromosome. Arrows indicate the direction of transcription and length of the messenger RNA for each gene or gene cluster. Symbols to the left of the light blue bar indicate untranscribed DNA sites and genes for which the direction of transcription is not known. For example, *attTn7* is an insertion site for the transposon Tn7. [From M. K. B. Berlyn, K. B. Low, and K. E. Rudd. 1996. In *Escherichia coli and Salmonella: Cellular and Molecular Biology,* 2nd ed. (F. C. Neidhardt, R. Curtiss III, J. L. Ingraham, E. C. C. Lin, K. B. Low, B. Magasanik, W. Reznikoff, M. Riley, M. Schaechter, and H. E. Umbarger, eds.) Washington, DC: American Society for Microbiology.]

unknown. The symbol *oriC* represents the origin of DNA replication of the *E. coli* chromosomes.

F' Plasmids

Occasionally, F is excised from Hfr DNA by an exchange between the same sequences used in the integration event. However, in some cases, the excision process is not a precise reversal of integration. Instead, breakage and reunion take place between nonhomologous sequences at the boundary of F and nearby chromosomal DNA (Figure 8.13). Aberrant excision creates a plasmid containing a fragment of chromosomal DNA, which is called an **F' plasmid** ("F prime"). By the use of Hfr strains having different origins of transfer, F' plasmids with chromosomal segments from many regions of the chromosome have been isolated. These elements are extremely useful because they render any recipient cell diploid for the region of the chromosome carried by the plasmid. These diploid regions make possible dominance tests and gene-dosage tests (studies of the effects on gene expression of increasing the number of copies of a gene). Because only a part of the genome is diploid, cells containing an F' plasmid are **partial diploids,** also called **merodiploids.** Examples of genetic analysis using F' plasmids will be offered in Chapter 11 in a discussion of the *E. coli lac* genes.

Greater detail of part of the *E. coli* map for the 2 minutes between minutes 84 and 86 is shown in Figure 8.12. Genes and gene clusters transcribed as a unit are indicated by the arrows pointing in the direction of transcription, shown to the right of the blue line. To the left of the line are either regions that are not transcribed or genes in which the direction of transcription is

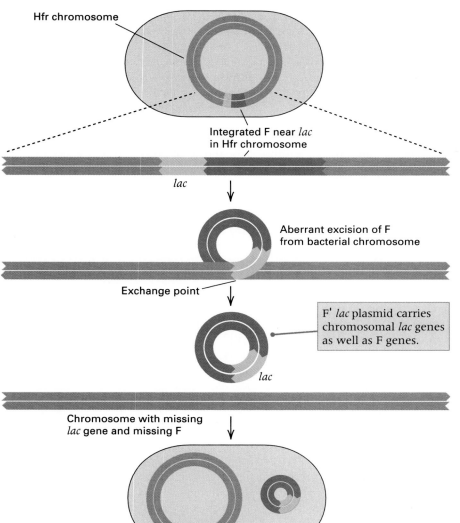

Hfr chromosome

Integrated F near *lac* in Hfr chromosome

lac

Aberrant excision of F from bacterial chromosome

Exchange point

F' *lac* plasmid carries chromosomal *lac* genes as well as F genes.

lac

Chromosome with missing *lac* gene and missing F

8.5 Transduction

In the process of **transduction,** a bacterial DNA fragment is transferred from one bacterial cell to another by a phage particle containing the bacterial DNA. Such a particle is called a **transducing phage.** Two types of transducing phages are known—generalized and specialized. A **generalized transducing phage** produces some particles that contain only DNA obtained from the host bacterium, rather than phage DNA; the bacterial DNA fragment can be derived from *any* part of the bacterial chromosome. A **specialized transducing phage** produces particles that contain both phage and bacterial genes linked in a single phage DNA molecule, and the bacterial genes are obtained from a *particular* region of the bacterial chromosome. In this section, we consider *E. coli* phage P1, a well-studied generalized transducing phage. Specialized transducing particles will be discussed in Section 8.7.

During infection by P1, the phage makes a nuclease that cuts the bacterial DNA into fragments. Single fragments of bacterial DNA comparable in size to P1 DNA are occasionally packaged into phage particles in place of P1 DNA. The positions of the nuclease cuts in the host chromosome are random, so a transducing particle may contain a fragment derived from any region of the host DNA. A large population of P1 phages will contain a few particles carrying any bacterial gene. On the average, any particular gene is present in roughly one transducing particle per 10^6 viable phages. When a transducing particle adsorbs to a bacterium, the bacterial DNA contained in the phage head is injected into the cell and becomes available for recombination with the homologous region of the host chromosome. A typical P1 transducing particle contains from 100 kb to 115 kb of bacterial DNA.

Let us now examine the events that follow infection of a bacterium by a generalized transducing particle obtained, for example, by growth of P1 on wildtype *E. coli* containing a *leu*+ gene (Figure 8.14). If such a particle adsorbs to a bacterial cell of *leu*− genotype and injects the DNA that it contains into the cell, then the cell survives because the phage head contained only bacterial genes and no phage genes. A recombination event exchanging the *leu*+ allele carried by the phage for the *leu*− allele carried by the host converts the genotype of the host cell from *leu*− into *leu*+. In such an experiment, typically about 1 *leu*− cell in 10^6 becomes *leu*+. Such frequencies are easily detected on selective growth medium. For example, if the infected cell is placed on solid medium that lacks leucine, it is able to multiply and a *leu*+ colony forms. A colony does not form unless recombination inserted the *leu*+ allele.

The fragment of bacterial DNA contained in a transducing particle is large enough to include about 50 genes, so transduction provides a valuable tool for genetic linkage studies of short regions of the bacterial genome. Consider a population of P1 prepared from a *leu*+ *gal*+ *bio*+ bacterium. This sample contains particles able to transfer any of these alleles to another cell; that is, a *leu*+ particle can transduce a *leu*− cell to *leu*+, or a *gal*+ particle can transduce a *gal*− cell to *gal*+. Furthermore, if a *leu*− *gal*− culture is infected with phage, both *leu*+ *gal*− and *leu*− *gal*+ bacteria are produced. However, *leu*+ *gal*+ colonies do not arise because the *leu* and *gal* genes are too far apart to be included in the same DNA fragment (Figure 8.15A).

The situation is quite different with a recipient cell with genotype *gal*− *bio*−, because the *gal* and *bio* genes are so closely linked that both genes are sometimes present in a single DNA fragment carried in a transducing particle—namely, a *gal*+-*bio*+ particle (Figure 8.15B). However, not all *gal*+ transducing particles also include *bio*+, nor do all *bio*+ particles include *gal*+. The probability of both markers being in a single particle—and hence the probability of simultaneous transduction of both markers (**cotransduction**)—depends on how close to each other the genes are. The closer they are, the greater the frequency of cotransduction. Cotransduction of the *gal*+-*bio*+ pair can be detected by plating infected cells on the appropriate growth medium. If *bio*+ transductants are selected by spreading the infected cells on a glucose-containing medium that lacks biotin, then both *gal*+ *bio*+ and *gal*− *bio*+ colonies will grow. If these colonies are tested for the *gal* marker, then 12 percent are found to be *gal*+ *bio*+ and the rest *gal*− *bio*+; similarly, if *gal*+ transductants are selected, then about 12 percent are found to be *gal*+ *bio*+. In other words, the **frequency of cotransduction** of *gal* and *bio* is 12 percent, which means that 12 percent of all transducing particles that contain one gene also include the other.

Studies of cotransduction can be used to map closely linked genetic markers by means of three-factor crosses analogous to those described in Chapter 4. That is, P1 is grown on wildtype bacteria and used to transduce cells that carry a mutation of each of three closely linked genes. Cotransductants containing various pairs of wildtype alleles are examined. The gene located in the middle can be identified because its wildtype allele is almost always cotransduced with the wildtype alleles of the genes that flank it. For example, in Figure 8.15B, a genetic marker located between *gal*+ and *bio*+ will almost always be present in *gal*+ *bio*+ transductants.

How is the frequency of cotransduction between genes related to their map distance in minutes in the standard *E. coli* map? The theoretical relation depends on the size of a molecule in a transducing phage relative to the size of the entire chromosome. For bacteriophage P1, the distance between

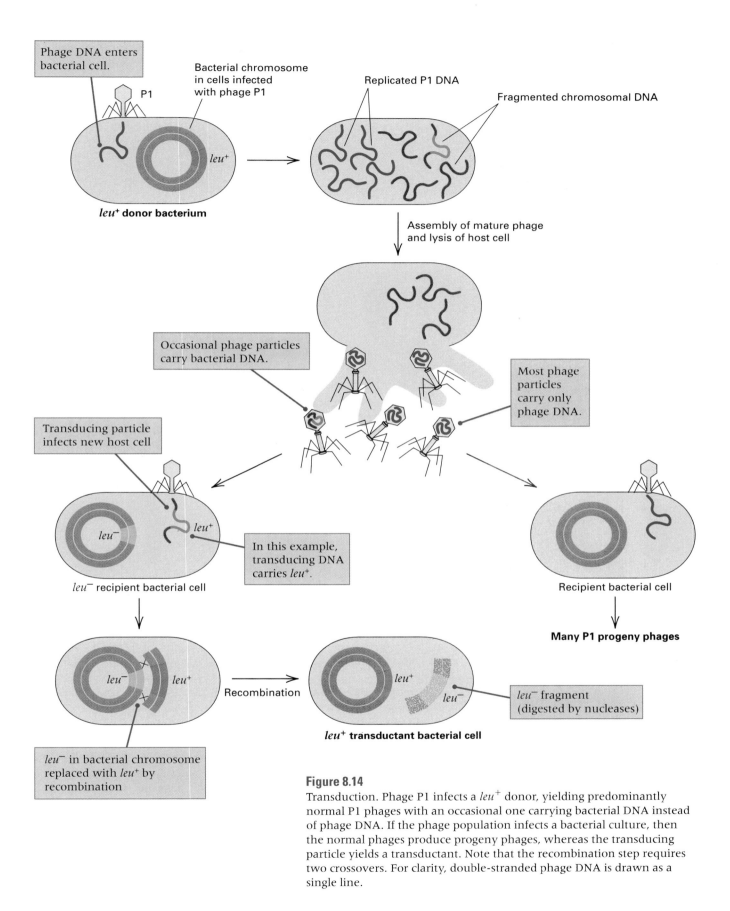

Figure 8.14

Transduction. Phage P1 infects a *leu*⁺ donor, yielding predominantly normal P1 phages with an occasional one carrying bacterial DNA instead of phage DNA. If the phage population infects a bacterial culture, then the normal phages produce progeny phages, whereas the transducing particle yields a transductant. Note that the recombination step requires two crossovers. For clarity, double-stranded phage DNA is drawn as a single line.

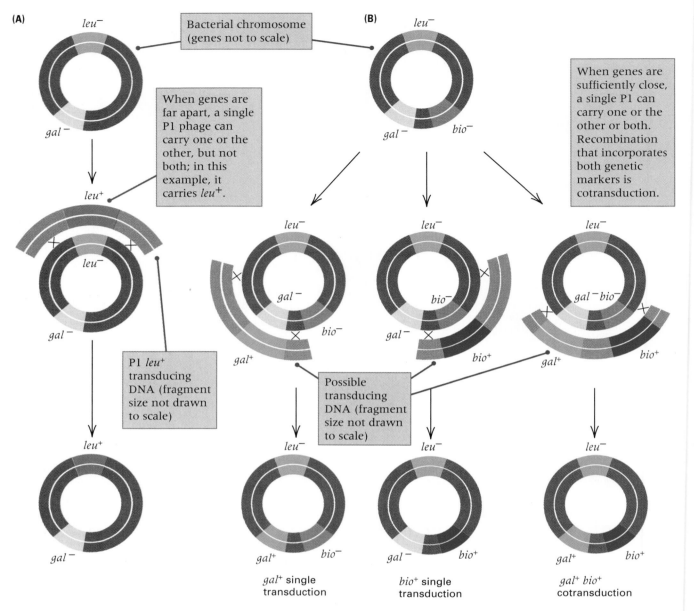

Figure 8.15
Demonstration of linkage of the *gal* and *bio* genes by cotransduction. (A) A P1 transducing particle carrying the *leu*⁺ allele can convert a *leu*⁻ *gal*⁻ cell into a *leu*⁺ *gal*⁻ genotype (but cannot produce a *leu*⁺ *gal*⁺ genotype). (B) The transductants that could be formed by three possible types of transducing particles—one carrying *gal*⁺, one carrying *bio*⁺, and one carrying the linked alleles *gal*⁺ *bio*⁺. The third type results in cotransduction. The distance between *gal* and *bio*, relative to that between *leu* and *gal*, is greatly exaggerated, and the size of the DNA fragment in a transducing particle, relative to the size of the bacterial chromosome, is not drawn to scale.

markers in minutes on a conjugational map is related to the frequency of cotransduction approximately as follows:

Map distance in minutes =
$$2 - 2 \times (\text{Cotransduction frequency})^{1/3}$$

The implication of this equation is that the frequency of cotransduction falls off very rapidly as the distance between the markers increases. In the formula, the shortest distance for which the cotransduction frequency equals 0 is 2 minutes; hence any markers separated by more than 2 minutes will not be cotransduced by phage P1. For *gal* and *bio*, which have a cotransduction frequency of about 12 percent, the distance equals 1 minute (see Figure 8.11).

Is a Bacteriophage an "Organism"?

Alfred D. Hershey and Raquel Rotman 1948
Washington University,
St. Louis, Missouri
*Genetic Recombination Between Host-range
and Plaque-type Mutants of Bacteriophage
in Single Bacterial Cells*

Are bacteriophages "organisms"? Well . . . yes and no. If by the term organism *you mean a cell or group of cells that function together as a whole to maintain life and its activities, then bacteriophages are not organisms. By* organism *you might also mean "anything resembling a living thing in its complexity of structure and function." In this sense bacteriophages are organisms. Even though they require a bacterial host to survive and reproduce, their reproductive cycle conforms in all respects to that of cellular organisms, including DNA as their genetic material, semiconservative replication of DNA, heritable mutations, and, as shown first in this report, a process of recombination. Hershey and Rotman were quite unsure whether phage recombination resulted from physical exchange of DNA molecules or whether, when two molecules recombined, both reciprocal products of the exchange were produced. Later research showed that the exchanges are reciprocal and that they are physical exchanges.*

We have shown that any two of several mutants of the bacterial virus T2 interact with each other, in bacterial cells infected with both, to give rise to wildtype and double mutant genetic recombinants. . . . In principle, the experimental technique we have to describe is very similar to genetic crossing. One starts with a pair of mutants, each corresponding to a mutant haploid germ cell differing from wildtype by a different unit change. Bacterial cells are infected with both members of the pair, and during viral growth the pair interact to produce viral progeny corresponding to germ cells of a new generation, but now including some individuals differing from wildtype

---

The analogy to other genetic recombination is obvious, and it is natural to look for a common mechanism.

---

by both unit changes, and other individuals differing from wildtype not at all. The analogy to other genetic recombination is obvious, and it is natural to look for a common mechanism. . . . The crosses between *h* and *r* mutants yield the linkage system shown below.

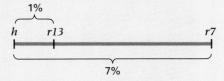

In the diagram, *h* refers to bacterial host range and *r13* and *r7* to two different, rapidly lysing mutants. The data show that

the *h* locus is very closely linked to *r13* (less than 1 percent of wildtype), and that the linkage relation between *r13* and *r7* is 6 percent. . . . The results further show that the two recombinants [wildtype and double mutant] appear in equal numbers in any one cross, and that pairs of reverse crosses yield equal numbers of recombinants. It is these relations that increase the resemblance to simple types of Mendelian segregation. . . . A hypothesis is proposed according to which one visualizes genetic interaction not between two viral particles, but between two sets of independently multiplying chromosome-like structures. Genetic exchange occurs either by reassortment of these structures, or by something like crossing-over between homologous pairs, depending on the structural relation between the genetic factors concerned. The interpretation made brings the linkage relations into superficial agreement with the requirements of linear structure, but there is little evidence that the genetic exchanges are reciprocal, and accordingly little evidence that they are material exchanges.

Source: Genetics 34: 44–71

8.6
Bacteriophage Genetics

The life cycles of phages fit into two distinct categories—the lytic and the lysogenic cycles. In the **lytic cycle,** phage nucleic acid enters a cell and replicates repeatedly, the bacterium is killed, and hundreds of phage progeny result (see Figure 8.3). All phage species can undergo a lytic cycle; a phage capable *only* of lytic growth is called

virulent. In the alternative, **lysogenic cycle,** no progeny particles are produced, the bacterium survives, and a phage DNA molecule is transmitted to each bacterial daughter cell. In most cases, transmission of this kind is accomplished by integration of the phage chromosome into the bacterial chromosome. A phage capable of such a life cycle is called **temperate.**

Several phage particles can infect a single bacterium, and the DNA of each particle can replicate. In a multiply infected cell in which a lytic cycle is underway, there are genetic exchanges between phage DNA molecules. If the infecting particles carry different mutations as genetic markers, then recombinant phages result. Recombination in phage may be **general recombination,** which means that it can take place anywhere along two homologous DNA molecules. General recombination will be considered in this section. In temperate phages there is a second type of recombination called **site-specific recombination** because it takes place only between a particular site in the phage and a particular site in the bacterial

chromosome. The end result of site-specific recombination is not the production of recombinant phage chromosomes but the joining of phage and bacterial DNA molecules. This process is described in Section 8.7.

Plaque Formation and Phage Mutants

Phages are easily detected because in a lytic cycle, an infected cell breaks open (**lysis**) and releases phage particles to the growth medium. The formation of plaques can be observed as outlined in Figure 8.16A. A large number of bacteria (about 10^8) are

(A)

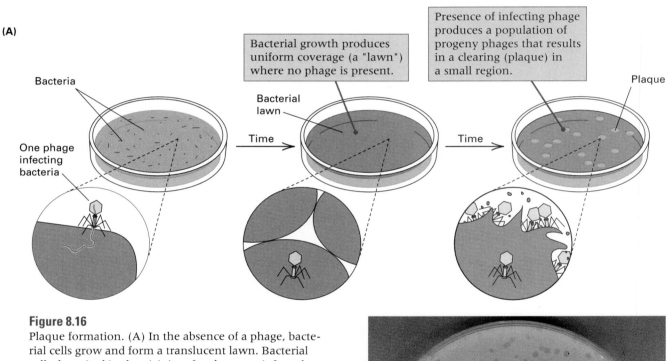

Bacteria

One phage infecting bacteria

Bacterial growth produces uniform coverage (a "lawn") where no phage is present.

Bacterial lawn

Time

Presence of infecting phage produces a population of progeny phages that results in a clearing (plaque) in a small region.

Plaque

Time

Figure 8.16

Plaque formation. (A) In the absence of a phage, bacterial cells grow and form a translucent lawn. Bacterial cells deposited in the vicinity of a phage are infected and lyse. Progeny phages diffusing outward from the original site infect other cells and cause their lysis. Because of phage infection and lysis, no bacteria can grow in a small region around the site of each phage particle originally present in the medium. The area devoid of bacteria remains transparent and is called a plaque. (B) Large plaques in a lawn of *E. coli* formed by infection with a mutant of bacteriophage λ. Each plaque results from an initial infection by a single bacteriophage.

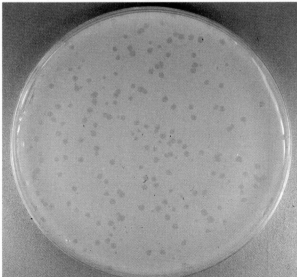

(B)

placed on a solid medium. After a period of growth, a continuous turbid layer of bacteria results. If a phage is present at the time the bacteria are placed on the medium, it adsorbs to a cell, and shortly afterward, the infected cell lyses and releases many phages. Each of these progeny adsorbs to a nearby bacterium, and after another lytic cycle, these bacteria in turn release phages that can infect still other bacteria in the vicinity. (Progeny phage remain close to their site of origin because their size prevents diffusion in agar medium.) These cycles of infection continue, and after several hours, the phages destroy all of the bacteria in a localized area, giving rise to a clear, transparent region—a **plaque**—in the otherwise turbid layer of confluent bacterial growth. Phages can multiply only in growing bacterial cells, so exhaustion of nutrients in the growth medium limits phage multiplication and the size of the plaque. Because a plaque is a result of an initial infection by one phage particle, the number of individual phages originally present on the medium can be counted.

The genotypes of phage mutants can be determined by studying the plaques. In some cases, the appearance of the plaque is sufficient. For example, phage mutations that decrease the number of phage progeny from infected cells often yield smaller plaques. Large plaques can be produced by mutants that cause premature lysis of infected cells, so that each round of infection proceeds more quickly (Figure 8.16B). Another type of phage mutation can be identified by the ability or inability of the phage to form plaques on a particular bacterial strain.

Genetic Recombination in Virulent Bacteriophages

If two phage particles with different genotypes infect a single bacterium, then some phage progeny are genetically recombinant. Figure 8.17 shows plaques from progeny of a mixed infection with *E. coli* phage T4 mutants. The r^- (*rapid lysis*) allele results in large plaques, and the h^- (*host range*) allele results in clear plaques. The cross is written as

$$r^- \, h^+ \text{ (large turbid plaque)} \times$$
$$r^+ \, h^- \text{ (small clear plaque)}$$

Four plaque types can be seen in Figure 8.17. Two—the large turbid plaque and the small clear plaque—correspond to the phenotypes of the parental phages. The other two phenotypes—the large clear plaque

Figure 8.17
A phage cross is performed by infecting host cells with both parental types of phage simultaneously. This example shows the progeny of a cross between T4 phages of genotypes $r^- \, h^+$ and $r^+ \, h^-$ when both parental phages infect cells of *E. coli*. The arrowheads point to plaques formed from progeny phages of the indicated genotypes. [Courtesy of Leslie Smith and John W. Drake.]

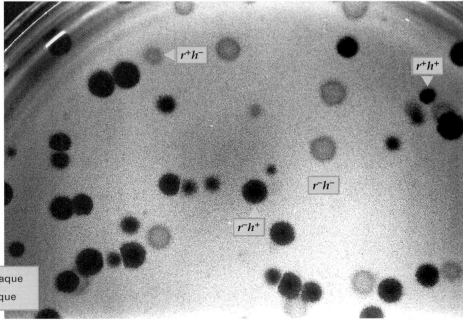

r^+ = small plaque	h^+ = turbid plaque
r^- = large plaque	h^- = clear plaque

Figure 8.18

Circular genetic map of phage T4. Each arc connects three or four markers that were mapped in the crosses indicated. The circular map results when all data are considered together. [After G. Streisinger, R. S. Edgar, and G. H. Denhardt. 1964. *Proc. Natl. Acad. Sci. USA* 51: 775.]

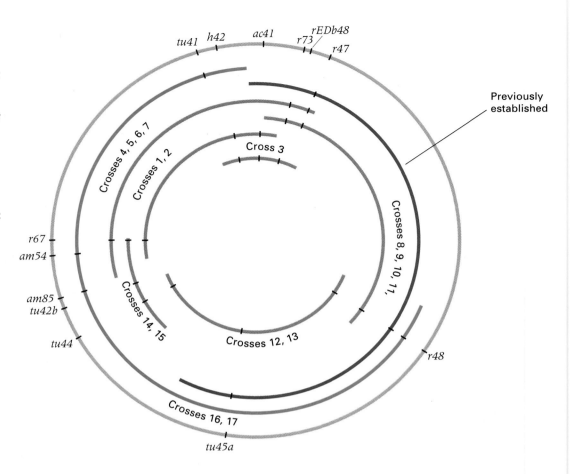

and the small turbid plaque—are recombinants that correspond to the genotypes $r^- h^-$ and $r^+ h^+$, respectively. When many bacteria are infected, approximately equal numbers of reciprocal recombinant types are usually found among the progeny phage. In an experiment like that in Figure 8.17, in which each of the four genotypes yields a different phenotype of plaque morphology, the number of each of the genotypes can be counted by examining each of the plaques that is formed. The recombination frequency, expressed as a percentage, is defined as

$$\text{Recombination frequency} = \frac{\text{Number of recombinant phage}}{\text{Total number of phage}} \times 100$$

Recombination frequencies can be used to estimate map distances, just as they are in eukaryotes. Early mapping experiments indicated that mutations in T4 mapped in three separate clusters. However, all three clusters showed linkage to one another. In elegant experiments with three-point crosses, George Streisinger and colleagues demonstrated in 1964 that the genetic map for T4 phage is actually circular.

In each cross, they mapped three or four genetic markers with respect to one another and proceeded systematically through the entire T4 genome, eventually demonstrating all of the linkages shown in Figure 8.18. Many additional genes were identified and mapped later by other researchers (Figure 8.19), and the results were fully consistent with the circular map. In Figure 8.19, the regions indicated in the innermost circle are the three clusters of T4 markers that had initially been identified and mapped. The outermost circle in Figure 8.19 presents a much larger set of markers

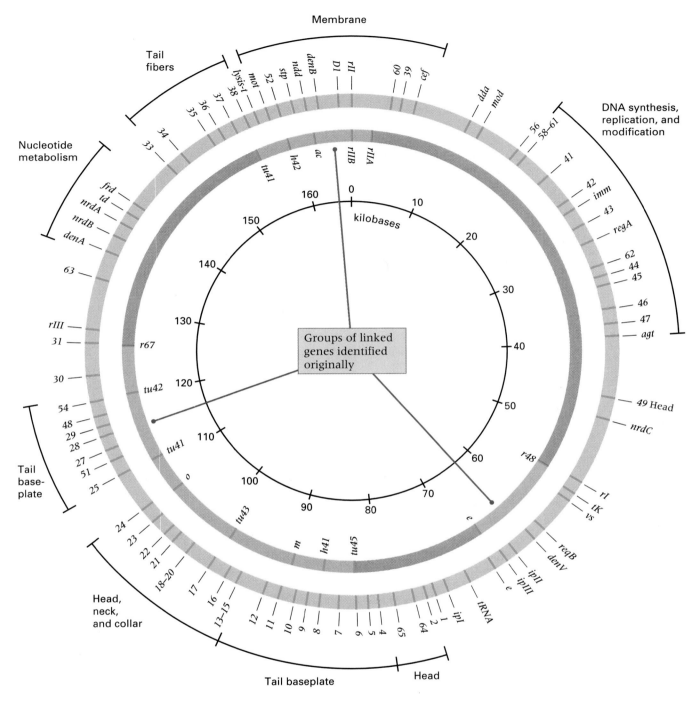

Figure 8.19
Circular genetic map of bacteriophage T4 with additional markers. The middle circle contains the markers mapped and used in the earliest T4 experiments. The units along the innermost circle are kilobases; 0 kb is at a fixed point between the *rIIA* and *rIIB* cistrons. The outer circle contains many genes defined by so-called conditional mutations, which cause a mutant phenotype under one condition (for example, at a high temperature) but not under other conditions (for example, at a low temperature). Many of the genes show clustering according to their functions. [After W. B. Wood and H. R. Revel. 1976. *Bacteriol. Rev.* 40:847.]

Seymour Benzer 1955
Purdue University,
West Lafayette, Indiana
Fine Structure of a Genetic Region in Bacteriophage

Just as every fan of "Star Wars" can identify the lovable robot Artoo-Detoo (a.k.a. R2-D2), every geneticist can identify the gene rII. The rII gene in bacteriophage T4 was the first experimental example of genetic fine structure. Benzer used the special property that rII *mutants cannot grow on* E. coli *strain K but can grow on strain B to examine recombination between different nucleotides within the* rII *gene. He demonstrated that the* rII *gene was divisible by recombination. (It is now known that, in principle, recombination can take place between any adjacent nucleotides.) But if the gene can be subdivided by recombination, then what is a gene, anyway? If two different mutations can undergo recombination whether or not they are in the same gene, then how can one decide, experimentally, whether two different mutations are, or are not, alleles? Benzer realized that the key experimental operation in the definition of allelism was not recombination but rather the complementation test. This is a rare paper with two great ideas in it: recombination within a gene, and the use of the complementa-tion test to determine experimentally whether two different mutations are, or are not, alleles of the same gene.*

The phenomenon of genetic recombination provides a powerful tool for separating mutations and discerning their positions along a chromosome. When it comes to very close neighboring mutations, a difficulty arises, since the closer two mutations lie to one another, the smaller is the probability that recombination between them will occur. Therefore, failure to observe recombinant types ordinarily does not justify the conclusion that the two mutations are inseparable. . . . A high degree of resolution can best be achieved if there is available a selective feature for the detection of small proportions of recombinants. Such a feature is offered by the case of the *rII* mutants of T4 bacteriophage described in this paper. The wildtype phage produces plaques on either of two bacterial hosts, *Escherichia coli* strain B or strain K, while a mutant of the *rII* group produces plaques only on B. Therefore, if a cross is made between two different *rII* mutants any wildtype recombinants which arise, even in proportions as low as 10^{-8}, can be detected by plating on strain K. . . . In this way, a series of eight *rII* mutants of T4 have been crossed with each other. The results of these crosses are given in the figure below.

The distances are only roughly additive; there is some systematic deviation in the sense that a long distance tends to be smaller than the sum of its component shorter ones, [which is accounted for by multiple recombination events]. . . . Thus, while all *rII* mutants in this set fall into a small portion of the phage linkage map, it is possible to seriate [order] them unambiguously, and their positions *within* the region are well scattered. . . . *Test for allelism.* The functional relatedness of two closely linked mutations causing similar defects may be tested by constructing diploid heterozygotes containing the two mutations. . . . The *trans* form, containing one of the mutations in each chromosome, may or may not produce the wild phenotype. If it does [complementation], it is concluded that the two mutations in question are located in separate functional units. [They are alleles of different genes.] . . . In order to characterize a unit of genetic "function," it is necessary to define what function is meant. . . . On the basis of phenotype tests of *trans* configuration heterozygotes [complementation tests], the *rII* region can be subdivided into two functionally separable segments. Each segment may have the "function" of specifying the sequence of amino acids in a polypeptide chain.

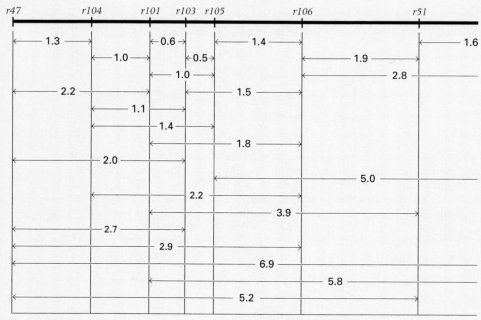

Source: *Proceedings of the National Academy of Sciences of the USA* 41: 344–354

establishing the overall circularity of the genetic map beyond doubt. The T4 genetic map in Figure 8.19 also indicates that genes in T4 show extensive clustering according to the function for which they are required. For example, there is a large cluster of genes for DNA replication in the upper right quadrant, and there is a cluster of genes for phage head components near the bottom.

In view of the circular nature of the T4 genetic map, it came as quite a surprise to find that the DNA molecule in a T4 phage particle is a single *linear* molecule. This discovery was completely unexpected and at first seemed inconsistent with the genetic data. However, the discrepancy was resolved by the finding that the very ends of the phage T4 DNA are duplicated, or have **terminal redundancy.** Because of the redundancy, each molecule is about 2 percent longer than would be expected. As the DNA is replicating inside the cell, recombination between each of the duplicated ends of one T4 genome with homolo-gous sequences in other T4 genomes results in products much longer than can be contained in a T4 head (Figure 8.20). Such concatenated molecules are formed because recombination in the T4 genome is very frequent, averaging about 20 recombination events per chromosome. When the DNA is packaged, it is enzymatically cleaved into "headful" packages consisting of about 102 percent of the minimum length of the T4 genome; hence the duplication at the ends. Because of the headful packaging mechanism, each T4 DNA molecule is terminally redundant. Moreover, in a whole population of DNA molecules present in a phage sample, except for the short terminal redundancy, each of the molecules is also a **circular permutation** of the others. (A molecule is a circular permutation of another if the other sequence can be created by joining its ends and cleaving at an internal position.) Because each T4 molecule is a circular permutation, different molecules begin at different points in the DNA sequence, but they

Figure 8.20
Terminal redundancy and circular permutation of the phage T4 linear DNA molecule. A long concatenate, or series of molecules linked together, is formed inside the cell by genetic recombination. Each successive headful of DNA is slightly longer than unit size and contains a terminal redundancy. Except for the terminal redundancy, the molecules are also circular permutations of each other.

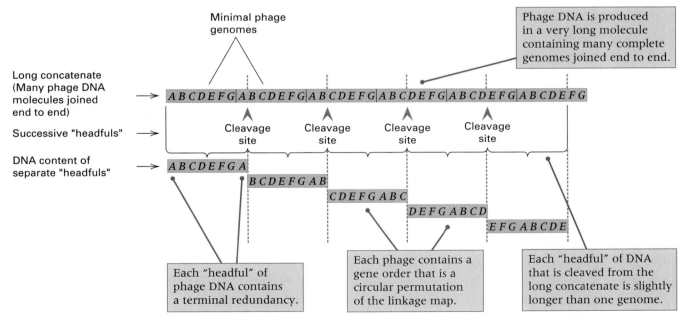

always incorporate a little more than one complete genome. The properties of terminal redundancy and circular permutation account for the circular genetic map.

Fine Structure of the *rII* Gene in Bacteriophage T4

In Section 4.7, we described experiments with *Drosophila* that showed that recombination within a gene could take place. A genotype heterozygous for two different mutant alleles of the same gene could, at low frequency, generate recombinant chromosomes carrying either both mutations or neither mutation. These experiments gave the first indication of intragenic recombination and the first evidence that genes have fine structure. Other studies of genes were performed in the years that followed this experiment, but none could equal the fine-structure mapping of the *rII* gene in bacteriophage T4 carried out by Seymour Benzer. Using novel genetic mapping techniques that reduced the number of required crosses from more than half a million to several thousand, Benzer succeeded in mapping 2400 independent mutations in the *rII* locus of phage T4.

Wildtype T4 bacteriophage is able to multiply in *E. coli* strains B and K12(λ) and gives small ragged plaques. Mutations in the *rII* gene of T4 yield large round plaques on strain B but are completely unable to propagate in strain K12(λ). If *E. coli* cells of strain B are all infected with two different *rII* mutants, then recombination between the mutants can be detected, even if the frequency is very low, by taking advantage of the inability of *rII* mutants to grow on K12(λ). Plating the progeny phages on K12(λ) selects for growth of the *rII*$^+$ recombinant progeny, because only these recombinants can grow. Furthermore, because very large numbers of progeny phage can be examined (numbers of 10^{10} bacteriophages/ml are not unusual), even very low frequencies of recombination can be detected. Typical results for three different *rII* mutations might yield a map like the following:

where the numbers above the line designate *rII* alleles and the numbers below are map distances given as the percent of *rII*$^+$, namely

$$\frac{\text{Map}}{\text{Distance}} = \frac{\text{Number of plaques per phage on strain K12 (λ)}}{\text{Number of plaques per phage on strain B}} \times 100$$

Some mutations failed to recombine with several mutations, each of which recombined with the others. These were interpreted to be deletion mutations, because they prevented recombination with two or more "point" mutations known to be at different sites in the gene. Each deletion eliminated a part of the bacteriophage genome, including a region of the *rII* gene. The use of deletions greatly simplified the ordering and mapping of thousands of mutations.

Figure 8.21 depicts the array of deletion mutations used for mapping. Deletion mapping is based on the presence or absence of recombinants; each cross yields a yes-or-no answer and so avoids many of the ambiguities of genetic maps based on frequencies of recombination. In any cross between an unknown "point" mutation (for example, a simple nucleotide substitution) and one of the deletions, the presence of wildtype progeny means that the point mutation is outside of the region missing in the deletion. (The reciprocal product of recombination, which carries the deletion plus the point mutation, is not detected in these experiments.) On the other hand, if the point mutation is present in the region missing in the deletion, then wildtype recombinant progeny cannot be produced. Because each cross clearly reveals whether a particular mutation is within the region missing in the deletion, deletion mapping also substantially reduces the amount of work needed to map a large number of mutations.

The series of crosses that would be made in order to map a particular *rII* mutation is presented in Figure 8.22, in which the large intervals A1 through A6 plus B are the same as those in Figure 8.21. To illustrate the method, suppose that a particular mutation being examined is located in the region denoted A4. This mutation would

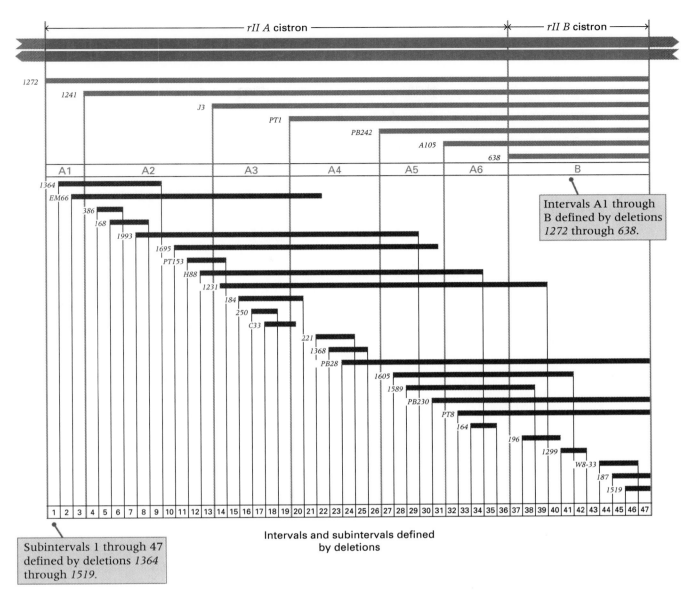

Figure 8.21
The array of deletion mutations used to divide the *rII* locus of bacteriophage T4 into 7 regions and 47 smaller subregions. The extent of each deletion is indicated by a horizontal bar. Any deletion endpoint used to establish a boundary between regions or subregions is indicated with a vertical line. [After S. Benzer. 1961. *Proc. Natl. Acad. Sci. USA* 47:403.]

fail to yield wildtype recombinants in crosses with the large deletions *r1272*, *r1241*, *rJ3*, and *rPT1*, but it would yield wildtype recombinant progeny in crosses with the large deletions *rPB242*, *rA105*, and *r638*. Conversely, any mutation that yielded the same pattern of outcomes in crosses with the large deletions would be assigned to region A4. Still finer resolution of the

genetic map within region A4 is made possible by the set of deletions shown at the bottom of Figure 8.22, the endpoints of which define seven subregions (a through g) within A4. For example, a mutation in region A4 that yields wildtype recombinants with the deletion *r1368* but not with *r221* would be assigned to the c subregion. At the finest level of resolution, mutations

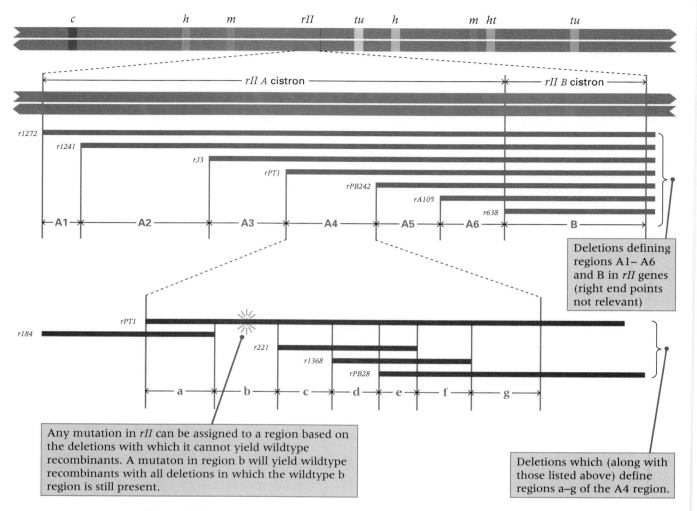

Figure 8.22

The location of *rII* in relation to other genetic markers in a linear version of the phage T4 genetic map. Major subdivisions of the *rII* region are defined by the left ends of seven deletion mutations. All seven deletions extend through the right end of the *rII* region. Genetic material deleted is indicated in red. Further subdivision of the A4 region is made possible by the additional deletions shown at the bottom of the figure. For simplicity, only the deleted regions are shown.

within a subregion are ordered by crosses among themselves. In phage T4, mutant sites that are very close can be separated by recombination, because on the average, 1 percent recombination corresponds to a distance of about 100 base pairs. Hence, any two mutations that fail to recombine can be assigned to the same site within the gene. The genetic map generated for a large number of independent *rII* mutations is given in Figure 8.23.

The *rII* mutation and mapping studies were important because they gave experimental support to these conclusions:

- Genetic exchange can take place within a gene and probably between any pair of adjacent nucleotides.

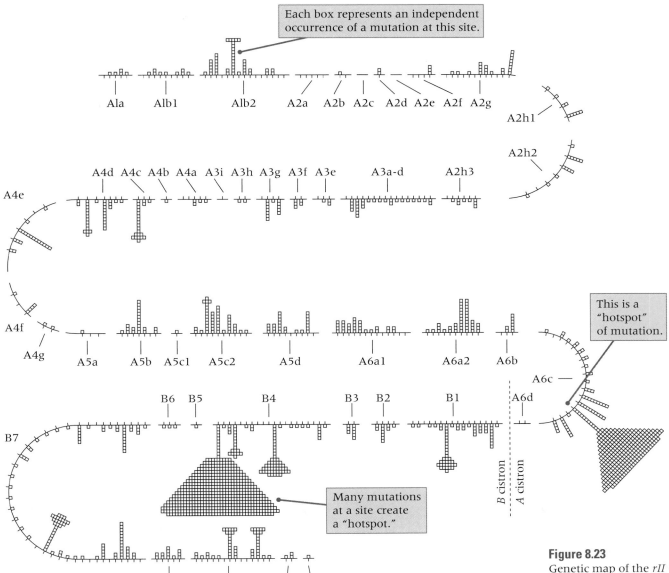

Each box represents an independent occurrence of a mutation at this site.

This is a "hotspot" of mutation.

Many mutations at a site create a "hotspot."

Figure 8.23
Genetic map of the *rII* locus of phage T4. Each small square indicates a separate, independent occurrence of a mutation at the indicated site. The arrangement of sites within each A or B segment is arbitrary. [After S. Benzer. 1961. *Proc. Natl. Acad. Sci. USA* 47:403.]

• Mutations are not produced at equal frequencies at all sites within a gene. For example, the 2400 *rII* mutations were located at only 304 sites. One of these sites accounted for 474 mutations (Figure 8.23); a site that shows such a high frequency of mutation is called a **hotspot** of mutation. At sites other than hotspots, mutations were recovered only once or a few times.

The *rII* analysis was also important because it helped to distinguish experimentally between three distinct meanings of the word *gene*. Most commonly, the word *gene* refers to a unit of function. Physically, this corresponds to a protein-coding segment of DNA. Benzer assigned the term **cistron** to this unit of function, and the term is still occasionally used. The unit of function is normally defined experimentally by a complementation test (see Sections 2.2 and 4.8), and indeed, two units of function, the *rIIA* cistron and the *rIIB* cistron, were defined by complementation. The limits of *rIIA* and *rIIB* are shown in Figures 8.21 and 8.23. The complementation between *rIIA* and *rIIB* is observed when two types of T4 phage, one with a mutation in *rIIA* and the other with a mutation in *rIIB*, are used simultaneously to infect *E. coli* strain K12(λ). The multiply infected cells produce normal numbers of phage progeny, most of which carry either the parental *rIIA* mutation or the parental *rIIB* mutation. However, rare recombination between the mutant sites results in recombinant progeny phage that are *rII*$^+$. In contrast, when K12(λ) is simultaneously infected with two phage having different *rIIA* mutations, or two phage having different *rIIB* mutations, no progeny phage are produced.

Besides the meaning of function, clarified by use of the term *cistron*, the term *gene* has two other distinct meanings: (1) the unit of genetic transmission that participates in recombination and (2) the unit of genetic change or mutation. Physically, both the recombinational and the mutational units correspond to the individual nucleotides in a gene. Despite potential ambiguity, the term *gene* is still the most important word in genetics, and in most cases, the shade of meaning intended is clear from the context.

8.7
Genetic Recombination in Temperate Bacteriophages

Temperate bacteriophages have two alternative life cycles—a lytic cycle and a lysogenic cycle. The lytic cycle is depicted in Figure 8.3. A temperate phage, such as *E. coli* phage λ, when reproducing in its lytic cycle, undergoes general recombination, much as phage T4 does. A map of the phage λ genome is depicted in Figure 8.24; it is linear rather than circular. The DNA molecule in the λ phage particle is also linear. Unlike the DNA molecules in T4 phage, however, every phage λ DNA molecule has identical ends. Indeed, the ends are single-stranded and complementary in sequence so that they can pair, forming a circular molecule.

The single-stranded ends are called **cohesive ends** to indicate their ability to undergo base pairing. The packaging of DNA in phage λ does not follow a headful mechanism, like T4. Rather, the λ packaging process recognizes specific sequences that are cleaved to produce the cohesive ends.

The complete DNA sequence of the genome of λ phage has been determined, and many of the genes and gene products and their functions have been identified. The map of genes in Figure 8.24 shows where each gene is located along the DNA molecule, scaled in kilobase pairs (kb), rather than in terms of its position in the genetic map. However, a genetic map with coordinates in map units (centimorgans, or percent recombination) has been placed directly below the molecular scale. Comparing the scales reveals that the frequency of recombination is not uniform along the molecule. For example, the 5 map units between genes *H* and *I* span about 3.1 kb, whereas the 5 map units between the genes *int* and *cIII* span about 4.8 kb.

The λ map is also interesting for what it implies about the evolution of the phage genome. The genes in λ show extensive clustering by function. The left half of the map consists entirely of genes whose products (head and tail proteins) are required for assembly of the phage structure, and within this region, the head genes and the tail genes themselves form subclusters. The right half of the λ genome also shows several gene clusters, which include genes for DNA replication, recombination, and lysis. The genes are clustered not only by function but also according to the time at which their products are synthesized. For example, the N gene acts early; genes *O* and *P* are active later; and genes *Q, S, R,* and the head-tail cluster are expressed last. The transcription patterns for mRNA synthesis are thus very simple and efficient. There are only two rightward transcripts, and all late genes except for *Q* are transcribed into the same mRNA.

Lysogeny

The lysogenic cycle comes about when the λ DNA molecule becomes integrated into the bacterial chromosomes, where it is passively replicated as part of the bacterial

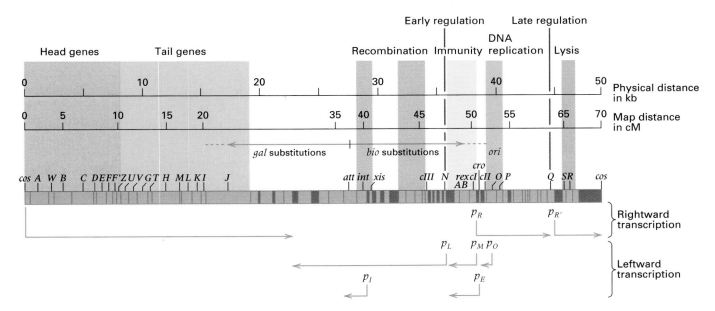

Figure 8.24

Molecular and genetic maps of bacteriophage λ. The scale of the molecular map is length in kilo-base pairs (kb); genetic distances are given in centimorgans (cM), equivalent in this case to percent recombination. Clusters of genes with related functions are indicated by the pastel colored boxes. The positions of the regulatory genes *N* and *Q* are indicated by vertical lines at about 35 and 45 kb, respectively. Promoters are indicated by the letter *p*. The direction of transcription and length of transcript are indicated by arrows. Many of the genes known through mutant phenotypes are identified. Coding regions of unidentified function are represented as unlabeled, light-colored rectangles; dark-colored regions indicate sequences unlikely to code for proteins. The λ *att* (attachment site) is the site of recombination for integration of the λ prophage. The origin of replication (*ori*) is the region denoted *O*. The extents of possible substitutions of λ DNA with *E. coli* DNA in the specialized transducing phages λ*dgal* and λ*dbio* are indicated by arrows just above the gene map; such transducing phages are discussed in the section on specialized transduction.

chromosome; phage particles are not produced (Figure 8.25). The inserted DNA is called a **prophage,** and the surviving cell is called a **lysogen.** A strain lysogenic for λ has the symbol (λ) appended to its name. For example, the strain *E. coli* K12(λ) that made possible the fine-structure analysis of the *rII* region of phage T4 was a K12 strain that had become lysogenic for λ.

As noted, the cohesive ends of the λ DNA molecule are single-stranded, with 12 unpaired bases at either end. The ends are complementary, however. Upon entering the cell, the complementary ends anneal to form a nicked circle, and ligation seals the nicks (Figure 8.26). Circularization, which takes place early in both the lytic and lysogenic cycles, is a necessary event in both cycles: for DNA replication in the lytic

mode, for prophage integration in the lysogenic cycle. In about 75 percent of infected cells, the circular molecule replicates, and the lytic cycle ensues. However, in about 25 percent of infected cells, the circular λ molecule and the circular *E. coli* DNA molecule interact and undergo a single, site-specific recombination event in which the phage DNA becomes incorporated into the bacterial chromosome. Because λ can exist either as an autonomous genetic element (in the lytic cycle) or as an integrated element in the chromosome (in lysogeny), λ, like the F factor, is classified as an episome.

The positions of the site-specific recombination in the bacterial and phage DNA are called the **bacterial** and **phage attachment sites,** respectively. Each attachment site consists of three segments. The central

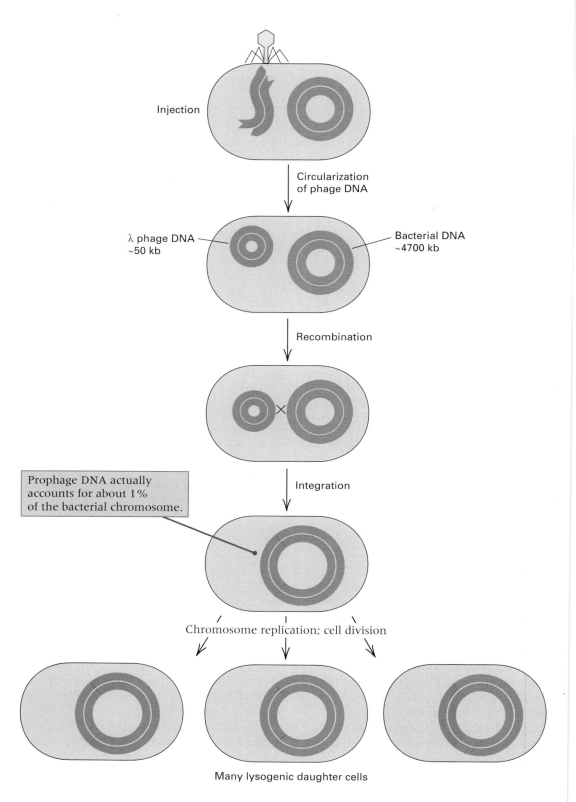

Injection

Circularization
of phage DNA

λ phage DNA
~50 kb

Bacterial DNA
~4700 kb

Recombination

Prophage DNA actually
accounts for about 1%
of the bacterial chromosome.

Integration

Chromosome replication; cell division

Many lysogenic daughter cells

Figure 8.25
The general mode of lysogenization by integration of phage DNA into the bacterial chromosome.
Some genes (those needed to establish lysogeny) are expressed shortly after infection and are then
turned off. The inserted brown DNA is the prophage. For clarity, the phage DNA is drawn much
larger than to scale; the size of phage λ DNA is actually about 1 percent of the size of the *E. coli*
genome.

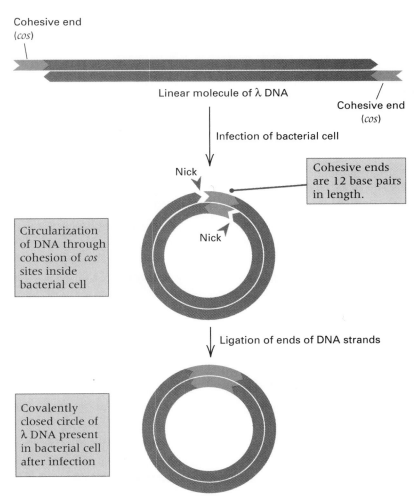

Cohesive end
(*cos*)

Linear molecule of λ DNA

Cohesive end
(*cos*)

Infection of bacterial cell

Nick

Cohesive ends
are 12 base pairs
in length.

Nick

Circularization
of DNA through
cohesion of *cos*
sites inside
bacterial cell

Ligation of ends of DNA strands

Covalently
closed circle of
λ DNA present
in bacterial cell
after infection

Figure 8.26
A diagram of a linear λ DNA molecule showing the cohesive ends (complementary single-stranded ends). Circularization by means of base pairing between the cohesive ends forms an open (nicked) circle, which is converted into a covalently closed (uninterrupted) circle by sealing (ligation) of the single-strand breaks. The length of the cohesive ends is 12 base pairs in a total molecule of approximately 50 kb.

segment has the same nucleotide sequence in both attachment sites and is the region in which the recombination actually takes place. The phage attachment site is denoted *POP'* (*P* for phage), and the bacterial attachment site is denoted *BOB'* (*B* for bacteria). A comparison of the genetic maps of the phage and the prophage indicates that *POP'* is located near the middle of the linear form of the phage DNA molecule (see Figure 8.24). A phage protein, **integrase,** catalyzes a site-specific recombination event; the integrase recognizes the phage and bacterial attachment sites and causes the physical exchange that results in integration of the λ DNA molecule into the bacterial DNA. The geometry of the exchange is shown in Figure 8.27. As a result of the recombination event, the genetic map of the prophage is not the same as the map of the phage; it is a circular permutation of it that arises from

the central location of λ *att* (the *POP'* site) and the circularization of the DNA.

The correct model for prophage integration was first suggested by Allan Campbell in 1962. The model was confirmed by bacterial crosses of lysogens with nonlysogens, as well as by transduction by phage P1. Campbell found that the order of genes in the integrated prophage was

N   *mi R A m6*   J
_____

where *m6* is a mutation affecting head formation and *mi* is one affecting lysis. In contrast, the gene order of genes in the free phage as determined by general recombination is

A m6   *J att N*   *mi R*
_____

Figure 8.27

The geometry of integration and excision of phage λ. The phage attachment site is *POP'*. The bacterial attachment site is *BOB'*. The prophage is flanked by two hybrid attachment sites denoted *BOP'* and *POB'*.

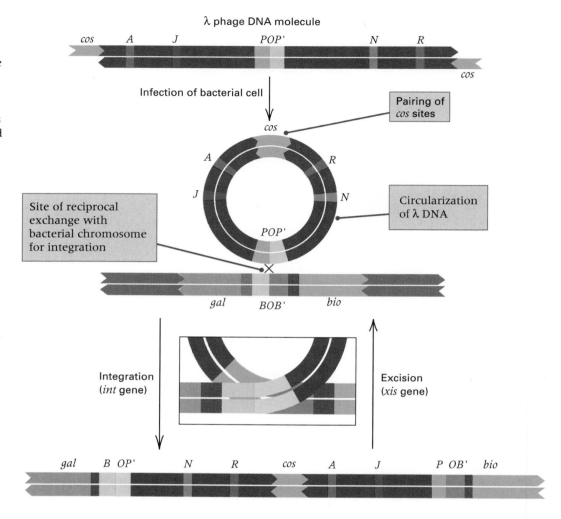

λ phage DNA molecule

Infection of bacterial cell

Pairing of *cos* sites

Circularization of λ DNA

Site of reciprocal exchange with bacterial chromosome for integration

Integration (*int* gene)

Excision (*xis* gene)

The prophage map is thus a circular permutation of the map of the free phage. The prophage is inserted into the *E. coli* chromosome between the genes *gal* and *bio*, as indicated in Figure 8.28. An additional finding that confirmed the physical insertion of λ was that the integrated prophage increased the distance between *gal* and *bio* so that these markers could no longer be cotransduced by phage P1. The distance between *gal* and *bio* in a λ lysogen is about 2 minutes, compared to 1 minute in a nonlysogen.

When a cell is lysogenized, the phage genes become part of the bacterial chromosome, so it might be expected that the phenotype of the bacterium would change. But most phage genes in a prophage are kept in an inactive state by a **repressor** protein, the product of one of the phage genes. The repressor protein is synthesized initially by the infecting phage and then continually by the prophage. The gene that codes for the repressor is frequently the only prophage gene that is expressed in lysogens. If a lysogen is infected with a phage of the same type as the prophage—for example, λ infecting a λ lysogen—then the repressor present within the cell from the prophage prevents expression of the genes of the infecting phage. This resistance to infection by a phage identical with the prophage, which is called **immunity,** is the usual criterion for determining whether a bacterial cell contains a particular prophage. For example, λ will not form

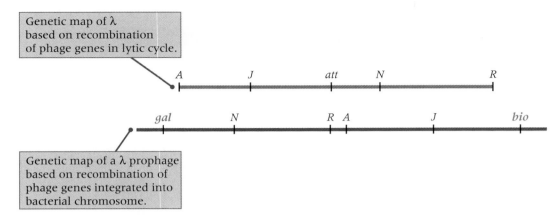

Genetic map of λ based on recombination of phage genes in lytic cycle.

A   J   att   N   R

gal   N   R A   J   bio

Genetic map of a λ prophage based on recombination of phage genes integrated into bacterial chromosome.

Figure 8.28
The map order of genes in phage λ as determined by phage recombination (lytic cycle) and in the prophage (prophage order). The genes have been selected arbitrarily to provide reference points. Bacterial DNA is labeled in red letters.

plaques on bacteria that contain a λ prophage.

A lysogenic cell can replicate nearly indefinitely without the release of phage progeny. However, the prophage can sometimes become activated to undergo a lytic cycle in which the usual number of phage progeny are produced. This phenomenon is called **prophage induction,** and it is initiated by damage to the bacterial DNA. The damage sometimes happens spontaneously but is more often caused by some environmental agent, such as chemicals or radiation. The ability to be induced is advantageous for the phage because the phage DNA can escape from a damaged cell. The biochemical mechanism of induction is complex and will not be discussed, but the excision of the phage is straightforward.

Excision is another site-specific recombination event that reverses the integration process. Excision requires the phage enzyme integrase plus an additional phage protein called **excisionase.** Genetic evidence and studies of physical binding of purified excisionase, integrase, and λ DNA indicate that excisionase binds to integrase and thereby enables the latter to recognize the prophage attachment sites *BOP'* and *POB'*; once bound to these sites, integrase makes cuts in the *O* sequence and recreates the *BOB'* and *POP'* sites. This reverses the integration reaction, causing excision of the prophage (Figure 8.27).

Specialized Transducing Phage

When a bacterium lysogenic for phage λ is subjected to DNA damage that leads to induction, the prophage is usually excised from the chromosome precisely. However, in about 1 cell per 10^6 to 10^7 cells, an excision error is made (Figure 8.29), and a chance breakage in two nonhomologous sequences takes place—one break within the prophage and the other in the bacterial DNA. The free ends of the excised DNA are then joined to produce a DNA circle capable of replication. The sites of breakage are not always located so as to produce a length of DNA that can fit in a λ phage head, and the DNA may be too large or too small. Sometimes, however, a molecule forms that can replicate and be packaged. In λ lysogens, the prophage lies between the *gal* and *bio* genes, and because the aberrant cut in the host DNA can be either to the right or to the left of the prophage, particles can arise that carry either the *bio* genes (cut at the right) or the *gal* genes (cut at the left). The resulting phage are called λ*bio* and λ*gal* transducing particles. These are **specialized transducing phages** because they can transduce only certain bacterial genes (*gal* or *bio*), in contrast with the P1-type generalized transducing particles, which can transduce any gene.

Often the specialized transducing phages are defective, because essential λ

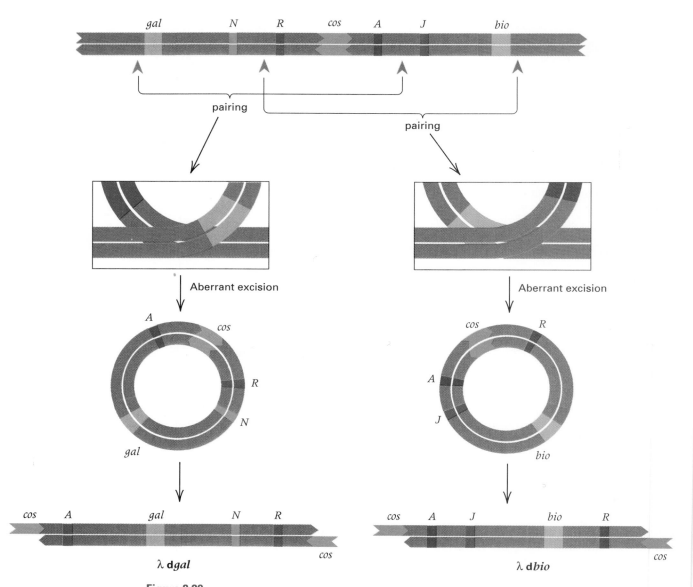

Figure 8.29
Aberrant excision leading to the production of specialized λ transducing phages. (A) Formation of a *gal* transducing phage (λ*dgal*). (B) Formation of a *bio* transducing phage (λ*dbio*).

genes are replaced by bacterial genes during formation of the *gal*⁺-bearing or *bio*⁺-bearing λ molecules. The phages are thus called λ*dgal* (defective, *gal*⁺-transducing) or λ*dbio* (defective, *bio*⁺-transducing). They are unable by themselves to infect *E. coli* productively. Addition of a wildtype helper phage both allows production of a lysate rich in λ*dgal* or λ*dbio* transducing phage and formation of a double lysogen. Presumably, integration of the wildtype λ provides hybrid attachment sites POB' and

BOP' that allow integration of λ*dbio* or λ*dgal* by homologous recombination between identical hybrid attachment sites.

8.8
Transposable Elements

DNA sequences that are present in a genome in multiple copies and that have the capability of occasional movement, or **transposition,** to new locations in the genome were described in Section 6.8. These *transposable elements* are widespread in living organisms.

Many transposable elements in bacteria have been extensively studied. The bacterial elements first discovered—called **insertion sequences** or **IS elements**—are small and do not contain any known host genes. Like many transposable elements in eukaryotes, these bacterial elements possess inverted-repeat sequences at their termini, and most code for a **transposase** protein required for transposition and one or more additional proteins that regulate the rate of transposition. The DNA organization of the insertion sequence IS*50* is diagrammed in Figure 8.30A. Insertion sequences are instrumental in the origin of Hfr bacteria from F$^+$ cells (Section 8.4), because the F plasmid normally integrates through genetic exchange between insertion sequences present in F and homologous copies present at various sites in the bacterial chromosome.

Other transposable elements in bacteria contain one or more bacterial genes that can be shuttled between different bacterial hosts by transposing into bacterial plasmids (such as the F plasmid), which are capable of conjugational transfer. These gene-containing elements are called **transposons,** and their length is typically several kilobases of DNA, but a few are much longer. Some transposons have composite structures with the bacterial genes sandwiched between insertion sequences, as is the case with the Tn*5* element illustrated in Figure 8.30B, which terminates in two IS*50* elements in inverted orientation. Transposons are usually designated by the abbreviation Tn followed by an italicized number (for example, Tn*5*). When it is necessary to refer to genes carried in such an element, the usual designations for the genes are used. For example, Tn*5* (*neo-r ble-r str-r*) contains genes for resistance to three different antibiotics: neomycin, bleomycin, and streptomycin. Such genes provide markers, making it easy to detect transposition of the composite element, as is shown in Figure 8.31. An F' *lac*$^+$ plasmid is transferred by conjugation into a bacterial

Figure 8.30

Transposable elements in bacteria. (A) Insertion sequence IS*50*. The element is terminated by short, nearly perfect inverted-repeat sequences, the terminal nine base pairs of which are indicated. IS*50* contains a region that codes for the transposase and for a repressor of transposition. The coding regions are identical in the region of overlap, but the repressor is somewhat shorter because it begins at a different place. (B) Composite transposon Tn*5*. The central sequence contains genes for resistance to neomycin, *neo-r;* bleomycin, *ble-r,* and streptomycin, *str-r;* it is flanked by two copies of IS*50* in inverted orientation. The left-hand element (IS*50L*) contains mutations and is nonfunctional, so the transposase and repressor are made by the right-hand element (IS*50R*).

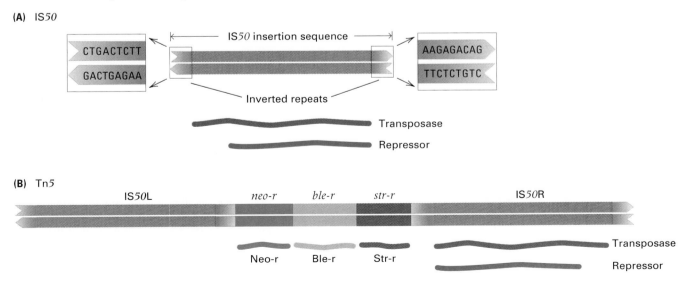

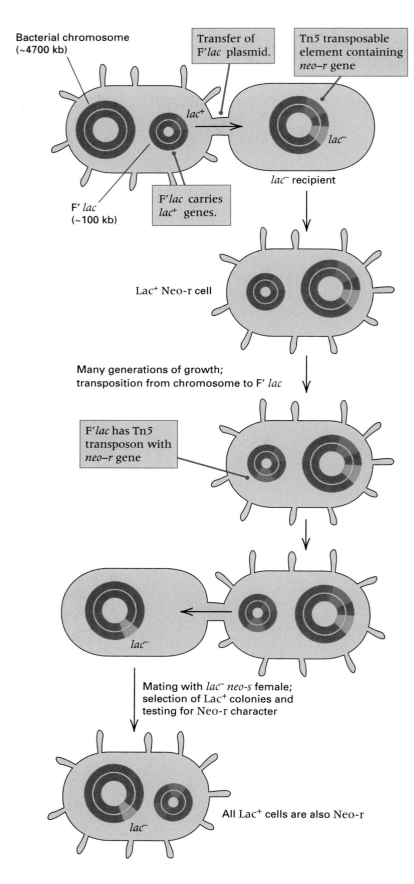

Bacterial chromosome
(~4700 kb)

Transfer of
F'*lac* plasmid.

Tn5 transposable
element containing
neo–r gene

lac⁺

lac⁻

lac⁻ recipient

F' *lac*
(~100 kb)

F'*lac* carries
lac⁺ genes.

Lac⁺ Neo-r cell

Many generations of growth;
transposition from chromosome to F' *lac*

F'*lac* has Tn5
transposon with
neo–r gene

lac⁻

Mating with *lac*⁻ *neo-s* female;
selection of Lac⁺ colonies and
testing for Neo-r character

lac⁻

lac⁻

All Lac⁺ cells are also Neo-r

cell containing a transposable element that carries the *neo-r* (neomycin-resistance) gene. The bacterial cell is allowed to grow, and in the course of multiplication, transposition of the transposon into the F' plasmid occasionally takes place in a progeny cell. Transposition yields an F' plasmid containing both the *lac*⁺ and *neo-r* genes. In a subsequent mating to a Neo-s Lac⁻ cell, the *lac*⁺ and *neo-r* markers are transferred together and so are genetically linked.

In nature, sequential transposition of transposons containing *different* antibiotic-resistance genes into the same plasmid results in the evolution of plasmids that confer resistance to multiple antibiotics. These multiple-resistance plasmids are called **R plasmids.** The evolution of R plasmids is promoted by the use (and regrettable overuse) of antibiotics, which selects for resistant cells because, in the presence of antibiotics, resistant cells have a growth advantage over sensitive cells. The presence of multiple antibiotics in the environment selects for multiple-drug resistance. Serious clinical complications result when plasmids resistant to multiple drugs are transferred to bacterial **pathogens,** or agents of disease. Infections with some pathogens containing R factors are extremely difficult to treat because the pathogen is resistant to all known antibiotics.

When transposition takes place, the transposable element can be inserted in any one of a large number of positions. The existence of multiple insertion sites can be shown when a wildtype lysogenic *E. coli* culture is infected with a temperate phage

Figure 8.31

An experiment demonstrating the transposition of transposon Tn5, which contains a neomycin-resistance gene, from the chromosome to an F' plasmid containing the bacterial gene for lactose utilization (F' *lac*). The bacterial chromosome is *lac*⁻ and the cells are unable to grow on lactose unless the F' *lac* is present. After the transposition, the F' *lac* plasmid also contains Tn5, indicated by the linkage of the *neo-r* gene to the F' factor. Note that transposition to the F' plasmid does not eliminate the copy of the transposon in the chromosome.

that is identical to the prophage (immunity prevents the phage from killing the cell) but carries a transposable element with an antibiotic-resistance marker. Transposition events can be detected by the production of antibiotic-resistant bacteria that contain new mutations resulting from insertion of the transposon into a gene in the bacterial chromosome. For example, a *lac⁺ leu⁺ neo-s* culture of *E. coli* infected with a *neo-r* transposon can yield both *lac⁻ neo-r* and *leu⁻ neo-r* mutants. If many hundreds of Neo-r bacterial colonies are examined and tested for a variety of nutritional requirements and for the ability to utilize different sugars as a carbon source, then colonies can usually be found bearing a mutation in almost any gene that is examined. This observation indicates that potential insertion sites for transposons are scattered throughout the chromosomes of *E. coli* and other bacterial species.

The end result of the transposition process is the insertion of a transposable element between two base pairs in a recipient DNA molecule. During the insertion process, most transposition events also create a duplication of a short (2–12 nucleotide pairs) sequence of host DNA, which after transposition flanks the insertion site (Section 6.8). The insertion of a transposable element does not require DNA sequence homology or the use of most of the enzymes of homologous recombination, because transposition is not inhibited even in cells in which the major enzyme systems for homologous recombination have been eliminated. Direct nucleotide sequence analysis of many transposable elements and their insertion sites confirms the absence of DNA sequence homology in the recipient with any sequence in the transposable element.

Transposons in Genetic Analysis

Transposons can be employed in a variety of ways in bacterial genetic analysis. Three features make them especially useful for this purpose.

1. Transposons can insert at a large number of potential target sites that are essentially random in their distribution throughout the genome.

2. Many transposons code for their own transposase and require only a small number of host genes for mobility.

3. Transposons contain one or more genes for antibiotic resistance that serve as genetic markers for selection.

Genetic analysis using transposons is particularly important in bacterial species that do not have readily exploited systems for genetic manipulation or large numbers of identified and mapped genes. Many of these species are bacterial pathogens, and transposons can be used to identify and manipulate the disease factors.

The use of a *neo-r* transposable element to identify a particular disease gene in a pathogenic bacterial species is diagrammed in Figure 8.32. In this example, the transposon is introduced into the pathogenic bacterium by a mutant bacteriophage that cannot replicate in the pathogenic host. A variety of other methods of introduction also are possible. After introduction of the transposon and selection on medium containing the antibiotic, the only resistant cells are those in which the transposon inserted into the bacterial chromosome, because the phage DNA in which it was introduced is incapable of replication. Any of a number of screening methods are then used to identify cells in which a particular disease gene became inactivated (nonfunctional) as a result of the transposon inserting into the gene. This method is known as **transposon tagging,** because the gene with the insertion is tagged (marked) with the antibiotic-resistance phenotype of the transposon.

Once the disease gene has been tagged by transposon insertion, it can be used in many ways (for example, in genetic mapping) because the phenotype resulting from the presence of the tagged gene is antibiotic resistance, which is easily identified and selected. The lower part of Figure 8.32 shows how the transposon tag is used to transfer the disease gene into *E. coli*. In the first step, DNA from the pathogenic strain that contains the tagged gene is purified, cut into fragments of suitable size, and inserted into a small plasmid capable of replication in *E. coli*. (Details of these genetic engineering methods are discussed

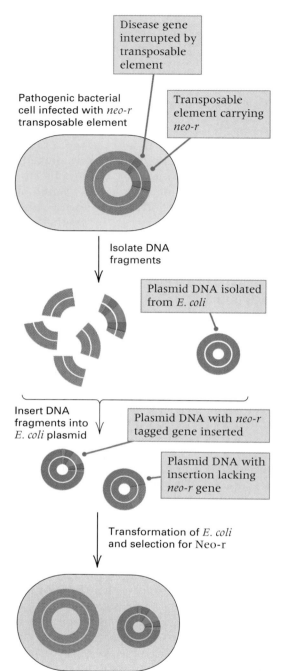

Disease gene interrupted by transposable element

Transposable element carrying *neo-r*

Pathogenic bacterial cell infected with *neo-r* transposable element

Isolate DNA fragments

Plasmid DNA isolated from *E. coli*

Insert DNA fragments into *E. coli* plasmid

Plasmid DNA with *neo-r* tagged gene inserted

Plasmid DNA with insertion lacking *neo-r* gene

Transformation of *E. coli* and selection for Neo-r

Figure 8.32
Transposon tagging making use of a *neo-r* transposon to mutate a disease gene by inserting into it. The *neo-r* resistance marker is then used to select plasmids containing the disease gene after transfer into *E. coli*, where the gene can be analyzed more safely and easily.

in Chapter 9.) The result is a heterogeneous collection of plasmids, most containing DNA fragments other than the one of interest. However, plasmids that do contain the fragment of interest also contain the transposon tag and so are able to confer antibiotic resistance on the host. The heterogeneous collection of plasmids is introduced into *E. coli* cells by transformation, and the cells are grown on solid medium containing the antibiotic. The only cells that survive contain the desired plasmid, which contains the transposon along with DNA sequences from the desired gene. The transposon-tagged gene is said to be **cloned** in *E. coli*.

The cloning of genes in *E. coli* greatly facilitates further genetic and molecular studies, such as DNA sequencing. The disease gene usually poses no hazard when it is present in *E. coli* for several reasons: (1) More than one gene in the pathogen is usually required to cause disease; and (2) the cloned transposon-tagged gene is inactivated by the insertion; and (3) the disease gene cloned in *E. coli* is usually incomplete or, if complete, is often inactive in this host. In cases in which some danger might result from cloning, potential problems are minimized by using special kinds of *E. coli* strains and following procedures that ensure containment in the laboratory.

Chapter Summary

DNA can be transferred between bacteria in three ways: transformation, transduction, and conjugation. In transformation, free DNA molecules, obtained from donor cells, are taken up by recipient cells; by a recombinational mechanism, a single-stranded segment becomes integrated into the recipient chromosome, replacing a homologous segment.

In conjugation, donor and recipient cells pair and a single strand of DNA is transferred by rolling-circle replication from

the donor cell to the recipient. Transfer is mediated by the transfer genes of the F plasmid. When F is present as a free plasmid, it becomes established in the recipient as an autonomously replicating plasmid; if F is integrated into the donor chromosome—that is, if the donor is an Hfr cell—then only part of the donor chromosome is usually transferred, and it can be maintained in the recipient only after an exchange event. About 100 minutes is required to transfer the entire *E. coli* chromosome, but the mating cells usually break apart before transfer is complete. DNA transfer starts at a particular point in the Hfr chromosome (the site of integration of F) and proceeds linearly. The times at which donor markers first enter recipient cells—the times of entry—can be arranged in order, yielding a map of the bacterial genome. This map is circular because of the multiple sites at which an F plasmid integrates into the bacterial chromosome. Occasionally, F is excised from the chromosome in an Hfr cell; aberrant excision, in which one cut is made at the end of F and the other cut is made in the chromosome, gives rise to an F' plasmid, which can transfer bacterial genes. The term *episome* applies to DNA elements, such as the F factor, that can be maintained within the cell either as an autonomously replicating molecule or by integration into the chromosome.

In transduction, a generalized transducing phage infects a donor cell, fragments the host DNA, and packages fragments of host DNA into phage particles. The transducing particles contain no phage DNA but can inject donor DNA into a recipient bacterium. By genetic exchange, the transferred DNA can replace homologous DNA of the recipient, producing a recombinant bacterium called a transductant.

When bacteria are infected with several phages, genetic exchange can take place between phage DNA molecules, generating recombinant phage progeny. Measurement of recombination frequency yields a genetic map of the phage. A common feature of these maps is clustering of phage genes with related function. The genetic map of bacteriophage T4, which infects cells of *E. coli*, is circular, but the genome consists of a linear DNA molecule. The circularity of the genetic map results from a packaging mechanism in which successive "headfuls" of DNA, each slightly longer than the minimum size required to contain all genes, are cleaved from a long concatemer of phage genomes. In a population of phage, the DNA molecules are related as circular permutations, and every DNA molecule has a short duplication at each end. Bacteriophage T4 undergoes efficient recombination. The fine-structure genetic analysis of the *rII* region led to the discovery of hotspots of mutation and to clarification of the term *gene* as meaning (1) the unit of function (cistron)

defined by a complementation test, (2) the unit of genetic transmission that participates in recombination, or (3) the unit of genetic change or mutation.

In contrast to virulent phage, such as T4, temperate phages (including phage λ) possess mechanisms for recombining phage and bacterial DNA. A bacterium that contains integrated phage DNA is called a lysogen, the integrated phage DNA is a prophage, and the overall phenomenon is lysogeny. Integration results from site-specific recombination between particular sequences, called attachment sites, in the bacterial and phage DNA. The bacterial and phage attachment sites are not identical but have a short common sequence in which the recombination takes place. Temperate phages circularize their DNA after infection. Because the phage has a single attachment site, which is not terminally located, the order of the prophage genes is a circular permutation of the order of the genes in the phage particle. The integrated prophage DNA is stable, but if the bacterial DNA is damaged, then prophage induction is initiated, the phage DNA is excised, and the lytic cycle of the phage ensues. Defective excision of a prophage yields a specialized transducing phage capable of transducing bacterial genes on either side of the phage attachment site.

Most bacteria possess transposable elements, which are capable of moving from one part of the DNA to another. Transposition does not require sequence homology. Bacterial insertion sequences (IS) are small transposable elements that code for their own transposase and that have inverted-repeat sequences at the ends. Some larger transposons have a composite structure consisting of a central region, often containing one or more antibiotic-resistance genes, flanked by two IS sequences in inverted orientation. Accumulation of such transposons in plasmids gives rise to multiple-drug-resistant R plasmids, which confer resistance to several chemically unrelated antibiotics. Transposable elements are useful in genetic analysis because they can create mutations by inserting within a host gene and interrupting its continuity. Such mutations make possible transposon tagging for genetic analysis of bacterial pathogens and other species.

Key Terms

antibiotic-resistant mutant	episome	lytic cycle
attachment site	excisionase	merodiploid
auxotroph	F plasmid	minimal medium
bacterial attachment site	F' plasmid	nonselective medium
bacteriophage	generalized transduction	nucleoid
carbon-source mutant	Hfr strain	nutritional mutant
circular permutation	hotspot	partial diploid
cistron	immunity	pathogen
clone	insertion sequence	phage
cloned gene	integrase	phage attachment site
cohesive end	interrupted-mating technique	phage repressor
conjugation	IS element	plaque
cotransduction	lysis	plasmid
cotransformation	lysogen	prophage
counterselected marker	lysogenic cycle	prophage induction

prototroph
R plasmid
selected marker
selective medium
site-specific recombination
specialized transduction
temperate phage

terminal redundancy
time of entry
transducing phage
transduction
transformation
transposase
transposition

transposon
transposon tagging
virulent phage

Review the Basics

- What are three phenotypes of mutations that are particularly useful for genetic analysis in bacteria.

- What is the difference between an F plasmid and an F' plasmid?

- What is the difference between a selected marker and a counterselected marker? Why are both necessary in analyzing the progeny of an Hfr × F⁻ mating?

- Distinguish between generalized and specialized transduction.

- Transfer of DNA in conjugation is always accompanied by DNA replication. What is the mode of replication and in which cell does it take place?

- In a mating between Hfr and F⁻ cells, why does the recipient cell usually remain F⁻?

- What prevents a bacterial cell that is lysogenic for some temperate bacteriophage from being reinfected with a phage of the same type ?

Guide to Problem Solving

Problem 1: In an Hfr × F⁻ cross in *E. coli*, the Hfr genotype is $a^+ b^+$ str-s and the F⁻ genotype is $a^- b^-$ str-r. Recombinants of genotype a^+ str-r are selected, and almost all of them prove to be b^+. How can this result be explained?

Answer: Because most colonies that are a^+ are also b^+, it is clear that *a* and *b* are close together in the bacterial chromosome.

Problem 2: Consider the following Hfr × F⁻ cross.

Hfr genotype: $a^+ b^+ c^+$ str-s
F⁻ genotype: $a^- b^- c^-$ str-r

The order of gene transfer is *a b c*, with *a* transferred at 9 minutes, *b* at 11 minutes, *c* at 30 minutes, and *str-s* at 40 minutes. Recombinants are selected by plating on a medium that lacks particular nutrients and contains streptomycin. Which of the following statements are true, and why? (*Note:* Genetic markers introduced before a selected marker, and not closely linked to it, are incorporated into the recipient genome about 50 percent of the time.)

(a) a^+ str-r colonies ≈ b^+ str-r colonies.

(b) a^+ str-r colonies > c^+ str-r colonies.

(c) b^+ str-r colonies < c^+ str-r colonies.

(d) Among colonies selected for
$$b^+ \text{ str-r}, a^+ b^+ \text{ str-r} \approx a^- b^+ \text{ str-r.}$$

(e) Among colonies selected for
$$c^+ \text{ str-r}, a^+ c^+ \text{ str-r} \approx a^- c^+ \text{ str-r.}$$

(f) Among colonies selected for
$$a^+ c^+ \text{ str-r}, a^+ b^+ c^+ \text{ str-r} < a^+ b^- c^+ \text{ str-r.}$$

Answer:

(a) True, because *a* and *b* are only 2 minutes apart, and if chromosome transfer has gone far enough to include *a*, it will usually also include *b*.

(b) True, because *a* enters long before *c*.

(c) False, because *b* enters long before *c*.

(d) False, because *a* and *b* are only 2 minutes apart, and strains selected for b^+ will usually also contain a^+.

(e) True, because *a* is transferred long before *c*, so the probability of incorporation of a^+ is about 50 percent.

(f) False, because *a* and *b* are only 2 minutes apart, and strains selected for a^+ will usually also contain b^+.

Problem 3: A temperate bacteriophage has the gene order *a b c d e f g h*, whereas the order of genes in the prophage present in the bacterial chromosome is *g h a b c d e f*. What information does this give you about the location of the attachment site in the phage?

Answer: The gene order in the prophage is circularly permuted with respect to the gene order in the phage, and the attachment site defines the site of breakage and rejoining with the bacterial chromosome. Because adjacent genes f and g are separated by prophage integration, the attachment site must be located between f and g.

Problem 4: An Hfr str-s strain that transfers genes in alphabetical order

$$a^+ \ b^+ \ c^+ \dots x^+ \ y^+ \ z^+$$

with a transferred early and z a terminal marker, is mated with $F^- \ z^-$ str-r cells. The mating mixture is agitated violently 15 minutes after mixing to break apart conjugating cells and then plated on a medium that lacks nutrient Z and contains streptomycin. The z gene is far from the str gene. The yield of z^+ str-r colonies is about one per 10^7 Hfr cells. What are two possible modes of origin and genotypes of such a colony? How could the two possibilities be distinguished?

Answer: The time of 15 minutes is too short to allow the transfer of a terminal gene. Thus, one explanation for a z^+ str-r colony is that aberrant excision of the F factor occurred in an Hfr cell, yielding an F' plasmid carrying terminal markers. When the F' z^+ plasmid is transferred into an F^- recipient, the genotype of the resulting cell would be F' z^+/z^- str-r. Alternatively, a z^+ str-r colony can originate by reverse mutation of $z^- \to z^+$ in the F^- recipient. The possibilities can be distinguished because the F' z^+/z^- str-r cell is able to transfer the z^+ marker to other cells.

Analysis and Applications

8.1 In the specialized transduction of a $gal^- \ bio^-$ strain of *E. coli* using bacteriophage λ from a $gal^+ \ bio^+$ lysogen, what medium would select for gal^+ transductants without selecting for bio^+?

8.2 Given that bacteriophage λ has a genome size of 50 kilobase pairs, what is the approximate genetic length of λ prophage in minutes? (*Hint:* There are 100 minutes in the entire *E. coli* genetic map.)

8.3 How could you obtain a lysate of bacteriophage P1 in which some of the transducing particles contained bacteriophage λ?

8.4 A temperate bacteriophage has gene order *A B C att D E F*. What is the gene order in the prophage?

8.5 Why are λ specialized transducing particles generated only by inducing a lysogen to produce phages rather than by lytic infection?

8.6 If leu^+ str-r recombinants are desired from the cross Hfr leu^+ str-s × $F^- \ leu^-$ str-r, on what kind of medium should the matings pairs be plated? Which are the selected and which the counterselected markers?

8.7 How many plaques can be formed by a single bacteriophage particle? A bacteriophage adsorbs to a bacterium in a liquid growth medium. Before lysis, the infected cell is added to a suspension of cells that are plated on solid medium to form a lawn. How many plaques will result?

8.8 Phage T2 (a relative of T4) normally forms small, clear plaques on a lawn of *E. coli* strain B. Another strain of *E. coli*, called B/2, is unable to adsorb T2 phage particles, and no plaques are formed. T2*h* is a host-range mutant capable of adsorbing to *E. coli* B and to B/2, and it forms normal-looking plaques. If *E. coli* B and the mutant B/2 are mixed in equal proportions and used to generate a lawn, how can plaques made by T2 and T2*h* be distinguished by their appearance?

8.9 If 10^6 phages are mixed with 10^6 bacteria and all phages adsorb, what fraction of the bacteria remain uninfected?

8.10 An Hfr strain transfers genes in alphabetical order. When tetracycline sensitivity is used for counterselection, the number of h^+ tet-r colonies is 1000-fold lower than the number of h^+ str-r colonies found when streptomycin sensitivity is used for counterselection. Suggest an explanation for the difference.

8.11 An Hfr strain transfers genes in the order *a b c*. In an Hfr $a^+ \ b^+ \ c^+$ str-s × $F^- \ a^- \ b^- \ c^-$ str-r mating, do all b^+ str-r recombinants receive the a^+ allele? Are all b^+ str-r recombinants also a^+? Why or why not?

8.12 If the genes in a bacterial chromosome are in alphabetical order and an Hfr cell transfers genes in the order *g h i . . . d e f*, what types of F' plasmids could be derived from the Hfr?

8.13 In an Hfr $lac^+ \ met^+$ str-s × $F^- \ lac^- \ met^-$ str-r mating, *met* is transferred much later than *lac*. If cells are plated on minimal medium containing glucose and streptomycin, what

GeNETics on the web will introduce you to some of the most important sites for finding genetic information on the Internet. To complete the exercises below, visit the Jones and Bartlett home page at

http://www.jbpub.com/genetics

Select the link to Genetics: Principles and Analysis and then choose the link to GeNETics on the web. You will be presented with a chapter-by-chapter list of highlighted keywords.

GeNETics EXERCISES

To view a list of keywords corresponding to the bold terms in the four GeNETics Exercises below, visit the GeNETics on the Web home page and choose the link for Chapter 8, then select "GeNETics Exercises." Clicking on a keyword will transfer you to a web site that contains information necessary to complete the exercise. Each exercise suggests a specific, written eport that makes use of the information available at that site.

1. Strains of E. coli as well as an updated **genetic map** are available at this keyword site. Browse the Working Map and find

the gal genes, the bacterial lambda attachment site (called attLAM), and the bio genes around 17 minutes. If assigned to do so, use the WebServer at the site to find out how many bacterial attachment sites (att) have been identified; make a list of the phage that have attachment sites in the E. coli chromosome.

2. A listing of available **Hfr** strains can be found by searching at this site. How many entries are retrieved? Clicking on any entry— say, 3310 Paris Hfr—will open a link to more detailed information. The PO# will give you the origin of transfer of the Hfr (in this case, 96.9 minutes) and the direction of transfer (1 = clockwise, 0 = counterclockwise). Check out the "classic" Hfr strains, Hfr Hayes and Hfr Cavalli. If assigned to do so, choose any other 10 Hfr strains, and draw a diagram of the E. coli genetic map showing the name, origin, and direction of transfer of each of the strains you selected.

3. The keyword **genes and metabolism** will link you with an encyclopedia of many of the metabolic pathways present in this model organism. You can choose any of an extensive list of pathways and also view the molecular structures of the intermediates. If assigned to do so, outline the metabolic pathway for the

---

fraction of the cells are expected to be lac^+? If cells are plated on minimal medium containing lactose, methionine, and streptomycin, what fraction of the cells are expected to be met^+?

8.14 In the mating

Hfr met^- his^+ leu^+ trp^+ × F$^-$ met^+ his^- leu^- trp^-

the met marker is known to be transferred very late. After a short time, the mating cells are interrupted and the cell suspensions plated on four different growth media. The amino acids in the growth media and the number of colonies observed on each are as follows:

histidine + tryptophan	250 colonies
histidine + leucine	50 colonies
leucine + tryptophan	500 colonies
histidine	10 colonies

What is the purpose of the met^- mutation in the Hfr strain? What is the order of transfer of the genes? Why is the number of colonies so small for the medium containing only histidine?

8.15 Bacterial cells of genotype pur^- pro^+ his^+ were transduced with P1 bacteriophage grown on bacteria of genotype

pur^+ pro^- his^-. Transductants containing pur^+ were selected and tested for the unselected markers pro and his. The numbers of pur^+ colonies with each of four genotypes are as follows:

pro^+ his^+	102
pro^- his^+	25
pro^+ his^-	160
pro^- his^-	1

What is the gene order?

8.16 For generalized transduction, the theoretical relationship between map distance in minutes and frequency of cotransduction is

map distance = $2 - 2 \times$ (cotransduction frequency)$^{1/3}$

(a) What map distance in minutes corresponds to 75 percent cotransduction? To 50 percent cotransduction? To 25 percent cotransduction?

(b) What is the cotransduction frequency between genetic markers separated by half a minute? One minute? Two minutes? Greater than two minutes?

(c) If genes a, b, and c are in alphabetical order and the cotransduction frequencies are 30 percent for a−b and

synthesis of the vitamin biotin, and sketch the molecular structure of biotin.

4. A highly annotated genetic map of **bacteriophage** λ can be found at this keyword site. You can get detailed information about any of the phage genes. To find the location of, for example, *int* and *xis,* select *GeneList* and then the genes. From the main menu, choose DNA and then GenBank report to access the entire annotated lambda DNA sequence. Using this report, find the nucleotide sequences corresponding to the site, *attP,* at which the phage recombines with the bacterial chromosome. Where is this site in relation to that of *int* and *xis*? If assigned to do so, draw a diagram of a portion of phage DNA, to the correct scale, showing the locations of the open reading frames for *int* and *xis* and the location of *attP.*

MUTABLE SITE EXERCISES

The Mutable Site Exercise changes frequently. Each new update includes a different exercise that makes use of genetics resources available on the World Wide Web. Select the **Mutable Site** for Chapter 8, and you will be linked to the current exercise that relates to the material presented in this chapter.

PIC SITE

The Pic Site showcases some of the most visually appealing genetics sites on the World Wide Web. To visit the showcase genetics site, select the **Pic Site** for Chapter 8.

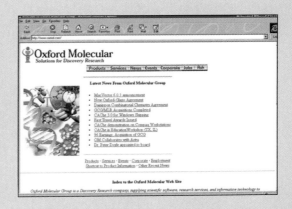

10 percent for *b–c,* what is the expected cotransduction frequency between *a* and *c*?

8.17 An ampicillin-resistant (Amp-r) strain of *E. coli* isolated from nature is infected with λ, and a suspension of phage progeny is obtained. This phage suspension is used to infect a culture of laboratory *E. coli* that is lysogenic for λ, sensitive to the antibiotic, and unable to undergo homologous recombination. Because the λ repressor is present in the lysogen, infecting λ molecules cannot replicate and are gradually diluted out of the culture by growth and continued division of the cells. However, rare Amp-r cells are found among the lysogens. Explain how they might have arisen.

8.18 On continued growth of a culture of a λ lysogen, rare cells become nonlysogenic by spontaneous loss of the prophage. Such loss is a result of prophage excision without subsequent development of the phage. The lysogen is said to have been "cured." A λ phage that carries a gene for tetracycline resistance is used to lysogenize a Tet-s cell. At a later time, a cell is isolated that is Tet-r but no longer contains a λ prophage. Explain how the cured cell could remain Tet-r.

8.19 A time-of-entry experiment is carried out in the mating

$$\text{Hfr } a^+ \ b^+ \ c^+ \ d^+ \ str\text{-}s \times \text{F}^- \ a^- \ b^- \ c^- \ d^- \ str\text{-}r$$

where the spacing between the genes is equal. The data in the accompanying table are obtained. What are the times of entry of each gene? Suggest one possible reason for the low frequency of d^+ *str-r* recombinants.

Time of mating in minutes	Number of recombinants of indicated genotype per 100 Hfr			
	a^+ *str-r*	b^+ *str-r*	c^+ *str-r*	d^+ *str-r*
0	0.01	0.006	0.008	0.0001
10	5	0.1	0.01	0.0004
15	50	3	0.1	0.001
20	100	35	2	0.001
25	105	80	20	0.1
30	110	82	43	0.2
40	105	80	40	0.3
50	105	80	40	0.4
60	105	81	42	0.4
70	103	80	41	0.4

8.20 You have reason to believe that a number of *rII* mutants of bacteriophage T4 are deletion mutations. You cross them in all possible combinations in *E. coli* strain B and plate them on *E. coli* strain K12(λ) to determine whether r^+ recombinants are formed. The formation of r^+ recombinants indicates that the mutations can recombine and so, if they are deletions, they must be nonoverlapping. The results are given in the accompanying table, in which *a* through *f* indicate an *rII* mutation and + indicates the formation of r^+ recombinant progeny in the cross.

	a	b	c	d	e	f
a	−	−	−	−	−	−
b		−	−	+	+	−
c			−	+	−	+
d				−	−	+
e					−	+
f						−

Assemble a deletion map for these mutations, using a line to indicate the DNA segment that is deleted in each mutant.

8.21 Seven *rII* mutants, *t* through *z*, thought to carry point mutations, are crossed to the six deletion mutants in Challenge Problem 8.20 and scored for the ability to produce r^+ recombinants. The results are given in the accompanying table.

	a	b	c	d	e	f
t	−	−	−	+	+	+
u	−	+	−	+	+	+
v	+	+	+	−	+	+
w	+	+	+	+	+	−
x	−	+	+	+	+	−
y	−	−	+	+	+	−
z	−	+	+	−	−	+

Using the deletion end points to define genetic intervals along the *rII* gene, position each point mutation within an interval. Refine the deletion end points if required by the data.

8.22 You cross the *rII* point mutations *t* through *z* in Challenge Problem 8.21 in all possible combinations in K12(λ) to assess complementation. The results are shown in the accompanying complementation matrix.

	t	u	v	w	x	y	z
t	−	−	+	−	−	−	+
u		−	+	−	−	−	+
v			+	+	+	+	−
w				−	−	−	+
x					−	−	+
y						−	+
z							−

The mutations *v* and *z* complement deletion *c* but not deletion *e*. The mutations *v* and *z* also fail to complement a mutant known to be defective for *rIIB* function but not for *rIIA* function. Which mutants are in the *rIIA* cistron and which in the *rIIB* cistron? Where does the boundary between the two genes lie? Assemble a complementation matrix for the deletion mutants, filling in all the squares if possible.

8.23 Bacteriophage 363 can be used for generalized transduction of genetic markers in *E. coli*. François Jacob used phage 363 to demonstrate that the lysogenic state of bacteriophage λ can be genetically transferred from one strain to another so long as the experiment is carried out at 20°C, a temperature at which λ phage do not multiply. Jacob studied four strains of bacteria:

A: $thr^- \ lac^- \ gal^-$

B: $thr^- \ lac^- \ gal^-$ (λ)

C: $thr^+ \ lac^+ \ gal^+$

D: $thr^+ \ lac^+ \ gal^+$ (λ)

In these designations, the symbol (λ) means that the strain is lysogenic for phage λ. Jacob made two kinds of crosses. Phage grown on strain D were used to transduce strain A, and phage grown on strain C were used to transduce strain B. The results are given in the table below.

Give a genetic explanation of these results. What genetic element is being transduced to give or to remove lysogeny?

Problem 8.23:

Donor strain	Recipient strain	Thr$^+$ Colonies tested	Thr$^+$ Colonies lysogenic	Lac$^+$ Colonies tested	Lac$^+$ Colonies lysogenic	Gal$^+$ Colonies tested	Gal$^+$ Colonies lysogenic
D	A	400	0	400	0	400	24
C	B	400	400	400	400	400	368

Adelberg, E. A., ed. 1966. *Papers on Bacterial Genetics.* Boston: Little, Brown.

Campbell, A. 1976. How viruses insert their DNA into the DNA of the host cell. *Scientific American,* December.

Clowes, R. D. 1975. The molecules of infectious drug resistance. *Scientific American,* July.

Davies, J. 1995. Vicious circles: Looking back on resistance plasmids. *Genetics* 139: 1465.

Drlica, K., and M. Riley. 1990. *The Bacterial Chromosome.* Washington, DC: American Society for Microbiology.

Edgar, R. S., and R. H. Epstein. 1965. The genetics of a bacterial virus. *Scientific American,* February.

Hopwood, D. A., and K. E. Chater, eds. 1989. *Genetics of Bacterial Diversity.* New York: Academic Press.

Losick, R., and D. Kaiser. 1997. Why and how bacteria communicate. *Scientific American,* February.

Low, K. B., and R. Porter. 1978. Modes of genetic transfer and recombination in bacteria. *Annual Review of Genetics* 12: 249.

Neidhardt, F. C., R. Curtiss III, J. L. Ingraham, E. C. C. Lin, K. B. Low, B. Magasanik, W. S. Reznikoff, M. Riley, M. Schaechter, and H. E. Umbarger, eds. 1996. *Escherichia coli and Salmonella typhimurium: Cellular and Molecular Biology* (2 volumes). 2d ed. Washington, DC: American Society for Microbiology.

Novick, R. P. 1980. Plasmids. *Scientific American,* December.

Robertson, B. D., and T. F. Meyer. 1992. Genetic variation in pathogenic bacteria. *Trends in Genetics* 8: 422.

Zinder, N. 1958. Transduction in bacteria. *Scientific American,* November.

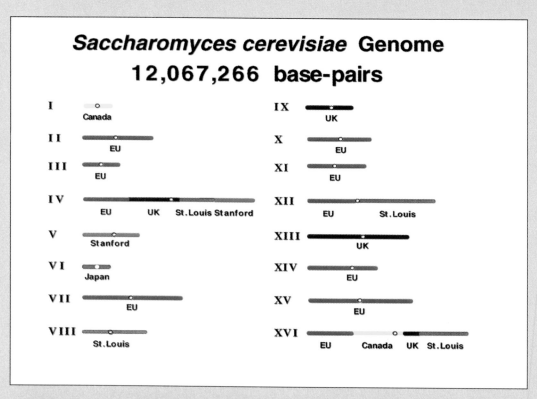

Who sequenced what? Shown are the locations of the research groups whose collaborative effort resulted in the complete sequence of the genome of the yeast, *Saccharomyces cerevisiae*—the first eukaryotic genome sequenced in its entirety. EU stands for a consortium of laboratories in Europe, and UK is the United Kingdom. [Courtesy of H. Mark Johnston.]

CHAPTER

9

Genetic Engineering and Genome Analysis

PRINCIPLES

- In recombinant DNA (gene cloning), DNA fragments are isolated, inserted into suitable vector molecules, and introduced into bacteria or yeast cells where they are replicated.
- Recombinant DNA is widely used in research, medical diagnostics, and the manufacture of drugs and other commercial products.
- In reverse genetics, gene function is analyzed by introducing an altered gene carrying a designed mutation into the germ line of an organism.
- Specialized methods for manipulating and cloning large fragments of DNA—up to one million base pairs—have made it possible to develop physical maps of the DNA in complex genomes.
- Large-scale automated DNA sequencing has resulted in the complete sequence of the genomes of several species of bacteria, the yeast *Saccharomyces cerevisiae* (the first eukaryote completely sequenced), and most of the genome of the nematode *Caenorhabditis elegans*. Large-scale sequencing projects are underway in many other organisms of genetic or commercial interest.

CONNECTIONS

CONNECTION: Hello, Dolly!
Ian Wilmut, Anagelika E. Schnieke, Jim McWhir, Alex J. Kind, and Keith H. S. Campbell 1997
Viable offspring derived from fetal and adult mammalian cells

CONNECTION: YAC-ity YAC
David T. Burke, Georges F. Carle and Maynard V. Olson 1987
Cloning of large segments of exogenous DNA into yeast by means of artificial chromosome vectors

Genetic analysis is a powerful experimental approach for understanding the mechanisms and evolution of biological processes. Genetics has played an equally important role in the development of techniques for the deliberate manipulation of biological systems to create organisms with novel genotypes and phenotypes. These organisms may be created for experimental studies or for economic reasons, such as to obtain superior varieties of domesticated animals and crop plants or to create organisms that produce molecules used in the treatment of human disease. The traditional foundations of genetic manipulation have been mutation and recombination. These processes are essentially random; selective procedures, often quite complex, are required to identify organisms with the desired characteristics among the many other types of organisms produced. Since the 1970s, techniques have been developed in which the genotype of an organism can instead be modified in a directed and predetermined way. This approach is called **recombinant DNA** technology, **genetic engineering,** or **gene cloning.** It entails isolating DNA fragments, joining them in new combinations, and introducing the recombined molecules back into a living organism. Selection of the desired genotype is still necessary, but the probability of success is usually many orders of magnitude greater than with earlier procedures. The basic technique is quite simple: Two DNA molecules are isolated and cut into fragments by one or more specialized enzymes, and then the fragments are joined together *in any desired combination* and introduced back into a cell for replication and reproduction. Recombinant DNA procedures complement the traditional methods of plant and animal breeding in supplying new types of genetic variation for improvement of the organisms.

Much of the current interest in genetic engineering is motivated by its many practical applications. Among these are: (1) the isolation of a particular gene, part of a gene, or region of a genome; (2) the production of a particular RNA or protein molecule in quantities formerly unobtainable; (3) improved efficiency in the production of biochemicals (such as enzymes and drugs) and commercially important organic chemicals; (4) the creation of organisms with particular desirable characteristics (for example, plants that require less fertilizer and animals that exhibit faster growth rates or increased resistance to disease); and (5) potentially, the correction of genetic defects in higher organisms, including human beings. Some specific examples will be considered later in this chapter.

9.1 Restriction Enzymes and Vectors

The technology of recombinant DNA relies heavily on restriction enzymes, which are discussed in Section 5.7 with reference to their site-specific DNA cleavage in the generation of DNA fragments with defined ends that can be separated by gel electrophoresis, isolated, sequenced, or otherwise manipulated. In recombinant DNA, after the DNA has been cleaved by a restriction enzyme at specific sites, the restriction fragments are ligated into a vector molecule capable of replication in a living host organism, typically *E. coli* or yeast. In this section, we take a closer look at the special features of restriction enzymes that make them so useful in DNA cloning. Some of the most popular types of vectors used in recombinant DNA are also examined.

Production of Defined DNA Fragments

DNA fragments are usually obtained by the treatment of DNA samples with restriction enzymes. A **restriction enzyme** is a type of nuclease that cleaves a DNA duplex wherever the DNA molecule contains a particular short sequence of nucleotides matching the **restriction site** of the enzyme (see Section 5.7). Most restriction sites consist of four or six nucleotides, within which the restriction enzyme makes a break in each DNA strand. Each break creates a free 3' hydroxyl (3'-OH) and a free 5' phosphate (5'-P) at the site of breakage. Several hundred restriction enzymes, each with a different restriction site, have been isolated from microorganisms. Most restriction sites are symmetrical in the sense that the recognition sequence is identical in both strands of the DNA duplex. For example, the restriction enzyme *Eco*RI, isolated from *Escherichia coli*, has the restriction site 5'-GAATTC-3' and cuts the strand between the G and the A. The sequence of the other strand is 3'-CTTAAG-5', which is identical but is written with the 3' end at the left. The term **palindrome** is used to denote this type of symmetry.

Soon after restriction enzymes were discovered, observations with the electron microscope indicated that the fragments produced by many restriction enzymes could spontaneously form circles. The circles could be made linear again by heating; however, if after circularization they were treated with *E. coli* DNA ligase, which joins 3'-OH and 5'-P groups (Section 5.6), then the ends became covalently joined. This observation was the first evidence for three important features of restriction enzymes:

- Restriction enzymes make breaks in palindromic sequences.

- The breaks need not be directly opposite one another in the two DNA strands.

- Enzymes that cleave the DNA strands asymmetrically generate DNA fragments with complementary ends.

These properties are illustrated for *Eco*RI in Figure 9.1.

Most restriction enzymes are like *Eco*RI in that the cuts in the DNA strands are staggered, producing single-stranded ends called **sticky ends** that can adhere to each other because they contain complementary nucleotide sequences. Some restriction

Figure 9.1
Circularization of DNA fragments produced by a restriction enzyme. The red arrowheads indicate the *Eco*RI cleavage sites.

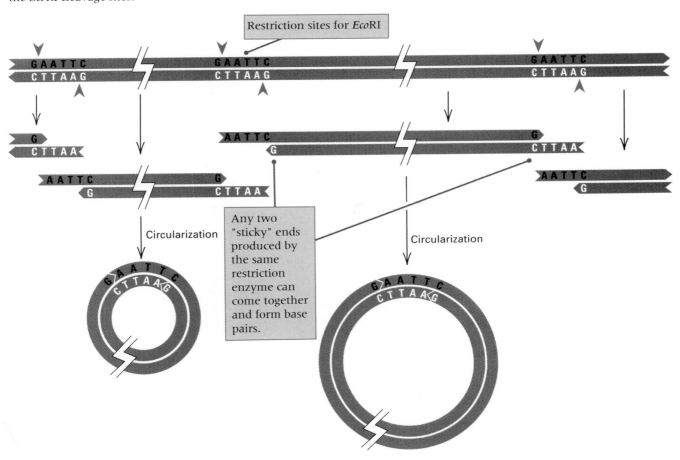

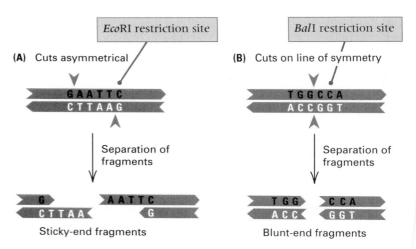

Figure 9.2

Two types of cuts made by restriction enzymes. The red arrowheads indicate the cleavage sites. (A) Cuts made in each strand at an equal distance from the center of symmetry of the restriction site. In this example with the enzyme *Eco*RI, the resulting molecules have complementary 5' overhangs. Other restriction enzymes produce fragments with complementary 3' overhangs. (B) Cuts made in each strand at the center of symmetry of the restriction site. The products have blunt ends. The specific enzyme in this example is *Bal*I.

enzymes, including *Eco*RI, leave a single-stranded overhang at the 5' end (Figure 9.2A); others leave a 3' overhang. A number of restriction enzymes cleave both DNA strands at the center of symmetry, forming **blunt ends.** Figure 9.2B shows the blunt ends produced by the enzyme *Bal*I. Blunt ends also can be ligated by DNA ligase. However, whereas ligation of sticky ends recreates the original restriction site, any blunt end can join with any other blunt end and not necessarily create a restriction site.

Most restriction enzymes recognize their restriction sequence without regard to the source of the DNA:

> DNA fragments obtained from one organism will have the same sticky ends as the fragments from another organism if they have been produced by the same restriction enzyme.

This principle is one of the foundations of recombinant DNA technology.

Because most restriction enzymes recognize a unique sequence, the number of cuts made in the DNA of an organism by a particular enzyme is limited. For example, an *E. coli* DNA molecule contains 4.7×10^6 base pairs, and any enzyme that cleaves a six-base restriction site will cut the molecule into about a thousand fragments because, with equal and random frequencies of each of the four nucleotides, a six-base restriction site is expected to occur, on average, every $4^6 = 4096$ base pairs. Mammalian nuclear DNA, because the genome is much larger than in *E. coli*, would be cut into about a million fragments. These large numbers still are small compared with the number that would be produced by random cuts. Of special interest are cleavage products from smaller DNA molecules, such as viral or plasmid DNA, which may have from only one to ten sites of cutting (or possibly no sites) for any particular enzyme. Plasmids containing a single site for a particular enzyme are especially valuable, as we will see shortly.

Recombinant DNA Molecules

A **vector** is a DNA molecule into which a DNA fragment can be cloned and which can replicate in a suitable host organism. In genetic engineering, a particular DNA

Figure 9.3

An example of cloning. A fragment of DNA from any organism is joined to a cleaved plasmid. The recombinant plasmid is then used to transform a bacterial cell, and thereafter, the foreign DNA is present in all progeny bacteria. The bacterial host chromosome is not drawn to scale. It is typically about 1000 times larger than the plasmid.

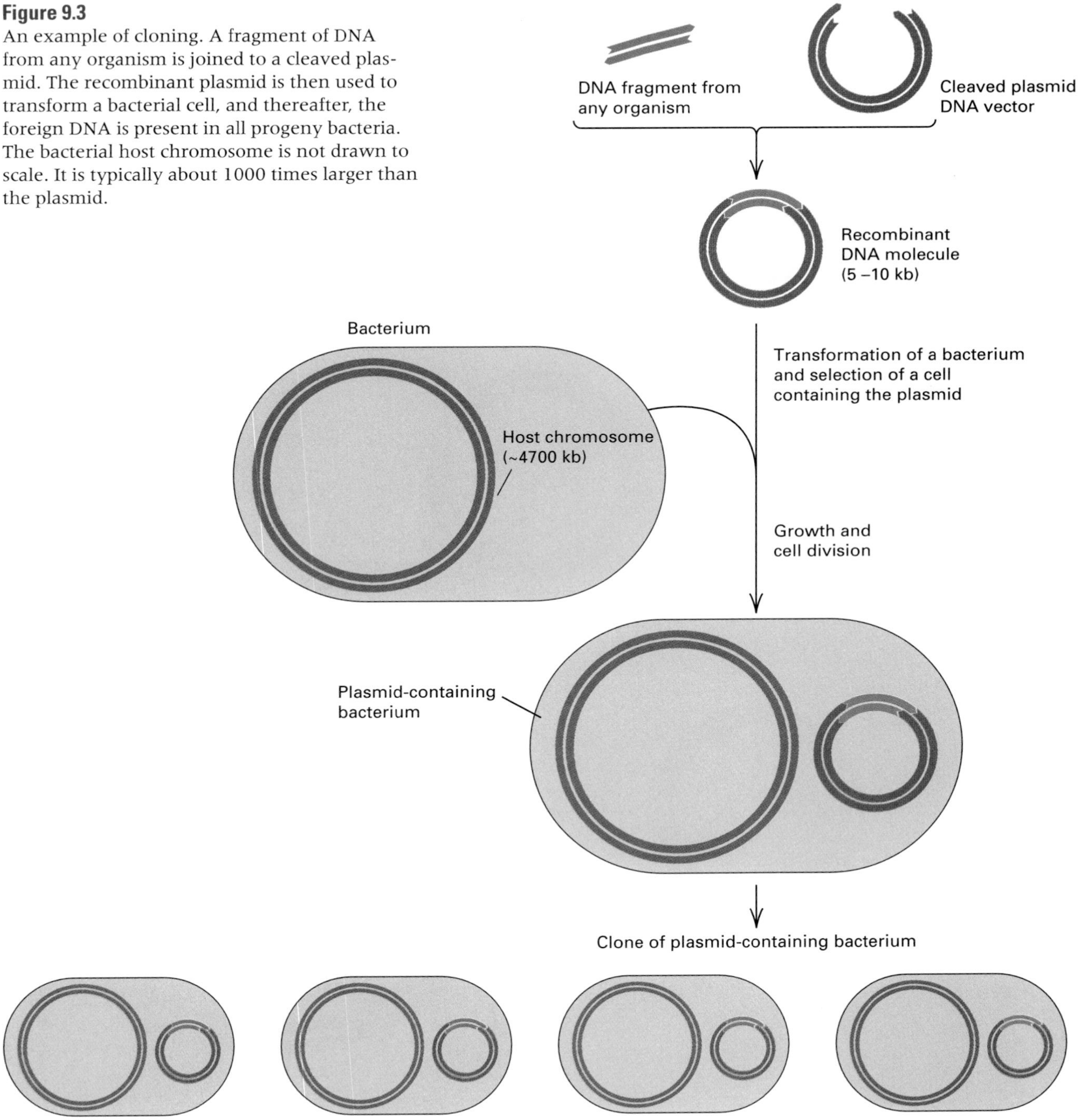

DNA fragment from any organism

Cleaved plasmid DNA vector

Recombinant DNA molecule (5 –10 kb)

Bacterium

Host chromosome (~4700 kb)

Transformation of a bacterium and selection of a cell containing the plasmid

Growth and cell division

Plasmid-containing bacterium

Clone of plasmid-containing bacterium

segment of interest is joined to a vector DNA molecule. By a transformation procedure very similar to that used by Avery, MacLeod, and McCarty in proving that DNA is the genetic material (Chapter 1), this recombinant molecule is placed in a cell in which replication can take place (Figure 9.3). When a stable transformant has been isolated, the genes or DNA sequences linked to the vector are said to be **cloned.** In the following section, several types of vectors are described.

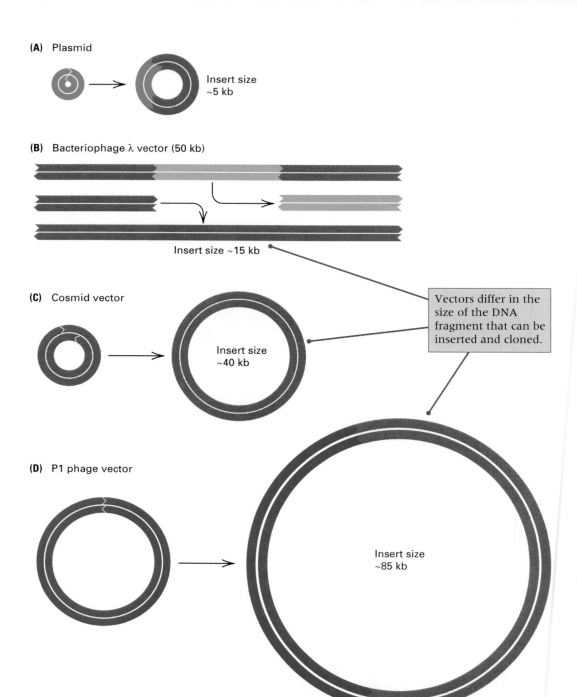

(A) Plasmid

Insert size
~5 kb

(B) Bacteriophage λ vector (50 kb)

Insert size ~15 kb

(C) Cosmid vector

Insert size
~40 kb

Vectors differ in the
size of the DNA
fragment that can be
inserted and cloned.

(D) P1 phage vector

Insert size
~85 kb

Figure 9.4
Common cloning vectors for use with *E. coli*, drawn approximately to scale. (A) Plasmid vectors
are ideal for cloning relatively small fragments of DNA. (B) Bacteriophage λ vectors contain
convenient restriction sites for removing the middle section of the phage and replacing it with the
DNA of interest. (C) Cosmid vectors are useful for cloning DNA fragments up to about 40 kb; they
can replicate as plasmids but contain the cohesive ends of phage λ and so can be packaged in
phage particles. (D) Vectors based on the bacteriophage P1 are used for cloning DNA fragments of
about 85 kb.

Plasmid, Lambda, Cosmid, and P1 Vectors

The most generally useful vectors have three properties:

- The vector DNA can be introduced into a host cell.

- The vector contains a replication origin so that it can replicate inside the host cell.

- Cells containing the vector can usually be selected in a straightforward manner, most conveniently through a selectable phenotype, such as antibiotic resistance, conferred on the host by genes present in the vector.

At present, the most commonly used vectors are *E. coli* plasmids and derivatives of the bacteriophages λ and M13. Other plasmids and viruses also have been developed for cloning into cells of animals, plants, and bacteria. Recombinant DNA can be detected in host cells by means of genetic features or particular markers made evident in the formation of colonies or plaques. Plasmid and phage DNA can be introduced into cells by a **transformation** procedure in which cells gain the ability to take up free DNA by exposure to a $CaCl_2$ solution. Recombinant DNA can also be introduced into cells by a kind of electrophoretic procedure called **electroporation.** After introduction of the DNA, the cells containing the recombinant DNA are plated on a solid medium. If the added DNA is a plasmid, then colonies that consist of bacterial cells containing the recombinant plasmid are formed, and the transformants can usually be detected by the phenotype that the plasmid confers on the host cell. For example, plasmid vectors typically include one or more genes for resistance to antibiotics, and plating the transformed cells on a selective medium with antibiotic prevents all but the plasmid-containing cells from growing (Section 8.1). Alternatively, if the vector is phage DNA, the infected cells are plated in the usual way to yield plaques. Variants of these procedures are used to transform animal or plant cells with suitable vectors,

but the technical details may differ considerably.

Four types of vectors commonly used for cloning into *E. coli* are illustrated in Figure 9.4. Plasmids (Figure 9.4A) are most convenient for cloning relatively small DNA fragments (5–10 kb). Somewhat larger fragments can be cloned with bacteriophage λ (Figure 9.4B). The wildtype phage is approximately 50 kb in length, but the central portion of the genome is not essential for lytic growth and can be removed and replaced with donor DNA. After the donor DNA has been ligated in place, the recombinant DNA is packaged into mature phage *in vitro,* and the phage is used to infect bacterial cells. However, to be packaged into a phage head, the recombinant DNA must be neither too large nor too small, which means that the donor DNA must be roughly the same size as the portion of the λ genome that was removed. Most λ cloning vectors accept inserts ranging in size from 12 to 20 kb. Still larger DNA fragments can be inserted into cosmid vectors (Figure 9.4C). These vectors can exist as plasmids, but they also contain the cohesive ends of phage λ (Section 8.6), which enables them to be packaged into mature phages. The size limitation on cosmid inserts usually ranges from 40 to 45 kb. A vector for cloning even larger fragments of DNA is a vector derived from the bacteriophage P1 (Figure 9.4D). The genome of the wildtype P1 phage is approximately 100 kb, but only about 15 kb are required for the genome to replicate and persist as a plasmid inside a suitable host bacterium. Therefore, P1 vectors that contain only the minimal 15 kb of essential DNA can be used to clone DNA fragments averaging approximately 85 kb.

9.2 Cloning Strategies

In genetic engineering, the immediate goal of an experiment is usually to insert a *particular* fragment of chromosomal DNA into a plasmid or a viral DNA molecule. One strategy for isolating DNA from a known gene is *transposon tagging,* which was discussed in Chapter 8. In transposon tagging, the target gene to be cloned is first inactivated by means of a transposon insertion. Because the insertion inactivates the gene, the mutant phenotype and the position of the mutation in the genetic map indicate that the correct gene has been hit. Because the transposon contains an antibiotic-resistance gene, the resistance affords a direct selection for any vector molecules that contain the transposon. The DNA flanking the transposon in the recombinant clone must necessarily derive from the target gene.

Although transposon tagging is an important approach in gene cloning, the method is feasible only in organisms whose genetics is already highly developed, such as maize and *Drosophila*. Transposon tagging is not possible in many organisms of interest, including human beings. Even in organisms in which the technique is available, particular genes may not be clonable with the approach because the gene contains no target sites for the transposon. It was therefore necessary to develop a set of alternative strategies for cloning genes. These are considered next.

Joining DNA Fragments

The circularization of restriction fragments having terminal single-stranded, "sticky" ends that have complementary bases was described in Section 9.1. Because a particular restriction enzyme produces fragments with *identical* sticky ends, without regard for the source of the DNA, fragments from DNA molecules isolated from two different organisms can be joined, as shown in Figure 9.5. In this example, the restriction enzyme *Eco*RI is used to digest DNA from any organism of interest and to cleave a bacterial plasmid that contains only one *Eco*RI restriction site. The donor DNA is digested into many fragments (one of which is shown) and the plasmid into a single linear fragment. When the donor fragment and the linearized plasmid are mixed, recombinant molecules can form by base pairing between the complementary single-stranded ends. At this point, the DNA is treated with DNA ligase to seal the joints, and the donor fragment becomes permanently joined in a combination that may never have existed before. The ability

to join a donor DNA fragment of interest to a vector is the basis of the recombinant DNA technology.

Joining sticky ends does not always produce a DNA sequence that has functional genes. For example, consider a linear DNA molecule that is cleaved into four fragments—A, B, C, and D—whose sequence in the original molecule was A B C D. Reassembly of the fragments can occasionally yield the original molecule, but if B and C have the same pair of sticky ends, then molecules with different arrangements of the fragments are also formed (Figure 9.6), including arrangements in which one or more of the restriction fragments are inverted in orientation (indicated in the figure by the reversed letters). Restriction fragments from the vector can join together in the wrong order, but this potential problem can be eliminated by the use of a vector that has only one cleavage site for a particular restriction enzyme. When a circular molecule has only one cleavage site for a restriction enzyme, cleavage with the restriction enzyme opens the circle at this single point, and any other DNA fragment of interest, if it has complementary ends, can be inserted at the point of cleavage. Many plasmids with single cleavage sites are available (most have been created by genetic engineering). Many vectors contain unique sites for several different restriction enzymes, but generally only one enzyme is used at a time.

DNA molecules that lack sticky ends also can be joined. A direct method uses the DNA ligase made by *E. coli* phage T4. This enzyme differs from other DNA ligases in that it not only heals single-stranded breaks in double-stranded DNA but can also join molecules with blunt ends.

Insertion of a Particular DNA Molecule into a Vector

In the cloning procedure described so far, a collection of fragments obtained by digestion with a restriction enzyme can be made to anneal with a cleaved vector molecule, yielding a large number of recombinant molecules containing different fragments of donor DNA. However, if a particular gene or segment of DNA is to be cloned, then the

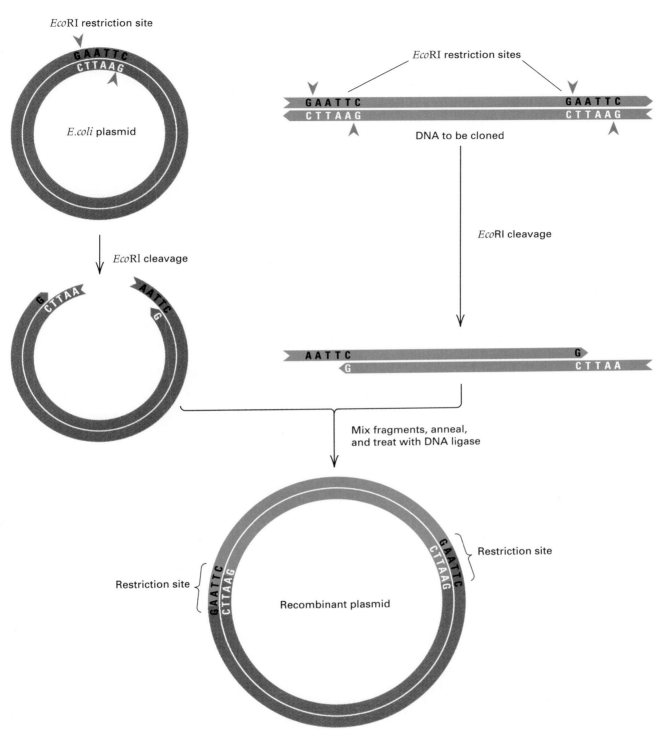

Figure 9.5
Construction of recombinant DNA plasmids containing fragments derived from a donor organism, by the use of a restriction enzyme (in this example *Eco*RI) and cohesive-end joining. Short red arrowheads indicate cleavage sites.

Figure 9.6

Fragments produced by a restriction enzyme can be rejoined in arbitrary ways because the ends of the cleavage sites are compatible. In this example, a DNA molecule is cut into four different fragments (A, B, C, and D) by a restriction enzyme. A few examples of how these can be rejoined are shown. Even among the rejoined molecules that have A and D at the ends and two fragments in the middle, many have one fragment represented twice and the other not at all, or they have a fragment present in the reverse orientation. Only occasionally does a rejoining recreate the original order A B C D.

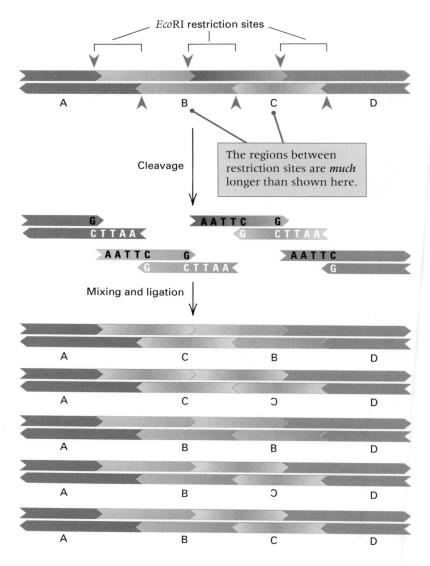

recombinant molecule possessing that particular segment must be isolated from among all the recombinant molecules that contain donor DNA. The simplest method is direct selection of the desired clone because of a phenotypic attribute it conveys. Indirect kinds of selection will be described shortly.

As an example of direct selection for the recovery of recombinant molecules containing a particular gene, we may consider the cloning of a *leu*⁺ gene from a bacterial cell. In this case, one would use a Leu⁻ host, which cannot grow in medium lacking leucine. After transformation with recombinant vectors, a few host cells will contain the *leu*⁺ gene. These transformants will each give rise to a Leu⁺ colony on a medium lacking leucine.

Direct selection for recombinant clones, such as in the Leu⁺ example, is not always possible. If the clone of interest is either very rare or very difficult to detect, it is often preferable to purify the DNA fragment that contains the gene of interest before joining it with the vector. Two methods of fragment isolation have already been discussed:

- Amplification by means of the **polymerase chain reaction (PCR),** discussed in Section 5.8, in which short oligonucleotide primers homologous to the ends of the fragment to be isolated are used repeatedly to replicate the fragment of interest. After a sufficient number of rounds of replication, the resulting solution contains predominantly replicas of the targeted DNA fragment.

- Purification of a restriction fragment known to contain the gene of interest (Section 5.7). For example, if a gene in a virus is known to be contained in a particular restriction fragment, then this fragment can be isolated from a gel after electrophoresis and joined to an appropriate vector.

Some genes in higher eukaryotes, such as human beings, are exceedingly long, as we shall see in Chapter 10. Any DNA fragment containing the complete gene is likely to contain multiple cleavage sites for any restriction enzyme and is also likely to be too long to be amplified by PCR. However, such large genes usually produce messenger RNA (mRNA) molecules of modest size. If the purpose of cloning the gene is to produce the protein product of the gene, then all the necessary information is contained in the mRNA. How to clone a DNA molecule whose sequence corresponds to a particular mRNA is described next.

The Use of Reverse Transcriptase: cDNA and RT-PCR

Some specialized animal cells make only one or a very small number of proteins in large amounts. In these cells, the cytoplasm contains a very high abundance of specific mRNA molecules, which constitute a large fraction of the total mRNA synthesized. Consequently, mRNA samples can usually be obtained that consist predominantly of a single mRNA species. An example is the chicken gene for the egg white protein ovalbumin, which is highly expressed in the oviduct of adult hens. In the cloning of genes whose products are major cellular proteins, the purified mRNA can serve as a starting point for creating a collection of recombinant plasmids, many of which contain coding sequences of the gene of interest.

Cloning from mRNA molecules depends on an unusual polymerase, **reverse transcriptase,** which can use a single-stranded RNA molecule as a template and synthesize a complementary strand of DNA called **complementary DNA,** or **cDNA.** Like other DNA polymerases, reverse transcriptase requires a primer. The stretch of A nucleotides usually found at the 3' end of eukaryotic mRNA serves as a convenient priming site, because the primer can be an oligonucleotide consisting of poly-T (Figure 9.7). Like any other single-stranded DNA molecule, the single strand of DNA produced from the RNA template can fold back upon itself at the extreme 3' end to form a "hairpin" structure that includes a very short double-stranded region consisting of a few base pairs. The 3' end of the hairpin serves as a primer for second-strand synthesis. The second strand can be synthesized either by DNA polymerase or by reverse transcriptase itself. Reverse transcriptase is the source of the second strand in RNA-based viruses that use reverse transcriptase, such as the human immunodeficieny virus (HIV). Conversion into a conventional double-stranded DNA molecule is achieved by cleavage of the hairpin by a nuclease.

In the reverse transcription of an mRNA molecule, the resulting full-length cDNA contains an uninterrupted coding sequence for the protein of interest. As we will see in Chapter 10, eukaryotic genes often contain DNA sequences, called *introns,* that are initially transcribed into RNA but are removed in the production of the mature mRNA. Because the introns are absent from the mRNA, the cDNA sequence is not identical with that in the genome of the original donor organism. However, if the purpose of forming the recombinant DNA molecule is to identify the coding sequence or to synthesize the gene product in a bacterial cell, then cDNA formed from

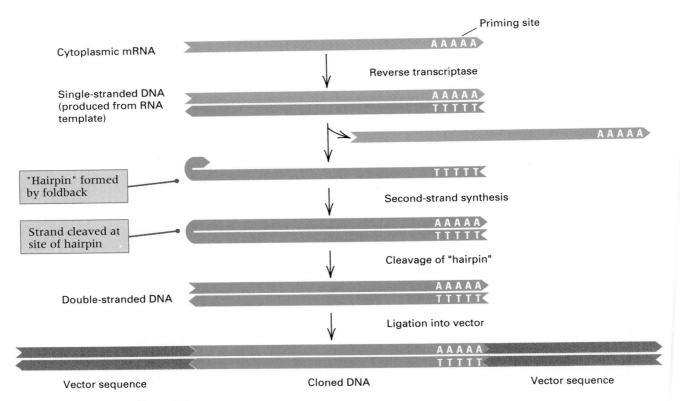

Priming site

Cytoplasmic mRNA

A A A A A

Reverse transcriptase

Single-stranded DNA
(produced from RNA
template)

A A A A A
T T T T T

A A A A A

T T T T T

"Hairpin" formed
by foldback

Second-strand synthesis

Strand cleaved at
site of hairpin

A A A A A
T T T T T

Cleavage of "hairpin"

Double-stranded DNA

A A A A A
T T T T T

Ligation into vector

A A A A A
T T T T T

Vector sequence   Cloned DNA   Vector sequence

Figure 9.7
Reverse transcriptase produces a single-stranded DNA complementary in sequence to a template
RNA. In this example, a cytoplasmic mRNA is copied. As indicated here, most eukaryotic mRNA
molecules have a tract of consecutive A nucleotides at the 3' end, which serves as a convenient
priming site. After the single-stranded DNA is produced, a foldback at the 3' end forms a hairpin
that serves as a primer for second-strand synthesis. After the hairpin is cleaved, the resulting
double-stranded DNA can be ligated into an appropriate vector either immediately or after PCR
amplification. The resulting clone contains the entire coding region for the protein product of
the gene.

processed mRNA is the material of choice
for cloning. The joining of cDNA to a vector
can be accomplished by available procedures for joining blunt-ended molecules
(Figure 9.7).

Genes that are not highly expressed are
represented by mRNA molecules whose
abundance ranges from low to exceedingly
rare. The cDNA molecules produced from
such rare RNAs will also be rare. The efficiency of cloning rare cDNA molecules can
be markedly increased by PCR amplification prior to ligation into the vector. The
only limitation on the procedure is the
requirement that enough DNA sequence
be known at both ends of the cDNA for
appropriate oligonucleotide primers to be
designed. PCR amplification of the cDNA
produced by reverse transcriptase is called

reverse transcriptase PCR (RT-PCR).
The resulting amplified molecules contain
the coding sequence of the gene of interest
with very little contaminating DNA.

Detection of Recombinant Molecules

When a vector is cleaved by a restriction
enzyme and renatured in the presence of
many different restriction fragments from a
particular organism, many types of molecules result, including such examples as a
self-joined circular vector that has not
acquired any fragments, a vector containing one or more fragments, and a molecule
consisting only of many joined fragments.
To facilitate the isolation of a vector con-

taining a particular gene, some means is needed to ensure (1) that the vector does indeed possess an inserted DNA fragment, and (2) that the fragment is in fact the DNA segment of interest. This section describes several useful procedures for detecting chimeric vectors.

In the use of transformation to introduce recombinant plasmids into bacterial cells, the initial goal is to isolate bacteria that contain the plasmid from a mixture of plasmid-free and plasmid-containing cells. A common procedure is to use a plasmid possessing an antibiotic-resistance marker and to grow the transformed bacteria on a medium that contains the antibiotic: Only cells that contain plasmid can form a colony. An example of a state-of-the-art cloning vector is the pBluescript plasmid illustrated in Figure 9.8A. The entire plasmid is 2961 base pairs. Different regions contribute to its utility as a cloning vector:

- The plasmid origin of replication is derived from the *E. coli* plasmid ColE1. The ColE1 is a high-copy-number plasmid, and its origin of replication enables pBluescript and its recombinant derivatives to exist in approximately 300 copies per cell.

- The ampicillin-resistance gene allows for selection of transformed cells in medium containing ampicillin.

Figure 9.8

(A) Diagram of the cloning vector pBluescript II. It contains a plasmid origin of replication, an ampicillin-resistance gene, a multiple cloning site (polylinker) within a fragment of the *lacZ* gene from *E. coli*, and a bacteriophage origin of replication. (B) Sequence of the multiple cloning site showing the unique restriction sites at which the vector can be opened for the insertion of DNA fragments. The numbers 657 and 759 refer to the position of the base pairs in the complete sequence of pBluescript. [Courtesy of Stratagene Cloning Systems, La Jolla, CA.]

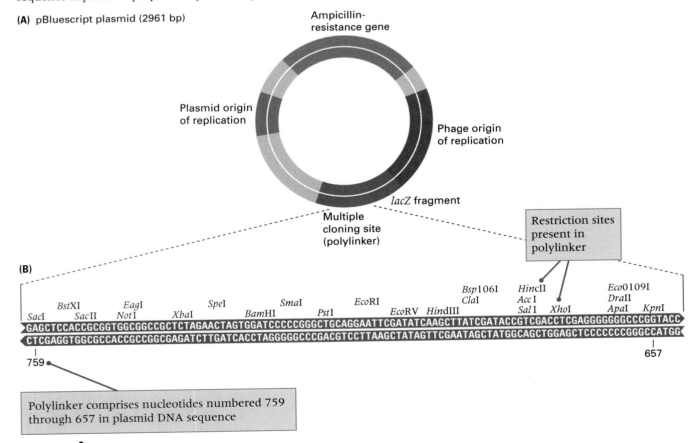

- The cloning site is called a **multiple cloning site (MCS)** or **polylinker** because it contains unique cleavage sites for many different restriction enzymes and enables many types of restriction fragments to be inserted. In pBluescript, the MCS is a 108-bp sequence that contains cloning sites for 23 different restriction enzymes (Figure 9.8B).

- The detection of recombinant plasmids is by means of a region containing the *lacZ* gene from *E. coli*. When the *lacZ* region is interrupted by a fragment of DNA inserted into the MCS, the recombinant plasmid yields Lac$^-$ cells. Nonrecombinant plasmids do not contain a DNA fragment in the MCS and yield Lac$^+$ colonies. The Lac$^+$ or Lac$^-$ phenotypes can be distinguished by color when the cells are grown on a special β-galactoside compound called X-gal, which releases a deep blue dye when cleaved. On medium containing X-gal, Lac$^+$ colonies contain nonrecombinant plasmids and are a deep blue, whereas Lac$^-$ colonies contain recombinant plasmids and are white.

- The bacteriophage origin of replication is from the single-stranded DNA phage f1. When cells containing a recombinant plasmid are infected with an f1 helper phage, the f1 origin enables a single strand of the inserted fragment, starting with *lacZ*, to be packaged in progeny phage. This feature is very convenient because it yields single-stranded DNA for sequencing. The plasmid shown in Figure 9.8A is the SK($+$) variety. There is also an SK($-$) variety in which the f1 origin is in the opposite orientation and packages the complementary DNA strand.

All good cloning vectors have an efficient origin of replication, at least one unique cloning site for the insertion of DNA fragments, and a second gene whose interruption by inserted DNA yields a phenotype indicative of a recombinant plasmid. Once a **library,** or large set of clones, has been obtained in a particular vector, the next problem is how to identify the particular recombinant clones that contain the gene of interest.

Screening for Particular Recombinants

The procedure of **colony hybridization** makes it possible to detect the presence of any gene for which DNA or RNA labeled with radioactivity or some other means is available (Figure 9.9). Colonies to be tested are transferred (*lifted*) from a solid medium onto a nitrocellulose or nylon filter by gently pressing the filter onto the surface. A part of each colony remains on the agar medium, which constitutes the reference plate. The filter is treated with sodium hydroxide (NaOH), which simultaneously breaks open the cells and denatures the DNA. The filter is then saturated with labeled DNA or RNA, complementary to the gene being sought, and the cellular DNA is renatured. The labeled nucleic acid used in the hybridization is called the **probe.** After washing to remove unbound probe, the positions of the bound probe identify the desired colonies. For example, with radioactively labeled probe, the desired colonies are located by means of autoradiography. A similar assay is done with phage vectors, but in this case plaques are lifted onto the filters.

If transformed cells can synthesize the protein product of a cloned gene or cDNA, then immunological techniques may allow the protein-producing colony to be identified. In one method, the colonies are transferred as in colony hybridization, and the transferred copies are exposed to a labeled antibody directed against the particular protein. Colonies to which the antibody adheres are those that contain the gene of interest.

Positional Cloning

Because of the large number of polymorphisms in DNA sequence that are found among members of natural populations of virtually every species, the genetic map of many species is covered almost completely with a dense concentration of molecular markers. The most common type of molecular marker is a *simple tandem repeat polymor-*

phism, or *STRP*, in which the polymorphism consists of variation in the number of copies of a simple repeating sequence. STRPs and similar types of molecular markers were examined in Section 4.4 in the context of the human genetic map. In the molecular map of the human genome, there is an average of approximately 700 kb between adjacent markers.

In a species in which there is a high density of molecular markers in the genetic map, the genetic map itself often supplies the information needed for cloning a gene. The procedure for cloning a mutation in any gene of interest has four steps.

1. Determine the genetic map position of the mutant gene as precisely as possible using molecular markers linked to it.

2. For molecular markers flanking the mutant gene, isolate clones that contain the marker sequences.

3. Verify that the DNA inserts in the isolated clones are mutually overlapping in such a way as to cover the entire region between the flanking markers without any gaps. Such a mutually overlapping set of clones is called a **contig** because adjacent clones are contiguous ("touching"). If the initial set of clones does include one or more gaps, complete the contig by screening the clone library with additional probes derived from the ends of the clones flanking each gap.

4. Identify the clone or clones in the contig that contain the gene of interest. This usually entails determining the nucleotide sequence of all or part of the contig in wildtype and mutant chromosomes to identify the site of the mutation.

Note that the cloning procedure outlined in these steps requires no knowledge of what the gene of interest does. All the method requires is that a mutation in the gene be mapped rather precisely with respect to molecular markers, which, in turn, are used to isolate a contig of clones covering the relevant region. Use of the genetic map to clone genes in this manner

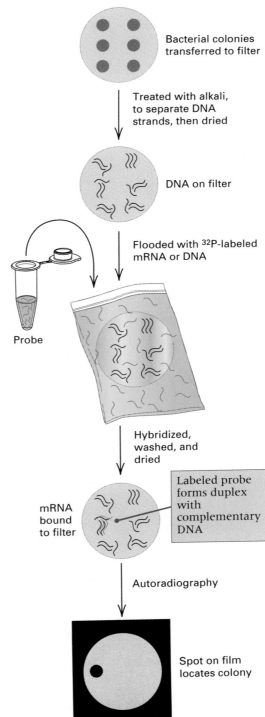

Figure 9.9
Colony hybridization. The reference plate, from which the colonies were obtained, is not shown.

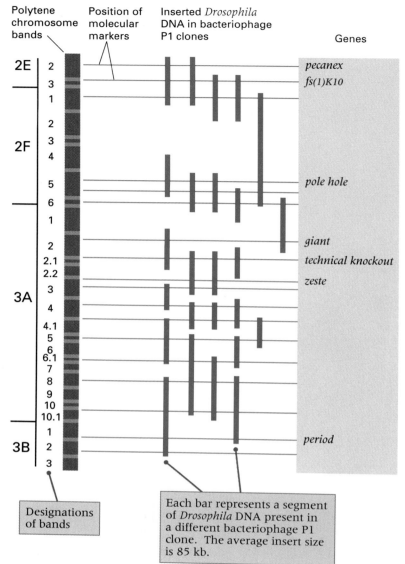

Polytene chromosome bands

Position of molecular markers

Inserted *Drosophila* DNA in bacteriophage P1 clones

Genes

pecanex
fs(1)K10

pole hole

giant
technical knockout
zeste

period

Designations of bands

Each bar represents a segment of *Drosophila* DNA present in a different bacteriophage P1 clone. The average insert size is 85 kb.

Figure 9.10

Positional cloning in *Drosophila* is made possible by a set of P1 clones that covers virtually the entire genome. At the right is illustrated a set of P1 clones covering a small region of the X chromosome. (Most of the X chromosome is also covered with cosmid contigs.) The P1 clones contain sequences for a set of molecular markers (red lines) that have known locations on the genetic map as well as in the polytene chromosomes of the salivary glands (left). The locations of several cloned genes in the region are shown. Any new mutation in the region is cloned by first determining its location with respect to flanking molecular markers and then identifying the gene within the P1 clones that contain the flanking markers. [After D. L. Hartl, D. I. Nurminsky, R. W. Jones, and E. R. Lozovskaya, *Proc. Natl. Acad. Sci. USA* 1994. 91: 6824.]

is called **positional cloning** or **map-based cloning.** It is one of the most important approaches to cloning genes in all species that have a dense genetic map, and it is particularly important in human molecular genetics.

In many well-studied organisms, contigs of clones are already available that cover almost the entire genome. These organisms include bacteria, such as *E. coli;* fungi, including *S. cerevisiae;* the nematode *C. elegans;* the fruit fly *D. melanogaster;* plants, such as *A. thaliana;* the mouse *Mus musculus;* human beings *Homo sapiens;* and many others. The utility of such contigs is illustrated for a small part of the *Drosophila* genome in Figure 9.10. On the left is a diagram of the banding pattern in the polytene salivary gland chromosomes across the region 2E2 through 3B3. (The conventions for naming the bands are discussed in Section 7.6.) These 26 bands comprise approximately 650 kb of DNA. To the right of the bands, the red lines represent the positions of molecular markers in the region, spaced at an average distance of about 30 kb. The blue bars depict P1 clones known to contain sequences of the molecular markers. Each clone contains an insert of approximately 85 kb. The red lines coincide with the positions of the molecular markers, and the points of intersection with the clones show the presence of each sequence within each clone. Seven cloned genes in the region are indicated. In the cloning of any new mutant gene that maps within the region, the location of the mutation with respect to the markers immediately identifies the clone or clones in which the gene is located.

9.3 Site-Directed Mutagenesis

Once a gene has been identified and cloned, the next step is usually to sequence the gene and ascertain whether similar sequences are already present in databases from other organisms. Similarity in sequence with a known gene sometimes gives an important clue to the function of a previously unknown gene. Other clues to function can be determined by deliberately introducing mutations into the gene and putting the gene back into the organism to observe the resulting phenotype. Methods by which a gene can be introduced into the germ line of a living organism are examined in Section 9.4. In this section we deal with the method of mutagenesis itself.

Virtually any desired mutation can be introduced into a cloned gene by direct manipulation of the DNA. An important method for producing changes in one or a few nucleotides is called **oligonucleotide site-directed mutagenesis.** Application of this method for the substitution of a single nucleotide in a gene is outlined in Figure 9.11. The first step is to isolate a single-stranded DNA molecule from a plasmid that contains the target gene and two antibiotic-resistance markers—in this example, kanamycin resistance (*kan-r*) and ampicillin resistance (*amp-r*). One of the genes is deliberately chosen to have a single-nucleotide substitution that inactivates the gene, which in this example is the *amp-r* gene. Hence, the double-stranded form of the plasmid would have the phenotype Kan-r Amp-s because of the mutation in *amp*. The single-stranded molecule is annealed with two short oligonucleotides (usually from 20 to 50 nucleotides), as shown in Figure 9.11A. One oligonucleotide is complementary to the target gene except for a single-nucleotide mismatch. The other oligonucleotide is complementary to a region of the *amp-s* allele except for a single-nucleotide mismatch that reverses the *amp* mutation. After annealing of the oligonucleotides, which act as primers, the complementary strand of the DNA is synthesized by the addition of DNA polymerase and DNA ligase. The oligonucleotide primers are incorporated into the new strand (Figure 9.11B).

The next step is transformation, in which the DNA is introduced into a bacterial cell. The circular plasmid DNA contains two mismatches, one to correct the antibiotic-resistance mutation and the other in the target gene, each of which is theoretically susceptible to repair by enzymes present in the cell. The mismatch repair enzymes work by cleaving one of the mismatched nucleotides out of the molecule, widening the resulting gap somewhat, and resynthesizing across the gap from the remaining template strand. In *E. coli*, the

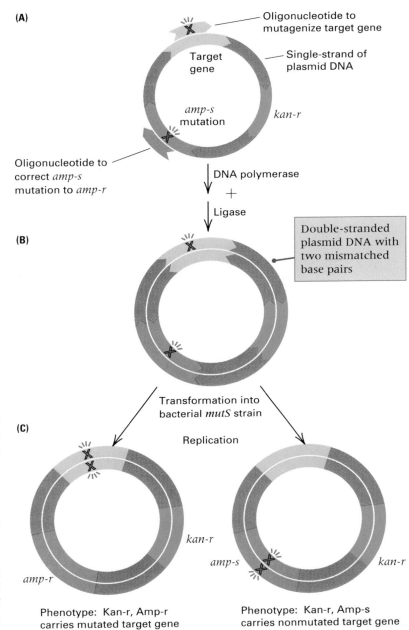

Figure 9.11
Method for introducing specific nucleotide substitutions into a DNA molecule. (A) A single strand of a plasmid is annealed with two oligonucleotides, one containing a mismatch that creates the desired mutation in the target gene and the other containing a mismatch that corrects a preexisting mutation in an ampicillin-resistance gene. (B) The second DNA strand, which incorporates the oligonucleotides, is synthesized by DNA polymerase and the gaps are closed by ligase. (C) After transformation and one round of DNA replication, the mutated strand yields a plasmid that confers ampicillin resistance and that also contains the desired mutation in the target gene. The *mutS* mutation in the host knocks out the mismatch repair system.

possibility of repair is eliminated by using a *mutS* strain for transformation, because *mutS* mutants lack one of the key enzymes necessary for mismatch repair (Chapter 13). After one round of replication, the original molecule segregates into two daughter molecules, each without mismatches (Figure 9.11C). One daughter molecule contains the mutations introduced by the oligonucleotides, which create the desired mutation in the target gene and reverse the *amp* mutation. The other daughter molecule contains neither mutation and so has the unmutated target and the *amp-s* mutation. Therefore, if the transformed cells are plated on medium containing both kanamycin and ampicillin, only those cells that contain descendants of the mutagenized strand can survive, because only these cells are Kan-r as well as Amp-r. Furthermore, in a typical experiment of this kind, more than 90 percent of

the Kan-r Amp-r cells also contain the desired mutation in the target gene.

Many genes are too large to be contained completely in a plasmid of the size used for site-directed mutagenesis. In such cases, only a fragment of the target gene is mutated, and then the mutated fragment is isolated from the plasmid and used to replace the nonmutant homologous fragment from the wildtype gene. Site-directed mutagenesis of part of the wildtype gene for the *E. coli* enzyme alkaline phosphatase is shown in Figure 9.12. The nucleotide substitution in the oligonucleotide is shown in red. The amino acid sequence coded by the wildtype and the mutant differ in a single amino acid, with serine replacing cysteine. This particular cysteine residue was thought to be critical in protein folding. As expected, the cysteine is critical, and the mutant enzyme was found to be completely nonfunctional.

Figure 9.12

Production of a specific mutation in the *E. coli* alkaline phosphatase by the use of oligonucleotide site-directed mutagenesis. The mismatch in the oligonucleotide is in red. The mutant protein contains a cysteine → serine replacement, which renders the enzyme nonfunctional.

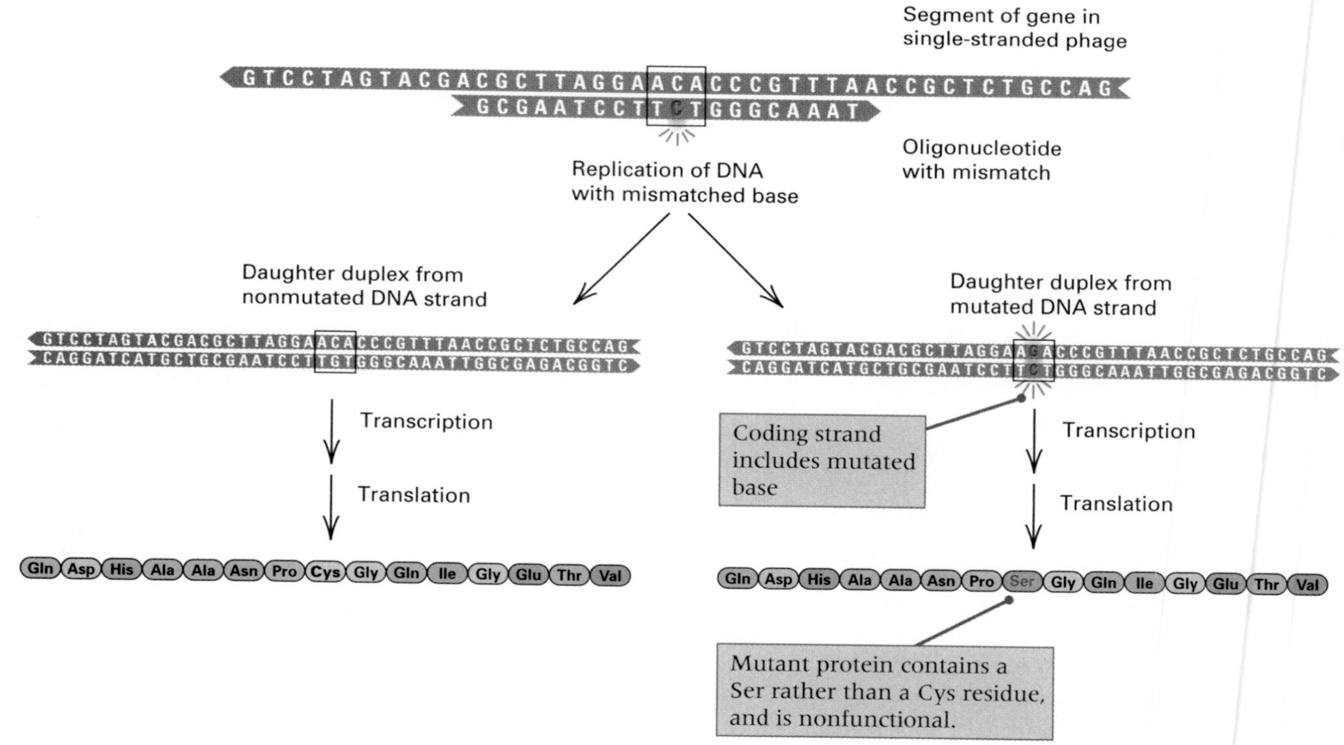

9.4 Reverse Genetics

Genetics has traditionally relied on mutation to provide the raw material needed for analysis. The customary procedure has been to use a mutant phenotype to recognize a mutant gene and then to identify the wildtype allele and its normal function. This approach has proved highly successful, as evidenced by numerous examples throughout this book. But the approach also has its limitations. For example, it may prove difficult or impossible to isolate mutations in genes that duplicate the functions of other genes or that are essential for the viability of the organism. As we saw in the previous section, site-directed mutagenesis opens up a quite different approach in which virtually any type of mutation can be created in almost any gene. Because the mutations are predetermined, a very fine level of resolution is possible in defining promoter and enhancer sequences that are necessary for transcription, the sequences necessary for messenger RNA production, and particular amino acids that are essential for protein function. This approach to genetic analysis is often called **reverse genetics** because it reverses the usual flow of study: Instead of starting with a mutant phenotype and trying to identify the nature of the mutation, reverse genetics starts with a known mutation and determines its effect on the phenotype. Of course, once a mutation has been created in the laboratory, there remains the problem of how to put it back into a living organism. The following sections describe some of the relevant techniques.

Germ-Line Transformation in Animals

Reverse genetics can be carried out in most organisms that have been extensively studied genetically, including the nematode *Caenorhabditis elegans, Drosophila,* the mouse, and many domesticated animals and plants. In nematodes, the basic procedure is to manipulate the DNA of interest in a plasmid that also contains a selectable genetic marker that will alter the phenotype of the transformed animal. The DNA is injected directly into the reproductive organs and sometimes spontaneously becomes incorporated into the chromosomes in the germ line. The result of transformation is observed and can be selected in the progeny of the injected animals because of the phenotype conferred by the selectable marker.

A somewhat more elaborate procedure is necessary for germ-line transformation in *Drosophila.* The usual method makes use of a 2.9-kb transposable element (Section 6.8) called the **P element,** which consists of a central region coding for transposase flanked by 31 base-pair inverted repeats (Figure 9.13A). A genetically engineered derivative of this P element, called *wings clipped,* can make functional transposase but cannot itself transpose because of deletions introduced at the ends of the inverted repeats (Figure 9.13B). For germ-line transformation, the vector is a plasmid containing a P element that includes, within the inverted repeats, a selectable genetic marker (usually one affecting eye color), as well as a large internal deletion that removes much of the transposase-coding region. By itself, this P element cannot transpose because it makes no transposase, but it can be mobilized by the transposase produced by the wings-clipped or other intact P elements.

In *Drosophila* transformation, any DNA fragment of interest is introduced between the ends of the deleted P element. The resulting plasmid and a different plasmid containing the wings-clipped element are injected into the region of the early embryo containing the germ cells. The DNA is taken up by the germ cells, and the wings-clipped element produces functional transposase (Figure 9.13B). This mobilizes the engineered P vector and results in its transposition into an essentially random location in the genome. Transformants are detected among the progeny of the injected flies because of the eye color or other genetic marker included in the P vector. Integration into the germ line is typically very efficient: From 10 to 20 percent of the injected embryos that survive and are fertile yield one or more transformed progeny. However, the efficiency decreases with the size of the DNA fragment in the P element, and the effective upper limit is approximately 20 kb.

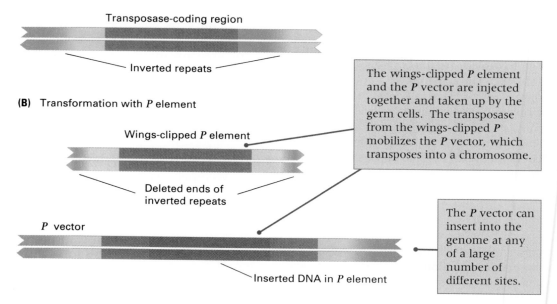

(A) Complete *P* element

Transposase-coding region

Inverted repeats

(B) Transformation with *P* element

Wings-clipped *P* element

Deleted ends of inverted repeats

P vector

Inserted DNA in *P* element

The wings-clipped *P* element and the *P* vector are injected together and taken up by the germ cells. The transposase from the wings-clipped *P* mobilizes the *P* vector, which transposes into a chromosome.

The *P* vector can insert into the genome at any of a large number of different sites.

Figure 9.13
Transformation in *Drosophila* mediated by the transposable element *P*. (A) Complete *P* element containing inverted repeats at the ends and an internal transposase-coding region. (B) Two-component transformation system. The vector component contains the DNA of interest flanked by the recognition sequences needed for transposition. The wings-clipped component is a modified *P* element that codes for transposase but cannot transpose itself because critical recognition sequences are deleted.

Transformation of the germ line in mammals can be carried out in several ways. The most direct is the injection of vector DNA into the nucleus of fertilized eggs, which are then transferred to the uterus of foster mothers for development. The vector is usually a modified retrovirus. **Retroviruses** have RNA as their genetic material and code for a reverse transcriptase that converts the retrovirus genome into double-stranded DNA that becomes inserted into the genome in infected cells (Section 7.7). Genetically engineered retroviruses containing inserted genes undergo the same process. Animals that have had new genes inserted into the germ line in this or any other manner are called **transgenic** animals.

Another method of transforming mammals uses **embryonic stem cells** obtained from embryos a few days after fertilization (Figure 9.14). Although embryonic stem cells are not very hardy, they can be isolated and then grown and manipulated in culture; mutations in the stem cells can be selected or introduced using recombinant DNA vectors. The mutant stem cells are introduced into another developing embryo and transferred into the uterus of a foster mother (Figure 9.14A), where they become incorporated into various tissues of the embryo and often participate in forming the germ line. If the embryonic stem cells carry a genetic marker, such as a gene for black coat color, then mosaic animals can be identified by their spotted coats (Figure 9.14B). Some of these animals, when mated, produce black offspring (Figure 9.14C), which indicates that the embryonic stem cells had become incorpo-

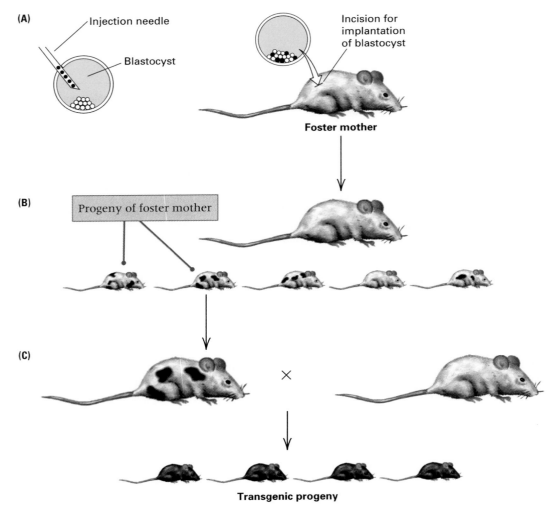

Figure 9.14
Transformation of the germ line in the mouse using embryonic stem cells. (A) Stem cells obtained from an embryo of a black strain are isolated and, after genetic manipulation in culture, injected into the embryo of a white strain, which is then introduced into the uterus of a foster mother. (B) The resulting offspring are often mosaics that contain cells from both the black and the white strains. (C) If cells from the black strain colonize the germ line, then the offspring of the mosaic animal will be black. [After M. R. Capecchi. 1989. *Trends Genet.*, 5: 70.]

rated into the germ line. In this way, mutations introduced into the embryonic stem cells while they were in culture may become incorporated into the germ line of living animals. The method in Figure 9.14 has been used to create strains of mice with mutations in genes associated with such human genetic diseases as cystic fibrosis. These strains serve as mouse models for studying the disease and for testing new drugs and therapeutic methods.

The procedure for introducing mutations into specific genes is called **gene targeting.** The specificity of gene targeting comes from the DNA sequence homology needed for homologous recombination. Two examples are illustrated in Figure 9.15, where the DNA sequences present in

Hello, Dolly!

Ian Wilmut, Anagelika E. Schnieke,
Jim McWhir, Alex J. Kind, and
Keith H. S. Campbell 1997
Roslin Institute, Roslin, Midlothian,
Scotland
*Viable Offspring Derived from Fetal
and Adult Mammalian Cells*

The Scottish Finn Dorset ewe known as "Dolly" is the first mammal to have been cloned from the nucleus of a cell taken from an adult mammal. The experiment created a press sensation. The President of the United States said that he would be against cloning people. A professor of public health said, "I don't think reasonable, rational people would want to clone themselves, but an eccentric millionaire might want to leave his money to a clone. There's no way of stopping the super-rich from going offshore to clone themselves." Some people were concerned for the animals. A spokesperson for People for the Ethical Treatment of Animals was quoted as saying, "It is time that society learned to respect our fellows, not exploit them for every fool thing." (Note in the excerpt that follows that the experiment was carried out with the prior approval of the appropriate Animal Welfare Committee.) Many important ethical issues are raised by the possibility of cloning humans. There are also important issues concerning the status of cloned animals. Should the clone of a racehorse that won the Triple Crown com-

pete under the same rules as other racehorses? Should the clones of champion show dogs and cats compete under the same rules? Who should decide?

Transfer of a single nucleus at a specific stage of development, to an enucleated unfertilized egg, provided an opportunity to investigate whether cellular differentiation to that stage involved irreversible genetic modification. The first offspring to develop

> **The fact that a lamb was derived from an adult cell confirms that differentiation of that cell did not involve the irreversible modification of genetic material required for development to term.**

from a differentiated cell were born after nuclear transfer from an embryo-derived cell that had been induced to become quiescent. Using the same procedure, we now report the birth of live lambs from adult mammary gland, fetus, and embryo. The fact that a lamb was derived from an adult cell confirms that differentiation of that cell did not involve the irreversible modification of genetic material required for development to term. . . . If the recipient cytoplasm is prepared by enucleation of an oocyte at

metaphase II, it is only possible to avoid chromosomal damage and maintain normal ploidy by transfer of diploid nuclei. Our studies with cultured cells suggest that there is an advantage if cells are quiescent. . . . Together our results indicate that nuclei from a wide range of cell types should prove to be totipotent after enhancing opportunities for reprogramming by using appropriate combinations of these cell-cycle stages. In turn, the dissemination of the genetic improvement obtained within elite selection herds will be enhanced by limited replication of animals with proven performance by nuclear transfer of cells derived from adult animals. . . . The lamb born after nuclear transfer from a mammary gland cell is, to our knowledge, the first mammal to develop from a cell derived from an adult tissue. . . . This is consistent with the generally accepted view that mammalian differentiation is almost all achieved by systematic, sequential changes in gene expression brought about by interactions between the nucleus and the changing cytoplasmic environment. . . . These experiments were conducted under the Animals (Scientific Procedure) Act 1986 [United Kingdom] and with the approval of the Roslin Institute Animal Welfare Experiments Committee.

Source: Nature 385: 810–813

gene-targeting vectors are shown as looped configurations paired with homologous regions in the chromosome prior to recombination. The targeted gene is shown in pink. In Figure 9.15A, the vector contains the targeted gene interrupted by an insertion of a novel DNA sequence, and homologous recombination results in the novel sequence becoming inserted into the targeted gene in the genome. In Figure 9.15B, the vector contains only flanking sequences, not the targeted gene, so homologous recombination results in replacement of the targeted gene with an

unrelated DNA sequence. In both cases, cells with targeted gene mutations can be selected by including an antibiotic-resistance gene, or other selectable genetic marker, in the sequences that are incorporated into the genome through homologous recombination.

Genetic Engineering in Plants

A procedure for the transformation of plant cells makes use of a plasmid found in the soil bacterium *Agrobacterium tumefaciens* and related species. Infection of susceptible

(A) Vector with DNA inserted into target gene

(B) Vector with DNA replacing target gene

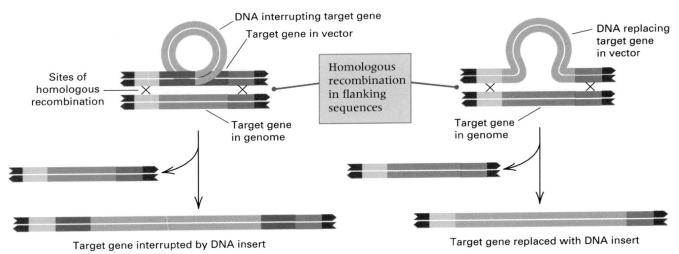

Target gene interrupted by DNA insert

Target gene replaced with DNA insert

Figure 9.15
Gene targeting in embryonic stem cells. (A) The vector (top) contains the target sequence (red) interrupted by an insertion. Homologous recombination introduces the insertion into the genome. (B) The vector contains DNA sequences flanking the targeted gene. Homologous recombination results in replacement of the targeted gene with an unrelated DNA sequence. [After M. R. Capecchi. 1989. *Trends Genet.*, 5: 70.]

plants with this bacterium results in the growth of tumors known as **crown gall tumors** at the entry site, usually a wound. Susceptible plants comprise about 160,000 species of flowering plants, known as the dicots, and include the great majority of the most common flowering plants.

The *Agrobacterium* contains a large plasmid of approximately 200 kb called the **Ti plasmid,** which includes a smaller (~25 kb) region known as the **T DNA** flanked by 25 base-pair direct repeats (Figure 9.16A). The *Agrobacterium* causes a profound change in the metabolism of infected cells because of transfer of the T DNA into the plant genome. The T DNA contains genes coding for proteins that stimulate division of infected cells, hence causing the tumor, and also coding for enzymes that convert the amino acid arginine into an unusual derivative, generally **nopaline** or **octopine** (depending on the particular type of *Ti* plasmid), that the bacterium needs in order to grow. The transfer functions are present not in the T DNA itself but in another region of the plasmid called the *vir* (stands for *virulence*) region of about 40 kb that includes six genes necessary for transfer.

Transfer of T DNA into the host genome is similar in some key respects to bacterial conjugation, which we examined in Chapter 8. In infected cells, transfer begins with the formation of a nick that frees one end of the T DNA (Figure 9.16A), which peels off the plasmid and is replaced by rolling-circle replication (Figure 9.16B). The region of the plasmid that is transferred is delimited by a second nick at the other end of the T DNA, but the position of this nick is variable. The resulting single-stranded T DNA is bound with molecules of a single-stranded binding protein (SSBP) and is transferred into the plant cell and incorporated into the nucleus. There it is integrated into the chromosomal DNA by a mechanism that is still unclear (Figure 9.16C). Although the SSBP has certain similarities in amino acid sequence to the *recA* protein from *E. coli*, which plays a key role in homologous recombination, it is clear that integration of the T DNA does not require homology.

Use of T DNA in plant transformation is made possible by engineered plasmids in which the sequences normally present in T DNA are removed and replaced with those to be incorporated into the plant genome

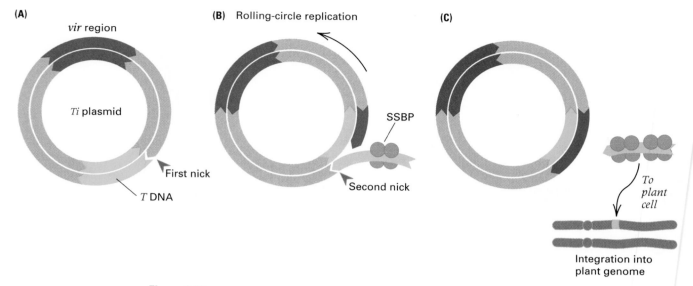

(A) vir region

Ti plasmid

First nick

T DNA

(B) Rolling-circle replication

SSBP

Second nick

(C)

To plant cell

Integration into plant genome

Figure 9.16

Transformation of a plant genome by *T* DNA from the *Ti* plasmid. (A) A nick forms at the 5' end of the *T* DNA. (B) Rolling-circle replication elongates the 3' end and displaces the 5' end, which is stabilized by single-stranded binding protein (SSBP). A second nick terminates replication. (C) The SSBP-bound *T* DNA is transferred to a plant cell and inserts into the genome.

along with a selectable marker. A second plasmid contains the *vir* genes and permits mobilization of the engineered T DNA. In infected tissues, the *vir* functions mobilize the T DNA for transfer into the host cells and integration into the chromosome. Transformed cells are selected in culture by the use of the selectable marker and then grown into mature plants in accordance with the methods described in Section 7.3.

9.5
Applications of Genetic Engineering

Engineered male sterility in plants is just one example of how recombinant DNA technology has revolutionized modern biology. At the present time, recombinant DNA technology is used mainly in (1) the efficient production of useful proteins, (2) the creation of cells capable of synthesizing economically important molecules, (3) the generation of DNA and RNA sequences as research tools or in medical diagnosis, (4) the manipulation of the genotype of organisms such as plants, and (5) the potential correction of genetic defects in animals,

including human patients (gene therapy). Some examples of these applications follow.

Giant Salmon with Engineered Growth Hormone

In many animals, the rate of growth is controlled by the amount of growth hormone produced. Transgenic animals with a growth-hormone gene under the control of a highly active regulatory sequence are often larger than their normal counterparts. An example of a highly active sequence that can drive gene expression is the regulatory region of a gene for **metallothionein.** The metallothioneins are proteins that bind heavy metals. They are ubiquitous in eukaryotic organisms and are often encoded in a family of related genes. Human beings, for example, have more than ten metallothionein genes that can be separated into two major groups according to their sequences. The regulatory region of a metallothionein gene accelerates transcription of any gene to which it is attached in response to heavy metals or steroid hormones. When DNA constructs consisting of a rat growth-hormone gene under metallothionein control are used to produce transgenic mice, the resulting animals grow about twice as large as normal mice.

The effect of another growth-hormone construct is shown in Figure 9.17. The fish are coho salmon at 14 months of age. Those on the left are normal, whereas those on the right are transgenic animals that contain a salmon growth-hormone gene driven by a metallothionein regulatory region. Both the growth-hormone gene and the metallothionein gene were cloned from the sockeye salmon. As an index of scale, the largest transgenic fish on the right has a length of about 42 cm. On average, the transgenic fish are 11 times heavier than their normal counterparts; the largest transgenic fish was 37 times the average weight of the nontransgenic animals. Not only do the transgenic salmon grow faster and become larger than normal salmon, they also mature faster.

Figure 9.17
Normal coho salmon (left) and genetically engineered coho salmon (right) containing a sockeye salmon growth-hormone gene driven by the regulatory region from a metallothionein gene. The transgenic salmon average 11 times the weight of the nontransgenic fish. The smallest fish on the left is about 4 inches long. [Courtesy of R. H. Devlin; see R. H. Devlin, T. Y. Yesaki, C. A. Biagi, E. M. Donaldson, P. Swanson, and W.-K. Chan. 1994. *Nature* 371: 209.]

Engineered Male Sterility with Suicide Genes

Another practical application of genetic engineering is in the production of male-sterile plants. Compared with their inbred parents, hybrid plants are usually superior in numerous respects, including higher yield and increased disease resistance (Chapter 16). Male sterility is important in hybrid seed production, because it promotes efficient hybridization between the inbred lines. Cytoplasmic male sterility is widely used in corn breeding (Chapter 14), but analogous mutations are not available in many crop plants. To be optimally useful, male sterility should also be reversible by fertility restorers so that the inbred lines can be propagated.

A genetically engineered system of male sterility and fertility restoration has been introduced into a number of plant species, including the oilseed rape, *Brassica napus,* which is a major source of vegetable oil. (Canola oil comes from the oilseed rape.) The genetic engineering of male sterility and fertility restoration makes use of two engineered genes. The basis of the sterility is an extracellular RNA nuclease called **barnase,** which is produced by the bacterium *Bacillus amyloliquefaciens.* The barnase nuclease is an extremely potent cellular toxin. In the use of barnase to produce male sterility, the coding sequence of the bacterial gene is fused with the regulatory sequence of a gene, *TA29,* that has a tissue-specific expression in the tapetal cell layer that surrounds the pollen sacs in the anther (Figure 9.18A). When the artificial *TA29-barnase* gene is transformed into the genome by the use of T DNA from *Agrobacterium,* the resulting plants are male-sterile because of the destruction of tapetal cell RNA by the barnase enzyme and the resulting lethality of cells in the tapetum (Figure 9.18B).

Fertility restoration makes use of another protein, called **barstar,** also produced by *B. amyloliquefaciens.* Barstar is an intracellular protein that protects against the lethal effects of barnase by forming a stable, enzymatically inactive complex in the cytoplasm. Hence, barstar is the bacterial cell's self-defense against its own barnase. Plants transformed with an artificial *TA29-barstar* gene are healthy and fertile; they merely produce barstar in the tapetum because of the tapetum-specific expression of the *TA29* regulatory region. However, when the male-sterile *TA29-barnase* plants are crossed with those carrying *TA29-barstar*, the resulting genotype that combines *TA29-barnase* with *TA29-barstar* is male-fertile (Figure 9.18C). The reason for the restoration of fertility is that the barstar protein combines with the barnase nuclease and renders it ineffective.

The macroscopic appearance and the microscopic appearance of oilseed rape plants of various genotypes are shown in Figure 9.19. The wildtype phenotype is shown in parts A and B, and numerous pollen grains (PG) are apparent in the cross section of the anther. Parts C and D show the flower and anthers of plants carrying the *TA29-barnase;* the anthers are small, shriveled, and devoid of pollen. Parts E and F show the flower and anthers of plants whose genotypes include *TA29-barnase* plus *TA29-barstar;* the phenotype is virtually indistinguishable from wildtype.

Other Commercial Opportunities

As we have seen, organisms with novel phenotypes can be produced by genetic engineering, sometimes by combining the features of different organisms. Additional examples include a species of marine bacte-

Figure 9.18

Engineered genetic male sterility and fertility restoration. (A) Pollen grains (PG) develop within a thin layer of tapetal cells (T). (B) Expression of the barnase RNA nuclease under the control of the tapetum-specific regulatory region of gene *TA29* destroys the tapetum and renders the plants male-sterile. (C) Expression of barstar (a barnase inhibitor) in tapetal cells inactivates barnase and restores fertility. [Courtesy of Robert B. Goldberg.]

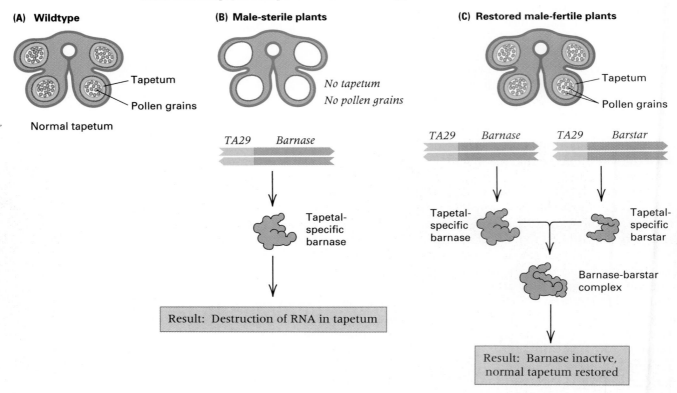

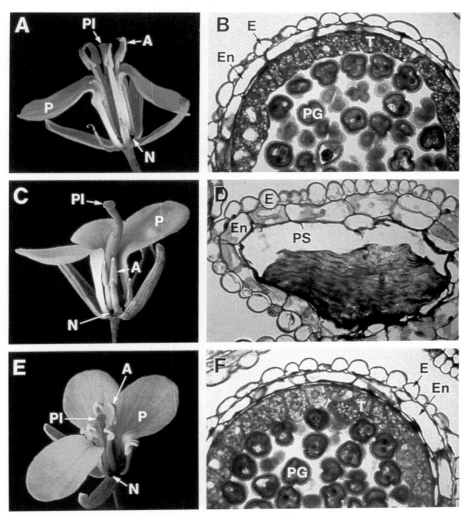

Figure 9.19

Flowers (A, C, and E) and cross sections of anthers (B, D, and F) of oilseed rape plants. The genotypes are as follows: A and B, wildtype; C and D, *TA29-barnase;* E and F, *TA29-barnase + TA29-barstar*. Note that the fertility-restored plants (E and F) are virtually indistinguishable from wildtype (A and B). The labeled parts of the flowers are A, anther; P, petal; Pl, pistil; and N, nectary. The labeled parts of the cross sections are E, epidermis; En, endothecium; PG, pollen grain; PS, pollen sac; and T, tapetum. [From C. Mariani, V. Gossele, M. De Beuckeleer, M. De Block, R. B. Goldberg, W. De Greef and J. Leemans. 1992. *Nature* 357: 384.]

ria that has been equipped with genes from other bacterial species for the metabolism of some components of petroleum, yielding an organism used in cleaning up oil spills in the oceans. Many biotechnology companies are at work designing bacteria that can synthesize industrial chemicals or degrade industrial wastes. Bacteria have also been created that are able to compost waste more efficiently and to fix nitrogen to improve the fertility of soil. A great deal of effort is currently being expended to create organisms that can convert biological waste into alcohol. A number of medicinally or commercially important molecules are routinely produced in genetically engineered cells, including human insulin and growth hormones produced in bacteria.

Altering the genotypes of plants (Figure 9.16) is an important application of recombinant DNA technology. It is possible to transfer genes from one plant species to another. The transferred genes affect yield, hardiness, or disease resistance. There are also attempts to alter the surface structure of the roots of grains, such as wheat, by introducing genes from legumes (peas, beans) to give the grains the ability that legumes possess to establish root nodules of nitrogen-fixing bacteria. If successful, this would eliminate the need to add nitrogenous fertilizers to soils where grains are grown.

The first engineered recombinant plant of commercial value was developed in 1985. An economically important herbicide is **glyphosate,** a weed killer that inhibits an enzyme in a metabolic pathway found only in plants and microorganisms. The pathway is the *shikimic acid pathway,* and glyphosate is a competitive inhibitor of the enzyme *EPSP synthase* (stands for *5-enolpyruvylshikimic acid-3-phosphate synthase)*. Glyphosate is the active ingredient in the world's largest-selling herbicide, Roundup, which is used on more than 100 crops. Treatment requires that the compound be applied directly to the green foliage of the weeds because it is also toxic to the crop plant itself. However, the target gene of glyphosate is also present in the bacterium *Salmonella typhimurium.* A resistant form of the gene was obtained by mutagenesis and growth of *Salmonella* in the presence of glyphosate. Then the gene was cloned in *E. coli* and recloned in the T DNA of *Agrobacterium.* Transformation of plants with the glyphosate-resistance gene has yielded varieties of maize, cotton, tobacco, and other plants that are resistant to the herbicide. Thus fields of these crops can be sprayed with glyphosate at any stage of growth of the crop. The weeds are killed but the crop is unharmed.

Genetic engineering can also be used to control insect pests. For example, the black cutworm causes extensive crop damage and is usually combated with noxious insecticides. The bacterium *Bacillus thuringiensis* produces a protein that is lethal to the black cutworm, but the bacterium does not normally grow in association with the plants that are damaged by the worm. However, the gene coding for the lethal protein has been introduced into the soil bacterium *Pseudomonas fluorescens,* which lives in association with maize and soybean roots. Inoculation of soil with the engineered *P. fluorescens* helps control the black cutworm and reduces crop damage.

Uses in Research

Modern molecular genetics could not exist without recombinant DNA technology. It is an essential research tool. Reverse genetics makes it possible to isolate and alter a gene at will and introduce it back into a living cell or even the germ line. Besides saving time and labor, reverse genetics enables mutants to be constructed that cannot be formed in any other way. An example is the formation of double mutants of animal viruses, which naturally undergo recombination at such a low frequency that mutations can rarely be recombined by genetic crosses. The greatest impact of recombinant DNA on basic research has been in the study of eukaryotic gene regulation and development, and many of the principles of regulation and development summarized in Chapters 11 and 12 were determined by using these methods.

Production of Useful Proteins

Among the most important applications of genetic engineering is the production of large quantities of particular proteins that are otherwise difficult to obtain (for example, proteins that are present in only a few molecules per cell or that are produced in only a small number of cells or only in human cells). The method is simple in principle. A DNA sequence coding for the desired protein is cloned in a vector adjacent to an appropriate regulatory sequence. This step is usually done with cDNA because cDNA has all the coding sequences spliced together in the right order. Using a vector with a high copy number ensures that many copies of the coding sequence will be present in each bacterial cell, which can result in synthesis of the gene product at concentrations ranging from 1 to 5 percent of the total cellular protein. In practice, the production of large quantities of a protein in bacterial cells is straightforward, but there are often problems that must be overcome because, in the bacterial cell, which is a prokaryote, the eukaryotic protein may be unstable, may not fold properly, or may fail to undergo necessary chemical modification. Many important proteins are currently produced in bacterial cells, including human growth hormone, blood-clotting factors, and insulin. Patent offices in Europe and the United States have already issued well over 1000 patents for the clinical use of the products of genetically engineered human genes. Figure 9.20 gives a breakdown of the number of patents issued relative to clinical application.

Genetic Engineering with Animal Viruses

The genetic engineering of animal cells often makes use of retroviruses because their reverse transcriptase makes a double-stranded DNA copy of the RNA genome, which then becomes inserted into the chromosomes of the cell. DNA-to-RNA transcription occurs only after the DNA copy is inserted. The infected host cell survives the infection, retaining the retroviral DNA in its genome. These features of retroviruses make them convenient vectors for the genetic manipulation of animal cells, including those of birds, rodents, monkeys, and human beings.

Genetic engineering with retroviruses allows the possibility of altering the genotypes of animal cells. Because a wide variety of retroviruses are known, including many that infect human cells, genetic defects may be corrected by these procedures in the future. However, many retroviruses contain a gene that results in uncontrolled growth of the infected cell, thereby causing a tumor. When retrovirus vectors are used for genetic engineering, the tumor-causing gene is first deleted. The deletion also provides the space needed for the incorporation of foreign DNA sequences. The recombinant DNA procedure employed with retroviruses consists of synthesis in the laboratory of double-stranded DNA from the viral RNA, through use of reverse transcriptase. The DNA is then cleaved with a restriction enzyme and, by means of the techniques already described, foreign DNA is inserted. Trans-

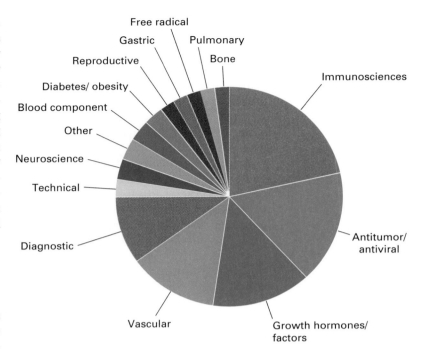

Figure 9.20
Relative number of patents issued for various clinical applications of the products of genetically engineered human genes. [Data from *Nature* 1996, 380: 388.]

formation yields cells with the recombinant retroviral DNA permanently inserted into the genome. In this way, the genotype of the cells can be altered.

Attempts are currently underway to assess the potential of genetic engineering in the treatment of cystic fibrosis. This disorder results from a mutation in a gene for chloride transport in the lungs, pancreas, and other glands, leading to abnormal secretions and the accumulation of a thick, sticky, honey-like mucus in the lungs. Among its chief symptoms are recurrent respiratory infections. A genetically engineered form of the common cold virus is being used in clinical trials in an attempt to introduce the nonmutant form of the cystic fibrosis gene directly into the lungs of affected patients. This is an example of **gene therapy.** The exciting potential of gene therapy lies in the possibility of correcting genetic defects—for example, restoring the ability to make insulin persons with diabetes or correcting immunological deficiencies in patients with defective immunity. However, a number of major problems stand in the way of gene therapy becoming widely used. At this time, there is no completely reliable way to ensure that a gene will be inserted only into the appropriate target cell or target tissue, and there is no completely reliable means of regulating the expression of the inserted genes.

A major breakthrough in disease prevention has been the development of synthetic vaccines. Production of certain vaccines is difficult because of the extreme hazards of working with large quantities of the active virus—for example, the human immunodeficiency virus (HIV1) that causes acquired immune deficiency syndrome (AIDS). The danger is minimized by cloning and producing viral antigens in a nonpathogenic organism. Vaccinia virus, the agent used in smallpox vaccination, has been very useful for this purpose. Viral antigens are often on the surface of virus particles, and some of these antigens can be engineered into the coat of vaccinia. For example, engineered vaccinia with surface antigens of hepatitis B, influenza virus, and vesicular stomatitis virus (which kills cattle, horses, and pigs) have produced useful vaccines in animal tests. A surface antigen of *Plasmodium falciparum*, the parasite that causes malaria, has also been placed in the vaccinia coat, a development that may ultimately lead to an antimalaria vaccine.

Diagnosis of Hereditary Diseases

Probes derived by recombinant DNA methods are widely used in prenatal detection of disease, including cystic fibrosis, Huntington disease, sickle-cell anemia, and hundreds of other genetic disorders. In some cases, probes derived from the gene itself are used; in other cases, molecular markers, such as STRPs, that are genetically linked to the disease gene are employed (Section 4.4). If the disease gene itself, or a region close to it in the chromosome, differs from the normal chromosome in the positions of one or more cleavage sites for restriction enzymes, then these differences can be detected with Southern blots using cloned DNA from the region as the probe (Section 5.7) or in DNA amplified by the polymerase chain reaction (Section 5.8).

The genotype of the fetus can be determined directly. These techniques are very sensitive and can be carried out as soon as tissue from the fetus (or even from the embryonic membranes) can be obtained.

9.6 Analysis of Complex Genomes

Gene structure and function are understood in considerable detail because of the manipulative capability of recombinant DNA technology combined with genetic analysis. Most recombinant DNA techniques are limited to the manipulation of DNA fragments smaller than about 40 kb. Although this range includes most cDNAs and many genes, some genes are much larger than 50 kb and so must be analyzed in smaller pieces. However, recent advances in recombinant DNA technology have made it possible for very large DNA molecules to be cloned and analyzed. These methods are suitable not only for the study of very large genes, but also for the analysis of entire genomes. This section discusses complex genomes and the recombinant DNA technology that has made their analysis possible.

Sizes of Complex Genomes

The simplest genomes are those of small bacterial viruses whose DNA is smaller than 10 kb—for example, the *E. coli* phage ϕX174 whose genome contains 5386 base pairs. Toward the larger end of the spectrum is the human genome, consisting of 3×10^9 base pairs. The range is so large that comparisons are often made in terms of information content, with the base pairs of DNA analogous to the letters in a book. The analogy is apt, because the range of sizes is from a few pages of text to an entire library.

The book analogy is used to compare genome sizes in Figure 9.21. The volumes are about the size of a big-city telephone book: 1500 pages per volume with 25,000 characters (nucleotides) per page. In these terms, the 50-kb genome of bacteriophage λ fills two pages, and the 4.7-Mb genome of *E. coli* needs about 200 pages. Among eukaryotes, the size range for organisms of

Phage λ
50 kb
2 pages

Escherichia coli
(bacteria)
4.7 Mb
200 pages

Saccharomyces cerevisiae
(yeast)
12.5 Mb
500 pages

Caenorhabditis elegans
(nematode)
Arabidopsis thaliana
(plant)
100 Mb
3 volumes

Drosophila melanogaster
(fruit fly)
165 Mb
5 volumes

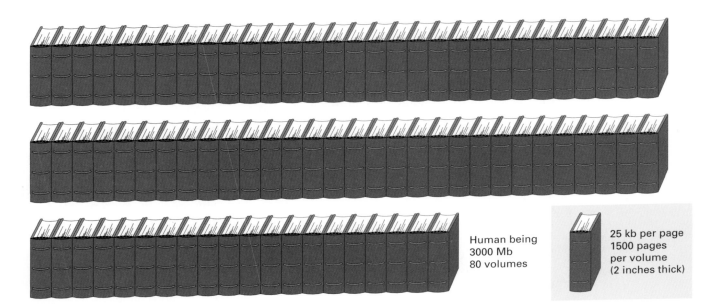

Human being
3000 Mb
80 volumes

25 kb per page
1500 pages
per volume
(2 inches thick)

Figure 9.21
Relative sizes of genomes if they were printed at 25,000 characters per page and bound in 1500-page volumes. One volume would contain about as many characters as a telephone book 2.5 inches thick. The *E. coli* genome would require about 200 pages, yeast 500 pages, and so forth.

genetic interest varies considerably. Yeast would take up one-third of a volume (genome size 12.5 Mb), the nematode *Caenorhabditis elegans* and the flowering plant *Arabidopsis thaliana* 3 volumes (100 Mb) each, *Drosophila melanogaster* 5 volumes (165 Mb), and human beings 80 volumes (3000 Mb). To put the matter in practical terms, let us calculate how many cosmids would be necessary to clone an entire genome in a cosmid library in which the average insert size is 40 kb. To have a good chance of containing most sequences of interest, a genomic library must contain several copies of the genome. For example, for a genomic library to have a 95 percent

chance of containing any given single-copy sequence, the library should consist of enough clones that the DNA fragments in the clones, added together, would equal 3 haploid genomes (3 **genome equivalents**).

The rationale for the 3 genome equivalents is that in a library of n clones, each containing a small fraction f of the haploid genome, the probability that any single-copy DNA fragment will be included in a particular clone is f, assuming that the DNA fragment is small in relation to the insert size of the clones and that there is no bias toward or against cloning the fragment of interest. The probability that none of the n

clones contains the fragment is therefore $(1 - f)^n$, so the probability that at least one clone contains the fragment is $1 - (1 - f)^n$, which we want to be greater than 95 percent. Hence,

$$1 - (1 - f)^n \geq 0.95 \quad \text{or} \quad (1 - f)^n < 0.05$$

Taking natural logarithms of both sides of the right-hand inequality and simplifying yields

$$n \geq \ln(0.05)/\ln(1 - f).$$

Now, $\ln(0.05) \approx -3$, and when f is small, $\ln(1 - f) = -f$ to a very close approximation, as can be seen by substituting in any small number for f (for example, $f = 0.001$). Hence the inequality becomes $n \geq -3/-f$ or $nf \geq 3$. But what is nf? It is the fraction of the genome present in each clone multiplied by the total number of clones or, in other words, the number of haploid genome equivalents present in the library. (To verify that you understand this argument, you should try to show that 4.6 genome equivalents are needed to ensure with 99 percent confidence that any fragment will be present.)

Taking 3 genome equivalents as needed for 95 percent coverage of the single-copy sequences, this coverage of yeast would require a little less than 1000 clones in a cosmid vector averaging 40-kb inserts. Three genome equivalents of nematode or *A. thaliana* DNA would require 7500 clones; that of *Drosophila* DNA, 12,000 clones; and that of human DNA, 225,000 clones. The larger genomes require a formidable number of clones. An alternative to larger numbers of clones is larger fragments within the clones. Clones that contain large fragments of DNA can be made and analyzed by the procedures described in the following sections.

Manipulation of Large DNA Fragments

Section 6.4 included an examination of pulsed-field gel electrophoresis, a procedure by which DNA fragments exceeding several megabases can be separated. In only a few organisms are any chromosomes as small as several megabases, however. In *Drosophila*, the smallest wildtype chromosome is about 6 Mb, and even the

smallest rearranged and deleted chromosome is 1 Mb. Most chromosomes in higher eukaryotes are very much larger. The smallest human chromosome is chromosome 21, and its long arm alone is approximately 42 Mb. Therefore, even with the ability to separate large DNA fragments by electrophoresis, it is necessary to be able to cut the DNA in the genome into fragments of manageable size. This can be done with a class of restriction enzymes, each of which cleaves at a restriction site consisting of eight bases rather than the usual six or four. For example, the restriction sites of the eight-cutter enzymes *Not*I and *Sfi*I are

*Not*I   5'-GC*GGCCGC-3'

*Sfi*I   5'-GGCCNNNN*NGGCC-3'

Both enzymes cleave double-stranded DNA at the positions of the restriction sites, and the asterisks denote the position at which the backbone is cut in each DNA strand. (The N's in the *Sfi*I restriction site mean that any nucleotide can be present at this site.) In a genome with equal proportions of the four nucleotides and random nucleotide sequences, the average size of both *Not*I and *Sfi*I fragments is $4^8 = 65,536$ nucleotide pairs, or about 66 kb. Many genomes are relatively A + T-rich, and the average *Not*I and *Sfi*I fragment size is larger than 66 kb. In vertebrate genomes, there is a bias against long runs of G's and C's, so many *Not*I and *Sfi*I fragments are considerably larger than 66 kb. In any case, the use of eight-cutter restriction enzymes allows complex DNA molecules to be cleaved into a relatively small number of large fragments that can be separated, cloned, and analyzed individually.

Cloning of Large DNA Fragments

Large DNA molecules can be cloned intact in bacterial cells with the use of specialized vectors that can accept large inserts. An example is a vector derived from the bacteriophage P1 (Figure 9.4D), which is used to clone DNA fragments averaging approximately 85 kb. DNA fragments in the appropriate size range can be produced by breaking larger molecules into fragments of the desired size by physical means, by treatment with restriction enzymes that have infrequent cleavage sites (for example, *Not*I or *Sfi*I), or by treatment with ordinary restriction enzymes under conditions in which only a fraction of the restriction sites are cleaved (**partial digestion**). Cloning the large molecules consists of mixing the large fragments of source DNA with the vector, ligation with DNA ligase, introduction of the recombinant molecules into bacterial cells, and selection for the clones of interest. These methods are generally similar to those described in Section 9.2 for the production of recombinant molecules containing small inserts of cloned DNA.

DNA fragments as large as 1 Mb can be cloned intact in yeast cells with the use of special vectors for creating **yeast artificial chromosomes,** or **YACs.** The general structure of a YAC vector is diagrammed at the upper left in Figure 9.22. The YAC vector contains four types of genetic elements: (1) a cloning site, (2) a yeast centromere

Figure 9.22

Cloning large DNA fragments in yeast artificial chromosomes (YACs). The vector (upper left) contains sequences that allow replication and selection in both *E. coli* and yeast, a yeast centromere, and telomeres from *Tetrahymena*. In producing the YAC clones, the vector is cut by two restriction enzymes (A and B) to free the chromosome arms. These arms are ligated to the ends of large fragments of source DNA, and yeast cells are transformed. Many ligation products are possible, but only those that consist of source DNA flanked by the left and right vector arms form stable artificial chromosomes in yeast.

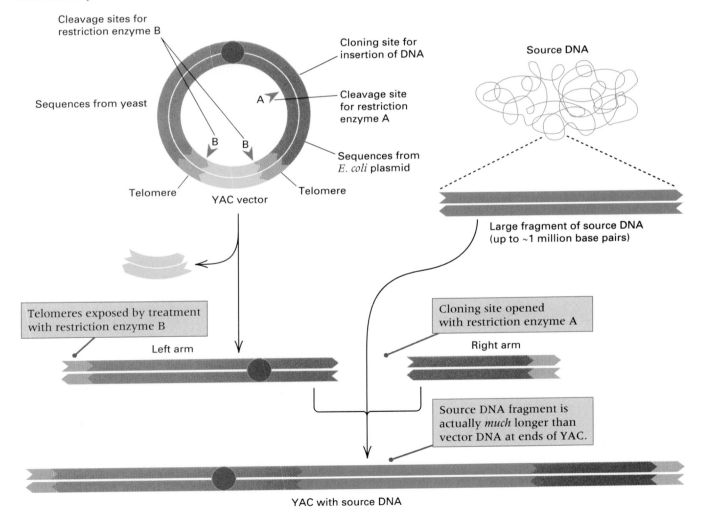

and genetic markers that are selectable in yeast, (3) an *E. coli* origin of replication and genetic markers that are selectable in *E. coli*, and (4) a pair of telomere sequences from *Tetrahymena*. Therefore, a YAC vector is a **shuttle vector** that can replicate and be selected in both *E. coli* and yeast.

Use of the YAC vector in cloning is also illustrated in Figure 9.22. The circular YAC vector is isolated after growth in *E. coli* and cleaved with two different restriction enzymes—one that cuts only at the cloning site (denoted A) and one that cuts near the tip of each of the telomeres (denoted B). Discarding the segment between the telomeres results in the two telomere-bearing fragments shown, which form the arms of the yeast artificial chromosomes. Ligation of a mixture containing source DNA and YAC vector arms results in a number of possible products. However, transformation of yeast cells and selection for the genetic markers in the YAC arms yields only the ligation product shown in Figure 9.22, in which a fragment of source DNA is inserted at a site within the right arm of the YAC vector. This product is recovered because it is the only true chromosome possessing a single centromere and two telomeres. Products that contain two YAC left arms are dicentric, those with two YAC right arms are acentric, and both of these types of products are genetically unstable (Section

7.1). YACs that have donor DNA inserted at the cloning site can be identified because the inserted DNA interrupts a yeast gene present at the cloning site and renders it nonfunctional. Yeast cells containing YACs with inserts of particular sequences of donor DNA can be identified in a number of ways, including colony hybridization (Figure 9.9), and these cells can be grown and manipulated in order to isolate and study the YAC insert. Figure 9.23 shows a region of the *Drosophila* salivary gland chromosomes that hybridizes with a YAC containing an insert of 300 kb. The entire genome of *Drosophila* could be contained in only 550 YAC clones of this size.

Physical Mapping

The development of methods for isolating and cloning large DNA fragments has stimulated major efforts to map and sequence the human genome, which is the principal goal of an effort termed the **human genome project.** The project also aims to map and sequence the genomes of a number of model genetic organisms, including *E. coli*, yeast, *Arabidopsis thaliana*, the nematode *C. elegans*, *Drosophila*, and the laboratory mouse. The first stage in the analysis of complex genomes is usually the production of a **physical map,** which is a diagram of the genome depicting the physical loca-

Figure 9.23

Hybridization *in situ* between *Drosophila* DNA cloned into a yeast artificial chromosome and the giant salivary gland chromosomes. The yeast artificial chromosome contains DNA sequences derived from numerous adjacent salivary bands—in this example, the bands in regions 52B through 52E in the right arm of chromosome 2. On average, each salivary band contains about 20 kb of DNA, but the bands vary widely in DNA content.

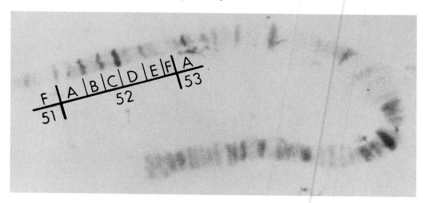

YAC-ity YAC

David T. Burke, Georges F. Carle, and Maynard V. Olson 1987
Washington University,
St. Louis, Missouri
Cloning of Large Segments of Exogenous DNA into Yeast by Means of Artificial Chromosome Vectors

Technological innovations often open up whole new areas of investigation by making novel types of experiments possible. A case in point is the development of methods for cloning large fragments of DNA, pioneered by the development of yeast artificial chromosomes (YACs). About the origin of the method, Olson recalls: "David Burke [then a graduate student] was in his third year of a molecular genetics project that was going well, when he did what all students are told not to do—start another project." The new approach was a great stimulus for genome analysis—the study of the organization of DNA in complex genomes. Large-fragment DNA cloning made it possible to obtain a physical map of the DNA in the entire human genome and to correlate the genetic map based on homologous recombination with the physical map based on the analysis of cloned DNA fragments. The result has been great activity in human gene mapping, including genetic mapping of mutations that cause inherited diseases, and also rapid progress in isolating the DNA of the mutant genes and identifying their normal functions.

Standard recombinant DNA techniques, . . . whose capacities for exogenous DNA range up to 50 kilobase pairs (kb), are well suited to the analysis and manipulation of genes from organisms in which the genetic information is tightly packed. It is increasingly apparent, however, that many of the functional genetic units in higher organisms span enormous tracts of DNA. For example, . . . recent estimates of the size of the gene that is defective in Duchenne's muscular dystrophy suggest that this single genetic locus, whose protein-coding func-

---

We report here the development of a high-capacity cloning system that is based on the in vitro construction of linear DNA molecules that can be transformed into yeast, where they are maintained as artificial chromosomes.

---

tion could be fulfilled by as little as 15 kb of DNA, actually covers more than a million base pairs. . . . We report here the development of a high-capacity cloning system that is based on the in vitro construction of linear DNA molecules that can be transformed into yeast, where they are maintained as artificial chromosomes. . . . The vector incorporates all necessary functions into a single plasmid that can replicate in *Escherichia coli*. This plasmid, called a "yeast artificial chromosome" (YAC) vector, supplies a cloning site within a gene whose interruption is phenotypically visible, an autonomous-replication sequence with properties expected of a replication origin, a yeast centromere, selectable markers on both sides of the centromere, and two sequences that seed telomere formation in yeast. . . . An initial test of the vector system involved cloning human DNA into the YAC vector. . . . A number of clones were analyzed to determine whether or not the artificial chromosomes that had been produced had the expected structure. . . . The test cases appear to be propagated as faithful copies of the source DNA. . . . Further experience with the YAC cloning system will be required to assess such issues as the stability of the clones, the extent to which the source DNA is randomly sampled, and the biological activity of the cloned DNA. Nevertheless, there are grounds for optimism that YAC vectors could even offer important advantages over standard cloning systems in these areas. . . . The demonstration of the basic feasibility of generating large recombinant DNA's in vitro and transforming them into easily manipulated host cells may stimulate experimentation with other combinations of vectors and hosts. There is a strong incentive to develop such systems since they are directed toward the major remaining gap in our ability to dissect the genomes of higher organisms.

Science 236: 806–812

---

tions of various landmarks along the DNA. The landmarks in a physical map usually consist of the locations of particular DNA sequences, such as coding regions or sequences present in particular cloned DNA fragments. If the landmarks are the locations of the cleavage sites for restriction enzymes, then the physical map is also a restriction map (Section 5.7). More useful landmarks are the positions of molecular markers, such as STRPs (simple tandem repeat polymorphisms), that have also been located in the genetic map (Section 4.4). Molecular markers serve to unify the genetic map and the physical map of an organism. The utility of a physical map is that it affords a single framework for organizing and integrating diverse types of genetic information, including the positions of chromosome bands, chromosome breakpoints, mutant genes, transcribed regions, and DNA sequences.

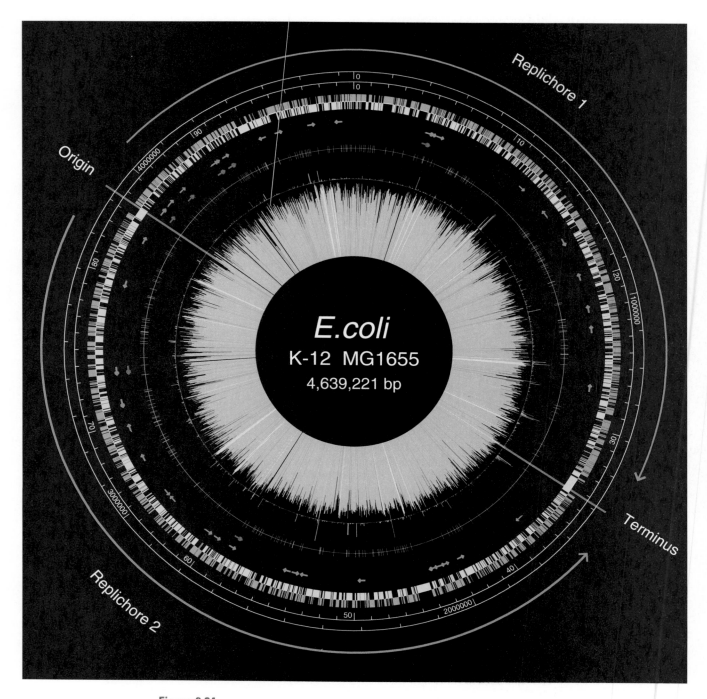

Figure 9.24

Diagram of the DNA sequence organization of *Escherichia coli* strain K-12. The coordinates are given in base pairs as well as in minutes on the genetic map. The coding sequences are shown as gold and yellow bars, which are transcribed in a clockwise (gold) or counterclockwise (yellow) direction. Green and red arrows denote genes for transfer RNAs or for ribosomal RNAs, respectively. The gold rays of the "sunburst" are proportional to the degree of randomness of codon usage in the coding sequences. Genes with the longest rays use the codons in the genetic code almost randomly. The origin and terminus of DNA replication are indicated. Bidirectional replication creates two "replichores." The peaks on the circle immediately outside the sunburst indicate coding sequences with high similarity to previously described bacteriophage proteins. [Courtesy of Frederick R. Blattner and Guy Plunkett III. From F. R. Blattner et al. 1997. *Science* 277: 1453.]

The Genome of *E. coli*

A diagram of the genome of *E. coli,* based on the complete DNA sequence, is illustrated in Figure 9.24. The coordinates of the circle are given in minutes on the genetic map (0–100) as well as in base pairs. The "replichores" are the two halves of the circle replicated bidirectionally starting from the origin. The gold bars on the outside denote genes whose transcription is from left to right, the yellow bars on the inside denote genes transcribed from right to left. The green arrows show the positions of tRNA genes, red arrows rRNA genes. The circle just inside the red arrows shows the positions of a 40-base-pair repetitive sequence of unknown function that is present 581 times. The rays of the yellow "sunburst" in the middle show the usage of codons among all of the coding sequences. The length of each ray is proportional to the degree to which codons are used randomly. Short rays indicate genes with a highly biased usage of codons, which is usually associated with a high level of gene expression. This strain has a genome of 4.6 megabases.

The Human Genome

A more complex type of physical map is illustrated in Figure 9.25. The map covers a small part of human chromosome 16, and it illustrates how the physical map is used to organize and integrate several different levels of genetic information. A map of the metaphase banding pattern of chromosome 16 is shown across the top. (The entire chromosome contains about 95 Mb of DNA.) Beneath the cytogenetic map, the somatic cell hybrid map shows the locations of chromosome breakpoints observed in cultured hybrid cells that contain only a part of chromosome 16. The genetic linkage map depicts the locations of various genetic markers studied in pedigrees; most of the genetic markers are molecular markers, such as STRPs (Section 4.4). One region of the genetic linkage map shows the location of a YAC clone, which has been assigned a physical location in the cytogenetic map by *in situ* hybridization. The large DNA insert in the YAC clone is

also represented in a set of overlapping cosmid or P1 clones; these clones define a contig covering a contiguous region of the genome without any gaps.

The various levels of the physical map in Figure 9.25 are connected by a special type of genetic marker shown at the bottom: a **sequence-tagged site,** or **STS.** An STS marker is a DNA sequence, present once per haploid genome, that can be amplified with a suitable pair of oligonucleotide primers by means of the polymerase chain reaction (PCR), described in Section 5.8. Hence an STS marker defines a unique site in the genome whose presence in a cloned DNA fragment can be detected by PCR amplification. In Figure 9.25, for example, the STS marker is present in two clones of the cosmid (or P1) contig, as well as in the YAC clone. Furthermore, the STS can be positioned on the somatic cell hybrid map by carrying out the PCR reaction with DNA from hybrid cells that contain rearranged or deleted chromosomes, and the amplified PCR product (or a clone containing the STS) can be used in *in situ* chromosome hybridization to localize the STS on the cytogenetic map. Therefore, STS markers are a type of genetic marker that can be used to integrate different types of information in a physical map. At present, 94 percent of the human genome is covered by a set of 16,494 YAC clones interconnected by 10,850 STS markers. The YAC clones define 377 contigs, averaging 8 Mb in size, and there is an average spacing between adjacent STS markers of 276 kb. About half of the STS markers have also been located on the human genetic linkage map.

Although an STS may consist of any type of sequence present once per haploid genome, some genome projects rely extensively on STS markers derived from cDNAs, because these markers represent coding regions that are likely to be of greater long-term interest than single-copy sequences that are noncoding. Intensive study of human cDNAs has identified a substantial proportion of the estimated 80,000 or so genes in the human genome. The most ambitious cDNA project to date included sequencing about 300,000 partial cDNAs, among them cDNAs obtained from cDNA libraries prepared from 37 distinct

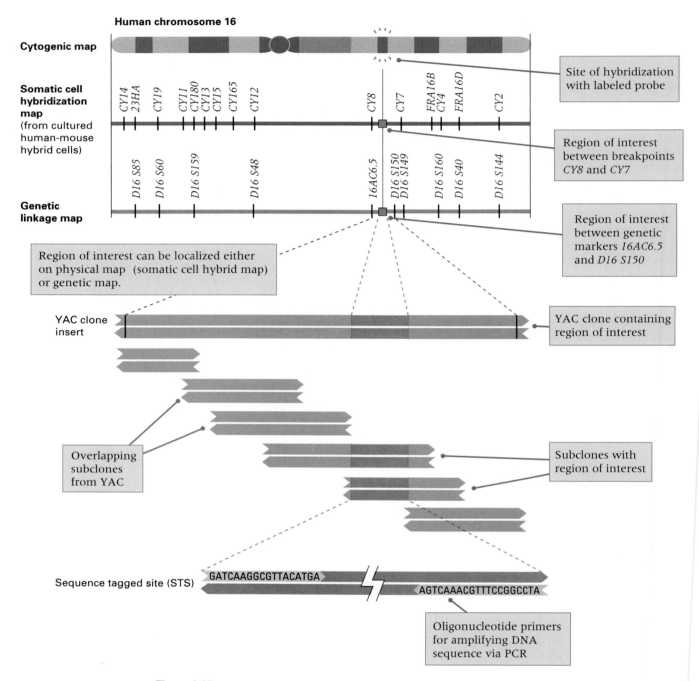

Human chromosome 16

Cytogenic map

Somatic cell hybridization map (from cultured human-mouse hybrid cells)

CY14
23HA
CY19
CY11
CY180
CY13
CY15
CY165
CY12
CY8
CY7
FRA16B
CY4
FRA16D
CY2

Genetic linkage map

D16 S85
D16 S60
D16 S159
D16 S48
16AC6.5
D16 S150
D16 S149
D16 S160
D16 S40
D16 S144

Site of hybridization with labeled probe

Region of interest between breakpoints *CY8* and *CY7*

Region of interest between genetic markers *16AC6.5* and *D16 S150*

Region of interest can be localized either on physical map (somatic cell hybrid map) or genetic map.

YAC clone insert

YAC clone containing region of interest

Overlapping subclones from YAC

Subclones with region of interest

Sequence tagged site (STS)

GATCAAGGCGTTACATGA

AGTCAAACGTTTCCGGCCTA

Oligonucleotide primers for amplifying DNA sequence via PCR

Figure 9.25

Integrated physical map of a small part of human chromosome 16. The map contains information about (1) the banding pattern of the metaphase chromosome (cytogenetic map), (2) the position of a particular sequence in the chromosome derived from *in situ* hybridization, (3) the locations of chromosome breakpoints in cultured cells (somatic cell hybrid map), (4) the positions of genetic markers analyzed by means of recombination in pedigrees (genetic linkage map), (5) the inserted DNA (blue) present in a YAC clone, and (6) a set of overlapping cosmid or P1 clones forming a contig (coverage of a contiguous region of the genome without any gaps). The various levels of the map are integrated by sequence-tagged sites (STSs), sequences present once per haploid genome that can be amplified with the polymerase chain reaction. [After an illustration in *Human Genome: 1991–92 Program Report,* United States Department of Energy.]

human organs and tissues. The total DNA sequence obtained was 83 million base pairs. Computer matching among the cDNA sequences revealed 87,983 distinct sequences, many of which could be assigned a function on the basis of similarity with already known genes from human beings or from other organisms. Figure 9.26 gives a breakdown of the cDNA sequences by type of function. Approximately 40 percent of human genes are implicated in basic energy metabolism, cell structure, homeostasis, or cell division; a further 22 percent are concerned with RNA and protein synthesis and processing; and 12 percent are associated with signaling and communication between cells. Figure 9.27 summarizes the results of examining the tissue-specific cDNA libraries. For each organ or tissue type, the first number is the total number of cDNA clones sequenced, and the number in parentheses is the number of distinct cDNA sequences found among the total from that organ or tissue type.

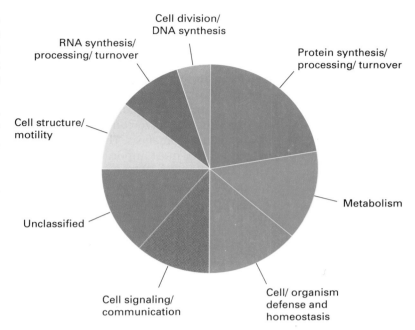

Figure 9.26

Classification of cDNA sequences by function. The chart is based on over 13,000 distinct, randomly selected human cDNA sequences. [Data courtesy of Craig Venter and the Institute for Genomic Research.]

Figure 9.27

Classification of cDNA sequences by organ or tissue type. In each category, the initial number is the total number of cDNA clones examined. The number in parentheses is the number of distinct sequences found per organ or tissue type. [Data from M. D. Adams and 84 other authors. 1995. *Nature* 377 (Suppl.): 3.]

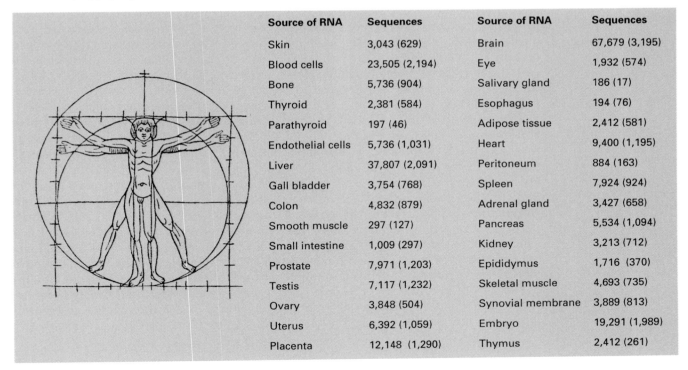

Source of RNA	Sequences	Source of RNA	Sequences
Skin	3,043 (629)	Brain	67,679 (3,195)
Blood cells	23,505 (2,194)	Eye	1,932 (574)
Bone	5,736 (904)	Salivary gland	186 (17)
Thyroid	2,381 (584)	Esophagus	194 (76)
Parathyroid	197 (46)	Adipose tissue	2,412 (581)
Endothelial cells	5,736 (1,031)	Heart	9,400 (1,195)
Liver	37,807 (2,091)	Peritoneum	884 (163)
Gall bladder	3,754 (768)	Spleen	7,924 (924)
Colon	4,832 (879)	Adrenal gland	3,427 (658)
Smooth muscle	297 (127)	Pancreas	5,534 (1,094)
Small intestine	1,009 (297)	Kidney	3,213 (712)
Prostate	7,971 (1,203)	Epididymus	1,716 (370)
Testis	7,117 (1,232)	Skeletal muscle	4,693 (735)
Ovary	3,848 (504)	Synovial membrane	3,889 (813)
Uterus	6,392 (1,059)	Embryo	19,291 (1,989)
Placenta	12,148 (1,290)	Thymus	2,412 (261)

Genome Evolution in the Grass Family

The discovery of unexpectedly regular relationships among the genomes of several cereal grasses in the Family Gramineae must be numbered among the extraordinary results that have come from applications of genome analysis. The cereal grasses are among our most important crop plants. They include rice, wheat, maize, millet, sugar cane, sorghum, and other cereals. The genomes of grass species vary enormously in size. The smallest, at 400 Mb, is found in rice; the largest, at 17,000 Mb, is found in wheat. Although some of the difference in genome size results from the fact that wheat is an allohexaploid (Chapter 7) whereas rice is a diploid, a far more important factor is the large variation from one species to the next in types and amount of repetitive DNA sequences present. Each chromosome in wheat contains approximately 25 times as much DNA as each chromosome in rice. For comparison, maize has a genome size of 2500 Mb; it is intermediate in size among the grasses and approximately the same size as the human genome.

In spite of the large variation in chromosome number and genome size in the grass family, there are a number of genetic and physical linkages between single-copy genes that are remarkably conserved amid a background of very rapidly evolving repetitive DNA sequences. In particular, each of the conserved regions can be identified in all the grasses and referred to a similar region in the rice genome. The situation is as depicted in Figure 9.28. The rice chromosome pairs are numbered R1 through R12, and the conserved regions within each chromosome are indicated by lowercase letters, for example, R1a and R1b. In each of the other species, each chromosome pair is diagrammed according to the arrangement of segments of the rice genome that contain single-copy DNA sequences homologous to those in the corresponding region of the chromosome of the species in question. For example, the wheat monoploid chromosome set is designated W1 through W7. One region of W1 contains single-copy sequences that are homologous to those in rice segment R5a, another contains single-copy sequences that are homologous to those in rice segment R10, and still another contains single-copy sequences that are homologous to those in rice segment R5b. The genomes of the other grass species can be aligned with those of rice as shown. Each of such conserved genetic and physical linkages is called a **synteny group.** Synteny groups are found in other species comparisons as well. For example, many synteny groups are shared between the human and the mouse genomes. The human-mouse synteny groups are often useful in identifying the mouse homolog of a human gene.

Relative to the synteny groups in Figure 9.28, note that the maize genome has a repetition of segments, indicated by the connecting lines. The relationships confirm what some maize geneticists had long suspected, that maize is a complete, very ancient tetraploid with the complication that the two complete genomes are rearranged relative to each other. Furthermore, most of the larger chromosomes (1, 2, 3, 4 and 6) comprise one of the genomes, and most of the short chromosomes (5, 7, 8, 9, and 10) comprise the other.

The synteny groups among the grass genomes are shown in a different format in Figure 9.29. In this case, the segments are formed into a circle in the same order in which they are aligned in the hypothetical ancestral chromosome (Figure 9.28G). There is no evidence that the ancestral cereal chromosome was actually a circle. It seems highly unlikely that it was anything other than a normal linear chromosome. However, the value of the circular diagram is that it shows the arrangement of the synteny groups in all of the grass genomes simultaneously, which makes comparisons much easier. Because of the synteny groups in the genomes, homologous genes can often be identified by location alone. For example, both wheat and maize have dwarfing mutations in which the mutant plants are insensitive to the plant hormone gibberellin. A line through the positions of these mutations in the wheat (Triticeae)

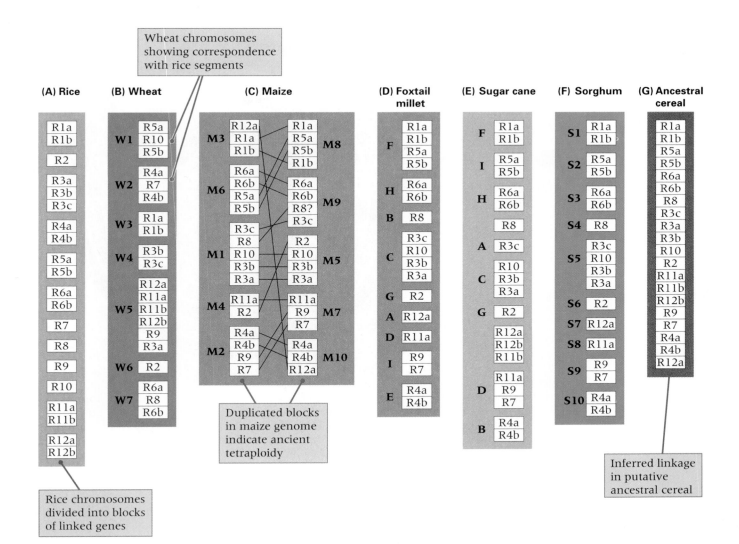

Figure 9.28
Conserved linkages (synteny groups) between the rice genome (A) and that of other grass species: wheat (B), maize (C), foxtail millet (D), sugar cane (E), and sorghum (F). Part (G) depicts the inferred or "reconstructed" order of segments in a hypothetical ancestral cereal genome consisting of a single chromosome pair. For each extant species, the hypothetical ancestral chromosome can be cleaved at different points to yield groups of blocks corresponding to the arrangement of the segments in the chromosomes of the species. [Courtesy of Graham Moore. From G. Moore, K. M. Devos, Z. Wang, and M. D. Gale. 1995. *Current Opinion Genet. Devel.* 5: 737.]

and maize chromosomes passes through rice block 3b, which probably contains the homologous gene. Similarly, rice block 4 contains a gene for *liguless* that aligns with similar mutations in the chromosomes of barley (again Triticeae) and maize, which indicates that the mutations are almost certainly in homologous genes in all three species. Such relationships between genes based on position in the physical maps affords an important method of positional cloning in all species in the grass family.

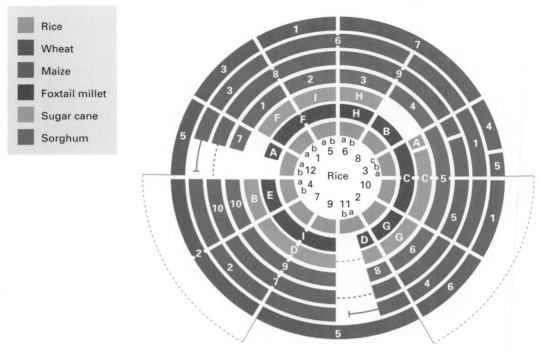

Rice

Wheat

Maize

Foxtail millet

Sugar cane

Sorghum

Figure 9.29
Circular arrangement of synteny groups in the cereal grasses makes simultaneous comparisons possible. The thin dashed lines indicate connections between blocks of genes. A number of transpositions of genetic segments are not shown. In some cases, a region within a synteny group is inverted; these inversions are not shown. The circular diagram is for convenience only; there is no indication that the ancestral grass chromosome was actually circular. [Courtesy of Graham Moore. From G. Moore, K. M. Devos, Z. Wang, and M. D. Gale. 1995. *Current Opinion Genet. Devel.* 5: 737.]

9.7
Large-Scale DNA Sequencing

Large-scale DNA sequencing is well under way in a number of model organisms. The 12.5-Mb genome of the yeast *Saccharomyces cerevisiae* is the first eukaryotic genome to have been sequenced in its entirety. Some of the important conclusions are summarized next.

Complete Sequence of the Yeast Genome

A summary of the analysis of the 666,448 nucleotide pairs in the sequence of yeast chromosome XI is illustrated in Figure 9.30. Like other yeast chromosomes, chromosome XI has a high density of coding regions. A coding region includes an **open reading frame (ORF),** which is a region of sequence containing an uninterrupted run of amino-acid-coding triplets (codons) with no "stop" codons that would terminate translation. An ORF of greater than random length is likely to code for a protein of some kind. The coding region associated with an ORF also includes the flanking regulatory sequences. In chromosome XI, approximately 72 percent of the sequence is present within 331 coding regions averaging 2 kb. The average length of ORF codes for a sequence of 488 amino acids, but the longest ORF codes for the protein dynein with 4092 amino acids. Worthy of note in the chromosome XI sequence is the low number of introns (sequences that are transcribed but removed from the RNA in producing the mRNA); only about 2 percent of the genes have introns. The chromosome also includes sequences for 16 transfer-RNA genes used in protein translation as well as representatives of the transposable elements δ and σ.

The yeast sequence contains evidence of an ancient duplication of the entire genome, although only a small fraction of the genes are retained in duplicate. Protein pairs derived from the ancient duplication make up 13 percent of all yeast proteins. These include pairs of cytoskeletal proteins, ribosomal proteins, transcription factors, glycolytic enzymes, cyclins, proteins of the secretory pathway, and protein kinases. When yeast protein sequences are compared with mammalian protein sequences in GenBank (the international repository of protein sequences of all organisms), a mammalian homolog is found for about 31 percent of yeast proteins. This is a minimum estimate of homology, because the mammalian sequences available for comparison are only a small fraction of those present in mammalian genomes. The homologous proteins are of many types. Examples include proteins that catalyze metabolic reactions, subunits of RNA polymerase, transcription factors, translation initiation and elongation factors, enzymes of DNA synthesis and repair, nuclear-pore proteins, and structural proteins and enzymes of mitochondria and peroxisomes. In some instances the similarities are known to reflect conservation of function, because in more than 70 cases, a human amino-acid coding sequence will substitute for a yeast sequence. These include coding sequences for cyclins, DNA ligase, the *RAS* proto-oncogenes, translation initiation factors, and proteins involved in signal transduction.

The most common method of cloning human disease genes is positional cloning, which means cloning by map position (Figure 9.25). Usually nothing is known about the gene except that, when defective, it results in disease. The first clue to function often comes by recognizing homology to a yeast gene. Striking examples are the human genes that cause hereditary nonpolyposis colon cancer and Werner's syndrome, a disease associated with premature aging. In cells of patients with hereditary nonpolyposis colon cancer, short repeated DNA sequences are unstable. These findings stimulated studies of stability of repeated DNA sequences in yeast mutants, which revealed that repeated DNA sequences are unstable in yeast cells that are deficient in the repair of mismatched nucleotides in DNA, including *msh*2 and *mlh*1 mutants (Chapter 13). The prediction that the cancer genes might also encode proteins for mismatch repair was later verified when the cancer genes were cloned. Cells of patients with Werner's syndrome of premature aging show a limited life span in culture. The human gene encodes a protein highly similar to a DNA helicase encoded by *SGS1* in yeast. The *sgs1* mutant yeast cells show accelerated aging and a reduced lifespan, as well as other cellular phenotypes, including relocation of proteins from telomeres to the nucleolus and nucleolar fragmentation. Examples like these demonstrate how research on model organisms can have direct applicability to human health and disease.

Automated DNA Sequencing

Large-scale sequencing studies of other eukaryotic genomes are also well advanced and, judging by the insights gained from yeast sequences, can be expected to be of considerable value. Most eukaryotic chromosomes are much larger than those of yeast, so complete DNA sequencing is not feasible without the use of instruments that partly automate the process. The principle behind automated DNA sequencing is shown in Figure 9.31. Figure 9.31A illustrates a conventional sequencing gel obtained from the dideoxy procedure described in Section 5.9; each lane contains the products of DNA synthesis carried out in the presence of a small amount of a dideoxy nucleotide (dideoxy-G, -A, -T or -C), which, when incorporated into a growing DNA strand, terminates further elongation. The products of each reaction are separated by electrophoresis in individual lanes, and the gel is placed in contact with photographic film so that radioactive atoms present in one of the normal nucleotides will darken the

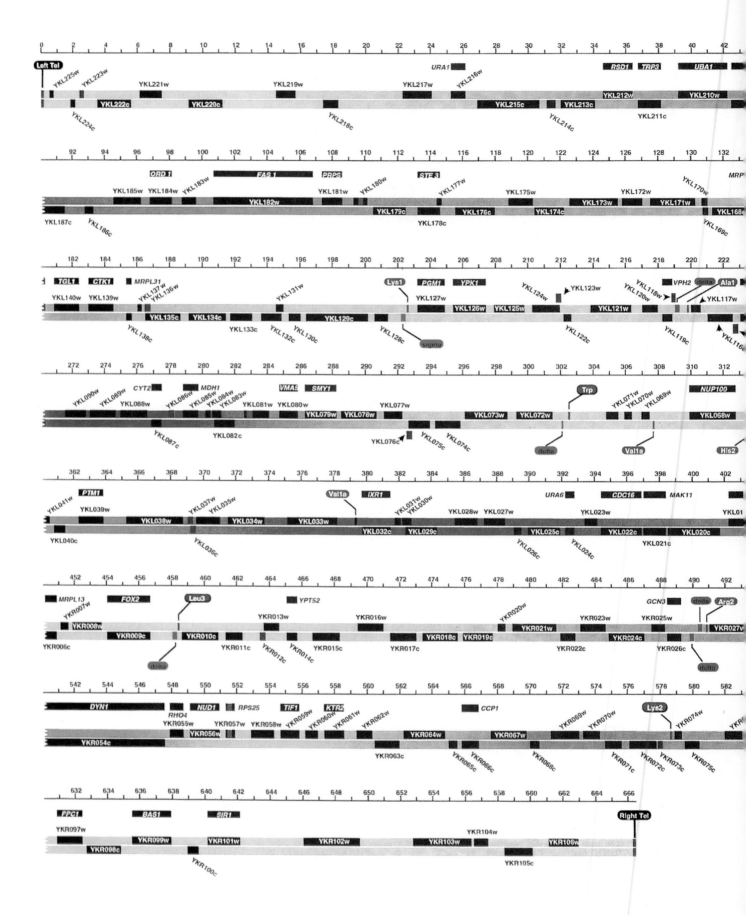

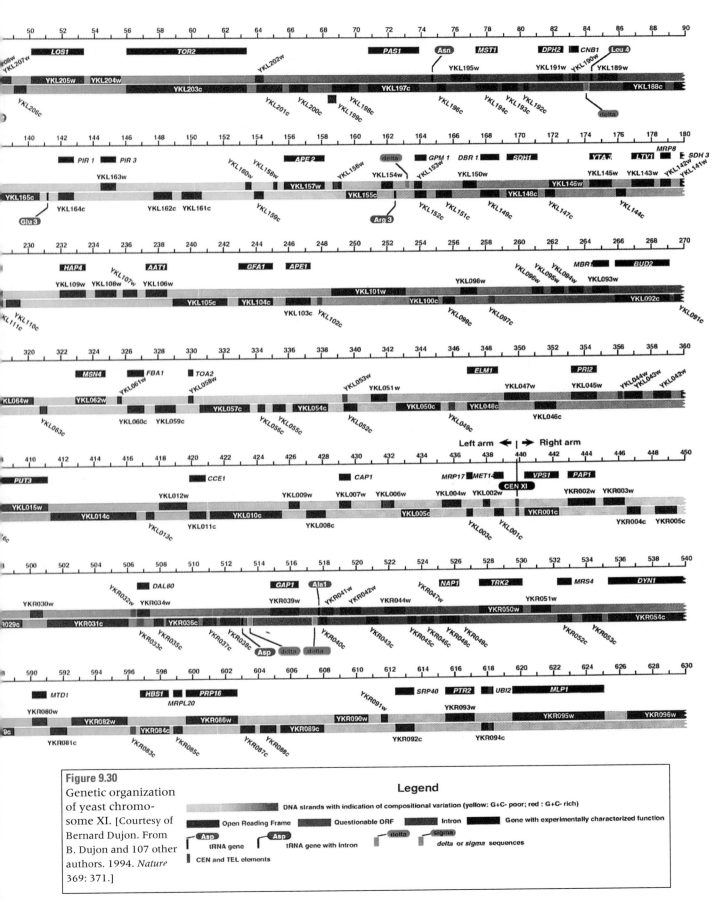

Figure 9.30
Genetic organization of yeast chromosome XI. [Courtesy of Bernard Dujon. From B. Dujon and 107 other authors. 1994. *Nature* 369: 371.]

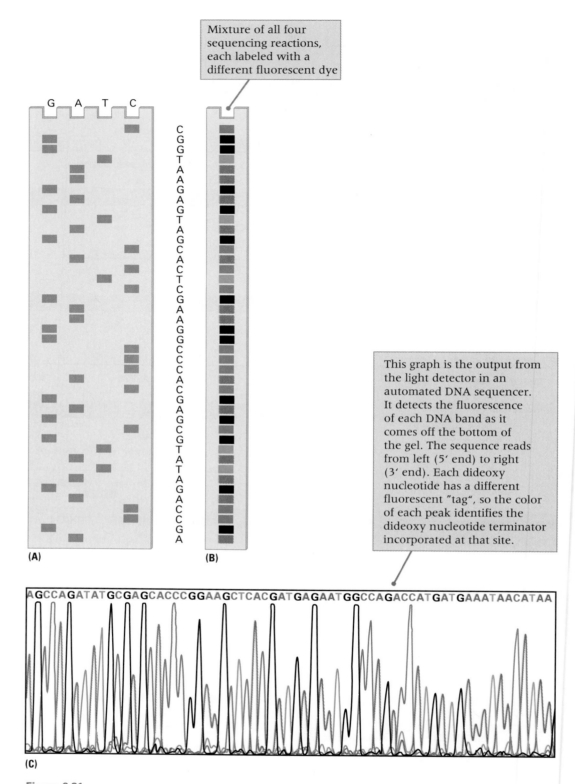

Figure 9.31

Automated DNA sequencing. (A) Conventional sequencing gel obtained from the dideoxy procedure (Section 5.9). The DNA sequence can be determined directly from photographic film according to the positions of the bands. (B) Banding pattern obtained when each of the terminating nucleotides is labeled with a different fluorescent dye and the bands are separated in the same lane of the gel. (C) Trace of the fluorescence pattern obtained from the gel in part B by automated detection of the fluorescence of each band as it comes off the bottom of the gel during continued electrophoresis.

film and reveal a band at the position to which each incomplete DNA strand migrated in electrophoresis. After the film is developed, the DNA sequence is read from the pattern of bands, as shown by the sequence at the right of the gel.

In automated DNA sequencing, illustrated in Figure 9.31B, the nucleotides that terminate synthesis are labeled with different fluorescent dyes (G, black; A, green; T, red; C, blue). Because the colors distinguish the products of DNA synthesis that terminate with each nucleotide, the products of all the synthesis reactions can be put together in the same tube and separated by electrophoresis in a single lane. In principle, the sequence could again be read directly from the gel, as shown in letters at the left of Figure 9.31B. However, a substantial improvement in efficiency is accomplished by continuing the electrophoresis until each band, in turn, drops off the bottom of the gel. As each band comes off the bottom of the gel, the fluorescent dye that it contains is excited by laser light, and the color of the fluorescence is read automatically by a photocell and recorded in a computer. Figure 9.31C is a trace of the fluorescence pattern that would emerge at the bottom of the gel in Figure 9.31B after continued electrophoresis. The nucleotide sequence is read directly from the colors of the alternating peaks along the trace. When used to maximum capacity, an automated sequencing instrument can ideally generate as much as 20 Mb of nucleotide sequence per year. The actual amount of finished sequence is considerably smaller, because in a sequencing project, each DNA strand needs to be sequenced completely for the sake of minimizing sequencing errors, and some troublesome regions need to be sequenced several times.

Chapter Summary

Recombinant DNA technology makes it possible to modify the genotype of an organism in a directed, predetermined way by enabling different DNA molecules to be joined into novel genetic units, altered as desired, and reintroduced into the organism. Restriction enzymes play a key role in the technique because they can cleave DNA molecules within particular base sequences. Many restriction enzymes generate DNA fragments with complementary single-stranded ends, which can anneal and be ligated together with similar fragments from other DNA molecules. The carrier DNA molecule used to propagate a desired DNA fragment is called a vector. The most common vectors are plasmids, phages, viruses, and yeast artificial chromosomes (YACs). Transformation is an essential step in the propagation of recombinant molecules because it enables the recombinant DNA molecules to enter host cells, such as those of bacteria, yeast, or mammals. If the recombinant molecule has its own replication system or can use the host replication system, then it can replicate. Plasmid vectors become permanently established in the host cell; phages can multiply and produce a stable population of phages carrying source DNA; retroviruses can be used to establish a gene in an animal cell; and YACs include source DNA within an artificial chromosome that contains a functional centromere and telomeres.

Recombinant DNA can also be used to transform the germ line of animals or to genetically engineer plants. These techniques form the basis of reverse genetics, in which genes are deliberately mutated in specified ways and introduced back into the organism to determine the effects on phenotype. Reverse genetics is routine in genetic analysis in bacteria, yeast, nematodes, *Drosophila*, the mouse, and other organisms. In *Drosophila*, transformation employs a system of two vectors based on the transposable *P* element. One vector contains sequences that produce the *P* transposase; the other contains the DNA of interest between the inverted repeats of *P* and other sequences needed for mobilization by transposase and insertion into the genome. Germ-line transformation in the mouse makes use of retrovirus vectors or embryonic stem cells. Dicot plants are transformed with T DNA derived from the *Ti* plasmid found in species of *Agrobacterium*, whose virulence genes promote a conjugation-like transfer of T DNA into the host plant cell, where it is integrated into the chromosomal DNA.

Practical applications of recombinant DNA technology include the efficient production of useful proteins, the creation of novel genotypes for the synthesis of economically important molecules, the generation of DNA and RNA sequences for use in medical diagnosis, the manipulation of the genotype of domesticated animals and plants, the development of new types of vaccines, and the potential correction of genetic defects (gene therapy). Production of eukaryotic proteins in bacterial cells is sometimes hampered by protein instability, inability to fold properly, or failure to undergo necessary chemical modification. These problems are often eliminated by production of the protein in yeast or mammalian cells.

Cloning in bacteriophage P1 vectors or yeast artificial chromosomes (YACs) allows very large DNA molecules to be isolated and manipulated and has stimulated a major effort to analyze complex genomes, such as those in nematodes, *Drosophila*, the mouse, and human beings. These efforts include the development of detailed physical maps that integrate many levels of genetic information, such as the positions of sequence-tagged sites or the positions and lengths of contigs. Among the important discoveries of genome analysis is that of the relationships among cereal grass genomes. Although the genomes differ enormously in size from one species to the next, largely because of the abundance and types of repetitive DNA sequences, cross-hybridization of single-copy sequences demonstrates that the genomes can be brought into register by postulating rearrangements of 20 segments, or synteny groups, found in rice, the smallest of the genomes. Large-scale DNA sequencing of the yeast genome has revealed an unexpectedly high density of genetic information. Many coding regions have previously unknown functions and yield no recognizable phenotype when mutated. Large-scale genomic sequencing is also underway in other organisms through the application of automated DNA-sequencing machines.

Key Terms

blunt end	library	restriction enzyme
cDNA	map-based cloning	restriction site
cloning	multiple cloning site	retrovirus
cohesive end	oligonucleotide site-directed	reverse genetics
colony hybridization assay	mutagenesis	reverse transcriptase
complementary DNA	open reading frame	reverse transcriptase PCR
contig	P1 bacteriophage	sequence-tagged site (STS)
crown gall tumor	palindrome	shuttle vector
electroporation	partial digestion	sticky end
embryonic stem cell	physical map	synteny group
gene cloning	polylinker	T DNA
gene targeting	polymerase chain reaction (PCR)	*Ti* plasmid
gene therapy	positional cloning	transformation
genetic engineering	primer oligonucleotides	transgenic animal
genome equivalent	probe	vector
human genome project	*P* transposable element	yeast artificial chromosome (YAC)
insertional inactivation	recombinant DNA technology	

Review the Basics

- What is recombinant DNA?

- What features are essential in a bacterial cloning vector?

- What is the reaction catalyzed by the enzyme reverse transcriptase? How is this enzyme used in recombinant DNA technology?

- What is a transgenic organism?

- In the context of genome analysis, what is a YAC? What feature makes YACs useful in the analysis of complex genomes?

- What is a physical map? How is a physical map related to a genetic map?

Guide to Problem Solving

Problem 1: In genetic engineering, when we are expressing eukaryotic gene products in bacterial cells, why is it necessary.

(a) to use cDNA instead of genomic DNA?

(b) to fuse the cDNA with a bacterial promoter?

Answer:

(a) Most eukaryotic genes have introns, which cannot be removed by bacterial cells.

(b) Eukaryotic promoters have a different sequence than prokaryotic promoters and are not normally recognized in bacterial cells.

Problem 2: What is the average distance between restriction sites for each of the following restriction enzymes? Assume that the DNA substrate has a random sequence with equal amounts of each base. The symbol N stands for any nucleotide, R for any purine (A or G), and Y for any pyrimidine (T or C).

(a) TCGA   *Taq*I

(b) GGTACC   *Kpn*I

(c) GTNAC   *Mae*III

(d) GGNNCC   *Nla*IV

(e) GRCGYC   *Acy*I

Answer: The average distance between restriction sites equals the reciprocal of the probability of occurrence of the restriction site. You must therefore calculate the probability of occurrence of each restriction site in a random DNA sequence.

(a) The probability of the sequence TCGA is $1/4 \times 1/4 \times 1/4 \times 1/4 = (1/4)^4 = 1/256$, so 256 bases is the average distance between *Taq*I sites.

(b) By the same reasoning, the probability of a *Kpn*I site is $(1/4)^6 = 4096$, so 4096 bases is the average distance between *Kpn*I sites.

(c) The probability of N (any nucleotide at a site) is 1, so the probability of the sequence GTNAC equals $1/4 \times 1/4 \times 1 \times 1/4 \times 1/4 = (1/4)^4 = 1/256$; therefore, 256 is the average distance between *Mae*III sites.

(d) The same reasoning yields the average distance between *Nla*IV sites as $(1/4 \times 1/4 \times 1 \times 1 \times 1/4 \times 1/4)^{-1} = 256$ bases.

(e) The probability of an R (A or G) at a site is $1/2$, and the probability of a Y (T or C) at a site is $1/2$. Hence the probability of the sequence GRCGYC is $1/4 \times 1/2 \times 1/4 \times 1/4 \times 1/2 \times 1/4 = 1/1024$, so there is an average of 1024 bases between *Acy*I sites.

Problem 3: What fundamental structural components are necessary for yeast artificial chromosomes to be stable in yeast cells?

Answer: Yeast artificial chromosomes need a centromere for segregation in cell division and a telomere at each end to stabilize the tips.

Analysis and Applications

9.1 The euchromatic part of the *Drosophila* genome that is highly replicated in the banded salivary gland chromosomes is approximately 110 Mb (million base pairs) in size. The salivary gland chromosomes contain approximately 5000 bands. For ease of reference, the salivary chromosomes are divided into about 100 approximately equal, numbered sections (1–100), each of which consists of six lettered subdivisions (A–F). On average, how much DNA is in a salivary gland band? In a lettered subdivision? In a numbered section? How do these compare with the size of the DNA insert in a 200-kb (kilobase pair) YAC? With the size of the DNA insert in an 80-kb P1 clone?

9.2 Restriction enzymes generate one of three possible types of ends on the DNA molecules that they cleave. What are the three possibilities?

9.3 Are the ends of different restriction fragments produced by a particular restriction enzyme always the same? Must opposite ends of each restriction fragment be the same? Why?

9.4 Will the sequences 5'-GGCC-3' and 3'-GGCC-5' in a double-stranded DNA molecule be cut by the same restriction enzyme?

9.5 In cloning into bacterial vectors, why is it useful to insert DNA fragments to be cloned into a restriction site inside an antibiotic-resistance gene? Why is another gene for resistance to a second antibiotic also required?

9.6 How frequently would the restriction enzymes *Taq*I (restriction site TCGA) and *Mae*III (restriction site GTNAC, in which N is any nucleotide) cleave double-stranded DNA molecules containing random sequences of

(a) 1/6 A, 1/6 T, 1/3 G, and 1/3 C?

(b) 1/3 A, 1/3 T, 1/6 G, and 1/6 C?

9.7 If the genomic and cDNA sequences of a gene are compared, what information does the cDNA sequence give you that is not obvious from the genomic sequence? What information does the genomic sequence contain that is not in the cDNA?

9.8 What might prevent a cloned eukaryotic gene from yielding a functional mRNA in a bacterial host? Assuming that these problems are overcome, why might the desired protein still not be produced?

9.9 When DNA isolated from phage J2 is treated with the enzyme *Sal*I, eight fragments are produced with sizes of 1.3, 2.8, 3.6, 5.3, 7.4, 7.6, 8.1, and 11.4 kilobase pairs. However, if J2 DNA is isolated from infected cells, only seven fragments are found, with sizes of 1.3, 2.8, 7.4, 7.6, 8.1, 8.9, and 11.4 kb. What form of the intracellular DNA can account for these results?

9.10 Phage X82 DNA is cleaved into six fragments by the enzyme *Bgl*I. A mutant is isolated with plaques that look quite different from the wildtype plaque. DNA isolated from

GeNETics on the web will introduce you to some of the most important sites for finding genetic information on the Internet. To complete the exercises below, visit the Jones and Bartlett home page at

http://www.jbpub.com/genetics

Select the link to *Genetics: Principles and Analysis* and then choose the link to *GeNETics on the web.* You will be presented with a chapter-by-chapter list of highlighted keywords.

GeNETics EXERCISES

Select the highlighted keyword in any of the exercises below, and you will be linked to a web site containing the genetic information necessary to complete the exercise. Each exercise suggests a specific, written report that makes use of the information available at the site. This report, or an alternative, may be assigned by your instructor.

1. The keyword genome analysis will lead you to an informative introduction to some of the methods used in genome research. Read the discussion of top-down and bottom-up physical mapping. If assigned to do so, write a brief summary of each type of mapping, and list some of the advantages and limitations of each.

2. Most people are very surprised to learn how many organisms have had their genomes sequenced either completely or in large part. An extensive list of genome sequencing projects is maintained at this keyword site. If assigned to do so, make a list of microbial genomes whose sequences are completely known and which are available in public databases.

3. What are the social implications of modern genetics? Some groups are worried because the application of genetic technologies poses ethical and legal issues of the foreknowledge of one's health

as well as issues of genetic privacy and insurability. Others are optimistic that the technologies will yield great benefits for medicine and society. The debate continues. This keyword site will connect you to resources that will enable you to learn more about these issues. If assigned to do so, choose one controversial ethical or legal issue related to modern genetic technologies, and write a 250-word paper defining the issue and the opposing views.

MUTABLE SITE EXERCISES

The Mutable Site Exercise changes frequently. Each new update includes a different exercise that makes use of genetics resources available on the World Wide Web. Select the Mutable Site for Chapter 9, and you will be linked to the current exercise that relates to the material presented in this chapter.

PIC SITE

The Pic Site showcases some of the most visually appealing genetics sites on the World Wide Web. To visit the showcase genetics site, select the Pic Site for Chapter 9.

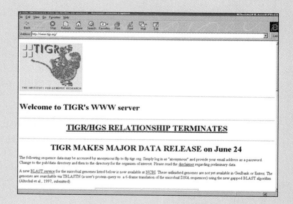

---

the mutant is cleaved into only five fragments. What possible genetic changes can account for the difference in the number of restriction fragments?

9.11 The wildtype allele of a bacterial gene is easily selected by growth on a special medium. Repeated attempts to clone the gene by digestion of cellular DNA with the enzyme *Eco*RI are unsuccessful. However, if the enzyme *Hin*dIII is used, then clones are easily found. Explain.

9.12 A *kan-r tet-r* plasmid is treated with the restriction enzyme *Bgl*I, which cleaves the *kan* (kanamycin) gene. The DNA is annealed with a *Bgl*I digest of *Neurospora* DNA and then used to transform *E. coli.*

(a) What antibiotic would you put in the growth medium to ensure that each colony has the plasmid?

(b) What antibiotic-resistance phenotypes will be found among the resulting colonies?

(c) Which phenotype will contain *Neurospora* DNA inserts?

9.13 A *lac+ tet-r* plasmid is cleaved in the *lac* gene with a restriction enzyme. The enzyme has a four-base restriction site and generates fragments with a two-base single-stranded overhang. The cutting site in the *lac* gene is in the codon for the second amino acid in the chain, a site that can tolerate any amino acid without loss of function. After cleavage, the single-stranded ends are converted to blunt ends with DNA polymerase I, and then the ends are joined by blunt-end ligation to recreate a circle. A *lac⁻ tet-s* bacterial strain is transformed with the DNA, and tetracycline-resistant bacteria are selected. What is the Lac phenotype of the colonies?

9.14 You want to introduce the human insulin gene into a bacterial host in hopes of producing a large amount of human insulin. Should you use the genomic DNA or the cDNA? Explain your reasoning.

Challenge Problem

9.15 Plasmid pBR607 DNA is a double-stranded circle of 4 kilobase pairs. This plasmid carries two genes whose protein products confer resistance to tetracycline (Tet-r) and ampicillin (Amp-r) in host bacteria. The DNA has a single site for each of the following restriction enzymes: *Eco*RI, *Bam*HI, *Hin*dIII, *Pst*I, and *Sal*I. Cloning DNA into the *Eco*RI site does not affect resistance to either drug. Cloning DNA into the *Bam*HI, *Hin*dIII, and *Sal*I sites abolishes tetracycline resistance. Cloning into the *Pst*I site abolishes ampicillin resistance. Digestion with the following mixtures of restriction enzymes yields fragments with the sizes listed below. Indicate the positions of the *Pst*I, *Bam*HI, *Hin*dIII, and *Sal*I cleavage sites on a restriction map, relative to the *Eco*RI cleavage site.

Enzyme mixture	Fragment size (kb)
*Eco*RI + *Pst*I	0.70, 3.30
*Eco*RI + *Bam*HI	0.30, 3.70
*Eco*RI + *Hin*dIII	0.08, 3.92
*Eco*RI + *Sal*I	0.85, 3.15
*Eco*RI + *Bam*HI + *Pst*I	0.30, 0.70, 3.00

Further Reading

Azpirozleehan, R., and K. A. Feldmann. 1997. T-DNA insertion mutagenesis in *Arabidopsis*: Going back and forth. *Trends in Genetics* 13: 152.

Bishop, J. E., and M. Waldholz. 1990. *Genome*. New York: Simon and Schuster.

Blaese, R. M. 1997. Gene therapy for cancer. *Scientific American,* June.

Botstein, D., A. Chervitz, and J. M. Cherry. 1997. Yeast as a model organism. *Science* 277: 1259.

Capecchi, M. R. 1994. Targeted gene replacement. *Scientific American,* March.

Chilton, M.-D. 1983. A vector for introducing new genes into plants. *Scientific American,* June.

Cohen, S. N. 1975. The manipulation of genes. *Scientific American, July.*

Cooke, H. 1987. Cloning in yeast: An appropriate scale for mammalian genomes. *Trends in Genetics* 3: 173.

Curtiss, R. 1976. Genetic manipulation of microorganisms: Potential benefits and hazards. *Annual Review of Microbiology* 30: 507.

Dujon, B. 1996. The yeast genome project: What did we learn? *Trends in Genetics* 12: 263.

Felgner, P. L. 1997. Nonviral strategies for gene therapy. *Scientific American,* June.

Friedmann, T. 1997. Overcoming the obstacles to gene therapy. *Scientific American,* June.

Gasser, C. S., and R. T. Fraley. 1992. Transgenic crops. *Scientific American,* June.

Gossen, J., and J. Vigg. 1993. Transgenic mice as model systems for studying gene mutations in vivo. *Trends in Genetics* 9: 27.

Havukkala, I. J. 1996. Cereal genome analysis using rice as a model. *Current Opinion in Genetics & Development* 6: 711.

Houdebine, L. M., ed. 1997. *Transgenic Animals: Generation and Use.* New York: Gordon and Breach.

Mariani, C., V. Gossele, M. De Beuckeleer, M. De Block, R. B. Goldberg, W. De Greef, and J. Leemans. 1992. A chimaeric ribonuclease-inhibitor gene restores fertility to male sterile plants. *Nature* 357: 384.

Meisler, M. H. 1992. Insertional mutation of "classical" and novel genes in transgenic mice. *Trends in Genetics* 8: 341.

Rennie, J. 1994. Grading the gene tests. *Scientific American,* June.

Sambrook, J., E. F. Fritsch, and T. Maniatis. 1989. *Molecular Cloning: A Laboratory Manual.* 2d ed. Cold Spring Harbor, NY: Cold Spring Harbor Laboratory.

Smith, D. H. 1979. Nucleotide sequence specificity of restriction enzymes. *Science* 205: 455.

Sternberg, N. L. 1992. Cloning high molecular weight DNA fragments by the bacteriophage P1 system. *Trends in Genetics* 8: 11.

Tanksley, S. D., and S. R. McCouch. 1977. Seed banks and molecular maps: Unlocking genetic potential from the wild. *Science* 277: 1063.

Watson, J. D. 1995. *Recombinant DNA.* 2nd ed. New York: Freeman.

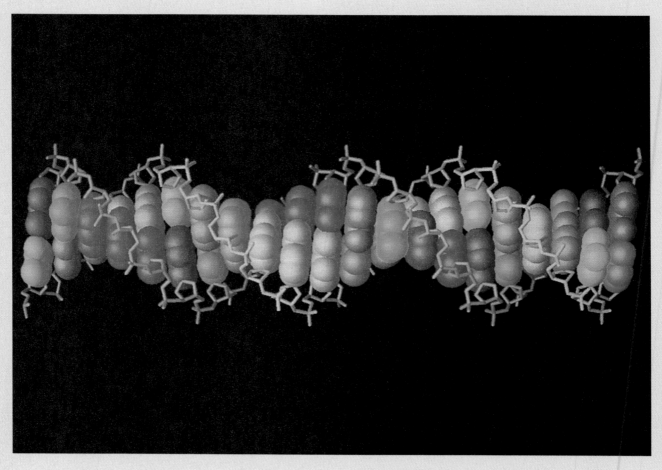

Structure of the DNA double helix showing the major (wide) and minor (narrow) grooves. Each base is shown in a dark or light color according to whether it is present on the template strand transcribed in RNA synthesis (light brown backbone) or on the nontemplate strand (blue backbone). The deoxyribose sugars are the pentagonal shapes, each adjacent pair connected by a phosphodiester bond (P—O—P). The base colors are brown (G), blue (C), red (A), and green (T). [Courtesy of Antony M. Dean.]

Gene Expression

PRINCIPLES

- In gene expression, information in the base sequence of DNA is used to dictate the linear order of amino acids in a polypeptide by means of an RNA intermediate.
- Transcription of an RNA from one strand of the DNA is the first step in gene expression.
- In eukaryotes, the RNA transcript is modified and may undergo splicing to make the messenger RNA.
- The messenger RNA is translated on ribosomes in groups of three bases (codons), each specifying an amino acid through an interaction with molecules of transfer RNA, each "charged" (chemically bonded) with one amino acid.
- Ribosomes are particles consisting of special types of RNA (ribosomal RNA) and numerous proteins. Each transfer RNA molecule contains a region of three bases that recognizes one (in some cases more than one) codon by base pairing. Each transfer RNA also has a particular amino acid attached at one end that corresponds to the amino acid encoded by the codon (or codons) with which the transfer RNA binds.
- Almost all organisms use the same genetic code, but exceptions are found in certain protozoa and in the genetic codes of mitochondrial and other organelle DNA.

CONNECTIONS

CONNECTION: One Gene, One Enzyme

George W. Beadle and Edward L. Tatum 1941

Genetic control of biochemical reactions in Neurospora

CONNECTION: Messenger *Light*

Sydney Brenner, François Jacob and Matthew Meselson 1961

An unstable intermediate carrying information from genes to ribosomes for protein synthesis

CONNECTION: Uncles and Aunts

Francis H. C. Crick, Leslie Barnett, Sydney Brenner, and R. J. Watts-Tobin, 1961

General nature of the genetic code for proteins

arlier chapters have been concerned with genetic analysis—with genes as units of genetic information, their relation to chromosomes, and the chemical structure and replication of the genetic material. In this chapter, we shift our perspective and consider the processes by which the information contained in genes is converted into molecules that determine the properties of cells and viruses. The transfer of genetic information from DNA into protein constitutes **gene expression.** The information transfer is accomplished by a series of events in which the sequence of bases in DNA is first copied into an RNA molecule and then the RNA is used, either directly or after some chemical modification, to determine the amino acid sequence of a protein molecule. The principal steps in gene expression can be summarized as follows:

1. RNA molecules are synthesized enzymatically by *RNA polymerase,* which uses the base sequence of a segment of a single strand of DNA as a template in a polymerization reaction similar to that used in replicating DNA. The overall process by which the segment corresponding to a particular gene is selected and an RNA molecule is made is called **transcription.**

2. In eukaryotes, the RNA usually undergoes chemical modification in the nucleus called **processing.**

3. Protein molecules are then synthesized by the use of the base sequence of a processed RNA molecule to direct the sequential joining of amino acids in a particular order, and so the amino acid sequence is a direct consequence of the base sequence. The production of an amino acid sequence from an RNA base sequence is called **translation,** and the protein made is called the **gene product.**

10.1
Proteins and Amino Acids

Proteins are the molecules responsible for catalyzing most intracellular chemical reactions (enzymes), for regulating gene expression (regulatory proteins), and for determining many features of the structures of cells, tissues, and viruses (structural proteins). A protein is composed of one or more chains of amino acids. Each of these chains is a series of covalently joined amino acids that constitute a **polypeptide.** The 20 different amino acids commonly found in polypeptides can be joined in any number and in any order. Because the number of amino acids in a polypeptide usually ranges from 100 to 1000, an enormous number of different protein molecules can be formed from the 20 common amino acids.

Each amino acid contains a carbon atom (the α carbon) to which is attached one carboxyl group ($-COOH$), one amino group ($-NH_2$), and a side chain commonly called an **R group** (Figure 10.1). The R groups are generally chains or rings of carbon atoms bearing various chemical groups. The simplest side chains are those of glycine ($-H$) and of alanine ($-CH_3$). For reference, the chemical structures of all 20 amino acids are shown in Figure 10.2. For each amino acid, the R group is indicated by a gold rectangle.

Polypeptide chains are formed when the carboxyl group of each amino acid becomes joined with the amino group of the next amino acid in line; the resulting chemical bond is an ordinary covalent bond called a **peptide bond** (Figure 10.3A). Thus the basic unit of a protein is a polypeptide chain in which α-carbon atoms alternate with peptide groups to form a backbone that has an ordered array of side chains (Figure 10.3B).

The two ends of every polypeptide molecule are distinct. One end has a free $-NH_2$ group and is called the **amino terminus;** the other end has a free $-COOH$ group and is the **carboxyl terminus.** Polypeptides

Figure 10.1
The general structure of an amino acid.

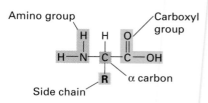

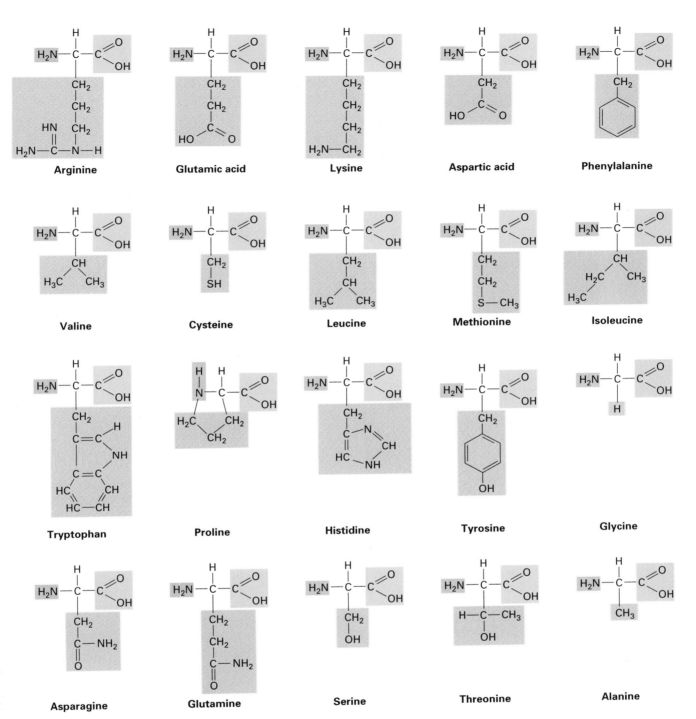

Figure 10.2

Chemical structures of the amino acids. Note that proline does not have the general structure shown in Figure 10.1 because it lacks a free amino group.

are synthesized by adding successive amino acids to the carboxyl end of the growing chain. Conventionally, the amino acids of a polypeptide chain are numbered starting at the amino terminus.

Most polypeptide chains are highly folded, and a variety of three-dimensional shapes have been observed. The manner of folding is determined primarily by the sequence of amino acids—in particular, by

Figure 10.3

Properties of a polypeptide chain. (A) Formation of a dipeptide by reaction of the carboxyl group of one amino acid (left) with the amino group of a second amino acid (right). A molecule of water (HOH) is eliminated to form a peptide bond (red line). (B) A tetrapeptide showing the alternation of α-carbon atoms (black) and peptide groups (blue). The four amino acids are numbered below.

(A)

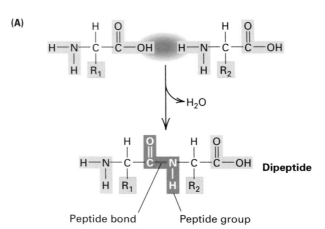

Dipeptide

Peptide bond   Peptide group

(B)

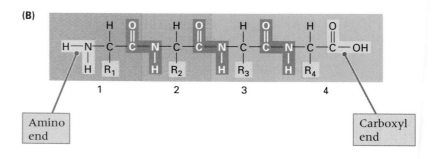

1   2   3   4

Amino end   Carboxyl end

noncovalent interactions between the side chains—so each polypeptide chain tends to fold into a unique three-dimensional shape as it is being synthesized. In some cases, protein folding is assisted by interactions with other proteins in the cell called **chaperones.** Generally speaking, the molecules fold so that amino acids with charged side chains tend to be on the surface of the protein (in contact with water) and those with uncharged side chains tend to be internal. Specific folded configurations also result from hydrogen bonding between peptide groups. Two fundamental polypeptide structures are the α helix and the β sheet (Figure 10.4). The α helix, represented as a coiled ribbon in Figure 10.4, is formed by interactions between neighboring amino acids that twist the backbone into a right-handed helix in which the N−H in each peptide group is hydrogen-bonded with the C−O in the peptide group located four amino acids further along the helix. In contrast, the β sheet, represented as parallel "flat" ribbons in Figure 10.4, is formed by interactions between amino acids in distant parts of the polypeptide chain; the back-

bones of the polypeptide chains are held flat and rigid (forming a "sheet"), because alternate N−H groups in one polypeptide backbone are hydrogen-bonded with alternate C−O groups in the polypeptide backbone of the adjacent chain. In each polypeptide backbone, alternate C−O and N−H groups are free to form hydrogen bonds with their counterparts in a different polypeptide backbone on the opposite side, so a β sheet can consist of multiple aligned segments in the same (or different) polypeptide chains. Other types of interactions also are important in protein folding—for example, covalent bonds may form between the sulfur atoms of pairs of cysteines in different parts of the polypeptide. However, the rules of folding are so complex that, except for the simplest proteins, the final shape of a protein cannot usually be predicted from the amino acid sequence alone.

Many protein molecules consist of more than one polypeptide chain. When this is the case, the protein is said to contain **subunits.** The subunits may be identical or different. For example, hemoglobin, the

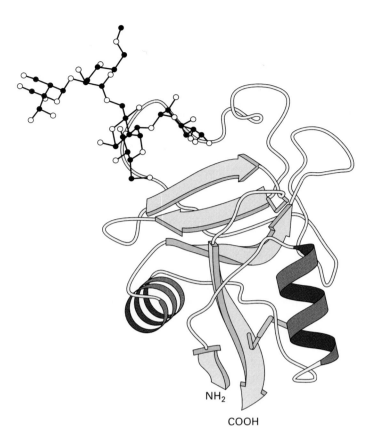

NH₂

COOH

Figure 10.4

A "ribbon" diagram of the path of the backbone of a polypeptide, showing the ways in which the polypeptide is folded. Arrows represent parallel β sheets, each of which is held to its neighboring β sheet by hydrogen bonds. Helical regions are shown as coiled ribbons. The polypeptide chain in this example is a mannose-binding protein. The stick figure at the upper left shows a molecule of mannose bound to the protein. [Adapted from William I. Weis, Kurt Drickamer, and Wayne A. Hendrickson. 1992. *Nature,* 360: 127.]

oxygen carrier of blood, consists of four subunits: two copies of each of two different polypeptides, which are designated the α chain and the β chain. (The use of α and β as names of the polypeptide chains has nothing to do with the α helices and β sheets that are found within the polypeptides.)

10.2
Relations Between Genes and Polypeptides

It took half a century to find out that genes control the structure of proteins. In the early 1900s, Archibald Garrod suggested that hereditary human diseases, such as phenylketonuria, result from inborn errors of metabolism (Chapter 1). Support for this idea came in the 1940s when George Beadle and Edward Tatum demonstrated, using *Neurospora crassa*, that genes govern the ability of the fungus to synthesize amino acids, purines, and vitamins. That genes control metabolism by determining protein structure was demonstrated when

it was shown, in the early 1950s, that the allele for sickle-cell anemia brings about a change in the charge of the hemoglobin molecule by causing substitution of an uncharged valine for a negatively charged glutamic acid at residue number 6 in the β-globin chain.

Most genes contain the information for the synthesis of only one polypeptide chain. Furthermore, the *sequence* of nucleotides in a gene determines the *sequence* of amino acids in a polypeptide. This point was first proved by studies of the tryptophan synthase gene *trpA* in *E. coli*, a gene in which many mutations had been obtained and accurately mapped. The effects of numerous mutations on the amino acid sequence of the enzyme were determined by directly analyzing the amino acid sequences of the wildtype and mutant enzymes. Each mutation was found to result in a single amino acid substituting for the wildtype amino acid in the enzyme; more important, *the order of the mutations in the genetic map was the same as the order of the affected amino acids in the polypeptide chain* (Figure 10.5). This attribute of genes and

Figure 10.5
Correlation of the positions of mutations in the genetic map of the E. coli trpA gene with positions of amino acid replacements in the TrpA protein.

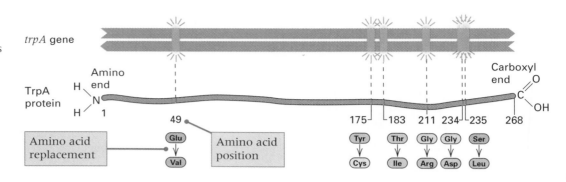

polypeptides is called **colinearity,** which means that the sequence of base pairs in DNA determines the sequence of amino acids in the polypeptide in a colinear, or point-to-point, manner. Colinearity is found almost universally in prokaryotes. However, we will see later that, in eukaryotes, noninformational DNA sequences interrupt the continuity of most genes; in these genes, the order of mutations along a gene (but not their spacing) correlates with the respective amino acid substitutions.

What Are the Minimal Genetic Functions Needed for Life?

The bacterium *Mycoplasma genitalium* belongs to a large group of bacteria, called mycoplasmas, that lack a cell wall. Mycoplasmas are free-living organisms that are parasites on a wide range of plant and animal hosts, including human beings. *M. genitalium,* which exists in parasitic association with ciliated epithelial cells of the genital and respiratory tracts of primates, is thought to have the smallest genome of all self-replicating organisms. The entire sequence of the genome has been determined, which enables us to see what constitutes a minimal functional gene set for a cell. The *M. genitalium* genome is a circular DNA molecule 580 kb in length (only about 3.5 times larger than that of the bacteriophage T4), and it encodes 471 genes.

The entire gene set of *M. genitalium* is depicted in Figure 10.6. The cellular processes in which these gene products participate are summarized in Table 10.1. A substantial fraction of the genome is devoted to macromolecular syntheses

Table 10.1 Summary of functions of 471 genes of *Mycoplasma genitalium*

Function	Number of Genes	Percent*	Function	Number of Genes	Percent*
DNA replication	32	10	Nucleoside & nucleotide synthesis	19	6
Transcription	12	4	Salvage (degradative pathways)	10	
Translation	101	32	Other metabolism		
Cell envelope	17	5	(lipids, cofactors, amino acids, intermediary metabolism)	18	6
Transport of small molecules	34	11			
Energy metabolism	31	10	Regulatory	7	2
ATP-proton force generation	8		Other	27	8
Glycolysis	10		Hypothetical or unknown	152	
Cell processes					
(cell division, secretion, stress response)	21	7			

*Percent among all genes with identified functions. Data from C. M. Fraser, J. D. Gocayne, O. White, M. D. Adams, R. A. Clayton, R. D. Fleischmann, and 23 other authors, *Science*, 1995. 270: 397.

Figure 10.6

Arrangement of coding sequences in *M. genitalium* as determined from the complete DNA sequence of the genome. The genes are color-coded according to the function of the gene product. Each arrowhead denotes the direction of transcription. [Figure design by O. White, courtesy of C. M. Fraser, J. D. Gocayne, O. White, M. D. Adams, R. A. Clayton, R. D. Fleischmann, and 23 other authors, 1995. *Science* 270: 397.]

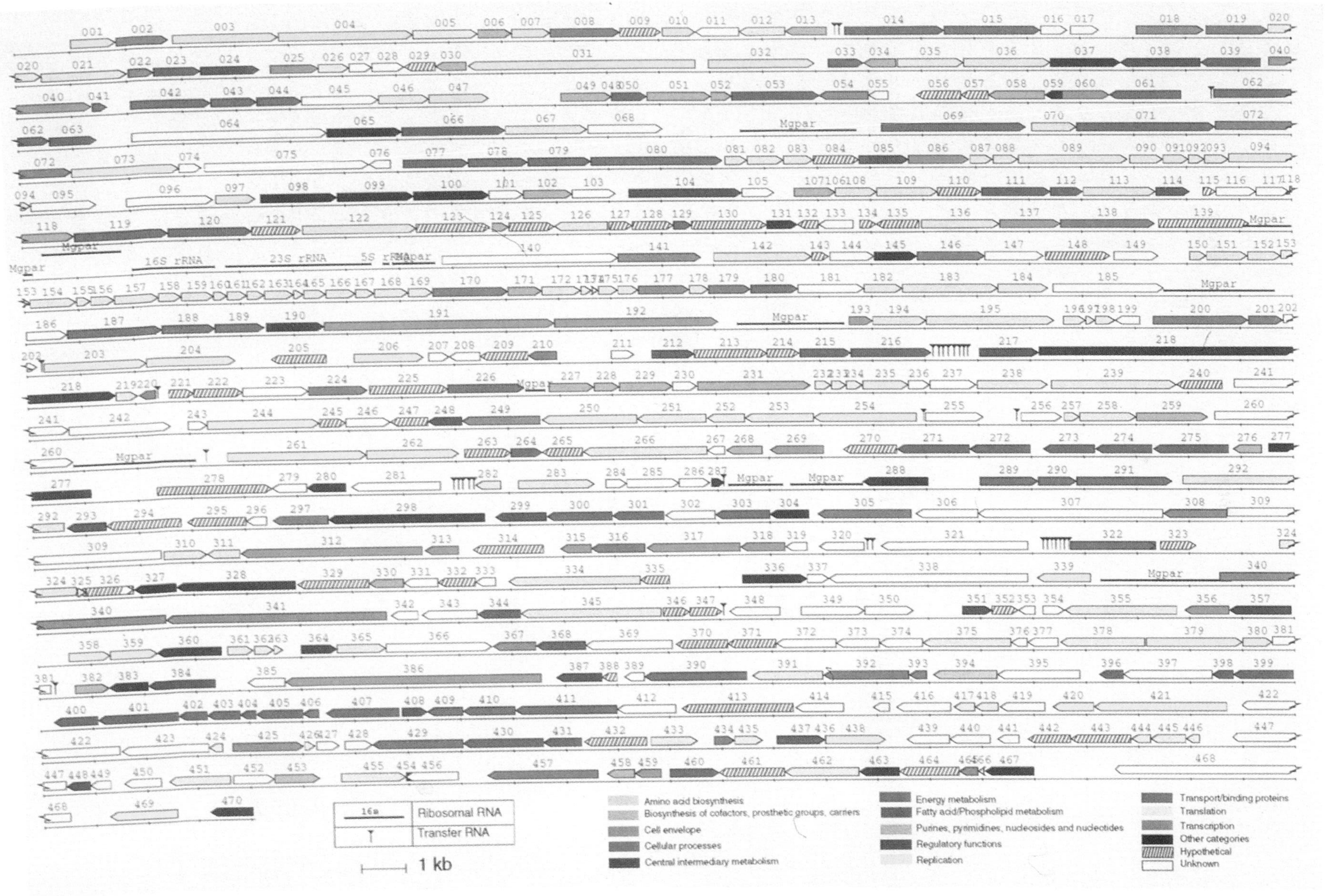

(DNA, RNA, protein), cell processes, and energy metabolism. There are very few genes for biosynthesis of small molecules. However, genes that encode proteins for salvaging and/or for transporting small molecules make up a substantial fraction of the total, which underscores the fact that the bacterium is parasitic. The remaining genes are largely devoted to formation of the cellular envelope and evasion of the immune system of the host.

10.3 Transcription

The first step in gene expression is the synthesis of an RNA molecule copied from the segment of DNA that constitutes the gene. The basic features of the production of RNA are described in this section.

General Features of RNA Synthesis

The essential chemical characteristics of the enzymatic synthesis of RNA resemble those of DNA synthesis (Chapter 5).

1. The precursors in the synthesis of RNA are the four ribonucleoside 5'-triphosphates—adenosine triphosphate (ATP), guanosine triphosphate (GTP), cytidine triphosphate (CTP), and uridine triphosphate (UTP). They differ from the DNA precursors only in that the sugar is ribose rather than deoxyribose and the base uracil (U) replaces thymine (T) (Figure 10.7).

2. In the synthesis of RNA, a sugar-phosphate bond is formed between the 3'-hydroxyl group of one nucleotide and the 5'-triphosphate of the next nucleotide in line (Figure 10.8A and B). This is the same chemical bond as in the synthesis of DNA, but the enzyme is different. The enzyme used in transcription is **RNA polymerase** rather than DNA polymerase.

3. The linear order of bases in an RNA molecule is determined by the sequence of bases in the DNA template. Each base added to the growing end of the RNA chain is chosen for its ability to base-pair with the DNA template strand. Thus the bases C, T, G, and A in a DNA strand cause G, A, C, and U, respectively, to be added to the growing end of an RNA molecule.

4. Nucleotides are added only to the 3'-OH end of the growing chain; as a result, the 5' end of a growing RNA molecule bears a triphosphate group. Note that the 5'-to-3' direction of RNA chain growth is the same as that in DNA synthesis.

A significant difference between DNA polymerase and RNA polymerase is that *RNA polymerase is able to initiate chain growth without a primer.* Furthermore,

> Each RNA molecule produced in transcription derives from a single strand of DNA, because in any particular region of the DNA, only one strand serves as a template for RNA synthesis.

The implications of this statement are shown in Figure 10.8C.

The synthesis of RNA can be described as consisting of four discrete stages.

1. *Promoter recognition* RNA polymerase binds to DNA within a base sequence from

Figure 10.7
Differences in the structures of ribose and deoxyribose and in those of uracil and thymine.

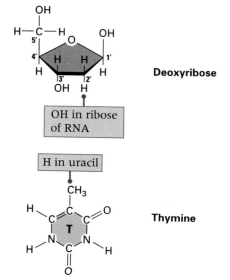

Deoxyribose

OH in ribose of RNA

H in uracil

Thymine

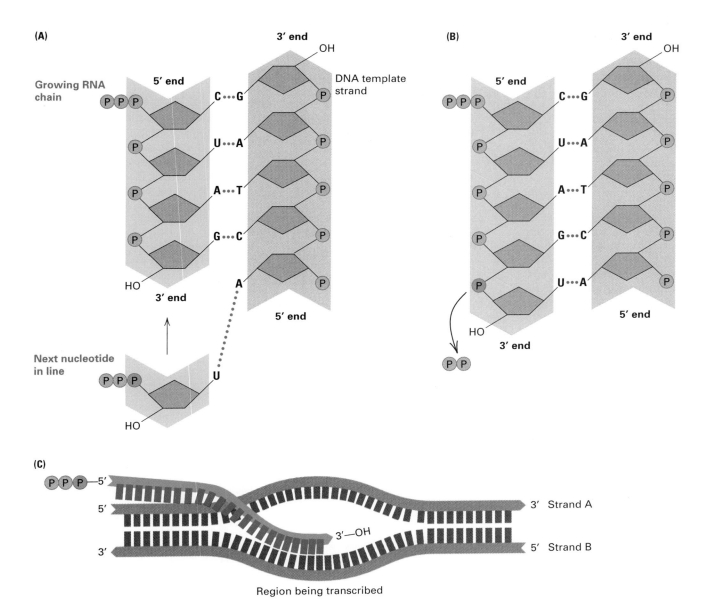

Figure 10.8

RNA synthesis. (A) The polymerization step in RNA synthesis. The incoming nucleotide forms hydrogen bonds (red dots) with a DNA base. The −OH group in the growing RNA chain reacts with the orange P in the next nucleotide in line (B). (C) Geometry of RNA synthesis. RNA is copied from only one strand of a segment of a DNA molecule—in this example, strand B—without the need for a primer. In this region of the DNA, RNA is not copied from strand A. However, in a different region (for example, in a different gene) strand A might be copied rather than strand B. Because RNA elongates in the 5'-to-3' direction, its synthesis moves along the DNA template in the 3'-to-5' direction; that is, the RNA molecule is antiparallel to the DNA strand being copied.

20 to 200 bases in length called a **promoter.** Many promoter sequences have been isolated and their base sequences determined. Although there is substantial sequence variation among promoter regions (in part corresponding to different strengths of the promoters in binding with the RNA polymerase) certain sequence patterns or "motifs" are quite frequent. Two such patterns often found in promoter regions in *E. coli* are illustrated in Figure 10.9. Each pattern is defined by a **consensus sequence** of bases determined from the actual sequences by majority rule: Each

One Gene, One Enzyme

George W. Beadle and Edward L. Tatum 1941
Stanford University, Stanford, California
Genetic Control of Biochemical Reactions in Neurospora

How do genes control metabolic processes? The suggestion that genes control enzymes was made very early in the history of genetics, most notably by the British physician Archibald Garrod in his 1908 book Inborn Errors of Metabolism. *But the precise relationship between genes and enzymes was still uncertain. Perhaps each enzyme is controlled by more than one gene, or perhaps each gene contributes to the control of several enzymes. The classic experiments of Beadle and Tatum showed that the relationship is usually remarkably simple: One gene codes for one enzyme. The pioneering experiments united genetics and biochemistry, and for the "one gene–one enzyme" concept, Beadle and Tatum were awarded a Nobel Prize in 1958 (Joshua Lederberg shared the prize for his contributions to microbial genetics). Because we now know that some enzymes contain polypeptide chains encoded by two (or occasionally more) different genes, a more accurate statement of the principle is "one gene, one polypeptide." Beadle and Tatum's experiments also demonstrate the importance of choosing the right organism.* Neurospora *had been introduced as a genetic organism only a few years earlier, and Beadle and Tatum realized that they could take advantage of the ability of this organism to grow on a simple medium composed of known substances.*

From the standpoint of physiological genetics the development and functioning of an organism consist essentially of an integrated system of chemical reactions controlled in some manner by genes. . . . In investigating the roles of genes, the physiological geneticist usually attempts to determine the physiological and biochemical bases of already known hereditary traits. . . . There are, however, a number of limitations inherent in this approach. Perhaps the most serious of these is that

---

These preliminary results appear to us to indicate that the approach may offer considerable promise as a method of learning more about how genes regulate development and function.

---

the investigator must in general confine himself to the study of non-lethal heritable characters. Such characters are likely to involve more or less non-essential so-called "terminal" reactions. . . . A second difficulty is that the standard approach to the problem implies the use of characters with visible manifestations. Many such characters involve morphological variations, and these are likely to be based on systems of biochemical reactions so complex as to make analysis exceedingly difficult. . . . Considerations such as those just outlined have led us to investigate the general problem of the genetic control of development and metabolic reactions by reversing the ordinary procedure and, instead of attempting to work out the chemical bases of known genetic characters, to set out to determine if and how genes control known biochemical reac-

tions. The ascomycete *Neurospora* offers many advantages for such an approach and is well suited to genetic studies. Accordingly, our program has been built around this organism. The procedure is based on the assumption that x-ray treatment will induce mutations in genes concerned with the control of known specific chemical reactions. If the organism must be able to carry out a certain chemical reaction to survive on a given medium, a mutant unable to do this will obviously be lethal on this medium. Such a mutant can be maintained and studied, however, if it will grow on a medium to which has been added the essential product of the genetically blocked reaction. . . . Among approximately 2000 strains [derived from single cells after x-ray treatment], three mutants have been found that grow essentially normally on the complete medium and scarcely at all on the minimal medium. One of these strains proved to be unable to synthesize vitamin B_6 (pyridoxine). A second strain turned out to be unable to synthesize vitamin B_1 (thiamine). A third strain has been found to be unable to synthesize para-aminobenzoic acid. . . . These preliminary results appear to us to indicate that the approach may offer considerable promise as a method of learning more about how genes regulate development and function. For example, it should be possible, by finding a number of mutants unable to carry out a particular step in a given synthesis, to determine whether only one gene is ordinarily concerned with the immediate regulation of a given specific chemical reaction.

Source: Proceedings of the National Academy of Sciences of the USA 27: 499–506

---

base in the consensus sequence is the base most often observed at that position in actual sequences. Any particular sequence may resemble the consensus sequence very well or very poorly.

The consensus sequences in the promoter regions in *E. coli* are TTGACA, cen-

tered approximately 35 base pairs upstream from the transcription start site (+1), and TATAAT, centered approximately 10 base pairs upstream from the +1 site. The -10 sequence, which is called the **TATA box,** is similar to sequences found at corresponding positions in many eukaryotic promoters. The positions of the promoter sequences determine where the RNA polymerase begins synthesis, and an A or G is often the first nucleotide in the transcript.

The strength of the binding of RNA polymerase to different promoters varies greatly, which causes differences in the extent of expression from one gene to another. Most of the differences in promoter strength result from variations in the −35 and −10 promoter elements and in the spacing between them. Promoter strength among *E. coli* genes differs by a factor of 10^4, and most of the variation can be attributed to the promoter sequences themselves. In general, the more closely the promoter elements resemble the consensus sequence, the stronger the promoter.

Mutations that change the base sequence in a promoter can alter the strength of the promoter; changes that result in less resemblance to the consensus sequence lower the strength, whereas those with greater resemblance to the consensus increase the strength. Furthermore, there are promoters that differ greatly from the consensus sequence in the −35 region. These promoters typically require accessory proteins to activate transcription by RNA polymerase. In eukaryotes, in addition to the promoter sequences, there are also other DNA sequences called *enhancers* that interact with the promoter to determine the level of transcription.

2. *Chain initiation* After the initial binding step, the RNA polymerase "melts" (locally denatures) the DNA double helix, causing the strand that is to be transcribed to separate from its partner strand and become accessible to the polymerase. The RNA polymerase then initiates RNA synthesis at a nearby transcription start site, denoted the +1 site in Figure 10.9. The first nucleoside triphosphate is placed at this site, and synthesis proceeds in a 5'-to-3' direction.

3. *Chain elongation* RNA polymerase moves progressively along the transcribed DNA strand, adding nucleotides to the growing RNA chain. Only one DNA strand, the **template strand,** is transcribed.

4. *Chain termination* RNA polymerase reaches a chain-termination sequence, and both the newly synthesized RNA molecule and the polymerase are released. Two kinds of termination events are known: those that are self-terminating and depend only on the base sequence in the DNA template, and those that require the presence of a termination protein. In self-termination, which is the most common case, transcription stops when the polymerase encounters a particular sequence of bases in the transcribed DNA strand that is able to fold back upon itself to form a hairpin loop. An example of such a terminator found in *E. coli* is shown in Figure 10.10. The hairpin

Figure 10.9
Base sequences in promoter regions of several genes in *E. coli*. The consensus sequences located 10 and 35 nucleotides upstream from the transcription start site (+1) are indicated. Promoters vary tremendously in their ability to promote transcription. Much of the variation in promoter strength results from differences between the promoter elements and the consensus sequences at −10 and −35.

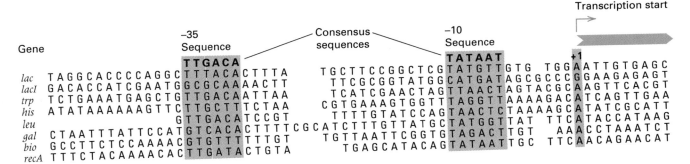

(A) DNA

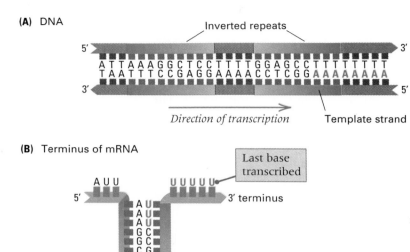

Inverted repeats

5′   3′
ATTAAAGGCTCCTTTTGGAGCCTTTTTTTT
TAATTTCCGAGGAAAACCTCGGAAAAAAAA
3′   5′

Direction of transcription   Template strand

(B) Terminus of mRNA

Last base transcribed

3′ terminus

Figure 10.10

(A) Base sequence of the transcription-termination region for the set of tryptophan-synthesizing genes in *E. coli*. The inverted repeat sequences (blue) are characteristic of termination sites. (B) The 3′ terminus of the RNA transcript, folded to form a stem-and-loop structure. The sequence of U's found at the end of the transcript in this and many other prokaryotic genes is in red. The RNA polymerase, not shown here, terminates transcription when the loop forms in the transcript.

loop alone is not enough for termination of transcription; the run of U's at the end of the hairpin is also necessary.

Initiation of a second round of transcription need not await completion of the first, because the promoter becomes available once RNA polymerase has polymerized from 50 to 60 nucleotides. For a rapidly transcribed gene; such reinitiation occurs repeatedly, and a gene can be cloaked with numerous RNA molecules in various degrees of completion. The micrograph in Figure 10.11 shows a region of the DNA of the newt *Triturus* that contains tandem repeats of a particular gene. Each gene is associated with growing RNA molecules. The shortest RNA molecules are at the promoter end of the gene; the longest are near the gene terminus.

The existence of promoters was first demonstrated in genetic experiments with *E. coli* by the isolation of particular Lac⁻

mutations, denoted p^-, that eliminate activity of the *lac* gene but *only when the mutations are adjacent to the gene in the same DNA molecule*. The need for the coupled genetic configuration, also called the *cis* configuration, can be seen by examining a cell with two copies of the gene *lacZ*—for example, a cell containing an F′ *lacZ* plasmid, which contains *lacZ* in the bacterial chromosome as well as *lacZ* in the F′ plasmid. Transcription of the *lacZ* gene enables the cell to synthesize the enzyme β-galactosidase. Table 10.2 shows that a wild-type *lacZ* gene (*lacZ*⁺) is inactive when it and a p^- mutation are present in the same DNA molecule (either in the chromosome or in an F′ plasmid); this can be seen by comparing entries 4 and 5. Analysis of the RNA shows that, in a cell with the genotype p^- *lacZ*⁺, the *lacZ*⁺ gene is not transcribed, whereas if the genotype is p^+ *lacZ*⁻, a mutant RNA is produced. The p^- mutations are called **promoter mutations.**

Figure 10.11

Electron micrograph of part of the DNA of the newt *Triturus viridescens* containing tandem repeats of genes being transcribed into ribosomal RNA. The thin strands forming each feather-like array are RNA molecules. A gradient of lengths can be seen for each rRNA gene. Regions in the DNA between the thin strands are spacer DNA sequences, which are not transcribed. [Courtesy of Oscar Miller.]

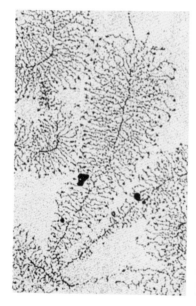

Table 10.2 Effect of promoter mutations on transcription of the *lacZ* gene

Genotype	Transcription of *lacZ*$^+$ gene
1. $p^+ lacZ^+$	Yes
2. $p^- lacZ^+$	No
3. $p^+ lacZ^+ / p^+ lacZ^-$	Yes
4. $p^- lacZ^+ / p^+ lacZ^-$	No
5. $p^+ lacZ^+ / p^- lacZ^-$	Yes

Note: lacZ$^+$ is the wildtype gene; *lacZ*$^-$ is a mutant that produces a nonfunctional enzyme.

Mutations have also been instrumental in defining the transcription-termination region. Mutations have been isolated that create a new termination sequence upstream from the normal one. When such a mutation is present, an RNA molecule is made that is shorter than the wildtype RNA. Other mutations eliminate the terminator, resulting in a longer transcript.

The best understood RNA polymerase is that of the bacterium *E. coli*. This enzyme consists of five protein subunits and can be easily seen by electron microscopy (Figure 10.12). In *E. coli*, all transcription is catalyzed by this enzyme. Eukaryotic cells have three distinct RNA polymerases, denoted I, II, and III, each of which makes a particular class of RNA molecule. RNA polymerase I catalyzes synthesis of all ribosomal RNA species except 5S RNA; RNA polymerase III catalyzes synthesis of 5S and all of the transfer RNAs. The RNA polymerase II is the enzyme responsible for the synthesis of all RNA transcripts that contain information specifying amino acid sequences. These transcripts are called messenger RNA molecules, which are discussed in the next section. RNA polymerase II also catalyzes synthesis of most small nuclear RNAs involved in RNA splicing, which is discussed in Section 10.4.

Messenger RNA

Amino acids do not bind directly to DNA. Consequently, intermediate steps are needed for arranging the amino acids in a polypeptide chain in the order determined by the DNA base sequence. This process begins with transcription of the base sequence of the template strand of DNA into the base sequence of an RNA molecule. In prokaryotes, this RNA molecule, which is called **messenger RNA,** or **mRNA,** is

Figure 10.12
E. coli RNA polymerase molecules bound to DNA. [Courtesy of Robley Williams.]

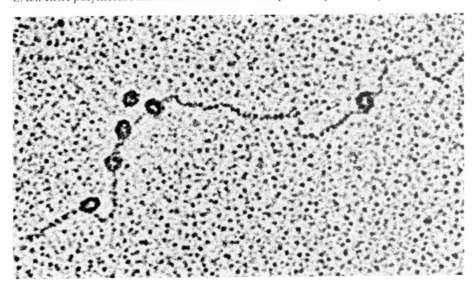

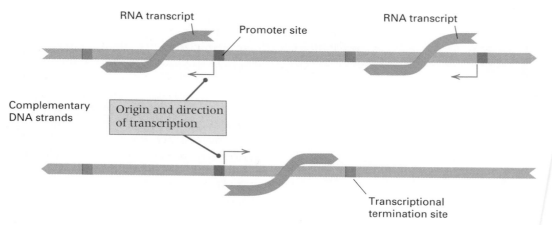

Figure 10.13

A typical arrangement of promoters (green) and termination sites (red) in a segment of a DNA molecule. Promoters are present in both DNA strands. Termination sites are usually located such that transcribed regions do not overlap.

used directly in polypeptide synthesis. In eukaryotes, the RNA molecule is generally processed before it becomes mRNA. The amino acid sequence is then determined by the base sequence in mRNA by the protein-synthesizing machinery of the cell.

Not all base sequences in an mRNA molecule are translated into the amino acid sequences of polypeptides. For example, translation of an mRNA molecule rarely starts exactly at one of its ends and proceeds to the other end; instead, initiation of polypeptide synthesis may begin many nucleotides downstream from the 5' end of the mRNA. The untranslated 5' segment of RNA is called a **leader** and in some cases contains regulatory sequences that affect the rate of protein synthesis. The leader is followed by a **coding sequence,** also called an **open reading frame,** or **ORF,** which specifies the order in which the amino acids are present in the polypeptide chain. A typical coding sequence in an mRNA molecule is between 500 and 3000 bases long (depending on the number of amino acids in the protein). Like the leader sequence, the 3' end of an mRNA molecule following the coding sequence is not translated.

The template strand of each gene is only one of the two DNA strands present in the gene, but which DNA strand is the template can differ from gene to gene along a DNA molecule. That is, except in some small viruses, *not all mRNA molecules are transcribed from the same DNA strand.* Thus in an extended segment of a DNA molecule, mRNA molecules would be seen growing in either of two directions (Figure 10.13), depending on which DNA strand functions as a template.

In prokaryotes, most mRNA molecules are degraded within a few minutes after synthesis. In eukaryotes, a typical lifetime is several hours, although some last only minutes, and others persist for days. In both kinds of organisms, the degradation enables cells to dispose of molecules that are no longer needed. The short lifetime of prokaryotic mRNA is an important factor in regulating gene activity (Chapter 11).

10.4 RNA Processing

The process of transcription is very similar in prokaryotes and eukaryotes, but there are major differences in the relation between the transcript and the mRNA used for polypeptide synthesis. In prokaryotes, the immediate product of transcription (the **primary transcript**) is mRNA; by contrast, *the primary transcript in eukaryotes must be converted into mRNA.* This conversion, which is called **RNA processing,** usually consists of two types of events: modification of the ends and excision of untranslated sequences embedded *within* coding sequences. These events are illustrated diagrammatically in Figure 10.14.

Each end of a eukaryotic transcript is processed. The 5' end is altered by the addition of a modified guanosine, 7-methyl guanosine, in an uncommon 5'-to-5' (instead of 3'-to-5') linkage; this terminal group is called a **cap.** The 3' terminus of a eukaryotic mRNA molecule is usually modified by the addition of a polyadenosine sequence (the **poly-A tail**) of as many as 200 nucleotides. The 5' cap is necessary for the mRNA to bind with the ribosome to begin protein synthesis, and the poly-A tail helps to determine mRNA stability.

A second important feature peculiar to the primary transcript in eukaryotes, also shown in Figure 10.14, is the presence of segments of RNA, called **introns** or **intervening sequences,** that are excised from the primary transcript. Accompanying the excision of introns is a rejoining of the coding segments (**exons**) to form the mRNA molecule. The excision of the introns and the joining of the exons is called **RNA splicing.** The mechanism of RNA splicing is illustrated schematically in Figure 10.15. Figure 10.15A shows the consensus sequence found at the 5' (**donor**) end and at the 3' (**acceptor**) end of most introns. The symbols are N, any nucleotide; R, any purine (A or G); Y, any pyrimidine (C or U); and S, either A or C. In the first step of

Figure 10.14
A schematic drawing showing the production of eukaryotic mRNA. The primary transcript is capped before it is released from the DNA. MeG denotes 7-methylguanosine (a modified form of guanosine), and the two asterisks indicate two nucleotides whose riboses are methylated. The 3' end is usually modified by the addition of consecutive adenines. Along the way, the introns are excised. These reactions take place within the nucleus.

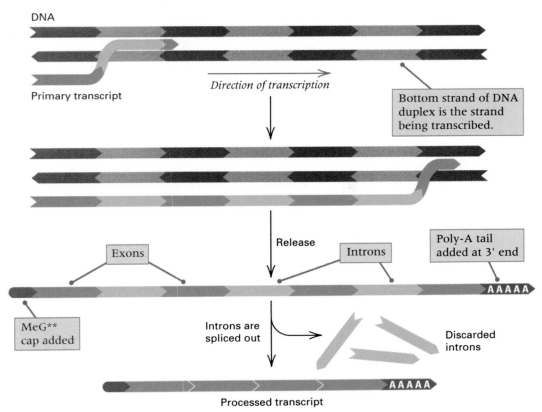

DNA

Primary transcript

Direction of transcription

Bottom strand of DNA duplex is the strand being transcribed.

Release

Exons

Introns

Poly-A tail added at 3' end

AAAAA

MeG** cap added

Introns are spliced out

Discarded introns

AAAAA

Processed transcript

splicing, the 2'−OH of the adenosine (A) at the branch site, which is located a short distance upstream from a run of prymidines (Y) near the acceptor site, attacks the phosphodiester bond at the donor splice site junction. The attack results in cleavage at the donor splice site and formation of a branched molecule (Figure 10.15B) known as a **lariat** because it has a loop and a tail. The A−G linkage at the "knot" of the lariat is unusual in being 2'-to-5' (instead of the usual 3'-to-5'). In the final step of splicing (Figure 10.15C), the 3'−OH of the guanosine of the donor exon attacks the phosphodiester bond at the acceptor splice site, freeing the lariat intron and joining the donor and acceptor exons together. The lariat intron is rapidly degraded into individual nucleotides by nucleases.

RNA splicing takes place in nuclear particles known as **spliceosomes.** These abundant particles are composed of protein and several types of specialized small RNA molecules ranging from 100 to 200 bases in length. The specificity of splicing comes from the small RNAs, some of which contain sequences that are complementary to the splice junctions, but numerous spliceosome proteins also are required for splicing. One model for the process is illustrated in Figure 10.16, in which U1, U2, and U5 are designations for three different types of small nuclear RNAs. The ends of the intron are brought together by U1, which forms base pairs with nucleotides in the intron at both the 5' and the 3' ends. The ends of the exons are brought together by U5, which forms base pairs with nucleotides in the exons at both the donor splice site and the acceptor splice site. The black arrow in Figure 10.16 indicates the initial attack of the branch site A on the donor splice site

Figure 10.15

A schematic diagram showing removal of one intron from a primary transcript. The A nucleotide at the branch site attacks the terminus of the 5' exon, cleaving the exon-intron junction and forming a loop connected back to the branch site. The 5' exon is later brought to the site of cleavage of the 3' exon, a second cut is made, and the exon termini are joined and sealed. The loop is released as a lariat-shaped structure that is degraded. Because the loop includes most of the intron, the loop of the lariat is usually very much longer than the tail.

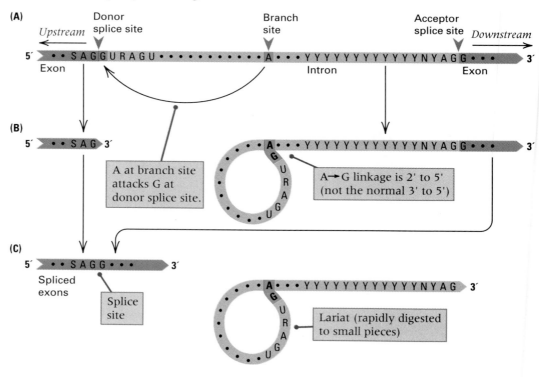

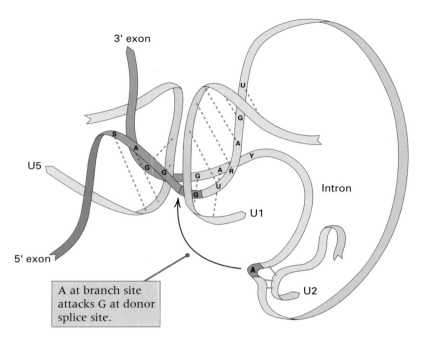

3' exon

U5

5' exon

A at branch site
attacks G at donor
splice site.

Intron

U1

U2

Figure 10.16

Model for RNA splicing. U1, U2, and U5 are small nuclear RNA molecules.
The exons are shown in dark green, and the intron is shown in light
green. The base pairs that form with nucleotides in the small nuclear
RNAs are indicated by dashed (U1 and U5) or solid (U2) lines between the
RNA strands. The arrow shows the first step, in which the adenosine of
the branch site, held in place by U2, attacks the phosphodiester bond at
the splice donor site, resulting in cleavage at the splice donor site and formation of a lariat structure. The process of splicing also requires additional
small nuclear RNAs, as well as numerous proteins. Together, these components form the spliceosome. [Modified from J. A. Steitz. 1992. *Science*,
257: 888.]

junction. Note that the branch site is held
in place by U2.

Introns are also present in some genes
in organelles, but the mechanisms of their
excision differ from those of introns in
nuclear genes because organelles do not
contain spliceosomes. In one class of
organelle introns, the intron contains a
sequence coding for a protein that participates in removing the intron that codes for
it. The situation is even more remarkable
in the splicing of a ribosomal RNA precursor in the ciliate *Tetrahymena*. In this case,
the splicing reaction is intrinsic to the folding of the precursor; that is, the RNA precursor is *self-splicing* because the folded

precursor RNA creates its own RNA-splic-
ing activity. The self-splicing *Tetrahymena*
RNA was the first example found of an
RNA molecule that could function as an
enzyme in catalyzing a chemical reaction;
such enzymatic RNA molecules are called
ribozymes.

The existence and the positions of
introns in a particular primary transcript
are readily demonstrated by renaturing the
transcribed DNA with the fully processed
mRNA molecule. The DNA-RNA hybrid
can then be examined by electron
microscopy. An example of adenovirus
mRNA (fully processed) and the corresponding DNA are shown in Figure 10.17.

The DNA copies of the introns appear as single-stranded loops in the hybrid molecule, because no corresponding RNA sequence is available for hybridization.

The number of introns per RNA molecule varies considerably from one gene to the next. For example, 2 introns are present in the primary transcript of human α-globin, and 52 introns occur in collagen RNA. Furthermore, within a particular RNA molecule, the introns are widely distributed and have many different sizes (Figure 10.18). In human beings and other mammals, most introns range in size from 100 to 10,000 base pairs, and in the processing of a typical primary transcript, the amount of discarded RNA ranges from about 50 percent to nearly 90 percent of the primary transcript. Genes in lower eukaryotes, such as yeast, nematodes, and fruit flies, generally have fewer introns than genes in mammals, and the introns tend to be much smaller.

Most introns appear to have no function in themselves. An artificial gene that lacks a particular intron usually functions normally. In those cases in which an intron seems to be required for function, it is usually not because the interruption of the gene is necessary, but because the intron happens to include certain nucleotide sequences that regulate the timing or tissue specificity of transcription. The implication is that many mutations in introns, including small deletions and insertions, should have essentially no effect on gene function, and this is the case. Moreover, the nucleotide sequence of a particular intron is found to undergo changes (including small deletions and insertions) extremely rapidly in the course of evolution, and this lack of sequence conservation is another indication that most of the nucleotide sequences present within introns are not important.

Mutations that affect any of the critical splicing signals do have important consequences, because they interfere with the splicing reaction. Two possible outcomes are illustrated in Figure 10.19. In Figure10.19A, the intron with the mutated splice site fails to be removed, and it remains in the processed mRNA. The result

Figure 10.17

(A) An electron micrograph of a DNA-RNA hybrid obtained by annealing a single-stranded segment of adenovirus DNA with one of its mRNA molecules. The loops are single-stranded DNA. (B) An interpretive drawing. RNA and DNA strands are shown in red and blue, respectively. Four regions do not anneal, creating three single-stranded DNA segments that correspond to the introns and the poly-A tail of the mRNA molecule. [Electron micrograph courtesy of Tom Broker and Louise Chow.]

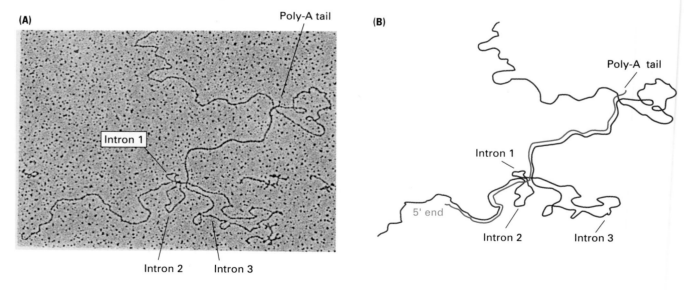

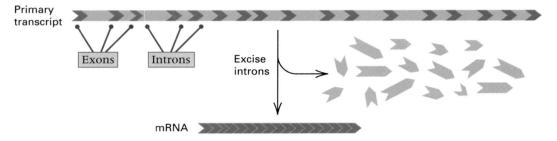

Figure 10.18
A diagram of the primary transcript and the processed mRNA of the conalbumin gene. The 16 introns, which are excised from the primary transcript, are shown in light green. The exons range in size from 29–331 bp (average 138 bp); the introns range in size from 124–1313 bp (average 512 bp). Approximately 75 percent of the primary transcript consists of introns.

is the production of a mutant protein with a normal sequence of amino acids up to the splice site but an abnormal sequence afterward. Most introns are long enough that, by chance, they contain a stop sequence that terminates protein synthesis, and once a stop is encountered, the protein grows no further. A second kind of outcome is shown in Figure10.19B. In this case, splicing does occur, but at an alternative splice site. (The example shows the alternative site downstream from the mutation, but alternative

Figure 10.19
Possible consequences of mutation in the donor splice site of an intron. (A) No splicing occurs, and the entire intron remains in the processed transcript. (B) Splicing occurs at a downstream cryptic splice site, and only the upstream part of the original intron still remains in the processed transcript. Neither outcome results in a normal protein product.

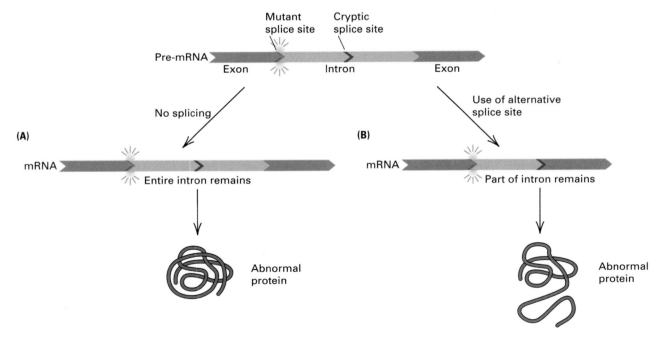

sites can also be upstream.) The alternative site is called a **cryptic splice site** because it is not normally used. The cryptic splice site is usually a poorer match with the consensus sequence and is ignored when the normal splice site is available. The result of using the alternative splice site is again an incorrectly processed mRNA and a mutant protein. In some splice-site mutations, both outcomes (Figure 10.19A and B) can occur: Some transcripts leave the intron unspliced, whereas others are spliced at cryptic splice sites.

Although introns are not usually essential in regulating gene expression, they may play a role in gene evolution. In some cases, the exons in a gene code for segments of the completed protein that are relatively independent in their folding characteristics. For example, the central exon of the β-globin gene codes for the segment of the protein that folds around an iron-containing molecule of heme. Relatively autonomous folding units in proteins are known as folding **domains,** and the correlation between exons and domains found in some genes suggests that the genes were originally assembled from smaller pieces. In some cases, the ancestry of the exons can be traced. For example, the human gene for the low-density lipoprotein receptor that participates in cholesterol regulation shares exons with certain blood-clotting factors and epidermal growth factor. The model of protein evolution through the combination of different exons is called the **exon-shuffle** model. The mechanism for combining exons from different genes is not known. Although some genes support the model, in other genes the boundaries of the folding domains do not coincide with exons.

The evolutionary origin of introns is unknown. On the one hand, introns may be an ancient feature of gene structure. The existence of self-splicing RNAs means that introns could have existed long before the evolution of the spliceosome mechanism, and therefore some introns may be as old as the genes themselves. Furthermore, the finding that some genes have introns in the same places in both plants and animals suggests that the introns may have been in place before plants and animals became separate lineages. If introns are ancient,

then exon shuffling might have been important early in evolution by creating new genes with novel combinations of exons. It has even been suggested that all forms of early life had introns in their genes and that today's prokaryotes, which lack introns, lost their introns in their evolution. On the other hand, it has also been argued that introns arose relatively late in evolution and became inserted into already existing genes, particularly in vertebrate genomes.

10.5 Translation

The synthesis of every protein molecule in a cell is directed by an mRNA originally copied from DNA. Protein production includes two kinds of processes: (1) information-transfer processes in which the RNA base sequence determines an amino acid sequence, and (2) chemical processes in which the amino acids are linked together. The complete series of events is called **translation.**

The main ingredients necessary for translation are as follows:

- *Messenger RNA* Messenger RNA is needed to bring the ribosomal subunits together (described below) and to provide the coding sequence of bases that determines the amino acid sequence in the resulting polypeptide chain.

- *Ribosomes* These components are particles on which protein synthesis takes place. They move along an mRNA molecule and align successive transfer RNA molecules; the amino acids are attached one by one to the growing polypeptide chain by means of peptide bonds. Ribosomes consist of two subunit particles. In *E. coli,* their sizes are 30S (the small subunit) and 50S (the large subunit). The counterparts in eukaryotes are 40S and 60S. (The S stands for *Svedberg unit,* which measures the rate of sedimentation of a particle in a centrifuge and so is an indicator of size.) Together, the small and large particles form a functional ribosome. An electron

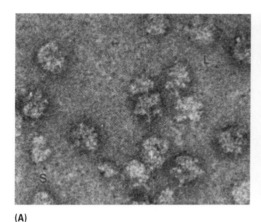
(A)

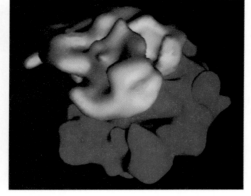
(B)

Figure 10.20

Ribosomes. (A) An electron micrograph of 70S ribosomes from *E. coli*. The 70S ribosome consists of one small subunit of size 30S and one large subunit of size 50S. (B) A three-dimensional model of the *E. coli* 70S ribosome based on high-resolution electron microscopy. The 30S subunit is in light green, and the 50S subunit is in dark blue. [A, courtesy of James Lake; B, courtesy of J. Frank, A. Verschoor, Y. Li, J. Zhu, R. K. Lata, M. Radermacher *et al.* 1995. *Biochemistry and Cell Biology.* 73: 357.]

micrograph and a model of an *E. coli* ribosome are shown in Figure 10.20.

- *Transfer RNA, or tRNA* The sequence of amino acids in a polypeptide is determined by the base sequence in the mRNA by means of a set of adaptor molecules, the tRNA molecules, each of which is attached to a particular amino acid. Each group of three adjacent bases in the mRNA forms a **codon** that binds to a particular group of three adjacent bases in the tRNA (an **anticodon**), bringing the attached amino acid into line for addition to the growing polypeptide chain.

- *Aminoacyl tRNA synthetases* This set of enzymes catalyzes the attachment of each amino acid to its corresponding tRNA molecule. A tRNA attached to its amino acid is called an **aminoacylated tRNA** or a **charged tRNA.**

- *Initiation, elongation, and termination factors* Polypeptide synthesis can be divided into three stages—(1) initiation, (2) elongation, and (3) termination. Each stage requires specialized proteins.

In prokaryotes, all of the components for translation are present throughout the cell; in eukaryotes, they are located in the cytoplasm, as well as in mitochondria and chloroplasts.

In overview, the process of translation is that an mRNA molecule binds to a ribosome. The aminoacylated tRNAs are brought along sequentially, one by one, to the ribosome that is translating the mRNA molecule. Peptide bonds are made between successively aligned amino acids, each time joining the amino group of the incoming amino acid to the carboxyl group of the amino acid at the growing end. Finally, the chemical bond between the last tRNA and its attached amino acid is broken, and the completed polypeptide is removed.

Initiation

The main features of the initiation step in polypeptide synthesis are the binding of mRNA to the small subunit of the ribosome and the binding of a charged tRNA bearing the first amino acid (Figure 10.21A). The ribosome includes three sites for tRNA molecules. They are called the **E (exit)** site, the **P (peptidyl)** site, and the **A (amino-acyl)** site. In the initiation of translation,

(A)

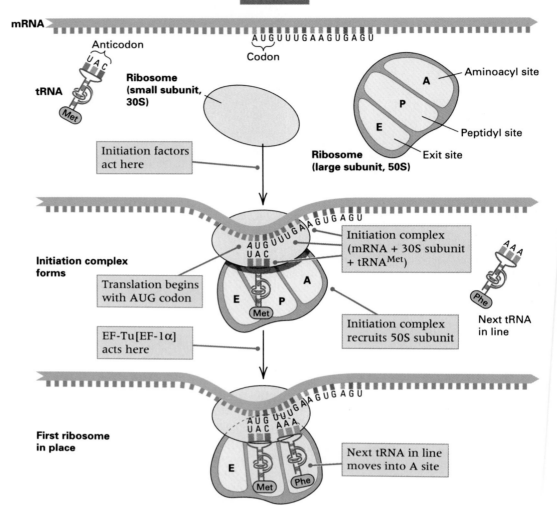

Figure 10.21 (*above and facing page*)
Initiation of protein synthesis. (A) The initiation complex—consisting of the mRNA, one 30S ribosomal subunit, and tRNA^Met—recruits a 50S ribosomal subunit in which the tRNA^Met occupies the P (peptidyl) site of the ribosome. A second charged tRNA (in this example, tRNA^Phe) joins the complex in the A (aminoacyl) site. (B) First steps in elongation. The methionine is transferred from the tRNA^Met onto the amino group of phenylalanine (the attacking group), resulting in cleavage of the bond between methionine and its tRNA and peptide bond formation in a concerted reaction catalyzed by peptidyl transferase in the 50S subunit. Then the ribosome shifts one codon along the mRNA to the next in line.

two initiation factors (IF-1 and IF-3) interact with the 30S subunit at the same time that another initiation factor (IF-2) binds with a special initiator tRNA charged with methionine. (In prokaryotes, the initiator tRNA actually carries formylmethionine, yielding tRNA^fMet.) These components come together and combine with an mRNA. In prokaryotes, the mRNA binding is facilitated by hydrogen bonding between the 16S RNA present in the 30S subunit and the **ribosome-binding site** of the mRNA; in eukaryotes, the 5' cap on the mRNA is instrumental. Together, the 30S + tRNA^Met + mRNA complex recruits a 50S subunit, in which the tRNA^Met is positioned in the P site and aligned with the AUG initiation codon, forming the 70S initiation complex (Figure 10.21A). The tRNA binding is accomplished by hydrogen bonding between the AUG codon in the mRNA and the three-base **anticodon** in

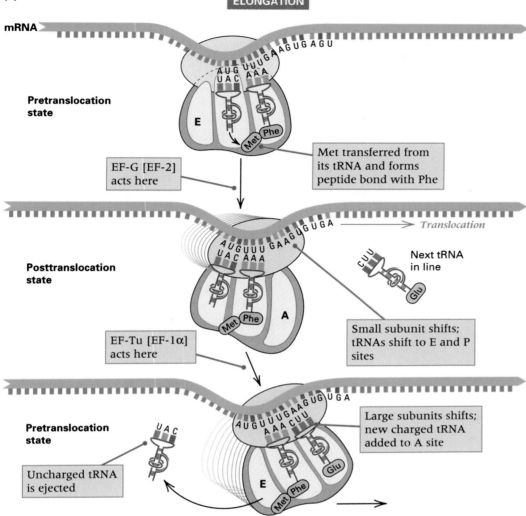

ELONGATION

mRNA

Pretranslocation state

E

Met Phe

EF-G [EF-2] acts here

Met transferred from its tRNA and forms peptide bond with Phe

Translocation

Posttranslocation state

Next tRNA in line

Glu

Met Phe

A

EF-Tu [EF-1α] acts here

Small subunit shifts; tRNAs shift to E and P sites

Pretranslocation state

UAC

Uncharged tRNA is ejected

Large subunits shifts; new charged tRNA added to A site

E

Met Phe

Glu

the tRNA. In the assembly of the completed ribosome, the initiation factors dissociate from the complex.

Elongation

The elongation stage of translation consists of three processes: bringing each new aminoacylated tRNA into line, forming the new peptide bond to elongate the polypeptide, and moving the ribosome along the mRNA so that the codons can be translated successively. The first step in elongation is illustrated in Figure 10.21B. A key role is played by the elongation factor EF-Tu, although a second protein, EF-Ts, is also required. (The eukaryotic counterpart of

EF-Tu is called EF-1α.) The EF-Tu, bound with guanosine triphosphate (EF-Tu-GTP), brings the next aminoacylated tRNA into the A site on the 50S subunit, which in this example is tRNA^{Phe}. This processes requires the hydrolysis of GTP to GDP, and once the GDP is formed, the EF-Tu-GDP has low affinity for the ribosome and diffuses away, becoming available for reconversion into EF-Tu-GTP.

Once the A site is filled, a **peptidyl transferase** activity catalyzes a concerted reaction in which the bond connecting the methionine to the tRNA^{Met} is transferred to the amino group of the phenylalanine, forming the first peptide bond. Peptidyl transferase activity is not due to a single

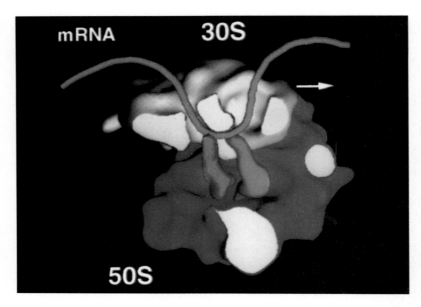

Figure 10.22
A model of a 70S ribosome with some parts cut away to show the orientation of the mRNA relative to the 30S and 50S subunits. The P and A tRNA sites are indicated in red and dark green, respectively. This is the pretranslocation state, in which the E site is unoccupied. [Courtesy of J. Frank, A. Verschoor, Y. Li, J. Zhu, R. K. Lata, M. Radermacher *et al.* 1995. *Biochemistry and Cell Biology* 73: 357.]

molecule but requires several components of the 50S subunit, including several proteins and an RNA component (called 23S) of the 50S subunit. Some evidence indicates that the actual catalysis is carried out by the 23S RNA, which would suggest that 23S is an example of a ribozyme at work.

Figure 10.22 shows a cutaway view of a 70S ribosome and the bound tRNA molecules in the P site and the A site. The 30S subunit, in light green at the top, binds the mRNA and moves along it in the direction indicated by the arrow. The 50S subunit, in blue at the bottom, contains the tRNA binding sites in the P and A sites. The P site is in red, to the left, and the A site is in dark green, to the right.

Note in Figure 10.21B that the relative positions of the 30S and 50S ribosomal subunits are shifted from one panel to the next. The configuration of the subunits in the top panel is called the **pretranslocation state.** In the middle panel, the 30S subunit shifts one codon to the right. This event is called *translocation*. After translocation, the ribosome is said to be in the *posttranslocation state*. In the next step of polypeptide synthesis, shown in the bot-

tom panel in Figure 10.21B, the 50S subunit shifts one step over to the right, which reconfigures the ribosome back into the pretranslocation state.

The term **translocation,** as applied to protein synthesis, means the movement of the 30S subunit one codon further along the mRNA. With each successive translocation of the 30S subunit, one more amino acid is added to the growing polypeptide chain. The entire cycle of charged tRNA addition, peptide bond formation, and translocation is elongation.

The repetitive steps in elongation are outlined in Figure 10.23. Starting with a ribosome in the pretranslocation state of Figure 10.23A, the elongation factor EF-G binds with the ribosome. (The eukaryotic counterpart of EF-G is called EF-2.) Like EF-Tu, the EF-G comes on in the form EF-G-GTP and, in fact, binds to the same ribosomal site as EF-Tu-GTP. Hydrolysis of the GTP to GDP yields the energy to shift the tRNAs in the P and A sites to the E and P sites, respectively, as well as to translocate the 30S subunit one codon along the mRNA (red arrow). The ribosome is thereby converted to the **posttranslocation state** (Figure 10.23B), and the EF-G-GDP is released.

At this stage, EF-Tu-GTP comes into play again, and four events happen, as indicated in Figure 10.23C:

- The next aminoacylated tRNA is brought into line (in this case, tRNA^{Val}).

- The uncharged tRNA is ejected from the E site.

- In a concerted reaction, the bond connecting the growing polypeptide chain to the tRNA in the P site is transferred to the amino group of the amino acid in the A site, forming the new peptide bond.

- The ribosome transitions to the *pretranslocation* state.

Also, the EF-Tu-GDP is released, making room for the EF-G-GTP, whose function is shown in Figure 10.23D:

- *Translocation* of the 30S ribosome one codon further along the mRNA and return of the ribosome to its *posttranslocation* state.

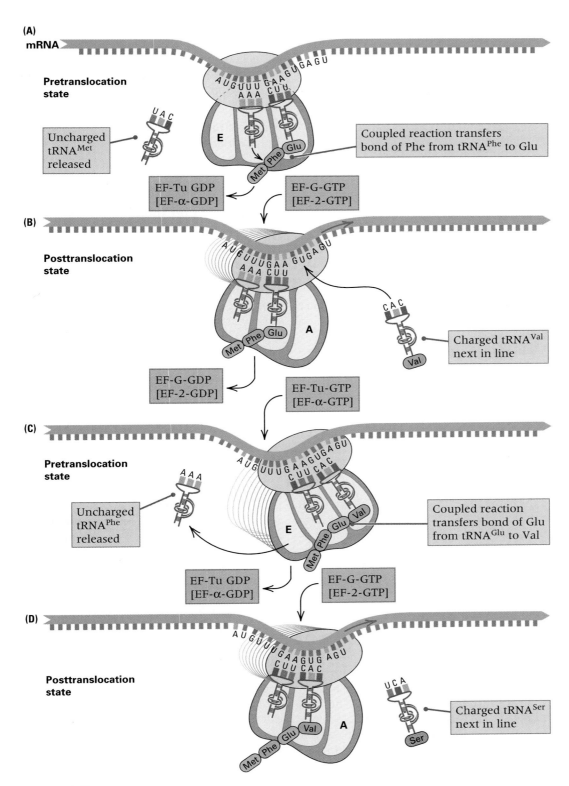

Figure 10.23
Elongation cycle in protein synthesis. (A) Pretranslocation state. (B) Posttranslocation state, in which an uncharged tRNA occupies the E site and the polypeptide is attached to the tRNA in the P site. (C) The function of EF-Tu is to release the uncharged tRNA and bring the next charged tRNA into the A site. A peptide bond is formed between the polypeptide and the amino acid held in the A site, in this case Val. Simultaneously, the 50S subunit is shifted relative to the 30S subunit, forming the pretranslocation state. (D) The function of EF-G is to translocate the 30S ribosome to the next codon, once again generating the posttranslocation state.

After these steps, the EF-G-GDP is released for regeneration into EF-G-GTP for use in another cycle. The translocation state of the ribosome is not depicted separately in Figure 10.23 because it happens so rapidly. In effect, the ribosome shuttles between the pretranslocation and posttranslocation states.

You will note that Figure 10.23D is essentially identical to Figure 10.23B except that the ribosome is one codon farther to the right and the polypeptide is one amino acid greater in length. Hence the ribosome is again available for EF-Tu-GTP to start the next round of elongation. Polypeptide elongation may therefore be considered as a cycle of events repeated again and again. The steps $B \rightarrow C \rightarrow B \rightarrow C$ (or, equivalently, $C \rightarrow D \rightarrow C \rightarrow D$) are carried out repeatedly until a termination codon is encountered.

In Figure 10.23D, for example, the configuration of the ribosome is such that the tRNA^{Gly} and tRNA^{Val} are occupying the E and P sites, respectively. The aminoacylated tRNA that corresponds to the codon AGU is brought into line, which is tRNA^{Ser}, and the bond connecting the polypeptide to tRNA^{Val} is transferred to the amino group of Ser, creating a new peptide bond and elongating the polypeptide by one amino acid. At the same time, the ribosome is converted to the pretranslocation state in preparation for translocation. The elongation cycle happens relatively fast. Under optimal conditions, *E. coli* synthesizes a polypeptide at the rate of about 20 amino acids per second; in eukaryotes, the rate of elongation is about 15 amino acids per second.

Termination

The elongation steps of protein synthesis are carried out repeatedly until a stop codon for termination is reached. The stop codons are UAA, UAG, and UGA. No tRNA exists that can bind to a stop codon, so the tRNA holding the polypeptide remains in the P site (Figure 10.24). Specific release factors act to cleave the polypeptide from the tRNA to which it is attached as well as to disassociate the 70S ribosome from the mRNA, after which the individual 30S and 50S subunits are recycled to initiate translation of another mRNA. The release factor RF-1 recognizes the stop codons UAA and UAG, whereas release factor RF-2 recognizes UAA and UGA. A third release factor, RF-3, is also required for translational termination.

Monocistronic and Polycistronic mRNA

The process of selecting the correct AUG initiation codon is of some importance in understanding many features of gene expression. In prokaryotes, mRNA molecules commonly contain information for the amino acid sequences of several different polypeptide chains; such a molecule is called a **polycistronic mRNA.** (*Cistron* is a term often used to mean a base sequence that encodes a single polypeptide chain.) In a polycistronic mRNA, each polypeptide coding region is preceded by its own ribosome-binding site and AUG initiation codon. After the synthesis of one polypeptide is finished, the next along the way is translated (Figure 10.25). The genes contained in a polycistronic mRNA molecule often encode the different proteins of a metabolic pathway. For example, in *E. coli*, the ten enzymes needed to synthesize histidine are encoded by one polycistronic mRNA molecule. The use of polycistronic mRNA is an economical way for a cell to regulate the synthesis of related proteins in a coordinated manner. For example, in prokaryotes, the usual way to regulate the synthesis of a particular protein is to control the synthesis of the mRNA molecule that codes for it (Chapter 11). With a polycistronic mRNA molecule, the synthesis of several related proteins can be regulated by a single signal, so that appropriate quantities of each protein are made at the same time; this is termed **coordinate regulation.**

In eukaryotes, the 5' terminus of an mRNA molecule binds to the ribosome, after which the mRNA molecule slides along the ribosome until the AUG codon nearest the 5' terminus is in contact with the ribosome. Then protein synthesis begins. There is no mechanism for initiating polypeptide synthesis at any AUG other than the first one encountered. Eukaryotic mRNA is always monocistronic (Figure 10.25).

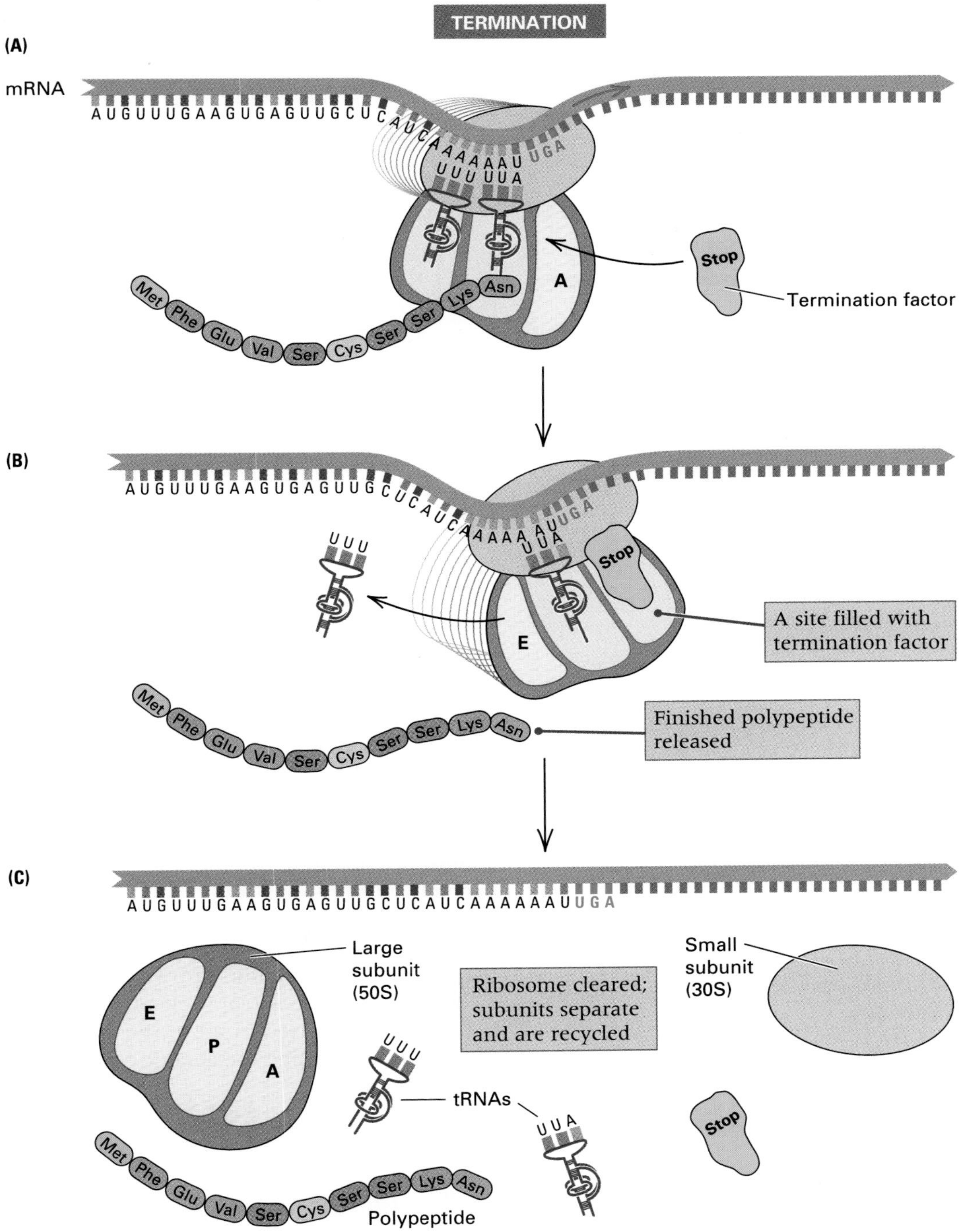

Figure 10.24
Termination of protein synthesis. When a stop codon is reached, no tRNA can bind to that site, which causes the release of the newly formed polypetide and the remaining bound tRNA.

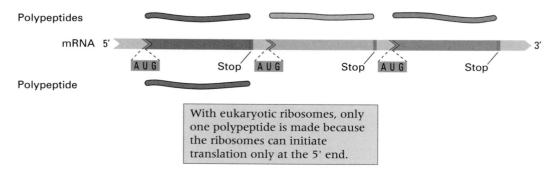

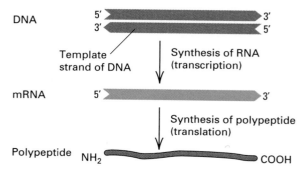

Figure 10.25
Different products are translated from a three-cistron mRNA molecule by the ribosomes of prokaryotes and eukaryotes. The prokaryotic ribosome translates all of the genes, but the eukaryotic ribosome translates only the gene nearest the 5' terminus of the mRNA. Translated sequences are shown in purple, yellow, and orange; stop codons in red; the ribosome binding sites in green; and the spacer sequences in light green.

The definitive feature of translation is that it proceeds in a particular direction along the mRNA and the polypeptide:

The mRNA is translated from an initiation codon to a stop codon in the 5'-to-3' direction. The polypeptide is synthesized from the amino end toward the carboxyl end by the addition of amino acids, one by one, to the carboxyl end.

For example, a polypeptide with the sequence

$$NH_2-Met-Pro- \cdots -Gly-Ser-COOH$$

Figure 10.26
Direction of synthesis of RNA with respect to the coding strand of DNA, and of synthesis of protein with respect to mRNA.

would start with methionine, and serine would be the last amino acid added to the chain. The directions of synthesis are illustrated schematically in Figure 10.26.

In writing nucleotide sequences, by convention, we place the 5' end at the left and, in writing amino acid sequences, we place the amino end at the left. Polynucleotides are generally written so that both synthesis and translation proceed from left to right, and polypeptides are written so that synthesis proceeds from left to right. This convention is used in all of the following sections concerning the genetic code.

10.6
The Genetic Code

The four bases in DNA—A, T, G, and C—are sufficient to specify the 20 amino acids in proteins because each codon is three bases in length. Each sequence of three adjacent bases in mRNA is a codon that specifies a particular amino acid (or chain termination). The **genetic code** is the list of all codons and the amino acid that each one encodes. Before the genetic code was determined experimentally, it was reasoned that if all codons were assumed to

Messenger *Light*

Sydney Brenner,[1] François Jacob,[2] and Matthew Meselson[3] 1961

[1]Cavendish Laboratory, Cambridge, England.
[2]Institute Pasteur, Paris, France.
[3]California Institute of Technology, Pasadena, California.

An Unstable Intermediate Carrying Information from Genes to Ribosomes for Protein Synthesis

Brenner and Jacob were guest investigators at the California Institute of Technology in 1961. At that time there was great interest in the mechanisms by which genes code for proteins. One possibility, which seemed reasonable at the time, was that each gene produced a different type of ribosome, differing in its RNA, which in turn produced a different type of protein. François Jacob and Jacques Monod had recently proposed an alternative, which was that the informational RNA ("messenger RNA") is actually an unstable molecule that breaks down rapidly. In this model, the ribosomes are nonspecific protein-synthesizing centers that synthesize different proteins according to specific instructions they receive from the genes through the messenger RNA. The key to the experiment is density-gradient centrifugation, which can separate macromolecules made "heavy" or "light" according to their content of ^{15}N or ^{14}N, respectively. (This technique is described in Chapter 5.) The experiment is a purely biochemical proof of an issue absolutely critical for genetics—that genes code for proteins through the

intermediary of a relatively short-lived messenger RNA.

A large amount of evidence suggests that genetic information for protein structure is encoded in deoxyribonucleic acid (DNA) while the actual assembling of amino acids into proteins occurs in cytoplasmic ribonucleoprotein particles called ribosomes. The

---

> **The results also suggest that the messenger RNA may be large enough to code for long polypeptide chains**

---

fact that proteins are not synthesized directly on genes demands the existence of an intermediate information carrier. . . . Jacob and Monod have put forward the hypothesis that ribosomes are non-specialized structures which receive genetic information from the gene in the form of an unstable intermediate or "messenger." We present here the results of experiments on phage-infected bacteria which give direct support to this hypothesis. . . . When growing bacteria are infected with T2 bacteriophage, synthesis of DNA stops immediately, to resume 7 minutes later, while protein synthesis continues at a constant rate; in all likelihood, the protein is genetically determined by the phage. . . . Phage-infected bacteria therefore provide a situation in which the synthesis of a protein is suddenly switched from bacterial to phage

control. . . . It is possible to determine experimentally [whether an unstable messenger RNA is produced] in the following way: Bacteria are grown in heavy isotopes so that all cell constituents are uniformly labelled "heavy." They are infected with phage and transferred immediately to a medium containing light isotopes so that all constituents synthesized after infection are "light." The distribution of new RNA and new protein, labelled with radioactive isotopes, is then followed by density gradient centrifugation of purified ribosomes. . . . We may summarize our findings as follows: (1) After phage infection no new ribosomes can be detected. (2) A new RNA with a relatively rapid turnover is synthesized after phage infection. This RNA, which has a base composition corresponding to that of the phage DNA, is added to pre-existing ribosomes, from which it can be detached in a cesium chloride gradient by lowering the magnesium concentration. (3) Most, if not all, protein synthesis in the infected cell occurs in pre-existing ribosomes. . . . The results also suggest that the messenger RNA may be large enough to code for long polypeptide chains. . . . It is a prediction of the messenger RNA hypothesis that the messenger RNA should be a simple copy of the gene, and its nucleotide composition should therefore correspond to that of the DNA. This appears to be the case in phage-infected cells. . . . If this turns out to be universally true, interesting implications for the coding mechanisms will be raised.

Source: Nature 190: 576–581

---

have the same number of bases, then each codon would have to contain at least three bases. Codons consisting of pairs of bases would be insufficient, because four bases can form only $4^2 = 16$ pairs; triplets of bases would suffice, because four bases can form $4^3 = 64$ triplets. In fact, the genetic code is a triplet code, and all 64 possible codons carry information of some sort. Most amino acids are encoded by more than one codon. Furthermore, in the trans-

lation of mRNA molecules, the codons do not overlap but are used sequentially (Figure 10.27).

Genetic Evidence for a Triplet Code

Although theoretical considerations suggested that each codon must contain at least three letters, codons having more than three letters could not be ruled out.

Start ————————————————————————— Stop

5' NNNAUGAGUCAGUGGGUCAGUCAGUCAGUCUAANNNN 3'

Direction of reading of codons in translation

Figure 10.27
Bases in an RNA molecule are read sequentially in the 5'-to-3' direction, in groups of three.

The first widely accepted proof for a triplet code came from genetic experiments using *rII* mutants of bacteriophage T4 that had been induced by replication in the presence of the chemical *proflavin*. These experiments were carried out in 1961 by Francis Crick and collaborators. Proflavin-induced mutations typically resulted in total loss of function. Because proflavin is a large planar molecule, it was suspected that it caused mutations that inserted or deleted a base pair by interleaving between base pairs in the double helix. Analysis of the properties of these mutations led directly to the deduction that the code is read three nucleotides at a time from a fixed point; in other words, there is a **reading frame** to each mRNA. Mutations that delete or add a base pair shift the reading frame and are called **frameshift mutations.** Figure 10.28 illustrates the profound effect of a frameshift mutation on the amino acid sequence of the polypeptide produced from the mRNA of the mutant gene.

The genetic analysis of the structure of the code began with an *rII* mutation called FC0, which was arbitrarily designated (+), as if it had an inserted base pair. (It could also arbitrarily have been designated (−),

as if it had a deleted base pair. Calling it (+) was a lucky guess, however, because when FC0 was sequenced, it did turn out to have a single-base insertion.) If FC0 has a (+) insertion, then it should be possible to revert the FC0 allele to "wildtype" by deletion of a nearby base. Selection for r^+ revertants was carried out by isolating plaques formed on a lawn of an *E. coli* strain K12 that was lysogenic for phage λ. The basis of the selection is that *rII* mutants are unable to propagate in K12(λ). Analysis of the revertants revealed that each still carried the original FC0 mutation along with a second (suppressor) mutation that reversed the effects of the FC0 mutation. The suppressor mutations could be separated by recombination from the original mutation by crossing each revertant to wildtype; each suppressor mutation proved to be an *rII* mutation that, by itself, would cause the *r* (rapid lysis) phenotype. If FC0 had an inserted base, then the suppressors should all result in deletion of a base pair; hence each suppressor of FC0 was designated (−). Three such revertants and their consequences for the translational reading frame, illustrated using ordinary three-letter words, are illustrated in Figure 10.29. The (−) mutations are designated $(-)_1$, $(-)_2$, and $(-)_3$, and those parts of the mRNA translated in the correct reading frame are indicated in green.

Each of the individual (−) suppressor mutations could, in turn, be used to select other "wildtype" revertants, with the expectation that these revertants would carry new suppressor mutations of the (+) variety, because the (−)(+) combination should yield a phage able to form plaques on K12(λ).

Various double mutant combinations were made. Usually any (+) (−) combination, or any (−)(+) combination, resulted in a wildtype phenotype, whereas (+)(+) and (−)(−) double mutant combinations always resulted in the mutant phenotype. The truly telling result came when triple mutants were made. Usually, the (+)(+)(+) and (−)(−)(−) triple mutants yielded the wildtype phenotype!

The phenotypes of the various (+) and (−) combinations were interpreted in terms of a reading frame. The initial FC0 mutation, a +1 insertion, shifts the reading

Figure 10.28
The change in the amino acid sequence of a protein caused by the addition of an extra base, which shifts the reading frame. A deleted base also shifts the reading frame.

mRNA from original DNA

A G C C A C U U A G A C A A A C U A

Ser His Leu Asp Lys Leu

mRNA from DNA in which a base has been added

A G C A C A C U U A G A C A A A C U A

Ser Thr Leu Arg Gln Thr

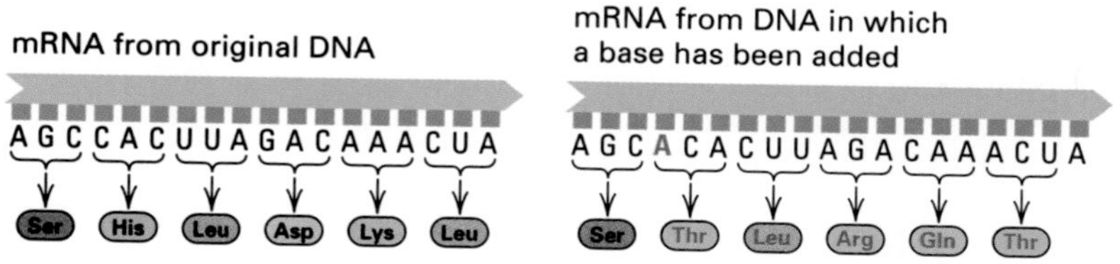

frame, resulting in incorrect amino acid sequence from that point on and thus a nonfunctional protein (Figure 10.29). Deletion of a base pair nearby will restore the reading frame, although the amino acid sequence encoded between the two mutations will be different and incorrect. In (+)(+) or (−)(−) double mutants, the reading frame is shifted by two bases; the protein made is still nonfunctional. However, in the (+)(+)(+) and (−)(−)(−) triple mutants, the reading frame is restored, though all amino acids encoded within the region bracketed by the outside mutations are incorrect; the protein made is one amino acid longer for (+)(+)(+) and one amino acid shorter for (−)(−)(−) (Figure 10.29).

The genetic analysis of the (+) and (−) mutations strongly supported the following conclusions:

- Translation of an mRNA starts from a fixed point.

- There is a single reading frame maintained throughout the process of translation.

- Each codon consists of three nucleotides.

Crick and his colleagues also drew other inferences from these experiments. First, in the genetic code, most codons must function in the specification of an amino acid. Second, each amino acid must be specified by more than one codon. They reasoned that if each amino acid had only one codon, then only 20 of the 64 possible codons could be used for coding amino acids. In this case, most frameshift mutations should have affected one of the remaining 44 "noncoding" codons in the

Figure 10.29

Interpretation of the *rII* frameshift mutations showing that combinations of appropriately positioned single-base insertions (+) and single-base deletions (−) can restore the correct reading frame (green). The key finding was that a combination of three single-base deletions, as shown in the bottom line, also restores the correct reading frame (green). Two single-base deletions do not restore the reading frame. These classic experiments gave strong genetic evidence that the genetic code is a triplet code.

Phage type	Insertion/deletion	Translational reading frame of mRNA
Wildtype sequence		THE BIG BOY SAW THE NEW CAT EAT THE HOT DOO ⋯
+1 insertion	(+)	THE BIG BOY SAW TTH ENE WCA TEA TTH EHO TDO G
Revertant 1	(−)₁ (+)	THE BIG OYS AWT THE NEW CAT EAT THE HOT DOG ⋯
Revertant 2	(+) (−)₂	THE BIG BOY SAW TTH ENE WCA TEA THE HOT DOG ⋯
Revertant 3	(+) (−)₃	THE BIG BOY SAW TTH ENE WAT EAT THE HOT DOG ⋯
(−) deletion number 1	(−)₁	THE BIG OYS AWT HEN EWC ATE ATT HEH OTD OG ⋯
(−) deletion number 2	(−)₂	THE BIG BOY SAW THE NEW CAT EAT HEH OTD OG ⋯
(−) deletion number 3	(−)₃	THE BIG BOY SAW THE NEW ATE ATT HEH OTD OG ⋯
Double (−) mutant	(−)₁ (−)₂	THE BIG OYS AWT HEN EWC ATE ATH EHO TDO G ⋯
Triple (−) mutant	(−)₁ (−)₂ (−)₃	THE BIG OYS AWT HEN EWA TEA THE HOT DOG ⋯

Uncles and Aunts

Francis H. C. Crick, Leslie Barnett, Sydney Brenner, and R. J. Watts-Tobin, 1961
Cavendish Laboratory,
Cambridge, England
General Nature of the Genetic Code for Proteins

No other paper affords a better demonstration of the power of mutational analysis in the hands of clever researchers. The issue was this: How many bases are needed to code for one amino acid in a protein? Crick et al. answered the question by using single-base insertions and deletions in the rIIB cistron (Seymour Benzer's name for a region of DNA that codes for a single polypeptide chain). In the laboratory they referred to the (+) and (−) mutations as "uncles" and "aunts" because it was not known, in any particular case, whether a mutation was truly an insertion or truly a deletion. The paper is written as if FCO were an insertion mutation, solely for the sake of simplicity. (This guess later proved to be right.) The presentation is unusual in another respect also. The paper starts by assuming its conclusion in order to describe the results, and then demonstrates how the conclusion was arrived at. Try to imagine writing it in the usual way. Any such effort would be bound to be less clear. (Incidentally, the suggestion that the use of synthetic polynucleotides would solve the coding problem within a year was not far wrong. By 1968, Marshall W. Nirenberg, Robert W. Holley, and Har Gobind Khorana would be awarded a Nobel Prize for their deciphering of the genetic code table.)

In this article we report genetic experiments which suggest that the genetic code [is one in which] a group of three bases (or, less likely, a multiple of three bases) codes for one amino acid. . . . Our genetic experiments have been carried out on the *B* cistron of the *rII* region of the bacteriophage T4. . . . We report here our work on the mutant *FCO*. This mutant was originally produced by the action of proflavin, . . . which we have previously argued acts as a mutagen because it adds or deletes a base

---

Fortunately, we have convincing evidence that the coding ratio is in fact 3 or a multiple of 3.

---

or bases. . . . If an acridine mutant is produced by, say, adding a base, it should revert to "wildtype" by deleting a base. Our work on *FCO* shows that it usually reverts not by reversing the original mutation but by producing a second mutation at a nearby point on the genetic map. . . . A genetic map of 18 suppressors of *FCO* shows that they scatter over a region about, say, one-tenth of the size of the *B* cistron. . . . In all we have isolated about eighty independent *rIIB* mutants, all suppressors of *FCO*, or suppressors of suppressors, or suppressors of suppressors of suppressors. . . . Although we have no direct evidence that the *B* cistron produces a polypeptide chain (probably through an RNA intermediate), in what follows we shall assume this to be so. To fix ideas, we imagine that the string of nucleotides is read, triplet by triplet, from a starting point on the left of the *B* cistron. We now suppose that, for example, the mutant *FCO* was produced by the insertion of an additional base in the wildtype sequence. This addition of a base at the *FCO* site will mean that the reading of all the triplets to the right of *FCO* will be shifted along one base, and will therefore be incorrect. Thus the amino-acid sequence of the protein which the *B* cistron is presumed to produce will be completely altered from that point onwards. This explains why the function of the gene is lacking. . . . We now postulate that a suppressor of *FCO* (for example, *FC1*) is formed by deleting a base. Thus when the *FC1* mutation is present by itself, all triplets to the right of *FC1* will be read incorrectly and thus the function will be absent. However, when both mutations are present in the same piece of DNA, then although the reading of triplets between *FCO* and *FC1* will be altered, the original reading will be restored to the rest of the gene. . . . So far we have spoken as if the evidence supported a triplet code, but this was simply for illustration. . . . Fortunately, we have convincing evidence that the coding ratio is in fact 3 or a multiple of 3. This we have obtained by constructing triple mutants of the form (+ with + with +) or (− with − with −). One must be careful not to make shifts across the "unacceptable" region of *rIIB*, but this we can avoid by a proper choice of mutants. . . . It is possible by various devices, either chemical or enzymatic, to synthesize polyribonucleotides with defined or partially defined sequences. If these will produce specific polypeptides, the coding problem is wide open for experimental attack . . . and the genetic code may well be solved within a year.

Source: Nature 192: 1227–1232

reading frame, and hence a nearby frameshift of the opposite polarity mutation should not have suppressed the original mutation. Consequently, the code was deduced to be **degenerate,** which means that more than one codon can specify a particular amino acid.

Elucidation of the Base Sequences of the Codons

Polypeptide synthesis can be carried out in *E. coli* cell extracts obtained by breaking cells open. Various components can be isolated, and a functioning protein-

synthesizing system can be reconstituted by mixing ribosomes, tRNA molecules, mRNA molecules, and various protein factors. If radioactive amino acids are added to the extract, then radioactive polypeptides are made. Synthesis continues for only a few minutes because mRNA is rapidly degraded by nucleases in the mixture. The elucidation of the genetic code began with the observation that when the degradation of mRNA was allowed to go to completion and the synthetic polynucleotide polyuridylic acid (poly-U) was added to the mixture as an mRNA molecule, a polypeptide consisting only of phenylalanine (Phe−Phe−Phe−···) was synthesized. From this simple result, and knowledge that the code is a triplet code, it was concluded that UUU must be a codon for the amino acid phenylalanine. Variations on this basic experiment identified other codons. For example, when a long sequence of guanines was added at the terminus of the poly-U, the polyphenylalanine was terminated by a sequence of glycines, indicating that GGG is a glycine codon (Figure 10.30). A trace of leucine or

tryptophan was also present in the glycine-terminated polyphenylalanine. Incorporation of these amino acids was directed by the codons UUG and UGG at the transition point between U and G. When a single guanine was added to the terminus of a poly-U chain, the polyphenylalanine was terminated by leucine. Thus UUG is a leucine codon, and UGG must be a codon for tryptophan. Similar experiments were carried out with poly-A, which yielded polylysine, and with poly-C, which produced polyproline.

Other experiments led to a complete elucidation of the code. Three codons

$$\text{UAA} \qquad \text{UAG} \qquad \text{UGA}$$

were found to be stop signals for translation, and one codon, **AUG,** which encodes methionine, was shown to be the initiation codon. AUG also codes for internal methionines, but it uses a different tRNA to do so.

A Summary of the Code

The *in vitro* translation experiments, which used components isolated from the bacterium *E. coli,* have been repeated with components obtained from many species of bacteria, yeast, plants, and animals. The standard genetic code deduced from these experiments is considered to be nearly universal because the same codon assignments can be made for nuclear genes in almost all organisms that have been examined. However, some minor differences in codon assignments are found in certain protozoa and in the genetic codes of organelles.

The standard code is shown in Table 10.3. Note that four codons—the three stop codons and the start codon—are signals. Altogether, 61 codons specify amino acids, and in many cases several codons direct the insertion of the same amino acid into a polypeptide chain. This feature confirms the inference from the *rII* frameshift mutations that the genetic code is redundant (degenerate). In a redundant genetic code, some amino acids are encoded by two or more different codons. In the actual genetic code, all amino acids except tryptophan and methionine are specified by more than one codon. The redundancy is not random. For example, with the exception of serine, leucine, and arginine, all

Figure 10.30
Polypeptide synthesis using

$$\text{UUUU} \ldots \text{UUGGGGGGG}$$

as an mRNA in three different reading frames, showing the reasons for the incorporation of glycine, leucine, and tryptophan.

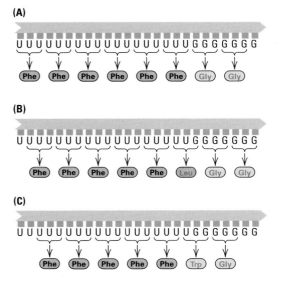

(A)

(B)

(C)

Table 10.3 The standard genetic code

First position (5' end)	Second position				Third position (3' end)
	U	C	A	G	
U	UUU Phe ⎫F UUC Phe ⎬ UUA Leu ⎫L UUG Leu ⎭	UCU Ser ⎫ UCC Ser ⎬S UCA Ser ⎪ UCG Ser ⎭	UAU Tyr ⎫Y UAC Tyr ⎬ UAA Stop UAG Stop	UGU Cys ⎫C UGC Cys ⎬ UGA Stop UGG Trp W	U C A G
C	CUU Leu ⎫ CUC Leu ⎬L CUA Leu ⎪ CUG Leu ⎭	CCU Pro ⎫ CCC Pro ⎬P CCA Pro ⎪ CCG Pro ⎭	CAU His ⎫H CAC His ⎬ CAA Gln ⎫Q CAG Gln ⎭	CGU Arg ⎫ CGC Arg ⎬R CGA Arg ⎪ CGG Arg ⎭	U C A G
A	AUU Ile ⎫ AUC Ile ⎬I AUA Ile ⎭ AUG Met M	ACU Thr ⎫ ACC Thr ⎬T ACA Thr ⎪ ACG Thr ⎭	AAU Asn ⎫N AAC Asn ⎬ AAA Lys ⎫K AAG Lys ⎭	AGU Ser ⎫S AGC Ser ⎬ AGA Arg ⎫R AGG Arg ⎭	U C A G
G	GUU Val ⎫ GUC Val ⎬V GUA Val ⎪ GUG Val ⎭	GCU Ala ⎫ GCC Ala ⎬A GCA Ala ⎪ GCG Ala ⎭	GAU Asp ⎫D GAC Asp ⎬ GAA Glu ⎫E GAG Glu ⎭	GGU Gly ⎫ GGC Gly ⎬G GGA Gly ⎪ GGG Gly ⎭	U C A G

Note: Each amino acid is given its conventional abbreviation in both the single-letter and three-letter format. The codon AUG, which codes for methionine (boxed) is usually used for initiation. The codons are conventinally written with the 5' base on the left and the 3' base on the right.

codons that correspond to the same amino acid are in the same box of Table 10.3; that is, *synonymous codons usually differ only in the third base*. For example, GGU, GGC, GGA, and GGG all code for glycine. Moreover, in all cases in which two codons code for the same amino acid, the third base is either A or G (both purines) or T or C (both pyrimidines).

The codon assignments shown in Table 10.3 are completely consistent with all chemical observations and with the amino acid sequences of wildtype and mutant proteins. In virtually every case in which a mutant protein differs by a single amino acid from the wildtype form, the amino acid substitution can be accounted for by a single base change between the codons corresponding to the two different amino acids. For example, substitution of glutamic acid by valine, which occurs in sickle-cell hemoglobin, results from a change from GAG to GUG in codon six of the β-globin mRNA.

Mutations that change the nucleotide sequence of a gene may differ in their con-

sequences, and a special terminology is used to describe them. A **missense** mutation results in the replacement of one amino acid by another. For example, the change from GAG to GTG in the DNA of the gene for β-globin results in the replacement of glutamic acid by valine in the sickle-cell hemoglobin molecule; the mutation is therefore a missense mutation. In contrast, a **silent mutation** is one that does not change the amino acid sequence. Silent mutations often result from changes in the third codon position; for example, a mutation that changes an AAA codon into an AAG codon is silent because both codons specify lysine. A most interesting class of mutations consists of changes that convert a codon that specifies an amino acid into a chain-terminating codon. A mutation of this type is called a **nonsense mutation,** and it results in premature termination of the polypeptide chain. An example of a nonsense mutation is found in the β-globin gene, in which a mutation from AAG to TAG in the seventeenth codon results in a truncated polypeptide only 16 amino acids in length. This mutation is one of several types associated with the disease β-thalassemia.

Transfer RNA and Aminoacyl-tRNA Synthetase Enzymes

The decoding operation by which the base sequence within an mRNA molecule becomes translated into the amino acid sequence of a protein is accomplished by aminoacylated, or charged, tRNA molecules, each of which is linked to the correct amino acid by an aminoacyl-tRNA synthetase.

The tRNA molecules are small, single-stranded nucleic acids ranging in size from about 70 to 90 nucleotides. Like all RNA molecules, they have a 3'-OH terminus, but the opposite end terminates with a 5'-monophosphate rather than a 5'-triphosphate, because tRNA molecules are cut from a larger primary transcript. Internal complementary base sequences form short double-stranded regions, causing the molecule to fold into a structure in which open loops are connected to one another by double-stranded stems (Figure 10.31). In two dimensions, a tRNA molecule is drawn as a

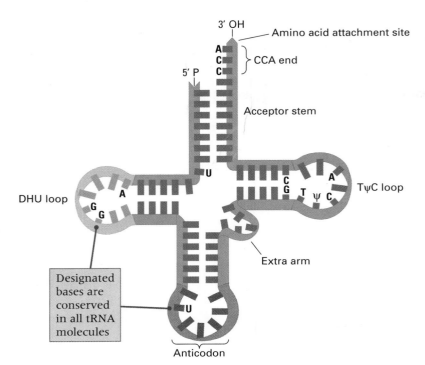

Figure 10.31

A tRNA cloverleaf configuration. The heavy black letters indicate a few bases that are conserved in the sequence of all tRNA molecules. The labeled loop regions are those found in all tRNA molecules. DHU refers to a base, dihydrouracil, found in one loop; the Greek letter Ψ is a symbol for the unusual base pseudouridine.

planar cloverleaf. Its three-dimensional structure is more complex, as is shown in Figure 10.32, in which part A shows a skeletal model of a yeast tRNA molecule that carries phenylalanine and part B is an interpretive drawing. Note how the TψC loop and the DHU loop are in close proximity. When viewed from either side, the folded structure roughly resembles the diagrammatic representations of the tRNAs used in Figures 10.21, 10.23, and 10.24.

Particular regions of each tRNA molecule are used in the decoding operation. One region is the anticodon sequence, which consists of three bases that can form base pairs with a codon sequence in the mRNA. No normal tRNA molecule has an anticodon complementary to any of the stop codons UAG, UAA, and UGA, which is why these codons are stop signals. A second critical site is at the 3' terminus of the tRNA molecule, where the amino acid attaches. A specific aminoacyl-tRNA synthetase matches the amino acid with the anticodon. At least one, and usually only one, aminoacyl-synthetase exists for each amino acid. To make the correct attachment, the synthetase must be able to distinguish one tRNA molecule from another. The necessary distinction is provided by recognition regions that encompass many parts of the tRNA molecule.

Figure 10.33 shows the three-dimensional structure of the seryl-tRNA synthetase complexed with its tRNA. On binding with the tRNA, a part of the protein makes contact with the variable part of the TψC loop of the tRNA and guides the acceptor stem into the active pocket of the enzyme. These interactions depend primarily on recognition of the shape of the tRNA^{Ser} through contacts with the backbone and only secondarily on interactions that are specific to the anticodon.

Redundancy and Wobble

Several features of the genetic code and of the decoding system suggest that something is missing in the explanation of codon-anticodon binding. First, the code is highly redundant. Second, the identity of the third base of a codon is often unimportant. In some cases, any nucleotide will do; examples include proline (Pro), threonine (Thr), and glycine (Gly). In other cases, either purine (A or G) or either pyrimidine (U or C) in the third position codes for the same amino acid; examples include histidine (His), glutamine (Gln), and tyrosine (Tyr).

(A)

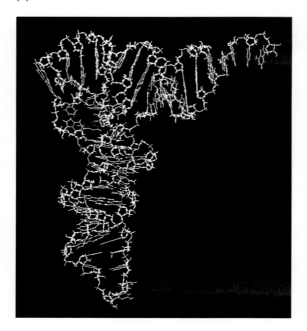

(B)

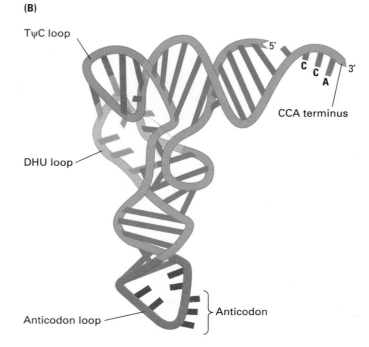

TψC loop

DHU loop

5'

C
C
A

3'

CCA terminus

Anticodon loop

Anticodon

Figure 10.32

Yeast phenylalanine tRNA (called tRNA^Phe). (A) A skeletal model. (B) A schematic diagram of the three-dimensional structure of yeast tRNA^Phe. [Courtesy of Sung-Hou Kim.]

Third, the number of distinct tRNA molecules that have been isolated from a single organism is less than the number of codons; because all codons are used, the anticodons of some tRNA molecules must be able to pair with more than one codon. Experi-

ments with several purified tRNA molecules showed this to be the case.

To account for these observations, the **wobble** concept was advanced in 1966 by Francis Crick. He proposed that the first two bases in a codon form base pairs with

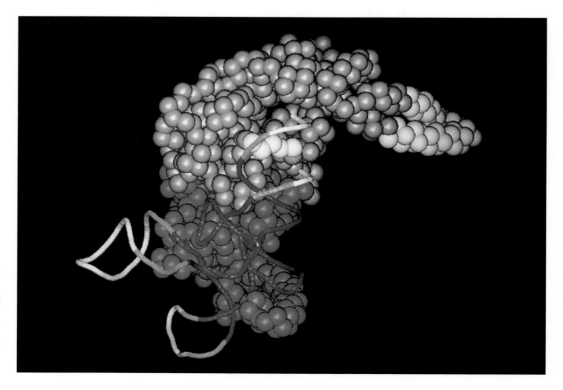

Figure 10.33

Three-dimensional structure of seryl-tRNA synthetase (solid spheres) complexed with its tRNA. Note that there are many points of contact between the enzyme and the tRNA. The molecules are from *Thermus thermophilus*. [Courtesy of Stephen Cusack. From V. Biou, A. Yaremchuk, M. Tukalo, and S. Cusack, 1996. *Science* 263:1404.]

the tRNA anticodon according to the usual rules (A−U and G−C) but that the base at the 5' end of the anticodon is less spatially constrained than the first two and can form hydrogen bonds with more than one base at the 3' end of the codon. He suggested the pairing rules given in Table 10.4. Evidence has confirmed the wobble concept and indicates that the pairings given in Table 10.4 are largely true for *E. coli*.

On the other hand, analysis of tRNAs in the yeast *Saccharomyces cerevisiae* has indicated that wobble is more restricted in yeast than in *E. coli*. Table 10.5 summarizes the wobble rules for *E. coli* and yeast. The yeast rules may hold for other eukaryotes as well. In yeast, single tRNAs can recognize the pairs of related codons ending in U or C. However, separate tRNAs are needed for codons that end in A or G. Thus at least three tRNAs are required for amino acids such as proline and glycine, which are specified by a set of four codons. A total of 46 tRNAs are needed to decode mRNA molecules in yeast. As noted, there are no tRNAs corresponding to the stop codons.

Nonsense Suppression

As described in Section 10.6, mutations can occur that result in premature chain termination during translation. For whimsical historical reasons, the stop codons are referred to as the amber (UAG), ochre (UAA), and UGA codons. A remarkable observation was that some phage mutants bearing nonsense mutations in the gene encoding the major head protein were able to propagate in some bacterial strains but not in others. How could this happen? On further analysis, it turned out that the strains able to support growth of the phage nonsense mutants carry suppressor mutations that act by changing the way the mRNA is read, not by changing the nucleotide sequence of the phage gene. The suppressor mutations proved to be mutant tRNA genes. These suppressor tRNA genes act by reading a stop codon as though it were a signal for a specific amino acid. The amino acid is inserted at that position and translation continues. So long as the inserted amino acid is compatible with the function of the protein, the effects of the original mutations are suppressed and plaques can be produced. In *E. coli*,

Table 10.4 Allowed pairings due to wobble

First base in anticodon (5' position)	Allowed bases in third codon position (3' position)
A	U
C	G
U	A or G
G	C or U
I	A or C or U

"amber suppressors" recognize and suppress only amber mutations because the anticodon of the altered tRNA pairs only with the UAG codon; on the other hand, "ochre suppressors" suppress both ochre and amber mutations because the mutant tRNA can recognize and suppress both UAG and UAA. In yeast there is more specificity because of the more stringent wobble rules: Ochre suppressors suppress only ochre mutations, and amber suppressors suppress only amber mutations.

We can illustrate nonsense suppression by examining a chain-termination codon formed by mutation of the tyrosine codon UAC to the stop codon UAG (Figure 10.34A and B). Such a mutation can be suppressed by a mutant leucine tRNA molecule. In *E. coli*, tRNA^{Leu} has the anticodon 3'-AAC-5', which pairs with the codon 5'-UUG-3'. A suppressor mutation in the tRNA^{Leu} gene produces an altered tRNA

Table 10.5 Wobble rules for tRNAs of *E. coli* and *Saccharomyces cerevisiae*

Third position of codon	First position of anticodon	
	E. coli	Yeast (*S. cerevisiae*)
U	A, G, or I	G or I
C	G or I	G or I
A	U or I	U*
G	C or U	C

Note: I indicates inosine, which is structurally similar to adenosine except that the −NH$_2$ is replaced with −OH. U* indicates a modified uridine.

Source: Data from C. Guthrie and J. Abelson. 1982. *The Molecular Biology of the Yeast Saccharomyces: Metabolism and Gene Expression*, edited by J. N. Strathern, E. W. Jones, and J. R. Broach, Cold Spring Harbor Laboratory Press, Cold Spring Harbor, NY, p. 487.

(A) Wildtype gene

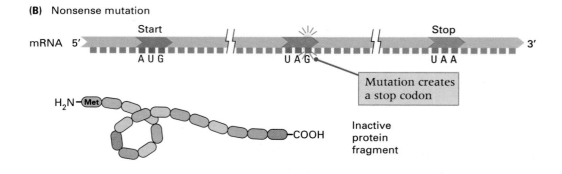

mRNA 5' ... Start AUG ... UAC ... Stop UAA ... 3'

H₂N—Met ... Tyr ... —COOH

Active
protein

(B) Nonsense mutation

mRNA 5' ... Start AUG ... UAG ... Stop UAA ... 3'

Mutation creates
a stop codon

H₂N—Met ... —COOH

Inactive
protein
fragment

(C) Nonsense suppression (tRNA suppression)

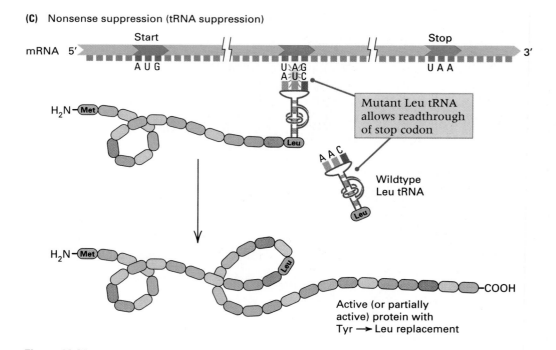

mRNA 5' ... Start AUG ... UAG / AUC ... Stop UAA ... 3'

Mutant Leu tRNA
allows readthrough
of stop codon

H₂N—Met ... Leu

AAC
Wildtype
Leu tRNA
Leu

H₂N—Met ... Leu ... —COOH

Active (or partially
active) protein with
Tyr → Leu replacement

Figure 10.34

The mechanism of suppression by a suppressor tRNA molecule. (A) The wildtype gene. (B) A UAC → UAG chain-termination mutation leads to an inactive, prematurely terminated protein. (C) A mutation in the tRNA^{Leu} gene produces an altered tRNA molecule, which has a codon complementary to a UAG stop codon but can still be charged with leucine. This tRNA molecule allows the protein to be completed but with a leucine at the site of the original tyrosine. Suppression will be achieved if the substitution restores activity to the protein.

with the anticodon AUC; this tRNA molecule is still charged with leucine but responds to the stop codon UAG rather than to the normal leucine codon UUG. Thus, in a cell that contains this suppressor tRNA, the mutant protein is completed and suppression occurs as long as the mutant protein can tolerate a replacement of leucine for tyrosine (Figure 10.34C). Many suppressor tRNA molecules of this type have been observed. Each suppressor is effective against only some nonsense mutations, because the resulting amino acid replacement may not yield a functional protein.

In *E. coli*, there are three classes of tRNA suppressors: those that suppress only UAG (amber suppressors), those that suppress both UAA and UAG (ochre suppressors), and those that suppress only UGA. They share the following properties:

1. The original mutant gene still contains the mutant base sequence (UAG in Figure 10.34).

2. The suppressor tRNA suppresses all chain-termination mutations with the same stop codon, provided that the amino acid inserted is an acceptable amino acid at the site.

3. A cell can survive the presence of a tRNA suppressor *only* if the cell contains two or more copies of the same tRNA gene. Taking the example in Figure 10.34, if only one tRNA^{Leu} gene were present in the genome and if it were mutated, then the normal leucine codon UUG would no longer be read as a sense codon, and all polypeptide chains would terminate wherever a UUG codon occurred. However, *multiple copies of most tRNA genes exist,* so if one copy is mutated to yield a suppressor tRNA, a normal copy nearly always remains.

4. Any chain-termination codon can be translated by a suppressor tRNA mutation that recognizes that codon. For example, translation of UAG by insertion of an amino acid would prevent termination of all wildtype mRNA reading frames terminating in UAG.

However, the anticodon of the suppressor tRNA usually binds rather weakly to the stop codon, so the stop codon often results in termination anyway.

Suppressor tRNA mutations are very useful in genetic analysis because they allow nonsense mutations to be identified through their ability to be suppressed. This is important because nonsense alleles usually result in a truncated and completely inactive protein and so are considered true loss-of-function alleles. Suppressor tRNAs have been widely used for genetic analysis in prokaryotes, yeast, and even nematodes, because their other harmful effects are tolerated by the organism. In higher organisms, suppressor tRNAs have such severe harmful effects that they are of limited usefulness.

The Sequence Organization of a Typical Prokaryotic mRNA Molecule

Most prokaryotic mRNA molecules are **polycistronic,** which means that they contain sequences specifying the synthesis of several proteins (Figure 10.25). A polycistronic mRNA molecule must therefore possess a series of start and stop codons for use in translation. If an mRNA molecule encodes three proteins, then the minimal coding requirement would be the sequence

AUG (start)/protein 1/stop—
   AUG/protein 2/stop—AUG/protein 3/stop

The stop codons might be UAA, UAG, or UGA. Actually, such an mRNA molecule is probably never so simple in that (1) the leader sequence preceding the first start signal may be several hundred bases long and (2) spacer sequences containing ribosome binding sites are usually present between one stop codon and the next start codon.

10.7
Overlapping Genes

The idea that two or more reading frames might exist within a single coding segment of DNA was not considered for many years. The reason is that a mutation in a gene that overlaps another gene often produces defects in both gene products, but double mutations are very rare. Furthermore, the existence of overlapping reading frames was thought to place severe constraints on the amino acid sequences of two proteins translated from the same part of an mRNA molecule. However, because the code is highly redundant, the constraints are not so rigid.

If genes contained multiple reading frames, then a single DNA segment would be utilized with maximal efficiency. However, a disadvantage is that evolution might be more difficult because random mutations would rarely improve the function of both proteins. Nevertheless, some cases of **overlapping genes** have been found. Most examples are in transposable elements or in small viruses in which there is a premium on packing the largest amount of genetic information into a small DNA molecule. Some of the best examples of overlapping genes occur in the *E. coli* phage ϕX174.

Phage ϕX174 contains a single strand of DNA consisting of 5386 nucleotides. If a single reading frame were used, at most 1795 amino acids could be encoded in the sequence, and with an average protein size of about 400 amino acids, only 4 or 5 proteins could be made. However, ϕX174 makes 11 proteins containing a total of more than 2300 amino acids. This paradox was resolved when it was shown that translation occurs in several reading frames from three mRNA molecules (Figure 10.35). For example, the sequence for protein B is contained totally in the sequence for protein A' but is translated in a different reading frame; similarly, the sequence for protein E is included within the sequence for protein D. Protein K is initiated near the end of gene *A'*, includes the base sequence of gene *B*, and terminates in gene *C;* synthesis is not in phase with either gene *A'* or gene *C*. Of note is protein A', which is formed by initiating translation within the mRNA for protein A using the same reading frame, so that it terminates at the same stop codon as protein A. Thus, the amino acid sequence of A' is identical with a segment of protein A. In total, six different proteins obtain some or all of their primary structure from shared base sequences in ϕX174.

10.8
Complex Translation Units

In most prokaryotes and eukaryotes, the unit of translation is almost never simply one ribosome traversing an mRNA molecule. Rather, it is a more complex structure, of which there are several forms. Two examples are given in this section.

After about 25 amino acids have been joined together in a polypeptide chain, an AUG initiation codon is completely free of the ribosome, and a second initiation complex can form. The overall configuration is that of two ribosomes moving along the mRNA at the same speed. When the second ribosome has moved along a distance similar to that traversed by the first, a third ribosome can attach to the initiation site. The process of movement and reinitiation continues until the mRNA is covered with ribosomes at a density of about one ribosome per 80 nucleotides. This large translation unit is called a **polysome,** and this is the usual form of the translation unit. An electron micrograph of a polysome is shown in Figure 10.36.

Figure 10.35

Physical map of *E. coli* phage ϕX174. The red arrows show the start points for synthesis of the major mRNA transcripts, and the uppercase letters indicate the regions from which different protein products are translated. The solid black regions are untranslated spacers.

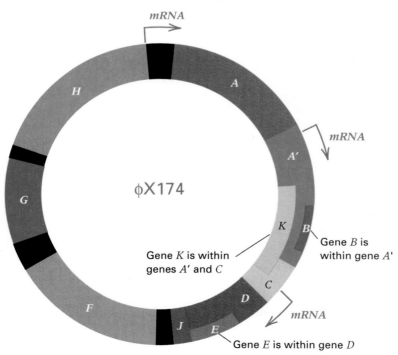

An mRNA molecule being synthesized has a free 5' terminus. Because translation takes place in the 5'-to-3' direction, the mRNA is synthesized in a direction appropriate for immediate translation. That is, the ribosome-binding site (in prokaryotes) and the 5' terminus (in eukaryotes) is transcribed first, followed in order by the initiating AUG codon, the region encoding the amino acid sequence, and finally the stop codon. Thus in prokaryotes, in which no nuclear envelope separates the DNA and the ribosome, the initiation complex can form before the mRNA is released from the DNA. This allows the simultaneous occurrence, or **coupling,** of transcription and translation. Figure 10.37 shows an electron micrograph of a DNA molecule with a number of attached mRNA molecules, each associated with ribosomes (Figure 10.37A), and an interpretation (Figure 10.37B). Transcription of DNA is beginning in the upper left part of the micrograph. The lengths of the polysomes increase with distance from the transcription initiation site, because the mRNA is farther from that site and hence of greater length because the process of transcription has been going on for a longer time. *Coupled transcription and translation does not take place in eukaryotes,* because the mRNA is synthesized and processed in the nucleus and later transported through the nuclear envelope to the cytoplasm where the ribosomes are located.

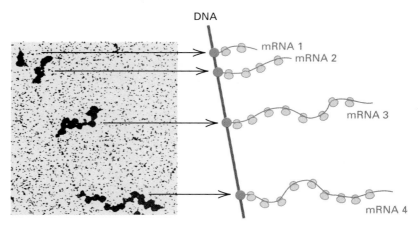

Figure 10.37
Visualization of transcription and translation. The photograph shows transcription of a section of the DNA of *E. coli* and translation of the nascent mRNA. The dark spots are ribosomes, which coat the mRNA. An interpretation of the electron micrograph is at the right. Each mRNA has ribosomes attached along its length. The large red dots are the RNA polymerase molecules; they are too small to be seen in the photo. The length of each mRNA is equal to the distance that each RNA polymerase has progressed from the transcription-initiation site. [Electron micrograph courtesy of O. L. Miller, B. A. Hamkalo, and C. A. Thomas. 1977. *Science* 169: 392.]

10.9
The Overall Process of Gene Expression

In this chapter, the main features of the process of gene expression have been described. The mechanisms of gene expressions are complex. Nonetheless, the basic process is a simple one:

> A base sequence in a DNA molecule is converted into a complementary base sequence in an intermediate molecule (mRNA), and then the base sequence in the mRNA is converted into an amino acid sequence of a polypeptide chain using tRNA molecules, each charged with the correct amino acid.

Both of these steps, which have a multitude of substeps, utilize the simplest of principles: (1) The rules of base pairing provide the base sequence of the mRNA, and (2) a two-ended molecule (tRNA), with an amino acid attached at one end and able to base-pair with RNA bases at the other, translates each set of three bases into one amino acid. Various recognition regions are needed to ensure that the correct base

Figure 10.36
Electron micrograph of *E. coli* polysomes. [Courtesy of Barbara Hamkalo.]

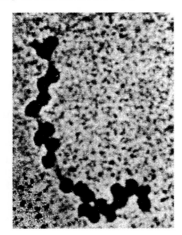

sequence is read and that the correct amino acid is put in the appropriate position in the protein. As is always the case when the information in nucleic acid molecules is used, base sequences provide the information for the first process. That is, specific sequences in the DNA are recognized as the beginning (promoter) and end (transcription-termination site) of a gene, and these sequences are recognized by an enzyme (RNA polymerase) that makes the copy of the gene that is used by the protein-synthesizing machinery. To ensure that the correct amino acid sequence is assembled, one codon (AUG) is used to tell the system where to start reading, and a stop codon defines the end of the polypeptide chain. Particular recognition sites in tRNA enable the aminoacyl tRNA synthetase enzymes to connect amino acids to the tRNA molecules with the correct anticodons.

An essential feature of the entire process of gene expression is that both DNA and RNA are scanned by molecules that move in a single direction. That is, RNA polymerase moves along the DNA as it polymerizes nucleotides, and the ribosome and the mRNA move with respect to one another as different amino acids are brought in for covalent linking.

Chapter Summary

The flow of information from a gene to its product is from DNA to RNA to protein. The properties of the different protein products of genes are determined by the sequence of amino acids of the polypeptide chain and by the way in which the chain is folded. Each gene is usually responsible for the synthesis of a single polypeptide.

Gene expression begins with the enzymatic synthesis of an RNA molecule that is a copy of one strand of the DNA segment corresponding to the gene. This process is called transcription and is carried out by the enzyme RNA polymerase. This enzyme joins ribonucleoside triphosphates by the same chemical reaction used in DNA synthesis. RNA polymerase differs from DNA polymerase in that a primer is not needed to initiate synthesis. Transcription is initiated when RNA polymerase binds to a promoter sequence. Each promoter consists of several subregions, of which two are the polymerase binding site and the polymerization start site. Polymerization continues until a termination site is reached. The product of transcription is an RNA molecule. In prokaryotes, this molecule is used directly as messenger RNA (mRNA) in polypeptide synthesis. In eukaryotes, the RNA is processed: Noncoding sequences called introns are removed, the exons are spliced together, and the termini are modified by formation of a 5' cap and usually by addition of a poly-A tail at the 3' end.

After mRNA is formed, polypeptide chains are synthesized by translation of the mRNA molecule. Translation is the successive reading of the base sequence of an mRNA molecule in groups of 3 bases called codons. There are 64 codons; 61 correspond to the 20 amino acids, of which 1 (AUG) is a start codon. The remaining 3 codons (UAA, UAG, and UGA) are stop codons. The code is highly redundant: Many amino acids have several codons. The codons in mRNA are recognized by tRNA molecules, which contain a 3-base sequence complementary to a codon and called an anticodon. When used in polypeptide synthesis, each tRNA molecule possesses a terminally bound amino acid (aminoacylated, or charged, tRNA). The correct amino acid is attached to each tRNA species by specific enzymes called aminoacyl-tRNA synthetases.

Polypeptides are synthesized on particles called ribosomes. Synthesis begins with the formation of a 30S ribosomal subunit + charged tRNA + mRNA complex, which recruits a 50S subunit to complete the mature 70S ribosome. (In eukaryotes, the 40S and 60S subunits come together to form the 80S ribosome.) Next, charged tRNA molecules are successively brought to the A site on the 50S ribosome by elongation factor EF-Tu (EF-1α in eukaryotes). These are hydrogen-bonded to the mRNA in the 30S subunit by a codon-anticodon interaction, and the 50S subunit is shifted to the pretranslocation state. As each charged tRNA is brought aboard, its amino acid is attached by a peptide bond to the growing polypeptide chain. Translocation of the 30S ribosome one codon farther along the tRNA is the function of elongation factor EF-G (EF-2 in eukaryotes), converting the ribosome to the posttranslation state and shifting the uncharged tRNA to the E site and the polypetidyl tRNA (the one carrying the incomplete polypeptide) to the P site.

The elongation process continues until a stop codon in the mRNA is reached. No tRNAs for the stop codons exist. Instead, specific release factors cleave the complete polypeptide from the last polypetidyl tRNA and free the ribosome components for reuse in translation.

Several ribosomes can translate an mRNA molecule simultaneously, forming a polysome. In prokaryotes, translation often begins before synthesis of mRNA is completed; in eukaryotes, this does not occur because mRNA is made in the nucleus, whereas the ribosomes are located in the cytoplasm. Prokaryotic mRNA molecules are often polycistronic, encoding several different polypeptides. Translation proceeds sequentially along the mRNA molecule from the start codon nearest the ribosome-binding site, terminating at stop codons and reinitiating at the next start codon. This is not possible in eukaryotes, because only the AUG site nearest the 5' terminus of the mRNA can be used to initiate polypeptide synthesis; thus eukaryotic mRNA is monocistronic.

Key Terms

amino terminus
aminoacyl-tRNA synthetase
aminoacylated tRNA
anticodon
AUG
cap
carboxyl terminus
chain elongation
chain initiation
chain termination
chaperone
charged tRNA
coding sequence
codon
colinearity
consensus sequence
coordinate regulation
coupled transcription-translation
cryptic splice site
degenerate code
exon
exon shuffle
folding domain
frameshift mutation
gene expression

gene product
genetic code
inosine (I)
intervening sequence
intron
lariat structure
leader
messenger RNA (mRNA)
missense mutation
monocistronic mRNA
mRNA
nonsense mutation
open reading frame (ORF)
overlapping genes
peptide bond
peptidyl transferase
poly-A tail
polycistronic mRNA
polypeptide chain
polysome
pretranslocation state
posttranslocation state
primary transcript
promoter
protein subunit

R group
reading frame
ribosome
ribosome-binding site
ribozyme
RNA polymerase
RNA processing
RNA splicing
silent mutation
splice acceptor
splice donor
spliceosome
start codon
stop codon
TATA box
template strand
transcription
transfer RNA (tRNA)
translation
translocation
triplet code
tRNA
uncharged tRNA]
wobble

Review the Basics

- Is the DNA strand that serves as the template for RNA polymerase transcribed in the 5' → 3' or the 3' → 5' direction? Which end of the mRNA molecule is translated first? Which end of the polypeptide encoded in the mRNA is synthesized first?

- What is the difference between the reaction catalyzed by DNA polymerase and that catalyzed by RNA polymerase?

- What are the principal characteristics of the standard genetic code?

- What is a primary transcript and how does a primary transcript differ from mRNA in prokaryotes? In eukaryotes?

- Give an example of overlapping genes. Why do you suppose overlapping genes are unusual except in certain phages and viruses in which there is a premium on small genome size?

- What are the roles of the different types of RNA molecules that are necessary for protein synthesis?

- How do prokaryotes and eukaryotes differ in the mechanism for selecting an AUG codon as a start for polypeptide synthesis?

- What is the consequence when an incorrect nucleotide is inserted into the new DNA strand during replication if it is not corrected by the proofreading function of DNA polymerase or other repair mechanisms prior to the next replication? What is the consequence when an incorrect nucleotide is inserted into an RNA molecule during transcription?

- What is a polysome and what is its role in polypeptide synthesis?

- Which of the following is the mechanism by which polypeptide chain termination takes place: (1) mRNA synthesis stops at a chain-termination codon; (2) the tRNA corresponding to a chain-termination codon cannot be charged with an amino acid; (3) chain-termination codons have no tRNA molecules that bind with them, but they interact with specific release-factor proteins instead.

Guide to Problem Solving

Problem 1: The following is the nucleotide sequence of a strand of DNA.

```
TACGTCTCCAGCGGAGATCTTTTCCGGTCGCAACTGAGGTTGATC
```

The strand is transcribed from left to right and codes for a small peptide.

(a) Which end is the 3' end and which the 5' end?

(b) What is the sequence of the complementary DNA strand?

(c) What is the sequence of the transcript?

(d) What is the amino acid sequence of the peptide?

Chapter 10

GeNETics on the web will introduce you to some of the most important sites for finding genetic information on the Internet. To complete the exercises below, visit the Jones and Bartlett home page at

http://www.jbpub.com/genetics

Select the link to *Genetics: Principles and Analysis* and then choose the link to *GeNETics on the web.* You will be presented with a chapter-by-chapter list of highlighted keywords.

GeNETics EXERCISES

Select the highlighted keyword in any of the exercises below, and you will be linked to a web site containing the genetic information necessary to complete the exercise. Each exercise suggests a specific, written report that makes use of the information available at the site. This report, or an alternative, may be assigned by your instructor.

1. The ribosomal RNA genes of *E. coli* are clustered together in seven transcriptional units, each called an *operon*. The RNA coding genes are greater than 99 percent identical from one operon to the next. Each 23S ribosomal RNA (2904 nucleotides in length) is encoded in an *rrl* gene; each 5S ribosomal RNA (120 nucleotides in length) is encoded in an *rrf* gene. One 23S and one 5S molecule are included in each 50S (large) ribosomal subunit. The 30S (small) ribosomal subunit contains one molecule of 16S RNA (1542 nucleotides in length), which is encoded in any of the *rrs* genes. Search at the keyword site for Name *rrn* and Type *operon.* Follow the links to learn the map position of each ribosomal RNA operon and the direction of transcription. You will see that some other genes are also included in the ribosomal RNA operons. What are these genes? Does their inclusion make any sense from the standpoint of translation? If assigned to do so, draw a map of the *E. coli* chromosome showing the name and location of each rRNA operon and the direction of transcription.

2. More about the various forms of RNA polymerase, including three-dimensional structural representations, can be found at this site. What common shape is found in the RNA polymerase holo-

enzyme from *E. coli* and PolII from yeast? If assigned to do so, write one paragraph describing the difference in subunit composition between the *E. coli* holoenzyme and the core enzyme.

3. The gene *CCA1* in yeast is critical for formation of a mature transfer RNA molecule ready for charging with the correct amino acid. Search this keyword site for the gene *CCA1* to find out what it does. If assigned to do so, diagram an immature tRNA showing what reaction is catalyzed by the *cca1* gene product. Locate the gene on the genetic map and retrieve its DNA and amino acid sequence.

MUTABLE SITE EXERCISES

The Mutable Site Exercise changes frequently. Each new update includes a different exercise that makes use of genetics resources available on the World Wide Web. Select the Mutable Site for Chapter 10, and you will be linked to the current exercise that relates to the material presented in this chapter.

PIC SITE

The Pic Site showcases some of the most visually appealing genetics sites on the World Wide Web. To visit the showcase genetics site, select the Pic Site for Chapter 10.

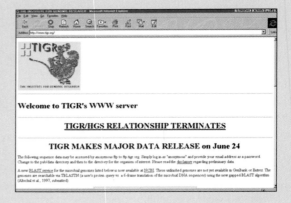

(e) In which direction along the transcript does translation occur?

(f) Which is the amino ($-NH_2$) and which the carboxyl ($-COOH$) end of the peptide?

Answer: Because the DNA strand is transcribed from left to right, and the first nucleotides incorporated into an RNA transcript form its 5' end, the 3' end of the DNA strand must be at the left. The original strand (answer to problem 1(a)), the complementary strand (answer to problem 1(b)), the transcript (answer to problem 1(c)), and the peptide (answer to problem 1(d)) are shown in the accompanying table.

(e) Translation goes from 5'-to-3' along the mRNA, so in this example, translation also goes from left to right.

(f) The amino end of the peptide is synthesized first (Met) and the carboxyl end last (Asn).

Problem 2: Using the sequence given in Problem 1, determine what effects the following mutations would have on the peptide produced. Each of the mutations affects the TTTT in the middle of the strand.

(a) TTTT → TCTT (substitution of C for T)

(b) TTTT → TATT (substitution of A for T)

(c) TTTT → TTTTT (one-base insertion)

(d) TTTT → TTT (one-base deletion)

Answer: In problems of this type, it is best to derive the sequence of the mutant mRNA and then the amino acid sequence of the peptide. The sequence alignments are shown in the table below.

(a) The mutation changes a GAA codon into a GAG codon, which still codes for glutamic acid, so no change in the peptide results.

(b) The mutation changes a GAA codon into a GAU codon, which changes glutamic acid into aspartic acid in the peptide.

(c) The insertion introduces a new nucleotide into the mRNA and shifts the reading frame downstream from the site of the mutation. In this case, the normal sequence starting with Lys is replaced with Lys-Gly-Gln-Arg, and the UGA stop codon following the Arg codon results in premature termination of the peptide.

(d) The deletion again shifts the reading frame downstream from the site of the mutation. In this case, translation continues because a stop codon is not reached immediately, but the entire amino acid sequence downstream from the mutation is altered.

Problem 3: What anticodon sequence would pair with the codon 5'-AUG-3', assuming only Watson-Crick base pairing?

Answer: With Watson-Crick base pairing only, the base pairing is the usual A with U and G with C. However, the orientations of the codon and anticodon are antiparallel, so the anticodon sequence is 3'-UAC-5'.

Problem 4: What amino acids can be present at the site of a UAA codon that is suppressed by a suppressor tRNA created by a mutant base in the anticodon, assuming only Watson-Crick base pairing?

Answer: The nonmutant tRNA must be able to pair with the codon at two sites, and the mutant base in the anticodon allows the third site to pair as well. Therefore, the amino acids that can be inserted into the suppressed site are those whose codons differ from UAA in a single base. These codons are AAA (Lys), CAA (Gln), GAA (Glu), UUA (Leu), UCA (Ser), UAC (Tyr), and UAU (Tyr).

Answer to Problem 1(a)–1(d)

(a) 3'-TACGTCTCCAGCGGAGACCTTTTCCGGTCGCAACTGAGGTTGATC-5'

(b) 5'-ATGCAGAGGTCGCCTCTGGAAAAGGCCAGCGTTGACTCCAACUAG-3'

(c) 5'-AUGCAGAGGUCGCCUCUGGAAAAGGCCAGCGUUGACUCCAACUAG-3'

(d) MetGlnArgSerProLeuGluLysAlaSerValAspSerAsn

Sequence Alignments for Answer to Problem 2(a)–2(d)

(a) DNA   3'-TACGTCTCCAGCGGAGACCTCTTCCGGTCGCAACTGAGGTTGATC-5'

mRNA 5'-AUGCAGAGGUCGCCUCUGGAGAAGGCCAGCGUUGACUCCAACUAG-3'

Peptide   MetGlnArgSerProLeuGluLysAlaSerValAspSerAsn

(b) DNA   3'-TACGTCTCCAGCGGAGACCTATTCCGGTCGCAACTGAGGTTGATC-5'

mRNA 5'-AUGCAGAGGUCGCCUCUGGAUAAGGCCAGCGUUGACUCCAACUAG-3'

Peptide   MetGlnArgSerProLeuAspLysAlaSerValAspSerAsn

(c) DNA   3'-TACGTCTCCAGCGGAGACCTTTTTCCGGTCGCAACTGAGGTTGATC-5'

mRNA 5'-AUGCAGAGGUCGCCUCUGGAAAAAGGCCAGCGUUGACUCCAACUAG-3'

Peptide   MetGlnArgSerProLeuGluLysGlyGlnArg

(d) DNA   3'-TACGTCTCCAGCGGAGACCTTTCCGGTCGCAACTGAGGTTGATC-5'

mRNA 5'-AUGCAGAGGUCGCCUCUGGAAAGGCCAGCGUUGACUCCAACUAG-3'

Peptide   MetGlnArgSerProLeuGluArgProAlaLeuThrProThr

Analysis and Applications _____

10.1 What are the translation initiation and stop codons in the genetic code? In a random sequence of four ribonucleotides, what is the probability that any three adjacent nucleotides will be a start codon? A stop codon? In an mRNA molecule of random sequence, what is the average distance between stop codons?

10.2 A part of the coding strand of a DNA molecule that codes for the 5' end of an mRNA has the sequence 3'-TTTTACGGGAATTAGAGTCGCAGGATG-5'. What is the amino acid sequence of the polypeptide encoded by this region, assuming that the normal start codon is needed for initiation of polypeptide synthesis?

10.3 Poly-U codes for polyphenylalanine. If a G is added to the 5' end of the molecule, the polyphenylalanine has a different amino acid at the amino terminus, and if a G is added to the 3' end, there is a different amino acid at the carboxyl terminus. What are the amino acids?

10.4 The synthetic polymer, poly-A, is used as an mRNA molecule in an *in vitro* protein-synthesizing system that does not need a special start codon. Polylysine is synthesized. A single guanine nucleotide is added to one end of the poly-A. The resulting polylysine has a glutamic acid at the amino terminus. Was the G added to the 3' or the 5' end of the poly-A?

10.5 What polypeptide products are made when the alternating polymer GUGU . . . is used in an *in vitro* protein-synthesizing system that does not need a start codon?

10.6 What polypeptide products are made when the alternating polymer GUCGUC . . . is used in an *in vitro* protein-synthesizing system that does not need a start codon?

10.7 Some codons in the genetic code were determined experimentally by the translation of random polymers. If a ribonucleotide polymer is synthesized that contains 3/4 A and 1/4 C in random order, which amino acids would the resulting polypeptide contain, and in what frequencies?

10.8 How many different sequences of nine ribonucleotides would code for the amino acids Met−His−Thr? For Met−Arg−Thr? Using the symbol Y for any pyrimidine, R for any purine, and N for any nucleotide, what are the sequences?

10.9 At one time, it was considered that the genetic code might be one in which the codons overlapped. For example, with a two-base overlap, the codons in the mRNA sequence CAUCAU would be translated as CAU AUC UCA CAU rather than as CAU CAU. How is this hypothesis affected by the observation that mutant proteins usually differ from the wildtype protein by a single amino acid?

10.10 What codons could pair with the anticodon 5'-IAU-3'? (I stands for inosine.) What amino acid would be incorporated?

10.11 Two possible anticodons could pair with the codon UGG, but only one is actually used. Identify the possible anticodons, and explain why one of them is not used.

10.12 Two *E. coli* genes, *A* and *B,* are known from mapping experiments to be very close to each other. A deletion mutation is isolated that eliminates the activity of both *A* and *B.* Neither the A nor the B protein can be found in the mutant, but a novel protein is isolated in which the amino-terminal 30 amino acids are identical to those of the *B* gene product and the carboxyl-terminal 30 amino acids are identical to those of the *A* gene product.

(a) With regard to the 5'-to-3' orientation of the nontranscribed DNA strand, is the order of the genes *A B* or *B A*?

(b) Can you make any inference about the number of bases deleted?

10.13 The nontranscribed sequence at the beginning of a gene reads

$$5'-ATGCATCCGGGCTCATTAGTCT . . . -3'$$

Two mutations are studied. Mutation X has an insertion of a G immediately after the underlined G, and mutation Y has a deletion of the red A. What is the amino acid sequence of each of the following?

(a) the wildtype polypeptide

(b) the polypeptide in mutant X

(c) the polypeptide in mutant Y

(d) the polypeptide in a recombinant organism containing both mutations

10.14 The amino terminus of a wildtype enzyme in yeast has the amino acid sequence

Met−Leu−His−Tyr−Met−Gly−Asp−Tyr−Pro

A mutant, X, is found that contains an inactive enzyme with the sequence Met−Gly−Asp−Tyr−Pro at the amino terminus and the wildtype sequence at the carboxyl terminus. A second mutant, Y, also lacks enzyme activity, but there is no trace of a full-length protein. Instead, mutant Y makes a short peptide containing just three amino acids. What single-base changes can account for the features of mutation X and mutation Y? What is the sequence of the tripeptide produced by mutant Y?

10.15 Protein synthesis occurs with high fidelity. In prokaryotes, incorrect amino acids are inserted at the rate of approximately 10^{-3} (that is, one incorrect amino acid per 1000 translated). What is the probability that a polypeptide of 300 amino acids has exactly the amino acid sequence specified in the mRNA?

10.16 A DNA fragment containing a particular gene is isolated from a eukaryotic organism. This DNA fragment is mixed with the corresponding mRNA isolated from the organism, denatured, renatured, and observed by electron microscopy. Heteroduplexes of the type shown in the accompanying figure are observed. How many introns does this gene contain?

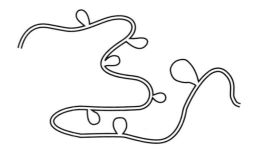

10.17 If the DNA molecule shown here is transcribed from left to right, what are the sequence of the mRNA and the amino acid sequence? What are the sequence of the mRNA and the amino acid sequence if the segment in red is inverted?

```
5'-AGACTTCAGGCTCAACGTGGT-3'
3'-TCTGAAGTCCGAGTTGCACCA-5'
```

10.18 You are given the nontemplate-strand nucleotide sequence of a part of an exon of an active gene.

```
5'-TAACGTATGCTTGACCTCCAAGCAATCGATGCCAGCTCAAGG-3'
```

Assuming the standard genetic code, what is the amino acid sequence in the polypeptide chain? What tells you that you have identified the correct reading frame?

Challenge Problems

10.19 In performing an evolutionary analysis, biologists often consider the 6-fold degenerate serine (Ser) as two separate amino acids—a 4-fold and a 2-fold degenerate class—even though the amino acid is the same. Taking into account what you know about translation and the genetic code, why does it make sense to do this for serine but not for other 6-fold degenerate amino acids?

10.20 For two different frameshift mutations in the second codon of a gene, the amino terminal sequences of the mutant proteins are

Mutant 1:   Met—Lys—UAG
Mutant 2:   Met—Ile—Val—UAA

Mutant 1 has a single-nucleotide addition, and mutant 2 has a single-nucleotide deletion. Furthermore, the first five amino acids of the wildtype protein are known to be Met-(Asn, Val, Ser, Lys), where the parentheses mean that the order of the amino acids is unknown. Using the information provided by the frameshift mutations, determine the first five codons in the wildtype gene as well as the nature of each frameshift mutation.

Further Reading

Barrell, B. G., A. T. Bankier, and J. Drouin. 1979. A different genetic code in human mitochondria. *Nature* 282: 189.

Beadle, G. W. 1948. Genes of men and molds. *Scientific American,* September.

Bird, R. C., ed. *Nuclear Structure and Gene Expression.* New York: Academic Press.

Blumenthal, T. 1995. *Trans*-splicing and polycistronic transcription in *Caenorhabditis elegans. Trends in Genetics* 11: 132.

Chambon, P. 1981. Split genes. *Scientific American,* May.

Crick, F. H. C. 1962. The genetic code. *Scientific American,* October.

Crick, F. H. C. 1966. The genetic code. *Scientific American,* October.

Crick, F. H. C. 1979. Split genes and RNA splicing. *Science* 204: 264.

Haseltine, W. A. 1997. Discovering genes for new medicines. *Scientific American,* March.

Henkin, T. M. 1996. Control of transcription termination in prokaryotes. *Annual Review of Genetics* 30: 35.

Hill, W. E., and A. Dahlberg, eds. 1990. *The Ribosome: Structure, Function, and Evolution.* Washington, DC: American Society for Microbiology.

Jackson, R . J., and M. Wickens. 1997. Translational controls impinging on the 5'-untranslated region and initiation factor proteins. *Current Opinion in Genetics & Development* 7: 233.

Kim, J. L., D. B. Nikolov, and S. K. Burley. 1993. Co-crystal structure of TBP recognizing the minor groove of a TATA element. *Nature* 3656: 520.

Kim, Y., J. H. Geiger, S. Hahn, and P. B. Sigler. 1993. Crystal structure of a yeast TBP/TATA-box complex. *Nature* 365: 512.

Lee, M. S., and P. A. Silver. 1997. RNA movement between the nucleus and the cytoplasm. *Current Opinion in Genetics & Development* 7: 212.

Neidhardt, F. C., R. Curtiss III, J. L. Ingraham, E. C. C. Lin, K. B. Low, B. Magasanik, W. S. Reznikoff, M. Riley, M. Schaechter, and H. E. Umbarger, eds. 1996. Escherichia coli *and* Salmonella typhimurium: *Cellular and Molecular Biology* (2 volumes). 2d ed. Washington, DC: American Society for Microbiology.

Nirenberg, M. 1963. The genetic code. *Scientific American,* March.

Rhodes, D., and A. Klug. 1993. Zinc fingers. *Scientific American,* February.

Ross, J. 1996. Control of messenger RNA stability in higher eukaryotes. *Trends in Genetics* 12: 171.

Steitz, J. A. 1992. Splicing takes a Holliday. *Science* 257: 888.

Taylor, J. H., ed. 1965. *Selected Papers on Molecular Genetics.* New York: Academic Press.

Wickens, M., P. Anderson, and R. J. Jackson. 1997. Life and death in the cytoplasm: Messages from the 3' end. *Current Opinion in Genetics & Development* 7: 220.

Yanofsky, C. 1967. Gene structure and protein structure. *Scientific American,* May.

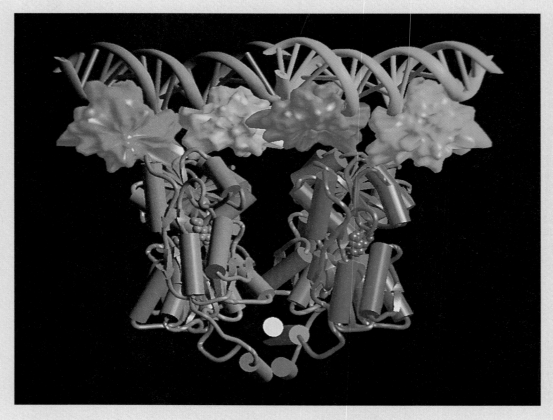

In *E. coli,* production of the enzymes necessary for growth on the sugar lactose is controlled by a transcriptional repressor protein, composed of four identical polypeptide subunits, which interacts with DNA sequences comprising the *lac* operator. This model shows how the operator sequences (red and blue double helices running across the top) are contacted by the repressor subunits shown below. [Courtesy of Thomas A. Steitz.]

CHAPTER *11*

Regulation of Gene Activity

CHAPTER OUTLINE

PRINCIPLES

- Genes can be regulated at any level, including transcription, RNA processing, translation, and post-translation.
- Control of transcription is an important mechanism of gene regulation.
- Transcriptional control can be negative ("on unless turned off") or positive ("off unless turned on"); many genes include regulatory regions for both types of regulation.
- Most genes have multiple, overlapping regulatory mechanisms that operate at more than one level, from transcription through post-translation.
- In prokaryotes, the genes coding for the enzymes in a metabolic pathway are often clustered in the genome and controlled jointly by a regulatory protein that binds with an "operator" region at the 5' end of the cluster. This type of gene organization is known as an operon.
- In eukaryotes, genes are not organized into operons. Genes at dispersed locations in the genome are coordinately controlled by one or more "enhancer" DNA sequences located near each gene that interact with transcriptional activator proteins that enable transcription of each nearby gene to occur.

CONNECTIONS

CONNECTION: Operator? Operator?
François Jacob, David Perrin, Carmen Sanchez,
and Jacques Monod 1960
The operon: A group of genes whose expression is coordinated by an operator

CONNECTION: Sex-Change Operations
James B. Hicks, Jeffrey N. Strathern, and Ira
Herskowitz 1977
The cassette model of mating-type interconversion

Not all genes are expressed continuously. The level of gene expression may differ from one cell type to the next or according to stage in the cell cycle. For example, the genes for hemoglobin are expressed at high levels only in precursors of the red blood cells. The activity of genes varies according to the functions of the cell. A vertebrate animal, such as a mouse, contains approximately 200 different types of cells with specialized functions. With minor exceptions, all cell types contain the same genetic complement. The cell types differ only in which genes are active. In general, the synthesis of particular gene products is controlled by mechanisms collectively called **gene regulation.**

In many cases, gene activity is regulated at the level of transcription, either through signals originating within the cell itself or in response to external conditions. For example, many gene products are needed only on occasion, and transcription can be regulated in an on-off manner that enables such products to be present only when external conditions demand. However, the flow of genetic information is regulated in other ways also. Control points for gene expression include the following:

1. *DNA rearrangements,* in which gene expression changes depending on the position of DNA sequences in the genome.

2. *Transcriptional regulation* of the synthesis of RNA transcripts by controlling initiation or termination.

3. *RNA processing,* or regulation through RNA splicing or alternative patterns of splicing.

4. *Translational control* of polypeptide synthesis.

5. *Stability of mRNA,* because mRNAs that persist in the cell have longer-lasting effects than those that are degraded rapidly.

6. *Post-translational control,* which includes a great variety of mechanisms that affect enzyme activity, activation, stability, and so on.

The regulatory systems of prokaryotes and eukaryotes are somewhat different from each other. Prokaryotes are generally free-living unicellular organisms that grow and divide indefinitely as long as environmental conditions are suitable and the supply of nutrients is adequate. Their regulatory systems are geared to provide the maximum growth rate in a particular environment, except when such growth would be detrimental. Prokaryotes can also use the coupling between transcription and translation (Chapter 10) for regulation, but the absence of introns eliminates RNA splicing as a possible control point.

The requirements of multicellular eukaryotes are different from those of prokaryotes. In a developing organism, not only must a cell grow and divide, but the progeny cells must also undergo considerable changes in morphology and biochemistry and then each maintain its altered state. Furthermore, during embryonic development, most eukaryotic cells are challenged less by the environment than are bacteria in that the composition and concentration of the growth medium does not change drastically with time. Finally, in an adult organism, growth and cell division in most cell types have stopped, and each cell needs only to maintain itself and its specialized characteristics.

In this chapter, we consider the basic mechanisms of the regulation of transcription and RNA processing. The examples we use are those in which the regulation is well understood.

11.1
Transcriptional Regulation in Prokaryotes

In bacteria and phages, on-off gene activity is often controlled through transcription. Synthesis of a particular mRNA takes place only when the gene product is needed, and when the gene product is not needed, mRNA synthesis occurs at greatly reduced levels. In discussing transcription, we use the term *off* for convenience, but remember that this usually means "very low." In bacteria, few examples are known of a system being switched completely off. When transcription is in the "off" state, a basal level of gene expression almost always remains, often averaging one transcriptional event or fewer per cell generation; hence there is very little synthesis of the gene product. Extremely low levels of expression are also found in certain classes of genes in eukaryotes, including many genes that participate in embryonic development. Regulatory mechanisms other than the on-off type also are known in both prokaryotes and eukaryotes; in these examples, the level of expression of a gene may be modulated in gradations from high to low according to conditions in the cell.

In bacterial systems, when several enzymes act in sequence in a single metabolic pathway, usually either all or none of these enzymes are produced. This **coordinate regulation** results from control of the synthesis of one or more polycistronic mRNA molecules encoding all of the gene products that function in the same metabolic pathway. This type of regulation is not found in eukaryotes because eukaryotic mRNA is monocistronic, as we saw in Chapter 10.

Several mechanisms of regulation of transcription are common. The particular one used often depends on whether the enzymes being regulated act in degradative or biosynthetic metabolic pathways. For example, in a multistep degradative (catabolic) system, the availability of the molecule to be degraded helps determine whether the enzymes in the pathway will be synthesized. In the presence of the molecule, the enzymes of the degradative (catabolic) pathway are synthesized; in its absence, they are not. Such a system, in which the presence of a small molecule results in enzyme synthesis, is said to be **inducible.** The small molecule is called the **inducer.** The opposite situation is often found in the control of the synthesis of enzymes that participate in biosynthetic (anabolic) pathways; in these cases, the final product of the pathway is frequently the regulatory molecule. In the presence of the final product, the enzymes of the biosynthetic pathway are not synthesized; in its absence, they are synthesized. Such a system, in which the presence of a small molecule results in failure to synthesize enzymes, is said to be **repressible.** The small molecule that participates in the regulation is called the **co-repressor.**

The molecular mechanisms for each of the regulatory patterns vary quite widely but usually fall into one of two major categories—**negative regulation** and **positive regulation.** In a negatively regulated system (Figure 11.1A), a **repressor** protein

Figure 11.1

The distinction between negative and positive regulation. (A) In negative regulation, the "default" state of the gene is one in which transcription takes place. The binding of a repressor protein to the DNA molecule prevents transcription. (B) In positive regulation, the default state is one in which transcription does not take place. The binding of a transcriptional activator protein stimulates transcription. A single genetic element may be regulated both positively and negatively; in such a case, transcription requires the binding of the transcriptional activator and the absence of repressor binding.

(A) Negative regulation

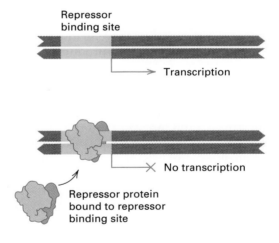

(B) Positive regulation

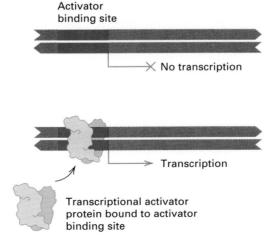

present in the cell prevents transcription. In an inducible system that is negatively regulated, the repressor protein acts by itself to prevent transcription. The inducer antagonizes the repressor, allowing the initiation of transcription. In a repressible system, an **aporepressor** protein combines with the co-repressor molecule to form the functional repressor, which prevents transcription. In the absence of the co-repressor, the aporepressor is unable to prevent transcription. On the other hand, in a positively regulated system (Figure 11.1B), mRNA synthesis only takes place if a regulatory protein binds to a region of the gene that activates transcription. Such a protein is usually referred to as a **transcriptional activator.** Negative and positive regulation are not mutually exclusive, and some systems are both positively and negatively regulated, utilizing two regulators to respond to different conditions in the cell. Negative regulation is more common in prokaryotes, positive regulation in eukaryotes.

A degradative system may be regulated either positively or negatively. In a biosynthetic pathway, the final product usually negatively regulates its own synthesis; in the simplest type of negative regulation, absence of the product increases its synthesis (through production of the necessary enzymes), and presence of the product decreases its synthesis (through repression of enzyme synthesis). Even in a system in which a single protein molecule (not necessarily an enzyme), is translated from a monocistronic mRNA molecule, the protein may be **autoregulated,** which means that the protein regulates its own transcription. In negative autoregulation, the protein inhibits transcription, and high concentrations of the protein result in less transcription of the mRNA that codes for the protein. In positive autoregulation, the protein stimulates transcription: As more protein is made, transcription increases to the maximum rate. Positive autoregulation is a common way for weak induction to be amplified. Only a weak signal is necessary to get production of the protein started, but then the positive autoregulation stimulates the production to the maximum level.

The next two sections are concerned with several systems of regulation in prokaryotes. These serve as an introduction to the remainder of the chapter, which deals with regulation in eukaryotes.

11.2
Lactose Metabolism and the Operon

Metabolic regulation was first studied in detail in the system in *E. coli* responsible for degradation of the sugar lactose, and most of the terminology used to describe regulation has come from genetic analysis of this system.

Lac⁻ Mutants

In *E. coli*, two proteins are necessary for the metabolism of lactose. They are the enzyme **β-galactosidase,** which cleaves lactose (a β-galactoside) to yield galactose and glucose; and a transporter molecule, **lactose permease,** which is required for the entry of lactose into the cell. The existence of two different proteins in the lactose-utilization system was first shown by a combination of genetic experiments and biochemical analysis.

First, hundreds of mutants unable to use lactose as a carbon source, designated Lac⁻ mutants, were isolated. Some of the mutations were in the *E. coli* chromosome and others were in an F' *lac*, a plasmid carrying the genes for lactose utilization. By performing F' × F⁻ matings, investigators constructed partial diploids with the genotypes F' *lac⁻/lac⁺* and F' *lac⁺/lac⁻*. (The genotype of the plasmid is given to the left of the slash and that of the chromosome to the right.) It was observed that all of these diploids always had a Lac⁺ phenotype (that is, they made β-galactosidase and permease); thus none produced an inhibitor that prevented functioning of the *lac* genes. Other partial diploids were then constructed in which both the F' *lac* plasmid and the chromosome carried a *lac⁻* allele. These were tested for the Lac⁺ phenotype, with the result that all of the mutants initially isolated could be placed into two complementation groups, called *lacZ* and *lacY*, a result that implies that the *lac* system consists of at least two genes. Complementation is indicated by the observation that the partial diploids

$$F'\ lacY^-\ lacZ^+/lacY^+\ lacZ^-$$

and

$$F'\ lacY^+\ lacZ^-/lacY^-\ lacZ^+$$

had a Lac$^+$ phenotype, producing both β-galactosidase and permease. However, the genotypes

$$F'\ lacY^-\ lacZ^+/lacY^-\ lacZ^+$$

and

$$F'\ lacY^+\ lacZ^-/lacY^+\ lacZ^-$$

had the Lac$^-$ phenotype because they were unable to synthesize the permease and the β-galactosidase, respectively. Hence the *lacZ* gene codes for the β-galactosidase and the *lacY* gene for the permease. (A third gene that participates in lactose metabolism was later discovered; it was not included among the early mutants because it is not essential for growth on lactose.) A final important result—that the *lacY* and *lacZ* genes are adjacent—was deduced from a high frequency of cotransduction observed in genetic mapping experiments.

Inducible and Constitutive Synthesis and Repression

The on-off nature of the lactose-utilization system is evident in the following observations:

1. If a culture of Lac$^+$ *E. coli* is growing in a medium that does not include lactose or any other β-galactoside, then the intracellular concentrations of β-galactosidase and permease are exceedingly low: roughly one or two molecules per bacterial cell. However, if lactose is present in the growth medium, then the number of each of these molecules is about 10^3-fold higher.

2. If lactose is added to a Lac$^+$ culture growing in a lactose-free medium (also lacking glucose, a point that will be discussed shortly), then both β-galactosidase and permease are synthesized nearly simultaneously, as shown

in Figure 11.2. Analysis of the total mRNA present in the cells before and after the addition of lactose shows that almost no *lac* mRNA (the polycistronic mRNA that codes for β-galactosidase and permease) is present before lactose is added and that the addition of lactose triggers synthesis of *lac* mRNA.

These two observations led to the view that transcription of the lactose genes is **inducible transcription** and that lactose is an *inducer* of transcription. Some analogs of lactose are also inducers, such as a sulfur-containing analog denoted IPTG (isopropylthiogalactoside), which is convenient for experiments because it induces but is not cleaved by β-galactosidase, so this inducer is stable in the cell whether or not the β-galactosidase enzyme is present.

Figure 11.2
The "on-off" nature of the *lac* system. The *lac* mRNA appears soon after lactose or another inducer is added; β-galactosidase and permease appear at nearly the same time but are delayed with respect to mRNA synthesis because of the time required for translation. When lactose is removed, no more *lac* mRNA is made, and the amount of *lac* mRNA decreases because of the degradation of mRNA already present. Both β-galactosidase and permease are stable proteins: their amounts remain constant even when synthesis ceases. However, their concentration per cell gradually decreases as a result of repeated cell divisions.

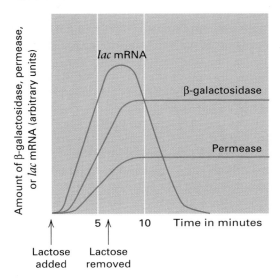

Mutants were also isolated in which *lac* mRNA was synthesized (and hence β-galactosidase and permease produced) in both the presence and the *absence* of an inducer. The mutants that eliminated regulation provided the key to understanding induction; because of their constant synthesis, the mutants were termed **constitutive.** Mutants were also obtained that failed to produce *lac* mRNA (and hence β-galactosidase and permease) even when the inducer was present. These uninducible mutants fell into two classes, *lacI*^{s} and *lacP*$^-$. The characteristics of the mutants are shown in Table 11.1 and discussed in the following sections.

The Repressor

In Table 11.1, genotypes 3 and 4 show that *lacI*$^-$ mutations are recessive. In the absence of inducer, a *lacI*$^+$ cell does not make *lac* mRNA, whereas the mRNA is made in a *lacI*$^-$ mutant. These results suggest that

> The *lacI* gene is a regulatory gene whose product is the repressor protein that keeps the system turned off. Because the repressor is necessary to shut off mRNA synthesis, regulation by the repressor is negative regulation.

A *lacI*$^-$ mutant lacks the repressor and hence is constitutive. Wildtype copies of the repressor are present in a *lacI*$^+$/*lacI*$^-$ partial diploid, so transcription is repressed. It is important to note that the single *lacI*$^+$ gene prevents synthesis of *lac* mRNA from both the F' plasmid and the chromosome. Therefore, the repressor protein must be diffusible within the cell to shut off mRNA synthesis from both DNA molecules present in a partial diploid.

On the other hand, genotypes 7 and 8 indicate that the *lacI*^{s} mutations are dominant and act to shut off mRNA synthesis from both the F' plasmid and the chromosome, whether or not the inducer is present (the superscript in *lacI*^{s} signifies *super-repressor.*) The *lacI*^{s} mutations result in repressor molecules that fail to recognize and bind the inducer and thus permanently shut off *lac* mRNA synthesis.

Genetic mapping experiments placed the *lacI* gene adjacent to the *lacZ* gene and established the gene order *lacI lacZ lacY*. How the *lacI* repressor prevents synthesis of *lac* mRNA will be explained shortly.

The Operator Region

Entries 1 and 2 in Table 11.1 show that *lacO*^{c} mutants are dominant. However, the dominance is evident only in certain combinations of *lac* mutations, as can be seen by examining the partial diploids shown in entries 5 and 6. Both combinations are Lac$^+$ because a functional *lacZ* gene is present. However, in the combination shown in entry 5, synthesis of β-galactosidase is inducible even though a *lacO*^{c} mutation is present. The difference between the two combinations in entries 5 and 6 is that in entry 5, the *lacO*^{c} mutation is present in the same DNA molecule as the *lacZ*$^-$ mutation, whereas in entry 6, *lacO*^{c} is contained in the same DNA molecule as *lacZ*$^+$. The key feature of these results is that

> A *lacO*^{c} mutation causes constitutive synthesis of β-galactosidase only when the *lacO*^{c} and *lacZ*$^+$ alleles are contained in the same DNA molecule.

The *lacO*^{c} mutation is said to be ***cis-dominant,*** because only genes in the *cis* configuration (in the same DNA molecule as that containing the mutation) are expressed in dominant fashion. Confirmation of this conclusion comes from an important biochemical observation: The mutant enzyme coded by the *lacZ*$^-$ sequence is synthesized constitutively in a *lacO*^{c} *lacZ*$^-$/*lacO*$^+$ *lacZ*$^+$ partial diploid (entry 5), whereas the wildtype enzyme (coded by the *lacZ*$^+$ sequence) is synthesized only if an inducer is added. All *lacO*^{c} mutations are located between the *lacI* and *lacZ* genes; hence the gene order of the four genetic elements of the *lac* system is

> *lacI   lacO   lacZ   lacY*

An important feature of all *lacO*^{c} mutations is that they cannot be complemented (a characteristic feature of all *cis*-dominant mutations); that is, a *lacO*$^+$ allele cannot alter the constitutive activity of a *lacO*^{c}

mutation. This observation implies that the *lacO* region does not encode a diffusible product and must instead define a site in the DNA that determines whether synthesis of the product of the adjacent *lacZ* gene is inducible or constitutive. The *lacO* region is called the **operator.** In the next section, we will see that the operator is in fact a *binding site* in the DNA for the repressor protein.

The Promoter Region

Entries 11 and 12 in Table 11.1 show that *lacP⁻* mutations, like *lacOᶜ* mutations, are *cis*-dominant. The *cis*-dominance can be seen in the partial diploid in entry 11. The genotype in entry 11 is uninducible, in contrast to the partial diploid of entry 12, which is inducible. The difference between the two genotypes is that in entry 11, the *lacP⁻* mutation is in the same DNA molecule with *lacZ⁺*, whereas in entry 12, the *lacP⁻* mutation is combined with *lacZ⁻*. This observation means that a wildtype *lacZ⁺* remains inexpressible in the presence of *lacP⁻*; no *lac* mRNA is transcribed from that DNA molecule. The *lacP⁻* mutations map between *lacI* and *lacO,* and the order

of the five genetic elements of the *lac* system is

$$lacI \quad lacP \quad lacO \quad lacZ \quad lacY$$

As expected because of the *cis*-dominance of *lacP⁻* mutations, they cannot be complemented; that is, a *lacP⁺* allele on another DNA molecule cannot supply the missing function to a DNA molecule carrying a *LacP⁻* mutation. Thus *lacP,* like *lacO,* must define a site that determines whether synthesis of *lac* mRNA will take place. Because synthesis does not occur if the site is defective or missing, *lacP* defines an essential site for mRNA synthesis. The *lacP* region is called the **promoter.** It is a site at which RNA polymerase binding takes place to allow initiation of transcription.

The Operon Model of Transcriptional Regulation

The genetic regulatory mechanism of the *lac* system was first explained by the **operon model** of François Jacob and Jacques Monod, which is illustrated in Figure 11.3. (The figure uses the abbreviations *i, o, p, z, y,* and *a* for *lacI, lacO, lacP, lacZ, lacY,*

Table 11.1 Characteristics of partial diploids containing several combinations of *lacI, lacO* and *lacP* alleles

Genotype	Synthesis of *lac* mRNA	Lac phenotype
1. F' *lacOᶜ lacZ⁺/lacO⁺ lacZ⁺*	Constitutive	+
2. F' *lacO⁺ lacZ⁺/lacOᶜ lacZ⁺*	Constitutive	+
3. F' *lacI⁻ lacZ⁺/lacI⁺ lacZ⁺*	Inducible	+
4. F' *lacI⁺ lacZ⁺/lacI⁻ lacZ⁺*	Inducible	+
5. F' *lacOᶜ lacZ⁻/lacO⁺ lacZ⁺*	Inducible	+
6. F' *lacOᶜ lacZ⁺/lacO⁺ lacZ⁻*	Constitutive	+
7. F' *lacIˢ lacZ⁺/lacI⁺ lacZ⁺*	Uninducible	−
8. F' *lacI⁺ lacZ⁺/lacIˢ lacZ⁺*	Uninducible	−
9. F' *lacP⁻ lacZ⁺/lacP⁺ lacZ⁺*	Inducible	+
10. F' *lacP⁺ lacZ⁺/lacP⁻ lacZ⁺*	Inducible	+
11. F' *lacP⁺ lacZ⁻/lacP⁻ lacZ⁺*	Uninducible	−
12. F' *lacP⁺ lacZ⁺/lacP⁻ lacZ⁻*	Inducible	+

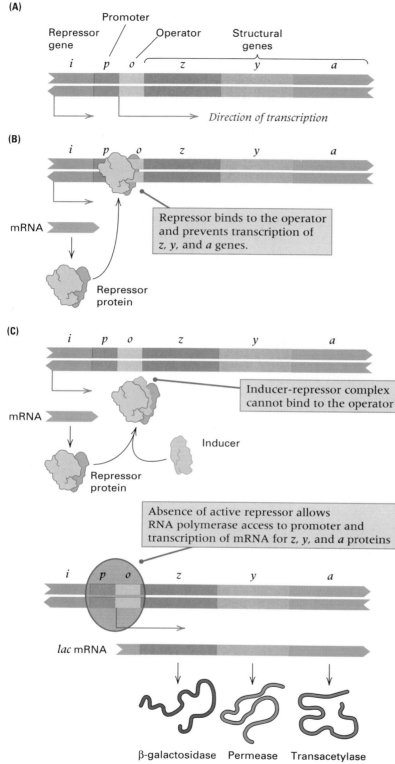

(A)

Repressor gene

Promoter

Operator

Structural genes

i   *p*   *o*   *z*   *y*   *a*

Direction of transcription

(B)

i   *p*   *o*   *z*   *y*   *a*

mRNA

Repressor binds to the operator and prevents transcription of *z*, *y*, and *a* genes.

Repressor protein

(C)

i   *p*   *o*   *z*   *y*   *a*

mRNA

Inducer-repressor complex cannot bind to the operator

Repressor protein

Inducer

Absence of active repressor allows RNA polymerase access to promoter and transcription of mRNA for *z*, *y*, and *a* proteins

i   *p*   *o*   *z*   *y*   *a*

lac mRNA

β-galactosidase   Permease   Transacetylase

Figure 11.3

(A) A map of the *lac* operon, not drawn to scale. The *p* and *o* sites are actually much smaller than the other regions and together comprise only 83 base pairs. (B) A diagram of the *lac* operon in the repressed state. (C) A diagram of the *lac* operon in the induced state. The inducer alters the shape of the repressor so that the repressor can no longer bind to the operator. The common abbreviations *i, p, o, z, y,* and *a* are used instead of *lacI, lacO,* and so on. The *lacA* gene is not essential for lactose utilization.

lacY, and *lacA.*) The operon model has the following features:

1. The lactose-utilization system consists of two kinds of components—*structural genes* (*lacZ* and *lacY*), which encode proteins needed for the transport and metabolism of lactose, and *regulatory elements* (the repressor gene *lacI,* the promoter *lacP,* and the operator *lacO*).

2. The products of the *lacZ* and *lacY* genes are coded by a single polycistronic mRNA molecule. (A third protein, encoded by *lacA,* is also translated from the mRNA. This protein is the enzyme transacetylase; it is used in the metabolism of certain β-galactosides other than lactose and will not be of further concern here.) The linked structural genes, together with *lacP* and *lacO,* constitute the **lac operon.**

3. The promoter mutations (*lacP⁻*) eliminate the ability to synthesize *lac* mRNA.

4. The product of the *lacI* gene is a repressor, which binds to a unique sequence of DNA bases constituting the operator.

5. When the repressor is bound to the operator, initiation of transcription of *lac* mRNA by RNA polymerase is prevented.

6. Inducers stimulate mRNA synthesis by binding to and inactivating the repressor. In the presence of an inducer, the operator is not bound with the repressor, and the promoter is available for the initiation of mRNA synthesis.

Note that regulation of the operon requires that the *lacO* operator either overlap or be adjacent to the promoter of the structural genes, because binding with the repressor prevents transcription. Proximity of *lacI* to *lacO* is not strictly necessary, because the *lacI* repressor is a soluble protein and is therefore diffusible throughout the cell. The presence of inducer has a profound effect on the DNA binding properties of the repressor; the inducer-repressor complex has an affinity for the operator that is approximately 10^3 smaller than that of the repressor alone.

The ratio of the numbers of copies of β-galactosidase, permease, and transacetylase

Operator? Operator?

François Jacob, David Perrin, Carmen Sanchez, and Jacques Monod, Institute Pasteur, Paris, France. 1960
The Operon: A Group of Genes Whose Expression Is Coordinated by an Operator (original in French)

How is gene expression controlled? Before Jacob and Monod and their collaborators addressed this question experimentally, it was all a matter of speculation. Prior to this report, the researchers had previously discovered the i (lacI) gene that controls expression of the β-galactosidase (z) and permease (y) genes needed for lactose utilization. They also had strong reason to believe that lacI produces a regulatory protein. How does the regulatory protein work? Here they give evidence that it works by directly binding to a DNA "operator" adjacent to the genes it regulates. Furthermore, the z and y genes are adjacent and are controlled coordinately by the same "operator" upstream from z. The discovery was immediately recognized as fundamental. Jacob and Monod, along with André Lwoff, were awarded the Nobel Prize in 1965. We now know that coordinate regulation via operons is restricted to bacteria. However, the underlying principle—that regulatory genes often control their target genes by direct binding to DNA—is valid for all organisms.

The analysis of different bacterial systems leads to the conclusion that, in the synthesis of certain proteins, there is a dual genetic determination involving two types of genes with distinct functions: one (the gene for structure) is responsible for the structure of the protein molecule, and the other (the regulatory gene) governs the expression of the former through the intermediary action of a repressor. The regulatory genes that have so far been identified show the remarkable property of exercising a *coordinated effect,* each governing the expression of several genes for structure, closely linked together, and corresponding to enzyme proteins belonging to the *same biochemical pathway.* To explain this effect, it seems necessary to invoke a new type of genetic entity, called an "operator," which

---

It seems necessary to invoke a new type of genetic entity, called an "operator," which would be (a) adjacent to the group of genes and would control their activity; and (b) would be sensitive to the repressor produced by a particular regulatory gene.

---

would be (a) adjacent to the group of genes and would control their activity; and (b) would be sensitive to the repressor produced by a particular regulatory gene. In the presence of the repressor, the expression of the group of genes would be inhibited through the mediation of the repressor. This hypothesis leads to some distinctive predictions concerning mutations that could affect the structure of the operator. (1) Certain mutations affecting an operator would be manifested by the loss of the capacity to synthesize the proteins determined by the group of linked genes "coordinated" by that operator. . . . (2) Other mutations, for example involving a loss of sensitivity (affinity) of the operator for the corresponding repressor, would be manifested by the constitutive synthesis of the protein determined by the coordinated genes. . . . We have studied certain mutations affecting the metabolism of lactose in *Escherichia coli* that act simultaneously on the synthesis of β-galactosidase [the product of the z gene] and galactoside permease [the product of the y gene]. . . . The *i* gene is the regulatory gene synthesizing a repressor specific for the system. The genes *i, z* and *y* are closely linked. . . . Constitutive mutants (o^c) have now been isolated. [In partially diploid genotypes] only the allele of *z* or *y* that is *cis* with respect to o^c is constitutively expressed. . . . Other mutants have been isolated that have lost the ability to synthesize both the permease and the β-galactosidase. . . . These mutants are recessive. . . . Genetic analysis shows that these mutations (o^o) are extremely closely linked to the o^c mutations and that the order of the *lac* region is *i–o–z–y*. . . . The remarkable properties of the o^c and o^o mutations are inexplicable according to the "classical" concept of the genes for structure [*z* and *y*] and distinguish them equally from mutations affecting the regulatory gene *i*. On the other hand, they conform to the predictions arising from the hypothesis of the operator.

Source: Comptes Rendus des Séances de l'Academie des Sciences 250: 1727–1729. Translated in E. A. Adelberg, 1966. *Papers on Bacterial Genetics.* Boston: Little Brown.

---

is 1.0 : 0.5 : 0.2 when the operon is induced. These differences are partly due to the order of the genes in the mRNA: Downstream cistrons are less likely to be translated owing to failure of reinitiation when an upstream cistron has finished translation.

The operon model is supported by a wealth of experimental data and explains many of the features of the *lac* system, as well as numerous other negatively regulated genetic systems in prokaryotes. One aspect of the regulation of the *lac* operon—the effect of glucose—has not yet been discussed. Examination of this feature indicates that the *lac* operon is also subject to positive regulation, as we will see in the next section.

Positive Regulation of the Lactose Operon

The function of β-galactosidase in lactose metabolism is to form glucose by cleaving lactose. (The other cleavage product, galactose, also is ultimately converted into glucose by the enzymes of the galactose operon.) If both glucose and lactose are present in the growth medium, activity of the *lac* operon is not needed. In fact, in the presence of glucose, no β-galactosidase is formed until virtually all of the glucose in the medium has been consumed. The lack of synthesis of β-galactosidase is a result of the lack of synthesis of *lac* mRNA. No *lac* mRNA is made in the presence of glucose, because in addition to an inducer to inactivate the *lacI* repressor, another element is needed for initiating *lac* mRNA synthesis; the activity of this element is regulated by the concentration of glucose.

The inhibitory effect of glucose on expression of the *lac* operon is indirect. The small molecule *cyclic adenosine monophosphate* **(cAMP),** shown in Figure 11.4, is widely distributed in animal tissues, and in multicellular eukaryotic organisms, in which it is important in mediating the action of many hormones. It is also present in *E. coli* and many other bacteria, where it has a different function. Cyclic AMP is synthesized by the enzyme *adenyl cyclase,* and the concentration of cAMP is regulated indirectly by glucose metabolism. When bacteria are growing in a medium containing glucose, the cAMP concentration in the cells is quite low. In a medium containing glycerol or any carbon source that cannot

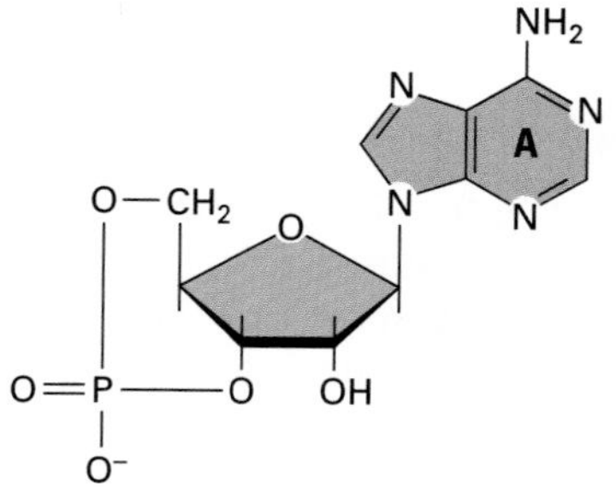

Table 11.2 Concentration of cyclic AMP in cells growing in media with the indicated carbon sources

Carbon source	cAMP concentration
Glucose	Low
Glycerol	High
Lactose	High
Lactose + glucose	Low
Lactose + glycerol	High

enter the biochemical pathway used to metabolize glucose (the glycolytic pathway), or when the bacteria are otherwise starved of an energy source, the cAMP concentration is high (Table 11.2). Glucose levels help regulate the cAMP concentration in the cell, and *cAMP regulates the activity of the lac operon* (as well as that of several other operons that control degradative metabolic pathways).

E. coli (and many other bacterial species) contain a protein called the *cyclic AMP receptor protein* **(CRP),** which is encoded by a gene called *crp.* Mutations of either the *crp* or the adenyl cyclase gene prevent synthesis of *lac* mRNA, which indicates that both CRP function and cAMP are required for *lac* mRNA synthesis. CRP and cAMP bind to one another, forming a complex denoted **cAMP-CRP,** which is an active regulatory element in the *lac* system. The requirement for cAMP-CRP is independent of the *lacI* repression system, because *crp* and adenyl cyclase mutants are unable to make *lac* mRNA even if a *lacI⁻* or a *lacOᶜ* mutation is present. The reason is that the cAMP-CRP complex must be bound to a base sequence in the DNA in the promoter region in order for transcription to occur (Figure 11.5). Unlike the repressor, which is a *negative* regulator, the cAMP-CRP complex is a *positive* regulator. The positive and negative regulatory systems of the *lac* operon are independent of each other.

Experiments carried out *in vitro* with purified *lac* DNA, *lac* repressor, cAMP-CRP, and RNA polymerase have established two further points:

Figure 11.4
Structure of cyclic AMP.

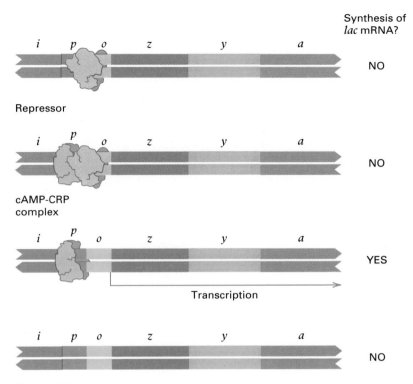

Synthesis of
lac mRNA?

NO

Repressor

NO

cAMP-CRP
complex

YES

Transcription

NO

Figure 11.5
Four regulatory states of the *lac* operon. The *lac* mRNA is synthesized only
if cAMP-CAP is present and the repressor is absent.

1. In the absence of the cAMP-CRP complex, RNA polymerase binds only weakly to the promoter, but its binding is stimulated when cAMP-CRP is also bound to the DNA. The weak binding rarely leads to initiation of transcription, because the correct interaction between RNA polymerase and the promoter does not occur.

2. If the repressor is bound to the operator, then RNA polymerase cannot stably bind to the promoter.

These results explain how lactose and glucose function together to regulate transcription of the *lac* operon. The relationship of these elements to one another, to the start of transcription, and to the base sequence in the region is depicted in Figure 11.6.

A great deal is also known about the three-dimensional structure of the regulatory states of the *lac* operon. Figure 11.7

shows that there is actually a 93-base-pair loop of DNA that forms in the operator region when it is in contact with the repressor. This loop corresponds to the *lac* operon region −82 to +11 (numbered as in Figure 11.6). The DNA region in red corresponds, on the right-hand side, to the operator region centered at +11 and, on the left-hand side, to a second repressor-binding site immediately upstream and adjacent to the CRP binding site. The *lac* repressor tetramer (violet) is shown bound to these sites. The DNA loop is formed by the region between the repressor-binding sites and includes, in medium blue, the CRP binding site, to which the CAP protein (dark blue) is shown bound. The DNA regions in green are the −10 and −35 sites in the *lacP* promoter indicated in Figure 11.6. In this configuration, the *lac* operon is not transcribed. Removal of the repressor opens up the loop and allows transcription to occur.

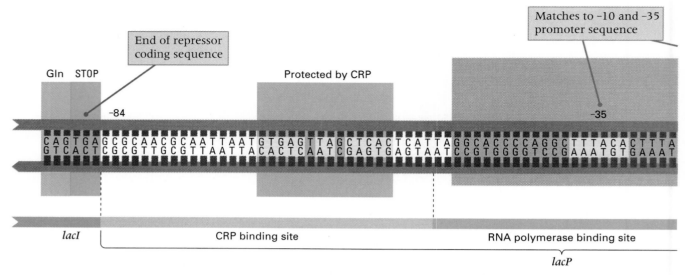

End of repressor
coding sequence

Gln STOP

Protected by CRP

–84

–35

CAGTGAGCGCAACGCAATTAATGTGAGTTAGCTCACTCATTAGGCACCCCAGGCTTTACACTTTA
GTCACTCGCGTTGCGTTAATTACACTCAATCGAGTGAGTAATCCGTGGGGGTCCGAAATGTGAAAT

lacI

CRP binding site

RNA polymerase binding site

lacP

Figure 11.6 (*above and facing page*)
The base sequence of the control region of the *lac* operon. Sequences protected from DNase digestion by binding of the stipulated proteins are indicated in the upper part. The end of the *lacI* gene is

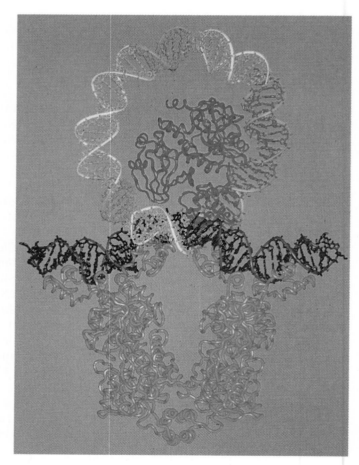

Figure 11.7
Structure of the *lac* operon
repression loop. The *lac* repressor,
shown in violet, binds to two
DNA regions (red) consisting of
the symmetrical operator region
indicated in Figure 11.6 and a
second region immediately
upstream from the CRP binding
site. Within the loop is the CRP
binding site (medium blue),
shown bound with CAP protein
(dark blue). The −10 and −35
promoter regions are in green.
[Courtesy of Mitchell Lewis;
from M. Lewis, G. Chang,
N. C. Horton, M. A. Kercher,
H. C. Pace, M. A. Schumacher,
R. G. Brennan, and P. Lu. 1996.
Science 271: 1247.]

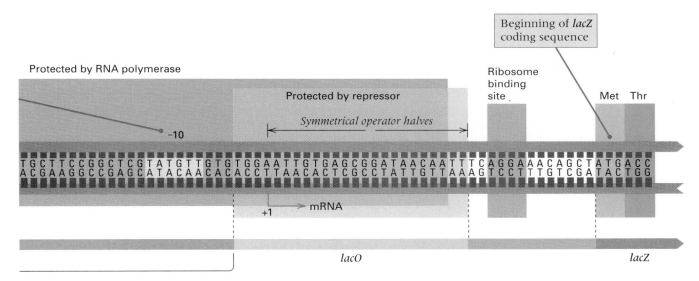

Protected by RNA polymerase

Beginning of *lacZ* coding sequence

Protected by repressor

Symmetrical operator halves

Ribosome binding site

Met Thr

-10

```
TGCTTCCGGCTCGTATGTTGTGTGTGGAATTGTGAGCGGATAACAATTTCAGGAAACAGCTATGACC
ACGAAGGCCGAGCATACAACACACCTTAACACTCGCCTATTGTTAAAGTCCTTTGTCGATACTGG
```

+1 → mRNA

lacO   *lacZ*

shown at the extreme left; the ribosome binding site is the site at which the ribosome binds to the *lac* mRNA. The consensus sites for CRP binding and for RNA polymerase promoter binding are indicated along the bottom.

11.3
Regulation of the Tryptophan Operon

The tryptophan (*trp*) operon of *E. coli* contains structural genes for enzymes that synthesize the amino acid tryptophan. This operon is regulated in such a way that when adequate tryptophan is present in the growth medium, transcription of the operon is repressed; however, when the supply of tryptophan is insufficient, transcription takes place. Regulation in the *trp* operon is similar to that of the *lac* operon because mRNA synthesis is regulated negatively by a repressor. However, it differs from regulation of *lac* in that tryptophan acts as a co-repressor, which stimulates binding of the repressor to the *trp* operator to shut off synthesis. The *trp* operon is a repressible rather than an inducible operon, although both the *lac* and the *trp* operons are negatively regulated. Furthermore, because the *trp* operon codes for a set of biosynthetic enzymes rather than degradative enzymes, neither glucose nor cAMP-CRP functions in regulation of the *trp* operon.

A simple on-off system, as in the *lac* operon, is not optimal for a biosynthetic

pathway. For example, a situation may arise in which some tryptophan is present in the growth medium, but the amount is not enough to sustain optimal growth. Under these conditions, it is advantageous to synthesize tryptophan, but at less than the maximum possible rate. Cells adjust to this situation by means of a regulatory mechanism in which *the amount of transcription in the derepressed state is determined by the concentration of tryptophan in the cell*. This regulatory mechanism is found in many operons responsible for amino acid biosynthesis.

Tryptophan is synthesized in five steps, each requiring a particular enzyme. The genes coding for these enzymes are adjacent and in the same linear order in the *E. coli* chromosome as the order in which the enzymes function in the biosynthetic pathway. The genes are called *trpE*, *trpD*, *trpC*, *trpB*, and *trpA*, and the enzymes are translated from a single polycistronic mRNA molecule. The *trpE* coding region is the first one translated. Upstream (on the 5' side) of *trpE* are the promoter, the operator, and two regions called the *leader* and the *attenuator*, which are designated *trpL* and *trpa* (not *trpA*), respectively (Figure 11.8). The repressor gene, *trpR*, is located quite far from this operon.

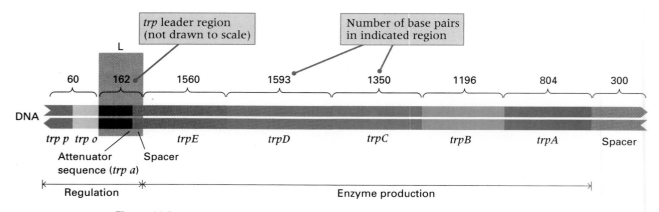

Figure 11.8

The *E. coli trp* operon. For clarity, the regulatory region is enlarged with respect to the coding region. The actual size of each region is indicated by the numbers of base pairs. Region *L* is the leader.

The regulatory protein of the *trp* operon is the product of the *trpR* gene. Mutations in either this gene or the operator cause constitutive initiation of transcription of *trp* mRNA, as in the *lac* operon. The *trpR* gene product is called the *trp* **aporepressor.** It does not bind to the operator unless it is first bound to tryptophan; that is, the aporepressor and the tryptophan molecule join together to form the active *trp* repres-

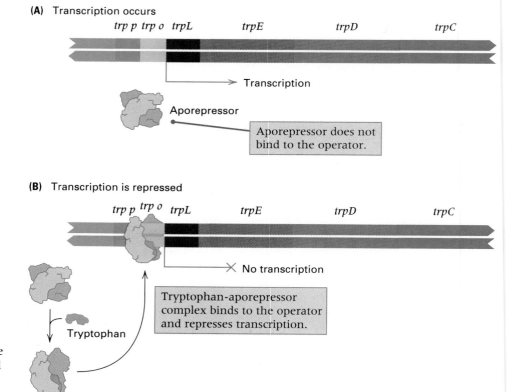

Figure 11.9

Regulation of the *E. coli trp* operon. (A) By itself, the *trp* aporepressor protein does not bind to the operator, and transcription occurs. (B) In the presence of sufficient tryptophan, the combination of aporepressor and tryptophan forms the active repressor that binds to the operator, and transcription is repressed.

sor, which binds to the operator. The reaction scheme is outlined in Figure 11.9. When there is not enough tryptophan, the aporepressor adopts a three-dimensional conformation unable to bind with the *trp* operator, and the operon is transcribed (Figure 11.9A). On the other hand, when tryptophan is present at high enough concentration, some molecules bind with the aporepressor and cause it to change conformation into the active repressor. The active repressor binds with the *trp* operator and prevents transcription (Figure 11.9B). Thus only when tryptophan is present in sufficient amounts is the active repressor molecule formed. This is the basic on-off regulatory mechanism.

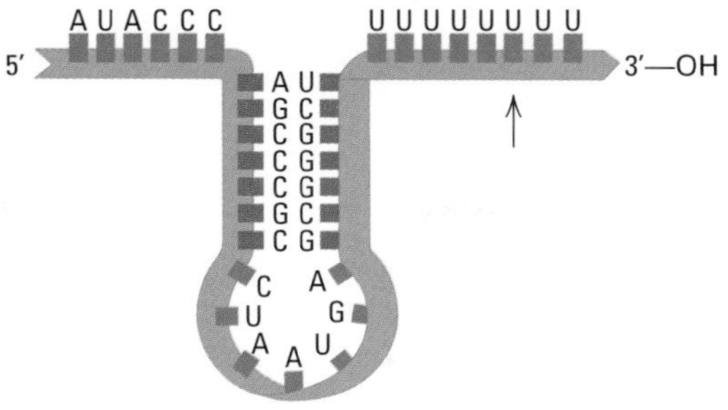

Attenuation

In the *on* state, a still more sensitive regulation of transcription is exerted by the internal concentration of tryptophan. This type of regulation is called **attenuation,** and it uses translation to control transcription. In the presence of even small concentrations of intracellular tryptophan, translation of part of the leader region of the mRNA immediately after its synthesis results in termination of transcription before the first structural gene of the operon is transcribed.

Attenuation results from interactions between DNA sequences present in the leader region of the *trp* transcript. In wild-type cells, transcription of the *trp* operon is often initiated. However, in the presence of even small amounts of tryptophan, most of the mRNA molecules terminate in a specific 28-base region within the leader sequence. The result of termination is an RNA molecule containing only 140 nucleotides that stops short of the genes

Figure 11.10
The terminal region of the *trp* attenuator sequence. The arrow indicates the final uridine in attenuated RNA. Nonattenuated RNA continues past that base. The bases in red letters form the hypothetical stem sequence that is shown.

coding for the *trp* enzymes. The 28-base region in which termination occurs is called the **attenuator.** The base sequence of this region (Figure 11.10) contains the usual features of a termination site, including a potential stem-and-loop configuration in the mRNA followed by a sequence of eight uridines.

The leader sequence, shown in Figure 11.11, contains several notable features.

1. An AUG codon and a downstream UGA stop codon in the same reading frame defining a region that codes for a polypeptide consisting of only 14 amino acids, which is called the **leader polypeptide.**

2. Two adjacent tryptophan codons that are located in the leader polypeptide at positions 10 and 11. We will see the significance of these repeated codons shortly.

Figure 11.11
The sequence of bases in the *trp* leader mRNA, showing the leader polypeptide, the two tryptophan codons (red letters), and the beginning of the TrpE protein. The numbers 23 and 91 are the numbers of bases in the sequence that, for clarity, are not shown.

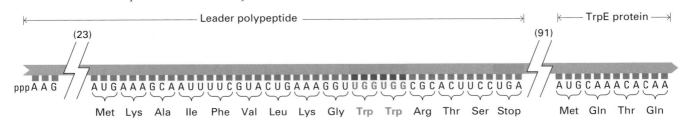

(A)

(B)

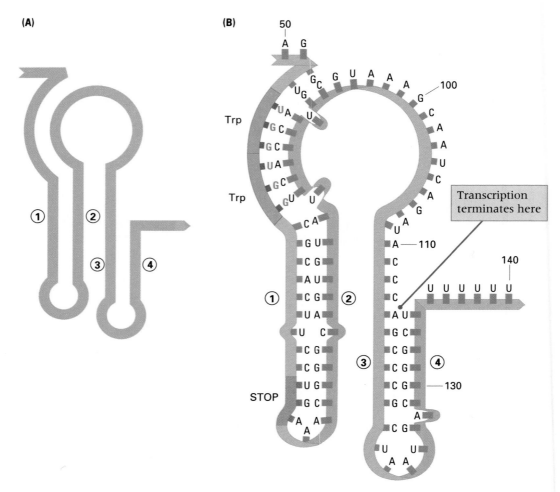

Figure 11.12

(A) Diagram of the transcript of the *trp* leader region, showing the proposed foldback structure in which a sequence of bases in region 1 can base-pair with a sequence in region 2 and a sequence of bases in region 3 can base-pair with a sequence in region 4. (B) Details of the structure. Note the two Trp codons in the 1–2 loop.

3. Four segments of the leader RNA—denoted in Figure 11.12 as regions 1, 2, 3, and 4—that are capable of base-pairing with each other. In one configuration, region 1 pairs with region 2, and region 3 with region 4. The details of this configuration are shown in Figure 11.12. When pairing takes place in this configuration, transcription is terminated at the run of uridines preceding nucleotide 140. This type of pairing occurs in purified *trp* leader mRNA.

4. An alternative type of pairing can also take place, in which region 2 pairs with region 3. The potential for this type of base pairing is apparent in Figure 11.12B in the nearly complementary sequence of bases present in regions 2 and 3.

Through the alternative modes of base pairing (essentially either 3–4 or 2–3), the sequence organization of the *trp* leader mRNA makes possible regulation of tran-

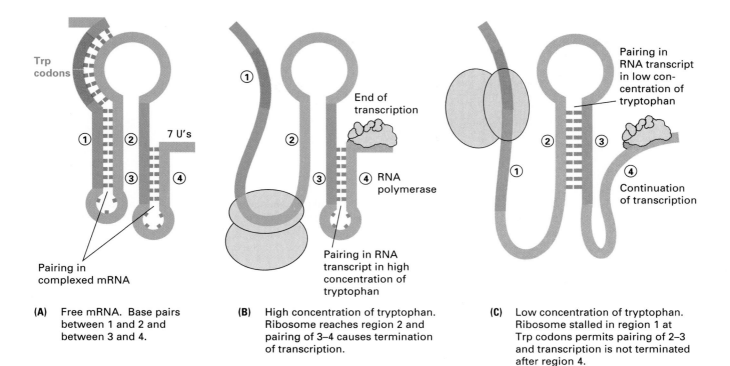

(A) Free mRNA. Base pairs between 1 and 2 and between 3 and 4.

(B) High concentration of tryptophan. Ribosome reaches region 2 and pairing of 3–4 causes termination of transcription.

(C) Low concentration of tryptophan. Ribosome stalled in region 1 at Trp codons permits pairing of 2–3 and transcription is not terminated after region 4.

Figure 11.13
The explanation for attenuation in the *E. coli trp* operon. The tryptophan codons in part A are those highlighted in red in Figure 11.11.

scription through translation of the leader polypeptide. The mechanism is shown in Figure 11.13. As the leader region is transcribed, translation of the leader polypeptide is initiated. Because there are two tryptophan codons in the coding sequence, the translation of the sequence is sensitive to the concentration of charged $tRNA^{Trp}$. If the supply of tryptophan is adequate for translation, the ribosome passes through the Trp codons and into region 2 (Figure 11.13B). Because the presence of a ribosome eliminates the possibility of base pairing in a region of about 10 bases on each side of the codons being translated, the presence of a ribosome in region 2 prevents its becoming paired with region 3. In this case, region 3 pairs with region 4 and forms the terminator shown in Figure 11.13B (and in detail in Figure 11.12B), and transcription is terminated at the run of uridines that follows region 4.

On the other hand, when the level of charged $tRNA^{Trp}$ is insufficient to support translation, the translation of the leader peptide is stalled at the tryptophan codons (Figure 11.13C). The stalling prevents the ribosome from proceeding into region 2, which is then free to pair with region 3. Pairing of regions 2 and 3 prevents formation of the terminator structure, so the complete *trp* mRNA molecule is made, including the coding sequences for the structural genes.

In summary, attenuation is a fine-tuning mechanism of regulation superimposed on the basic negative control of the *trp* operon:

When charged tryptophan tRNA is present in amounts that support translation of the leader polypeptide, transcription is terminated, and the *trp* enzymes are not synthesized. When the level of charged tryptophan

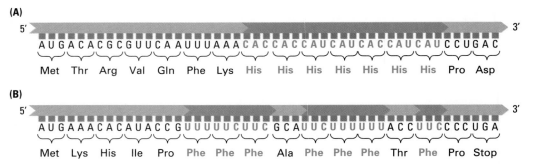

Figure 11.14
Amino acid sequence of the leader peptide and base sequence of the corresponding segment of mRNA from the histidine operon (A) and the phenylalanine operon (B). The repetition of these amino acids is emphasized in red letters.

tRNA is too low, transcription is not terminated, and the *trp* enzymes are made. At intermediate concentrations, the fraction of transcription initiation events that result in completion of *trp* mRNA depends on how frequently translation is stalled, which in turn depends on the intracellular concentration of charged tryptophan tRNA.

Many operons responsible for amino acid biosynthesis (for example, the leucine, isoleucine, phenylalanine, and histidine operons) are regulated by attenuators that function by forming alternative paired regions in the transcript. In the histidine operon, the coding region for the leader polypeptide contains seven adjacent histidine codons (Figure 11.14A). In the phenylalanine operon, the coding region for the leader polypeptide contains seven phenylalanine codons divided into three groups (Figure 11.14B). This pattern, in which codons for the amino acid produced by enzymes of the operon are present at high density in the leader peptide mRNA, is characteristic of operons in which attenuation is operative. Through these codons, the cell monitors the level of aminoacylated tRNA charged with the amino acid that is the end product of each amino acid biosynthetic pathway. Note that

Attenuation cannot take place in eukaryotes because transcription and translation are uncoupled; transcription takes place in the nucleus and translation in the cytoplasm.

Regulation of the *lac* and *trp* operons exemplifies some of the important mechanisms that control transcription of genes in prokaryotes. In the following section, we will see that similar mechanisms are used in the control of genes in bacteriophages.

11.4
Regulation in Bacteriophage λ

When Jacob and Monod proposed the operon model and negative regulation by repression, they suggested that the model could account not only for regulation in inducible and repressible operons for metabolic enzymes but also for the lysogenic cycle of temperate bacteriophages. They proposed that λ bacteriophage was kept quiescent and prevented from replicating within bacterial lysogens by a repressor. This explanation ultimately proved to be correct, although the biochemical route to achievement of the repressed, lysogenic state is more complicated than was initially thought.

When λ bacteriophage infects *E. coli*, each infected cell can undergo one of two possible outcomes: (1) a lytic infection, resulting in lysis and production of phage particles, or (2) a lysogenic infection, resulting in integration of the λ molecule into the *E. coli* chromosome and formation of a lysogen. Because of this dichotomy, λ normally produces turbid (not completely clear) plaques on a lawn of *E. coli*. The initial infection and lysis do produce a cleared region in the bacterial lawn, but a few lysogens grow within the cleared region, partially repopulating the cleared zone and producing a turbid plaque.

Mutations in regulatory genes in λ were first identified in phage mutants that give clear rather than turbid plaques. The mutants proved to fall into four classes: λ*vir*, *cI⁻*, *cII⁻*, and *cIII⁻*. The characteristics of these mutants are shown in Table 11.3. The genetic positions of the *cI*, *cII*, and *cIII* regions are shown in the simplified genetic map of λ bacteriophage in Figure 11.15, in which the genes are grouped by functional categories. Recall that, upon infection, the λ DNA molecule circularizes, bringing the *R* and *A* genes adjacent to one another.

Among the mutants in Table 11.3, the "clear" mutants proved to be analogous to *lacI* and *lacO* mutants in *E. coli*. The λ*vir* mutant is dominant to the wildtype λ⁺ in mixed infection, as indicated by the combination of infecting phage designated 1 in Table 11.3. This combination of phage carries out a productive infection and prevents lysogeny by the wildtype λ⁺. The λ*vir* mutant is therefore analogous to the *lacOᶜ* mutation. However, λ*vir* proves to be a double mutant, bearing mutations in two different operators, O_L and O_R, as depicted in Figure 11.16. The *cI⁻* mutations are recessive, as can be seen at entry 2 in Table 11.3. These *cI⁻* mutations are analogous to *lacI⁻* mutations in that the *cI⁺* gene encodes the λ repressor, which is diffusible.

Table 11.3 Characteristics of mixed infections containing several combinations of λ*vir*, *cI⁻* *cII⁻*, and *cIII⁻* mutants

Infecting phages	Clear or turbid plaques
1. λ*vir* + λ⁺	Clear
2. *cI⁻* + *cI⁺*	Turbid
3. *cII⁻* + *cII⁺*	Turbid
4. *cIII⁻* + *cIII⁺*	Turbid

The *cII⁺* and *cIII⁺* genes encode not for repressor but rather for proteins needed in establishing lysogeny. The *cII⁻* and *cIII⁻* mutations are also recessive in mixed infections (entries 3 and 4 in Table 11.3).

The molecular basis on which the decision between the lysogenic and the lytic cycle is determined is summarized in Figure 11.16. Upon infection of *E. coli* by λ, the λ molecule circularizes, and RNA polymerase binds at P_L and P_R and initiates transcription of the *N* and *cro* genes. N protein acts to prevent termination of the transcripts from P_L and P_R, allowing production of cII protein; cII protein activates transcription at P_E and P_I, thus allowing production of the cI and int proteins. The cI protein shuts down further transcription from P_L and P_R and stimulates transcription at P_M, increasing its own synthesis. Lysogeny is achieved if the concentration of cI protein reaches levels high enough to prevent transcription from P_L and P_R and to allow int protein to catalyze site-specific recombination between the circular λ molecule and the *E. coli* chromosome at their respective attachment (*att*) sites.

Figure 11.15

Genetic map of λ bacteriophage. The map is drawn to emphasize the functional organization of genes within the phage genome and to draw attention to the regulatory features. For a more detailed map, see Figure 8.24.

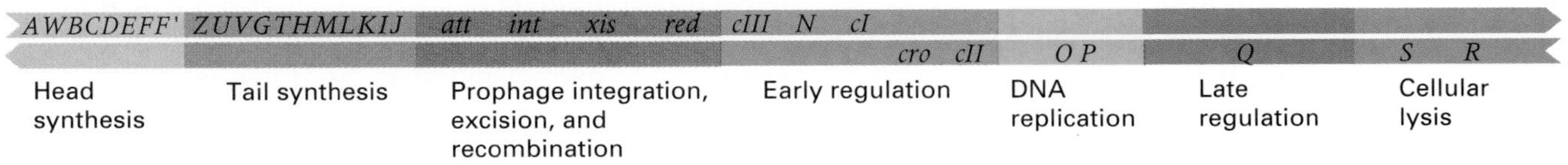

A W B C D E F F'	*Z U V G T H M L K I J*	*att* *int* *xis* *red*	*cIII* *N* *cI*				
				cro *cII*	*O P*	*Q*	*S R*
Head synthesis	Tail synthesis	Prophage integration, excision, and recombination	Early regulation	DNA replication	Late regulation	Cellular lysis	

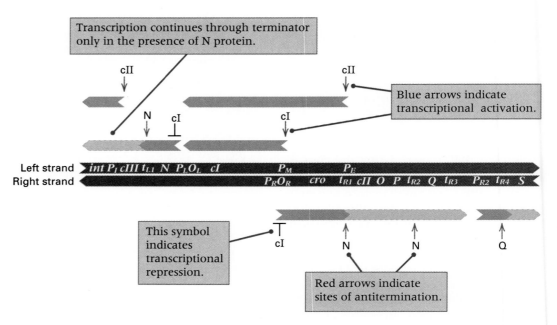

Figure 11.16

Genetic and transcriptional map of the control region of bacteriophage λ as expressed in the early stages of a lysogenic infection. The green arrows show the origin, direction, and extent of transcription. Light green arrows indicate portions of transcripts that are synthesized as a result of antitermination activity of the N or Q proteins. The sites of antitermination activity are indicated with red arrows pointing to the interior of a transcript. Blue arrows pointing to the origin of a transcript indicate transcriptional activation by cI or cII proteins; the sites of transcriptional repression by cI protein are indicated.

The alternative pathway to lysogeny, which leads to lytic development, takes place when the cro protein dominates. The cro protein also can bind to O_R and, in doing so, blocks transcription from P_M. If this occurs, then the concentration of repressor cannot rise to the levels required to block transcription from P_L and P_R. Transcription will continue from P_R and P_{R2}, and N and Q proteins will prevent termination in the rightward (and also leftward) transcripts. Because the λ DNA molecule is in a circular configuration, rightward transcription moves through genes S and R and thence through the head and tail genes (A through J, see Figures 11.15 and 11.16). The production of proteins needed for cellular lysis and formation of phage particles ensues, followed by phage assembly and cellular lysis to release phage.

In λ regulation, the cro and cI proteins compete for binding to O_L and O_R (each operator has three subsites that participate in the competition, but for our purposes this level of detail is unnecessary). If cro wins, the lytic cycle results in that cell; if cI wins, a lysogen is formed. Hence

> The cI and cro proteins function as a genetic switch; cI turns on lysogeny and cro turns on the lytic cycle.

The details that determine whether cI or cro controls the fate of a particular infection are quite complex. This complexity is apparent in Figure 11.17, where the major regulatory components and their interactions are shown in a form analogous to a wiring diagram used by electrical engineers. The key promoters are indicated in yellow, the proteins are encircled, and the interactions are shown in red. The multiple feedback loops and interactions are evident. The lesson from Figure 11.17 is that over the course of hundreds of millions of years, the regulation of even apparently

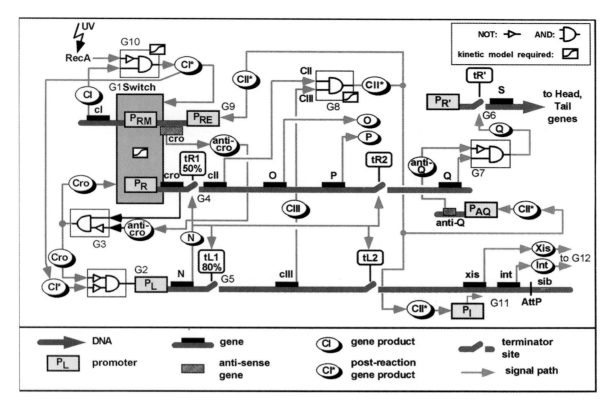

Figure 11.17
Genetic "circuit" determining the phage λ lysis-versus-lysogeny decision. The λ early promoters are shown in yellow, major protein components in circles, and the various regulatory interactions as red arrows. (Proteins CI, CII, and CIII are called cI, cII, and cIII in Figure 11.16.) The orientation of the operons is for convenience in representation and does not reflect their orientation in the genome. [Courtesy of Lucy Shapiro. From H. H. McAdams and L. Shapiro. 1995. *Science* 269: 650.]

"simple" systems such as lysogeny in phage λ may evolve great complexity, ultimately comprising layer upon layer of checks and balances.

11.5
Regulation in Eukaryotes

Eukaryotic cells and organisms have different needs for regulation than prokaryotic cells. At the cellular level, eukaryotic cells are compartmentalized and can sequester and mobilize small molecules intracellularly, which can serve to damp environmental change. At the organismal level, multicellular eukaryotes have elaborate developmental programs and numerous specialized cell types. Within the organism, the environment of the cells may not change drastically in time. During development of the organism, cells differentiate for the following reasons:

- As a result of sequential changes in gene activity that are programmed in the genome

- In response to molecular signals released by other cells

- In response to physical contact with other cells

- In response to changes in the external environment

After cells have differentiated, they remain genetically quite stable, producing particular substances either at a constant rate or in response to external stimuli such as hormones, nutrient concentrations, or temperature changes.

The great complexity of multicellular eukaryotes requires a wide variety of genetic regulatory mechanisms. On the whole, these mechanisms are not understood as thoroughly as are those in prokaryotes. However, many important examples of different types of mechanisms have been studied in animals as diverse as mammals (especially the mouse), birds (usually the chicken), amphibians (toads of the genus *Xenopus*), insects (*Drosophila*), nematode worms (*Caenorhabditis elegans*), echinoderms (the sea urchin), and ciliates (*Tetrahymen*a), as well as in yeast and other fungi. These examples reveal the general features of eukaryotic gene regulation that are discussed in the following sections.

Differences in Genetic Organization of Prokaryotes and Eukaryotes

Numerous differences exist between prokaryotes and eukaryotes with regard to transcription and translation, and in the spatial organization of DNA, as described in Chapters 6 and 10. Here are some of those most relevant to regulation:

1. In a eukaryote, usually only a single type of polypeptide chain can be translated from a completed mRNA molecule. Thus polycistronic mRNA of the type seen in prokaryotes is not found in eukaryotes.

2. The DNA of eukaryotes is bound to histones, forming chromatin, and to numerous nonhistone proteins. Only a small fraction of the DNA is bare. In bacteria, some proteins are present in the folded chromosome, but most of the DNA is free.

3. A significant fraction of the DNA of eukaryotes consists of moderately or highly repetitive nucleotide sequences. Some of the repetitive sequences are repeated in tandem copies, but others are not. Bacteria contain little repetitive DNA other than duplicated rRNA (ribosomal RNA) and tRNA genes and a few transposable elements.

4. A large fraction of eukaryotic DNA is untranslated; most of the nucleotide sequences do not code for proteins. Unicellular eukaryotes, such as yeast, are exceptions to this generalization, as are "lower" multicellular eukaryotes, such as *Drosophila* and *C. elegans*, and even certain vertebrates, such as the pufferfish, *Fugu rubripes*, with its relatively small (for a vertebrate) genome of 400 Mb.

5. Some eukaryotic genes are expressed and regulated by the use of mechanisms for rearranging certain DNA segments in a controlled way and for increasing the number of specific genes when needed.

6. Genes in eukaryotes are split into exons and introns, and the introns must be removed in the processing of the RNA transcript before translation begins.

7. In eukaryotes, mRNA is synthesized in the nucleus and must be transported through the nuclear envelope to the cytoplasm, where it is utilized. Bacterial cells do not have a nucleus separated from the cytoplasm.

We shall see in the following sections how some of these features are incorporated into particular modes of regulation.

11.6 Alteration of DNA

Some genes in eukaryotes are regulated by alteration of the DNA. For example, certain sequences may be amplified or rearranged in the genome, or the bases may be chemically modified. Some of the alterations are reversible, but others permanently change the genome of the cells. However, the permanent changes take place only in somatic cells, so they are not genetically transmitted to the offspring through the germ line.

Gene Dosage and Gene Amplification

Some gene products are required in much larger quantities than others. One means of maintaining particular ratios of certain gene products (other than by differences in transcription and translation efficiency, as discussed earlier) is by **gene dosage.** For example, if two genes, *A* and *B*, are transcribed at the same rate and the translation efficiencies are the same, then 20 times as much of product A can be made as of product B if there are 20 copies of gene *A* per copy of gene *B*. The histone genes exemplify a gene-dosage effect: To synthesize the huge amount of histone required to form chromatin, most cells contain hundreds of times as many copies of histone genes as of genes required for DNA replication. In this case, the high expression is automatic because the repeated genes are part of the normal chromosome complement.

In some cases, gene dosage is increased temporarily by a process called **gene amplification,** in which the number of genes increases in response to some signal. An example of gene amplification is found in the development of the oocytes of the toad *Xenopus laevis*. The formation of an egg from its precursor, the oocyte, is a complex process that requires a huge amount of protein synthesis. To achieve the necessary rate, a very large number of ribosomes are needed. Ribosomes contain molecules of rRNA, and the number of rRNA genes in the genome is insufficient to produce the required number of ribosomes for the oocyte in a reasonable period of time. In the development of the oocyte, the number of rRNA genes increases by about 4000-fold. The precursor to the oocyte, like all somatic cells of the toad, contains about 600 rRNA-gene (rDNA) units; after amplification, about 2×10^6 copies of each unit are present. This large amount enables the oocyte to synthesize 10^{12} ribosomes, which are required for the protein synthesis that occurs later during early development of the embryo, at a time when no ribosomes are being formed.

Before amplification, the 600 rDNA units are arranged in tandem. During amplification, which occurs over a 3-week period in which the oocyte develops from a precursor cell, the rDNA no longer consists of a single contiguous DNA segment containing 600 rDNA units but instead forms a large number of small circles and replicating rolling circles. The rolling-circle replication accounts for the increase in the number of copies of the genes. The precise mechanism of excision of the circles from the chromosome and formation of the rolling circles is not known.

When the oocyte is mature, no more rRNA needs to be synthesized until well after fertilization and into early development, at which time 600 copies are sufficient. The excess rDNA serves no purpose and is slowly degraded by intracellular enzymes. Following fertilization, the chromosomal DNA replicates and mitosis ensues, occurring repeatedly as the embryo develops. During this period, the extra chromosomal rDNA does not replicate; degradation continues, and by the time several hundred cells have formed, none of this extra rDNA remains. Amplification of rRNA genes during oogenesis occurs in many organisms, including insects, amphibians, and fish.

Some protein-coding genes also undergo amplification. For example, in *Drosophila* females, the genes that produce chorion proteins (a component of the sac that encloses the egg) are amplified in follicle cells just before maturation of the egg. The amplification enables the cells to produce a large amount of protein in a short time. In some cases, amplification occurs in abnormal regulation. For example, a gene called N-*myc* is frequently amplified in human tumor cells in the disease neuroblastoma, and the degree of amplification is correlated with progress of the disease and tendency of the tumor to spread. N-*myc* is the normal cellular counterpart of a viral oncogene. (Oncogenes are discussed in Section 7.8).

Programmed DNA Rearrangements

Rearrangement of DNA sequences in the genome is an unusual but important mechanism by which some genes are regulated. An example is the phenomenon known as **mating-type interconversion** in yeast. As we saw in Chapter 4, yeast has two mating types, denoted **a** and α. Mating between haploid **a** and haploid α cells produces the **a**α diploid, which can undergo meiosis to produce four-spored asci that contain haploid **a** and α spores in the ratio 2 : 2. If a single yeast spore of either the **a** or the α genotype is cultured in isolation from other spores, then mating between progeny cells would not be expected because the progeny cells would have the mating type of the original parent. However, *S. cerevisiae* has a mating system called **homothallism,** in which some cells undergo a conversion into the opposite mating type that allows matings between cells in what would otherwise be a pure culture of one mating type or the other.

The outlines of mating-type interconversion are shown in Figure 11.18. An original haploid spore (in this example, α) undergoes germination to produce two progeny cells. Both the mother cell (the original parent) and the daughter cell have mating-type α, as expected from a normal mitotic division. However, in the next cell division, a switching (interconversion) of mating type takes place in both the mother cell and its *new* progeny cell, in which the original α mating type is replaced with the **a** mating type. After this second cell division is complete, the α and **a** cells are able

to undergo mating because they now are of opposite mating types. Fusion of the nuclei produces the **a**α diploid, which undergoes mitotic divisions and later sporulation to again produce **a** and α haploid spores.

The genetic basis of mating-type interconversion is DNA rearrangement as outlined in Figure 11.19. The gene that controls mating type is the *MAT* gene in chromosome III, which can have either of two allelic forms, **a** or α. If the allele in a haploid cell is *MAT***a,** then the cell has mating-type **a;** if the allele is *MAT*α, then the cell has mating-type α. However, both genotypes normally contain both **a** and α genetic information in the form of unexpressed **cassettes** present in the same chromosome. The *HML*α cassette contains the α DNA sequence about 200 kb away from the *MAT* gene, and the *HMR***a** cassette contains the **a** DNA sequence about 150 kb away from *MAT* on the other side. (Figure 11.19 shows the relative positions of the genes in the chromosome.) When mating-type interconversion occurs, a specific endonuclease, encoded by the *HO* gene elsewhere in the genome, is produced and cuts both strands of the DNA in the *MAT* region. The double-stranded break initiates a process in which genetic information in the unexpressed cassette that contains the opposite mating type becomes inserted into *MAT*. In this process, the DNA sequence in the donor cassette is duplicated, so the mating type becomes converted, but the same genetic information is retained in unexpressed form in the cassette. The terminal regions of *HML*, *MAT*, and *HMR* are identical (illustrated in light blue and dark blue in Figure 11.19), and these regions are critical in making possible recognition of the regions for interconversion. The unique part of the α region is 747 base pairs in length; that of the **a** region is 642 base pairs long. The molecular details of the conversion process are similar to those of the double-strand gap mechanism of recombination, which is discussed in Chapter 13.

Figure 11.19 illustrates two sequential mating-type interconversions. In the first, an α cell (containing the *MAT*α allele) undergoes conversion into **a,** using the DNA sequence contained in the *HMR***a** cassette. The converted cell has the genotype

Figure 11.18

Mating-type switching in the yeast *Saccharomyces cerevisiae*. Germination of a spore (in this example, one of mating type α) forms a mother cell and a bud that grows into a daughter cell. In the next division, the mother cell and its new daughter cell switch to the opposite mating type (in this case, **a**). The result is two α and two **a** cells. Cells of opposite mating type can fuse to form **a**α diploid zygotes. In a similar fashion, germination of an **a** spore is accompanied by switching to the α mating type.

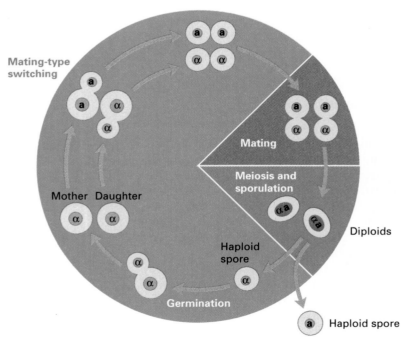

*MAT***a.** In a later generation, a descendant **a** cell may become converted into mating-type α, using the unexpressed DNA sequence contained in *HML*α. This cell has the genotype *MAT*α. Mating-type switches can occur repeatedly in the lineage of any particular cell.

Antibodies and Antibody Variability

Another important example of programmed DNA rearrangement takes place in vertebrates in cells that form the immune system. In this case, the precursor cells contain numerous DNA sequences that can serve as alternatives for various regions in the final gene. In the maturation of each cell, a combination of the alternatives is created by DNA cutting and rejoining, producing a great variety of possible genes that enable the immune system to recognize and attack most bacteria and viruses.

It has been estimated that a normal mammal is capable of producing more than 10^8 different antibodies, each of which can combine specifically with a particular antigen. Antibodies are proteins, and each unique antibody has a different amino acid sequence. If antibody genes were conventional in the sense that each gene codes for a single polypeptide, then mammals would need more than 10^8 genes for the production of antibodies. This is considerably more genes than are present in the entire genome. In fact, mammals use only a few hundred genes for antibody production, and the huge number of different antibodies derives from remarkable events that take place in the DNA of certain somatic cells. These events are discussed in this section.

Although an individual organism is capable of producing a vast number of different antibodies, only a fraction of them are synthesized at any one time. Antibodies are produced by a type of white blood cell called a B cell. Each B cell can produce a single type of antibody, but the antibody is not secreted until the cell has been stimulated by the appropriate antigen. Once stimulated, the B cell undergoes successive mitoses and eventually produces a clone of identical cells that secrete the antibody. Moreover, antibody secretion may continue even if the antigen is no longer pres-

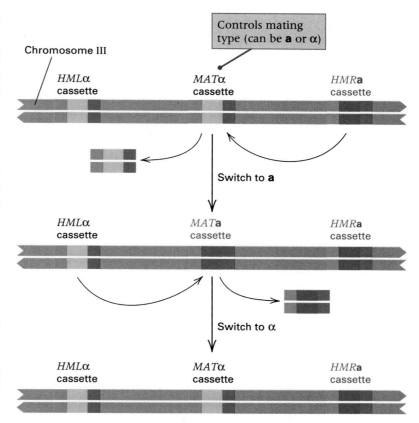

Figure 11.19
Genetic basis of mating-type interconversion. The mating type is determined by the DNA sequence present at the *MAT* locus. The *HML* and *HMR* loci are cassettes that contain unexpressed mating-type genes, either α or **a.** In the interconversion from α to **a,** the α genetic information present at *MAT* is replaced with the **a** genetic information from *HMR***a.** In the switch from **a** to α, the **a** genetic information at *MAT* is replaced with the α genetic information from *HML*α.

ent. In this manner, organisms produce antibodies only to the antigens to which they have been exposed.

The five distinct classes of antibodies known are designated IgG, IgM, IgA, IgD, and IgE (Ig stands for **immunoglobulin**). These classes serve specialized functions in the immune response and exhibit certain structural differences. However, each contains two types of polypeptide chains differing in size: a large one called the **heavy (H) chain** and a small one called the **light (L) chain.**

Immunoglobulin G (IgG) is the most abundant class of antibodies and has the simplest molecular structure. Its molecular organization is illustrated in Figure 11.20.

Sex-Change Operations

James B. Hicks, Jeffrey N. Strathern, and Ira Herskowitz, University of Oregon, Eugene, Oregon 1977

The Cassette Model of Mating-type Interconversion

*Mating in yeast requires cells of opposite mating types, **a** and α, to come together and fuse. Both **a** and α cells release signaling substances into the medium that prepare the opposite cell type for mating. In the **a**α diploid cell, genes specific for the diploid phase of the life cycle are expressed, and those specific for the haploid phase of the life cycle are turned off. Remarkably, yeast cells can change their mating type. In homothallic cells, the switch can take place in every generation. In heterothallic cells, it takes place at a frequency of 10^{-6}. This paper proposed a very bold hypothesis, later confirmed experimentally, that all yeast cells contain, in addition to the information for the expressed mating type at the* MAT *locus, both* α *and* **a** *genetic information in unexpressed cassettes at* HML *and* HMR *genetically linked to, but distinct from,* MAT. *The* HO *gene that distinguishes homothallic from heterothallic yeast codes for a site-specific endonuclease that cleaves within the mating-type locus and initiates the information-transfer process from either* HMLα *or* HMR**a**. *This is the physical basis of mating-type interconversion.*

Studies of mating-type interconversion in the yeast *Saccharomyces cerevisiae* have led us to propose a new mechanism of gene control involving mobile genes. . . . The mating-type locus of *S. cerevisiae* exists in two states, **a** or α, which control the ability of yeast cells to mate and sporulate. . . . It is clear that the **a** and α alleles are distinct entities, as they are codominant—an **a**/α diploid differs from a [rarely formed] **a**/**a** or α/α diploid. . . . The **a** allele

---

> **The mating-type locus is viewed as analogous to a playback head of a tape recorder which can give expression to whatever cassette of information is plugged into it.**

---

thus is not simply the absence of α, and the α allele is not simply the absence of **a**. . . . In homothallic strains, changes to opposite mating types occur frequently, as often as every generation. These strains carry a dominant nuclear gene (*HO*), unlinked to the mating-type locus. . . . *HO* cells that have sustained a change in mating type are fully capable of continuing to change mating type. However, when the *HO* gene is removed by genetic crosses, the new mating type is stable. . . . Cells with a defect at

the α mating-type locus can be converted to functional **a** cells. However, these **a** cells are then observed to switch to become functional α cells. In other words, a functional α mating-type locus can be restored through the mating-type interconversion process. We explain this recovery by proposing that yeast cells contain an additional copy (or copies) of the mating-type locus information. Specifically, we propose that yeast cells contain a silent (unexpressed) copy of **a** information and a silent copy of α information and that the *HO* gene activates this information by inserting it (or a copy) into the mating-type locus. Genetic studies have revealed the existence of two loci [*HML*α and *HMR***a**] in addition to *HO* which are necessary for mating-type interconversion and which we propose are the silent α and **a** information. . . . To summarize, we propose that cell type in *S. cerevisiae* is regulated by a locus [*MAT*] into which various blocs of information can be inserted. The mating-type locus is viewed as analogous to a playback head of a tape recorder which can give expression to whatever cassette of information is plugged into it.

Source: DNA Insertion Elements, Plasmids, and Episomes, eds. Ahmad I. Bukhari, James A. Shapiro, and Sankar L. Adhya. Cold Spring Harbor, New York: Cold Spring Harbor Laboratory, pp. 457–462.

An IgG molecule consists of two heavy and two light chains held together by disulfide bridges (two joined sulfur atoms) and has the overall shape of the letter Y. The sites on the antibody that carry its specificity and combine with the antigen are located in the upper half of the arms above the fork of the Y. Each IgG molecule with a different antigen specificity has a different amino acid sequence for the heavy and light chains in this part of the molecule. These specificity regions are called the **variable regions** (blue pointers in Figure 11.20) of the heavy and light chains. The remaining regions of the polypeptide are the **con-stant regions,** which are called constant because they have virtually the same amino acid sequence in all IgG molecules.

Initial understanding of the genetic mechanisms responsible for variability in the amino acid sequences of antibody polypeptide chains came from cloning a gene for the light chain of IgG. The critical observation was made by comparing the nucleotide sequence of the gene in embryonic cells or germ cells with that in mature antibody-producing cells. In the genome of a B cell that was actively producing the antibody, the DNA segments corresponding to the constant and variable regions of the light chain were found to be very close together, as expected of DNA that codes for different parts of the same polypeptide. However, in embryonic cells, these same DNA sequences were located far apart. Similar results were obtained for the variable and constant regions of the heavy chains: Segments encoding these regions were close together in B cells but widely separated in embryonic cells.

Extensive DNA sequencing of the genomic region that codes for antibody proteins revealed not only the reason for the different gene locations in B cells and germ cells but also the mechanism for the origin of antibody variability. Cells in the germ line contain a small number of genes corresponding to the constant region of the light chain, which are close together along the DNA. Separated from them, but on the same chromosome, is another cluster consisting of a much larger number of genes that correspond to the variable region of the light chains. In the differentiation of a B cell, one gene for the constant region is spliced (cut and joined) to one gene for the variable region, and this splicing produces a complete light-chain antibody gene. A similar splicing mechanism yields the constant and variable regions of the heavy chains.

The formation of a finished antibody gene is slightly more complicated than this description implies, because light-chain genes consist of three parts and heavy-chain genes consist of four parts. Gene splicing in the origin of a light chain is illustrated in Figure 11.21. For each of two parts of the variable region, the germ line contains multiple coding sequences called the **V (variable)** and **J (joining)** regions. In the differentiation of a B cell, a deletion makes possible the joining of one of the V regions with one of the J regions. The DNA joining process is called **combinatorial joining** because it can create many combinations of the V and J regions. When transcribed, this joined V-J sequence forms the 5' end of the light-chain RNA transcript. Transcription continues on through the DNA region coding for the constant (C) part of the gene. RNA splicing subsequently attaches the C region, creating the light-chain mRNA.

Combinatorial joining also takes place in the genes for the antibody heavy chains.

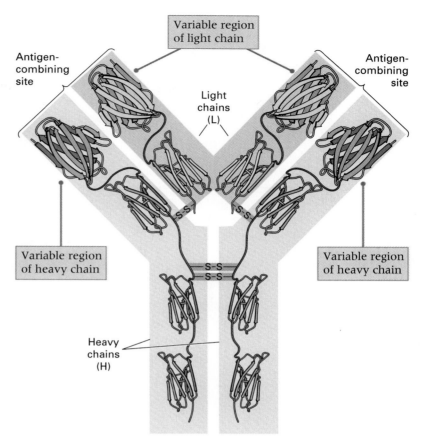

Figure 11.20
Structure of the immunoglobulin G (IgG) molecule showing the light chains (L, shaded blue) and heavy chains (H, shaded yellow). Variable and constant regions are indicated.

In this case, the DNA splicing joins the heavy-chain counterparts of V and J with a third set of sequences, called D (for diversity), located between the V and J clusters.

The amount of antibody variability that can be created by combinatorial joining is calculated as follows: In mice, the light chains are formed from combinations of about 250 V regions and 4 J regions, giving $250 \times 4 = 1000$ different chains. For the heavy chains, there are approximately 250 V, 10 D, and 4 J regions, producing $250 \times 10 \times 4 = 10,000$ combinations. Because any light chain can combine with any heavy chain, there are at least $1000 \times 10,000 = 10^7$ possible types of antibodies. The number of DNA sequences used for antibody production is quite small (about 500), but the number of possible antibodies is very large.

The value of 10^7 different antibody types is an underestimate, because there

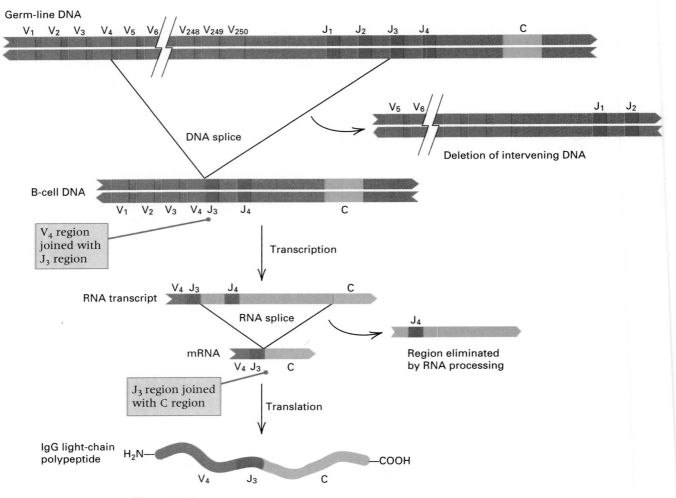

Figure 11.21

Formation of a gene for the light chain of an antibody molecule. One variable (V) region is joined with one randomly chosen J region by deletion of the intervening DNA. The remaining J regions are eliminated from the RNA transcript during RNA processing.

are two additional sources of antibody variability:

1. The junction for V-J (or V-D-J) splicing in combinatorial joining can be between different nucleotides of a particular V-J combination in light chains (or a particular V-D or D-J combination in heavy chains). The different splice junctions can generate different codons in the spliced gene. For example, a particular combination of V and J sequences can be spliced in five different ways. At the splice junction, the V sequence contains the nucleotides CATTTC, and the J sequence contains CTGGGTG. The splicing event determines the codons for amino acids 97,

98, and 99 in the completed antibody light chain, and it can occur in any of the following ways, the last two of which result in altered amino acid sequences:

Spliced sequence	97	98	99
CACTGGGTG	His	Trp	Val
CATTGGGTG	His	Trp	Val
CATTGGGTG	His	Trp	Val
CATTTGGTG	His	Leu	Val
CATTTCGTG	His	Phe	Val

In this manner, variability in the junction of V-J joining can result in polypeptides that differ in amino acid sequence.

2. The V regions are susceptible to a high rate of *somatic mutation,* which occurs in B-cell development. These mutations allow different B-cell clones to produce different polypeptide sequences, even if they have undergone exactly the same V-J joining. The mechanism for this high mutation rate is unknown.

Gene Splicing in the Origin of T-Cell Receptors

Immunity is also mediated by a different type of white blood cell called a T cell. A T cell carries an antigen receptor on its surface that combines with an antigen, stimulating the T cell to respond. Like the antibody molecules produced in B cells, the T-cell receptors are highly variable in amino acid sequence, enabling the T cells to respond to many antigens. Although the polypeptide chains in T-cell receptors are different from those in antibody molecules, they have a similar organization in that they are formed from the aggregation of two pairs of polypeptide chains. A particular T cell may carry either of two types of receptors. The majority carry the $\alpha\beta$ receptor, composed of polypeptide chains designated α and β, and the rest carry the $\gamma\delta$ receptor, composed of chains designated γ and δ. Each receptor polypeptide includes a variable region and a constant region. As their variability and similarity in organization suggest, the T-cell receptor genes are formed by somatic rearrangement of components analogous to those of the V, D, J, and C regions in the B cells. For example, in the mouse, the β chain of the T-cell receptor is spliced together from one each of approximately 20 V regions, 2 D regions, 12 J regions, and 2 C regions. Note that there are far fewer V regions for T-cell receptor genes than there are for antibody genes, yet T-cell receptors seem able to recognize just as many foreign antigens as B cells. The extra variation results from a higher rate of somatic mutation in the T-cell receptor genes.

DNA Methylation

In most eukaryotes, a small proportion of the cytosine bases are modified by the addition of a methyl (CH_3) group to the number-5 carbon atom (Figure 11.22). The cytosines are incorporated in their normal, unmodified form in the course of DNA replication, but they are modified later by an enzyme called a **DNA methylase.** Cytosines are modified preferentially in 5'-CG-3' dinucleotides. When a CG dinucleotide that is methylated in both strands undergoes DNA replication, the result is two daughter molecules, each of which contains one parental strand with a methylated CG and one daughter strand with an unmethylated CG. The DNA methylase recognizes the half-methylation in these molecules and methylates the cytosines in the daughter strands. Methylation of CG dinucleotides in the sequence CCGG can easily be detected by the use of the restriction enzymes *Msp*I and *Hpa*II. Both enzymes cleave the sequence CCGG. However, *Msp*I cleaves regardless of whether the interior C is methylated, whereas *Hpa*II cleaves only unmethylated DNA. Therefore, *Msp*I restriction sites that are not cleaved by *Hpa*II are sites at which the interior C is methylated (Figure 11.23).

Many eukaryotic genes have CG-rich regions upstream of the coding region, providing potential sites for methylation that may affect transcription. A number of observations suggest that high levels of methylation are associated with genes for which the rate of transcription is low. One example is the inactive X chromosome in mammalian cells, which is extensively methylated. Another example is the *Ac* transposable element in maize. Certain *Ac* elements lose activity of the transposase gene without any change in DNA sequence. These elements prove to have heavy methylation in a region particularly rich in the CG dinucleotides. Return to normal activity of the methylated *Ac* elements coincides with loss of methylation through the action of demethylating enzymes in the nucleus.

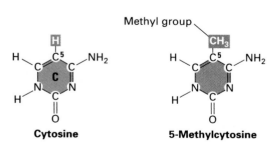

Figure 11.22
Structures of cytosine and 5-methylcytosine.

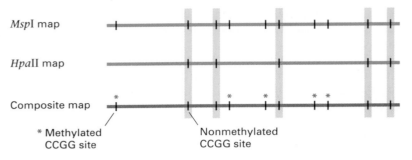

MspI map

HpaII map

Composite map

* Methylated CCGG site

Nonmethylated CCGG site

Figure 11.23
Detection of methylated cytosines in CCGG sequences by means of restriction enzymes. The enzyme *MspI* cleaves all CCGG sites regardless of methylation, whereas *HpaII* cleaves only nonmethylated sites. The positions of the methylated sites are determined by comparing the restriction maps.

Although there is a correlation between methylation and gene inactivity, it is possible that heavy methylation is a result of gene inactivity rather than a cause of it. However, treatment of cells with the cytosine analog *azacytidine* reverses methylation and can restore gene activity. For example, some clones of rat pituitary tumor cells express the gene for prolactin, whereas other related clones do not. The gene is methylated in the nonproducing cells but is not methylated in the producers. Reversal of methylation in the nonproducing cells with azacytidine results in prolactin expression. On the other hand, not all organisms exhibit methylation. For example, *Drosophila* DNA is not methylated. In organisms in which the DNA is methylated,

methylation increases susceptibility to certain kinds of mutations, which are discussed in Chapter 13.

11.7 Transcriptional Regulation in Eukaryotes

Many eukaryotic genes code for essential metabolic enzymes or cellular components and are expressed constitutively at relatively low levels in all cells; they are called **housekeeping genes.** The expression of other genes differs from one cell type to the next or among different stages of the cell cycle; these genes are often regulated at the level of transcription. In prokaryotes, the levels of expression in induced and uninduced cells may differ by a thousandfold or more. Such extreme levels of induction are uncommon in eukaryotes, except for some genes in lower eukaryotes such as yeast. Most eukaryotic genes are induced by factors ranging from 2 to 10.

In this section, we consider some components of transcriptional regulation in eukaryotes.

Galactose Metabolism in Yeast

We will introduce transcriptional regulation in eukaryotes by examining the control of galactose metabolism in yeast and comparing it with the *lac* operon in *E. coli*. The first steps in the biochemical pathway for galactose degradation are illustrated in Figure 11.24. Three enzymes, encoded by

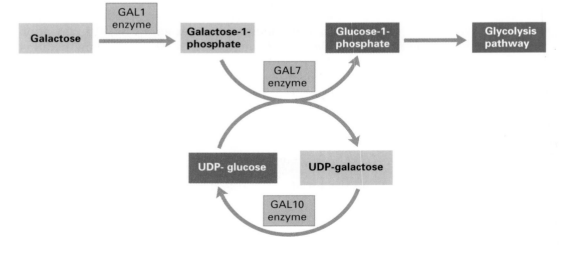

Figure 11.24
Metabolic pathway by which galactose is converted to glucose-1-phosphate in the yeast *Saccharomyces cerevisiae.*

Figure 11.25

The linked *GAL* genes of *Saccharomyces cerevisiae*. Arrows indicate the transcripts produced. The *GAL1* and *GAL10* transcripts come from divergent promoters, *GAL7* from its own promoter.

the genes *GAL1*, *GAL7*, and *GAL10*, are required for conversion of galactose to glucose-1-phosphate. These three genes are tightly linked, as shown in Figure 11.25. Despite the tight linkage of the three genes, the genes are not part of an operon; the mRNAs are monocistronic. The *GAL1* and *GAL10* mRNAs are synthesized from divergent promoters lying between the genes, and *GAL7* mRNA is synthesized from its own promoter. The mRNAs are synthesized only when galactose is present as inducer; the genes are thus inducible.

Constitutive and uninducible mutants have been observed. In two types of mutants, *gal80* and *GAL81*^{c}, the mutants synthesize *GAL1*, *GAL7* and *GAL10* mRNAs constitutively. Another type of mutant, *gal4*, does not synthesize the mRNAs whether or not galactose is present; it is uninducible. The characteristics of the mutants are shown in Table 11.4. The terms *cis* and *trans* are of no help in interpreting these results, for the regulatory genes are unlinked to the genes they regulate: *GAL1*, *GAL7* and *GAL10* are on chromosome II, *GAL80* is on chromosome XIII, and *GAL4* and *GAL81* are on chromosome XVI.

The *gal80* mutation is recessive (entry 1 of Table 11.4). Thus, superficially, it behaves like a *lacI⁻* mutation. The wildtype *GAL80* allele does indeed encode a protein, called a "repressor," that is a negative regulator of transcription. However, the GAL80 protein acts not by binding to an operator but by binding to, and inactivating, a transcriptional activator protein. The activator is the product of the *GAL4* gene.

The wildtype *GAL4* allele encodes a protein that is required for transcription of the three *GAL* genes. The *gal4* mutation is therefore recessive. In the absence of the GAL4 protein, the *GAL* genes are all unin-

ducible. The GAL4 protein is a positive regulatory protein that activates transcription of the three *GAL* genes. However, it does so by activating transcription of three different mRNAs starting at three different sites upstream of each of the activated genes. The GAL4 protein bound with its target site in the DNA is shown in Figure 11.26, in which the GAL4 protein (a dimer) is shown in blue and the DNA molecule in red. The small yellow spheres represent ions of zinc, which are essential components in the DNA binding.

The GAL80 "repressor" protein acts by binding to GAL4 protein and sequestering it so that GAL4 is not free to activate transcription. The inducer (galactose) eliminates the ability of GAL80 protein to bind to GAL4 freeing the GAL4 protein to activate transcription.

The constitutive mutation, *GAL81*^{c}, is dominant; it results in constitutive synthesis of all three mRNAs. However, because it does not map near *GAL1*, *GAL7*, and *GAL10* and is a single mutation, it cannot be an operator mutation comparable to *lacO*^{c}. The *GAL81*^{c} mutation does not define a separate regulatory gene but instead maps to a position within the *GAL4* gene. The mutation gives rise to a GAL4 protein that no longer binds to GAL80 protein. Hence the GAL4 protein produced by the *GAL81*^{c} allele cannot be sequestered by GAL80 and is able to activate transcription in the absence of galactose, whether or not wildtype GAL4 protein is also present.

The main point of these comparisons is that the superficial similarity between the constitutive and uninducible mutations in the *lac* operon of *E. coli* and the *GAL* genes of yeast are not indicative of similar molecular regulatory mechanisms. However, some physiological similarities remain in

Table 11.4 Characteristics of diploids containing various combinations of *gal80*, *gal4*, and *GAL81*^{c} mutations

Genotype	Synthesis of GAL1, GAL7, and GAL10 mRNAs	Gal phenotype
1. *gal80 GAL1 / GAL80 GAL 1*	Inducible	+
2. *gal4 GAL1 / GAL4 GAL 1*	Inducible	+
3. *GAL81*^{c} *GAL1 / GAL 81 GAL1*	Constitutive	+

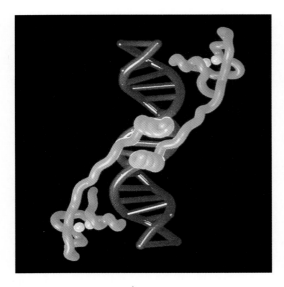

Figure 11.26

Three-dimensional structure of the GAL4 protein (blue) bound to DNA (red). The protein is composed of two polypeptide subunits held together by the coiled regions in the middle. The DNA-binding domains are at the extreme ends, and each physically contacts three base pairs in the major groove of the DNA. The zinc ions in the DNA-binding domains are shown in yellow. [Courtesy of Dr. Stephen C. Harrison. See also R. Marmorstein, M. Carey, M. Ptashne, and S. C. Harrison. 1992. DNA recognition by GAL4: Structure of a protein-DNA complex. *Nature* 356: 408–414.]

that, in both prokaryotes and eukaryotes, the genes for a particular metabolic (or developmental) pathway are expressed in a coordinated manner in response to a signal. The principle at work is that alternative molecular mechanisms can be employed to achieve similar ends.

Yeast Mating Type

As mentioned earlier, the mating type of a yeast cell is controlled by the allele of the *MAT* gene that is present (refer to Figure 11.19 for the genetic basis of mating-type interconversion in yeast). Both *MAT***a** (mating-type **a**) and *MAT*α (mating-type α) express a set of haploid-specific genes. They differ in that *MAT***a** expresses a set of **a**-specific genes and *MAT*α expresses a set of α-specific genes. The haploid-specific genes that cells of both mating types express include *HO*, which encodes the HO endonuclease used in mating-type inter-

conversion, and *RME1*, which encodes a repressor of meiosis-specific genes. The functions of the mating-type-specific genes include (1) secretion of a mating peptide that arrests cells of the opposite mating type before DNA synthesis and prepares them for cell fusion, and (2) production of a receptor for the mating peptide secreted by the opposite mating type. Therefore, when **a** and α cells are in proximity, they prepare each other for mating and undergo fusion.

Regulation of mating type is at the level of transcription according to the regulatory interactions diagrammed in Figure 11.27. These regulatory interactions were originally proposed on the basis of the phenotypes of various types of mutants, and most of the details have been confirmed by direct molecular studies. The symbols **a**sg, αsg, and *hsg* represent the **a**-specific genes, the α-specific genes, and the haploid-specific genes, respectively; each set of genes is represented as a single segment (lack of a "sunburst" indicates that transcription does not take place). In a cell of mating-type **a** (Figure 11.27A), the *MAT***a** region is transcribed and produces a polypeptide called **a**1. By itself, **a**1 has no regulatory activity, and in the absence of any regulatory signal, **a**sg and *hsg* are transcribed, but not αsg. In a cell of mating-type α (Figure 11.27B), the *MAT*α region is transcribed, and two regulatory proteins denoted α1 and α2 are produced: α1 is a *positive regulator* of the α-specific genes, and α2 is a *negative regulator* of the **a**-specific genes. The result is that αsg and *hsg* are transcribed, but transcription of **a**sg is turned off. Both α1 and α2 bind with particular DNA sequences upstream from the genes that they control.

In the diploid (Figure 11.27C), both *MAT***a** and *MAT*α are transcribed, but the only polypeptides produced are **a**1 and α2. The reason is that the **a**1 and α2 polypeptides combine to form a negative regulatory protein that represses transcription of the α1 gene in *MAT*α and of the haploid-specific genes. The α2 polypeptide acting alone is a negative regulatory protein that turns off **a**sg. Because α1 is not produced, transcription of αsg is not turned on. In sum, then, the αsg are not turned on because α1 is absent, the **a**sg are turned off because α2 is present, and the *hsg* are turned off by the α2/**a**1 complex. This ensures that meiosis

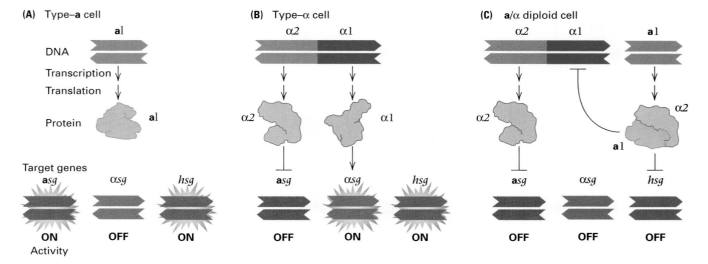

Figure 11.27

Regulation of mating type in yeast. The symbols **a**sg, αsg, and *hsg* denote sets of **a**-specific genes, α-specific genes, and haploid-specific genes, respectively. Sets of genes represented with a "sunburst" are *on,* and those unmarked are *off.* (A) In an **a** cell, the **a**1 peptide is inactive, and the sets of genes manifest their basal states of activity (**a**sg and *hsg on* and αsg *off*), so the cell is an **a** haploid. (B) In an α cell, the α2 peptide turns the **a**sg *off* and the α1 peptide turns the αsg *on,* so the cell is an α haploid. (C) In an **a**/α diploid, the α2 and **a**1 peptides form a complex that turns the *hsg off,* the α2 peptide turns the **a**sg *off,* and the αsg manifest their basal activity of *off,* so physiologically the cell is non-**a,** non-α, and nonhaploid (that is, it is a normal diploid).

can occur (because expression of *RME1* is turned off) and that mating type switching ceases (because the HO endonuclease is absent). Thus the homothallic **a**α diploid is stable and able to undergo meiosis. The result is that the **a**α diploid does not transcribe either the mating-type-specific set of genes or the haploid-specific genes.

The repression of transcription of the haploid-specific genes mediated by the **a**1/α2 protein is an example of negative control of the type already familiar from the *lac* and *trp* systems in *E. coli.* The interesting twist in the yeast example is that the α2 protein has a regulatory role of its own in repressing transcription of the **a**-specific genes. Why does the α2 protein, on its own, not repress the haploid-specific genes as well? The answer lies in the specificity of its DNA binding. By itself, the α2 protein has low affinity for the target sequences in the haploid-specific genes. However, the **a**1/α2 heterodimer has both high affinity and high specificity for the target DNA sequences in the haploid-specific genes. The three-dimensional structure of the **a**1/α2 protein in complex with target DNA

is shown in Figure 11.28. Upon binding, the **a**1/α2 complex produces a pronounced 60° bend in the DNA molecule, which may play a role in transcriptional repression.

Transcriptional regulation of the mating-type genes includes negative control (**a**-specific genes and haploid-specific genes) and positive control (α-specific genes). Although the regulation of transcription in eukaryotes is both positive and negative, positive regulation is more usual. The regulatory proteins required are the subject of the next section.

Transcriptional Activator Proteins

The α1 protein that functions in the activation of the α-specific genes is an example of a **transcriptional activator protein,** which must bind with an upstream DNA sequence in order to prepare a gene for transcription. We have already seen an example of another transcriptional activator in the case of the GAL4 protein (Figure 11.26). Some transcriptional activator proteins work by direct interaction with one or more proteins present in large complexes

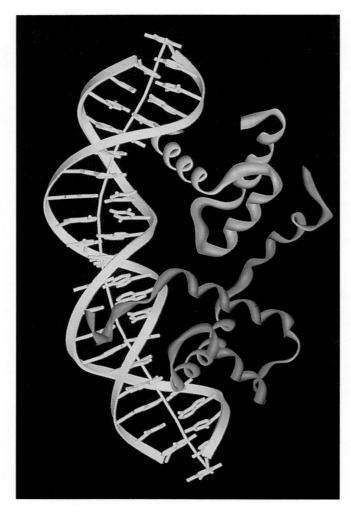

Figure 11.28
Structure of the **a**1/α2 protein bound with DNA. The **a**1 subunit is shown in blue, the α2 subunit in red. Contact with the DNA target results in a sharp bend in the DNA. [Courtesy of Cynthia Wolberger. From T. Li, M. R. Stark, A. D. Johnson, and C. Wolberger. 1995. *Science* 270: 262.]

of proteins needed for transcription, among them RNA polymerase II (PolII), and attract the transcription complexes to the promoter of the gene to be activated. Other transcriptional activator proteins may initiate transcription by an already assembled transcription complex. In either case, the activator proteins are essential for the transcription of genes that are positively regulated.

Many transcriptional activator proteins can be grouped into categories on the basis of characteristics that their amino acid sequences share. For example, one category has a **helix-turn-helix** motif, which consists of a sequence of amino acids forming a pair of α-helices separated by a bend; the helices are so situated that they can fit neatly into the groves of a double-stranded DNA molecule. The helix-turn-helix motif is the basis of the DNA-binding ability, although the sequence specificity of the binding results from other parts of the protein. The α2 protein that regulates yeast mating type has a helix-turn-helix motif, as do many other transcriptional activator proteins in both prokaryotes and eukaryotes.

A second large category of transcriptional activator proteins includes a DNA-binding motif that is called a **zinc finger** because the folded structure incorporates a zinc ion. An example is the GAL4 transcriptional activator protein in yeast. The protein functions as a dimer composed of two identical GAL4 polypeptides oriented with their zinc-binding domains at the extreme ends (Figure 11.26 shows the zinc ions in yellow). The DNA sequence recognized by the protein is a symmetrical sequence, 17 base pairs in length, which includes a CCG triplet at each end that makes direct contact with the zinc-containing domains.

A more detailed illustration of the DNA-binding domain of the GAL4 protein is shown in Figure 11.29. Each of two zinc ions (Zn^{2+}) is chelated by bonds with four cysteine residues in characteristic positions at the base of a loop that extends for an additional 841 amino acids beyond those shown. The amino acids marked by a red asterisk are the sites of mutations that result in mutant proteins unable to activate transcription. Replacements at amino acid positions 15 (Arg → Gln), 26 (Pro → Ser), and 57 (Val → Met) are particularly interesting because they provide genetic evidence that zinc is necessary for DNA binding. In particular, the mutant phenotypes can be rescued by extra zinc in the growth medium because the molecular defect reduces the ability of the zinc finger part of the molecule to chelate zinc; extra zinc in the medium overcomes the defect and restores the ability of the mutant activator protein to attach to its particular binding sites in the DNA.

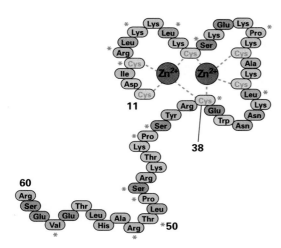

Figure 11.29
DNA-binding domain present in the GAL4 transcriptional activator protein in yeast. The four cysteine residues bind a zinc ion and form a peptide loop called a zinc finger. The zinc finger is a common motif found in DNA-binding proteins. The amino acids marked by a red asterisk have been identified by mutations as sites at which amino acid replacements can abolish the DNA-binding activity of the protein. The result is that the target genes cannot be activated.

Hormonal Regulation

Among the known regulators of transcription in higher eukaryotes, the **hormones**—small molecules or polypeptides that are carried from hormone-producing cells to target cells—have perhaps been studied in most detail. One class of hormones consists of small molecules synthesized from cholesterol; these steroid hormones include the principal sex hormones. Many of the steroid hormones act by turning on the transcription of specific sets of genes. If a hormone regulates transcription, then it must somehow signal the DNA. The signaling mechanism for the steroid hormone cortisol is outlined in Figure 11.30. The hormone penetrates a target cell through diffusion, because steroids are hydrophobic (nonpolar) molecules that pass freely through the cell membrane into the cytoplasm. There it encounters a receptor molecule that is complexed with another protein called Hsp82, which functions to mask the receptor. Once cortisol binds to the receptor, it liberates the receptor from Hsp82, and the hormone-receptor complex migrates to the nucleus where it binds to its DNA target sequences to activate transcription. Nontarget cells do not contain the receptors and so are unaffected by the hormone.

A well-studied example of induction of transcription by a hormone is stimulation of the synthesis of ovalbumin in the chicken oviduct by the steroid sex hormone **estrogen.** When hens are injected with estrogen, oviduct tissue responds by synthesizing ovalbumin mRNA. This synthesis continues as long as estrogen is administered. When the hormone is withdrawn,

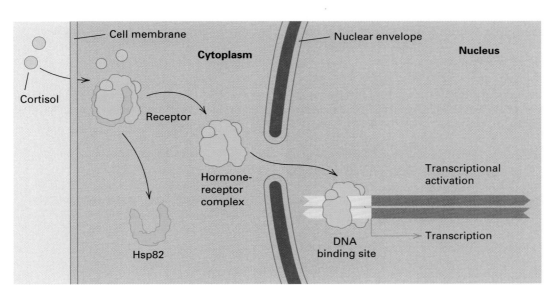

Figure 11.30
A schematic diagram showing how a steroid hormone reaches a DNA molecule and triggers transcription by binding with a receptor to form a transcriptional activator. Entry into the cytoplasm is by passive diffusion.

the rate of synthesis decreases. Before injection of the hormone, and 60 hours after the injections have stopped, no ovalbumin mRNA is detectable. When estrogen is given to hens, only the oviduct synthesizes mRNA because other tissues lack the hormone receptor.

Transcriptional Enhancers

Hormone receptors and other transcriptional activator proteins bind with particular DNA sequences known as **enhancers.** Enhancer sequences are typically rather short (usually fewer than 20 base pairs) and are found at a variety of locations around the gene affected. Most enhancers are upstream of the transcriptional start site (sometimes many kilobases away), others are in introns within the coding region, and a few are even located at the 3' end of genes. One of the most thoroughly studied enhancers is in the mouse mammary tumor virus and determines transcriptional activation by the glucocorticoid steroid hormone. The consensus sequence of the enhancer is AGAQCAGQ, in which Q stands for either A or T. The virus contains five copies of the enhancer positioned throughout its genome (Figure 11.31), providing five binding sites for the hormone-receptor complex that activates transcription.

Enhancers are essential components of gene organization in eukaryotes because they enable genes to be transcribed only when proper transcriptional activators are present. Some enhancers respond to molecules outside the cell—for example, steroid hormones that form receptor-hormone complexes. Other enhancers respond to

molecules that are produced inside the cell (for example, during development), and these enhancers enable the genes under their control to participate in cellular differentiation or to be expressed in a tissue-specific manner. Many genes are under the control of several different enhancers, so they can respond to a variety of different molecular signals, both external and internal.

Figure 11.32 illustrates several of the genetic elements found in a typical eukaryotic gene. The transcriptional complex binds to the promoter (*P*) to initiate RNA synthesis. The coding regions of the gene (exons) are interrupted by one or more intervening sequences (introns) that are eliminated in RNA processing. Transcription is regulated by means of enhancer elements (numbered 1 through 6) that respond to different molecules that serve as induction signals. The enhancers are located both upstream and downstream of the promoter, and some (in this example, enhancer 1) are present in multiple copies.

Many enhancers stimulate transcription by means of **DNA looping,** which refers to physical interactions between relatively distant regions along the DNA. The mechanism is illustrated in Figure 11.33. The factors necessary for transcription include a transcriptional activator protein that interacts with at least one protein subunit present in one or more large, multisubunit protein complexes. The protein factors in these complexes are known as **general transcription factors** because they are associated with the transcription of many different genes. The general transcription factors in eukaryotes have been highly con-

Figure 11.31

Positions of enhancers (orange) in the mouse mammary tumor virus that allow transcription of the viral sequence to be induced by glucocorticoid steroid hormone. LTR stands for long terminal repeat, a DNA sequence found at both ends of the virus.

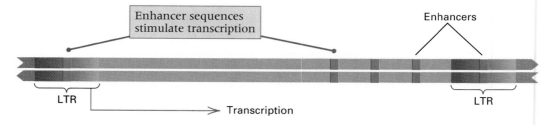

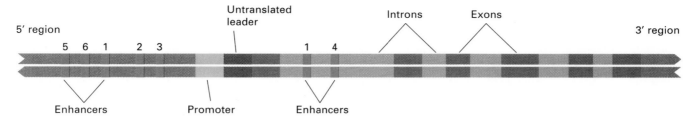

Figure 11.32
Schematic diagram of the organization of a typical gene in a higher eukaryote. Scattered throughout the sequence, but tending to be concentrated near the 5' region, are a number of different enhancer sequences. The enhancers are binding sites for transcriptional activator proteins that make possible temporal and tissue-specific regulation of the gene.

served in evolution. One of the complexes is TFIID, which includes a **TATA-box-binding protein (TBP)** that binds with the promoter in the region of the TATA box. In addition to TBP, the TFIID complex may also include a number of other proteins, called **TBP-associated factors (TAFs),** that act as intermediaries through which the effects of the transcriptional activator are transmitted. (Not all of the TBP is found in association with TAFs.) Transcription also requires an **RNA polymerase holoenzyme,** which consists of PolII (itself composed of 12 protein subunits) combined with at least 9 other protein subunits. In yeast these subunits include the transcription factors TFIIB, TFIIF, and TFIIH, as well as other proteins. Other general transcription factors have also been identified (for example, TFIIA and TFIIE), but it is not known whether these are components of larger complexes or whether they join the transcriptional apparatus as it is being assembled at the promoter.

Illustrated in Figure 11.33 is a mechanism of transcriptional activation called activation by **recruitment.** The key players, shown in Figure 11.33A, are the transcriptional activator protein and the TFIID and RNA polymerase holoenzyme complexes. The actual structures of TFIID and the holoenzyme complexes are not known, but for concreteness they are shown as multisubunit aggregates. To activate transcription (Figure 11.33B), the transcriptional activator protein binds to an enhancer in the DNA and to one of the TAF subunits in the TFIID complex. This inter-

action attracts ("recruits") the TFIID complex to the region of the promoter (Figure 11.33C). Attraction of the TFIID to the promoter also recruits the holoenzyme (Figure 11.33D) as well as any remaining general transcription factors, and in this manner the transcriptional complex is assembled for transcription to begin.

Experimental evidence for transcriptional activation by recruitment has come from studies of a number of artificial proteins created by fusing a DNA binding domain with one of the protein subunits in TFIID. Such fusion proteins act as transcriptional activators wherever they bind to DNA (provided that a promoter is nearby), because the TFIID is "tethered" to the DNA binding domain and so the "recruitment" of TFIID is automatic. Similarly, fusion proteins that are tethered to a subunit of the holoenzyme can recruit the holoenzyme to the promoter. In this case, TFIID and the remaining general transcription factors are also attracted to the promoter, and the transcriptional complex is assembled. These experiments suggest that a transcriptional activator protein can activate transcription by interacting with subunits of either the TFIIA complex or the holoenzyme.

As Figure 11.33 suggests, the fully assembled transcription complex in eukaryotes is a very large structure. A real example, taken from early development in *Drosophila*, is shown in Figure 11.34. In this case, the enhancers, located a considerable distance upstream from the gene to be activated, are bound by the transcriptional

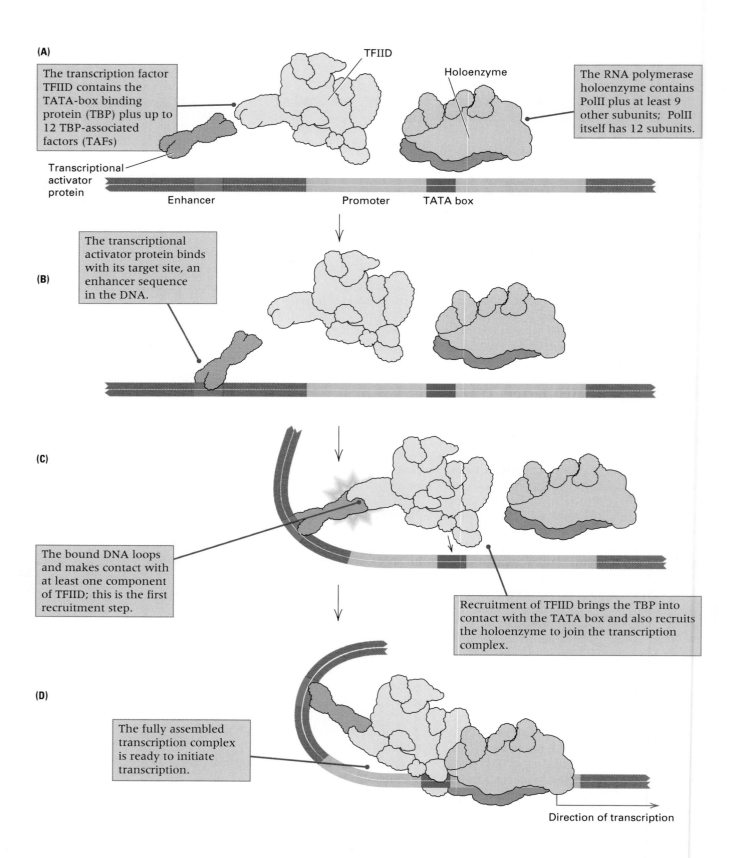

(A)

The transcription factor TFIID contains the TATA-box binding protein (TBP) plus up to 12 TBP-associated factors (TAFs)

TFIID

Holoenzyme

The RNA polymerase holoenzyme contains PolII plus at least 9 other subunits; PolII itself has 12 subunits.

Transcriptional activator protein

Enhancer   Promoter   TATA box

(B)

The transcriptional activator protein binds with its target site, an enhancer sequence in the DNA.

(C)

The bound DNA loops and makes contact with at least one component of TFIID; this is the first recruitment step.

Recruitment of TFIID brings the TBP into contact with the TATA box and also recruits the holoenzyme to join the transcription complex.

(D)

The fully assembled transcription complex is ready to initiate transcription.

Direction of transcription

Figure 11.33 (*facing page*)
Transcriptional activation by recruitment. (A) Relationship between enhancer and promoter and the protein factors that bind to them. (B) Binding of the transcriptional activator protein to the enhancer. (C) Bound transcriptional activator protein makes physical contact with a subunit in the TFIID complex, which contains the TATA-box-binding protein, and attracts ("recruits") the complex to the promoter region. (D) The PolII holoenzyme and any remaining general transcription factors are recruited by TFIID, and the transcription complex is fully assembled and ready for transcription. In the cell, not all of the PolII is found in the holoenzyme, and not all of the TBP is found in TFIID. In this illustration, transcription factors other than those associated with TFIID and the holoenzyme are not shown.

activator proteins BCD and HB, which are products of the genes *bicoid (bcd)* and *hunchback (hb)*, respectively; these transcriptional activators function in establishing the anterior-posterior axis in the embryo. (Early *Drosophila* development is discussed in Chapter 12.) Note the position of the TATA box in the promoter of the gene. The TATA box binding is the function of the TBP. The functions of a number of other components of the transcription complex have also been identified. For example, the TFIIH contains both helicase and kinase activity to melt the DNA and to phosphorylate RNA polymerase II. Phosphorylation allows the polymerase to leave the promoter and elongate mRNA. The looping of the DNA effected by the transcriptional activators is an essential feature of the activation process. Transcriptional activation in eukaryotes is a complex process, especially when compared to the prokaryotic RNA polymerase, which consists of a core $\alpha_2\beta\beta'$ tetramer and a σ factor.

The versatility of some enhancers results from their ability to interact with two different promoters in a competitive fashion; that is, at any one time, the enhancer can stimulate one promoter or the other, but not both. An example of this mechanism is illustrated in Figure 11.35, in which P1 and P2 are alternative promoters

Figure 11.34

An example of transcriptional activation during *Drosophila* development. The transcriptional activators in this example are bicoid protein (BCD) and hunchback protein (HB). The numbered subunits are TAFs (TBP-associated factors) that, together with TBP (TATA-box-binding protein) correspond to TFIID. BCD acts through a 110-kilodalton TAF, and HB, through a 60-kilodalton TAF. The transcriptional activators act via enhancers to cause recruitment of the transcriptional apparatus. The fully assembled transcription complex includes TBP and TAFs, RNA polymerase II, and general transcription factors TFIIA, TFIIB, TFIIE, TFIIF, and TFIIH.

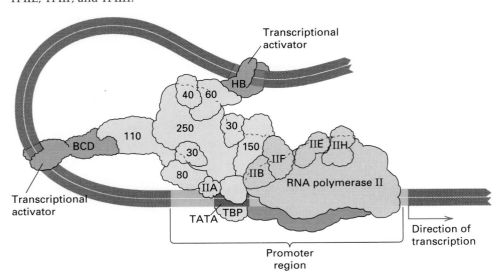

Figure 11.35

Genetic switching regulated by competition for an enhancer. Promoters *P1* and *P2* compete for a single enhancer located between them. When complexed with an appropriate transcriptional activator protein, the enhancer binds preferentially with promoter *P1* (A) or promoter *P2* (B). Binding to the promoter recruits the transcription complex. If either promoter is mutated or deleted, then the enhancer binds with the alternative promoter. The location of the enhancer relative to the promoters is not critical.

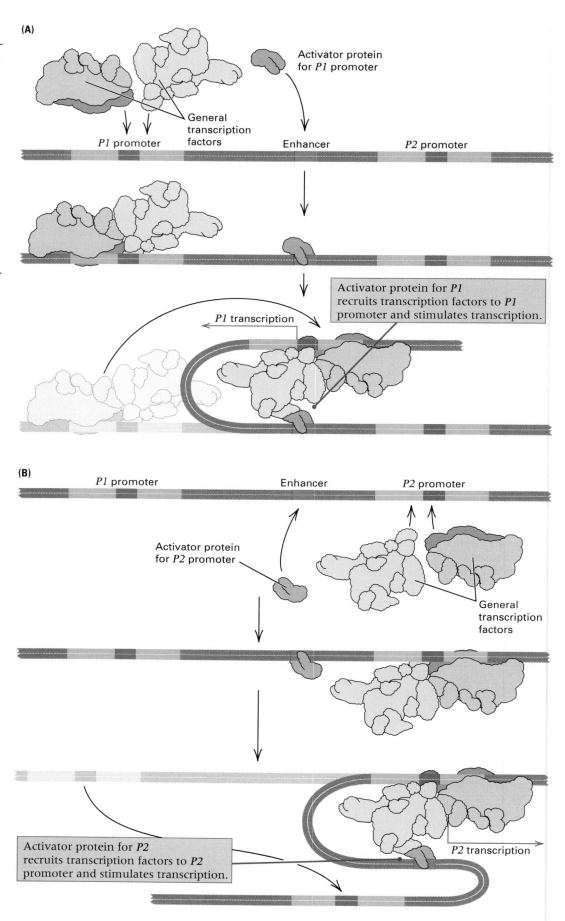

(A)

Activator protein for *P1* promoter

General transcription factors

P1 promoter   Enhancer   *P2* promoter

P1 transcription

Activator protein for *P1* recruits transcription factors to *P1* promoter and stimulates transcription.

(B)

P1 promoter   Enhancer   *P2* promoter

Activator protein for *P2* promoter

General transcription factors

Activator protein for *P2* recruits transcription factors to *P2* promoter and stimulates transcription.

P2 transcription

that compete for an enhancer located between them (Figure 11.35). When complexed with a transcriptional activator specific for promoter P1, the enhancer binds preferentially with promoter P1 and stimulates transcription (Figure 11.35A). When complexed with a different transcriptional activator specific for promoter P2, the enhancer binds preferentially with promoter P2 and stimulates transcription from it (Figure 11.35B). In this way, competition for the enhancer serves as a sort of switch mechanism for the expression of the P1 or P2 promoter. This regulatory mechanism is present in chickens and results in a change from the production of embryonic β-globin to that of adult β-globin in development. In this case, the embryonic globin gene and the adult gene compete for a single enhancer, which in the course of development changes its preferred binding from the embryonic promoter to the adult promoter. In human beings, enhancer competition appears to control the developmental switch from the fetal γ-globin to that of the adult β-globin polypeptide chains. In persons in whom the β-globin promoter is deleted or altered in sequence and unable to bind with the enhancer, there is no competition for the enhancer molecules, and the γ-globin genes continue to be expressed in adult life when normally they would not be transcribed. Adults with these types of mutations have fetal hemoglobin instead of the adult forms. The condition is called high-F disease because of the persistence of fetal hemoglobin, but the clinical manifestations are very mild.

The Logic of Combinatorial Control

Because the genome of a complex eukaryote contains many possible enhancers that respond to different signals or cellular conditions, in principle each gene could be regulated by its own distinct combination of enhancers. This is called **combinatorial control,** and it is a powerful means of increasing the complexity of possible regulatory states by employing several simple regulatory states in combination.

To consider a simple example, suppose that a gene has a single binding site for only one type of regulatory molecule. Then there are only two regulatory states (call them $+$ or $-$), which reflect whether or not the binding site is occupied. If the gene has single binding sites for each of two different regulatory molecules, then there are four possible regulatory states: $++$, $+-$, $-+$, and $--$. Single binding sites for three different regulatory molecules yield eight combinatorial states, and in general, n different types of regulatory molecules yield 2^n states. If transcription occurs according to the particular pattern of which binding sites are occupied, then a small number of types of regulatory molecules can result in a large number of different patterns of regulation. For example, because eukaryotic cells contain approximately 200 different cell types, each gene would theoretically need as few as $8 +/-$ types of binding sites to specify whether it should be on or off in each cell type, because $2^8 = 256$.

The actual regulatory situation is certainly more complex than the naive calculation based on $+/-$ switches would imply, because genes are not merely on or off; their level of activity is modulated. For example, a gene may have multiple binding sites for an activator protein, and this allows the level of gene expression to be adjusted according to the number of binding sites that are occupied. Furthermore, each cell type is programmed to respond to a variety of conditions both external and internal, so the total number of regulatory states must be considerably greater than the number of cell types. On the other hand, the $+/-$ example demonstrates that combinatorial control is extremely powerful in multiplying the number of regulatory possibilities, and therefore a large number of regulatory states does not necessarily imply a hopelessly complex regulatory apparatus.

Enhancer-Trap Mutagenesis

The ability of enhancers to regulate transcription is the basis of the **enhancer trap,** a genetically engineered transposable element designed to detect tissue-specific expression resulting from insertion of the element near enhancers. A simplified diagram of an enhancer trap based on the transposable *P* element in *Drosophila* is shown in Figure 11.36. The element (Figure 11.36A) contains a weak promoter, unable to initiate transcription without the aid of an enhancer, linked to the β-galactosidase gene from *E. coli*; also shown are the inverted-repeat sequences necessary for transposition. When the enhancer trap transposes and inserts at a random position in the genome, no transcription occurs if the insertion is not near an enhancer (Figure 11.36B). On the other hand, if the insertion is near an enhancer, then the β-galactosidase gene will be transcribed in whatever tissues stimulate the enhancer. Any tissue-specific expression can be detected by the use of staining reagents specific for β-galactosidase; one commonly used stain turns bright blue in the presence of the enzyme. When used in this manner, the β-galactosidase gene is called a **reporter gene** because its expression reveals ("reports") that the gene has been activated.

For example, the enhancer trap has been used to identify genes expressed only in the eye. When the enhancer trap inserts into the genome and comes under the control of eye-specific enhancers, the β-galactosidase is expressed in the eye and nowhere else. The eyes, and only the eyes, of flies with such insertions stain blue. Insertions of the enhancer trap provide an important method for identifying genes that are expressed in particular cells or tissues, because the insertions do not necessarily disrupt the function of the normal gene. Furthermore, the presence of DNA sequences from the *P* element at the site of insertion provides a molecular tag with which to clone the gene.

Alternative Promoters

Some eukaryotic genes have two or more promoters that are active in different cell types. The different promoters result in different primary transcripts that contain the same protein-coding regions. An example from *Drosophila* is shown in Figure 11.37. The gene codes for alcohol dehydrogenase, and its organization in the genome is shown in Figure 11.37; there are three

Figure 11.36

Identification of enhancers by genetic means. (A) The enhancer trap, consisting of a transposable element that contains a reporter gene with a weak promoter. (B) If insertion is at a site that lacks a nearby enhancer, then the reporter gene cannot be activated. (C) If insertion is near an enhancer, then the reporter gene is transcribed in a temporally regulated or tissue-specific manner determined by the type of enhancer.

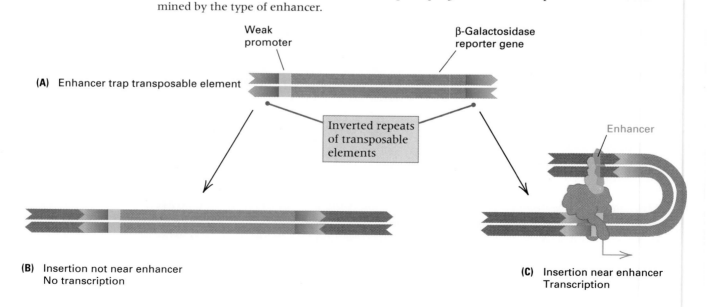

(A) Enhancer trap transposable element

Weak promoter

β-Galactosidase reporter gene

Inverted repeats of transposable elements

Enhancer

(B) Insertion not near enhancer
No transcription

(C) Insertion near enhancer
Transcription

(A) Gene structure

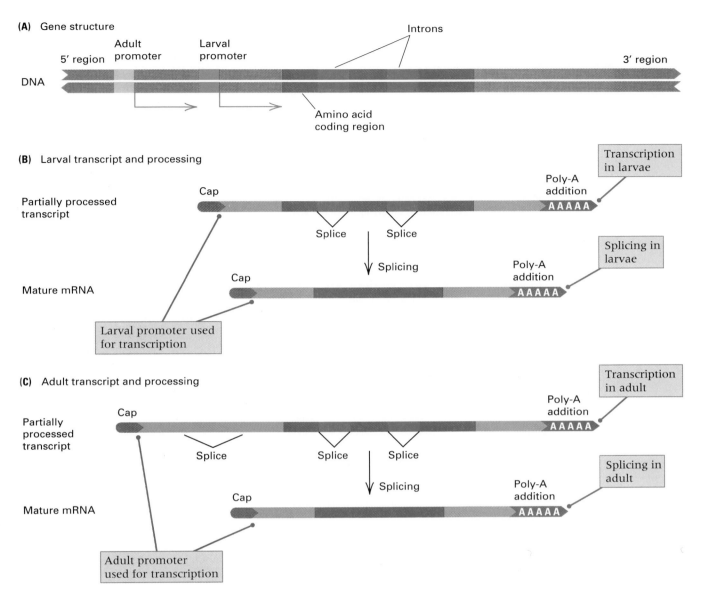

Figure 11.37

Use of alternative promoters in the gene for alcohol dehydrogenase in *Drosophila*. (A) The overall gene organization includes two introns within the amino acid coding region. (B) Transcription in larvae uses the promoter nearest the 5' end of the coding region. (C) Transcription in adults uses a promoter farther upstream, and much of the larval leader sequence is removed by splicing.

protein-coding regions interrupted by two introns. Transcription in larvae (Figure 11.37A) uses a different promoter from that used in transcription in adults (Figure 11.37B). The adult transcript has a longer 5' leader sequence, but most of this sequence is eliminated in splicing. Alternative promoters make possible the independent regulation of transcription in larvae and adults.

11.8
Alternative Splicing

Even when the same promoter is used to transcribe a gene, different cell types can produce different quantities of the protein (or even different proteins) because of differences in the mRNA produced in processing. The reason is that the same transcript can be spliced differently from one cell type to the next. The different splicing patterns may include exactly the same protein-coding exons, in which case the protein is identical but the rate of synthesis differs because the mRNA. molecules are not translated with the same efficiency. In other cases, the protein-coding part of the transcript has a different splicing pattern in each cell type, and the resulting mRNA molecules code for proteins that are not identical even though they share certain exons.

In the synthesis of α-amylase in the mouse, different mRNA molecules are produced from the same gene because of different patterns of intron removal in RNA processing. The mouse salivary gland produces more of the enzyme than the liver, although the same coding sequence is transcribed. In each cell type, the same primary transcript is synthesized, but two different splicing patterns are used. The initial part of the primary transcript is shown in Figure 11.38. The coding sequence begins 50 base pairs inside exon 2 and is formed by joining

exon 3 and subsequent exons. In the salivary gland, the primary transcript is processed such that exon S is joined with exon 2 (that is, exon L is removed as part of introns 1 and 2). In the liver, exon L is joined with exon 2, and exon S is removed along with intron 1 and the leader L. The exons S and L become alternative 5' leader sequences of the amylase mRNA, and the alternative mRNAs are translated at different rates.

In chicken skeletal muscle, two different forms of the muscle protein myosin are made from the same gene. A different primary transcript is made from each of two promoters, and these transcripts are processed differently to form mRNA molecules that encode distinct forms of the protein. In *Drosophila*, the myosin RNA is processed in four different ways; the precise mode depends on the stage of development of the fly. One class of myosin is found in pupae and another in the later embryo and larval stages. How the mode of processing is varied is not known.

Figure 11.38

Production of distinct amylase mRNA molecules by different splicing events in cells of the salivary gland and liver of the mouse. The leader and the introns are distinguished by color from the exons. Exons S and L form part of the untranslated 5' end of the mRNA in salivary glands and liver, respectively. The coding sequence begins at the AUG codon in exon 2. (A) Splicing in the liver. Exon L is joined with exons 2, 3, and 4. (B) Splicing in the salivary gland. Exon S is joined with exons 2, 3, and 4.

(A)

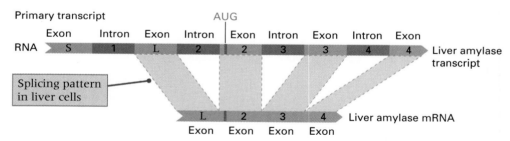

(B)

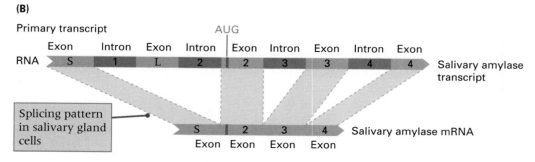

11.9
Translational Control

In bacteria, most mRNA molecules are translated about the same number of times, with little variation from gene to gene. In eukaryotes, translation is sometimes regulated. The principal types of translational control are

1. Inability of an mRNA molecule to be translated unless a molecular signal is present

2. Regulation of the lifetime of a particular mRNA molecule

3. Regulation of the overall rate of protein synthesis

4. Aborted translation of the principal open reading frame because of the presence of smaller open reading frames upstream in the mRNA

In this section, examples of each of the first three modes of regulation will be presented.

An important example of translational regulation is that of **masked mRNA.** Unfertilized eggs are biologically static, but shortly after fertilization, many new proteins must be synthesized—among them, for example, the proteins of the mitotic apparatus and the cell membranes. Unfertilized sea urchin eggs can store large quantities of mRNA for many months in the form of mRNA-protein particles made during formation of the egg. This mRNA is translationally inactive, but within minutes after fertilization, translation of these molecules begins. Here, the timing of translation is regulated; the mechanisms for stabilizing the mRNA, for protecting it against RNases, and for activation are unknown.

Translational regulation of another type occurs in mature unfertilized eggs. These cells need to maintain themselves but do not have to grow or undergo a change of state. Thus the rate of protein synthesis in eggs is generally low. This is not a consequence of an inadequate supply of mRNA but apparently results from failure to form the ribosome-mRNA complex.

The synthesis of some proteins is regulated by direct action of the protein on the mRNA. For instance, the concentration of one type of antibody molecule is kept constant by self-inhibition of translation; that is, the antibody molecule itself binds specifically to the mRNA that encodes it and thereby inhibits initiation of translation.

A dramatic example of translational control is the extension of the lifetime of silk fibroin mRNA in the silkworm. During cocoon formation, the silk gland of the silkworm predominantly synthesizes a single type of protein, silk fibroin. Because the worm takes several days to construct its cocoon, it is the total amount (not the rate) of fibroin synthesis that must be great; the silkworm achieves this in two ways. First, the silk gland cells become highly polyploid, accumulating thousands of copies of each chromosome. Second, each fibroin gene synthesizes an mRNA molecule that has a very long lifetime. Transcription of the fibroin gene is initiated at a strong promoter, and about 10^4 fibroin mRNA molecules are made in a period of several days. (This synthesis is under transcriptional regulation.) A typical eukaryotic mRNA molecule has a lifetime of about 3 hours before it is degraded. However, fibroin mRNA survives for several days, during which each mRNA molecule is translated repeatedly to yield 10^5 fibroin molecules. Thus each gene is responsible for the synthesis of 10^9 protein molecules in 4 days. Altogether, the silk gland makes about 10^{15} molecules of fibroin in this period. If the lifetime of the mRNA were not extended, then either 25 times as many genes would be needed or synthesis of the required fibroin would take about 100 days.

Another example of an mRNA molecule with an extended lifetime is the mRNA that encodes casein, the major protein of milk, in mammary glands. When the hormone prolactin is received by the gland, the lifetime of casein mRNA increases. Synthesis of the mRNA also continues, so the overall rate of production of casein is markedly increased by the hormone. When the body no longer supplies prolactin, the concentration of casein mRNA decreases because the RNA is degraded more rapidly, and lactation terminates.

11.10
Is There a General Principle of Regulation?

With microorganisms, whose environment frequently changes drastically and rapidly, a general principle of regulation is that bacteria make what they need when it is required and in appropriate amounts. Although such extraordinary efficiency is common in prokaryotes, it is rare in eukaryotes. For example, efficiency appears to be violated by the large amount of DNA in the genome that has no protein-coding function and by the large amount of intron RNA that is discarded.

What is abundantly clear is that there is no universal regulatory mechanism. Many control points are possible, and different genes are regulated in different ways. Furthermore, evolution has not always selected for simplicity in regulatory mechanisms; it sometimes just settles for something that works. If a cumbersome regulatory mechanism were to arise, then it would in time evolve, be refined, and become more effective, but it would not necessarily become simpler. On the whole, regulatory mechanisms include a variety of seemingly *ad hoc* processes, each of which has stood the test of time primarily because it works.

Chapter Summary

Most cells do not synthesize molecules that are not needed. There are important opportunities for controlling gene expression in transcription, RNA processing, and translation. Gene expression can also be regulated through the stability of mRNA, and the activity of proteins can be regulated in a variety of ways after translation. The processes of gene regulation generally differ between prokaryotes and eukaryotes.

In bacteria, the synthesis of most proteins is regulated by controlling the rate of transcription of the genes that code for the proteins. The concentration of a few proteins is autoregulated, usually by direct binding of the protein at or near its promoter.

The synthesis of degradative enzymes needed only on occasion, such as the enzymes required to metabolize lactose, is typically regulated by an off-on mechanism. When lactose is present, the genes that code for the enzymes required to metabolize lactose are transcribed; when lactose is absent, transcription does not occur. Lactose metabolism is negatively regulated. Two enzymes that are needed to degrade lactose—permease, required for the entry of lactose into bacteria, and β-galactosidase, the enzyme that does the degrading—are encoded in a single polycistronic mRNA molecule, *lac* mRNA. Immediately adjacent to the promoter for *lac* mRNA is a regulatory sequence of bases called an operator. A repressor protein is made by a tightly linked gene, and this protein binds tightly to the operator, thereby preventing RNA polymerase from initiating transcription at the promoter. Lactose is an inducer of transcription because it can bind to the repressor and prevent the repressor from binding to the operator. Therefore, in the presence of lactose, there is no active repressor, and the *lac* promoter is always accessible to RNA polymerase. The operator, the promoter, and the structural genes are adjacent to one another and together constitute the *lac* operon. Repressor mutations have been isolated that inactivate the repressor protein, and operator mutations are known that prevent recognition of the operator by an active repressor; such mutations cause continuous production of *lac* mRNA and are said to be constitutive.

When lactose is cleaved by β-galactosidase, the products are glucose and galactose. Glucose is metabolized by enzymes that are made continuously; galactose is broken down by enzymes of the inducible galactose operon. When glucose is present in the growth medium, the enzymes for degrading lactose and other sugars are unnecessary. The general mechanism for preventing transcription of many sugar-degrading operons is as follows: High concentrations of glucose suppress the synthesis of the small molecule cyclic AMP (cAMP). Initiation of transcription of many sugar-degrading operons requires the binding of a protein, called CRP, to a specific region of the promoter. This binding takes place only after CRP has first bound cAMP and formed a cAMP-CRP complex. Only when glucose is absent is the concentration of cAMP sufficient to produce cAMP-CRP and hence to permit transcription of the sugar-degrading operons. Thus, in contrast with a repressor, which must be removed before transcription can begin (negative regulation), cAMP is a positive regulator of transcription.

Biosynthetic enzyme systems exemplify the repressible type of negative transcriptional control. In the synthesis of tryptophan, transcription of the genes encoding the *trp* enzymes is controlled by the concentration of tryptophan in the growth medium. When excess tryptophan is present, it binds with the *trp* aporepressor to form the active repressor that prevents transcription. The *trp* operon is also regulated by attenuation, in which transcription is initiated continually but the transcript forms a hairpin structure that results in premature termination. The frequency of termination of transcription is determined by the availability of charged tryptophan tRNA; with decreasing concentrations of tryptophan, termination occurs less often and the enzymes for tryptophan synthesis are made, thereby increasing the concentration of tryptophan. Attenuators also regulate operons for the synthesis of other amino acids.

Bacteriophage λ adopts either the quiescent lysogenic state or the lytic cycle as a result of competition between two repressors, cI and cro. If cI repressor dominates, the λ genome becomes integrated into the bacterial chromosome. A lysogen is formed, in which cI continues to be synthesized and prevents transcription of all other λ genes. If cro repressor dominates, cI repressor is no longer synthesized, and the lytic cycle ensues. Although there are elements of positive transcriptional regulation in the regulatory circuitry, the predominant mode of regulation is negative.

Eukaryotes employ a variety of genetic regulatory mechanisms, occasionally including changes in DNA. .The number of copies of a gene may be increased by DNA amplification; programmed rearrangements of DNA may occur; and in some cases, gene inactivity coincides with the methylation of cytosine bases.

Many genes in eukaryotes are regulated at the level of transcription. Although both negative and positive regulation occur, positive regulation is typical. In the control of yeast mating type by the *MAT* locus, the **a**-specific genes and the haploid-specific genes are negatively regulated and the α-specific genes are positively regulated. In general, positive regulation is effected through transcriptional activator proteins that contain particular structural motifs that bind to DNA—for example, the helix-turn-helix motif or the zinc finger motif. Hormones also can regulate transcription. Steroid hormones bind with receptor proteins to form transcriptional activators.

Transcriptional activators bind to DNA sequences known as enhancers, which are usually short sequences that may be present at a variety of positions around the genes that they regulate. In the recruitment model of transcriptional activation, a transcriptional activator protein interacts directly with one or more protein components of the RNA polymerase holoenzyme and attracts the holoenzyme to the promoter along with other protein complexes, such as TFIID. The fully assembled transcriptional apparatus in eukaryotes consists of a complex assemblage of RNA polymerase II, TATA-box-binding protein (TBP), other general transcription factors (TFIIA, TFIIB, and so forth), and TBP-associated factors (TAFs).

In genetic analysis, enhancers can be identified by the use of transposable elements containing genes that come under the control of enhancers near the sites of insertion. There are many types of enhancers responsive to different transcriptional activators. Combining different types of enhancers provides a large number of possible types of regulation.

Some genes contain alternative promoters used in different tissues; other genes use a single promoter, but the transcripts are spliced in different ways. Alternative splicing can result in mRNA molecules that are translated with different efficiencies, or even in different proteins if there is alternative splicing of the protein-coding exons. Regulation can also be at the level of translation—for example, through masked mRNAs or through factors that affect mRNA stability.

Key Terms

aporepressor
attenuation
attenuator
autoregulation
β-galactosidase
cAMP-CRP complex
cassette
cis-dominant
combinatorial control
combinatorial joining
constant region
constitutive mutant
coordinate regulation
co-repressor
cyclic adenosine monophosphate (cAMP)
cyclic AMP receptor protein (CRP)
DNA looping
DNA methylase
effector

enhancer
enhancer trap
estrogen
gene amplification
gene dosage
general transcription factor
gene regulation
heavy chain
helix-turn-helix
homothallism
hormone
housekeeping genes
immunoglobulin
inducer
inducible
inducible transcription
lac operon
lactose permease
leader polypeptide
light chain

masked mRNA
mating-type interconversion
methylation
negative regulation
operator
operon model
permease
polyprotein
positive regulation
recruitment
reporter gene
repressible transcription
repressor
RNA polymerase holoenzyme
TATA-box-binding protein (TBP)
TBF-associated factor (TAF)
transcriptional activator protein
variable region
zinc finger

GeNETics on the web will introduce you to some of the most important sites for finding genetic information on the Internet. To complete the exercises below, visit the Jones and Bartlett home page at

http://www.jbpub.com/genetics

Select the link to *Genetics: Principles and Analysis* and then choose the link to *GeNETics on the web*. You will be presented with a chapter-by-chapter list of highlighted keywords.

GeNETics EXERCISES

Select the highlighted keyword in any of the exercises below, and you will be linked to a web site containing the genetic information necessary to complete the exercise. Each exercise suggests a specific, written report that makes use of the information available at the site. This report, or an alternative, may be assigned by your instructor

1. Look up lactose degradation at this site for a diagram of the chemical reaction catalyzed by β-galactosidase. Examine the pathway in more detail to see the molecular structures of the substrate and the products of the enzymes. Click on *lacZ* to get more information about the gene itself. If assigned to do so, follow the *Unification link* and pursue further links until you find the amino acid sequence of the β-galactosidase polypeptide (each active enzyme contains four of these polypeptide chains). List, in order, the links you followed to find this information, as well as the number of amino acids in the polypeptide and its molecular weight. Note also which protein database contains this sequence and the entry number.

2. More about the genes that control mating type in *Saccharomyces cerevisiae* can be retrieved at this site by examining the map of chromosome 3 and clicking on *HML, MAT,* and *HMR.* See if you can follow the links to the Entrez Protein database to find the amino acid sequence of each of the regulatory proteins **a**1, α1, and α2. If assigned to do so, write a short paragraph summarizing

Review the Basics

*   Distinguish between positive regulation and negative regulation of transcription, and give an example of each.

*   What is an operon and what is the significance of an operon for gene expression?

*   An operon containing genes that encode the enzymes in a metabolic pathway for the synthesis of an amino acid includes a short open reading frame in the leader sequence. This short open reading frame contains multiple codons for the amino acid synthesized by the pathway. What does this observation suggest about regulation of the operon?

*   In yeast, which genes control the transcription of haploid-specific genes?

*   What is a transcriptional enhancer? What distinguishes it from a promoter?

*   What is alternative splicing and what is its significance in gene regulation?

*   Distinguish between a repressor and an aporepressor. Give an example of each.

*   What is autoregulation? Distinguish between positive and negative autoregulation. Which would be used to amplify a weak induction signal? Which to prevent overproduction?

*   What term describes a gene that is expressed continuously?

Guide to Problem Solving

Problem 1: A gene *R* codes for a protein that is a negative regulator of transcription of a gene *S*. Is gene *S* transcribed in an R^- mutant? How does the situation differ if the product of *R* is a positive regulator of *S* transcription?

Answer: A negative regulator of transcription is needed to turn transcription off; hence, in the R^- mutant, transcription of *S* is constitutive. The opposite happens in positive regulation. A positive regulator of transcription is needed to turn transcription on, so if the product of *R* is a positive regulator, transcription does not occur in R^- cells.

Problem 2: For each of the following partial diploid genotypes of *E. coli*, state whether β-galactosidase is made and whether its synthesis is inducible or constitutive. The convention for writing partial diploid genotypes is to put the plasmid genes at the left of the slash and the chromosomal genes at the right.

(a) $lacZ^+ \ lacY^-/lacZ^- \ lacY^+$

(b) $lacO^c \ lacZ^- \ lacY^+/lacZ^+ \ lacY^-$

(c) $lacP^- \ lacZ^+/lacO^c \ lacZ^-$

what each entry says about the function of the regulatory molecule.

3. Gene splicing in the origin of **human antibodies** is described at this site, which also offers much other information about the genetics of this system. If assigned to do so, write a one-page summary of the genetics of the immunoglobulin heavy-chain family, and discuss specifically the role of sequence homology and unequal crossing-over in the evolution of the gene families.

MUTABLE SITE EXERCISES

The Mutable Site Exercise changes frequently. Each new update includes a different exercise that makes use of genetics resources available on the World Wide Web. Select the **Mutable Site** for Chapter 11, and you will be linked to the current exercise that relates to the material presented in this chapter.

PIC SITE

The Pic Site showcases some of the most visually appealing genetics sites on the World Wide Web. To visit the showcase genetics site, select the **Pic Site** for Chapter 11.

(d) $lacI^+$ $lacP^-$ $lacZ^+/lacI^-$ $lacZ^+$

(e) $lacI^+$ $lacP^-$ $lacY^+/lacI^-$ $lacY^-$

Answer: The location of the genes does not matter because they function in the same manner whether present in a plasmid or in a chromosome.

(a) In this partial diploid, there are no *cis*-dominant mutations. The plasmid operon is $lacZ^+$ and can make the enzyme, whereas the chromosomal operon is $lacZ^-$ and cannot. The cell will produce the enzyme from the plasmid gene as long as the operon can be turned on. Turning it on requires the presence of functional regulatory elements, which is the case in this example. (Recall that if a gene is not listed, for example *lacI*, it is assumed to be wildtype.) It is also necessary that the inducer be able to enter the cell, which it can in this example because the chromosomal genotype can supply the *lacY* permease. Thus for this partial diploid, β-galactosidase can be made, and its synthesis is inducible.

(b) This partial diploid has the *cis*-dominant mutation $lacO^c$ in the plasmid, so the genes in the plasmid operon are always expressed. However, the *lacZ* gene in the plasmid makes a defective enzyme. The chromosome has a functional *lacZ* gene from which active enzyme can be made, but its synthesis is under the control of a normal operator (because the operator genotype is not indicated, it is wildtype). Hence, enzyme synthesis must be inducible. Thus the partial diploid makes a defective enzyme constitutively and a normal enzyme inducibly, so the overall phenotype is that the cell can be induced to make β-galactosidase.

(c) A promoter mutation, which is *cis*-dominant, is in the plasmid, which means that no *lac* mRNA can be made from this operon. The chromosome contains a *cis*-dominant $lacO^c$ mutation that drives constitutive synthesis of an mRNA, but the enzyme is defective. Thus there is no way for the cell to make active β-galactosidase.

(d) The plasmid genotype contains a promoter mutation, so no mRNA can be produced from the *lacZ* structural gene. However, the *lacI* gene in the plasmid has its own promoter, so *lac* repressor molecules are present in the cell. The chromosomal genotype alone would make enzyme constitutively because of the *lacI* mutation, but the presence of the functional *lacI* product made by the plasmid means that any synthesis that occurs must be induced. The chromosomal operon can provide both β-galactosidase and permease, so β-galactosidase is inducible in this genotype.

(e) This genotype differs from that in part (d) by the presence of a $lacY^-$ mutation in the chromosome. Again, the plasmid contributes only *lacI* repressor to the cell, so any synthesis of the enzyme must be inducible. However, the chromosomal genotype is $lacY^-$. Because the $lacY^+$ allele in the plasmid cannot be expressed, no inducer can enter the cell, so the cell is unable to make any enzyme.

Problem 3: With regard to mating type in yeast, what phenotypes would each of the following haploid cells exhibit?

(a) a duplication of the mating-type gene giving the genotype *MATa*/*MAT*α

(b) a deletion of the *HML*α cassette in a *MATa* cell

Answer:

(a) The haploid cell expresses both **a** and α, so the **a**-specific genes, the α-specific genes, and the haploid-specific genes are all inactive. The phenotype is similar to the **a**α diploid. It will not mate, and if it attempts to sporulate, it will self-destruct.

(b) The phenotype is that of a normal **a** haploid, but mating-type switching to α cannot take place.

Analysis and Applications

11.1 The metabolic pathway for glycolysis is responsible for the degradation of glucose and is one of the fundamental energy-producing systems in living cells. Would you expect the enzymes in this pathway to be regulated? Why or why not?

11.2 Among mammals, the reticulocyte cells in the bone marrow extrude their nuclei in the process of differentiation into red blood cells. Yet the reticulocytes and red blood cells continue to synthesize hemoglobin. Suggest a mechanism by which hemoglobin synthesis can continue for a long period of time in the absence of the hemoglobin genes.

11.3 Several eukaryotes are known in which a single effector molecule regulates the synthesis of different proteins coded by distinct mRNA molecules—say, X and Y. At what level in the process of gene expression does regulation occur in each of the following situations?

(a) Neither nuclear nor cytoplasmic RNA can be found for either X or Y.

(b) Nuclear but not cytoplasmic RNA can be found for both X and Y.

(c) Both nuclear and cytoplasmic RNA can be found for both X and Y, but none of it is associated with polysomes.

11.4 Is it necessary for the gene that codes for the repressor of a bacterial operon to be near the structural genes? Why or why not?

11.5 Consider a eukaryotic transcriptional activator protein that binds to an enhancer sequence and promotes transcription. What change in regulation would you expect from a duplication in which several copies of the enhancer were present instead of just one?

11.6 The following questions pertain to the *lac* operon in *E. coli.*

(a) Which proteins are regulated by the repressor?

(b) How does binding of the *lac* repressor to the *lac* operator prevent transcription?

(c) Is production of the *lac* repressor constitutive or induced?

11.7 The permease of *E. coli* that transports the α-galactoside melibiose can also transport lactose, but it is temperature-sensitive: Lactose can be transported into the cell at 30°C but not at 37°C. In a strain that produces the melibiose permease constitutively, what are the phenotypes of *lacZ⁻* and *lacY⁻* mutants at 30°C and 37°C?

11.8 How do inducers enable transcription to occur in a bacterial operon under negative transcriptional control?

11.9 When glucose is present in an *E. coli* cell, is the concentration of cyclic AMP high or low? Can a mutant with either an inactive adenyl cyclase gene or an inactive *crp* gene synthesize β-galactosidase? Does the binding of cAMP-CRP to DNA affect the binding of a repressor?

11.10 The operon allows a type of coordinate regulation of gene activity in which a group of enzymes with related functions are synthesized from a single polycistronic mRNA. Does this imply that all proteins in the polycistronic mRNA are made in the same quantity? Explain.

11.11 Both repressors and aporepressors bind molecules that are substrates or products of the metabolic pathway encoded by the genes in an operon. Generally speaking, which binds the substrate of a metabolic pathway and which the product?

11.12 Is an attenuator a region of DNA that, like an operator, binds with a protein? Is RNA synthesis ever initiated at an attenuator?

11.13 How do steroid hormones induce transcription of eukaryotic genes?

11.14 In regard to mating-type switching in yeast, what phenotype would you expect from a type α cell that has a deletion of the *HMR**a*** cassette?

11.15 What mating-type phenotype would you expect of a haploid cell of genotype *MAT**a**'*, where the prime denotes a mutation that renders the **a**1 protein inactive? What mating type would you expect in the diploid *MAT**a**'/MATα*?

11.16 What mating-type phenotype would you expect from a diploid cell of genotype *MAT**a**/MATα* with a mutation in α in which the α2 gene product functions normally in turning off the **a**-specific genes but is unable to combine with the **a**1 product?

11.17 If a wildtype *E. coli* strain is grown in a medium without lactose or glucose, how many proteins are bound to the *lac* operon? How many are bound if the cells are grown in glucose?

11.18 A mutant strain of *E. coli* is found that produces both β-galactosidase and permease constitutively (that is, whether lactose is present or not).

(a) What are two possible genotypes for this mutant?

(b) A second mutant is isolated that produces no active β-galactosidase at any time but produces permease if lactose is present in the medium. What is the genotype of this mutant?

(c) A partial diploid is created from the mutants in parts (a) and (b): When lactose is absent, neither enzyme is made, and when lactose is present, both enzymes are made. What is the genotype of the mutant in part (a)?

11.19 A $lacI^+$ $lacO^+$ $lacZ^+$ $lacY^+$ Hfr strain is mated with an F^- $lacI^-$ $lacO^+$ $lacZ^-$ $lacY^-$ strain. In the absence of any inducer in the medium, β-galactosidase is made for a short time after the Hfr and F^- cells have been mixed. Explain why it is made and why only for a short time.

11.20 What amino acids can be inserted at the site of the UGA codon that is suppressed by a suppressor tRNA?

11.21 An *E. coli* mutant is isolated that is simultaneously unable to utilize a large number of sugars as sources of carbon. However, genetic analysis shows that the operons responsible for metabolism of each sugar are free of mutations. What genotypes of this mutant are possible?

Challenge Problems

11.22 The histidine operon is negatively regulated and contains ten structural genes for the enzymes needed to synthesize histidine. The repressor protein is also coded within the operon—that is, in the polycistronic mRNA molecule that codes for the other proteins. Synthesis of this mRNA is controlled by a single operator regulating the activity of a single promoter. The co-repressor of this operon is tRNA^{His}, to which histidine is attached. This tRNA is not coded by the operon itself. A collection of mutants with the following defects is isolated. Determine whether the histidine enzymes would be synthesized by each of the mutants and whether each mutant would be dominant, *cis*-dominant only, or recessive to its wildtype allele in a partial diploid.

(a) The promoter cannot bind with RNA polymerase.

(b) The operator cannot bind the repressor protein.

(c) The repressor protein cannot bind with DNA.

(d) The repressor protein cannot bind histidyl-tRNA^{His}.

(e) The uncharged tRNA^{His} (that is, without histidine attached) can bind to the repressor protein.

11.23 The attenuator of the histidine operon contains seven consecutive histidines. The relevant part of the attentuator coding sequence is

5'-AAACACCACCATCATCACCATCATCCTGAC-3'

A mutation occurs in which an additional A nucleotide is inserted immediately after the red A. What amino acid sequences are coded by the wildtype and mutant attenuators? What phenotype would you expect of the mutant?

Further Reading

Cohen, S., and G. Jürgens. 1991. *Drosophila* headlines. *Trends in Genetics* 7: 267.

Gellert, M. 1992. V(D)J recombination gets a break. *Trends in Genetics* 8: 408.

Gralla, J. D. 1996. Activation and repression of *E. coli* promoters. *Current Opinion in Genetics & Development* 6: 526.

Guarente, L. 1984. Yeast promoters: Positive and negative elements. *Cell* 36: 799.

Holliday, R. 1989. A different kind of inheritance. *Scientific American,* June.

Khoury, G., and P. Gruss. 1983. Enhancer elements. *Cell* 33: 83.

Klar, A. J. S. 1992. Developmental choices in mating-type interconversion in fission yeast. *Trends in Genetics* 8: 208.

Laird, P. W., and R. Jaenisch. 1996. The role of DNA methylation in cancer genetics and epigenetics. *Annual Review of Genetics* 30: 441.

Maniatis, T., and M. Ptashne. 1976. A DNA operator-repressor system. *Scientific American,* January.

Marmorstein, R., M. Carey, M. Ptashne, and S. C. Harrison. 1992. DNA recognition by GAL4: Structure of a protein-DNA complex. *Nature* 356: 408.

Miller, J., and W. Reznikoff, eds. 1978. *The Operon.* Cold Spring Harbor, New York: Cold Spring Harbor Laboratory.

Neidhardt, F. C., R. Curtiss III, J. L. Ingraham, E. C. C. Lin, K. B. Low, B. Magasanik, W. S. Reznikoff, M. Riley, M. Schaechter, and H. E. Umbarger, eds. 1996. Escherichia coli *and* Salmonella typhimurium: *Cellular and Molecular Biology* (2 volumes). 2d ed. Washington, DC: American Society for Microbiology.

Novina, C. D., and A. L. Roy. 1996. Core promoters and transcriptional control. *Trends in Genetics* 12: 351.

Ptashne, M. 1992. *A Genetic Switch.* 2d ed. Cambridge, MA: Cell Press.

Ptashne, M., and A. Gann. 1997. Transcriptional activation by recruitment. *Nature* 386: 569.

Struhl, K. 1995. Yeast transcriptional regulatory mechanisms. *Annual Review of Genetics* 29: 651.

Tijan, R. 1995. Molecular machines that control genes. *Scientific American,* February.

Ullman, A., and A. Danchin. 1980. Role of cyclic AMP in regulatory mechanisms of bacteria. *Trends in Biochemical Sciences* 5: 95.

Yanofsky, C. 1981. Attenuation in the control of expression of bacterial operons. *Nature* 289: 751.

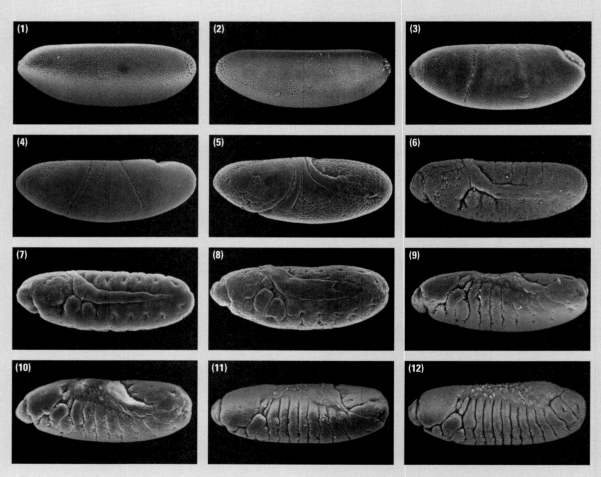

Early development in *Drosophila*, represented by scanning electron micrographs of 12 stages. Each embryo is arranged with the anterior end at the left and the ventral side down. The stages are arranged chronologically from 1 through 12. The first five stages are the early cleavage divisions, ending with the cellular blastoderm (stage 5) at 3.25 hours after fertilization. Gastrulation (stages 6–8) is followed by formation of the characteristic pattern of segments along the body, ending with stage 12 at 9 hours after fertilization. Compare stages 1–5 with Figure 12.21 and stage 12 with Figure 12.23. [Courtesy of Thomas C. Kaufman and Rudi Turner.]

The Genetic Control of Development

PRINCIPLES

- In animals cells, maternal gene products in the oocyte control the earliest stages of development, including the establishment of the main body axes.
- Developmental genes are often controlled by gradients of gene products, either within cells or across parts of the embryo.
- Regulation of developmental genes is hierarchical—genes expressed early in development regulate the activities of genes expressed later.
- Regulation of developmental genes is combinatorial—each gene is controlled by a combination of other genes.
- Many of the fundamental processes of pattern formation appear to be similar in animals and plants.

CONNECTIONS

CONNECTION: Distinguished Lineages
John E. Sulston, E. Schierenberg, J. G. White, and J. N. Thomson 1983
The embryonic cell lineage of the nematode Caenorhabditis elegans

CONNECTION: Embryo Genesis
Christiane Nüsslein-Volhard and Eric Wieschaus 1980
Mutations affecting segment number and polarity in Drosophila

Understanding gene regulation is central to understanding how an organism as complex as a human being develops from a fertilized egg. In development, genes are expressed according to a prescribed program to ensure that the fertilized egg divides repeatedly and that the resulting cells become specialized in an orderly way to give rise to the fully differentiated organism. The genotype determines not only the events that take place in development, but also the temporal order in which the events unfold.

Genetic approaches to the study of development often make use of mutations that alter developmental patterns. These mutations interrupt developmental processes and make it possible to identify

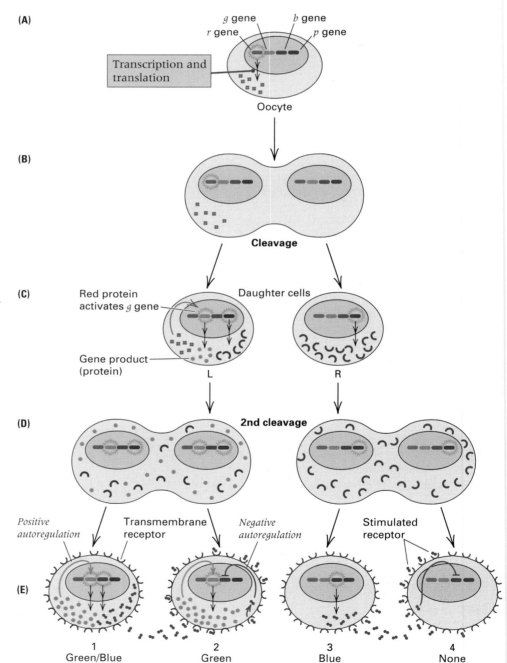

Figure 12.1
Hypothetical example illustrating some regulatory mechanisms that result in differences in gene expression in development. (A) The *r* gene is expressed in the oocyte. (B) Polarized presence of the *r* gene product in the egg. (C) Presence of the *r* gene product in cell L stimulates transcription of the *g* gene. The *p* gene codes for a transmembrane receptor expressed in both L and R cells. (D) The *g* gene has positive autoregulation and so continues to be expressed in cells 1 and 2. (E) Expression of the *b* gene in one cell represses its expression in the neighboring cell by stimulation of the *p* gene transmembrane receptor. The result of these mechanisms is that cells 1 through 4 differ in their combinations of *b* gene and *g* gene activity.

factors that control development and to study the interactions among them. This chapter demonstrates how genetics is used in the study of development. Many of the examples come from the nematode *Caenorhabditis elegans* and the fruit fly *Drosophila melanogaster*, because these organisms have been studied intensively from the standpoint of developmental genetics.

The key process in development is **pattern formation,** which means the emergence, from cell division and differentiation of the fertilized egg, of spatially organized and specialized cells that form the embryo. One might make an analogy between pattern formation in development and a pattern formed by fitting the pieces of a jigsaw puzzle together, but the analogy is not quite right. In a jigsaw puzzle, the pattern exists independently of the shape and position of the pieces; the cuts in the pattern are superimposed on a preexisting picture that is merely reassembled. In biological development, the emergence of the image on each piece is *caused by* the shape and position of the piece. It is remarkable also that the picture changes throughout life as the organism undergoes growth and aging.

12.1
Genetic Determinants of Development

Conceptually, the relationship between genotype and development is straightforward:

> The genotype contains a developmental program that unfolds and results in the expression of different sets of genes in different types of cells.

In other words, development consists of a program that results in the specific expression of some genes in one cell type and not in another. As development unfolds, cell types progressively appear that differ qualitatively in the genes that are expressed. Whether a particular gene is expressed may depend on the presence or absence of a particular transcription factor, a change in

chromatin structure, or the synthesis of a particular receptor molecule. These and other molecular mechanisms define the pattern of interactions by which one gene controls another. Collectively, these regulatory interactions ultimately determine the fate of the cell. Moreover, many interactions that are important in development make use of more general regulatory elements. For example, developmental events often depend on regional differences in the concentration of molecules within a cell or within an embryo, and the activity of enhancers may be modulated by the local concentration of these substances.

The formation of an embryo is also affected by the environment in which development takes place. Although genotype and environment work together in development, the genotype determines the developmental potential of the organism. Genes determine whether a developing embryo will become a nematode, a fruit fly, a chicken, or a mouse. However, the expression of the genetically determined developmental potential is also influenced by the environment—in some cases, very dramatically. Fetal alcohol syndrome is one example in which an environmental agent, chronic alcohol poisoning, affects various aspects of fetal growth and development.

Development includes many examples in which genes are selectively turned on or off by the action of regulatory proteins that respond to environmental signals. The identification of genetic regulatory interactions that operate during development is an important theme in developmental genetics. The implementation of different regulatory interactions means that genetically identical cells can become qualitatively different in the genes that are expressed.

Several common mechanisms by which genes become activated or repressed at particular stages in development or in particular tissues are illustrated in the hypothetical example in Figure 12.1. In this example, the developmentally regulated genes are denoted *r* (red), *g* (green), *b* (blue), and *p* (purple). The initial event is transcription of the *r* gene in the oocyte (Figure 12.1A). Unequal partitioning of the *r* gene product into one region of the egg establishes a polarity or regionalization of the egg with respect to presence of the r gene product (Figure 12.1B). When cleavage takes place (Figure 12.1C), the polarized cell produces daughter cells either with (L, left) or without (R, right) the r gene product. The r gene product is a transcriptional activator of the g gene, so the g gene product is expressed in cell L but not in cell R; hence, even at the two-cell stage, the daughter cells may differ in gene expression as a result of the initial polarization of the egg. Also in the two-cell stage, the p gene is expressed in both daughter cells. This gene codes for a **transmembrane receptor** containing amino acid sequences that span the cell membrane.

The presence of receptor molecules in the cell membrane provides an important mechanism for signaling between cells in development, as illustrated in this example with the regulation of the b gene in Figure 12.1D. In the absence of the transmembrane receptor, the b gene would be expressed in all four cells. However, expression of the b gene in one cell inhibits its expression in the neighboring cell because of the transmembrane receptor. The mechanism is that the product of the b gene stimulates the receptor of neighboring cells, and stimulation of the receptor represses transcription of b. The process in which gene expression in one cell inhibits expression of the same or differing genes in neighboring cells is known as *lateral inhibition.*

At the same time that b-gene expression is regulated by lateral inhibition, the initial activation of the g gene in cell L is retained in daughter cells 1 and 2 because the g gene product has a positive autoregulatory activity and stimulates its own transcription. Positive autoregulation is one way in which cells can amplify weak or transient regulatory signals into permanent changes in gene expression. The result of lateral inhibition of b expression and g autoregulation is that cells 1 through 4 in Figure 12.1E, though genetically identical, are developmentally different because they express different combinations of the g and b genes. Specifically, cell 1 has g and b activity, cell 2 has g activity only, cell 3 has b activity only, and cell 4 has neither.

Among the most intensively studied developmental genes are those that act early in development, before tissue differentiation, because these genes establish the overall pattern of development.

12.2 Early Embryonic Development in Animals

The early development of the animal embryo establishes the basic developmental plan for the whole organism. Fertilization initiates a series of mitotic **cleavage** divisions in which the embryo becomes multicellular. There is little or no increase in overall size compared with the egg, because the cleavage divisions are accompanied by little growth and merely partition the fertilized egg into progressively smaller cells. The cleavage divisions form the **blastula,** which is essentially a ball of about 10^4 cells containing a cavity (Figure 12.2). Completion of the cleavage divisions is followed by the formation of the **gastrula** through an infolding of part of the blastula wall and extensive cellular migration. In the reorganization of cells in the gastrula, the cells become arranged in several distinct layers. These layers establish the basic body plan of an animal. In higher plants, as we shall see later, the developmental processes differ substantially from those in animals.

A fertilized animal egg has full developmental potential because it contains the genetic program in its nucleus and macromolecules present in the oocyte cytoplasm that are necessary for giving rise to an entire organism. Full developmental potential is maintained in cells produced by the early cleavage divisions. However, the developmental potential is progressively channeled and limited in early embryonic development. Cells within the blastula usually become committed to particular developmental outcomes, or **fates,** that limit the differentiated states possible among descendant cells.

Autonomous Development and Intercellular Signaling

Two principal mechanisms progressively restrict the developmental potential of cells within a lineage.

- Developmental restriction may be **autonomous,** which means that it is determined by genetically programmed changes in the cells themselves.

- Cells may respond to **positional information,** which means that developmental restrictions are imposed by the position of cells within the embryo. Positional information may be mediated by signaling interactions between neighboring cells or by gradients in concentration of morphogens. A **morphogen** is a molecule that participates directly in the control of growth and form during development.

The distinction between autonomous development and development that depends on positional information is illustrated in Figure 12.3. In the normal embryo (Figure 12.3A), cells 1 and 2 have different fates either because of autonomous developmental programs or because they respond to positional information near the anterior and posterior ends. When the cells are transplanted (Figure 12.3B), autonomous development is indicated if the developmental fate of cells 1 and 2 is unchanged in spite of their new locations; in this case, the embryo has anterior and posterior reversed. When development depends on positional information, the transplanted cells respond to their new locations, and the embryo is normal.

Restriction of developmental fate is usually studied by transplanting cells of the embryo to new locations to determine whether they can substitute for the cells that they displace. Alternatively, individual cells are isolated from early embryos and cultured in laboratory dishes to study their developmental potential. In some eukaryotes, such as the soil nematode *Caenorhabditis elegans,* many lineages develop autonomously according to genetic programs that are induced by interactions with neighboring cells very early in embryonic development. Subsequent cell interactions also are important, and each stage in development is set in motion by the successful completion of the preceding stage. Figure 12.4 illustrates the first three cell

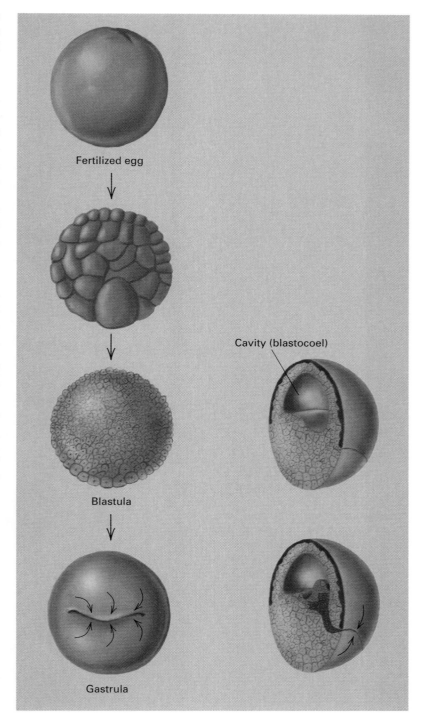

Fertilized egg

Cavity (blastocoel)

Blastula

Gastrula

Figure 12.2

Early development of the animal embryo. The cleavage divisions of the fertilized egg result in first a clump of cells and then a hollow ball of cells (the blastula). Extensive cell migrations form the gastrula and establish the basic body plan of the embryo.

(A) Normal embryo

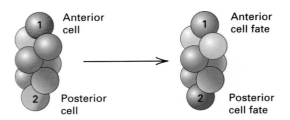

(B) Cell transplantation

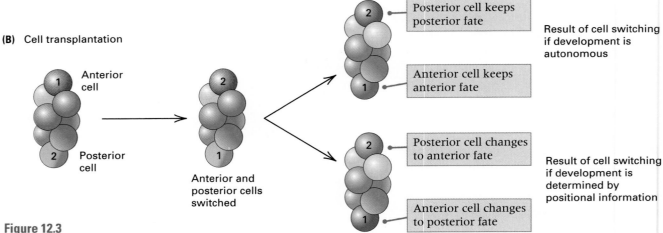

Figure 12.3
Distinction between autonomous determination and positional information. (A) Cells 1 and 2 differentiate normally as shown. (B) Transplantation of the cells to reciprocal locations. If the transplanted cells differentiate autonomously, then they differentiate as they would in the normal embryo, but in their new locations. If they differentiate in response to positional information (signals from neighboring cells), then their new positions determine their fate.

Figure 12.4
Early cell divisions in *C. elegans* development. (A) Spatial organization of cells. (B) Lineage relationships of the cells. The transmission of the polar granules illustrates cell-autonomous development. The arrows denote cell-to-cell signaling mechanisms that determine developmental fate. (From Wade Roush. 1996. *Science* 272: 1871.)

divisions in the development of *C. elegans,* which result in eight embryonic cells that differ in their genetic activity and developmental fate. The determination of cell fate in these early divisions is in part autonomous and in part a result of interactions between cells. Figure 12.4B shows the lineage relationships between the cells. Cell-autonomous mechanisms are illustrated by the transmission of cytoplasmic particles called polar granules from the cells P0 to P1 to P2 to P3. Normal segregation of the polar granules is a function of microfilaments in the cytoskeleton. Cell-signaling mechanisms are illustrated by the effects of P2 on EMS and on ABp. The EMS fate is determined by the activity of the *mom-2* gene in P2. The P2 cell also produces a signaling molecule, APX-1, which determines the fate of ABp through the cell-surface receptor GLP-1. (The specific mechanisms that determine early cell fate in *C. elegans* strongly resemble some of the general mechanisms outlined in Figure 12.1.) In contrast to *C. elegans,* in which many developmental decisions are cell-autonomous, in *Drosophila* and *Mus* (the mouse), regulation by cell-to-cell signaling is more the rule than the exception. The use of cell signaling to regulate development provides a sort of insurance that helps to overcome the death of individual cells that might happen by accident during development.

One special case of cell-to-cell signaling is **embryonic induction,** in which the development of a major embryonic structure is determined by a signal sent from neighboring cells. An example is found in early development of the sea urchin, *Strongylocentrotus purpuratus,* in which, after the fourth cleavage division, specialized cells called micromeres are produced at the ventral pole of the embryo(Figure 12.5A and B). After additional cell proliferation, the region of the embryo immediately above the micromeres folds inward to form a tube, the archenteron, that eventually develops into the stomach, intestine, and related structures. If micromere cells from a 16-cell embryo are removed and transplanted into the dorsal region of another embryo at the 8-cell stage (Figure 12.5C), then the next cleavage results in an embryo with two sets of micromeres (Figure 12.5D), the normal set being at the ventral end and the transplanted set at the dorsal

end. In these embryos, as development proceeds, an archenteron forms in the normal position from cells lying above the ventral micromeres. However, cells lying beneath the upper, transplanted micromeres also form an archenteron (Figure 12.5E), which indicates that the transplanted micromeres are capable of inducing the development of archenteron in the cells with which they come into contact.

Figure 12.5

Embryonic induction of the archenteron by micromere cells in the sea urchin. (A) Normal embryo at the 8-cell stage. (B) Normal embryo at the 16-cell stage showing micromeres on ventral side. (C) Transplanted 8-cell embryo with micromeres from another embryo placed at the dorsal end. (D) Transplanted embryo after the fourth cleavage division with two sets of micromeres. (E) The result of the transplant is an embryo with two archenterons. The doughnut-shaped objects in this embryo are the developing stomachs. [Photomicrograph, courtesy of Eric H. Davidson and Andrew Ransick. From A. Ransick and E. H. Davidson. 1993. *Science* 259: 1134.]

(A) Eight-cell stage   **(B)** Sixteen-cell stage

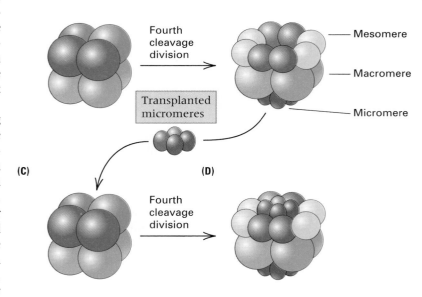

(E)

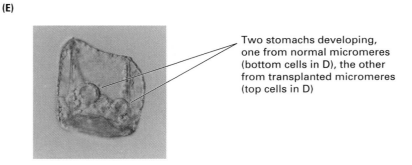

Two stomachs developing, one from normal micromeres (bottom cells in D), the other from transplanted micromeres (top cells in D)

Early Development and Activation of the Zygote Genome

In most animals, the earliest events in embryonic development do not depend on genetic information in the cell nucleus of the zygote. For example, fertilized frog eggs with the nucleus removed are still able to carry out the cleavage divisions and form rudimentary blastulas. Similarly, when gene transcription in sea urchin or amphibian embryos is blocked shortly after fertilization, there is no effect on the cleavage divisions or on blastula formation, but gastrulation does not take place. The reason why the genes in the zygote are not needed in the early stages of development is that the oocyte cytoplasm produced by the mother contains all the necessary macromolecules.

Following the period early in development in which the genes in the zygote nucleus are not needed, the embryo becomes dependent on the activity of its own genes. In mice, and probably in all mammals, the zygotic genes are needed much earlier than in lower vertebrates. The shift from control by the maternal genome to control by the zygote genome begins in the two-cell stage of development, when proteins coded by the zygote nucleus are first detectable. Inhibitors of transcription stop development of the mouse embryo when they are applied at any time after the first cleavage division. However, even in mammals, the earliest stages of development are greatly influenced by the cytoplasm of the oocyte, which determines the planes of the initial cleavage divisions and other events that ultimately affect cell fate.

Early activation of the zygote nucleus in mammals may be necessary because, in gamete formation, certain genes undergo a process of *imprinting* that prevents their expression in either the egg or the sperm nuclei that unite to form the zygote nucleus. Imprinting of a gene is thought to be associated with methylation of the DNA in the gene (Chapter 11). Very few genes—probably fewer than ten in the mouse—are subject to imprinting. Among these are the gene for an insulin-like growth factor (*Igf2*), which is imprinted during oogenesis, and the gene for the *Igf2* transmembrane receptor (*Igf2-r*), which is imprinted during spermatogenesis. Therefore, expression of both *Igf2* and *Igf2-r* in the embryo requires activation of the sperm nucleus for *Igf2* and of the egg nucleus for *Igf2-r*. Imprinting also affects some genes in the human genome and has been implicated in some unusual aspects of the genetic transmission of the fragile-X syndrome of mental retardation (Chapter 7).

Composition and Organization of Oocytes

The oocyte is a diploid cell during most of oogenesis. The cytoplasm of the oocyte is extensively organized and regionally differentiated (Figure 12.6). This spatial differentiation ultimately determines the different developmental fates of cells in the blastula. The cytoplasm of the animal egg has two essential functions:

1. Storage of the molecules needed to support the cleavage divisions and the rapid RNA and protein synthesis that take place in early embryogenesis.

2. Organization of the molecules in the cytoplasm to provide the positional

Figure 12.6
The animal oocyte is highly organized internally, which is revealed by the visible differences between the dorsal (dark) and ventral (light) parts of these *Xenopus* eggs. The regional organization of the oocyte determines many of the critical events in early development. [Courtesy of Michael W. Klymkowsky.]

information that results in differences between cells in the early embryo.

To establish the proper composition and organization of the oocyte requires the participation of many genes within the oocyte itself and gene products supplied by adjacent helper cells of various types (Figure 12.7). Numerous maternal genes are transcribed in oocyte development, and the mature oocyte typically contains an abundance of transcripts that code for proteins needed in the early embryo. Some of the maternal mRNA transcripts are stored complexed with proteins in special ribonucleoprotein particles that cannot be translated, and release of the mRNA, enabling it to be translated, does not happen until after fertilization.

Developmental instructions in oocytes are determined in part by the presence of distinct types of molecules at different positions within the cell and in part by gradients of morphogens that differ in concentration from one position to the next. Although the oocyte contains the products of many genes, only a small number of genes are expressed exclusively in the formation of the oocyte. Most genes expressed in oocyte formation are also important in the development of other tissues or at later times in development. Therefore, it is not only gene products but also the spatial organization of the gene products within the cell that give the oocyte its unique developmental potential.

12.3
Genetic Control of Cell Lineages

The mechanisms that control early development can be studied genetically by isolating mutations with early developmental abnormalities and altered cell fates. In most organisms, it is difficult to trace the lineage of individual cells in development because the embryo is not transparent, the cells are small and numerous, and cell migrations are extensive. The **lineage** of a cell refers to the ancestor-descendant relationships among a group of cells. A cell lineage can be illustrated with a **lineage diagram,** the sort of cell pedigree in Figure 12.8 that

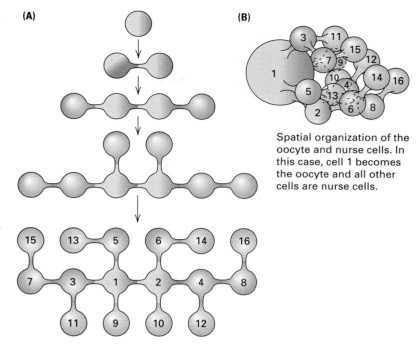

Spatial organization of the oocyte and nurse cells. In this case, cell 1 becomes the oocyte and all other cells are nurse cells.

Figure 12.7
Pattern of cell divisions in the development of the *Drosophila* oocyte. (A) The cells surrounding the oocyte are connected to it and to each other by cytoplasmic bridges. These cells synthesize products transported into the oocyte and contribute to its regional organization. (B) Geometrical organization of the cells. [From R. C. King. 1965. *Genetics,* Oxford University Press.]

Figure 12.8
Hypothetical cell-lineage diagram. Different terminally differentiated cell fates are denoted W, X, and Y. One cell in the lineage (A.aa) undergoes programmed cell death. The lowercase letters *a* and *p* denote anterior and posterior daughter cells. For example, cell A.ap is the posterior daughter of the anterior daughter of cell A.

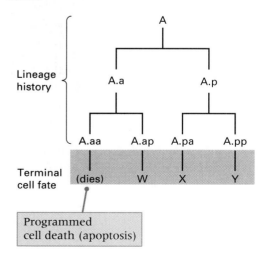

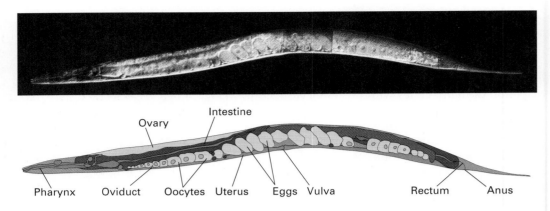

Figure 12.9

The soil nematode *Caenorhabditis elegans*. This organism offers several advantages for the genetic analysis of development, including the fact that all individuals of both sexes exhibit an identical pattern of cell lineages in the development of the somatic cells. DNA sequencing of the 100-megabase genome is nearing completion. [Photograph courtesy of Tim Schedl.]

shows each cell division and that indicates the terminal differentiated state of each cell. Figure 12.8 is another lineage diagram of a hypothetical cell A in which the cell fate is either programmed cell death or one of the terminally differentiated cell types designated W, X, and Y. The letter symbols are the kind normally used for cells in nematodes, in which the name denotes the cell lineage according to ancestry and position in the embryo. For example, the cells A.a and A.p are the anterior and posterior daughters of cell A, respectively; and A.aa and A.ap are the anterior and posterior daughters of cell A.a.

Genetic Analysis of Development in the Nematode

The soil nematode *Caenorhabditis elegans* (Figure 12.9) is popular for genetic studies because it is small, is easy to culture, and has a short generation time with a large number of offspring. The worms are grown on agar surfaces in petri dishes and feed on bacterial cells such as *E. coli*. Because they are microscopic in size, as many as 10^5 animals can be contained in a single petri dish. Sexually mature adults of *C. elegans* are capable of laying more than 300 eggs within a few days. At 20°C, it requires about 60 hours for the eggs to hatch,

undergo four larval molts, and become sexually mature adults.

Nematodes are diploid organisms with two sexes. In *C. elegans*, the two sexes are the hermaphrodite and the male. The hermaphrodite contains two X chromosomes (XX), produces both functional eggs and functional sperm, and is capable of self-fertilization. The male produces only sperm and fertilizes the hermaphrodites. The sex-chromosome constitution of *C. elegans* consists of a single X chromosome; there is no Y chromosome, and the male karyotype is XO.

The transparent body wall of the worm has made possible the study of the division, migration, and death or differentiation of all cells present in the course of development. Nematode development is unusual in that the pattern of cell division and differentiation is virtually identical from one individual to the next; that is, cell division and differentiation are highly stereotyped. The result is that both sexes show the same geometry in the number and arrangement of somatic cells. The hermaphrodite contains exactly 959 somatic cells, and the male contains exactly 1031 somatic cells. The complete developmental history of each somatic cell is known at the cellular and ultrastructural level, including the *wiring diagram*, which describes all the

Distinguished Lineages

John E. Sulston,[1] E. Schierenberg,[2] J. G. White,[1] and J. N. Thomson[1] 1983

[1]Medical Research Council Laboratory for Molecular Biology, Cambridge, England
[2]Max-Planck Institute for Experimental Medicine, Gottingen, Germany

The Embryonic Cell Lineage of the Nematode Caenorhabditis elegans

The data produced in this landmark study form the basis for interpreting developmental mutants in the nematode worm. This long paper offers voluminous data, and it is presently available through the Internet. During embryogenesis, 671 cells are generated; 113 of these, or 17%, undergo programmed cell death. What is the reason for such a high proportion of programmed cell deaths? The embryonic lineage is highly invariant—the same from one organism to the next. Why is there not more developmental flexibility, as is found in most other organisms? These issues are addressed in this excerpt, in which the emphasis is on the historical background and motivation of the study, the big picture of development, and interpretation of the lineage in terms of the evolution of the nematode. The technique of Nomarski microscopy is a modern invention also called differential interference contrast microscopy. When light passes through living material, it changes phase according to the refractive index of the material. Adjacent parts of a cell or organism that differ in refractive index cause different changes in phase. When two sets of waves combine after passing through an object, the difference in phase creates an interference pattern that yields an image of the object. The major advantage of Nomarski microscopy is that it can be used to observe living tissue.

This report marks the completion of a project begun over one hundred years ago—namely the determination of the entire cell lineage of a nematode. Nematode embryos were attractive to nineteenth-century biologists because of their simplicity and the reproducibility of their development, and considerable progress was made in determining their lineages by the use of fixed specimens. By the technique of Nomarski microscopy, which is nondestructive and yet produces high resolution, cells can be followed in living larvae. The use of living material lends a previously unattainable continuity and certainty to the observations, and has permitted the origin and fate of every cell in one nematode species [*Caenorhabditis elegans*] to be determined. Thus, not only are the broad relationships between tissues now known unambiguously but also the detailed pattern of cell fates is clearly revealed. . . . The lineage is of

---

The nematode belongs to an ancient phylum, and its cell lineage is a piece of frozen evolution. In the course of time, new cell types were generated from precursors selected not so much for their intrinsic properties as for the accident of their position in the embryo

---

significance both for what it can tell us immediately about relationships between cells and also as a framework into which future observations can be fitted. . . . The embryonic cell lineage is essentially invariant. The patterns of division, programmed cell death, and terminal differentiation are constant from one individual to another, and no great differences are seen in timing. . . . We shall use the term "sublineage" as an abbreviation for the more descriptive, but cumbersome, phrase "intrinsically determined sublineage"— namely, a fragment of the lineage which is thought to be generated by a programme within its precursor cell. . . . Two of the available criteria for postulating the existence of a sublineage are: (1) the

generation of the same lineage, giving rise to the same cell fates, from a series of precursors of diverse origin and position; (2) evidence for cell autonomy within the lineage, obtained from laser ablation experiments or the study of mutants. . . . The large number of programmed cell deaths, and their reproducibility, is evident from the lineage. The most likely reason for the occurrence of most of them is that, because of the existence of sublineages, unneeded cells are frequently generated along with needed ones. . . . Perhaps the most striking findings are firstly the complexity and secondly the cell autonomy of the lineages. . . . With hindsight, we can rationalize both this complexity and this rigidity. The nematode belongs to an ancient phylum, and its cell lineage is a piece of frozen evolution. In the course of time, new cell types were generated from precursors selected not so much for their intrinsic properties as for the accident of their position in the embryo. . . . Cell-cell interactions that were initially necessary for developmental decisions may have been gradually supplanted by autonomous programmes that were fast, economical, and reliable, the loss of flexibility being outweighed by the gain in efficiency. On this view, the perverse assignments, the cell deaths, the long-range migration—all the features which could, it seems, be eliminated from a more efficient design—are so many developmental fossils. These are the places to look for clues both to the course of evolution and to the mechanisms by which the lineage is controlled today.

Source: Developmental Biology 100: 64–119

interconnections among cells in the nervous system.

Nematode development is largely autonomous, which means that in most cells, the developmental program unfolds automatically with no need for interactions with other cells. However, in the early embryo, some of the developmental fates are established by interactions among the cells. In later stages of development of these cells, the fates established early are reinforced by still other interactions between cells. Worm development also provides important examples of the effects of intercellular signaling on determination (for example, those shown in Figure 12.4).

Mutations That Affect Cell Lineages

Many mutations that affect cell lineages have been studied in nematodes, and they reveal several general features by which genes control development.

- The division pattern and fate of a cell are generally affected by more than one gene and can be disrupted by mutations in any of them.

- Most genes that affect development are active in more than one type of cell.

Figure 12.10
Transformation mutations cause cells to undergo abnormal terminal differentiation. (A) The wildtype lineage. (B) A mutant lineage in which the cell A.ap differentiates into a Y-type cell rather than the normal W-type cell.

- Complex cell lineages often include simpler, genetically determined lineages within them; these components are called *sublineages* because they are expressed as an integrated pattern of cell division and terminal differentiation.

- The lineage of a cell may be triggered autonomously within the cell itself or by signaling interactions with other cells.

- Regulation of development is controlled by genes that determine the different sublineages that cells can undergo and the individual steps within each sublineage.

The next section deals with some of the types of mutations that affect cell lineages and development.

Types of Lineage Mutations

Mutations can affect cell lineages in two major ways. One is through nonspecific metabolic blocks—for example, in DNA replication—that prevent the cells from undergoing division or differentiation. The other is through specific molecular defects that result in patterns of division or differentiation that are characteristic of cells normally found elsewhere in the embryo or at a different time in development. From the standpoint of genetic analysis of development, the latter class of mutants is the more informative, because the mutant genes must be involved in developmental processes rather than in general "housekeeping" functions found in all cells. *C. elegans* provides several examples of each of the following general types of lineage mutations.

Mutations that cause cells to undergo developmental fates characteristic of other types of cells are called **transformation mutations.** For example, in the wildtype cell lineage shown in Figure 12.10A, the cell designated A.ap differentiates into cell-type W. In the transformation mutant (Figure 12.10B), cell A.ap differentiates into cell-type Y. Transformation mutations have the consequence that one or more differentiated cell types are missing from the embryo and are replaced with extra,

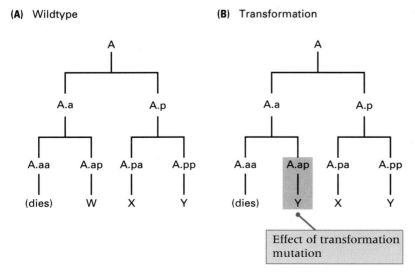

(A) Wildtype

(B) Transformation

supernumerary copies of other cell types. In this example, the missing cell type is W, and the extra cell type is Y. In the nematode, the mutation *unc-86* (*unc* = *uncoordinated*) provides an example in which the supernumerary cells become neurons. Supernumerary cells can impair development by interfering with the normal developmental process in nearby cells. In Section 12.4, we will see that transformation mutations in nematodes are analogous to a class of mutations in *Drosophila* called *homeotic* mutations, in which certain cells undergo developmental fates that are normally characteristic of other cells. Indeed, the wildtype *unc-86* gene codes for a transcription factor that contains a DNA-binding domain resembling that found in the proteins of the *Drosophila* homeotic genes.

Developmental programs often require sister cells or parent and offspring cells to adopt different fates. Mutations that cause sister cells or parent-offspring cells to fail to become differentiated from each other are called **segregation mutations** because the factors that govern the daughter cells' fates fail to segregate (become separated) in the daughter cells. Two examples based on the wildtype lineage in Figure 12.8 are illustrated in Figure 12.11. In the mutation in Figure 12.11A, the sister cells A.a and A.p give rise to sublineages in which the anterior derivative undergoes programmed cell death and the posterior derivative adopts fate W. This pattern contrasts with the wildtype situation in Figure 12.8, in which A.a and A.p give rise to different sublineages. In Figure 12.11B, the daughter cell A.pp adopts the fate of its parent cell, A.p, in that A.pp divides, the anterior daughter A.ppa adopts fate X, and the posterior daughter A.ppp divides again. Continuation of this pattern results in a group of supernumerary cells of type X.

Figure 12.11

Two types of segregation mutations, which cause related cells to fail to become differentiated from each other (see Figure 12.8 for the wildtype lineage). (A) Aberrant sister-cell segregation, in which the mutant cell A.p adopts the same lineage as its sister cell A.a instead of undergoing its normal fate. (B) Aberrant parent-offspring segregation, in which the mutant cell A.pp, like its parent A.p, undergoes cell division, the anterior daughter of which differentiates into an X-type cell. In both of these types of segregation mutations, the result of abnormal segregation is the absence of certain cell types and the presence of supernumerary copies of other cell types.

(A) Aberrant sister-cell segregation

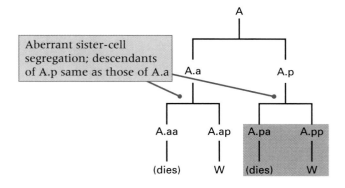

(B) Aberrant parent-offspring segregation

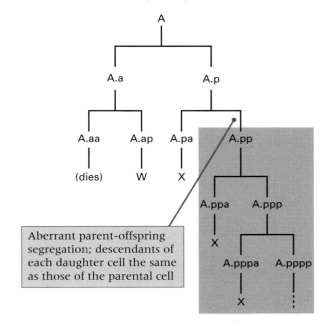

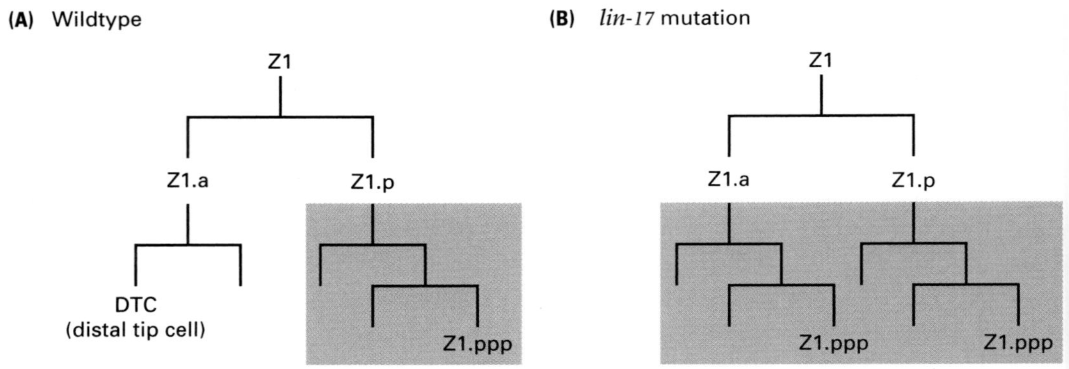

(A) Wildtype

Z1

Z1.a   Z1.p

DTC
(distal tip cell)   Z1.ppp

(B) *lin-17* mutation

Z1

Z1.a   Z1.p

Z1.ppp   Z1.ppp

Figure 12.12
Aberrant sister-cell segregation caused by the *lin-17* mutation in *C. elegans*. (A) Z1.a and Z1.p normally undergo different fates, and the Z1.a lineage includes a distal tip cell (DTC). (B) In the *lin-17* segregation mutant, Z1.a and Z1.p have the same fate, and the distal tip cell is absent.

Mutations in the *lin-17* gene in *C. elegans* result in sister-cell segregation defects. In males, a gonadal precursor cell, Z1, produces daughter cells that are different in that the sister cells Z1.a and Z1.p have different lineages and fates (Figure 12.12A). In *lin-17* mutants, the developmental determinants do not segregate, and the sister cells have the same lineage and fate as the normal Z1.p (Figure 12.12B).

When mutant cells are unable to execute their normal developmental fates, the mutation is an **execution mutation.** The *lin-11* gene in *C. elegans* provides an exam-

ple of an execution mutation. In the normal development of the vulva in the hermaphrodite, a lineage designated the 2° lineage gives rise to four cells in the spatial pattern N-T-L-L (Figure] 12.13A), in which N indicates a cell that does not divide, T a cell that divides in a transverse plane relative to the orientation of the larva, and L a cell that divides in a longitudinal plane. In *lin-11* mutants, the 2° lineage is not executed, and the four cells divide in the abnormal pattern L-L-L-L (Figure 12.13B).

The process of *programmed cell death*, technically known as **apoptosis,** is an

Figure 12.13
Example of an execution mutation. (A) Wildtype development of the 2° lineage in the vulva of *C. elegans*. The letter N denotes no further cell division, and T and L denote cell division in either a transverse (T) or a longitudinal (L) plane with respect to the embryo. (B) The *lin-11* mutation results in failure to execute the N and T sublineage. All cells divide in a longitudinal plane.

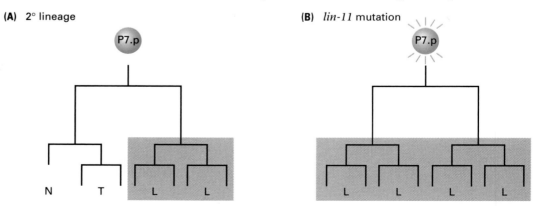

(A) 2° lineage

P7.p

N   T   L   L

(B) *lin-11* mutation

P7.p

L   L   L   L

important feature of normal development in many organisms. Apoptosis is a completely normal process in which, at the appropriate time in development, a cell commits suicide. In many cases, the signaling molecules that determine this fate have been identified (a number are known to be transcription factors). Failure of programmed cell death often results in specific developmental abnormalities. For example, compared with the wildtype lineage in Figure 12.14A, the lineage in Figure 12.14B is abnormal in that cell A.aa fails to undergo apoptosis and, instead, differentiates into the cell-type V. Phenotypically, when apoptosis fails and the surviving cells differentiate into recognizable cell types, the result is the presence of supernumerary cells of that type. For example, with mutations in the *ced-3* gene (*ced = cell death abnormal*) in *C. elegans,* a particular cell that normally undergoes programmed cell death survives and often differentiates into a supernumerary neuron. There are exactly 113 programmed cell deaths in the development of the *C. elegans* hermaphrodite. None of these deaths is essential. Mutants that cannot execute programmed cell death are viable

and fertile but are slightly impaired in development and in some sensory capabilities. On the other hand, mutants in *Drosophila* that fail to execute apoptosis are lethal, and in mammals, including human beings, failure of programmed cell death results in severe developmental abnormalities or, in some instances, leukemia or other forms of cancer.

The events of development are coordinated in time, so mutations that affect the timing of developmental events are of great interest. Mutations that affect timing are called **heterochronic mutations.** An example is shown in Figure 12.14C. In comparison with the wildtype situation in Figure 12.14A, the heterochronic mutant cell A.a delays cell division until the daughter cells A.aa and A.ap become contemporaneous with X and Y. Therefore, the heterochronic mutant in Figure 12.14C is *retarded* in that developmental events are normal but delayed. Heterochronic mutations can also be *precocious* in the expression of developmental events at times earlier than normal. For example, in certain heterochronic mutants in *C. elegans,* specific sublineages that normally develop in males

Figure 12.14

Apoptosis (programmed cell death) and heterochrony. (A) A wildtype lineage. (B) Failure of apoptosis, in which cell A.aa does not die but instead differentiates into a V-type cell. In some cases in which programmed cell death fails, the surviving cells differentiate into identifiable types. (C) Heterochronic mutants are abnormal in the timing of events in development. Retarded mutants undergo normal events too late, and precocious mutants undergo normal events too early. The example shown here is a retarded mutant in which cell A.a delays division until its products (A.aa and A.ap) are contemporaneous with the X and Y cells derived from A.p.

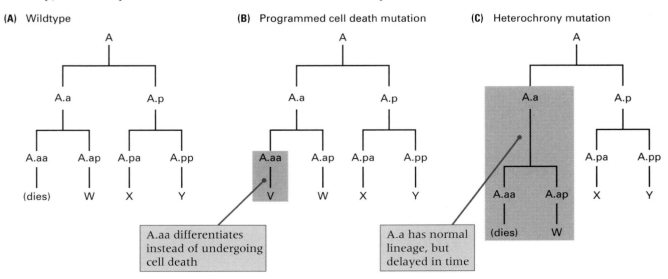

only at the fourth larval molt develop precociously in the mutant in the second or third larval molt.

The *lin-12* Developmental-Control Gene

Control genes that cause cells to diverge in developmental fate are not always easy to recognize. For example, an execution mutation may identify a gene that is necessary for the expression of a particular developmental fate, but the gene may not actually control or determine the developmental fate of the cells in which it is expressed. This possibility complicates the search for genes that control major developmental decisions.

Genes that control decisions about cell fate can sometimes be identified by the unusual characteristic that dominant or recessive mutations have opposite effects;

that is, if alternative alleles of a gene result in opposite cell fates, then the product of the gene must be both necessary and sufficient for expression of the fate. Identification of possible regulatory genes in this way excludes the large number of genes whose functions are merely necessary, but not sufficient, for the expression of cell fate. Recessive mutations in genes controlling development often result from **loss of function** in that the mRNA or the protein is not produced; loss-of-function mutations are exemplified by nonsense mutations that cause polypeptide chain termination in translation (Chapter 10). Dominant mutations in developmental-control genes often result from **gain of function** in that the gene is overexpressed or is expressed at the wrong time.

In *C. elegans,* only a small number of genes have dominant and recessive alleles that affect the same cells in opposite ways. Among them is the *lin-12* gene, which controls developmental decisions in a number of cells. One example concerns the cells denoted Z1.ppp and Z4.aaa in Figure 12.15. These cells lie side by side in the embryo, but they have quite different lineages (cell P_0 is the zygote). Normally, one of the cells differentiates into an *anchor cell* (AC), which participates in development of the vulva, and the other differentiates into a *ventral uterine precursor cell* (VU) (Figure 12.16A). Either Z1.ppp or Z4.aaa may become the anchor cell with equal likelihood.

Direct cell-cell interaction between Z1.ppp and Z4.aaa controls the AC-VU decision. If either cell is burned away (ablated) by a laser microbeam, then the remaining cell differentiates into an anchor cell (Figure 12.16B). This result implies that the preprogrammed fate of both Z1.ppp and Z4.aaa is that of an anchor cell. When either cell becomes committed to the anchor-cell fate, its contact with the other cell elicits the latter's fate as the ventral uterine precursor cell. As noted, recessive and dominant mutations of *lin-12* have opposite effects. Mutations in which *lin-12* activity is lacking or is greatly reduced are denoted *lin-12(0)*. These mutations are recessive, and in the mutants, both Z1.aaa and Z4.aaa become anchor cells (Figure 12.16C). In contrast, *lin-12(d)* mutations are those in which *lin-12* activity is overex-

Figure 12.15

Complete lineage of Z1.ppp and Z4.aaa in *C. elegans.* P_0 represents the zygote, and the dashed lines indicate three cell divisions not shown. In the normal development of the vulva, Z1.ppp and Z4.aaa are equally likely to differentiate into the anchor cell. Whichever cell remains differentiates into a ventral uterine precursor cell.

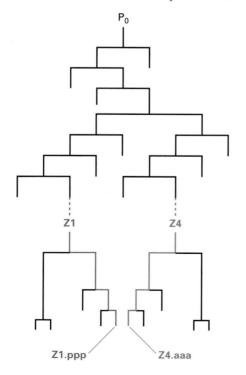

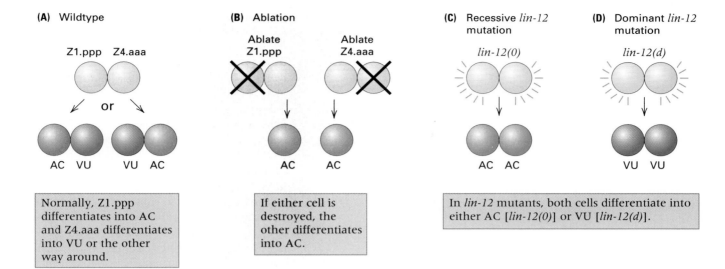

(A) Wildtype

Z1.ppp Z4.aaa

or

AC VU   VU AC

Normally, Z1.ppp
differentiates into AC
and Z4.aaa differentiates
into VU or the other
way around.

(B) Ablation

Ablate
Z1.ppp

Ablate
Z4.aaa

AC   AC

If either cell is
destroyed, the
other differentiates
into AC.

(C) Recessive *lin-12*
mutation

lin-12(0)

AC AC

(D) Dominant *lin-12*
mutation

lin-12(d)

VU VU

In *lin-12* mutants, both cells differentiate into
either AC [*lin-12(0)*] or VU [*lin-12(d)*].

Figure 12.16

Control of the fates of Z1.ppp and Z4.aaa in vulval development. For the complete lineage of these
cells, see Figure 12.15. (A) In wildtype cells, each cell has an equal chance of becoming the anchor
cell (AC); the other becomes a ventral uterine precursor cell (VU). (B) If either cell is destroyed
(ablated) by a laser beam, then the other differentiates into the anchor cell. (C) Genetic control of
cell fate by the *lin-12* gene. With recessive loss-of-function mutations [*lin-12(0)*], both cells become
anchor cells. (D) With dominant gain-of-function mutations [*lin-12(d)*], both cells become ventral
uterine precursor cells.

pressed. These mutations are dominant or
partly dominant, and in the mutants, both
Z1.aaa and Z4.ppp become ventral uterine
precursor cells (Figure 12.16D).

The effects of *lin-12* mutations suggest
that the wildtype gene product is a receptor
of a developmental signal. The molecular
structure of the *lin-12* gene product is typical of a transmembrane receptor protein,
and it shares domains with other proteins
that are important in developmental control (Figure 12.17). The transmembrane
region separates the LIN-12 protein into an
extracellular part (the amino end) and an
intracellular part (the carboxyl end). The
extracellular part contains 13 repeats of a
domain found in a mammalian peptide
hormone, epidermal growth factor (EGF),
as well as in the product of the *Notch* gene
in *Drosophila*, which controls the decision
between epidermal and neural cell fates.
Nearer the transmembrane region, the
amino end contains three repeats of a
cysteine-rich domain also found in the
Notch gene product. Inside the cell, the carboxyl part of the LIN-12 protein contains

six repeats of a domain also found in the
genes *cdc10* and *SWI6*, which control cell
division in two species of yeast.

The anchor cell expresses a signaling
gene, called *lin-3*, that illustrates another
case in which either loss-of-function or
gain-of-function alleles have opposite
effects on phenotype. In the anchor cell,
the gene *lin-3* controls the fate of certain
cells in the development of the vulva.
Figure 12.18 illustrates five precursor cells,
P4.p through P8.p, that participate in the
development of the vulva. Each precursor
cell has the capability of differentiating into
one of three fates, called the 1°, 2°, and 3°
lineages, which differ in whether descendant cells remain in a syncytium (S) or
divide longitudinally (L), transversely (T),
or not at all (N). The precursor cells normally differentiate as shown in Figure
12.18, giving five lineages in the order 3°-
2°-1°-2°-3°. The vulva itself is formed from
the 1° and 2° cell lineages. The spatial
arrangement of some of the key cells is
shown in the photograph in Figure 12.19.
The black arrow indicates the anchor cell,

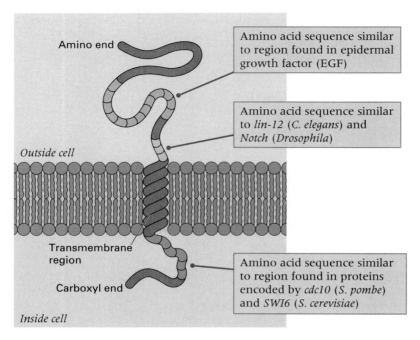

Figure 12.17
The structure of the LIN-12 protein is that of a receptor protein containing a transmembrane region and various types of repeated units that resemble those in epidermal growth factor (EGF) and other developmental control genes.

and the white lines show the pedigrees of 12 cells. The four cells in the middle derive from P6.p and the four on each side derive from P5.p and P7.p.

The important role of the *lin-3* gene product (LIN-3) is suggested by the opposite phenotypes of loss-of-function and gain-of-function alleles. Loss of LIN-3 results in the complete absence of vulval development. Conversely, overexpression of LIN-3 results in excess vulval induction. LIN-3 is a typical example of an interacting molecule, or **ligand,** that interacts with an EGF-type transmembrane receptor. In this case, the receptor is located in cell P6.p and is the product of the gene *let-23*. The LET-23 protein is a tyrosine-kinase receptor that, when stimulated by the LIN-3 ligand, stimulates a series of intracellular signaling events that ultimately results in the synthesis of transcription factors that determine the 1° fate. Among the genes that are induced is a gene for yet another ligand, which stimulates receptors on the cells P5.p and P7.p, causing these cells to adopt the 2° fate (blue horizontal arrows in Figure 12.18). The evidence for horizontal signaling is found in genetic mosaic worms in which the LET-23 receptor is missing in some or all of P5.p through P7.p. If the receptor is missing in all three cell types, none of the cells adopts its normal fate.

Figure 12.18
Determination of vulval differentiation by means of intercellular signaling. Cells P4.p through P8.p in the hermaphrodite give rise to lineages in the development of the vulva. The three types of lineages are designated 1°, 2°, and 3°. The 1° lineage is induced in P6.p by the ligand LIN-3 produced in the anchor cell (AC), which stimulates the LET-23 receptor tyrosine kinase in P6.p. The P6.p cell, in turn, produces a ligand that stimulates receptors in P5.p and P7.p to induce the 2° fate. On the other hand, the 3° fate is the default or baseline condition, which P4.p and P8.p adopt normally and all cells adopt in the absence of AC.

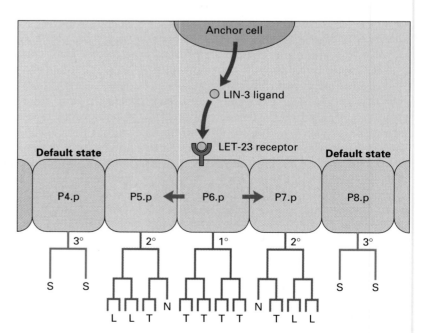

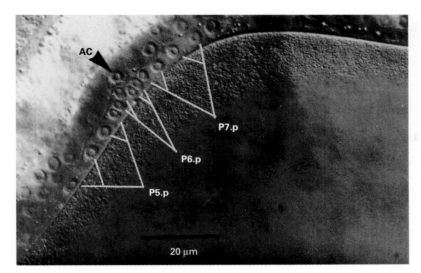

Figure 12.19

Spatial organization of cells in the vulva, including the anchor cell (black arrowhead) and the daughter cells produced by the first two divisions of P5.p through P7.p (white tree diagrams). The length of the scale bar equals 20 micrometers. [Courtesy of G. D. Jongeward, T. R. Clandinin, and P. W. Sternberg. 1995. *Genetics* 139: 1553.]

However, if the receptor is present in P6.p but absent in P5.p and P7.p, all three cell types differentiate as they should, which implies that receipt of the LIN-3 signal is necessary for 1° determination and that a stimulated P6.p is necessary for 2° determination.

In vulval development, the adoption of the 3° lineages by the P4.p and P8.p cells is determined not by a positive signal, but by the lack of a signal, because in the absence of the anchor cell, all of the cells P4.p through P8.p express the 3° lineage. Thus development of the 3° lineage is the uninduced or *default* state, which means that the 3° fate is preprogrammed into the cell and must be overridden by another signal if the cell's fate is to be altered.

12.4
Development in *Drosophila*

Many important insights into developmental processes have been gained from genetic analysis in *Drosophila*. In 1995, the pioneering work of Christiane Nüsslein-Volhard, Eric Wieschaus, and Edward B. Lewis was recognized with the awarding of the Nobel Prize in Physiology or Medicine.

The developmental cycle of *D. melanogaster*, summarized in Figure 12.20, includes egg, larval, pupal, and adult stages. Early development includes a series of cell divisions, migrations, and infoldings that result in the gastrula. About 24 hours after fertilization, the first-stage larva, composed of about 10^4 cells, emerges from the egg. Each larval stage is called an *instar*. Two successive larval molts that give rise to the second and third instar larvae are followed by pupation and a complex metamorphosis that gives rise to the adult fly composed of more than 10^6 cells. In wildtype strains reared at 25°C, development requires from 10 to 12 days.

Early development in *Drosophila* takes place within the egg case (Figure 12.21A). The first nine mitotic divisions occur in rapid succession without division of the cytoplasm and produce a cluster of nuclei within the egg (Figure 12.21B). The nuclei migrate to the periphery, and the germ line is formed from about 10 **pole cells** set off at the posterior end (Figure 12.21C); the pole cells undergo two additional divisions and are reincorporated into the embryo by invagination. The nuclei within the embryo undergo four more mitotic divisions without division of the cytoplasm, forming the **syncytial blastoderm**, which contains about 6000 nuclei (Figure 12.21D). Cellularization of the blastoderm takes place from about 150 to 180 minutes after fertilization by the synthesis of membranes that separate the nuclei. The **blastoderm** formed by cellularization (Figure 12.21E) is a flattened hollow ball of cells that corresponds to the blastula in other animals.

The experimental destruction of patches of cells within a *Drosophila* blastoderm results in localized defects in the larva and adult. The spatial correlation between the position of the cells destroyed and the type

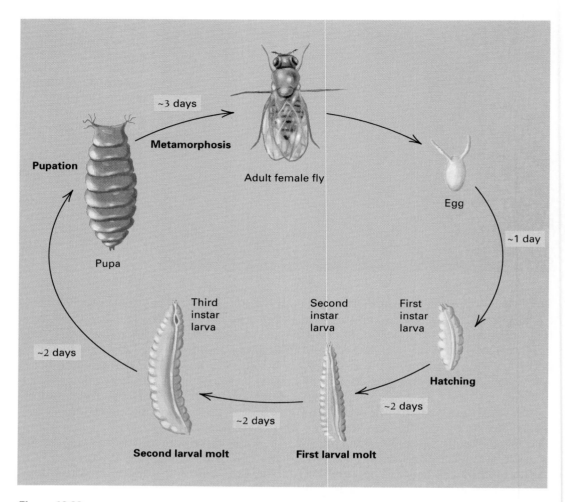

Figure 12.20
Developmental program of *Drosophila melanogaster*. The durations of the stages are at 25°C.

of defects results in a **fate map** of the blastoderm, which specifies the cells in the blastoderm that give rise to the various larval and adult structures (Figure 12.22). Use of genetic markers in the blastoderm has made possible further refinement of the fate map. Cell lineages can be genetically marked during development by inducing recombination between homologous chromosomes in mitosis, resulting in genetically different daughter cells (Chapter 4). Much like the cells in the early blastula of *Caenorhabditis,* cells in the blastoderm of *Drosophila* have predetermined developmental fates, with little ability to substitute in development for other, sometimes even adjacent, cells.

Evidence that blastoderm cells in *Drosophila* have predetermined fates comes from experiments in which cells from a genetically marked blastoderm are implanted into host blastoderms. Blastoderm cells implanted into the equivalent regions of the host become part of the normal adult structures. However, blastoderm cells implanted into different regions develop autonomously and are not integrated into host structures.

Because of the relatively high degree of determination in the blastoderm, genetic analysis of *Drosophila* development has tended to focus on the early stages of development when the basic body plan of the embryo is established and key regulatory

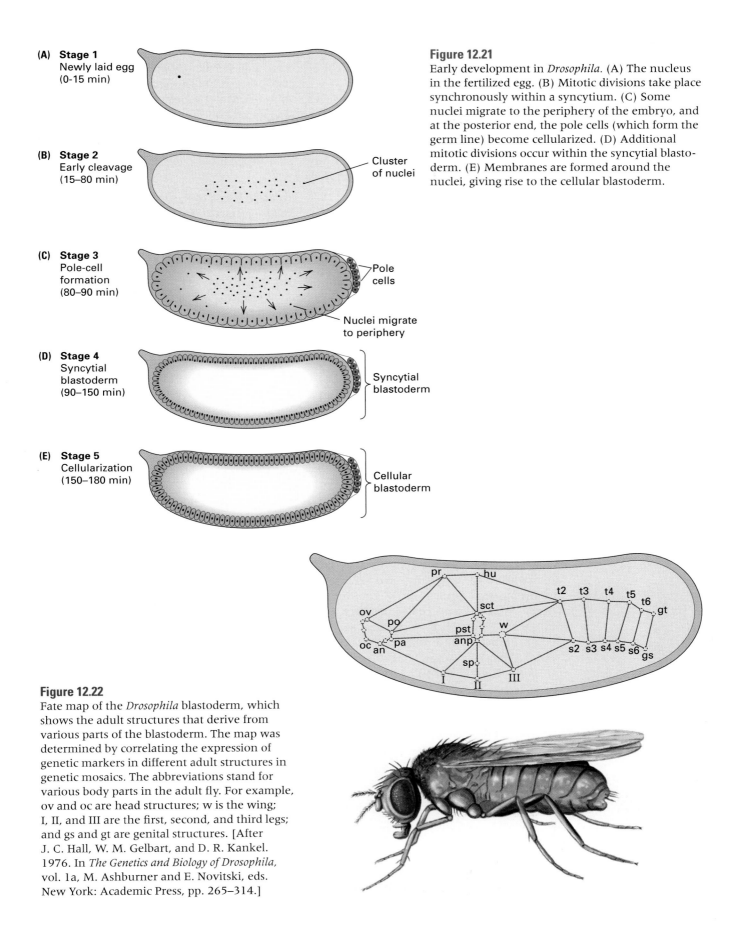

(A) Stage 1
Newly laid egg
(0-15 min)

(B) Stage 2
Early cleavage
(15–80 min)

Cluster of nuclei

(C) Stage 3
Pole-cell formation
(80–90 min)

Pole cells

Nuclei migrate to periphery

(D) Stage 4
Syncytial blastoderm
(90–150 min)

Syncytial blastoderm

(E) Stage 5
Cellularization
(150–180 min)

Cellular blastoderm

Figure 12.21
Early development in *Drosophila*. (A) The nucleus in the fertilized egg. (B) Mitotic divisions take place synchronously within a syncytium. (C) Some nuclei migrate to the periphery of the embryo, and at the posterior end, the pole cells (which form the germ line) become cellularized. (D) Additional mitotic divisions occur within the syncytial blastoderm. (E) Membranes are formed around the nuclei, giving rise to the cellular blastoderm.

Figure 12.22
Fate map of the *Drosophila* blastoderm, which shows the adult structures that derive from various parts of the blastoderm. The map was determined by correlating the expression of genetic markers in different adult structures in genetic mosaics. The abbreviations stand for various body parts in the adult fly. For example, ov and oc are head structures; w is the wing; I, II, and III are the first, second, and third legs; and gs and gt are genital structures. [After J. C. Hall, W. M. Gelbart, and D. R. Kankel. 1976. In *The Genetics and Biology of Drosophila*, vol. 1a, M. Ashburner and E. Novitski, eds. New York: Academic Press, pp. 265–314.]

processes become activated. The following sections summarize the genetic control of these early events.

Maternal-Effect Genes and Zygotic Genes

Early development in *Drosophila* requires translation of maternal mRNA molecules present in the oocyte. Blockage of protein synthesis during this period arrests the early cleavage divisions. Expression of the zygote genome is also required, but the timing varies. Blockage of transcription of the zygote genome at any time after the ninth cleavage division prevents formation of the blastoderm.

Because the earliest stages of *Drosophila* development are programmed in the oocyte, mutations that affect oocyte composition or structure can upset development of the embryo. Genes that function in the mother and are needed for development of the embryo are called **maternal-effect genes,** and developmental genes that function in the embryo are called **zygotic genes.**The interplay between the two types of genes is as follows:

> The zygotic genes interpret and respond to the positional information laid out in the egg by the maternal-effect genes.

Mutations in maternal-effect genes are easy to identify because homozygous females produce eggs that are unable to support normal embryonic development, whereas homozygous males produce normal sperm. Therefore, reciprocal crosses give dramatically different results. For example, a recessive maternal-effect mutation, *m*, will yield the following results in crosses:

$$m/m \; ♀ \times +/+ \; ♂ \rightarrow +/m \text{ progeny}$$
(abnormal development)

$$+/+ \; ♀ \times m/m \; ♂ \rightarrow +/m \text{ progeny}$$
(normal development)

The $+/m$ progeny of the reciprocal crosses are genetically identical, but development is upset when the mother is homozygous m/m.

The reason why maternal-effect genes are needed in the mother is that the maternal-effect genes establish the polarity of the *Drosophila* oocyte even before fertilization

takes place. They are active during the earliest stages of embryonic development, and they determine the basic body plan of the embryo. Maternal-effect mutations provide a valuable tool for investigating the genetic control of pattern formation and for identifying the molecules that are important in morphogenesis.

Genetic Basis of Pattern Formation in Early Development

The *Drosophila* embryo features 14 superficially similar repeating units visible as a pattern of stripes along the main trunk (Figure 12.23). The stripes can be recognized externally by the bands of *denticles,* which are tiny, pigmented, tooth-like projections from the surface of the larva. The 14 stripes in the embryo correspond to the segments in the larva that forms from the embryo. Each **segment** is defined morphologically as the region between successive indentations formed by the sites of muscle attachment in the larval cuticle. The designations of the segments are indicated in Figure 12.23. There are three head segments (C1–C3), three thoracic segments (T1–T3), and eight abdominal segments (A1–A8.). In addition to the segments, another type of repeating unit is also important in development. These repeating units are called **parasegments;** each parasegment consists of the posterior region of one segment and the anterior region of the adjacent segment. Parasegments have a transient existence in embryonic development. Although they are not visible morphologically, they are important in gene expression because the patterns of expression of many genes coincide with the boundaries of the parasegments rather than with the boundaries of the segments.

The early stages of pattern formation are determined by genes that are often called **segmentation genes** because they determine the origin and fate of the segments and parasegments. There are four classes of segmentation genes, which differ in their times and patterns of expression in the embryo.

1. The *coordinate genes* determine the principal coordinate axes of the embryo: the anterior-posterior axis, which defines

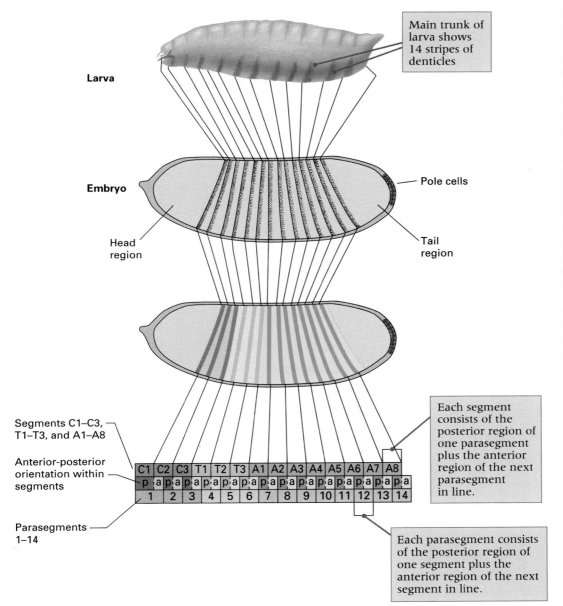

Main trunk of larva shows 14 stripes of denticles

Larva

Embryo

Pole cells

Head region

Tail region

Segments C1–C3, T1–T3, and A1–A8

Anterior-posterior orientation within segments

Parasegments 1–14

C1	C2	C3	T1	T2	T3	A1	A2	A3	A4	A5	A6	A7	A8
p a	p a	p a	p a	p a	p a	p a	p a	p a	p a	p a	p a	p a	p a
1	2	3	4	5	6	7	8	9	10	11	12	13	14

Each segment consists of the posterior region of one parasegment plus the anterior region of the next parasegment in line.

Each parasegment consists of the posterior region of one segment plus the anterior region of the next segment in line.

Figure 12.23
Segmental organization of the *Drosophila* embryo and larva. The segments are defined by successive indentations formed by the sites of muscle attachment in the larval cuticle. The parasegments are not apparent morphologically but include the anterior and posterior regions of adjacent segments. The distinction is important, because the patterns of expression of segmentation genes are more often correlated with the parasegment boundaries than with the segment boundaries.

the front and rear, and the dorsalventral axis, which defines the top and bottom.

2. The *gap genes* are expressed in contiguous groups of segments along the embryo (Figure 12.24A), and they establish the next level of spatial organization. Mutations in gap genes result in the absence of contiguous body segments, so gaps appear in the normal pattern of structures in the embryo.

3. The *pair-rule genes* determine the separation of the embryo into discrete segments (Figure 12.24B). Mutations in

pair-rule genes result in missing pattern elements in alternate segments. The reason for the two-segment periodicity of pair-rule genes is that the genes are expressed in a zebra-stripe pattern along the embryo.

4. The *segment-polarity genes* determine the pattern of anterior-posterior development within each segment of the embryo (Figure 12.24C). Mutations in segment-polarity genes affect all segments or parasegments in which the normal gene is active. Many segment-polarity mutations have the normal number of segments, but part of each

Figure 12.24

Patterns of expression of different types of segmentation genes. (A) The gap genes are expressed in a set of contiguous segments. (B) The pair-rule genes are expressed in alternating segments. (C) The segment-polarity genes are expressed in each segment and determine the anterior-posterior pattern of differentiation within each parasegment.

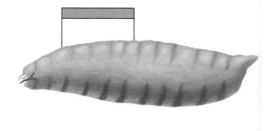

(A) Gap genes affect contiguous groups of segments

(B) Pair-rule genes affect alternating segments

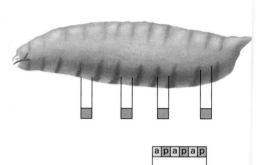

(C) Segment-polarity genes affect anterior-posterior polarity within each segment

segment is deleted and the remainder is duplicated in mirror-image symmetry.

Evidence for the existence of the four classes of segmentation genes—coordinate genes, gap genes, pair-rule genes, and segment-polarity genes—is presented in the following sections.

Coordinate Genes

The **coordinate genes** are maternal-effect genes that establish early polarity through the presence of their products at defined positions within the oocyte or through gradients of concentration of their products. The genes that determine the anterior-posterior axis can be classified into three groups according to the effects of mutations in them, as illustrated in Figure 12.25.

1. The first group of coordinate genes includes the *anterior genes,* which affect the head and thorax. The key gene in this class is *bicoid.* Mutations in *bicoid* produce embryos that lack the head and thorax and

occasionally have abdominal segments in reverse polarity duplicated at the anterior end. The *bicoid* phenotype resembles that produced by certain kinds of surgical manipulations. For example, when *Drosophila* eggs are punctured and small amounts of cytoplasm allowed to escape, loss of cytoplasm from the anterior end results in embryos in which some posterior structures develop in place of the head. Similarly, replacement of anterior cytoplasm with posterior cytoplasm by injection yields embryos with two mirror-image abdomens and no head.

The *bicoid* gene product is a transcription factor for genes determining anterior structures. Because the *bicoid* mRNA is localized in the anterior part of the early-cleavage embryo, these genes are activated primarily in the anterior region. The *bicoid* mRNA is produced in nurse cells (the cells surrounding the oocyte in Figure 12.7) and exported to a localized region at the anterior pole of the oocyte. The protein product is less localized and, during the syncytial cleavages, forms an anterior-posterior concentration gradient with the maximum at

Embryo Genesis

**Christiane Nüsslein-Volhard
and Eric Wieschaus 1980**
European Molecular Biology
Laboratory, Heidelberg, Germany
*Mutations Affecting Segment Number
and Polarity in* Drosophila

*Nüsslein-Volhard and Wieschaus were
exceptionally bold in supposing that the
molecular mechanisms governing a
process as complex as early embryonic
development could be understood by the
genetic and molecular analysis of mutations. The phenotype of such mutants is
superficially identical: The embryo dies.
The* Drosophila *genetic map was already
littered with mutations classified collectively as "recessive lethals." These were
generally considered as not amenable to
further analysis because, in any particular
case, the search for the specific defect
was regarded as a needle-in-a-haystack
problem. Nüsslein-Volhard and Wieschaus
ignored most of the existing mutants. They
set out to acquire systematically a new set
of recessive-lethal mutants, each showing
a specific and characteristic type of defect
in the formation of organized patterns in
the early embryo. Their first efforts,
reported in this paper, yielded a number of
mutations in each of three major classes of
genes concerned with development. The
paper sparked an enormous interest in*
Drosophila *developmental genetics. Today,
a typical Annual Drosophila Research
Conference includes approximately 500*

*presentations (mainly posters) dealing
with aspects of* Drosophila *development.
Nüsslein-Volhard and Wieschaus were
awarded a Nobel Prize in 1995. They
shared it with Edward B. Lewis for his
pioneering genetic studies of the homeotic
genes.*

The construction of complex form from
similar repeating units is a basic feature of
spatial organization in all higher animals.
Very little is known for any organism about
the genes involved in this process. In
Drosophila, the metameric [repeating]
nature of the pattern is most obvious in the
thoracic and abdominal segments of the
larval epidermis and we are attempting to

---

*In Drosophila, it would seem
feasible to identify all
genetic components involved
in the complex process of
embryonic pattern formation*

---

identify all loci required for the establishment of this pattern. . . . We have undertaken a systematic search for mutations
that affect the segmental pattern. We
describe here mutations at 15 loci which
show one of three novel types of pattern
alteration: pattern duplication (segment
polarity mutants; six loci), pattern deletion
in alternating segments (pair-rule mutants;

six loci) and deletion of a group of adjacent
segments (gap mutants; three loci). . . .
Segment polarity mutants have the normal
number of segments. However, in each
segment a defined fraction of the normal
pattern is deleted and the remainder is present as a mirror-image duplication. The
duplicated part is posterior to the 'normal'
part and has reversed polarity. . . . In pair-
rule mutants homologous parts of the pattern are deleted in every other segment.
Each of the six loci is characterized by its
own pattern of deletions. . . . One of the
striking features of the [segment polarity
and pair-rule] classes is that the alteration
in the pattern is repeated at specific intervals along the antero-posterior axis of the
embryo. No such repeated pattern is found
in mutants of the gap class and instead a
group of up to eight adjacent segments is
deleted. . . . The lack of a repeated pattern
suggests that the loci are involved in
processes in which position along the
antero-posterior axis of the embryo is
defined by unique values. . . . The majority
of mutants described here have been isolated in systematic searches for mutations
affecting the segmentation pattern. These
experiments are still incomplete. . . . In
Drosophila, it would seem feasible to identify all genetic components involved in the
complex process of embryonic pattern
formation.

Source: Nature 287: 795–801

---

the anterior tip of the embryo (Figure
12.26). The bicoid protein is a principal
morphogen in determining the blastoderm
fate map. The protein is a transcriptional
activator that contains a helix-turn-helix
motif for DNA binding (Chapter 11). Genes
affected by the bicoid protein contain multiple upstream binding domains that consist
of nine nucleotides resembling the consensus sequence 5'-TCTAATCCC-3'. Binding
sites that differ by as many as two base
pairs from the consensus sequence can
bind the bicoid protein with high affinity,
and sites that contain four mismatches bind
with low affinity. The combination of high-
and low-affinity binding sites determines
the concentration of bicoid protein needed
for gene activation; genes with many high-
affinity binding sites can be activated at low
concentrations, but those with many low-
affinity binding sites need higher concentrations. Such differences in binding
affinity mean that the level of gene expression can differ from one regulated gene to

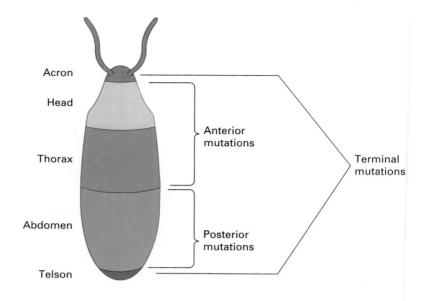

Figure 12.25

Regional differentiation of the early *Drosophila* embryo along the anterior-posterior axis. Mutations in any of the classes of genes shown result in elimination of the corresponding region of the embryo.

Acron

Head

Anterior mutations

Thorax

Terminal mutations

Abdomen

Posterior mutations

Telson

the next along the bicoid concentration gradient. One of the important genes activated by *bicoid* is the gap gene *hunchback*. Five other genes in the anterior class are known, and they code for cellular components necessary for *bicoid* localization.

2. The second group of coordinate genes includes the *posterior genes*, which affect the abdominal segments (Figure 12.25). Some of the mutants also lack pole cells. One of

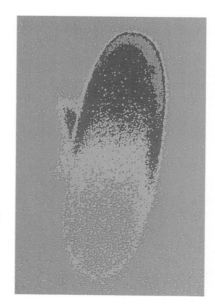

Figure 12.26

A gradient of gene expression resembling that of *bicoid* in the early *Drosophila* embryo. In this photograph, the intensity of the fluorescent signal has been pseudocolored so that the region of highest expression is pink and the region of lowest expression is green. [Courtesy of James Langeland, Stephen Paddock, and Sean Carroll.]

the posterior mutations, *nanos,* yields embryos with defective abdominal segmentation but normal pole cells, abnormalities that resemble those produced by surgical removal of the posterior cytoplasm. The phenotype does not result merely from a generalized disruption of development at the posterior end, because the pole cells—as well as a posterior structure called the telson, which normally develops between the pole cells and the abdomen—are not affected in either *nanos* or the surgically manipulated embryos. The *nanos* mRNA is localized tightly to the posterior pole of the oocyte, and the gene product is a repressor of translation. Among the genes whose mRNA is not translated in the presence of nanos protein is the gene *hunchback*. Hence *hunchback* expression is controlled jointly by the bicoid and nanos proteins; bicoid protein activates transcription in an anterior-posterior gradient, and nanos protein represses translation in the posterior region.

3. The third group of coordinate genes includes the *terminal genes*, which simultaneously affect the most anterior structure (the acron) and the most posterior structure (the telson) (Figure 12.25). The key gene in this class is *torso*, which codes for a transmembrane receptor that is uniformly distributed throughout the embryo in the

early developmental stages. The torso receptor is activated by a signal released only at the poles of the egg by the nurse cells in that location (Figure 12.7). The torso receptor is a tyrosine kinase that initiates cellular differentiation by means of phosphorylation of specific tyrosine residues in one or more target proteins, among them a *Drosophila* homolog of the vertebrate oncogene *D-raf*.

Apart from the three sets of genes that determine the anterior-posterior axis of the embryo, a fourth set of genes determines the dorsal-ventral axis. The morphogen for dorsal-ventral determination is the product of the gene *dorsal*, which is present in a pronounced ventral-to-dorsal gradient in the late syncytial blastoderm. The dorsal protein is a transcription factor related to the avian oncogene *v-rel*. An additional 16 other genes are known to affect dorsal-ventral determination. Mutations in these genes eliminate ventral and lateral pattern elements. In many cases, the mutant embryos can be rescued by the injection of wildtype cytoplasm, no matter where the wildtype cytoplasm is taken from or where it is injected. Examples include the genes called *snake, gastrulation-defective,* and *easter.* All three genes code for proteins called *serine proteases.* Serine proteases are synthesized as inactive precursors that require a specific cleavage for activation. They often act in a temporal sequence, which means that activation of one enzyme in the pathway is necessary for activation of the next enzyme in line (Figure 12.27). About half the clotting factors in human blood are serine proteases. The serial activation of the enzymes results in a **cascade effect** that greatly amplifies an initial signal. Each step in the cascade multiplies the signal produced in the preceding step.

Figure 12.27

Amplification of a signal by a cascade of activation. The number of activated components at each step increases exponentially. This is a simplified example with a threefold amplification at each step. The primed symbols denote inactive enzyme forms; the unprimed symbols denote active forms.

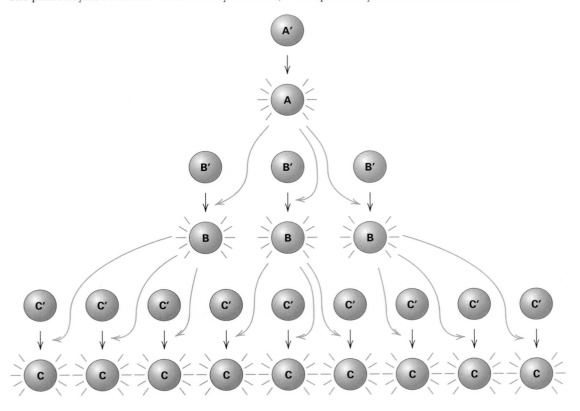

Gap Genes

The main role of the coordinate genes is to regulate the expression of a small group of approximately six gap genes along the anterior-posterior axis. The genes are called **gap genes** because mutations in them result in the absence of pattern elements derived from a group of contiguous segments (Figure 12.24A). Gap genes are zygotic genes. The gene *hunchback* serves as an example of the class because *hunchback* expression is controlled by offsetting effects of *bicoid* and *nanos*. Transcription of *hunchback* is stimulated in an anterior-to-posterior gradient by the bicoid transcription factor, but posterior *hunchback* expression is prevented by translational repression because of the posteriorly localized nanos protein. In the early *Drosophila* embryo in Figure 12.28, the gradient of *hunchback* expression is indicated by the green fluorescence of an antibody specific to the *hunchback* gene product. The superimposed red fluorescence results from antibody specific to the product of *Krüppel*, another gap gene. The region of overlapping gene expression appears in yellow. The products of both *hunchback* and *Krüppel* are transcription factors of the zinc finger type (Chapter 11). Other gap genes also are transcription factors. Together, the gap genes have a pattern of regional specificity and partly overlapping domains of expression that enable them to act in combinatorial fashion to control the next set of genes in the segmentation hierarchy, the pair-rule genes.

Pair-Rule Genes

The coordinate and gap genes determine the polarity of the embryo and establish broad regions within which subsequent development takes place. As development proceeds, the progressively more refined organization of the embryo is correlated with the patterns of expression of the segmentation genes. Among these are the **pair-rule genes,** in which the mutant phenotype has alternating segments absent or malformed (Figure 12.24B). Approximately eight pair-rule genes have been identified. For example, mutations of the pair-rule gene *even-skipped* affect even-numbered segments, and those of another pair-rule gene, *odd-skipped,* affect odd-numbered segments. The function of the pair-rule genes is to give the early *Drosophila* larva a segmented body pattern with both repetitiveness and individuality of segments. For example, there are eight abdominal segments that are repetitive in that they are regularly spaced and share several common features, but they differ in the details of their differentiation.

One of the earliest pair-rule genes expressed is *hairy*, whose pattern of expression is under both positive and negative regulation by the products of *hunchback*, *Krüppel,* and other gap genes. Expression of *hairy* yields seven stripes (Figure 12.29). The striped pattern of pair-rule gene expression is typical, but the stripes of expression of one gene are usually slightly out of register with those of another. Together with the continued regional expression of the gap genes, the combinatorial patterns of gene expression in the embryo are already complex and linearly differentiated. Figure 12.30 shows an embryo stained for the products of three genes: *hairy* (green), *Krüppel* (red), and *giant* (blue). The regions of overlapping expression appear as color mixtures—orange, yellow, light green, or purple. Even at the early stage in Figure 12.30, there is a unique combinatorial pattern of gene expression in every segment and parasegment. The complexity of combinatorial control can be appreciated by considering that the expression of the *hairy* gene in stripe 7 depends on a promoter element smaller than 1.5 kb that contains a series of binding sites for the protein products of the genes *caudal, hunchback, knirps, Krüppel ,* *tailless, huckbein, bicoid*, and perhaps still other proteins yet to be identified. The combinatorial patterns of gene expression of the pair-rule genes define the boundaries of expression of the segment-polarity genes, which function next in the hierarchy.

Segment-Polarity Genes

Whereas the pair-rule genes determine the body plan at the level of segments and parasegments, the **segment-polarity**

genes create a spatial differentiation within each segment. Approximately 14 segment-polarity genes have been identified. The mutant phenotype has repetitive deletion of pattern along the embryo (Figure 12.24C) and usually a mirror-image duplication of the part that remains. Among the earliest segment-polarity genes expressed is *engrailed,* whose stripes of expression approximately coincide with the boundaries of the parasegments and so divide each segment into anterior and posterior domains (Figure 12.31).

Expression of the segment-polarity genes finally establishes the early polarity and linear differentiation of the embryo into segments and parasegments. The regulatory interactions within the hierarchy of segmentation genes are illustrated in Figure 12.32. These interactions govern the activities of the second set of developmental genes, the *homeotic genes,* which control the pathways of differentiation in each segment or parasegment.

Figure 12.28
An embryo of *Drosophila,* approximately 2.5 hours after fertilization, showing the regional localization of the *hunchback* gene product (green), the Krüppel gene product (red), and their overlap (yellow). [Courtesy of James Langeland, Stephen Paddock, and Sean Carroll.

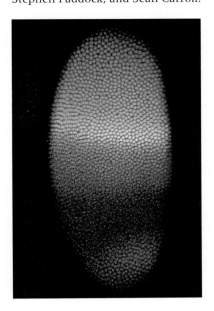

Figure 12.29
Characteristic seven stripes of expression of the gene *hairy* in a *Drosophila* embryo approximately 3 hours after fertilization. [Courtesy of James Langeland, Stephen Paddock, and Sean Carroll.]

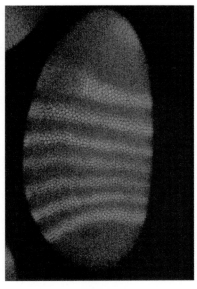

Figure 12.30
Combined patterns of expression of *hairy* (green), *Krüppel* (red), and *giant* (blue) in a *Drosophila* embryo approximately 3 hours after fertilization. Already considerable linear differentiation is apparent in the patterns of gene expression. [Courtesy of James Langeland, Stephen Paddock, and Sean Carroll.]

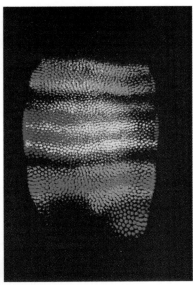

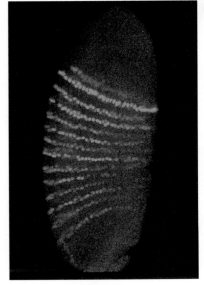

Figure 12.31
Expression of the segment-polarity gene *engrailed* partitions the early *Drosophila* embryo into 14 regions. These eventually differentiate into three head segments, three thoracic segments, and eight abdominal segments. [Courtesy of James Langeland, Stephen Paddock, and Sean Carroll.]

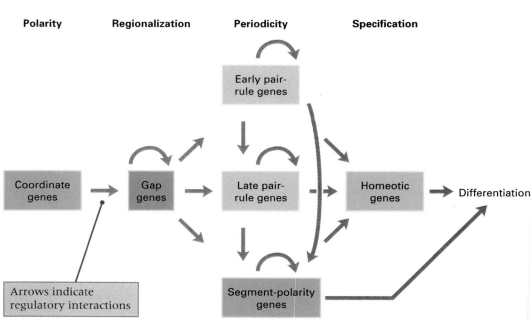

Figure 12.32

Hierarchy of regulatory interactions among genes controlling early development in *Drosophila*. Each gene is controlled by a unique combination of other genes. The terms *polarity, regionalization, periodicity,* and *specification* refer to the major developmental determinations that are made in each time interval.

Homeotic Genes

As with most other insects, the larvae and adults of *Drosophila* have a segmented body plan consisting of a head, three thoracic segments, and eight abdominal segments (Figure 12.33). Metamorphosis makes use of about 20 structures called **imaginal disks** present inside the larvae (Figure

Figure 12.33

Relationship between larval and adult segmentation in *Drosophila*. Each of the three thoracic segments in the adult carries a pair of legs. The wings develop on the second thoracic segment (T2) and the halteres (flight balancers) on the third thoracic segment (T3).

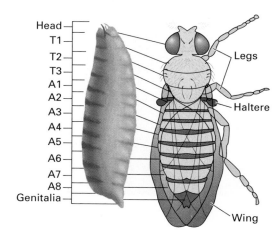

12.34). Formed early in development, the imaginal disks ultimately give rise to the principal structures and tissues in the adult organism. Examples of imaginal disks are the pair of wing disks (one on each side of the body), which give rise to the wings and related structures; the pair of eye-antenna disks, which give rise to the eyes, antennae, and related structures; and the genital disk, which gives rise to the reproductive apparatus. During the pupal stage, when many larval tissues and organs break down, the imaginal disks progressively unfold and differentiate into adult structures. The morphogenic events that take place in the pupa are initiated by the hormone ecdysone, secreted by the larval brain.

Cell determination in *Drosophila* also takes place within bounded units called **compartments.** Cells in the body segments and imaginal discs do not migrate across the boundaries between compartments. For example, the *Drosophila* wing disk includes five compartment boundaries, and most body segments include one

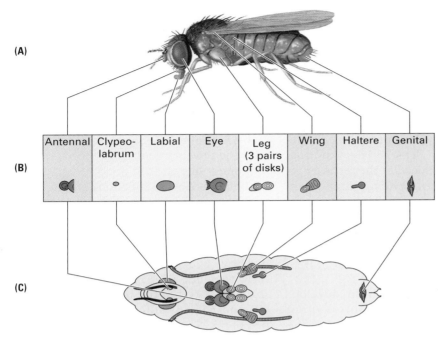

(A)

(B)

Antennal	Clypeo-labrum	Labial	Eye	Leg (3 pairs of disks)	Wing	Haltere	Genital

(C)

Figure 12.34

(A) Structures of the adult *Drosophila* larva and the adult structures derived from them. (B) Larval locations of the nine pairs of imaginal disks and one genital disk. (C) General morphology of the disks late in larval development.

boundary that divides the segment into anterior and posterior halves. The evidence for compartments comes from genetic marking of individual cells by means of mitotic recombination and observation of the positions of their descendants. Within each compartment, neighboring groups of cells not necessarily related by ancestry undergo developmental determination together.

As in the early embryo, overlapping patterns of gene expression and combinatorial control guide later events in *Drosophila* development. The expression patterns of two genes in the wing imaginal disk are shown in Figure 12.35. The expression of *apterous* is indicated in green, that of *vestigial* in red. Regions of overlapping expression are various shades of orange and yellow. Patterns of gene expression in imaginal disks are highly varied. Some genes are expressed in patterns with radial symmetry—for example, in alternating sectors or in concentric rings. The varied and overlapping patterns of expression ultimately yield the exquisitely fine level of cellular and morphological differentiation observed in the adult animal.

Among the genes that transform the periodicity of the *Drosophila* embryo into a body plan with linear differentiation are two small sets of **homeotic genes** (Figure

12.32). Mutations in homeotic genes result in the transformation of one body segment into another, which is recognized by the misplaced development of structures that are normally present elsewhere in the embryo. One class of homeotic mutation is

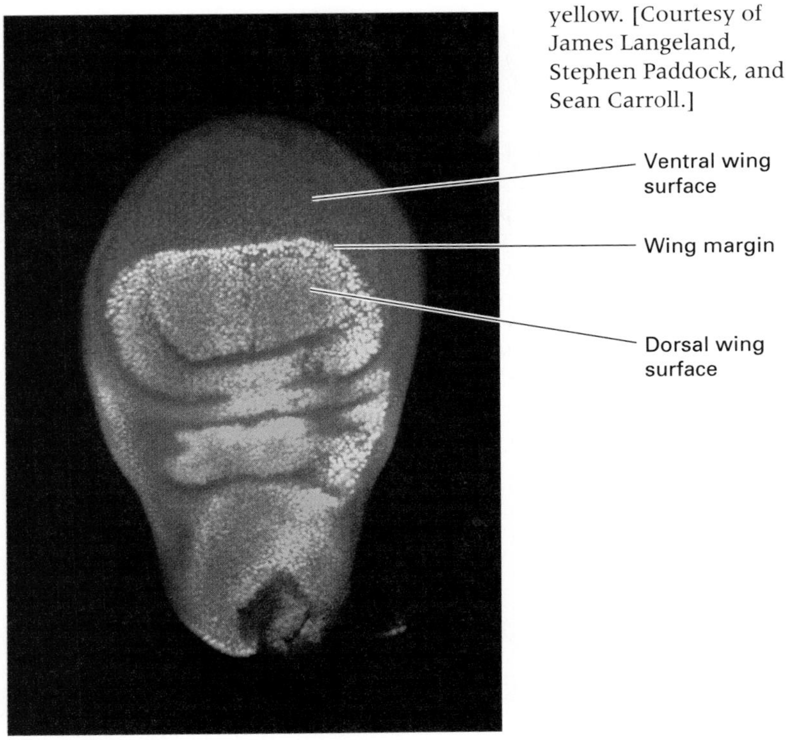

Figure 12.35

Expression of two genes that affect wing development in the wing imaginal disk. Expression of *apterous* is in green; that of *vestigial* is in red. Regions of overlapping expression are orange and yellow. [Courtesy of James Langeland, Stephen Paddock, and Sean Carroll.]

Ventral wing surface

Wing margin

Dorsal wing surface

illustrated by *bithorax*, which causes transformation of the anterior part of the third thoracic segment into the anterior part of the second thoracic segment, with the result that the halteres (flight balancers) are transformed into an extra pair of wings (Figure 12.36). The other class of homeotic mutation is illustrated by *Antennapedia*, which results in transformation of the antennae into legs. The normal *Antennapedia* gene specifies the second thoracic segment. *Antennapedia* mutations are dominant gain-of-function mutations in which the gene is overexpressed in the dorsal part of the head, transforming it into the second thoracic segment, with the antennae becoming transformed into a pair of misplaced legs.

Homeotic genes act within developmental compartments to control other genes concerned with such characteristics as rates of cell division, orientation of mitotic spindles, and the capacity to differentiate bristles, legs, and other features. Homeotic genes are also important in restricting the activities of groups of structural genes to definite spatial patterns. The homeotic genes represented by *bithorax* and *Antennapedia* are in fact gene clusters. The cluster containing *bithorax* is designated BX-C (stands for *bithorax*-complex), and that containing *Antennapedia* is called ANT-C (stands for *Antennapedia*-complex). Both gene clusters were initially discovered through their homeotic effects in adults. Later they were shown to affect the identity of larval segments. The BX-C is primarily concerned with the development of larval segments T3 through A8 (Figure 12.37) and has its principal effects in T3 and A1. The ANT-C is primarily concerned with the development of the head (H) and of thoracic segments T1 and T2.

Deletion (loss of function) and duplication (gain of function) of genes in the homeotic complexes help to define their functions. For example, deletion of the entire BX-C complex (Figure 12.37) results in a larva with a normal head (H), first thoracic segment (T1), and second thoracic segment (T2), but the remaining segments (T3 and A1–A8) differentiate in the manner of T2. Therefore, the function of the BX-C genes is to shift development progressively to more posterior types of segments. The BX-C region extends across approximately 300 kb of DNA yet contains only three essential coding regions. The rest of the region appears to consist of a complex series of enhancers and other regulatory elements that function to specify segment identity by activating the different coding regions to different degrees in particular parasegments.

Figure 12.36

A) Wildtype *Drosophila* showing wings and halteres (the pair of knob-like structures protruding posterior to the wings). (B) A fly with four wings produced by mutations in the *bithorax* complex. The mutations convert the third thoracic segment into the second thoracic segment, and the halteres that are normally present on the third thoracic segment become converted into the posterior pair of wings. [Courtesy of E. B. Lewis.]

(A)

Haltere

(B)

(A)

(B)

Wildtype larva

Larva with deletion
of BX-C
(*bithorax* complex)

Figure 12.37

Segmentation patterns in *Drosophila* larvae. (A) A wildtype larva. (B) A mutant. The wildtype larva has three thoracic segments (T1–T3) and eight abdominal segments (A1–A8), in addition to head (H) and genital (G) segments. The mutant larva has a genetic deletion of most of the *bithorax* (BX-C) complex. The segments H, T1, and G are normal. All other segments develop as T2 segments.

The homeotic genes are transcriptional activators of other genes. Most homeotic genes contain one or more copies of a characteristic sequence of about 180 nucleotides called a **homeobox,** which is also found in key genes concerned with the development of embryonic segmentation in organisms as diverse as segmented worms, frogs, chickens, mice, and human beings. Homeobox sequences are present in exons; they code for a protein-folding domain that includes a helix-turn-helix DNA-binding motif (Chapter 11), as well as other transcriptional activation components that are not so well understood.

12.5
Genetic Control of Development in Higher Plants

Reproductive and developmental processes in plants differ significantly from those in other eukaryotes. For example, plants have an alternation of generations between the diploid sporophyte and the haploid gametophyte, and the plant germ line is not established in a discrete location during embryogenesis but rather at many locations in the adult organism. In a corn plant, for example, each ear contains germ-line cells that undergo meiosis to form the pollen and ovules.

In animals, as we have seen, most of the major developmental decisions are made early in life, in embryogenesis. In higher plants, however, differentiation takes place almost continuously throughout life in regions of actively dividing cells called **meristems** in both the vegetative organs (root, stem, and leaves) and the floral organs (sepal, petal, pistil, and stamen). The shoot and root meristems are formed during embryogenesis and consist of cells that divide in distinctive geometric planes and at different rates to produce the basic morphological pattern of each organ system. The floral meristems are established by a reorganization of the shoot meristem after embryogenesis and eventually differentiate

Figure 12.38
The ability of plant development to adjust to perturbations is illustrated by this tree. Encroached on by a fence, it eventually incorporates the fence into the trunk. [Courtesy of Robert Pruitt.]

into floral structures characteristic of each particular species. One important difference between animal and plant development is that

> In higher plants, as groups of cells leave the proliferating region of the meristem and undergo further differentiation into vegetative or floral tissue, their developmental fate is determined almost entirely by their position relative to neighboring cells.

The critical role of positional information in development of higher plants stands in contrast to animal development, in which cell lineage often plays a key role in determinating cell fate.

The plastic or "indeterminate" growth patterns of higher plants are the result of continuous production of both vegetative and floral organ systems. These patterns are conditioned largely by day length and the quality and intensity of light. The plasticity of plant development gives plants a remarkable ability to adjust to environmental insults. Figure 12.38 shows a tree that, over time, adjusts to the presence of a nearby fence by engulfing it into the trunk.

Higher plants can also adjust remarkably well to a variety of genetic aberrations. For example, transgenic plants of *Arabidopsis thaliana* (a member of the mustard family) have been created that either overexpress or underexpress cyclin B. Overexpression of cyclin B results in an accelerated rate of cell division; underexpression of cyclin B results in a decelerated rate. Plants with the faster rate of cell division contain more cells and are somewhat larger than their wildtype counterparts, but otherwise they look completely normal. Likewise, plants with the decreased rate of cell division have less than half the normal number of cells, but they grow at almost the same rate and reach almost the same size as wildtype plants, because as the number of cells decreases, each individual cell gets larger. One of the important consequences of plants being able to adjust to abnormal growth conditions is that plant cells rarely undergo a transformation into proliferative cancer cells, as often happens in animals. The common plant tumors are produced only as a result of complex interactions with pathogens such as *Agrobacterium* (Chapter 9).

Flower Development in *Arabidopsis*

Genetic analysis of *Arabidopsis* has revealed some important principles in the genetic determination of floral structures. As is typical of flowering plants, the flowers of *Arabidopsis* are composed of four types of organs arranged in concentric rings or whorls. Figure 12.39 illustrates the geometry, looking down at a flower from the top. From outermost to innermost, the whorls are designated 1, 2, 3, and 4 (Figure 12.39A). In the development of the flower, each whorl gives rise to a different floral organ (Figure 12.39B). Whorl 1 yields the sepals (the green, outermost floral leaves), whorl 2 the petals (the white, inner floral leaves), whorl 3 the stamens (the male organs, which form pollen), and whorl 4 the carpels (which fuse to form the ovary).

Mutations that affect floral development fall into three major classes, each with a characteristic phenotype (Figure 12.40). Compared with the wildtype flower (panel A), one class lacks sepals and petals (panel B), another class lacks petals and stamens (panel C), and the third class lacks stamens and carpels (panel D). On the basis of crosses between homozygous mutant organisms, these classes of mutants can be assigned to four complementation groups, each of which defines a different gene (Chapter 2). Each gene and the phenotype of a plant homozygous for a recessive mutation in the gene are shown in Table 12.1. The phenotype lacking sepals and petals is caused by mutations in the gene *ap2 (apetala-2)*. The phenotype lacking stamens and petals is caused by a mutation in either of two genes, *ap3 (apetala-3)* and *pi (pistillata)*. The phenotype lacking stamens and carpels is caused by mutations in the gene *ag (agamous)*. Each of these genes has been cloned and sequenced. They are all transcription factors. The transcription factors encoded by *ap3, pi,* and *ag* are members of what is called the *MADS box* family of transcription factors; each member of this family contains a sequence of 58 amino acids in which common features can be identified. MADS box transcription factors are very common in plants but are also found, less frequently, in animals.

Combinatorial Determination of the Floral Organs

The role of the *ap2, ap3, pi,* and *ag* transcription factors in the determination of floral organs can be inferred from the phenotypes of the mutations. The logic of the inference is based on the observation (see Table 12.1) that mutation in any of the genes eliminates two floral organs that arise from adjacent whorls. This pattern suggests that *ap2* is necessary for sepals and petals, *ap3* and *pi* are both necessary for petals and stamens, and *ag* is necessary for stamens and carpels. Because the mutant phenotypes are caused by loss-of-function alleles of the genes, it may be inferred that *ap2* is expressed in whorls 1 and 2, that *ap3* and *pi* are expressed in whorls 2 and 3, and that *ag* is expressed in whorls 3 and 4. The overlapping patterns of expression are shown in Table 12.2.

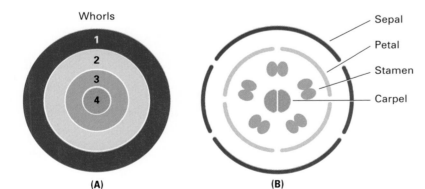

(A)   **(B)**

Figure 12.39
(A) The organs of a flower are arranged in four concentric rings, or whorls. (B) Whorls 1, 2, 3, and 4 give rise to sepals, petals, stamens, and carpels, respectively.

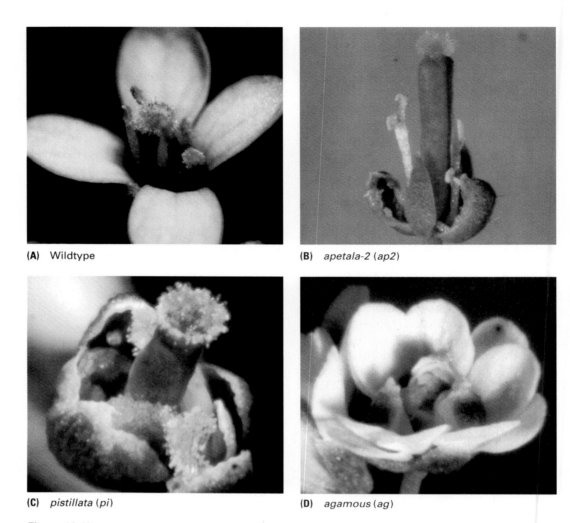

(A) Wildtype

(B) *apetala-2* (*ap2*)

(C) *pistillata* (*pi*)

(D) *agamous* (*ag*)

Figure 12.40
Phenotypes of the major classes of floral mutations in *Arabidopsis*. (A) The wildtype floral pattern consists of concentric whorls of sepals, petals, stamens, and carpels. (B) The homozygous mutation *ap2* (*apetala-2*) results in flowers missing sepals and petals. (C) Genotypes that are homozygous for either *ap3* (*apetala-3*) or *pi* (*pistillala*) yield flowers that have sepals and carpels but lack petals and stamens. (D) The homozygous mutation *ag* (*agamous*) yields flowers that have sepals and petals but lack stamens and carpels. [Courtesy of Elliot M. Meyerowitz and John Bowman. Part B from Elliot M. Meyerowitz. 1994. The genetics of flower development. *Scientific American*, 271: 56 (November 1994).]

Table 12.1 Floral development in mutants of *Arabidapsis*

Genotype	Whorl			
	1	2	3	4
wildtype	sepals	petals	stamens	carpels
ap2/ap2	carpels	stamens	stamens	carpels
ap3/ap3	sepals	sepals	carpels	carpels
pi/pi	sepals	sepals	carpels	carpels
ag/ag	sepals	petals	petals	sepals

Table 12.2 Domains of expression of genes determining floral development

Whorl	Genes expressed	Determination
1	*ap2*	sepal
2	*ap2* + *ap3* and *pi*	petal
3	*ap3* and *pi* + *ag*	stamen
4	*ag*	carpel

The model of gene expression in Table 12.2 suggests that floral development is controlled in combinatorial fashion by the four genes. Sepals develop from tissue in which only *ap2* is active; petals are evoked by a combination of *ap2*, *ap3*, and *pi*; stamens are determined by a combination of *ap3*, *pi* and *ag*; and carpels derive from tissue in which only *ag* is expressed. This model is illustrated graphically in Figure 12.41.

You may have noted already that the model in Table 12.2 does not account for all the phenotypic features of the *ap2* and *ag* mutations in Table 12.1. In particular, according to the combinatorial model in Table 12.2, the development of carpels and stamens from whorls 1 and 2 in homozygous *ap2* plants would require expression of *ag* in whorls 1 and 2. Similarly, the development of petals and sepals from whorls 3 and 4 in homozygous *ag* plants

Figure 12.41

Control of floral development in *Arabidopsis* by the overlapping expression of four genes. The sepals, petals, stamens, and carpels are floral organ systems that form in concentric rings or whorls. The developmental identity of each concentric ring is determined by the genes *ap2*, *ap3* and *pi*, and *ag*, each of which is expressed in two adjacent rings. Gene *ap2* is expressed in the outermost two rings, *ap3* and *pi* in the middle two, and *ag* in the inner two. Therefore, each whorl has a unique combination of active genes.

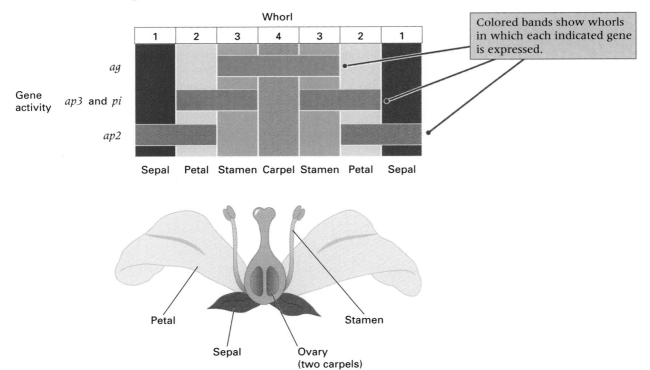

would require expression of *ap2* in whorls 3 and 4. This discrepancy can be explained if it is assumed that *ap2* expression and *ag* expression are mutually exclusive: If the presence of the *ap2* transcription factor, *ag* is repressed, and in the presence of the *ag* transcription factor, *ap2* is repressed. If this were the case, then, in *ap2* mutants, *ag* expression would spread into whorls 1 and 2, and, in *ag* mutants, *ap2* expression would spread into whorls 3 and 4. This additional assumption, enables us to explain the phenotypes of the single and even double mutants.

With the additional assumption we have made about *ap2* and *ag* interaction, the model in Table 12.2 fits the data, but is the model true? For these genes, the patterns of gene expression, assayed by *in situ* hybridization of RNA in floral cells with labeled probes for each of the genes, fits the patterns in Table 12.2. In particular, *ap2* is expressed in whorls 1 and 2, *ap3* and *pi* in whorls 2 and 3, and *ag* in whorls 3 and 4. (The *ap2* gene is also expressed in nonfloral tissue, but its role in these tissues is unknown.) Furthermore, the seemingly *ad hoc* assumption about *ap2* and *ag* expression being mutually exclusive turns out to be true. In *ap2* mutants, *ag* is expressed in whorls 1 and 2; reciprocally, in *ag* mutants, *ap2* is expressed in whorls 3 and 4. It is also known how *ap3* and *pi* work together. The active transcription factor that corresponds to these genes is a dimeric protein composed of Ap3 and Pi polypeptides. Each component polypeptide, in the absence of the other, remains inactive in the cytoplasm. Together, they form an active dimeric transcription factor that migrates into the nucleus.

Figure 12.42
The homozygous triple mutant *ap2 pi ag* lacks all of the transcription factors needed for floral development. Hence the "flowers" lack all of the wildtype floral structures. They have no sepals, petals, stamens, or carpels. Without the floral genetic determinants, the flowers consist entirely of leaves arranged in concentric whorls. [Courtesy of Elliot M. Meyerowitz and John Bowman.]

Given the critical role of the Ap2, Ap3/Pi, and Ag transcription factors in floral determination, it might be speculated that triple mutants lacking all three types of transcription factors would have very strange flowers. The phenotype of the *ap2 pi ag* triple mutant is shown in the photograph in Figure 12.42. The flowers have none of the normal floral organs. They consist merely of leaves arranged in concentric whorls.

Chapter Summary

The genotype determines the developmental potential of the embryo. By means of a developmental program that results in different sets of genes being expressed in different types of cells, the genotype controls the developmental events that take place and their temporal order. Mutations that interrupt developmental processes identify genetic factors that control development.

The early development of the animal embryo establishes the basic developmental plan for the whole organism. The earliest events in embryonic development depend on the correct spatial organization of numerous constituents present in the oocyte. Developmental genes that are needed in the mother for proper oocyte formation and zygotic development are maternal-effect genes. Genes that are required in the zygote nucleus are zygotic genes. Fertilization of the oocyte initiates a series of mitotic cleavage divisions that form the "hollow ball" blastula, which rapidly undergoes a restructuring into the gastrula. Accompanying these early morphological events are a series of molecular events that determine the developmental fates that cells undergo. Execution of a developmental state may be autonomous (genetically programmed), or it may require positional information supplied by neighboring cells or the local concentration of one or more morphogens.

The soil nematode *Caenorhabditis elegans* is used widely in studies of cell lineages because many lineages in the organism undergo virtually autonomous development and the developmental program is identical from one organism to the next. Most lineages are affected by many genes, including genes that control the sublineages into which the lineage can differentiate. Mutations that affect cell lineages define several types of developmental mutants: (1) execution mutants, which prevent a developmental program from being carried out; (2) transformation mutants, in which a wrong developmental program is executed; (3) segregation mutants, in which developmental determinants fail to segregate normally between sister cells or mother and daughter cells; (4) apoptosis (programmed cell death) mutants, in which cells fail to undergo a normal developmental program that normally leads to their death; and (5) heterochronic mutants, in which execution of a normal developmental program is either precocious or retarded in time.

Genes that control key points in development can often be identified by the unusual feature that recessive alleles (ideally, loss-of-function mutations) and dominant alleles (ideally, gain-of-function mutations) have opposite effects on phenotype. For example, if loss of function results in failure to execute a developmental program in a particular anatomical position, then gain of function should result in execution of the program in an abnormal location. The *lin-12* gene in *C. elegans* is an important developmental control gene identified by these criteria. The *lin-12* gene controls the developmental "decision" of a pair of cells whether to become anchor cells or ventral uterine precursor cells. The gene codes for a transmembrane receptor protein that shares some domains with the mammalian peptide hormone epidermal growth factor, shares other domains with the *Notch* protein in *Drosophila* (which controls epidermal or neural commitment), and shares still other domains with proteins that control the cell cycle. In mutations that inactivate *lin-12*, both precursors develop into anchor cells; in mutations in which *lin-12* is overexpressed, both become ventral uterine precursor cells.

Early development in *Drosophila* includes the formation of a syncytial blastoderm by early cleavage divisions without cytoplasmic division, the setting apart of pole cells that form the germ line at the posterior of the embryo, the migration of most nuclei to the periphery of the syncytial blastoderm, cellularization to form the cellular blastoderm, and determina-

tion of the blastoderm fate map at or before the cellular blastoderm stage. Metamorphosis into the adult fly makes use of about 20 imaginal disks present in the larva that contain developmentally committed cells that divide and develop into the adult structures. Most imaginal disks include several discrete groups of cells or compartments separated by boundaries that progeny cells do not cross.

Early development in *Drosophila* to the level of segments and parasegments requires four classes of segmentation genes: (1) coordinate genes that establish the basic anterior-posterior and dorsal-ventral aspect of the embryo, (2) gap genes for longitudinal separation of the embryo into regions, (3) pair-rule genes that establish an alternating on/off striped pattern of gene expression along the embryo, and (4) segment-polarity genes that refine the patterns of gene expression within the stripes and determine the basic layout of segments and parasegments. For example, the *bicoid* gene is a maternal-effect gene in the coordinate class that controls anterior-posterior differentiation of the embryo and determines position on the blastoderm fate map. The *bicoid* gene product is a transcriptional activator protein that is present in a concentration gradient from anterior to posterior along the embryo. Genes that are regulated by *bicoid* have different sensitivities to its concentration, depending on the number and types of high-affinity and low-affinity binding sites that they contain. The segmentation genes can regulate themselves, other members of the same class, and genes of other classes farther along the hierarchy. Together, the segmentation genes control the homeotic genes that initiate the final stages of developmental specification.

Mutations in homeotic genes result in the transformation of one body segment into another. For example, *bithorax* causes transformation of the anterior part of the third thoracic segment into the anterior part of the second thoracic segment, and *Antennapedia* results in transformation of the dorsal part of the head into the second thoracic segment. Homeotic genes act within developmental compartments to control other genes concerned with such characteristics as rates of cell division, orientation of mitotic spindles, and the capacity to differentiate bristles, legs, and other features. Both the *bithorax* complex (BX-C) and the *Antennapedia* complex (ANT-C) are clusters of genes, each containing homeoboxes that are characteristic of developmental-control genes in many organisms and that code for DNA-binding and other protein domains.

Developmental processes in higher plants differ significantly from those in animals in that developmental decisions continue throughout life in the meristem regions of the vegetative organs (root, stem, and leaves) and the floral organs (sepal, petal, pistil, and stamen). However, genetic control of plant development is mediated by transcription factors analogous to those in animals. For example, control of floral development in *Arabidopsis* is through the combinatorial expression of four genes in each of a series of four concentric rings, or whorls, of cells that eventually form the sepals, petals, stamens, and carpels.

Key Terms

apoptosis
autonomous determination
blastoderm
blastula
cascade
cell fate
cleavage division
compartment
coordinate gene
embryonic induction
execution mutation
fate map
gain-of-function mutation
gap gene
gastrula
heterochronic mutation

homeobox
homeotic gene
imaginal disk
ligand
lineage
lineage diagram
loss-of-function mutation
maternal-effect genes
meristem
morphogen
pair-rule gene
parasegment
pattern formation
pole cell
positional information
programmed cell death

segmentation gene
segment-polarity gene
segregation mutation
syncytial blastoderm
transformation mutation
transmembrane receptor
zygotic gene

Review the Basics

- What is meant by *polarity* in the mature oocyte?

- What is a loss-of-function mutation? What is a gain-of-function mutation? In developmental genetics, what is the significance of the observation that loss-of-function alleles and gain-of-function alleles of the same gene have opposite phenotypes?

- How does knowledge of the complete cell lineage of nematode development demonstrate the importance of programmed cell death (apoptosis) in development?

- Among genes that control embryonic development in *Drosophila*, distinguish between coordinate genes, gap genes, pair-rule genes, and segment-polarity genes. Generally speaking, what is the temporal order of expression of these classes of genes?

- What is a homeotic mutation? Give an example from *Drosophila*. Do homeotic mutations occur in organisms other than *Drosophila?*

- Do plants have a germ line in the same sense as animals? Explain.

- What does it mean to say that a cell in an organism is *totipotent?* Are all cells in a mature higher plant totipotent?

- What is the genetic basis of the developmental determination of sepals, petals, stamens, and carpels in floral development in *Arabidopsis?*

Guide to Problem Solving

Problem 1: What is the logic behind the following principle of developmental genetics: "For a gene affecting cell determination, if loss-of-function alleles eliminate a particular cell fate, and gain-of-function alleles induce the fate, then the product of the gene must be both necessary and sufficient for expression of the fate."

Answer: A gene product is necessary for cell fate if its absence prevents normal expression of the fate. Hence the effect of loss-of-function alleles means that the gene product is necessary for the fate. On the other hand, a gene product is sufficient for cell determination if its presence at the wrong time, in the wrong tissues, or in the wrong amount results in

expression of the cell fate. The fact that gain-of-function mutations induce a particular cell fate means that the gene product is sufficient to trigger the fate.

Problem 2: A mutation m is called a recessive maternal-effect lethal if eggs from homozygous mm females are unable to support normal embryonic development, irrespective of the genotype of the zygote. What kinds of crosses are necessary to produce mm females?

Answer: Although the eggs from mm females are abnormal, those from m^+m females allow normal embryonic development because the wildtype m^+ allele supplies the cytoplasm of the oocyte with sufficient products for embryogenesis. Therefore, homozygous *mm* females can be obtained by crossing heterozygous m^+m females with either m^+m or *mm* males. The first cross produces 1/4 *mm* females, the second 1/2 *mm*.

Problem 3: The *bicoid* mutation in *Drosophila* results in absence of head structures in the early embryo. A similar phenotype results when cytoplasm is removed from the anterior end of a *Drosophila* embryo. What developmental effect would you expect in each of the following circumstances?

(a) A *bicoid* embryo was injected in the anterior end with some cytoplasm taken from the anterior end of a wildtype embryo.

(b) A wildtype *Drosophila* embryo was injected in the middle with some cytoplasm taken from the anterior end of another wildtype embryo.

Answer: The experimental results imply that the anterior cytoplasm, including the *bicoid* gene product, induces development of head structures.

(a) Anterior cytoplasm from a wildtype embryo should be able to rescue *bicoid,* so head structures should be formed.

(b) Anterior cytoplasm injected into the middle of the embryo should induce head structures in that location, so the embryo should have head structures at the anterior end and also in the middle.

Analysis and Applications

12.1 Distinguish between a developmental fate determined by autonomous development and one determined by positional information. What two types of surgical manipulation are used to distinguish between the processes experimentally?

12.2 What are lineage mutations and how are they detected?

12.3 What is the *Drosophila* blastoderm, and what is the blastoderm fate map? Does the existence of a fate map imply that *Drosophila* developmental decisions are autonomous?

12.4 What is the result of a maternal-effect lethal allele in *Drosophila*? If an allele is a maternal-effect lethal, how can a fly be homozygous for it?

12.5 What is the principal consequence of a failure in programmed cell death?

12.6 With regard to its effects on cell lineages, what kind of mutation is a homeotic mutation in *Drosophila*?

12.7 A cell, A, in *Caenorhabditis elegans* normally divides and the daughter cells differentiate into cell types B and C. What developmental pattern would result from a mutation in cell A that prevents sister-cell differentiation? From a mutation in cell A that prevents parent-offspring segregation?

12.8 A mutation is found in which the developmental pattern is normal but slow. Does this qualify as a heterochronic mutation? Explain.

12.9 Why is transcription of the zygote nucleus dispensable in *Drosophila* early development but not in early development of the mouse?

12.10 Distinguish between a loss-of-function mutation and a gain-of-function mutation. Can the same gene undergo both types of mutations? Can the same allele have both types of effects?

12.11 A particular gene is necessary, but not sufficient, for a certain developmental fate. What is the expected phenotype of a loss-of-function mutation in the gene? Is the allele expected to be dominant or recessive?

12.12 A particular gene is sufficient for a certain developmental fate. What is the expected phenotype of a gain-of-function mutation in the gene? Is the allele expected to be dominant or recessive?

12.13 What is the phenotype of a *Drosophila* mutation in the *gap* class? Of a mutation in the *pair-rule* class?

12.14 The drug actinomycin D prevents RNA transcription but has little direct effect on protein synthesis. When fertilized sea urchin eggs are immersed in a solution of the drug, development proceeds to the blastula stage, but gastrulation does not take place. How would you interpret this finding?

12.15 A mutation in the axolotl designated *o* is a maternal-effect lethal because embryos from *oo* females die at gastrulation, irrespective of their own genotype. However, the embryos can be rescued by injecting oocytes from *oo* females with an extract of nuclei from either o^+o^+ or o^+o eggs. Injection of cytoplasm is not as effective. Suggest an explanation for these results.

12.16 The nuclei of brain cells in the adult frog normally do not synthesize DNA or undergo mitosis. However, when transplanted into developing oocytes, the brain cell nuclei behave as follows: (a) In rapidly growing premeiosis oocytes, they synthesize RNA. (b) In more mature oocytes, they do not synthesize DNA or RNA, but their chromosomes condense and they begin meiosis. How would you explain these results?

Chapter 12

GeNETics on the web will introduce you to some of the most important sites for finding genetic information on the Internet. To complete the exercises below, visit the Jones and Bartlett home page at

http://www.jbpub.com/genetics

Select the link to *Genetics: Principles and Analysis* and then choose the link to *GeNETics on the web.* You will be presented with a chapter-by-chapter list of highlighted keywords.

GeNETics EXERCISES

Select the highlighted keyword in any of the exercises below, and you will be linked to a web site containing the genetic information necessary to complete the exercise. Each exercise suggests a specific, written report that makes use of the information available at the site. This report, or an alternative, may be assigned by your instructor.

1.   Browse the *C. elegans* database, ACeDB, to learn more about *lin-12* and other developmental-control genes in this organism. Choose *Locus* and then the alphabetical sublist containing *lin-12*. Read the entry for *lin-12*. If assigned to do so, pick any other of the *lin* mutations, and write a short description of its phenotype and genetic map position.

2.   Great scanning electron microscopic images of *Drosophila* embryogenesis can be found at this site. Examine the process of gastrulation from the dorsal and lateral views. Note the prominence of the pole cells. If assigned to do so, pick one of the mutants (*bcd* or *ftz*), and sketch a mutant and wildtype embryo at approximately the same stage of development, pointing out the differences.

3.   *Arabidopsis* is an important model organism for genetic studies of plant development. Use this site to find the genetic map positions of *ap2, ap3, pi,* and *ag.* If assigned to do so, show the rela-

Challenge Problems

12.17 The autosomal gene *rosy* (*ry*) in *Drosophila* is the structural gene for the enzyme xanthine dehydrogenase (XDH), which is necessary for wildtype eye pigmentation. Flies of genotype *ry/ry* lack XDH activity and have rosy eyes. The X-linked gene *maroonlike* (*mal*) is also necessary for XDH activity, and *mal/mal; ry$^+$/ry$^+$* females and *mal/Y; ry$^+$/ry$^+$* males also lack XDH activity; they have maroonlike eyes. The cross *mal$^+$/mal; ry/ry* / × *mal/Y; ry$^+$/ry$^+$* produces *mal/mal; ry$^+$/ry* females and *mal/Y; ry$^+$/ry* males that have wildtype eye color even though they lack active XDH enzyme. Suggest an explanation.

12.18 You wish to demonstrate that during segmentation of the *Drosophila* embryo, normal pair-rule patterns of expression require normal expression of the gap-genes, whereas

gap gene expression does not require pair-rule expression. You have the following four mutations available:

(a) A mutation in the zygotic-effect gap gene *knirps* (*kni*).

(b) A mutation in the zygotic-effect pair-rule gene *fushi tarazu* (*ftz*).

(c) A transgene consisting of a reporter gene (*lacZ*) fused to the enhancer elements of *kni*.

(d) A transgene consisting of a reporter gene (*lacZ*) fused to the enhancer elements of *ftz*.

Describe the strains you would need and how you would use them to show that *kni* is epistatic to *ftz*. You do not need to give details of the crosses.

Further Reading

Avery, L., and S. Wasserman. 1992. Ordering gene function: The interpretation of epistasis in regulatory hierarchies. *Trends in Genetics* 8: 312.

Capecchi, M. R., ed. 1989. *The Molecular Genetics of Early Drosophila and Mouse Development.* Cold Spring Harbor, NY: Cold Spring Harbor Laboratory.

Chater, K., A. Downie, B. Drobak and C. Martin. 1995. Alarms and diversions: The biochemistry of development. *Trends in Genetics* 11: 79.

De Robertis, E. M., G. Oliver, and C. V. E. Wright. 1990. Homeobox genes and the vertebrate body plan. *Scientific American,* July.

Duke, R. C., D. M. Ojcius, and J. D. E. Young. 1996. Cell suicide in health and disease. *Scientific American,* December.

Gaul, U., and H. Jäckle. 1990. Role of gap genes in early *Drosophila* development. *Advances in Genetics* 27: 239.

tive positions of these genes on the genetic map of the entire genome.

4. Check out this site for further illustrations of floral development, including sites of expression of some of the major genes. If assigned to do so, find which of the genes involved in floral development is expressed in all four whorls and describe the hypotheses put forward to explain this unexpected observation.

MUTABLE SITE EXERCISES

The Mutable Site Exercise changes frequently. Each new update includes a different exercise that makes use of genetics resources available on the World Wide Web. Select the Mutable Site for Chapter 12, and you will be linked to the current exercise that relates to the material presented in this chapter.

PIC SITE

The Pic Site showcases some of the most visually appealing genetics sites on the World Wide Web. To visit the showcase genetics site, select the Pic Site for Chapter 12.

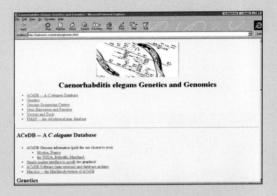

Grunert, S., and D. St. Johnston. 1996. RNA localization and the development of asymmetry during *Drosophila* oogenesis. *Current Opinion in Genetics & Development* 6: 395.

Irish, V. 1987. Cracking the *Drosophila* egg. *Trends in Genetics* 3: 303.

Kaufman, T. C., M. A. Seeger, and G. Olsen. 1990. Molecular and genetic organization of the Antennapedia gene complex of *Drosophila melanogaster*. *Advances in Genetics* 27: 309.

Kennison, J. A. 1995. The Polycomb and Trithorax group proteins of *Drosophila*: Trans-regulators of homeotic gene function. *Annual Review of Genetics* 29: 289.

Kornfeld, K. 1997. Vulval development in *Caenorhabditis elegans*. *Trends in Genetics* 13: 55.

Lawrence, P. A. 1992. *The Making of a Fly: The Genetics of Animal Design*. Oxford, England: Blackwell.

Nüsslein-Volhard, C. 1996. Gradients that organize embryo development. *Scientific American*, August.

McCall, K., and H. Steller. 1997. Facing death in the fly: Genetic analysis of apoptosis in *Drosophila*. *Trends in Genetics* 13: 222.

Meyerowitz, E. M. 1996. Plant development: Local control, global patterning. *Current Opinion in Genetics & Development* 6: 475.

Morisato, D., and K. V. Anderson. 1995. Signaling pathways that establish the dorsal-ventral pattern of the *Drosophila* embryo. *Annual Review of Genetics* 29: 371.

Riverapomar, R., and H. Jäckle. 1996. From gradients to stripes in *Drosophila* embryogenesis: Filling in the gaps. *Trends in Genetics* 12: 478.

Sternberg, D. W. 1990. Genetic control of cell type and pattern formation in *Caenorhabditis elegans*. *Advances in Genetics* 27: 63.

Weigel, D. 1995. The genetics of flower development: From floral induction to ovule morphogenesis. *Annual Review of Genetics* 29: 19.

Wieschaus, E. 1996. Embryonic transcription and the control of developmental pathways. *Genetics* 142: 5.

Wolpert, L. 1996. One hundred years of positional information. *Trends in Genetics* 12: 359.

Wood, W. B., ed. 1988. *The Nematode Caenorhabditis elegans*. Cold Spring Harbor, NY: Cold Spring Harbor Laboratory.

Wright, T. R. F., ed. 1990. *Genetic Regulatory Hierarchies in Development*. New York: Academic Press.

This hummingbird is a rare, mutant form of the ruby-throated humming-bird found in the eastern United States. It has no pigment in its plumage. The round tail is a sign that this bird is a female. [Courtesy of Steve and Dave Maslowski.]

CHAPTER 13

Mutation, DNA Repair, and Recombination

CHAPTER OUTLINE

PRINCIPLES

- Substitution of one base for another is an important mechanism of spontaneous mutation. A single base substitution in a coding region may result in an amino acid replacement; a single-base deletion or insertion results in a shifted reading frame.
- Transposable element insertion is also an important mechanism of spontaneous mutation.
- Mutations can be induced by various agents, including highly reactive chemicals and x rays.
- Most mutagens are also carcinogens.
- Cells contain enzymatic pathways for the repair of different types of damage to DNA. Among the most important repair systems is mismatch repair of duplex DNA, in which a nucleotide that contains a mismatched base is excised and replaced with the correct nucleotide.
- Recombination between homologous DNA molecules includes the invasion of a duplex by one or both strands from another, forming a heteroduplex whose junction can migrate until it is finally resolved by breakage and reunion. Mismatch repair in the heteroduplex region can lead to 3 : 1 or 1 : 3 segregation of alleles; these types of aberrant segregation can be observed directly in organisms in which all four products of meiosis remain together, such as in fungal asci.

CONNECTIONS

CONNECTION: X-Ray Daze
Hermann J. Muller 1927
Artificial transmutation of the gene

CONNECTION: Replication Slippage in Unstable Repeats
Micheline Strand, Tomas A. Prolla, R. Michael Liskay, and Thomas D. Petes 1993
Destabilization of tracts of simple repetitive DNA in yeast by mutations affecting DNA mismatch repair

In preceding chapters, numerous examples were presented in which the information contained in the genetic material had been altered by mutation. A **mutation** is any heritable change in the genetic material. In this chapter, we examine the nature of mutations at the molecular level. You will learn how mutations are created, how they are detected phenotypically, and the means by which many mutations are corrected by special DNA repair enzymes almost immediately after they occur. You will see that mutations can be induced by radiation and a variety of chemical agents that produce strand breakage and other types of damage to DNA. The breakage and repair of DNA serve as an introduction to the process of recombination at the DNA level.

13.1
General Properties of Mutations

Mutations can happen at any time and in any cell. The phenotypic effects can range from minor alterations that are detectable only by biochemical methods to drastic changes in essential processes that cause, at one extreme, unrestrained cell proliferation (cancer) or, at the other extreme, the death of the cell or organism. Mutations that produce clearly defined effects are regularly used, when genetic phenomena are studied in the laboratory, but most mutations are not of this type. The effect of a mutation is determined by the type of cell containing the mutant allele, by the stage in the life cycle or development of the organism that the mutation affects, and, in diploid organisms, by the dominance or recessiveness of the mutant allele. A recessive mutation is usually not detected until a later generation when two heterozygous genotypes mate. Dominance does not complicate the expression of mutations in bacteria and haploid eukaryotes.

Mutations can be classified in a variety of ways. In multicellular organisms, one distinction is based on the type of cell in which the mutation first occurs: those that arise in cells that ultimately form gametes are **germ-line mutations,** all others are **somatic mutations.** A somatic mutation yields an organism that is genotypically,

and for many dominant mutations phenotypically, a mixture of normal and mutant tissue. Because reproductive cells are not affected, such a mutant allele will not be transmitted to the progeny and may not be detected or be recoverable for genetic analysis. In higher plants, however, somatic mutations can often be propagated by vegetative means (without going through seed production), such as grafting or the rooting of stem cuttings. This process has been the source of valuable new varieties such as the 'Delicious' apple and the 'Washington' navel orange.

Among the mutations that are most useful for genetic analysis are those whose effects can be turned on or off at will. These are called **conditional mutations** because they produce changes in phenotype in one set of environmental conditions (called the **restrictive conditions**) but not in another (called the **permissive conditions**). A **temperature-sensitive mutation,** for example, is a conditional mutation whose expression depends on temperature. Usually, the restrictive temperature is high, and the organism exhibits a mutant phenotype above this critical temperature. The permissive temperature is lower, and under permissive conditions the phenotype is wildtype or nearly wildtype. Temperature-sensitive mutations are frequently used to block particular steps in biochemical pathways in order to test the role of the pathways in various cellular processes, such as DNA replication.

An example of temperature sensitivity is found in the ordinary Siamese cat, with its black-tipped paws, ears, and tail. In this breed of cat, the biochemical pathway leading to black pigmentation is temperature-sensitive and inactivated at normal body temperature. Consequently, the pigment is not present in the hair over most of the body. The tips of the legs, ears, and tail are cooler than the rest of the body, so the pigment is deposited in the hair in these areas.

Mutations can also be classified by other criteria, such as the kinds of alterations in the DNA, the kinds of phenotypic effects produced, and whether the mutational events are spontaneous in origin or were induced by exposure to a known **mutagen** (a mutation-causing agent). Such classifications are often useful in discussing aspects of the mutational process. *Spontaneous* usually means that the event that caused a mutation is unknown, and **spontaneous mutations** are those that take place in the absence of any known mutagenic agent. The properties of spontaneous mutations and of **induced mutations** will be described in later sections.

13.2
The Molecular Basis of Mutation

All mutations result from changes in the nucleotide sequence of DNA or from deletions, insertions, or rearrangement of DNA sequences in the genome. Some types of major rearrangements in chromosomes were discussed in Chapter 7. In this section, we discuss mutations whose molecular basis can be specified.

Base Substitutions

The simplest type of mutation is a **base substitution,** in which a nucleotide pair in a DNA duplex is replaced with a different nucleotide pair. For example, in an A $\rightarrow$ G substitution, an A is replaced with a G in one of the DNA strands. This substitution temporarily creates a mismatched G—T base pair; at the very next replication, the mismatch is resolved as a proper G—C base pair in one daughter molecule and as a proper A—T base pair in the other daughter molecule. In this case, the G—C base pair is the mutant and the A—T base pair is nonmutant. Similarly, in an A $\rightarrow$ T substitution, an A is replaced with a T in one strand, creating a temporary T—T mismatch, which is resolved by replication as T—A in one daughter molecule and A—T in the other. In this example, the T—A base pair is mutant and the A—T base pair is nonmutant. The T—A and the A—T are not equivalent, as can be seen by considering the nucleotide context. If the original unmutated DNA strand has the sequence 5'-GAC-3', for example, then the mutant strand has the sequence 5'-GTC-3' (which we have written as T—A), and the nonmutant strand has the sequence 5'-GAC-3' (which we have written as A—T).

Some base substitutions replace one pyrimidine base with the other or one purine base with the other. These are called **transition mutations.** The possible transition mutations are

$$T \rightarrow C \quad \text{or} \quad C \rightarrow T$$
$$(\text{pyrimidine} \rightarrow \text{pyrimidine})$$

$$A \rightarrow G \quad \text{or} \quad G \rightarrow A$$
$$(\text{purine} \rightarrow \text{purine})$$

Other base substitutions replace a pyrimidine with a purine or the other way around. These are called **transversion mutations.** The possible transversion mutations are

$$T \rightarrow A, \ T \rightarrow G, \ C \rightarrow A, \quad \text{or} \quad C \rightarrow G$$
$$(\text{pyrimidine} \rightarrow \text{purine})$$

$$A \rightarrow T, \ A \rightarrow C, \ G \rightarrow T, \quad \text{or} \quad G \rightarrow C$$
$$(\text{purine} \rightarrow \text{pyrimidine})$$

Altogether, there are four possible transitions and eight possible transversions. Therefore, if base substitutions were strictly random, then one would expect a 1 : 2 ratio of transitions to transversions. However,

Spontaneous base substitutions are biased in favor of transitions. Among spontaneous base substitutions, the ratio of transitions to transversions is approximately 2 : 1.

Examination of the genetic code (Table 10.2) shows that the bias toward transitions has an important consequence for base substitutions in the third position of codons. In all codons with a pyrimidine in the third position, it does not matter which pyrimidine is present; likewise, in most codons that end in a purine, either purine will do. This means that most transition mutations in the third codon position do not change the amino acid that is encoded. Such mutations change the nucleotide sequence without changing the amino acid sequence; these are called **silent mutations** or **silent substitutions** because they are not detectable as changes in phenotype.

Mutational changes in nucleotides that are outside of coding regions can also be silent. In noncoding regions, which include introns and the DNA between genes, the precise nucleotide sequence is often not critical. These sequences can undergo base substitutions, small deletions or additions, insertions of transposable elements, and other rearrangements, and yet the mutations may have no detectable effect on phenotype. On the other hand, some noncoding sequences do have essential functions—for example, promoters, enhancers, transcription termination signals, and intron splice junctions. Mutations in these sequences often do have phenotypic effects.

Most base substitutions in coding regions do result in changed amino acids; these are called **missense mutations.** A change in the amino acid sequence of a protein may alter the biological properties of the protein. The classic example of a phenotypic effect of a single amino acid change is that responsible for the human hereditary disease **sickle-cell anemia,** which we discussed in Chapter 1. This condition, characterized by a change in shape of the red blood cells into an elongate form that blocks capillaries, results from the replacement of a glutamic acid by a valine at position 6 in the β-hemoglobin chain.

On the other hand, an amino acid replacement does not always create a mutant phenotype. For instance, replacement of one amino acid for another with the same charge (say, lysine for arginine) may in some cases have no effect on either protein structure or phenotype. Whether the substitution of a similar amino acid for another produces an effect depends on the precise role of that particular amino acid in the structure and function of the protein. Any change in the active site of an enzyme usually decreases enzymatic activity.

Figure 13.1 illustrates the nine possible codons that can result from a single base substitution in the UAU codon for tyrosine. One mutation is silent (box), and six are missense mutations that change the amino acid inserted in the polypeptide at this position. The other two mutations create a stop codon resulting in premature termination of translation and production of a truncated polypeptide. A base substitution that creates a new stop codon is called a **nonsense mutation.** Because nonsense mutations cause premature chain termination, the remaining polypeptide fragment is almost always nonfunctional.

Insertions and Deletions

The genomes of most higher eukaryotes contain, at a very large number of locations, tandem repeats of any of a number of short nucleotide sequences; a particularly prevalent repeat in the human genome is that of the dinucleotide CA. A repeat of this type is called a *simple tandem repeat polymorphism,* or *STRP* (Chapter 4). In most organisms in which STRPs are found, the number of copies of the repeat often differs from one chromosome to the next. Hence, populations are usually highly polymorphic for the number of repeating units. The high level of polymorphism makes these repeats very useful in such applications as linkage mapping (Chapter 4), DNA typing (Chapter 15), and family studies to localize genes that influence multifactorial traits (Chapter 16).

STRPs are usually polymorphic because they are susceptible to errors in replication or recombination that change the number of repeats or the length of the run. For

Figure 13.1

The nine codons that can result from a single base change in the tyrosine codon UAU. Blue arrows indicate transversions, gray arrows, transitions. Tyrosine codons are in boxes. Two possible stop ("nonsense") codons are shown in red. Altogether, the codon UAU allows for six possible missense mutations, two possible nonsense mutations, and one silent mutation.

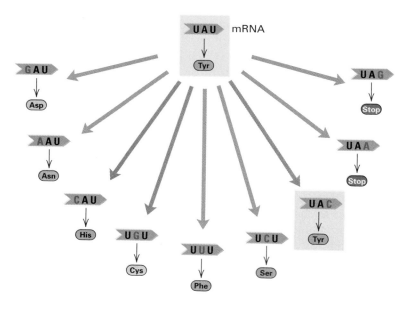

example, a run of consecutive CA dinucleotides may have extra copies added or a few deleted. Any short sequence repeated in tandem a number of times may gain or lose a few copies because of these types of errors, although the rate of change in the number of repeats depends, in some unknown manner, on the sequence in question as well as on its location in the genome. The umbrella term generally used to describe the processes leading to a change in the number of copies of a short repeating unit is **replication slippage.** One specific process that can result in such additions or deletions is unequal crossing-over, discussed in Chapter 7 (see Figure 7.20).

An increase in the number of repeating units in an STRP is genetically equivalent to an insertion, and a decrease in the number of copies is equivalent to a deletion. Nonrepeating DNA sequences are also subject to insertion or deletion. The phenotypic consequences of insertion or deletion mutations depend on their location. In nonessential regions, no effects may be seen. STRPs are usually present in noncoding regions. When insertions or deletions happen in regulatory or coding regions, however, their effects may be significant.

When they take place in coding regions, small insertions or deletions add or delete amino acids from the polypeptide, provided that the number of nucleotides added or deleted is an exact multiple of three (the length of a codon). Otherwise, the insertion or deletion shifts the phase in which the ribosome reads the triplet codons and, consequently, alters all of the amino acids downstream from the site of the mutation. As noted in Chapter 10, such mutations are called *frameshift* mutations because they shift the reading frame of the codons in the mRNA. A common type of frameshift mutation is a single-base addition or deletion. The consequences of a frameshift can be illustrated by the insertion of an adenine in this simple mRNA sequence:

Leu	Leu	Leu	Leu	
··· CUG	CUG	CUG	CUG	···
··· CUG	CAU	GCU	GCU	G···
Leu	His	Ala	Ala	

Because of the frameshift, all the amino acids downstream from the insertion are different from the original. Any addition or deletion that is not a multiple of three nucleotides will produce a frameshift. Unless it is very near the carboxyl terminus of a protein, a frameshift mutation results in the synthesis of a nonfunctional protein.

Transposable-Element Mutagenesis

All organisms contain multiple copies of several different kinds of transposable elements, which are DNA sequences capable of readily changing their positions in the genome. The structure and function of transposable elements were examined in Chapter 6 (eukaryotic transposable elements), Chapter 8 (prokaryotic transposable elements), and Chapter 9 (in connection with genetic engineering). Transposable elements are also important agents of mutation. For example, in some genes in *Drosophila,* approximately half of all spontaneous mutations that have visible phenotypic effects result from insertions of transposable elements.

Among the ways in which transposable elements can cause mutations are the mechanisms illustrated in Figure 13.2. Most transposable elements are present in nonessential regions of the genome and usually do little or no harm. When an element transposes, however, it can insert into an essential region and disrupt that region's function. Figure 13.2A shows the result of transposition into a coding region of DNA (an exon). The insert interrupts the coding region. Because most transposable elements contain coding regions of their own, either transcription of the transposable element interferes with transcription of the gene into which it is inserted or transcription of the gene terminates within the transposable element. Even if transcription proceeds through the element, the phenotype will be mutant because the coding region then contains incorrect sequences.

Another mechanism of transposable-element mutagenesis results from recombination (Figure 13.2B and C). Transposable elements can be present in multiple copies, often with two or more in the same chromosome. If the copies are near enough

Figure 13.2

Some types of mutations resulting from transposable elements. (A) Insertion into the coding region of a gene eliminates gene function. (B) Crossing-over between two copies of a transposable element present in the same orientation in a single chromatid results in the formation of a single "hybrid" copy and the deletion of the DNA sequence between the copies. (C) Unequal crossing-over between homologous elements present in the same orientation in different chromatids results in products that contain either a duplication or a deletion of the DNA sequences between the elements. Crossing-over takes place at the four-strand stage of meiosis (Chapter 4), but in parts B and C, only the two strands participating in the crossover are shown.

(A) Interruption of coding region

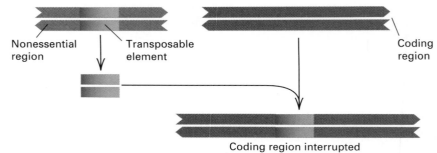
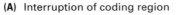

(B) Intrachromosomal crossing-over

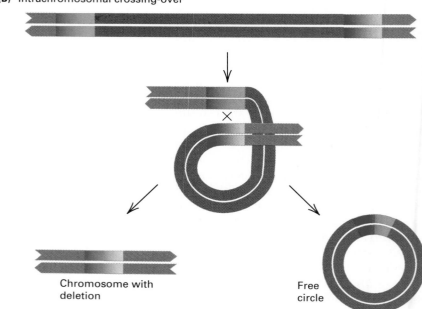

(C) Unequal crossing-over

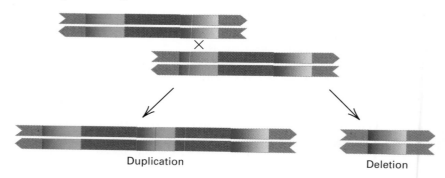

together, they can pair during synapsis and undergo crossing-over, with the result that the region of DNA between the elements is deleted. Figure 13.2B shows what happens when elements located within the same chromosome undergo pairing and crossing-over. Alternatively, the downstream element in one chromosome can pair with the upstream element in the homologous chromosome, diagrammed in Figure 13.2C. An exchange within the paired elements results in one chromatid that is missing the region between the elements (right) and one chromatid in which the region is duplicated (left).

13.3
Spontaneous Mutations

Mutations are statistically random events. There is no way of predicting when, or in which cell, a mutation will take place. However, every gene mutates spontaneously at a characteristic rate, so it is possible to assign probabilities to particular mutational events. Hence there is a definite probability that a specified gene will mutate in a particular cell and, likewise, a definite probability that a mutant allele of a specified gene will appear in a population of a designated size. The various kinds of mutational alterations in DNA differ substantially in complexity, so their probabilities of occurrence are quite different. The mutational process is also random in the sense that whether a particular mutation happens is unrelated to any adaptive advantage it may confer on the organism in its environment. A potentially favorable mutation does not arise *because* the organism has a need for it. The experimental basis for this conclusion will be presented in the next section.

The Nonadaptive Nature of Mutation

The concept that mutations are spontaneous, statistically random events unrelated to adaptation was not widely accepted until the late 1940s. Before that time, it was believed that mutations occurred in bacterial populations *in response to* particular selective conditions. The basis for this belief was the observation that when antibiotic-sensitive bacteria are spread on a solid growth medium containing the antibiotic, some colonies form that consist of cells having an inherited resistance to the drug. The initial interpretation of this observation (and similar ones) was that these adaptive variations were *induced* by the selective agent itself.

Several types of experiments showed that adaptive mutations take place spontaneously and hence were present at low frequency in the bacterial population even *before* it was exposed to the antibiotic. One experiment utilized a technique developed by Joshua and Esther Lederberg called **replica plating** (Figure 13.3). In this procedure, a suspension of bacterial cells is spread on a solid medium. After colonies have formed, a piece of sterile velvet mounted on a solid support is pressed onto the surface of the plate. Some bacteria from each colony stick to the fibers, as shown in Figure 13.3A. Then the velvet is pressed onto the surface of fresh medium, transferring some of the cells from each colony, which give rise to new colonies that have positions identical with those on the first plate. Figure 13.3B shows how this method was used to demonstrate the spontaneous origin of phage T1-r mutants. A master plate containing about 10^7 cells growing on nonselective medium (lacking phage) was replica-plated onto a series of plates that had been spread with about 10^9 T1 phages. After incubation for a time sufficient for colony formation, a few colonies of phage-resistant bacteria appeared in the same positions on each of the selective replica plates. This meant that the T1-r cells that formed the colonies must have been transferred from corresponding positions on the master plate. Because the colonies on the master plate had never been exposed to the phage, the mutations to resistance must have been present, by chance, in a few original cells not exposed to the phage.

The replica-plating experiment illustrates the principle that

> Selective techniques merely select mutants that preexist in a population.

(A) **The transfer process**

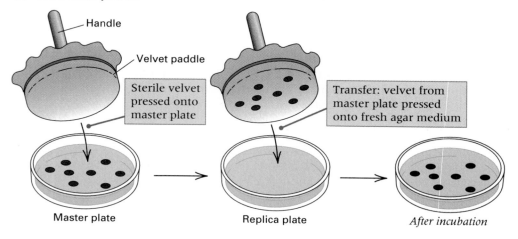

- Handle
- Velvet paddle

Sterile velvet pressed onto master plate

Transfer: velvet from master plate pressed onto fresh agar medium

Master plate

Replica plate

After incubation

(B) **Replica plating**

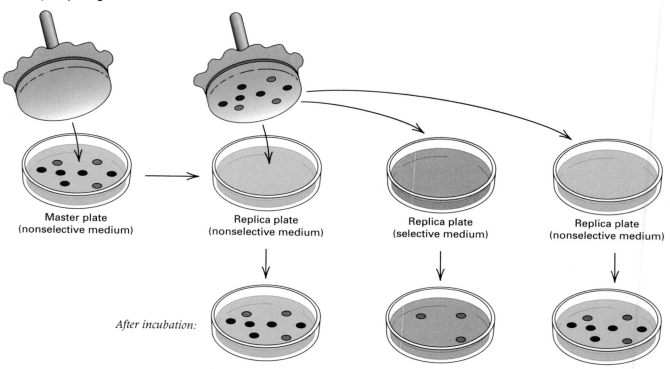

Master plate (nonselective medium)

Replica plate (nonselective medium)

Replica plate (selective medium)

Replica plate (nonselective medium)

After incubation:

Figure 13.3

Replica plating. (A) In the transfer process, a velvet-covered disk is pressed onto the surface of a master plate in order to transfer cells from colonies on that plate to a second medium. (B) For the detection of mutants, cells are transferred onto two types of plates containing either a nonselective medium (on which all form colonies) or a selective medium (for example, one spread with T1 phages). Colonies form on the nonselective plate in the same pattern as on the master plate. Only mutant cells (for example, T1-r) can grow on the selective plate; the colonies that form correspond to certain positions on the master plate. Colonies consisting of mutant cells are shown in red.

This principle is the basis for understanding how natural populations of rodents, insects, and disease-causing bacteria become resistant to the chemical substances used to control them. A familiar example is the high level of resistance to insecticides, such as DDT, that now exists in many insect populations. It is the result of selection for spontaneous mutations affecting behavioral, anatomical, and enzymatic traits that enable the insect to avoid or resist the chemical. Similar problems are encountered in controlling plant pathogens. For example, the introduction of a new variety of a crop plant that is resistant to a particular strain of disease-causing fungus results in only temporary protection against the disease. The resistance inevitably breaks down because of the occurrence of spontaneous mutations in the fungus that enable it to attack the new plant genotype. Such mutations have a clear selective advantage, and the mutant alleles rapidly become widespread in the fungal population.

Measurement of Mutation Rates

Spontaneous mutations tend to be rare, and the methods used to estimate the frequency with which they arise require large populations and often special techniques. The **mutation rate** is the probability that a gene undergoes mutation in a single generation or in forming a single gamete. Measurement of mutation rates is important in population genetics, in studies of evolution, and in analysis of the effect of environmental mutagens.

One of the earliest techniques to be developed for measuring mutation rates is the **ClB method.** ClB refers to a special X chromosome of *Drosophila melanogaster*; it has a large inversion (C) that prevents the recovery of crossover chromosomes in the progeny from a female heterozygous for the chromosome (as described in Section 7.6); a recessive lethal (l); and the dominant marker Bar (B), which reduces the normal round eye to a bar shape. The presence of a recessive lethal in the X chromosome means that males with that chromosome and females homozygous for chromosome cannot survive. The technique is designed to detect mutations arising in a normal X chromosome.

In the *ClB* procedure, females heterozygous for the *ClB* chromosome are mated with males that carry a normal X chromosome (Figure 13.4). From the progeny produced, females with the Bar phenotype are selected and then individually mated with normal males. (The presence of the Bar phenotype indicates that the females are heterozygous for the *ClB* chromosome and the normal X chromosome from the male parent.) A ratio of two females to one male is expected among the progeny from such a cross, as shown in the lower part of the illustration. The critical observation in determining the mutation rate is the fraction of males produced in the second cross. Because the *ClB* males die, all of the males in this generation must contain an X chromosome derived from the X chromosome of the initial normal male (top row of illustration). Furthermore, it must be a nonrecombinant X chromosome because of the inversion (C) in the *ClB* chromosome. Occasionally, the progeny include no males, and this means that the X chromosome present in the original sperm underwent a mutation somewhere along its length to yield a new recessive lethal. The method provides a quantitative estimate of the rate at which mutation to a recessive lethal allele occurs *at any of the large number of genes* in the X chromosome. About 0.15 percent of the X chromosomes acquire new recessive lethals in spermatogenesis; that is, the mutation rate is 1.5×10^{-3} recessive lethals per X chromosome per generation. Note that the *ClB* method tells us nothing about the mutation rate for a particular gene, because the method does not reveal how many genes on the X chromosome would cause lethality if they were mutant. Since the time that the *ClB* method was devised, a variety of other methods have been developed for determining mutation rates in *Drosophila* and other organisms. Of significance is the fact that mutation rates vary widely from one gene to another. For example, the yellow-body mutation in *Drosophila* occurs at a frequency of 10^{-4} per gamete per generation, whereas mutations to streptomycin resistance in *E. coli*. occur at a frequency of 10^{-9} per cell per generation. Furthermore, within a single organism, the frequency can vary enormously, ranging in *E. coli*,

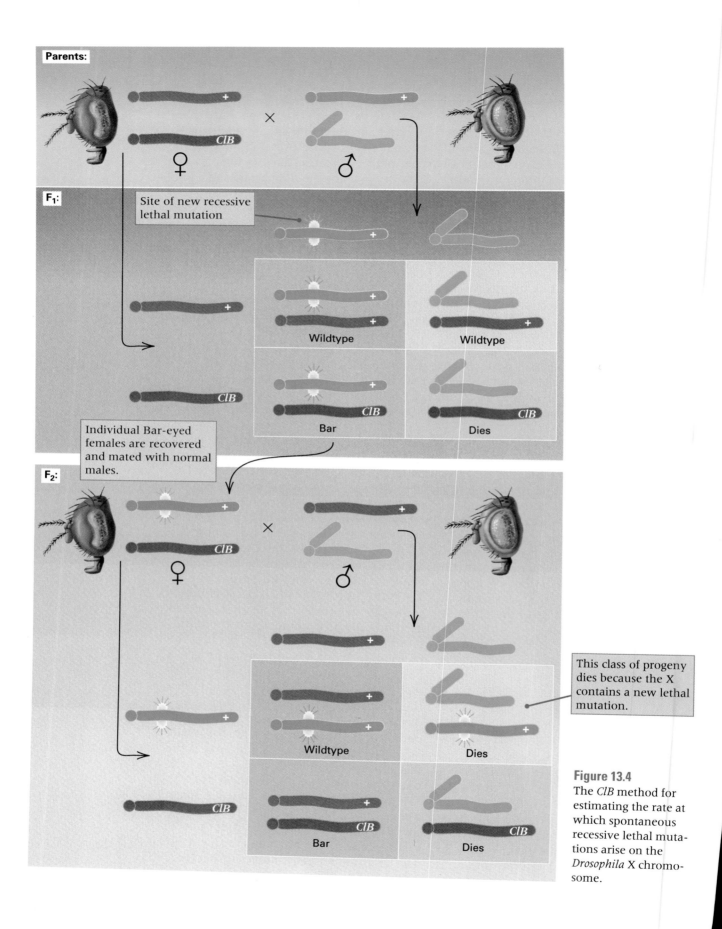

Parents:

♀ × ♂

F₁:

Site of new recessive lethal mutation

Individual Bar-eyed females are recovered and mated with normal males.

Wildtype | Wildtype
Bar | Dies

F₂:

♀ × ♂

Wildtype | Dies
Bar | Dies

This class of progeny dies because the X contains a new lethal mutation.

Figure 13.4
The *ClB* method for estimating the rate at which spontaneous recessive lethal mutations arise on the *Drosophila* X chromosome.

from 10^{-5} for some genes to 10^{-9} for others.

Hot Spots of Mutation

Certain DNA sequences are called mutational **hot spots** because they are more likely to undergo mutation than others. Mutational hot spots include monotonous runs of a single nucleotide and tandem repeats of short sequences (STRPs), which may gain or lose copies by any of a variety of mechanisms. Hot spots are found at many sites throughout the genome and within genes. For genetic studies of mutation, the existence of hot spots means that a relatively small number of sites accounts for a disproportionately large fraction of all mutations. For example, in the analysis of the *rII* gene in bacteriophage T4 discussed in Section 8.6, one extreme hot spot accounted for approximately 20 percent of all the mutations obtained.

Sites of cytosine methylation are usually highly mutable, and the mutations are usually GC → AT transitions. In many organisms, including bacteria, maize, and mammals (but not *Drosophila*), about 1 percent of the cytosine bases are methylated at the carbon-5 position, yielding 5-methylcytosine instead of ordinary cytosine (Figure 13.5). Special enzymes add the methyl groups to the cytosines in certain target sequences of DNA. In DNA replication, the 5-methylcytosine pairs with guanine and replicates normally. The genetic function of cytosine methylation is not entirely clear, but regions of DNA high in cytosine methylation tend to have reduced gene activity. Examples include the inactive X chromosome in female mammals (Chapter 7), genes that are imprinted during mammalian gametogenesis (Chapter 11), and inactive copies of certain transposable elements in maize.

Cytosine methylation is an important contributor to mutational hot spots, as illustrated in Figure 13.5. Both cytosine and 5-methylcytosine are subject to occasional loss of an amino group. For cytosine, this loss yields uracil. Because uracil pairs with adenine instead of guanine, replication of a molecule containing a GU base pair would ultimately lead to substitution

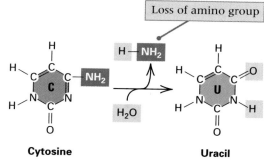

(A)

Cytosine   Uracil

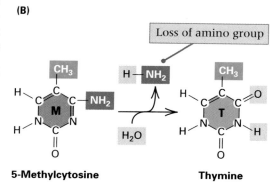

(B)

5-Methylcytosine   Thymine

Figure 13.5
(A) Spontaneous loss of the amino group of cytosine to yield uracil. (B) Spontaneous loss of the amino group of 5-methylcytosine to yield thymine.

of an AT pair for the original GC pair (by the process GU → AU → AT in successive rounds of replication). However, cells possess a special repair enzyme that specifically removes uracil from DNA. The repair process is shown in Figure 13.6. Figure 13.6A shows deamination of cytosine leading to the presence of a uracil-containing base. Repair is initiated by an enzyme called **DNA uracil glycosylase,** which recognizes the incorrect GU base pair and cleaves the offending uracil from the deoxyribose sugar to which it is attached (Figure 13.6B). It is known how this enzyme works. It scans along duplex DNA until a uracil nucleotide is encountered, at which point a specific arginine residue in the enzyme intrudes into the DNA through the minor groove. The intrusion compresses the DNA backbone flanking the uracil and results in the flipping out of the

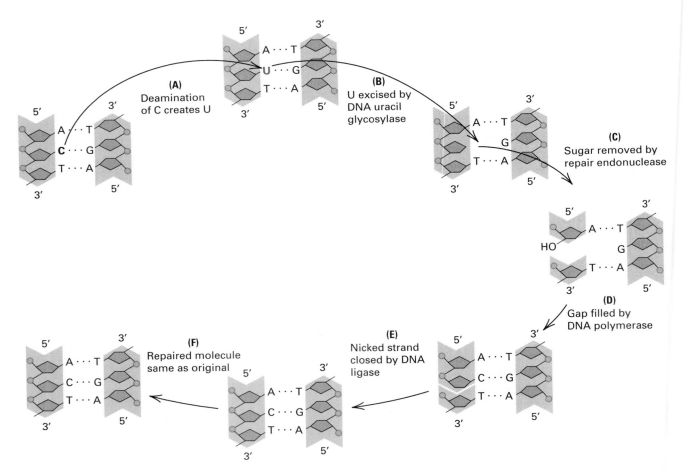

Figure 13.6

Mechanism by which uracil-containing nucleotides are formed in DNA and removed. (A) Deamination of cytosine produces uracil. (B) Uracil is cleaved from the deoxyribose sugar by DNA uracil glycosylase. (C) The deoxyribose sugar that does not contain a base is removed from the DNA backbone by another enzyme. (D) The gap in the DNA strand is filled by DNA polymerase, using the ungapped strand as template. (E) The nick remaining at the end of the gap-filling process is closed by DNA ligase. (F) The repair process restores the original sequence.

uracil base so that it sticks out of the helix. The flipping out makes it accessible to cleavage by the uracil glycosylase. After the uracil is removed, another enzyme, called the **AP endonuclease,** cleaves the base-less sugar from the DNA (Figure 13.6C), leaving a single-stranded gap that is repaired by one of the DNA polymerases (Figure 13.6D). The gap filling leaves one nick remaining in the repaired strand, which is closed by DNA ligase (Figure 13.6E). The net result of the repair is that the original C → U conversion rarely leads to mutation because the U is removed and replaced with C (Figure 13.6F). Unfortunately, loss of the amino group of 5-methylcytosine yields thymine (Figure 13.5B), which is a normal DNA base and hence is not removed by the DNA uracil glycosylase. Thus the original GC pair becomes a GT pair and, in the next round of replication, gives GC (wildtype) and AT (mutant) DNA molecules.

13.4
Induced Mutations

The first evidence that external agents could increase the mutation rate was presented in 1927 by Hermann Muller, who showed that x rays are mutagenic in

Drosophila. (He later was awarded the Nobel Prize for this work.) Since then, a large number of physical agents and chemicals have been shown to increase the mutation rate. The use of these mutagens, several of which will be discussed in this section, is a means of greatly increasing the number of mutants that can be isolated. Because some environmental contaminants are mutagenic, as are numerous chemicals found in tobacco products, mutagens are also of great importance in public health.

Base-Analog Mutagens

A **base analog** is a molecule sufficiently similar to one of the four DNA bases that it can be incorporated into a DNA duplex in the course of normal replication. Such a substance must be able to pair with a base in the template strand. Some base analogs are mutagenic because they are more prone to mispairing than are the normal nucleotides. The molecular basis of the mutagenesis can be illustrated with 5-bromouracil (Bu), a commonly used base analog that is efficiently incorporated into the DNA of bacteria and viruses.

The base 5-bromouracil is an analog of thymine, and the bromine atom is about the same size as the methyl group of thymine (Figure 13.7A). If cells are grown in a medium containing 5-bromouracil, a thymine is sometimes replaced by a 5-bromouracil in the replication of the DNA, resulting in the formation of an A-Bu pair at a normal AT site (Figure 13.7B). The subsequent mutagenic activity of the incorporated 5-bromouracil stems from a rare shift in the configuration of the base. This shift is influenced by the bromine atom and happens in 5-bromouracil more frequently than it does in thymine. In the mutagenic configuration, 5-bromouracil pairs preferentially with guanine (Figure

Figure 13.7

Mispairing mutagenesis by 5-bromouracil. (A) Structures of thymine and 5-bromouracil. (B) A base pair between adenine and the keto form of 5-bromouracil. (C) A base pair between guanine and the rare enol form of 5-bromouracil. One of 5-bromouracil's hydrogen atoms changes position to create the keto form.

(A)

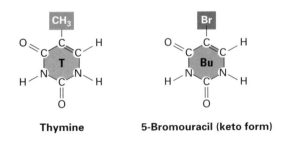

Thymine   5-Bromouracil (keto form)

(B) A-Bu base pair

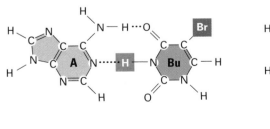

Adenine   5-Bromouracil (keto form)

(C) G-Bu base pair

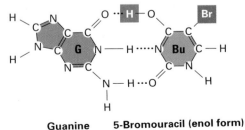

Guanine   5-Bromouracil (enol form)

13.7C), so in the next round of replication, one daughter molecule has a GC pair at the altered site, yielding an AT → GC transition (Figure 13.8).

Other experiments suggest that 5-bromouracil is also mutagenic in another way. The concentration of the nucleoside triphosphates in most cells is regulated by the concentration of thymidine triphosphate (dTTP). This regulation results in appropriate relative amounts of the four triphosphates for DNA synthesis. One part of this complex regulatory process is inhibition of the synthesis of deoxycytidine triphosphate (dCTP) by excess dTTP. The 5-bromouracil nucleoside triphosphate also inhibits production of dCTP. When 5-bromouracil is added to the growth medium, dTTP continues to be synthesized by cells at the normal rate, whereas the synthesis of dCTP is significantly reduced. The ratio of dTTP to dCTP then becomes quite high, and the frequency of misincorporation of T opposite G increases. Although repair systems can remove some incorrectly incorporated thymine, in the presence of 5-bromouracil the rate of misincorporation can exceed the rate of correction. An incorrectly incorporated T pairs with A in the next round of DNA replication, yielding a GC → AT transition in one of the daughter molecules. Thus 5-bromouracil usually induces transitions in both directions: AT → GC by the route in Figure 13.7, and GC → AT by the misincorporation route.

Just as a base analog is incorporated into DNA in place of a normal base, a **nucleotide analog** can be incorporated in place of a normal nucleotide. Modern DNA-sequencing methods are based on the use of dideoxy nucleotide analogs to terminate strand elongation at predetermined sites (Chapter 5). Dideoxy nucleotide analogs are also used in the clinical treatment of various diseases, including cancer and viral infections (Section 5.9). One of the most important drugs in the treatment of AIDS is the dideoxy analog AZT (Figure 13.9), which slows down replication of the HIV virus because the polymerase of the

Figure 13.8

One mechanism for mutagenesis by 5-bromouracil (Bu). An AT → GC transition is produced by the incorporation of 5-bromouracil in DNA replication, forming an A−Bu pair. In the mutagenic round of replication, the Bu (in its rare enol form) pairs with G. In the next round of replication, the G pairs with C, completing the transition mutation.

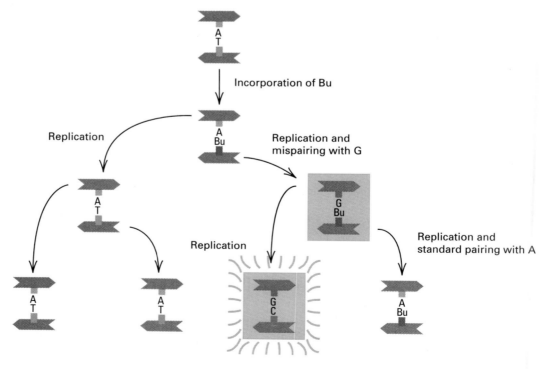

virus incorporates AZT in preference to the normal deoxythymidine.

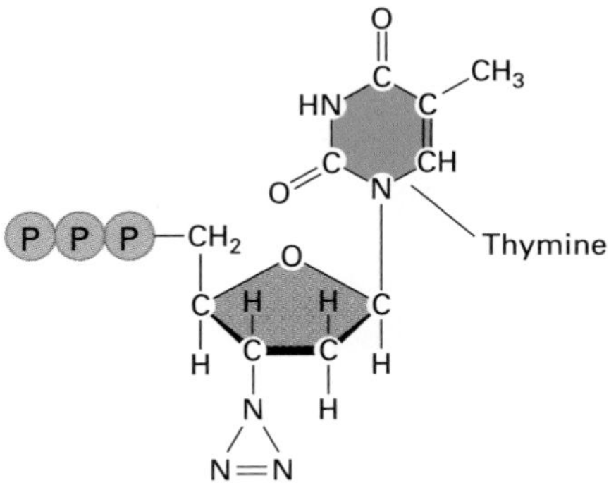

Chemical Agents That Modify DNA

Many mutagens are chemicals that react with DNA and change the hydrogen-bonding properties of the bases. These mutagens are active on both replicating and nonreplicating DNA, in contrast to the base analogs, which are mutagenic only when DNA replicates. Several of these chemical mutagens, of which nitrous acid is a well-understood example, are highly specific in the changes they produce. Others—for example, the alkylating agents—react with DNA in a variety of ways and produce a broad spectrum of effects.

Nitrous acid (HNO_2) acts as a mutagen by converting amino (NH_2) groups of the bases adenine, cytosine, and guanine into keto ($=O$) groups. This process of *deamination* alters the hydrogen-bonding specificity of each base. Deamination of adenine

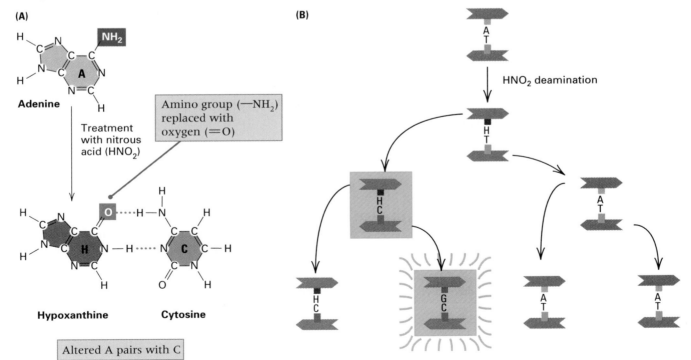

Figure 13.9
Chemical structure of AZT, a nucleotide analog used in the clinical treatment of AIDS.

3'-azido-2',3'-deoxythymidine
(AZT)

yields a base, hypoxanthine, that pairs with cytosine rather than thymine, resulting in an AT → GC transition (Figure 13.10). As discussed earlier, deamination of cytosine yields uracil (Figure 13.5A), which pairs with adenine instead of guanine, producing a GC → AT transition.

In genetic research, chemical mutagens such as nitrous acid (and many others) are

Figure 13.10
Nitrous acid mutagenesis. (A) Conversion of adenine into hypoxanthine, which pairs with cytosine. (The cytosine → uracil conversion is shown in Figure 13.5.) (B) Production of an AT → GC transition. In the mutagenic round of replication, the hypoxanthine (H) pairs with C. In the next round of replication, the C pairs with G, completing the transition mutation.

(A)

Adenine

Amino group (—NH$_2$) replaced with oxygen (=O)

Treatment with nitrous acid (HNO$_2$)

Hypoxanthine   Cytosine

Altered A pairs with C

(B)

HNO$_2$ deamination

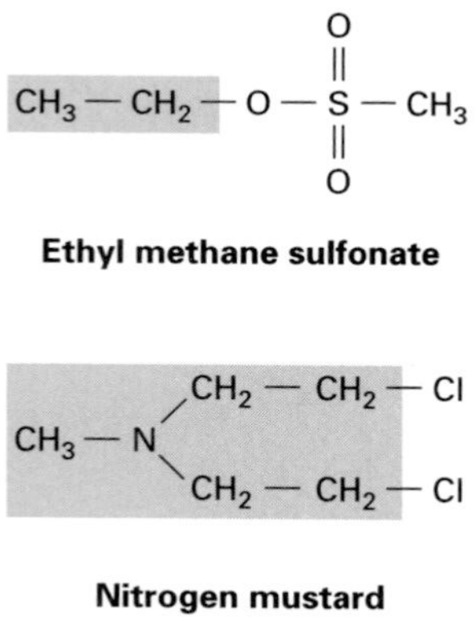

$$CH_3 - CH_2 - O - \overset{\displaystyle O}{\underset{\displaystyle O}{\overset{\|}{\underset{\|}{S}}}} - CH_3$$

Ethyl methane sulfonate

$$CH_3 - N \overset{\displaystyle CH_2 - CH_2 - Cl}{\underset{\displaystyle CH_2 - CH_2 - Cl}{}}$$

Nitrogen mustard

Figure 13.11
The chemical structures of two highly mutagenic alkylating agents; the alkyl groups are shown in red.

Figure 13.12
Deoxyribose-purine bonds are somewhat unstable and prone to undergo spontaneous reaction with water (hydrolysis), which results in loss of a purine base from the DNA (depurination). (A) Part of a DNA molecule prior to depurination. The bond between the labeled G and the deoxyribose to which it is attached is about to be hydrolyzed. (B) Hydrolysis of the bond releases the G purine, which diffuses away from the molecule and leaves a hydroxyl (−OH) in its place in the depurinated DNA.

(A) Normal DNA duplex

(B) DNA with one site of depurination

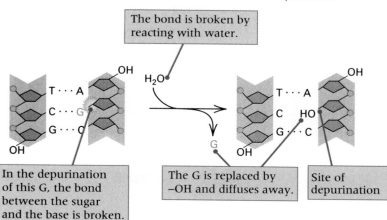

The bond is broken by reacting with water.

In the depuration of this G, the bond between the sugar and the base is broken.

The G is replaced by −OH and diffuses away.

Site of depurination

exceedingly useful in prokaryotic systems but are not particularly useful as mutagens in higher eukaryotes because the chemical conditions necessary for reaction are not easily obtained. However, alkylating agents are highly effective in eukaryotes, as well as prokaryotes. **Alkylating agents,** such as ethyl methane sulfonate (EMS) and nitrogen mustard (the structures of which are shown in Figure 13.11), are potent mutagens that have been used extensively in genetic research. These agents add to the DNA bases various chemical groups that either alter their base-pairing properties or cause structural distortion of the DNA molecule. Alkylation of either guanine or thymine causes mispairing, leading to the transitions AT → GC and GC → AT. EMS reacts less readily with adenine and cytosine.

Another phenomenon that results from the alkylation of guanine is **depurination,** or loss of the alkylated base from the DNA molecule by breakage of the bond joining the purine nitrogen with the deoxyribose. The process is illustrated in Figure 13.12. Depurination is not always mutagenic, because the loss of the purine can be repaired by excision of the deoxyribose (a function of the AP endonuclease discussed in Section 13.3) and gap repair by DNA polymerase. If, however, the replication fork reaches the apurinic site before repair has taken place, then replication almost always inserts an adenine nucleotide in the daughter strand opposite the apurinic site. After another round of replication, the original GC pair becomes a TA pair, an example of a transversion mutation.

DNA is susceptible to spontaneous depurination also. In air, the rate of spontaneous depurination is approximately 3×10^{-9} depurinations per purine nucleotide per minute. This rate is at least tenfold greater than any other single source of spontaneous degradation. At this rate, the half-life of a purine nucleotide exposed to air is about 300 years. It is for this reason that reports of PCR amplification of ancient DNA from any material exposed to the elements for more than a few thousand years should be treated with some skepticism. There are, however, a few

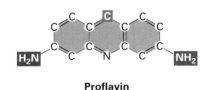

Proflavin

Figure 13.13
The structure of proflavine, an acridine derivative. Other mutagenic acridines have additional
atoms on the NH₂ group and on the C of the
central ring. Hydrogen atoms are not shown.

substances in which DNA that may be present degrades at a slower rate than normal.
For example, the DNA in organisms pre-
served in amber may be more resistant to
degradation.

Misalignment Mutagenesis

The **acridine** molecules, of which
proflavin is an example (Figure13.13), are
planar, three-ringed molecules whose
dimensions are roughly the same as those
of a purine-pyrimidine pair. These substances insert between adjacent base pairs
of DNA, a process called **intercalation.**
The effect of the intercalation of an acridine
molecule is to cause the adjacent base pairs
to move apart by a distance roughly equal
to the thickness of one pair (Figure 13.14).
When DNA that contains intercalated
acridines is replicated, the template and
daughter DNA strands can misalign, particularly in regions where nucleotides are
repeated. The result is that a nucleotide can
be either added or deleted in the daughter
strand. In a coding region, the result of a
single-base addition or deletion is a
frameshift mutation (Section 13.2).

Ultraviolet Irradiation

Ultraviolet (UV) light produces both mutagenic and lethal effects in all viruses and
cells. The effects are caused by chemical
changes in the bases resulting from absorption of the energy of the light. The major
products formed in DNA after UV irradiation are covalently joined pyrimidines
(**pyrimidine dimers**), primarily thymine
(Figure 13.15A), that are adjacent in the
same polynucleotide strand. This chemical
linkage brings the bases closer together,
causing a distortion of the helix (Figure
13.15B), which blocks transcription and
transiently blocks DNA replication.

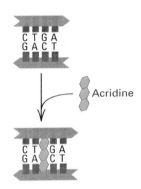

Figure 13.14
Separation of two base
pairs caused by inter-
calation of an acridine
molecule.

Figure 13.15
(A) Structural view of the formation of a thymine dimer. Adjacent
thymines in a DNA strand that have been subjected to ultraviolet (UV)
irradiation are joined by formation of the bonds shown in red. Other
types of bonds between the thymine rings also are possible. Although not
drawn to scale, these bonds are considerably shorter than the spacing
between the planes of adjacent thymines, so the double-stranded structure becomes distorted. The shape of each thymine ring also changes. (B)
The distortion of the DNA helix caused by two thymines moving closer
together when joined in a dimer.

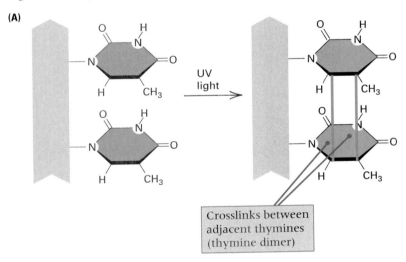

Crosslinks between
adjacent thymines
(thymine dimer)

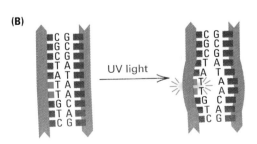

Pyrimidine dimers can be repaired in ways discussed later in this chapter.

In human beings, the inherited autosomal recessive disease **xeroderma pigmentosum** is the result of a defect in a system that repairs ultraviolet-damaged DNA. Persons with this disease are extremely sensitive to sunlight, and this sensitivity results in excessive skin pigmentation and the development of numerous skin lesions that frequently become cancerous. Removal of pyrimidine dimers does not occur in the DNA of cells cultured from patients with the disease, and the cells are sensitive to much lower doses of ultraviolet light than are cells from normal persons.

Even in normal people, excessive exposure to the UV rays in sunlight increases the risk of skin cancer. Ozone, a form of oxygen (O_3) present high in the atmosphere (at a height of 15 to 30 kilometers), absorbs UV rays and ameliorates, to some extent, the damaging effects of sunlight. The thinning of the ozone layer (the "ozone hole"), which is especially pronounced at certain latitudes, greatly increases the effective exposure to UV and correspondingly increases the rate of skin cancer. The thinning of the ozone layer can be attributed to the destructive effects of chlorofluorocarbons (CFCs) used as pressurizers in aerosol spray cans and as refrigerants, when these agents ascend to the upper atmosphere.

Ionizing Radiation

Ionizing radiation includes x rays and the particles and radiation released by radioactive elements (α and β particles and γ rays). When x rays were first discovered late in the nineteenth century, their power to pass through solid materials was regarded as a harmless entertainment and a source of great amusement:

> By 1898, personal x rays had become a popular status symbol in New York. The *New York Times* reported that "there is quite as much difference in the appearance of the hand of a washerwoman and the hand of a fine lady in an x-ray picture as in reality." The hit of the exhibition season was Dr. W. J. Morton's full-length portrait of *"the x-ray lady,"* a "fashionable woman who had evi-

dently a scientific desire to see her bones." The portrait was said to be a "fascinating and coquettish" picture, the lady having agreed to be photographed without her stays and corset, the better to satisfy the "longing to have a portrait of well-developed ribs." Dr. Morton said women were not afraid of x rays: "After being assured that there is *no danger* they take the rays without fear."

> The titillating possibility of using x rays to see through clothing or to invade the privacy of locked rooms was a familiar theme in popular discussions of x rays and in cartoons and jokes. Newspapers carried advertisements for *"x-ray proof underclothing"* for those seeking to protect themselves from x-ray inspection.

> The luminous properties of radium soon produced a full-fledged radium craze. A famous woman dancer performed *radium dances* using veils dipped in fluorescent salts containing radium. *Radium roulette* was popular at New York casinos, featuring a "roulette wheel washed with a radium solution, such that it glowed brightly in the darkness; an unseen hand cast the ball on the turning wheel and sparks marked its course as it bounded from pocket to glimmery pocket." A patent was issued for a process for making women's gowns luminous with radium, and Broadway producer Florenz Ziegfeld snapped up the rights for his stage extravaganzas.

> Even while the unrestrained use of x rays and radium was growing, evidence was accumulating that the new forces might not be so benign after all. Hailed as tools for fighting cancer, they could also cause cancer. Doctors using x rays were the first to learn this bitter lesson. (Quoted from S. Hilgartner, R. C. Bell, and R. O'Connor. 1982. *Nukespeak.* Sierra Club Books, San Francisco.)

Doctors were indeed the first to learn the lesson. Many suffered severe x-ray burns or required amputation of overexposed hands or arms. Many others died from radiation poisoning or from radiation-induced cancer. By the mid-1930s, the number of x-ray deaths had grown so large that a monument to the "x-ray martyrs" was erected in a hospital courtyard in Germany. Yet the full hazards of x-ray exposure were not widely appreciated until the 1950s.

When ionizing radiation interacts with water or with living tissue, highly reactive ions called **free radicals** are formed. The free radicals react with other molecules, including DNA, which results in the carcinogenic and mutagenic effects. The intensity of a beam of ionizing radiation can be described quantitatively in several ways. There are, in fact, a bewildering variety of units in common use (Table 13.1). Some of the units (becquerel, curie) deal with the number of disintegrations emanating from a material, others (roentgen) with the number of ionizations the radiation produces in air, still others (gray, rad) with the amount of energy imparted to material exposed to the radiation, and some (rem, sievert) with the effects of radiation on living tissue. The types of units have proliferated through the years in attempts to encompass different types of radiation, including nonionizing radiation, in a common frame of reference. The units in Table 13.1 are presented only for ease of interpreting the multitude of units found in the literature on radiation genetics. They need not be memorized.

Genetic studies of ionizing radiation support the following general principle:

> Over a wide range of x-ray doses, the frequency of mutations induced by x rays is proportional to the radiation dose.

One type of evidence supporting this principle is the frequency with which X-chromosome recessive lethals are induced in *Drosophila* (Figure 13.16). The mutation

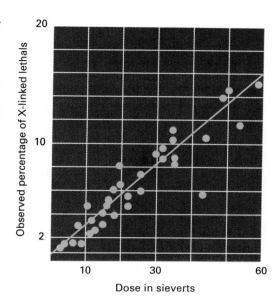

Figure 13.16
The relationship between the percentage of X-linked recessive lethals in *D. melanogaster* and x-ray dose, obtained from several experiments. The frequency of spontaneous X-linked lethal mutations is 0.15 percent per X chromosome per generation.

rate increases linearly with increasing x-ray dose. For example, an exposure of 10 sieverts increases the frequency from the spontaneous value of 0.15 percent to about 3 percent. The mutagenic and lethal effects of ionizing radiation at low to moderate doses

Table 13.1 Units of radiation

Unit (abbreviation)	Magnitude
Becquerel (Bq)*	1 disintegration/second = 2.7×10^{-11} Ci
Curie (Ci)	3.7×10^{10} disintegrations/second = 3.7×10^{10} Bq
Gray (Gy)*	1 joule/kilogram = 100 rad
Rad (rad)	100 ergs/gram = 0.01 Gy
Rem (rem)	damage to living tissue done by 1 rad = 0.01 Sv
Roentgen (R)	produces 1 electrostatic unit of charge per cubic centimeter of dry air under normal conditions of pressure and temperature. (By definition, 1 electrostatic unit repels with a force of 1 dyne at a distance of 1 centimeter.)
Sievert (Sv)	100 rem

*Units officially recognized by the International System of Units as defined by the General Conference on Weights and Measures.

X-Ray Daze

Hermann J. Muller 1927
University of Texas, Austin, Texas
Artificial Transmutation of the Gene

Mutagenesis, which means the deliberate induction of mutations, plays an important role in genetics because it makes possible the identification of genes that control biological processes, such as development and behavior. For the demonstration that x rays could be used to induce new mutations, Muller was awarded a Nobel Prize in 1946. His discovery also had important practical implications in regard to x rays (and other mutagenic agents) present in the environment. After World War II ended with the nuclear bombing of Hiroshima and Nagasaki, Muller became a leader in the fight against the above-ground testing of nuclear bombs because of the release of radioactivity into the atmosphere. He also condemned the casual use of x rays, such as for the fitting of children's shoes, and he campaigned for the engineering of safer x-ray machines for medical use. Muller's prose could be turgid at times, thanks to his use of subordinate clauses. This excerpt captures the style.

Most modern geneticists will agree that gene mutations form the chief basis of organic evolution, and therefore of most of the complexities of living things. Unfortunately for the geneticists, however, the study of these mutations, and, through them, of the genes themselves, has heretofore been very seriously hampered by the extreme infrequency of their occurrence under ordinary conditions, and by the general unsuccessfulness of attempts to modify decidedly, and in a sure and detectable way, this sluggish "natural" mutation rate. . . . On theoretical grounds, it has

---

Treatment of the sperm with relatively heavy doses of x-rays induces the occurrence of true "gene mutations" in a high proportion of the treated germ cells

---

appeared to the present writer that radiations of short wave length should be especially promising for the production of mutational changes, and for this and other reasons a series of experiments concerned with this problem has been undertaken during the past year on the fruit fly, *Drosophila melanogaster.* . . . It has been found quite conclusively that treatment of the sperm with relatively heavy doses of x-rays induces the occurrence of true "gene mutations" in a high proportion of the treated germ cells. Several hundred mutants have been obtained in this way in a short time and considerably more than a hundred have been followed through three, four or more generations. They are (nearly all of them, at any rate) stable in their inheritance. . . . Regarding the types of mutations produced, it was found that the recessive lethals greatly outnumbered the nonlethals producing a visible morphological abnormality. . . . In addition to gene mutations, it was found that there is also caused by x-ray treatment a high proportion of rearrangements in the linear order of the genes. . . . The transmuting action of x-rays on the genes is not confined to the sperm cells, for treatment of the unfertilized females causes mutations about as readily as treatment of the males. . . . In conclusion, the attention of those working along classical genetic lines may be drawn to the opportunity, afforded them by the use of x-rays, of creating in their chosen organisms a series of artificial races for use in the study of genetics. If, as seems likely on genetic considerations, the effect is common to most organisms, it should be possible to produce, "to order," enough mutations to furnish respectable genetic maps.

Source: Science 66: 84–87

---

result primarily from damage to DNA. Three types of damage in DNA are produced by ionizing radiation: single-strand breakage (in the sugar-phosphate backbone), double-strand breakage, and alterations in nucleotide bases. The single-stand breaks are usually efficiently repaired, but the other damage is responsible for mutation and lethality. In eukaryotes, ionizing radiation also results in chromosome breaks. Although systems exist for repairing the breaks, the repair often leads to translocations, inversions, duplications, and deletions. In human cells in culture, a dose of 0.2 sievert results in an average of one visible chromosome break per cell.

Although the effects of extremely low levels of radiation are extremely difficult to measure because of the background of spontaneous mutation, most experiments support the following principle:

There appears to be no threshold exposure below which mutations are not induced. Even very low doses of radiation induce some mutations.

Ionizing radiation is widely used in tumor therapy. The basis for the treatment is the increased frequency of chromosomal breakage (and the consequent lethality) in cells undergoing mitosis compared with cells in interphase. Tumors usually contain many more mitotic cells than most normal tissues, so more tumor cells than normal cells are destroyed. Because not all tumor cells are in mitosis at the same time, irradiation is carried out at intervals of several days to allow interphase tumor cells to enter mitosis. Over a period of time, most tumor cells are destroyed.

Table 13.2 gives representative values of doses of ionizing radiation received by human beings in the United States in the course of a year. The unit of measure is the millisievert, which equals 0.1 rem. The exposures in Table 13.2 are on a yearly basis, so over the course of a generation, the total exposure is approximately 100 millisieverts. Note that, with the exception of diagnostic x rays, which yield important compensating benefits, most of the total radiation exposure comes from natural sources, particularly radon gas. Less than 20 percent of the average radiation exposure comes from artificial sources. Nevertheless, there are dangers inherent in any exposure to ionizing radiation, particularly in an increased risk of leukemia and certain others cancers in the exposed persons. In regard to increased genetic diseases in future generations the risk that a small amount of additional radiation will have mutagenic effects is low enough that most geneticists are currently more concerned about the effects of the many mutagenic (as well as carcinogenic) chemicals that are introduced into the environment from a variety of sources.

The National Academy of Sciences of the United States regularly updates the estimated risks of radiation exposure. The latest estimates are summarized in Table 13.3. The message is that an additional 10 millisieverts of radiation per generation (about a 10 percent increase in the annual exposure) is expected to cause a relatively modest increase in diseases that are wholly or in part due to genetic factors. The most common conditions in the table are heart disease and cancer. No estimate for the radiation-induced increase is given for either of these traits because the genetic contribution to the total is still very uncertain.

Genetic Effects of the Chernobyl Nuclear Accident

The city of Chernobyl in the Ukraine has become a symbol for nuclear disaster. On April 26, 1986, a nuclear power plant near the city exploded, heavily contaminating the immediate area and sending clouds of radioactive debris over long distances. It was the largest publicly acknowledged nuclear accident in history. The meltdown is estimated to have released between 50 and 200 million curies of radiation, along with a wide variety of heavy-metal and chemical pollutants. Iodine-131 and cesium-137 were the principal radioactive contaminants. People living in the area were evacuated almost immediately, but many were heavily exposed to radiation.

Table 13.2 Annual exposure of human beings in the United States to various forms of ionizing radiation

Source	Dose (in millisieverts*)
Natural radiation	
Radon gas	2.06
Cosmic rays	0.27
Natural radioisotopes in the body	0.39
Natural radioisotopes in the soil	0.28
Total natural radiation	3.00
Other radiation sources	
Diagnostic x rays	0.39
Radiopharmaceuticals	0.14
Consumer products (x rays from TV, radioisotopes in clock dials) and building materials	0.10
Fallout from weapon tests	<0.01
Nuclear power plants	<0.01
Total from non-natural sources	0.63
Total from all sources	3.63

*One millisievert (mSv) equals 1/1000 sievert, or 0.1 rem.
Source: From National Research Council, Committee on the Biological Effects of Ionizing Radiations, Health Effects of Exposure to Low Levels of Ionizing Radiation (*BEIR V*) National Academy Press, 1990.

Within a short time there was a notable increase in the frequency of thyroid cancer in children, and other health effects of radiation exposure were also detected.

At first, little attention was given to possible genetic effects of the Chernobyl disaster, because relatively few people were acutely exposed and because the radiation dose to people outside the immediate area was considered too small to worry about. In the district of Belarus, some 200 kilometers north of Chernobyl, the average exposure to iodine-131 was estimated at approximately 0.185 sievert per person. This is a fairly high dose, but the exposure was brief because the half-life of iodine-131 is only 8 days. Radiation from the much longer-lived (30-year half-life) cesium-137 was estimated at less than 5 millisieverts per year. On the basis of data from laboratory animals and studies of the survivors of the Hiroshima and Nagasaki atomic bombs, little detectable genetic damage was expected from these exposures.

Nevertheless, 10 years after the meltdown, studies of people living in Belarus did indicate a remarkable increase in the mutation rate. The observations focused on STRP polymorphisms because of their intrinsically high mutation rate due to replication slippage and other factors. Each STRP was examined in parents and their offspring to detect any DNA fragments that increased or decreased in size, indicative of an increase or decrease in the number of repeating units contained in the DNA fragment. Five STRP polymorphisms were studied, with the results summarized in Figure 13.17. Two of the loci (blue dots) showed no evidence for an increase in the mutation rate. This is the expected result. The unexpected finding was that three of the loci (red dots) did show a significant increase, and the level of increase was consistent with an approximate doubling of the mutation rate. At the present time, it is still not known whether the increase was detected because replication slippage is

Table 13.3 Estimated genetic effects of an additional 10 millisieverts per generation

Type of disorder	Current incidence per million liveborn	Additional cases per million liveborn per 10 mSv per generation	
		First generation	At equilibrium
Autosomal dominant			
Clinically severe	2500	5–20	25
Clinically mild	7500	1–15	75
X-linked	400	<1	<5
Autosomal recessive	2500	<1	Very slow increase
Chromosomal			
Unbalanced translocations	600	<5	Very little increase
Trisomy	3800	<1	<1
Congenital abnormalities (multifactorial)	20,000–30,000	10	10–100
Other multifactorial disorders			
Heart disease	600,000	Unknown	Unknown
Cancer	300,000	Unknown	Unknown
Others	300,000	Unknown	Unknown

Source: Health Effects of Exposure to Low Levels of Ionizing Radiation (BEIR V), National Research Council, Washington, D.C.

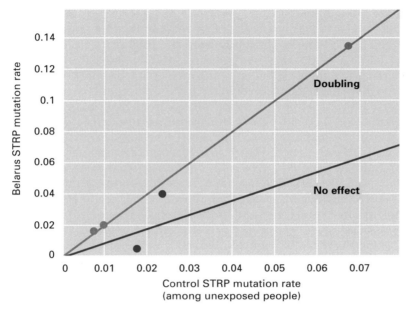

Figure 13.17
Mutation rates of five STRP polymorphisms among people of Belarus exposed to radiation from Chernobyl and among unexposed British people. Three of the loci (red dots) show evidence of an approximately twofold increase in the mutation rate. [Data from Y. E. Dubrova, V. N. Nesterov, N. G. Krouchinsky, V. A. Ostapenko, R. Neumann, D. L. Neil, and A. J. Jeffreys, *Nature* 1996. 380: 183.]

more sensitive to radiation than other types of mutations or because the effective radiation dose sustained by the Belarus population was much higher than originally thought.

The finding of increased mutation among the people of Belarus is paralleled by observations carried out in two species of voles, small rodents related to hamsters and gerbils, that live in the immediate vicinity of Chernobyl. These animals live in areas still highly contaminated by radiation, heavy metals, and chemical pollutants. The animals eat food that is so heavily contaminated that they themselves become extremely radioactive. Yet they survive and reproduce in this harsh environment.

Among the voles of Chernobyl, the rate of mutation in the cytochrome *b* gene present in mitochondrial DNA was estimated at 2×10^{-4} mutations per nucleotide per generation. This rate is at least 200 times higher than that found among animals living in a relatively uncontaminated area just 30 kilometers to the southwest of Chernobyl. It is so high that, on average, each mitochondrial DNA molecule of approximately 17 kb undergoes three new mutations in each generation. It is as yet unknown whether this high rate of mutation is due to radiation exposure, to nonradioactive pollutants, or to the synergistic effects of both.

13.5
Mechanisms of DNA Repair

Spontaneous damage to DNA in human cells takes place at a rate of approximately 1 event per billion nucleotide pairs per minute (or, per nucleotide pair, at a rate of 1×10^{-9} per minute). This may seem like quite a small rate, but it implies that every 24 hours, in every human cell, the DNA is damaged at approximately 10,000 different sites.

Fortunately for us, and for all living organisms, much of the damage done to DNA by spontaneous chemical reactions in the nucleus, chemical mutagens, and radiation can, under ordinary circumstances, be repaired. We have already seen (Section 13.3) one example of repair in the manner in which DNA uracil glycosylase and the AP endonuclease act in tandem first to remove and then to replace uracil nucleotides that get into DNA, such as by the deamination of cytosine. In this section, we introduce a number of other processes that are important in the repair of aberrant or damaged DNA.

Mismatch Repair

Base-substitution mutations usually start as a single mismatched (mispaired) base in a double-stranded DNA molecule—for example, a GT base pair resulting from spontaneous loss of an amino group from 5-methylcytosine (Figure 13.5). However, many of these mismatched bases need not persist through DNA replication to be resolved. The mismatched bases can be detected and corrected by enzymes that carry out **mismatch repair,** a process of excision and resynthesis illustrated in Figure 13.18. When a mismatched base is detected, one of the strands is cut in two places and a region around the mismatch is removed. The excised region is usually quite precise. In prokaryotes, the cleavage sites are 5 nucleotides to the 3' end and 8 nucleotides to the 5' end of the damage, yielding an excised fragment of 13 nucleotides. In eukaryotes, the cleavage sites are 5 nucleotides to the 3' end and 24 nucleotides to the 5' end, yielding an excised fragment of 29 nucleotides. After excision, DNA polymerase fills in the gap by using the remaining strand as template,

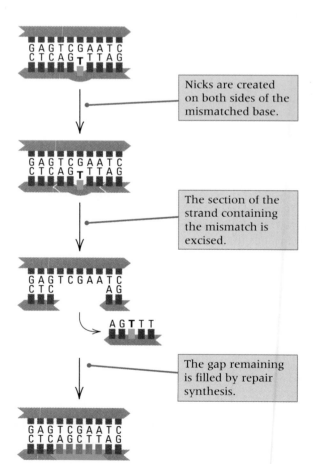

Figure 13.18
Mismatch repair consists of the excision of a segment of a DNA strand that contains a base mismatch, followed by repair synthesis. The length of the excised strand is typically 13 nucleotides in prokaryotes and 29 nucleotides in eukaryotes. Either strand can be excised and corrected, but in newly synthesized DNA, methylated bases in the template strand often direct the excision mechanism to the newly synthesized strand that contains the incorrect nucleotide.

Nicks are created on both sides of the mismatched base.

The section of the strand containing the mismatch is excised.

The gap remaining is filled by repair synthesis.

Replication Slippage in Unstable Repeats

Micheline Strand,[1] Tomas A. Prolla,[2] R. Michael Liskay,[2] and Thomas D. Petes[1]
1993

[1]University of North Carolina, Chapel Hill, North Carolina
[2]Yale University, New Haven, Connecticut

Destabilization of Tracts of Simple Repetitive DNA in Yeast by Mutations Affecting DNA Mismatch Repair

One of the most important principles of genetics is that many fundamental processes are carried out in much the same way in many different organisms. Basic cellular processes are similar in all eukaryotic cells, and the basic mechanisms of DNA replication and repair are fundamentally the same in eukaryotes and prokaryotes. The similarity of mismatch repair among organisms motivated this paper. The authors report that yeast cells deficient in mismatch repair are very deficient in their ability to replicate faithfully tracts of short repeating sequences. Because this type of instability also characterizes certain human hereditary colon cancers, the authors correctly predicted that the mutant genes causing the colon cancer would be genes involved in mismatch repair.

The genomes of all eukaryotes contain tracts of DNA in which a single base or a small number of bases is repeated. Expansions of such tracts have been associated with several human disorders including the fragile-X syndrome. In addition, simple repeats are unstable in certain forms of colorectal cancer, suggesting a

---

The instability of poly(GT) tracts in yeast is increased by mutations in the mismatch repair genes

---

defect in DNA replication of repair. . . . One mechanism [for changes in repeat number] is DNA polymerase slippage: during replication of the tract, the primer and template strand transiently dissociate and then reassociate in a misaligned configuration. If the unpaired bases are in the primer strand, continued synthesis results in an elongation of the tract whereas unpaired bases in the template strand result in a deletion. . . . The most common simple repeat in most eukaryotes is poly$(GT)_{10-30}$. We examined the genetic control of the stability of poly(GT) tracts. . . . Three mutations affect-

ing mismatch repair in yeast were examined, *pms1* and *mlh1* (homologs of *MutL*) and *msh2* (a homolog of *MutS*). . . . In two different genetic backgrounds using two different instability assay systems, all three mutations lead to 100- to 700-fold elevated levels of tract instability. . . . Yeast contains two DNA polymerases that have 3'–5' exonuclease ("proofreading") activities. In strains with mutations in the nuclease portion of these genes, neither affects tract stability as dramatically as mutations that reduce mismatch repair. . . . The results show that the instability of poly(GT) tracts in yeast is increased by mutations in the mismatch repair genes but is relatively unaffected by mutations affecting the proofreading functions of DNA polymerase. These results indicate that DNA polymerase *in vivo* has a very high rate of slippage on templates containing simple repeats, but most of the errors are corrected by cellular mismatch repair systems. Thus, the instability of simple repeats observed for some human diseases may be a consequence of either an increased rate of DNA polymerase slippage or a decreased efficiency of mismatch repair.

Source: Nature 365: 274–276

---

thereby eliminating the mismatch. The mismatch-repair system also corrects most single-base insertions or deletions.

If the DNA strand that is removed in mismatch repair is chosen at random, then the repair process sometimes creates a mutant molecule by cutting the strand that contains the correct base and using the mutant strand as template. However, in newly synthesized DNA, there is a way to prevent this from happening because the daughter strand is less methylated than the parental strand. The mismatch-repair system recognizes the degree of methylation of a strand and *preferentially excises nucleotides from the undermethylated strand.*

This helps ensure that incorrect nucleotides incorporated into the daughter strand in replication will be removed and repaired. The daughter strand is always the undermethylated strand because its methylation lags somewhat behind the moving replication fork, whereas the parental strand was fully methylated in the preceding round of replication.

The mismatch-repair system, like the proofreading function of DNA polymerase (Chapter 5) and all other biochemical systems, is not perfect. Some spontaneous base substitutions still arise as a result of the incorporation of errors that escape correction. Figure 13.19 summarizes the

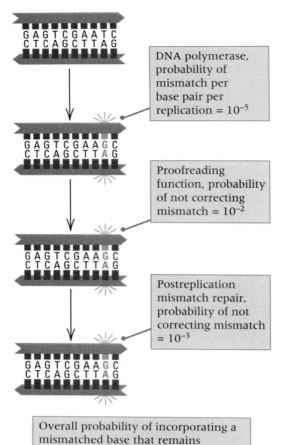

GAGTCGAATC
CTCAGCTTAG

DNA polymerase, probability of mismatch per base pair per replication = 10^{-5}

GAGTCGAAGC
CTCAGCTTAG

Proofreading function, probability of not correcting mismatch = 10^{-2}

GAGTCGAAGC
CTCAGCTTAG

Postreplication mismatch repair, probability of not correcting mismatch = 10^{-3}

GAGTCGAAGC
CTCAGCTTAG

Overall probability of incorporating a mismatched base that remains uncorrected = $10^{-5} \times 10^{-2} \times 10^{-3} = 10^{-10}$ per base pair per replication

Figure 13.19
Summary of rates of error in DNA polymerization, proofreading, and postreplication mismatch repair. The initial rate of nucleotide misincorporation is 10^{-5} per base pair per replication. The proofreading function of DNA polymerase corrects 99 percent of these errors, and of those that remain, postreplication mismatch repair corrects 99.9 percent. The overall rate of misincorporated nucleotides that are not repaired is 10^{-10} per base pair per replication.

The mechanism of mismatch repair has been studied extensively in bacteria. Mutants defective in the process were identified as having high rates of spontaneous mutation. The products of two genes, *mutL* and *mutS*, recognize and bind to a mismatched base pair. This triggers the excision of a tract of nucleotides from the newly synthesized strand. Experiments carried out in yeast by Thomas Petes and colleagues revealed that simple sequence repeats (tandem repeats of short nucleotide sequences) are a thousandfold less stable in mutants deficient in mismatch repair than in wildtype yeast. By that time it was already known that some forms of human hereditary colorectal cancer result in decreased stability of simple repeats, and the yeast researchers suggested that these high-risk families might be segregating for an allele causing a mismatch-repair deficiency. Within less than two years, scientists in several laboratories had identified four human genes homologous to *mutL* or *mutS*, any one of which, when mutated, results in hereditary nonpolyposis colorectal cancer (HNPCC). Most cases of this type of cancer may be caused by mutations in one of these four mismatch-repair genes.

Photoreactivation

Various enzymes can recognize and catalyze the direct reversal of specific DNA damage. A classic example is the reversal of UV-induced pyrimidine dimers by **photoreactivation,** in which an enzyme breaks the bonds that join the pyrimidines in the dimer and thereby restores the original

relevant error rates in *E. coli.* The rate of initial misincorporation is about 10^{-5} per nucleotide per replication. This consists of two components: chemical mispairing of the bases, which takes place at a rate of about 10^{-2} per nucleotide, and selection of the incorrect nucleotide by the DNA polymerase, which occurs at a rate of about 10^{-3} per nucleotide. Once there is an initial mismatch (probability 10^{-5}), the chance that the mismatch escapes the proofreading function is approximately 10^{-2}, and the chance that it escapes the postreplication mismatch-repair system is approximately 10^{-3}. Overall, the probability of incorporating a mismatched base that remains uncorrected by either mechanism is $10^{-5} \times 10^{-2} \times 10^{-3} = 10^{-10}$ per base pair per replication.

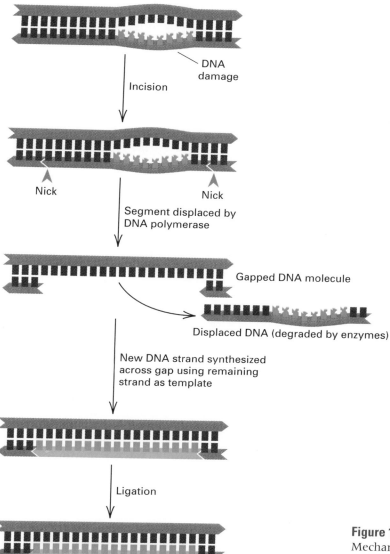

DNA damage

Incision

Nick   Nick

Segment displaced by DNA polymerase

Gapped DNA molecule

Displaced DNA (degraded by enzymes)

New DNA strand synthesized across gap using remaining strand as template

Ligation

Figure 13.20
Mechanism of excision repair of DNA damage.

ses. The enzyme binds to the dimers in e dark but then utilizes the energy of e light to cleave the bonds. Another portant example is the reversal of gua- e methylation in O^6-methyl guanine by ethyl transferase enzyme, which other- would pair like adenine in replication.

sion Repair

sion repair is a ubiquitous multistep natic process by which a stretch of a ged DNA strand is removed from a molecule and replaced by resynthe-

sis using the undamaged strand as a template. Mismatch repair of single mispaired bases is one important type of excision repair. The overall process of excision repair is illustrated in Figure 13.20. The DNA damage can be due to anything that produces a distortion in the duplex molecule; one example is a pyrimidine dimer. In excision repair, a repair endonuclease recognizes the distortion produced by the DNA damage and makes one or two cuts in the sugar-phosphate backbone, several nucleotides away from the damage on either side. A 3'-OH group is produced at the 5' cut,

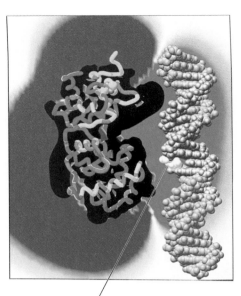

Damaged base
in DNA duplex

Figure 13.21
Model of endonuclease III from *E. coli* (left) in close proximity with a region of 15 base pairs (purple) of DNA (right) that it contacts in carrying out its function. The region of binding is asymmetrical around a damaged base (white), which the enzyme removes while producing a nick at the same site in the DNA backbone. The blue cloud indicates a region of strong electrostatic attraction; the red cloud indicates a region of electrostatic repulsion. [Courtesy of John A. Tainer. Computer graphic model by Michael Pique, Cindy Fisher, and John A. Tainer. From C.-F. Kuo, D. E. McRee, C. L. Fisher, S. F. O'Handley, R. P. Cunningham and J. A Tainer. 1992. *Science* 258:434.]

which DNA polymerase uses as a primer to synthesize a new strand while displacing the DNA segment that contains the damage. The final step of the repair process is the joining of the newly synthesized segment to the contiguous strand by DNA ligase.

Figure 13.21 shows a computer model based on the crystal structure of the repair enzyme endonuclease III from *E. coli*. The purple region in the DNA (right) is a region of 15 base pairs contacted by the enzyme in carrying out the repair of the damaged base (white). The enzyme removes the damaged base and introduces a single-strand nick at the site from which the damaged base was removed. The endonuclease is unusual in containing a cluster of four iron and four sulfur atoms, which are represented by the

cluster of yellow and red spheres near the top of the enzyme.

Postreplication Repair

Sometimes DNA damage persists rather than being reversed or removed, but its harmful effects may be minimized. This often requires replication across damaged areas, so the process is called **postreplication repair**. For example, when DNA polymerase reaches a damaged site (such as a pyrimidine dimer), it stops synthesis of the strand. However, after a brief time, synthesis is reinitiated beyond the damage and chain growth continues, producing a gapped strand with the damaged spot in the gap.

The gap can be filled by strand exchange with the parental strand that has the same polarity, and the secondary gap produced in that strand can be filled by repair synthesis (Figure 13.22). The products of this exchange and resynthesis are two intact single strands, each of which can then serve in the next round of replication as a template for the synthesis of an undamaged DNA molecule.

The SOS Repair System

SOS repair, which is found in *E. coli*. and related bacteria, is a complex set of processes that includes a bypass system that allows DNA replication to take place across pyrimidine dimers or other DNA distortions, but at the cost of the fidelity of replication. Even though intact DNA strands are formed, the strands are often defective. SOS repair is said to be *error prone*. A significant feature of the SOS repair system is that it is not always active but is induced by DNA damage. Once activated, the SOS system allows the growing fork to advance across the damaged region, adding nucleotides that are often incorrect because the damaged template strand cannot be replicated properly. The proofreading system of DNA polymerase is relaxed to allow polymerization to proceed across the damage, despite the distortion of the helix.

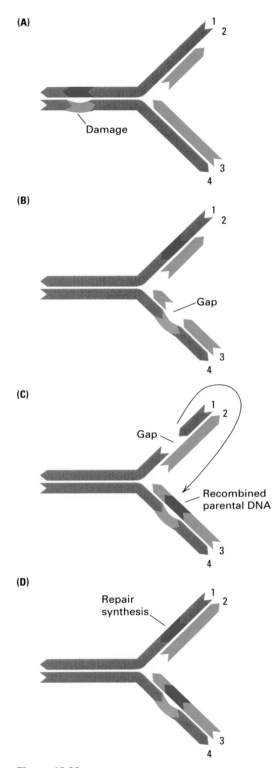

(A)

Damage

(B)

Gap

(C)

Gap

Recombined
parental DNA

(D)

Repair
synthesis

Figure 13.22

Postreplication repair. (A) A molecule with
DNA damage in strand 4 is being replicated. (B)
By reinitiation of synthesis beyond the damage,
a gap is formed in strand 3. (C) A segment of
parental strand 1 is excised and inserted in
strand 3. (D) The gap in strand 1 is next filled in
by repair synthesis.

13.6
Reverse Mutations and
Suppressor Mutations

In most of the mutations we have considered so far, a wildtype (normal) gene is
changed into a form that results in a
mutant phenotype, an event called a **forward mutation.** Mutations are frequently
reversible, and an event that restores the
wildtype phenotype is called a **reversion.**
A reversion may result from a **reverse
mutation,** an exact reversal of the alteration in base sequence that occurred in the
original forward mutation, restoring the
wildtype DNA sequence. A reversion may
also result from the occurrence, at some
other site in the genome, of a second mutation that in any of several ways compensates for the effect of the original mutation.
Reversion by the exact reversal mechanism
is infrequent. The second-site mechanism
is much more common, and a mutation of
this kind is called a **suppressor mutation.**
A suppressor mutation can occur at a different site in the gene containing the mutation that it suppresses (*intra*genic
suppression) or in a different gene in either
the same chromosome or a different chromosome (*inter*genic suppression). Most
suppressor mutations do not fully restore
the wildtype phenotype, for reasons that
will be apparent in the following discussion
of the two kinds of suppression.

Intragenic Suppression

Reversion of frameshift mutations usually
results from **intragenic suppression;** the
mutational effect of the addition (or deletion) of a nucleotide pair in changing the
reading frame of the mRNA is rectified by a
compensating deletion (or addition) of a
second nucleotide pair at a nearby site in
the gene (Section 10.6). A second type of
intragenic suppression is observed when
loss of activity of a protein, caused by one
amino acid change, is at least partly
restored by a change in a second amino
acid. An example of this type of suppression in the protein product of the *trpA* gene
in *E. coli* is shown in Figure 13.23. The A
polypeptide, composed of 268 amino acids,
is one of two polypeptides that make up
the enzyme tryptophan synthase in *E. coli*.

Figure 13.23

Simplified model for the effect of mutation and intragenic suppression on the folding and activity of the A protein of tryptophan synthase in *E. coli*. (A) The wildtype protein in which two critical regions (orange segments) of the polypeptide chain interact. (B) Disruption of proper folding of the polypeptide chain by a substitution of amino acid 210 prevents the critical regions from interacting. (C) Suppression of the effect of the original forward mutation by a subsequent change in amino acid 174. The structure of the region that contains this amino acid is altered, bringing the critical regions together again.

(A) Wildtype A polypeptide

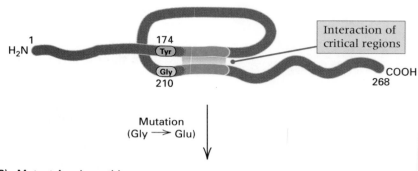

Mutation
(Gly ⟶ Glu)

(B) Mutant A polypeptide

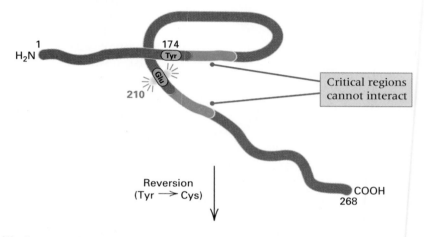

Reversion
(Tyr ⟶ Cys)

(C) Revertant A polypeptide

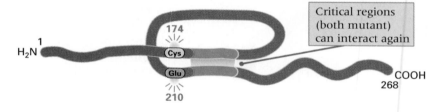

The mutation shown, one of many that inactivate the enzyme, is a change of amino acid number 210 from glycine in the wildtype protein to glutamic acid. This glycine is not in the active site; rather, the inactivation is caused by a change in the folding of the protein, which indirectly affects the active site. The activity of the mutant protein is partly restored by a second mutation, in which amino acid number 174 changes from tyrosine to cysteine. A protein in which only the tyrosine has been replaced by cysteine is inactive, again because of a change in the shape of the protein. This change at the second site, found repeatedly in different reversions of the original mutation, restores activity because the two regions containing amino acids 174 and 210 interact fortuitously to produce a protein with the correct folding. This type of reversion can usually be taken to mean that the two regions of the protein interact, and studies of the amino acid sequences of mutants and revertants are often informative in elucidating the structure of proteins.

Intergenic Suppression

Intergenic suppression refers to a mutational change *in a different gene* that eliminates or suppresses the mutant phenotype. Many intergenic suppressors are very specific in being able to suppress the effects of mutations in only one or a few other genes. For example, phenotypes resulting from the accumulation of intermediates in a metabolic pathway as a consequence of mutational inactivation of one step in the pathway can sometimes be suppressed by mutations in other genes that function earlier in the pathway, which reduces the concentration of the intermediate. These suppressors act only to suppress mutations that affect steps further along the pathway, and they suppress all alleles of these genes.

Other intergenic suppressors are more general in being able to suppress mutations in many functionally unrelated genes—and usually in only a subset of alleles of these genes. For example, certain intergenic suppressors in *Drosophila* affect transcription factors that control the transcription of particular families of transposable elements. Alleles that are mutant because they contain a transposable element whose transcription interferes with that of the gene can be suppressed by an intergenic suppressor that prevents transcription of the inserted transposable element.

In prokaryotes, the best-understood type of intergenic suppression is found in some tRNA genes, mutant forms of which can suppress the effects of particular mutant alleles of many other genes. These suppressor mutations change the anticodon sequence in the tRNA and thus the specificity of mRNA codon recognition by the tRNA molecule. Mutations of this type were first detected in certain strains of *E. coli*, which were able to suppress particular phage-T4 mutants that failed to form plaques on standard bacterial strains. These strains were also able to suppress particular alleles of numerous other genes in the bacterial genome. The suppressed mutations were, in each case, nonsense mutations—those in which a stop codon (UAA, UAG, or UGA) had been introduced within the coding sequence of a gene, with the result that polypeptide synthesis was prematurely terminated and only an amino-terminal fragment of the polypeptide was synthesized. Because they suppressed nonsense mutations, the suppressors were called nonsense suppressors.

Nonsense suppression takes place when the anticodon of one of the normal tRNA molecules is mutated so that it pairs with a nonsense codon and inserts an amino acid, allowing translation to continue (Chapter 10). Such mutations are not lethal because most tRNA genes are present in several copies, so nonmutated copies of the tRNA are also present to allow correct translation of the codon.

Reversion as a Means of Detecting Mutagens and Carcinogens

In view of the increased number of chemicals present as environmental contaminants, tests for the mutagenicity of these substances have become important. Furthermore, most carcinogens are also mutagens, so mutagenicity provides an initial screening for potential hazardous agents. One simple method of screening large numbers of substances for mutagenicity is a reversion test that uses nutritional mutants of bacteria. In the simplest type of reversion test, a compound that is a potential mutagen is added to solid growth medium, a known number of a mutant bacterium is plated, and the number of revertant colonies is counted. An increase in the reversion frequency significantly greater than that obtained in the absence of the test compound identifies the substance as a mutagen. However, simple tests of this type fail to demonstrate the mutagenicity of a large number of potent carcinogens. The explanation for this failure is that many substances are not directly mutagenic (or carcinogenic); rather, they require a conversion into mutagens by enzymatic reactions that take place in the livers of animals and that have no counterpart in bacteria. The normal function of these enzymes is to protect the organism from various naturally occurring harmful substances by converting them into soluble nontoxic substances that can be disposed of in the urine. However, when the enzymes encounter certain human-made and natural compounds, they convert these substances, which may not be harmful in themselves, into mutagens or carcinogens. The enzymes of liver cells, when added to the bacterial growth medium, activate the compounds and allow their mutagenicity to be recognized. The addition of liver extract is one step in the Ames test for carcinogens and mutagens.

In the **Ames test** histidine-requiring (His^-) mutants of the bacterium *Salmonella typhimurium*, containing either a base substitution or a frameshift mutation, are tested for reversion to His^+. In addition, the bacterial strains have been made more sensitive to mutagenesis by the incorporation of several mutant alleles that inactivate the excision-repair system and that make the cells more permeable to foreign molecules. Because some mutagens act only on replicating DNA, the solid medium used contains enough histidine to support a few rounds of replication but not enough to permit the formation of a visible colony. The medium also contains the potential mutagen to be tested and an extract of rat liver. If the test substance is a mutagen or is converted into a mutagen, then some colonies form. A quantitative analysis of reversion frequency can also be carried out by incorporating various amounts of the potential mutagen in the medium. The reversion frequency generally depends on the concentration of the substance being tested and, for a known carcinogen or mutagen, correlates roughly with its carcinogenic potency in animals.

The Ames test has been used with thousands of substances and mixtures (such as industrial chemicals, food additives, pesticides, hair dyes, and cosmetics), and numerous unsuspected substances have been found to stimulate reversion in this test. A high frequency of reversion does not necessarily indicate that the substance is definitely a carcinogen but only that it has a high probability of being so. As a result of these tests, many industries have reformulated their products. For example, the cosmetics industry has changed the formulation of many hair dyes and cosmetics to render them nonmutagenic. Ultimate proof of carcinogenicity is determined by testing for tumor formation in laboratory animals. However, only a few percent of the substances known from animal experiments to be carcinogens fail to increase the reversion frequency in the Ames test.

13.7
Recombination

Genetic recombination may be regarded as a process of breakage and repair between two DNA molecules. In eukaryotes, the process takes place early in meiosis after each molecule has replicated, and with respect to genetic markers, it results in two molecules of the parental type and two recombinants (Chapter 4). For genetic studies of recombination, fungi such as yeast or *Neurospora* are particularly useful because all four products of meiosis are contained in a four-spore or eight-spore ascus (Section 4.5). Most asci from heterozygous *Aa* individuals contain a ratio of

$2\,A : 2\,a$ in four-spored asci,

or

$4\,A : 4\,a$ in eight-spored asci

because of normal Mendelian segregation. Occasionally, however, aberrant ratios are also found, such as

$3\,A : 1\,a$ or $1\,A : 3\,a$ in four-spored asci

and

$5\,A : 3\,a$ or $3\,A : 5\,a$ in eight-spored asci

Different types of aberrant ratios can also occur. The aberrant asci are said to result from **gene conversion** because it appears as though one allele has "converted" the other allele into a form like itself. Gene conversion is frequently accompanied by recombination between genetic markers on either side of the conversion event, even when the flanking markers are tightly linked. This implies that gene conversion can be one consequence of the recombination process. The accepted explanation of gene conversion is that it results from mismatch repair in a small segment of duplex DNA containing one strand from each parental molecule (Figure 13.24). The region of hybrid DNA is called a **heteroduplex region.**

Molecular models of recombination are quite complex because heteroduplexes may or may not lead to recombination between flanking genetic markers. The models also strive to account for all the types of aberrant asci that are observed and for their relative frequencies. We will consider three models that differ in the type of DNA breakage that initiates the process.

The Holliday Model

In the **Holliday model** of recombination, each of the participating DNA duplexes (Figure 13.25A) undergoes a symmetrically positioned single-stranded nick in strands of the same polarity (Figure 13.25B). The nicked strands unwind in the region of the nicks, switch pairing partners, and form a heteroduplex region in each molecule (Figure 13.25C). Ligation anchors the ends of the exchanged strands in place, and further unwinding increases the length of the heteroduplex region (termed *branch migration*, Figure 13.25D). Two more nicks and rejoinings are necessary to resolve the structure in Figure 13.25D. As the

Figure 13.24

Mismatch repair resulting in gene conversion. Only a small part of the heteroduplex region is shown. Each mismatched heteroduplex can be repaired to give an *A* allele or an *a* allele. The patterns of repair leading to 3 : 1 and 1 : 3 segregation are shown.

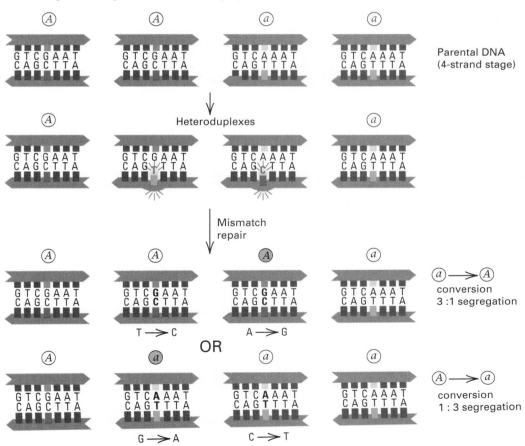

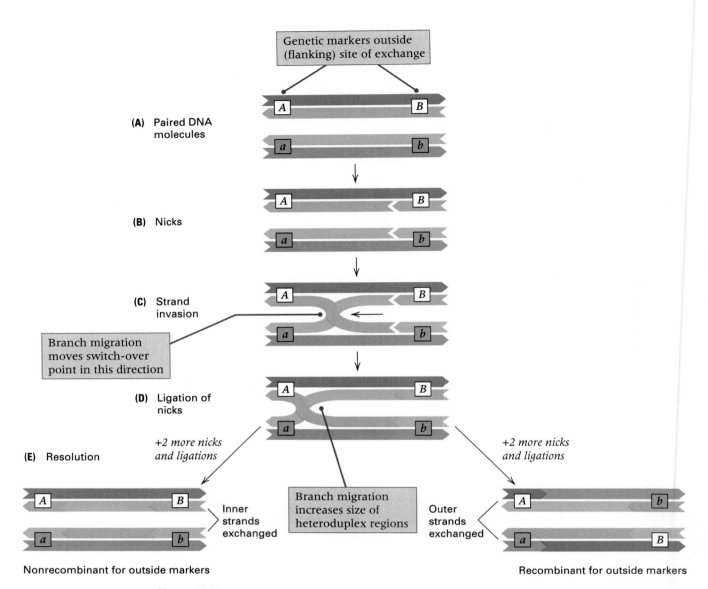

(A) Paired DNA molecules

Genetic markers outside (flanking) site of exchange

A   B

a   b

(B) Nicks

A   B

a   b

(C) Strand invasion

A   B

a   b

Branch migration moves switch-over point in this direction

(D) Ligation of nicks

A   B

a   b

Branch migration increases size of heteroduplex regions

(E) Resolution

+2 more nicks and ligations

+2 more nicks and ligations

A   B

a   b

Inner strands exchanged

Outer strands exchanged

A   b

a   B

Nonrecombinant for outside markers

Recombinant for outside markers

Figure 13.25
Holliday model of recombination. (A) Paired DNA molecules. (B) The exchange process is initiated by single-stranded breaks in strands of the same polarity. (C) Strands exchange pairing partners (referred to as "invasion"). (D) Ligation and branch migration. (E) Exchange is resolved either by the breaking and rejoining of the inner strands (resulting in molecules that are nonrecombinant for outside markers) or by the breaking and rejoining of the outer strands (resulting in recombinant molecules). Either type of resolution gives a heteroduplex region of the same length in each participating molecule.

configuration is drawn, these breaks can be in the inner strands, resulting in molecules that are *nonrecombinant* for markers flanking the heteroduplex region, or they can be in the outer strands, resulting in molecules that are *recombinant* for outside markers (Figure 13.25E).

Of the two possibilities for resolution of the structure in Figure 13.25D, one results

in recombination of the outside markers and the other does not. These outcomes are equally likely because there is no topological difference between the inner strands and the outer strands. To understand why, note that an end-for-end rotation of the lower duplex in Figure 13.25D gives the configuration in Figure 13.26. This type of configuration is known as a **Holliday**

structure. Drawing the structure in this manner makes it clear that breakage and rejoining of the east-west strands or the north-south strands are topologically equivalent. However, north-south resolution results in recombination of the outside markers (*A b* and *a B* gametes), whereas east-west resolution gives nonrecombinants (*A B* and *a b* gametes).

An electron micrograph of a Holliday structure formed between two DNA duplexes is shown in Figure 13.27A. This is the physical structure of DNA corresponding to the diagram in Figure 13.26. An interpretation of the structure that preserves the double-helical structure of duplex DNA is shown in Figure 13.27B. A pair of genetic markers—*A, a* and *B, b*—flanking the exchange are shown. One of the duplexes that participates in the exchange carries the alleles *A B*, and the other duplex carries *a b*. Near the site of exchange, each of the double-stranded molecules contains a short heteroduplex region. The key feature of the structure is that there are four short, single-stranded regions where the duplexes come together. The Holliday structure can be resolved either by breakage and reunion in the single-stranded regions on the top and bottom (called north and south in Figure 13.26) or by breakage and reunion in the single-stranded regions on the left and right (called east and west in Figure 13.26). The left-right breakage and rejoining result in two duplexes that are nonrecombinant for the outside markers (*A B* and *a b*), whereas the top-bottom breakage and rejoining result in two duplexes that are recombinant for the outside markers (*A b* and *a B*).

Several types of enzymes are required for recombination. In *E. coli*, the exchange of strands is mediated by the recA protein. Another enzyme, called the recBCD nuclease because it contains three protein subunits (B, C, and D), acts as an endonuclease, an exonuclease, and a helicase; recBCD has been implicated in strand exchange and branch migration. The PolI DNA polymerase also promotes strand displacement in branch migration, and DNA ligase is necessary to connect the strands with a covalent bond. Finally, the resolution of the Holliday structure is an enzymatic function, carried out by the

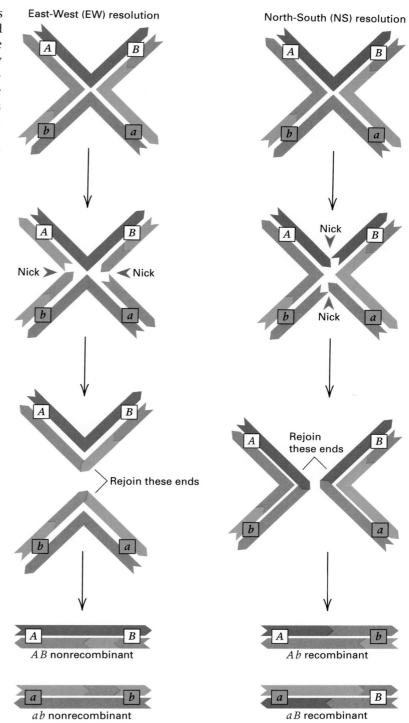

Figure 13.26
The cross-shaped configuration shows the topological equivalence of outer-strand and inner-strand resolution of the Holliday structure in Figure 13.25D. East-west resolution corresponds to inner-strand breakage and rejoining; north-south resolution corresponds to outer-strand breakage and rejoining.

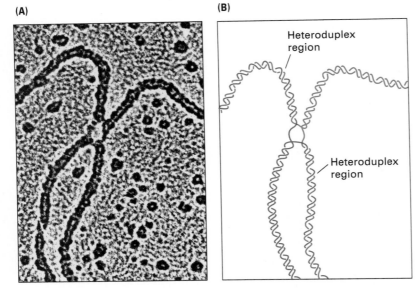

(A)

(B)

Heteroduplex region

Heteroduplex region

Figure 13.27

(A) Electron micrograph of a Holliday structure. Note the short, single-stranded region of DNA on each side of the oval-shaped structure where the duplexes come together. (B) Interpretative drawing of a Holliday structure in terms of double-helical molecules. Note the single-stranded regions where the duplexes come together, as well as the short heteroduplex region in two of the duplexes.

Holliday junction-resolving enzyme, which is denoted RuvC.

An important feature of the Holliday model is that heteroduplex regions are found in the region of the exchange. Mismatch repair in this region can result in gene conversion, and the equal likelihood of north-south and east-west resolution (the RuvC function) implies that gene conversion will frequently be accompanied by recombination of flanking markers.

Asymmetrical Single-Strand Break Model

Observed patterns of gene conversion indicate that the heteroduplex region is not always the same length in the DNA duplexes participating in an exchange. This is a difficulty for the Holliday model, because in its original form, the heteroduplex regions are symmetric. However, in the single-strand break and repair model (Figure 13.28), the heteroduplex regions can be asymmetrical. Starting with a single-strand nick (Figure 13.28A), progressive unwinding and DNA synthesis (red strand)

liberate one DNA strand from a duplex. This strand can "invade" the other duplex (Figure 13.28B), resulting in displacement of the original strand and formation of a **D loop (displacement loop)**. Nucleases degrade the D loop, and the invading strand is ligated in place (Figure 13.28C). Branch migration (Figure 13.28D) forms a second heteroduplex region, but the heteroduplex regions in the participating duplexes are of different lengths. As with the Holliday model, the configuration in Figure 13.28D can be resolved by RuvC in either the east-west or the north-south direction, yielding nonrecombinant or recombinant products, respectively.

Double-Strand Break Model

In yeast, duplex DNA molecules with double-stranded breaks are very efficient in initiating recombination with a homologous unbroken duplex. A model for this process is shown in Figure 13.29. The initiating molecule will typically contain a gap with missing sequences and strands of unequal length at the ends because of exonuclease activity. The ends are free to invade a homologous duplex, forming a D loop (Figure 13.29B), which increases in extent until it bridges the gap (Figure 13.29C). Repair synthesis and ligation in both duplexes gives the configuration in Figure 13.29D, which can be resolved in four ways because it contains two Holliday-type junctions, each of which can be resolved in east-west or north-south orientation. Two resolutions yield nonrecombinant duplexes and two yield recombinant duplexes, so genetic recombination is a frequent outcome.

Although each of the three models of recombination accounts for the principal features of gene conversion and recombination, each falls short of completely explaining all of the details and the aberrations that are occasionally found. For example, the Holliday model predicts an excess of two-strand double recombination, which is not observed. All three models may represent alternative pathways of recombination whose frequencies depend on whether the initiating event consists of a pair of single-stranded nicks or a double-stranded gap.

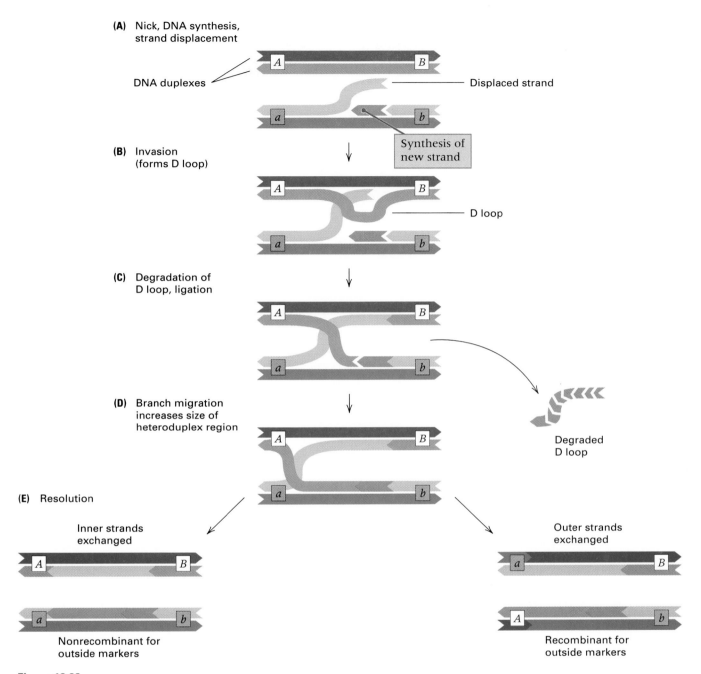

(A) Nick, DNA synthesis, strand displacement

DNA duplexes

Displaced strand

Synthesis of new strand

(B) Invasion (forms D loop)

D loop

(C) Degradation of D loop, ligation

Degraded D loop

(D) Branch migration increases size of heteroduplex region

(E) Resolution

Inner strands exchanged

Outer strands exchanged

Nonrecombinant for outside markers

Recombinant for outside markers

Figure 13.28

Model of recombination initiated by a break in one strand of a pair of DNA molecules (the layout is like that in Figure 13.25). (A) A freed DNA strand is replaced by new synthesis and invades the partner molecule. (B) Displacement of a strand in the partner molecule forms a D (displacement) loop. (C) The D loop is degraded, and free DNA ends are ligated. (D) Branch migration increases the length of the heteroduplex region. (E) Resolution occurs in either of two topologically equivalent ways. In this model, the heteroduplex regions are not of equal length.

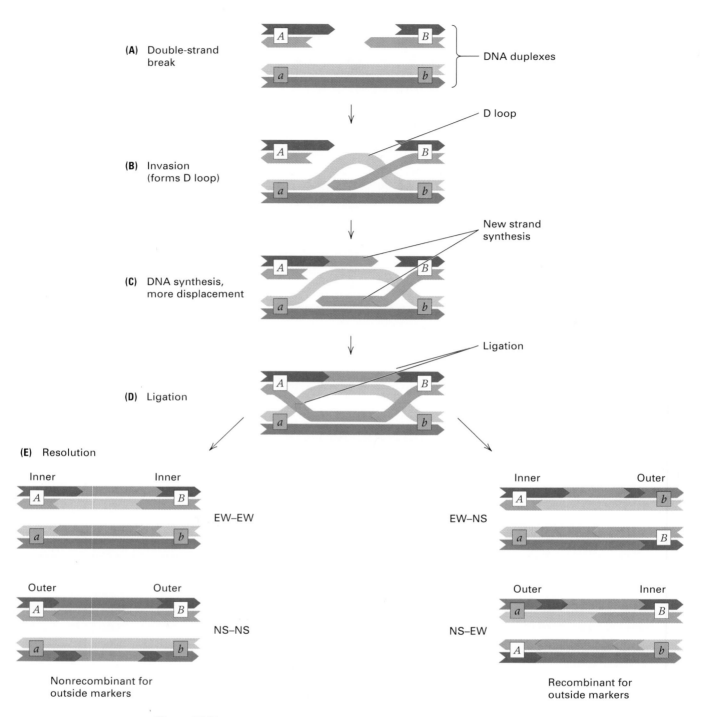

(A) Double-strand break

(B) Invasion (forms D loop)

(C) DNA synthesis, more displacement

(D) Ligation

(E) Resolution

DNA duplexes

D loop

New strand synthesis

Ligation

Inner — Inner — EW–EW

Outer — Outer — NS–NS

Nonrecombinant for outside markers

Inner — Outer — EW–NS

Outer — Inner — NS–EW

Recombinant for outside markers

Figure 13.29

Model of recombination initiated by a double-stranded break (the layout is like that in Figure 13.25). (A) The double-stranded break is trimmed back asymmetrically by nucleases. (B) One of the strands invades the partner molecule, forming a displacement loop. (C) The invading strand is elongated by new synthesis, and the displacement loop serves as a template for new synthesis in the other duplex. (D) Free DNA ends are ligated. (E) Two Holliday junctions are formed, which can be resolved in any one of four topologically equivalent ways, two of which result in recombination of outside makers. In this model, the heteroduplex regions are unequal in length, and all of them contain some newly synthesized DNA

Mutations can be classified in a variety of ways in terms of (1) how they come about, (2) the nature of the chemical change, or (3) how they are expressed. Conditional mutations cause a change in phenotype under restrictive but not permissive conditions. For example, temperature-sensitive mutations are expressed only above a particular temperature. Spontaneous mutations are of unknown origin, whereas induced mutations result from exposure to chemical reagents or radiation. All mutations ultimately result from changes in the sequence of nucleotides in DNA, including base substitutions, insertions, deletions, and rearrangements. Base-substitution mutations are single-base changes that may be silent (for example, in a noncoding region or a synonymous codon position), may change an amino acid in a polypeptide chain (a missense mutation), or may cause chain termination by production of a stop codon (a nonsense mutation). A transition is a base-substitution mutation in which a pyrimidine is substituted for another pyrimidine or a purine for another purine. In a transversion, a pyrimidine is substituted for a purine, or the other way around. A mutation may consist of the insertion or deletion of one or more bases; in a coding region, if the number of bases is not a multiple of three, the mutation is a frameshift.

Spontaneous mutations are random. Favorable mutations do not take place in response to harsh environmental conditions. Rather, the environment selects for particular mutations that arise by chance and happen to confer a growth advantage. Spontaneous mutations often arise by errors in DNA replication that fail to be corrected either by the proofreading system or by the mismatch-repair system. The proofreading system removes an incorrectly incorporated base immediately after it is added to the growing end of a DNA strand. The mismatch-repair system removes incorrect bases at a later time. Methylation of parental DNA strands and delayed methylation of daughter strands allows the mismatch-repair system to operate selectively on the daughter DNA strand.

Mutations can be induced chemically by direct alteration of DNA—for example, by nitrous acid, which deaminates bases. Base analogs are incorporated into DNA in replication. They undergo mispairing more often than do the normal bases, which gives rise to transition mutations. The base analog 5-bromouracil is an example of such a mutagen. Alkylating agents often cause depurination; this type of mutation is usually repaired, but often an adenine is inserted opposite a depurination site, which results in a transversion. Acridine molecules intercalate (interleaf) between base pairs of DNA and cause misalignment of parental and daughter strands in DNA replication, giving rise to frameshift mutations, usually of one or two bases. Ionizing radiation results in oxidative free radicals that cause a variety of alterations in DNA, including double-stranded breaks. Although the amount of genetic damage from normal background and other sources of radiation is believed to be low, studies of persons exposed to fallout from the Chernobyl nuclear melt-

down indicate an increased mutation rate, at least for simple tandem repeat polymorphisms (STRPs).

A variety of systems exist for repairing damage to DNA. In excision repair, a damaged stretch of a DNA strand is excised and replaced with a newly synthesized copy of the undamaged strand. In photoreactivation, there is direct cleavage of the pyrimidine dimers produced by ultraviolet radiation. Damage to DNA that results in the formation of O^6-methyl guanine is repaired by a special methyl transferase. In all of these systems, the correct template is restored. Postreplication repair is an exchange process in which gaps in one daughter strand produced by aberrant replication across damaged sites are filled by nondefective segments from the parental strand of the other branch of the newly replicated DNA. Thus a new template is produced by this system. The SOS repair system in bacteria is error prone (mutagenic) because it facilitates replication across stretches in the template DNA that are too damaged to serve as a normal template.

Mutant organisms sometimes revert to the wildtype phenotype. Reversion is normally due to an additional mutation at another site. Reversion by a second mutation is called suppression, and the secondary mutations are called suppressor mutations. These can be intragenic or intergenic. In intragenic suppression, a mutation in one region of a protein alters the folding of the protein, and a change in another amino acid causes correct folding again. Intergenic suppression takes place through a variety of mechanisms. For example, in nonsense suppression, a chain-termination mutation (a stop codon) is read by a mutant tRNA molecule with an anticodon that can hydrogen-bond with the stop codon, and an amino acid is inserted at the site of the stop codon.

The Ames test measures reversion as an indicator of mutagens and carcinogens; it uses an extract of rat liver, which in mammals occasionally converts intrinsically harmless molecules into mutagens and carcinogens.

Genetic recombination is intimately connected with DNA repair because the process always includes breakage and rejoining of DNA strands. Three current models of recombination differ in the types of DNA breaks that initiate the process and in predictions about the types and frequencies of gene-conversion events. In gene conversion, one allele becomes converted into a homologous allele, which is detected by aberrant segregation in fungal asci, such as 3 : 1 or 1 : 3. Gene conversion is the outcome of mismatch repair in heteroduplexes. All recombination models include the creation of heteroduplexes in the region of the exchange and predict that 50 percent of gene conversions will be accompanied by recombination of genetic markers flanking the conversion event. The Holliday model assumes that single-strand nicks occur at homologous positions in the participating duplexes. Other models assume initiation with a single-strand nick or a double-strand break. All three processes may be operative.

Key Terms

acridine
alkylating agent
Ames test
base analog
base-substitution mutation
ClB method
conditional mutation
depurination
displacement loop
D loop
DNA uracil glycosylase
excision repair
forward mutation
free radical
gene conversion
germ-line mutation
heteroduplex region
Holliday junction-resolving enzyme

Holliday model
Holliday structure
hot spot
induced mutation
intercalation
intergenic suppression
intragenic suppression
ionizing radiation
mismatch repair
missense mutation
mutagen
mutation
mutation rate
nitrous acid
nonsense mutation
nonsense suppression
nucleotide analog
permissive condition

photoreactivation
postreplication repair
pyrimidine dimer
replica plating
replication slippage
restrictive condition
reverse mutation
reversion
sickle-cell anemia
silent mutation
somatic mutation
SOS repair
spontaneous mutation
suppressor mutation
temperature-sensitive mutation
transition mutation
transversion mutation
xeroderma pigmentosum

Review the Basics

• What is a spontaneous mutation? An induced mutation?

• What is meant by the term *mutation rate*?

• How was the technique of replica plating used to demonstrate that antibiotic-resistance mutations are present at low frequency in populations of bacteria even in the absence of the antibiotic?

• What is mismatch repair and how does it happen? Name one human disease associated with a defective mismatch-repair system.

• What features of DNA polymerase could qualify it as a repair enzyme?

• What is a hot spot of mutation? Why might a mutational hot spot coincide with the site of a methylated cytosine in the DNA?

• Explain the following statement: "There is no threshold dose for the mutagenic effect of radiation."

• What unusual types of tetrads provide evidence that a region of heteroduplex DNA is created in the process of recombination?

Guide to Problem Solving

Problem 1: The molecule 2-aminopurine (Ap) is an analog of adenine that pairs with thymine. It also occasionally pairs with cytosine. What types of mutations will be induced by 2-aminopurine?

Answer: In problems of this sort, first note the base that is replaced, and then follow what can happen in subsequent rounds of DNA replication when the base analog pairs normally or abnormally. In this example, the 2-aminopurine (Ap) is incorporated in place of adenine opposite thymine, forming an Ap–T base pair. In the following rounds of replication, the Ap will usually pair with T, but when it occasionally pairs with C, it forms an Ap–C pair. In the next round of

replication, the Ap again pairs with T (forming an Ap–T pair), but the C pairs with G (forming a mutant G–C pair). Hence the type of mutation induced by 2-aminopurine is a change from an A–T pair to a G–C pair, which is a transition mutation.

Problem 2: Twenty spontaneous white-eye (*w*) mutants of *Drosophila* are examined individually to determine their reversion frequencies. Of these, 12 revert spontaneously at a frequency of 10^{-3} to 10^{-5}. Another 6 do not revert spontaneously at detectable frequency but can be induced to revert at frequencies ranging from 10^{-5} to 10^{-6} by treatment with alkylating agents. The remaining 2 cannot be made to revert

GeNETics on the web will introduce you to some of the most important sites for finding genetic information on the Internet. To complete the exercises below, visit the Jones and Bartlett home page at

http://www.jbpub.com/genetics

Select the link to *Genetics: Principles and Analysis* and then choose the link to *GeNETics on the web*. You will be presented with a chapter-by-chapter list of highlighted keywords.

GeNETics EXERCISES

Select the highlighted keyword in any of the exercises below, and you will be linked to a web site containing the genetic information necessary to complete the exercise. Each exercise suggests a specific, written report that makes use of the information available at the site. This report, or an alternative, may be assigned by your instructor.

1. Search this database for **HNPCC** (hereditary nonpolyposis colorectal cancer) to learn more about the homologs of *E. coli mutS* and *mutL* involved in human colon cancer. Check out each of the entries to learn whether the genetic basis is a mutation in a *mutS* homolog, a *mutL* homolog, or neither. If assigned to do so, choose either HNPCC Type 1 or HNPCC Type 2, and write a 150-word report on how the genetic basis of the disease was discovered.

2. Some inherited defects in **DNA repair** are included in this database of pediatric disorders. Find the list of disorders, and follow the links to ataxia-telangiectasia, Bloom syndrome, Fanconi anemia, and xeroderma pigmentosum. If assigned to do so, pick any one of these conditions, and write a 250-word report on its incidence, genetic basis, clinical features, and treatment.

3. You may view an animated **Holliday structure** in the process of formation and resolution at this keyword site. If assigned to do so, draw key stages of the process using the colors in the animation; draw similar pictures showing how the Holliday structure is resolved without an exchange of genetic markers flanking the heteroduplex region.

MUTABLE SITE EXERCISES

The Mutable Site Exercise changes frequently. Each new update includes a different exercise that makes use of genetics resources available on the World Wide Web. Select the **Mutable Site** for Chapter 13, and you will be linked to the current exercise that relates to the material presented in this chapter.

PIC SITE

The Pic Site showcases some of the most visually appealing genetics sites on the World Wide Web. To visit the showcase genetics site, select the **Pic Site** for Chapter 13.

under any conditions. Answer the following questions, and explain your answers.

(a) Which class of mutants would you suspect to result from the insertion of transposable elements?

(b) Which class of mutants are probably missense mutations?

(c) Which class of mutants are probably deletions?

Answer:

(a) Insertions of transposable elements in eukaryotes are often characterized by genetic instability and high reversion rates. Therefore, the twelve mutants with the highest reversion frequencies are probably transposable-element insertions.

(b) The fact that alkylating agents can induce the second class of mutations to revert suggests that they are missense mutations. Although they do not revert spontaneously, this is probably because the exact reversal of a base substitution is an exceedingly rare spontaneous event.

(c) Deletion mutations cannot revert under any conditions because there is too much missing genetic information to be restored. Therefore, the class of mutations that does not revert probably consists of deletions.

13.1 Occasionally, a person is found who has one blue eye and one brown eye or who has a sector of one eye a different color from the rest. Can these phenotypes be explained by new mutations? If so, in what types of cells must the mutations occur?

13.2 A mutation is isolated that cannot be induced to revert. What types of molecular changes might be responsible?

13.3 There are 12 possible substitutions of one nucleotide pair for another (for example, A → G). Which changes are transitions and which transversions?

13.4 Do all nucleotide substitutions in the second position of a codon necessarily produce an amino acid replacement?

13.5 Do all nucleotide substitutions in the first position of a codon necessarily produce an amino acid replacement? Do all of them necessarily produce a nonfunctional protein?

13.6 In a coding sequence made up of equal proportions of A, U, G, and C, what is the probability that a random nucleotide substitution will result in a chain-termination codon?

13.7 If a mutagen produces frameshift mutations, what kinds of mutations will it induce to revert?

13.8 If a deletion eliminates a single amino acid in a protein, how many base pairs are missing?

13.9 Does the nucleotide sequence of a mutation tell you anything about its dominance or recessiveness?

13.10 What are two ways in which pyrimidine dimers are removed from DNA in *E. coli*?

13.11 This problem illustrates how conditional mutations can be used to determine the order of genetically controlled steps in a developmental pathway. A certain organ undergoes development in the sequence of stages A → B → C, and both gene X and gene Y are necessary for the sequence to proceed. A conditional mutation X' is sensitive to heat (the gene product is inactivated at high temperatures), and a conditional mutation Y' is sensitive to cold (the gene product is inactivated at cold temperatures). The double mutant X'/X' Y'/Y' is created and reared at either high or low temperatures. How far would development proceed in each of the following cases at the high temperature and at the low temperature?

(a) Both X and Y are necessary for the A → B step.

(b) Both X and Y are necessary for the B → C step.

(c) X is necessary for the A → B step, and Y for the B → C step.

(d) Y is necessary for the A → B step, and X for the B → C step.

13.12 A Lac⁺ culture of *E. coli* gives rise to a Lac⁻ strain that results from a UGA nonsense mutation in the *lacZ* gene. The Lac⁻ strain gives rise to Lac⁺ colonies at the rate of 10^{-8} per cell per generation, and 90 percent of the latter are caused by suppressor tRNA molecules. What is the rate of production of suppressor mutations in the original Lac⁺ culture?

13.13 If a gene in a particular chromosome has a probability of mutation of 5×10^{-5} per generation, and if the allele in a particular chromosome is followed through successive generations.

(a) What is the probability that the allele does not undergo a mutation in 10,000 consecutive generations?

(b) What is the average number of generations before the allele undergoes a mutation?

13.14 In the mouse, a dose of approximately 1 gray (Gy) of x rays produces a rate of induced mutation equal to the rate of spontaneous mutation. Taking the spontaneous mutation rate into account, what is the total mutation rate at 1 Gy? What dose of x rays will increase the mutation rate by 50 percent? What dose will increase the mutation rate by 10 percent?

13.15 Among several hundred missense mutations in the gene for the A protein of tryptophan synthase in *E. coli*, fewer than 30 of the 268 amino acid positions are affected by one or more mutations. Explain why the number of positions affected by amino acid replacements is so low.

13.16 Human hemoglobin C is a variant in which a lysine in the β-hemoglobin chain is substituted for a particular glutamic acid. What single-base substitution can account for the hemoglobin-C mutation?

13.17 How many amino acids can substitute for tyrosine by a single-base substitution? (Do not assume that you know which tyrosine codon is being used.)

13.18 Which of the following amino acid replacements would be expected with the highest frequency among mutations induced by 5-bromouracil? (1) Met → Leu, (2) Met → Lys, (3) Leu → Pro, (4) Pro → Thr, (5) Thr → Arg.

13.19 Dyes are often incorporated into a solid medium to determine whether bacterial cells can utilize a particular sugar as a carbon source. For instance, on eosin-methylene blue (EMB) medium containing lactose, a Lac⁺ cell yields a purple colony and a Lac⁻ cell yields a pink colony. If a population of Lac⁺ cells is treated with a mutagen that produces Lac⁻ mutants and the population is allowed to grow for many generations before the cells are placed on EMB-lactose medium, a few pink colonies are found among a large number of purple ones. However, if the mutagenized cells are plated on the medium immediately after exposure to the mutagen, some colonies appear that are *sectored* (half purple and half pink). Explain how the sectored colonies arise.

13.20 Mutant alleles of two genes, *a* and *b*, in a haploid organism interact as follows: Most ab^+ and a^+b combinations are mutant, but a few *ab* combinations are wildtype. Each *a* allele that is suppressed requires a particular *b* allele, and each *b* allele that is suppressed requires a particular *a* allele. However, most *a* and *b* alleles are not suppressible by any alleles of the other gene. What kind of interaction between the *a* and *b* gene products can explain these results?

13.21 The following technique has been used to obtain mutations in bacterial genes that are otherwise difficult to isolate. Suppose a mutation is desired in a gene, *a*, and suppose a mutation, b^-, in a closely linked gene is available. The method is to grow phage P1 on a b^+ strain to obtain $a^+ b^+$ transducing particles. The P1 phages containing the transducing particles are exposed to a mutagen, such as hydroxylamine, and the b^- strain is infected with the phages. What step or steps would you carry out next to isolate an a^- mutation? On what genetic phenomenon does this procedure depend?

Challenge Problems

13.22 A *Neurospora* strain that is unable to synthesize arginine (and that therefore requires this amino acid in the growth medium) produces a revertant colony able to grow in the absence of arginine. A cross is made between the revertant and a wildtype strain. What proportion of the progeny from this cross would be arginine-independent if the reversion occurred by each of the following mechanisms?

(a) a precise reversal of the nucleotide change that produced the original *arg*⁻ mutant allele

(b) a mutation to a suppressor of the *arg*⁻ allele occurring in another gene located in a different chromosome

(c) a suppressor mutation occurring in another gene located 10 map units away from the *arg*⁻ allele in the same chromosome

13.23 Which of the following 8-spore asci might arise from gene conversion at the *m* locus? Describe how these asci might arise.

Ascus type	A	B	C	D	E	F
Spore 1	+	+	+	+	+	+
Spore 2	+	+	+	+	+	+
Spore 3	+	*m*	+	+	*m*	*m*
Spore 4	+	*m*	*m*	*m*	*m*	*m*
Spore 5	*m*	*m*	*m*	+	+	*m*
Spore 6	*m*	*m*	*m*	*m*	+	*m*
Spore 7	*m*	+	*m*	*m*	*m*	*m*
Spore 8	*m*	+	*m*	*m*	*m*	*m*

Further Reading

Ames, B. W. 1979. Identifying environmental chemicals causing mutations and cancer. *Science* 204: 587.

Bhatia, P. K., Z. G. Wang, and E. C. Friedberg. 1996. DNA repair and transcription. *Current Opinion in Genetics & Development* 6: 146.

Cameriniotero, R. D., and P. Hsieh. 1995. Homologous recombination proteins in prokaryotes and eukaryotes. *Annual Review of Genetics* 29: 509.

Chu, G., and L. Mayne. 1996. Xeroderma pigmentosum, Cockayne syndrome and trichothiodystrophy: Do the genes explain the diseases? *Trends in Genetics* 12: 187.

Cox, E. C. 1997. *mutS*, proofreading, and cancer. *Genetics* 146: 443.

Crow, J. F., and C. Denniston. 1985. Mutation in human populations. *Advances in Human Genetics* 14: 59.

De la chapelle, A., and P. Peltomaki. 1995. Genetics of hereditary colon cancer. *Annual Review of Genetics* 29: 329.

Drake, J. W. 1991. Spontaneous mutation. *Annual Review of Genetics* 25: 125.

Eggleston, A. K., and S. C. West. 1996. Exchanging partners: Recombination in *E. coli*. *Trends in Genetics* 12: 20.

Hoeijmakers, J. H. J. 1993. Nucleotide excision repair I: From *E. coli* to yeast. *Trends in Genetics* 9: 173.

Jackson, S. P. 1996. The recognition of DNA damage. *Current Opinion in Genetics & Development* 6: 19.

Shcherbak, Y. M. 1996. Confronting the nuclear legacy. I. Ten years of the Chernobyl era. *Scientific American*, April.

Singer, B., and J. T. Kusmierek. 1982. Chemical mutagenesis. *Annual Review of Biochemistry* 51: 655.

Sommer, S. S. 1995. Recent human germ-line mutation: Inferences from patients with hemophilia B. *Trends in Genetics* 11: 141.

Weinberg, R. A. 1996. How cancer arises. *Scientific American*, September.

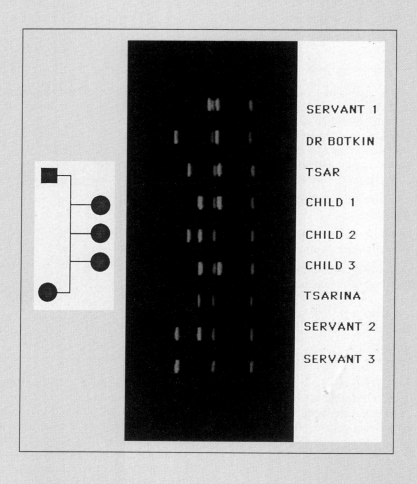

DNA typing confirmed the suspected identities of the skeletons of six adults and three children found in a common, shallow, unmarked grave near Ekaterinburg, Russia, in July 1991. The red bands across the gel are size markers that are the same in all lanes. The blue bands are from DNA markers that are polymorphic in the human population. Sex was identified separately by PCR amplification of genes in the X and Y chromosomes. The identities of the Tsar, Tsarina, and Dr. Botkin were inferred from DNA typing of living relatives. Because the three children, all girls, share one band from the Tsar and one from the Tsarina for this DNA marker, as well as four other markers, they are undoubtedly three of the four daughters. The remains had been buried since 1918. The skeletons of the fourth daughter and of the son affected with hemophilia have not been found. [Courtesy of Peter D. Gill.]

Extranuclear Inheritance

PRINCIPLES

- Some inherited traits are determined by DNA molecules present in organelles, principally mitochondria and chloroplasts. A hallmark of such traits is uniparental inheritance, either maternal or paternal.
- Progeny resembling the mother is characteristic of both maternal inheritance and maternal effects. Although maternal inheritance results from the transmission of non-nuclear genetic factors, maternal effects result from nuclear genes expressed in the mother that determine the phenotype of the progeny.
- Mitochondria, photosynthetic plastids, and certain other cellular organelles are widely believed to have originated through independent episodes of symbiosis in which a prokaryotic cell, the progenitor of the present organelle, became an intracellular symbiont of the proto-eukaryotic cell.
- Evidence from DNA sequence variation in human mitochondrial DNA suggests that the mitochondrial DNA present in all modern human beings originated from a single female who lived approximately 150,000 to 300,000 years ago, probably in Africa.

CONNECTIONS

CONNECTION: *Chlamydomonas* **Moment**
Ruth Sager and Zenta Ramanis 1965
Recombination of nonchromosomal genes in Chlamydomonas

CONNECTION: **A Coming Together**
Lynn Margulis (formerly Lynn Sagan) 1967
The origin of mitosing cells

Most traits in eukaryotes show Mendelian inheritance because they are determined by genes in the nucleus that segregate in meiosis. Less common are traits determined by factors located outside the nucleus. These include traits determined by the DNA in mitochondria and in chloroplasts—the self-replicating cytoplasmic **organelles** specialized for respiration and for photosynthesis, respectively. Each of these organelles contains its own DNA that codes for numerous proteins and RNA molecules. The DNA is replicated within the organelles and transmitted to daughter organelles as they form. The organelle genetic system is separate from that of the nucleus, and traits determined by organelle genes show patterns of inheritance quite different from the familiar Mendelian ratios.

Many organisms also contain intracellular parasites or symbionts, including cytoplasmic bacteria, viruses, and other elements. In some cases, inherited traits of the infected organism are determined by such cytoplasmic entities. Organelle heredity and other examples of the diverse phenomena that make up **extranuclear,** or **cytoplasmic, inheritance** are presented in this chapter.

14.1
Recognition of Extranuclear Inheritance

Other than a non-Mendelian pattern of inheritance, no criterion is universally applicable to distinguish extranuclear from nuclear inheritance. In higher organisms, the transmission of a trait through only one parent, **uniparental inheritance,** usually indicates extranuclear inheritance. Genetic transmission of cytoplasmic factors, such as mitochondria or chloroplasts, is determined by maternal and paternal contributions at the time of fertilization, by mechanisms of elimination from the zygote, and by the irregular sorting of cyto-

Figure 14.1
Maternal inheritance of human mitochondrial DNA. (A) Pattern of DNA fragments obtained when mitochondrial DNA is digested with the restriction enzyme *Hae*II. The DNA type at the left includes a fragment of 8.6 kb (red). The DNA type at the right contains a cleavage site for *Hae*II within the 8.6-kb fragment, which results in smaller fragments of 4.5 kb and 4.1 kb (blue). (B) Pedigree showing maternal inheritance of the DNA pattern with the 8.6-kb fragment (red symbols). The mitochondrial DNA type is transmitted only through the mother. [After D. C. Wallace. 1989. *Trends in Genetics,* 5: 9.]

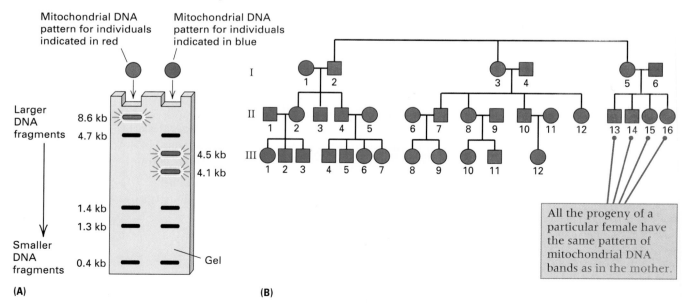

plasmic elements during cell division. Uniparental transmission through the mother constitutes **maternal inheritance;** uniparental transmission through the father is **paternal inheritance.**

Mitochondrial Genetic Diseases

In higher animals, mitochondria are typically inherited through the mother because the egg is the major contributor of cytoplasm to the zygote. Therefore, mitochondria usually show maternal inheritance in which the progeny of a mutant mother and a normal father are mutant, whereas the progeny of a normal mother and a mutant father are normal. A typical pattern of maternal inheritance of mitochondria is shown in the human pedigree in Figure 14.1. When human mitochondrial DNA is cleaved with the restriction enzyme *Hae*II, the cleavage products include either one fragment of 8.6 kb or two fragments of 4.5 kb and 4.1 kb (Figure 14.1A). The pattern with two smaller fragments is typical, and it results from the presence of a *Hae*II cleavage site within the 8.6-kb fragment. Maternal inheritance of the 8.6-kb mitochondrial DNA fragment is indicated in Figure 14.1B, because males (I-2, II-7, and II-10) do not transmit the pattern to their progeny, whereas females (I-3, I-5, and II-8) transmit the pattern to all of their progeny. Although the mutation in the *Hae*II site yielding the 8.6-kb fragment is not associated with any disease, a number of other mutations in mitochondrial DNA do cause diseases and have similar patterns of mitochondrial inheritance. Most of these conditions decrease the ATP-generating capacity of the mitochondria and affect the function of muscle and nerve cells, particularly in the central nervous system, leading to blindness, deafness, or stroke. Table 14.1 summarizes some of the human diseases caused by mitochondrial mutations. Many of these conditions are lethal in the absence of some normal mitochondria, and there is variable expressivity because of differences in the proportions of normal and mutant mitochondria among affected persons.

An example of a human pedigree for inheritance of a mitochondrially inherited disease is shown in Figure 14.2A. The dis-

Table 14.1 Phenotypes associated with some mitochondrial mutations

Nucleotide changed	Mitochondrial component affected	Phenotype[a]
3460	ND1 of Complex I[b]	LHON
11778	ND4 of Complex I	LHON
14484	ND6 of Complex I	LHON
8993	ATP6 of Complex V[b]	NARP
3243	tRNA$^{Leu(UUR)c}$	MELAS, PEO
3271	tRNA$^{Leu(UUR)}$	MELAS
3291	tRNA$^{Leu(UUR)}$	MELAS
3251	tRNA$^{Leu(UUR)}$	PEO
3256	tRNA$^{Leu(UUR)}$	PEO
5692	tRNA^{Asn}	PEO
5703	tRNA^{Asn}	PEO, myopathy
5814	tRNA^{Cys}	Encephalopathy
8344	tRNA^{Lys}	MERRF
8356	tRNA^{Lys}	MERRF
9997	tRNA^{Gly}	Cardiomyopathy
10006	tRNA^{Gly}	PEO
12246	tRNA$^{Ser(AGY)c}$	PEO
14709	tRNA^{Glu}	Myopathy
15923	tRNA^{Thr}	Fatal infantile multisystem disorder
15990	tRNA^{Pro}	Myopathy

[a]LHON Leber's hereditary optic neuropathy
 NARP Neurogenic muscle weakness, ataxia, retinitis pigmentosa
 MERRF Myoclonic epilepsy and ragged-red fiber syndrome
 MELAS Mitochondrial myopathy, encephalopathy, lactic acidosis, stroke-like episodes
 PEO Progressive external ophthalmoplegia
[b]Complex I is NADH dehydrogenase. Complex V is ATP synthase.
[c]In tRNA$^{Leu(UUR)}$, the R stands for either A or G; in tRNA$^{Ser(AGY)}$, the Y stands for either T or C.

ease is myoclonic epilepsy associated with ragged-red muscle fibers (MERRF), a disease of the central nervous system and skeletal muscle. The cause of the disease is a mutation in a tRNA^{Lys} gene encoded in mitochondrial DNA (mtDNA). Figure 14.3 shows the organization of genes in human mtDNA. The tRNA^{Lys} mutation is very pleiotropic because it affects synthesis of all proteins encoded in mitochondrial DNA in all cells. Like many mitochondrial mutations, MERRF is lethal in the absence of normal mitochondria. All of the individuals

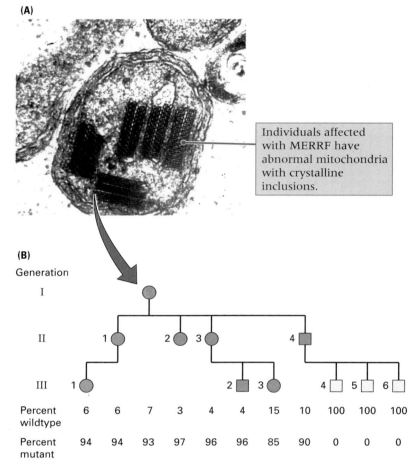

(A)

(B)
Generation

I

II

III

	1		2	3			4			

Percent wildtype: 6 6 7 3 4 4 15 10 100 100 100

Percent mutant: 94 94 93 97 96 96 85 90 0 0 0

Individuals affected with MERRF have abnormal mitochondria with crystalline inclusions.

Figure 14.2
Inheritance of myoclonic epilepsy with ragged-red fiber disease (MERFF) in humans. (A) Electron micrograph of an abnormal MERRF mitochondrion containing paracrystalline inclusions. (B)The pedigree shows inheritance of MERFF in one family and the percentage of the mitochondria in each person found to be wildtype or mutant. [Micrograph courtesy of D. C. Wallace, from J. M. Shoffner, M. T. Lott, A. M. S. Lezza, P. Seibel, S. W. Ballinger, and D. C. Wallace. 1990. *Cell* 61: 931.]

indicated by red symbols in the pedigree manifest some symptoms of the disease, which include hearing loss, seizures, fatigue, dementia, tremors, and jerkiness of movement. The affected individuals in the pedigree vary considerably in the symptoms they express. This variable expressivity is due to differences in the proportions of normal and mutant mitochondria in the cells of affected persons. Those most severely affected have the highest proportion of mutant mitochondria; those least affected have the lowest, as indicated in

Figure 14.2B. An electron micrograph of an abnormal mitochondrion from one of the affected patients is shown in Figure 14.2A. It contains numerous paracrystalline inclusion bodies formed from abnormal protein aggregates.

Consideration of the phenotypes caused by the mutations listed in Table 14.1, and the variable expressivity seen in pedigrees of mitochondrial diseases, indicate that mutations in different mitochondrial genes can result in similar phenotypes and that mutations in the same mitochondrial gene can result in different phenotypes. Some of the variation appears to arise from differences among affected persons in the ratio of mutant to wildtype mitochondria, as seen in Figure 14.2B. Another source of variation results from differences in the structural and metabolic consequences of each individual mutation.

Heteroplasmy

The condition in which two or more genetically different types of mitochondria (or other organelle) are present in the same cell is known as **heteroplasmy.** Heteroplasmy of mitochondria is unusual among animals. For example, a typical human cell contains from 1000 to 10,000 mitochondria, and all of them are normally genetically identical.

The rarity of heteroplasmy made it possible to identify the remains of the last Russian Tsar, Nicholas II, who, together with his family and servants, was secretly executed by a Bolshevik firing squad on the night of July 16, 1918. Their bodies were hidden in a shallow grave and remained undisturbed until they were discovered in 1992. DNA analysis of the exhumed bone samples thought to be from members of the murdered family resulted in identification of nine bodies, five of whom were related. Three were female siblings and another was the mother, the putative Tsarina; all had mitochondrial DNA sequences similar to that of the present Duke of Edinburgh, the maternal grandnephew of the Tsarina. The male relative of the three female siblings was thought to be the father, the Tsar Nicholas. He proved to be heteroplasmic for mitochondrial DNA. At the time of the

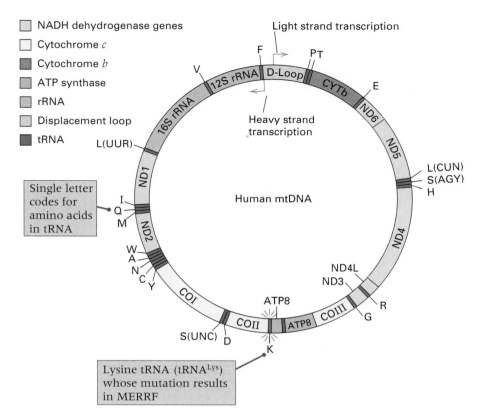

NADH dehydrogenase genes
Cytochrome *c*
Cytochrome *b*
ATP synthase
rRNA
Displacement loop
tRNA

Light strand transcription

Human mtDNA

Single letter codes for amino acids in tRNA

Heavy strand transcription

Lysine tRNA (tRNA^Lys) whose mutation results in MERRF

Figure 14.3

Genes in human mitochondrial DNA. The tRNA genes are indicated by the one-letter amino acid symbols; hence tRNA^Lys is denoted K. The positions of these and other genes in the mitochondrial DNA are indicated by color according to the key at the upper left. The arrows indicate the promoters for transcription of the heavy (HSP) and light (LSP) strands. [From N-G. Larsson and D. A. Clayton. 1995. *Ann. Rev. Genet.* 29: 151.]

1992 report, no reference DNA sample was available to confirm identification of the Tsar. In 1994, however, the body of the Tsar's brother, Georgij, who had died of tuberculosis in 1899, was exhumed. The mitochondrial DNA sequence of Georgij Romanov not only matched that of the putative Tsar but was also heteroplasmic at the identical base pair, providing powerful evidence supporting identification of the remains as those of the Tsar. The two brothers differed in the ratios of the two mitochondrial types: Georgij Romanov had about 62 percent T and 38 percent C at the heteroplasmic base pair, whereas Nicholas had 28 percent T and 72 percent C. The heteroplasmy is not present in two living members of this matrilineage, who are separated by two or three additional genera-

tions from the brothers. This suggests that random segregation of mitochondria within the lineage has finally resulted in homoplasmy.

Maternal Inheritance and Maternal Effects

A maternal pattern of inheritance usually indicates extranuclear inheritance, but not always. The difficulty is in distinguishing between *maternal inheritance* and *maternal effects*. The influence of the mother's nuclear genotype on the phenotype of the progeny is referred to as a **maternal effect.** Such effects may be mediated either through substances present in the egg that affect early development or through the effects of nurture. Examples of

(A)   (B)

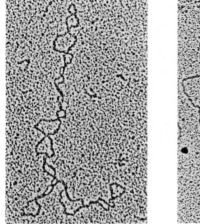
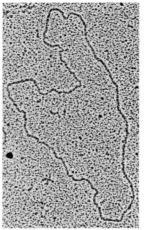

Figure 14.4
DNA from rat liver mitochondria. (A) Native supercoiled configuration. (B) Relaxed configuration produced by a nick in the DNA strand in the process of isolation. The size of each molecule is 16,298 nucleotide pairs. [Courtesy of David Wolstenholme.]

developmental and nurturing effects are, respectively, the intrauterine environment of female mammals and their ability to produce milk. The distinction between maternal inheritance and maternal effects is as follows:

1. In *maternal inheritance,* the hereditary determinants of a trait are extranuclear, and genetic transmission is only through the maternal cytoplasm; Mendelian segregation is not observed.

2. In a *maternal effect,* the nuclear genotype of the mother determines the phenotype of the progeny. The hereditary determinants are nuclear genes transmitted by both sexes, and in suitable crosses the trait undergoes Mendelian segregation.

In higher plants, organelle inheritance is often uniparental, but the particular type of uniparental inheritance depends on the organelle and the organism. For example, among angiosperms, mitochondria typically show maternal inheritance, but chloroplasts may be inherited maternally (the predominant mode), paternally, or

from both parents, depending on the species. In conifers, chloroplast DNA is usually inherited paternally, but a few offspring result from maternal transmission. The redwood *Sequoia sempervirens* shows paternal transmission of both mitochondrial and chloroplast DNA. Whatever the pattern of cytoplasmic transmission, most species have a few exceptional progeny, which indicates that the predominant mode of transmission is usually not absolute.

14.2
Organelle Heredity

As noted earlier, mitochondria are respiratory organelles, and chloroplasts are photosynthetic organelles. The DNA of these organelles is usually in the form of supercoiled circles of double-stranded molecules (Figure 14.4). The chloroplast DNA in most plants ranges in size from 120 to 160 kb. Mitochondrial genomes are usually smaller and have a greater size range. For example, the mitochondrial genome in mammals is about 16.5 kb, that in *D. melanogaster* is about 18.5 kb, and that in some higher plants is more than 100 kb. The mitochondrial genomes of higher plants are exceptional in consisting of two or more circular DNA molecules of different sizes. Several copies of the genome are present in each organelle. There are typically from 2 to 10 copies in mitochondria and from 20 to 100 copies in chloroplasts. In addition, most cells contain multiple copies of each organelle.

Organelle genes code for DNA polymerases that replicate the organelle DNA. They also code for other components essential to their function or replication, but many organelle components are determined by nuclear genes. These components are synthesized in the cytoplasm and transported into the organelle. Mitochondrial DNA contains a relatively small number of genes. For example, both the human mitochondrial genome and that of yeast contain approximately 40 genes. Chloroplast genomes are generally larger than those of mitochondria and contain more genes. Many genes in mitochondria and chloroplasts code for the ribosomal

RNA and transfer RNA components used in protein synthesis (Chapter 10). Figure 14.5 shows the organization of genes in the 120-kb chloroplast genome of the liverwort *Marchantia polymorpha*.

RNA Editing

The mRNAs of mitochondria and chloroplasts often are modified post-transcriptionally in processes called **RNA editing.** The mechanisms of modification and the predominant patterns of modification differ between plants and animals. In most land plants, except algae and mosses, mRNAs may be extensively modified by conversion of cytosine residues to uracils. Because of RNA modification, an organelle gene may carry a CGG codon for arginine, but the mRNA will carry a UGG codon for tryptophan. The modification appears to be accomplished by deamination of selected cytosine residues in the mRNA, leaving uracil residues in their places. In the mitochondria of the trypanosome protozoans that cause sleeping sickness, uridine residues are inserted and/or deleted as templated by a guide RNA in a process that resembles mRNA splicing (Chapter 10). Whatever the mechanism, the outcome is an mRNA whose sequence may differ substantially from that of the DNA from which it was transcribed.

The Genetic Codes of Organelles

Translation in organelles follows the standard decoding process in which the organelle tRNA molecules translate codon by codon along the mRNA. However, the genetic code in mitochondria may differ somewhat from that used to encode proteins in prokaryotes, in archaea, and in nuclear genes of eukaryotes. Human mitochondria depart from the standard code in three principal ways: (1) The UGA codon is not a stop codon but encodes tryptophan, (2) the AGA and AGG codons are stop codons rather than arginine codons, and (3) AUA encodes methionine rather than isoleucine. The genetic code in yeast mitochondria resembles that in humans in that UGA is a tryptophan codon; however, in yeast, AUA is an isoleucine codon, AGA

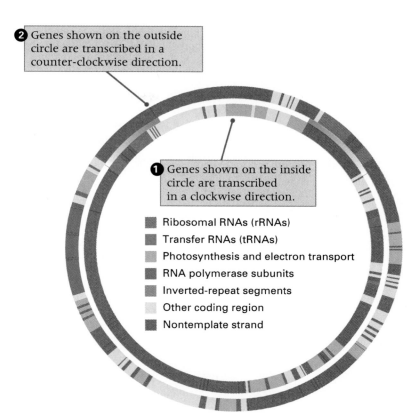

Figure 14.5

Organization of genes in the chloroplast genome of the liverwort *Marchantia polymorpha*. The large inverted-repeat segments (green) that include the genes coding for ribosomal RNA (light brown) are a feature of many other chloroplast genomes. The red lines show the locations of the genes for transfer RNA. Yellow bars are the genes for components of the photosynthetic and electron transport systems; purple bars are the genes for subunits of RNA polymerase. Genes shown inside the circle are transcribed in a clockwise direction; those outside are transcribed counterclockwise. The length of the entire molecule is 121,024 nucleotide pairs. [Data from K. Ohyama, H. Fukuzawa, T. Kohchi, H. Shirai, T. Sano, S. Sano, K. Umesono, Y. Shiki, M. Takeuchi, Z. Chang, S. Aota, H. Inokuchi, and H. Ozeki. 1986. *Nature* 322: 752.]

and AGG are arginine codons (as in the standard code), and the four codons of the form CUN code for threonine rather than leucine. Although the mitochondrial genetic code may differ from the standard genetic code from one animal species to the next, some are identical to the standard code. On the other hand, plant species employ the standard genetic code in both mitochondria and chloroplasts.

Because the structure and function of mitochondria and chloroplasts depend on both nuclear and organelle genes, it is sometimes difficult to distinguish these contributions. In addition, numerous mitochondria or chloroplasts are present in a zygote, but only one of them may contain a mutant allele. The result is that traits determined by organelle genes exhibit a pattern of determination and transmission very different from simple Mendelian traits determined entirely by nuclear genes.

In the following sections, we briefly consider some examples of traits determined by chloroplast and mitochondrial genes.

Leaf Variegation in Four-O'clock Plants

Chloroplast transmission accounts for the unusual pattern of inheritance observed with leaf variegation in a strain of the four-o'clock plant, *Mirabilis jalapa*. Variegation refers to the appearance, in the leaves and stems of plants, of white regions that result from the lack of green chlorophyll (Figure 14.6). On the variegated plants, some branches are completely green, some completely white, and others variegated. Branches of all three types produce flowers, and these flowers can be used to perform nine possible crosses (Table 14.2). From each cross, the seeds are collected and planted, and the phenotypes of the progeny are examined. The results of the crosses are summarized in Table 14.2. Two significant observations are that

1. Reciprocal crosses yield different results. For example, the cross

   green ♀ × white ♂

   yields green plants, whereas the cross

   white ♀ × green ♂

   yields white plants (which usually die shortly after germination because they lack chlorophyll and hence photosynthetic activity).

2. The phenotype of the branch bearing the female parent in each case determines the phenotype of the progeny (compare columns 1 and 3 in the table). The male makes no contribution to the phenotype of the progeny.

Furthermore, when flowers present on either the green or the variegated progeny plants are used in subsequent crosses, the

Figure 14.6
Leaf variegation in *Mirabilis jalapa*. The sectors result from cytoplasmic segregation of normal (green) and defective (white) chloroplasts.

patterns of transmission are identical with those of the original crosses. These observations suggest direct transmission of the trait through the mother, or maternal inheritance.

The genetic explanation for the variegation is as follows:

1. Green color depends on the presence of chloroplasts, and pollen contains no chloroplasts.

2. Segregation of chloroplasts into daughter cells is determined by cytoplasmic division and is somewhat irregular.

3. The cells of white tissue contain mutant chloroplasts that lack chlorophyll.

These points enable us to draw the model shown in Figure 14.7. Variegated plants germinate from embryos that are

Table 14.2 Crosses and progeny phenotypes in variegated four-o'clock plants

Phenotype of branch bearing egg parent	Phenotype of branch bearing pollen parent	Phenotype of progeny
white	white	white
white	green	white
white	variegated	white
green	white	green
green	green	green
green	variegated	green
variegated	white	variegated, green, or white
variegated	green	variegated, green, or white
variegated	variegated	variegated, green, or white

Figure 14.7

Genetic model for leaf variegation. Branches that are all green, all white, and variegated form on the same plant. Flowers form on all three kinds of branches. The insets show the chloroplast composition of cells in each type of branch. Cells in all-green or all-white branches contain only green or white chloroplasts, respectively, whereas cells in variegated branches are heteroplasmic.

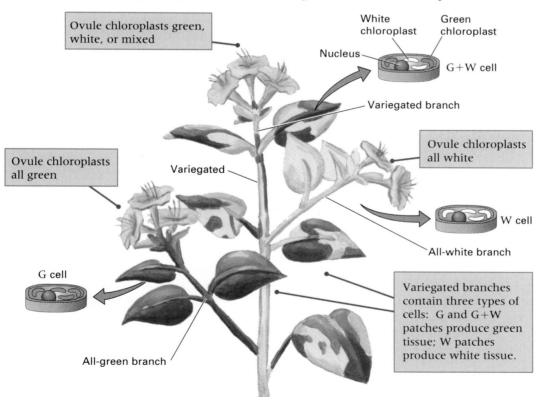

Chlamydomonas Moment

Ruth Sager and Zenta Ramanis 1965
Columbia University,
New York, New York
*Recombination of Nonchromosomal Genes
in* Chlamydomonas

Before the widespread use of polymorphic DNA markers in mitochondria and chloroplasts, the study of extranuclear inheritance was hindered in most organisms not only by uniparental inheritance of cellular organelles but also by the lack of mutant phenotypes showing extranuclear transmission. The alga Chlamydomonas *proved to be suitable for studies of extranuclear inheritance. Mutants showing nonchromosomal inheritance could be obtained, and strains were discovered in which the inheritance of nonchromosomal genes was biparental, making it possible to examine segregation of nonchromosomal alleles. The compelling evidence for the existence of genetic systems apart from genes in the nucleus was the discovery of intragenic recombination in nonchromosomal genes. In the matings described here, the parental genotypes for the nonchromosomal genes were ac_1 sd (slow growth on acetate, streptomycin requiring) and ac_2 sr (acetate requiring, streptomycin resistant). The intragenic recombinant products were either wildtype ac^+ (acetate independent) versus the double mutant ac_1 ac_2 or they were wildtype ss (streptomycin sensitive) versus the double mutant sd sr. The authors leave no doubt about their opinion (which turned out to be correct) that the nonchro-mosomal genetic systems are due to DNA in the cellular organelles.*

The analysis of nonchromosomal heredity has made slow progress, often under severe attack, [in part because of] the difficulty in obtaining mutations of nonchromosomal genes. . . . Within the past few years, a concerted attack has been made on this problem with the alga *Chlamydomonas.* . . . Two findings have been of key importance in our investigation. First, streptomycin was developed as a mutagen for nonchromosomal (NC) genes. . . . Second, the existence of an exceptional class of zygotes was dis-

---

**We view our results
as providing evidence of
the nucleic acid nature
of a nonchromosomal
genetic system.**

---

covered that transmits NC genes from both parents to the progeny, in contrast to the standard pattern of maternal inheritance. With this material it was established that the NC genes are particulate in nature, genetically autonomous, and stable. . . . In this paper we describe what is to our knowledge the first evidence of recombination of nonchromosomal genes. . . . The occurrence of linked recombination demonstrates that NC genes, like the chromosomal ones, are capable of close pairing and precise reciprocal exchange. . . . The salient features of our results are the following: (1) Segregation of NC genes did not occur during meiosis. . . . (2) NC gene segregation began at the first or second mitotic division after meiosis, with pure clones arising so long as any heterozygous cells remained. (3) The average segregation ratio is 1 : 1 although the progeny of individual zygotes gave ratios significantly different from 1 : 1. (4) The acetate alleles segregated independently of the streptomycin alleles. (5) Novel classes of progeny were recovered that were phenotypically indistinguishable from wildtype ac^+ and from wildtype ss. New mutant phenotypes were also observed with new levels of acetate requirement and new levels of streptomycin resistance and dependence. [These were the reciprocal, double-mutant types.] . . . From existing knowledge, we infer that intragenic recombination, in which the wildtype gene is reconstituted, requires a close intermolecular pairing and exchange of the sort only known to occur between nucleic acids. Consequently, we view our results as providing evidence of the nucleic acid nature of a nonchromosomal genetic system. These genetic findings have been paralleled by the discovery of DNA of high molecular weight in chloroplasts and mitochondria. It is a reasonable surmise that these organelle DNA's carry primary genetic information.

Source: Proceedings of the National Academy of Sciences of the USA 53: 1053–1061

---

heteroplasmic for green and white (mutant) chloroplasts. Random segregation of the chloroplasts in cell division results in some branches derived from cells that contain only green chloroplasts and some branches derived from cells that contain only mutant chloroplasts. The flowers on green branches produce ovules with normal chloroplasts, so all progeny are green. Flowers on white branches produce ovules with only mutant chloroplasts, so only white progeny are formed. The ovules formed by flowers on variegated branches are of three types: those with only chlorophyll-containing chloroplasts, those with only mutant chloroplasts, and those

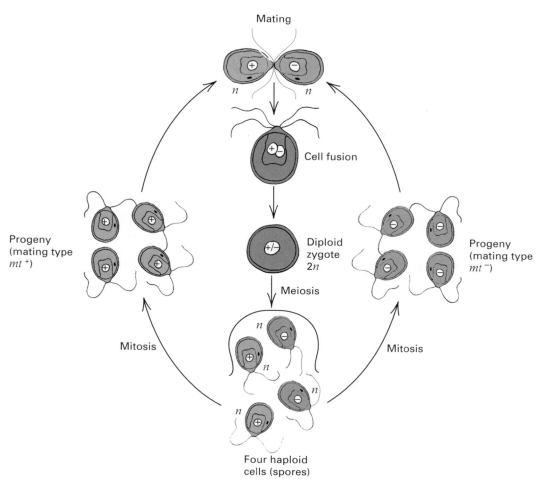

Mating

Cell fusion

Diploid
zygote
2n

Meiosis

Four haploid
cells (spores)

Mitosis

Mitosis

Progeny
(mating type
mt^+)

Progeny
(mating type
mt^-)

Figure 14.8
Life cycle of *Chlamydomonas reinhardtii.*

that are heteroplasmic. The heteroplasmic class again yields variegated plants.

Drug Resistance in *Chlamydomonas*

Cytoplasmic inheritance has been studied extensively in the unicellular green alga *Chlamydomonas*. Cells of this organism have a single large chloroplast that contains from 75 to 80 copies of a double-stranded DNA molecule approximately 195 kb in size. The life cycle of *Chlamydomonas* is shown in Figure 14.8. There are two mating types, mt^+ and mt^-. These cells are of equal size and appear to contribute the same amount of cytoplasm to the zygote. Cells of opposite mating type fuse to form a diploid zygote, which undergoes meiosis to form a tetrad of four haploid cells. The cells can be grown on solid growth medium to form visible clusters of cells (colonies) that can indicate the genotype. Reciprocal crosses between certain antibiotic-resistant strains (mutant) and antibiotic-sensitive strains (wildtype) yield different results. For example, for streptomycin resistance (*str-r*),

$$str\text{-}r\ mt^+ \times str\text{-}s\ mt^- \quad \text{yields only}$$
$$str\text{-}r \text{ progeny}$$

$$str\text{-}s\ mt^+ \times str\text{-}r\ mt^- \quad \text{yields only}$$
$$str\text{-}s \text{ progeny}$$

This is a clear example of uniparental inheritance of the streptomycin resistance or sensitivity, because the resistance or sensitivity comes from the mt^+ parent. In contrast, the mt alleles behave in strictly Mendelian fashion, yielding 1 : 1 progeny ratios, as expected of nuclear genes.

A large number of antibiotic-resistance markers with uniparental inheritance have been examined in *Chlamydomonas*. In each case, the antibiotic resistance has been traced to the chloroplast. Haploid *Chlamydomonas* contains only one chloroplast. If the chloroplast could be derived from the mt^+ or mt^- cell with equal probability, then uniparental inheritance would not be observed because the chloroplast could come from either parent. However, studies in which DNA markers are used to distinguish the chloroplasts of the two mating types have shown that the chloroplast of the mt^- parent is preferentially lost after mating, which accounts for the fact that the antibiotic-resistance phenotype of the mt^+ parent is the one transmitted to the progeny.

About 5 percent of the progeny from *Chlamydomonas* crosses are heteroplasmic. Both parental chloroplasts are retained, although they ultimately segregate in later cell divisions. The heteroplasmic cells have been valuable for genetic analysis, because recombination can take place between the DNA in the two chloroplasts. Analysis of the recombination frequencies for the various phenotypes has made possible the construction of fairly detailed genetic maps of the chloroplast genome.

An interesting inference can be made from the study of *Chlamydomonas* mutants that are resistant to myxothiazol and mucidin, which are known to be inhibitors of the cytochrome b of mitochondria. The mutations that confer resistance are therefore in the mitochondrial DNA. As before, reciprocal crosses can be made to demonstrate the pattern of inheritance of the resistance trait. When crosses are made between antibiotic-resistant (*MUD2*) and antibiotic-sensitive (*mud2*) strains (the *mud2* strains are wildtype), the results are as follows:

$$MUD2\ mt^+ \times mud2\ mt^-\ \text{yields only}$$
$$mud2\ \text{(sensitive) progeny}$$

$$mud2\ mt^+ \times MUD2\ mt^-\ \text{yields only}$$
$$MUD2\ \text{(resistant) progeny}$$

In each cross, the tetrads yield 2 mt^+ : 2 mt^- spores. However, all spore progeny have the resistance phenotype of the mt^- parent. Inheritance of resistance to mucidin or myxothiazol therefore shows uniparental inheritance, because the phenotype is the same as that of the mt^- parent. Resistance to myxothiazol or mucidin results from a change in the mitochondrial DNA, so we can also infer that the meiotic progeny must receive their mitochondrial DNA from the mt^- parent, because the resistance comes from this parent. In contrast, as indicated by the streptomycin example above, they receive their chloroplast DNA from the mt^+ parent. It is presently unknown whether there is any biological significance in the fact that in *Chlamydomonas*, meiotic progeny inherit mitochondria from one parent and chloroplasts from the other parent, or whether the patterns are merely a matter of chance evolutionary accident.

Respiration-Defective Mitochondrial Mutants

Very small colonies of the yeast *Saccharomyces cerevisiae* are occasionally observed when the cells are grown on solid medium that contains glucose. These are called **petite mutants.** Microscopic examination indicates that the cells are of normal size. Physiological studies show that the cells grow normally during early stages of colony formation when they are fermenting the glucose to ethanol but cease to grow thereafter because of a defect in oxygen-requiring respiration that is needed for further metabolism of the ethanol. The colonies are quite small, because the yield of ATP is much higher during respiration than during fermentation. There are several types of petite mutants. We will discuss two of them.

Genetic analysis of crosses between petites and wildtype distinguishes the two major types of petites illustrated in Figure 14.9. One type, called **segregational petites,** exhibits the typical segregation of a simple Mendelian recessive. In a cross between a segregational petite and

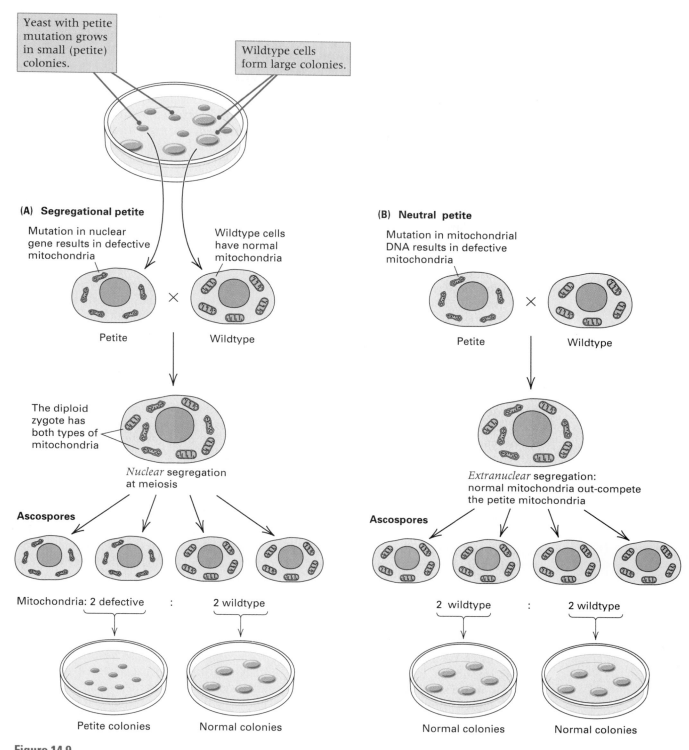

Figure 14.9

Behavior of petite mutations in genetic crosses. The brown circles represent petite cells. The brown nucleus in part A contains an allele resulting in petite colonies. The blue cells have normal mitochondria.

wildtype, the diploid zygotes are normal, and when the diploids undergo meiosis, half of the ascospores in an ascus produce petite colonies and half form wildtype colonies. The 1 : 1 ratio (actually 2 : 2) indicates that these petites are the result of a nuclear mutation and that the allele for petite is recessive.

The second type, **neutral petites,** is quite different: In a cross with wildtype, all diploid zygotes produce wildtype ascospores (4 : 0 segregation). The same pattern is found when the progeny from such a cross are backcrossed with neutral petites. In both cases, the phenotype that is inherited is that of the normal parent, so the inheritance is uniparental. The explanation for these results is that the majority of neutral petites lack most or all mitochondrial DNA, in which many of the genes determining oxidative respiration are encoded. When a neutral-petite cell mates with a wildtype cell, the cytoplasm of the latter is the source of the normal mitochondrial DNA in the resulting progeny spores.

Petites arise at a frequency of roughly 10^{-5} per generation. The neutral petite phenotype is the result of large deletions in mitochondrial DNA. For unknown reasons, yeast mitochondria often fuse and fragment, which may cause occasional deletions of DNA. The production of neutral petites is apparently a result of the segregation of aberrant mitochondrial DNA from normal DNA and the further sorting out of mutant mitochondria in the course of cell division.

Cytoplasmic Male Sterility in Plants

An example of extranuclear inheritance that is important in agriculture is **cytoplasmic male sterility** in plants, a condition in which the plant does not produce functional pollen, but the female reproductive organs and fertility are normal. This type of sterility is used extensively in the production of hybrid corn seed, because it circumvents the need for manual "detasseling" of the plants to be used as females to produce the hybrid. Because the tassel is the pollen-bearing organ, the detasseled female plants cannot produce pollen and so cannot fertilize themselves. The detasseled female plants must therefore use pollen from the still-tasseled male plants used to produce the hybrid.

The genetically transmitted male sterility is not controlled by nuclear genes but is transmitted through the egg cytoplasm from generation to generation. The pattern of inheritance of cytoplasmic male sterility in corn is summarized in Figure 14.10. The key observation is that repeated backcrossing of a male-sterile variety, using pollen from a male-fertile variety, does not restore male fertility. The resulting plants, in which nearly all nuclear genes from the male-sterile variety have been replaced with those from the male-fertile variety, remain male-sterile. Because a small amount of functional pollen is produced by the male-sterile variety, it is also possible to carry out the reciprocal cross:

male-fertile ♀ × male-sterile ♂

In this case, the progeny are fully male-fertile. The difference between the reciprocal crosses means that male sterility or fertility in these varieties is maternally inherited.

Cytoplasmic male sterility in maize results from rearrangements in mitochondrial DNA. For example, one DNA rearrangement fuses two mitochondrial genes and creates a novel protein that causes male sterility. Although cytoplasmic male sterility is maternally inherited, certain nuclear genes called **fertility restorers** can suppress the male-sterilizing effect of the cytoplasm. Different types of cytoplasmic male sterility can be classified on the basis of their response to restorer genes. For example, in one case, restoration of fertility requires the presence of dominant alleles of two different nuclear genes, and plants with male-sterile cytoplasm and both restorer alleles produce normal pollen. In another case, suppression requires the dominant allele of a different restorer gene, but this gene acts only in the gametophyte; hence plants with male-sterile cytoplasm that are heterozygous for the restorer allele produce a 1 : 1 ratio of normal to aborted pollen grains.

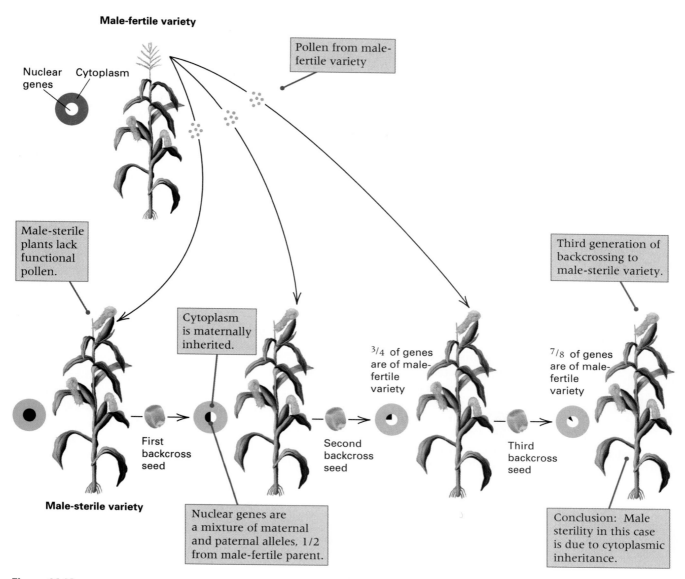

Male-fertile variety

Nuclear genes   Cytoplasm

Pollen from male-fertile variety

Male-sterile plants lack functional pollen.

Third generation of backcrossing to male-sterile variety.

Cytoplasm is maternally inherited.

$^3/_4$ of genes are of male-fertile variety

$^7/_8$ of genes are of male-fertile variety

First backcross seed

Second backcross seed

Third backcross seed

Male-sterile variety

Nuclear genes are a mixture of maternal and paternal alleles, 1/2 from male-fertile parent.

Conclusion: Male sterility in this case is due to cytoplasmic inheritance.

Figure 14.10
Diagram of an experiment demonstrating maternal inheritance of male sterility (orange cytoplasm) in maize.

14.3
The Evolutionary Origin of Organelles

A widely accepted theory holds that organelles originally evolved from prokaryotic cells that lived inside primitive eukaryotic cells in symbiosis, a mutually beneficial interaction. This is the **endosymbiont theory.** Mitochondria and chloroplasts are thought to have originated from bacteria and cyanobacteria (formerly called blue-green algae), respectively. The archaea, which comprise the third major type of life form (the other two are eubacteria and eukaryotes), are good candidates for the primitive hosts of endosymbionts that ultimately evolved into eukaryotes, including animals, plants, and fungi. They share some features with eubacteria, including their prokaryotic cytological organization. However, in their macromolecular syntheses—replication, transcription, and translation—the archaea resemble the eukaryotes.

One line of evidence for the endosymbiont theory is that mitochondrial and chloroplast organelles of eukaryotes share numerous features with prokaryotes that distinguish them from eukaryotic cells. Among these common features:

1. The genomes are composed of circular DNA that is not extensively complexed with histone-like proteins.

2. The genomes are organized with functionally related genes close together and often expressed as a single unit.

3. The ribosome particles on which protein synthesis takes place have major subunits whose size is similar in organelles and in prokaryotes but differs from that in cytoplasmic ribosomes in eukaryotic cells.

4. The nucleotide sequences of the key RNA constituents of ribosomes are similar in chloroplasts, cyanobacteria such as *Anacystis nidulans,* and even bacteria such as *E. coli.*

The genomes of today's organelles are small compared with those of bacteria and cyanobacteria. For example, chloroplast genomes are only from 3 to 5 percent as large as the genomes of cyanobacteria. Organelle evolution was probably accompanied by major genome rearrangements, some genes being transferred into the nuclear genome of the host and others being eliminated. The transfer of organelle genes into the nucleus differed from one lineage to the next. For example, the gene for one subunit of the mitochondrial ATP synthase is located in the mitochondrial genome in yeast but in the nuclear genome in *Neurospora.*

14.4
The Cytoplasmic Transmission of Symbionts

In eukaryotes, a variety of cytoplasmically transmitted traits result from the presence of bacteria and viruses living in the cytoplasm of certain cells. One of the best known is the killer phenomenon found in certain strains of the protozoan *Paramecium aurelia.* Killer strains of *Paramecium* release to the surrounding medium a substance that is lethal to many other strains of the protozoan. The killer phenotype requires the presence of cytoplasmic bacteria referred to as **kappa particles,** whose maintenance is dependent on a dominant nuclear gene, *K.* The kappa particles are cells of *Caedobacter taeniospiralis,* which produce the killer substance. Why killer strains are immune to the substance has not yet been determined.

Paramecium is a diploid protozoan that undergoes sexual exchange through a mating process called **conjugation** (Figure 14.11A). Initially, each cell has two diploid micronuclei. Many protozoans contain micronuclei (small nuclei) and a macronucleus (a large nucleus with specialized functions), but only the micronuclei are relevant to the genetic processes described here. When two cells come into contact for conjugation, the two micronuclei in each cell undergo meiosis, forming eight micronuclei in each cell. Seven of the micronuclei and the macronucleus disintegrate, leaving each cell with one micronucleus, which then undergoes a mitotic division. The cell membrane between the two cells breaks down slightly, and the cells exchange a single micronucleus; then the nuclei fuse to form a diploid nucleus. After nuclear exchange, the two new cells (the exconjugants) are genotypically identical. In this sequence of events, only the micronuclei are exchanged, without any mixing of cytoplasm.

Single cells of *Paramecium* also occasionally undergo an unusual nuclear phenomenon called **autogamy** (Figure 14.11B). Meiosis takes place, and again seven of the micronuclei disintegrate. The surviving nucleus then undergoes mitosis and nuclear fusion to recreate the diploid state.

(A) Conjugation

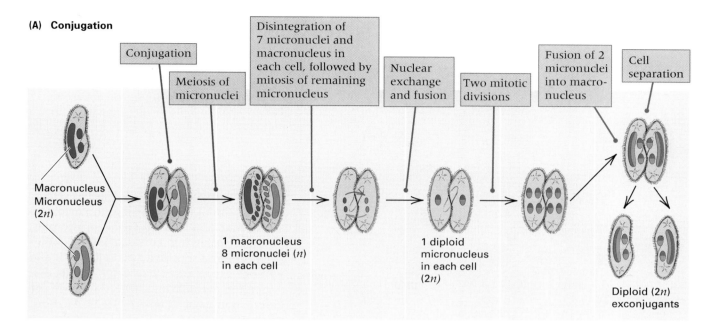

Conjugation

Meiosis of micronuclei

Disintegration of 7 micronuclei and macronucleus in each cell, followed by mitosis of remaining micronucleus

Nuclear exchange and fusion

Two mitotic divisions

Fusion of 2 micronuclei into macronucleus

Cell separation

Macronucleus
Micronucleus
(2n)

1 macronucleus
8 micronuclei (n)
in each cell

1 diploid micronucleus in each cell (2n)

Diploid (2n) exconjugants

(B) Autogamy

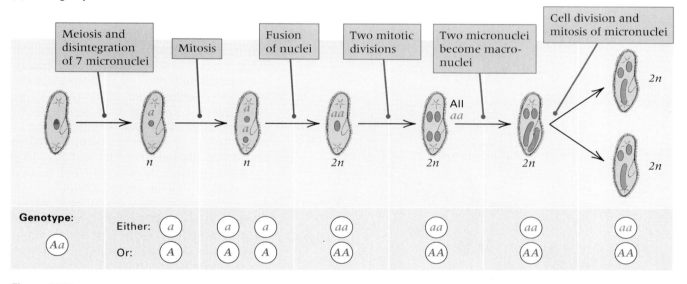

Meiosis and disintegration of 7 micronuclei

Mitosis

Fusion of nuclei

Two mitotic divisions

Two micronuclei become macronuclei

Cell division and mitosis of micronuclei

n

n

2n

2n

All
aa

2n

2n

2n

2n

2n

Genotype:

Aa

Either: a | a a | aa | aa | aa | aa

Or: A | A A | AA | AA | AA | AA

Figure 14.11

(A) Conjugation in *Paramecium* results in reciprocal fertilization and the formation of two genetically identical exconjugant cells. (B) Autogamy in *Paramecium*. The alleles in the heterozygous *Aa* cell segregate in meiosis, with the result that the progeny cells become homozygous for either *A* or *a*. The production of *aa* progeny cells is shown in the diagram. The initial steps resemble those in conjugation in part A: The micronuclei undergo meiosis, after which seven of the products and the macronucleus degenerate.

The genetic importance of autogamy is that even if the initial cell is heterozygous, the newly formed diploid nucleus becomes homozygous because it is derived from a single haploid meiotic product. In a population of cells undergoing autogamy, the surviving micronucleus is selected randomly, so for any pair of alleles, half of the new cells contain one allele and half contain the other.

Conjugation usually allows an exchange *only* of micronuclei at the surface of

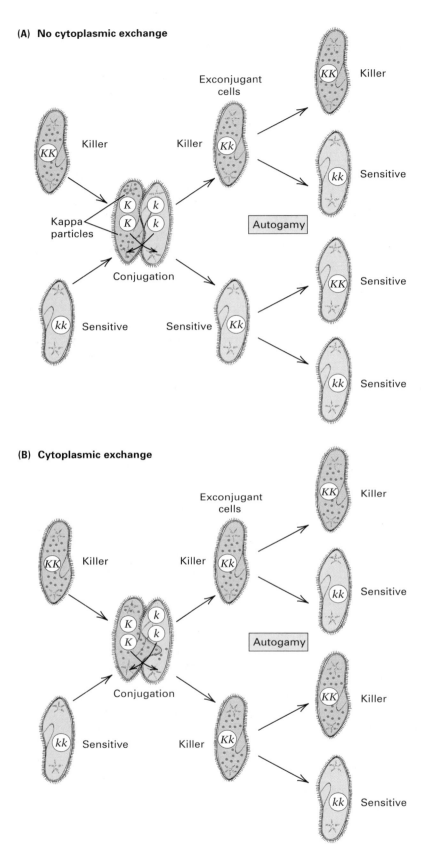

(A) No cytoplasmic exchange

Killer

Exconjugant
cells

Killer

KK Killer

Kappa
particles

Killer

Kk

Kk Killer

KK Killer

kk Sensitive

Autogamy

Conjugation

kk Sensitive

Sensitive

Kk

Sensitive

KK Sensitive

kk Sensitive

(B) Cytoplasmic exchange

Killer

Exconjugant
cells

Killer

KK Killer

KK Killer

Kk

Kk Killer

KK Killer

kk Sensitive

Autogamy

Conjugation

kk Sensitive

Killer

Kk

Killer

KK Killer

kk Sensitive

Figure 14.12

Crosses between killer (*KK*) and sensitive (*kk*) *Paramecium.* (A) The result when no cytoplasmic exchange takes place in conjugation. (B) The result when cytoplasmic exchange occurs. Cells containing kappa are shown in yellow. The genotype and phenotype of each cell are indicated.

contact between the two cells. Occasionally, small amounts of cytoplasm also are exchanged. A comparison of the phenotypes following conjugation of killer (*KK*) cells and sensitive (*kk*) cells, with and without cytoplasmic exchange, yields evidence for cytoplasmic inheritance of the killer phenotype (Figure 14.12). Without cytoplasmic mixing (Figure 14.12A), the expected 1 : 1 ratio of killer to sensitive exconjugant cells is observed; with cytoplasmic mixing (Figure 14.12B), both exconjugants are killer cells because each cell has kappa particles derived from the cytoplasm of the killer parent. Conjugation is eventually followed by autogamy of each of the exconjugants (Figure 14.12). Autogamy of *Kk* killer cells results in an equal number of *KK* and *kk* cells, but the *kk* cells cannot maintain the kappa particle and so become sensitive.

Another example of a symbiont-related cytoplasmic effect is a condition in *Drosophila* called **maternal sex ratio,** which is characterized by the production of almost no male progeny. The daughters of sex-ratio females pass on the trait, whereas the occasional sons that are produced do not. Cytoplasm taken from the eggs of sex-ratio females transmits the condition when injected into the female embryos from unaffected cultures of the same or other *Drosophila* species. A species of bacteria has been isolated from the cytoplasm of sex-ratio females; when these bacteria are allowed to infect other *Drosophila* females, these acquire the sex-ratio trait. The causative agent of the sex-ratio condition is not the bacteria themselves but rather a virus that multiplies in them. When released by the bacterial cells, this virus kills most male *Drosophila* embryos. Why female embryos are not killed by the virus is still a mystery.

14.5
Maternal Effect in Snail-Shell Coiling

There are also some examples in which maternal transmission takes place, but the maternal transmission is not mediated by cytoplasmic organelles or symbionts but rather through the cytoplasmic transmission of the products of nuclear genes. An example is the determination of the direction of coiling of the shell of the snail *Limnaea peregra*. The direction of coiling, as viewed by looking into the opening of the shell, may be either to the right (dextral coiling) or to the left (sinistral coiling). Reciprocal crosses between homozygous strains give the following results:

dextral ♀ × sinistral ♂ → all F$_1$ dextral
sinistral ♀ × dextral ♂ → all F$_1$ sinistral

In these crosses, the F$_1$ snails have the same genotype, but the reciprocal crosses give different results in that the direction of coiling is the same as that of the mother. This result is typical of traits with maternal inheritance. However, in this case, all F$_2$ progeny from both crosses exhibit dextral coiling, a result that is inconsistent with cytoplasmic inheritance. The F$_3$ generation provides the explanation.

The F$_3$ generation, obtained by self-fertilization of the F$_2$ snails (the snail is her-maphroditic), indicates that the inheritance of coiling direction depends on nuclear genes rather than extranuclear factors. Three-fourths of the F$_3$ progeny have dex-trally coiled shells and one-fourth have sinistrally coiled shells (Figure 14.13). This is a typical 3 : 1 ratio, which indicates Mendelian segregation in the F$_2$. Dextral coiling ($+/+$ and $s/+$) is dominant to sinis-tral coiling (s/s). The reason why reciprocal crosses give different results, and why the production of the 3 : 1 Mendelian ratio is delayed for a generation, is as follows:

The coiling phenotype of an individual snail is determined by the genotype of its mother.

This means that the direction of shell coil-ing is not a case of maternal inheritance but

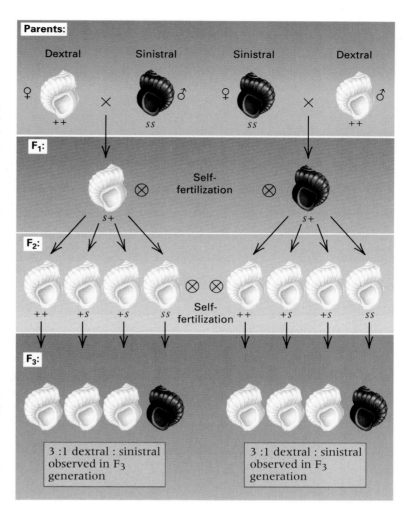

Figure 14.13
Inheritance of the direction of shell coiling in the snail *Limnaea.* Sinistral coiling is determined by the recessive allele *s*, dextral coiling by the wild-type + allele. However, the direction of coiling is determined by the nuclear genotype of the mother, not by the genotype of the snail itself or by extranuclear inheritance. The F$_2$ and F$_3$ generations can be obtained by self-fertilization because the snail is hermaphroditic and can undergo either self-fertilization or cross-fertilization.

rather a case of maternal effect (Section 14.1). Cytological analysis of developing eggs has provided the explanation: The genotype of the mother determines the ori-entation of the spindle in the initial mitotic division after fertilization, and this in turn controls the direction of shell coiling of the offspring (Figure 14.14). This classic exam-ple of maternal effect indicates that more than a single generation of crosses is needed to provide conclusive evidence of extranuclear inheritance of a trait.

(A) Sinistral coiling

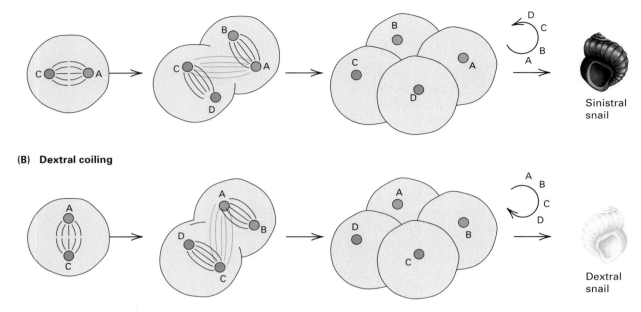

Sinistral snail

(B) Dextral coiling

Dextral snail

Figure 14.14

The direction of shell coiling in *Limnaea* is determined by the orientation of the mitotic spindles during the second cleavage division in the zygote. The orientation is predetermined in the egg cytoplasm as a result of the mother's genotype. The spiral patterns established by the cleavage result in (A) sinistral (leftward) or (B) dextral (rightward) coiling.

14.6
In Search of
Mitochondrial "Eve"

Mitochondrial DNA has a number of features that make it useful for studying the genetic relationships among organisms in natural populations, including human populations. First, it is maternally inherited and does not undergo genetic recombination, and hence the DNA molecule in any mitochondrion derives from a single ancestral molecule. Second, mammalian mitochondrial DNA evolves considerably faster than that of nuclear genes; for example, differences in mitochondrial DNA sequences accumulate among human lineages at a rate of approximately 1 change per mitochondrial lineage every 1500 to 3000 years.

To put this rate of evolution into perspective, consider the people of Australia and New Guinea, who have been relatively isolated genetically from each other since an aboriginal people colonized these areas approximately 30,000 years ago (Australia) and 40,000 years ago (New Guinea). The total time of separation between the populations is therefore 70,000 years (30,000 years in Australia and 40,000 years in New Guinea). If one nucleotide change accumulates every 1500 to 3000 years, the total number of differences in the mitochondrial DNA of Australian aborigines and New Guineans is expected to range from 23 nucleotides (calculated as 70,000/3000) to 47 nucleotides (calculated as 70,000/1500). This example shows how the rate of mitochondrial DNA evolution can be used to predict the number of differences between populations. In practice, the calculation is done the other way around, and the observed number of differences between pairs of populations is used to estimate the rate of mitochondrial evolution.

Nucleotide differences in mitochondrial DNA have been used to reconstruct the probable historical relationships among human populations. Figure 14.15 is an example that includes people from five native populations: African, Asian, Australian, Caucasian (European), and

A Coming Together

Lynn Margulis (a.k.a. Lynn Sagan) 1967
Boston University, Boston,
Massachusetts
The Origin of Mitosing Cells

By the 1960s certain features of the genetic apparatus of cellular organelles were already known to resemble those in prokaryotes. A number of scientists had suggested that the similarities may reflect an ancient ancestral relationship. Lynn Margulis took these ideas a giant step further by proposing a comprehensive hypothesis of the origin of the eukaryotic cell as a series of evolved symbiotic relationships. The main players are the mitochondria, the kinetosomes (basal bodies) of the flagella and cilia, and the photosynthetic plastids. All are hypothesized to have originated as prokaryotic symbionts. When the paper was submitted for publication, the anonymous reviewers were so strongly negative that it was almost rejected. Through the years, specific predictions of the hypothesis have been experimentally verified. A few revisions have been necessary; for example, current evidence suggests that the earliest symbiont was kinetosome-bearing, not a protomitochondrion. However, no major inconsistencies have been uncovered. Today the serial endosymbiont theory is widely accepted.

This paper presents a theory of the origin of the discontinuity between eukaryotic and prokaryotic cells. Specifically, the mitochondria, the basal bodies [kinetosomes] of the flagella, and the photosynthetic plastids can be considered to have derived from free-living cells, and the eukaryotic cell is the result of the evolution of ancient symbioses. Although these ideas are not new, in this paper they have been synthesized in such a way as to be consistent with recent data on the biochemistry and cytology of subcellular organelles. . . . Many aspects of this theory are verifiable by modern techniques of molecular biology. . . .

Prokaryotic cells containing DNA, synthesizing protein on ribosomes, and using messenger RNA as intermediate between DNA and protein, are ancestral to all extant cellular life. Such cells arose under reducing conditions of the primitive terrestrial atmosphere. . . . Eventually, a population of cells arose using photoproduced ATP with water as the source of hydrogen atoms in

---

The eukaryotic cell is the result of the evolution of ancient symbioses.

---

the reduction of CO_2 for the production of cell material. This led to the formation of gaseous oxygen as a by-product of photosynthesis. . . . The continued production of free oxygen resulted in a crisis. . . . It is suggested that the first step in the origin of eukaryotes from prokaryotes was related to survival in the new oxygen-containing atmosphere: an aerobic prokaryotic microbe (the protomitochondrion) was ingested into the cytoplasm of a heterotrophic anaerobe. This symbiosis became obligate and resulted in the first aerobic amoeboid [without, at this stage, mitosis]. . . . Some of the amoeboids ingested certain motile prokaryotes. The genes of the parasite coded for its characteristic morphology, $(9 + 2)$ fibrils in cross section. This parasite also became symbiotic, forming primitive amoeboflagellates . . . that could actively pursue their own food. . . . The replicating nucleic acid of the endosymbiont genes (which determines its characteristic $9 + 2$ structure of sets of microtubules) was eventually utilized to form the chromosomal centromeres and centrioles of eukaryotic mitosis and to distribute newly synthesized host nuclear chromatin to host daughters. . . . Eukaryotic plant cells acquired photosynthesis by symbiosis with photosynthetic prokaryotes (protoplastids), which themselves evolved from organisms related to cyanobacteria. . . . In order to document this theory, the following, at the very least, are required: (1), the theory must be consistent with geological and fossil records; (2), each of the three symbiotic organelles (the mitochondria, the $(9 + 2)$ homologues, and the plastids) must demonstrate general features characteristic of cells originating in hosts as symbionts. None may have features that conflict with such an origin; (3), predictions based on this account of the origin of eukaryotes must be verified. The rest of this paper [24 pages] discusses the evidence for this theory in terms of assumptions based on molecular biology which can be made concerning evolutionary mechanisms.

Source: Journal of Theoretical Biology 14: 225–274

New Guinean. The diagram is *not* a pedigree, although it superficially resembles one. Rather, the diagram is a hypothetical **phylogenetic tree,** which depicts the lines of descent that connect the mitochondrial DNA sequences. The "true" phylogenetic tree connecting the mitochondrial DNA sequences is unknown, so the tree illustrated should be regarded as an approximation (or an "estimate") of the true tree. It is approximate because it is inferred from analysis of the mitochondrial DNA, and this analysis requires a number of assumptions, including, for example, uniformity of rates of evolution in different branches. Nevertheless, the tree in Figure 14.15 has two interesting features: (1) Each population contains multiple mitochondrial types connected to the tree at widely dispersed positions, and (2) one of the primary branches issuing from the inferred common ancestor is composed entirely of

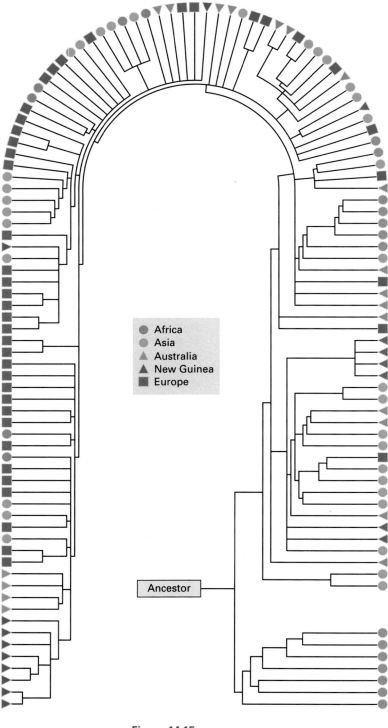

Figure 14.15
Possible phylogenetic tree of human mitochondria based on analysis of presence or absence of restriction sites for 12 different restriction enzymes in mitochondrial DNA from 134 persons. [From R. L. Cann, M. Stoneking, and A. C. Wilson. 1987. *Nature* 325: 31.]

Legend:
● Africa
● Asia
▲ Australia
▲ New Guinea
■ Europe

Ancestor

Africans. These features can be explained most readily by postulating that the mitochondrial DNA type of the common ancestor was African. Furthermore, because the inferred common ancestor links mitochondrial DNA types that differ at an average of 95 nucleotides, the time since the existence of the common ancestor can be estimated as between $95 \times 1500 = 142,500$ years and $95 \times 3000 = 285,000$ years. Therefore, the common ancestor of all human mitochondrial DNA was a woman who lived in Africa between 142,500 and 285,000 years ago. This woman has been given the name "Eve" by the popular press, although she was not the only female alive at the time. Nevertheless, Eve's mitochondrial DNA became the common ancestor of all human mitochondrial DNA that exists today, because during the intervening years, all other mitochondrial lineages except hers became extinct (probably by chance).

The hypothesis of an African mitochondrial Eve has been very controversial. The difficulty is not with postulating a common ancestor of human mitochondrial DNA: It is a mathematical certainty that all human mitochondrial DNA must, eventually, trace back to a single ancestral woman. This principle is true because there is differential reproduction, and therefore, each mitochondrial lineage has a chance of becoming extinct in every generation. Hence, given enough time, all lineages except one must eventually become extinct. Was the common ancestor of all mitochondria an African? When did she live? These are the questions in dispute. Later analysis of the same data suggested that other trees were almost as good as the one in Figure 14.15 and that some trees were even better. The main criterion for comparing hypothetical trees based on DNA sequences is the total number of mutational steps in all the branches together: Trees with fewer mutational steps are generally regarded as better. Therefore, the finding of better trees than the one in Figure 14.15 was a serious blow to the hypothesis of an African Eve, because none of the better trees had an earliest branch composed exclusively of Africans. However, much of the uncertainty is intrinsic in the data themselves. Another hypothetical phylogenetic tree, with fewer mutational steps than that in

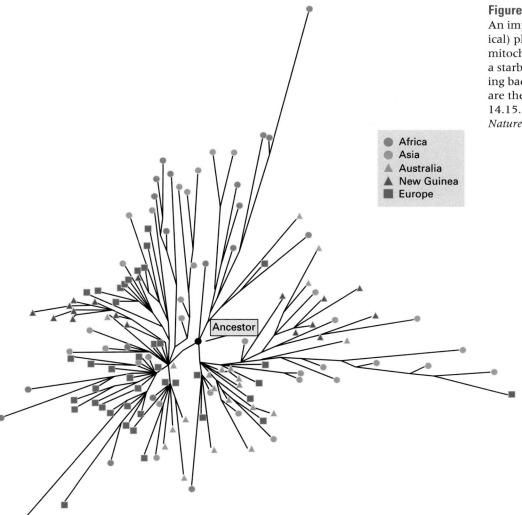

Figure 14.16
An improved (but still hypothetical) phylogenetic tree of human mitochondrial DNA presented as a starburst with the reader looking back in time. The symbols are the same as those in Figure 14.15. [From C. Wills. 1992. *Nature* 356:389.]

- ● Africa
- ● Asia
- ▲ Australia
- ▲ New Guinea
- ■ Europe

Figure 14.15, is shown in Figure 14.16. This tree is drawn with the lengths of the branches proportional to the number of mutational steps, and it resembles a starburst with the viewer looking back in time. The difficulty of identifying the earliest common ancestor is evident from this style of representation. Although there is an early branch (at a position near one o'clock in the starburst) composed primarily of Africans, this branch also includes two Asians, one Caucasian, and one Australian aborigine. On the other hand, this is no proof that the mitochondrial Eve was *not* an African, and considerable evidence of other kinds supports Africa as the cradle of human evolution and colonization. However, the evidence for postulating an African Eve solely on the basis of current mitochondrial DNA data is inconclusive.

Chapter Summary

The cytoplasm of eukaryotic cells contains complex organelles, the most prevalent of which are mitochondria, found in both plant and animal cells, and chloroplasts, found in plant cells. Mitochondria are responsible for respiratory metabolism. Chloroplasts are responsible for photosynthesis. Both mitochondria and chloroplasts contain DNA molecules that encode a variety of proteins. The genes contained in mitochondrial and chloroplast DNA are not present in chromosomal DNA. Hence, certain traits determined by these genes are not inherited according to Mendelian principles but show extranuclear (cytoplasmic) inheritance instead. Extranuclear inheritance is also observed for traits that are determined by bacteria and viruses transmitted in the cytoplasm from one generation to the next. In eukaryotic microorganisms, cytoplasmic mixing usually takes place in zygote formation, and both parents contribute cytoplasmic particles to the zygote. However, in multicellular eukaryotes, the egg contains a large amount of cytoplasm and the sperm very little, so cytoplasmic transmission results in maternal inheritance.

Progeny resembling the mother is characteristic of both maternal inheritance and maternal effects. Maternal inheritance results from the transmission of extranuclear factors by the mother. Maternal effects occur when the nuclear genotype of the mother determines the phenotype of the offspring—for example, through the molecular organization of the egg or by nurturing ability. The direction of shell coiling in snails results from a maternal effect on the orientation of the spindle in the first mitotic division in the zygote.

Variegation in the four-o'clock plant is caused by mutations in chloroplast DNA that eliminate chlorophyll. Variegated plants contain patches of green and white derived from wildtype cells and from cells with mutant chloroplasts, respectively. Inheritance of the variegated phenotype depends on the chloroplast content of the egg because pollen does not contain chloroplasts. The inheritance of resistance to certain antibiotics in *Chlamydomonas* is also cytoplasmic and depends on mutations in the chloroplast.

Mitochondrial defects account for the cytoplasmic inheritance of some petite mutations in yeast. Petites produce very small colonies because the defective mitochondria are unable to carry out aerobic metabolism. Petite cells can use only anaerobic metabolism, which is so inefficient that the cells grow very slowly, thereby producing small colonies. There are two major types of petites. Segregational petites result from a mutation in a nuclear gene and show strictly Mendelian inheritance. Neutral petites show non-Mendelian inheritance (4 : 0 segregation in ascospores) and result from large deletions in mitochondrial DNA. Male sterility in maize also results from defective mitochondria that contain certain types of deletions or rearrangements in the DNA. Male sterility is very useful in plant breeding, because it makes it possible to cross-pollinate plants without removing immature anthers from the recipient plants.

The killer phenotype in *Paramecium* exemplifies cytoplasmic inheritance through a symbiotic bacterium that produces a product that is toxic to many other paramecia. Cells containing the bacterium are resistant to the killer substance. *Paramecium* mates by two cells coming in contact and exchanging haploid nuclei. After the nuclei exchange, the cells separate. Normally, the exchange takes place without any cytoplasmic mixing, in which case all segregating traits are inherited according to Mendelian principles. Sometimes cytoplasmic mixing occurs, and in this case one sees cytoplasmic inheritance. Maternal sex ratio in *Drosophila*, expressed as a deficiency of males among progeny, is another phenomenon determined by the presence of a symbiotic bacterium. In this case, the bacterium contains a virus that inhibits the development of male embryos.

Human mitochondrial DNA types have been organized into a hypothetical phylogenetic tree in which one branch issuing from the inferred common ancestor is composed entirely of mitochondrial DNA from Africans. The woman whose mitochondrial DNA is thought to be the common ancestor of all human mitochondria has been named "Eve." The all-African branch in the phylogenetic tree and the number of nucleotide differences observed in mitochondrial DNA have been interpreted to mean that the mitochondrial Eve was present in a woman who lived in Africa between 142,500 and 285,000 years ago. On the other hand, additional analyses of the same data have yielded hypothetical phylogenetic trees that have fewer mutational steps overall and that do not contain exclusively African branches. Hence, the evidence that suggests a mitochondrial Eve who lived in Africa is inconclusive.

Key Terms

autogamy	heteroplasmy	paternal inheritance
conjugation	kappa particle	petite mutant
cytoplasmic inheritance	maternal effect	phylogenetic tree
cytoplasmic male sterility	maternal inheritance	RNA editing
endosymbiont theory	maternal sex ratio	segregational petite
extranuclear inheritance	neutral petite	symbiosis
fertility restorer	organelle	uniparental inheritance

Review the Basics

- What are the normal functions of mitochondria and chloroplasts?

- What eukaryotic cells can survive without mitochondria? Can plant cells survive without chloroplasts?

- What is an inherited mitochondrial disease?

- Distinguish between maternal inheritance and maternal effect.

- What is the endosymbiont theory of the origin of certain cellular organelles?

- What is meant by variegation?

- How does leaf variegation in four-o'clock plants demonstrate heteroplasmy of chloroplasts in certain cell lineages and homoplasmy in others?

- Of what practical use is cytoplasmic male sterility in maize breeding?

- What does the occurrence of segregational petites in yeast imply about the relationship between the nucleus and mitochondria?

- What is meant by the term *Eve* with reference to human mitochondrial DNA? Do you think *Eve* is a suitable term for this concept? Why or why not?

Guide to Problem Solving

Problem 1: Two strains of maize exhibit male sterility. One is cytoplasmically inherited; the other is Mendelian. What crosses could you make to identify which is which?

Answer: Cytoplasmic male sterility is transmitted through the female, so outcrosses (crosses to unrelated strains) using the cytoplasmic male-sterile strain as the female parent will yield male-sterile progeny. Repeated outcrosses using the strain with Mendelian male sterility will result in either no male sterility (if the sterility gene is recessive) or 50 percent sterility (if it is dominant).

Problem 2: A snail of the species *Limnaea peregra* in which the shell coils to the left undergoes self-fertilization, and all of the progeny coil to the right. What is the genotype of the parent? If the progeny are individually self-fertilized, what coiling phenotypes are expected?

Answer: In this species, shell coiling is determined by the genotype of the mother. Because the coiling of the mother is leftward, its own mother had the genotype *ss*, so the individual must carry at least one *s* allele. Because the progeny coil to the right, the mother's genotype must be s^+. The progeny genotypes are $1/4$ s^+s^+, $1/2$ s^+s, and $1/4$ ss. When self-fertilized, the s^+s^+ and s^+s individuals yield rightward-coiled progeny, and the ss individuals yield leftward-coiled progeny.

Problem 3: Consider an inherited trait in *Drosophila* with the following characteristics: (1) It is transmitted from father to son to grandson. (2) It is not expressed in females. (3) It is not transmitted through females.

(a) Is the trait likely to result from extranuclear inheritance?

(b) Can male-limited expression explain the data?

(c) What hypothesis of Mendelian inheritance can explain the data?

(d) What observations could be made to confirm the Mendelian hypothesis? (*Hint*: Think about the consequences of nondisjunction.)

Answer:

(a) The trait is very unlikely to result from extranuclear inheritance. First, it would be very unusual for extranuclear inheritance in animals to be transmitted through the male. Second, if it were transmitted through sperm, both sexes should be affected.

(b) The trait is also unlikely to have male-limited expression. Although the observation that only males are affected is consistent with this hypothesis, a trait with male-limited expression could be transmitted through females.

(c) The father → son → grandson pattern of inheritance parallels that of the Y chromosome. Therefore, the data fit a Y-linked model of inheritance.

(d) Nondisjunction in an XY male results in XXY zygotes (which in *Drosophila* are female) and XO zygotes (which in *Drosophila* are male). Therefore, if the trait is Y-linked, then XXY females resulting from nondisjunction should be affected, and XO males resulting from nondisjunction should be unaffected. Furthermore, the affected XXY females should give rise to affected sons (XY) and some affected daughters (XXY).

Problem 4: The rate of mitochondrial DNA evolution is given as 1 nucleotide change per mitochondrial lineage every 1500 to 3000 years. Given that human mitochondrial DNA is approximately 16,500 nucleotides, what is the rate of change per nucleotide per million years?

Answer: 16,500 nucleotides for 1500 years equals 2.5×10^7 nucleotide-years, and 16,500 nucleotides for 3000 years equals 5.0×10^7 nucleotide-years. The rate per nucleotide is therefore between $1/2.5 \times 10^7 = 4 \times 10^{-8}$ per year and $1/5.0 \times 10^7 = 2 \times 10^{-8}$ per year. That is, it is between 0.02 and 0.04 per million years, which is usually expressed as 2 percent to 4 percent nucleotide changes per million years.

Analysis and Applications

14.1 A mutant plant is found with yellow instead of green leaves. Microscopic and biochemical analyses show that the cells contain very few chloroplasts and that the plant manages to grow by making use of other pigments. The plant does not grow very well, but it does breed true. The inheritance of yellowness is studied in the following crosses:

$$\text{green } \delta \times \text{yellow } \varphi \rightarrow \text{all yellow}$$

$$\text{yellow } \delta \times \text{green } \varphi \rightarrow \text{all green}$$

What do these results suggest about the type of inheritance?

14.2 For the phenotype in the preceding problem, suppose that variegated progeny are found in the crosses at a frequency of about 1 per 1000 progeny. How might this be explained?

Chapter 14

GeNETics on the web will introduce you to some of the most important sites for finding genetic information on the Internet. To complete the exercises below, visit the Jones and Bartlett home page at

http://www.jbpub.com/genetics

Select the link to *Genetics: Principles and Analysis* and then choose the link to *GeNETics on the web.* You will be presented with a chapter-by-chapter list of highlighted keywords.

GeNETics EXERCISES

Select the highlighted keyword in any of the exercises below, and you will be linked to a web site containing the genetic information necessary to complete the exercise. Each exercise suggests a specific, written report that makes use of the information available at the site. This report, or an alternative, may be assigned by your instructor.

1. The complete, fully annotated sequence of **human mitochondrial DNA** can be found at this site, along with the sequences of many other mitochondrial genomes. Find the coordinates defining the complete nucleotide sequence of the lysine tRNA implicated in the MERRF type of myoclonic epilepsy, and determine whether the tRNA sequence matches the strand given or its complement. If assigned to do so, draw the self-paired structure of the tRNA in a "cloverleaf" configuration.

2. Genetic maps of the *Chlamydomonas* chloroplast and mitochondrion can be found by choosing the *Linkage group* option at this site. From the maps, deduce the approximate size of each molecule in kilobase pairs. If assigned to do so, determine how many genes are in each organelle genome; use the search engine to find out what the mitochondrial *S* and *L* genes encode as well as the function of the chloroplast *tufA* gene.

3. Search this maize site for **cytoplasmic male sterility,** and follow the links to the *cms* types of cytoplasmic male sterility. If assigned to do so, make a list of which *cms* types can be suppressed by fertility restorers and which types confer susceptibility to fungal infection.

MUTABLE SITE EXERCISES

The Mutable Site Exercise changes frequently. Each new update includes a different exercise that makes use of genetics resources available on the World Wide Web. Select the **Mutable Site** for Chapter 14, and you will be linked to the current exercise that relates to the material presented in this chapter.

PIC SITE

The Pic Site showcases some of the most visually appealing genetics sites on the World Wide Web. To visit the showcase genetics site, select the **Pic Site** for Chapter 14.

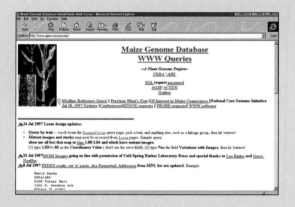

14.3 What kinds of progeny result from the following crosses with *Mirabilis*?

(a) green ♀ × white ♂
(b) white ♀ × green ♂
(c) variegated ♀ × green ♂
(d) green ♀ × variegated ♂

14.4 In the maternally inherited leaf variegation in *Mirabilis*, how does the F_2 of the cross green ♀ × white ♂ differ from that which would be expected if white leaves were due to a conventional X-linked recessive gene in animals?

14.5 Assuming that chloroplasts duplicate and segregate randomly in cell division, what is the probability that both replicas of a single mutant chloroplast will segregate to the same daughter cell during mitosis? After two divisions, what is the probability that at least one daughter cell will lack the mutant chloroplast?

14.6 If *a* and *b* are chloroplast markers in *Chlamydomonas*, what genotypes of progeny are expected from the cross $a^+ b^- mt^+ \times a^- b^+ mt^-$?

14.7 What is the phenotype of a diploid yeast produced by crossing a segregational petite with a neutral petite?

14.8 An antibiotic-resistant haploid strain of yeast is isolated. Mating with a wildtype (antibiotic-sensitive) strain produces a diploid that, when grown for some generations and induced to undergo sporulation, produces tetrads containing either four antibiotic-resistant spores or four antibiotic-sensitive spores. What can you conclude about the inheritance of the antibiotic resistance?

14.9 What are the possible genotypes of the progeny of an *Aa Paramecium* that undergoes autogamy?

14.10 Equal numbers of *AA* and *aa Paramecium* undergo conjugation in random pairs. What are the expected matings? What genotypes and frequencies are expected after the conjugating pairs separate? What genotypes and frequencies are expected after the cells undergo autogamy?

14.11 Corn plants of genotype *AA bb* with cytoplasmic male sterility and normal *aa BB* plants are planted in alternate rows, and pollination is allowed to occur at random. What progeny genotypes would be expected from the two types of plants?

14.12 Females of a certain strain of *Drosophila* are mated with wildtype males. All progeny are female. There are three possible explanations: (1) The females are homozygous for an autosomal allele that produces lethal male embryos (that is, a maternal effect). (2) The females are homozygous for an X-linked allele that is lethal in males. (3) The females carry a cytoplasmically inherited factor that kills male embryos. A cross is carried out between F_1 females and wildtype males, and only female progeny are produced. What mode of inheritance does this result imply?

14.13 A rightward-coiled snail, when self-fertilized, has all leftward-coiled progeny. What is the genotype of the rightward-coiled parent? What was the genotype of the parent's mother?

Challenge Problems

14.14 There are an average of 78 nucleotide differences in mitochondrial DNA between two Africans, an average of 58 differences between two Asians, and an average of 40 differences between two Caucasians. Assuming a rate of evolution of mitochondrial DNA of 1 nucleotide difference per 1500 to 3000 years, what is the estimated average age of the common ancestor of the mitochondrial DNA among Africans? Among Asians? Among Caucasians?

14.15 The illustration shows the locations of restriction sites A through N in the mitochondrial DNA molecules of five species of crabs. The presence of a restriction site is represented by a short vertical line. Assuming that the proportion of shared restriction sites between two molecules is an indication of their closeness of relationship, draw a diagram showing the inferred relationships among the species.

Further Reading

André, C., A. Levy, and V. Walbot. 1992. Small repeated sequences and the structure of plant mitochondrial genomes. *Trends in Genetics* 8: 128.

Cann, R. L., M. Stoneking, and A. C. Wilson. 1987. Mitochondrial DNA and human evolution. *Nature* 325: 31.

Gillham, N. 1978. *Organelle Heredity.* New York: Raven.

Grivell, L. A. 1983. Mitochondrial DNA. *Scientific American,* March.

Jacobs, H. T., and D. M. Lonsdale. 1987. The selfish organelle. *Trends in Genetics* 3: 337.

Larsson, N. G., and D. A. Clayton. 1995. Molecular genetic aspects of human mitochondrial disorders. *Annual Review of Genetics* 29: 151.

Laughnan, J. R., and S. Gabay-Laughnan. 1983. Cytoplasmic male sterility in maize. *Annual Review of Genetics* 17: 27.

Palmer, J. D. 1997. The mitochondrion that time forgot. *Nature* 387: 454.

Preer, J. R., Jr. 1971. Extrachromosomal inheritance: Hereditary symbionts, mitochondria, chloroplasts. *Annual Review of Genetics* 5: 361.

Preer, J. R., Jr. 1997. Whatever happened to *Paramecium* genetics? *Genetics* 145: 217.

Rochaix, J. D. 1995. *Chlamydomonas reinhardtii* as the photosynthetic yeast. *Annual Review of Genetics* 29: 209.

Stoneking, M., and H. Soodyall. 1996. Human evolution and the mitochondrial genome. *Current Opinion in Genetics & Development* 6: 731.

Sturtevant, A. H. 1923. Inheritance of the direction of coiling in *Limnaea. Science* 58: 269.

Tattersall, I. 1997. Out of Africa again . . . and again. *Scientific American,* April.

Thorne, A. G., and M. H. Wolpoff. 1992. The multiregional evolution of humans. *Scientific American,* April.

Vilà, C., P. Savolainen, J. E. Maldonado, I. R. Amorim, J. E. Rice, R. L. Honeycutt, K. A. Crandall, J. Lundeberg, and R. K. Wayne. 1997. Multiple and ancient origins of the domestic dog. *Science* 276, 1687.

Wallace, D. C. 1993. Mitochondrial diseases: Genotype versus phenotype. *Trends in Genetics* 9: 128.

Wilson, A. C., and R. L. Cann. 1992. The recent African genesis of humans. *Scientific American,* April.

This spectacular white baby emperor penguin is a mutant discovered by wildlife researchers in snow-covered sea ice in the western Ross Sea of Antarctica. Other than its black eyes, the bird is completely white, including its bill and feet. [Courtesy of Gerald Kooyman.]

Population Genetics and Evolution

PRINCIPLES

- Many genes in natural populations are polymorphic—they have two or more common alleles. Genetic polymorphisms can be used as genetic markers in pedigree studies and for individual identification (DNA typing).
- The Hardy-Weinberg principle asserts that, with random mating, the frequencies of the genotypes AA, Aa, and aa in a population are expected to be p^2, $2pq$, and q^2, respectively, where p and q are the allele frequencies of A and a. One implication of this principle is that the great majority of rare, harmful recessive alleles are present in heterozygous carriers.
- Inbreeding means mating between relatives. Through inbreeding, replicas of an allele present in a common ancestor may be passed down both sides of a pedigree and come together in an inbred organism. Such alleles are said to be identical by descent. Because inbreeding can result in alleles that are identical by descent, inbreeding results in an excess of homozygous genotypes relative to the frequencies of genotypes expected with random mating.
- Mutation and migration both introduce new alleles into populations. Because mutation rates are typically small, mutation is a weak force for changing allele frequencies.
- Natural selection and random genetic drift are the usual causes of change in allele frequency—the former in a systematic direction, the latter randomly. Selection can result in a stable genetic polymorphism if the heterozygous genotype has the highest fitness. Selection against recurrent harmful mutations leads to a mutation-selection balance in which the allele frequency of the harmful recessive remains constant generation after generation.

CONNECTIONS

CONNECTION: **A Yule Message from Dr. Hardy**
Godfrey H. Hardy 1908
Mendelian proportions in a mixed population

CONNECTION: **Be Ye Son or Nephew?**
Alec J. Jeffreys, John F. Y. Brookfield, and Robert Semeonoff 1985
Positive identification of an immigration test-case using human DNA fingerprints

In the genetic analyses we have examined so far, matings have been deliberately designed by geneticists. In most organisms, matings are not under the control of the investigator, and often the familial relationships among individuals are unknown. This situation is typical of studies of organisms in their natural habitat. Most organisms do not live in discrete family groups but exist as part of populations of individuals with unknown genealogical relationships. At first, it might seem that classical genetics with its simple Mendelian ratios could have little to say about such complex situations, but this is not the case. The principles of Mendelian genetics can be used both to interpret data collected in natural populations and to make predictions about the genetic composition of populations. Application of genetic principles to entire populations of organisms constitutes the subject of **population genetics.**

15.1
Allele Frequencies and Genotype Frequencies

The term **population** refers to a group of organisms of the same species living within a prescribed geographical area. Sometimes the area is large, as when we refer, say, to the population of sparrows in North America; it may even include the entire earth, as in reference to the "human population." More commonly, the area is considered to be of a size within which individuals in the population are likely to find mates. Many geographically widespread species are subdivided into more or less distinct breeding groups, called **subpopulations,** that live within limited geographical areas. Each subpopulation is a **local population.** The complete set of genetic information contained within the individuals in a population is called the **gene pool.** The gene pool includes all alleles present in the population. This section begins with an analysis of a local population with respect to a phenotype determined by two alleles.

Allele Frequency Calculations

The genetic composition of a population can often be described in terms of the frequencies, or relative abundances, in which alternative alleles are found. The concepts will be illustrated by using the human MN blood groups because this genetic system is exceptionally simple. There are three possible phenotypes—M, MN, and N—corresponding to the combinations of M and N antigens that can be present on the surface of red blood cells. These antigens are unrelated to ABO and other red-cell antigens. The M, MN, and N phenotypes correspond to three genotypes of one gene: *MM, MN,* and *NN,* respectively. In a study of a British population, a sample of 1000 people yielded 298 M, 489 MN, and 213 N phenotypes. From the one-to-one correspondence between genotype and phenotype in this system, the genotypes can be directly inferred to be

$$298\ MM \qquad 489\ MN \qquad 213\ NN$$

These numbers contain a surprising amount of information about the population—for example, whether there may be mating between relatives. One of the goals of population genetics is to be able to interpret this information. First, note that the sample contains two types of data: (1) the number of each of the three genotypes, and (2) the number of individual M and N alleles. Furthermore, the 1000 genotypes represent 2000 alleles of the gene because each genome is diploid. These alleles break down in the following way:

298 *MM* persons = 596 *M* alleles		
489 *MN* persons = 489 *M* alleles	+	489 *N* alleles
213 *NN* persons =		426 *N* alleles
Totals = 1085 *M* alleles	+	915 *N* alleles

Usually, it is more convenient to analyze the data in terms of relative frequency than in terms of the observed numbers. For genotypes, the **genotype frequency** in a population is the proportion of organisms that have the particular genotype. For each allele, the **allele frequency** is the proportion of all alleles that are of the specified type. In the *MN* example, the genotype frequencies are obtained by dividing the observed numbers by the total sample size—in this case, 1000. Therefore, the genotype frequencies are

$$0.298 \; MM \qquad 0.489 \; MN \qquad 0.213 \; NN$$

Similarly, the allele frequencies are obtained by dividing the observed number of each allele by the total number of alleles (in this case, 2000), so

$$\text{Allele frequency of } M = 1085/2000$$
$$= 0.5425$$

$$\text{Allele frequency of } N = 915/2000$$
$$= 0.4575$$

Note that the genotype frequencies add up to 1.0, as do the allele frequencies; this is a consequence of their definition in terms of proportions, which must add up to 1.0 when all of the possibilities are taken into account. *Allele and genotype frequencies must always be between 0 and 1.* A population with an allele frequency of 1.0 for some allele is said to be **fixed** for that allele. If an allele frequency becomes 0, then the allele is said to be **lost.**

Allele frequencies can be used to make inferences about matings in a population and to predict the genetic composition of future generations. Allele frequencies are often more useful than genotype frequencies because genes, not genotypes, form the bridge between generations. Genes rarely undergo mutation in a single generation, so they are relatively stable in their transmission from one generation to the next. Genotypes are not permanent; they are disrupted in the processes of segregation and recombination that take place in each reproductive cycle. We know from simple Mendelian considerations what types of gametes must be produced from the *MM*, *MN*, and *NN* genotypes:

Consequently, the *M*-bearing gametes in the population consist of all the gametes from *MM* individuals and half the gametes from *MN* individuals. Likewise, the *N*-bearing gametes consist of all the gametes from *NN* individuals and half the gametes from *MN* individuals. Therefore, in terms of gametes, the population produces the following frequencies:

0.298 *MM* ⟶ 0.298 *M* ⎱ 0.5425 *M* gametes
0.489 *MM* ⟵ 0.2445 *M* ⎰
⟶ 0.2445 *N* ⎱ 0.4575 *N* gametes
0.213 *NN* ⟶ 0.213 *N* ⎰

Note that the allele frequencies among *gametes* equal the allele frequencies among *adults* calculated earlier, which must be true whenever each adult in the population produces the same number of functional gametes.

Enzyme Polymorphisms

An important method for studying genes in natural populations is **electrophoresis,** or the separation of charged molecules in an electric field (Section 5.7). This method is commonly used to study genetic variation in proteins coded by alternative alleles and to study the relative sizes of homologous DNA fragments produced by a restriction enzyme.

In protein electrophoresis, protein-containing tissue samples are placed near the edge of a gel and a voltage is applied for several hours. All charged molecules in the sample move in response to the voltage, and a variety of techniques can be used to locate particular substances. For example, an enzyme can be located by staining the gel with a reagent that is converted into a colored product by the enzyme; wherever the enzyme is located, a colored band appears.

Figure 15.1, a photograph of a gel stained to reveal the enzyme phosphoglucose isomerase in tissue samples from 16 mice, illustrates typical raw data from an electrophoretic study. The pattern of bands varies from one mouse to the next, but only three patterns are observed. Samples

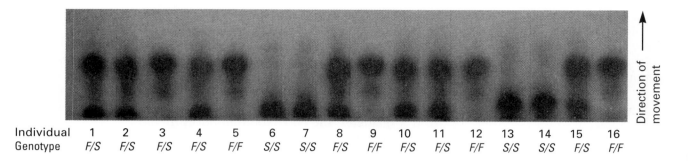

Individual	1	2	3	4	5	6	7	8	9	10	11	12	13	14	15	16
Genotype	F/S	F/S	F/S	F/S	F/F	S/S	S/S	F/S	F/F	F/S	F/S	F/F	S/S	S/S	F/S	F/F

Direction of movement

Figure 15.1

Electrophoretic mobility of glucose phosphate isomerase in a sample of 16 mice. [Courtesy of S. E. Lewis and F. M. Johnson.]

from mice 1, 2, 4, 8, 10, 11, and 15 have two bands: one that moves fast (upper band) and one that moves slowly (lower band). A second pattern is seen for the samples from mice 3, 5, 9, 12, and 16, in which only the fast band appears. The samples from mice 6, 7, 13, and 14 show a third pattern, in which only the slow band appears. Electrophoretic analysis of many enzymes in a wide variety of organisms has shown that mobility variation almost always has a simple genetic basis: Each electrophoretic form of the enzyme contains one or more polypeptides with a genetically determined *amino acid replacement* that changes the electrophoretic mobility of the enzyme. Alternative forms of an enzyme coded by alleles of a single gene are known as **allozymes.** Alleles coding for allozymes are usually codominant, which means that heterozygotes express the allozyme corresponding to each allele. Therefore, in Figure 15.1, mice with only the fast allozyme are *FF* homozygotes, those with only the slow allozyme are *SS* homozygotes, and those with both fast and slow allozymes are *FS* heterozygotes. The 16 genotypes in Figure 15.1 are consequently

$$5 \; FF \qquad 7 \; FS \qquad 4 \; SS$$

and the allele frequency of *F* in this small sample is

$$\text{Allele frequency of } F = [(2 \times 5) + 7]/(2 \times 16)$$
$$= 0.53$$

Because only two alleles are represented in Figure 15.1, the allele frequency of *S* must equal $1 - 0.53 = 0.47$, which can also be calculated directly as

$$\text{Allele frequency of } S = [7 + (2 \times 4)]/(2 \times 16)$$
$$= 0.47$$

Many genes in natural populations are **polymorphic,** which means that they have two or more relatively frequent alleles. Among plants and vertebrate animals, approximately 20 percent of enzyme genes are polymorphic (Table 15.1). The implication of this finding is that genetic variation among organisms is very common in natural populations of most organisms. Furthermore, among plants and vertebrate animals, the average proportion of genes that are heterozygous within an organism is about 0.05, which again emphasizes the prevalence of alternative alleles in populations. Invertebrate animals such as *Drosophila* have even more genetic variation.

DNA Polymorphisms

Polymorphisms can also be detected directly in DNA molecules. One method uses a *probe* DNA that is generally obtained from plasmid or phage vectors into which DNA fragments from the organism of interest have been cloned. The probe DNA is complementary in sequence to a segment of DNA in the organism of interest, but its presence in clones allows the probe to be

Table 15.1 Detection of genetic variation through use of protein electrophoresis

	Number of species examined	Proportion of protein-coding loci that are polymorphic	Proportion of heterozygous loci (average) ± standard deviation
Plants	15	0.26 ± 0.17	0.07 ± 0.07
Invertebrates	93	0.40 ± 0.20	0.11 ± 0.07
Vertebrates	135	0.17 ± 0.12	0.05 ± 0.04

Source: From Nevo, E. 1978. "Genetic variation in natural populations: Patterns and theory." *Theor. Pop. Biol.* 43: 121–177

obtained in large quantities and pure form. In the *Southern blot* procedure, in which genomic DNA is digested with a restriction enzyme and fragments of different sizes separated by electrophoresis, radioactive probe DNA hybridizes only with DNA fragments that contain complementary sequences and so identifies these restriction fragments (Section 5.7).

When this procedure is used to study natural populations, it is common to find restriction fragments complementary to a probe that differ in size among individuals. These differences result from differences in the locations of restriction sites along the DNA. Each region of the genome that hybridizes with the probe generates a restriction fragment. The restriction fragments that derive from corresponding positions in homologous chromosomes are analogous to alleles in that they segregate in meiosis: The offspring of a heterozygote may inherit a chromosome with either fragment size, but not both. Moreover, allelic restriction fragments of different sizes are codominant because both fragments are detected in heterozygotes; this is the best kind of situation for genetic studies, because genotypes can be inferred directly from phenotypes.

One type of variation in restriction fragment length is illustrated in Figure 15.2A. A single nucleotide change knocks out a restriction site, so the restriction fragment that is complementary to the probe stretches farther downstream to the next restriction site. Therefore, hybridization with probe DNA will reveal a shorter restriction fragment produced from allele 1 than from allele 2. As with any other type

of allelic variation, an organism may be homozygous for allele 1, homozygous for allele 2, or heterozygous. Restriction fragments of different sizes produced from the alleles of a gene constitute a *restriction fragment length polymorphism,* or *RFLP,* of the type discussed in Section 4.4. RFLPs are extremely common in the human genome and in the genomes of most other organisms; hence they provide an extensive set of highly polymorphic, codominant genetic markers scattered throughout the genome, which can be studied with Southern blotting procedures identical except for the use of different restriction enzymes and different types of probe DNA. Figure 15.2B illustrates the segregation of alleles of a restriction fragment polymorphism within a human pedigree.

A second type of variation in DNA consists of a short sequence of nucleotides repeated in tandem a number of times, such as 5'-TGTGTGTGTGTG-3'. The number of repeats may differ from one chromosome to the next, so the number of repeats can be used to identify different chromosomes. A polymorphism of this type is called a *simple tandem repeat polymorphism,* or *STRP* (Section 4.4). Although there are many places in the genome where an STRP of TGs (or any other short repeat) may occur, each place is flanked by sequences that are unique in the genome. By means of oligonucleotide primers specific for the unique flanking sequences, any one of the STRPs in the genome can be amplified in the polymerase chain reaction, and the number of repeats in any chromosome can be determined by examining the length of the fragment that was amplified.

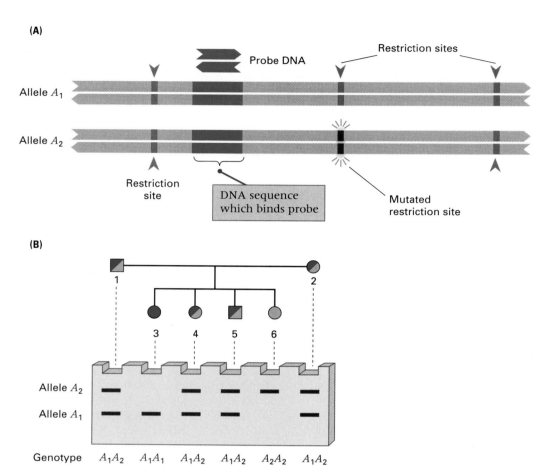

(A)

Probe DNA

Restriction sites

Allele A_1

Allele A_2

Restriction site

DNA sequence which binds probe

Mutated restriction site

(B)

1   2

3   4   5   6

Allele A_2

Allele A_1

Genotype A_1A_2 A_1A_1 A_1A_2 A_1A_2 A_2A_2 A_1A_2

Figure 15.2

Restriction fragment size variation in DNA. (A) Nucleotide substitution in a DNA molecule (highlighted in A_2) eliminates a restriction site. The result is that in the region detected by the probe DNA, alleles 1 and 2 differ in the size of a restriction fragment. (B) Segregation of a restriction fragment length polymorphism like that in part A. In the pedigree, both parents (individuals 1 and 2) are heterozygous A_1A_2. Expected progeny resulting from segregation are A_1A_1, A_1A_2, and A_2A_2 in the proportions 1 : 2 : 1. The diagram of the electrophoresis gel shows the expected patterns of bands in a Southern blot. Homozygous individuals numbered 3 (A_1A_1) and 6 (A_2A_2) yield a single restriction fragment that hybridizes with the probe DNA. The heterozygous individuals show both bands.

15.2 Systems of Mating

The transmission of genetic material from one generation to the next is analyzed in terms of alleles rather than genotypes because genotypes are broken up in each generation by the process of segregation and recombination. When gametes unite in the process of fertilization, the genotypes of the next generation are formed. The genotype frequencies in the zygotes of the prog-

eny generation are determined by the frequencies with which the various types of parental gametes come together, and these frequencies are, in turn, determined by the **mating system** in which the genotypes pair as mates. Three types of mating systems predominate in natural populations: random mating, assortative mating, and inbreeding.

The mating system called **random mating** means that mating pairs are formed without regard to genotype. With random mating, a mating between any two genotypes takes place in proportion to the relative frequencies of the genotypes in the population. A population can undergo random mating with respect to some traits but, at the same time, nonrandom mating with respect to other traits. In a type of mating system called **assortative mating,** mating pairs come together on the basis of their degree of similarity in phenotype. In **positive assortative mating,** organisms tend to mate with others that are similar in phenotype. Positive assortative mating takes place for many traits in human populations. In the United States, these traits include skin color and height, traits for which mates tend to resemble each other more than would be expected by chance. In **negative assortative mating,** the mating pairs are more dissimilar in phenotype than would be expected by chance. Negative assortative mating is not so common as positive assortative mating, but a famous example is found in some species of primroses (*Primula*) and their relatives. In most populations of these species, two types of flowers are found in approximately equal proportions (Figure 15.3). One type, known as *pin*, has a long style and short stamens. The anthers are the organs that produce pollen and are located atop the stamens; the stigma is the organ that receives pollen and is located atop the style. Pin flowers produce pollen low in the flowers and receive pollen high in the flowers. In the other type of plant, known as *thrum*, the relative positions of stigma and anthers are reversed. Consequently, insect pollinators that work low in the flowers pick up mainly pin pollen and deposit it mainly on thrum stigmas, whereas pollinators that work high in the flowers pick mainly thrum pollen and deposit it mainly on pin stigmas. This behavior promotes negative assortative mating, which is further enhanced because pin pollen grows better in thrum stigmas, and vice versa. The genetic basis of the flower type resides in a single genetic region: pin is homozygous *ss,* and thrum is heterozygous *Ss.*

Figure 15.3

Pin and thrum forms of flowers of *Primula officalis.* Negative assortative mating is promoted by the anatomy of these flowers. Pollinators that work low in the flowers tend to pick up pin pollen and deposit it on thrum stigmas, and pollinators that work high in the flowers tend to do the reverse.

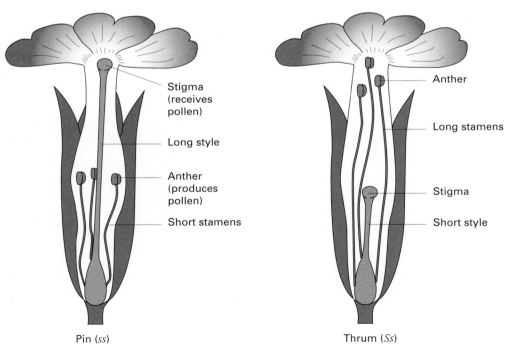

Pin (*ss*)

Stigma (receives pollen)

Long style

Anther (produces pollen)

Short stamens

Thrum (*Ss*)

Anther

Long stamens

Stigma

Short style

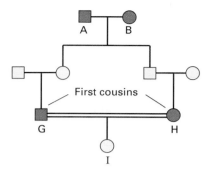

Figure 15.4

An inbreeding pedigree in which individual I is the result of a mating between first cousins.

quency is particularly simple because *the random mating of individuals is equivalent to the random union of gametes.* Conceptually, we may imagine all the gametes of a population as present in a large container. To form zygote genotypes, pairs of gametes are withdrawn from the container at random. To be specific, consider the alleles M and N in the MN blood groups, whose allele frequencies are p and q, respectively (remember that $p + q = 1$). The genotype frequencies expected with random mating can be deduced from the following diagram:

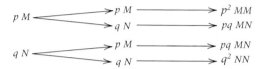

In this diagram, the gametes at the left represent the sperm and those in the middle the eggs. The genotypes that can be formed with two alleles are shown at the right, and with random mating, the frequency of each genotype is calculated by multiplying the allele frequencies of the corresponding gametes. However, the genotype MN can be formed in two ways—the M allele could have come from the father (top part of diagram) or from the mother (bottom part of diagram). In each case, the frequency of the MN genotype is pq; considering both possibilities, we find that the frequency of MN is $pq + pq = 2pq$. Consequently, the overall genotype frequencies expected with random mating are

$$MM: p^2 \qquad MN: 2pq \qquad NN: q^2 \qquad (1)$$

The frequencies p^2, $2pq$, and q^2 result from random mating for a gene with two alleles; they constitute what is called the **Hardy-Weinberg principle** (after Godfrey Hardy and Wilhelm Weinberg, each of whom derived it independently of the other in 1908). Sometimes the Hardy-Weinberg principle is demonstrated by a Punnett square, as illustrated in Figure 15.5. Such a square is completely equivalent to the tree diagram used above. Although the Hardy-Weinberg principle is exceedingly simple, it has a number of important implications that are not obvious. These are described in the following sections.

A third important type of mating system is **inbreeding,** mating between relatives. In human pedigrees, a mating between relatives is often called a **consanguineous mating.** Figure 15.4 represents an example of inbreeding. In this case, the female I is the offspring of a first cousin mating (G with H). The closed loop in the pedigree is diagnostic of inbreeding, and the individuals designated A and B are called *common ancestors* of I, because they are ancestors of both of I's parents. Because A and B are common ancestors, a particular allele present in A (or in B) could, by chance, be transmitted in inheritance down both sides of the pedigree, to meet again in the formation of I. This possibility, which is the most important characteristic and consequence of inbreeding, will be discussed later in this chapter.

Random Mating and the Hardy-Weinberg Principle

In random mating, organisms form mating pairs independently of genotype; each type of mating pair is formed as often as would be expected by chance encounters. Random mating is by far the most prevalent mating system for most species of animals and plants, except for plants that regularly reproduce through self-fertilization.

With random mating, the relationship between allele frequency and genotype fre-

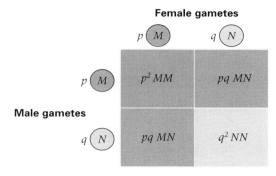

Female gametes

p (M) q (N)

Male gametes

p (M) | p^2 MM | pq MN
q (N) | pq MN | q^2 NN

Figure 15.5

A cross-multiplication square (Punnett square) showing the result of the random union of male and female gametes, which is equivalent to the random mating of individuals.

Implications of the Hardy-Weinberg Principle

One important implication of the Hardy-Weinberg principle is that *the allele frequencies remain constant from generation to generation*. Consider a gene with two alleles, *A* and *a*, that have frequencies p and q, respectively ($p + q = 1$). With random mating, the frequencies of genotypes *AA*, *Aa*, and *aa* among zygotes are p^2, $2pq$, and q^2, respectively. Assuming equal *viability* (ability to survive) among the genotypes, these frequencies equal those among adults. If all of the genotypes are equally fertile, then the frequency, p', of allele *A* among gametes of the next generation will be

$$p' = p^2 + 2pq/2 = p(p + q) = p$$

(because $p + q = 1$). This argument shows that the frequency of allele *A* remains constant at the value of p through the passage of one (or any number of) complete generations. This principle depends on certain assumptions, of which the most important are

1. Mating is random; there are no subpopulations that differ in allele frequency.

2. Allele frequencies are the same in males and females.

3. All the genotypes are equal in viability and fertility; *selection* does not operate.

4. Mutation does not occur.

5. Migration into the population is absent.

6. The population is sufficiently large that the frequencies of alleles do not change from generation to generation because of chance.

A Test for Random Mating

As a practical application of the Hardy-Weinberg principle, consider again the MN blood groups among British people, which we discussed in Section 15.1. The frequencies of alleles *M* and *N* among 1000 adults were 0.5425 and 0.4575, respectively; assuming random mating, the *expected* genotype frequencies can be calculated from Equation (1) as

MM: $(0.5425)^2 = 0.2943$
MN: $2(0.5425)(0.4575) = 0.4964$
NN: $(0.4575)^2 = 0.2093$

Because the total number of people in the sample is 1000, the expected numbers of these genotypes are 294.3 MM, 496.4 MN, and 209.3 NN. The *observed* numbers are 298 MM, 489 MN, and 213 NN. Goodness of fit between observed numbers and expected numbers would normally be determined by means of the χ^2 test described in Chapter 3. We will not apply the test here because the agreement between expected and observed numbers is quite good. The χ^2 test yields a probability of about 0.67, which means that the hypothesis of random mating can account for the data. On the other hand, the χ^2 test detects only deviations that are rather large, so a good fit to Hardy-Weinberg frequencies should not be overinterpreted, because:

It is entirely possible for one or more assumptions of the Hardy-Weinberg principle to be violated, including the assumption of random mating, and still not produce deviations from the expected genotype frequencies that are large enough to be detected by the χ^2 test.

The use of a χ^2 test to determine departures from random mating may seem somewhat circular inasmuch as the data

A Yule Message from Dr. Hardy

Godfrey H. Hardy 1908
Trinity College, Cambridge, England
Mendelian Proportions in a Mixed Population

An early argument against Mendelian inheritance was that dominant traits, even harmful ones, should eventually come to have a 3 : 1 ratio in any population. This is obviously not the case. There are many dominant traits in human beings that are rare now and have always been rare, such as brachydactyly (short fingers), which is the specific issue in this paper. Hardy, a mathematician, realized that with random mating, and assuming no differences in survival or fertility, the frequencies of dominant and recessive genotypes would have no innate tendency to change of their own accord. The form of the principle as written by Hardy is that if the genotype frequencies of AA, Aa, and aa are P, 2Q, and R in any generation, then with random mating, the frequencies will be $(P + Q)^2 : 2(P + Q)(Q + R) : (Q + R)^2$ in the next and subsequent generations. The allele frequencies p and q of A and a are not stated explicitly, so the familiar form of the principle as stated today—$p^2 : 2pq : q^2$—does not appear as such in Hardy's paper. But it is there implicitly, because $p = P + Q$ and $q = Q + R$.

I am reluctant to intrude in a discussion concerning matters of which I have no expert knowledge, and I should have expected the very simple point which I wish to make to have been familiar to biologists. However, some remarks of Mr. Udny Yule suggest that it may still be worth making. . . . Mr. Yule is reported to have suggested, as a criticism of the Mendelian position, that if brachydactyly is dominant "in the course of time one would expect, in the absence of counteracting forces, to get

---

There is not the slightest foundation for the idea that a dominant character should show a tendency to spread, or that a recessive should tend to die out.

---

three brachydactylous persons to one normal." . . . It is not difficult to prove, however, that such an expectation would be quite groundless. Suppose that A, a is a pair of Mendelian characters, A being dominant, and that in any given generation the numbers of pure dominants (AA), heterozygotes (Aa), and pure recessives (aa) are as P : 2Q : R. Finally, suppose that the numbers are fairly large, so that the mating may be

regarded as random, that the sexes are evenly distributed among the three varieties, and that they are all equally fertile. A little mathematics of the multiplication-table type is enough to show that in the next generation the numbers will be as

$$(P + Q)^2 : 2(P + Q)(Q + R) : (Q + R)^2$$

or as $P_1 : 2Q_1 : R_1$, say. The interesting question is—in what circumstances will this distribution be the same as that in the generation before? It is easy to see that the condition for this is $Q^2 = PR$. And since $Q_1^2 = P_1R_1$, whatever the values of P, Q, and R may be, the distribution will in any case continue unchanged after the second generation. . . . In a word, there is not the slightest foundation for the idea that a dominant character should show a tendency to spread, or that a recessive should tend to die out. . . . I have, of course, considered only the very simplest hypothesis possible. Hypotheses other than that of purely random mating will give different results, and, of course, if, as appears to be the case sometimes, the character is not independent of sex, or has an influence on fertility, the whole question may be greatly complicated. But such complications seem to be irrelevant to the simple issue raised by Mr. Yule's remarks.

Source: Science 28: 49–50

---

themselves are used in calculating the expected genotype frequencies. However, the use of the data is exactly compensated for by the simple expedient of deducting one degree of freedom for every parameter estimated from the data. In the MN case, there would ordinarily be two degrees of freedom, because there are three classes of data (the numbers of M, MN, and N phenotypes). The allele frequency p was estimated from the data, so we deduct one degree of freedom, which leaves one degree of freedom in the final χ^2 test. (No deduction is made for q, because $q = 1 - p$.)

To verify that the χ^2 test as employed is not circular, consider the following observed numbers of genotypes in the flower *Phlox cuspidata* for a gene coding for phosphoglucomutase: *Pgm^a Pgm^a*, 15; *Pgm^a Pgm^b*, 6; and *Pgm^b Pgm^b*, 14. You should verify for yourself that the allele frequencies of *Pgm^a* and *Pgm^b* are 0.51 and 0.49, respectively, and that the expected numbers of the genotypes with random mating are 9.1, 17.5, and 8.4, respectively. The χ^2 value is 15.1, which has one degree of freedom because there are three classes of data (the observed number of each of the three genotypes) and one parameter (the allele frequency of either Pgm^a or Pgm^b) had to be estimated from the data. The corresponding probability is about 0.0002. The probability associated with the χ^2 value is less than 0.05 (in this case, very much less), so these data must be judged significantly different from the Hardy-Weinberg expectations. The reason, as we will soon see, is inbreeding: In this plant, some of the ovules are fertilized by self-fertilization.

Frequency of Heterozygotes

Another important implication of the Hardy-Weinberg principle is that *for a rare allele, the frequency of heterozygotes far exceeds the frequency of the rare homozygote.* For example, when the frequency of the rarer allele is $q = 0.1$, the ratio of heterozygotes to homozygotes equals

$$2pq/q^2 = 2p/q = 2(0.9)/(0.1)$$

or approximately 20; when $q = 0.01$, this ratio is about 200; and when $q = 0.001$, it is about 2000. In other words,

> When an allele is rare, there are many more heterozygotes than there are homozygotes for the rare allele.

The reason for this perhaps unexpected relationship is shown in Figure 15.6, which plots the frequencies of homozygotes and heterozygotes with random mating. Note that at allele frequencies near 0 or 1, the frequency of the heterozygous genotype goes to 0 as a linear function, whereas that of the rarer, homozygous genotype goes to 0 more rapidly.

One practical implication of this principle is seen in the example of cystic fibrosis, an inherited secretory disorder of the pancreas and lungs. Cystic fibrosis is one of the

Figure 15.6

Graphs of p^2, $2pq$, and q^2. If the allele frequencies are between 1/3 and 2/3, then the heterozygote is the genotype with the highest frequency.

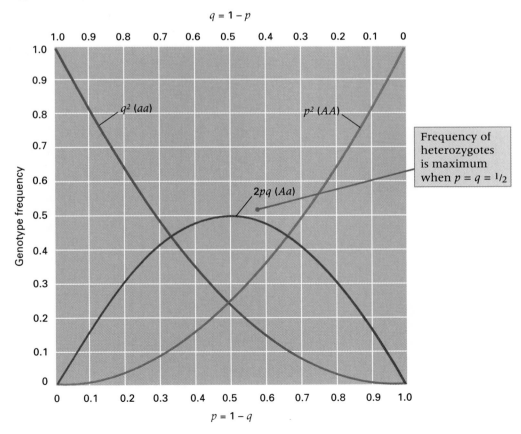

most common recessively inherited severe disorders among Caucasians; it affects about 1 in 1700 newborns. In this case, the heterozygotes cannot readily be identified by phenotype, so a method of calculating allele frequencies that is different from the gene-counting method used earlier must be employed. This method is straightforward because, with random mating, the frequency of recessive homozygotes must correspond to q^2. In the case of cystic fibrosis,

$$q^2 = 1/1700 = 0.00059 \text{ or}$$
$$q = (0.00059)^{1/2} = 0.024$$

and consequently,

$$p = 1 - q = 1 - 0.024 = 0.976$$

The frequency of heterozygotes that carry the allele for cystic fibrosis is calculated as

$$2pq = 2(0.976)(0.024) = 0.047 = 1/21$$

This calculation implies that for cystic fibrosis, although only 1 person in 1700 is affected with the disease (homozygous), about 1 person in 21 is a carrier (heterozygous). The calculation should be regarded as approximate because it is based on the assumption of Hardy-Weinberg genotype frequencies. Nevertheless, considerations like these are important in predicting the outcome of population screening for the

detection of carriers of harmful recessive alleles, which is essential in evaluating the potential benefits of such screening programs.

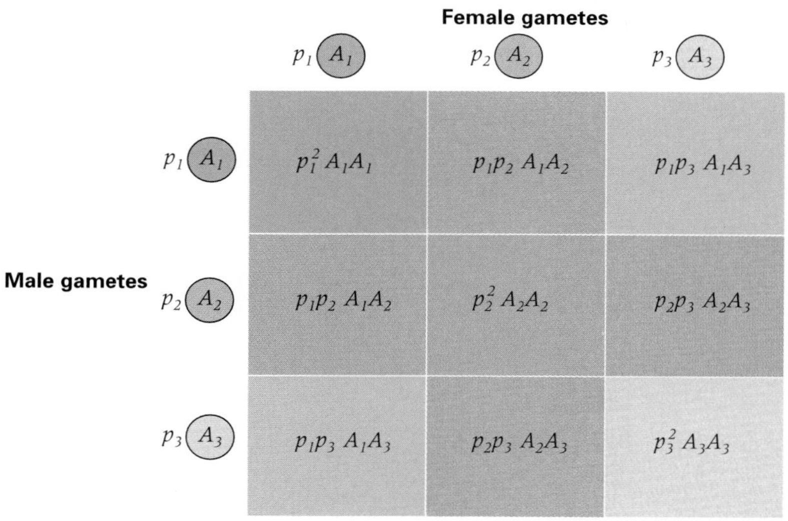

Multiple Alleles

Extension of the Hardy-Weinberg principle to multiple alleles of a single autosomal gene can be illustrated by a three-allele case. Figure 15.7 shows the results of random mating in which three alleles are considered. The alleles are designated A_1, A_2, and A_3, where the uppercase letter represents the gene and the subscript designates the particular allele. The allele frequencies are p_1, p_2, and p_3, respectively. With three alleles (as with any number of alleles), the allele frequencies of all alleles must sum to 1; in this case, $p_1 + p_2 + p_3 = 1.0$. As in Figure 15.5, the entry in each square is obtained by multiplying the frequencies of the alleles at the corresponding margins; any homozygote (such as A_1A_1) has a random-mating frequency equal to the square of the corresponding allele frequency (in this case, p_1^2). Any heterozygote (such as A_1A_2) has a random-mating frequency equal to twice the product of the corresponding allele frequencies (in this case, $2p_1p_2$). The extension to any number of alleles is straightforward:

Frequency of any homozygote =
    square of allele frequency

Frequency of any heterozygote =
    2 × product of allele frequencies   (2)

Multiple alleles determine the human ABO blood groups (Chapter 2). The gene has three principal alleles, designated I^A, I^B, and I^O. In one study of 3977 Swiss people, the allele frequencies were found to be 0.27 I^A, 0.06 I^B, and 0.67 I^O. Applying the rules for multiple alleles, we can expect the genotype frequencies resulting from random mating to be

Figure 15.7
Punnett square showing the results of random mating with three alleles.

$$
\begin{aligned}
I^A I^A &= (0.27)^2 = 0.0729 \\
I^A I^O &= 2(0.27)(0.67) = 0.3618
\end{aligned} \quad \text{Type A} = 0.4347
$$

$$
\begin{aligned}
I^B I^B &= (0.06)^2 = 0.0036 \\
I^B I^O &= 2(0.06)(0.67) = 0.0804
\end{aligned} \quad \text{Type B} = 0.0840
$$

$$
\begin{aligned}
I^O I^O &= (0.67)^2 = 0.4489 \quad \text{Type O} = 0.4489 \\
I^A I^B &= 2(0.27)(0.06) = 0.0324 \quad \text{Type AB} = 0.0324
\end{aligned}
$$

(A)

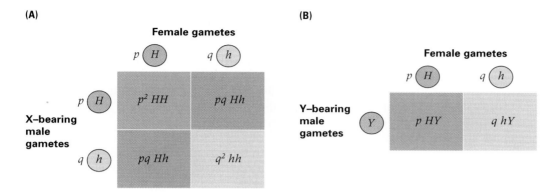

(B)

Figure 15.8
The results of random mating for an X-linked gene. (A) Genotype frequencies in females. (B) Genotype frequencies in males.

Because both I^A and I^B are dominant to I^O, the expected frequency of blood-group *phenotypes* is that shown at the right. Note that the majority of A and B phenotypes are heterozygous for the I^O allele.

X-Linked Genes

Random mating for two X-linked alleles (*H* and *h*) is illustrated in Figure 15.8. The principles are the same as those considered earlier, but male gametes carrying the X chromosome (Figure 15.8A) must be distinguished from those carrying the Y chromosome (Figure 15.8B). When the male gamete carries an X chromosome, the Punnett square is exactly the same as that for the two-allele autosomal gene in Figure 15.5. However, because the male gamete carries an X chromosome, all the offspring in question are female. Consequently, among females, the genotype frequencies are

$$HH: p^2 \qquad Hh: 2pq \qquad hh: q^2$$

When the male gamete carries a Y chromosome, the outcome is quite different (Figure 15.8B). All the offspring are male, and each has only one X chromosome, which is inherited from the mother. Therefore, each male receives only one copy of each X-linked gene, and the genotype frequencies among males are the same as the allele frequencies: *H* males with frequency *p*, and *h* males with frequency *q*.

An important implication of Figure 15.8 is that if *h* is a rare recessive allele, then many more males will exhibit the trait than females because the frequency of affected females (q^2) will be much smaller than the frequency of affected males (*q*). For example, the common form of X-linked color blindness in human beings affects about 1 in 20 males among Caucasians, so $q = 1/20 = 0.05$. The expected frequency of color-blind females is calculated as $q^2 = (0.05)^2 = 0.0025$, or about 1 in 400.

15.3
DNA Typing and Population Substructure

In principle, no two human beings are genetically identical. The only exceptions are identical twins, identical triplets, and so on. Each human genotype is unique because so many genes in human populations are polymorphic. One practical application of the theoretical principle of genetic uniqueness has come through the study of DNA polymorphisms. Small samples of human material from an unknown person (for example, material left at the scene of a crime) often contain enough DNA that the genotype can be determined for a number of polymorphisms and matched against those present among a group of suspects. Typical examples of crime-scene evidence include blood, semen, hair roots, and skin

cells. Even a small number of cells is sufficient, because predetermined regions of
DNA can be amplified by the polymerase
chain reaction (Chapter 5).

If a suspect's DNA contains sequences
that are clearly not present in the crime-
scene sample, then the sample must have
originated from a different person. On the
other hand, if a suspect's DNA *does* match
that of the crime-scene sample, then the
suspect could be the source. The strength of
the DNA evidence depends on the number
of polymorphisms that are examined and
on the number of alleles present in the
population. The greater the number of
polymorphisms that match, especially if
they are highly polymorphic, the stronger
the evidence linking the suspect to the
sample taken from the scene of the crime.
The use of polymorphisms in DNA to link
suspects with samples of human material is
called **DNA typing.** Although there is
some controversy over certain technical

aspects of the procedures, DNA typing is
generally regarded as the most important
innovation in criminal investigation since
the development of fingerprinting more
than a century ago. One of the controversies involves the application of population
genetics to evaluate the significance of a
matching DNA type.

Figure 15.9 illustrates the type of polymorphism commonly used in DNA typing.
The restriction fragments that correspond
to each allele differ in length because they
contain different numbers of units repeated
in tandem. For this reason, each polymorphism of this type is called a *VNTR,* which
stands for *variable number of tandem repeats*
(Section 4.4). VNTRs are of value in DNA
typing because many alleles are possible,
owing to the variable number of repeating
units. Although many alleles may be present in the population as a whole, each
person can have no more than two alleles
for each VNTR polymorphism. An example

Figure 15.9
Allelic variation resulting from a variable number of units repeated in tandem in a nonessential
region of a gene. The probe DNA detects a restriction fragment for each allele. The length of the
fragment depends on the number of repeating units present.

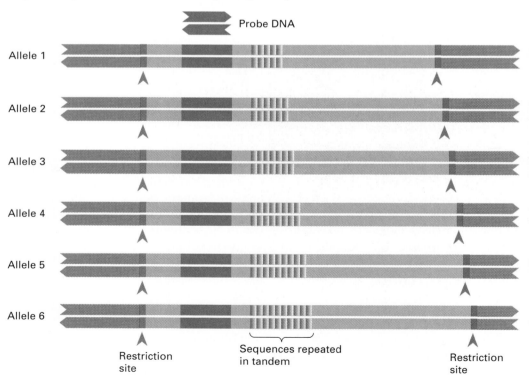

Be Ye Son or Nephew?

Alec J. Jeffreys, John F. Y. Brookfield, and Robert Semeonoff 1985
University of Leicester, Leicester, England
Positive Identification of an Immigration Test-Case Using Human DNA Fingerprints

One of the first practical applications of DNA typing was this immigration case. A boy from Ghana was refused permission to rejoin his mother in the United Kingdom because it was not clear whether he was the woman's son or nephew. One difficulty in the case was that the father was in Ghana and unavailable for DNA typing. As you will see, there was enough genetic information available from the mother's other children to infer the genotype of the father and to reach a conclusion about the parentage of the boy. The term "minisatellites" used in the excerpt refers to types of simple tandem repeat polymorphisms that are abundant in the human genome and highly variable ("hypervariable") from one person to the next. These features make minisatellites suitable for use in DNA typing.

The human genome contains a set of minisatellites, each of which consists of tandem repeats of a DNA segment. . . . [Analysis of] many hypervariable minisatellites simultaneously produces DNA fingerprints . . . suitable for individual identification in, for example, forensic science and paternity testing. . . . They can also be used to resolve immigration disputes arising from lack of proof of family relationships. . . . The case in question concerned a Ghanian boy born in the United Kingdom who emigrated to Ghana to join his father and subsequently returned to the United Kingdom to be reunited with his mother, brother and two sisters. However,

---

At the request of the family's solicitor, we therefore carried out a DNA fingerprint analysis to determine the maternity of the boy

---

there was evidence to suggest that a substitution might have occurred, either for an unrelated boy, or for a son of a sister of the mother; she has several sisters, all of whom live in Ghana. As a result the returning boy was not granted residence in the United Kingdom. . . . At the request of the family's solicitor, we therefore carried out a DNA fingerprint analysis to determine the

maternity of the boy. . . . DNA fingerprints from blood DNA samples were taken from the mother M, brother B, sisters S1 and S2 and the boy X in dispute. . . . Although the father was unavailable, most of his DNA fingerprint could be reconstructed from paternal-specific DNA fragments present in at least one of B, S1 or S2 but absent from M. Of 39 paternal fragments so identified, approximately half were present in the DNA fingerprint of X. Since DNA fragments are seldom shared between unrelated individuals, this suggests very strongly that X has the same father as B, S1 and S2. . . . There remained 40 DNA fragments in X, all of which were present in M. This in turn provides strong evidence that M is the mother of X, and therefore that X, B, S1 and S2 are true sibs. . . . How certain is this identification? . . . The odds that M is a sister of X's true mother but by chance contains all of X's maternal-specific bands are 6×10^{-6}. We therefore conclude that, beyond any reasonable doubt, M must be the true mother of X. This evidence . . . was provided to the immigration authorities, who dropped the case against X and granted him residence in the United Kingdom.

Source: Nature 317:818–819

---

of a VNTR used in DNA typing is shown in Figure 15.10. The lanes in the gel labeled M contain multiple DNA fragments of known size to serve as molecular-weight markers. Each numbered lane 1–9 contains DNA from a single person. Two typical features of VNTRs are to be noted:

1. Most people are heterozygous for VNTR alleles that produce restriction fragments of different sizes. Heterozygosity is indicated by the presence of two distinct bands. In Figure 15.10, only the person numbered 1 appears to be homozygous for a particular allele.

2. The restriction fragments from different people cover a wide range of sizes. The

variability in size indicates that the population as a whole contains many VNTR alleles.

The reason why VNTRs are useful in DNA typing is also evident in Figure 15.10: Each of the nine people tested has a different pattern of bands and thus could be identified uniquely by means of this VNTR. On the other hand, the uniqueness of each person in Figure 15.10 is due in part to the high degree of polymorphism of the VNTR and in part to the small sample size. If more people were examined, then pairs that matched in their VNTR types by chance would certainly be found. Table 15.2 shows the frequencies of alleles for the VNTR designated *D1S80* in major population groups

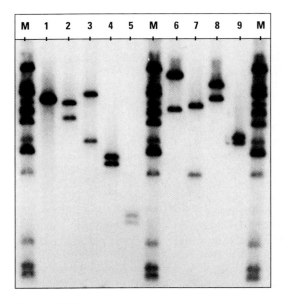

Figure 15.10

Genetic variation in a VNTR used in DNA typing. Each numbered lane contains DNA from a single person; the DNA has been cleaved with a restriction enzyme, separated by electrophoresis, and hybridized with radioactive probe DNA. The lanes labeled M contain molecular-weight markers. [Courtesy of R. W. Allen.]

VNTR and then to multiply the expected frequencies across each VNTR in order to calculate the expected frequency of the multilocus genotype. The factor p^2 for homozygous genotypes is usually modified, because when a single band is observed (as in person 1 in Figure 15.10), it is uncertain whether the genotype is really homozygous or is heterozygous with one DNA band not observed (perhaps because of running off the end of the electrophoresis gel). Thus the factor for single-band patterns is taken as $2p$ rather than the p^2 in Equation (2) because $2p$ is always greater than p^2 for VNTR alleles.

With the assumptions of random mating and independence of VNTRs, if a DNA match included the allele A_i of VNTR A (a single-banded pattern); alleles B_j and B_k of VNTR B; and alleles C_l and C_m of VNTR C, then the expected frequency of the DNA type in the population as a whole would be calculated as

$$2 \times \text{Freq}(A_i) \times 2 \times \text{Freq}(B_j) \times \\ \text{Freq}(B_k) \times 2 \times \text{Freq}(C_l) \times \text{Freq}(C_m) \quad (3)$$

where $\text{Freq}(A_i)$ is the frequency of the allele A_i, $\text{Freq}(B_j)$ is that of allele B_j, and so on. Because human populations can differ in their allele frequencies, the calculation would be carried out using allele frequencies among Caucasians for white suspects, using those among Blacks for black suspects, and using those among Hispanics for Hispanic suspects.

Differences Among Populations

Equation (3) makes a number of assumptions about human populations: (1) that the Hardy-Weinberg principle holds for each locus, (2) that each locus is statistically independent of the others so that the multiplication across loci is justified, and (3) that the only level of population substructure that is important is that of race. The term **population substructure** refers to the extent to which a larger population is subdivided into smaller subpopulations that may differ in allele frequencies from one subpopulation to the next.

Critics of the method of calculation are worried about differences in the frequencies of VNTR alleles among different

in the United States. The allele with 24 repeats has a frequency of 0.378 in Caucasians, so with random mating, more that 14 percent of the population would be homozygous for this allele. The same allele also occurs quite frequently in the other population groups. Because of the possibility of chance matches between VNTR types, applications of DNA typing are usually based on at least three—and preferably more—highly polymorphic loci.

Suppose a person is found whose DNA type for three VNTR loci matches that of a sample from the scene of a crime. How is the significance of this match to be evaluated? The significance of the match depends on the likelihood of its happening by chance, and hence matches of rare DNA types are more significant than matches of common DNA types. If random mating proportions can be assumed, and it is also assumed that each VNTR is independent of the others, then it is reasonable to use Equation (2) to calculate the expected frequency of the particular genotype for each

Table 15.2 Allele frequencies for *D1S80* among U.S. population groups

Repeat number	Caucasian	Hispanic	African American	Asian
14	0	0	0	0
15	0	0.001	0	0
16	0.001	0.010	0.002	0.034
17	0.002	0.009	0.028	0.025
18	0.237	0.224	0.073	0.152
19	0.003	0.005	0.003	0.022
20	0.018	0.013	0.032	0.007
21	0.021	0.028	0.115	0.034
22	0.038	0.024	0.081	0.017
23	0.012	0.009	0.014	0.017
24	0.378	0.315	0.234	0.230
25	0.046	0.072	0.045	0.027
26	0.020	0.007	0.006	0
27	0.007	0.016	0.008	0.047
28	0.063	0.078	0.130	0.076
29	0.052	0.055	0.053	0.042
30	0.008	0.039	0.009	0.123
31	0.072	0.053	0.054	0.093
32	0.006	0.005	0.007	0.012
33	0.003	0.004	0.004	0.005
34	0.001	0.006	0.086	0.005
35	0.003	0	0.002	0.005
36	0.004	0.011	0.001	0.005
37	0.001	0.004	0	0.007
38	0	0	0	0
39	0.003	0.004	0.003	0.005
40	0	0	0	0
41	0	0.002	0.002	0.007
>41	0.001	0.006	0.007	0.002
Sample size	718	409	606	204

Source: Data from B. Budowle, et al. 1995. *Journal of Forensic Science* 40:38

human subpopulations. For example, in Figure 15.11, each bar shows the range in frequency (minimum to maximum) of each allele of *D1S80* observed among approximately 30 subpopulations. For some alleles, the range is very great; for example, alleles with 21, 30, 31, or 34 repeats are very rare in some subpopulations but relatively common in others. Even the common alleles, with 18 and 24 repeats, can vary substantially in frequency from one subpopulation to the next. Critics

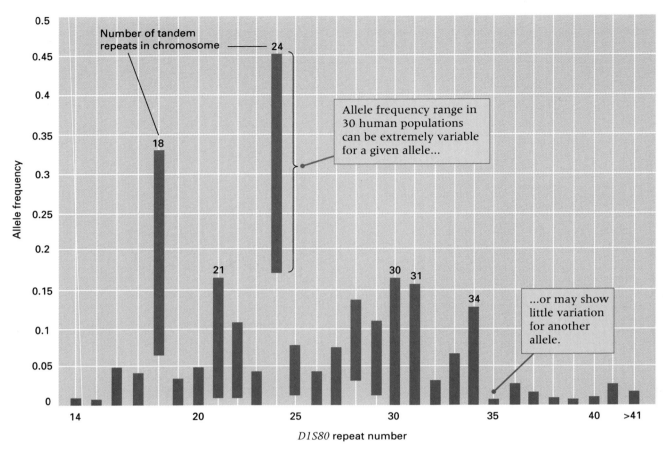

Figure 15.11
Range of allele frequencies found among human subpopulations for the VNTR *D1S80*. [Data from B. Budowle et al. *J. Forensic Science* 1995. 40:38.]

argue that DNA matches across multiple VNTRs may be much more common among persons within a particular ethnic group than among persons drawn at random from the population as a whole, and so calculations of genotype frequency should be based on the ethnic group of the accused person, not on the race as a whole. Others argue that population substructure has a relatively minor effect on the final outcome of the calculation and that what matters most is not a high degree of accuracy but rather a general sense of whether a particular multilocus genotype is rare or common. The underlying issue is very serious: how to maintain a proper balance between the need to protect the rights of accused persons and the desire to get rapists and murderers off the streets.

DNA Exclusions

Issues of population genetics apply only when DNA types match. DNA types that do *not* match can exclude innocent suspects irrespective of any considerations about population substructure. For example, if the DNA type of a suspected rapist does not match the DNA type of semen taken from the victim, then the suspect could not be the perpetrator. Another example, which illustrates the use of DNA typing in paternity testing, is given in Figure 15.12. The numbers 1 and 2 designate different cases, in each of which a man was accused of fathering a child. In each case, DNA was obtained from the mother (M), the child (C), and the accused man (A). The pattern of bands obtained for one VNTR is shown.

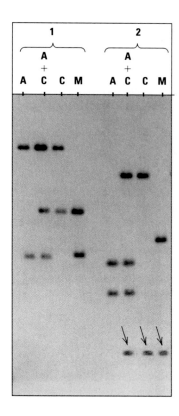

Figure 15.12

Use of DNA typing in paternity testing. The sets of lanes numbered 1 and 2 contain DNA samples from two different paternity cases. In each case, the lanes contain DNA fragments from the following sources: M, the mother; C, the child; A, the accused father. The lanes labeled A + C contain a mixture of DNA fragments from the accused father and the child. The arrows in case 2 point to bands of the same size present in lanes M, C, and A + C. Note that the accused male in case 2 could not be the father because neither of his bands is shared with the child. [Courtesy of R. W. Allen.]

The lanes labeled A + C contain a mixture of DNA from the accused man and the child. In case 1, the lower band in the child was inherited from the mother and the upper band from the father; because the upper band is the same size as one of those in the accused, the accused man could have contributed this allele to the child. This finding does not prove that the accused man is the father; it says only that he cannot be excluded on the basis of this particular gene. (On the other hand, if a large enough number of genes are studied, and the man cannot be excluded by any of the genes, then it makes it more likely that he really is the father.) Case 2 in Figure 15.12 is an exclusion. The band denoted by the arrows is the band inherited by the child from the mother. The other band in the child does not match either of the bands in the accused, so the accused man could not be the biological father. (In theory, mutation could be invoked to explain the result, but this is extremely unlikely.)

15.4 Inbreeding

Inbreeding means mating between relatives, such as first cousins. The principal consequence of inbreeding is that the frequency of heterozygous offspring is smaller than it is with random mating. This effect is seen most dramatically in repeated self-fertilization, which happens naturally in certain plants. Consider a hypothetical population consisting exclusively of *Aa* heterozygotes. With self-fertilization, each plant would produce offspring in the proportions 1/4 *AA*, 1/2 *Aa*, and 1/4 *aa*, which means that one generation of self-fertilization reduces the proportion of heterozygotes from 1 to 1/2. In the second generation, only the heterozygous plants can again produce heterozygous offspring, and only half of their offspring will again be heterozygous. Heterozygosity is therefore reduced to 1/4 of what it was originally. Three generations of self-fertilization reduce the heterozygosity to

$1/4 \times 1/2 = 1/8$, and so on. The remainder of this section demonstrates how the reduction in heterozygosity because of inbreeding can be expressed in quantitative terms.

The Inbreeding Coefficient

Repeated self-fertilization is a particularly intense form of inbreeding, but weaker forms of inbreeding are qualitatively similar in that they also lead to a reduction in heterozygosity. A convenient measure of the effect of inbreeding is based on the reduction in heterozygosity. Suppose that H_I is the frequency of heterozygous genotypes in a population of inbred individuals. The most widely used measure of inbreeding is called the **inbreeding coefficient,** symbolized F, which is defined as the proportionate reduction in H_I compared with the value of $2pq$ that would be expected with random mating; in symbols,

$$F = (2pq - H_I)/2pq \qquad (4)$$

Equation (4) can be rearranged as

$$H_I = 2pq(1 - F)$$

The homozygous genotypes increase in frequency as the heterozygotes decrease in frequency, and overall, the genotype frequencies in the inbred population are

$$
\begin{aligned}
AA: &\quad p^2(1 - F) + pF \\
Aa: &\quad 2pq(1 - F) \\
aa: &\quad q^2(1 - F) + qF \qquad (5)
\end{aligned}
$$

Equation (5) is a modification of the Hardy-Weinberg principle made to take inbreeding into account. When $F = 0$ (no inbreeding), the genotype frequencies are the same as those given in the Hardy-Weinberg principle in Equation (1): p^2, $2pq$, and q^2. At the other extreme, when $F = 1$ (complete inbreeding), the inbred population consists entirely of AA and aa genotypes in the frequencies p and q, respectively.

As an illustration of the use of Equation (5), consider again the *Pgm* gene in *Phlox cuspidata,* the genotype frequencies of which are not at all in agreement with the Hardy-Weinberg principle (Section 15.2).

In this species, about 78 percent of the plants undergo self-fertilization. It can be shown that the inbreeding coefficient corresponding to this amount of self-fertilization is $F = 0.64$, and the expected genotype frequencies must be modified to take this inbreeding into account. The frequencies of alleles Pgm^a and Pgm^b are 0.51 and 0.49, respectively, so by Equation (5) for inbreeding, the expected genotype frequencies are

$Pgm^a\,Pgm^a$: $(0.51)^2(1 - 0.64) + (0.51)(0.64) = 0.420$
$Pgm^a\,Pgm^b$: $2(0.51)(0.49)(1 - 0.64) = 0.180$
$Pgm^b\,Pgm^b$: $(0.49)^2(1 - 0.64) + (0.49)(0.64) = 0.400$

In a sample size of 35, the observed and expected numbers of the three genotypes were

$Pgm^a\,Pgm^a$: 15, versus $0.420 \times 35 = 14.7$
$Pgm^a\,Pgm^b$: 6, versus $0.180 \times 35 = 6.3$
$Pgm^b\,Pgm^b$: 14, versus $0.400 \times 35 = 14$

In this example, the agreement with the inbreeding model is excellent.

Allelic Identity by Descent

The approach to inbreeding based on genotype frequencies is useful for considering natural populations. However, in human genetics, as well as in animal and plant breeding, a geneticist may wish to calculate the inbreeding coefficient of a particular individual whose pedigree is known in order to determine the probability of each possible genotype for the individual. This section introduces a concept that is particularly suited to pedigree calculations.

The diagnostic criterion of inbreeding is a closed loop in the pedigree. Such a closed loop is often called a **path.** Figure 15.13A is an inbreeding pedigree in which individual I is the product of a mating between half sibs, and the path (closed loop) from I through the common ancestor A is evident. A streamlined version of this pedigree useful for inbreeding calculations is depicted in Figure 15.13B; individuals of both sexes are shown as circles, diagonal lines trace alleles transmitted from parent to offspring, and ancestors that did not contribute to the inbreeding of I (ancestors B and C) are left out altogether. The contrasting dots inside the circles represent particular alleles present in the corresponding individuals, and the depicted pattern of inheritance of the alleles represents the principal consequence of inbreeding and illustrates why closed loops are important. Specifically, individuals D and E both received the blue allele from A and received a white allele from B and C, respectively. The blue alleles in D and E have a very special relationship; they both originate by replication of the blue allele in A and therefore are said to be **identical by descent.** In the interval of the relatively few generations involved in most pedigrees, alleles that are identical by descent can be assumed to be identical in nucleotide sequence, because mutation can usually be ignored in so few generations. Individual I is shown as having received the blue allele from both D and E, and because the alleles in question are identical by descent, this individual must be homozygous for the allele inherited from the common ancestor. Without the closed loop in the pedigree, identity by descent of the alleles in I would be much less probable. The concept of identity by descent provides an alternative definition of the inbreeding coefficient, which can be shown to be completely equivalent to the definition in terms of heterozygosity given earlier. Specifically,

> The inbreeding coefficient of an individual is the probability that the individual carries alleles that are identical by descent.

Calculation of the Inbreeding Coefficient from Pedigrees

Figure 15.13C illustrates how the probability of identity by descent can be calculated in a pedigree, and the alleles are drawn as dots on the diagonal lines to emphasize their pattern of transmission. The double-headed arrows indicate which pairs of alleles must be identical by descent in order for I to have alleles that are identical by descent. The values of 1/2 represent the probability that a diploid genotype will transmit the allele inherited from a particular parent; this value must be 1/2 because of Mendelian segregation. The term $(1/2)(1 + F_A)$ can be explained by

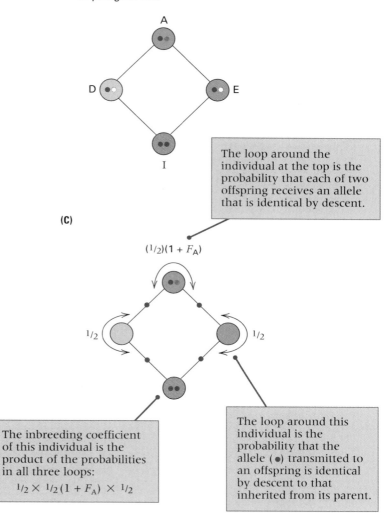

The loop around the individual at the top is the probability that each of two offspring receives an allele that is identical by descent.

$(1/2)(1 + F_A)$

The inbreeding coefficient of this individual is the product of the probabilities in all three loops:

$1/2 \times 1/2(1 + F_A) \times 1/2$

The loop around this individual is the probability that the allele (•) transmitted to an offspring is identical by descent to that inherited from its parent.

Figure 15.13
(A) An inbreeding pedigree in which individual I results from a mating between half siblings. (B) The same pedigree redrawn to show gametes in the closed loop leading to individual I. The colored dots represent alleles of a particular gene. (C) Calculation of the probability that the alleles indicated are identical by descent.

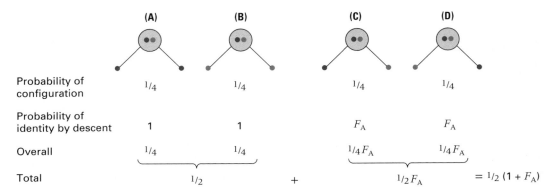

	(A)	(B)	(C)	(D)
Probability of configuration	$1/4$	$1/4$	$1/4$	$1/4$
Probability of identity by descent	1	1	F_A	F_A
Overall	$1/4$	$1/4$	$1/4 F_A$	$1/4 F_A$
Total		$1/2$	$+$	$1/2 F_A$ $= 1/2 (1 + F_A)$

Figure 15.14

A diagram illustrating the reason why $(1/2)(1 + F_A)$ is the probability that two different gametes from the same individual, A, carry alleles that are identical by descent.

looking at the situation in Figure 15.14, which shows four possible pairs of gametes that A can transmit, each equally likely. In cases A and B, the transmitted alleles are certainly identical by descent because they are products of DNA replication of the same allele (blue or red). In cases C and D, the transmitted alleles have contrasting colors, but they can still be identical by descent if A is already inbred. However, the probability that A carries alleles that are identical by descent is, by definition, the inbreeding coefficient of A, and this is designated F_A. Altogether, the probability that A transmits two alleles that are identical by descent is the sum of the probabilities of the four eventualities in Figure 15.14, and this sum equals $(1/2)(1 + F_A)$. Matters are somewhat simpler when A is not itself inbred, because in this case $F = 0$, and the corresponding probability is simply $1/2$.

Returning now to Figure 15.13C, the alleles in I can be identical by descent only if all three identities indicated by the loops are true, and this probability is the product of the three numbers shown. That is, the inbreeding coefficient of I, symbolized F_I, is

$$F_I = (1/2) \times (1/2)(1 + F_A) \times (1/2)$$
$$= (1/2)^3(1 + F_A)$$

For a more remote common ancestor, the formula becomes

$$F_I = (1/2)^n(1 + F_A) \qquad (6)$$

in which n is the number of individuals in the path traced from one of the parents of I back through the common ancestor and forward again to the other parent of I.

Most pedigrees of interest are more complex than the one in Figure 15.13, but the inbreeding coefficient of any individual can still be calculated by using Equation (6). Application to complex pedigrees requires several additional rules:

1. If there are several distinct paths that can be traced through a common ancestor, then each path provides a contribution to the inbreeding coefficient calculated as in Equation (6).

2. If there are several common ancestors, then each is considered in turn in order to calculate its separate contribution to the inbreeding.

3. The overall inbreeding coefficient is then calculated as the sum of all these separate contributions of the common ancestors.

The extension of Equation (6) to complex pedigrees may be written as

$$F_I = \Sigma(1/2)^n(1 + F_A) \qquad (7)$$

in which Σ denotes the summation of all paths through all common ancestors.

Application of Equation (7) can be illustrated in the pedigree in Figure 15.15. Each path and its contribution to the inbreeding coefficient of I is shown below (the red letter indicates the common ancestor).

EBD	$(1/2)^3(1 + F_B)$
ECD	$(1/2)^3(1 + F_C)$
ECABD	$(1/2)^5(1 + F_A)$
EBACD	$(1/2)^5(1 + F_A)$

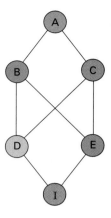

Figure 15.15

A complex pedigree in which the inbreeding of individual I results from several independent paths through three different common ancestors.

The inbreeding coefficient of I is the sum of these numbers. If none of the common ancestors is inbred (that is, if $F_A = F_B = F_C = 0$), then

$$F_I = (1/2)^3 + (1/2)^3 + (1/2)^5 + (1/2)^5 = 5/16$$

The value 5/16 is to be interpreted to mean that the probability that individual I will be heterozygous at any particular locus is smaller, by the proportion 5/16, than the probability that a noninbred individual will be heterozygous at the same locus. Equivalently, the 5/16 could be interpreted to mean that individual I is heterozygous at 5/16 fewer loci in the whole genome than a noninbred individual.

Effects of Inbreeding

The effects of inbreeding differ according to the normal mating system of an organism. At one extreme, in regularly self-fertilizing plants, inbreeding is already so intense and the organisms are so highly homozygous that additional inbreeding has virtually no effect. However,

> In most species, inbreeding is harmful, and much of the effect is due to rare recessive alleles that would not otherwise become homozygous.

Among human beings, inbreeding is usually uncommon because of social conven-

tions and laws, although in small isolated populations (isolated villages, religious communities, aboriginal groups) it does occur, mainly through matings between relatives more distant than second cousins ("remote relatives"). The most common type of close inbreeding is between first cousins. The effect is always an increase in the frequency of genotypes that are homozygous for rare, usually harmful recessives. For example, among American whites, the frequency of albinism among offspring of matings between nonrelatives is approximately 1 in 20,000; among offspring of first-cousin matings, the frequency is approximately 1 in 2000. The reason for the increased risk may be understood by comparing the genotype frequencies of homozygous recessives in Equation (5) for inbreeding and Equation (1) for random mating. Among the offspring of a mating between first cousins, $F = 1/16$, or 0.062. Therefore, the frequency of homozygous recessives produced by first-cousin mating will be

$$q^2(1 - 0.062) + q(0.062)$$

whereas among the offspring of nonrelatives, the frequency of homozygous recessives is q^2. For albinism, $q = 0.007$ (approximately), and the calculated frequencies are 5×10^{-4} for the offspring of first cousins and 5×10^{-5} for the offspring of nonrelatives. The increased frequency with first-cousin mating is the principal consequence of inbreeding.

15.5 Genetics and Evolution

Evolution refers to changes in the gene pool resulting in the progressive adaptation of populations to their environment. Evolution occurs because genetic variation exists in populations and because there is a natural selection favoring organisms that are best adapted to the environment. Genetic variation and natural selection are population phenomena, so they are conveniently discussed in terms of allele frequencies.

Four processes account for most of the changes in allele frequency in populations. They form the basis of cumulative change in the genetic characteristics of populations, leading to the descent with modification that characterizes the process of evolution. Although the point has yet to be proved, most evolutionary biologists believe that these same processes, when carried out continuously over geological time, can also account for the formation of new species and higher taxonomic categories. These process are

1. **Mutation,** the origin of new genetic capabilities in populations by means of spontaneous heritable changes in genes.

2. **Migration,** the movement of individuals among subpopulations within a larger population.

3. **Natural selection,** which results from the differing abilities of individuals to survive and reproduce in their environment.

4. **Random genetic drift,** the random, undirected changes in allele frequency that occur by chance in all populations, but particularly in small ones.

Some of the implications of these processes for population genetics are considered in the following sections.

15.6
Mutation and Migration

Mutation is the ultimate source of genetic variation. It is an essential process in evolution, but it is a relatively weak force for changing allele frequencies, primarily because typical mutation rates are too low. Moreover, most newly arising mutations are harmful to the organism. Although some mutations may be **selectively neutral,** which means they do not affect the ability of the organism to survive and reproduce, only a very few mutations are favorable for the organism and contribute to evolution. The low mutation rates that are observed are thought to have evolved as a compromise: The mutation rate is high enough to generate the favorable muta-

tions that a species requires to continue evolving, but it is not so high that the species suffers excessive genetic damage from the preponderance of harmful mutations.

Migration is similar to mutation in that new alleles can be introduced into a local population, although the alleles derive from another subpopulation rather than from new mutations. In the absence of migration, the allele frequencies in each local population can change independently, so local populations can undergo considerable genetic differentiation. Genetic differentiation among subpopulations means that there are differing frequencies of common alleles among the local populations or that some local populations possess certain rare alleles not found in others. The accumulation of genetic differences among subpopulations can be minimized if the subpopulations exchange individuals (undergo migration). In fact, only a relatively small amount of migration among subpopulations (on the order of just a few migrant individuals in each local population in each generation) is usually sufficient to prevent the accumulation of high levels of genetic differentiation. On the other hand, genetic differentiation can occur in spite of migration if other evolutionary forces, such as natural selection for adaptation to the local environments, are sufficiently strong.

Irreversible Mutation

To appreciate the nature of the force of mutation, consider an allele A and a completely equivalent mutant allele a. (The mutational change in a could be, for example, a nucleotide substitution at a nonessential site in an intron or a change from one synonymous codon to another, so the protein product is unchanged.) Suppose allele A undergoes mutation to a at a rate of μ per A allele per generation. What this rate means is that in any generation, the probability that a particular A allele mutates to a is μ. Equivalently, we could say that in any generation, the proportion of A alleles that do *not* mutate to a equals $1 - \mu$. If we let p and q represent the allele frequencies of A and a, respectively, then the allele frequency of A changes according to the rule that the allele frequency of A in any generation equals the allele frequency of A in the previous generation times the proportion of A alleles that failed to mutate in that generation. (Here we are assuming that reverse mutation of a to A can be neglected.) In symbols, the rule implies that

$$p_n = p_{n-1}(1 - \mu)$$

where the subscripts n and $n - 1$ denote any generation n and the previous generation, respectively. Hence the allele frequency of A in generations 0, 1, 2, 3, and so on changes according to

$$p_1 = p_0(1 - \mu)$$
$$p_2 = p_1(1 - \mu)$$
$$p_3 = p_2(1 - \mu)$$

Substituting p_1 from the first equation into the second, we obtain

$$p_2 = p_1(1 - \mu) = p_0(1 - \mu)^2$$

Substituting this into the equation for p_3 yields

$$p_3 = p_2(1 - \mu) = p_0(1 - \mu)^3$$

The general rule follows by consecutive substitutions:

$$p_n = p_0(1 - \mu)^n \qquad (8)$$

where p_n represents the allele frequency of A in the nth generation.

If we assume that allele A is fixed in the population at time $n = 0$, then $p_0 = 1$, and Equation (8) becomes

$$p_n = (1 - \mu)^n \qquad (9)$$

For realistic values of the mutation rate, Equation (9) implies a very slow rate of change in allele frequency. The pattern of change is illustrated by the lower curve in Figure 15.16, which assumes a mutation rate of $\mu = 1 \times 10^{-5}$ per generation. With this value of μ, the allele frequency of A decreases by half its present value every 69,314 generations. To put this rate in human terms, assuming 20 years per generation, it would require $69,314 \times 20 = 1,368,280$ years, or approximately 1.4 million years, to reduce the allele frequency of A by half. For example, if a human allele A were fixed at the time of *Homo erectus*, 1.4 million years ago, and A underwent mutation to a at a rate of 1×10^{-5} per generation, then 50 percent of the alleles present in the modern human population would still be of type A.

Figure 15.16
Effects of irreversible mutation (bottom curve) and reversible mutation (top curve) on allele frequency. In this example, the forward mutation rate is $\mu = 1 \times 10^{-5}$ and the reverse mutation rate is $\mu = 1 \times 10^{-6}$.

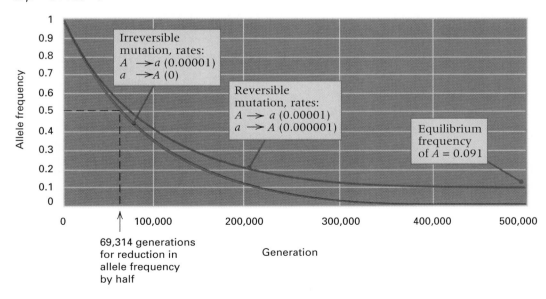

Reversible Mutation

When the possibility of reverse mutation is taken into account, then **forward mutation** of A to a is offset by **reverse mutation** of a to A. Eventually, an **equilibrium** of allele frequency is reached at which the allele frequencies become stable and no longer change, even though there is continuous mutation. The reason for the equilibrium is that at this point, each new A allele that is lost by forward mutation is replaced by a new A allele created by reverse mutation.

It is quite straightforward to deduce the equilibrium frequencies when mutation is reversible. As before, let A and a have allele frequencies p and q, respectively. Suppose that the rate of forward mutation (A → a) is μ and that of reverse mutation (a → A) is ν. In any generation, the number of A alleles lost to forward mutation is proportional to $p\mu$, and the number of new A alleles created by reverse mutation is proportional to $q\nu$. At equilibrium, the loss of A must equal the gain of a; hence $p\mu = q\nu$. Substituting $q = 1 - p$, we obtain

$$p\mu = (1 - p)\nu = \nu - p\nu$$

and therefore, at equilibrium,

$$\hat{p} = \frac{\nu}{\mu + \nu} \qquad (10)$$

where the symbol $\hat{p}$ is used, by convention, to mean the equilibrium value of p. As a check of the correctness of Equation (10), note that $\hat{q} = \mu/(\mu + \nu)$ and so $\hat{p}\mu = \hat{q}\nu$ as required.

In Figure 15.16, the upper curve gives the change in allele frequencies for reversible mutation with $\mu = 1 \times 10^{-5}$ and $\nu = 1 \times 10^{-6}$. The equilibrium in this example is

$$\hat{p} = \frac{1 \times 10^{-6}}{} = \frac{1}{11}$$

The time required to reach equilibrium is again very long. When $\mu = 1 \times 10^{-5}$ and $\nu = 1 \times 10^{-6}$, it requires 63,013 generations for any specified values of p and q to go halfway to their equilibrium values. (The half-time of 69,314 generations derived earlier is different because of the effect of reverse mutation.)

15.7 Natural Selection

The driving force of adaptive evolution is *natural selection,* which is a consequence of hereditary differences among organisms in their ability to survive and reproduce in the prevailing environment. Since it was first proposed by Charles Darwin in 1859, the concept of natural selection has been revised and extended, most notably by the incorporation of genetic concepts. In its modern formulation, the occurrence of natural selection rests on three premises.

1. In all organisms, more offspring are produced than can possibly survive and reproduce. (This part of the theory Darwin borrowed from Thomas Malthus.)

2. Organisms differ in their ability to survive and reproduce, and some of these differences are due to genotype.

3. In every generation, the genotypes that promote survival in the prevailing environment (favored genotypes) are present in excess among individuals of reproductive age, and hence the favored genotypes contribute disproportionately to the offspring of the next generation.

By means of this process, the alleles that enhance survival and reproduction increase in frequency from generation to generation, and the population becomes progressively better able to survive and reproduce in the environment. This progressive genetic improvement in populations constitutes the process of evolutionary **adaptation.**

Selection in a Laboratory Experiment

Selection is easily studied experimentally in bacterial populations because of the short generation time (about 30 minutes). Figure 15.17 shows the result of competition between two bacterial genotypes, A and B. Genotype A is the superior competitor under the particular conditions. In the experiment, the competition was allowed to continue for 290 generations, in which time the proportion of A genotypes (p) increased from 0.60 to 0.9995 and that of B genotypes decreased from 0.40 to 0.0005. The data points give a satisfactory fit to an equation of the form

$$\frac{p_n}{q_n} = \left(\frac{p_0}{q_0}\right)\left(\frac{1}{w}\right)^n \qquad (11)$$

in which p_0 and q_0 are the initial frequencies of A and B (in this case 0.6 and 0.4, respectively), p_n and q_n are the frequencies after n generations of competition, and w is a measure of the competitive ability of B when competing against A under the conditions of the experiment. Equation (11) can be derived in a similar manner as Equation (8) for mutation. Because the relative rates of survival and reproduction of strains A and B are in the ratio $1 : w$, the ratio of frequencies in any generation, p_n/q_n, must equal the ratio of frequencies in the previous generation, p_{n-1}/q_{n-1}, times $1/w$. Therefore,

$$(p_1/q_1) = (p_0/q_0) \times (1/w)$$
$$(p_2/q_2) = (p_1/q_1) \times (1/w)$$
$$(p_3/q_3) = (p_2/q_2) \times (1/w)$$

and so on. Equation (11) is derived by repeated substitution of each equation into the next.

The value of w is called the **relative fitness** of the B genotype relative to the A genotype under these particular conditions. The data in Figure 15.17 fit Equation (11) for a value of $w = 0.958$, which generates the smooth curve. Relative fitness measures the comparative contribution of each parental genotype to the pool of offspring genotypes produced in each generation. In this example, for each offspring cell produced by an A genotype, a B genotype produces an average of 0.958 offspring cell.

In population genetics, relative fitnesses are usually calculated with the most favored genotype (A in this case) taken as the standard with a fitness of 1.0. However, the selective disadvantage of a genotype is often of greater interest than its relative fitness, because some of the formulas are simplified. The selective disadvantage of a disfavored genotype is called the **selection coefficient** against the genotype, and it is calculated as the difference between the fitness of the standard (taken as 1.0) and the relative fitness of the genotype in question. In the case at hand, the selection coefficient against B, denoted s, is

$$s = 1.000 - 0.958 = 0.042 \qquad (12)$$

In words, the selective disadvantage of strain B is 4.2 percent per generation. When the fitnesses are known, Equation (11) also enables us to predict the allele frequencies in any future generation, given the original frequencies. Alternatively, it can be used to calculate the number of generations required for selection to change the allele frequencies from any specified initial values to any later ones. For example, from the relative fitnesses of A and B just estimated, we can calculate the number of generations required to change the frequency of A from 0.1 to 0.8. In this example, $p_0/q_0 = 0.1/0.9$, $p_n/q_n = 0.8/0.2$, and $w = 0.958$. A little manipulation of Equation (11) gives

$$n = [\log(0.1/0.9) - \log(0.8/0.2)]/\log(0.958)$$
$$= 83.5 \text{ generations}$$

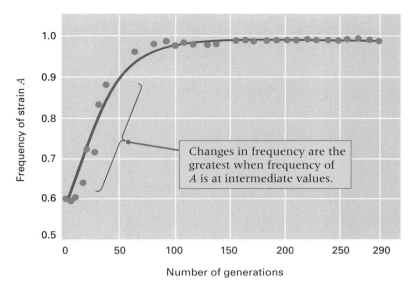

Figure 15.17
Increase in frequency of a favored strain of *E. coli* resulting from selection in a continuously growing population. The *y*-axis is the number of *A* cells at any time divided by the total number of cells ($A + B$). Note that the changes in frequency are the greatest when the frequency of the favored strain, A, is at intermediate values.

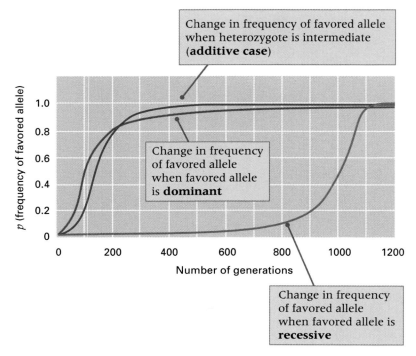

Change in frequency of favored allele when heterozygote is intermediate (**additive case**)

Change in frequency of favored allele when favored allele is **dominant**

Change in frequency of favored allele when favored allele is **recessive**

Figure 15.18

Theoretically expected change in frequency of an allele favored by selection in a diploid organism undergoing random mating when the favored allele is dominant, when the favored allele is recessive, and when the heterozygote is exactly intermediate in fitness (additive). In each case, the selection coefficient against the least fit homozygous genotype is 5 percent. The data are plotted in terms of p, the allele frequency of the beneficial allele. These curves demonstrate that selection for or against a recessive allele is very inefficient when the recessive allele is rare.

Selection in Diploids

Selection in diploids is analogous to that in haploids, but dominance and recessiveness create additional complications. Figure 15.18 shows the change in allele frequencies for both a favored dominant and a favored recessive in a random-mating diploid population. The striking feature of the figure is that the frequency of the favored dominant allele changes very slowly when the allele is common, and the frequency of the favored recessive allele changes very slowly when the allele is rare. The reason is that, with random mating, rare alleles are found much more frequently in heterozygotes than in homozygotes. With a favored dominant at high frequency, most of the recessive alleles are present in heterozygotes, and the heterozygotes are not exposed to selection and hence do not contribute to change in allele frequency. Conversely, with a favored recessive at low frequency, most of the favored alleles are in heterozygotes, and again the heterozygotes are not exposed to selection and do not contribute to change in allele frequency. The principle is quite general:

Selection for or against a recessive allele is very inefficient when the recessive allele is rare.

The inefficiency of selection against rare recessive alleles has an important practical implication. There is a widely held belief that medical treatment to save the lives of persons with rare recessive disorders will cause a deterioration of the human gene pool because of the reproduction of persons who carry the harmful genes. This belief is unfounded. With rare alleles, the proportion of homozygous genotypes is so small that reproduction of the homozygous genotypes contributes a negligible amount to change in allele frequency. Considering their low frequency, it matters little whether homozygous genotypes reproduce or not. Similar reasoning applies to eugenic proposals to "cleanse" the human gene pool of harmful recessives by preventing the reproduction of affected persons. People with severe genetic disorders rarely reproduce anyway, and even when they do, they have essentially no effect on allele frequency. *The largest reservoir of harmful recessive alleles is in the genomes of phenotypically normal carrier heterozygotes.*

Components of Fitness

Data like those in Figure 15.17 are essentially unobtainable in higher organisms because of their long generation time. Observations extending over 290 generations would take 12 years with *Drosophila* and 6000 years with humans. How, then, is one to estimate fitness? The usual approach is to divide fitness into its component parts and estimate these separately. The two major components of the fitness of a genotype are usually considered to be

- The **viability,** which means the probability that a newly formed zygote of the specified genotype survives to reproductive age

- The **fertility,** which means the average number of offspring produced by an individual of the specified genotype during the reproductive period

Together, viability and fertility measure the contribution of each genotype to the offspring of the next generation.

Table 15.3 is an example of experiments carried out to estimate the viability of two genotypes coding for allozymes of the enzyme alcohol dehydrogenase in *Drosophila.* The genotypes are homozygous for either the *F* or the *S* allele. The tabulated numbers are averages obtained in seven separate experiments in which larvae were placed together in competition in culture vials. The ratio of surviving larvae to those introduced originally is the probability of survival of each genotype. As noted at the bottom of the table, these numbers can be converted into relative viabilities in two ways, using either *SS* or *FF* as the standard. Because of the convention that the larger of the two relative values should be 1.0, the situation can be summarized either by saying that the viability of *FF* relative to *SS* is 0.94 or that the selective disadvantage of *FF* relative to *SS* with respect to viability is 0.06.

Estimates of overall fitness based on viability and fertility components may be used for predicting changes in allele frequency that would be expected in experimental or natural populations. However, the fitness of a genotype may depend on the environment and on other factors such as population density and the frequencies of other genotypes. Consequently, fitnesses estimated in a particular laboratory situation may have little relevance to predicting allele frequency changes in natural populations unless other factors are comparable.

Selection-Mutation Balance

Natural selection usually acts to minimize the frequency of harmful alleles in a population. However, the harmful alleles can never be totally eliminated, because mutation of wildtype alleles continually creates new harmful mutations. With the forces of selection and mutation acting in opposite directions, the population eventually attains a state of equilibrium at which new mutations exactly offset selective eliminations. In determining the allele frequencies at equilibrium, it makes a great deal of difference whether the harmful allele is completely recessive or has, instead, a small effect (partial dominance) on the fitness of the heterozygous carriers. To deduce the equilibrium frequencies, we will symbolize the wildtype allele as A and the harmful allele as a and represent the allele frequencies as p and q, respectively.

Table 15.3 Relative viability of alcohol dehydrogenase genotypes in *Drosophila*

Experimental data and relative viability	Genotype	
	FF	*SS*
Experimental data		
Number of larvae introduced per vial	200	100
Average number of survivors per vial	132.72	70.43
Probability of survival	0.6636	0.7043
Relative viability		
With *SS* as standard	0.6636/0.7043 = 0.94	0.7043/0.7043 = 1.0
With *FF* as standard	0.6636/0.6636 = 1.0	0.7043/0.6636 = 1.06

In the case of a complete recessive, the fitnesses of AA, Aa, and aa genotypes can be written as $1 : 1 : 1 - s$, where s represents the selection coefficient against the aa homozygote and measures the fraction of aa genotypes that fail to survive or reproduce. For a recessive lethal, for example, $s = 1$. When q is very small (as it will be near equilibrium), the number of a alleles eliminated by selection will be proportional to $q^2 s$, assuming Hardy-Weinberg proportions. At the same time, the number of new a alleles introduced by mutation will be proportional to μ. At equilibrium, the selective eliminations must balance the new mutations, and so $q^2 s = \mu$, or

$$\hat{q} = \sqrt{\frac{\mu}{s}} \qquad (13)$$

As before, $\hat{q}$ means the equilibrium value. To apply Equation (13) to a specific example, consider the recessive allele for Tay-Sachs disease, which in most non-Jewish populations has a frequency of $q = 0.001$. Because the condition is lethal, $s = 1$. Assuming that $q = 0.001$ represents the equilibrium frequency and that the allele has no effect on the fitness of the heterozygous carriers, the mutation rate required to account for the observed frequency would be

$$\mu = q^2 s = (0.001)^2 \times 1.0 = 1 \times 10^{-6}$$

This example shows how considerations of selection-mutation balance can be used in estimating human mutation rates.

If a harmful allele shows partial dominance in having a small detrimental effect on the fitness of the heterozygous carriers, then the fitnesses of AA, Aa, and aa can be written as $1 : 1 - hs : 1 - s$, in which hs is the selection coefficient against the heterozygous carriers. The parameter h is called the **degree of dominance.** When $h = 1$, the harmful allele is completely dominant; when $h = 0$, it is completely recessive; and when $h = 1/2$, the fitness of the heterozygous genotype is exactly the average of those of the homozygous genotypes. For most harmful alleles that show partial dominance, the value of h is considerably smaller than $1/2$.

With random mating, when an allele is rare (as a harmful allele will be at equilibrium), the frequency of heterozygous genotypes far exceeds the frequency of those that are homozygous for the harmful recessive. This means that most of the alleles that are eliminated by selection are eliminated because of the selection against the heterozygous genotypes, because there are so many more of them. For each Aa genotype that fails to survive or reproduce, half of the alleles that are lost are a, so the number of alleles lost to selection in any generation is proportional to $2pq \times hs \times 1/2$. On the other hand, the number of new a alleles created by $A \rightarrow a$ mutation equals μ. At equilibrium, the loss due to selection must balance the gain from mutation, so $2pq \times hs \times 1/2 = \mu$, or

$$\hat{p}\hat{q}hs = \mu$$

Because a harmful allele will be rare at equilibrium, especially if it shows partial dominance, $\hat{p}$ will be so close to 1 that it can be replaced with 1 in the above expression. Therefore, the equilibrium value for the frequency of a partially dominant harmful allele is

$$\hat{q} = \frac{\mu}{hs} \qquad (14)$$

Note that Equation (14) does not include a square root, which makes a substantial difference in the allele frequency. This effect can be seen by comparing the curves in Figure 15.19 for a harmful allele that is lethal when homozygous ($s = 1$). The top curve is for a complete recessive ($h = 0$), and the other curves refer to various degrees of partial dominance; for example, $h = 1/2$ means that the heterozygous genotype has a fitness exactly intermediate between the homozygous genotypes. Even a small amount of partial dominance has a substantial effect on reducing the equilibrium allele frequency.

Heterozygote Superiority

So far we have considered examples of selection in which the fitness of the heterozygote is intermediate between those of the homozygotes (or possibly equal in fitness to one homozygote). In these cases, the allele associated with the most fit homozygote eventually becomes fixed, unless the selection is opposed by mutation. In this section we consider the possi-

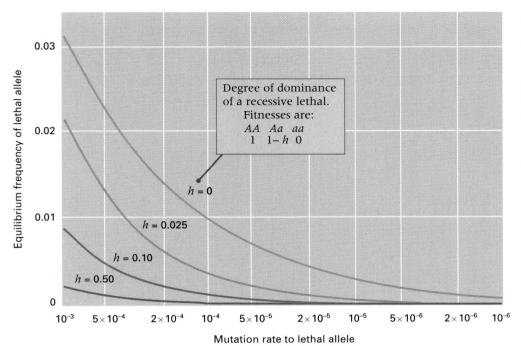

Figure 15.19
Effect of partial dominance on the equilibrium frequency of a harmful allele that is lethal when homozygous. The degree of dominance is measured by h. For example, a value of $h = 0.025$ means that the relative fitnesses of AA and Aa are $1 : 0.975$.

bility that the fitness of the heterozygote is greater than that of both homozygotes; this case is called **overdominance** or **heterozygote superiority.**

When there is overdominance, neither allele can be eliminated by selection, because in each generation, the heterozygotes produce more offspring than the homozygotes. The selection in favor of heterozygous genotypes keeps both alleles in the population. Eventually there is equilibrium in which the allele frequency no longer changes. With overdominance, the fitnesses of AA, Aa and aa may be written as $1 - s : 1 : 1 - t$, where s and t are the selection coefficients against AA and aa, respectively, relative to Aa. Assuming random mating, in any generation, the proportion of A alleles eliminated by selection is $p^2s/p = ps$, and the proportion of a alleles eliminated by selection is $q^2t/q = qt$. At equilibrium, the selective eliminations of A must balance the selective eliminations of a, and hence $ps = qt$, or

$$p = \frac{t}{s + t} \qquad (15)$$

With overdominance, the allele frequencies always go to equilibrium, but the rate

of approach depends on the magnitudes of s and t. The equilibrium is attained most rapidly when there is strong selection against the homozygotes.

Overdominance does not appear to be a particularly common form of selection in natural populations. However, there are several well-established cases, the best known of which involves the sickle-cell hemoglobin mutation. The initial hint of a connection between sickle-cell anemia and falciparum malaria was the substantial overlap of the geographical distribution of sickle-cell anemia and the geographical distribution of malaria (Figure 15.20).

On the basis of the rates at which people of different genotypes are infected with malaria, and the mortality of infected people, the relative fitnesses of the genotypes AA, AS and SS have been estimated as $0.86 : 1 : 0$. (S represents the sickle-cell mutation.) In the absence of intensive medical treatment, the homozygous SS genotype is lethal as a result of severe anemia. On the other hand, there is overdominance because AS heterozygotes are more resistant to malarial infection than either of the homozygous genotypes.

From these fitnesses, we have $s = 0.24$ and $t = 1$, so the predicted equilibrium

(A)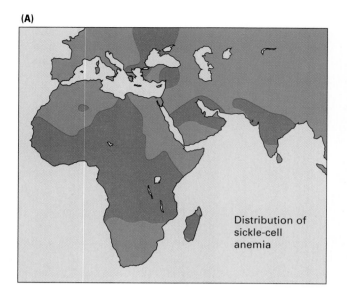

(B)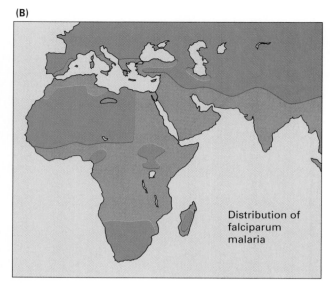

Figure 15.20

Geographic distribution of (A) sickle cell anemia and (B) falciparum malaria in the 1920s, before extensive malaria-control programs were launched. Shades of brown indicate the areas in question.

frequency of *A* from Equation (15) equals 1/(0.24 + 1) = 0.81. The predicted equilibrium frequency of *S* is therefore 1 − 0.806 = 0.19. This value is reasonably close to that observed in many parts of West Africa.

15.8
Random Genetic Drift

The process of **random genetic drift** comes about because populations are not infinitely large (as we have been assuming all along) but rather are finite, or limited in size. The breeding individuals of any one generation produce a potentially infinite pool of gametes. In the absence of fertility differences, the allele frequencies among gametes will equal the allele frequencies among adults. However, because of the finite size of the population, only a few of these gametes will participate in fertilization and be represented among zygotes of the next generation. In other words, there is a process of *random sampling* that takes place in going from one generation to the next. Because there is variation among

samples due to chance, the allele frequencies among gametes and zygotes may differ. Changes in allele frequency that come about because of the generation-to-generation sampling in finite populations are the cause of random genetic drift.

Consider a population consisting of exactly *N* diploid individuals in each generation. At each autosomal locus, there will be exactly 2*N* alleles. Suppose that *i* of these are of type *A* and the remaining 2*N* − *i* are of type *a*. (The allele frequency of *A*, designated *p*, therefore equals *i*/2*N*.) The number of *A* alleles in the next generation cannot be specified with certainty, because it will differ from case to case as a result of random genetic drift. On the other hand, it is possible to specify a *probability* for each possible number of *A* alleles in the next generation. This probability is given by the terms of the binomial distribution (Chapter 3). The probability *P*(*k*|*i*) that there will be exactly *k* copies of *A* in the next generation, given that there are exactly *i* copies among the parents, is given by

$$P(k|i) = \frac{(2N)!}{k!\,(2N-k)!}p^k q^{2N-k} \quad (16)$$

In Equation (16), $p = i/2N$, $q = 1 - p$, and k can equal 0, 1, 2, and so on, up to and including 2N. The part of the equation with factorials is the number of possible orders in which the k copies of allele A can be chosen. (The A alleles could be the first k chosen, the last k chosen, alternating with a, and so on.) The powers of p and q represents the probability that any particular order will be realized. For any specified values of N and i, one can use Equation (16) along with a table of random numbers (or, better yet, a personal computer) to calculate an actual value for k.

A concrete example illustrates the essential features of random genetic drift. Table 15.4 shows the result of random genetic drift in 12 subpopulations (labeled a through l), each initially consisting of eight diploid individuals and containing eight copies of allele A and eight copies of allele a. Within each subpopulation, mating

Table 15.4 Effects of random genetic drift. Each entry is the number of A alleles in a subpopulation of size 16.

Generation	\multicolumn{12}{c}{Population designation}	Average $\bar{p}$
	a	b	c	d	e	f	g	h	i	j	k	l	
0	8	8	8	8	8	8	8	8	8	8	8	8	0.500
1	11	7	9	8	8	6	8	7	8	6	11	10	0.516
2	10	9	10	8	8	8	6	7	6	7	13	11	0.536
3	7	11	6	5	12	5	8	5	8	7	14	9	0.505
4	8	11	7	4	12	5	8	8	7	4	15	6	0.495
5	8	8	5	3	13	2	8	12	9	5	15	6	0.490
6	11	5	3	1	13	3	10	13	6	7	15	3	0.469
7	11	8	4	3	11	1	8	14	3	7	15	2	0.453
8	14	7	4	3	10	1	9	14	3	9	16	1	0.474
9	15	5	3	5	7	0	12	14	2	11	16	0	0.469
10	16	6	5	9	8	0	9	13	3	10	16	0	0.495
11	16	1	5	11	6	0	10	13	3	10	16	0	0.474
12	16	0	5	12	6	0	9	13	2	9	16	0	0.458
13	16	0	3	12	7	0	9	13	1	11	16	0	0.458
14	16	0	5	15	7	0	9	11	1	12	16	0	0.479
15	16	0	3	14	7	0	8	12	3	13	16	0	0.479
16	16	0	2	14	9	0	11	14	2	14	16	0	0.510
17	16	0	1	15	6	0	12	14	2	13	16	0	0.495
18	16	0	1	15	4	0	13	15	5	13	16	0	0.510
19	16	0	1	16	2	0	14	16	6	14	16	0	0.526
20	16	0	1	16	2	0	15	16	9	16	16	0	0.557
21	16	0	2	16	3	0	15	16	10	16	16	0	0.573
	A	a		A		a		A		A	A	a	
	fixed	fixed		fixed		fixed		fixed		fixed	fixed	fixed	

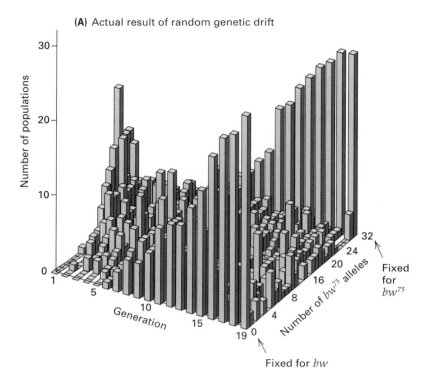

(A) Actual result of random genetic drift

Number of populations / Generation / Number of *bw^{75}* alleles

Fixed for *bw^{75}*

Fixed for *bw*

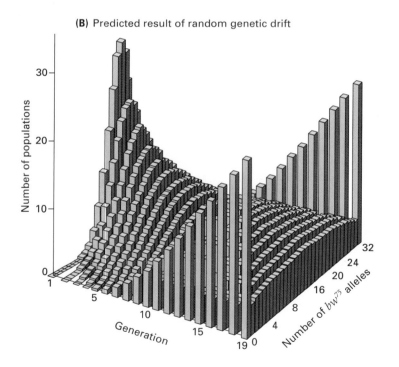

(B) Predicted result of random genetic drift

Number of populations / Generation / Number of *bw^{75}* alleles

Figure 15.21

(A) Random genetic drift in 107 experimental populations of *Drosophila melanogaster,* each consisting of 8 females and 8 males. (B) Theoretical expectation of the same situation, calculated from the binomial distribution. [Data in part A from P. Buri, 1956, *Evolution* 10:367. Graphs from D. L. Hartl and A. G. Clark, 1989, *Principles of Population Genetics,* Sunderland, MA: Sinauer Associates.]

is random. A computer program generating values of *k* with Equation (16) was used to calculate the number of *A* alleles in each subpopulation in each successive generation. These numbers are shown in the vertical columns, and the dispersion of allele frequencies resulting from random genetic drift is apparent. These changes in allele frequency would be less pronounced, and would require more generations, in larger populations than in the very small populations illustrated here, but the overall effect would be the same. Because of the prominent role of 2*N* in Equation (16), the dispersion of allele frequency resulting from random genetic drift depends on population size; the smaller the population, the greater the dispersion and the more rapidly it takes place.

In Table 15.4, the principal effect of random genetic drift is evident in the first generation: The allele frequencies have begun to spread out over a wider range. By the seventh generation, the spreading is extreme, and the number of *A* alleles ranges from 1 to 15. This speading out means that the allele frequencies among the subpopulations become progressively more different. In general:

> Random genetic drift causes differences in allele frequency among subpopulations; it is a major cause of genetic differentiation among subpopulations.

Although allele frequencies among subpopulations spread out over a wide range because of random genetic drift, the *average* allele frequency among subpopulations remains approximately constant. This point is illustrated by the column headed $\overline{p}$ in Table 15.4. The average allele frequency stays close to 0.5, its initial value. If an infinite number of subpopulations were being considered instead of the 12 subpopulatons in Table 15.4, then the average allele frequency would be exactly 0.5 in every generation. This principle implies that in spite of the random drift of allele frequency in individual subpopulations, the average allele frequency among a large number of subpopulations remains constant and equal to the average allele frequency among the original subpopulations.

After a sufficient number of generations of random genetic drift, some of the subpopulations become fixed for *A* and others for *a*. Because we are excluding the occurrence of mutation, a population that becomes fixed for an allele remains fixed thereafter. After 21 generations in Table 15.4, only four of the populations (*c, e, g,* and *i*) are still segregating; eventually, these too will become fixed. Because the average allele frequency of *A* remains constant, it follows that a fraction p_0 of the populations will ultimately become fixed for *A* and a fraction $1 - p_0$ will become fixed for *a*. (The symbol p_0 represents the allele frequency of *A* in the initial generation.) Therefore,

> With random genetic drift, the probability of ultimate fixation of a particular allele is equal to the frequency of the allele in the original population.

In Table 15.4, five of the fixed populations are fixed for *A* and three for *a*, which is not very different from the equal numbers expected theoretically with an infinite number of subpopulations.

An example of random genetic drift in small experimental populations of *Drosophila* exhibiting the characteristics pointed out in connection with Table 15.4 is illustrated in Figure 15.21A. The figure is based on 107 subpopulations, each initiated with eight bw^{75}/bw females (bw = brown eyes) and eight bw^{75}/bw males and maintained at a constant size of 16 by randomly choosing eight males and eight females from among the progeny of each generation. Note how the allele frequencies among subpopulations spread out because of random genetic drift and how subpopulations soon begin to be fixed for either bw^{75} or bw. Although the data are somewhat rough because there are only 107 subpopulations, the overall pattern of genetic differentiation has a reasonable resemblance to that expected from the theory based on the binomial distribution (Figure 15.21B).

If random genetic drift were the only force at work, then all alleles would become either fixed or lost and there would be no polymorphism. On the other hand, many factors can act to retard or prevent the effects of random genetic drift, of which the following are the most important: (1) large population size; (2) mutation and migration, which impede fixation because alleles lost by random genetic drift can be reintroduced by either process; and (3) natural selection, particularly those modes of selection that tend to maintain genetic diversity, such as heterozygote superiority.

Chapter Summary

Population genetics is the application of Mendel's laws and other principles of genetics to populations of organisms. A subpopulation, or local population, is a group of organisms of the same species living within a geographical region of such size that most matings are between members of the group. In most natural populations, many genes are polymorphic in that they have two or more common alleles. One of the goals of population genetics is to determine the nature and causes of genetic variation in natural populations.

The relationship between the relative proportions of particular alleles (allele frequencies) and genotypes (genotype frequencies) is determined in part by the frequencies with which particular genotypes form mating pairs. In random mating, mating pairs are independent of genotype. When a population undergoes random mating for an autosomal gene with two alleles, the frequencies of the genotypes are given by the Hardy-Weinberg principle. If the alleles of the gene are *A* and *a,* and their allele frequencies are *p* and *q*, respectively, then the Hardy-Weinberg principle states that the genotype frequencies with random mating are *AA*, p^2; *Aa*, $2pq$; and *aa*, q^2. These are often good approximations for genotype frequencies within subpopulations. An important implication of the Hardy-Weinberg principle is that rare alleles are found much more frequently in heterozygotes than in homozygotes ($2pq$ versus q^2).

Inbreeding means mating between relatives, and the extent of inbreeding is measured by the inbreeding coefficient, *F*. The main consequence of inbreeding is that an allele present in a common ancestor may be transmitted to both parents of an inbred individual in a later generation and become homozygous in the inbred offspring. The replicas of an ancestral allele are said to be identical by descent. The inbreeding coefficient of an inbred organism can be deduced directly from the pedigree of inbreeding by calculating the probability of identity by descent along every possible path of descent through each common ancestor. Among inbred individuals, the frequency of heterozygous genotypes is smaller, and that of homozygous genotypes greater, than it would be with random mating.

Evolution is the progressive increase in the degree to which a species becomes genetically adapted to its environment. A principal mechanism of evolution is natural selection, in which individuals superior in survival or reproductive ability in the prevailing environment contribute a disproportionate share of genes to future generations, thereby gradually increasing the frequency of the favorable alleles in the whole population. However, at least three other processes also can change allele frequency: mutation (heritable change in a gene), migration (movement of individuals among subpopulations), and random genetic drift (resulting from restricted population size). Spontaneous mutation rates are generally so low that the effect of mutation on changing allele frequency is minor, except for rare alleles. Migration can have significant effects on allele frequency because migration rates may be very large. The main effect of migration is the tendency to equalize allele frequencies among the local populations that exchange migrants. Selection occurs through differences in viability (the probability of survival of a genotype) and in fertility (the probability of successful reproduction).

Populations maintain harmful alleles at low frequencies as a result of a balance between selection, which tends to eliminate the alleles, and mutation, which tends to increase their frequencies. When there is selection-mutation balance, the allele frequency at equilibrium is usually greater if the allele is completely recessive than if it is partially dominant.

This difference arises because selection is quite ineffective in affecting the frequency of a completely recessive allele when the allele is rare, owing to the almost exclusive appearance of the allele in heterozygotes.

A few examples are known in which the heterozygous genotype has a greater fitness than either of the homozygous genotypes (heterozygote superiority). Heterozygote superiority results in an equilibrium in which both alleles are maintained in the population. An example is sickle-cell anemia in regions of the world where falciparum malaria is endemic. Heterozygous persons have an increased resistance to malaria and only a mild anemia, a circumstance that results in greater fitness than that of either homozygote.

Random genetic drift is a statistical process of change in allele frequency in small populations, resulting from the inability of every individual to contribute equally to the offspring of successive generations. In a subdivided population, random genetic drift results in differences in allele frequency among the subpopulations. In an isolated population, barring mutation, an allele will ultimately become either fixed or lost because of random genetic drift. The probability of ultimate fixation of an allele as a result of random genetic drift is equal to the frequency of the allele in the initial population.

Key Terms

adaptation
allele frequency
allozyme
assortative mating (positive and negative)
consanguineous mating
degree of dominance
DNA typing
electrophoresis
equilibrium
evolution
fertility
fixed allele
forward mutation
gene pool
genotype frequency
Hardy-Weinberg principle

heterozygote superiority
identity by descent
inbreeding
inbreeding coefficient
local population
lost allele
mating system
migration
mutation
natural selection
overdominance
path in a pedigree
polymorphic gene
population
population genetics
population substructure
random genetic drift

random mating
relative fitness
reverse mutation
selection coefficient
selection-mutation balance
selectively neutral mutation
subpopulation
viability
VNTR (variable number of tandem repeats)

Review the Basics

• With regard to allele frequency, what does it mean to say that an allele is fixed? That an allele is lost?

• What is the Hardy-Weinberg principle and how does it apply to multiple alleles?

• Name four evolutionary processes that can change allele frequencies in natural populations. How do allele frequencies change in the absence of these processes?

• Distinguish between inbreeding and assortative mating.

• What is the role of mutation in the evolutionary process? Why is mutation a weak force for changing allele frequencies?

• What is the meaning of the term *fitness* in population genetics? Give an example of an organism with high fitness and an example of an organism with low fitness.

Chapter 15

GeNETics on the web will introduce you to some of the most important sites for finding genetic information on the Internet. To complete the exercises below, visit the Jones and Bartlett home page at

http://www.jbpub.com/genetics

Select the link to *Genetics: Principles and Analysis* and then choose the link to *GeNETics on the web.* You will be presented with a chapter-by-chapter list of highlighted keywords.

GeNETics EXERCISES

Select the highlighted keyword in any of the exercises below, and you will be linked to a web site containing the genetic information necessary to complete the exercise. Each exercise suggests a specific, written report that makes use of the information available at the site. This report, or an alternative, may be assigned by your instructor.

1. This **Hardy-Weinberg** site has a clever demonstration of the principle in a game that makes use of several decks of playing cards. If assigned to do so, write a one-paragraph description of the game, explaining why it works and why there have to be at least 25 players.

2. A colorful introduction to **DNA typing** is located at this site. The *how-is-it-done* link illustrates the procedure step by step. If assigned to do so, use the information at this site to write a one-page report on the applications of DNA typing and identify some of the problems in interpretation.

3. Search this keyword site for **Pingelap disease** to identify a human population in which from 4 to 9 percent of the population is blind from infancy. If assigned to do so, write one paragraph describing how random genetic drift might explain the high incidence of the condition in this population.

MUTABLE SITE EXERCISES

The Mutable Site Exercise changes frequently. Each new update includes a different exercise that makes use of genetics resources available on the World Wide Web. Select the **Mutable Site** for Chapter 15, and you will be linked to the current exercise that relates to the material presented in this chapter.

PIC SITE

The Pic Site showcases some of the most visually appealing genetics sites on the World Wide Web. To visit the showcase genetics site, select the **Pic Site** for Chapter 15.

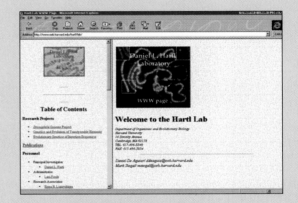

• What is the fitness of an organism that dies before the age of reproduction? What is the fitness of an organism that is sterile?

• Why, in a random-mating population, does the allele frequency of a rare recessive allele change slowly under selection?

• What is a mutation-selection balance? How does the equilibrium allele frequency of a harmful allele depend on the degree of dominance?

• What is random genetic drift and why does it occur? In the absence of any counteracting forces, what is the ultimate effect of random genetic drift on allele frequency? Are random changes in allele frequency from one generation to the next greater in small or large populations, and why?

Problem 1: A sample of 300 plants from a population is examined for the electrophoretic mobility of an enzyme that varies according to the genotype determined by two alleles, *F* and *S*, of a single gene. The results are 7 plants with genotype *FF*, 106 with genotype *FS*, and 187 with genotype *SS*. What are the allele frequencies of *F* and *S*? What are the expected numbers of the three genotypes, assuming random mating?

Answer: The allele frequencies are determined by counting the alleles. The 7 *FF* plants represent 14 *F* alleles, the 106 *FS* plants represent 106 *F* and 106 *S* alleles, and the 187 *SS* plants represent 374 *S* alleles, for a total of 300 $\times$ 2 = 600 alleles altogether. The allele frequency p of *F* is $(14 + 106)/600 = 0.2$, and the allele frequency q of *S* is $(106 + 374)/600 = 0.8$. As a check on the calculations, note that the allele frequencies sum to unity, as they should. For the second part of the question, the expected genotype frequencies with random mating are p^2 *FF*, $2pq$ *FS*, and q^2 *SS*, so the expected numbers are

$$FF: \quad (0.2)^2 \times 300 = 12$$
$$FS: \quad 2(0.2)(0.8) \times 300 = 96$$
$$SS: \quad (0.8)^2 \times 300 = 192$$

As a check on these calculations, note that $12 + 96 + 192 = 300$.

Problem 2: Excessive secretion of male sex hormones results in premature sexual maturation in males and masculization of the sex characters in females. This disorder is called the adrenogenital syndrome, and in Switzerland there is an autosomal recessive form of the disease that affects about one in 5000 newborns.

(a) Assuming random mating, what is the allele frequency of the recessive?

(b) What is the frequency of heterozygous carriers?

(c) What is the expected frequency of the condition among the offspring of first cousins?

Answer:

(a) Set $q^2 = 1/5000$, which implies that
$$q = (1/5000)^{1/2} = 0.014.$$

(b) The frequency of heterozygotes is $2pq$, in which $p = 1 - 0.014 = 0.986$. Thus
$$2pq = 2 \times 0.014 \times 0.986 = 0.028 = 1/36$$

that is, almost 3 percent of the population are carriers, even though only about 0.02 percent of the population are affected.

(c) With first-cousin mating, the inbreeding coefficient $F = 1/16 = 0.062$, and the expected proportion of affected people is

$$q^2(1 - F) + qF = (0.014)^2(0.938) + (0.014)(0.062)$$
$$= 0.001$$

In other words, there is about a five-fold greater risk of homozygosity for this allele among the offspring of first cousins.

Problem 3: Warfarin is a rat killer that acts by hindering blood coagulation. Continued use of the poison has resulted in the evolution of resistance in most rat populations because of selection favoring a resistance mutation *R*. The normal, sensitive allele may be denoted *S*. In the absence of warfarin, the relative fitnesses of *SS*, *RS*, and *RR* are in the ratio 1.00 : 0.77 : 0.46, and in the presence of warfarin, the ratio of fitnesses is 0.68 : 1.00 : 0.37.

(a) With regard to persistence or loss of the *R* and *S* alleles, what is the expected result of the continued use of the chemical?

(b) What would happen if warfarin were no longer used?

(c) Under what circumstances might the *R* allele become fixed?

Answer:

(a) In the presence of warfarin, there is heterozygote superiority, so continued use would be expected to result in a stable equilibrium in which both *R* and *S* persisted in the population.

(b) In the absence of warfarin, the genotype *SS* is favored and *RS* is intermediate in fitness, so allele *S* would become fixed and *R* would become lost.

(c) The allele *R* could become fixed in a small population where, because of chance fluctuations in allele frequency, the *S* allele might become lost. (However, small populations would also increase the chance that the *R* allele might be lost.)

Analysis and Applications

15.1 If the genotype *AA* is an embryonic lethal and the genotype *aa* is fully viable but sterile, what genotype frequencies would be found in adults in an equilibrium population containing the *A* and *a* alleles? Is it necessary to assume random mating?

15.2 For an X-linked recessive allele maintained by mutation-selection balance, would you expect the equilibrium frequency of the allele to be greater or smaller than that of an autosomal recessive allele, assuming that the relative fitnesses are the same in both cases? Why?

15.3 How many *A* and *a* alleles are present in a sample of organisms that consists of 10 *AA*, 15 *Aa*, and 4 *aa* individuals? What are the allele frequencies in this sample?

15.4 Allozyme phenotypes of alcohol dehydrogenase in the flowering plant *Phlox drummondii* are determined by codominant alleles of a single gene. In one sample of 35 plants, the following data were obtained:

Genotype	AA	AB	BB	BC	CC	AC
Number	2	5	12	10	5	1

What are the frequencies of the alleles *A, B,* and *C* in this sample?

15.5 The accompanying illustration shows the gel patterns observed with a probe for a VNTR (variable number of tandem repeats) locus among four pairs of parents (1–8) and four children (A–D). One child comes from each pair of parents.

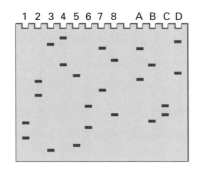

(a) Why does each person have two bands?

(b) How would it be possible for a person to have only one band?

(c) What type of dominance is illustrated by the VNTR alleles?

(d) Which pairs of people who have the DNA in lanes 1–8 are the parents of each child?

15.6 DNA from 100 unrelated people was digested with the restriction enzyme *Hin*dIII, and the resulting fragments were separated and probed with a sequence for a particular gene. Four fragment lengths that hybridized with the probe were observed—namely 5.7, 6.0, 6.2, and 6.5 kb—where each fragment defines a different restriction-fragment allele. The accompanying illustration shows the gel patterns observed; the number of individuals with each gel pattern is shown across the top. Estimate the allele frequencies of the four restriction-fragment alleles.

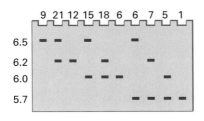

15.7 Which of the following genotype frequencies of *AA*, *Aa*, and *aa*, respectively, satisfy the Hardy-Weinberg principle?

(a) 0.25, 0.50, 0.25

(b) 0.36, 0.55, 0.09

(c) 0.49, 0.42, 0.09

(d) 0.64, 0.27, 0.09

(e) 0.29, 0.42, 0.29

15.8 If the frequency of a homozygous dominant genotype in a randomly mating population is 0.09, what is the frequency of the dominant allele? What is the combined frequency of all the other alleles of this gene?

15.9 Hartnup disease is an autosomal-recessive disorder of intestinal and renal transport of amino acids. The frequency of affected newborn infants is about 1 in 14,000. Assuming random mating, what is the frequency of heterozygotes?

15.10 A randomly mating population of dairy cattle contains an autosomal-recessive allele causing dwarfism. If the frequency of dwarf calves is 10 percent, what is the frequency of heterozygous carriers of the allele in the entire herd? What is the frequency of heterozygotes among nondwarf cattle?

15.11 In certain grasses, the ability to grow in soil contaminated with the toxic metal nickel is determined by a dominant allele.

(a) If 60 percent of the seeds in a randomly mating population are able to germinate in contaminated soil, what is the frequency of the resistance allele?

(b) Among plants that germinate, what proportion are homozygous?

15.12 In Caucasians, the M-shaped hairline that recedes with age (sometimes to nearly complete baldness) is due to an allele that is dominant in males but recessive in females. The frequency of the baldness allele is approximately 0.3. Assuming random mating, what frequencies of the bald and nonbald phenotypes are expected in males and females?

15.13 In a Pygmy group in Central Africa, the frequencies of alleles determining the ABO blood groups were estimated as 0.74 for I^O, 0.16 for I^A, and 0.10 for I^B. Assuming random mating, what are the expected frequencies of ABO genotypes and phenotypes?

15.14 Among 35 individuals of the flowering plant *Phlox roemariana*, the following genotypes were observed for a gene determining electrophoretic forms of the enzyme phosphoglucose isomerase: 2 *AA*, 13 *AB*, 20 *BB*.

(a) What are the frequencies of the alleles *A* and *B*?

(b) Assuming random mating, what are the expected numbers of the genotypes?

15.15 If an X-linked recessive trait is present in 2 percent of the males in a population with random mating, what is the frequency of the trait in females? What is the frequency of carrier females?

15.16 In a population of *Drosophila*, an X-linked recessive allele causing yellow body color is present in genotypes at frequencies typical of random mating; the frequency of the recessive allele is 0.20. Among 1000 females and 1000 males, what are the expected numbers of the yellow and wildtype phenotypes in each sex?

15.17 How does the frequency of heterozygotes in an inbred population compare with that in a randomly mating population with the same allele frequencies?

15.18 Which of the following genotype frequencies of *AA*, *Aa*, and *aa*, respectively, are suggestive of inbreeding?

(a) 0.25, 0.50, 0.25

(b) 0.36, 0.55, 0.09

(c) 0.49, 0.42, 0.09

(d) 0.64, 0.27, 0.09

(e) 0.29, 0.42, 0.29

15.19 Galactosemia is an autosomal-recessive condition associated with liver enlargement, cataracts, and mental retardation. Among the offspring of unrelated individuals, the frequency of galactosemia is 8.5×10^{-6}. What is the expected frequency among the offspring of first cousins

$(F = 1/16)$ and among the offspring of second cousins $(F = 1/64)$?

15.20 Self-fertilization in the annual plant *Phlox cuspidata* results in an average inbreeding coefficient of $F = 0.66$.

(a) What frequencies of the genotypes for the enzyme phosphoglucose isomerase would be expected in a population with alleles *A* and *B* at frequencies 0.43 and 0.57, respectively?

(b) What frequencies of the genotypes would be expected with random mating?

15.21 Two strains of *Escherichia coli*, *A* and *B*, are inoculated into a chemostat in equal frequencies and undergo competition. After 40 generations, the frequency of the *B* strain is 35 percent. What is the fitness of strain *B* relative to strain *A*, under the particular experimental conditions, and what is the selection coefficient against strain *B*?

15.22 Determine the ultimate fate of the *A* allele with random mating when the relative fitnesses of *AA*, *Aa*, and *aa* are

(a) 1.00, 0.98, 0.96

(b) 1.00, 1.02, 1.02

(c) 1.00, 1.02, 0.98

(d) 1.00, 1.01, 1.02

15.23 For an autosomal gene with two alleles in a randomly mating population, what is the probability that a heterozygous female will have a heterozygous offspring?

Challenge Problems

15.24 In a randomly mating population that contains *n* equally frequent alleles, what is the expected frequency of heterozygotes?

15.25 Human genetic polymorphisms in which there are a variable number of tandem repeats (VNTR) are often used in DNA typing. The accompanying table shows the alleles of four VNTR loci and their estimated average frequencies in genetically heterogeneous Caucasian populations. (The allele 9.3 of the *HUMTHO1* locus has a partial copy of one of the

tandem repeats.) Using Equation (3), estimate

(a) the probability that an individual will be of genotype (6, 9.3) for *HUMTHO1*, (10, 11) for *HUMFES*, (6, 6) for *D12S67*, and (18, 24) for *D1S80*.

(b) The probability that an individual will be of genotype (8, 10) for *HUMTHO1*, (8, 13) for *HUMFES*, (1, 2) for *D12S67*, and (19, 19) for *D1S80*.

VNTR locus

HUMTHO1		HUMFES		D12S67		D1S80	
Repeats	Freq.	Repeats	Freq.	Repeats	Freq.	Repeats	Freq.
6	0.230	8	0.007	1	0.015	18	0.293
7	0.160	10	0.321	2	0.005	19	0.011
8	0.105	11	0.373	3	0.058	20	0.021
9	0.193	12	0.233	4	0.078	21	0.032
9.3	0.310	13	0.066	5	0.118	22	0.043
10	0.002			6	0.324	23	0.016
				7	0.196	24	0.335
				8	0.127	25	0.037
				9	0.059	26	0.016
				10	0.020	28	0.059
						29	0.059
						30	0.016
						31	0.043

Allison, A. C. 1956. Sickle cells and evolution. *Scientific American,* August.

Ayala, F. 1978. The mechanisms of evolution. *Scientific American,* September.

Bittles, A. H., W. M. Mason, J. Greene, and N. A. Rao. 1991. Reproductive behavior and health in consanguineous marriages. *Science* 252: 789.

Bodmer, W. F., and L. L. Cavalli-Sforza. 1976. *Genetics, Evolution, and Man.* New York: Freeman.

Cavalli-Sforza, L. L. 1974. The genetics of human populations. *Scientific American,* September.

Crow, J. F., and M. Kimura. 1970. An *Introduction to Population Genetics Theory.* New York: Harper & Row.

Dobzhansky, T. 1941. *Genetics and the Origin of the Species.* New York: Columbia University Press.

Falconer, D. S. and T. F. C. Mackay. 1996. *Introduction of Quantitative Genetics.* 4th ed. Essex, England: Longman.

Harris, H., and D. A. Hopkinson. 1972. Average heterozygosity in man. *Journal of Human Genetics* 36: 9.

Hartl, D. L. 1988. *A Primer of Population Genetics.* 2d ed. Sunderland, MA: Sinauer.

Hartl, D. L., and A. G. Clark. 1997. *Principles of Population Genetics.* 3rd ed. Sunderland, MA: Sinauer.

Hoffmann, A. A., C. M. Sgro, and S. H. Lawler. 1995. Ecological population genetics: The interface between genes and the environment. *Annual Review of Genetics* 29: 349.

Kimura, M. 1983. *The Neutral Theory of Molecular Evolution.* Cambridge, England: Cambridge University Press.

Kline, J., N. Takahata, and F. J. Ayala. 1993. MHC polymorphism and human origins. *Scientific American,* November.

Lewontin, R. C. 1974. *Genetic Basis of Evolutionary Change.* New York: Columbia University Press.

Li, W.-H., and D. Graur. 1991. *Fundamentals of Molecular Evolution.* Sunderland, MA: Sinauer.

Nei, M. 1987. *Molecular Evolutionary Genetics.* New York: Columbia University Press.

Neufeld, D. J., and N. Colman. 1990. When science takes the witness stand. *Scientific American,* May.

Potts, W. K., and S. K. Wakeland. 1993. Evolution of MHC genetic diversity: A tale of incest, pestilence and sexual preference. *Trends in Genetics* 9: 408.

Stebbins, G. L. 1977. *Processes of Organic Evolution.* 3d ed. Englewood Cliffs, NJ: Prentice-Hall.

Weir, B. S. 1996. *Genetic Data Analysis II.* Sunderland, MA: Sinauer.

White, R., and J.-M. Lalouel. 1988. Chromosome mapping with DNA markers. *Scientific American,* February.

Wilson, A. C., and R. L. Cann. 1992. The recent African genesis of humans. *Scientific American,* April.

Wright, S. 1978. *Evolution and the Genetics of Populations* (4 volumes). Chicago: University of Chicago Press.

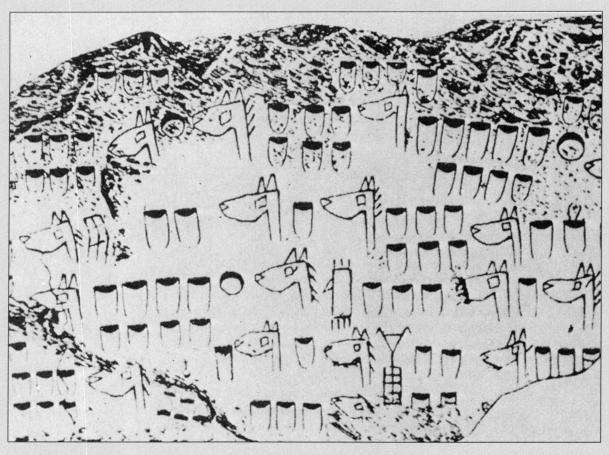

A stone carving found in Asia, about 4000 years old, showing a herd of domesticated horses. The markings indicate that the breeders kept records of desirable traits in their horses and may have used these traits as the basis of artificial selection. [Courtesy of Dorsey Stuart.]

Quantitative Genetics and Multifactorial Inheritance

PRINCIPLES

- Multifactorial traits are determined by multiple genetic and environmental factors acting together.
- The relative contributions of genotype and environment to a trait are measured by the variance due to genotype (genotypic variance) and the variance due to environment (environmental variance).
- Correlations between relatives are used to estimate various components of variation, such as genotypic variance, additive variance, and dominance variance.
- Additive variance accounts for the parent-offspring correlation; dominance variance accounts for the sib-sib correlation over and above that expected from the additive variance.
- Narrow-sense heritability is the ratio of additive (transmissible) variance to the total phenotypic variance; it is widely used in plant and animal breeding.
- Genes that affect quantitative traits can be identified and genetically mapped using genetic polymorphisms, especially highly polymorphic DNA sequences.

CONNECTIONS

CONNECTION: The Supreme Law of Unreason
Francis Galton 1989
Natural Inheritance

CONNECTION: Human Gene Map
Jeffrey C. Murry and 26 other investigators 1994
A comprehensive human linkage map with centimorgan density

Earlier chapters have emphasized traits in which differences in phenotype result from alternative genotypes of a single gene. Examples include green versus yellow peas, red eyes versus white eyes in *Drosophila*, normal versus sickle-cell hemoglobin, and the ABO blood groups. These traits are particularly suited for genetic analysis through the study of pedigrees because of the small number of genotypes and phenotypes and because of the simple correspondence between genotype and phenotype. However, many traits of importance in plant breeding, animal breeding, and medical genetics are influenced by *multiple* genes as well as by the effects of environment. These are known as **multifactorial traits** because of the multiple genetic and environmental factors implicated in their causation. With a multifactorial trait, a single genotype can have many possible phenotypes (depending on the environment), and a single phenotype can include many possible genotypes. Genetic analysis of such complex traits requires special concepts and methods, which are introduced in this chapter.

16.1
Quantitative Inheritance

Multifactorial traits are often called **quantitative traits** because the phenotypes in a population differ in the quantity of a characteristic rather than in type. Height is a quantitative trait. Heights are not found in discrete categories but differ merely in quantity from one person to the next. The opposite of a quantitative trait is a *discrete* trait, in which the phenotypes differ in kind—for example, brown eyes versus blue eyes.

Quantitative traits are typically influenced not only by the alleles of two or more genes but also by the effects of environment. Therefore, with a quantitative trait, the phenotype of an organism is potentially influenced by

- **Genetic factors** in the form of alternative genotypes of one or more genes.

- **Environmental factors** in the form of conditions that are favorable or unfa-

vorable for the development of the trait. Examples include the effect of nutrition on the growth rate of animals, and the effects of fertilizer, rainfall, and planting density on the yield of crop plants.

With some quantitative traits, differences in phenotype result largely from differences in genotype, and the environment plays a minor role. With other quantitative traits, differences in phenotype result largely from the effects of environment, and genetic factors play a minor role. However, most quantitative traits fall between these extremes, and both genotype and environment must be taken into account in their analysis.

In a genetically heterogeneous population, many genotypes are formed by the processes of segregation and recombination. Variation in genotype can be eliminated by studying inbred lines, which are homozygous for most genes, or the F_1 progeny from a cross of inbred lines, which are uniformly heterozygous for all genes in which the parental inbreds differ. In contrast, complete elimination of environmental variation is impossible, no matter how hard the experimenter may try to render the environment identical for all members of a population. With plants, for example, small variations in soil quality or exposure to the sun will produce slightly different environments, sometimes even for adjacent plants. Similarly, highly inbred *Drosophila* still show variation in phenotype (for example, in body size) brought about by environmental differences among animals within the same culture bottle. Therefore, traits that are susceptible to small environmental effects will never be uniform, even in inbred lines.

Most traits that are important in plant and animal breeding are quantitative traits. In agricultural production, one economically important quantitative trait is yield—for example, the size of the harvest of corn, tomatoes, soybeans, or grapes per unit area. In domestic animals, important quantitative traits include meat quality, milk production per cow, egg laying per hen, fleece weight per sheep, and litter size per sow. Important quantitative traits in human genetics include infant growth rate, adult weight, blood pressure, serum cholesterol, and length of life. In evolutionary studies, fitness is the preeminent quantitative trait.

Most quantitative traits cannot be studied by means of the usual pedigree methods because the effects of segregation of alleles of one gene may be concealed by effects of other genes, and environmental effects may cause identical genotypes to have different phenotypes. Therefore, individual pedigrees of quantitative traits do not fit any simple pattern of dominance, recessiveness, or X linkage. Nevertheless, genetic effects on quantitative traits can be assessed by comparing the phenotypes of relatives who, because of their familial relationship, must have a certain proportion of their genes in common. Such studies utilize many of the concepts of population genetics discussed in Chapter 15.

Three categories of traits are frequently found to have quantitative inheritance. They are described in the following section.

Continuous, Meristic, and Threshold Traits

Most phenotypic variation in populations is not manifested in a few easily distinguished categories. Instead, the traits vary continuously from one phenotypic extreme to the other, with no clear-cut breaks in between. Height is again a prime example of such a trait. Other examples include milk production in cattle, growth rate in poultry, yield in corn, and blood pressure in human beings. Such traits are called **continuous traits** because there is a continuous gradation from one phenotype to the next, from minimum to maximum, with no clear categories. Weight is an example of such a trait, because the weight of an organism can fall anywhere along a continuous scale of weights. The distinguishing characteristic of continuous traits is that the phenotype of an individual organism can fall anywhere on a continuous scale of measurement, and so the number of possible phenotypes is virtually unlimited.

Two other types of quantitative traits are not continuous:

Meristic traits are traits in which the phenotype is determined by counting. Some examples are number of skin ridges forming the fingerprints, number of kernels on an ear of corn, number of eggs laid by a hen, number of bristles on the abdomen of a fly, and number of puppies in a litter. Another example of a meristic trait is the number of ears on a stalk of corn, which typically has the value 1, 2, 3, or 4 ears on a given stalk.

Threshold traits are traits that have only two, or a few, phenotypic classes, but their inheritance is determined by the effects of multiple genes together with the environment. Examples of threshold traits are twinning in cattle and parthenogenesis (development of unfertilized eggs) in turkeys. In a threshold trait, each organism has an underlying and not directly observable predisposition to express the trait, such as a predisposition for a cow to give birth to twins. If the underlying predisposition is high enough (above a "threshold"), the cow will actually give birth to twins; otherwise she will give birth to a single calf. In many threshold-trait disorders, the phenotypic classes are "affected" versus "not affected." Examples of threshold-trait disorders in human beings include adult-onset diabetes, schizophrenia, and many congenital abnormalities, such as spina bifida. Threshold traits can be interpreted as continuous traits by imagining that each individual has an underlying risk or *liability* toward manifestation of the condition. A liability above a certain cutoff, or *threshold,* results in expression of the condition; a liability below the threshold results in normality. The liability of an individual to expressing a threshold trait cannot be observed directly, but inferences about liability can be drawn from the incidence of the condition among individuals and their relatives. The manner in which this is done is discussed in Section 16.6.

Distributions

The **distribution** of a trait in a population is a description of the population in terms of the proportion of individuals that have each of the possible phenotypes. Characterizing the distribution of some traits is straightforward because the number of phenotypic classes is small. For example, the distribution of progeny in a certain pea cross may consist of 3/4 green seeds and 1/4 yellow seeds, and the distribution of ABO blood groups among Greeks consists of 42 percent O, 39 percent A, 14 percent B, and 5 percent AB. However, with continuous traits, the large number of possible phenotypes makes such summaries impractical. Often, it is convenient to reduce the number of phenotypic classes by grouping similar phenotypes together. Data for an example pertaining to the distribution of height among 4995 British women are given in Table 16.1 and in Figure 16.1. You can imagine each bar in the graph in Figure 16.1 being built, step by step, as each of the women is measured, by placing a small square along the *x*-axis at the location corresponding to the height of that woman. As sampling proceeds, the squares begin to pile up in certain places, leading ultimately to the bar graph shown.

Displaying a distribution completely, in either tabular form (Table 16.1) or graphical form (Figure 16.1), is always helpful but often unnecessary; frequently, a description of the distribution in terms of two major features—the *mean* and the *variance*—is sufficient. To discuss the mean and the variance in quantitative terms, we will need some symbols. In Table 16.1, the height intervals are numbered from 1 (53–55 inches) to 11 (73–75 inches). The

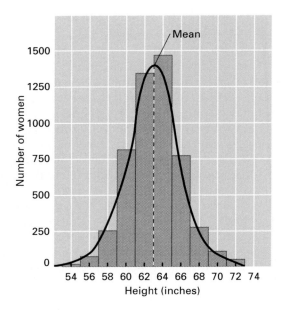

Figure 16.1
Distribution of height among 4995 British women and the smooth normal distribution that approximates the data.

symbol x_i designates the midpoint of the height interval numbered *i*; for example, $x_1 = 54$ inches, $x_2 = 56$ inches, and so on. The number of women in height interval *i* is designated f_i; for example, $f_1 = 5$ women, $f_2 = 33$ women, and so forth. The total size of the sample, in this case 4995, is denoted as *N*. The mean and variance serve to characterize the distribution of height among these women as well as the distribution of many other quantitative traits.

- The **mean,** or average, is the peak of the distribution. The mean of a population is estimated from a sample of individuals from the population, as follows:

$$\bar{x} = \frac{\Sigma f_i x_i}{N} \qquad (1)$$

in which $\bar{x}$ is the estimate of the mean and Σ symbolizes summation over all classes of data (in this example, summation over all 11 height intervals). In Table 16.1, the mean height in the sample of women is 63.1 inches.

- The **variance** is a measure of the spread of the distribution and is estimated in terms of the squared *deviation* (differ-

Table 16.1 Distribution of height among British women

Interval number (*i*)	Height interval (inches)	Midpoint (*x_i*)	Number of women (*f_i*)
1	53–55	54	5
2	55–57	56	33
3	57–59	58	254
4	59–61	60	813
5	61–63	62	1340
6	63–65	64	1454
7	65–67	66	750
8	67–69	68	275
9	69–71	70	56
10	71–73	72	11
11	73–75	74	4
		Total *N* =	4995

ence) of each observation from the mean. The variance is estimated from a sample of individuals as follows:

$$s^2 = \frac{\Sigma f_i(x_i - \bar{x})^2}{N - 1} \qquad (2)$$

in which s^2 is the estimated variance and x_i, f_i, and N are as in Table 16.1. Note that $(x_i - \bar{x})$ is the difference from the mean of each height category and that the denominator is the total number of individuals minus 1. The variance describes the extent to which the phenotypes are clustered around the mean, as shown in Figure 16.2. A large value implies that the distribution is spread out, and a small value implies that it is clustered near the mean. From the data in Table 16.1, the variance of the population of British women is estimated as $s^2 = 7.24$ in^2.

A quantity closely related to the variance—the **standard deviation** of the dis-tribution—is defined as the square root of the variance. For the data in Table 16.1, the estimated standard deviation s is obtained from Equation 2 as

$$s = (s^2)^{1/2} = (7.24 \text{ in}^2)^{1/2} = 2.29 \text{ inches}$$

The standard deviation has the useful feature of having the same units of dimension as the mean—in this example, inches.

When the data are symmetrical, or approximately symmetrical, the distribution of a trait can often be approximated by a smooth arching curve of the type shown in Figure 16.1. The arch-shaped curve is called the **normal distribution.** Because the normal curve is symmetrical, half of its area is determined by points that have values greater than the mean and half by points with values less than the mean, and thus the proportion of phenotypes that exceed the mean is 1/2. One important characteristic of the normal distribution is that it is completely determined by the value of the mean and the variance.

Figure 16.2

Graphs showing that the variance of a distribution measures the spread of the distribution around the mean. The area under each curve covering any range of phenotypes equals the proportion of individuals having phenotypes within that range.

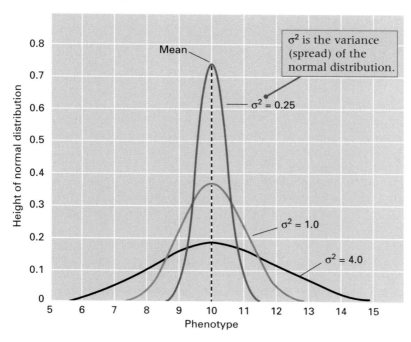

The Supreme Law of Unreason

Francis Galton 1989
42 Rutland Gate, South Kensington,
London, England

Francis Galton pioneered the application of statistics to biological problems. He was captivated by the properties of the normal distribution. This form of distribution is also known by several other names, including the "bell curve," the "Gaussian distribution," and the "law of the frequency of errors." The name Gaussian *comes from Carl Friedrich Gauss, a German mathematician who studied the distribution in the early 1800s. The distribution was popularized by the Belgian mathematician and astronomer Adolphe Quetelet (1796–1874). He showed that various human measurements conformed to the normal distribution, including the heights of French draftees and the chest girths of Scottish soldiers. From this conformity, Quetelet developed the concept of the* homme moyen, *or "average person," as the mean of the distribution. Deviations from the* homme moyen *were regarded as "errors" of the same sort that arise in astronomy when the magnitude of any celestial quantity is estimated. In this concept, variation around the mean is a regrettable fact of an imperfect world. It was Galton who realized that in biology, variation is an intrinsic result of differences in genotype and environment among individuals. He regarded*

biological variation as a subject demanding study in its own right, and to him we owe such fundamental concepts as correlation and regression. Why is the normal distribution so often found in practice? A theorem important in statistics, called the Central Limit Theorem, states that under quite general conditions, the sum of a large number of independent random quantities is bound to converge to a normal distribu-

---

The law would have been personified by the Greeks if they had known of it.

---

tion. In quantitative genetics, the independent quantities are multiple genes and multiple environmental factors, which often do interact to yield a normal distribution of a trait.

It is difficult to understand why statisticians commonly limit their inquiries to averages, and do not revel in more comprehensive views. Their souls seem dull to the charm of variety. . . . An average is but a solitary fact, whereas if a single other fact [the standard deviation] be added to it, an entire normal distribution, which nearly corresponds to the observed one, starts potentially into

existence. . . . Some people hate the very name of statistics, but I find it full of beauty and interest. Measurements, whenever they are not brutalized, but delicately handled, and are warily interpreted, their power of dealing with complicated phenomena is extraordinary. They are the only tools by which an opening can be cut through the formidable thicket of difficulties that bar the path of those who pursue the science of humanity. . . . I know of scarcely anything so apt to impress the imagination as the wonderful form of cosmic order expressed by the "law of frequency of error" [the normal distribution]. Whenever a large sample of chaotic elements is taken in hand and marshaled in the order of their magnitude, this unexpected and most beautiful form of regularity proves to have been latent all along. The law would have been personified by the Greeks if they had known of it. It reigns with serenity and complete self-effacement amidst the wildest confusion. The larger the mob and the greater the apparent anarchy, the more perfect is its sway. It is the supreme law of unreason.

Source: Natural Inheritance. Macmillan Publishers, London

The mean and standard deviation (square root of the variance) of a normal distribution provide a great deal of information about the distribution of phenotypes in a population, as is illustrated in Figure 16.3. Specifically, for a normal distribution,

1. Approximately 68 percent of the population have a phenotype within 1 standard deviation of the mean (in the symbols of Figure 16.3, between $\mu - \sigma$ and $\mu + \sigma$).

2. Approximately 95 percent lie within 2 standard deviations of the mean (between $\mu - 2\sigma$ and $\mu + 2\sigma$).

3. Approximately 99.7 percent lie within 3 standard deviations of the mean (between $\mu - 3\sigma$ and $\mu + 3\sigma$).

Applying these rules to the data in Figure 16.1, in which the mean and standard deviation are 63.1 and 2.69 inches, reveals that approximately 68 percent of the women are expected to have heights in the range

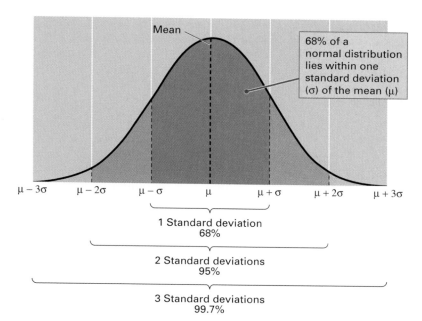

Figure 16.3

Features of a normal distribution. The proportions of individuals lying within 1, 2, and 3 standard deviations of the mean are approximately 68 percent, 95 percent, and 99.7 percent, respectively. In this normal distribution, the mean is symbolized μ and the standard deviation σ.

Labels on figure: Mean; 68% of a normal distribution lies within one standard deviation (σ) of the mean (μ)

$\mu - 3\sigma$ $\quad$ $\mu - 2\sigma$ $\quad$ $\mu - \sigma$ $\quad$ μ $\quad$ $\mu + \sigma$ $\quad$ $\mu + 2\sigma$ $\quad$ $\mu + 3\sigma$

1 Standard deviation
68%

2 Standard deviations
95%

3 Standard deviations
99.7%

from $63.1 - 2.69$ inches to $63.1 + 2.69$ inches (that is, 60.4–65.8), and approximately 95 percent are expected to have heights in the range from $63.1 - 2 \times 2.69$ inches to $63.1 + 2 \times 2.69$ inches (that is, 57.7–68.5).

Real data frequently conform to the normal distribution. Normal distributions are usually the rule when the phenotype is determined by the cumulative effect of many individually small independent factors. This is the case for many multifactorial traits.

16.2
Causes of Variation

In considering the genetics of multifactorial traits, an important objective is to assess the relative importance of genotype and environment. In some cases in experimental organisms, it is possible to separate genotype and environment with respect to their effects on the mean. For example, a plant breeder may study the yield of a series of inbred lines grown in a group of environments that differ in planting density or amount of fertilizer. It would then be possible (1) to compare yields of the same genotype grown in different environments and thereby rank the *environments*

relative to their effects on yield, or (2) to compare yields of different genotypes grown in the same environment and thereby rank the *genotypes* relative to their effects on yield.

Such a fine discrimination between genetic and environmental effects is not usually possible, particularly in human quantitative genetics. For example, with regard to the height of the women in Figure 16.1, environment could be considered favorable or unfavorable for tall stature only in comparison with the mean height of a genetically identical population reared in a different environment. This reference population does not exist. Likewise, the genetic composition of the population could be judged as favorable or unfavorable for tall stature only in comparison with the mean of a genetically different population reared in an identical environment. This reference population does not exist either.

Without such standards of comparison, it is impossible to distinguish genetic from environmental effects on the mean. However, it is still possible to assess genetic versus environmental contributions to the *variance*, because instead of comparing the means of two or more populations, we can compare the phenotypes of individuals within the *same* population. Some of the differences in phenotype result from differences in genotype and others from differences in environment, and it is often possible to separate these effects.

In any distribution of phenotypes, such as the one in Figure 16.1, four sources contribute to phenotypic variation:

1. Genotypic variation.

2. Environmental variation.

3. Variation due to genotype-environment interaction.

4. Variation due to genotype-environment association.

Each of these sources of variation is discussed in the following sections.

Genotypic Variance

The variation in phenotype caused by differences in genotype among individuals is termed **genotypic variance.** Figure 16.4 illustrates the genetic variation expected among the F_2 generation from a cross of two inbred lines that differ in genotype for three unlinked genes. The alleles of the three genes are represented as Aa, Bb, and Cc, and the genetic variation in the F_2 caused by segregation and recombination is

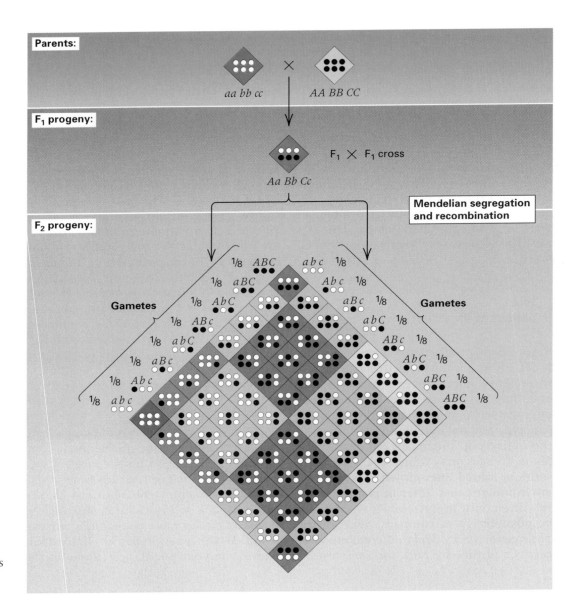

Figure 16.4
Segregation of three independent genes affecting a quantitative trait. Each uppercase allele in a genotype contributes one unit to the phenotype.

evident in the differences in color. Relative to a meristic trait (a trait whose phenotype is determined by counting, such as ears per stalk in corn), if we assume that each uppercase allele is favorable for the expression of the trait and adds one unit to the phenotype, whereas each lowercase allele is without effect, then the *aa bb cc* genotype has a phenotype of 0, and the *AA BB CC* genotype has a phenotype of 6. There are seven possible phenotypes (0–6) in the F_2 generation. The distribution of phenotypes in the F_2 generation is shown in the colored bar graph in Figure 16.5. The normal distribution approximating the data has a mean of 3 and a variance of 1.5. In this case, we are assuming that *all* of the variation in phenotype in the population results from differences in genotype among the individuals.

Figure 16.5 also includes a bar graph with diagonal lines that represent the theoretical distribution when the trait is determined by 30 unlinked genes segregating in a randomly mating population, grouped into the same number of phenotypic classes as the 3-gene case. We assume that 15 of the genes are nearly fixed for the favorable allele and 15 are nearly fixed for the unfavorable allele. The contribution of each favorable allele to the phenotype has been chosen to make the mean of the distribution equal to 3 and the variance equal to 1.5. Note that the distribution with 30 genes is virtually identical to that with 3 genes and that both are approximated by the same normal curve. If such distributions were encountered in actual research, the experimenter would not be able to distinguish between them. The key point is that

> Even in the absence of environmental variation, the distribution of phenotypes, by itself, provides no information about the number of genes influencing a trait and no information about the dominance relations of the alleles.

However, the number of genes influencing a quantitative trait is important in determining the potential for long-term genetic improvement of a population by means of artificial selection. For example, in the 3-gene case in Figure 16.5, the best possible genotype would have a phenotype of 6, but in the 30-gene case, the best possible

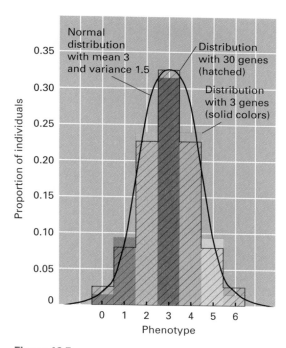

Figure 16.5

The colored bar graph is the distribution of phenotypes determined by the segregation of the 3 genes illustrated in Figure 16.4. The bar graph with the diagonal lines is the theoretical distribution expected from the segregation of 30 independent genes. Both distributions are approximated by the same normal distribution (black curve).

genotype (homozygous for the favorable allele of all 30 genes) would have a phenotype of 30.

Later in this chapter, some methods for estimating how many genes affect a quantitative trait will be presented. All the methods depend on comparing the phenotypic distributions in the F_1 and F_2 generations of crosses between nearly or completely homozygous lines.

Environmental Variance

The variation in phenotype caused by differences in environment among individuals is termed **environmental variance.** Figure 16.6 is an example showing the distribution of seed weight in edible beans. The mean of the distribution is 500 mg, and the standard deviation is 95 mg. All of the beans in this population are genetically identical and homozygous because they are highly inbred. Therefore, in this population, *all* of the phenotypic variation in seed

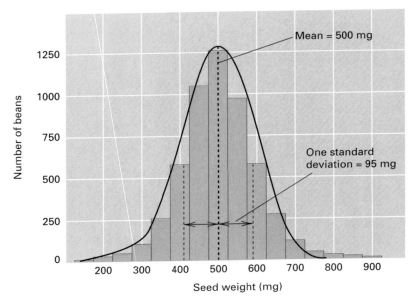

Figure 16.6

Distribution of seed weight in a homozygous line of edible beans. All variation in phenotype among individuals results from environmental differences.

weight results from environmental variance. A comparison of Figure 16.5 and 16.6 demonstrates the following principle:

> The distribution of a trait in a population provides no information about the relative importance of genotype and environment. Variation in the trait can be entirely genetic, it can be entirely environmental, or it can reflect a combination of both influences.

Genotypic and environmental variance are seldom separated as clearly as in Figures 16.5 and 16.6, because usually they work together. Their combined effects are illustrated for a simple hypothetical case in Figure 16.7. Figure 16.7A is the distribution of phenotypes for three genotypes assumed not to be influenced by environment. As depicted, the trait can have one of three distinct and nonoverlapping phenotypes determined by the effects of two additive alleles. The genotypes are in random-mating proportions for an allele frequency of 1/2, and the distribution of phenotypes has mean 5 and variance 2. Because it results solely from differences in genotype, this variance is *genotypic variance,*

which is symbolized σ_g^2. Figure 16.7B is the distribution of phenotypes in the presence of environmental variation, illustrated for each genotype separately. Each distribution corresponds to the one in Figure 16.6, and the variance is 1. Because this variance results solely from differences in environment, it is *environmental variance,* which is symbolized σ_e^2. When the effects of genotype and environment are combined in the same population, each genotype is affected by environmental variation, and the distribution shown in Figure 16.7C of the figure results. The variance of this distribution is the **total variance** in phenotype, which is symbolized σ_t^2. Because we are assuming that genotype and environment have separate and independent effects on phenotype, we expect σ_t^2 to be greater than either σ_g^2 or σ_e^2 alone. In fact,

$$\sigma_t^2 = \sigma_g^2 + \sigma_e^2 \qquad (3)$$

In words, Equation (3) states that

> When genetic and environmental effects contribute independently to phenotype, the total variance equals the sum of the genotypic variance and the environmental variance.

Equation (3) is one of the most important relations in quantitative genetics. How it can be used to analyze data will be explained shortly. Although the equation serves as an excellent approximation in very many cases, it is valid in an exact sense only when genotype and environment are independent in their effects on phenotype. The two most important departures from independence are discussed in the next section.

Genotype-Environment Interaction and Genotype-Environment Association

In the simplest cases, environmental effects on phenotype are additive, and each environment adds or detracts the same amount from the phenotype, independent of the genotype. When this is not true, environmental effects on phenotype differ according to genotype, and a **genotype-environment interaction (G-E interaction)** is said to be present. In some cases,

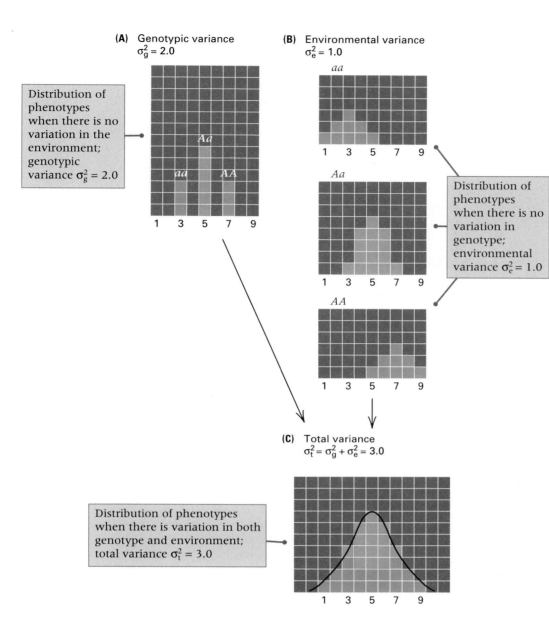

(A) Genotypic variance
$\sigma_g^2 = 2.0$

Distribution of phenotypes when there is no variation in the environment; genotypic variance $\sigma_g^2 = 2.0$

Aa

aa   *AA*

1 3 5 7 9

(B) Environmental variance
$\sigma_e^2 = 1.0$

aa

1 3 5 7 9

Aa

1 3 5 7 9

AA

1 3 5 7 9

Distribution of phenotypes when there is no variation in genotype; environmental variance $\sigma_e^2 = 1.0$

(C) Total variance
$\sigma_t^2 = \sigma_g^2 + \sigma_e^2 = 3.0$

Distribution of phenotypes when there is variation in both genotype and environment; total variance $\sigma_t^2 = 3.0$

1 3 5 7 9

Figure 16.7
The combined effects of genotypic and environmental variance. (A) Population affected only by genotypic variance, σ_g^2. (B) Populations of each genotype affected only by environmental variance, σ_e^2. (C) Population affected by both genotypic and environmental variance, in which the total phenotypic variance, σ_t^2, equals the sum of σ_g^2 and σ_e^2.

G-E interaction can even change the relative ranking of the genotypes, so a genotype that is superior in one environment may become inferior in another.

An example of genotype-environment interaction in maize is illustrated in Figure 16.8. The two strains of corn are hybrids formed by crossing different pairs of inbred lines, and their overall means, averaged across all of the environments, are approximately the same. However, the strain designated A clearly outperforms B in the stressful (negative) environments, whereas the performance is reversed when the envi-

ronment is of high quality. (Environmental quality is judged on the basis of soil fertility, moisture, and other factors.) In some organisms, particularly plants, experiments like those illustrated in Figure 16.8 can be carried out to determine the contribution of G-E interaction to the total observed variation in phenotype. In other organisms, particularly human beings, the effect cannot be evaluated separately.

Interaction of genotype and environment is common and is very important in both plants and animals. Because of interaction, no one plant variety will

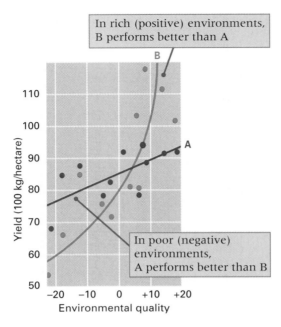

In rich (positive) environments, B performs better than A

In poor (negative) environments, A performs better than B

Figure 16.8
Genotype-environment interaction in maize. Strain A is superior when environmental quality is low (negative numbers), but strain B is superior when environmental quality is high. [Data from W. A. Russell. 1974. *Annual Corn & Sorghum Research Conference*, 29: 81.]

outperform all others in all types of soil and climate, and therefore plant breeders must develop special varieties that are suited to each growing area.

When the different genotypes are not distributed at random in all the possible environments, there is **genotype-environment association (G-E association).** In these circumstances, certain genotypes are preferentially associated with certain environments, which may either increase or decrease the phenotype of these genotypes compared with what would result in the absence of G-E association. An example of deliberate genotype-environment association can be found in dairy husbandry, in which some farmers feed each of their cows in proportion to its level of milk production. Because of this practice, cows with superior genotypes with respect to milk production also have a superior environment in that they receive more feed. In plant breeding, genotype-environment association can often be eliminated or minimized by appropriate

randomization of genotypes within the experimental plots. In other cases, human genetics again being a prime example, the possibility of G-E association cannot usually be controlled.

16.3
Analysis of Quantitative Traits

Equation (3) can be used to separate the effects of genotype and environment on the total phenotypic variance. Two types of data are required: (1) the phenotypic variance of a genetically uniform population, which provides an estimate of σ_e^2 because a genetically uniform population has a value of $\sigma_g^2 = 0$, and (2) the phenotypic variance of a genetically heterogeneous population, which provides an estimate of $\sigma_g^2 + \sigma_e^2$. An example of a genetically uniform population is the F_1 generation from a cross between two highly homozygous strains, such as inbred lines. An example of a genetically heterogeneous population is the F_2 generation from the same cross. If the environments of both populations are the same, and if there is no G-E interaction, then the estimates may be combined to deduce the value of σ_g^2.

To take a specific numerical illustration, consider variation in the size of the eyes in the cave-dwelling fish, *Astyanax*, all individuals being reared in the same environment. The variances in eye diameter in the F_1 and F_2 generations from a cross of two highly homozygous strains were estimated as 0.057 and 0.563, respectively. Written in terms of the components of variance, these are

$$F_2: \quad \sigma_t^2 = \sigma_g^2 + \sigma_e^2 = 0.563$$

$$F_1: \quad \sigma_e^2 = 0.057$$

The estimate of genotypic variance σ_g^2 is obtained by subtracting the second equation from the first; that is,

$$\sigma_g^2 = 0.563 - 0.057 = 0.506$$

because

$$(\sigma_g^2 + \sigma_e^2) - \sigma_e^2 = \sigma_g^2$$

Hence, the estimate of σ_g^2 is 0.506, whereas that of σ_e^2 is 0.057. In this example, the genotypic variance is much greater than the environmental variance, but this is not always the case.

The next section shows what information can be obtained from an estimate of the genotypic variance.

The Number of Genes Affecting a Quantitative Trait

When the number of genes influencing a quantitative trait is not too large, knowledge of the genotypic variance can be used to estimate the number of genes. All we need are the means and variances of two phenotypically divergent strains and their F_1, F_2, and backcrosses. In ideal cases, the data appear as in Figure 16.9, in which P_1 and P_2 represent the divergent parental strains (for example, inbred lines). The points lie on a triangle, with increasing variance according to the increasing genetic heterogeneity (genotypic variance) of the populations. If the F_1 and backcross means lie exactly between their parental means, then these means will lie at the midpoints along the sides of the triangle, as shown in Figure 16.9. This finding implies that the alleles affecting the trait are *additive*; that is, for each gene, the phenotype of the heterozygote is the average of the phenotypes of the corresponding homozygotes. In such a simple situation, it can be shown that the number, n, of genes contributing to the trait is

$$n = \frac{D^2}{8\sigma_g^2} \qquad (4)$$

in which D represents the difference between the means of the original parental strains, P_1 and P_2. This equation can be verified by applying it to the ideal case in Figure 16.7. In this case, the parental strains would be the homozygous genotypes *AA*, with a mean phenotype of 7, and *aa*, with a mean phenotype of 3. Consequently, $D = 7 - 3 = 4$. The genotypic variance is given in Figure 16.7 as $\sigma_g^2 = 2$. Substituting D and σ_g^2 into Equation 4, we obtain $n = 16/(8 \times 2) = 1$, which is correct because there is only one gene, with alleles *A* and *a*, that affects the trait.

Applied to actual data, Equation (4) requires several assumptions that are not necessarily valid. In addition to the assumption that all generations are reared in the same environment, the theory also makes the genetic assumptions that (1) the alleles of each gene are additive, (2) the genes contribute equally to the trait, (3) the genes are unlinked, and (4) the original parental strains are homozygous for alternative alleles of each gene. However, when the assumptions are invalid, then the calculated n is smaller than the actual number of genes affecting the trait. The estimated number is a minimum because almost any departure from the genetic assumptions leads to a smaller genotypic variance in the F_2 generation and so, for the same value of D, would yield a larger value of n in Equation (4). This is why the estimated n is the *minimum* number of genes that can account for the data. For the cave-dwelling *Astyanax* fish discussed in the preceding section, the parental strains had average phenotypes of 7.05 and 2.10, giving $D = 4.95$. The estimated value of $\sigma_g^2 = 0.506$, and so the minimum number of genes affecting eye diameter is $n = (4.95)^2/(8 \times 0.506) = 6.0$. Therefore, at least six different genes affect the diameter of the eye of the fish.

The number of genes that affect a quantitative trait is important because it

Figure 16.9

Means and variances of parents (P), backcross (B), and hybrid (F) progeny of inbred lines for an ideal quantitative trait affected by unlinked and completely additive genes. [After R. Lande. 1981. *Genetics*, 99: 541.]

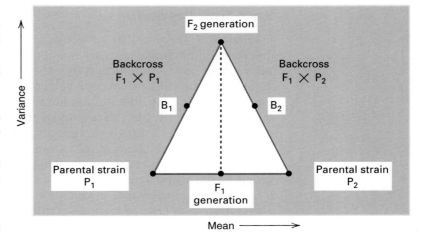

influences the amount by which a population can be genetically improved by selective breeding. With traits determined by a small number of genes, the potential for change in a trait is relatively small, and a population consisting of the best possible genotypes may have a mean value that is only 2 or 3 standard deviations above the mean of the original population. However, traits determined by a large number of genes have a large potential for improvement. For example, after a population of the flour beetle *Tribolium* was bred for increased pupa weight, the mean value for pupa weight was found to be 17 standard deviations above the mean of the original population. Determination of traits by a large number of genes implies that

> Selective breeding can create an improved population in which the value of *every* individual greatly exceeds that of the *best* individuals that existed in the original population.

This principle at first seems paradoxical, because in a large enough population, every possible genotype should be created at some low frequency. The resolution of the paradox is that real populations subjected to selective breeding typically consist of a few hundred organisms (at most), and hence many of the theoretically possible genotypes are never formed because their frequencies are much too rare. As selection takes place, and the allele frequencies change, these genotypes become more common and allow the selection of superior organisms in future generations.

Broad-Sense Heritability

Estimates of the number of genes that determine quantitative traits are frequently unavailable because the necessary experiments are impractical or have not been carried out. Another attribute of quantitative traits, which requires less data to evaluate, makes use of the ratio of the genotypic variance to the total phenotypic variance. This ratio of σ_g^2 to σ_t^2 is called the **broad-sense heritability,** symbolized as H^2, and it measures the importance of genetic variation, relative to environmental variation,

in causing variation in the phenotype of a trait of interest. Broad-sense heritability is a ratio of variances, specifically

$$H^2 = \frac{\sigma_g^2}{\sigma_t^2} = \frac{\sigma_g^2}{\sigma_g^2 + \sigma_e^2} \qquad (5)$$

Substitution of the data for eye diameter in *Astyanax,* in which $\sigma_g^2 = 0.506$ and $\sigma_g^2 + \sigma_e^2 = 0.563$, into Equation (5) yields $H^2 = 0.506/0.563 = 0.90$ for the estimate of broad-sense heritability. This value implies that 90 percent of the variation in eye diameter in this population results from differences in genotype among the fish.

Knowledge of heritability is useful in the context of plant and animal breeding, because heritability can be used to predict the magnitude and speed of population improvement. The broad-sense heritability defined in Equation (5) is used in predicting the outcome of selection practiced among clones, inbred lines, or varieties. Analogous predictions for randomly bred populations utilize another type of heritability, different from H^2, which we will discuss shortly. Broad-sense heritability measures how much of the total variance in phenotype results from differences in genotype. For this reason, H^2 is often of interest in regard to human quantitative traits.

Twin Studies

In human beings, twins would seem to be ideal subjects for separating genotypic and environmental variance, because **identical twins,** which arise from the splitting of a single fertilized egg, are genetically identical and are often strikingly similar in such traits as facial features and body build. **Fraternal twins,** which arise from two fertilized eggs, have the same genetic relationship as ordinary siblings, and therefore only half of the genes in either twin are identical with those in the other. Theoretically, the variance between members of an identical-twin pair would be equivalent to σ_e^2, because the twins are genetically identical; whereas the variance between members of a fraternal-twin pair would include not only σ_e^2 but also part of the genotypic variance (approximately

$\sigma_g^2/2$, because of the identity of half of the genes in fraternal twins). Consequently, both σ_g^2 and σ_e^2 could be estimated from twin data and combined as in Equation (5) to estimate H^2. Table 16.2 summarizes estimates of H^2 based on twin studies of several traits.

Unfortunately, twin studies are subject to several important sources of error, most of which increase the similarity of identical twins, so the numbers in Table 16.2 should be considered very approximate and probably too high. Four of the potential sources of error are

1. Genotype-environment interaction, which increases the variance in fraternal twins but not in identical twins.

2. Frequent sharing of embryonic membranes between identical twins, resulting in more similar intrauterine environments than those of fraternal twins.

3. Greater similarity in the treatment of identical twins by parents, teachers, and peers, resulting in a decreased environmental variance in identical twins.

4. Different sexes in half of the pairs of fraternal twins, in contrast with the same sex of identical twins.

These pitfalls and others imply that data from human twin studies should be interpreted with caution and reservation.

16.4 Artificial Selection

The practice among breeders of choosing a select group of organisms from a population to become the parents of the next generation is termed **artificial selection.** When artificial selection is carried out either by choosing the best organisms in a species that reproduces asexually or by choosing the best subpopulation among a series of subpopulations, each propagated by close inbreeding (such as self-fertilization), the broad-sense heritability makes possible an assessment of how rapidly progress can be achieved. Broad-sense heritability is important in this context because, with clones, inbred lines, or varieties, superior genotypes can be perpetuated without disruption of favorable gene combinations by Mendelian segregation. An example is the selection of superior varieties of plants that are propagated asexually by means of cuttings or grafts. Because there is no sexual reproduction, each plant has exactly the same genotype as its parent.

In sexually reproducing populations that are genetically heterogeneous, broad-sense heritability is not relevant in predicting progress resulting from artificial selection, because superior genotypes must necessarily be broken up by the processes of segregation and recombination. For example, if the best genotype is heterozygous for

Table 16.2 Broad-sense heritability, in percent, based on twin studies

Trait	Heritability (H^2)	Trait	Heritability (H^2)
Longevity	29	Verbal ability	63
Height	85	Numerical ability	76
Weight	63	Memory	47
Amino acid excretion	72	Sociability index	66
Serum lipid levels	44	Masculinity index	12
Maximum blood lactate	34	Temperament index	58
Maximum heart rate	84		

each of two unlinked loci, *Aa Bb,* then because of segregation and independent assortment, among the progeny of a cross between parents with the best genotypes— *Aa Bb* × *Aa Bb*—only 1/4 will have the same favorable *Aa Bb* genotype as the parents. The rest of the progeny will be genetically inferior to the parents. For this reason, to the extent that high genetic merit may depend on particular combinations of alleles, each generation of artificial selection results in a slight setback in that the offspring of superior parents are generally not quite so good as the parents themselves. Progress under selection can still be predicted, but the prediction must make use of another type of heritability, the narrow-sense heritability, which is discussed in the next section.

Narrow-Sense Heritability

Figure 16.10 illustrates a typical form of artificial selection and its result. The organism is *Nicotiana longiflora* (tobacco), and the trait is the length of the corolla tube (the corolla is a collective term for all the petals of a flower). Figure 16.10A shows the distribution of phenotypes in the parental generation, and Figure 16.10B shows the distribution of phenotypes in the offspring generation. The parental generation is the population from which the parents were chosen for breeding. The type of selection is called **individual selection,** because each member of the population to be selected is evaluated according to its own individual phenotype. The selection is practiced by choosing some arbitrary level of phenotype—called the **truncation point**—that determines which individuals will be saved for breeding purposes. All individuals with a phenotype above the threshold are randomly mated among themselves to produce the next generation.

Phenotypic Change with Selection: A Prediction Equation

In evaluating progress through individual selection, three distinct phenotypic means are important. In Figure 16.10, these means are symbolized as M, $M*$ and M' and are defined as follows:

1. M is the mean phenotype of the entire population in the parental generation, including both the selected and the nonselected individuals.

2. $M*$ is the mean phenotype among those individuals selected as parents (those with a phenotype above the truncation point).

3. M' is the mean phenotype among the progeny of selected parents.

The relationship among these three means is given by

$$M' = M + h^2(M* - M) \qquad (6)$$

where the symbol h^2 is the **narrow-sense heritability** of the trait in question.

Later in this chapter, a method for estimating narrow-sense heritability from the similarity in phenotype among relatives will be explained. In Figure 16.10, h^2 is the only unknown quantity, so it can be estimated from the data themselves. Rearranging Equation (6) and substituting the values for the means from Figure 16.10, we find that

$$h^2 = \frac{M' - M}{M* - M} = \frac{77 - 70}{81 - 70} = 0.64$$

In a manner analogous to the way in which total phenotypic variance can be split into the sum of the genotypic variance and the environmental variance (Equation (3), the genotypic variance can be split into parts resulting from the additive effects of alleles, dominance effects, and effects of interaction between alleles of different genes. The difference between the broad-sense heritability, H^2, and the narrow-sense heritability, h^2, is that H^2 includes all of these genetic contributions to variation, whereas h^2 includes only the additive effects of alleles. From the standpoint of animal or plant improvement, h^2 is the heritability of interest, because

The narrow sense heritability, h^2, is the proportion of the variance in phenotype that can be used to predict changes in population mean with individual selection according to Equation (6).

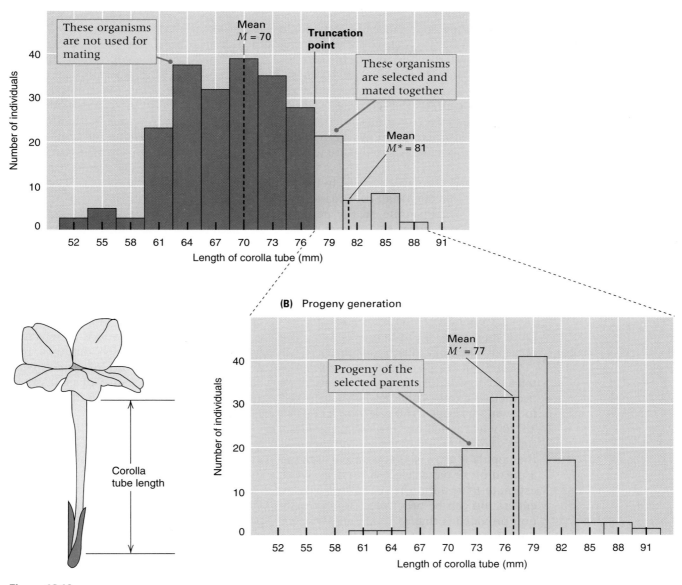

(A) Parental generation

These organisms are not used for mating

Mean
M = 70

Truncation point

These organisms are selected and mated together

Mean
M* = 81

Number of individuals

40

30

20

10

0

52   55   58   61   64   67   70   73   76   79   82   85   88   91

Length of corolla tube (mm)

Corolla tube length

(B) Progeny generation

Mean
M' = 77

Progeny of the selected parents

Number of individuals

40

30

20

10

0

52   55   58   61   64   67   70   73   76   79   82   85   88   91

Length of corolla tube (mm)

Figure 16.10
Selection for increased length of corolla tube in tobacco. (A) Distribution of phenotypes in the parental generation. The symbol *M* denotes the mean phenotype of the entire population, and *M** denotes the mean phenotype of the organisms chosen for breeding (organisms with a phenotype that exceeds the truncation point). (B) Distribution of phenotypes among the offspring bred from the selected parents. The symbol *M'* denotes the mean.

The distinction between the broad-sense heritability and the narrow-sense heritability can be appreciated intuitively by considering a population in which there is a rare recessive gene. In such a case, most homozygous recessive genotypes come from matings between heterozygous carriers. Some such kindreds have more than one affected offspring. Hence, affected siblings can resemble each other more than they resemble their parents. For example, if *a* is a recessive allele, the mating *Aa* × *Aa*, in which both parents show the dominant phenotype, may yield two offspring that are *aa*, which both show the recessive phenotype; in this case, the offspring are more similar to each other than to either of the parents. It is the dominance of the wildtype allele that causes this paradox, contributing to the broad-sense heritability of the trait but not to the narrow-sense heritability. The narrow-sense heritability includes only those genetic effects that contribute to the resemblance between parents and their offspring, because narrow-sense heritability measures how similar offspring are to their parents.

In general, the narrow-sense heritability of a trait is smaller than the broad-sense heritability. For example, in the parental generation in Figure 16.10, the broad-sense heritability of corolla tube length is $H^2 = 0.82$. The two types of heritability are equal only when the alleles affecting the trait are additive in their effects; with additive effects, each heterozygous genotype shows a phenotype that is exactly intermediate between the phenotypes of the respective homozygous genotypes.

Equation (6) is of fundamental importance in quantitative genetics because of its predictive value. This can be seen in the following example. The selection in Figure 16.10 was carried out for several generations. After two generations, the mean of the population was 83, and parents having a mean of 90 were selected. By use of the estimate $h^2 = 0.64$, the mean in the next generation can be predicted. The information provided is that $M = 83$ and $M^* = 90$. Therefore, Equation (6) implies that the predicted mean is

$$M' = 83 + (0.64)(90 - 83) = 87.5$$

This value is in good agreement with the observed value of 87.9.

Long-Term Artificial Selection

Artificial selection is analogous to natural selection in that both types of selection cause an increase in the frequency of alleles that improve the selected trait (or traits). The principles of natural selection discussed in Chapter 15 also apply to artificial selection. For example, artificial selection is most effective in changing the frequency of alleles that are in an intermediate range of

frequency $(0.2 < p < 0.8)$. Selection is less effective for alleles with frequencies outside this range and is least effective for rare recessive alleles. For quantitative traits, including fitness, the total selection is shared among all the genes that influence the trait, and the selection coefficient affecting each allele is determined by (1) the magnitude of the effect of the allele, (2) the frequency of the allele, (3) the total number of genes affecting the trait, (4) the narrow-sense heritability of the trait, and (5) the proportion of the population that is selected for breeding.

The value of heritability is determined both by the magnitude of effects and by the frequency of alleles. If all favorable alleles were fixed ($p = 1$) or lost ($p = 0$), then the heritability of the trait would be 0. Therefore, the heritability of a quantitative trait is expected to decrease over many generations of artificial selection as a result of favorable alleles becoming nearly fixed. For example, ten generations of selection for less fat in a population of Duroc pigs decreased the heritability of fatness from 73 to 30 percent because of changes in allele frequency that resulted from the selection.

Population improvement by means of artificial selection cannot continue indefinitely. A population may respond to selection until its mean is many standard deviations different from the mean of the original population, but eventually the population reaches a **selection limit** at which successive generations show no further improvement. Progress may stop because all alleles affecting the trait are either fixed or lost, and so the narrow-sense heritability of the trait becomes 0. However, it is more common for a selection limit to be reached because natural selection counteracts artificial selection. Many genes that respond to artificial selection as a result of their favorable effect on a selected trait also have indirect harmful effects on fitness. For example, selection for increased size of eggs in poultry results in a decrease in the number of eggs, and selection for extreme body size (large or small) in most animals results in a decrease in fertility. When one trait (for example, number of eggs) changes in the course of selection for a different trait (for example, size of eggs), then the unselected trait is said to have undergone a **correlated response** to selection. Correlated response of fitness is typical in long-term artificial selection. Each increment of progress in the selected trait is partially offset by a decrease in fitness because of correlated response; eventually, artificial selection for the trait of interest is exactly balanced by natural selection against the trait. Thus a selection limit is reached, and no further progress is possible without changing the strategy of selection.

Inbreeding Depression and Heterosis

Inbreeding can have harmful effects on economically important traits such as yield of grain and egg production. This decline in performance is called **inbreeding depression,** and it results principally from rare harmful recessive alleles becoming homozygous because of inbreeding (Chapter 15). The degree of inbreeding is measured by the inbreeding coefficient F discussed in Chapter 15. Figure 16.11 is an example of inbreeding depression in yield of corn, in which the yield decreases linearly as the inbreeding coefficient increases.

Most highly inbred strains suffer from many genetic defects, as might be expected from the uncovering of deleterious recessive alleles. One would also expect that if two different inbred strains were crossed, then the F_1 would show improved features, because a harmful recessive allele inherited from one parent would be likely to be covered up by a normal dominant allele from the other parent. In many organisms, the F_1 generation of a cross between inbred lines is superior to the parental lines, and the phenomenon is called **heterosis,** or **hybrid vigor.** The phenomenon, which is widely used in the production of corn and other agricultural products, yields genetically identical hybrid plants with traits that are sometimes more favorable than those of the ancestral plants from which the inbreds were derived. The features that most commonly distinguish hybrid plants from their inbred parents are their rapid growth, larger size, and greater yield.

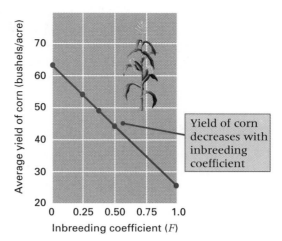

Figure 16.11
Inbreeding depression for yield in corn. [Data from N. Neal. 1935. *J. Amer. Soc. Agronomy,* 27: 666]

Furthermore, the F_1 plants have a fairly uniform phenotype (because $\sigma_g^2 = 0$). Genetically heterogeneous crops with high yields or certain other desirable features can also be produced by traditional plant-breeding programs, but growers often prefer hybrid plants because of their uniformity. For example, uniform height and time of maturity facilitate machine harvesting, and plants that all bear fruit at the same time accommodate picking and shipping schedules.

Hybrid varieties of corn are used almost exclusively in the United States for commercial crops. A farmer cannot plant the seeds from his crop because the F_2 generation consists of a variety of genotypes, most of which do not show hybrid vigor. The production of hybrid seeds is a major industry in corn-growing sections of the United States.

16.5 Correlation Between Relatives

Quantitative genetics relies extensively on similarity among relatives to assess the importance of genetic factors. Particularly in the study of complex behavioral traits in human beings, genetic interpretation of familial resemblance is not always straightforward because of the possibility of non-genetic, but nevertheless familial, sources of resemblance. However, the situation is usually less complex in plant and animal breeding, because genotypes and environments are under experimental control.

Covariance and Correlation

Genetic data about families are frequently pairs of numbers: pairs of parents, pairs of twins, or pairs consisting of a single parent and a single offspring. An important issue in quantitative genetics is the degree to which the numbers in each pair are associated. The usual way to measure the association is to calculate a statistical quantity called the **correlation coefficient** between the members of each pair.

The correlation coefficient among relatives is based on the covariance in phenotype among them. Much as the variance describes the tendency of a set of measurements to vary (Equation 2), the covariance describes the tendency of pairs of numbers to vary together (co-vary). Calculation of the covariance is similar to calculation of the variance in Equation (2) except that the squared deviation term $(x_i - \bar{x})^2$ is replaced with the product of the deviations of the pairs of measurements from their respective means—that is, $(x_i - \bar{x})(y_i - \bar{y})$.

For example, $(x_i - \bar{x})$ could be the deviation of a particular father's height from the overall father mean, and $(y_i - \bar{y})$ could be the deviation of his son's height from the overall son mean. In symbols, let f_i be the number of pairs of relatives with phenotypic measurements x_i and y_i. Then the estimated **covariance (Cov)** of the trait among the relatives is

$$Cov(x,y) = \frac{\Sigma f_i(x_i - \bar{x})(y_i - \bar{y})}{N - 1} \quad (7)$$

where N is the total number of pairs of relatives studied.

The **correlation coefficient** (r) of the trait between the relatives is calculated from the covariance as follows:

$$r = \frac{Cov(x,y)}{s_x s_y} \quad (8)$$

where s_x and s_y are the standard deviations of the measurements, estimated from Equation 2. The correlation coefficient can range from -1.0 to $+1.0$. A value of $+1.0$ means perfect association. When $r = 0$, x and y are not associated.

The Geometrical Meaning of a Correlation

Figure 16.12 shows a geometrical interpretation of the correlation coefficient. Just as the distribution of a single variable can be built up, step by step, by placing small squares corresponding to each individual along the x-axis, so too can the joint distribution of two variables (for example, height

Figure 16.12

Distribution of a quantitative trait in parents and offspring when there is no correlation (A) or a high correlation (B) between them. [From R. R. Sokal and F. J. Rohlf. 1969. *Biometry*. San Francisco: Freeman.]

(A)

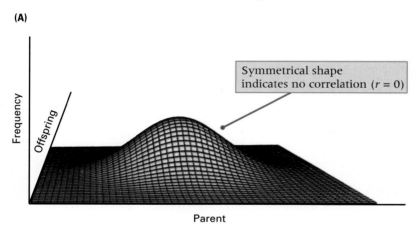

Symmetrical shape indicates no correlation ($r = 0$)

(B)

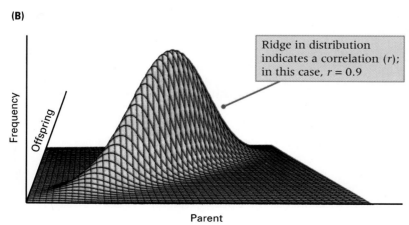

Ridge in distribution indicates a correlation (r); in this case, $r = 0.9$

of fathers and height of sons) be built up, step by step, by placing small cubes on the x-y plane at a position determined by each pair of measurements. As these cubes pile up, the joint distribution assumes the form of an inverted bowl, cross sections of which are themselves normal distributions. When there is no correlation between the variables, the surface is completely symmetrical (Figure 16.12A), which means that the distribution of y is completely independent of that of x. However, when the distributions are correlated, the distribution of y varies according to the particular value of x, and the joint distribution becomes asymmetrical with a ridge built up in the x-y plane (Figure 16.12B).

Normally, there are not enough data to form a full three-dimensional joint distrib-

ution of x and y as depicted in Figure 16.12. Instead, the data are in the form of individual points, sparse enough to be plotted on a scatter diagram like those in Figure 16.13. In these diagrams, the clustering of points around each line corresponds to the ridge of points in Figure 16.13. The higher the ridge, the more clustered the points are around the line.

Estimation of Narrow-Sense Heritability

Covariance and correlation are important in quantitative genetics, because the correlation coefficient of a trait between individuals with various degrees of genetic relationship is related fairly simply to the narrow-sense or broad-sense heritability, as

Figure 16.13
Scatter diagrams of x and y variables with various degrees of correlation r. [From R. R. Sokal and F. J. Rohlf. 1969. *Biometry*. San Francisco: Freeman.]

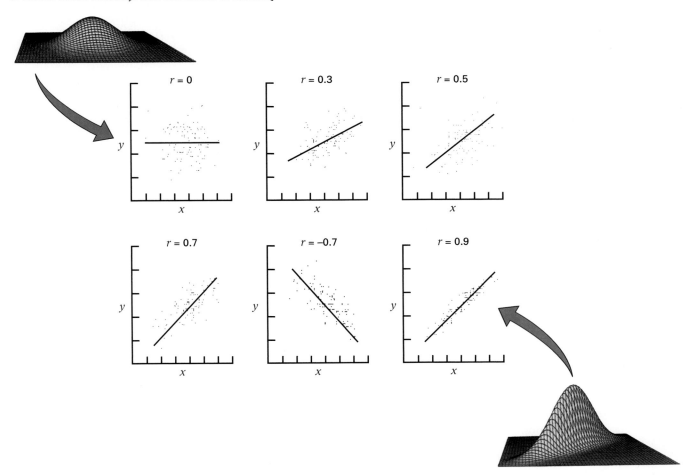

Table 16.3 Theoretical correlation coefficient in phenotype between relatives

Degree of relationship	Correlation coefficient*
Offspring and one parent	$h^2/2$
Offspring and average of parents	$h^2/2$
Half siblings	$h^2/4$
First cousins	$h^2/8$
Monozygotic twins	H^2
Full siblings	$\sim H^2/2$

*Contributions from interactions among alleles of different genes have been ignored. For this and other reasons, H^2 correlations are approximate

shown in Table 16.3. The table gives the theoretical values of the correlation coefficient for various pairs of relatives; h^2 represents the narrow-sense heritability and H^2 the broad-sense heritability. For parent-offspring, half-sibling, or first-cousin pairs, narrow-sense heritability can be estimated directly by multiplication. Specifically, h^2 can be estimated as twice the parent-offspring correlation, four times the half-sibling correlation, or eight times the first-cousin correlation.

With full siblings, identical twins, and double first cousins, the correlation coefficient is related to the broad-sense heritability, H^2, because phenotypic resemblance depends not only on additive effects but also on dominance. In these relatives, dominance contributes to resemblance because the relatives can share *both* of their alleles as a result of their common ancestry, whereas parents and offspring, half siblings, and first cousins can share at most a single allele of any gene because of common ancestry. Therefore, to the extent that phenotype depends on dominance effects, full siblings can resemble each other more than they resemble their parents.

The potentially greater resemblance between siblings than between parents and offspring may be understood by considering an autosomal recessive trait caused by a rare recessive allele. In a randomly mating population, the probability that an offspring will be affected, if the mother is affected, is q (the allele frequency), which

corresponds to the random-mating probability that the sperm giving rise to the offspring carries the recessive allele. In contrast, the probability that both of two siblings will be affected, if one of them is affected, is 1/4, because most of the matings that produce affected individuals will be between heterozygous parents. When the trait is rare, the parent-offspring resemblance will be very small (as q approaches 0), whereas the sibling-sibling resemblance will be substantial. This discrepancy is entirely a result of dominance, and it arises only because, at any particular locus, full siblings may have the same genotype.

16.6
Heritabilities of Threshold Traits

Application of the concepts of individual selection can be used to understand the meaning of heritability in the context of threshold traits. The analogy is illustrated in Figure 16.14. Threshold traits depend on an underlying quantitative trait, called the **liability,** which refers to the degree of genetic risk or predisposition to the trait. Only those individuals with a liability greater than a certain threshold develop the trait. Figure 16.14A shows the distribution of liability in a population, where liability is measured on an arbitrary scale so that the mean liability (M) equals 0 and the variance in liability equals 1. Affected individuals are in the right-hand area of the distribution. Using various characteristics of the normal distribution, we can use the observed proportion of affected individuals to calculate the position of the threshold and the value of M^*, which in this context is the mean liability of affected individuals.

The distribution of liability among offspring of affected individuals is shown in Figure 16.14B. It is shifted slightly to the right, which implies that, relative to the population as a whole, a greater proportion of individuals with affected parents have liabilities above the threshold and so will be affected. The observed proportion of affected individuals in the offspring generation can be used to calculate M'. The important point is that M^* and M' can be obtained from the observed proportions corresponding to the shaded red areas in

(A) Parental generation

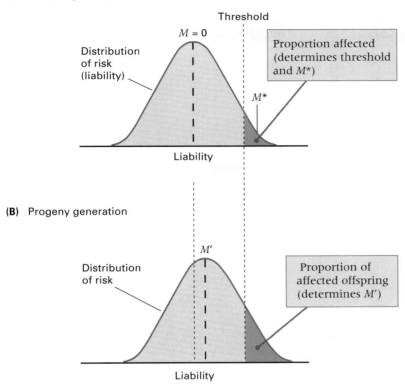

Threshold

$M = 0$

Distribution
of risk
(liability)

Proportion affected
(determines threshold
and $M*$)

$M*$

Liability

(B) Progeny generation

M'

Distribution
of risk

Proportion of
affected offspring
(determines M')

Liability

Figure 16.14
Interpretation of a threshold trait in terms of an underlying continuous distribution of liability. Individuals with a liability greater than the threshold are affected. M (arbitrarily set equal to 0), $M*$, and M' are the mean liabilities of the parental generation, of affected individuals, and of the offspring of affected individuals, respectively.

Figure 16.14. With $M = 0$, Equation (6) can be used to calculate the narrow-sense heritability of liability of the trait in question.

Examples of narrow-sense heritability of the liability to important congenital abnormalities are given in Table 16.4. How these heritabilities translate into risk is illustrated in Figure 16.15, along with observed data for the most common congenital abnormalities in Caucasians. The theoretical curves correspond to the relationships among the incidence of the trait in the general population, the type of inheritance, and the risk among first-degree relatives of affected individuals. (First-degree relatives of an affected individual include parents, offspring, and siblings.) Simple Mendelian inheritance produces the highest risks, but these traits, as a group, tend to be rare. The most common traits are threshold traits, and with a few exceptions, their risks tend to be moderate or low, corresponding to the heritabilities indicated.

Table 16.4 Narrow-sense heritability of liability for congenital abnormalities

Trait	Description	Population frequency (%)	Narrow-sense Heritability (%)
Cleft lip	Upper lip not completely formed	0.10	76
Spina bifida	Exposed spinal cord	0.50	60
Club foot	Abnormally formed foot	0.10	68
Pyloric stenosis	Obstructed stomach	0.30	75
Dislocation of hip	Hip joint mispositioned	0.06	70
Hydrocephalus	Fluid accumulation around brain	0.05	76
Celiac disease	Inability to digest fats, starches, and sugars	0.05	45
Hypospadias	Abnormally formed penis	0.30	75
Atrial septal defect	Hole between upper chambers of heart	0.07	70
Patent ductus arteriosus	Hole between aorta and pulmonary artery; blood bypasses lungs	0.05	60

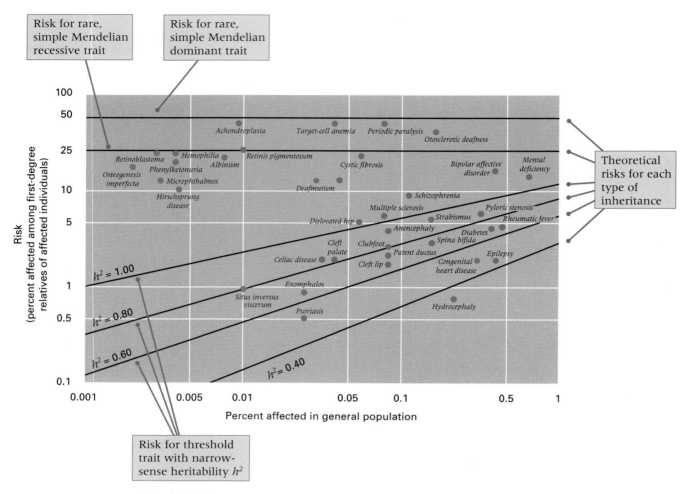

Risk for rare, simple Mendelian recessive trait

Risk for rare, simple Mendelian dominant trait

Theoretical risks for each type of inheritance

Risk for threshold trait with narrow-sense heritability h^2

Figure 16.15

Recurrence risks of common abnormalities. Diagonal lines are the theoretical risks for threshold traits with the indicated values of the narrow-sense heritability of liability. Horizontal lines are the theoretical risks for simple dominant or recessive traits.

16.7
Linkage Analysis of Quantitative-Trait Loci

Genes that affect quantitative traits cannot usually be identified in pedigrees, because their individual effects are obscured by the segregation of other genes and by environmental variation. Even so, genes that affect quantitative traits can be localized if they are genetically linked with genetic markers, such as simple tandem repeat polymorphisms (STRPs), discussed in Chapter 4, because the effects of the genotype affecting the quantitative trait will be correlated with the genotype of the linked STRP. A

gene that affects a quantitative trait is a **quantitative-trait locus (QTL).** Locating QTLs in the genome is important to the manipulation of genes in breeding programs and to the cloning and study of genes in order to identify their functions.

STRPs result from variation in number, n, of repeating units of a simple-sequence repeat, such as $(CA)_n$. The number of repeats can be determined by PCR amplification of the STRP using oligonucleotide primers to flanking unique-sequence DNA. The size of the amplified DNA fragment is a function of the number of repeats in the STRP. Such STRP markers are abundant, are distributed throughout the genome,

Human Gene Map

Jeffrey C. Murry and 26 other investigators
1994
University of Iowa and 9 other
research institutions
*A Comprehensive Human Linkage Map
with Centimorgan Density*

Assessed by the distance between genetic markers, the human genetic map is one of the most dense genetic maps of any organism. At present, the human map includes more than 5000 genetic markers. The assembly of the human genetic map represents a major achievement in international science. It was pieced together by hundreds of investigators working in many countries throughout the world. This paper is one example of a progress report illustrating several important features. First, the human genetic map was made possible by researchers studying a single, very extensive set of families assembled by CEPH (Centre d'Étude du Polymorphisme Humain, Paris) so that their data could be collated. Second, a very large proportion of the markers in the human map are detected by means of the polymerase chain reaction (PCR) to amplify short tandem repeats (STRPs) that are abundant and highly polymorphic in the human genome. Third, the data are so extensive that they are not presented in the paper itself but rather in a large wall chart and (more conveniently) at a site on the Internet. Fourth, the authors point out the utility of the genetic map in determining the chromosomal locations of mutations that cause disease, using the methods of QTL mapping, as a first step in identifying the genes themselves. Finally, as is clear from the closing passages of the excerpt, the authors are well aware of the implications of modern human genetics for ethics, law, and social policy.

For the first time, humans have been presented with the capability of understanding their own genetic makeup and how it contributes to morbidity of the individual and the species. Rapid scientific advances have made this possible, and developments in molecular biology, genetics, and computing, coupled with a cooperative and interactive biomedical community, have accelerated the progress of investigation into human inherited disorders. A primary

---

> **It should be emphasized that the new opportunities and challenges for biomedical research provided by these marker-dense maps also create an urgency to face the parallel challenges in the areas of ethics, law, and social policy.**

---

engine driving these advances has been the development and use of human gene maps that allow the rapid positional assignment of an inherited disorder as a starting point for gene identification and characterization. . . . The genetic maps were based on genotypes generated from DNA samples obtained from the CEPH reference pedigree set. Since this material is publicly available, individual research groups can add their own genetic markers in the future. . . . Although only markers genotyped on CEPH reference families are included in the maps, the list is extensive. The CEPH database contains blood group markers and protein polymorphisms as well as a variety of DNA-based markers including RFLPs and short tandem repeat polymorphisms (STRPs). There has been an

emphasis on PCR-based markers. . . . The 3617 STRP markers alone provide about one marker every 1×10^6 base pairs (bp) throughout the genome. Once an initial localization [of a new gene] is identified by means of a set of genome-wide genetic maps, subsequent steps can be undertaken to increasingly narrow the region to be searched and eventually to select a specific interval on which studies of physical reagents, such as cloned DNA fragments, can be carried out. . . . Once the physical reagents are in hand, gene and mutation searches are possible with a number of strategies, . . . but considerable problems and difficulties still remain. . . . Public access to databases is an especially important feature of the Human Genome Project. They are easy to use and facilitate rapid communication of new findings and can be updated efficiently. We have made primary data available electronically on the World Wide Web at http://gdbwww.gdb.org/. . . . Finally, it should be emphasized that the new opportunities and challenges for biomedical research provided by these marker-dense maps also create an urgency to face the parallel challenges in the areas of ethics, law, and social policy. Our ability to distinguish individuals for forensic purposes, identify genetic predispositions for rare and common inherited disorders, and to characterize, if present, the underlying nature of the genetic components of normal trait variability such as height, intelligence, sexual preference, or personality type, has never been greater. Although technically feasible, whether these maps should be used for these ends should be resolved after open dialogue to review those implications and devise policy to deal with the as yet unpredictable outcomes.

Source: Science 265: 2049–2054

---

and often have multiple codominant alleles; thus they are ideally suited for linkage studies of quantitative traits. In STRP studies, as many widely scattered STRPs as possible are monitored, along with the quantitative trait, in successive generations of a genetically heterogeneous population. Statistical studies are then carried out to identify which STRP alleles are the best predictors of phenotype of the quantitative trait because their presence in a genotype is consistently accompanied by superior

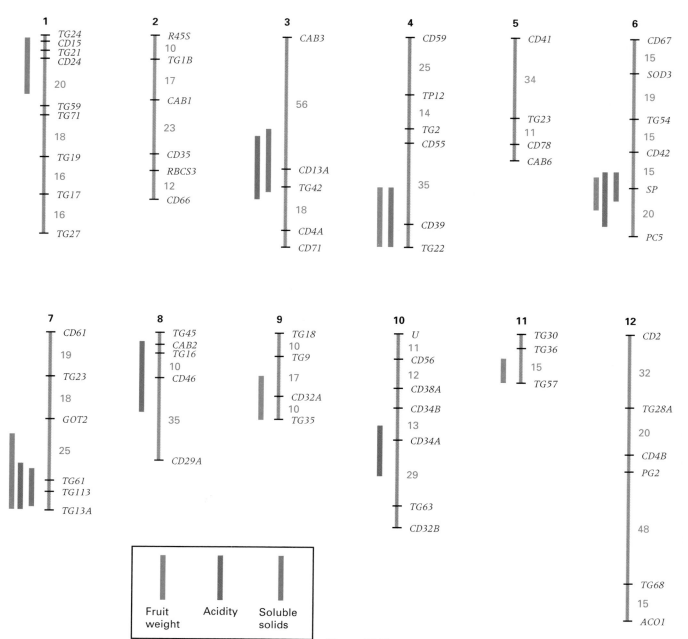

Figure 16.16

Location of QTLs for several quantitative traits in the tomato genome. The genetic markers are shown for each of the 12 chromosomes. The numbers in red are distances in map units between adjacent markers, but only map distances of 10 or greater are indicated. The regions in which the QTLs are located are indicated by the bars—green bars, QTLs for fruit weight; blue bars, QTLs for content of soluble solids; and dark red bars, QTLs for acidity (pH). The data are from crosses between the domestic tomato (*Lycopersicon esculentum*) and a wild South American relative with small, green fruit (*Lycopersicon chmielewskii*). The photograph is of fruits of wild tomato. Note the small size in comparison with the coin (a U.S. quarter). The F$_1$ generation was backcrossed with the domestic tomato, and fruits from the progeny were assayed for the genetic markers and each of the quantitative traits. [Data from A. H. Paterson, et al. 1988. *Nature*, 335: 721. Photograph courtesy of Steven D. Tanksley.]

performance with respect to the quantitative trait. These STRPs identify regions of the genome that contain one or more genes that have important effects on the quantitative trait, and the STRPs can be used to trace the segregation of the important regions in breeding programs and even as starting points for cloning genes with particularly large effects.

An example of genetic mapping of quantitative-trait loci for several quantitative traits in tomatoes is illustrated in Figure 16.16. More than 300 highly polymorphic genetic markers have been mapped in the tomato genome, with an average spacing between markers of 5 map units. The chromosome maps in Figure 16.16 show a subset of 67 markers that were segregating in crosses between the domestic tomato and a wild South American relative. The average spacing between the markers is 20 map units. Backcross progeny of the cross $F_1 \times$ domestic tomato were tested for the genetic markers, and the fruits of the backcross progeny were assayed for three quantitative traits—fruit weight, content of soluble solids, and acidity. Statistical analysis of the data was carried out in order to detect marker alleles that were associated with phenotypic differences in any of the traits; a significant association indicates genetic linkage between the marker gene and one or more QTLs affecting the trait. A total of six QTLs affecting fruit weight were detected (green bars), as well as four QTLs affecting soluble solids (blue bars) and five QTLs affecting acidity (dark red bars). Although additional QTLs with smaller effects undoubtedly remained undetected in these types of experiments, the effects of the mapped QTLs are substantial: The mapped QTLs account for 58 percent of the total phenotypic variance in fruit weight, 44 percent of the phenotypic variance in soluble solids, and 48 percent of the phenotypic variance in acidity. The genetic markers linked to the QTLs with substantial effects make it possible to trace the transmission of the QTLs in pedigrees and to manipulate the QTLs in breeding programs by following the transmission of the linked marker genes. Figure 16.16 also indicates a number of chromosomal regions containing QTLs for two or more of the traits—for example, the QTL regions on chromosomes 6 and 7, which affect all three traits. Because the locations of the QTLs can be specified only to within 20 to 30 map units, it is unclear whether the coincidences result from (1) pleiotropy, in which a single gene affects several traits simultaneously, or (2) independent effects of multiple, tightly linked genes. However, the locations of QTLs for different traits coincide frequently enough so that, in many cases, pleiotropy is likely to be the explanation.

Chapter Summary

Many traits that are important in agriculture and human genetics are determined by the effects of multiple genes and by the environment. Such traits are multifactorial traits (also known as quantitative traits), and their analysis is known as quantitative genetics. There are three types of multifactorial traits: continuous, meristic, and threshold traits. Continuous traits are expressed according to a continuous scale of measurement, such as height. Meristic traits are traits that are expressed in whole numbers, such as the number of grains on an ear of corn. Threshold traits have an underlying risk and are either expressed or not expressed in each individual; an example is diabetes. The genes affecting quantitative traits are no different from those affecting simple Mendelian traits, and the genes can have multiple alleles, partial dominance, and so forth. When several genes affect a trait, the pattern of genetic transmission need not fit a simple Mendelian pattern, because the effects of one gene can be obscured by other genes or the environment. However, the number of genes can be estimated and many of them can be mapped using linkage to simple tandem repeat polymorphisms.

Many quantitative and meristic traits have a distribution that approximates the bell-shaped curve of a normal distribution. A normal distribution can be completely described by two quantities: the mean and the variance. The standard deviation of a distribution is the square root of the variance. In a normal distribution, approximately 68 percent of the individuals have a phenotype within 1 standard deviation of the mean, and approximately 95 percent of the individuals have a phenotype within 2 standard deviations of the mean.

Variation in phenotype of multifactorial traits among individuals in a population derives from four principal sources: (1) variation in genotype, which is measured by the genotypic variance; (2) variation in environment, which is measured by the environmental variance; (3) variation resulting from the interaction between genotype and environment (G-E interaction); and (4) variation resulting from nonrandom association of genotypes and environments (G-E association). The ratio of genotypic variance to the total phenotypic variance of a trait is called the broad-sense heritability; this quantity is useful in predicting the outcome of artificial selection among clones, inbred lines, or varieties. When artificial selection is carried out in a randomly mating population, the narrow-sense heritability is used for prediction. The value of the narrow-sense heritability can be determined from the correlation in phenotype among groups of relatives.

One common type of artificial selection is called truncation selection, in which only those individuals that have a phenotype above a certain value (the truncation point) are saved for breeding the next generation. Artificial selection usually results in improvement of the selected population. The general principle is that the deviation of the progeny mean from the mean of the previous generation equals the narrow-sense heritability times the deviation of the parental mean from the mean of the parental generation: $M' - M = h^2(M^* - M)$. When selection is carried out for many generations, progress often slows or ceases because (1) some of the favorable genes become nearly fixed in the population, which decreases the narrow-sense heritability, and (2) natural selection may counteract the artificial selection.

A gene that affects a quantitative trait is a quantitative-trait locus, or QTL. The locations of QTLs in the genome can be determined by genetic mapping with respect to simple tandem repeat polymorphisms or other types of genetic markers. The mapping of QTLs aids in the manipulation of genes in breeding programs and in the identification of the genes.

Key Terms

artificial selection	genotype-environment interaction	multifactorial trait
broad-sense heritability	genotypic variance	narrow-sense heritability
continuous trait	heterosis	normal distribution
correlated response	hybrid vigor	quantitative trait
correlation coefficient	identical twins	quantitative-trait locus (QTL)
covariance	inbreeding depression	selection limit
distribution	individual selection	standard deviation
environmental variance	liability	threshold trait
fraternal twins	mean	truncation point
genotype-environment association	meristic trait	variance

Review the Basics

• What is a quantitative trait? How do a continuous trait, a meristic trait, and a threshold trait differ?

• How can a trait be affected both by genetic factors and by environmental factors? Give an example.

• What is the genotypic variance of a quantitative trait? What is the environmental variance?

• In terms of genotypic variance, what is special about an inbred line or about the F_1 progeny of a cross between two inbred lines?

• Why are correlations between relatives of interest to the quantitative geneticist?

• Distinguish between broad-sense heritability and narrow-sense heritability. Which type of heritability is relevant to individual truncation selection?

• Distinguish between the variance due to genotype-enviroment interaction and the variance due to genotype-environment association. Which type of variance is more easily controlled by the experimentalist?

• How are the variance and the standard deviation related? When does the variance of a distribution of phenotypes equal zero?

Guide to Problem Solving

Problem 1: The following data are the values of a quantitative trait measured in a sample of 100 individuals taken at random from a population.

(a) Estimate the mean, variance, and standard deviation in the population.

(b) Assuming a normal distribution, what proportion of individuals in the entire population would be expected to have a phenotype in the range of 93 to 107? Above 114? (For purposes of this problem, round off the value of the standard deviation to the nearest whole number.)

Phenotype	Number of individuals
86–90	10
91–95	16
96–100	26
101–105	29
106–110	10
111–115	9

Answer: The data are tabulated in ranges, but for purposes of calculation it is better to retabulate the data as the midpoints of the ranges, as is shown in the next table.

Chapter 16

GeNETics on the web will introduce you to some of the most important sites for finding genetic information on the Internet. To complete the exercises below, visit the Jones and Bartlett home page at

http://www.jbpub.com/genetics

Select the link to *Genetics: Principles and Analysis* and then choose the link to *GeNETics on the web*. You will be presented with a chapter-by-chapter list of highlighted keywords.

GeNETics EXERCISES

Select the highlighted keyword in any of the exercises below, and you will be linked to a web site containing the genetic information necessary to complete the exercise. Each exercise suggests a specific, written report that makes use of the information available at the site. This report, or an alternative, may be assigned by your instructor.

1. This keyword will open a link to a table of frequencies of the most common types of **birth defects**. The overall frequency of these conditions is about one in 25 births. Most of the genetically determined conditions are multifactorial. If assigned to do so, make a bar graph showing the relative incidence of each of the types of abnormality.

2. The trait of **parthogenesis** in turkeys is an example of a quantitative threshold trait influenced by genetic and environmental factors. At this site you can learn about parthenogenesis and its role in commercial turkey production. If assigned to do so, prepare a table listing the genetic and environmental factors affecting parthenogenesis that have so far been identified.

3. One of the founders of evolutionary genetics was **Sewall Wright**, who related his strong interest in animal breeding to the principles of quantitative genetics. His paper on *the relation of live-stock breeding to theories of evolution* is a good summary of his thoughts and is well worth reading. If assigned to do so, write a 250-word paper describing his shifting-balance theory of evolution and the four very diverse research projects that led him to it.

MUTABLE SITE EXERCISES

The Mutable Site Exercise changes frequently. Each new update includes a different exercise that makes use of genetics resources available on the World Wide Web. Select the Mutable Site for Chapter 16, and you will be linked to the current exercise that relates to the material presented in this chapter.

PIC SITE

The Pic Site showcases some of the most visually appealing genetics sites on the World Wide Web. To visit the showcase genetics site, select the Pic Site for Chapter 16.

Phenotype	Number of individuals
88	10
93	16
98	26
103	29
108	10
113	9

(a) The mean is estimated as $(88 \times 10 + 93 \times 16 + 98 \times 26 + 103 \times 29 + 108 \times 10 + 113 \times 9)/100 = 10,000/100 = 100$. The variance is estimated as $[(88 - 100)^2 \times 10 + (93 - 100)^2 \times 16 + (98 - 100)^2 \times 26 + (103 - 100)^2 \times 29 + (108 - 100)^2 \times 10 + (113 - 100)^2 \times 9]/99 = 4750/99 = 47.98$. The standard deviation is then estimated as $(47.98)^{1/2} = 6.93$.

(b) Rounding off the standard deviation to 7, the range of 93 to 107 equals the mean plus or minus one standard deviation (100 ± 7), and approximately 68 percent of the population is expected to be within this range. To estimate the proportion with a phenotype above 114, note that this is two standard deviations above the mean. Because approximately 95 percent of the population have phenotypes within the range 100 ± 14 (two standard deviations), the remaining 5 percent will have phenotypes either greater than 114 or less than 86. Because the normal distribution is symmetrical, half of the 5 percent (or 2.5 percent) will be expected to have phenotypes greater than 114.

Problem 2: A genetically heterogeneous population of wheat has a variance in the number of days to maturation of 40, whereas two inbred populations derived from it have a variance in the number of days to maturation of 10.

(a) What is the genotypic variance, σ_g^2, the environmental variance, σ_e^2, and the broad-sense heritability, H^2, of days to maturation in this population?

(b) If the inbred lines were crossed, what would be the predicted variance in days to maturation of the F_1 generation?

Answer:

(a) The total variance, σ_t^2, in the genetically heterogeneous population is the sum $\sigma_g^2 + \sigma_e^2 = 40$. The variance in the inbred lines equals the environmental variance (because $\sigma_g^2 = 0$ in a genetically homogeneous population); hence $\sigma_e^2 = 10$. Therefore, σ_g^2 in the heterogeneous population equals $40 - 10 = 30$. The broad-sense heritability H^2 equals $\sigma_g^2/(\sigma_g^2 + \sigma_e^2) = 30/40 = 75$ percent.

(b) If the inbred lines were crossed, the F_1 generation would be genetically uniform, and consequently the predicted variance would be $\sigma_e^2 = 10$.

Problem 3: A breeder aims to increase the height of maize plants by artificial selection. In a population with an average plant height of 150 cm, plants averaging 180 cm are selected and mated at random to give the next generation. If the narrow-sense heritability of plant height in this population is 40 percent, what is the expected average plant height among the progeny generation?

Answer: Use Equation (6) with $M = 150$, $M^* = 180$, and $h^2 = 0.40$. Then the predicted mean is calculated as $M' = 150 + 0.40 \times (180 - 150) = 162$ cm.

Problem 4: In maize, data for the logarithm of the percent oil content in the kernels lie approximately on a triangle like that depicted in Figure 16.9. Two inbred lines with average values of 0.513 and 1.122 are crossed. The variance of the trait in the F_1 generation is 0.0003. The F_1 plants are self-fertilized, and the variance of the trait in the F_2 generation is 0.0030. Estimate the minimum number of genes affecting oil content.

Answer: The minimum number of genes is estimated as $n = D^2/8\sigma_g^2$, where D is the difference between inbred lines and σ_g^2 is the genotypic variance. In this problem, $D = 1.122 - 0.513 = 0.609$. The variance of the F_2 generation equals $\sigma_g^2 + \sigma_e^2$ and that of the F_1 generation equals σ_e^2 (because the F_1 generation is genetically uniform). Therefore, $\sigma_g^2 = 0.0030 - 0.0003 = 0.0027$. The minimum number of genes affecting the oil content of kernels is estimated as $(0.609)^2/[8 \times 0.0027] = 17.2$.

Analysis and Applications

16.1 Two varieties of corn, A and B, are field-tested in Indiana and North Carolina. Strain A is more productive in Indiana, but strain B is more productive in North Carolina. What phenomenon in quantitative genetics does this example illustrate?

16.2 A distribution has the feature that the standard deviation is equal to the variance. What are the possible values for the variance?

16.3 The following questions pertain to a normal distribution.

(a) What term applies to the value along the x-axis that corresponds to the peak of the distribution?

(b) If two normal distributions have the same mean but different variances, which is the broader?

(c) What proportion of the population is expected to lie within one standard deviation of the mean? Within two standard deviations?

16.4 Distinguish between the broad-sense heritability of a quantitative trait and the narrow-sense heritability. If a population is fixed for all genes that affect a particular quantitative trait, what are the values of the narrow-sense and broad-sense heritabilities?

16.5 When we compare a quantitative trait in the F_1 and F_2 generations obtained by crossing two highly inbred strains, which set of progeny provides an estimate of the environmental variance? What determines the variance of the other set of progeny?

16.6 For the difference between the domestic tomato, *Lycopersicon esculentum*, and the wild South American relative, *Lycopersicon chmielewskii*, the environmental variance σ_e^2 accounts for 13 percent of the total phenotypic variance σ_t^2 of fruit weight, 9 percent of σ_t^2 of soluble-solid content, and 11 percent of σ_t^2 of acidity. What are the broad-sense heritabilities of these traits?

16.7 Ten female mice had the following numbers of liveborn offspring in their first litters: 11, 9, 13, 10, 9, 8, 10, 11, 10, 13. Considering these females as representative of the total population from which they came, estimate the mean, variance, and standard deviation of size of the first litter in the entire population.

16.8 Values of IQ score are distributed approximately according to a normal distribution with mean 100 and standard deviation 15. What proportion of the population has a value above 130? Below 85? Above 85?

16.9 The data in the adjoining table pertain to milk production over an 8-month lactation among 304 two-year-old Jersey cows.

Pounds of milk produced	Number of cows
1000–1500	1
1500–2000	0
2000–2500	5
2500–3000	23
3000–3500	60
3500–4000	58
4000–4500	67
4500–5000	54
5000–5500	23
5500–6000	11
6000–6500	2

The data have been grouped by intervals, but for purposes of computation, each cow may be treated as though the milk production were equal to the midpoint of the interval (for example, 1250 for cows in the 1000–1500 interval).

(a) Estimate the mean, variance, and standard deviation in milk yield.

(b) What range of yield would be expected to include 68 percent of the cows?

(c) Round the limits of this range to the nearest 500 pounds, and compare the observed with the expected numbers of animals.

(d) Do the same calculations as in parts (b) and (c) for the range expected to include 95 percent of the animals.

16.10 In the F_2 generation of a cross of two cultivated varieties of tobacco, the number of leaves per plant was distributed according to a normal distribution with mean 18 and standard deviation 3. What proportion of the population is expected to have the following phenotypes?

(a) between 15 and 21 leaves

(b) between 12 and 24 leaves

(c) fewer than 15 leaves

(d) more than 24 leaves

(e) between 21 and 24 leaves

16.11 Two highly homozygous inbred strains of mice are crossed and the 6-week weight of the F_1 progeny determined.

(a) What is the magnitude of the genotypic variance in the F_1 generation?

(b) If all alleles affecting 6-week weight were additive, what would be the expected mean phenotype of the F_1 generation compared with the average 6-week weight of the original inbred strains?

16.12 In a cross between two cultivated inbred varieties of tobacco, the variance in leaf number per plant is 1.46 in the F_1 generation and 5.97 in the F_2 generation. What are the genotypic and environmental variances? What is the broad-sense heritability in leaf number?

16.13 Two inbred lines of *Drosophila* are crossed, and the F_1 population has a mean number of abdominal bristles of 20 and a standard deviation of 2. The F_2 generation has a mean of 20 and a standard deviation of 3. What are the environmental variance, the genetic variance, and the broad-sense heritability of bristle number in this population?

16.14 In an experiment with weight gain between ages 3 and 6 weeks in mice, the difference in mean phenotype between two strains was 17.6 grams and the genotypic variance was estimated as 0.88. Estimate the minimum number of genes affecting this trait.

16.15 Estimate the minimum number of genes affecting fruit weight in a population of the domestic tomato produced by crossing two inbred strains. When average fruit weight is expressed as the logarithm of fruit weight in grams, the inbred lines have average fruit weights of −0.137 and 1.689. The F_1 generation had a variance of 0.0144, and the F_2 generation had a variance of 0.0570.

16.16 A flock of broiler chickens has a mean weight gain of 700 g between ages 5 and 9 weeks, and the narrow-sense heritability of weight gain in this flock is 0.80. Selection for increased weight gain is carried out for five consecutive generations, and in each generation the average of the parents is 50 g greater than the average of the population from which the parents were derived. Assuming that the heritability of the trait remains constant at 80 percent, what is the expected mean weight gain after the five generations?

16.17 To estimate the heritability of maze-learning ability in rats, a selection experiment was carried out. From a population in which the average number of trials necessary to learn the maze was 10.8, with a variance of 4.0, animals were selected that managed to learn the maze in an average of 5.8 trials. Their offspring required an average of 8.8 trials to learn the maze. What is the estimated narrow-sense heritability of maze-learning ability in this population?

16.18 A replicate of the population in the preceding problem was reared in another laboratory under rather different conditions of handling and other stimulation. The mean number of trials required to learn the maze was still 10.8, but the variance was increased to 9.0. Animals with a mean learning time of 5.8 trials were again selected, and the mean learning time of the offspring was 9.9.

(a) What is the heritability of the trait under these conditions?

(b) Is this result consistent with that in Problem 16.17 and how can the difference be explained?

16.19 In terms of the narrow-sense heritability, what is the theoretical correlation coefficient in phenotype between first cousins who are the offspring of monozygotic twins?

16.20 If the correlation coefficient of a trait between first cousins is 0.09, what is the estimated narrow-sense heritability of the trait?

16.21 A representative sample of lamb weights at the time of weaning in a large flock is shown to the right. If the narrow-sense heritability of weaning weight is 20 percent, and the half of the flock consisting of the heaviest lambs is saved for breeding for the next generation, what is the best estimate of the average weaning weight of the progeny? (*Note*: If a normal distribution has mean μ and standard deviation σ, then the mean of the upper half of the distribution is given by $\mu + 0.8\sigma$.)

81	81	83	101	86
65	68	77	66	92
94	85	105	60	90
94	90	81	63	58

Challenge Problems

16.22 Consider a trait determined by n unlinked, additive genes with two alleles of each gene, in which each favorable allele contributes one unit to the phenotypic value and each unfavorable allele contributes nothing. What is the value of the genotypic variance?

16.23 Two varieties of chickens differ in the average number of eggs laid per hen per year. One variety yields an average of 300 eggs per year, the other an average of 270 eggs per year. The difference is due to a single quantitative-trait locus (QTL) located between two restriction fragment length polymorphisms. The genetic map is

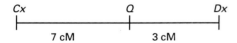

In this diagram, *Cx* and *Dx* are the locations of the restriction fragment length polymorphisms, and *Q* is the location of the QTL affecting egg production. The map distances are given in centimorgans (cM). The *Cx* alleles are denoted *Cx4.3* and *Cx2.1* according to the size of the restriction fragment produced by digestion with a particular restriction enzyme, and the *Dx* alleles are denoted *Dx3.2* and *Dx1.1*. The allele *Q* is associated with the higher egg production and *q* with the lower egg production, with the *Qq* genotype having an annual egg production equal to the average of the *QQ* (300 eggs) and *qq* (270 eggs) genotypes. A cross of the following type is made:

$$\frac{Cx4.3 \quad Q \quad Dx1.1}{Cx2.1 \quad q \quad Dx3.2} \times \frac{Cx4.3 \quad q \quad Dx1.1}{Cx2.1 \quad q \quad Dx3.2}$$

Female offspring are classified into the following four types:

(a) *Cx4.3 Dx1.1/Cx2.1 Dx3.2*

(b) *Cx4.3 Dx3.2/Cx2.1 Dx3.2*

(c) *Cx2.1 Dx1.1/Cx2.1 Dx3.2*

(d) *Cx2.1 Dx3.2/Cx2.1 Dx3.2*

What is the expected average egg production of each of these genotypes? You may assume complete chromosome interference across the region.

16.24 The pedigree shown to the right includes information about the alleles of three genes that are segregating.

- Gene 1 has four alleles denoted *A, B, C,* and *D*. Each allele is identified according to the electrophoretic mobility of a DNA restriction fragment that hybridizes with a probe for gene 1. The pattern of bands observed in each individual in the pedigree is shown in the diagram of the gel below the pedigre. When an allele is homozygous, the band is twice as thick as otherwise, because homozygous genotypes yield twice as many restriction fragments corresponding to the allele than do homozygotes. The small table at the upper right of the pedigree gives the frequencies of the gene 1 alleles in the population as a whole.

- Gene 2 has four alleles denoted *E, F, G,* and *H*. These also are identified by the electrophoretic mobility of restriction fragments. For each person in the pedigree, the genotype with respect to gene 2 is indicated. The population frequencies of the gene-2 alleles are also given in the table at the upper right.

- Gene 3 is associated with a rare genetic disorder. Affected people are indicated by the brown symbols in the pedigree. The genotype associated with the disease has complete penetrance. There are two alleles of gene 3, which are denoted *Q* and *q*.

(a) What genetic hypothesis can explain the pattern of affected and nonaffected individuals in the pedigree?

(b) If the male IV-2 fathers a daughter, what is the probability that the daughter will be affected?

(c) Assuming random mating, if the male IV-2 fathers a daughter, what is the probability that the daughter will have the genotype *HF* for gene 2?

(d) Identify the genotype of each person in the pedigree with respect to genes 1, 2, and 3. Wherever possible, specify the linkage phase (coupling or repulsion) of the alleles present in each individual.

(e) Examine the data for evidence of linkage. Are any of the pairs of genes linked? On the basis of this pedigree, what is the estimated recombination frequency between gene 1 and gene 2? Between gene 2 and gene 3?

Pedigree for Problem 16.24:

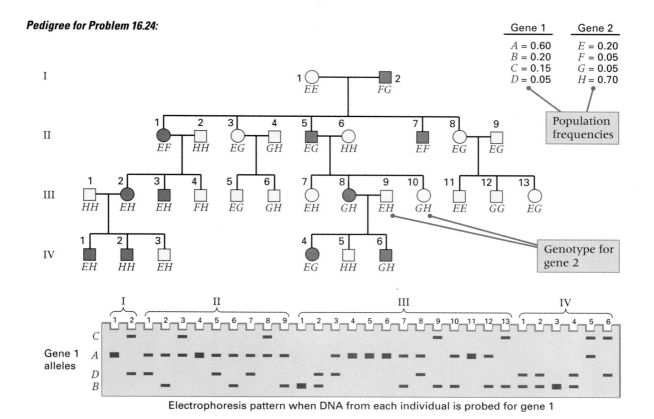

Electrophoresis pattern when DNA from each individual is probed for gene 1

Further Reading

Black, D. M., and E. Solomon. 1993. The search for the familial breast/ovarian cancer gene. *Trends in Genetics* 9: 22.

Bouchard, T. J., Jr., D. T. Lykken, M. McGue, N. L. Segal, and A. Tellegen. 1990. Sources of human psychological differences: The Minnesota study of twins reared apart. *Science* 250: 223.

Comprehensive Human Genetic Linkage Center: J. C. Murray, K. H. Buetow, J. L. Weber, S. Ludwigsen, T. Scherpbier-Heddema, F. Manion, J. Quillen, V. C. Schffield, S. Sunden, G. M. Duyk; Généthon: J. Weissenbach, G. Gyapay, C. Dib, J. Morrissette, G. M. Lathrop, A. Vignal; University of Utah: R. White, N. Matsunami, S. Gerken, R. Melis, H. Albertson, R. Plaetke, S. Odelberg; Yale University: D. Ward; Centre d'Etude du Polymorphisme Humain (CEPH): J. Dausset, D. Cohen, H. Cann. 1994. A comprehensive human linkage map with centimorgan density. *Science* 265: 2049.

Crow, J. F. 1993. Francis Galton: Count and measure, measure and count. *Genetics* 135: 1.

East, E. M. 1910. Mendelian interpretation of inheritance that is apparently continuous. *American Naturalist* 44: 65.

Falconer, D. S., and T. F. C. Mackay. 1996. *Introduction of Quantitative Genetics*, 4th ed. Essex, England: Longman.

Feldman, M. W., and R. C. Lewontin. 1975. The heritability hang-up. *Science* 190: 1163.

Greenspan, R. J. 1995. Understanding the genetic construction of behavior. *Scientific American,* April.

Haley, C. S. 1996. Livestock QTLs: Bringing home the bacon. *Trends in Genetics* 11: 488.

Hartl, D. L., and A. G. Clark. 1997. *Principles of Population Genetics*. 3rd ed. Sunderland, MA: Sinauer.

Lander, E. S., and N. J. Schork. 1994. Genetic dissection of complex traits. *Science* 265: 2037.

Paterson, A. H., E. S. Lander, J. D. Hewitt, S. Peterson, S. E. Lincoln, and S. D. Tanksley. 1988. Resolution of quantitative traits into Mendelian factors by using a complete linkage map of restriction fragment length polymorphisms. *Nature* 335: 721.

Pirchner, F. 1983. *Population Genetics in Animal Breeding*. 2d ed. New York: Plenum.

Risch, N., and H. Zhang. 1995. Extreme discordant sib pairs for mapping quantitative trait loci in humans. *Science* 268: 1584.

Stigler, S. M. 1995. Galton and identification by fingerprints. *Genetics* 140: 857.

Stuber, C. W. 1996. Mapping and manipulating quantitative traits in maize. *Trends in Genetics* 11: 477.

Tanksley, S. D. and S. R. McCouch. 1997. Seed banks and molecular maps: Unlocking genetic potential from the wild. *Science* 277:1063.

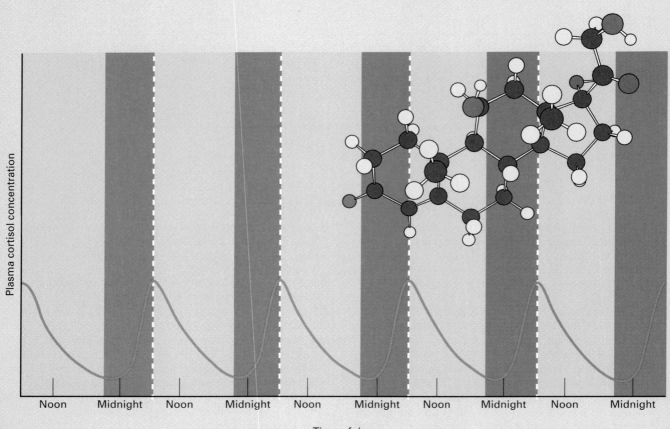

Circadian rhythms with a period of approximately 24 hours affect many physiological and neurological traits. This example shows a circadian rhythm in the production of cortisol in human beings. The highest levels occur at night, just before waking. N means noontime and M midnight. [Adapted with permission from G. A. Hedge, H. D. Colby, and R. L. Goodman. 1987. *Clinical Endocrine Physiology.* Philadelphia: Saunders.]

Genetics of Biorhythms and Behavior

PRINCIPLES

- Behavior includes all changes that an organism undergoes in response to its environment.
- Some behaviors are completely determined genetically. An example is chemotaxis in *E. coli*, which results from control of counterclockwise or clockwise rotation of the flagella by means of intracellular signals transmitted in response to attractants or repellents in the medium.
- Other behaviors are determined by innate biological rhythms, such as the circadian rhythm of roughly 24 hours. The molecular genetics of the *Drosophila* circadian rhythm involves cyclic transcriptional control of genes such as *per* (*period*) and *tim* (*timeless*).
- Learning in laboratory animals is affected by genetic as well as environmental factors. Genetic factors are evidenced by the success of artificial selection in increasing or decreasing maze-learning ability in laboratory rats. In this case, there is also a strong genotype-environment interaction.
- Some human behavioral differences are due to simple Mendelian factors, such as the ability to detect the chemical phenylthiocarbamide. More complex human behaviors are affected by genetic and environmental factors and the interactions between them.
- Heritability estimates of human behavioral traits are often inflated because of cultural transmission; these values, however estimated, cannot be used to infer the genetic or environmental causation of differences among human racial or ethnic groups.

CONNECTIONS

CONNECTION: **A Trip to the Zoo**
Joan Fisher Box 1978
R. A. Fisher: The Life of a Scientist

All organisms interact with their environments. The bacterium *E. coli* responds to the presence of lactose (milk sugar) in its environment by synthesizing new enzymes that allow the lactose to be brought into the cell and metabolized. Sunflowers respond to sunlight by tracking the sun across the sky. Canada geese respond to signs of autumn by beginning their long migrations. You respond to touching a hot stove by involuntarily jerking your hand away. **Behavior** in its broadest sense includes all changes that an organism undergoes in response to its environment. The underlying genetics and molecular biology of some seemingly complex behaviors may be relatively simple. For example, the enzymes for utilizing lactose in *E. coli* are induced by the straightforward molecular mechanism discussed in Chapter 11. At the other extreme of complexity, behaviors such as learning and memory in humans beings are currently beyond explanation by genetics and molecular biology. Between the bacterial cell's response to lactose and a human child's ability to learn a language are immeasurable gradations of behavioral complexity. The field of **behavior genetics** deals with the study of observable genetic differences that affect behavior.

Because behavior genetics is carried out in a wide variety of organisms, from bacteria to human beings, virtually all techniques of genetic analysis are applicable at one level or another. The induction and analysis of specific mutations affecting behavior has proved fruitful in behavioral studies in *E. coli* and *Drosophila*. In human behavior, the most common techniques of genetic analysis are pedigree studies and quantitative genetics. Quantitative genetics is essential in human behavior genetics, because most variation in human behavior-particularly variation within the range considered "normal"-is multifactorial in origin; most normal human behavior is influenced by both genetic and environmental factors.

This chapter summarizes examples of behavioral genetic analysis. Its emphasis is on the types of genetic approaches that have been developed for the study of behavior in different organisms. The examples start with *E. coli* and end with human beings.

17.1
Chemotaxis in Bacteria

Chemotaxis is the tendency of organisms to move toward certain chemical substances (attractants) and away from others (repellents). Such behaviors are widespread among living organisms. Among human beings, who is not attracted to the aroma of freshly baked bread? Some organisms rely on chemical attractants for finding mates; one example is the female gypsy moth, whose pheromones can attract males across long distances. Other organisms, such as the skunk, rely on chemical repellents for self-defense. Even relatively simple creatures, such as bacteria, have a chemotactic response. In *E. coli*, the chemotactic response can be observed by inserting a capillary tube filled with an attractant or repellent into a test tube containing bacterial cells in agar dilute enough that the bacteria can swim (Figure 17.1). If the capillary contains a neutral chemical that neither attracts nor repels, the bacterial cells are indifferent to it and remain uniformly dispersed throughout the medium (part A). If the capillary contains an attractant, the cells move toward the mouth of

Figure 17.1
Chemotactic behavior of *E. coli* in a soft agar solution with an inserted capillary tube containing (A) a neutral chemical, (B) an attractant, or (C) a repellent. The cells swim toward the attractant, swim away from the repellent, and are indifferent to the neutral chemical.

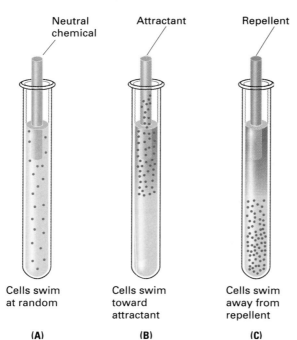

Neutral chemical

Attractant

Repellent

Cells swim at random

Cells swim toward attractant

Cells swim away from repellent

(A)

(B)

(C)

the capillary and eventually swim up into it (part B). If the capillary contains a repellent, the cells swim away from it (part C). Most attractants are nutritious substances such as sugars and amino acids, whereas most repellents are harmful substances or excretory products such as acids or alcohols. Hence, the chemotactic response is important in helping bacterial cells move toward favorable conditions or away from harmful ones.

Cells of *E. coli* propel themselves by means of hollow filaments, called **flagella,** that twirl individually like propellers on a motorboat. A typical cell has six flagella, each approximately 7 μm long (Figure 17.2), which is roughly three times the length of the cell itself. At the base of each flagellum is a molecular "motor" that rotates the flagellum either counterclockwise or clockwise. Counterclockwise rotation twists the flagella into a single bundle that propels the cell forward (Figure 17.3A), whereas clockwise rotation causes the flagella to splay out and work at cross

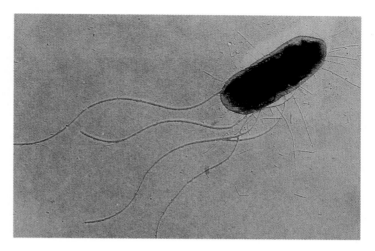

Figure 17.2
Flagella (long fibers) emanate from random places from the sides of an *E. coli* cell. When the flagella rotate counterclockwise, they form a bundle and propel the cell forward; when they rotate clockwise, they splay apart and cause the cell to tumble. The shorter, thinner objects projecting from the cell are a different type of fiber that aids in adherence to surfaces. Magnification × 27,600. [Courtesy of Howard C. Berg.]

Figure 17.3
Each flagellum of *E. coli* is rotated by a molecular "motor" at the base. (A) If the flagella are rotated counterclockwise, they form a coherent bundle that propels the cell forward (swimming). (B) If the flagella are rotated clockwise, they spread apart and the cell tumbles, causing an unpredictable change in direction. (C) After tumbling, a return to counterclockwise rotation causes the cell to dart off in a new direction.

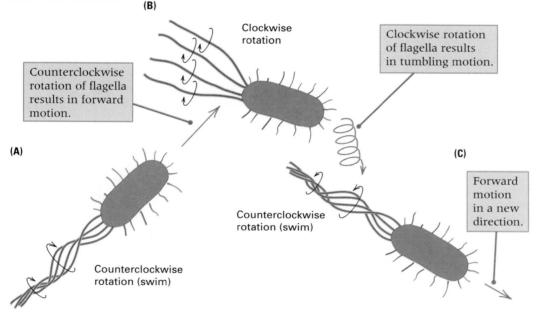

(B)
Clockwise rotation

Counterclockwise rotation of flagella results in forward motion.

Clockwise rotation of flagella results in tumbling motion.

(A)

Counterclockwise rotation (swim)

Counterclockwise rotation (swim)

(C)

Forward motion in a new direction.

purposes (Figure 17.3B). Consequently, a cell that switches flagellar rotation from counterclockwise to clockwise stops swimming forward and begins to tumble, thereby changing its orientation in the medium. A return to counterclockwise rotation of the flagella then sends the cell off swimming again in a new direction (Figure 17.3C). In the absence of attractants or repellents, cells trace a kind of random walk, darting forward in some direction for a second or two, then tumbling for a fraction of a second, and then darting off in perhaps a quite different direction (Figure 17.4). In the presence of an attractant, periods of swimming toward the attractant become longer, and periods of moving in the other direction become shorter. The random walk consequently becomes biased toward the direction of the attractant, and the cells eventually swim and tumble their way to the source of the attractant (Figure 17.1). Repellents act in the reverse manner, and the cells gradually disperse from them.

Mutations Affecting Chemotaxis

Mutations affecting chemotaxis have revealed the genetic components of the chemotactic response. The phenotypes of chemotaxis mutations fall into five very general categories, which are summarized in Table 17.1.

1. **Specific nonchemotaxis** means that mutant cells show no response to a particular attractant or repellent. For example, galactose is an attractant. A mutant phenotype that fails to respond to galactose, but that is otherwise completely wildtype, shows a nonchemotaxis that is specific for galactose. A mutation in a different gene results in a phenotype of specific nonchemotaxis in which mutant cells fail to be attracted to glucose.

2. **Multiple nonchemotaxis** refers to a phenotype with no chemotactic response to a group of chemically unrelated substances. For example, mutations in the gene designated *tsr* cause the mutants to fail to respond to the amino acid attractant serine or to the repellents indole, leucine, and acetate. Mutants that are multiply nonchemotactic fall into four complementation groups and thus define four genes, which are called *tsr, tar, trg,* and *tap.*

3. **General nonchemotaxis** is a failure to respond to any attractant or repellent. Although such mutants have normal flagella, as well as the apparatus necessary for switching flagellar rotation between clockwise and counterclockwise, they are unable to control the flagellar rotation in response to attractants or repellents. Mutations for general nonchemotaxis can be grouped into six complementation groups, each of

Figure 17.4
Three-dimensional path of a single *E. coli* cell followed automatically in a computerized tracking microscope. The track starts at the position of the red dot. In the 30-second tracking period, the cell made 26 swims (straight paths) and 26 tumbles (changes of direction). The swims were at a speed of about 20 μm per second. The cube has dimensions of 130 μm on each side. [After H. C. Berg. 1975. *Scientific American* 233: 36.]

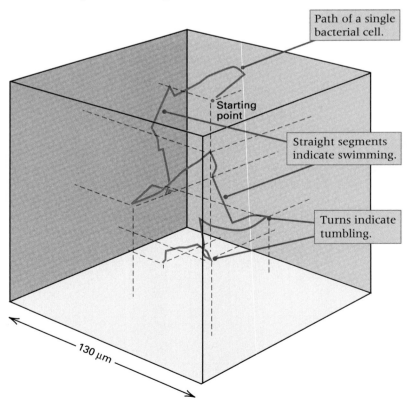

Path of a single bacterial cell.

Starting point

Straight segments indicate swimming.

Turns indicate tumbling.

130 μm

Table 17.1 Major categories of nonchemotactic mutants

Type of mutant	Genes	Phenotype
1. Specific nonchemotaxis	Galactose-binding protein, glucose transporter, etc.	No response to a particular attractant (e.g., galactose)
2. Multiply nonchemotactic	*tsr, tar, trg, tap*	No response to attractants that stimulate a particular receptor
3. Generally nonchemotactic	*cheA, cheW, cheR, cheB, cheZ, cheY*	No response to any attractant
4. Motor components	*fliG, fliM, fliN*	Continuous tumbling or continuous swimming
5. Flagellar components	Genes needed for synthesis of flagella	No motility

which defines a gene that is essential for any kind of chemotaxis. These genes are designated *cheA, cheB, cheR, cheW, cheY,* and *cheZ* (Table 17.1).

4. Mutations in the **motor components** of the flagella cause nonchemotactic behavior because of defects in the apparatus that controls switching between clockwise and counterclockwise flagellar rotation. These mutants continuously swim or continuously tumble, depending on the particular type of mutation. Figure 17.5 is the track of a mutant cell that swims continuously. Mutations in the motor components define three genes necessary for switching between swimming and tumbling, which are designated *fliG, fliM,* and *fliN*.

5. **Nonmotile mutants** are nonchemotactic for the trivial reason that they are unable either to swim or to tumble. Nonmotile mutations define a large number of different genes that are necessary for the synthesis or function of the flagella. Figure 17.6 is a diagram showing some of the gene products that are structural components of the flagellum.

Analysis of mutations has been critical to our understanding of the process of chemotaxis. For example, the correlation between the direction of rotation of the flagella and the swimming or tumbling behavior of the cells was determined by observation of the two types of mutations in the motor components of the flagella

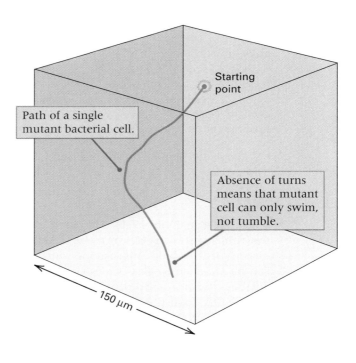

Starting point

Path of a single mutant bacterial cell.

Absence of turns means that mutant cell can only swim, not tumble.

150 μm

Figure 17.5
Swimming path of an *E. coli* cell with a mutation in one of the flagellar motor genes, making the cell unable to tumble. The track starts at the position of the red dot. Compare this pattern with that of the wildtype cell in Figure 17.4. In a 7-second tracking period, the cell swam 225 μm without a tumble. The square is 150 μm on each side. [After H. C. Berg. 1972. *Nature* 239: 500.]

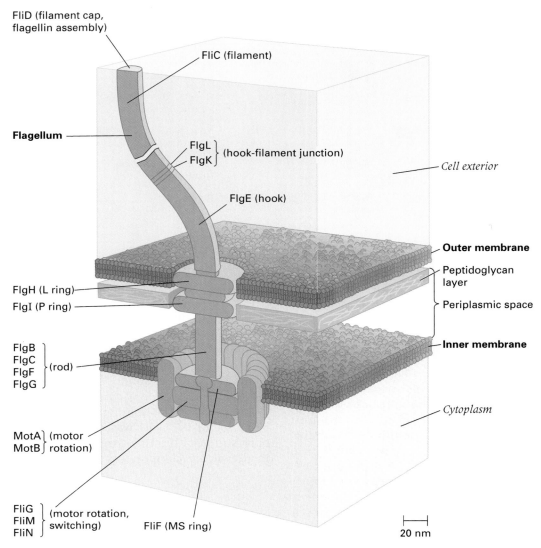

FliD (filament cap, flagellin assembly)

FliC (filament)

Flagellum

FlgL
FlgK } (hook-filament junction)

FlgE (hook)

FlgH (L ring)

FlgI (P ring)

FlgB
FlgC
FlgF
FlgG } (rod)

MotA } (motor
MotB } rotation)

FliG
FliM
FliN } (motor rotation, switching)

FliF (MS ring)

Cell exterior

Outer membrane

Peptidoglycan layer

Periplasmic space

Inner membrane

Cytoplasm

20 nm

Figure 17.6

Protein components of the flagellum of *E. coli*. The products of more than 40 genes necessary for flagellar synthesis and function have been identified. The flagellum is anchored to the outer and inner cell membranes and is rotated by means of a motor apparatus at the base. The products of the genes *fliG, fliM,* and *fliN,* denoted FliG, FliM, and FliN, respectively, are necessary for switching between counterclockwise and clockwise rotation. Mutations in these genes result in motile but nonchemotactic cells. Mutations in the genes for most of the other flagellar components result in loss of motility. [From R. M. Macnab and J. S. Parkinson. 1991. *Trends Genet.* 7: 196.]

(category 4 in Table 17.1). The flagella are too thin for their rotation to be observed directly in the light microscope, but they can be attached to glass slides coated with antiflagellar antibody proteins that combine tightly with the flagellar proteins and hold them to the slide. Under these conditions, the bacteria become tethered to the slide through their flagella, and the flagella remain fixed while the *body* of the cell rotates. The cell bodies are large enough that the direction of rotation can be observed directly, and the cell bodies rotate about the tethered flagella in the same direction as the flagella themselves would rotate if they were untethered (Figure

17.7). Observation of tethered cells has revealed that

Mutants that swim continuously rotate counterclockwise continuously; mutants that tumble continuously rotate clockwise continuously.

All of the genes required for chemotaxis in categories 2 through 4 have been cloned and sequenced, and their gene products have been identified. The complex behavior of chemotaxis is a relatively simple consequence of the various products of these genes and their interactions. The cellular components of chemotaxis and the mechanism of the response are the subjects of the next section.

The Cellular Components of Chemotaxis

The cellular components of chemotaxis are illustrated in the diagram of an *E. coli* cell in Figure 17.8. A mutational defect in any of these components results in abnormal chemotaxis, and the phenotypes of the major classes of mutants listed in Table 17.1 are indicated in the blue boxes. Mutations can affect components in the system at any of the following levels.

Wildtype chemotaxis requires a set of proteins called **chemosensors,** each of which binds with a particular attractant or repellent. Examples of chemosensors include the galactose-binding protein and the glucose transporter. More than 20 different chemosensors are known; about half of them bind with attractants and half with repellents. Some chemosensors, such as the galactose-binding protein, are present in the periplasmic space between the cytoplasmic membrane and the outer membrane. Other chemosensors, such as the glucose transporter, are tightly associated with the cytoplasmic membrane. The chemosensors for attractants are typically components of the molecular systems used for transporting the attractant into the cell. Mutations in the chemosensors result in a phenotype of specific nonchemotaxis (Table 17.1).

The next step in chemotaxis is interaction between the chemosensor and any of

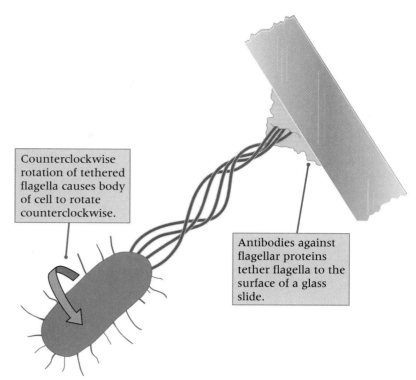

Counterclockwise rotation of tethered flagella causes body of cell to rotate counterclockwise.

Antibodies against flagellar proteins tether flagella to the surface of a glass slide.

Figure 17.7
A molecular motor at the base of a cell that rotates the flagellum in any direction (in this case, counterclockwise) will rotate the cell body in the same direction (in this case, counterclockwise) when the flagellum is tethered and unable to turn.

four **chemoreceptors,** which are the products of the genes *tsr, tar, trg,* and *tap.* Each chemoreceptor interacts with a particular subset of chemosensors. For example, the chemoreceptor Tsr (Figure 17.9), encoded by the gene *tsr,* interacts with the chemosensor for the attractant serine (an amino acid) and with chemosensors for the repellents leucine, indole, and acetate. Similarly, the chemoreceptor Trg, corresponding to the *trg* gene, responds to several different chemosensors, including those for galactose and ribose. Mutations in any of the chemoreceptor genes result in loss of responsiveness to all of the chemosensors with which the chemoreceptor interacts. Hence chemoreceptor mutants have a phenotype of multiple nonchemotaxis (category 2 in Table 17.1).

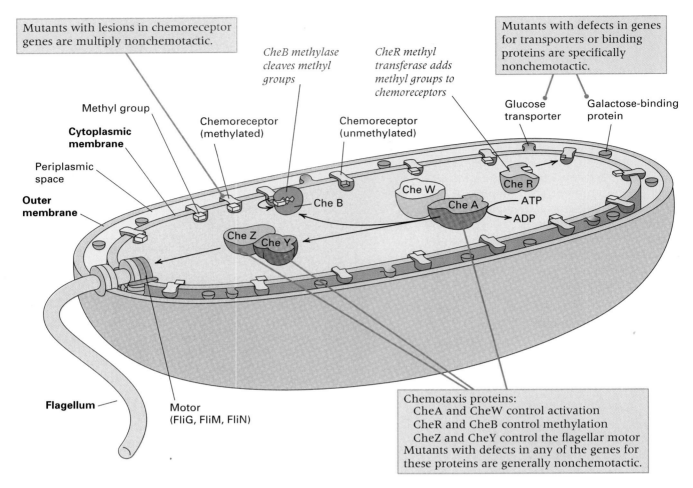

CheB methylase cleaves methyl groups

CheR methyl transferase adds methyl groups to chemoreceptors

Methyl group

Cytoplasmic membrane

Chemoreceptor (methylated)

Chemoreceptor (unmethylated)

Glucose transporter

Galactose-binding protein

Periplasmic space

Outer membrane

Che W

Che B

Che R

ATP

Che A

ADP

Che Z

Che Y

Flagellum

Motor (FliG, FliM, FliN)

Figure 17.8
Diagram of a cell of *E. coli* showing the interactions of the major types of proteins implicated in chemotaxis. The phenotypes associated with the major classes of mutations are indicated in the blue boxes.

All of the chemoreceptors feed molecular signals into a common pathway containing the proteins denoted CheA, CheB, CheR, CheW, CheY, and CheZ, illustrated in Figure 17.8; these proteins are the products of the genes *cheA, cheB, cheR, cheW, cheY,* and *cheZ,* respectively. The molecular nature of the chemotactic signals is described in the next section. For the present, the important point is that all of the chemoreceptors use the same pathway, and this explains why mutations in any of the genes in the common pathway result in a phenotype of general nonchemotaxis (category 3 in Table 17.1).

The target of the common pathway is the molecular motor at the base of the flagella. Switching between clockwise rotation (tumbling) and counterclockwise rotation (swimming) is controlled by the products of the genes *fliG, fliM,* and *fliN.*

Mutations in any of these genes result in continuous tumbling or continuous swimming (Table 17.1).

Once the chemotaxis signal is passed to the switching apparatus, successful attraction or repulsion requires functional flagella that are able to rotate. The flagella are composed of a number of essential components (Figure 17.6) in which mutational defects result in nonmotile cells (category 5 in Table 17.1).

Molecular Mechanisms in Chemotaxis

The molecular basis of the chemotactic response is mediated by chemical modification of some of the key components in the central pathway illustrated in Figure 17.8. For clarity, the chemoreceptors are shown distributed uniformly in the cytoplasmic

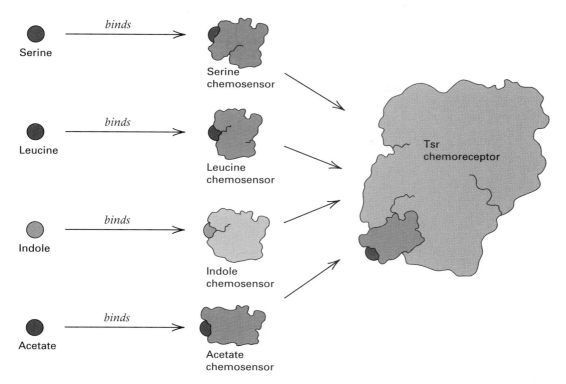

Figure 17.9
Pattern in which the Tsr chemoreceptor interacts with a number of different types of chemosensors.

membrane, and CheA and CheW in the cytoplasm. However, localization of key components of the chemotaxis apparatus by the use of fluorescent antibodies implies that they are concentrated at the poles of the cell (Figure 17.10). To understand how chemotaxis in *E. coli* works, let us first follow the normal course of events in the absence of attractants or repellents, as shown by the arrows in Figure 17.8.

The key player in regulating chemotaxis is CheA. This protein contains a region that catalyzes the transfer of a phosphate group from ATP to a specific histidine within the molecule (Figure 17.11A), yielding the phosphorylated form of CheA (denoted P-CheA). The formation of P-CheA is stimulated by the chemoreceptors acting in combination with CheW. Once transferred to P-CheA, the phosphoryl group on the histidine in P-CheA is readily transferred to either CheY or CheB, resulting in the phosphorylated forms P-CheY and P-CheB. The P-CheY form interacts directly with the switching apparatus that controls flagellar

Figure 17.10
Evidence that the chemoreceptor proteins are preferentially located at the poles of the cell. In this photograph, the green color results from fluorescent antibodies to Tsr combining with the Tsr protein where it is localized in the cell-namely, at the poles. Similar polar localization is also observed for CheA and CheW. [Courtesy of L. Shapiro, from J. R. Maddock and L. Shapiro. 1993. *Science* 259: 1717.]

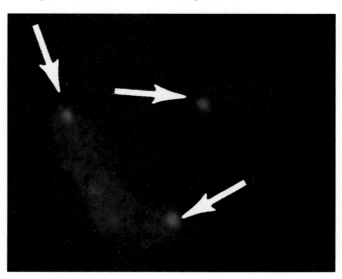

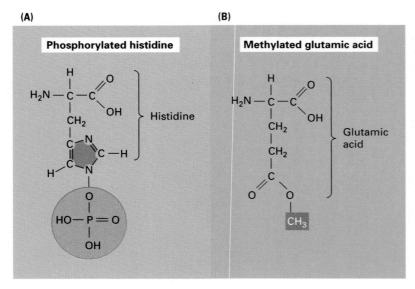

Figure 17.11
Protein modifications in the regulation of chemotaxis. (A) One particular histidine at amino acid position 48 in the CheA protein becomes phosphorylated (chemical group in yellow) at the position shown. (B) A number of glutamic acid residues in the chemoreceptor proteins become methylated (methyl group shown in red). Addition of the methyl groups is catalyzed by CheR, and removal is catalyzed by P-CheB.

rotation and causes the flagella to rotate clockwise (tumbling). As might be expected from this description, extreme mutations in *cheA* that cannot make P-CheA, and extreme mutations in *cheY* that cannot make P-CheY, result in a phenotype of continuous swimming because the clockwise rotation of the flagella necessary for tumbling cannot be stimulated.

If only CheA and CheY were at work, the cells could be expected to tumble continuously because of nearly constant stimulation of the rotational switch by P-CheY. However, continuous tumbling is prevented by P-CheB, which acts on the chemoreceptors to remove methyl groups from certain glutamic acid subunits that are methylated (Figure 17.11B). The methylated forms of the chemoreceptors stimulate production of P-CheA. Removal of the methyl groups by P-CheB inhibits the formation of P-CheA and thereby diminishes further tumbling. Continuous tumbling is also prevented by CheZ, which rapidly removes the phosphoryl group from P-CheY, so that each tumbling episode caused by P-CheY typically lasts for only a fraction

of a second. As might be expected, extreme *cheZ* mutations result in a phenotype of continuous tumbling because of the increased levels of P-CheY. Extreme *cheB* mutations also result in a phenotype of continuous tumbling, because when the chemoreceptors remain methylated, they stimulate additional production of P-CheA and therefore P-CheY.

As noted, methylation of the chemoreceptors stimulates greater production of P-CheA and enhances the tendency of the cells to tumble. Addition of the methyl groups is the function of CheR. In the chemoreceptor Tsr, for example, as many as four glutamic acid residues can be methylated. Methylation requires the amino acid methionine because a derivative of methionine (*S-adenosylmethionine*) is the donor of the methyl groups that CheR transfers. Because they accept methyl groups, the chemoreceptor proteins are often called **MCP** proteins, which stands for **methyl-accepting chemotaxis proteins.** Extreme mutations in *cheR* cannot methylate the receptors; therefore, the mutants swim continuously.

In the absence of any chemotactic stimulus, the basal level of phosphorylation of CheA results in enough P-CheY that the cells undergo repeated episodes of tumbling and reorientation in random directions; the tumbling episodes are terminated by the effects of CheZ and P-CheB, after which the cells swim for several seconds in the direction in which they happen to be oriented at the moment. The phosphorylation of CheA is inhibited by the binding of attractants to the chemoreceptors. Therefore, in the presence of an attractant, the episodes of tumbling become less frequent. Cells that, by chance, happen to be propelling themselves toward higher concentrations of the attractant undergo greater inhibition of tumbling, which results in longer runs of swimming toward the attractant. The upshot is that, in time, all the cells in the medium converge toward the attractant (Figure 17.1B). With repellents, the situation is the reverse: The repellent-excited chemoreceptors increase the rate of CheA phosphorylation, which results in more tumbling and, ultimately, divergence from the source of the repellent (Figure 17.1C).

A Trip to the Zoo

Joan Fisher Box 1978
Madison, Wisconsin

Ronald A. Fisher combined extraordinary mathematical skills with a deep love of biological observation. Shortly after the discovery that the human population is polymorphic for the ability to taste phenylthiocarbamide (PTC)-the substance is harmless, but tasters report that it is very bitter-Fisher and a small group of other geneticists decided to find out whether chimpanzees are also polymorphic for the taster trait. If such a polymorphism persists over long periods of evolutionary time, even through events of species formation and divergence, then the persistence is evidence that the polymorphism may be maintained by some sort of balancing selection in which, on the average, the heterozygous genotype is favored. With this in mind, and solutions of PTC in hand, the researchers paid a visit to the chimpanzees at the Edinburgh Zoo. The episode is recalled by Joan Fisher Box, Fisher's daughter and biographer.

At the end of August 1939 the International Genetical Congress met in Edinburgh. It was an uneasy session in anticipation of the outbreak of World War II. . . . Fisher and a number of others abandoned the confer-ence one afternoon and made their way in high spirits to the cage of the chimpanzees [at the Edinburgh Zoo]. The eight chimpanzees were given mugs containing first a small quantity of slightly sweetened water (2% sucrose), a pleasant mixture for which they were expected to return for more. They drank it up and came back. This time they were given a sugared mixture containing 6 parts per million of phenylthiocarbamide; there were one or two dubious

---

Obviously, there were tasters among the chimpanzees.

---

expressions but they returned again, and received a mixture of 50 parts per million of phenylthiocarbamide. This time there was no doubt about their reaction: one chimpanzee turned his back emphatically on his visitors; another poured his drink on the ground with disgust and went to the opposite side of the cage; one sat gazing reproachfully at his audience while the fluid dribbled from a lower lip, a picture of misery; a more spirited beast spat it out in a strong jet straight at his persecutors. The geneticists were much diverted by these exhibitions. Obviously, there were tasters among the chimpanzees. There were also nontasters: two of the eight animals tested returned to receive and to drink a mixture containing 400 parts per million and still returned for more. This result was astonishing. The small sample was inconclusive as to the exact proportion of tasters, but it suggested not only that chimpanzees were polymorphic for the taste factor in the same way as people but in approximately the same proportions.

Source: R. A. Fisher: The Life of a Scientist. John Wiley & Sons, New York (pp. 371–372)

It should be noted that the Animal Caretaker approved of these experiments and that the animals were not harmed. Today the experiments would need advance approval by an Animal Welfare Committee in response to a written application explaining the details of the experiment and making a case for its scientific value. Try having a discussion with a few fellow students by pretending that you are members of such a committee; the issue for debate is whether to approve an application for similar experiments with a different, but also harmless, chemical.

Sensory Adaptation

Methylation of the chemoreceptors by CheR is the basis of **sensory adaptation,** a process by which bacterial cells become accommodated to external stimuli. Sensory adaptation can be demonstrated by a simple experiment in which a solution of attractant is suddenly stirred into a suspension of cells. Because of the mixing, the concentration of the attractant is the same everywhere in the medium, but nevertheless tumbling is suppressed and each cell swims in whatever direction it happens to be oriented. After a few minutes, however, the cells adapt to the presence of the attractant and begin to tumble and swim in the normal, unstimulated manner. What happens is that the temporarily reduced level of P-CheB allows the activity of CheR to increase the overall level of methylation of the chemoreceptors. Methylation stimulates the production of P-CheA, which not only restores tumbling behavior through the production of P-CheY but also yields enough P-CheB methylase activity to result in an equilibrium level of methylation that is higher than it was before the attractant was added. The attractant still attracts the cells, but higher concentrations are necessary to be effective. For example, the Tsr chemoreceptor for serine contains four glutamic acid residues that are capable of being methylated. Each additional methyl group increases the level of serine required for excitation, so sensory adaptation to serine is graded in response to the existing level of serine. Should the level of an attractant suddenly decrease, the P-CheB methylase removes additional methyl groups and returns the chemoreceptor to its basal level of methylation.

The mixing experiment also demonstrates another important point about chemotaxis. Because bacterial cells swim toward increasing concentrations of an attractant, it may seem that the cells are comparing concentrations at various points in space-for example, between opposite ends of the cell. This would be very difficult because the bacterial cell is so small (approximately $1 \times 2 \; \mu$m). What the cells, in fact, do is compare the concentration that they experience at any time with the concentration of a moment ago. If the present concentration of an attractant is greater than that in the immediate past, the cell swims. If the present concentration is smaller, the cell tumbles. The comparison of concentrations at different points in *time*, rather than different points in *space*, explains why bacteria respond as they do to a substance suddenly dispersed uniformly in a liquid medium: If it is an attractant, each cell swims in a random direction; if it is a repellent, each cell tumbles.

17.2
Animal Behavior

The most complex behaviors are found in animals with well-developed nervous systems, such as insects, fish, reptiles, birds, and mammals. Although behaviors can differ among normal organisms in a species, the inheritance of these differences is usually through multiple genes. For example, inbred lines of mice differ in their nesting behavior, but the differences are due to multiple genes, no one of which has such a strong effect that it stands out above the rest. Complex behaviors are usually also influenced by environmental factors, such as learning and experience. Nevertheless, mutations in single genes can affect complex behaviors, so the genetic approach can be used to identify genes that are essential for the behavior. Because of the relative ease with which specific mutations can be obtained and analyzed in *D. melanogaster*, this organism is a favorite for research in behavior genetics. Behavioral mutants in *Drosophila* include mutations affecting the ability of the animal to walk or fly, the response to light, the ability to learn or to retain prior learning, sexual behavior or sexual responsiveness, and the length of cycle of the internal clock that regulates the animal's activities in synchrony with the cycle of day and night. The following sections illustrate the genetic approach to dissecting a particular complex behavior in *Drosophila*.

Circadian Rhythms

Many animal behaviors repeat at regular intervals. Biological rhythms synchronized with the 24-hour cycle of day and night caused by Earth's rotation are known as **circadian** rhythms. The natural rhythm of human sleep is an example of a circadian rhythm. The unpleasantness of jet lag after high-speed, long-distance air travel results from the tendency of the sleep rhythm to persist for several days even after the day-night cycle has undergone a major dislocation. It takes a traveler about one day to adjust for each time zone passed through, but it is usually easier to adjust to travel from East to West than from West to East. The jet-lag example, and many others, indicates that circadian rhythms are intrinsic to the metabolism of the organism, even though the rhythms eventually undergo adjustment to changes in the day-night cycle.

Among many behaviors in *Drosophila* that exhibit a circadian rhythm is **locomotor activity,** or movement from one place to another, with the peak period of activity during the night. An apparatus for automatic monitoring of locomotor activity is illustrated in Figure 17.12A. It consists of a narrow glass tube containing a single fly. Each end of the tube is illuminated with red light of a wavelength invisible to the fly, and the intensity of the light transmitted through the tube at each end is measured by a photocell. The photocells are connected to an event recorder in such a way that a single event is recorded whenever the light intensity on the two photo-

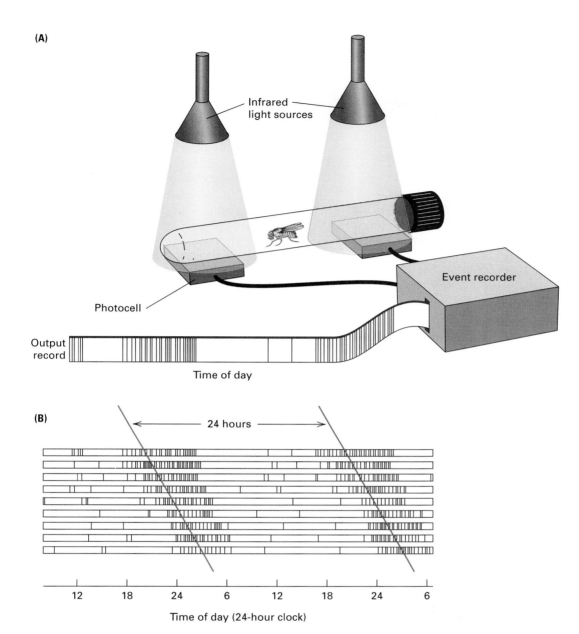

Figure 17.12

Locomotor activity in *Drosophila*. (A) The monitoring apparatus consists of a narrow tube illuminated with red light (invisible to the fly) at each end. The event recorder is adjusted to register an event whenever the light intensity on the two photocells becomes unequal. Hence, each vertical line in the output record indicates that the fly moved into one end of the tube or the other. (B) Output record of a single wildtype fly. Note that the periods of high activity (indicated by the dense vertical marks) have an approximately 24-hour periodicity. [Data from Y. Citri, H. V. Colot, A. C. Jacquier, Q. Yu, J. C. Hall, D. Baltimore, and M. Rosbash. 1987. *Nature* 346: 82.]

cells becomes unequal. Hence, in the output record, each vertical line indicates that the fly moved into one end of the tube or the other. The high density of vertical lines in the middle of the output record indicates a period of high locomotor activity. Data

from this apparatus are illustrated in Figure 17.12B, which shows the locomotor activity of a single wildtype fly over a period of 9 days. (Each horizontal line represents only 24 hours, because in order to maintain visual continuity, the activity pattern on

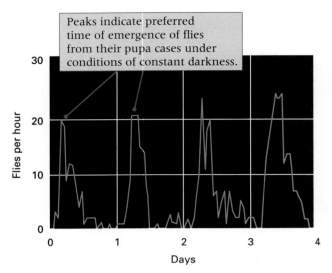

Peaks indicate preferred time of emergence of flies from their pupa cases under conditions of constant darkness.

Figure 17.13
Circadian rhythm in time of emergence of adult *Drosophila* from their pupa cases. Wildtype flies were grown as larvae under a cycle of 12 hours light : 12 hours dark and then shifted to constant darkness for formation of pupae and metamorphosis. The graph shows the number of adult flies emerging during each hour of a 4-day period. [Data from R. J. Konopka and S. Benzer. 1971. *Proc. Natl. Acad. Sci. USA* 68: 2112.]

the right-hand side of each line is duplicated on the left-hand side of the line below.) Before being monitored, the fly was maintained in a regular cycle of 12 hours light and 12 hours dark, but during the monitoring, there was no visible light. Nevertheless, a daily peak of locomotor activity is evident, and although the onset of activity shifts slightly from one day to the next, the periods of peak activity recur at regular intervals of approximately 24 hours.

Another example of circadian rhythm in *Drosophila* is the time of day at which the newly developed adults emerge from their pupa cases after undergoing metamorphosis. An example is shown in Figure 17.13, which plots the number of wildtype flies emerging each hour of a 4-day period. Periods of emergence peak at approximately 24-hour intervals, even though the pupae were in total darkness during metamorphosis. Abnormal emergence rhythms of a number of *Drosophila* mutants are shown in Figure 17.14. The mutant strains

were obtained by exposing adult flies to a chemical mutagen and mating them in such a way as to produce a series of populations, within each of which the males carried a replica of a single mutagen-treated X chromosome. Each mutant strain was maintained in total darkness during metamorphosis, like the wildtype strain in Figure 17.13, and the pattern of emergence was examined to identify abnormal circadian rhythms. The three mutants in Figure 17.14 were obtained from tests of 2000 mutagen-treated X chromosomes.

All three mutations in Figure 17.14 proved to be alleles of the same X-linked gene, called *period* (*per*). However, the mutant alleles have dramatically different circadian phenotypes, which suggests that the *per* gene is important in the timing of circadian rhythms. The allele *per^o* knocks out the circadian clock (Figure 17.14A). Males hemizygous for *per^o*, and females that are homozygous, are arrhythmic, so the adults are equally likely to emerge at any time of day. The alleles *per^S* and *per^L* result in phenotypes that still exhibit an emergence rhythm, but the period of the rhythm is altered. A short period of about 19 hours is characteristic of *per^S* (Figure 17.14B), and a long period of about 29 hours is characteristic of *per^L* (Figure 17.14C). In a manner paralleling their effects on adult emergence, the *per* alleles also affect other circadian rhythms. For example, in locomotor activity, the *per^o* mutation results in an arrhythmic phenotype, whereas *per^S* and *per^L* cause, respectively, phenotypes with short (19-hour) and long (29-hour) periods between successive peaks of locomotor activity.

Love-Song Rhythms in *Drosophila*

Fruit flies exhibit many types of behavioral rhythms that are not circadian, and it is remarkable that some of the noncircadian rhythms are also affected by mutations in *per*. An example is the *Drosophila* "love song," which consists of rhythmic patterns of wing vibrations that the male directs toward the female during courtship. The *Drosophila* courtship ritual is depicted in Figure 17.15. In part A, the male approaches the female from the front, ori-

(A) No preferred period of emergence

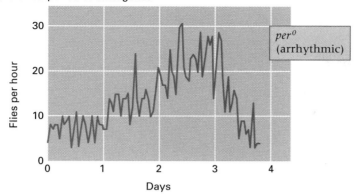

(B) Peaks of emergence spaced ~19 hours

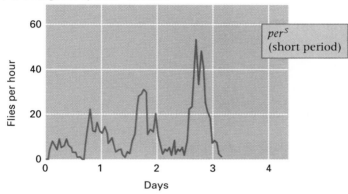

(C) Peaks of emergence spaced ~29 hours

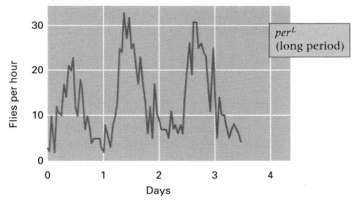

Figure 17.14

Mutant alleles of the *per* gene affecting the period of the adult emergence rhythm. (A) In *per⁰* flies, emergence is equally likely at any time of day, and because there is no circadian rhythm, the mutant is arrhythmic. (B) In *perˢ* flies, there is an emergence rhythm, but the period is shortened to about 19 hours. (C) In *perᴸ* flies, there is an emergence rhythm, but the period is lengthened to about 29 hours. [Data from R. J. Konopka and S. Benzer. 1971. *Proc. Natl. Acad. Sci. USA* 68: 2112.]

ented at an angle. In parts B and C, the male moves toward the rear, following the female if she moves, and repeatedly extends one of his wings, vibrating it rapidly up and down. In parts D and E, the male moves directly behind the female and touches her genitalia with his proboscis. In parts F and G, there is attempted copulation. More often than not, the attempt at copulation is unsuccessful, and the sequence of behaviors must be repeated several times until either copulation occurs or the encounter is broken off.

The *Drosophila* courtship song is defined by the pattern of wing vibrations in the stage of courtship depicted in Figure 17.15C. The wing vibrations are studied by the use of a small microphone connected to a device that converts the sound into electrical signals that can be displayed visually. A typical song burst is shown in Figure 17.16. The song burst is preceded by a

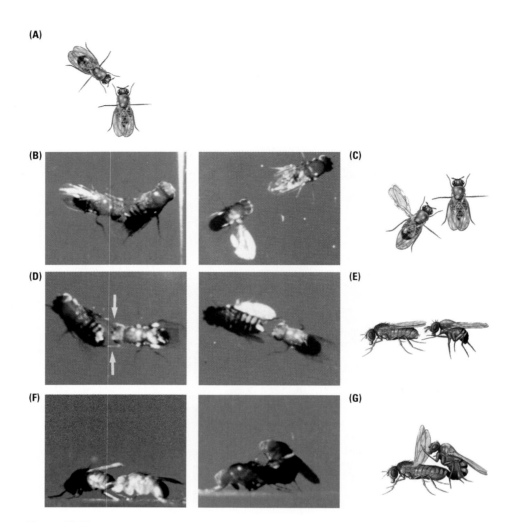

Figure 17.15

Courtship behavior in *Drosophila melanogaster.* (A) The male approaches the female and becomes oriented toward her, usually slightly to one side. (B) The male moves to a position slightly behind the female and extends the wing nearer her head (C), vibrating it rapidly up and down before returning the wing to its normal position (this process may be repeated several times). (D–F) The male moves to a position directly behind the female and touches her genitalia with his proboscis. (G) Attempted copulation. [Photographs courtesy of Jeffrey C. Hall; sketches from B. Burnet and K. Connolly, in *The Genetics of Behaviour,* edited by J. H. F. van Abeelen (Amsterdam: North-Holland, 1974), pp. 212–213.]

"hum" of extremely rapid vibrations, which serves to "warm up," or sensitize, the female. The "song" follows: a series of discrete pulses that vary in length from 15 to 100 milliseconds from one song burst to the next. A sequence of song bursts is diagrammed in Figure 17.17A. As the sequence progresses, the interpulse intervals become shorter, then longer, then shorter again. It is the rhythmic expansion and contraction of the interpulse intervals in the sequence of song bursts that defines the *Drosophila* courtship song. A plot of the interpulse intervals, as in Figure 17.17B, produces a smooth oscillating curve.

The effects of *per* alleles on the period of the courtship song are illustrated in Figure 17.18. The period observed with wildtype *D. melanogaster* is approximately 60 seconds. The *per*^*S* allele shortens the period to about 40 seconds, and the *per*^*L* allele lengthens the period to about 90 seconds.

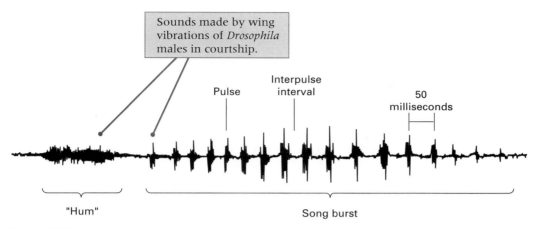

Figure 17.16

Typical song burst in *Drosophila* courtship. The burst is preceded by a "hum" of extremely rapid vibrations, followed by a series of discrete pulses. The time between each pulse is called an inter-pulse interval. [From C. P. Kyriacou, M. L. Greenacre, J. R. Thackeray, and J. C. Hall, in *Molecular Genetics of Biological Rhythms,* edited by M. W. Young (New York: Marcel Dekker, 1992), p. 172.]

Figure 17.17

Rhythm in the *Drosophila* courtship song composed of wing vibrations. (A) Acoustical recording of a single burst. (B) Pattern of pulses in a sequence of 11 consecutive bursts. Note that the interpulse intervals expand and contract during the sequence. A plot (yellow dots) of interpulse intervals yields a rhythmic pattern, which in *D. melanogaster* has a period of approximately 60 seconds. [Data from C. P. Kyriacou, M. L. Greenacre, J. R. Thackeray, and J. C. Hall, in *Molecular Genetics of Biological Rhythms,* edited by M. W. Young (New York: Marcel Dekker, 1992), p. 171.

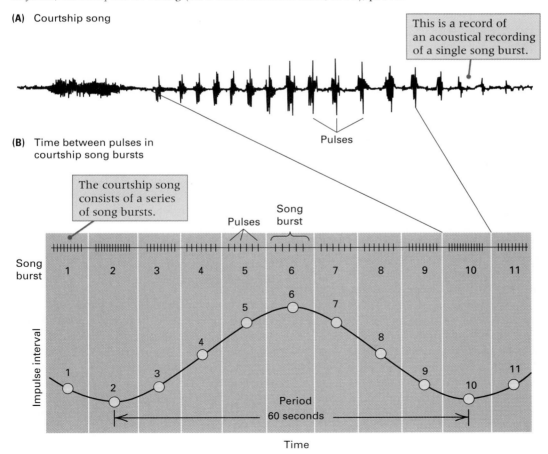

Figure 17.18

Effects of mutations in the *period* gene on the period of the *Drosophila* courtship song. (A) Wildtype flies have a courtship song with a period of approximately 60 seconds (see also Figure 17.17B). (B) Males carrying the *per^S* allele have a shorter period, approximately 40 seconds. (C) Males carrying the *per^L* allele have a longer period, approximately 90 seconds.

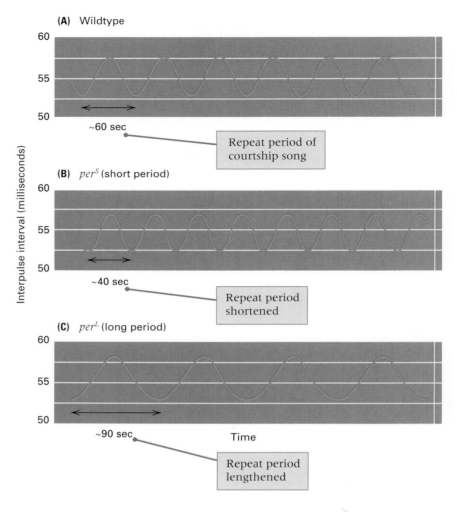

Therefore, the *per* gene affects the periods of both the courtship song and the circadian rhythm, even though the time scales are very different (60 seconds versus 24 hours). The *per* gene is also important in controlling differences in the period of the courtship song in different species. For example, in *Drosophila melanogaster* the period is about 60 seconds (Figure 17.17B), whereas in the closely related species *Drosophila simulans,* the period is about 35 seconds. The difference is apparently important in mate recognition: *D. melanogaster* females are maximally stimulated to mate when the courtship song has a 60-second period, whereas *D. simulans* females prefer the 35-second period. In any event, the role of the *per* gene in the differ-

ence between the species is indicated by the observation that transfer of the *per^+* gene from *D. simulans* to *D. melanogaster* results in *D. melanogaster* males that exhibit the rhythm of a typical *D. simulans* courtship song.

Molecular Genetics of the *Drosophila* Clock

Some years after the discovery of *per,* mutant screens in *Drosophila* turned up other mutations that were arrhythmic and still others with a shortened or lengthened period. Complementation tests showed that some of the new mutations were additional alleles of *per.* However, two of the mutations revealed a new complementation group that also controls the circadian clock. The gene defined by the new complementation group was designated *timeless* (*tim*). Unlike *per*, which is X-linked, *tim* is autosomal (chromosome 2).

Both *per* and *tim* have been cloned and sequenced, but each of the proteins, called PER and TIM, has an amino acid sequence that affords little insight into what the protein does. The cloned genes are important as probes, however, and show that the abundance of both *per* and *tim* mRNA cycle in a circadian manner and both peak after dark. The cyclic abundances of the mRNAs are illustrated by the dashed lines in Figure 17.19A. During the night, the abundance of the mRNAs plateaus and then begins to decrease until, by the next morning, both mRNAs are at low levels. The abundance of PER cycles also, with about an 8-hour time lag behind the *per* mRNA. The abundance of TIM cycles as well, but with a much shorter time lag behind its mRNA. The amount of TIM is at a low level in the daylight hours but increases rapidly to a peak after dark and diminishes thereafter.

If flies are reared in complete darkness, the TIM protein remains at a constant high level. When the lights are turned on, the TIM disappears within an hour. It has been noted that when wildtype *tim*$^+$ flies are reared in constant light, they become arrhythmic. The last observation suggests that TIM protein is destroyed by light.

In the circadian cycle, the PER protein is localized in the cytoplasm and gradually degraded as long as the level of TIM is low. After dark, when the amount of TIM begins to increase, there is a point at which PER moves into the nucleus. The movement of PER is associated with the increase in the level of TIM, and it has been suggested that TIM and PER molecules form a dimer that is transported into the nucleus. In any case, there is good evidence that both TIM and PER, once in the nucleus, exert negative feedback effects on the transcription of the genes that encode them.

Changes in the levels of *per* and *tim* mRNA and PER and TIM protein during the course of one day are shown in Figure 17.19A. At dawn, the levels of *per* and *tim* mRNA are low because transcription is repressed. When the sun comes up, the light destroys TIM, and both genes are derepressed. Transcription begins and the mRNA accumulates. Translation results in an accumulation of PER protein as well, but any TIM protein produced is immediately destroyed by the light. Not until the onset of darkness can TIM begin to accumulate, and it does so very rapidly because the amount of *tim* mRNA has already built up. Once TIM reaches an appreciable level, TIM and PER move into the nucleus and repress transcription. As the existing mRNA decays, no more is produced, and the protein levels drop as well until, in the morning, the situation is as it was 24 hours earlier.

The light sensitivity of TIM helps explain how the circadian clock is reset each day. No circadian rhythms have an exact 24-hour cycle. In the absence of light, the cycle time is either slightly shorter or slightly longer than 24 hours. To maintain synchrony with Earth's light-dark cycle, the circadian clock must be reset a small amount each day. The molecular changes that accompany the resetting are shown in Figures 17.19B and C. For clarity, both the "short-day" and "long-day" situations are exaggerated. Figure 17.19B shows what happens in a "short day," in which darkness comes on early; in this case, TIM accumulates at an earlier hour of the cycle than usual and so allows PER and TIM to enter the nucleus and repress transcription earlier than usual. The result is that the cycle is shortened (or, said another way, the clock advances). The response to a longer day is shown in Figure 17.19C. In this case, the longer day prevents TIM from accumulating at the normal time. It does not begin to accumulate until after dark, so PER and TIM enter the nucleus and repress transcription later than usual. The result is that the cycle is lengthened or, equivalently, the clock delayed.

Although the light sensitivity of TIM helps explain how the clock becomes reset each day to remain in synchrony with day length, it does not explain why, once the clock has been set, it continues to run even in complete darkness. Beyond the feedback repression of transcription of PER and TIM, the molecular mechanism by which the genes affect biological rhythms is not

Figure 17.19
Circadian cycling of *tim* and *per* mRNA and of TIM and PER protein. The TIM protein is destroyed by light and so does not accumulate until after dark. When sufficiently abundant in the cytoplasm, TIM and PER move into the nucleus and repress *tim* and *per* transcription. (A) Normal cycles. (B) A "short day" advances the clock by allowing TIM to accumulate sooner. (C) A "long day" delays the clock by delaying the accumulation of TIM.

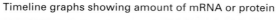

Timeline graphs showing amount of mRNA or protein

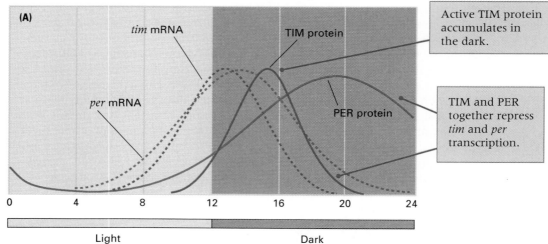

(A)

tim mRNA

per mRNA

TIM protein

PER protein

Active TIM protein accumulates in the dark.

TIM and PER together repress *tim* and *per* transcription.

0   4   8   12   16   20   24

Light   Dark

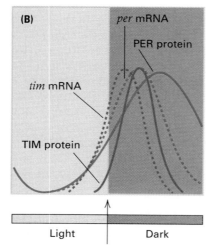

(B)

per mRNA

PER protein

tim mRNA

TIM protein

Light   Dark

Lights off early; short day advances clock

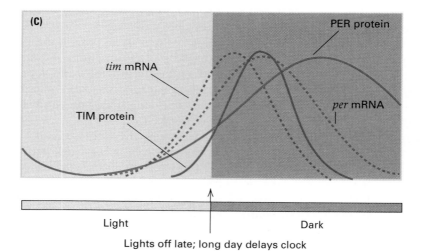

(C)

tim mRNA

TIM protein

PER protein

per mRNA

Light   Dark

Lights off late; long day delays clock

known. Much remains to be explored, including the mechanisms by which the proteins affect the physiology and behavior of the organism. For example, it is not known how PER affects the courtship song. In any case, *per* and *tim* are only part of a larger story of biological rhythms, because various rhythms can be uncoupled. For example, rearing flies in continuous bright light completely obliterates the circadian rhythm of emergence of adults illustrated in Figure 17.13, whereas the rhythm of the courtship song remains unaltered.

The Mammalian Clock, Prion Protein, and Mad Cow Disease

The TIM-PER mechanism in *Drosophila* is by no means universal in the control of circadian rhythms. Although circadian rhythms are extremely widespread, the mechanism of control differs from one species of organism to the next. For example, the fungus *Neurospora* has a circadian rhythm in growth rate. This rhythm is controlled by the gene *frq* (*frequency*) through its protein product FRQ. The cycling of FRQ is the opposite of that of TIM: The abundance of FRQ peaks in daylight and decreases at night. Although light controls the abundance of FRQ, the mechanism is through transcription of the *frq* gene, which is stimulated by light.

Mammals also have pronounced circadian rhythms that govern, among other physiological states, the sleep cycle. The phenomenon of jet lag results from the time needed to adjust the sleep cycle after a change of time zone. A key gene is called *CREM,* which encodes a transcriptional repressor protein, ICER, the level of which undergoes a circadian cycle in the pineal gland. At night, nerve input into the pineal gland from a region of the brain called the suprachiasmatic nucleus induces *CREM* transcription through a transcriptional activator protein called CREB. Induction of *CREM* causes an increase in the amount of *CREM* mRNA, and therefore of ICER protein, that is self-limiting and peaks during the night because ICER feedback represses *CREM* transcription. The regulatory system adjusts itself to changes in day length by altering the sensitivity of *CREM* to induction: In long-day situations, *CREM* is super-

sensitive to induction; in long-night situations, it is subsensitive to induction.

A number of mutations are known to affect the mammalian circadian rhythm. A mutation called *tau* in the hamster results in a short 20-hour cycle rather than the normal 23.5-hour cycle. In humans, there is a rare autosomal-dominant disease known as *familial fatal insomnia* (*FFI*); first appearing between the ages of about 35 and 60, the symptoms include progressive insomnia and degeneration of certain parts of the thalamus; the condition results in death about a year after onset. FFI is caused by a mutation in the gene for **prion protein,** located in chromosome 20pter-p12. The term *prion* stands for **protein infectious agent.** Prion proteins were originally discovered through their ability to act as infectious agents in the transmission of any of a number of serious neurological diseases. Produced initially as a precursor protein of 253 amino acids, the mature prion protein is found on the surface of neurons and glial cells in the brain. Its normal function is unknown. However, that it plays an important role in circadian regulation is supported by the observation that, in transgenic mice produced by genetic engineering (Section 9.4), a knockout mutation of the prion-protein gene results in altered circadian activity and sleep patterns resembling those in human FFI.

In human kindreds, other mutations in the prion-protein gene result in **Creutzfeldt-Jakob disease (CJD)** and **Gerstmann-Straussler disease (GSD).** Both of these conditions are marked by a mid-life age of onset and by progressive ataxia (loss of control of movement) and dementia (literally, "out of mind"), but their clinical symptoms differ somewhat. In both diseases, parts of the brain lose their normal histology and become full of holes (technically, *spongiform encephalopathy*).

Given the seemingly conventional genetics of FFI, CJD and GSD, each a manifestation of a different type of mutation in the prion-protein gene, it came as an almost unbelievable finding that prion-related diseases could also be transmitted from one organism to another by transfer of a protease-resistant form of the prion protein itself. For example, injection of a chimpanzee with prion protein from a

Table 17.2 Diseases Associated with Defective Prion Protein

	Organism	Disease
Inherited diseases	human	familial fatal insomnia (FFI)
	human	Creutzfeldt-Jakob disease (CJD)
	human	Gerstmann-Straussler disease (GSD)
"Infectious" diseases	human	kuru
	sheep and goat	scrapie
	cattle	mad cow disease

human CJD patient results in a progressive spongiform encephalopathy indistinguishable from the human condition. Other infectious diseases once attributed to transmission of a "slow virus" are now known to have prions as the active agent (Table 17.2). One example is **kuru,** discovered originally among the Fore people of Papua New Guinea who practiced ritual cannibalism as a rite of mourning in which the brain of the dead was handled and eaten, particularly by the women and children. Another example is **scrapie,** the equivalent prion-protein disease in sheep and goats, so called because the diseased animal is so intolerably itchy that it rubs and scrapes against any convenient object. A third example is "mad cow disease" (technically, *bovine spongiform encephalopathy*), whose identification as a prion-protein defect resulted in an eruption of near hysteria over the importation of British beef into Europe. Although the transmissible prion agent is thought to be a protease-resistant, insoluble form of the protein derived from the normal protease-sensitive protein, it is not clear whether the transmissible protein should be regarded as a truly infectious agent or merely as a cytotoxic metabolite.

17.3 Learning

Higher animals have sophisticated nervous systems and correspondingly sophisticated behavioral repertoires. Birds and mammals are especially adept at modifying their behavior in response to environmental cues and previous experiences. In other words, they can learn. Although learning of sorts can be demonstrated in organisms such as *Drosophila*, the ability is most highly developed in birds and mammals-particularly in human beings and higher primates. Partly because of the importance of learning in human beings, and partly because of its intrinsic interest, psychologists and geneticists have tried to assess the genetic contribution to learning ability in various organisms.

Artificial Selection for Learning Ability

Selection experiments for learning ability were some of the earliest studies in behavior genetics in animals. In these experiments, the animal is introduced into one end of a maze and must make its way to the other end to find a reward. One apparatus for testing the maze-learning ability of laboratory rats is illustrated in Figure 17.20A. For an animal that has previous experience in getting through the maze, the number of wrong turns ("errors") is a measure of the extent to which the animal has learned the layout of the maze. Artificial selection for better maze-learning ability is carried out by choosing superior learners as the parents in each generation; for poorer maze-learning ability, inferior learners are chosen as parents. The results of 21 generations of selection for better or poorer maze-learning ability are shown in Figure 17.20B. The initial population was a large, genetically heterogeneous strain of laboratory rats. In each generation, the rats were tested for

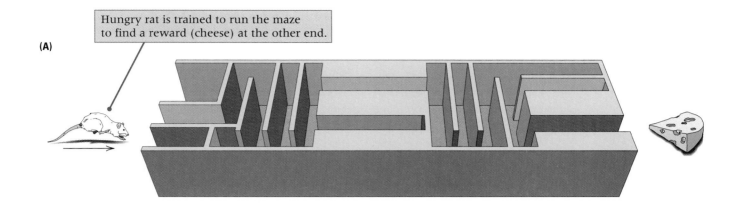

(A)

Hungry rat is trained to run the maze to find a reward (cheese) at the other end.

(B)

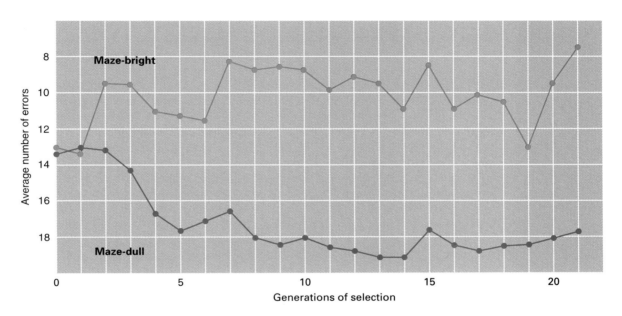

Figure 17.20
Maze-learning ability in laboratory rats. (A) A typical type of maze. Wrong turns are scored as "errors." (B) The results of 21 generations of artificial selection for maze-learning ability in laboratory rats. The "maze-bright" population was selected for better learning, and the "maze-dull" population for poorer learning. Each data point gives the average number of errors made by the animals in each population after the same number of learning trials. [Part A, after S. D. Porteus, *Porteus Maze Test: Fifty Years' Application* (Palo Alto: Pacific Books, 1965); part B, data from R. C. Tyron (1940) reported in G. E. McClearn, in *Psychology in the Making,* edited by L. Postman (New York: Knopf, 1962), p. 144.]

their ability to learn a maze. Those rats that performed best (that is, made the fewest errors after a given number of trials) were mated among themselves to form a "maze-bright" selected line, and in each generation, selection was carried out within this line for the brightest animals. Similarly, those rats that performed worst (made the most errors) were mated among them-selves to form a "maze-dull" line, and in each generation, the dullest of the dull were selected to perpetuate the line.

As Figure 17.20B shows, selection in both directions was successful. By generation 2, there was virtually no overlap in learning ability between the two lines; the dullest "brights" were about equal in maze-learning ability to the brightest "dulls."

However, beyond generation 7, almost no further progress was made in either direction. Crosses between the "bright" and "dull" lines produced an F_1 generation that was intermediate in maze-learning ability. Additional crosses revealed that maze learning is inherited as a typical multifactorial trait of the sort discussed in Chapter 16.

The experiments represented in Figure 17.20B have a number of technical deficiencies. First, there was no control experiment consisting of a randomly mated population, unselected for maze-learning ability, established from the same initial population as the selected lines and maintained with the same population size as the selected lines. An unselected control is necessary to detect possible systematic changes in the environment in the course of the selection. Second, the experiments in Figure 17.20B are unreplicated. A better set of experiments would have included at least two "bright" selected lines, at least two "dull" selected lines, and at least one unselected control line. The replication is necessary to assess the reproducibility of the result in independently selected populations. Nevertheless, in spite of the technical limitations, the experiments in Figure 17.20B are a classic in behavior genetics, because they do demonstrate hereditary variation affecting maze-learning ability in the rat population from which the selected lines were derived. In the absence of genetic variation, the divergence between the selected lines would be impossible. In quantitative terms, the response to selection in Figure 17.20B is of the magnitude that would be expected for a trait with a narrow-sense heritability of maze learning of 21 percent. (Narrow-sense heritability, discussed in Chapter 16, refers to the rate at which the mean of a trait can be changed by artificial selection. Roughly speaking, it is the proportion of the total variance in phenotype that is transmitted from parents to offspring.)

Genotype-Environment Interaction

The maze-bright and maze-dull rats in Figure 17.20B are also of interest in demonstrating the complexity of the response to selection. Many behavioral differences were observed between the selected populations. For example, the selected populations differed in their maze-learning ability depending on whether the rats were motivated by hunger or by escape from water. The maze-bright animals learned better than the maze-dull animals when motivated by hunger (the condition in which the selection was carried out), but the maze-dull rats actually learned better than the maze-bright rats when tested in a situation in which getting through the maze allowed them to escape from immersion in water. Evidently, motivation makes a big difference, even to rats. Furthermore, the maze-bright rats were more active than the maze-dull rats when in mazes, but they were less active than the maze-dull rats when in exercise wheels. The differences between the strains were varied and complex, and it is not clear which of the differences may have been related to maze learning in a causal sense and which were fortuitous differences. Indeed, it is not even clear that the maze-bright rats were in any sense "smarter" than the maze-dull rats. The maze-bright rats may simply have responded better to the specific maze learning test situation.

The populations in Figure 17.20B can also demonstrate *genotype-environment interaction*, a complication that may arise with any multifactorial trait (Chapter 16) but particularly with complex behavioral traits. Genotype-environment interaction means that the relative performance of different genotypes changes drastically according to the environment. In another set of maze-learning experiments, the bright and dull rats from the thirteenth generation of an experiment like that in Figure 17.20B were reared separately in three different types of environments and then tested for their maze-learning ability. The environments were

1. The normal environment.

2. An "enriched" environment having cage walls decorated with designs in luminous paint and containing ramps, mirrors, swings, polished balls, marbles, barriers, slides, tunnels, bells, teeter-totters, and springboards-all constructed so that they could be easily shifted to new positions.

3. A "restricted" environment having cages surrounded by drab, gray walls and containing only food and water pans.

The maze-learning ability of the rats reared in these three environments is summarized in Figure 17.21. In the normal environment (Figure 17.21A), the rats from the bright strain performed much better than those from the dull strain. This difference is expected, because the normal environment is the one in which the selection experiments were carried out. The genotype-environment interaction is observed in the enriched and the restricted environments. Although the enriched environment (Figure 17.21B) produces a dramatic improvement in the maze performance of the dull rats, it hardly affects the performance of the bright rats. Therefore, the difference in maze-learning ability is much diminished in the enriched environment. In the restricted environment (Figure 17.21C), the performance of both strains is very adversely affected, but the bright strain is affected more than the dull strain. The result is that in the restricted environment, the maze-learning ability of both strains is approximately equal. The principal conclusion from these experiments is that, though the maze-bright and maze-dull strains certainly differ in genetic factors affecting maze learning, the magnitude of the difference between the strains results as much from their rearing environment as from their difference in genotype. This example of genotype-environment interaction should be kept in mind as a cautionary tale in interpreting experiments on the genetics of human behavioral variation.

17.4
Genetic Differences
in Human Behavior

One can hardly imagine a subject more controversial than the inheritance of behavioral differences in human beings, particularly behavior that is considered socially undesirable. From the very beginning of modern genetics, commentators have been divided, some claiming the sim-

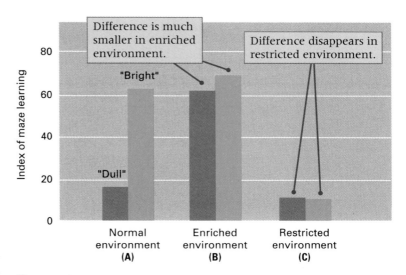

Figure 17.21
Maze-learning ability of maze-bright and maze-dull rats reared in three different environments. (A) In the normal environment, the bright rats are clearly better at learning the maze. (B) In an enriched environment, the difference is much smaller. (C) In a restricted environment, the difference disappears completely. [Data from R. M. Cooper and J. P. Zubek. 1958. *Canad. J. Psychol.* 12: 159.]

ple and direct inheritance of virtually all forms of socially unacceptable behavior-"feeble-mindedness," habitual drunkenness, criminality, prostitution, and so on-and others arguing for environmental causation of such behavior. Extreme hereditarian views are exemplified by the following statements:

It may be stated that feeble-mindedness is generally of the inherited, not the induced, type; and that the inheritance is generally recessive. (R. R. Gates, 1933)

In feeble-minded stocks, mental defect is interchangeable with imbecility, insanity, alcoholism, and a whole series of mental (and often physical) anomalies. (K. Pearson, 1931)

Such extreme hereditarian views were supported by studies of certain families with a high incidence of undesirable traits, the most famous being the Jukes family and the Kallikak family (both pseudonyms). Figure 17.22 is part of the Jukes pedigree published in 1902 and purporting to show the inheritance of "shiftlessness"

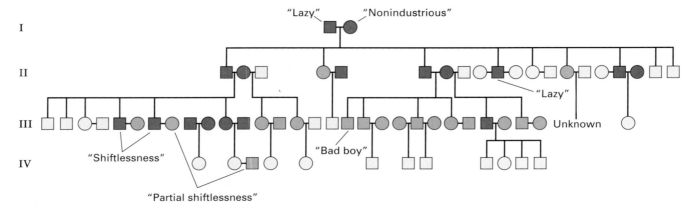

Figure 17.22

Part of the Jukes pedigree illustrating hereditarian prejudices of some early investigators. Red symbols purported to represent "shiftlessness" and pink symbols "partial shiftlessness." [After R. L. Dugdale, *The Jukes: A Study in Crime, Pauperism, Disease and Heredity* (New York: Putnam, 1902); and C. B. Davenport, *Heredity in Relation to Eugenics* (New York: Holt, 1911).]

and "partial shiftlessness." The red- and pink-shaded symbols are as in the original report, and they give the false impression of objectivity. However, the diagnosis of "shiftlessness" or "partial shiftlessness" is entirely subjective, as indicated by the descriptions of some of the individuals in the pedigree. Individual I-1 is described as "a lazy mulatto," I-2 as "a nonindustrious harlot, but temperate," II-10 as "lazy, in poorhouse," and III-18 as a "bad boy." Today this sort of analysis is rejected as subjective and prejudicial, but at the time it was taken quite seriously as an element in the "inheritance" of poverty.

Even if we accept the pedigree at face value, all that it implies is that no family exists outside of its environment and that social environments, like genes, tend to be perpetuated from generation to generation. Familial association of certain characteristics (for example, poverty) and certain types of behavior is not evidence that the traits are genetically inherited, rather than socially transmitted. A clear example is English grammar and pronunciation. Children tend to speak with the same quirks and idioms as their parents and other relatives, yet nobody would presume that English usage is genetically transmitted. Nevertheless, some early behaviorists were extreme hereditarians.

The opposite view was held by the extreme environmentalists, whose thinking is typified by the following example:

So let us hasten to admit-yes, there are heritable differences in form, in structure. . . . These differences are in the germ plasm and are handed down from parent to child. . . . But do not let these undoubted facts of inheritance lead us astray as they have some of the biologists. The mere presence of these structures tells us not one thing about function. . . . Our hereditary structure lies ready to be shaped in a thousand different ways-the same structure-depending on the way in which the child is brought up. . . . We have no real evidence of the inheritance of behavioral traits. (J. B. Watson, 1925)

Watson went on to add,

I should like to go one step further now and say, "Give me a dozen healthy infants, well formed, and my own specified world to bring them up in and I'll guarantee to take any one at random and train him to become any type of specialist I might select-doctor, lawyer, artist, merchant-chief and, yes, even beggar-man and thief, regardless of his ancestors." I am going beyond my facts and I admit it, but so have the advocates of the contrary. . . .

At the present time, environmentalist interpretations of human behavioral differences predominate. In the academic literature, you will not find any hereditarian views as extreme as those illustrated by the quotations from Gates and Pearson. However, apart from the use of the masculine pronoun, it is rather easy to find contemporary environmentalist statements that resemble the quotations from Watson. On the other hand, even though extreme hereditarian interpretations of human behavioral differences are strongly out of favor, most geneticists and psychologists are willing to concede the importance of both heredity *and* environment in human behavioral variation. The important questions involve the relative importance of "nature" (genetics) and "nurture" (environment) and how best to assess the situation experimentally.

Sensory Perceptions

If "behavior" is interpreted in its widest sense to include sensory perception and processing, then genetic variation within the normal range is easy to demonstrate. Two examples will serve to illustrate this point. About 70 percent of North American whites, and 97 percent of North American blacks, are able to taste the chemical **phenylthiocarbamide (PTC)** when it is present in filter paper. Phenotypically, these people are "tasters." The remaining 30 and 3 percent, respectively, are unable to detect PTC and therefore have the "nontaster" phenotype. PTC tasting is a

clear example of variation in sensory perception, and it is a simple Mendelian trait. The ability to taste PTC is due to a dominant allele, denoted *T*, located in chromosome 7q. The recessive allele is designated *t*. The genotypes *TT* and *Tt* are tasters of PTC, and *tt* genotypes are nontasters. Family data of the type illustrated in Table 17.3 provide an excellent fit to the simple two-allele hypothesis. However, a finer classification of people according to the *concentration* of PTC that can be tasted reveals subtler shades of tasting and nontasting ability that seem to have a multifactorial basis. This may account for the small number of unexpected taster offspring from nontaster × nontaster matings.

A second example of genetic variation in sensory perception concerns the ability to smell an odiferous substance, *methanethiol,* that some people excrete in the urine a few hours after eating asparagus. This trait is a good example of the difficulty of studying the genetics of human behavior. Simple as the trait may seem to be, there are two diametrically opposed explanations of its inheritance. One hypothesis is that the trait is a classic type of inborn error of metabolism in which about 40 percent of the population lacks an unidentified enzyme that results in the urinary excretion of methanethiol. The other hypothesis is that everyone excretes methanethiol in the urine but that the ability to smell the substance is a genetically determined hypersensitivity to the smell.

Table 17.3 Distribution of PTC taster and nontaster offspring

Mating	Number of families	Taster Observed	Taster Expected	Nontaster Observed	Nontaster Expected	Proportion of nontasters among offspring Observed	Proportion of nontasters among offspring Expected
Taster × taster	425	929	930	130	129	0.123	0.122
Taster × nontaster	289	483	495	278	266	0.366	0.349
Nontaster × nontaster	86	5	0	218	223	0.978	1.000

Note: The expected numbers are derived from the supposition of a dominant *T* and a recessive *t* allele with random mating. When the recessive has allele frequency *q*, it can be shown that the proportions expected in the right-hand column are $[q/(1 + q)]^2$, $q/(1 + q)$, and 1, respectively. In this case $q = 0.537$. Although the fit with the single-locus hypothesis is excellent, the five taster offspring in the nontaster × nontaster matings are unexplained. They could be due to misclassification of phenotype, to misassigned parentage, or perhaps to alleles at other loci that influence tasting ability. [From C. Stern, *Principles of Human Genetics,* 3d ed. (San Francisco: Freeman, 1973).]

About 10 percent of people tested are able to smell the substance at low concentration. An important observation is that those people who *can* smell it can detect its presence in everyone's urine after ingestion of asparagus. This observation supports the second hypothesis: Every person excretes the substance, but not every person is able to smell it.

Severe Mental Disorders

Examples of both genetic and environmental effects on mental function are easily enumerated. Down syndrome and the fragile-X syndrome are major genetic causes of mental retardation (Section 7.5). Another inherited form of mental retardation is **phenylketonuria,** an inborn error of metabolism caused by an autosomal recessive gene at 12q24.1 that results in a deficiency of the enzyme phenylalanine hydroxylase and an accumulation of the amino acid phenylalanine. The metabolic defect causes profound mental deficiency unless the dietary intake of phenylalanine is rigorously controlled. On the environmental side, severe mental impairment of genetically normal infants can result from prenatal infection with rubella virus (German measles), excessive exposure of the fetus to alcohol (*fetal alcohol syndrome*), insufficient oxygen or brain damage at birth, and many other factors. These examples illustrate that assignment of the primary cause of mental retardation to genetics or environment must be done on a case-by-case basis. As might be expected, in many instances, the precise cause is unassignable, so it remains unclear what overall proportion of mental retardation is genetic in origin and what proportion is environmental.

Genetic factors can also influence personality. Table 17.4 summarizes family and adoption data on the risk of schizophrenia, a classic form of "madness" characterized by delusions, auditory hallucinations, and other progressively worsening forms of dementia. In view of its severity, the trait is remarkably common, affecting about 1 person in 100. The genetic contribution to the trait is seen in the high concordance (46 percent) in identical twins, in the high risk among children with two affected parents, and in the substantial risk among adopted children when their biological mother is affected. Indeed, regarding schizophrenia, adopted children resemble their biological parents more than their adoptive parents. The genetic risk factors for schizophrenia have not been identified, but it is clear that there are multiple genetic factors and that the relative importance of the genetic factors differs from one kindred to the next. Genetic mapping studies in families with schizophrenia have implicated genes at least in chromosomes 5q, 6p, and 22q.

On the other hand, Table 17.4 also shows that environmental influences are important in schizophrenia. The environmental component is evidenced by the incomplete concordance of identical twins (when one twin is affected, there is a 46 percent chance that the other will also be affected) and by the increased risk in normal children adopted into families in which one partner later develops schizophrenia. Most psychiatric geneticists currently favor the view that genotype is an important predisposing factor toward schizophrenia but that environmental stress is an essential trigger for actual onset of the disease.

Severe mental retardation, senile dementia (such as Alzheimer disease), and personality disorders (such as schizophrenia) lie well outside the range of normal intellectual function. Hence the relative importance of genetic and environmental

Table 17.4 Risk of schizophrenia: family and adoption data

Relationship	Number of persons	Risk (%)
Familial data		
Identical twins, one affected	210	46
Same-sex fraternal twins, one affected	309	14
Siblings, one affected	9921	10
Children with one affected parent	1577	13
Children with two affected parents	134	46
Genetically unrelated controls	399	2
Adoption data		
Adopted child of affected mother	47	17
Adopted child of nonaffected mother	50	0
Normal child adopted by future schizophrenic	28	11
Biological parent of affected adoptee	66	12
Adopted parent of affected adoptee	63	2

Data are the combined results of various studies differing somewhat in diagnostic criteria. Data from I. I. Gottesman and J. Shields, *Schizophrenia: The Epigenetic Puzzle* (Cambridge, England: Cambridge University Press, 1982)

contributions to these conditions may not yield an accurate picture of behavioral differences among people. Among any random collection of people, there is enormous variation in personality, emotional traits, and intellectual abilities. How much of this variation can be attributed to genotypic differences among people and how much can be attributed to differences in their environments? This is the key issue in understanding how genes influence human behavioral traits.

Genetics and Personality

The possible role of hereditary factors in influencing human personality traits within the normal range of variation in personality is difficult to assess. These traits are affected by environmental as well as genetic factors. If the genetic factors consist of more than a handful of genes, each with a relatively small effect, then the genetic factors are very difficult to detect, even using the methods outlined in Chapter 16 for genetic mapping of quantitative-trait loci. Even though the genes cannot be identified individually, it is nevertheless possible to estimate how important genetic factors are, in the aggregate, by separating the observed variance in the trait into a genetic variance component (estimating the importance of genetic factors in causing the total variance in phenotype) and an environmental component (estimating the importance of environmental factors in causing the total variance in phenotype). The concepts of genetic and environmental variance were discussed in Chapter 16. One overall measure of the importance of genetic factors in determining the variance of a trait is the broad-sense heritability, which is the ratio obtained by dividing the variance in phenotype attributable to differences in genotype by the total variance in phenotype from all causes together. The uses and limitations of this concept are also discussed in Chapter 16.

The broad-sense heritability of personality traits has been estimated experimentally from studies of identical twins reared together and reared apart in comparison with fraternal twins and ordinary siblings reared together and reared apart. For each of five broad personality traits, the broad-sense heritability ranges from 30 percent to 50 percent (Figure 17.23). Each of these broad personality traits is expressed not in an all-or-none fashion but quantitatively. After examination by trained personnel, each person can be assigned a score on a continuous scale of measurement reflecting the degree to which the trait is expressed in the person. The traits are

- *Agreeableness: a trait associated with friendliness, likeability, and pleasantness* People who strongly express this trait are sympathetic, warm, kind, and good-natured, and they tend not to take advantage of others. The opposite end of the scale is associated with aggressiveness, pugnacity, coldness, and vindictiveness.

- *Conscientiousness: associated with conformity and dependability* People who are high in conscientiousness are well organized; they plan ahead; they are responsible and practical. People who score strongly negative for this trait are impulsive, careless, and undependable.

- *Extroversion: associated with social dominance* Extroverts are outgoing, decisive, and persuasive. At the other end of the scale are introverts, who tend to be reserved and withdrawn and to avoid attention.

- *Neuroticism: associated with anxiety and emotional instability* People high on the neuroticism scale tend to be nervous, self-conscious, tense, and prone to worry, and they try to avoid situations that they perceive as harmful. People at the opposite end of the scale are easygoing, emotionally stable, and not prone to high anxiety.

- *Openness: associated with receptiveness to novel ideas and stimuli* People high on the openness scale are curious, imaginative, insightful, and original. People at the negative end of the openness scale are described as shallow or lacking in curiosity.

A knowledge that the broad-sense heritability of each of these personality traits is from 30 percent to 50 percent tells us nothing about the number of genes that are

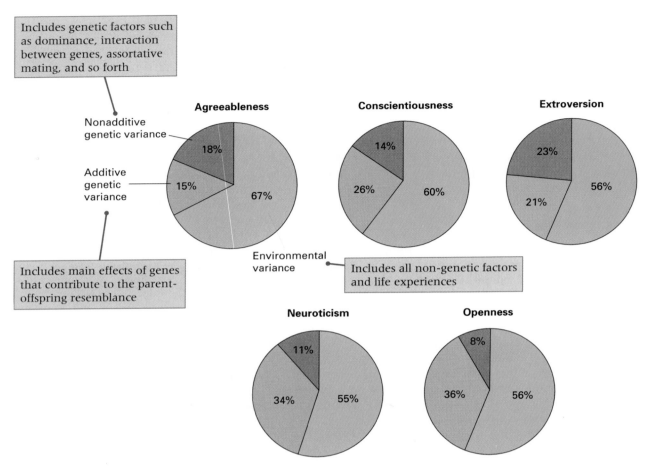

Includes genetic factors such as dominance, interaction between genes, assortative mating, and so forth

Nonadditive genetic variance

Additive genetic variance

Includes main effects of genes that contribute to the parent-offspring resemblance

Environmental variance

Includes all non-genetic factors and life experiences

Agreeableness
18%
15%
67%

Conscientiousness
14%
26%
60%

Extroversion
23%
21%
56%

Neuroticism
11%
34%
55%

Openness
8%
36%
56%

Figure 17.23
Percentages of the total phenotypic variance attributable to additive gene effects, nonadditive gene effects, and environmental effects for each of five broad categories of personality traits. The broad-sense heritability equals the sum of the additive and nonadditive percentages. [Data from T. J. Bouchard, Jr. 1994. *Science* 264: 1700.]

involved or the magnitude of the effects of the genes, let alone what the genes are at the molecular level. Because many of the genes affecting personality may have relatively small effects, they are likely to remain undetected in the usual types of pedigree studies for genetic mapping.

Another approach to the identification of genes that influence personality and behavior is through candidate genes. A **candidate gene** for a trait is a gene for which there is some *a priori* basis for suspecting that the gene may affect the trait. In human behavior genetics, for example, if a pharmacological agent is known to affect a personality trait, and the molecular target of the drug is known, then the gene that

codes for the target molecule and any gene whose product interacts with the drug or with the target molecule are all candidate genes for affecting the personality trait.

One example of the use of candidate genes in the study of human behavior genetics is in the identification of a naturally occurring genetic polymorphism that affects neuroticism and, in particular, the anxiety component of neuroticism. The neurotransmitter substance serotonin (5-hydroxytryptamine) is known to influence a variety of psychiatric conditions, such as anxiety and depression. Among the important components in serotonin action is the serotonin transporter protein. Neurons that release serotonin to stimulate other neurons also take it up again through the serotonin transporter. The uptake terminates the stimulation and also recycles the molecule for later use.

The serotonin transporter became a strong candidate gene for anxiety-related personality traits when it was discovered that the transporter is the target of a class of antidepressants known as selective serotonin reuptake inhibitors. The widely prescribed antidepressant Prozac is an example of such a drug.

Motivated by the strong suggestion that the serotonin transporter might be involved in anxiety-related traits, researchers looked for evidence of genetic polymorphisms affecting the transporter gene in human populations. Such a polymorphism was found in the promoter region of the transporter. About 1 kb upstream from the transcription start site is a series of 16 tandem repeats of a nearly identical DNA sequence of about 15 base pairs. This is the *l* (*long*) allele, which has an allele frequency of 57 percent among Caucasians. There is also an *s* (*short*) allele, in which three of the repeated sequences are not present. The *s* allele has an allele frequency of 43 percent. The genotypes *l l*, *l s*, and *s s* are found in the Hardy-Weinberg proportions of 32 percent, 49 percent, and 19 percent, respectively (Chapter 15). Further studies revealed that the polymorphism does have a physiological effect. In cells grown in culture, *l l* cells had approximately 50 percent more mRNA for the transporter than *l s* or *s s* cells, and *l l* cells had approximately 35 percent more membrane-bound transporter protein than *l s* or *s s* cells.

On the basis of these results, the researchers predicted that the serotonin transporter polymorphism would be found to be associated with anxiety-related personality traits. The predicted result was observed in a study of 505 people who were genotyped for the transporter polymorphism and whose personality traits were classified by means of a self-report questionnaire. Significant associations were found between the transporter genotype and the overall neuroticism score, and the highest correlations were with the anxiety-related traits "tension" and "harm avoidance." A comparison between the genotypes with respect to neuroticism score is shown in Figure 17.24. Because the *s* allele is dominant to *l* with respect to gene expression and personality score, the L

Figure 17.24

Distribution of neuroticism scores among different genotypes for the serotonin transporter polymorphism. The genotype L includes only *l/l*, whereas S includes *s/s* and *s/l*. The total is 163 subjects for the L genotype and 342 for the S genotypes. [From K.-P. Lesch, et al. 1996. *Science* 274: 1527.]

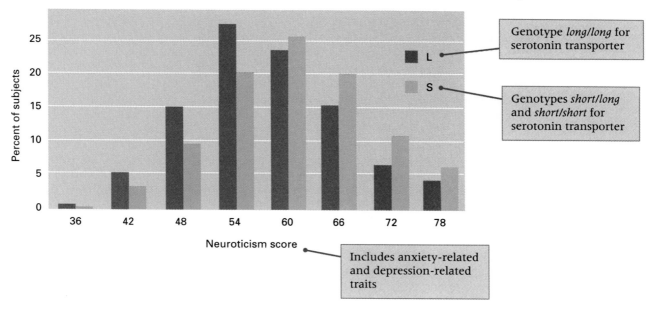

category includes only the *l l* genotype whereas the S category includes both *s s* and *l s*.

Note in Figure 17.24 that there is a great deal of overlap in the L and S distributions. The overlap means that the transporter polymorphism accounts for a relatively small proportion of the total variance in neuroticism score in the whole population. In quantitative terms, the transporter polymorphism accounts for 3 to 4 percent of the total variance in neuroticism score and 7 to 9 percent of the genotypic variance in neuroticism score. If all genes affecting neuroticism had the same magnitude of effect as the serotonin transporter polymorphism, approximately 10 to 15 genes would be implicated in the 45 percent of the variance that can be attributed to genetic factors.

Genetic and Cultural Effects in Human Behavior

Human beings exhibit many important behavioral traits for which no candidate genes have been identified and for which the effects of individual genes are probably too small to detect with genetic linkage studies. For these traits, one measure of the overall importance of genetic factors is the broad-sense heritability, which is the ratio of the genotypic variance in the trait to the total phenotypic variance in the trait.

Estimating the heritability of any human behavioral trait requires that special attention be given to cultural transmission of nongenetic factors that potentially affect the behavior. These familial factors increase the resemblance between relatives, and, unless properly taken into account, they inflate the apparent genetic heritability. Most conventional genetic studies of human behavior are biased in that they include cultural transmission in the estimate of heritability, treating the situation as though cultural differences were genetic in origin.

Another difficulty in applying the concepts of quantitative genetics to human behavior is that the concept of heritability is prone to being misinterpreted. The most common mistake is to interpret the heritability as applying to genetic differences *between* populations. For essentially the same reasons that identical twins reared in different environments may have different phenotypes, two genetically identical populations that differ in their environments may differ in average phenotypes. The error in this interpretation is that

> The heritability of a trait *within* groups tells one nothing about the genetic or environmental causes of a difference in the average of a trait *between* groups.

Both genetic and cultural factors are relevant to interpreting the variation in phenotypes within a specified population. Use of these quantities for comparison between populations is unjustified and may often lead to incorrect conclusions. Because the emotional overtones of human behavioral differences are so great, the central issue may become clearer by means of analogy. Instead of a human behavioral difference, one might just as well ask why farmer Brown's hens lay an average of 230 eggs per year but farmer Smith's average only 210. Lacking any knowledge of the genetic relationship between the flocks, we cannot specify the reason for the difference. This is true even though we might know with certainty that the genetic heritability within each flock is 30 percent, because heritability within flocks is irrelevant to the comparison. On the one hand, farmer Brown's hens could be genetically superior in egg laying. On the other hand, farmer Smith's chicken feed or husbandry could be inferior. In order to determine the reason for the difference, we would have to rear some of Brown's chickens on Smith's farm, and vice versa. If the egg laying of the transplanted hens did not change, then the difference between the two groups of hens would have been shown to be entirely genetic. If the egg laying of the transplanted hens did change, then the difference between the groups would have been shown to be environmental. Because experiments of this type cannot be carried out with humans, the issue of genetic versus environmental causes of behavioral differences among human groups is essentially unresolvable.

Chemotaxis refers to movement toward attractants and away from repellents. Cells of *E. coli* propel themselves by means of flagella, counterclockwise rotation of which forms a bundle that propels the cell forward and clockwise rotation of which causes tumbling. Chemotactic movements of *E. coli* are the result of a bias toward swimming when the cell is headed toward an attractant or away from a repellent. Most attractants for *E. coli* are nutrients, and most repellents are harmful substances.

Mutations affecting chemotaxis fall into a number of distinct categories. Mutants that are specifically nonchemotactic show no response to a particular attractant or repellent, usually because of a nonfunctional chemosensor protein responsible for binding the particular attractant or repellent. Mutants that are multiply nonchemotactic show no response to a group of chemically unrelated substances, usually because of a nonfunctional chemoreceptor that interacts with a particular subset of chemosensors. All of the chemoreceptors feed into a common biochemical pathway of chemotaxis. Mutants with defects in genes for the components of the common pathway are generally nonchemotactic and fail to respond to any attractant or repellent. Mutations that affect chemotaxis also may cause defects in the structural components of the flagella or in the motor components that control rotational switching.

The molecular basis of chemotaxis is mediated by phosphate transfer and methylation. Chemoreceptor proteins repeatedly initiate a chain of phosphorylations, resulting in clockwise flagellar rotation (tumbling). When a chemoreceptor becomes bound with a chemosensor, the phosphoryl groups are rapidly removed, restoring swimming, and the rate of phosphorylation is decreased, so the episodes of swimming get longer. There is also a feedback mechanism because phosphorylation results in removal of certain methyl groups from the chemoreceptors. Demethylation decreases the rate of phosphorylation and, hence, also promotes swimming. Methylation of the chemoreceptors is the basis of sensory adaptation, in which the cells become adjusted to a particular level of an attractant because each additional methyl group increases the level of attractant needed for excitation. Chemoreceptors are also called methyl-accepting chemotaxis proteins (MCPs) because of the role of methylation in their function.

Circadian rhythms are cycles synchronized with the 24-hour cycle of day and night. Several mutations in *Drosophila* alter the period of circadian rhythms and affect such behaviors as locomotor activity, the time of day at which the newly developed adults emerge from the pupa cases, and the courtship song, which refers to the rhythmic patterns of wing vibrations that the male directs toward the female during courtship. In *Drosophila,* circadian rhythms are controlled by two proteins, TIM and PER. TIM is the product of the gene *timeless* (*tim*), and PER is the product of the gene *period* (*per*). TIM is light-sensitive and so cannot accumulate during daylight hours. The mRNA of both genes accumulates in the daytime, as does PER protein. TIM begins to accumulate after nightfall, and when it is sufficiently abundant, enters the nucleus along with PER, where transcription of both *tim* and *per* is repressed, resulting in a progressive decrease in *tim* and *per* mRNA, as well as TIM and PER proteins, throughout the night.

Other organisms have different molecular mechanisms that control circadian rhythms. In *Neurospora,* transcription of the gene *frequency* (*frq*) is stimulated by light. In mammals, prion protein is implicated in the control of circadian rhythms. Mutations in the prion-protein gene result in inherited forms of familial fatal insomnia (FFI), Creutzfeldt-Jakob disease (CJD), and other conditions. In certain of these conditions, there is also produced a protease-resistant form of the prion protein, which can be transmitted from one organism to another as though it were an infectious agent. Transmissible prion protein has been implicated in kuru in human beings, scrapie in sheep and goats, and mad cow disease (bovine spongiform encephalopathy).

Selection experiments for maze-learning ability in laboratory rats demonstrate that a substantial amount of genetic variation affects the trait. The experiments also demonstrate the importance of genotype-environment interaction in complex behaviors. The maze-bright rats were superior to maze-dull rats in learning the maze when motivated by hunger (the condition of the selection), but the maze-dull rats were superior to the maze-bright rats in learning the maze when motivated by escape from water. Furthermore, although rats from the bright strain performed much better than those from the dull strain in a normal environment, the difference was much reduced in an enriched environment and virtually disappeared in a restricted environment.

Although a few differences in human behavior have a relatively simple genetic basis (for example, the ability to taste phenylthiocarbamide, or PTC), most human behavior is influenced by multiple genes and by environmental factors. For example, a genetic predisposition to schizophrenia is indicated by the increased risk in close relatives of affected persons, but environmental factors also are important, as shown by the increased risk in normal children adopted into families in which one partner later develops schizophrenia. With a behavioral trait that can be measured on a continuous scale, the heritability of the trait can be estimated by analysis of the correlation between relatives. Such heritability estimates often include cultural transmission in the "genetic" component of variation, so the heritability estimate is inflated. In any case, the heritability of a human behavioral trait applies only to variation in the trait within the relevant population; it provides no information about the genetic basis of possible differences in the trait among diverse racial or ethnic groups.

attractant
behavior
candidate gene
chemoreceptor
chemosensor
chemotaxis
circadian rhythm
courtship song

flagella
kuru
locomotor activity
methyl-accepting chemotaxis protein
    (MCP)
nontaster
phenylketonuria
phenylthiocarbamide (PTC)

prion protein
PTC tasting
repellent
sensory adaptation
taster

Review the Basics

- When cells of *E. coli* are placed in medium containing lactose, they begin to transcribe the lactose operon and to produce the proteins needed for lactose transport and degradation. Is this an example of behavior? Why or why not?

- How is rotation of the flagella related to chemotaxis in *E. coli?*

- What is the molecular basis of sensory adaptation in bacterial chemotaxis?

- How do the *Drosophila* mutations *period* and *timeless* demonstrate that circadian rhythms are under genetic control?

- What does it mean to say that the broad-sense heritability of a personality trait is 50 percent?

- What is a candidate gene for a trait and how are candidate genes identified?

- What is sensory adaptation? How does sensory adaptation in *E. coli* take place in the presence of a repellent?

Guide to Problem Solving

Problem 1: Three strains of *E. coli* were tested for chemotactic response to the amino acids serine and leucine. Strain 1 swims toward serine and away from leucine; strain 2 swims toward serine but shows no response to leucine; and strain 3 shows no response to either serine or leucine. Additional tests showed that each strain had a normal chemotactic response in the presence of other attractants and repellents. Which strain is wildtype, which specifically nonchemotactic, and which multiply nonchemotactic? Which genes and gene products are likely to be defective in the mutant strains?

Answer: Strain 1 is wildtype with respect to serine and leucine chemotaxis. Strain 2 is specifically nonchemotactic with respect to leucine; it is likely to have a mutation in the gene encoding the chemosensor for leucine. Strain 3 is multiply nonchemotactic. It is likely that strain 3 has a mutation in the gene for the chemoreceptor that is common to the chemotaxis pathway stimulated by both serine and leucine; alternatively, strain 3 may have two mutations-one in the serine chemosensor and the other in the leucine chemosensor.

Problem 2: What types of genetic crosses would you do, and what tests would you carry out, to determine whether per^0, per^S, and per^L are alleles of the same gene affecting locomotor activity?

Answer: A complementation test is necessary, which means that females heterozygous for pairs of the alleles need to be examined for monitoring locomotor activity. Therefore, cross per^0/per^0 females with per^S males and examine the locomotor activity of the per^0/per^S daughters. Likewise, cross per^0/per^0 females with per^L males and examine the locomotor activity of the per^0/per^L daughters. (The required female genotypes could also be produced by carrying out the reciprocal of each cross.)

The result of the cross is that per^0/per^S females have a short-period rhythm and per^0/per^L females have a long-period rhythm. Therefore, per^S is an allele of per^0, and per^L is an allele of per^0. By inference, per^S is an allele of per^L. (The direct test of the allelism between per^S and per^L by examining per^S/per^L heterozygotes is ambiguous, because per^S shortens the period and per^L lengthens it. The effects of the mutant alleles are opposite, with the result that the period of per^S/per^L heterozygotes is approximately normal.)

Problem 3: Studies of pedigrees, twins, and adoptions suggest that some cases of the psychiatric disorder called bipolar affective disorder have a genetic basis. It has been difficult to identify genetic markers associated with this disorder. For example, one group of researchers reported genetic linkage with markers on chromosome 11, only to report at a later date that additional data eliminated the statistical signifi-

Chapter 17

GeNETics on the web will introduce you to some of the most important sites for finding genetic information on the Internet. To complete the exercises below, visit the Jones and Bartlett home page at

http://www.jbpub.com/genetics

Select the link to *Genetics: Principles and Analysis* and then choose the link to *GeNETics on the web*. You will be presented with a chapter-by-chapter list of highlighted keywords.

GeNETics EXERCISES

Select the highlighted keyword in any of the exercises below, and you will be linked to a web site containing the genetic information necessary to complete the exercise. Each exercise suggests a specific, written report that makes use of the information available at the site. This report, or an alternative, may be assigned by your instructor.

1. In the control of the *Drosophila* circadian rhythm, it has been suggested that the TIM and PER proteins form a dimer that enables them to enter the nucleus together and act to repress transcription of the *tim* and *per* genes. The evidence for this model can be found by requesting a *full report* on the gene *tim* at the keyword site. If assigned to do so, write one paragraph describing the nature of the evidence.

2. You can learn more about mad cow disease and related diseases at this site. Particularly interesting is the link to *History*. If assigned to do so, write one paragraph identifying Victoria Rimmer and explaining her significance in the story.

3. A personality trait known as novelty seeking can be found by searching at this keyword site. As with anxiety-related traits, some of the genetic variation in novelty seeking has been traced to a

genetic polymorphism in a candidate gene. If assigned to do so, write a 200-word report about the candidate gene, specifying its function and why it seemed like a good candidate.

MUTABLE SITE EXERCISES

The Mutable Site Exercise changes frequently. Each new update includes a different exercise that makes use of genetics resources available on the World Wide Web. Select the Mutable Site for Chapter 17, and you will be linked to the current exercise that relates to the material presented in this chapter.

PIC SITE

The Pic Site showcases some of the most visually appealing genetics sites on the World Wide Web. To visit the showcase genetics site, select the Pic Site for Chapter 17.

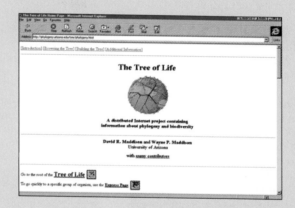

cance of their original results. Other researchers, studying different pedigrees, observed weak but suggestive linkage with markers on chromosomes other than chromosome 11. Give at least two reasons why difficulties in genetic analysis might be expected with psychiatric traits such as bipolar affective disorder.

Answer: First, psychiatric disorders are difficult phenotypes to identify and classify, and there is some subjectivity even when the examination is done by trained psychiatrists. Second, psychiatric disorders are complex, multifactorial traits that are likely to be influenced by several different

genes. Even a single pedigree may have genetic variation at more than one locus, and the occurrence and severity of the disorder may depend on the presence of particular combinations of alleles of several genes. (Distinct genes influencing the trait may also explain why different pedigrees may indicate genetic linkage to markers in different chromosomes.) Third, environmental factors, such as personal experiences, diet, and alcohol or drug abuse may influence the expression or severity of the trait. Misclassification of the phenotype, multiple genes influencing the trait, and environmental effects on expression-each of these complicates the detection of genetic linkage.

17.1 Why are strains of *E. coli* with mutations that affect chemoreceptors multiply nonchemotactic rather than specifically nonchemotactic or generally nonchemotactic?

17.2 CheW is necessary for autophosphorylation of CheA. Knowing this, what would you predict for the phenotype of a *cheW* deletion mutation?

17.3 Extreme *cheB* mutations have a phenotype of continuous tumbling but swim in response to very strong attractant stimuli. Why? Extreme *cheR* mutations have a phenotype of continuous swimming but tumble in response to very strong repellent stimuli. Why?

17.4 Why is phosphorylation of proteins important in bacterial chemotaxis? If the transfer of phosphate groups were inhibited, would the bacteria be specifically nonchemotactic, multiply nonchemotactic, or generally nonchemotactic? Explain.

17.5 Are mutations in the *per* gene of *Drosophila* more likely to affect the mating success of males or females? Explain.

17.6 How could you use an attached-X stock of *Drosophila* to produce a series of populations in each of which all of the males carry replicas of a single mutagen-treated X chromosome?

17.7 In an experiment to obtain mutants affecting courtship behavior in *D. melanogaster*, several strains with abnormal male courtship songs were recovered. Most of the strains had wing defects that were probably responsible for the song abnormalities, but one strain was morphologically normal. In males of this strain, the song starts out normally but eventually becomes louder than the normal courtship song, with pulses much longer than wildtype. The circadian rhythm of these flies is also abnormal. How would you determine whether this strain of flies has a mutation in the *per* gene? Is it possible that the mutant gene in these flies is different from *per*? Explain.

17.8 The mutation *transformer* (*tra*) in *Drosophila* affects the sexual phenotype of females (chromosomally XX), transforming the flies into phenotypic males. Although the males are sterile, their sexual behavior is that of normal males, and they readily court females and mate with them. What period of courtship song would you expect in each of the following genotypes?

(a) *per^S/per^S; tra/tra*

(b) *per^L/per^L; tra/tra*

(c) *per^S/per^L; tra/tra*

17.9 Adults of *D. melanogaster* may be attracted to light of 480 nanometers (nm), which is colored blue to blue-green, or to light of 370 nm, which is ultraviolet, depending on the intensity of the light and the physiological state of the flies. Mutations in several different genes can eliminate either the

response to 480-nm light or the response to 370-nm light. Describe these behavioral traits using categories of phototactic behavior analogous to the categories of chemotactic behavior described for *E. coli*. What types of molecules or cellular processes might be affected in the *Drosophila* mutants? (*Hint*: The perception of 370-nm light and that of 480-nm light make use of different visual pigments present in different photoreceptor cells.)

17.10 If extremely bright light of 480 nm is flashed into the eyes of wildtype adults of *D. melanogaster*, subsequent attempts to elicit physiological or behavioral responses to this wavelength of light meet with no response for several seconds after the flash. Would you regard this phenomenon as an example of sensory adaptation? Explain.

17.11 Suppose that the maze-learning experiments in rats described in the text were repeated using an inbred, genetically uniform strain of rats as the initial population. Would you expect to be able to select for maze-bright and maze-dull lines? Explain.

17.12 The detection of odors by mammals is mediated by odorant-binding proteins, membrane-bound olfactory receptors, and a host of cellular proteins that work together to stimulate olfactory neurons to respond to specific chemicals in the environment. Mutations in genes for the odorant-binding proteins or the olfactory-receptor proteins might affect the sense of smell and may affect an animal's ability to detect food, potential mates, dangerous predators, and so forth. How might the olfactory system affect the performance of rats in a maze-learning experiment where the goal is to find food at the end of the maze? If the initial population were known to have heterogeneity in olfactory receptors, would maze learning be an effective measure of the rat's "intelligence"? How might you control for this type of variability in a maze-learning experiment?

17.13 After several generations of selection for maze-learning ability in rats, the maze-bright rats were superior to maze-dull rats in learning the maze when motivated by hunger (the condition of the selection). However, the maze-dull rats were superior to the maze-bright rats in learning the maze when motivated by escape from water. Is this an example of genotype-environment interaction? Explain.

17.14 Although most behaviors are complex traits influenced by several genes, some behavioral differences among individuals can be ascribed to alleles of a single gene (for example, the *period* gene in *Drosophila*). Mice that are mutant for the *waltzer* gene run around in circles, shake their heads both horizontally and vertically, and are very irritable. Four matings among animals isolated from a small population containing *waltzer* yielded the following litters:

wildtype × waltzer	3 wildtype, 2 waltzer
wildtype × wildtype	7 wildtype
wildtype × wildtype	6 wildtype, 2 waltzer
waltzer × waltzer	5 waltzer

Judged on the basis of these data, do you think it is likely that the *waltzer* allele is dominant or recessive to the wildtype allele?

17.15 The mammalian visual system is very complex, but the ability to detect colors is under relatively simple genetic control. The genes that are defective in color blindness code for a family of photoreceptor proteins known as opsins. Normal people have four opsin genes that code for four different photoreceptor proteins: an opsin (called rhodopsin) constituting the major visual pigment, an opsin that is sensitive to blue light, an opsin that is sensitive to green light, and an opsin that is sensitive to red light. Under what conditions might a simple genetic trait, such as color blindness, affect a person's behavior?

17.16 Huntington disease is an autosomal-dominant neurodegenerative disease marked by motor disturbance, cognitive loss, and psychiatric disorders; it has a typical age of onset of between 30 and 40 years. The function of the wildtype gene product is not known. However, the gene shares an interesting feature with a number of other disease genes in human beings: It contains a region with several repeats of the trinucleotide CAG. Wildtype alleles of the gene typically have 11 to 34 CAG repeats; among mutant alleles, the number of CAG repeats ranges from 40 to 86. What molecular test could be used to detect the mutant Huntington disease allele in the child of a person who develops the disease at the age of 39? If the affected person's spouse is homozygous normal for the gene, what proportion of their children would be expected to be affected?

Challenge Problem

17.17 Recurring behavioral disorders were observed in some male members of a large pedigree extending over several generations. The males were mildly mentally retarded and, especially when under stress, were prone to repeated acts of aggression-including sex offenses, attempted murder, and arson. An X-linked gene, *MAO*, coding for the enzyme monoamine oxidase, which participates in the breakdown of neurotransmitters, was found to be defective in the affected men in this pedigree. Other researchers found abnormal levels of monoamine oxidase in some unrelated men with similar disorders, even though the *MAO* gene was not defective in these cases. Does this evidence support the hypothesis that defective monoamine oxidase is responsible for the behavioral disorder? Explain.

Further Reading

Lesch, K.-P., D. Bengel, A. Heils, S. Z. Sabol, B. D. Greenberg, S. Petri, J. Benjamin, C. R. Müller, D. H. Hamer, and D. L. Murphy. 1996. Association of anxiety-related traits with a polymorphism in the serotonin transporter gene regulatory region. *Science* 274: 1527.

Comprehensive Human Genetic Linkage Center: J. C. Murray, K. H. Buetow, J. L. Weber, S. Ludwigsen, T. Scherpbier-Heddema, F. Manion, J. Quillen, V. C. Schffield, S. Sunden, G. M. Duyk; Généthon: J. Weissenbach, G. Gyapay, C. Dib, J. Morrissette, G. M. Lathrop, A. Vignal; University of Utah: R. White, N. Matsunami, S. Gerken, R. Melis, H. Albertson, R. Plaetke, S. Odelberg; Yale University: D. Ward; Centre d'Etude du Polymorphisme Humain (CEPH): J. Dausset, D. Cohen, H. Cann. 1994. A comprehensive human linkage map with centimorgan density. *Science* 265: 2049.

Bodmer, W. F., and L. L. Cavalli-Sforza. 1976. *Genetics, Evolution, and Man.* San Francisco, CA: Freeman.

Bouchard, T. J., Jr., D. T. Lykken, M. McGue, N. L. Segal, and A. Tellegen. 1990. Sources of human psychological differences: The Minnesota study of twins reared apart. *Science* 250: 223.

Devor, E. J., and C. R. Cloninger. 1989. The genetics of alcoholism. *Annual Review of Genetics* 23: 19.

Dunlap, J. C. 1996. Genetic and molecular analysis of circadian rhythms. *Annual Review of Genetics* 30: 579.

Gottesman, I. 1997. Twins: En route to QTLs for cognition. *Science* 276: 1522.

Gottesman, I. I., and J. Shields. 1982. *Schizophrenia: The Epigenetic Puzzle.* Cambridge, England: Cambridge University Press.

Hall, J. C. 1995. Tripping along the trail to the molecular mechanisms of biological clocks. *Trends in Neurosciences* 18: 230.

Ho, D. Y., and R. M. Sapolsky. 1997. Gene therapy for the nervous system. *Scientific American,* June.

Iwasaki, K., and J. H. Thomas. 1997. Genetics in rhythm. *Trends in Genetics* 13: 111.

Kyriacou, C. P., M. L. Greenacre, J. R. Thackeray, and J. C. Hall. 1992. Genetic and molecular analysis of song rhythms in *Drosophila.* In *Molecular Genetics of Biological Rhythms,* ed. M. W. Young. New York, NY: Marcel Dekker.

Maddock, J. R., and L. Shapiro. 1993. Polar location of the chemoreceptor complex in the *Escherichia coli* cell. *Science* 259: 1717.

Mind and brain. 1992. *Scientific American,* September.

Parkinson, J. S., and D. F. Blair. 1993. Does *E. coli* have a nose? *Science* 259: 1701.

Pericakvance, M. A., and J. L. Haines. 1995. Genetic susceptibility to Alzheimer disease. *Trends in Genetics* 11: 504.

Smith, J. M. 1978. The evolution of behavior. *Scientific American,* September.

Tuite, M. F., and S. L. Lindquist. 1996. Maintenance and inheritance of yeast prions. *Trends in Genetics* 12: 467.

Tuomanen, E. 1993. Breaking the blood-brain barrier. *Scientific American,* February.

Youdim, M. B. H., and P. Riederer. 1997. Understanding Parkinson's disease. *Scientific American,* January.

Chapter 1

1.1   It could be considered either way. On the one hand, the deficiency of the enzyme G6PD is hereditary; on the other hand, the disease itself has an environmental trigger (eating broad beans). Both answers are too simple, because the disease actually results from an interaction of a particular genetic constitution with a particular factor in the environment.

1.2   Replication results in two daughter DNA duplexes, each identical in base sequence to the parental molecule (except for possible mutations in the sequence). Each of the daughter molecules contains one of the original intact parental strands.

1.3   The messenger RNA carries the genetic information in DNA (in the form of a specific base sequence) to the ribosome. The ribosome is the physical structure in the cell on which a polypeptide chain is formed. The transfer RNA molecules are adapters that enable each codon in the messenger RNA to specify the presence of a particular amino acid at the corresponding position in the polypeptide chain. There is only one type of ribosome, but there are many different types of transfer RNA (typically, more than one for each amino acid).

1.4   A mixture of heat-killed S cells and living R cells causes pneumonia in mice, but neither heat-killed S cells nor living R cells alone do so.

1.5   The substance was destroyed by DNA-degrading enzymes but not by protein-degrading enzymes.

1.6   The inside of the head is mainly DNA, and the rest of the phage is mainly protein. When a bacterial cell is infected, the head contents are transferred into the cell, but the outer "ghost" of the phage remains attached to the external surface of the cell.

1.7   A small amount of protein was also transferred into the bacterial cell during infection, and some of this was even recovered in the progeny phage. A die-hard advocate of protein as the genetic material could claim that the transmitted protein, though small in amount, was critical in the hereditary process.

1.8   Because the amount of A equals that of T, and the amount of G equals that of C, it can be inferred that the DNA in this bacteriophage is double-stranded; 56 percent of the base pairs are A—T pairs, and 44 percent are G—C pairs.

1.9   Because, in this case, the amount of A does not equal that of T, and the amount of G does not equal that of C, it can

be concluded that the DNA molecule is single-stranded, not double-stranded.

1.10   Because A pairs with T, and G pairs with C, the base composition of the other strand must be 24 percent T, 28 percent A, 22 percent C, and 26 percent G.

1.11   Yes, it is possible, because in some viruses the genetic material is RNA.

1.12   Because of A—T and G—C base pairing, the complementary strand has the sequence

3'-TAGCATACGTGAAATGGGCC-5'

Note that the paired strands have opposite 5'-to-3' polarity.

1.13   In this region the complementary strand has the sequence

3'-AGCAGCAGCAGCAGC-5'

1.14   The probability that four particular bases have the sequence 5'-GGCC-3' is $(1/4)^4 = 1/256$, so 256 base pairs is the average spacing between consecutive occurrences; similarly, the average spacing for the sequence 5'-GAATTC-3' is 4096 base pairs.

1.15   The corresponding region of RNA will contain no U (uracil), because U in RNA pairs with A in DNA.

1.16   The top strand is the RNA because it contains U instead of T; the mismatched base pair is the sixth from the right.

1.17   The initial part of the RNA transcript matches the DNA template starting at the tenth nucleotide. Hence the completed transcript has the sequence

5'-UAGCUACGAUCGCGUUGGA-3'

1.18   The complementary sequence is 3'-UAUGCUAU-5'.

1.19   The codon for phenylalanine must be 5'-UUU-3'.

1.20   The codon for leucine must be 5'-UUA-3'.

1.21   Three with *in vitro* translation:

5'-CGC/UUA/CCA/CAU/GUC/GCG/AAC/UCG-3'
5'-C/GCU/UAC/CAC/AUG/UCG/CGA/ACU/CG-3'
5'-CG/CUU/ACC/ACA/UGU/CGC/GAA/CUC/G-3'

With *in vivo* translation, there is only one

5'-CGCUUACCAC > AUG/UCG/CGA/ACU/CG-3'

where the symbol > means "start translation with next codon."

1.22 Six. Either DNA strand could be transcribed, and each transcript could be translated in any one of three reading frames.

1.23 The amino acids alternate, because with a triplet code, the codons alternate: 5'-UCU/CUC/UCU/CUC/UCU/CUC-3'. From this result we can conclude that Ser and Leu are encoded by 5'-UCU-3' and 5'-CUC-3'; we cannot specify which codon corresponds to which amino acid, because *in vitro* translation begins with either 5'-UCU-3' or 5'-CUC-3'.

1.24 The result means that an mRNA is translated in nonoverlapping groups of three nucleotides: the genetic code is a triplet code.

1.25 There are 64 possible codons and only 20 amino acids; hence some amino acids (most, in fact) are specified by two or more codons. A mutation that changes a codon for a particular amino acid into a synonymous codon for the same amino acid does not change the amino acid sequence.

Chapter 2

2.1 The strain or variety must be homozygous for all genes that affect the trait.

2.2 *Aa* yields *A* and *a* gametes, *Bb* yields *B* and *b* gametes, and *Aa Bb* yields *A B, A b, a B,* and *a b* gametes.

2.3 Multiply the number of possible gametes formed for each gene (one for each homozygous gene and two for each heterozygous gene). In the case of *AA Bb Cc Dd Ee*, there are a total of $1 \times 2 \times 2 \times 2 \times 2 = 16$ possible gametes. With *n* homozygous genes and *m* heterozygous genes, the number of possible gametes is $1^n \times 2^m$, which equals 2^m.

2.4 What Mendel apparently means is that the gametes consist of equal numbers of all possible combinations of the alleles present in the true-breeding parents; in other words, the cross *AA BB × aa bb* produces F_1 progeny of genotype *Aa Bb*, which yields the gametes *A B, A b, a B,* and *a b* in equal numbers. Segregation is illustrated by the 1 : 1 ratio of *A : a* and *B : b* gametes, and independent assortment is illustrated by the equal numbers of *A B, A b, a B,* and *a b* gametes.

2.5 The round seeds in the F_2 generation consist of 1/3 *AA* and 2/3 *Aa*. Only the latter produce *a*-bearing pollen, so the fraction of *aa* seeds expected is $2/3 \times 1/2 = 1/3$.

2.6 With dominance, there are two phenotypes, corresponding to the genotypes *RR* or *Rr* (dominant) and *rr* (recessive), in the ratio 3 : 1. With no dominance, there are three phenotypes, corresponding to the genotypes *RR, Rr,* and *rr,* in the ratio 1 : 2 : 1.

2.7 The production of a homozygous genotype would require that the pollen and egg contain identical self-sterility alleles, but this is impossible because pollen cannot function on plants that contain the same self-sterility allele.

2.8 1/2; 1/2. The probability argument is that each birth is independent of all the previous ones, so the number of girls and boys already born has no influence on the sexes of future children.

2.9 Because the probability of each child's being a girl is 1/2, and the two children are independent, the probability of two girls in a row is $(1/2) \times (1/2) = 1/4$. The probability of a girl and a boy (not necessarily in that order) equals $2 \times (1/4) \times (1/4) = 1/2$. The reason for the factor of 2 in this case is that the sexes may occur in either of two possible orders: girl-boy or boy-girl. Each of these has a probability of 1/4, so the total is $1/4 + 1/4 = 1/2$.

2.10 (a) The parent with the dominant phenotype must carry one copy of the recessive allele and hence must be heterozygous. **(b)** Because some of the progeny are homozygous recessive, both parents must be heterozygous. **(c)** The parent with the dominant phenotype could be either homozygous dominant or heterozygous; the occurrence of no homozygous recessive offspring encourages the suspicion that the parent may be homozygous dominant, but because there are only two offspring, heterozygosity cannot be ruled out.

2.11 An *A b* gamete can be formed only if the parent is *AA Bb*, which has probability 1/2, and in this case, 1/2 of the gametes are *A b*; overall, the probability of an *A b* gamete is $(1/2) \times (1/2) = 1/4$. *A B* gametes derive from *AA BB* parents with probability 1 and from *AA Bb* parents with probability 1/2; overall, the probability of an *A B* gamete is $(1/2) \times 1 + (1/2) \times (1/2) = 3/4$. (Alternatively, the probability of an *A B* gamete may be calculated as $1 - 1/4 = 3/4$, because *A b* and *A B* are the only possibilities.)

2.12 (a) Two phenotypic classes are expected for the *A, a* pair of alleles (*A—* and *aa*), two for the *B, b* pair (*B—* and *bb*), and three for the *R, r* pair (*RR, Rr,* and *rr*), yielding a total number of phenotypic classes of $2 \times 2 \times 3 = 12$. **(b)** The probability of an *aa bb* RR offspring is $1/4$ *(aa)* $\times 1/4$ *(bb)* $\times 1/4$ *(RR)* $= 1/64$. **(c)** Homozygosity may occur for either allele of each of the three genes, yielding such combinations as *AA BB rr, aa bb RR, AA bb rr,* and so forth. Because the probability of homozygosity for either allele is 1/2 for each gene, the proportion expected to be homozygous for all three genes is $(1/2) \times (1/2) \times (1/2) = 1/8$.

2.13 The probability of a heterozygous genotype for any one of the genes is 1/2, and so for all four together, the probability is $(1/2)^4 = 1/16$.

2.14 Because both parents have solid coats but produce some spotted offspring, they must be *Ss*. With respect to the *A, a* pair of alleles, the female parent (tan) is *aa*, and because there are some tan offspring, the genotype of the black male parent must be *Aa*. Thus the parental genotypes are *Ss aa* (solid tan female) and *Ss Aa* (solid black male).

2.15 Because one of the children is deaf (genotype *dd*, in which *d* is the recessive allele), both parents must be heterozygous *Dd*. The progeny genotypes expected from the mating *Dd* × *Dd* are 1/4 *DD*, 2/4 *Dd*, and 1/4 *dd*. The son is not deaf and hence cannot be *dd*. Among the nondeaf offspring, the genotypes *DD* and *Dd* are in the proportions 1 : 2, so their relative probabilities are 1/3 *DD* and 2/3 *Dd*. Therefore, the probability that the normal son is heterozygous is 2/3.

2.16 **(a)** Because the trait is rare, it is reasonable to assume that the affected father is heterozygous *HD/hd*, where *hd* represents the normal allele. Half of his gametes contain the *HD* allele, so the probability is 1/2 that the son received the allele and will later develop the disorder. **(b)** We do not know whether the son is heterozygous *HD/hd*, but the probability is 1/2 that he is; if he is heterozygous, half of his gametes will contain the *HD* allele. Therefore, the overall probability that his child has the *HD* allele is (1/2) × (1/2) = 1/4.

2.17 Both parents must be heterozygous (*Aa*) because each had an albino (*aa*) parent. Therefore, the probability of an albino child is 1/4, and the probability of two homozygous recessive children is (1/4) × (1/4) = 1/16. The probability that at least one child is an albino is the probability that exactly one is albino plus the probability that both are albinos. The probability that exactly one is an albino is 2 × (3/4) × (1/4) = 6/16, in which the factor of 2 comes from the fact that there may be two birth orders: normal-albino or albino-normal. Therefore, the overall probability of at least one albino child equals 1/16 + 6/16 = 7/16. Alternatively, the probability of at least one albino child may be calculated as 1 minus the probability that both are nonalbino, or $1 - (3/4)^2 = 7/16$.

2.18 Compatible transfusions are A donor with A or AB recipient, B donor with B or AB recipient, AB donor with AB recipient, and O donor with any recipient.

2.19 **(a)** The cross *RR BB* × *rr bb* yields F₁ progeny of genotype *Rr Bb*, which have red kernels. **(b)** Because there is independent segregation (independent assortment), the F₂ genotypes are *R− B−*, *R− bb*, *rr B−*, and *rr bb*. These genotypes are in the proportions 9 : 3 : 3 : 1, and they produce kernels that are red, brown, brown, and white, respectively. Therefore, the F₂ plants have red, brown, or white seeds in the proportions 9/16 : 6/16 : 1/16.

2.20 Because the genes segregate independently, they can be considered separately. For the *Cr, cr* pair of alleles, the genotypes of the zygotes are *Cp Cp*, *Cp cp*, and *cp cp* in the proportions 1/4, 1/2, and 1/4, respectively. However, the *Cp Cp* zygotes do not survive, so among the survivors the genotypes are *Cp cp* (creeper) and *cp cp* (noncreeper), in the proportions 2/3 and 1/3, respectively. For the *W, w* pair of alleles, the progeny genotypes are W− (white) and *ww* (yellow), in the proportions 3/4 and 1/4, respectively. Altogether, the phenotypes and proportions of the surviving offspring can be obtained by multiplying (2/3 creeper + 1/3 noncreeper) × (3/4 white + 1/4 yellow), which yields 6/12 creeper, white + 2/12 creeper, yellow + 3/12 noncreeper, white + 1/12 noncreeper, yellow. Note that the proportions sum to 12/12 = 1, which serves as a check.

2.21 Colored offspring must have the genotype *C− ii*; otherwise, color could not be produced (as in *cc*) or would be suppressed (as in *I−*). Among the F₂ progeny, 3/4 of the offspring have the *C−* genotype, and 1/4 of the offspring have the *ii* genotype. Because the genes undergo independent assortment, the overall expected proportion of colored offspring is (3/4) × (1/4) = 3/16.

2.22 The 9 : 3 : 3 : 1 ratio is that of the genotypes *A− B−*, *A− bb*, *aa B−*, and *aa bb*. The modified 9 : 7 ratio implies that all of the last three genotypes have the same phenotype. The F₁ testcross is between the genotypes *Aa Bb* × *aa bb*, and the progeny are expected in the proportions 1/4 *Aa Bb*, 1/4 *Aa bb*, 1/4 *aa Bb*, and 1/4 *aa bb*. Again, the last three genotypes have the same phenotype, so the ratio of phenotypes among progeny of the testcross is 1 : 3.

2.23 For the child to be affected, both III-1 and III-2 must be heterozygous *Aa*. For III-1 to be *Aa*, individual II-2 must be *Aa* and must transmit the recessive allele to III-1. The probability that II-2 is *Aa* equals 2/3, and then the probability of transmitting the *a* allele to III-1 is 1/2. Altogether, the probability that III-1 is a carrier equals (2/3) × (1/2) = 1/3. This is also the probability that III-2 is a carrier. Given that both III-1 and III-2 have genotype *Aa*, the probability of an *aa* offspring is 1/4. Overall, the probability that both III-1 and III-2 are carriers and that they have an *aa* child is (1/3) × (1/3) × (1/4) = 1/36. The numbers are multiplied because each of the events is independent. (The reason for the 2/3 is as follows: Because II-3 is affected, the genotypes of I-1 and I-2 must be *Aa*. Among nonaffected individuals from this mating (individuals II-2 and II-4), the ratios of *AA* and *Aa* are 1/4 : 2/4, so the probability of *Aa* is 2/3.)

2.24 For any individual offspring, the probabilities of black *B−* and white *bb* are 3/4 and 1/4, respectively. **(a)** The order white-black-white occurs with probability (1/4) × (3/4) × (1/4) = 3/64. The order black-white-black occurs with probability (3/4) × (1/4) × (3/4) = 9/64. Therefore, the probability of either the order white-black-white or the order black-white-black equals 3/64 + 9/64 = 3/16. The probabilities are summed because the events are mutually exclusive. **(b)** The probability of exactly two white and one black equals $3 × (1/4)^2 × (3/4)^1 = 9/64$. The factor of 3 is the number of birth orders of two white and one black (there are three, because the black offspring must be the first, second, or third), and the rest of the expression is the probability that any one of the birth orders occurs (in this case, 3/64).

2.25 The probability of all boys is $(1/2)^4 = 1/16$. The probability of all girls is also 1/16, so the probability of either all boys or all girls equals 1/16 + 1/16 = 1/8. The probability of equal numbers of the two sexes is $6 × (1/2)^4 = 3/8$. The factor of 6 is the number of possible birth orders of two boys and two girls (BBGG, BGBG, BGGB, GGBB, GBGB, and GBBG).

2.26 Black and splashed white are the homozygous genotypes and slate blue the heterozygote. The probabilities of each of these phenotypes from the mating between two heterozygotes are 1/4 black, 1/2 slate blue, and 1/4 splashed white. The probability of occurrence of the particular birth order black, slate blue, and splashed white is $(1/4) \times (1/2) \times (1/4) = 1/32$, but altogether there are six different orders in which exactly one of each type could be produced. (The black offspring could be first, second, or third, and in each case, the remaining phenotypes could occur in either of two orders.) Consequently, the overall probability of one of each phenotype, in any order, is $6 \times (1/32) = 3/16$.

2.27 The probability that one or more offspring are aa equals 1 minus the probability that all are $A-$. In the mating $Aa \times Aa$, the probability that all of n offspring will be Aa equals $(3/4)^n$, and the question asks for the value of n such that $1 - (3/4)^n \geq 0.95$. Solving this inequality yields $n \geq \log(1 - 0.95)/\log(3/4) = 10.4$. Therefore, 11 is the smallest number of offspring for which there is a greater than 95 percent chance that at least one will be aa. As a check, note that $(3/4)^{10} = 0.056$ and $(3/4)^{11} = 0.042$, so with 10 offspring, the chance of one or more aa equals 0.944, and with 11 offspring, the chance of one or more aa equals 0.958.

2.28 The ratio of probabilities that the sire is $AA : Aa$ is $1/3 : 2/3$. The ratio of the probabilities of producing n pups, all $A-$, for $AA : Aa$ sires is $1 : (1/2)^n$. Hence, for n $A-$ pups, the ratio of probabilities of $AA : Aa$ sires is $1/3 : (2/3)(1/2)^n$. Therefore, the probability that the sire is AA, given that he had a litter of n $A-$ pups, equals $(1/3)/[(1/3) + (2/3)(1/2)^n]$. This is the degree of confidence you can have that the sire is AA. For $n = 1$ to 15, the answer is shown in the accompanying table. It is interesting that 6 progeny are required for 95 percent confidence and 8 for 99 percent confidence.

1	0.5000	6	0.9697	11	0.9990
2	0.6667	7	0.9846	12	0.9995
3	0.8000	8	0.9922	13	0.9998
4	0.8889	9	0.9961	14	0.9999
5	0.9412	10	0.9981	15	0.9999

2.29 Make a Punnett square with the ratio of $D : d$ along each margin as $3/4 : 1/4$. **(a)** The progeny genotypes DD, Dd, and dd therefore occur in the proportions $9/16 : 6/16 : 1/16$. **(b)** If D is dominant, the ratio of dominant : recessive phenotypes is $15 : 1$. **(c)** Among the $D-$ genotypes, the ratio $DD : Dd$ is $9 : 6$. **(d)** If meiotic drive happens in only one sex, then the Punnett square has $3/4 : 1/4$ along one margin and $1/2 : 1/2$ along the other. The progeny genotypes DD, Dd, and dd are in the ratio $3/8 : 1/2 : 1/8$. The ratio of dominant : recessive phenotypes is $7 : 1$. Among $D-$ genotypes, the ratio $DD : Dd$ is $3 : 4$.

Chapter 3

3.1 Chromosome replication takes place in the S period, which is in the interphase stage of mitosis and the interphase

I stage of meiosis. (Note that chromosome replication does *not* take place in interphase II of meiosis.) Chromosomes condense and first become visible in the light microscope in prophase of mitosis and prophase I of meiosis.

3.2 Each chromosome present in telophase of mitosis is identical to one of a pair of chromatids present in the preceding metaphase that were held together at the centromere. In this example, there are 23 pairs of chromosomes present after telophase, which implies $23 \times 2 = 46$ chromosomes altogether. Therefore, at the preceding metaphase, there were $46 \times 2 = 92$ chromatids.

3.3 In leptotene, the chromosomes first become visible as *thin threads*. In zygotene, they become *paired threads* because of synapsis. The chromosomes continue to condense to become *thick threads* in pachytene. In diplotene, the sister chromatids become evident, and each chromosome is seen to be a *paired thread* consisting of the sister chromatids. (Early signs of the bipartite nature of each chromosome are evident in pachytene.) In diakinesis, the homologous chromosomes repulse each other and *move apart;* they would fall apart entirely were they not held together by the chiasmata.

3.4 The number of chromosomes. Each centromere defines a single chromosome; two chromatids are considered part of a single chromosome as long as they share a single centromere. As soon as the centromere splits in anaphase, each chromatid is considered a chromosome in its own right. Meiosis I reduces the chromosome number from diploid to haploid because the daughter nuclei contain the haploid number of chromosomes (each chromosome consisting of a single centromere connecting two chromatids.) In meiosis II, the centromeres split, so the numbers of chromosomes are kept equal.

3.5 **(a)** 20 chromosomes and 40 chromatids; **(b)** 20 chromosomes and 40 chromatids; **(c)** 10 chromosomes and 20 chromatids.

3.6 **(a)** For each centromere, there are two possibilities (for example, A or a); so altogether there are $2^7 = 128$ possible gametes. **(b)** The probability of a gamete's receiving a particular centromere designated by a capital letter is 1/2 for each centromere, and each centromere segregates independently of the others; hence the probability for all seven simultaneously is $(1/2)^7 = 1/128$.

3.7 The wheat parent produces gametes with 14 chromosomes and the rye parent gametes with 7, so the hybrid plants have 21 chromosomes.

3.8 The crisscross occurs because the X chromosome in a male is transmitted only to his daughters. The expression is misleading because any X chromosome in a female can be transmitted to a son or a daughter.

3.9 Because there is an affected son, the phenotypically normal mother must be heterozygous. This implies that the daughter has a probability of 1/2 of being heterozygous.

3.10 The *Bar* mutation is an X-linked dominant. Because the sexes are affected unequally, some association with the sex chromosomes is suggested. Mating (a) provides an important clue. Because a male receives his X chromosome from his mother, the wildtype phenotype of the sons suggests that the *Bar* gene is on the X chromosome. The fact that all daughters are affected is also consistent with X-linkage, provided that the *Bar* mutation is dominant. Mating (b) confirms the hypothesis, because the females from mating (a) would have the genotype *Bar/+* and so would produce the indicated progeny.

3.11 (a) Cross is $v/Y \times +/+$, in which Y represents the Y chromosome; progeny are $1/2 +/Y$ males (wildtype) and $1/2$ $v/+$ females (phenotypically wildtype, but heterozygous. **(b)** Mating is $v/v \times +/Y$; progeny are $1/2 v/Y$ males (vermilion) and $1/2 v/+$ females (phenotypically wildtype, but heterozygous). **(c)** Mating is $v/+ \times +/Y$; progeny are $1/4$ $v/+$ females (phenotypically wildtype), $1/4 ++/+$ females (wildtype), $1/4 v/Y$ males (vermilion), and $1/4 +/Y$ males (wildtype). **(d)** Mating is $v/+ \times v/$ Y; progeny are $1/4 v/v$ females (vermilion), $1/4 v/+$ females (wildtype), $1/4 v/Y$ males (vermilion), and $1/4 +/Y$ males (wildtype).

3.12 All the females will be v/v^+; *bw/bw* and will have brown eyes; all the males will be v/Y; *bw/bw* and will have white eyes.

3.13 Because the sex-chromosome situation is the reverse of that in mammals, female chickens are ZW and males ZZ, and females receive their Z chromosome from their father. Therefore, the answer is to mate a gold male (*ss*) with a silver female (*S*). The male progeny will all be *Ss* (silver plumage), the female progeny will be *s* (gold plumage), and these are easily distinguished.

3.14 (a) The mother is a heterozygous carrier of the mutation, so the probability that a daughter is a carrier is $1/2$. The probability that both daughters are carriers is $(1/2)^2 = 1/4$. **(b)** If the daughter is not heterozygous, the probability of an affected son is 0, and if the daughter is heterozygous, the probability of an affected son is $1/2$; the overall probability is therefore $1/2$ (that is, the chance that the daughter is a carrier) $\times$ $1/2$ (that is, the probability of an affected son if the daughter is a carrier) $= 1/4$.

3.15 Let *A* and *a* represent the normal and mutant X-linked alleles. Y represents the Y chromosome. The genotypes are as follows: I-1 is *Aa* (because there is an affected son), I-2 is *A*Y, II-1 is *A*Y, II-2 is *a*Y, II-3 is *Aa* (because there is an affected son), II-4 is *A*Y, and III-1 is *a*Y.

3.16 20, because in females there are 5 homozygous and 10 heterozygous genotypes, and in males there are 5 hemizygous genotypes.

3.17 The accompanying Punnett square shows the outcome of the cross. The X and Y chromosomes from the male are denoted in red, the attached-X chromosomes (yoked Xs) and Y from the female in black. The red X chromosome also con-

tains the *w* mutation. The zygotes in the upper left and lower right corners (shaded) fail to survive because they have three X chromosomes or none. The surviving progeny consist of white-eyed males (XY) and red-eyed females (XXY) in equal proportions. Attached-X inheritance differs from the typical situation in that the sons, rather than the daughters, receive their fathers' X chromosome.

3.18 The genotype of the female is *Cc*, where *c* denotes the allele for color blindness. In meiosis, the *c*-bearing chromatids undergo nondisjunction and produce an XX-bearing egg of genotype *cc*. Fertilization by a normal Y-bearing sperm results in an XXY zygote with genotype *cc*, which results in color blindness.

3.19 The female produces $1/2$ X-bearing eggs and $1/2$ O-bearing eggs ("O" means no X chromosome), and the male produces $1/2$ X-bearing sperm and $1/2$ Y-bearing sperm. Random combinations result in $1/4$ XX, $1/4$ XO, $1/4$ XY, and $1/4$ YO, and the last class dies because no X chromosome is present. Among the survivors, the expected progeny are $1/3$ XX females, $1/3$ XO females, and $1/3$ XY males.

3.20 (a) II-2 must be heterozygous, because she has an affected son, which means that half of her sons will be affected. The probability of a nonaffected child is therefore $3/4$, and the probability of two nonaffected children is $(3/4)^2 = 9/16$. **(b)** Because the mother of II-4 is heterozygous, the probability that II-4 is heterozygous is $1/2$. If she is heterozygous, half of her sons will be affected. Therefore, the overall probability of an affected child is $1/2 \times 1/4 = 1/8$.

3.21 (a) This question is like asking for the probability of the sex distribution GGGBBB (in that order), which equals $(1/2)^6 = 1/64$. **(b)** This question is like asking for the probability of either sex distribution GGGBBB or BBBGGG, which equals $2(1/2)^6 = 1/32$.

3.22 With 1 : 1 segregation, the expected numbers are 125 in each class, so the χ^2 value equals $(140 - 125)^2/125 + (110 - 125)^2/125 = 3.6$, and there is 1 degree of freedom because there are two classes of data. The associated *P* value is approximately 6 percent, which means that a fit at least as bad would be expected 6 percent of the time even with Mendelian segregation. Therefore, on the basis of these data, there is no justification for rejecting the hypothesis of 1 : 1 segregation. (On the other hand, the observed *P* value is close to 5 percent, which is the conventional level for rejecting the hypothesis, so the result should not inspire great confidence.)

3.23 The total number of progeny equals 800, and on the assumption of a 9 : 3 : 3 : 1 ratio, the expected numbers are 450, 150, 150, and 50, respectively.

The χ^2 equals $(462 - 450)^2/450 + (167 - 150)^2/150 + (127 - 150)^2/150 + (44 - 50)^2/50 = 6.49$ and there are 3 degrees of freedom (because there are four classes of data). The P value is approximately 0.09, which is well above the conventional rejection level of 0.05. Consequently, the observed numbers provide no basis for rejecting the hypothesis of a 9 : 3 : 3 : 1 ratio.

Chapter 4

4.1 Because the progeny are wildtype, the mutations complement each other, and both m_1 and m_2 must be heterozygous. This indicates that the genotype of the progeny is $m_1 +/+ m_2$, which implies that m_1 and m_2 are mutations in different genes.

4.2 (a) $A\,b$, $a\,B$, $A\,B$, and $a\,b$. (b) $A\,b$ and $a\,B$.

4.3 The nonrecombinant gametes are $A\,B$ and $a\,b$, and the recombinant gametes are $A\,b$ and $a\,B$. The nonrecombinant gametes are always more frequent than the recombinant gametes.

4.4 The gametes from the $A\,B/a\,b$ parent are $A\,B$ and $a\,b$; those from the other parent are $A\,b$ and $a\,B$. The offspring genotypes are expected to be $A\,B/A\,b$, $A\,B/a\,B$, $a\,b/A\,b$, and $a\,b/a\,B$, in equal numbers. The phenotypes are $A-\,B-$, $A-\,bb$, and $aa\,B-$, which are expected in the ratio 2 : 1 : 1.

4.5 The *cis* configuration is $a\,b/A\,B$.

4.6 There is no crossing-over in male *Drosophila*, so the gametes will be $A\,B$ and $a\,b$ in equal proportions. In the female, each of the nonrecombinant gametes $A\,B$ and $a\,b$ is expected with a frequency of $(1 - 0.05)/2 = 0.475$, and each of the recombinant gametes $A\,b$ and $a\,B$ is expected with a frequency of $0.05/2 = 0.025$.

4.7 (a) 17; (b) 1; (c) in diploids there is one linkage group per pair of homologous chromosomes, so the number of linkage groups is $42/2 = 21$.

4.8 Recombination at a rate of 6.2 percent implies that double crossovers (and other multiple crossovers) occur at a negligible frequency; in such a case, the distance in map units equals the frequency of recombination, so the map distance between the genes is 6.2 map units.

4.9 Each meiotic cell produces four products: namely, $A\,B$ and $a\,b$, which contain the chromatids that do not participate in the crossover, and $A\,b$ and $a\,B$, which contain the chromatids that do take part in the crossover. Therefore, the frequency of recombinant gametes is 50 percent. Because one crossover occurs in the region between the genes in every cell, the occurrence of multiple crossing-over does not affect the recombination frequency as long as the chromatids participating in each crossover are chosen at random.

4.10 The $aa\,bb$ genotype requires an $a\,b$ gamete from each parent. From the $A\,b/a\,B$ parent, the probability of an $a\,b$ gamete is $0.16 / 2 = 0.08$, and from the $A\,B/a\,b$ parent, the probability of an $a\,b$ gamete is $(1 - 0.16)/2 = 0.42$; therefore, the probability of an $aa\,bb$ progeny is $0.08 \times 0.42 = 0.034$, or 3.4 percent.

4.11 All the female gametes are $bz\,m$. The male gametes are bz^+M with frequency $(1 - 0.06)/2 = 47$ percent, $bz\,m$ with frequency 47 percent, bz^+m with frequency $0.06/2 = 3$ percent, and $bz\,M$ with frequency 3 percent. The progeny are therefore black males ($bz^+\,M/bz\,m$) with frequency 47 percent, bronze females ($bz\,m/bz\,m$) with frequency 47 percent, black females ($bz^+\,m/bz\,m$) with frequency 3 percent, and bronze males ($bz\,M/bz\,m$) with frequency 3 percent).

4.12 The most frequent gametes from the double heterozygous parent are the nonrecombinant gametes, in this case $Gl\,ra$ and $gl\,Ra$. The genotype of the parent was therefore $Gl\,ra/gl\,Ra$. The recombinant gametes are $Gl\,Ra$ and $gl\,ra$, and the frequency of recombination is calculated as the ratio $(6 + 3)/(88 + 103 + 6 + 3) = 4.5$ percent.

4.13 There is no recombination in the male, so the male gametes are $++$ and $b\,cn$, each with a probability of 0.5. A map distance of 8 units means that the recombination frequency between the genes is 0.08. The female gametes are $++$, $b\,cn$ [the nonrecombinants, each of which occurs at a frequency of $(1 - 0.08)/2 = 0.46$] and $+\,cn$ and $b\,+$ [the recombinants, each of which occurs at a frequency of $0.08/2 = 0.04$]. When the male and female gametes are combined at random, their frequencies are multiplied, and the progeny and their frequencies are as follows:

$++/++$	0.23	$++/b\,cn$	0.23
$b\,cn/++$	0.23	$b\,cn/b\,cn$	0.23
$b\,+/++$	0.02	$b\,+/b\,cn$	0.02
$+\,cn/++$	0.02	$+\,cn/b\,cn$	0.02

Note that $b\,cn/++$ is the same as $++/b\,cn$, so the overall frequency of this genotype is 0.46. With regard to phenotypes, the progeny are wildtype (73 percent), black cinnabar (23 percent), black (2 percent), and cinnabar (2 percent).

4.14 The most frequent classes of gametes are the nonrecombinants, and so the parental genotype was $A\,B\,c/a\,b\,C$. The least frequent gametes are the double recombinants, and the allele that distinguishes them from the nonrecombinants (in this case, C and c) is in the middle. Therefore, the gene order is $A-C-B$.

4.15 Comparison of the first three numbers implies that r lies between c and p, and the last two numbers imply that s is closer to r than to c. The genetic map is therefore $c-10-r-3-p-5-s$. From this map, you would expect the recombination frequency between c and p to be 13 percent and that between c and s to be 18 percent. The observed values are a little smaller because of double crossovers.

4.16 (a) The nonrecombinant gametes (the most frequent classes) are $y\ v^+\ sn$ and $y^+\ v\ sn^+$, and the double recombinants (the least-frequent classes) indicate that sn is in the middle. Therefore, the correct order of the genes is $y−sn−v$, and the parental genotype was $y\ sn\ v^+/y^+\ sn^+\ v$. The total progeny is 1000. The recombination frequency in the $y−sn$ interval is $(108 + 5 + 95 + 3)/1000 = 21.1$ percent, and the recombination frequency in the $sn−v$ interval is $(53 + 5 + 63 + 3)/1000 = 12.4$ percent. The genetic map is $y−21.1−sn−12.4−v$. **(b)** On the assumption of independence, the expected frequency of double crossovers is $0.211 \times 0.124 = 0.0262$, whereas the observed frequency is $(5 + 3)/1000 = 0.008$. The coincidence is the ratio $0.008/0.0262 = 0.31$, so the interference is $1 − 0.31 = 0.69$, or 69 percent.

4.17 The nonrecombinant gametes are $c\ wx\ Sh$ and $C\ Wx\ sh$, and the double recombinants indicate that sh is in the middle. The parental genotype was therefore $c\ Sh\ wx/C\ sh\ Wx$. The frequency of recombination in the $c−sh$ region is $(84 + 20 + 99 + 15)/6708 = 3.25$ percent, and that in the $sh−wx$ region is $(974 + 20 + 951 + 15)/6708 = 29.2$ percent. The genetic map is therefore $c−3.25−sh−29.2−wx$.

4.18 Let x be the observed frequency of double crossovers. Then single crossovers in the $a−b$ interval occur with frequency $0.15 − x$, and single crossovers in the $b−c$ interval occur with frequency $0.20 − x$. The observed recombination frequency between a and c equals the sum of the single-crossover frequencies (because double crossovers are undetected), which implies that $0.15 − x + 0.20 − x = 0.31$, or $x = 0.02$. The expected frequency of double crossovers equals $0.15 \times 0.20 = 0.03$, and the coincidence is therefore $0.02/0.03 = 0.67$. The interference equals $1 − 0.67 = 33$ percent.

4.19 b and st are unlinked, as are hk and st. The evidence is the $1 : 1 : 1 : 1$ segregation observed. For b and st, the comparisons are $243 + 10$ (black, scarlet) versus $241 + 15$ (black, nonscarlet) versus $226 + 12$ (nonblack, scarlet) versus $235 + 18$ (nonblack, nonscarlet). For hk and st, the comparisons are $226 + 10$ (hook, scarlet) versus $235 + 15$ (hook, nonscarlet) versus $243 + 12$ (nonhook, scarlet) versus $241 + 18$ (nonhook, nonscarlet). On the other hand, b and hk are linked, and the frequency of recombination between them is $(15 + 10 + 12 + 18)/1000 = 5.5$ percent.

4.20 The frequency of dpy-21 unc-34 recombinants is expected to be $0.24/2 = 0.12$ among both eggs and sperm, so the expected frequency of dpy-21 unc-34 / dpy-21 unc-34 zygotes is $0.12^2 = 1.44$ percent.

4.21 The genes must be linked, because with independent assortment, the frequency of dark eyes ($R−\ P−$) would be 9/16 in both experiments. The first experiment is a mating of $R\ P/r\ p \times r\ p/r\ p$. The dark-eyed progeny represent half the nonrecombinants, so the recombination frequency r can be estimated as $(1 − r)/2 = 628/(628 + 889)$, or $r = 0.172$. The second experiment is a mating of $R\ p/r\ P \times r\ p/r\ p$. In this case, the dark-eyed progeny represent half the recombinants, so r can be estimated as $r/2 = 86/(86 + 771)$, or $r = 0.201$. It is not uncommon for repeated experiments or crosses carried out in different ways to yield somewhat different estimates of the recombination frequency—in this problem, 0.17 and 0.20. The most reliable value is obtained by averaging the experimental values—in this case the average is $(0.171 + 0.201)/2 = 0.186$.

4.22 Consider row 1: there are $−$ entries for 3 and 7, and so 1, 3, and 7 form one complementation group. Row 2: all $+$, and so 2 is the only representative of a second complementation group. Row 3: mutation already classified. Row 4: there is a $−$ for 6 only and no other $−$ in column 4, and so 4 and 6 make up a third complementation group. Row 5: $−$ only for 9 and no other $−$ in column 5, and so 5 and 9 form a fourth complementation group. Rows 6 and 7: mutations already classified. Row 8: all $+$, and so 8 is the only member of a fifth complementation group. In summary, there are five complementation groups as follows: group 1 (mutations 1, 3, and 7); group 2 (mutation 2); group 3 (mutations 4 and 6); group 4 (mutations 5 and 9); group 5 (mutation 8).

4.23 Consider each gene in relation to first- and second-division segregation. Gene a gives 1766 asci with first-division segregation and 234 asci with second-division segregation; the frequency of second-division segregation is $234/(1766 + 234) = 0.117$, which implies that the distance between a and the centromere is $0.117/2 = 5.85$ percent. Gene b gives 1780 first-division and 220 second-division segregations, for a frequency of second-division segregation of $220/(1780 + 220) = 0.110$. The distance between b and the centromere is therefore $0.110/2 = 5.50$ map units. If we consider a and b together, there are 1986 PD asci, 14 TT asci, and no NPD. Because NPD $\ll$ PD, genes a and b are linked. The frequency of recombination between a and b is $[(1/2) \times 14]/2000 = 0.35$ percent. Comparing this distance with the gene-centromere distances calculated earlier results in the map $a−0.35−b−5.50$—centromere.

4.24 (a) On the assumption of independence, $(0.30 \times 0.10)/2 = 0.015$. (Division by 2 is necessary because $sh\ wx\ gl$ gametes represent only one of the two classes of double recombinants.) **(b)** If the interference is 60 percent, then the coincidence equals $1 − 0.60 = 0.40$. The observed frequency of doubles is therefore 40 percent as great as with independence, so the expected frequency of $sh\ wx\ gl$ gametes is $0.015 \times 0.40 = 0.006$.

4.25 Considered individually, gene c yields 90 first-division segregation asci and 10 $(1 + 9)$ second-division segregation asci, so the frequency of recombination between c and the centromere is $(10/100) \times 1/2 = 5$ percent. Gene v yields 79 first-division and 21 $(20 + 1)$ second-division asci, so the frequency of recombination between v and the centromere is $(21/100) \times 1/2 = 10.5$ percent. If the genes are taken together, the asci include 34 PD, 36 NPD, and 30 $(20 + 1 + 9)$ TT. Because NPD = PD, the genes are unlinked.

In summary, *c* and *v* are in different chromosomes, *c* being 5 map units from the centromere and *v* being 10.5 map units from the centromere.

4.26 Analysis of cross 1. Ascospore arrangements 1, 2, 4, and 5 result when there is a crossover between the gene and the centromere. Types 3 and 6 result when there is no crossing-over between the gene and the centromere. The map distance between the gene and its centromere is $(1/2)$ (Frequency of second-division segregation) $\times 100$ = $(1/2)[(6 + 6 + 6 + 6)/120] \times 100 = 10$ centimorgans. Analysis of cross 2. There is no crossing-over between the gene and its centromere. Analysis of cross 3. The proportion of second-division segregation (asci types 1, 2, 4, and 5) equals $80/120 = 67$ percent. This distribution of asci results when crossing-over is frequent enough to randomize the position of alleles with respect to their centromeres. The gene is said to be "unlinked" with its centromere.

4.27 For each pair of genes, classify the tetrads as PD, NPD, or TT, and tabulate the results as follows:

leu2-trp1	PD 230 + 235	= 465
	NPD 215 + 220	= 435
	TT 54 + 46	= 100
leu2-met14	PD 235 + 220	= 455
	NPD 230 + 215	= 445
	TT 54 + 46	= 100
trp1-met14	PD 235 + 215 + 46	= 496
	NPD 230 + 220 + 54	= 504
	TT	0

For *leu2-trp1*, PD = NPD; therefore, *leu2* and *trp1* are unlinked. Similarly, for *leu2-met14*, PD = NPD; these genes are also unlinked. Likewise, *trp1* and *met14* are unlinked because PD = NPD. However, for the *trp1-met14* gene pair, there are no tetratypes, which means that each gene is closely linked to its own centromere. Because they do not recombine with their centromeres, the segregation of *trp1* and *met14* can be used as genetic markers of centromere segregation; their presence marks the sister spores created by segregation in the first meiotic division. Using the *trp1* and *met14* markers in this fashion allows the remaining *leu2* gene to be mapped with respect to its centromere; all tetratype tetrads will result from crossing-over between *leu2* and its centromere because neither of the other two genes recombines with its centromere. The map distance between *leu2* and its centromere is therefore $(1/2)(100/1000) \times 100 = 5$ centimorgans.

4.28 For the *rad6-trp5* gene pair, PD = 188 + 206 = 140 + 154 + 154 + 109 = 797, NPD = 0, and TT = 105 + 92 + 3 + 2 + 1 = 203. Because PD >> NPD, the genes are linked. In this case, TT > 0 and NPD = 0, so the map distance is calculated as $(1/2)(203/1000) \times 100 = 10.15$ centimorgans. For the *trp5-leu1* gene pair, PD = 188 + 206 + 105 + 92 + 109 = 700, NPD = 0, and TT = 140 + 154 + 3 + 2 + 1 = 300; here again, PD >> NPD, TT > 0, and NPD = 0, so the map distance is calculated as $(1/2)(300/1000) \times 100 = 15$ centimorgans. For the *rad6-leu1* gene pair, PD = 188 + 206 + 109 = 503, NPD = 1, and TT = 105 + 92 + 140 + 154 + 3 + 2 = 496. In this case, PD >> NPD, but TT > 0 and NPD > 0. The map distance is calculated as $(1/2)[(496 + 6 \times 1)/1000] \times 100 = 25.1$ centimorgans. For the *leu1 − met14* gene pair, PD = 188 + 105 + 140 + 2 = 435, NPD = 206 + 92 + 154 + 3 + 1 = 456, and TT = 109. In this case PD = NPD, so *leu1* is not linked to *met14*. However, we know from the previous problem that *met14* is tightly linked with its centromere, and this information can be used to calculate the map distance between *leu1* and its centromere. Because TT = 109 and also TT < 2/3, the map distance between *leu1* and its centromere is $(1/2)(109/1000) \times 100 = 5.45$ centimorgans. For the *rad6-met14* gene pair, PD = 188 + 1 = 189, NPD = 206, and TT = 105 + 92 + 140 + 154 + 109 + 3 + 2 = 605. Here again, PD = NPD, which indicates that there is no linkage between the genes. Because *met14* is tightly linked to its centromere, and TT = 605 but TT < 2/3, the map distance between *rad6* and its centromere is $(1/2)(605/1000) \times 100 = 30.25$ centimorgans. For the *trp5 - met14* gene pair, PD = 188 + 105 = 293, NPD = 206 + 92 = 298, and TT = 140 + 154 + 109 + 3 + 2 + 1 = 409. Once again, PD = NPD, indicating no linkage between the genes. As before, we use *met14* as a centromere marker to calculate the distance between *trp5* and its centromere as $(1/2)(409/1000) \times 100 = 20.5$ centimorgans. The overall conclusion is that *rad6*, *trp5*, and *leu1* are all linked to each other; *met14* is tightly linked to its centromere but not to the other three genes. The genetic map based on these data is shown below:

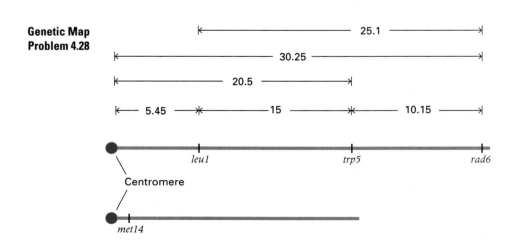

Genetic Map Problem 4.28

Chapter 5

5.1 The complementary strand is 5'-T-C-C-G-A-G-3'.

5.2 A nuclease is an enzyme capable of breaking a phosphodiester bond. An endonuclease can break any phosphodiester bond, whereas an exonuclease can only remove a terminal nucleotide.

5.3 Movement along the template strand is from the 3' end to the 5' end, because consecutive nucleotides are added to the 3' end of the growing chain. In double-stranded DNA, one strand is replicated continuously, and the other strand is replicated in short segments that are later joined.

5.4 DNA polymerase joins a 5'-triphosphate with a 3'-OH group, and the outermost two phosphates are released; the resulting phosphodiester bond contains one phosphate. DNA ligase joins a single 5'-P group with a 3'-OH group.

5.5 The enzyme has polymerizing activity, it is a 3' → 5' exonuclease (the proofreading function), and it is a 5' → 3' exonuclease.

5.6 Neither RNA polymerases nor primases require a primer to get synthesis started.

5.7 Smaller molecules move faster because they can penetrate the pores of the gel more easily.

5.8 **(a)** In rolling-circle replication, one parental strand remains in the circular part and the other is at the terminus of the branch. Therefore, only half of the parental radioactivity will appear in progeny. **(b)** If we are assuming no recombination between progeny DNA molecules, one progeny phage will be radioactive. If a single recombination occurs, two particles will be radioactive. If recombination is frequent and occurs at random, the radioactivity will be distributed among virtually all the progeny.

5.9 Rolling-circle replication must be initiated by a single-stranded break; θ replication does not need such a break.

5.10 **(a)** Because 18 percent is adenine, 18 percent is also thymine; so [T] = 18 percent. **(b)** Because [A] + [T] = 36 percent, [G] + [C] equals 64 percent, or [G] = [C] = 32 percent each. Overall, the base composition is [A] = 18 percent, [T] = 18 percent, [G] = 32 percent, and [C] = 32 percent; the [G] + [C] content is 64 percent.

5.11 **(a)** The distance between nucleotide pairs is 3.4 Å, or 0.34 nm. The length of the molecule is 34 μm, or 34 × 10^3 nm. Therefore, the number of nucleotide pairs equals $(34 \times 10^3)/0.34 = 10^5$. **(b)** There are ten nucleotide pairs per turn of the helix, so the total number of turns equals $10^5/10 = 10^4$.

5.12 **(a)** First, note that the convention for writing polynucleotide sequences is to put the 5' end at the left. Thus the sequence complementary to AGTC (which is 5'-AGTC-3') is 3'-TCAG-5', written in conventional format as GACT (that is, 5'-GACT-3'). Hence the frequency of CT is the same as that of

its complement AG, which is given as 0.15. Similarly, AC = 0.03, TC = 0.08, and AA = 0.10. **(b)** If DNA had a parallel structure, then the 5' ends of the strands would be together, as would the 3' ends, so the sequence complementary to 5'-AGTC-3' would be 5'-TCAG-3'. Thus AG = 0.15 implies that TC = 0.15, and similarly, the other observed nearest neighbors would imply that CA = 0.03, CT = 0.08, and AA = 0.10.

5.13 Remember that the chemical group at the growing end of the leading strand is always a 3'-OH group, because DNA polymerases can add nucleotides only to such a group. Because DNA is antiparallel, the opposite end of the leading strand is a 5'-P group. Therefore, **(a)** 3'-OH; **(b)** 3'-OH; **(c)** 5'-P.

5.14 Rolling-circle replication begins with a single-strand break that produces a 3'-OH group and a 5'-P group. The polymerase extends the 3'-OH terminus and displaces the 5'-P end. Because the DNA strands are antiparallel, the strand complementary to the displaced strand must be terminated by a 3'-OH group.

5.15 **(a)** The first base in the DNA that is copied is a T, so the first base in the RNA is an A. RNA grows by addition to the 3'-OH group, so the 5'-P end remains free. Therefore, the sequence of the eight-nucleotide primer is

5'-AGUCAUGC-3'

(b) The C at the 3' end has a free -OH; the A at the 5' end has a free *tri*phosphate. **(c)** Right to left, because this would require discontinuous replication of the strand shown.

5.16 **(a)** The cloned *Sal*I fragment is 20 kb in length, and there are no *Sal*I sites within it. **(b)** There is a single *Eco*RI site 7 kb from one end. **(c)** There are two *Hin*dIII sites within the fragment. **(d)** The *Hin*dIII sites must flank the *Eco*RI site, dividing the 7-kb *Eco*RI fragment into 3-kb and 4-kb subfragments and the 13-kb *Eco*RI fragment into 5-kb and 8-kb subfragments. The 3-kb and 8-kb fragments must be adjacent, because digestion with *Sal*I and *Hin*dIII alone produces an 11-kb fragment. Except for the left-to-right orientation, the inferred restriction map must be as follows:

5.17 The bands in the gel result from incomplete chains whose synthesis was terminated by incorporation of the dideoxynucleotide indicated at the top. The smaller fragments are at the bottom of the gel, the larger ones at the top. Because DNA strands elongate by addition to their 3' ends, the strand synthesized in the sequencing reactions has the sequence, from bottom to top, of

5'-ACTAGAGACCATGATCCTGTGATGAATAGC-3'

The template strand is complementary in sequence, and antiparallel, so its sequence is

3'-TGATCTCTGGTACTAGGACACTACTTATCG-5'

5.18 For each site, the probability equals the product of the probability of each of the nucleotide pairs in turn, 1/4 for any of the specified pairs and 1/2 for R-Y. The average spacing is the reciprocal of the probability. Hence for *Taq*I, the average distance between sites is $4^4 = 256$ base pairs; for *Bam*HI, it is $4^6 = 4096$ base pairs; and for *Hae*II, it is $4^4 \, 2^2 = 1024$ base pairs.

Chapter 6

6.1 The 1.3-kb insertion is probably a transposable element. (It is a transposable element, called *mariner*.)

6.2 Transposition in somatic cells could cause mutations that kill the host or decrease the ability of the host to reproduce, which reduces the chance of transmission of the transposable element to the next generation. Restriction of activity to the germ line lessens these potentially unfavorable effects.

6.3 Insertion is initiated by a staggered cut in the host DNA sequence, and the overhanging single-stranded ends are later filled in by a DNA polymerase. This process creates a direct repeat.

6.4 *E. coli*: 4,700,000 bp $\times$ 3.4 Å/bp $\times 10^{-7}$ mm/Å = 1.6 mm. Human beings: 3×10^9 bp $\times$ 3.4 Å/bp $\times 10^{-7}$ mm/Å = 1020 mm; in other words, there is approximately 1 *meter* of DNA in a human gamete.

6.5 DNA structure is a long, thin thread, and its total length in most cells greatly exceeds the diameter of a nucleus, even allowing for a large number of fragments per cell. This forces you to conclude that each DNA molecule must be folded back on itself repeatedly.

6.6 Matching of chromomeres, which are locally folded regions of chromatin. Polytene chromosomes do not divide because the cells in which they exist do not divide.

6.7 No, because the *E. coli* chromosome is circular. There are no termini.

6.8 The repeated sequences are located at the ends of the element. They may be in direct or inverted orientation, depending on the particular transposable element.

6.9 **(a)** There are ten base pairs for every turn of the double helix. With four turns of unwinding, $4 \times 10 = 40$ base pairs are broken. **(b)** Each twist compensates for the underwinding of one full turn of the helix; hence, there will be four twists.

6.10 The supercoiled form is in equilibrium with an underwound form having many unpaired bases. At the instant that a segment is single-stranded, S1 can attack it. Because a nicked circle lacks the strain of supercoiling, there is no tendency to have single-stranded regions, and so a nicked circle is resistant to S1 cleavage.

6.11 The rate of movement in the gel is determined by the overall charge and by the ability of the molecule to penetrate the pores of the gel. Supercoiled and relaxed circles migrate at different rates because they have different conformations.

6.12 **(a)** Renaturation is concentration-dependent because it requires complementary molecules to collide by chance before the formation of base pairs can occur. **(b)** Here, the lengths of the GC tracts are important. Molecule 2 has a long GC segment, which will be the last region to separate. Hence molecule 1 has the lower temperature for strand separation.

6.13 The hybrid molecules that can be produced from single-stranded DNA fragments from species A and B are AA, AB, BA, and BB, and these are expected in equal amounts. The molecules with a hybrid ($^{14}N/^{15}N$) density are AB and BA, and these account for 5 percent of the total. The same sequences can also renature as AA or BB which together must make up another 5 percent, so a total of 10 percent of the base sequences are common to the two species.

6.14 In a direct repeat, the sequence is repeated in the same 5'-to-3' orientation. In an inverted repeat, the sequence is repeated in the reverse orientation and on the opposite strand, which is necessary to preserve the correct 5'-to-3' polarity in view of the antiparallel nature of DNA strands. The accompanying figure shows a direct and inverted repeat of the example sequence. The dashed lines represent unspecified DNA sequences between the repeats.

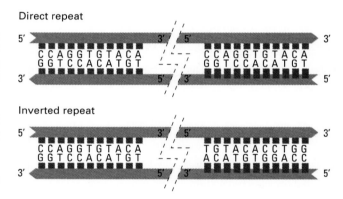

6.15 All such mutations are probably lethal. The fact that the amino acid sequence of histones is virtually the same in all organisms suggests that functional histones are extremely intolerant of amino acid changes.

6.16 The relatively large number of bands and the diverse locations in the genome among different flies suggests very strongly that the DNA sequence is a transposable element of some kind.

6.17 In the first experiment, the nucleosomes form at random positions in different molecules. In the second experiment, the protein binds to a unique sequence in all the molecules, and the nucleosomes form by sequential addition of histone octamers on both sides of the protein-binding site. Because nuclease attacks only in the linker regions, the cuts are made in small regions that are highly localized.

Chapter 7

7.1 **(a)** The Klinefelter karyotype is 47,XXY; hence one Barr body. **(b)** The Turner karyotype is 45,X; hence no Barr bodies. **(c)** The Down syndrome karyotype is 47,+21; people with this condition have one or zero Barr bodies, depending on whether they are female (XX) or male (XY). **(d)** Males with the karyotype 47,XYY have no Barr bodies. **(e)** Females with the karyotype 47,XXX have two Barr bodies.

7.2 An autopolyploid series is formed by the combination of identical sets of chromosomes. Therefore, the chromosome numbers must be even multiples of the monoploid number, or 10 (diploid), 20 (tetraploid), 30, 40, and 50.

7.3 Species S is an allotetraploid of A and B formed by hybridization of A and B, after which the chromosomes in the hybrid became duplicated. The univalent chromosomes in the S $\times$ A cross are the 12 chromosomes from B, and the univalent chromosomes in the S $\times$ B cross are the 14 chromosomes from A.

7.4 The chromosomes underwent replication with no cell division (endoreduplication), resulting in an autotetraploid.

7.5 The only 45-chromosome karyotype found at appreciable frequencies in spontaneous abortions is 45,X, and so 45,X is the probable karyotype. Had the fetus survived, it would have been a 45,X female with Turner syndrome.

7.6 Because the X chromosome in the 45,X daughter contains the color-blindness allele, the 45,X daughter must have received the X chromosome from her father through a normal X-bearing sperm. The nondisjunction must therefore have occurred in the mother, resulting in an egg cell lacking an X chromosome.

7.7 The mother has a Robertsonian translocation that includes chromosome 21. The child with Down syndrome has 46 chromosomes, including two copies of the normal chromosome 21 plus an additional copy attached to another chromosome (the Robertsonian translocation). This differs from the usual situation, in which Down syndrome children have trisomy 21 (karyotype 47,+21).

7.8 The inversion has the sequence $A\ B\ E\ D\ C\ F\ G$, the deletion $A\ B\ F\ G$. The possible translocated chromosomes are **(a)** $A\ B\ C\ D\ E\ T\ U\ V$ and $M\ N\ O\ P\ Q\ R\ S\ F\ G$ or **(b)** $A\ B\ C\ D\ E\ S\ R\ Q\ P\ O\ N\ M$ and $V\ U\ T\ F\ G$. One of these possibilities includes two monocentric chromosomes; the other includes a dicentric and an acentric. Only the translocation with two monocentrics is genetically stable.

7.9 The order is $a\ e\ d\ f\ c\ b$ or the other way around.

7.10 Genes b, a, c, e, d, and f are located in bands 1, 2, 3, 4, 5, and 6, respectively. The reasoning is as follows. Because deletion 1 uncovers any gene in band 1 but the other deletions do not, the pattern $-++++$ observed for gene b puts b in band 1. Deletions 1–3 uncover band 2 but deletions 4–5 do

not, so the pattern $---++$ observed for gene a localizes gene a to band 2. Genes in band 3 are uncovered by deletion 1–4 but not deletion 5, so the pattern $----+$ implies that gene c is in band 3. Band 4 is uncovered by deletions 3–5 but not deletions 1–2, which means that gene e, with pattern $++---$, is in band 4. Genes in band 5 are uncovered by deletions 4–5 but not deletions 1–3, so the pattern $+++--$ observed for gene d places it in band 5. Genes in band 6 are uncovered by deletion 5 but not deletions 1–4, so the pattern $++++-$ puts gene f in band 6.

7.11 The most probable explanation is an inversion. In the original strain the inversion is homozygous, so no problems arise in meiosis. The F_1 is heterozygous, and crossing-over within the inversion produces the dicentrics and acentrics. Because the crossover products are not recovered, the frequency of recombination is greatly reduced in the chromosome pair in which the inversion is heterozygous, so the inversion is probably in chromosome 6.

7.12 Starting with chromosome (c), compare the chromosomes pairwise to find those that differ by a single inversion. In this manner, you can deduce that inversion of the i-d-c region in (c) gave rise to (d), inversion of the h-g-c-d region in (d) gave rise to (a), and inversion of the c-d-e-f region in (a) gave rise to (b). Because you were told that (c) is the ancestral sequence, the evolutionary ancestry is (c) $\rightarrow$ (d) $\rightarrow$ (a) $\rightarrow$ (b).

7.13 The group of four synapsed chromosomes implies that a reciprocal translocation has taken place. The group of four includes both parts of the reciprocal translocation and their nontranslocated homologs.

7.14 Translocation heterozygotes are semisterile because adjacent segregation from the four synapsed chromosomes in meiosis produces aneuploid gametes, which have large parts of chromosomes missing or present in excess. The aneuploid gametes fail to function in plants, and in animals they result in zygote lethality. The translocation homozygote is fully fertile because the translocated chromosomes undergo synapsis in pairs, and segregation is completely regular. In the cross of translocation homozygote $\times$ normal homozygote, all progeny are expected to be semisterile.

7.15 In this male, there was a reciprocal translocation between the *Curly*-bearing chromosome and the Y chromosome. Because there is no crossing-over in the male, all Y-bearing sperm that give rise to viable progeny contain *Cy*, and all X-bearing sperm that give rise to viable progeny contain Cy^+.

7.16 In this male, the tip of the X chromosome containing y^+ became attached to the Y chromosome, giving the genotype y X / y^+ Y. The gametes are y X and y^+ Y, so all of the offspring are either yellow females or wildtype males.

7.17 The $A\ B\ c\ D\ E$ chromosome results from two-strand double crossing-over in the B–C and C–D regions. The double recombinant chromosome still has the inversion.

7.18 The unusual yeast strain is not completely haploid but is disomic for chromosome 2. Segregation of markers on this chromosome, like *his*7, is aberrant, but segregation of markers on the other chromosomes is completely normal.

7.19 The parental classes are wildtype and *brachytic, fine-stripe* and the recombinant classes are *brachytic, fine-stripe*$^+$ and *brachytic*$^+$, *fine-stripe*. The recombination frequency between *brachytic* and *fine-stripe* is therefore equal to $(17 + 1 + 6 + 8)/682 = 0.047$.

Because semisterility is associated with the presence of the translocation, determining the recombination frequency between each of the mutations and the translocation breakpoint follows the same logic as mapping any other kind of genetic marker. For the parental classes, the mutations and the translocation are on different chromosomes. Members of the recombinant classes are those plants that show both the mutation in question and semisterility or neither semisterility nor the mutation in question. The recombination frequency between *brachytic* and the translocation breakpoint is therefore $(17 + 25 + 19 + 8)/682 = 0.102$; the recombination frequency between *fine-stripe* and the translocation breakpoint is $(19 + 6 + 1 + 25)/682 = 0.075$.

Chapter 8

8.1 Minimal medium containing galactose and biotin; the galactose would require the presence of *gal*$^+$, and the biotin would allow both *bio*$^+$ and *bio*$^-$ to grow.

8.2 The *E. coli* genome contains approximately 4700 kilobase pairs, which implies approximately 47 kb per minute. Because the λ genome is 50 kb, the genetic length of the prophage is approximately 1 minute (more precisely, $47/50 = 0.94$ minutes).

8.3 Infect an *E. coli* λ lysogen with P1. The λ lysogen contains λ prophage, so some of the resulting P1 phage will include the part of the chromosome that contains the prophage.

8.4 After circularization and integration at the bacterial *att* site, the map is . . . *att D E F A B C att* . . . .

8.5 The specialized-transducing particles originate from aberrant prophage excision.

8.6 Plate on minimal medium that lacks leucine (selects for Leu$^+$) but contains streptomycin (selects for Str-r). The *leu*$^+$ allele is the selected marker, and *str-r* the counterselected marker.

8.7 One plaque per phage. One plaque, because the plating was done before lysis, so the progeny phage are confined to one tiny area.

8.8 The T2 plaques will be turbid because T2 fails to lyse the resistant bacteria. Plaques made by the T2*h* mutant will be clear because the mutant can lyse both the normal and the resistant cells.

8.9 The mean number of phage per cell equals 1. If you know the Poisson distribution, then it is apparent that the proportion of uninfected cells is $e^{-1} = 0.37$. If you do not know the Poisson distribution, then the answer can be calculated as follows. The probability that a bacterial cell escapes infection by a particular phage is $1 - 10^{-6} = 0.999999$, and the probability that it escapes infection by all one million phages is $P_0 = (0.999999)^{1000000}$. Therefore, $\ln P_0 = 1000000 \times \ln(0.999999) = -1$, so $P_0 = e^{-1} = 0.37$.

8.10 Apparently, *h* and *tet* are closely linked, so recombinants containing the *h*$^+$ allele of the Hfr tend also to contain the *tet-s* allele of the Hfr, and these recombinants are eliminated by the counterselection for *tet-r*.

8.11 All receive the *a*$^+$ allele, but whether or not a particular *b*$^+$ *str-r* recombinant contains *a*$^+$ depends on the positions of the genetic exchanges.

8.12 Depending on the size of the F', it could be F' *g*, F' *g h*, F' *g h i*, and so forth, or F' *f*, F' *f e*, F' *f e d*, and so forth.

8.13 The first selection is for Met$^+$ and so *lac*$^+$ must have been transferred. The probability that a marker that is transferred is incorporated into the recombinants is about 50 percent, so about 50 percent of the recombinants would be expected to be *lac*$^+$. The second selection is for Lac$^+$, and *met*$^+$ is a late marker. The great majority of mating pairs will spontaneously break apart before transfer of *met*$^+$, so the frequency of *met*$^+$ recombinants in this experiment will be close to zero.

8.14 The order 500 *his*$^+$, 250 *leu*$^+$, 50 *trp*$^+$ implies that the genes are transferred in the order *his leu trp*. The *met* mutation is the counterselected marker that prevents growth of the Hfr parent. The medium containing histidine selects for *leu*$^+$ *trp*$^+$, and the number is small because both genes must be incorporated by recombination.

8.15 The three possible orders are (1) *pur–pro–his*, (2) *pur–his–pro*, and (3) *pro–pur–his*. The predictions of the three orders follow. (1) Virtually all *pur*$^+$ *his*$^-$ transductants should be *pro*$^-$, but this is not true. (2) Virtually all *pur*$^+$ *pro*$^-$ transductants should be *his*$^-$, but this is not true. (3) Some *pur*$^+$ *pro*$^-$ transductants will be *his*$^-$, and some *pur*$^+$ *his*$^-$ transductants will be *pro*$^-$ (depending on where the exchanges occur). Order (3) is the only one that is not contradicted by the data.

8.16 (a) Use $m = 2 - 2d^{1/3}$, in which *m* is map distance in minutes and *d* is cotransduction frequency. With the values of *d* given, the map distances are 0.74, 0.41, and 0.18 minutes, respectively. **(b)** Use $d = [1 - (m/2)]^3$. With the values of *m* given, cotransduction frequencies are 0.42, 0.12, and 0, respectively. For markers greater than 2 minutes apart, the frequency of cotransduction equals zero. **(c)** Map distance *a–b* equals 0.66 minutes, *b–c* equals 1.07 minutes, and these are additive, giving 1.73 minutes for the distance *a–c*. Predicted cotransduction frequency for this map distance is 0.002.

8.17 It is probable that the *amp-r* gene in the natural isolate was present in a transposable element. Transposition into λ occurs in a small proportion of cases, and when these infect the laboratory strain, the transposable element can transpose into the chromosome before the infecting λ is lost.

8.18 The *tet-r* gene in the phage was included in a transposable element. Transposition occurred in the lysogen to another location in the *E. coli* chromosome, and when the prophage was lost, the antibiotic resistance remained.

8.19 Both the order of times of entry and the level of the plateaus imply that the order of gene transfer is *a b c d*. The times of entry are obtained by plotting the number of recombinants of each type versus time and extrapolating back to the time axis. The values are *a*, 10 minutes; *b*, 15 minutes; *c*, 20 minutes; *d*, 30 minutes. The low plateau value for the *d* gene probably results from close linkage with *str-s*, because in this case, the *d* allele in the Hfr strain would tend to be inherited along with the *str-r* allele.

8.20 The deletion map is shown in the accompanying illustration. A completely equivalent map can be drawn with the order of the genes reversed.

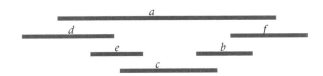

8.21 The genetic map shows the genetic intervals defined by the deletion endpoints and the locations of the mutations with respect to these intervals.

Intervals defined by
deletion endpoints

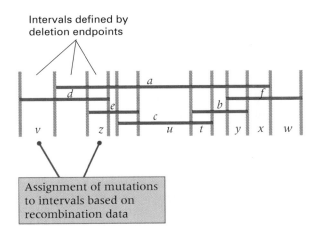

Assignment of mutations
to intervals based on
recombination data

8.22 The mutations *v* and *z* are in the *rIIB* cistron, whereas the mutations *t*, *u*, *w*, *x*, and *y* are in the *rIIA* cistron. As indicated in the genetic map, the boundary between the cistrons lies between the left end of the *c* deletion and the right end of the *e* deletion.

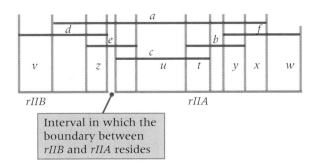

rIIB   rIIA

Interval in which the
boundary between
rIIB and *rIIA* resides

The matrix for complementation among the deletions must be as follows:

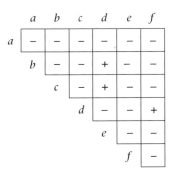

	a	b	c	d	e	f
a	−	−	−	−	−	−
b		−	−	+	−	−
c			−	+	−	−
d				−	−	+
e					−	
f						−

8.23 The λ prophage itself is being transduced as part of the chromosome. It maps next to *gal*. Hence when phage 363 is grown on strain D, 6 percent of the Gal$^+$ transductants are lysogenic, whereas none of the Thr$^+$ or Lac$^+$ transductants is lysogenic. The reciprocal cross gives the same basic result, only here λ is being removed from the chromosome and being replaced by phage-free DNA. Hence about 8 percent of the Gal$^+$ transductants are nonlysogenic. As expected from the close linkage of the phage λ attachment site and *gal*, no linkage is seen for the other two markers.

Chapter 9

9.1 The DNA content per average band equals about $1.1 \times 10^8/5000 = 22$ kb, that per average lettered subdivision equals approximately $1.1 \times 10^8/600 = 183$ kb, and that per average numbered section equals about $1.1 \times 10^8/100 = 1.1$ Mb. A YAC with a 200-kb insert contains the DNA equivalent to about 9 salivary bands, 1.1 lettered subdivisions, and 0.2 numbered section. An 80-kb P1 clone contains the DNA equivalent of about 4 salivary bands, 0.48 lettered subdivision, and 0.07 numbered section.

9.2 The ends may be blunt or they may be cohesive (sticky) with either a 3' overhang or a 5' overhang.

9.3 All restriction fragments have the same ends, because the restriction sites are all the same and they are cleaved in the same places. Opposite ends of the same fragment must also be identical, because restriction sites are palindromes.

9.4 No. The complement of 5'-GGCC-3' is 3'-CCGG-5'.

9.5 The insertion disrupts the gene, and sensitivity to the antibiotic shows that the vector has an inserted DNA sequence. The second antibiotic-resistance gene is needed to select transformants.

9.6 (a) The probability of a *Taq*I site is $1/6 \times 1/3 \times 1/3 \times 1/6 = 1/324$, so *Taq*I cleavage is expected every 324 base pairs; the probability of a *Mae*III site is $1/3 \times 1/6 \times 1 \times 1/6 \times 1/3 = 1/324$, so *Mae*III cleavage is expected every 324 base pairs. **(b)** The answers are the same as in part (a), because for both *Taq*I and *Mae*III sites, the number of A + T = the number of G + C.

9.7 Comparison of the genomic and cDNA sequences tells you where the introns are in the genomic sequence. The genomic sequence contains the introns; the cDNA does not.

9.8 Bacterial cells do not normally recognize eukaryotic promoters, and if transcription does occur, the transcript usually contains introns that the bacterial cell cannot remove. If these problems are overcome, the protein may still not be produced because the mRNA lacks a bacterial ribosome-binding site, because the protein might require post-translational processing, or because the protein might be unstable in bacterial cells and subject to degradation.

9.9 The finding is that the 3.6-kb and 5.3-kb fragments in the free phage are joined to form an 8.9-kb fragment in the intracellular form. This suggests that the the 3.6-kb and 5.3-kb fragments are terminal fragments of a linear molecule and that the intracellular form is circular.

9.10 A base substitution may destroy a restriction site and prevent two potential fragments from being separated, or a deletion may eliminate a restriction site.

9.11 The gene probably contains an *Eco*RI site that disrupts the gene in the process of cloning; the gene does not contain a *Hin*dIII site.

9.12 (a) The *tet-r* gene is not cleaved with *Bgl*I, so addition of tetracycline to the medium requires that the colonies be tetracycline-resistant (Tet-r) and hence contain the plasmid. **(b)** Cells with the phenotype Tet-r Kan-r or Tet-r Kan-s will form colonies. **(c)** Colonies with the phenotype Tet-r Kan-s contain inserts within the cleaved *kan* gene.

9.13 An insertion of two bases is generated within the codon, resulting in a frameshift. Therefore, all colonies should be Lac⁻.

9.14 You must use cDNA under a suitable promoter that functions in the prokaryotic system. Eukaryotic genomic DNA includes regulatory elements that will not work in bacteria and introns that cannot be spliced out of RNA in prokaryotic cells.

9.15 The restriction map is shown in the accompanying figure.

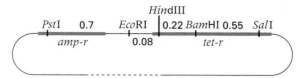

Rest of plasmid between *Pst*I and *Sal*I = 2.45 kb

Chapter 10

10.1 The start codon is AUG, which codes for methionine; the stop codons are UAA, UAG, and UGA. The probability of a start codon is 1/64, and the probability of a stop codon is 3/64. The average distance between stop codons is 64/3 = 21.3 codons.

10.2 The mRNA sequence is

5'-AAAAUGCCCUUAAUCUCAGCGUCCUAC-3'

and translation begins with the first AUG, which codes for methionine. The amino acid sequence coded by this region is Met-Pro-Leu-Ile-Ser-Ala-Ser-Tyr.

10.3 With a G at the 5' end, the first codon is GUU, which codes for valine; and with a G at the 3' end, the last codon is UUG, which codes for leucine.

10.4 Codons are read from the 5' end of the mRNA, and the first amino acid in a polypeptide chain is at the amino terminus. If the G were at the 5' end, the first codon would be GAA, coding for glutamic acid at the amino terminus of the polypeptide. If the G were at the 3' end, the last codon would be AAG, coding for arginine at the carboxyl terminus. Thus the G was added at the 5' end.

10.5 GUG codes for valine, and UGU codes for cysteine. Therefore, an alternating polypeptide containing only valine and cysteine would be made.

10.6 The artificial mRNA can be read in any one of three reading frames, depending on where translation starts. The reading frames are

GUC GUC GUC . . ., UCG UCG UCG . . ., and CGU CGU CGU . . .

The first codes for polyvaline, the second for polyserine, and the third for polyarginine.

10.7 The possible codons, their frequencies, and the amino acids coded are as follows:

AAA	$(3/4)^3 = 0.421$	Lys
AAC	$(3/4)^2(1/4) = 0.141$	Asn
ACA	$(3/4)^2(1/4) = 0.141$	Thr
CAA	$(3/4)^2(1/4) = 0.141$	Gln
CCA	$(3/4)(1/4)^2 = 0.047$	Pro
CAC	$(3/4)(1/4)^2 = 0.047$	His
ACC	$(3/4)(1/4)^2 = 0.047$	Thr
CCC	$(1/4)^3 = 0.015$	Pro

Therefore, the amino acids in the random polymer and their frequencies are lysine (42.1 percent), asparagine (14.1 percent), threonine (18.8 percent), glutamine (14.1 percent), histidine (4.7 percent), and proline (6.2 percent).

10.8 Methionine has 1 codon, histidine 2, and threonine 4, so the total number is $1 \times 2 \times 4 = 8$. The sequence AUGCAYACN encompasses them all. Because arginine has six codons, the possible number of sequences coding for Met-Arg-Thr is $1 \times 6 \times 4 = 24$. The sequences are AUGCGNACN (16 possibilities) and AUGAGRACN (8 possibilities).

10.9 In an overlapping code, a single base change affects more than one codon, so changes in more than one amino acid should often occur (but not always, because single amino acid changes can result from redundancy in the code). For this reason, an overlapping code was eliminated from consideration as a likely possibility.

10.10 Because the pairing of codon and anticodon is antiparallel, it is convenient to write the anticodon as 3'-UAI-5'. The first two codon positions are therefore 5'-AU-3', and the third is either A, U, or C (because each can pair with I). The possible codons are 5'-AUA-3', 5'-AUU-3', and 5'-AUC-3', all of which code for isoleucine.

10.11 In theory, the codon 5'-UGG-3' (tryptophan) could pair with either 3'-ACC-5' or 3'-ACU-5'. However, 3'-ACU-5' can also pair with 5'-UGA-3', which is a chain-termination codon. If this anticodon were used for tryptophan, an amino acid would be inserted instead of terminating the chain. Therefore, the other anticodon (3'-ACC-5') is the one used.

10.12 (a) The deletion must have fused the amino-coding terminus of the *B* gene with the carboxyl-coding terminus of the *A* gene. The nontranscribed strand must therefore be oriented 5'-*B*-*A*-3'. **(b)** The number of bases deleted must be a multiple of 3; otherwise, the carboxyl terminus would not have the correct reading frame.

10.13 (a) The nontranscribed strand is given, so the wildtype reading frame of the mRNA is

AU<u>G</u> CAU CCG GGC UCA UUA GUC U . . .

which codes for Met-His-Pro-Gly-Ser-Leu-Val- . . . **(b)** The mutant X mRNA reading frame is

AU<u>G</u> GCA UCC GGG CUC AUU AGU CU . . .

which codes for Met-Ala-Ser-Gly-Leu-Ile-Ser-Leu- . . . .

(c) The mutant Y mRNA reading frame is

AU<u>G</u> CAU CCG GGC UCU UAG UCU . . .

which codes for Met-His-Pro-Gly-Ser (the UAG is a termination codon). **(d)** The double-mutant mRNA reading frame is

AU<u>G</u> GCA UCC GGG CUC UUA GUC U . . .

which codes for Met-Ala-Ser-Gly-Leu-Leu-Val- . . . . Note that the double mutant differs from wildtype only in the region between the insertion and deletion mutations, in which the reading frame is shifted.

10.14 Mutation X can be explained by a single base change in the start codon for the first Met, in which case translation would begin with the second AUG codon. Mutation Y can be explained by a single base change that converts a codon for Tyr (UAU or UAC) into a chain-termination codon (UAA or UAG). The tripeptide sequence is Met-Leu-His.

10.15 The probability of a correct amino acid at a particular site is 0.999, so the probability of all 300 amino acids being correctly translated is $(0.999)^{300} = 0.74$, or approximately $3/4$.

10.16 Each loop represents an intron, so the total number is seven introns.

10.17 If transcribed from left to right, the transcribed strand is the bottom one, and the mRNA sequence is

AGA CUU CAG GCU CAA CGU GGU

which codes for Arg-Leu-Gln-Ala-Gln-Arg-Gly. To invert the molecule, the strands must be interchanged in order to preserve the correct polarity, and so the sequence of the inverted molecule is

5'-AGAACGTTGAGCCTGAAGGGT-3'
3'-TCTTGCAACTCGGACTTCCCA-5'

This codes for an mRNA with sequence

AGA ACG UUG AGC CUG AAG GGU

which codes for Arg-Thr-Leu-Ser-Leu-Lys-Gly.

10.18 The DNA nontemplate strand is shown as sequence (1) in the table below. The corresponding mRMA for this region is shown as sequence (2). Translation in all three possible reading frames yields (in the single-letter amino acid codes) the sequences shown in (3).

DNA nontemplate strand, mRNA region, and translation
Problem 10.18

(1)   5'-TAACGTATGCTTGACCTCCAAGCAATCGATGCCAGCTCAAGG-3'

(2)   5'-UAACGUAUGCUUGACCUCCAAGCAAUCGAUGCCAGCUCAAGG-3'

(3)   * R M L D L Q A I D A S S R
          N V C L T S K Q S M P A Q
          T Y A * P P S N R C Q L K

The sequence is from the middle of an exon, so you do not know the reading frame. However, two of the possible reading frames contain stop codons (*). Therefore, the only sequence without stop codons must be correct, because you know that the sequence comes from the middle of an exon of an active gene. Thus the correct amino acid sequence of the polypeptide is N V C L T S K Q S M P A Q.

10.19 Unlike other 6-fold degenerate amino acids (Arg, Leu), is not possible to change between Ser codons with a single mutation. In particular, to get from the 4-fold degenerate class (UCN, where N is any of four bases) to the 2-fold degenerate class (AGY, where Y is any pyrimidine) requires a change in both the first and the second codon positions. It is highly unlikely that two point mutations would occur simultaneously in the same codon, so an organism would have temporarily had an amino acid other than Ser when changing from one codon class to the other. It is believed that the two classes of Ser codons may have evolved independently.

10.20 Mutant 1 indicates the sequence AUGAARUAG, where R means any purine. Mutant 2 indicates the sequence AUGAU[U or C or A]GUNUAA, where N means any nucleotide. Because mutation 2 is a deletion in the second codon, the second codon in the wildtype gene must be either ?AU, A?U, or AU?, where the question mark indicates the deleted nucleotide; also, the third codon in the wildtype gene must be [U or C or A]GU, the fourth codon must be NUA, and the fifth must be A??. Similarly, because mutation 1 is an insertion in the second codon, the second codon in the wildtype gene must consist of three of the four bases AARU; also, the third codon in the wildtype gene must be AG?. The only consistent possibility for codon 2 is AAU, which codes for Asn. The third codon must be AGU, which codes for Ser. That leaves, for codon 4, the possibilities Val and Lys, and of these, only Val is coded by NUA, where N = G. Finally, the fifth codon must be AAR (Lys).

Hence the sequence of the nontranscribed strand of the wildtype gene, as nearly as can be determined from the data given, and separated into codons, is

5'-ATG-AAT-AGT-GTA-AAR-3'

Mutation 1 has a single nucleotide addition in codon 2 that yields a codon for Lys, so it must be an addition of a purine (R) at position 6 or of an A at either position 4 or position 5. Mutation 2 has a single nucleotide deletion that yields a codon for Ile, which means that the A at either position 3 or position 4 was deleted.

Chapter 11

11.1 The enzymes are expressed constitutively because glucose is metabolized in virtually all cells. However, the levels of enzyme are regulated to prevent runaway synthesis or inadequate synthesis.

11.2 The hemoglobin mRNA has a very long lifetime before being degraded.

11.3 (a) Transcription. (b) RNA processing. (c) Translation.

11.4 The repressor gene need not be near because the repressor is diffusible. The *trp* repressor is one example in which the repressor gene is located quite far from the structural genes.

11.5 The mutant gene should bind more of the activator protein or bind it more efficiently. Therefore, the mutant gene should be induced with lower levels of the activator protein or expressed at higher levels than the wildtype gene, or both.

11.6 (a) β-galactosidase (*lacZ* product), lactose permease (*lacY* product), and the transacetylase (*lacA* product). (b) The promoter region becomes inaccessible to RNA polymerase. (c) Repressor synthesis is constitutive.

11.7 At 37°C, both $lacZ^-$ and $lacY^-$ are phenotyipcally Lac$^-$; at 30°C, the $lacY^-$ strain is Lac$^+$, and the $lacZ^-$ strain is Lac$^-$.

11.8 Inducers combine with repressors and prevent them from binding with the operator regions of the respective operons.

11.9 The cAMP concentration is low in the presence of glucose. Both types of mutations are phenotypically Lac$^-$. The cAMP-CRP binding does not interfere with repressor binding because the DNA binding sites are distinct.

11.10 No; there is often a polar effect in which the relative amount of each protein synthesized decreases toward the 3' end of the mRNA.

11.11 With a repressor, the inducer is usually an early (often the first) substrate in the pathway, and the repressor is inactivated by combining with the inducer. With an aporepressor, the effector molecule is usually the product of the pathway, and the aporepressor is activated by the binding.

11.12 The attenuator is not a binding site for any protein, and RNA synthesis does not begin at an attenuator. An attenuator is strictly a potential termination site for transcription.

11.13 Steroid hormones pass through the cell membrane and combine with specific receptor proteins; thereupon they are transported to the nucleus and act as transcriptional activator proteins that stimulate transcription of their target genes.

11.14 The α cell would be unable to switch to the **a** mating type.

11.15 The **a**1 protein functions only in combination with the α2 protein in diploids; hence the *MATa'* haploid cell would appear normal. However, the diploid *MATa'/MATα* has the phenotype of an α haploid. This is because the genotype lacks an active **a**1-α2 complex to repress the haploid-specific genes as well as repressing α1, and the α1 gene product activates the α-specific genes.

11.16 The mutant diploid *MATa/MATα* has the phenotype of an α haploid. As in Problem 11.15, the genotype lacks an active **a**1-α2 complex to repress the haploid-specific genes as well as α1, and the α1 gene product activates the α-specific genes.

11.17 In the absence of lactose or glucose, two proteins are bound: the *lac* repressor and CRP-cAMP. In the presence of glucose, only the repressor is bound.

11.18 (a) The strain is constitutive and has either the genotype *lacI⁻* or *lacO^c*. **(b)** The second mutant must be *lacZ⁻* because no β-galactosidase is made; and because permease synthesis is induced by lactose, the mutant must have a normal repressor-operator system. Therefore, the genotype is *lacI⁺ lacO⁺ lacZ⁻ lacY⁺*. **(c)** Because the partial diploid is regulated, the operator in the operon with a functional *lacZ* gene must be wildtype. Hence the mutant in part (a) must be *lacI⁻*.

11.19 The *lac* genes are transferred by the Hfr cell and enter the recipient. No repressor is present in the recipient initially, so *lac* mRNA is made. However, the *lacI* gene is also transferred to the recipient, and soon afterward the repressor is made and *lac* transcription stops.

11.20 Amino acids with codons one step away from the nonsense codon, because their tRNA genes are the most likely to mutate to nonsense suppressors. For UGA, these are UGC (Cys), UGU (Cys), UGG (Trp), UUA (Leu), UCA (Ser), AGA (Arg), CGA (Arg), and GGA (Gly).

11.21 The mutant clearly lacks a general system that regulates sugar metabolism, and the most likely possibility is the cAMP-dependent regulatory system. Several mutations can prevent activity of this system. For example, the mutant either could make a defective CRP protein (which either fails to bind to the promoter or is unresponsive to cAMP) or could have an inactive adenyl cyclase gene and be unable to synthesize cAMP.

11.22 (a) No enzyme synthesis; the mutation is *cis*-dominant because lack of RNA polymerase binding to the promoter prevents transcription of the adjacent genes. **(b)** Enzymes synthesized; this one is *cis*-dominant because failure to bind the repressor cannot prevent constitutive transcription of the adjacent genes. **(c)** Enzymes synthesized; this mutation is recessive because the defective repressor can be complemented by a wildtype repressor in the same cell. **(d)** Enzymes synthesized; the mutation is recessive because the mutant repressor can be complemented by a wildtype repressor in the same cell. **(e)** No enzyme synthesis; this mutation is dominant because the repressor is always activated and all *his* transcription is prevented.

11.23 The wildtype mRNA reads

AAA CAC CAC CAU CAU CAC CAU CAU CCU GAC

which codes for Lys-His-His-His-His-His-His-His-Pro-Asp. The mutant mRNA reads AAA ACA CCA CCA UCA UCA CCA UCA UCC UGA C, which codes for Lys-Thr-Pro-Pro-Ser-Ser-Pro-Ser-Ser (the UGA is a normal stop codon). The expected phenotype of the mutant is His⁻ because of lack of attenuation by low levels of histidine. That is, translation of the attenuator occurs (preventing efficient transcription of the operon) even when the level of histidine within the cell is very low. However, low levels of proline or serine will prevent translation of the attenuator and allow transcription of the histidine operon.

Chapter 12

12.1 Autonomous developmental fates are controlled by genetically programmed changes within cells. Fates determined by positional information are controlled by interactions with morphogens or with other cells. Cell transplantation and cell ablation (destruction) are two kinds of surgical manipulation that help investigators distinguish between them.

12.2 Lineage mutations affect the developmental fates of one or more cells within a group of cells related by descent that normally undergo a defined pattern of development. Lineage mutations are detected by the occurrence of abnormal types or numbers of cells at characteristic positions within the embryo.

12.3 The *Drosophila* blastoderm is the flattened hollow ball of cells formed by cellularization of the syncytium formed by the early nuclear divisions. The fate map is a diagram of the blastoderm showing the locations of cells and their developmental fates. The existence of the fate map does not mean that development is autonomous, because many developmental decisions are made according to positional information in the blastoderm. In fact, most *Drosophila* developmental decisions are nonautonomous.

12.4 A female that is homozygous for a maternal-effect lethal produces defective eggs resulting in inviable embryos. However, the homozygous genotype itself is viable.

12.5 Cells normally destined to die adopt some other developmental fate, with the result that there are supernumerary copies of one or more cell types.

12.6 In a homeotic mutation, certain segments develop in a manner characteristic of that of other segments, and therefore the lineage effect is one of transformation.

12.7 A sister-cell segregation defect would lead to both daughter cells differentiating as B or to both differentiating as C. A parent-offspring segregation defect would cause both daughter cells to differentiate as A.

12.8 Heterochronic mutations affect the timing of developmental events relative to each other; hence a mutation that uniformly slows or accelerates development is not a heterochronic mutation.

12.9 Because the *Drosophila* oocyte contains all the transcripts needed to support the earliest stages of development; the mouse oocyte does not.

12.10 In a loss-of-function mutation, the genetic information in a gene is not expressed in some or all cells. In a gain-of-function mutation, the genetic information is expressed at inappropriate times or in inappropriate cells. Both mutations can occur in the same gene: Some alleles may be loss-of-function and others gain-of-function. However, a particular allele must be one or the other (or neither).

12.11 Because the gene is necessary, a loss-of-function mutation will prevent the occurrence of the proper developmental fate. The allele will be recessive, however, because a normal allele in the homologous chromosome will allow the developmental pathway to occur.

12.12 Because the gene is sufficient, a gain-of-function mutation will induce the developmental fate to occur. The allele will be dominant because a normal allele in the homologous chromosome will not prevent the developmental pathway from occurring.

12.13 Developmental abnormalities in gap mutants occur across contiguous regions, resulting in a gap in the pattern of normal development. Developmental abnormalities in pair-rule mutants are in even- or odd-numbered segments or parasegments, so the abnormalities appear in horizontal stripes along the embryo.

12.14 The proteins required for cleavage and blastula formation are translated from mRNAs present in the mature oocyte, but transcription of zygotic genes is required for gastrulation.

12.15 The material required for rescue is either mRNA transcribed from the maternal o^+ gene or a protein product of the gene that is localized in the nucleus.

12.16 The transplanted nuclei respond to the cytoplasm of the oocyte and behave like the oocyte nucleus at each stage of development.

12.17 XDH activity is needed only at an early stage in development for the eyes to have wildtype pigmentation. The data are explained by a maternal effect in which the mal^+/mal; ry/ry females in the cross transmit enought mal^+ gene product to allow brief activity of XDH in the ry^+/ry embryos. In theory, the mal^+ product could be a transcription factor that allows ry^+ transcription, but in fact the mal^+ gene product participates in the synthesis of a cofactor that combines with the ry^+ gene product to make an active XDH enzyme.

12.18 First, examine embryos of strains (c) and (d) to determine the wildtype expression patterns of *kni* and *ftz*. Next, cross strains (a) and (d) and strains (b) and (c). Intercross the F_1 heterozygotes from the cross (a) × (d), and examine the embryos to detect changes in *ftz* expression. Intercross the F_1 heterozygotes from the cross (b) × (c), and examine the embryos to detect changes in *kni* expression. Some of the F_2 embryos from the (a) × (d) cross will show abnormal expression of *ftz*, indicating that wildtype *kni* is required for proper *ftz* expression. The F_2 embryos of (b) × (c) will show normal expression of *kni*, indicating that *ftz* is not required for proper *kni* expression.

Chapter 13

13.1 Such people are somatic mosaics. This condition can be explained by somatic mutations in the pigmented cells of the iris of the eye or in their precursor cells.

13.2 Causes of a nonreverting mutation include (1) deletion of the gene, (2) multiple nucleotide substitutions in the gene, and (3) complex DNA rearrangement with more than one breakpoint in the gene. Deletions do not revert because the genetic information in the gene is missing, and the others do not revert because the nucleotide substitutions or DNA rearrangements are unlikely to be reversed precisely in one step.

13.3 Transitions are AT → GC, GC → AT, TA → CG and CG → TA (total four). Transversions are AT → TA, AT → CG, GC → CG, GC → TA, TA → AT, TA → GC, CG → GC, and CG → AT (total eight).

13.4 All nucleotide substitutions in the second codon position result in either an amino acid replacement or a terminator codon.

13.5 Both UUR and CUR code for leucine, and both CGR and AGR code for arginine (R stands for any purine, either A or G). Hence, in each of these codons, one of the possible substitutions in the first codon position does not result in an amino acid replacement. Even when an amino acid replacement does occur, the mutant protein may still be functional, because more than one amino acid at a given position in a protein molecule may be compatible with normal or nearly normal function.

13.6 Any single nucleotide substitution creates a new codon with a random nucleotide sequence, so the probability that it is a chain-termination codon (UAA, UAG, or UGA) is 3/64, or approximately 5 percent.

13.7 A frameshift mutagen will also revert frameshift mutations.

13.8 Exactly three. Otherwise, there would be a frameshift, and all downstream amino acids would be different.

13.9 No. Dominance and recessiveness are characteristics of the phenotype. They depend on the function of the gene and on which attributes of the phenotype are examined.

13.10 Photoreactivation and excision repair.

13.11 (a) Development arrests at A at both the high and low temperatures. (b) Development arrests at B at both the high and low temperatures. (c) Development arrests at A at the high temperature and arrests at B at the low temperature. (d) Development arrests at B at the high temperature and arrests at A at the low temperature.

13.12 The mutation frequency in the Lac$^-$ culture is at least $0.90 \times 10^{-8} = 9 \times 10^{-9}$. Because the mutation rate to a nonsense suppressor is independent of the presence of a suppressible mutation (in this case, in the Lac$^-$ mutant), the mutation rate in the original Lac$^+$ culture is also at least 9×10^{-9}. The qualifier *at least* is important because not all nonsense suppressors may suppress the mutation in the particular Lac$^-$ strain (that is, some suppressors may insert an amino acid that still results in a nonfunctional enzyme.)

13.13 (a) 5×10^{-5} is the probability of a mutation per generation, and so $1 - (5 \times 10^{-5}) = 0.99995$ is the probability of no mutation per generation. The probability of no mutations in 10,000 generations equals $(0.99995)^{10,000} = 0.61$. **(b)** The average number of generations until a mutation occurs is $1/(5 \times 10^{-5}) = 20,000$ generations.

13.14 At 1 Gy, the total mutation rate is double the spontaneous mutation rate. Because a dose of 1 Gy increases the rate by 100 percent, a dose of 0.5 Gy increases it by 50 percent and a dose of 0.1 Gy by 10 percent.

13.15 Amino acid replacements at many positions in the A protein do not eliminate its ability to function; only amino acid replacements at critical positions that do eliminate tryptophan synthase activity are detectable as Trp$^-$ mutants.

13.16 The possible lysine codons are AAA and AAG, and the possible glutamic acid codons are GAA and GAG. Therefore, the mutation results from an AT $\rightarrow$ GC transition in the first position of the codon.

13.17 Six amino acid replacements are possible because the codons UAY (Y stands for either pyrimidine, C or U) can mutate in a single step to codons for Phe (UUY), Ser (UCY), Cys (UGY), His (CAY), Asn (AAY), or Asp (GAY).

13.18 Single nucleotide substitutions that can account for the amino acid replacements are (1) Met (AUG) to Leu (UUG), (2) Met (AUG) to Lys (AAG), (3) Leu (CUN) to Pro (CCN), (4) Pro (CCN) to Thr (ACN), and (5) Thr (ACR) to Arg (AGR), where R stands for any purine (A or G) and N for any nucleotide. Substitutions (1), (2), (4), and (5) are transversion, and substitution (3) is a transition. Because 5-bromouracil primarily induces transitions, the Leu $\rightarrow$ Pro replacement is expected to be the most frequent.

13.19 The sectors are explained by a nucleotide substitution in only one strand of the DNA duplex. Replication of this heteroduplex yields one daughter DNA molecule with the wildtype sequence and another with the mutant sequence. Cell division produces one Lac$^+$ and one Lac$^-$ cell, located side by side. Because the cells do not move on the agar surface, the Lac$^+$ cells form a purple half-colony, and the Lac$^-$ cells form a pink half-colony.

13.20 The *a* and *b* gene products may be subunits of a multimeric protein. A few proteins with compensatory amino acid replacements can interact to give wildtype activity, but most mutant proteins cannot interact in this way with each other or with wildtype.

13.21 Some of the mutagenized P1 particles acquire mutations in the *a* gene and become $a^- b^+$. Therefore, the next step is to select b^+ transductants and screen among these for ones that are a^-. The procedure depends on close linkage.

13.22 (a) All progeny would be able to grow in the absence of arginine. **(b)** The genotype of the diploid is Su^+/Su^- arg^+/arg^-, where Su^+ indicates the ability to suppress, and Su^- indicates inability to suppress. The resulting haploid spores are 1/4 Su^+ arg^+, 1/4 Su^+ arg^-, 1/4 Su^- arg^+, and 1/4 Su^- arg^-. The first three are Arg$^+$ and the last Arg$^-$, so the ratio is 3 Arg$^+$: 1 Arg$^-$. **(c)** The diploid genotype is Su^+ arg^-/Su^- arg^+. The haploid spores are in the proportions 0.05 Su^+ arg^+, 0.45 Su^+ arg^-, 0.45 Su^- arg^+, and 0.05 Su^- arg^-. Therefore, 95 percent of the progeny are Arg$^+$ and 5 percent are Arg$^-$.

13.23 Asci C, D, and F might arise from gene conversion at the *m* locus as follows. Branch migration through *m* results in two regions of heteroduplex DNA. In ascus type C, one region is replicated before repair, leaving + and *m* alleles intact; the other region is repaired before replication (+ to *m*), yielding two *m* alleles (therefore a 3 : 5 ratio). In ascus type D, both heteroduplex regions are replicated before repair, leaving all alleles intact but changing their order in the ascus (therefore a 3 : 1 : 1 : 3 ratio). In ascus type F, both heteroduplex regions are repaired (+ to *m*) before replication, yielding four *m* alleles (therefore a 2 : 6 ratio).

Chapter 14

14.1 Inheritance is strictly maternal, so yellow is probably inherited cytoplasmically. A reasonable hypothesis is that the gene for yellow is contained in chloroplast DNA.

14.2 Pollen is not totally devoid of cytoplasm, and occasionally it contains a chloroplast that is transmitted to the zygote.

14.3 This trait is maternally inherited, so the phenotype is that of the mother. The progeny are **(a)** green, **(b)** white, **(c)** variegated, and **(d)** green.

14.4 With X-linked inheritance, half of the F$_2$ males (1/4 of the total progeny) would be white. The observed result is that all of the F$_2$ progeny are green.

14.5 If the daughter chloroplasts segregate at random, both of them will go to the same pole half the time. For all four products to contain a descendant chloroplast, the segregation has to be perfect in three divisions, which occurs with probability $(1/2)^3 = 1/8$. Therefore, the probability that at least one cell lacks a descendant chloroplast is $1 - 1/8 = 7/8$.

14.6 The chloroplast from the mt^- strain is preferentially lost, and the mt alleles segregate 1 : 1. The expected result is 1/2 $a^+ b^- mt^+$ and 1/2 $a^+ b^- mt^-$.

14.7 Petite, because neither parent contributes functional mitochondria.

14.8 Because every tetrad shows uniparental inheritance, it is likely that the trait is determined by factors transmitted through the cytoplasm, such as mitochondria. The 4 : 0 tetrads result from diploids containing only mutant factors, and the 0 : 4 tetrads result from diploids containing only wildtype factors.

14.9 In autogamy, one product of meiosis becomes a homozygous diploid nucleus, so half of the *Aa* cells that undergo autogamy become *AA* and half become *aa*.

14.10 The mating pairs are 1/4 *AA* × *AA*, 1/2 *AA* × *aa*, and 1/4 *aa* × *aa*. These produce only *AA*, *Aa*, and *aa* progeny, respectively, so the progeny are 1/4 *AA*, 1/2 *Aa*, and 1/4 *aa*. In autogamy, the *Aa* genotypes become homozygous for either *A* or *a*, so the expected result of autogamy is 1/2 *AA* and 1/2 *aa*.

14.11 The only functional pollen is *a B*. All the progeny of the *AA bb* plants are *Aa Bb*, and those of the *aa BB* plants are *aa BB*.

14.12 The F₁ females will be heterozygous for nuclear genes, and possibilities (1) and (2) imply that these females should have some viable sons. The fact that all progeny are female supports the hypothesis of cytoplasmic inheritance.

14.13 Because the progeny coiling is determined by the genotype of the mother, the rightward-coiling snail has genotype *ss*. To explain the rightward coiling, the mother of this parent snail would have had the + allele and must in fact have been +/*s*.

14.14 Among Africans, the average time traversed in tracing the ancestry from a present-day mitochondrial DNA back to a common ancestor and forward again to another present-day mitochondrial DNA is between 78 × 1500 years and 78 × 3000 years, or 117,000 to 234,000 years; the backward and forward times are equal, so the person having the common ancestor of the mitochondrial DNA lived between 117,000/2 = 58,500 years ago and 234,000/2 = 117,000 years ago. Among Asians, similar calculations result in a range of 43,500 to 87,000 years for the common ancestor of mitochondrial DNA. Among Caucasians, the calculated range is 30,000 to 60,000 years.

14.15 The accompanying table shows the percent of restriction sites shared by each pair of mitochondrial DNA molecules.

	1	2	3	4	5
1	100	0	80	10	0
2		100	0	10	100
3			100	11	0
4				100	10

The entries in the table may be explained by example. Molecules 3 and 4 together have 9 restriction sites, of which 1 is shared, so the entry in the table is 1/9 = 11 percent. The diagonal entries of 100 mean that each molecule shares 100 percent of the restriction sites with itself. The two most closely related molecules are 2 and 5 (in fact, they are identical), so these must be grouped together. The two next closest are 1 and 3, so these also must be grouped together. The 2–5 group shares an average of 10 percent restriction sites with molecule 4, and the 1–3 group shares an average of 10.5 percent restriction sites [that is, (10 + 11)/2] with molecule 4. Therefore, the 2–5 group and the 1–3 group are about equally distant from molecule 4. The diagram shows the inferred relationships among the species. Data of this type do not indicate where the "root" should be positioned in the tree diagram.

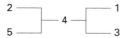

Chapter 15

15.1 Adults 2/3 *Aa* and 1/3 *aa*. It is not necessary to assume random mating, because the only fertile matings are *Aa* × *Aa*. These produce 1/4 AA, 1/2 *Aa*, and 1/4 *aa* zygotes, but the *AA* zygotes die, leaving 2/3 *Aa* and 1/3 *aa* adults.

15.2 The harmful recessive allele will have a smaller equilibrium frequency in the X-linked case, because the allele is exposed to selection in hemizygous males.

15.3 The 10 *AA* individuals have 20 *A* alleles, the 15 *Aa* have 15 *A* and 15 *a*, and the 4 *aa* have 8 *a*. Altogether there are 35 *A* and 23 *a* alleles, for a total of 58. The allele frequency of *A* is 35/58 = 0.603, and that of *a* is 23/58 = 0.397.

15.4 The numbers of alleles of each type are *A*, 2 + 2 + 5 + 1 = 10; *B*, 5 + 12 + 12 + 10 = 39; *C*, 10 + 5 + 5 + 1 = 21. The total number is 70, so the allele frequencies are *A*, 10/70 = 0.14; *B*, 39/70 = 0.56; and *C*, 21/70 = 0.30.

15.5 **(a)** Two bands indicate heterozygosity for two different VNTR alleles. **(b)** A person who is homozygous for a VNTR allele will have a single VNTR band. (An alternative explanation is also possible: Some VNTR alleles are present in small DNA fragments that migrate so fast that they are lost from the gel; a heterozygote for one of these alleles will have only one band—the one corresponding to the larger DNA fragment—because the band corresponding to the smaller DNA fragment will not be detected. **(c)** VNTR alleles are codominant because both are detected in heterozygotes. **(d)** Because none of the bands in the four pairs of parents have the same electrophoretic mobility, the parents of each child can be identified as the persons who share a single VNTR band with the child. Child A has parents 7 and 2; child B has parents 4 and 1; child C has parents 6 and 8; and child D has parents 3 and 5.

15.6 Allele 5.7 frequency = (2 × 9 + 21 + 15 + 6)/200 = 0.30; allele 6.0 frequency = (2 × 12 + 21 + 18 + 7)/200 = 0.35; allele 6.2 frequency = (2 × 6 + 15 +18 + 5)/200 =

0.25; allele 6.5 frequency = $(2 \times 1 + 6 + 7 + 5)/200 = 0.10$.

15.7 If the genotype frequencies satisfy the Hardy-Weinberg principle, they should be in the proportions p^2, $2pq$, and q^2. In each case, p equals the frequency of AA plus one-half the frequency of Aa, and $q = 1 - p$. The allele frequencies and expected genotype frequencies with random mating are **(a)** $p = 0.50$, expected: 0.25, 0.50, 0.25; **(b)** $p = 0.635$, expected: 0.403, 0.464, 0.133; **(c)** $p = 0.7$, expected : 0.49, 0.42, 0.09; **(d)** $p = 0.775$, expected: 0.601, 0.349, 0.051; **(e)** $p = 0.5$, expected: 0.25, 0.50, 0.25. Therefore, only (a) and (c) fit the Hardy-Weinberg principle.

15.8 $p^2 = 0.09$, and so $p = (0.09)^{1/2} = 0.30$. All other alleles have a combined frequency of $1 - 0.30 = 0.70$.

15.9 The frequency of the recessive allele is $q = (1/14,000)^{1/2} = 0.0085$. Hence $p = 1 - 0.0085 = 0.9915$, and the frequency of heterozygotes is $2pq = 0.017$ (about 1 in 60).

15.10 Because the frequency of homozygous recessives is 0.10, the frequency of the recessive allele q is $(0.10)^{1/2} = 0.316$. This means p $= 1 - 0.316 = 0.684$, so the frequency of heterozygotes is $2 \times 0.316 \times 0.684 = 0.43$. - Because the proportion 0.90 of the individuals are non-dwarfs, the frequency of heterozygotes among nondwarfs is $0.43/0.90 = 0.48$.

15.11 (a) Because the proportion 0.60 can germinate, the proportion 0.40 that cannot are the homozygous recessives. Thus the allele frequency q of the recessive is $(0.4)^{1/2} = 0.63$. This implies that $p = 1 - 0.63 = 0.37$ is the frequency of the resistance allele. **(b)** The frequency of heterozygotes is $2pq = 2 \times 0.37 \times 0.63 = 0.46$. The frequency of homozygous dominants is $(0.37)^2 = 0.14$, and the proportion of the surviving genotypes that are homozygous is $0.14/0.60 = 0.23$.

15.12 The bald and nonbald alleles have frequencies of 0.3 and 0.7, respectively. Bald females are homozygous and occur in the frequency $(0.3)^2 = 0.09$; nonbald females occur with frequency 0.91. Bald males are homozygous or heterozygous and occur with frequency $(0.3)^2 + 2(0.3)(0.7) = 0.51$; nonbald males occur with frequency 0.49.

15.13 $I^A I^A$, $(0.16)^2 = 0.026$; $I^A I^O$, $2(0.16)(0.74) = 0.236$; $I^B I^B$, $(0.10)^2 = 0.01$; $I^B I^O$, $2(0.10)(0.74) = 0.148$; $I^O I^O$, $(0.74)^2 = 0.548$; $I^A I^B$, $2(0.16)(0.10) = 0.032$. The phenotype frequencies are O, 0.548; A, $0.026 + 0.236 = 0.262$; B, $0.01 + 0.148 = 0.158$; AB, 0.032.

15.14 (a) There are 17 A and 53 B alleles, and the allele frequencies are A, $17/70 = 0.24$; B, $53/70 = 0.76$. **(b)** The expected genotype frequencies are AA, $(0.24)^2 = 0.058$; AB, $2(0.24)(0.76) = 0.365$; BB, $(0.76)^2 = 0.578$. The expected numbers are 2.0, 12.8, and 20.2, respectively.

15.15 The frequency in males implies that $q = 0.2$. The expected frequency of the trait in females is $q^2 = 0.0004$. The

expected frequency of heterozygous females is $2pq = 2(0.98)(0.02) = 0.0392$.

15.16 The frequency of the yellow allele y is 0.2. In females, the expected frequencies of wildtype $(+/+$ and $y/+)$ and yellow (y/y) are $(0.8)^2 + 2(0.8)(0.2) = 0.96$ and $(0.2)^2 = 0.04$, respectively. Among 1000 females, there would be 960 wildtype and 40 yellow. The phenotype frequencies are very different in males. In males, the expected frequencies of wildtype $(+/Y)$ and yellow (y/Y) are 0.8 and 0.2, respectively, where Y represents the Y chromosome. Among 1000 males, there would be 800 wildtype and 200 yellow.

15.17 The frequency of heterozygotes is smaller by the amount $2pqF$, in which F is the inbreeding coefficient.

15.18 Inbreeding is suggested by a deficiency of heterozygotes relative to the $2pq$ expected with random mating. The suggestive populations are (d) with expected heterozygote frequency 0.349 and (e) with expected heterozygote frequency 0.5.

15.19 The allele frequency is the square root of the frequency of affected offspring among unrelated individuals, or $(8.5 \times 10^{-6})^{1/2} = 2.9 \times 10^{-3}$. With inbreeding, the expected frequency of homozygous recessives is $q^2(1 - F) + qF$, in which F is the inbreeding coefficient. When $F = 1/16$, the expected frequency of homozygous recessives is $(2.9 \times 10^{-3})^2(1 - 1/16) + (2.9 \times 10^{-3})(1/16) = 1.9 \times 10^{-4}$; when $F = 1/64$, the value is 5.3×10^{-5}.

15.20 (a) Here $F = 0.66$, $p = 0.43$, and $q = 0.57$. The expected genotype frequencies are AA, $(0.43)^2(1 - 0.66) + (0.43)(0.66) = 0.347$; AB, $2(0.43)(0.57)(1 - 0.66) = 0.167$; BB, $(0.57)^2(1 - 0.66) + (0.57)(0.66) = 0.487$. **(b)** With random mating, the expected genotype frequencies are AA, $(0.43)^2 = 0.185$; AB, $2(0.43)(0.57) = 0.490$; BB, $(0.57)^2 = 0.325$.

15.21 Use Equation (11) with $n = 40$, $p_0/q_0 = 1$ (equal amounts inoculated), and $p_{40}/q_{40} = 0.35/0.65$. The solution for w is $w = 0.9846$, which is the fitness of strain B relative to strain A, and the selection coefficient against B is $s = 1 - 0.9846 = 0.0154$.

15.22 (a) A fixed; **(b)** A lost; **(c)** stable equilibrium; **(d)** a lost.

15.23 An Aa female produces $1/2$ A and $1/2$ a gametes, and these yield heterozygous offspring if they are fertilized by a and by A, respectively. The overall probability of a heterozygous offspring is $(1/2)q + (1/2)p = (1/2)(p + q) = 1/2$.

15.24 The frequency of each allele is $1/n$, and there are n possible homozygotes. Therefore, the total frequency of homozygotes is $n \times (1/n)^2 = 1/n$, which implies that the total frequency of heterozygotes is $1 - 1/n$.

15.25 The frequencies are calculated as in the Hardy-Weinberg principle except that, for apparently homozygous alleles, $2p$ is used instead of p^2. With regard to the genotype (6, 9.3) for *HUMTHO1*, (10, 11) for *HUMFES*, (6, 6) for

$D12S67$, and (18, 24) for $D1S80$, the calculation is $2 \times 0.230 \times 0.310 \times 2 \times 0.321 \times 0.373 \times 2 \times 0.324 \times 2 \times 0.293 \times 0.335 = 0.0043$, or about one in 230 people. With regard to the genotype (8, 10) for $HUMTHO1$, (8, 13) for $HUMFES$, (1, 2) for $D12S67$, and (19, 19) for $D1S80$, the calculation is $2 \times 0.105 \times 0.002 \times 2 \times 0.007 \times 0.066 \times 2 \times 0.015 \times 0.005 \times 2 \times 0.011 = 1.28 \times 10^{-12}$, or about one in 8×10^{11} people. The point is that DNA typing is at its most powerful in discriminating among people who carry rare alleles.

Chapter 16

16.1 Genotype-environment interaction.

16.2 Since the variance equals the square of the standard deviation, the only possibilities are that the variance equals either 1 or 0.

16.3 (a) The mean or average. (b) The one with the greater variance is broader. (c) 68 percent within one standard deviation of the mean, 95 percent within two.

16.4 Broad-sense heritability is the proportion of the phenotypic variance attributable to differences in genotype, or $H^2 = \sigma_g^2/\sigma_t^2$, which includes dominance effects and interactions between alleles. Narrow-sense heritability is the proportion of the phenotypic variance due only to additive effects, or σ_a^2/σ_t^2, which is used to predict the resemblance between parents and offspring in artificial selection. If all allele frequencies equal 1 or 0, both heritabilities equal zero.

16.5 The F_1 generation provides an estimate of the environmental variance. The variance of the F_2 generation is determined by the genotypic variance and the environmental variance.

16.6 For fruit weight, $\sigma_e^2/\sigma_t^2 = 0.13$ is given. Because $\sigma_t^2 = \sigma_g^2 + \sigma_e^2$, then $H^2 = \sigma_g^2/\sigma_t^2 = 1 - \sigma_e^2/\sigma_t^2 = 1 - 0.13 = 0.87$. Therefore, 87 percent is the broad-sense heritability of fruit weight. The values of H^2 for soluble-solid content and acidity are 91 percent and 89 percent, respectively.

16.7 Estimates are as follows: mean, $104/10 = 10.4$; variance, $24.4/(10 - 1) = 2.71$; standard deviation $(2.71)^{1/2} = 1.65$.

16.8 An IQ of 130 is two standard deviations above the mean, so the proportion greater equals $(1 - 0.95)/2 = 2.5$ percent; 85 is one standard deviation below the mean, so the proportion smaller equals $(1 - 0.68)/2 = 16$ percent; the proportion above 85 equals $1 - 0.16 = 84$ percent, which can be calculated alternatively as $0.50 + 0.68/2 = 0.84$.

16.9 (a) To determine the mean, multiply each value for the pounds of milk produced (using the midpoint value) by the number of cows, and divide by 304 (the total number of cows) to obtain a value of the mean of 4032.9. The estimated variance is 693,633.8, and the estimated standard deviation, which is the square root of the variance, is 832.8. (b) The range that includes 68 percent of the cows is the mean minus the standard deviation to the mean plus the standard deviation, or $4033 - 833 = 3200$ to $4033 + 833 = 4866$. (c) Rounding to the nearest 500 yields 3000–5000. The observed number in this range is 239; the expected number is $0.68 \times 304 = 207$. (d) For 95 percent of the animals, the range is within two standard deviations, or 2367–5698. Rounding off to the nearest 500 yields 2500–6000. The observed number is 296, which compares well to the expected number of $0.95 \times 304 = 289$.

16.10 (a) Because 15 to 21 is the range of one standard deviation from the mean, the proportion is 68 percent. (b) Because 12 to 24 is the range of two standard deviations from the mean, the proportion is 95 percent. (c) A phenotype of 15 is one standard deviation below the mean; because a normal distribution is symmetrical about the mean, the proportion with phenotypes more than one standard deviation below the mean is $(1 - 0.68)/2 = 0.16$. (d) Those with 24 leaves or greater are more than two standard deviations above the mean, so the proportion is $(1 - 0.95)/2 = 0.025$. (e) Those with 21 to 24 leaves are between one and two standard deviations above the mean, and the expected proportion is $0.16 - 0.025 = 0.135$ (because 0.16 have more than 21 leaves, and 0.25 have more than 24 leaves).

16.11 (a) The genotypic variance is 0 because all F_1 mice have the same genotype. (b) With additive alleles, the mean of the F_1 generation will equal the average of the parental strains.

16.12 Among the F_1, the total variance equals the environmental variance, which is given as 1.46. Among the F_2, the total variance, given as 5.97, is the sum of the environmental variance (1.46) and the genotypic variance. Thus the genotypic variance is $5.97 - 1.46 = 4.51$. The broad-sense heritability is the genotypic variance (4.51) divided by the total variance (5.97), or $4.51/5.97 = 0.76$.

16.13 From the F_1 data, $\sigma_e^2 = (2)^2 = 4$; from the F_2 data, $\sigma_g^2 + \sigma_e^2 = (3)^2 = 9$; hence $\sigma_g^2 = 9 - 4 = 5$, and so $H^2 = 5/9 = 56$ percent.

16.14 Using Equation (4), we find that the minimum number of genes affecting the trait is $D^2/8\sigma_g^2 = (17.6)^2/(8 \times 0.88) = 44$.

16.15 $\sigma_g^2 = 0.0570 - 0.0144 = 0.0426$ and $D = 1.689 - (-0.137) = 1.826$. The minimum number of genes is estimated as $n = (1.826)^2/(8 \times 0.0426) = 9.8$.

16.16 Use Equation (6) with $M* - M = 50$ grams in each generation of selection, with M initially 700 grams. The expected gain in each generation of selection is $0.8 \times 50 = 40$ grams, so after five generations, the expected mean weight is $700 + 5 \times 40 = 900$ grams.

16.17 Use Equation (6) with $M = 10.8$, $M* = 5.8$, and $M' = 8.8$. The narrow-sense heritability is estimated as $h^2 = (8.8 - 10.8)/(5.8 - 10.8) = 0.40$.

16.18 (a) In this case, $M = 10.8$, $M* = 5.8$, and $M' = 9.9$. The estimate of h^2 is $(9.9 - 10.8)/(5.8 - 10.8) = 0.18$. (b) The result is consistent with the earlier result because of the differences in the variance. In the previous case, the additive variance was $0.4 \times 4 = 1.6$. In the present case, the phenotypic variance is 9. If the additive variance did not change, then the expected heritability would be $1.6/9 = 0.18$, which is that observed. This problem has been idealized somewhat; in real examples, the additive variance would also be expected to change with a change in environment.

16.19 The offspring are genetically related as half-siblings, and Table 16.3 says that the theoretical correlation coefficient is $h^2/4$.

16.20 Table 16.3 gives the correlation coefficient between first cousins as $h^2/8$, and so $h^2 = 8 \times 0.09 = 72$ percent.

16.21 The mean weaning weight of the 20 individuals is 81 pounds, and the standard deviation is 13.8 pounds. The mean of the upper half of the distribution can be calculated by using the relationship given in the problem, which says that the mean of the upper half of the population is $81 + 0.8 \times 13.8 = 92$ pounds. Using Equation (6), we find that the expected weaning weight of the offspring is $81 + 0.2(92 - 81) = 83.2$ pounds.

16.22 One approach is to use Equation (4) with $D = 2n - 0 = 2n$, from which $n = (2n)^2/8\sigma_g^2$, or $\sigma_g^2 = n/2$. A more direct approach uses the fact that with unlinked, additive genes, the total genetic variance is the summation of the genetic variance resulting from each gene. With 1 gene, the genetic variance equals $1/2$, so with n genes, the total genetic variance equals $n/2$.

16.23 (a) This is a nonrecombinant class and so should have the genotype Q/q for the QTL affecting egg lay; the average should be $(300 + 270)/2 = 285$ eggs per hen. (b) This is a crossover class of progeny. In view of the relative distances of the two intervals of $0.07 : 0.03$, $7/10$ of the crossovers should occur in the region $Cx-Q$ and $3/10$ in the region $Q-Dx$. The expected genotypes with respect to the QTL are $7/10$ q/q (crossing-over in the $Cx-Q$ region) and $3/10$ Q/q (crossing-over in the $Q-Dx$ region). The average phenotype is expected to be $(7/10) \times 270 + (3/10) \times 285 = 274.5$ eggs per hen. (c) This is also a crossover class, but it is the reciprocal of that in the previous part. In this case, the expected genotypes with respect to the QTL are $7/10$ Q/q (crossing-over in the $Cx-Q$ region) and $3/10$ q/q (crossing-over in the $Q-Dx$ region). The expected average egg lay is $(7/10) \times 285 + (3/10) \times 270 = 280.5$ eggs per hen. (d) These are again nonrecombinants with QTL genotype q/q, and the expected egg lay is 270 eggs per hen.

16.24 (a) The rare genetic disease is transmitted as a simple Mendelian dominant. Genotype Q/q is affected and q/q is not. Because the disease is rare, the frequency of Q/Q homozygous genotypes is negligible. (b) Male IV-2 has the genotype Q/q, so the probability that he has an affected daugher is $1/2$. (c) Male IV-2 has the genotype H/H for gene

2; with random mating, the probability that his daughter will have the genotype H/F is equal to the probability that a randomly chosen egg contains the F allele, which is given in the population-frequency table as 0.05, or 5 percent. (d) The genotypes of the individuals in the pedigree are as shown in the accompanying table. At this stage in the analysis, we do not yet know anything about linkage, but we can, in some cases, specify the linkage phase in terms of which alleles were transmitted together in a single gamete. When known, the alleles transmitted in a single gamete are grouped together in parentheses. Also, the order in which the genes are listed is arbitrary. (The two columns on the right are referred to in the next part of the analysis.)

Genotypes of individuals in pedigree

I-1	$(q\,A\,E)/(q\,A\,E)$		
I-2	$Q/q\,;\,C/D\,;\,F/G$		
II-1	$(q\,A\,E)/(Q\,D\,F)$	NRC	
II-2	$(q\,A\,H)/(q\,B\,H)$		
II-3	$(q\,A\,E)/(q\,C\,G)$	NRC	
II-4	$(q\,A\,G)/(q\,A\,H)$		
II-5	$(q\,A\,E)/(Q\,D\,G)$	NRC	
II-6	$(q\,A\,H)/(q\,B\,H)$		
II-7	$(q\,A\,E)/(Q\,D\,F)$	NRC	
II-8	$(q\,A\,E)/(q\,C\,G)$	NRC	
II-9	$q/q\,;\,A/B\,;\,E/G$		
III-1	$(q\,B\,H)/(q\,B\,H)$		
III-2	$(q\,B\,H)/(Q\,D\,E)$	NRC	REC
III-3	$(q\,A\,H)/(Q\,D\,E)$	NRC	REC
III-4	$(q\,A\,H)/(q\,A\,F)$	NRC	REC
III-5	$(q\,A\,G)/(q\,A\,E)$		NRC
III-6	$(q\,A\,H)/(q\,A\,G)$		REC
III-7	$(q\,B\,H)/(q\,A\,E)$	NRC	NRC
III-8	$(q\,A\,H)/(Q\,D\,G)$	NRC	NRC
III-9	$q/q\,;\,B/C\,;\,E/H$		
III-10	$(q\,B\,H)/(q\,A\,G)$	NRC	REC
III-11	$(q\,A\,E)/(q\,A\,E)$		NRC
III-12	$(q\,B\,G)/(q\,A\,G)$		REC
III-13	$q/q\,;\,B/C\,;\,E/G$		Undeterminable
IV-1	$(q\,B\,H)/(Q\,D\,E)$	NRC	NRC
IV-2	$(q\,B\,H)/(Q\,D\,H)$	NRC	REC
IV-3	$(q\,B\,H)/(q\,B\,E)$	NRC	REC
IV-4	$(q\,B\,E)/(Q\,D\,G)$	NRC	NRC
IV-5	$(q\,C\,H)/(q\,A\,H)$	NRC	NRC
IV-6	$(q\,C\,H)/(Q\,D\,G)$	NRC	NRC

(e) There is evidence of linkage between genes 1 and 3. In particular, the allele Q is always transmitted with allele D, starting with male I-2 at the top of the pedigree. If we assume that this male has the genotype $(Q\,D)/(q\,C)$, then throughout the pedigree, there are 17 individuals in which recombination between genes 1 and 3 could have been detected. These cases are identified in the third column of the foregoing table. All 17 are nonrecombinant, so the recombination fraction between genes 1 and 3 in these data is 0/17. As far as one can discern from the pedigree alone, the Q and D alleles could be identical. However, the population-frequency table gives the allele frequency of D as 5 percent, and the text says that the Q allele is rare. Hence most D alleles in the population must be on chromosomes carrying q, so that D and Q are separable. Turning to genes 1 and 2, there are 17 people in whom recombination could be detected. These are identified in the fourth column in the foregoing table. Among the 17 cases, one is undeterminable, 8 are recombinant, and 8 are nonrecombinant. Genes 1 and 2 are therefore unlinked. They are probably on different chromosomes, but they could also be very far apart on the same chromosome.

Chapter 17

17.1 There are numerous chemoreceptors in *E. coli*, each of which interacts with a specific set of chemosensors for different small molecules. Mutations in a particular chemoreceptor affect the chemotactic response to all molecules whose chemosensors feed into the chemoreceptor. Hence cells with mutant chemosensors are multiply nonchemotactic.

17.2 The mutant would be unable to make P-CheA and, therefore, P-CheY, so the phenotype would be that of continuous swimming.

17.3 Extreme *cheB* mutants lack the CheB methylase and have high levels of chemoreceptor methylation; their phenotype is that of continuous tumbling, because the methylated chemoreceptors require high concentrations of attractant to be stimulated (which is why they swim in response to very strong attractant stimuli). Extreme *cheR* mutants cannot methylate the chemoreceptors and so are highly sensitive to stimuli; their phenotype is that of continuous swimming, but very strong repellent stimuli overcome the effects of under-methylation and result in tumbling.

17.4 Phosphate groups are transferred among several key components (CheA, CheY, and CheB) in the chemotaxis-signaling pathway. Stimulated chemoreceptors promote auto-phosphorylation of CheA, in which a phosphate group is transferred from ATP to a specific amino acid residue in cheA. The phosphate group can be transferred readily from P-CheA to CheB or CheY. The P-CheY promotes clockwise flagellar rotation and causes the bacteria to tumble. If phosphate transfers were inhibited, the lack of P-CheY would result in failure to switch from swimming behavior to tumbling behavior, and chemotaxis would be impossible. Because protein phosphorylation is common to all of the chemoreceptors, inhibition of phosphorylation would result in a generally nonchemotactic phenotype.

17.5 Male mating success is more likely to be affected by mutations in *per* because these mutations affect the courtship song, and it is the male who must sing acceptably to the female for successful courtship.

17.6 Mate a mutagen-treated male with an attached-X female and select the male progeny. Each male progeny carries a different, mutagen-treated X chromosome. Mate each of the male progeny with an attached-X female; the progeny of this mating consist of attached-X daughters and XY sons, and the X chromosome in each son is a replica of the single mutagen-treated X chromosome present in the father.

17.7 Either genetic or molecular tests would reveal whether the mutant allele is in *per*. The first step in genetic analysis is genetic mapping. If the mutant does not map near the chromosomal location of *per*, then it must be a different gene; if it does map near *per*, then a complementation test should be carried out with *per*^{0}. A molecular analysis would entail determining the DNA sequence of all or part of the *per* gene and comparing it with wildtype. If the sequences are identical, then the mutation must be in a gene different from *per*; if the sequences differ, then separate tests must be carried out to demonstrate that the mutant phenotype results from the difference in DNA sequence. The new mutation could easily be in a gene different from *per*, because courtship behavior is a complex trait influenced by many different genes.

17.8 (a) A short-period rhythm of approximately 40 seconds. **(b)** A long-period rhythm of approximately 90 seconds. **(c)** Because *per*^{S} shortens the period and *per*^{L} lengthens it, the *per*^{S}/*per*^{L} heterozygote would be expected to exhibit an intermediate period of approximately $(40 + 90)/2 = 65$ seconds, which is very close to the period of the wildtype courtship song.

17.9 A fly with a mutation that eliminates only one of the wavelength preferences would be classified as specifically nonphototactic. The mutation might affect the photopigments necessary for the detection of light of a particular wavelength, or it might affect the photoreceptor cells containing the particular visual pigment. A fly with a mutation that eliminates the phototactic response to both wavelengths is generally nonphototactic. The mutation might affect components of the visual pathway common to both 480-nm and 370-nm light—for example, a component common to all photoreceptor cells that is necessary for stimulation by light.

17.10 It is an example of sensory adaptation, because intense light of a given wavelength results in a temporarily decreased response to light of the same wavelength.

17.11 It is unlikely that an inbred strain would produce selected lines that differed significantly in maze-learning ability (or in any other trait), because genetic variation is essential for the success of artificial selection.

17.12 Variability in the sense of smell in the initial population might result in variability in the ability of the rats to detect the food at the end of the maze. Therefore, differences in maze-learning ability may result from differences in the ability of the animals to detect the food used to motivate them to run through the maze. In this example, maze-learning ability is not a good measure of intelligence. You could control for variation in the sense of smell by using odorless food, masking the smell of the food, smothering the entire maze with the smell of the food, or using a nonfood goal.

17.13 It is an example of genotype-environment interaction, because the relative performances of the two populations are reversed merely by changing the environment in which the test is carried out.

17.14 The *waltzer* allele is recessive to the wildtype allele.

17.15 Under conditions in which behavioral choices are strongly influenced by the ability to distinguish among different colors—for example, the ability to distinguish red and green in obeying traffic lights.

17.16 The test would take advantage of the size differences between the wildtype and mutant alleles. The DNA of parents and child would be digested with a restriction enzyme that cleaves the DNA at sites flanking the CAG repeat. The resulting fragments would be separated by size with electro-phoresis, and the fragments of interest would be identified by hybridization using a radioactive probe for the CAG-containing fragment. The mutant allele should be longer than the wildtype allele because of the greater number of CAG repeats. Because the mutant allele is dominant, any mating between heterozygous mutant and homozygous wildtype is expected to result in 50 percent affected offspring.

17.17 This example illustrates some of the complexities of behavior genetics in human beings. The correlation between the *MAO* gene and the behavioral disorder in the large pedigree is suggestive. However, single pedigrees may be misleading, because a correlation with a genetic marker may result from chance. Even if the disorder in this example is genetically determined, the cause need not be *MAO*. The disorder may result from a different gene that is genetically linked to *MAO*, or the large pedigree may also contain other genes influencing the disorder, and the *MAO* mutation may be only a contributing factor. The finding that unrelated men with similar behavioral disorders may have abnormal levels of monoamine oxidase is also suggestive. However, without data on the distribution of enzyme levels among normal men and among those with the behavioral disorder, it is impossible to evaluate the strength of the evidence.

Chapter 1 The Molecular Basis of Heredity and Variation

S1.1 Classify each of the following statements as true or false.

- **(a)** Each gene is responsible for only one visible trait.
- **(b)** Every trait is potentially affected by many genes.
- **(c)** The sequence of nucleotides in a gene specifies the sequence of amino acids in a protein encoded by the gene.
- **(d)** There is one-to-one correspondence between the set of codons in the genetic code and the set of amino acids encoded.

S1.2 Is the following statement correct? "Heredity is solely responsible for the manifestation of anemia." Explain.

S1.3 When a human trait is genetic in origin (for example, a disorder caused by an inborn error of metabolism), does this mean that the trait cannot be influenced by the environment?

S1.4 What is meant by the term "pleiotropy" and what is a "pleiotropic effect"? What is the relevance of sickle-cell anemia to this phenomenon?

S1.5 What facts known prior to Avery, MacLeod, and McCarty's experiment indirectly suggested that DNA might play a role in the genetic material?

S1.6 Prior to the experiments of Avery, MacLeod, and McCarty and of Hershey and Chase, why did most biologists and biochemists believe that proteins were probably the genetic material?

S1.7 Each of the following statements pertains to the experiments of Avery, MacLeod, and McCarty on *Streptococcus pneumoniae*. Classify each statement as true or false. Which single statement is a conclusion that can be drawn from the experiments themselves?

- **(a)** The polysaccharide capsule is required for virulence (the ability to infect and kill mice).
- **(b)** The polysaccharide capsule is not important for virulence.
- **(c)** DNA encodes the sequence of sugars in the polysaccharide capsule.
- **(d)** DNA is the polysaccharide capsule.
- **(e)** Proteins are not the molecules that carry the hereditary information for synthesis of the polysaccharide capsule.

S1.8 What critical finding in Hershey and Chase's experiment enabled them to interpret the results as a proof that the genetic material in bacteriophage T2 is DNA?

S1.9 Define the following terms: replication, transcription, translation, mutation, natural selection.

S1.10 What can we infer about the function of DNA from its structure?

S1.11 Is it correct to say that DNA is always the genetic material?

S1.12 One of the early models for the structure of DNA was the so-called tetranucleotide hypothesis. According to this hypothesis, one of each of the four nucleotides containing adenine, thymine, guanine, and cytosine joined together to form a covalently linked planar unit (the "tetranucleotide"); these units were, in turn, linked together to yield a repeating polymer of the tetranucleotide. Why could a molecule with this structure never serve as the genetic material?

S1.13 What are the major chemical components found in DNA?

S1.14 The repeating unit of DNA is:

- **(a)** a ribonucleotide.
- **(b)** an amino acid.
- **(c)** a nitrogenous base.
- **(d)** a sugar, deoxyribose.
- **(e)** none of the above.

S1.15 The two strands of a duplex DNA molecule are said to be complementary in base sequence. Which of the following statements are implied by the complementary nature of the strands?

- **(a)** The two strands have the same sequence of bases.
- **(b)** The base A in one strand is paired with a T in the other strand.
- **(c)** One strand consists of A's paired with T's; the other strand consists of G's paired with C's.
- **(d)** The base G in one strand is paired with a C in the other strand.
- **(e)** The two strands have the same sequence of bases when read in opposite directions.

S1.16 In the following statements about double-stranded DNA, square brackets indicate the number of molecules. For example, [A] means the number of molecules of the base adenine, and [deoxyribose] means the number of molecules of 2' deoxyribose. Classify each of the statements as true or false.

(a) [A] = [G]
(b) [A] = [C]
(c) [A] = [T]
(d) [A] + [G] = [T] + [C]
(e) [deoxyribose] = [phosphate]

S1.17 When the base composition of double-stranded DNA from *Micrococcus luteus* was determined, 37.5 percent of the bases were found to be cytosine. What is the percentage of adenine in the DNA of this organism?

S1.18 Which nucleic acid base is found only in DNA? Which is found only in RNA?

S1.19 What are three principal structural differences between RNA and DNA?

S1.20 What is the most likely reason for the participation of RNA intermediates in the process of information flow from DNA to protein?

S1.21 What are the consequences of the Glu → Val replacement at amino acid position 6 in the beta-globin chain?

S1.22 Using the *vermilion* and *cinnabar* mutations affecting eye color in *D. melanogaster* as an example, explain what is meant by an epistatic interaction between genes.

S1.23 Can all the existing diversity of life be explained as a result of adaptive evolution brought about by natural selection?

Chapter 2   Principles of Genetic Transmission

S2.1 Define the following terms: genotype, phenotype, allele, dominant, recessive.

S2.2 How many different alleles of a particular gene may exist in a population of organisms? How many alleles of a particular gene can be present in an individual organism?

S2.3 A breeder of Irish setters has a particularly valuable male show dog descended from a famous female named Rheona Didona, which was known to carry a harmful recessive gene for blindness caused by atrophy of the retina. Before she breeds the dog, she wants some assurance that it does not carry this harmful allele. How would you advise her to test the dog?

S2.4 Any of a large number of different recessive mutations cause profound hereditary deafness. Children homozygous for such mutations are often found in small, isolated communities in which matings tend to be within the group. In these instances, a mutation that originates in one person can be transmitted to several members of the population in future generations and become homozygous through a marriage

between group members who are carriers. A deaf man from one community marries a deaf woman from a different, unrelated community. Both of them have deaf parents. The marriage produces three children, all with normal hearing. How can you explain this result?

S2.5 Consider a human family with four children, and remember that each birth is equally likely to result in a boy or a girl.

(a) What proportion of sibships will include at least one boy?
(b) What fraction of sibships will have the gender order GBGB?

S2.6 Mendel's cross of homozygous round yellow with homozygous wrinkled green peas yielded the phenotypes 9/16 round yellow, 3/16 round green, 3/16 wrinkled yellow, and 1/16 wrinkled green progeny in the F_2 generation. If he had testcrossed the round green progeny individually, what proportion of them would have yielded some progeny with round yellow seeds?

(a) 1/4
(b) 1/3
(c) 1/2
(d) 3/4
(e) 0

S2.7 Intestinal lactase deficiency (ILD) is a common inborn error of metabolism in humans; affected persons are homozygous for a recessive allele of a gene on chromosome 2. A woman and her husband, both phenotypically normal, had a daughter with ILD. She married a normal man and had three sons. One had ILD and two were normal. Recognizing that each succeeding child's genotype is not influenced by the genotypes of older siblings, determine the probability that the couple's next child will have ILD.

(a) 0.25
(b) 0.50
(c) 0.75
(d) 1.00
(e) 0.0625

S2.8 Seborrheic keratosis is a rare hereditary skin condition due to an autosomal dominant mutation. Affected people have skin marked with numerous small, sharply margined, yellowish or brownish areas covered with a thin, greasy scale. A man with keratosis marries a normal woman, and they have three children. What is the chance that all three are normal? What is the chance that all three are affected?

S2.9 In *Drosophila*, the dominant allele *Cy* (*Curly*) results in curly wings. The cross *Cy+* × *Cy+* (where + represents the wildtype allele of *Cy*) results in a ratio of 2 curly : 1 wildtype F_1 progeny. The cross between the curly F_1 progeny also gives a ratio of 2 curly : 1 wildtype F_2 progeny. How can this result be explained?

S2.10 White Leghorn chickens are homozygous for a dominant allele, *C*, of a gene responsible for colored feathers, and also for a dominant allele, *I*, of an independently segregating gene that prevents the expression of *C*. The White Wyandotte breed is homozygous recessive for both genes *cc ii*. What proportion of the F₂ progeny obtained from mating White Leghorn × White Wyandotte F₁ hybrids would be expected to have colored feathers?

S2.11 The tailless trait in the Manx cat is determined by the alleles of a single gene. In the cross Manx × Manx, both tailless and tailed progeny are produced, in a ratio of 2 tailless : 1 tailed. All tailless progeny from this cross, when mated with tailed, produce a 1 : 1 ratio of tailless to tailed progeny.

(a) Is the allele for the tailless trait dominant or recessive?
(b) What genetic hypothesis can account for the 2 : 1 ratio of tailless : tailed and the result of the testcrosses with the tailless cats?

S2.12 Absence of the enzyme alpha-1-antitrypsin is associated with a very high risk of developing emphysema and early death due to damaged lungs. Enzyme deficiency is usually due to homozygosity for an allele denoted *PI-Z*. A phenotypically normal man whose father died of the emphysema associated with homozygosity for *PI-Z* marries a phenotypically normal woman with a negative family history for the trait. They consider having a child.

(a) Draw the pedigree as described above.
(b) If the frequency of heterozygous *PI-Z* in the population is 1 in 30, what is the chance that the first child will be homozygous *PI-Z*?
(c) If the first child is homozygous *PI-Z*, what is the probability that the second child will be either homozygous or heterozygous for the normal allele?

S2.13 Plants of the genotypes *Aa Bb cc* and *aa Bb Cc* are crossed. The three genes are on different chromosomes. What proportion of the progeny will have the phenotype *A− B− C−*, and what fraction of the *A− B− C−* plants will have the genotype *Aa Bb Cc*?

(a) 1/16; all
(b) 3/8; 2/3
(c) 3/16; 2/3
(d) 1/2; 1/2
(e) 1/2; all

S2.14 In the cross *Aa Bb Cc Dd* × *Aa Bb Cc Dd*, in which all four genes undergo independent assortment, what proportion of offspring are expected to be heterozygous for all four genes? What is the answer if the cross is *Aa Bb Cc Dd* × *Aa Bb cc dd*?

S2.15 In the cross *AA Bb Cc* × *Aa Bb cc*, where all three genes assort independently and each uppercase allele is dominant, what proportion of the progeny are expected to have the phenotype *A− bb cc*?

(a) 1/64
(b) 1/32
(c) 1/16

(d) 1/8
(e) 1/4

S2.16 In the cross *Aa Bb* × *Aa Bb*, what fraction of the progeny will be homozygous for one gene and heterozygous for the other if the two genes assort independently?

(a) 1/16
(b) 1/8
(c) 1/2
(d) 1/4
(e) 3/4

S2.17 The pedigree in the accompanying illustration shows the inheritance of coat color in a group of cocker spaniels. The coat colors and genotypes are indicated in the key.

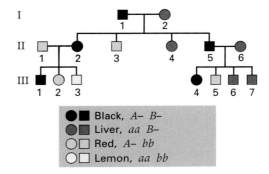

(a) Specify in as much detail as possible the genotype of each dog in the pedigree.
(b) What are the possible genotypes of animal III-4, and what is the probability of each genotype?
(c) If a single pup is produced from the mating of III-4 × III-3, what is the probability that the pup will be lemon?

S2.18 In terms of the type of epistasis indicated, would you say that a 27 : 37 ratio has more in common with a 9 : 7 ratio or with a 15 : 1 ratio? Explain your argument.

S2.19 A trihybrid cross *AA BB rr* × *aa bb RR* is made in a plant species in which *A* and *B* are dominant to their respective alleles but there is no dominance between *R* and *r*. Assuming independent assortment, consider the F₂ progeny from this cross.

(a) How many phenotypic classes are expected?
(b) What is the probability of the parental *aa bb RR* genotype?
(c) What proportion of the progeny would be expected to be homozygous for all three genes?

S2.20 Four babies accidentally become mixed up in a maternity ward. The ABO types of the four babies are known to be O, A, B, and AB, and those of the four sets of parents are shown below. One of the attendants insists that DNA typing will be necessary to straighten out the mess, but another claims that the ABO blood types alone are sufficient. To determine who is correct, try to match each baby with its correct set of parents.

(a) AB × O
(b) A × O
(c) A × AB
(d) O × O

S2.21 For each of the following F_2 ratios, some indicating epistasis, what phenotypic ratio would you expect in a test-cross of the F_1 dihybrid?

(a) 15 : 1
(b) 13 : 3
(c) 9 : 4 : 3
(d) 9 : 6 : 1
(e) 12 : 3 : 1
(f) 9 : 3 : 3 : 1
(g) 3 : 6 : 3 : 1 : 2 : 1
(h) 1 : 2 : 1 : 2 : 4 : 2 : 1 : 2 : 1

S2.22 In plants of the genus *Primula*, the *K* locus controls synthesis of a compound called malvidin. Two plants heterozygous at the *K* locus were crossed, producing the following distribution of progeny:

> 1010 Make malvidin
> 345 Do not make malvidin

You wish to determine if the *D* locus (located on a different chromosome from *K*), affects production of malvidin as well. True-breeding plants that make malvidin (*KK dd*) were crossed to true-breeding plants that do not make malvidin (*kk DD*). All F_1 plants failed to produce malvidin. F_1 heterozygotes were self-fertilized, and the F_2 progeny, assayed for malvidin synthesis, yielded the following distribution:

> 522 Make malvidin
> 2270 Do not make malvidin

(a) Write the genotypes and the corresponding phenotypes of all the F_2 progeny obtained.
(b) Does the *D* locus affect malvidin synthesis? What is the basis of your conclusion?

S2.23 From the complementation data in the table below, assign the ten mutations *a1* through *a10* to complementation groups. Use the complementation groups to deduce the expected results of the complementation tests indicated as missing data (denoted by the question marks).

Chapter 3   Genes and Chromosomes

S3.1 Classify each statement as true or false as it applies to mitosis.

(a) DNA synthesis takes place in prophase; chromosome distribution to daughter cells takes place in metaphase.
(b) DNA synthesis takes place in metaphase; chromosome distribution to daughter cells takes place in anaphase.
(c) Chromosome condensation takes place in prophase; chromosome distribution to daughter cells takes place in anaphase.
(d) Chromosomes pair in metaphase.

S3.2 Which of the following statements are true?

(a) A bivalent contains one centromere, two sister chromatids, and one homolog.
(b) A bivalent contains two centromeres, four sister chromatids, and two homologs.
(c) A bivalent contains two linear, duplex DNA molecules.
(d) A bivalent contains four linear, duplex DNA molecules.
(e) A bivalent contains eight linear, duplex DNA molecules.

S3.3 Which of the following statements correctly describes a difference between mitosis and meiosis?

(a) Meiosis includes two nuclear divisions, mitosis only one.
(b) Bivalents appear in mitosis but not in meiosis.
(c) Sister chromatids appear in meiosis but not in mitosis.
(d) Centromeres do not divide in meiosis; they do in mitosis.

Complementation Data
Problem S2.23

	a1	a2	a3	a4	a5	a6	a7	a8	a9	a10
a1	−	+	+	+	−	+	+	+	?	+
a2		−	+	?	+	+	+	−	?	+
a3			−	+	+	−	?	+	+	+
a4				−	+	+	?	−	+	+
a5					−	+	+	+	−	+
a6						−	−	+	+	+
a7							−	+	+	+
a8								−	+	+
a9									−	+
a10										−

S3.4 In which of the following situations can the result of segregation be observed phenotypically?

(a) In the haploid products of meiosis or the cells derived from them by mitosis, provided that these can grow and divide (as in certain fungi).
(b) In the testcross progeny of a heterozygous diploid organism.
(c) In the F_2 progeny of genetically different diploid parents.
(d) Statements (a), (b), and (c).
(e) Statements (a) and (b) only.

S3.5 Classify each of the following statements as true or false as it applies to the segregation of alleles for a genetic trait in a diploid organism.

(a) There is a 1 : 1 ratio of meiotic products.
(b) There is a 3 : 1 ratio of meiotic products.
(c) There is a 1 : 2 : 1 ratio of meiotic products.
(d) There is a 3 : 1 ratio of F_2 genotypes.
(e) All of the above.

S3.6 What is a chiasma? In what stage of meiosis are chiasmata observed?

S3.7 When in meiosis does the physical manifestation of independent assortment take place?

S3.8 In the life cycle of corn, *Zea mays*, how many haploid chromosome sets are there in cells of the following tissues?

(a) Megasporocyte
(b) Megaspore
(c) Root tip
(d) Endosperm
(e) Embryo
(e) Microspore

S3.9 Somatic cells of the red fox, *Vulpes vulpes*, normally have 38 chromosomes. What is the number of chromosomes present in the nucleus in a cell of this organism in each of the following stages? (For purposes of this problem, count each chromatid as a chromosome in its own right.)

(a) Metaphase of mitosis
(b) Metaphase I of meiosis
(c) Telophase I of meiosis
(d) Telophase II of meiosis

S3.10 The terms "reductional division" and "equational division" are literally correct in describing the two meiotic divisions with respect to division of the centromeres. To assess the applicability of the terms to alleles along a chromosome, consider a chromosome arm in which exactly one exchange event takes place in prophase I. Denote as *A* the region between the centromere and the position of the exchange and as *B* the region between the position of the exchange and the tip of the chromosome arm. If "reductional" is defined as resulting in products that are genetically different and "equational" is defined as resulting in products that are genetically identical, what can you say about the "reduc-

tional" or "equational" nature of the two meiotic divisions with respect to the regions *A* and *B*?

S3.11 What differences are there between the sexes in the pattern of transmission of genes located on the autosomes versus those on the sex chromosomes? (Assume that females are the homogametic sex and males the heterogametic sex.)

S3.12 What observations about *Drosophila* provided proof of the chromosomal theory of heredity?

S3.13 People with the chromosome constitution 47, XXY are phenotypically males. A normal woman whose father had hemophilia mates with a normal man and produces an XXY son who also has hemophilia. What kind of nondisjunction can explain this result?

S3.14 Which of the following types of inheritance have the feature that an affected male has all affected daughters but no affected sons?

(a) Autosomal recessive
(b) Autosomal dominant
(c) Y-linked
(d) X-linked recessive
(e) X-linked dominant

S3.15 Mendel studied the inheritance of phenotypic characters determined by seven pairs of alleles. It is an interesting coincidence that the pea plant also has seven pairs of chromosomes. What is the probability that, if seven loci are chosen at random in an organism that has seven chromosomes, each locus is in a different chromosome? (*Note*: Mendel's seven genes are actually located in only four chromosomes, but only two of the genes—tall versus short plant and smooth versus constricted pod—are close enough together that independent assortment would not be observed; he apparently never examined these two traits for independent assortment.)

S3.16 *Drosophila montana* has a somatic chromosome number of 12. Say the centromeres of the six homologous pairs are designated as A/a, B/b, C/c, D/d, E/e , and F/f.

(a) How many different combinations of centromeres can be produced in meiosis?
(b) What is the probability that a gamete will contain only centromeres designated by capital letters?

S3.17 How many different genotypes are possible for

(a) an autosomal gene with four alleles?

(b) an X-linked gene with four alleles?

S3.18 For an autosomal gene with *m* alleles, how many different genotypes are there?

S3.19 Consider the mating $Aa \times Aa$.

(a) What is the probability that a sibship of size 8 will contain no *aa* offspring?

(b) What is the probability that a sibship of size 8 will contain a perfect Mendelian ratio of 3 $A-$: 1 aa?

S3.20 Matings between two types of guinea pigs with normal phenotypes produce 64 offspring, 11 of which showed a hyperactive behavioral abnormality. The other 53 offspring were behaviorally normal. Assume that the parents were heterozygous.

(a) Do these data fit the model that hyperactivity is caused by a recessive allele of a single gene?
(b) Do they fit the model that hyperactivity is caused by simultaneous homozygosity for recessive alleles of each of two unlinked genes?

S3.21 In a χ^2 test with 1 degree of freedom, the value of χ^2 that results in significance at the 5 percent level ($P = 0.05$) is approximately 4. If a genetic hypothesis predicts a 1 : 1 ratio of two progeny types, calculate what deviations from the expected 1 : 1 ratio yield $P = 0.05$ (and rejection of the genetic hypothesis) when

(a) the total number of progeny equals 50
(b) the total number of progeny equals 100

Chapter 4   Gene Linkage and Chromosome Mapping

S4.1 After a large number of genes have been mapped, how many linkage groups will be found in each of the following?

(a) The bacterial cell *Salmonella typhimurium*, with a single, circular DNA molecule
(b) The haploid yeast *Saccharomyces cerevisiae*, with 16 chromosomes
(c) The common wheat *Triticum vulgare*, with 42 chromosomes per somatic cell

S4.2 Does physical distance always correlate with map distance?

S4.3 In genetic linkage studies, what is the definition of a map unit?

S4.4 Distinguish between the terms "chromosome interference" and "chromatid interference." Which is measured by the coincidence?

S4.5 What does the coefficient of coincidence measure?

(a) The frequency of double crossovers
(b) The ratio of double crossovers to single crossovers
(c) The ratio of single crossovers in two adjacent regions
(d) The ratio of double crossovers to the number expected if there were no interference
(e) The ratio of recombinant to parental progeny

S4.6 A coefficient of coincidence of 0.25 means that:

(a) the frequency of double crossovers is 1/4.

(b) the frequency of double crossovers is 1/4 of the number expected if there were no interference.
(c) there were four times as many single crossovers as double crossovers.
(d) there were four times as many single crossovers in one region as there were in an adjacent region.
(e) there were four times as many parental as recombinant progeny.

S4.7 In wheat, normal plant height requires the presence of either or both of two dominant alleles, A and B. Plants homozygous for both recessive alleles ($a\ b/a\ b$) are dwarfed but otherwise normal. The two genes are on the same chromosome and recombine with a frequency of 16 percent. From the cross $A\ b/a\ B \times A\ B/a\ b$, what is the expected frequency of dwarfed plants among the progeny?

S4.8 Two yeast strains were mated; one was *red* with mating type **a**; the other was *red*$^+$ with mating type α. The resulting diploids were induced to undergo meiosis, and the meiotic products of the diploid were as follows (total progeny = 100):

40	*red*$^+$	α
44	*red*	**a**
9	*red*$^+$	**a**
7	*red*	α

How many map units separate the *red* gene from that for mating type?

S4.9 In meiosis, under what conditions does second-division segregation of a pair of alleles take place?

S4.10 Explain why the most accurate measurement of map distance between two genes is obtained by summing shorter distances between the genes.

S4.11 In *D. melanogaster*, the recessive allele *ap* (*apterous*) drastically reduces the size of the wings and halteres (flight balancers), and the recessive allele *cl* (*clot*) produces dark maroon eye color. Both genes are located in the second chromosome. You have a friend taking a genetics laboratory course. To obtain the genetic distance between these genes, she performs a cross of a double-heterozygous female and a maroon-eyed, wingless male. She examines 25 flies in the F_1 generation and calculates that the frequency of recombination is 56 percent. What is wrong with this number? What is a plausible explanation for the results obtained?

S4.12 Your friend in Problem S4.11 logs onto FlyBase and learns that the *ap-cl* distance is 36 map units. She therefore decides to repeat the cross and to examine more F_1 flies. Because both genes are autosomal, she does not expect any difference between reciprocal crosses. Hence, in setting up the cross this time, she mates a double-heterozygous male with a maroon-eyed, wingless female. To her astonishment, among 1000 F_1 progeny she was unable to find even one recombinant progeny. How can this anomaly be explained?

S4.13 The genetic map of an X chromosome of *Drosophila melanogaster* has a length of 73.1 map units. The X chromosome of the related species, *Drosophila virilis*, is even longer—170.5 map units. In view of the fact that the frequency of recombination between two genes cannot exceed 50 percent, how is it possible for the map distance between genes at opposite ends of a chromosome to exceed 50 map units?

S4.14 In *D. virilis*, the mutations dusky body color (*dy*), cut wings (*ct*), and white eyes (*w*) are all recessive alleles located in the X chromosome. Females heterozygous for all three mutations were crossed to *dy ct w* males, and the following phenotypes were observed among the offspring.

wildtype	18
dusky	153
cut	6
white	76
cut, white	150
dusky, cut	72
dusky, cut, white	22
dusky, white	3

(a) What are the genotypes of two maternal X chromosomes?

(b) What is the map order of the genes?

(c) What are the two-factor recombination frequencies between each pair of genes?

(d) Are there as many observed double-crossover gametes as would be expected if there were no interference?

(e) What are the coefficient of coincidence and the interference across this region of the X chromosome?

S4.15 The genes considered in Problem S4.14 are also located on the X chromosome in *D. melanogaster*. However, their order and the distances between them are completely different because of chromosome rearrangements that have happened since the divergence of two species approximately 40 million years ago. In *D. melanogaster*, the genetic map of this region is *dy*−16.2−*ct*−18.5−*w*. If you performed a cross identical to that in Problem S4.14 with *D. melanogaster* and observed 500 progeny, what types of progeny would be expected and in what numbers? Assume that the interference between the genes is the same in *D. melanogaster* as in *D. virilis*, and round the expected number of progeny to the nearest whole number. For convenience, list the progeny types in the same order as in Problem S4.14.

S4.16 What experiment in *Drosophila* demonstrated that crossing-over takes place in the four-strand stage? How would the result of the experiment have been different if crossing-over took place in the two-strand stage?

S4.17 *Neurospora crassa* is a haploid ascomycete fungus. A mutant strain that requires arginine for growth (*arg⁻*) is crossed with a strain that requires tyrosine and phenylalanine (*tyr⁻ phe⁻*). Each requirement results from a single mutation. Tests on 200 randomly collected spores from the cross gave the following results, in which the + in any column indicates the presence of the wildtype allele of the gene.

arg	tyr	phe	Number of spores
+	−	−	43
−	+	+	42
+	−	+	43
−	+	−	42
+	+	−	8
−	−	+	9
+	+	+	6
−	−	−	7

(a) What are the three possible two-gene recombination frequencies?

(b) Is any linkage suggested by these data?

S4.18 In the case of independently assorting genes, why are parental ditype (PD) and nonparental ditype (NPD) tetrads formed in equal numbers? Why are nonparental ditype tetrads rare for linked genes?

S4.19 The yeast *Saccharomyces cerevisiae* has unordered tetrads. In a cross made to study the linkage relationships among three genes, the following tetrads were obtained. The cross was between a strain of genotype + *b c* and one of genotype *a* + +.

Tetrad type	Genotypes of spores in tetrads				Number of tetrads
1	*a* + +	*a* + +	+ *b c*	+ *b c*	135
2	*a b* +	*a b* +	+ + *c*	+ + *c*	124
3	*a* + +	*a* + *c*	+ *b* +	+ *b c*	59
4	*a b* +	*a b c*	+ + +	+ + *c*	70
5	*a* + *c*	*a* + *c*	+ *b* +	+ *b* +	1
6	*a b c*	*a b c*	+ + +	+ + +	1
				Total	390

(a) From these data, determine which (if any) of the genes are linked.

(b) For any linked genes, determine the map distances.

S4.20 Consider a strain of *Neurospora* that exhibits complete chromosomal interference, even across the centromere. Two genes, *A* and *B*, are located in the same chromosome but on different sides of the centromere. Gene *A* is 10 map units from the centromere, and gene *B* is 5 map units from the centromere. Calculate the expected percentage of each of the following types of asci.

(a) Parental ditype

(b) Nonparental ditype

(c) Tetratype

(d) First-division segregation for *A*, first for *B*

(e) First-division segregation for *A*, second for *B*

(f) Second-division segregation for *A*, first for *B*

(g) Second-division segregation for *A*, second for *B*

S4.21 You are a geneticist writing a set of test questions and wish to invent plausible numbers of the possible types of progeny in a three-point testcross to yield a frequency of recombination of 0.2 in "region I," a frequency of recombination of 0.3 in "region II," and an interference value of 2/3. Among 1000 total progeny, how many should be:

(a) nonrecombinant?
(b) recombinant in region I only?
(c) recombinant in region II only?
(d) recombinant in both regions (double crossovers)?
(e) In each class of progeny above, how should the total number be allocated between the two reciprocal products of recombination?

S4.22 What is the answer to Problem S4.21 in the general case when the frequency of recombination in region I is r_1, that in region II is r_2, and the coincidence equals c?

S4.23 In a testcross of a parent heterozygous for each of four linked genes, in which the order of the genes is known, the following numbers of progeny were obtained:

Nonrecombinant	486
Recombinant in region I only	61
Recombinant in region II only	143
Recombinant in region III only	226
Recombinant in regions I and II only	10
Recombinant in regions II and III only	45
Recombinant in regions I and III only	27
Recombinant in regions I and II and III	2

(a) What are the map distances in centimorgans of regions I, II, and III? (*Hint*: Solve the problem as in a three-point cross, but remember that recombinants in two or three regions are recombinant for each region separately.)
(b) Calculate the values of the interference for regions I + II, II + III, and I + III. Does the ranking of the interference values make intuitive sense?

S4.24 In a fungal organism with ordered tetrads, deduce whether one observes first-division segregation or second-division segregation when the region between a gene and the centromere undergoes

(a) a two-strand double crossover
(b) either of the two types of three-strand double crossover
(c) a four-strand double crossover

If the four types of double crossover are equally frequent, what proportion of asci formed from double crossing-over will show second-division segregation?

S4.25 Show that each of the following statements is true in a fungal organism with ordered tetrads,

(a) If a configuration of multiple crossovers will result in first-division segregation, then an additional crossover (distal to the existing ones) will convert the ascus to second-division segregation with probability 1.
(b) If a configuration of multiple crossovers will result in second-division segregation, then an additional crossover (distal to the existing ones) will convert the ascus to first-division segregation with probability 1/2.

Chapter 5   The Molecular Structure and Replication of the Genetic Material

S5.1 You have determined that one strand of a DNA double helix has the sequence 5'-AGCCTAG-3'. What is the sequence of the complementary strand?

S5.2 In the replication of a bacterial chromosome, does replication begin at a random point?

S5.3 The haploid genome of the wall cress, *Arabidopsis thaliana*, contains 100,000 kb and has five chromosomes. If a particular chromosome contains 10 percent of the DNA in the haploid genome, what is the approximate length of its DNA molecule in micrometers?

S5.4 For the chromosome of *A. thaliana* in Problem S5.3, estimate the time of replication, assuming that there is only one origin of replication (exactly in the middle), that replication is bidirectional, and that the rate of DNA synthesis is

(a) 1500 nucleotide pairs per second (typical of bacterial cells)
(b) 50 nucleotide pairs per second (typical of eukaryotic cells)

S5.5 In terms of the mechanism of DNA replication, explain why a linear chromosome without telomeres is expected to become progressively shorter in each generation.

S5.6 A friend brings you three samples of nucleic acid and asks you to determine each sample's chemical identity (whether DNA or RNA) and whether the molecules are double-stranded or single-stranded. You use powerful nucleases to degrade each sample to its constituent nucleotides, and then you determine the approximate relative proportions of nucleotides. The results of your assays are presented below. What can you tell your friend about the nature of these samples?

Assay results
Problem S5.6

Sample 1:	dGTP	14%	dCTP	15%	dATP	36%	dTTP	35%
Sample 2:	dGTP	12%	dCTP	36%	dATP	47%	dTTP	5%
Sample 3:	GTP	22%	CTP	47%	ATP	17%	UTP	14%

S5.7 What is meant by the statement that the DNA replication fork is asymmetrical?

S5.8 What is the role of each of the following proteins in DNA replication?

(a) DNA helicase
(b) Single-stranded DNA-binding protein
(c) DNA polymerase
(d) DNA ligase

S5.9 Which enzyme in DNA replication

(a) alters the helical winding of double-stranded DNA by causing a single-strand break, exchanging the relative position of the strands, and sealing the break?
(b) uses ribonucleoside triphosphates to synthesize short RNA primers?

S5.10 Describe why topoisomerase is necessary during replication of the *E. coli* chromosome.

S5.11 The accompanying drawing is of a replication fork. The ends of each strand are labeled with the letters a through h.

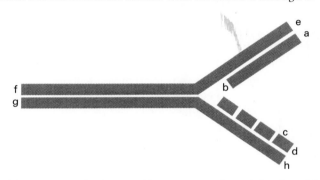

(a) Label the leading and lagging strands and, by the addition of arrowheads, show the direction of movement of the replication fork and the direction of synthesis of each DNA strand or fragment.
(b) Specify which of the letters a through h label 5′ ends and which label 3′ ends.

S5.12 Some "thermophilic" bacteria can live in water so hot that it approaches the boiling point. Which of the following DNA sequences would be more likely to be found in such an organism? Explain.

(a) 5′-TTATAAAATATATTTTTATAT-3′
       3′-AATATTTTATATAAAAATATA-5′
(b) 5′-CCCCCGCGCGGCCGGGCGCGCG-3′
       3′-GGGGGCGCGCCGGCCCGCGCGC-5′

S5.13 A 3.1-kilobase linear fragment of DNA was digested with *Pst*I and produced a 2-kb and a 1.1-kb fragment. When the same 3.1-kb fragment was cut with *Hin*dIII, it yielded a 1.5-kb fragment, a 1.3-kb fragment, and a 0.3-kb fragment. When the 3.1-kb molecule was cut with a mixture of the two enzymes, fragments of 1.5, 0.8, 0.5, and 0.3 kb resulted. Draw a map of the original 3.1-kb fragment, and label restriction sites and the distances between the sites.

S5.14 The X-linked recessive mutation *dusky* in *Drosophila* causes small, dark wings. In a stock of wildtype flies, you find a single male that has the dusky phenotype. In wildtype flies, the *dusky* gene is contained within an 8-kb *Xho*I restriction fragment. When you digest genomic DNA from the mutant male with *Xho*I and probe with the 8-kb restriction fragment on a Southern blot, the size of the labeled fragment is 10 kb. You clone the 10-kb fragment and use it as a probe for a polytene chromosome *in situ* hybridization in a number of different wildtype strains, and you notice that this fragment hybridizes to multiple locations along the polytene chromosomes. Each wildtype strain has a different pattern of hybridization. What do these data suggest about the origin of the *dusky* mutation that you isolated?

S5.15 A short region of an mRNA molecule is used as a primer in dideoxy sequencing. After transfer to a filter and autoradiography, the resulting film looks as follows, where the orientation is with the samples loaded at the top and the letters indicate the dideoxynucleotide present during synthesis.

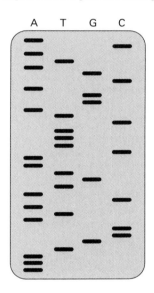

(a) Write the sequence implied by the bands in the gel from the 5′ end to the 3′ end.
(b) Is this the sequence of the elongated RNA strand or the sequence of its complementary DNA strand?

S5.16 For the following restriction endonucleases, calculate the average distance between restriction sites in an organism whose DNA has a random sequence and equal proportions of all four nucleotides. The symbol R means any purine (A or G) and Y means any pyrimidine (T or C), but an R—Y pair must be either A—T or G—C.

*Taq*I	5′-TCGA-3′
	3′-AGCT-5′
*Bam*HI	5′-GGATCC-3′
	3′-CCTAGG-5′
*Hae*II	5′-RGCGCY-3′
	3′-YCGCGR-5′

S5.17 For each of the restriction enzymes in Problem S5.16, calculate the expected number of restriction sites in the genome of a nematode *Caenorhabditis elegans*, assuming the sequence is approximately random. The genome size of this organism is 100,000 kb.

S5.18 In early studies of the properties of Okazaki fragments, it was shown that these fragments could hybridize to both strands of *E. coli* DNA. This was taken as an evidence that they are synthesized on both branches of the replication fork. However, one feature of the replication of *E. coli* DNA that was not known at the time invalidates this conclusion. What is this characteristic?

S5.19 In experiments that confirmed that DNA replication is semiconservative, *E. coli* DNA was labeled by growing cells for a number of generations in a medium containing "heavy" ^{15}N isotope and then observing the densities of the molecules during subsequent generations of growth in a medium containing a great excess of "light" ^{14}N. If DNA replication were conservative, what pattern of molecular densities would have been observed after one generation and after two generations of growth in the ^{14}N-containing medium?

S5.20 A covalently closed circular single-stranded DNA molecule is hybridized to a linear complementary strand that has a segment of 100 nucleotides missing. This duplex is exhaustively treated with various nucleases. List the final reaction products for each enzyme.

(a) An endonuclease that cuts only single-stranded DNA
(b) An exonuclease that degrades only single-stranded DNA
(c) An exonuclease that attacks only double-stranded DNA and degrades it in the $3' \rightarrow 5'$ direction (removing the shorter strand from a paired duplex when the partners are of unequal length)
(d) An endonuclease that attacks and degrades only double-stranded DNA
(e) A mixture of enzymes (b) and (d)
(f) A mixture of enzymes (a) and (c)
(g) A mixture of enzymes (c) and (d)

Chapter 6   The Molecular Organization of Chromosomes

S6.1 What are the principal molecular constituents of the nucleosome?

S6.2 Is it possible for a gene of average size to be present in a single nucleosome?

S6.3 A geneticist wishes to isolate mutations that affect the amino acid sequence of any of the histone genes H2, H3A, H3B, and H4. What is the likely phenotype of such a mutation?

S6.4 In spermatogenesis in many male animals, the normal somatic-cell histones are removed from the DNA and replaced with a still more highly basic (arginine-rich and lysine-rich) set of sperm-specific histones. Suggest one possible explanation.

S6.5 What are principal structural levels of chromosome organization?

S6.6 Why are centromeres and telomeres required for chromosomes to be genetically stable?

S6.7 The nematode *C. elegans* has holocentric chromosomes. If the chromosomes in a cell are fragmented into many pieces by treatment with x rays, how would you expect each fragment to behave in cell division?

S6.8 Name three features that distinguish euchromatin from heterochromatin.

S6.9 Explain why topoisomerase is necessary during replication of the *E. coli* chromosome.

S6.10 What is one possible reason why there is usually a single gene in the genome coding for a particular protein, whereas genes for rRNA and tRNA are repeated many times?

S6.11 The X-linked recessive mutation *singed* (*sn*) causes gnarled bristles in *Drosophila*. In a stock of wildtype flies, you find a single male that has the singed phenotype.

(a) Diagram a crossing scheme that you will use to generate a true-breeding line of *sn* flies. Note that, other than the single singed male, you have only wildtype flies available.
(b) In wildtype flies, the *sn* gene is contained within an 8-kb *Bam*HI restriction fragment. When you digest genomic DNA from the mutant line with *Bam*HI and probe with the 8-kb restriction fragment on a Southern blot, the size of the labeled fragment on the filter is 13 kb. What possible explanations could account for this finding?

S6.12 A genetically engineered strain of *Drosophila virilis* (strain 1) is homozygous for the recessive mutation *y* (yellow), an X-linked mutation that causes yellow body color. This strain also carries, at a site in the X chromosome, a genetically engineered transposable element called *Hermes* into which a copy of the y^+ gene was inserted; the engineered *Hermes* element is unable to produce the transposase protein needed for transposition of the element. Another genetically engineered strain, strain 2, is also homozygous *y* but contains a copy of *Hermes* that can produce the transposase in each homolog of chromosome 3 . A female of strain 1 is crossed with a male of strain 2. The phenotypically wildtype F_1 males are crossed with yellow females. About 5 percent of the F_2 progeny males have wildtype body color. Give two possible explanations of how these males might arise.

S6.13 The term "satellite DNA" was originally coined for highly repetitive simple sequences that, in equilibrium density-gradient centrifugation in CsCl, formed a distinct band (a "satellite" band) different from the bulk of the genomic DNA, because of a difference in density of the DNA fractions. You analyze the genome organization of a new species by the kinetics of DNA renaturation and find that 10 percent of DNA is represented by a highly repetitive simple sequence. Does this mean that, after equilibrium density-gradient centrifugation in CsCl, you will find a satellite band?

S6.14 DNA from species A that is labeled with ^{14}N and randomly fragmented is renatured with an equal concentration of DNA from species B that is labeled with ^{15}N and also randomly fragmented. The resulting mixture of molecules is centrifuged to equilibrium in CsCl. Ten percent of the total renatured DNA has a hybrid density. What fraction of the base sequences in the two species will be similar enough to renature?

S6.15 You are studying the denaturation of two unusual DNA molecules, both consisting of 99 percent AT base pairs. Both molecules have the same melting temperature. However, observations in the electron microscope indicate that the strands of one of the molecules do not separate completely until raised to a temperature 12°C higher than that required for complete separation of the strands of the other molecule. Suggest an explanation.

S6.16 A molecule of 5000 nucleotide pairs consists of the repeating sequence 5'-ACGTAG-3'. How would the melting temperature of this molecule compare to that of another molecule of the same length and base composition (50 percent G + C) but with a random sequence?

S6.17 A Cot analysis is carried out with DNA obtained from a particular plant species. The chloroplast DNA is readily detected as a rapidly reannealing fraction because of its abundance in the DNA sample. Plants of the same species are maintained in the dark for several weeks until they lose all their chlorophyll and the leaves become white. DNA from these plants has the same Cot curve as observed in plants maintained in the light. What does this result tell you about the loss of chlorophyll?

Chapter 7   Variation in Chromosome Number and Structure

S7.1 *Drosophila* and human beings have very different mechanisms of sex determination. Which of the following statements correctly describes the phenotype of an XXY organism in both species?

(a) Fertile male in *Drosophila*, sterile female in humans.
(b) Sterile female in *Drosophila*, fertile male in humans.
(c) Fertile female in *Drosophila*, fertile male in humans.
(d) Fertile female in *Drosophila*, sterile male in humans.
(e) Fertile female in *Drosophila*, fertile female in humans.

S7.2 What are the consequences of X-chromosome inactivation in mammals?

(a) The formation of Barr bodies
(b) A mutation

(c) A mosaic expression of different phenotypes in different cells of heterozygous X-linked genes in females
(d) A mosaic expression of heterozygous autosomal genes in females
(e) Turner syndrome

S7.3 Nondisjunction of chromosomes in meiosis may take place at the first or the second meiotic division. If nondisjunction of the XY pair of chromosomes took place at the first division in the formation of sperm, which of the following chromosomal types of sperm could be formed?

(a) XY
(b) YY
(c) XX
(d) X
(e) Y

S7.4 Which of the following human disorders can result from nondisjunction during the formation of the egg?

(a) Down syndrome
(b) Klinefelter syndrome
(c) XYY male

S7.5 Why are abnormalities in the number of sex chromosomes in mammals usually less severe in their phenotypic effects than abnormalities in the number of autosomes?

S7.6 What are the consequences of a single crossover within the inverted region of a pair of homologous chromosomes with the gene order is *A B C D* in one and *a c b d* in the other, in each of the following cases?

(a) The centromere is not included within the inversion.
(b) The centromere is included within the inversion.

S7.7 What role might inversions play in evolution?

S7.8 In human beings, trisomy 21 is the cause of Down syndrome. Ordinarily, only one child with Down syndrome is seen in a sibship. Occasionally, a sibship with multiple affected children is observed. In one such family, a pair of normal parents had seven children. Four were normal, and three had an atypical form of Down syndrome. In this atypical form, the slowed growth, mental handicap, transverse crease on the palm of the hand, and short broad hands were present, but the ear and heart defects often observed with Down syndrome were not present. One of the normal children gave rise to five offspring, two of whom had the same atypical form of Down syndrome. Provide a genetic explanation for this atypical form of Down syndrome and for its inheritance pattern.

S7.9 Colchicine is used to cause a doubling of chromosome number in many plant species. Starting with a diploid strain of watermelon (*Citrullus lanatus*), outline a series of crosses you would use to create a pentaploid strain of this species.

S7.10 Explain why, in *Drosophila*, triploid flies are much more likely to survive than trisomics.

S7.11 In a certain fertile tetraploid plant, a locus with alleles *A* and *a* is situated very near the centromere of its chromosome. If a tetraploid plant of genotype *AAaa* is crossed with a diploid plant of genotype *aa,* what genotypic ratios are expected? What phenotypic ratios are expected if *A* is dominant to *a*? (Assume that homologous chromosomes in the tetraploid form pairs at random and that all gametes produced by the tetraploid are diploid. Assume also that the dominant phenotype is expressed whenever at least one *A* allele is present.)

S7.12 Explain the apparent paradox that familial retinoblastoma is regarded as an example of a recessive oncogene (tumor suppressor gene) even though it is inherited in pedigrees as an autosomal dominant. Familial retinoblastomas are generally bilateral (that is, there are multiple independent tumors in both eyes of affected individuals), whereas sporadic retinoblastoma is generally limited to one eye and to fewer independent tumors. Account for these observations.

S7.13 Two strains of a plant called shepherd's purse (*Capsella bursa-pastoris*), a widespread lawn and roadside weed of the mustard family, are compared. In strain 1, the recombination frequency between genes *A* and *B* is 6 percent, whereas in strain 2, it is 50 percent. The F_1 progeny from a cross between these strains produce gametes with the following properties: About half are viable and contain only nonrecombinant chromosomes, and about half are nonviable. How can these results be explained?

S7.14 True-breeding tetrasomic organisms are much easier to create than true-breeding trisomic organisms. Suggest a reason for this observation.

S7.15 In plants, a cross between a diploid organism and a tetraploid organism of the same species is usually more fertile than that between a diploid organism and a triploid organism of the same species. Suggest a possible explanation.

S7.16 In *Drosophila melanogaster,* the genes for *brown eyes* (*bw*) and *humpy thorax* (*hy*) are about 12 map units distant on the same arm of chromosome 2. A paracentric inversion spans about one-third of the region but does not include these genes. What recombinant frequency between *bw* and *hy* would you expect in each of the following?

(a) Females that are homozygous for the inversion
(b) Females that are heterozygous for the inversion

S7.17 The sequence of genes in a normal human chromosome is *123 • 456789*, where the raised dot (•) represents the centromere. Aberrant chromosomes with the following arrangements of genes were observed.

    I.   *123 • 476589*
   II. *123 • 46789*
  III. *1654 • 32789*
  IV. *123 • 4566789*

(a) Identify each chromosome as a paracentric inversion, pericentric inversion, deletion, or duplication, and diagram how each rearrangement would pair with the normal homolog in meiosis.
(b) Draw the meiotic products of a crossover between genes 5 and 6 when chromosome I pairs with a wildtype chromosome.
(c) Draw the products of a crossover between genes 5 and 6 when chromosome III pairs with a wildtype chromosome.

S7.18 Semisterile tomato plants heterozygous for a reciprocal translocation between chromosomes 5 and 11 were crossed with chromosomally normal plants homozygous for the recessive mutant broad leaf on chromosome 11. When semisterile F_1 plants were crossed with the plants of broad-leaf parental type, the following phenotypes were found in the backcross progeny:

semisterile broad leaf	38
fertile broad leaf	242
semisterile normal leaf	282
fertile normal leaf	33

(a) What is the recombination frequency between the broad-leaf gene and the translocation breakpoint in chromosome 11?
(b) What ratio of phenotypes in the backcross progeny would have been expected if the broad-leaf gene had not been on the chromosome involved in the translocation?

S7.19 Semisterile tomato plants heterozygous for the same translocation referred to in the previous problem were crossed with chromosomally normal plants homozygous for the recessive mutants *wilty* and *leafy* in chromosome 5. When semisterile F_1 plants were crossed with plants of *wilty, leafy* parental genotype, the following phenotypes were found in the backcross progeny.

	Semisterile	Fertile	Total
wildtype	333	19	352
wilty	17	6	23
leafy	1	8	9
wilty and leafy	25	273	298

What is the map distance between the two genes and the map distance between each of the genes and the translocation breakpoint in chromosome 5?

S7.20 The accompanying diagrams depict wildtype chromosomes 2 and 3 from a certain plant. The filled circles represent the respective centromeres.

Chromosome 2   A   B   C   D   E   F   G   H

Chromosome 3   I   J   K   L   M   N   O   P   Q

(a) Show how these chromosomes would pair in meiosis in an individual heterozygous for an inversion that includes genes *E*, *F*, and *G* as well as heterozygous for a reciprocal translocation between chromosomes 2 and 3 with breakpoints between *B* and *C* in 2 and between *K* and *L* in 3.

(b) Suppose that the breakpoints of the above inversion are sufficiently close to genes *D* and *H* that recombination between the genes is completely eliminated when the inversion is heterozygous. A plant heterozygous for the inversion and the translocaton and of genotype *Dd Hh* is crossed with a plant with normal chromosomes and of genotype *dd hh*. The most frequent classes of offspring are *dd Hh* semisterile and *Dd hh* ertile. The rarest types are *Dd hh* semisterile and *dd Hh* fertile, which together have a combined frequency of 0.15. What is the recombination frequency between *D* and the translocation breakpoint?

S7.21 Six mutations of bacteriophage T4 (*a* through *f*) are isolated and crossed in all pairwise combinations to three deletion mutations (numbered 1–3) with the results shown below. In the table, R means that wildtype recombinants were observed and 0 means that there were no wildtype recombinants.

Point mutations	*a*	*b*	*c*	*d*	*e*	*f*
Deletions 1	0	0	R	R	R	R
Deletions 2	0	R	0	0	0	0
Deletions 3	0	0	R	0	0	0

In parallel with the above experiments, the mutations were tested in all possible pairwise combinations for complementation. The results are tabulated below, where + indicates complementation and − indicates no complementation.

	a	*b*	*c*	*d*	*e*	*f*
a	−	+	+	−	+	+
b	+	−	+	+	+	+
c	+	+	−	+	−	+
d	−	+	+	−	+	+
e	+	+	−	+	−	+
f	+	+	+	+	+	−

Assuming that complementation indicates that two mutations are in different genes, combine the deletion mapping and the complementation data to determine the most likely relative order of the mutations. Group the mutations into allelic classes, and determine the position and extent of the three deletions.

Chapter 8 The Genetics of Bacteria and Viruses

S8.1 Define the following terms:

(a) Transformation
(b) Conjugation
(c) Transduction
(d) Transposition
(e) Temperate phage

S8.2 Describe the differences between phage λ and F-plasmid integration into bacterial chromosome.

S8.3 Is it possible for a plasmid without any genes to exist?

S8.4 What process is responsible for the conversion of an F^+ bacterial cell into an Hfr cell?

S8.5 What is the major difference in bacteriophage between the lytic cycle and the lysogenic cycle?

S8.6 Why are specialized transducing particles of phage λ generated only when a lysogen is induced to produce phage rather than in the process of lytic infection?

S8.7 Why is generalized transduction not the preferred method for mapping two unknown mutants relative to one another on the bacterial chromosome?

S8.8 A temperate phage has the gene order *a b c d e f g h*, whereas the prophage of the same phage has the gene order *g h a b c d e f*. What information does this circular permutation give you about the location of the phage attachment site?

S8.9 Recombination in phage λ crosses reveals the order of certain genes to be

where *cos* represents the cohesive ends and *att* is the phage attachment site. What order of the genes would be found if the λ prophage were mapped by generalized P1 transduction of an *E. coli* strain lysogenic for phage λ?

S8.10 You are studying a biochemical pathway in *Neurospora* that leads to the production of substance A. You isolate a set of mutations, each of which is unable to grow on minimal medium unless it is supplemented with A. By performing appropriate matings, you group all the mutants into four complementation groups (genes) designated *a1*, *a2*, *a3*, and *a4*. You know beforehand that the biochemical pathway for the production of A includes four intermediates: B, C, D, and E. You test the nutritional requirements of your mutants by growing them on minimal medium supplemented with each of these intermediates in turn. The results are summarized in the accompanying table, where the plus signs indicate growth and the minus signs indicate failure to grow.

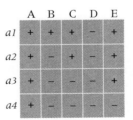

	A	B	C	D	E
a1	+	+	+	−	+
a2	+	−	+	−	+
a3	+	−	−	−	+
a4	+	−	−	−	−

Determine in what order the substances A, B, C, D, and E are most likely to participate in the biochemical pathway, and indicate the enzymatic steps by arrows. Label each arrow with the name of the gene that codes for the corresponding enzyme.

S8.11 A experiment was carried out in *E. coli* to map five genes around the chromosome using each of three different Hfr strains. The genetic markers were *bio, met, phe, his,* and *trp.* The Hfr strains were found to transfer the genetic markers at the times indicated in the accompanying table. Construct a genetic map of the *E. coli* chromosome that includes all five genetic markers, the genetic distances in minutes between adjacent gene pairs, and the origin and direction of transfer of each Hfr. Complete the missing entries in the table, indicated by the question marks.

Hfr1 markers	bio	met	phe	his
Time of entry	26	44	66	?

Hfr2 markers	phe	met	bio	trp
Time of entry	?	26	44	75

Hfr3 markers	phe	his	?	bio
Time of entry	6	27	35	?

S8.12 A first-year graduate student carried out transformation of an *E. coli* strain of genotype *trp⁻ azi⁻ phe⁻ gal⁻ pur⁻ his⁻ ser⁻* using a plasmid with the selectable markers *amp-r kan-r trp⁺ azi⁺ bio⁺ phe⁺ pur⁺ tet-r.* After transformation, he plated the cells on minimal medium containing galactose as a carbon source along with histidine and serine and added ampicillin and kanamycin to select for the transformants. After incubating the plates overnight, the student returned to the lab to discover that no colonies had formed! Then he realized that he had made a stupid mistake. What was it?

S8.13 The genes *A, B, G, H, I,* and *T* were tested in all possible pairs for cotransduction with bacteriophage P1. Only the following pairs were found able to be cotransduced: *G* and *H, G* and *I, T* and *A, I* and *B, A* and *H.* What is the order of the genes along the chromosome? Explain your logic.

S8.14 Bacteriophage P1 was grown on a wildtype strain of *E. coli,* and the resulting progeny phage were used to infect a strain with three mutations, *arg⁻, pro⁻,* and *his⁻,* each of which results in a requirement for an amino acid: arginine, proline, and histidine, respectively. Equal volumes of the infected culture were spread on plates containing minimal medium supplemented with various combinations of the amino acids, and the colonies appearing on each plate were counted. The results are shown here. In each row, the + signs indicate the amino acids present in the medium.

Plate number	Supplement			Number of colonies
	Arginine	Proline	Histidine	
1	+	+	0	1000
2	+	0	+	1000
3	0	+	+	1000
4	+	0	0	200
5	0	+	0	100
6	0	0	+	0

(a) For each of the six types of medium, list the donor markers that must be present to allow the transduced cell to form colonies.

(b) Using these results, construct a genetic map of the *arg, pro, his* region showing the relative order of the genes and the cotransduction frequencies between the markers.

S8.15 The order of genes in the λ phage virion is

A B C D E att int xis N cI O P Q S R

(a) Given that the bacterial attachment site, *att,* is between *gal* and *bio* in the bacterial chromosome, what is the prophage gene order?

(b) A new phage is discovered whose gene order is identical in the virion and in the prophage. What does this say about the location of the *att* site with respect to the termini of the phage chromosome?

(c) A wildtype λ lysogen is infected with another λ phage carrying a genetic marker, *Z,* located between *E* and *att.* The superinfection gives rise to a rare, doubly lysogenic *E. coli* strain that carries both λ and λ-Z prophage. Assuming that the second phage also entered the chromosome at an *att* site, diagram two possible arrangements of the prophages in the bacterial chromosome, and indicate the locations of the bacterial genes *gal* and *bio.*

S8.16 Four Hfr strains of *E. coli* transfer their genetic material in the following order. Construct a genetic map of the *E. coli* chromosome that includes all the markers listed here and the distances in minutes between adjacent gene pairs.

Hfr1 markers	mal	met	thi	thr	trp	
Time of entry	10	17	22	33	57	

Hfr2 markers	his	phe	arg	mal		
Time of entry	18	23	35	45		

Hfr3 markers	phe	his	bio	azi	thr	thi
Time of entry	6	11	33	48	49	60

Hfr4 markers	arg	thy	met	thr		
Time of entry	15	21	32	48		

S8.17 An F' factor consisting of the *E. coli* sex factor, F, and a segment of the bacterial chromosome that includes the wildtype tryptophan synthetase A gene (*trpA⁺*) was isolated. The F' *trpA* factor was introduced by conjugation into an F⁻ strain with a single nucleotide mutation in the *trpA* gene, and the resulting exconjugants were F' *trpA⁺/trpA⁻* partial diploids. These cells were streaked to obtain single colonies, and each colony was inoculated into liquid nutrient medium. When mixed with female cells, most of the resulting cultures transferred only the original F' *trpA* factor. However, a very small percentage of the cultures transferred a variable length of the bacterial chromosome, starting with the *trpA* gene. These cultures fell into two classes. Class 1 transferred *trpA⁺* early, whereas class 2 transferred *trpA⁻* early.

(a) Draw a diagram of the interaction between the F' factor and the bacterial chromosome, showing how the cells of class 1 are likely to have arisen.

(b) Draw a similar diagram showing the probable origin of cells of class 2.

Chapter 9 Genetic Engineering and Genome Analysis

S9.1 Define the following terms:

(a) Vector
(b) Restriction site
(c) Palindrome
(d) Cohesive end
(e) Blunt end
(f) Complementary DNA
(g) Insertional inactivation
(h) Colony hybridization assay
(i) Transgenic organism

S9.2 What are some of the principal practical applications of genetic engineering?

S9.3 Reverse transcriptase, like most enzymes that make DNA, requires a primer. Explain why, when cDNA is to be made for the purpose of cloning a eukaryotic gene, a convenient primer is a short sequence of poly(dT)? Why does this method not work with a prokaryotic messenger RNA?

S9.4 A DNA molecule has 23 occurrences of the sequence 5'-AATT-3' along one strand. How many times does the same sequence occur along the other strand?

S9.5 After doing a restriction digest with the enzyme *Sse*I, which has the recognition site 5'-CCTGCA/GG-3' (the slash indicates the position of the cleavage), you wish to separate the fragments in an agarose gel. In order to choose the proper concentration of agarose, you need to know the expected size of the fragments. Assuming equivalent amounts of each of the four nucleotides in the target DNA, what average-size fragment would you expect?

S9.6 How many clones are needed to establish a library of DNA from a species of monkey with a diploid genome size of 6×10^9 base pairs if (1) fragments whose sizes average 2×10^4 base pairs are used, and (2) one wishes 99 percent of the genomic sequences to be in the library? (*Hint*: If the genome is cloned at random with x-fold coverage, the probability that a particular sequence will be missing is e^{-x}.)

S9.7 The yeast *Saccharomyces cerevisiae* has 16 chromosomes. All of them can be separated by pulse field gel electrophoresis. Strains with the following chromosome rearrangements were examined by pulse field electrophoresis along with DNA from a wildtype control:

(a) A reciprocal translocation in which about one-third of chromosome IV (the largest chromosome) was exchanged with about one-third of chromosome I (the smallest chromosome).
(b) An insertional translocation in which two-thirds of one chromosome was inserted into another chromosome.
(c) A pericentric inversion in chromosome IV.

(d) A paracentric inversion in chromosome IV.

What band pattern would you expect to see in each case?

S9.8 As a laboratory exercise, a friend is required to express a cloned *Drosophila* gene in bacteria to obtain large amounts of the recombinant protein. She is given a recombinant phage containing a 12-kb fragment of *Drosophila* genomic DNA, which, when introduced into the germline of a mutant fly, is able to rescue the mutant phenotype. Hence, the 12-kb fragment contains the entire gene. The fragment also hybridizes to a 4.7-kb RNA on Northern blots. (A Northern blot is one in which RNA is separated by electrophoresis and the bands are transferred to a filter and probed with a labeled complementary sequence). The protein encoded by the gene is approximately 110 amino acids in length and can be isolated by standard methods. Your friend excises the 12-kb fragment out of the phage and inserts it into a bacterial expression vector (a plasmid designed to allow transcription of inserts). Although molecular analysis indicates that the DNA was inserted correctly and that the insert is being transcribed, she is unable to isolate any of the gene product. She turns to you for advice. What explanations would you suggest and what would you recommend she do to obtain the protein?

S9.9 Aliquots of a 7.8-kilobase (kb) linear piece of DNA are digested with the restriction enzymes *Pvu*II, *Hinc*II, *Cla*I, and *Ban*II, alone and in pairs. The digestion products are separated by gel electrophoresis, and the size of each fragment is determined by comparison to size standards. The fragment sizes obtained, in kilobase pairs, are given below. Draw a diagram of the 7.8-kb fragment, showing the location of the restriction sites for each enzyme.

(a) Digestion with *Pvu*II: 1.3 kb, 6.5 kb
(b) Digestion with *Hinc*II: 0.5 kb, 3.3 kb, 4 kb
(c) Digestion with *Cla*I: 1 kb, 6.8 kb
(d) Digestion with *Ban*II: 1.3 kb, 6.5 kb
(e) Digestion with *Pvu*II and *Cla*I: 1 kb, 5.5 kb, 1.3 kb
(f) Digestion with *Pvu*II and *Hinc*II: 0.5 kb, 0.8 kb, 2.5 kb, 4 kb
(g) Digestion with *Hinc*II and *Cla*I: 0.5 kb, 1 kb, 3 kb, 3.3 kb
(h) Digestion with *Hinc*II and *Ban*II: 0.5 kb, 0.8 kb, 2.5 kb, 4 kb
(i) Digestion with *Cla*I and *Ban*II: 1 kb, 1.3 kb, 5.5 kb

S9.10 A 3.3-kb size fragment created after the *Hinc*II digestion in the previous problem is inserted into a unique *Hinc*II site of the plasmid pUC18 and transformed into *E. coli*. Individual transformed colonies are isolated, and the recombinant plasmids from each colony are purified and mapped with the restriction enzyme *Pvu*II. To the surprise of a novice molecular geneticist, the plasmids do not all give the same pattern of restriction fragments on a gel. The plasmids fall into two distinct groups. Both groups yield the same number of bands in the gel, the aggregate size of which is the same, but two of the fragment sizes are different. Explain the reason for this difference.

S9.11 You have cloned a region of the genome of the nematode *C. elegans* that contains a mutation in a gene involved in controlling biorhythms. You make a restriction map of the cloned region and find that the following *Hin*dIII restriction map can be constructed:

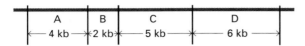

Now you perform a *Hin*dIII digest of genomic DNA from individuals homozygous for the mutation and also from those that are homozygous for the wildtype allele. You subject both of these digests to electrophoresis in an agarose gel and transfer the DNA to a nylon membrane. When the membrane is probed with fragment A, B, or D, the probe hybridizes to fragments of 4, 2, and 6 kb, respectively, in both the mutant and the wildtype strain. When the membrane is probed with fragment C, the probe hybridizes to 11 different fragments in the mutant genome (including one fragment of 5 kb) and to 10 fragments in the wildtype genome (but not one of 5 kb). What could account for both the mutation and the pattern of bands seen with the C probe?

S9.12 You are given a plasmid containing part of a gene of *D. melanogaster*. The gene fragment is 303 base pairs long. You would like to amplify it using the polymerase chain reaction (PCR). You design oligonucleotide primers 19 nucleotides in length that are complementary to the plasmid sequences immediately adjacent to both ends of the cloning site. What would be the exact size of the resulting PCR product?

S9.13 Suppose that you digest the genomic DNA of a particular organism with *Sau*3A (/GATC). Then you ligate the resulting fragments into a unique *Bam*HI (G/GATCC) cloning site of a plasmid vector. Would it be possible to isolate the cloned fragments from the vector using *Bam*HI? From what proportion of clones would it be possible?

S9.14 Digestion of a DNA molecule with *Hin*dIII yields two fragments of 2.2 kb and 2.8 kb. *Eco*RI cuts the molecule, creating 1.8 kb and 3.2 kb fragments. When treated with both enzymes, the same DNA molecule produces four fragments of 0.8 kb, 1.0 kb, 1.2 kb, and 2.0 kb. Draw a restriction map of this molecule.

S9.15 What are the main differences between cDNA and genomic DNA libraries?

S9.16 You wish to amplify a unique 100-bp sequence from the human genome (3×10^9 bp) using the polymerase chain reaction. At the end of the amplification, you want the amount of amplified DNA to represent not less than 99 percent of all the DNA present. Assume that the amplification is exponential.

(a) How many cycles of amplification are necessary to achieve the 99 percent goal?

(b) How would the number of cycles change if you were to amplify a fragment of 1000 bp?

S9.17 You are given a DNA fragment containing the *Drosophila* developmental gene *fushi tarazu* (*ftz*) along with the restriction map shown below. The restriction sites are *Hin*dIII (5'-A/AGCTT-3'), *Bcl*I (5'-T/GATCA-3'), *Eco*RI (5'-G/AATTC-3'), *Bam*HI (5'-G/GATCC-3').

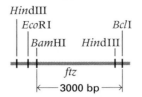

You are instructed to subclone the gene into a 3.8-kb plasmid vector having relevant restriction sites *Pst*I (5'-CTGCA/G-3'), *Bgl*II (5'-A/GATCT-3'), and *Bam*HI (5'-G/GATCC-3') positioned as follows:

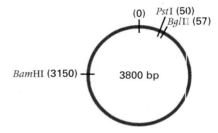

With what restriction enzymes should you cleave the *ftz*-containing region before ligating it with plasmid DNA that has been linearized with *Bgl*II?

S9.18 Once you have subcloned the *ftz* region as in the previous problem, you want to determine the orientation in which the fragment is inserted in the resulting plasmid. You cleave the *ftz*-containing plasmid with *Bam*HI (5'-G/GATCC-3'), *Pst*I (5'-CTGCA/G-3'), and *Hin*dIII (5'-A/AGCTT-3'), and observe fragments of sizes approximately 3 kb and 1 kb.

(a) Can you determine the orientation of the inserted fragment from this result?

(b) Would your answer change if the cloning site used in the plasmid were *Bam*HI rather than *Bgl*II? Explain.

S9.19 You are given two strains of *E. coli* containing plasmids whose restriction maps are

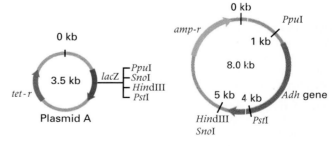

You are asked to subclone the gene for alcohol dehydrogenase (*Adh*) from plasmid B into plasmid A. Unsure of which

cloning procedure would be the best, you try two different approaches.

(a) You digest both plasmids to completion with *Ppu*I (A/TGCAT) and *Sno*I (G/TGCAT), isolate the 4-kb *Ppu*I-*Sno*I *Adh* gene fragment and the 3.5-kb *Ppu*I-*Sno*I vector A fragment, carry out a ligation reaction containing both isolated DNA fragments, transform *E. coli* with the resulting DNA, and plate the transformed cells on medium containing tetracycline. The plates also contain Xgal and IPTG, which makes it possible to identify cells that contain the original plasmid A, because they turn blue as a result of the functional *lacZ* gene. Colonies with an insertion that interrupts the *lacZ* gene are white because of insertional inactivation.

(b) You follow the same procedure except that you use *Hin*dIII (A/AGCCT) instead of *Sno*I. Using approach (a), you obtain many more blue colonies than using approach (b). Explain why this is so.

S9.20 You purify plasmid DNA from white colonies obtained in both approaches in the previous problem. Predict the sizes of the resulting fragments when the isolated plasmids are digested with *Pst*I.

Chapter 10   Gene Expression

S10.1 Describe the main features of the genetic code.

S10.2 If the triplet codons in the genetic code were read in an overlapping fashion, would there be any constraints on the amino acid sequences found in proteins?

S10.3 If DNA consisted of only two nucleotides (say, A and T) in any sequence, what is the minimal number of adjacent nucleotides that would be needed to specify uniquely each of the 20 amino acids?

S10.4 If DNA consisted of three possible base pairs (say, A—T, G—C, and X—Y), what is the minimal number of adjacent nucleotides that would be needed to specify uniquely each of the 20 amino acids?

S10.5 A DNA strand consists of any sequence of four kinds of nucleotides. Suppose there were only 16 different amino acids instead of 20. Which of the following statements would be correct descriptions of the minimal number of nucleotides necessary to create a genetic code?

(a) 1
(b) 2, provided that chain termination does not require a special codon
(c) 3, provided that chain termination does require a special codon
(d) 2, no matter how chain termination is accomplished
(e) statements (b) and (c)

S10.6 In the standard genetic code, synonymous codons often have the feature that the third nucleotide in the codon is either of the pyrimidines (T or C) or either of the purines (A or G) because of "wobble" in the base pairing at the third position of the codon. If this rule were followed consistently for all codons, would it be possible to have a triplet code that specifies 20 amino acids and one stop codon? Would there be any room for degeneracy?

S10.7 Which of the following statements is true of polypeptide chains and nucleotide chains?

(a) Both have a sugar-phosphate backbone.
(b) Both are chains consisting of 20 types of repeating units.
(c) Both types of molecules contain unambiguous and interconvertible triplet codes.
(d) Both types of polymers are linear and unbranched.
(e) None of the above.

S10.8 The notion that a strand of DNA serves as a template for transcription of an RNA that is translated into a polypeptide is known as the "central dogma" of gene expression. All three types of molecules have a polarity. In the DNA template and the RNA transcript, the polarity is determined by the free 3' and 5' groups at opposite ends of the polynucleotide chains; in a polypeptide, the polarity is determined by the free amino group (N terminal) and carboxyl group (C terminal) at opposite ends. Each statement below describes one possible polarity of the DNA template, the RNA transcript, and the polypeptide chain, respectively, in temporal order of use as a template or in synthesis. Which is correct?

(a) 5' to 3' DNA; 3' to 5' RNA; N terminal to C terminal.
(b) 3' to 5' DNA; 3' to 5' RNA; N terminal to C terminal.
(c) 3' to 5' DNA; 3' to 5' RNA; C terminal to N terminal.
(d) 3' to 5' DNA; 5' to 3' RNA; N terminal to C terminal.
(e) 5' to 3' DNA; 5' to 3' RNA; C terminal to N terminal.

S10.9 Which of the following features of a protein structure is most directly determined by the genetic code?

(a) Its shape
(b) Its secondary structure
(c) Its catalytic or structural role
(d) Its amino acid sequence
(e) Its subunit composition (quaternary structure)

S10.10 The structure below is that of a dipeptide composed of alanine and glycine.

(a) Which end, left or right, is the amino terminal end?
(b) Which amino acid is at the amino terminal end?
(c) Which end, left or right, is the carboxyl terminal end?
(d) Which amino acid is at the carboxyl terminal end?
(e) Which bond is the peptide bond?

S10.11 Summarize the studies in phage T4 that provided the first genetic evidence for a triplet code.

S10.12 An alternating polymer AGAG is used as a messenger RNA in an *in vitro* protein-synthesizing system that does not need a start codon. What types of polypeptide chain are expected?

S10.13 In a segment of DNA that contains overlapping genes, would you expect the two proteins to terminate at the same site or have the same number of amino acids? Explain.

S10.14 What unique feature of the 5' end of a eukaryotic mRNA distinguishes it from a prokaryotic mRNA?

S10.15 Consider the two codons AGA and AGG, which code for the amino acid arginine. A, C, U, G, or I is allowed in the 5' position of the anticodon.

(a) What is the minimum number of tRNA^{Arg} types needed to translate these codons?

(b) For each of the tRNA^{Arg} types in part (a), what is the anticodon?

S10.16 A wildtype protein has glutamic acid (Glu) at amino acid position 56. A mutation is found in which glutamic acid is replaced with a different amino acid. Among the reverse mutations that restore protein function, some have threonine (Thr) at the position, some have isoleucine (Ile), and some have glutamic acid. Assuming that the original mutation and all the revertants result from single nucleotide substitutions, determine the wildtype codon for position 56, that of the original mutation, and those of the revertants.

S10.17 What amino acid sequence would correspond to the following mRNA sequence?

5'-GCAUCAGAAUGUAGGGAGACA-3'

(If you use the conventional single-letter abbreviations for the amino acids, shown below, you will find a secret.)

Ala A	Arg R	Asn N	Asp D	Cys C	Gln Q	Glu E
Gly G	His H	Ile I	Leu L	Lys K	Met M	Phe F
Pro P	Ser S	Thr T	Trp W	Tyr Y	Val V	

S10.18 If the following double-stranded DNA molecule is transcribed from the bottom strand, does transcription start at the left or right end? What is the mRNA sequence? If translation starts at the first AUG, what is the polypeptide sequence?

5'-CGCTAGCATGGAGAACGACGAATTGAGCCCAGAAGCGTCATGAGG-3'
3'-GCCATCGTACCTCTTGCTGCTTAACTCGGGTCTTCGCAGTACTCC-5'

S10.19 The following table shows matching regions of the DNA, mRNA, tRNA, and amino acids encoded in a particular gene. The mRNA is shown with its 5' end at the left, and the tRNA anticodon is shown with its 3' end at the left. The vertical lines define the reading frame.

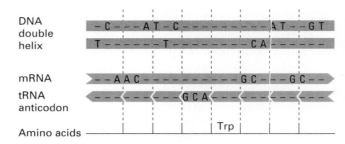

(a) Complete the nucleic acid sequences, assuming Watson-Crick pairing between each codon and anticodon.

(b) Is the DNA strand transcribed the top or the bottom strand?

(c) Translate the mRNA in all three reading frames.

(d) Specify the nucleic acid strand(s) whose sequence could be used as a probe:
  (i) In a Southern blot hybridization, in which the hybridization is carried out against genomic DNA.
  (ii) In a Northern blot hybridization, in which the hybridization is carried out against mRNA.

S10.20 Part of a DNA molecule containing a single short intron has the sequence

5'-ATAGCTCGACGGAAGGTCGACCAAATTTCGCAGGAGCTCGCT-3'
3'-TATCGAGCTGCCTTCCAGCTGGTTTAAAGCGTCCTCGAGCGA-5'

and the corresponding polypeptide has the sequence

Ile—Ala—Arg—Arg—Lys—Leu—Ala

Find the intron.

S10.21 A double mutation produced by recombination contains two single nucleotide frameshifts separated by about 20 base pairs. The first is an insertion and the second a deletion. The amino acid sequence of the wildtype and that of the mutant polypeptide in this part of the protein are

Wildtype: Lys—Lys—Tyr—His—Gln—Trp—Thr—Cys—Asn
Mutant: Lys—Gln—Ile—Pro—Pro—Val—Asp—Met—Asn

What are the original and the double-mutant mRNA sequences? Which nucleotide in the wildtype sequence is the frameshift addition? Which nucleotide in the double-mutant sequence is the frameshift deletion? (In working this problem, it will be convenient to use the conventional symbols Y for unknown pyrimidine, R for unknown purine, N for unknown nucleotide, and H for A, C, T.)

S10.22 The following RNA sequence is a fragment derived from a polycistronic messenger RNA including two genes.

5'-CUUAUGGAAGUAAACCCGAGC-3'

This fragment is known to include the initiation codon of one protein and the termination codon of the other.

(a) Determine the first four amino acids of the protein whose initiation codon is in the fragment.

(b) Determine the last two amino acids of the protein whose translation terminates in the fragment.

(c) What is unusual about these genes?

(d) What would be the consequence of a frameshift mutation that deleted the G from the initiation codon?

S10.23 A strain of *E. coli* carries a mutation that completely inactivates the enzyme encoded in the gene. Several revertants with partly or fully restored activity were selected and the amino acid sequence of the enzyme determined. The only differences found were at position 10 in the polypeptide chain.

> Revertant 1 had Thr.
> Revertant 2 had Glu.
> Revertant 3 had Met.
> Revertant 4 had Arg.

Assume that the initial mutation itself, as well as each revertant, resulted from a single nucleotide substitution.

(a) What amino acid is present at position 10 in the mutant protein?

(b) What codon in the messenger RNA would encode this amino acid?

S10.24 Make a sketch of a mature eukaryotic messenger RNA molecule hybridized to the transcribed strand of DNA of a gene that contains two introns, oriented with the promoter region of the DNA at the left. Clearly label the DNA and the mRNA. Use the following letters to label the location and/or boundaries of each segment. Some letters may be used several times, as appropriate; some, which are not applicable, may not be used at all.

(a) 5' end
(b) 3' end
(c) Promoter region
(d) Attenuator
(e) Intron
(f) Exon
(g) Polyadenylation signal
(h) Leader region
(i) Shine-Dalgarno sequence
(j) Translation start codon
(k) Translation stop codon
(l) 5' cap
(m) Poly-A tail

Chapter 11 The Regulation of Gene Activity

S11.1 Define the following terms:

(a) Enhancer
(b) Attenuator
(c) *Cis*-dominant mutation
(d) Autoregulation
(e) Combinatorial control
(f) Polycistronic mRNA
(g) Transcriptional activation by recruitment

S11.2 How do gene and genome organization differ in eukaryotes and prokaryotes?

S11.3 At what different levels can gene expression be regulated?

S11.4 Give some examples of gene regulation by alterations in the DNA.

S11.5 The regulation of transcription by attenuation cannot take place in eukaryotes. Why not?

S11.6 Why is the *lac* operon of *E. coli* not inducible in the presence of glucose?

S11.7 In eukaryotic organisms, the amounts of some proteins found in the cell are known to vary substantially with environmental conditions, whereas the amount of mRNA for these same proteins remains relatively constant. What mechanisms of regulation might be responsible for this phenomenon?

S11.8 Two genotypes of *E. coli* are grown and assayed for levels of the enzymes of the *lac* operon. Using the information provided here for these two genotypes, predict the enzyme levels for the other genotypes listed in (a) through (d). (The levels of activity are expressed in arbitrary units relative to those observed under the induced conditions.)

Genotype	Uninduced level		Induced level	
	Z	Y	Z	Y
$I^+ O^+ Z^+ Y^+$	0.1	0.1	100	100
$I^+ O^c Z^+ Y^+$	25	25	100	100

(a) $I^- O^+ Z^+ Y^+$
(b) F' $I^+ O^+ Z^- Y^-/I^- O^+ Z^+ Y^+$
(c) F' $I^+ O^+ Z^- Y^-/I^+ O^c Z^+ Y^+$
(d) F' $I^+ O^+ Z^- Y^-/I^- O^c Z^+ Y^+$

S11.9 Imagine a bacterial species in which the methionine operon is regulated only by an attenuator and there is no repressor. In its mode of operation, the methionine attenuator is exactly analogous to the *trp* attenuator of *E. coli*. The relevant portion of the attenuator sequence is

5'-AAA**A**TGATGATGATGATGATGATGGACGAT-3'

The translation start site is located upstream from this sequence, and the AUG codons in the region shown cannot be used for translation initiation. What phenotype (constitutive, wildtype, or *met*⁻) would you expect of each of the following types of mutations? Explain your reasoning.

(a) The boldface A is deleted.
(b) Both the boldface and the underlined A are deleted.
(c) The first three As in the sequence are deleted.

S11.10 How many proteins are bound to the *trp* operon when:

(a) Tryptophan and glucose are present?
(b) Tryptophan and glucose are absent?

(c) Tryptophan is present and glucose is absent?

S11.11 Why are mutations of the *lac* operator often called *cis*-dominant? Why are some constitutive mutations of the *lac* repressor (*lacI*) called *trans*-recessive? Can you think of a way in which a noninducible mutation in the *lacI* gene might be *trans*-dominant?

S11.12 An *E. coli* mutant strain synthesizes β-galactosidase whether or not the inducer is present. What genetic defect(s) might be responsible for this phenotype?

S11.13 Temperature-sensitive mutations in the *lacI* gene of *E. coli* render the repressor nonfunctional (unable to bind the operator) at 42°C but leave it fully functional at 30°C. Under which of the following conditions would β-galactosidase be produced?

(a) In the presence of lactose at 30°C
(b) In the presence of lactose at 42°C
(c) In the absence of lactose at 30°C
(d) In the absence of lactose at 42°C

S11.14 Many metabolic pathways are regulated by so-called feedback inhibition, in which the first enzyme in the pathway becomes inhibited by interacting with the final product of the entire pathway if the final product is present in a sufficient amount. Explain why a metabolic pathway should evolve in such a way that the first enzyme in the pathway is inhibited rather than the last enzyme in the pathway.

S11.15 Among cells of a *lacI⁺ lacO⁺ lacZ⁺ lacY⁺* strain of *E. coli*, a mutation was found in which the *lacY* gene underwent a precise inversion without any change in the coding region. However, the strain was found to be defective in the production of the permease. Why?

S11.16 Cristina, a young immunologist, told her friend Jorge, who studied molecular biology, that there are 10^8 different antibodies produced by human B cells. To her astonishment, Jorge strongly disagreed, saying that there are not enough genes in the human genome to code for such a great variety of immunoglobulins. Who is right? Is there any way to reconcile their views?

S11.17 You are engaged in the study of a gene, the transcription of which is activated in a particular tissue by a steroid hormone. From cultured cells isolated from this tissue, you are able to isolate mutant cells that no longer respond to the hormone. What are some possible explanations?

S11.18 A laboratory strain of *E. coli* produces β-galactosidase constitutively and permease inducibly, despite the fact that these two proteins are co-regulated in the *lac* operon in wild-type strains. What is the genotype of the *lac* operon in this strain?

Chapter 12   The Genetic Control of Development

S12.1 Define the following terms:

(a) Embryonic induction
(b) Gain-of-function mutation
(c) Fate map
(d) Pair-rule gene
(e) Segment-polarity gene
(f) Homeobox
(g) Homeotic mutation

S12.2 What is the relationship between maternal-effect genes and zygotic genes?

S12.3 In the early development of *C. elegans*, what is the role of the anchor cell in the differentiation of the vulva?

S12.4 When in the development of most metazoan animals is the polarity of the embryo determined?

S12.5 The same transmembrane receptor protein encoded by the *lin-12* gene is used in the determination of different developmental fates. What is the principal difference between the two types of target cells that develop differently in response to Lin-12 stimulation?

S12.6 Programmed cell death (apoptosis) is responsible in part for shaping many organs and tissues in normal development. If a group of cells in the duck leg primordium that are destined to die are transplanted from their normal leg site to another part of the embryo just prior to the time they would normally die, they still die on schedule. The same operation performed a few hours earlier rescues the cells, and they do not die. How can you explain this observation?

S12.7 Classify each of the following mutations:

(a) A mutation in *C. elegans* in which a cell that normally produces two distinct daughter cells gives rise to two identical cells.
(b) A mutation in *Drosophila* that causes an antenna to appear at the normal site of a leg.
(c) A lethal mutation in *Drosophila* that is responsible for abnormal gene expression in alternating segments of the embryo

S12.8 What causes induction of the morphogenic events in the *Drosophila* imaginal disks? When does this induction happen?

S12.9 If a small amount of cytoplasm is removed from the posterior end of the *Drosophila* embryo before blastoderm formation, the mature fly that develops from this embryo is normal in every way except it does not have germ cells. How can you explain this finding?

S12.10 How would you prove that a newly discovered mutation has a maternal effect?

S12.11 A recessive mutation *smooth* is found in a species of flower in which hairs (trichomes) normally present on the leaf surface are missing. Another mutation in the same species, called *hairy*, results in the presence of trichomes on the flower petals, where they are not normally formed. Cloning and sequencing indicate that these mutations are different alleles of the same gene. Molecular analysis reveals that *smooth* is a frameshift mutation near the start of the coding sequence and that *hairy* has an upstream insertion of a transposable element that allows transcription of the gene in petal tissue. What features of these observations suggest that the gene in question is an important gene for the regulation of trichome formation?

S12.12 Two classes of genes involved in segmentation of the *Drosophila* embryo are gap genes, which are expressed in one region of the developing embryo, and pair-rule genes, which are expressed in seven stripes. Homozygotes for mutations in gap genes lack a continuous block of larval segments; homozygous mutations in pair-rule genes lack alternating segments. You examine gene expression by mRNA *in situ* hybridization and find that (1) the embryonic expression pattern of gap genes is normal in all pair-rule mutant homozygotes, and (2) the pair-rule gene expression pattern is abnormal in all gap gene mutant homozygotes. What do these observations tell you about the temporal hierarchy of gap genes and pair-rule genes in the developmental pathway of segmentation?

S12.13 A paracentric inversion in *Drosophila* causes a dominant mutation that produces small wings. You mutagenize flies homozygous for this inversion with x rays and cross them to normal mates. Most progeny that arise have small wings, but a few rare progeny have normal wings. These rare progeny still have the original inversion but also contain a deletion spanning the more proximal of the two inversion breakpoints. (Proximal means closer to the centromere.) You correctly conclude from these observations that the dominant mutation is due to a genetic "gain of function" rather than to a reduction in gene activity. Explain how these results support this conclusion, and suggest an interpretation of how the gain-of-function phenotype arises.

S12.14 One approach to the identification of important developmental genes in the mouse is to focus on genes whose DNA sequences are similar to those that have significant developmental roles in other organisms. Suppose that you want to study the mouse homolog of the *lin-3* gene of *C. elegans*, which is important in vulva formation in the worm. Homozygotes for a null mutation of *lin-3* lack a vulva. You propose that a homolog of *lin-3* exists in the mouse, and you want to examine the phenotype of a mutant mouse that lacks a functional form of this gene. Arrange the following steps in the order in which you would perform them to generate a mouse homozygous for a null mutation of the gene.

(a) Identify an open reading frame in a mouse cDNA that is similar to the *C. elegans lin-3* open reading frame.

(b) Create a deletion in a mouse genomic clone, and insert a gene resistant to the drug G418.

(c) Breed chimeric mice to donor strains to identify heterozygotes for a null mutation.

(d) Inject embryonic stem cells into differentially marked mouse blastocysts.

(e) Probe a mouse wildtype genomic library with a labeled mouse cDNA.

(f) Probe a mouse cDNA library with a labeled *C. elegans lin-3* cDNA.

(g) Select for embryonic stem cells that can grow on G418.

(h) Intercross null mutant heterozygotes to obtain a homozygous null mouse.

(i) Identify chimeric mice.

(j) Inject the DNA of an engineered mutation in the mouse *lin-3* homolog into embryonic stem cells.

(k) Identify embryonic stem cells that have a null mutation in the mouse *lin-3* homolog.

S12.15 Edward B. Lewis, Christianne Nüsslein-Volhard, and Eric Weichaus shared a 1995 Nobel Prize in Physiology or Medicine for their work on the developmental genetics of *Drosophila*. In their screen for developmental genes, Nüsslein-Volhard and Weichaus initially identified 20 lines bearing maternal-effect mutations that produced embryos lacking anterior structures but having the posterior structures duplicated. When Nüsslein-Volhard mentioned this result to a colleague, he was astonished to hear that mutations in 20 genes could give rise to this phenotype. Explain why his astonishment was completely unfounded.

S12.16 The gene *bicoid* exemplifies a maternal-effect gene that, when mutated, produces embryos that lack anterior structures but have the posterior structures duplicated. The Bicoid protein is a morphogen that establishes the domain of expression of the gap gene *hunchback* in a concentration-dependent manner. The Bicoid protein is usually present in an anterior-to-posterior gradient with the highest concentration at the anterior end of the early embryo. Diagrammed below is the major domain of expression of *hunchback* in an embryo from a wildtype mother.

hunchback **expression pattern**

Anterior Posterior

What are the expected domains of *hunchback* expression in embryos from mothers with the following genotypes. Express your answers in the form of diagrams similar to the one here.

(a) Homozygous mutant for *bicoid*

(b) Homozygous for a duplication of a wildtype copy of *bicoid*

(c) Homozygous for transgene that causes bicoid to be localized at both poles of the embryo

Chapter 13 Mutation, DNA Repair, and Recombination

S13.1 Define the following terms:

(a) Gene conversion
(b) Mismatch repair
(c) Silent mutation
(d) Photoreactivation
(e) Transversion mutation

S13.2 Prior to the early 1940s, some bacteriologists maintained that the exposure of a bacterial population to an antibiotic or other selective agent induced mutations that enabled the bacteria to survive. Which of the following statements describes what Joshua and Esther Lederberg demonstrated with the replica plating technique?

(a) Streptomycin caused the formation of streptomycin-resistant bacteria.
(b) Streptomycin revealed the presence of streptomycin-resistant bacteria.
(c) Mutations are usually beneficial.
(d) Mutations are usually deleterious.
(e) None of the above.

S13.3 If base substitutions were random, what ratio of transversions to transitions would be expected?

S13.4 In the real world, spontaneous base substitutions are biased in favor of transitions. What is the consequence of this bias for base substitutions in the third position of codons?

S13.5 How many different mutations can result from a single base substitution in the unique tryptophan codon TGG? Classify each as silent, missense, or nonsense.

S13.6 What is the minimum number of single-nucleotide substitutions that would be necessary for each of the following amino acid replacements?

(a) Trp → Lys
(b) Tyr → Gly
(c) Met → His
(d) Ala → Asp

S13.7 Every human gamete contains, very approximately, 50,000 genes. If a mutation rate is between 10^{-5} and 10^{-6} new mutations per gene per generation, what percentage of all gametes will carry a gene that has undergone a spontaneous mutation?

S13.8 If a mutation rate is between 10^{-5} and 10^{-6} per gene per generation, what does this imply in terms of the number of mutations of a particular gene per million gametes per generation?

S13.9 If the spontaneous mutation rate of a gene is 10^{-6} mutations per generation, and a dose of ionizing radiation of 10 grays doubles the rate of mutation, what mutation rate would be expected with 1 gray of radiation? Assume that the rate of induced mutation is proportional to the radiation dose.

S13.10 In theory, deamination of both cytosine and 5-methylcytosine results in a mutation. However, nucleotide substitutions produced by loss of the amino group from cytosine are rarely observed, although those produced by loss of the amino group from 5-methylcytosine are relatively frequent. What is a possible explanation for this phenomenon?

S13.11 The mismatch repair system can detect mismatches that occur in DNA replication and can correct them after replication. In *E. coli*, how does the system use DNA methylation to determine which base in a pair is the incorrect one?

S13.12 If one of the many copies of the tRNA gene for Glu underwent a mutation in which the anticodon became CUA instead of CUC, what would be the result? (*Hint:* The key is that only one of many tRNA copies is mutated.)

S13.13 A strain of *E. coli* contains a mutant Leu tRNA. Instead of recognizing the codon 5'-CUG-3', as it does in nonmutant cells, the mutant tRNA now recognizes the codon 5'-GUG-3'. A missense mutation of another gene, affecting amino acid number 28 along the chain, is suppressed in cells with the mutant Leu tRNA.

(a) What are the anticodons of the wildtype and mutant Leu tRNA molecules?
(b) What kind of mutation is present in the mutant Leu tRNA gene?
(c) What amino acid would be inserted at position 28 of the mutant polypeptide chain if the missense mutation were not suppressed?
(d) What amino acid is inserted at position 28 when the missense mutation is suppressed?

S13.14 A mutation is selected after treatment with a particular mutagen, and organisms containing the mutation are then mutagenized to find revertants.

(a) Can a mutation caused by hydroxylamine be induced to revert to the wildtype sequence by re-exposure to hydroxylamine?
(b) Can a mutation caused by hydroxylamine be induced to revert by nitrous acid?

(Hydroxylamine reacts specifically with cytosine and causes GC R AT transitions; nitrous acid causes deamination of A and C).

S13.15 The tryptophan synthase of *E. coli* is a tetrameric enzyme formed by the polypeptide products of *trpB* and *trpA* genes. The wildtype subunit encoded by *trpB* has a glycine in position 38, whereas two mutants, B13 and B32, have Arg and Glu at this position, respectively. Plating these mutants on minimal medium sometimes yields spontaneous revertants to prototrophy. Four independent revertants of B13 have Ile, Thr, Ser, and Gly in position 38, and three independent revertants of B32 have Gly, Ala, and Val in the same position. What codon is in position 38 in the wildtype *trpB* gene?

S13.16 In the *rIIB* gene of bacteriophage T4, there is a small coding region near the beginning of the gene in which the particular amino acid sequence that is present is not essential for protein function. (Single-base insertions and deletions in this region were used to prove the triplet nature of the genetic code.) Under what condition would a +1 frameshift mutation at the beginning of this region not be suppressed by a −1 frameshift mutation at the end of the region?

S13.17 *E. coli* mutation *lacZ-1* was induced by acridine treatment, whereas *lacZ-2* was induced by 5-bromouracil. What kind of mutations are they likely to be? What sort of mutant β-galactosidase will these mutations produce?

S13.18 Is it correct to say that a Holliday structure and a chiasma are the same?

S13.19 When two mutations are closely linked (for example, two nucleotide substitutions in the same gene), a single gene conversion event may include both sites. This phenomenon is called co-conversion. How would you expect the frequency of co-conversion to depend on the distance between mutant nucleotide sites?

S13.20 In spite of gene conversion, among gametes chosen at random, one still observes the Mendelian segregation ratio of 1 : 1 in crosses. Why?

S13.21 You carry out a large-scale cross of genotypes $A\ m\ B \times a + b$ of two strains of the mold *Neurospora crassa* and observe a number of aberrant asci, some of which are shown below.

A m B	A m B	A m B	A m B	A m B
A m B	A m B	A m B	A m B	A m B
A m b	A m b	A m b	A m b	A + b
A + b	A m b	A + b	A m b	A + b
a m B	a m B	a + B	a m B	a + B
a + B	a + B	a + B	a m B	a + B
a + b	a + b	a + b	a + b	a + b
a + b	a + b	a + b	a + b	a + b
(1)	**(2)**	**(3)**	**(4)**	**(5)**

The colleague who provided the strains insists that they are both deficient in the same gene in the DNA mismatch repair pathway. The results of your cross seem to contradict this assertion.

(a) Which of the asci depicted above would you exhibit as evidence that your colleague is incorrect?

(b) Which ascus (or asci) would you exhibit as definitive evidence that DNA mismatch repair in these strains is not 100 percent efficient?

(c) Among all asci resulting from meioses in which heteroduplexes form across the *m* + region, what proportion of acsi will be tetratype for the *A* and *B* markers and also show 4 : 4 segregation of *m* : +, under the assumption that 30 percent of heteroduplexes in the region are not corrected, 40 percent are corrected to *m*, and 30 percent are corrected to +?

S13.22 In molecular evolutionary studies, biologists often regard serine as though it were two different amino acids, one type of "serine" corresponding to the fourfold-degenerate UCN codon and the other type of "serine" corresponding to the twofold-degenerate AGY codon. Why does it make sense to regard serine in this way but not the other amino acids encoded by six codons: leucine (UUR and CUN) and arginine (AGR and CGN)? (In this symbolism for the nucleotides, Y is any pyrimidine, R any purine, and N any nucleotide.)

Chapter 14   Extranuclear Inheritance

S14.1 Define the following terms:

(a) Maternal inheritance
(b) Maternal effect
(c) Cytoplasmic male sterility
(d) Heteroplasmy
(e) RNA editing
(f) Phylogenetic tree

S14.2 A woman with the mild form of a mitochondrial disease, Leber's hereditary optic neuropathy (LHON), has four children: two with severe disease expression, one unaffected, and one with the same level of expression as the mother. Explain this pattern of inheritance.

S14.3 How do mitochondrial and nuclear genomes interact at the functional level?

S14.4 A dwarfed corn plant is crossed as a female parent to a normal plant, and the F_1 progeny are found to be dwarfed. The F_1 plants are self-fertilized, and the F_2 plants are found to be normal. Each F_2 plant is self-fertilized, and 3/4 of F_3 generation are found to be normal whereas 1/4 of the plants are dwarfed. Which of the following hypotheses are consistent with these results? Explain your reasoning.

(a) Cytoplasmic inheritance
(b) Nuclear inheritance
(c) Maternal effect
(d) Cytoplasmic inheritance plus mutation in a nuclear gene

S14.5 A strain of yeast with a mutation *ery-r* conferring resistance to the antibiotic erycetin is crossed with a strain of opposite mating type with the genotype *ery-s*. After meiosis and sporulation, three tetrads are isolated. The spores have the following genotypes:

I	II	III
a *ery-r*	α *ery-s*	α *ery-r*
a *ery-r*	α *ery-s*	**a** *ery-r*
α *ery-r*	**a** *ery-s*	**a** *ery-r*
α *ery-r*	**a** *ery-s*	α *ery-r*

(a) What genetic hypothesis can explain these results?
(b) If an *ery-r* strain were used to generate petites, would you expect some of the petites to be sensitive to erycetin? Explain.

S14.6 In a fungal organism closely related to *Saccharomyces cerevisiae,* a strain resistant to three antibiotics is used to induce petite colonies. The antibiotics are erycetin, chlorpromanin, and paralysin, and the resistance genes are denoted *ery-r, chl-r,* and *par-r,* respectively. Each of the resistance genes is known to be present in mitochondrial DNA. Each row in the accompanying table is the result of an experiment in which petite colonies selected for loss of genetic marker A (one of the antibiotic resistances) were tested for simultaneous loss of genetic marker B (another of the antibiotic resistances).

Marker A	Marker B	Loss of marker A only	Loss of markers A and B
ery-r	chl-r	170	19
ery-r	par-r	54	81
chl-r	par-r	116	8

From these data, deduce the order of the genes and the relative distances between them in the mitochondrial DNA.

S14.7 Two haploid strains of *Neurospora* are isolated that have genetic defects in the ATPase complex. The mutants are designated $A1^-$ and $A2^-$. The crosses $A1^- \times A1^+$ and $A2^- \times A2^+$ were carried out, and one ascus was analyzed from each cross. The cross $A1^- \times A1^+$ yielded a 4 : 4 ratio of $A1^+ : A1^-$ ascospores, whereas the cross $A2^- \times A2^+$ yielded an 8 : 0 ratio of $A2^+ : A2^-$ ascospores.

(a) What difference between the mutations can explain these discordant results?

(b) What other types of tetrads would one expect from these crosses?

S14.8 Mothers addicted to cocaine who use the drug during pregnancy can give birth to babies who are also addicted. Suggest a mechanism for this maternal effect.

S14.9 If organelles were originally symbionts, why don't we call them symbionts instead of organelles?

S14.10 How would you prove the existence of RNA editing in mitochondria?

S14.11 What is controversial about the hypothesis of a mitochondrial Eve who lived in Africa? Is the controversial part the conclusion that there is a common ancestor of human mitochondrial DNA, or is it the inference of the geographical location of the common ancestor? Explain.

Chapter 15 Population Genetics and Evolution

S15.1 Define the following terms:

(a) Population
(b) Natural selection
(c) Random genetic drift
(d) Polymorphic gene
(e) Selectively neutral mutation
(f) Selection-mutation balance

S15.2 In a large, randomly mating herd of cattle, 16 percent of the newborn calves have a certain type of dwarfism because of a homozygous recessive allele. Assume Hardy-Weinberg frequencies.

(a) What is the frequency of the recessive dwarfing allele?
(b) What is the frequency of heterozygous carriers in the herd?
(c) Among the cattle that are nondwarfs, what is the frequency of carriers?

S15.3 A population of fish includes 1 percent albinos resulting from a homozygous recessive allele that eliminates the normal pigmentation. Assuming Hardy-Weinberg proportions, determine the frequency of heterozygous genotypes.

S15.4 Hereditary hemochromatosis, a defect in iron metabolism resulting in impaired liver function, arthritis, and other symptoms, is one of the most common genetic diseases in northern Europe. The disease is relatively mild and easily treated when recognized. About one person in 400 is affected. Assuming Hardy-Weinberg proportions, determine the expected frequency of carriers.

S15.5 An inherited defect in a certain receptor protein confers resistance to infection with HIV1 (human immunodeficiency virus 1, the cause of AIDS). Resistant people are homozygous recessive and are found at a frequency of 1 percent in the population. Assuming Hardy-Weinberg proportions, determine the frequency of heterozygous genotypes.

S15.6 Organisms were sampled from three different populations polymorphic for the alleles *A* and *a* and yielded these observed numbers of each of the possible genotypes:

(a) *AA*: 5   *Aa*: 60   *aa*: 75   Total: 140
(b) *AA*: 10   *Aa*: 50   *aa*: 80   Total: 140
(c) *AA*: 15   *Aa*: 40   *aa*: 85   Total: 140

In terms of the magnitude of the chi-square obtained in a test of goodness of fit to Hardy-Weinberg proportions, which of these samples is closest to Hardy-Weinberg proportions? Do any of the samples deviate significantly from Hardy-Weinberg proportions?

S15.7 In a population in Hardy-Weinberg equilibrium for a recessive allele at frequency q, what value of q is necessary for the ratio of carriers to affected organisms to equal each of the following?

(a) 10
(b) 100
(c) 1000

S15.8 A man is known to be a carrier of the cystic fibrosis allele. He marries a phenotypically normal woman. In the general population, the incidence of cystic fibrosis at birth is approximately one in 1700. Assume Hardy-Weinberg proportions.

(a) What is the probability that the wife is also a carrier?

(b) What is the probability that their first child will be affected?

S15.9 A man with normal parents whose brother has phenylketonuria marries a phenotypically normal woman. In the general population, the incidence of phenylketonuria at birth is approximately one in 10,000. Assume Hardy-Weinberg proportions.

(a) What is the probability that the man is a carrier?
(b) What is the probability that the wife is also a carrier?
(c) What is the probability that their first child will be affected?

S15.10 What genotype frequencies constitute the Hardy-Weinberg principle for a gene in a large, randomly mating allotetraploid population with two alleles at frequencies p and q?

S15.11 A large population includes 10 alleles of a gene. The ratio of the allele frequencies forms the arithmetic progression $1 : 2 : 3 : \ldots : 8 : 9 : 10$. The genotypes are in Hardy-Weinberg proportions.

(a) What ratio is formed by the frequencies of the homozygous genotypes?
(b) Among all heterozygous genotypes that carry the most common allele, what ratio is formed by the frequencies of the heterozygous genotypes?

S15.12 For an X-linked gene with two alleles in a large, randomly mating population, the frequency of carrier females equals one-half of the frequency of the males carrying the recessive allele. What are the allele frequencies?

S15.13 A trait due to a harmful recessive X-linked allele in a large, randomly mating population affects one male in 50. What is the frequency of carrier females? What is the expected frequency of affected females?

S15.14 In DNA typing, how many genes must be examined to prove that evidence left at a crime scene comes from a suspect? How many genes must be examined to prove that the evidence does not come from a suspect?

S15.15 The DNA type of a suspect and that of blood left at the scene of a crime match for each of n genes, and each gene is heterozygous and so yields two bands. The frequency of each of the bands in the general population is 0.1. Use the product rule to calculate:

(a) The number of genes for which the probability of a match by chance is less than one in a million.
(b) The number of genes for which the probability of a match by chance is less than one in ten billion.

S15.16 Two strains of bacteria, A and B, are placed into direct competition in a chemostat. A is favored over B. What is the value of the selection coefficient (s) if, after one generation, the ratio of the number of A cells to the number of B cells:

(a) Increases by 10 percent?
(b) Increases by 90 percent?
(c) Increases by a factor of 2?

S15.17 Two strains of bacteria, A and B, are placed into direct competition in a chemostat. A is favored over B. If the selection coefficient per generation is constant, what is its value if, in an interval of 100 generations:

(a) The ratio of A cells to B cells increases by 10 percent?
(b) The ratio of A cells to B cells increases by 90 percent?
(c) The ratio of A cells to B cells increases by a factor of 2?

S15.18 A strain of pathogenic bacteria undergoes cell division exactly two times as often as a nonpathogenic mutant strain of the same bacteria. What is the relative fitness of the pathogenic strain compared to the nonpathogenic strain? You may regard the generation time as the length of time for one division of a nonpathogenic cell.

S15.19 An allele A undergoes mutation to the allele a at the rate of 10^{-5} per generation. If a very large population is fixed for A (generation 0), what is the expected frequency of A in the following generation (generation 1)? What is the expected frequency of A in generation 2? Deduce the rule for the frequency of A in generation n.

S15.20 Large samples are taken from a single population of each of four related species of the flowering plant *Phlox*. The genotype frequencies for a gene with two alleles are estimated from each of the samples, and this yields the genotype frequencies in the accompanying table. For each species, calculate the allele frequencies and the genotype frequencies expected using the Hardy-Weinberg principle. Identify which of the populations give evidence of inbreeding, and for each apparently inbreeding population, calculate the inbreeding coefficient.

| Species | Genotype frequency | | |
	AA	*Aa*	*aa*
(a)	0.0720	0.2560	0.6720
(b)	0.0400	0.3200	0.6400
(c)	0.0080	0.3840	0.6080
(d)	0.0880	0.2240	0.6880

S15.21 The considerations in this problem illustrate the principle that for a rare, harmful, recessive allele, matings between relatives can account for a majority of affected individuals even though they account for a small proportion of all matings. Suppose that the frequency of the rare homozygous recessive genotype among the progeny of nonrelatives is one in 4 million and that first-cousin matings constitute 1 percent of all matings in the population.

(a) What is the frequency of the homozygous recessive among the offspring of first-cousin matings?
(a) Among all homozygous recessive genotypes, what proportion come from first-cousin matings?

S15.22 A diploid population consists of 50 breeding organisms, of which one organism is heterozygous for a new mutation. If the population is maintained at a constant size of 50, what is the probability that the new mutation will be lost by chance in the first generation as a result of random genetic drift? (You may assume that the species is monoecious.) What is the answer in the more general case in which the breeding population consists of N diploid organisms?

S15.23 A population segregating for the alleles A and a is sampled, and genotypes are found in the proportions 1/3 AA, |1/3 Aa, and 1/3 aa. These are not Hardy-Weinberg proportions. How big would the size of the sample have to be for the resulting P value to equal 0.05 in a chi-square test for goodness of fit?

S15.24 An observed sample of a gene with two alleles from a population conforms almost exactly to the genotype frequencies expected for a population with an inbreeding coefficient F and allele frequencies 0.2 and 0.8. What is the smallest value of F that will lead to rejection of Hardy-Weinberg proportions at the 5 percent level of significance in each of the following cases? (*Hint*: Set chi-square equal to 4.0.)

(a) A sample size of 200 organisms.
(b) A sample size of 500 organisms.
(c) A sample size of 1000 organisms.

Chapter 16 Quantitative Genetics

S16.1 Define the following terms:

(a) Multifactorial trait
(b) Quantitative-trait locus (QTL)
(c) Threshold trait
(d) Individual selection
(e) Broad-sense heritability
(f) Genotype-environment association

S16.2 A population of sweet clover includes 20 percent with two leaves, 70 percent with three leaves, and 10 percent with four leaves. What are the mean and variance in leaf number in this population?

S16.3 In an experimental population of the flour beetle *Tribolium castaneum*, the pupal weight is distributed normally with a mean of 2.0 mg and a standard deviation of 0.2 mg. What proportion of the population is expected to have a pupal weight between 1.8 and 2.2 mg? Between 1.6 and 2.4 mg? Would you expect to find an occasional pupa weighing 3.0 mg or more? Explain your answer.

S16.4 Tabulated below are the numbers of eggs laid by 50 hens over a 2-month period. The hens were selected at random from a much larger population. Estimate the mean, variance, and standard deviation of the distribution of egg number in the entire population from which the sample was drawn.

48	50	51	47	54	45	50	38	40	52
58	47	55	53	54	41	59	48	53	49
51	37	31	47	55	46	49	48	43	68
59	51	52	66	54	37	46	55	59	45
44	44	57	51	50	57	50	40	63	33

S16.5 Suppose that the 50 hens in the previous problem constituted the entire population of hens present in a flock instead of a sample from a larger flock. What are the mean, variance, and standard deviation of egg number in the population? Why are some of these numbers different from those calculated in the previous problem?

S16.6 Two homozygous genotypes of *Drosophila* differ in the number of abdominal bristles. In genotype AA, the mean bristle number is 20 with a standard deviation of 2. In genotype aa, the mean bristle number is 23 with a standard deviation of 3. Both distributions conform to the normal curve, in which the proportions of the population that have a phenotype within an interval defined by the mean ± 1, ± 1.5, ± 2, and ± 3 standard deviations are 68, 87, 95, and 99.7 percent, respectively.

(a) In genotype AA, what is the proportion of flies with a bristle number between 20 and 23?
(b) In genotype aa, what is the proportion of flies with a bristle number between 20 and 23?
(c) What proportion of AA flies have a bristle number greater than the mean of aa flies?
(d) What proportion of aa flies have a bristle number greater than the mean of AA flies?

S16.7 If the genotypes in the previous problem were crossed and the F_1 flies mated to produce an F_2 generation, would you expect the distribution of bristle number among F_2 flies to show segregation of the A and a alleles? Explain your answer.

S16.8 A meristic trait is determined by a pair of alleles at each of four unlinked loci: A, a and B, b and C, c and D, d. The alleles act in an additive fashion such that each capital letter adds one unit to the phenotypic score. Thus the phenotype ranges from a score of 0 (genotype $aa\ bb\ cc\ dd$) to 8 (genotype $AA\ BB\ CC\ DD$). What is the expected distribution of phenotype score in the F_2 generation of the cross $aa\ bb\ cc\ dd \times AA\ BB\ CC\ DD$? Draw a histogram of the distribution.

S16.9 With respect to the trait in the previous problem, what is the expected distribution of phenotypes in a randomly mating population in which the frequency of each allele indicated with a capital letter is 0.8 and in which the alleles of the different loci are combined at random to produce the genotypes?

S16.10 The F_1 generation of a cross between two inbred strains of maize has a standard deviation in mature plant height of 15 cm. When grown under the same conditions, the F_2 generation produced by the $F_1 \times F_1$ cross has a standard deviation in plant height of 25 cm. What are the genotypic variance and the broad-sense heritability of mature plant height in the F_2 population?

S16.11 Two inbred strains of *Drosophila* are crossed and the F_1 generation is brother-sister mated to produce a large, genetically heterogeneous F_2 population that is split into two parts. One part is reared at a constant temperature of 25°C, and the other part is reared under conditions in which the temperature fluctuates among 18°C, 25°C, and 29°C. The broad-sense heritability of body size at the constant temperature is 45 percent, but under fluctuating temperatures it is 15 percent. If the genotypic variance is the same under both conditions, by what factor does the temperature fluctuation increase the environmental variance?

S16.12 The narrow-sense heritability of withers height in a population of quarterhorses is 30 percent. (Withers height is the height at the highest point of the back, between the shoulder blades.) The average withers height in the population is 17 hands. (A "hand" is a traditional measure equal to the breadth of the human hand, now taken to equal 4 inches.) From this population, studs and mares with an average withers height of 16 hands are selected and mated at random. What is the expected withers height of the progeny? How does the value of the narrow-sense heritability change if withers height is measured in meters rather than hands?

S16.13 The narrow-sense heritability of adult stature in people is about 25 percent. What does this value imply about the correlation coefficient in adult height between father and son?

S16.14 Suppose that the narrow-sense heritability of adult stature in people equals 25 percent. A man whose height is 5 cm above the average height for men marries a woman whose average height equals that for women. Among their progeny, what is the expected average deviation from the mean adult height in the entire population?

S16.15 From a population of *Drosophila* with an average abdominal bristle number of 23.2, flies with an average bristle number of 27.6 were selected and mated at random to produce the next generation. The average bristle number among the progeny was 25.4. What is the realized heritability? Is this a narrow-sense or a broad-sense heritability? Explain your answer.

S16.16 In human beings, there is about a 90 percent correlation coefficient between identical twins in the total number of raised skin ridges forming the fingerprints. What is the broad-sense heritability of total fingerprint ridge count? What is the maximum value of the narrow-sense heritability?

S16.17 In human beings, the correlation coefficient between first cousins in the total fingerprint ridge count is 10 percent. On the basis of this value, what is the narrow-sense heritability of this trait?

S16.18 The narrow-sense heritability of 6-week body weight in mice is 25 percent. From a population whose average 6-week weight is 20 grams, individual mice weighing an average of 24 grams are selected and mated at random. What is the expected weight of the progeny mice in the next generation?

S16.19 In the mouse population in the problem above, the selection for greater 6-week weight is carried out for a total of five generations, and in each generation, parents are chosen with an average 6-week weight that is, on average, 4 grams heavier than the population mean in their generation. What is the expected average 6-week weight after the five generations of selection? Make a graph of average 6-week weight against generation number.

S16.20 Suppose that the selection for 6-week weight in the mouse population above were carried out in both directions for five generations. In each generation, the parents in the "up" selection are, on average, 4 grams heavier than the population mean, and the parents in the "down direction" are, on average, 4 grams lighter than the population mean.

(a) In each generation, how different are the expected means of the "up" and "down" populations?

(b) Draw a graph illustrating the expected course of selection in both populations.

(c) In the "down" population, is the realized heritability of 6-week weight expected to remain 25 percent indefinitely?

S16.21 Two inbred lines of maize differ in ear length. In one, the average length of ear is 30 cm, and in the other it is 15 cm. In both lines the variance in ear length is 2 cm^2. In the F_2 generation of a cross between the lines, the average ear length is 22.5 cm with a variance of 5 cm^2. What is the minimum number of genes required to account for these data?

S16.22 A threshold trait affects 2.5 percent of the organisms in a population. The underlying liability is a multifactorial trait that is normally distributed with mean 0 and variance 1.

(a) What is the value of the threshold, a liability above which causes the appearance of the condition?

(b) If the underlying liability were normally distributed with a mean of 1 and a variance of 4, what would be the corresponding value of the threshold?

(c) Is there any real difference between these two situations?

Chapter 17 Genetics of Biorhythms and Behavior

S17.1 Define the following terms:

(a) Specific nonchemotaxis

(b) Multiple nonchemotaxis

(c) General nonchemotaxis

(d) Sensory adaptation

(e) Circadian rhythm

(f) Genotype-enviroment interaction

S17.2 What is the target of the molecular pathway common to all forms of chemotaxis?

S17.3 What is the role of methylation in sensory adaptation? What is the expected phenotype of a mutant in *cheR* that cannot transfer methyl groups?

S17.4 What period of courtship song would you expect of *D. melanogaster* males hemizygous for the *per⁰* allele?

S17.5 What kind of an experiment would enable you to demonstrate that *per* controls the expression of other genes?

S17.6 How does the circadian system of *Drosophila* become reset each day so that the fly's intrinsic period becomes synchronized to the 24-hour day?

S17.7 You are studying two groups of laboratory mice raised in similar environments. One is a highly inbred strain, and the other is derived from a large randomly mating population. Which group would you expect to have a higher value of the broad-sense heritability of a particular behavioral trait? Explain your reasoning.

S17.8 Suppose that the broad-sense heritability of IQ score is 0.5. Can you conclude that genes are responsible for 50 percent of the IQ score of a particular person? Why or why not?

S17.9 Explain why twins and adoption studies are useful in genetic research on behavioral traits in human beings. Why are ordinary family studies not sufficient?

S17.10 For a given actual population size, how would the magnitude of random genetic drift differ in two animal populations, one with a strong social hierarchy and another lacking an established social structure?

S17.11 It has been reported that males of *Drosophila* with a genotype that is uncommon in the population have greater reproductive success than males with more common genotypes. What effect would a rare-male mating advantage have on the genetic composition of a population of flies?

S17.12 Humans have changed from a species living in small hunting populations to a species many members of which live in large and highly organized communities. Have those changes also led to changes in the relative contribution of different evolutionary factors (such as, for instance, random genetic drift) to the shaping of human genotypic structure?

S17.13 Consider a complex behavioral trait in which the environmental variance consists solely of the variance that results from errors in measurement. In other words, the behavioral trait is completely determined genetically but is difficult to measure because of random fluctuations in measurement. For a single measurement of the trait, let the environmental variance be represented as σ_e^2. If errors in measurement are the only cause of environmental variance, then it can be shown that the environmental variance for the mean of n independent measurements is σ_e^2/n. For such a trait with a broad-sense heritability of $H^2 = \sigma_g^2/(\sigma_g^2 + \sigma_e^2)$ for a single measurement, show that the broad-sense heritability of the mean of n measurements, H'^2, equals $nH^2/[1 + (n-1)H^2]$. (*Hint*: Find an expression for σ_e^2/σ_g^2 in terms of H^2.)

A

aberrant 4 : 4 segregation   The presence of equal numbers of alleles among the spores in a single ascus, yet with a spore pair in which the two spores have different genotypes.

acentric chromosome   A chromosome with no centromere. (p. 260)

acridine   A chemical mutagen that intercalates between the bases of a DNA molecule, causing single-base insertions or deletions. (p. 571)

acrocentric chromosome   A chromosome with the centromere near one end. (p. 261)

active site   The part of an enzyme at which substrate molecules bind and are converted into their reaction products.

acylated tRNA   A tRNA molecule to which an amino acid is linked. (p. 431)

adaptation   Any characteristic of an organism that improves its chance of survival and reproduction in its environment; the evolutionary process by which a species undergoes progressive modification favoring its survival and reproduction in a given environment. (p. 652)

addition rule   The principle that the probability that any one of a set of mutually exclusive events is realized equals the sum of the probabilities of the separate events. (p. 49)

additive variance   The magnitude of the genetic variance that results from the additive action of genes and that accounts for the genetic component of the parent-offspring correlation; the value that the genetic variance would assume if there were no dominance or interaction of alleles affecting the trait.

adenine (A)   A nitrogenous purine base found in DNA and RNA. (p. 9)

adenyl cyclase   The enzyme that catalyzes the synthesis of cyclic AMP. (p. 468)

adjacent segregation   Segregation of a heterozygous reciprocal translocation in which a translocated chromosome and a normal chromosome segregate together, producing an aneuploid gamete.

adjacent-1 segregation   Segregation from a heterozygous reciprocal translocation in which homologous centromeres go to opposite poles of the first-division spindle. (p. 289)

adjacent-2 segregation   Segregation from a heterozygous reciprocal translocation in which homologous centromeres go to the same pole of the first-division spindle. (p. 290)

agarose   A component of agar used as a gelling agent in gel electrophoresis; its value is that few molecules bind to it, so it does not interfere with electrophoretic movement. (p. 206)

agglutination   The clumping or aggregation, as of viruses or blood cells, caused by an antigen-antibody interaction. (p. 70)

albinism   Absence of melanin pigment in the iris, skin, and hair of an animal; absence of chlorophyll in plants. (p. 53)

alkaptonuria   A recessively inherited metabolic disorder in which a defect in the breakdown of tyrosine leads to excretion of homogentisic acid (alkapton) in the urine.

alkylating agent   An organic compound capable of transferring an alkyl group to other molecules. (p. 570)

allele   Any of the alternative forms of a given gene. (p. 40)

allele frequency   The relative proportion of all alleles of a gene that are of a designated type. (p. 628)

allopolyploid   A polyploid formed by hybridization between two different species. (p. 263)

allosteric protein   Any protein whose activity is altered by a change of shape induced by binding a small molecule.

allozyme   Any of the alternative electrophoretic forms of a protein coded by different alleles of a single gene. (p. 630)

α helix   A fundamental unit of protein folding in which successive amino acids form a right-handed helical structure held together by hydrogen bonding between the amino and carboxyl components of the peptide bonds in successive loops of the helix. (p. 414)

alpha satellite   Highly repetitive DNA sequences associated with mammalian centromeres. (p. 247)

alternate segregation   Segregation from a heterozygous reciprocal translocation in which both parts of the reciprocal translocation separate from both nontranslocated chromosomes in the first meiotic division. (p. 292)

amber codon   Common jargon for the UAG stop codon; an amber mutation is a mutation in which a sense codon has been altered to UAG. (p. 447)

Ames test   A bacterial test for mutagenicity; also used to screen for potential carcinogens. (p. 586)

amino acid   Any one of a class of organic molecules that have an amino group and a carboxyl group; 20 different amino acids are the usual components of proteins. (pp. 12, 412)

amino acid attachment site   The 3' terminus of a tRNA molecule at which an amino acid is attached.

aminoacylated tRNA   A tRNA covalently attached to its amino acid; charged tRNA. (p. 431)

aminoacyl site   The tRNA-binding site on the ribosome to which each incoming charged tRNA is initially bound. (p. 431)

A site   *See* **aminoacyl site**.

aminoacyl tRNA synthetase   The enzyme that attaches the correct amino acid to a tRNA molecule. (p. 431)

amino terminus   The end of a polypeptide chain at which the amino acid bears a free amino group ($-NH_2$). (p. 412)

amniocentesis   A procedure for obtaining fetal cells from the amniotic fluid for the diagnosis of genetic abnormalities. (p. 273)

amorph   A mutation in which the function of the gene product is completely abolished. (p. 66)

analog   *See* **base analog**.

anaphase   The stage of mitosis or meiosis in which chromosomes move to opposite ends of the spindle. In anaphase I of meiosis, homologous centromeres separate; in anaphase II, sister centromeres separate. (pp. 86, 94)

aneuploid   A cell or organism in which the chromosome number is not an exact multiple of the haploid number; more generally, aneuploidy refers to a condition in which particular genes or chromosomal regions are present in extra or fewer copies compared with wildtype. (p. 269)

annealing   The coming together of two strands of nucleic acid that have complementary base sequences to form a duplex molecule. (p. 200)

antibiotic-resistant mutant   A cell or organism that carries a mutation conferring resistance to an antibiotic. (p. 311)

antibiotic-resistant mutation   A mutation conferring resistance to one or more antibiotics. (p. 311)

antibody   A blood protein produced in response to a specific antigen and capable of binding with the antigen. (pp. 69, 483)

anticodon   The three bases in a tRNA molecule that are complementary to the three-base codon in mRNA. (pp.431, 432)

antigen   A substance able to stimulate the production of antibodies. (p. 69)

antigen presentation   An immune process in which a cell carrying an antigen bound to an MHC protein stimulates a responsive cell.

antigen-presenting cell   Any cell, such as a macrophage, expressing certain products of the major histocompatibility complex and capable of stimulating T cells that have appropriate antigen receptors.

antimorph   A mutation whose phenotypic effects are opposite to those of the wildtype allele. (p. 67)

antiparallel   The chemical orientation of the two strands of a double-stranded nucleic acid molecule; the 5'-to-3' orientations of the two strands are opposite one another. (p. 181)

AP endonuclease   An endonuclease that cleaves a DNA strand at any site at which the deoxyribose lacks a base. (p. 566)

apoptosis   Genetically programmed cell death, especially in embryonic development. (p. 524)

aporepressor   A protein converted into a repressor by binding with a particular molecule. (p. 462, 472)

Archaea   One of the three major classes of organisms; also called archaebacteria, they are unicellular microorganisms, usually found in extreme environments, that differ as much from bacteria as either group differs from eukaryotes. *See also* **Bacteria**. (p. 22)

artificial selection   Selection, imposed by a breeder, in which organisms of only certain phenotypes are allowed to breed. (p. 683)

ascospore   *See* **ascus**.

ascus   A sac containing the spores (ascospores) produced by meiosis in certain groups of fungi, including *Neurospora* and yeast. (p. 150)

assortative mating   Nonrandom selection of mating partners with respect to one or more traits; it is positive when like phenotypes mate more frequently than would be expected by chance and is negative when the reverse occurs. (p. 633)

ATP   Adenosine triphosphate, the primary molecule for storing chemical energy in a living cell.

attached-X chromosome   A chromosome in which two X chromosomes are joined to a common centromere; also called a compound-X chromosome. (p. 134)

attachment site   The base sequence in a bacterial chromosome at which bacteriophage DNA can integrate to form a prophage; either of the two attachment sites that flank an integrated prophage. (p. 341)

attenuation   *See* **attenuator**.

attenuator   A regulatory base sequence near the beginning of an mRNA molecule at which transcription can be terminated; when an attenuator is present, it precedes the coding sequences. (p. 473)

attractant   A substance that draws organisms to it through chemotaxis. (p. 704)

AUG   The usual initiation codon for polypeptide synthesis as well as the internal codon for methionine. (pp. 432, 443)

autogamy   Reproductive process in ciliates in which the genotype is reconstituted from the fusion of two identical haploid nuclei resulting from mitosis of a single product of meiosis; the result is homozygosity for all genes. (p. 614)

autonomous determination   Cellular differentiation determined intrinsically and not dependent on external signals or interactions with other cells. (p. 515)

autopolyploidy   Type of polyploidy in which there are more than two sets of homologous chromosomes. (p. 263)

autoradiography   A process for the production of a photographic image of the distribution of a radioactive substance in a cell or large cellular molecule; the image is produced on a photographic emulsion by decay emission from the radioactive material.

autoregulation   Regulation of gene expression by the product of the gene itself. (p. 462)

autosomes   All chromosomes other than the sex chromosomes. (p. 97)

auxotroph   A mutant microorganism unable to synthesize a compound required for its growth but able to grow if the compound is provided. (p. 312)

B

backcross   The cross of an F_1 heterozygote with a partner that has the same genotype as one of its parents. (p. 41)

Bacteria   One of the major kingdoms of living things; includes most bacteria. *See also* **Archaea.** (p. 22)

bacterial attachment site   *See* **attachment site**

bacterial transformation   *See* **transformation**

bacteriophage   A virus that infects bacterial cells; commonly called a phage. (pp. 6, 308)

band   In gel electrophoresis, a compact region of heavy staining due to the accumulation of molecules of a given size; in *Drosophila* genetics, one of the striations found in the giant polytene chromosomes in the larval salivary glands. (p. 206)

Barr body   A darkly staining body found in the interphase nucleus of certain cells of female mammals; consists of the condensed, inactivated X chromosome. (p. 277)

base   Single-ring (pyrimidine) or double-ring (purine) component of a nucleic acid. (p. 9)

base analog   A chemical so similar to one of the normal bases that it can be incorporated into DNA. (p. 567)

base composition   The relative proportions of the bases in a nucleic acid or of A + T versus G + C in duplex DNA. (p. 175)

base pair   A pair of nitrogenous bases, most commonly one purine and one pyrimidine, held together by hydrogen bonds in a double-stranded region of a nucleic acid molecule; commonly abbreviated bp, the term is often used interchangeably with the term *nucleotide pair*. The normal base pairs in DNA are AT and GC. (pp. 9, 13)

base-substitution mutation   Incorporation of an incorrect base into a DNA duplex. (p. 557)

B cells   A class of white blood cells, derived from bone marrow, with the potential for producing antibodies.

becquerel   A unit of radiation equal to 1 disintegration/second, abbreviated Bq; equal to 2.7×10^{-11} curie.

behavior   The observed response of an organism to stimulation or environmental change. (p. 704)

B-form DNA   The right-handed helical structure of DNA proposed by Watson and Crick. (p. 177)

β-galactosidase   An enzyme that cleaves lactose into its glucose and galactose constituents; produced by a gene in the *lac* operon. (p. 462)

β sheet   A fundamental structure formed in protein folding in which two polypeptide segments are held together in an antiparallel array by hydrogen bonds. (p. 414)

bidirectional replication   DNA replication proceeding in opposite directions from an origin of replication. *See also* θ **replication.** (p. 189)

binding site   A DNA or RNA base sequence that serves as the target for binding by a protein; the site on the protein that does the binding.

biochemical pathway   A diagram showing the order in which intermediate molecules are produced in the synthesis or degradation of a metabolite in a cell. (p. 19)

bivalent   A pair of homologous chromosomes, each consisting of two chromatids, associated in meiosis I. (p. 92)

blastoderm   Structure formed in the early development of an insect larva; the syncytial blastoderm is formed from repeated cleavage of the zygote nucleus without cytoplasmic division; the cellular blastoderm is formed by migration of the nuclei to the surface and their inclusion in separate cell membranes. (p. 529)

blastula   A hollow sphere of cells formed early in development. (p. 514)

blood-group system   A set of antigens on red blood cells resulting from the action of a series of alleles of a single gene, such as the ABO, MN, or Rh blood-group systems. (p. 68)

blunt ends   Ends of a DNA molecule in which all terminal bases are paired; the term usually refers to termini formed by a restriction enzyme that does not produce single-stranded ends. (pp. 204, 362)

branched pathway   A metabolic pathway in which an intermediate serves as a precursor for more than one product.

branch migration   In a DNA duplex invaded by a single strand from another molecule, the process in which the size of the heteroduplex region increases by progressive displacement of the original strand. (p. 138)

broad-sense heritability   The ratio of genotypic variance to total phenotypic variance. (p. 682)

C

cAMP-CRP complex   A regulatory complex consisting of cyclic AMP (cAMP) and the CAP protein; the complex is needed for transcription of certain operons. (p. 468)

candidate gene   A gene proposed to be involved in the genetic determination of a trait because of the role of the gene product in the cell or organism. (p. 732)

cap   A complex structure at the 5' termini of most eukaryotic mRNA molecules, having a 5'-5' linkage instead of the usual 3'-5' linkage. (p. 425)

CAP protein   Acronym for catabolite activator protein, the protein that binds cAMP and that regulates the activity of inducible operons in prokaryotes; also called CRP, for cAMP receptor protein. (p. 469)

carbon-source mutant   A cell or organism that carries a mutation preventing the use of a particular molecule or class of molecules as a source of carbon. (p. 312)

carboxyl terminus   The end of a polypeptide chain at which the amino acid has a free carboxyl group ($-COOH$). (p. 412)

carcinogen   A physical agent or chemical reagent that causes cancer.

carrier   A heterozygote for a recessive allele. (p. 53)

cascade   A series of enzyme activations serving to amplify a weak chemical signal. (p. 537)

cassette   In yeast, either of two sets of inactive mating-type genes that can become active by relocating to the *MAT* locus. (p. 482)

catabolite activator protein   *See* **CAP protein.**

cDNA   *See* **complementary DNA.**

cell cycle   The growth cycle of a cell; in eukaryotes, it is subdivided into G_1 (gap 1), S (DNA synthesis), G_2 (gap 2), and M (mitosis). (p. 83)

cell fate   The pathway of differentiation that a cell normally undergoes. (p. 514)

cell hybrid   Product of fusion of two cells in culture, often from different species.

cell lineage   The ancestor-descendant relationships of a group of cells in development. (p. 519)

cellular oncogene   A gene coding for a cellular growth factor whose abnormal expression predisposes to malignancy. *See also* **oncogene.** (p. 298)

centimorgan   A unit of distance in the genetic map equal to 1 percent recombination; also called a map unit. (p. 128)

central dogma   The concept that genetic information is transferred from the nucleotide sequence in DNA to the nucleotide sequence in an RNA transcript to the amino acid sequence of a polypeptide chain. (p. 12)

centromere   The region of the chromosome that is associated with spindle fibers and that participates in normal chromosome movement during mitosis and meiosis. (p. 86)

chain elongation   The process of addition of successive amino acids to the growing end of a polypeptide chain. (p. 421)

chain initiation   The process by which polypeptide synthesis is begun. (p. 421)

chain termination   The process of ending polypeptide synthesis and releasing the polypeptide from the ribosome; a chain-termination mutation creates a new stop codon, resulting in premature termination of synthesis of the polypeptide chain. (p. 421)

chaperone   A protein that assists in the three-dimensional folding of another protein. (p. 414)

Chargaff's rules   In double-stranded DNA, the amount of A equals that of T, and the amount of G equals that of C. (p. 176)

charged tRNA   A tRNA molecule to which an amino acid is linked; acylated tRNA. (p. 431)

chemoreceptor   Any molecule, usually a protein, that binds with other molecules and stimulates a behavioral response, such as chemotaxis, sensation, etc. *See also* **chemosensor.** (p. 709)

chemosensor   In bacteria, a class of proteins that bind with particular attractants or repellents. Chemosensors interact with chemoreceptors to stimulate chemotaxis. *See also* **chemotaxis.** (p. 709)

chemotaxis   Behavioral response in which organisms move toward chemicals that attract them and away from chemicals that repel them. (p. 704)

chiasma   The cytological manifestation of crossing-over; the cross-shaped exchange configuration between nonsister chromatids of homologous chromosomes that is visible in prophase I of meiosis. The plural can be either *chiasmata* or *chiasmas*. (p. 93)

chimeric gene   A gene produced by recombination, or by genetic engineering, that is a mosaic of DNA sequences from two or more different genes. (p. 285)

chi-square (χ^2)   A statistical quantity calculated to assess the goodness of fit between a set of observed numbers and the theoretically expected numbers. (p. 109)

chromatid   Either of the longitudinal subunits produced by chromosome replication. (p. 86)

chromatid interference   In meiosis, the effect that crossing-over between one pair of nonsister chromatids may have on the probability that a second crossing-over in the same chromosome will involve the same or different chromatids; chromatid interference does not generally occur. (p. 141)

chromatin   The aggregate of DNA and histone proteins that makes up a eukaryotic chromosome. (p. 228)

chromocenter   The aggregate of centromeres and adjacent heterochromatin in nuclei of *Drosophila* larval salivary gland cells. (p. 234)

chromomere   A tightly coiled, bead-like region of a chromosome most readily seen during pachytene of meiosis; the beads are in register in a polytene chromosome, resulting in the banded appearance of the chromosome. (pp. 92)

chromosome   In eukaryotes, a DNA molecule that contains genes in linear order to which numerous proteins are bound and that has a telomere at each end and a centromere; in prokaryotes, the DNA is associated with fewer proteins, lacks telomeres and a centromere, and is often circular; in viruses, the chromosome is DNA or RNA, single-stranded or double-stranded, linear or circular, and often free of bound proteins. (p. 2, 82)

chromosome complement   The set of chromosomes in a cell or organism. (p. 82)

chromosome interference   In meiosis, the phenomenon by which a crossing-over at one position inhibits the occurrence of another crossing-over at a nearby position. (p. 143)

chromosome map   A diagram showing the locations and relative spacing of genes along a chromosome. (p. 127)

chromosome painting   Use of differentially labeled, chromosome-specific DNA strands for hybridization with chromosomes to label each chromosome with a different color. (p. 264)

chromosome theory of heredity   The theory that chromosomes are the cellular objects that contain the genes.(p. 104)

circadian rhythm   A biological cycle with an approximate 24-hour period synchronized with the cycle of day and night. (p. 714)

circular permutation   A permutation of a group of elements in which the elements are in the same order but the beginning of the sequence differs. (p. 335)

cis configuration   The arrangement of linked genes in a double heterozygote in which both mutations are present in the same chromosome—for example, $a_1 a_2/++$; also called coupling. (pp. 126, 451)

cis-dominant mutation   A mutation that affects the expression of only those genes on the same DNA molecule as that containing the mutation. (p. 464)

cis heterozygote   *See* **cis configuration.**

cis-trans test   *See* **complementation test.**

cistron   A DNA sequence specifying a single genetic function as defined by a complementation test; a nucleotide sequence coding for a single polypeptide. (p. 339)

ClB method   A genetic procedure used to detect X-linked recessive lethal mutations in *Drosophila melanogaster;* so named because one X chromosome in the female parent is marked with an inversion (*C*), a recessive lethal allele (*l*), and the dominant allele for Bar eyes (*B*). (p. 563)

cleavage division   Mitosis in the early embryo. (p. 514)

clone   A collection of organisms derived from a single parent and, except for new mutations, genetically identical to that parent; in genetic engineering, the linking of a specific gene or DNA fragment to a replicable DNA molecule, such as a plasmid or phage DNA. (p. 308)

cloned gene   A DNA sequence incorporated into a vector molecule capable of replication in the same or a different organism. (pp. 205, 360)

cloning   The process of producing cloned genes. (p. 360)

coding region   The part of a DNA sequence that codes for the amino acids in a protein.

coding sequence   A region of a DNA strand with the same sequence as is found in the coding region of a messenger RNA, except that T is present in DNA instead of U. (p. 424)

coding strand   In a transcribed region of a DNA duplex, the strand that is not transcribed.

codominance   The expression of both alleles in a heterozygote. (p. 68)

codon   A sequence of three adjacent nucleotides in an mRNA molecule, specifying either an amino acid or a stop signal in protein synthesis. (pp. 14, 431)

coefficient of coincidence   An experimental value obtained by dividing the observed number of double recombinants by the expected number calculated under the assumption that the two events take place independently. (p. 143)

cohesive end   A single-stranded region at the end of a double-stranded DNA molecule that can adhere to a complementary single-stranded sequence at the other end or in another molecule. (p. 340)

colchicine   A chemical that prevents formation of the spindle in nuclear division. (p. 267)

colinearity   The linear correspondence between the order of amino acids in a polypeptide chain and the corresponding sequence of nucleotides in the DNA molecule. (p. 416)

colony   A visible cluster of cells formed on a solid growth medium by repeated division of a single parental cell and its daughter cells. (p. 3)

colony hybridization assay   A technique for identifying colonies containing a particular cloned gene; many colonies are transferred to a filter, lysed, and exposed to radioactive DNA or RNA complementary to the DNA sequence of interest, after which colonies that contain a sequence complementary to the probe are located by autoradiography. (p. 372)

color blindness In human beings, the usual form of color blindness is X-linked red-green color blindness. Unequal crossing-over between the adjacent red and green opsin pigment genes results in chimeric opsin genes causing mild or severe green-vision defects (deuteranomaly or deuteranopia, respectively) and mild or severe red-vision defects (protanomaly or protanopia, respectively). (p. 284)

combinatorial control Strategy of gene regulation in which a relatively small number of time- and tissue-specific positive and negative regulatory elements are used in various combinations to control the expression of a much larger number of genes. (p. 499)

combinatorial joining The DNA splicing mechanism by which antibody variability is produced, resulting in many possible combinations of V and J regions in the formation of a light-chain gene and many possible combinations of V, D, and J regions in the formation of a heavy-chain gene. (p. 485)

common ancestor Any ancestor of both one's father and one's mother. (p. 634)

compartment In *Drosophila* development, a group of descendants of a small number of founder cells with a determined pattern of development. (p. 540)

compatible Blood or tissue that can be transfused or transplanted without rejection. (p. 70)

complementary DNA A DNA molecule made by copying RNA with reverse transcriptase; usually abbreviated cDNA. (p. 369)

complementation The phenomenon in which two recessive mutations with similar phenotypes result in a wildtype phenotype when both are heterozygous in the same genotype; complementation means that the mutations are in different genes. (p. 55)

complementation group A group of mutations that fail to complement one another. (p. 58)

complementation test A genetic test to determine whether two mutations are alleles (are present in the same functional gene). (p. 55)

complex A term used to refer to an ordered aggregate of molecules, as in an enzyme-substrate complex or a DNA-histone complex.

compound-X chromosome *See* **attached-X chromosome.**

condensation The coiling process by which chromosomes become shorter and thicker during prophase of mitosis or meiosis. (p. 137)

conditional mutation A mutation that results in a mutant phenotype under certain (restrictive) environmental conditions but results in a wildtype phenotype under other (permissive) conditions. (p. 556)

congenital Present at birth.

conjugation A process of DNA transfer in sexual reproduction in certain bacteria; in *E. coli,* the transfer is unidirec-

tional, from donor cell to recipient cell. Also, a mating between cells of *Paramecium.* (p. 314, 614)

consanguineous mating A mating between relatives. (p. 634)

consensus sequence A generalized base sequence derived from closely related sequences found in many locations in a genome or in many organisms; each position in the consensus sequence consists of the base found in the majority of sequences at that position. (p. 419)

conserved sequence A base or amino acid sequence that changes very slowly in the course of evolution. (p. 229)

constant region The part of the heavy and light chains of an antibody molecule that has the same amino acid sequence among all antibodies derived from the same heavy-chain and light-chain genes. (p. 484)

constitutive mutation A mutation that causes synthesis of a particular mRNA molecule (and the protein that it encodes) to take place at a constant rate, independent of the presence or absence of any inducer or repressor molecule. (p. 464)

contig A set of cloned DNA fragments overlapping in such a way as to provide unbroken coverage of a contiguous region of the genome; a contig contains no gaps. (p. 373)

continuous trait A trait in which the possible phenotypes have a continuous range from one extreme to the other rather than falling into discrete classes. (p. 671)

coordinate gene Any of a group of genes that establish the basic anterior-posterior and dorsal-ventral axes of the early embryo. (pp. 532, 534)

coordinate regulation Control of synthesis of several proteins by a single regulatory element; in prokaryotes, the proteins are usually translated from a single mRNA molecule. (pp. 436, 461)

core particle The aggregate of histones and DNA in a nucleosome, without the linking DNA. (p. 230)

co-repressor A small molecule that binds with an aporepressor to create a functional repressor molecule. (p. 461)

correlated response Change of the mean in one trait in a population accompanying selection for another trait. (p. 687)

correlation coefficient A measure of association between pairs of numbers, equaling the covariance divided by the product of the standard deviations. (p. 688)

Cot Mathematical product of the initial concentration of nucleic acid (C_0) and the time (t). (p. 237)

Cot curve A graph of percent nucleic acid renaturation as a function of the initial concentration (C_0) multiplied by time. (p. 237)

cotransduction Transduction of two or more linked genetic markers by one transducing particle. (p. 326)

cotransformation Transformation in bacteria of two genetic markers carried on a single DNA fragment. (p. 313)

counterselected marker   A mutation used to prevent growth of a donor cell in an Hfr × F⁻ bacterial mating. (p. 319)

coupled transcription-translation   In prokaryotes, the translation of an mRNA molecule before its transcription is completed.

coupling   *See* **cis configuration.**

courtship song   In *Drosophila,* the regular oscillation in the time interval between pulses in successive episodes of wing vibration that the male directs toward the female during mating behavior.

covalent bond   A chemical bond in which electrons are shared.

covalent circle   *See* **covalently closed circle.**

covalently closed circle   A ring-shaped duplex DNA molecule whose ends are joined by covalent bonds. (p. 224)

covariance   A measure of association between pairs of numbers that is defined as the average product of the deviations from the respective means. (p. 688)

C region   *See* **constant region.**

crossing-over   A process of exchange between nonsister chromatids of a pair of homologous chromosomes that results in the recombination of linked genes. (p. 93)

crown gall tumor   Plant tumor caused by infection with *Agrobacterium tumefaciens.* (p. 381)

CRP protein   *See* **CAP protein.**

curie   A unit of radiation equal to 3.7×10^{10} disintegrations/second, abbreviated Ci; equal to 3.7×10^{10} becquerels.

cryptic splice site   A potential splice site not normally used in RNA processing unless a normal splice site is blocked or mutated. (p. 430)

cyclic adenosine monophosphate (cAMP)   *See* **cyclic AMP.**

cyclic AMP   Cyclic adenosine monophosphate, usually abbreviated cAMP, used in the regulation of cellular processes; cAMP synthesis is regulated by glucose metabolism; its action is mediated by the CAP protein in prokaryotes and by protein kinases (enzymes that phosphorylate proteins) in eukaryotes. (p. 468)

cyclic AMP receptor protein (CRP)   *See* **CAP protein.**

cyclin   One of a group of proteins that participates in controlling the cell cycle. Different types of cyclins interact with the p34 kinase subunit and regulate the G_1-S and G_2-M transitions. The proteins are called cyclins because their abundance rises and falls rhythmically in the cell cycle. (p. 84)

cytogenetics   The study of the genetic implications of chromosome structure and behavior.

cytokinesis   Division of the cytoplasm. (p. 83)

cytological map   Diagrammatic representation of a chromosome. (p. 234)

cytoplasmic inheritance   Transmission of hereditary traits through self-replicating factors in the cytoplasm—for example, mitochondria and chroloplasts. (p. 600)

cytoplasmic male sterility   Type of pollen abortion in maize with cytoplasmic inheritance through the mitochondria. (p. 612)

cytosine (C)   A nitrogenous pyrimidine base found in DNA and RNA. (p. 9)

cytotoxic T cell   Type of white blood cell that participates directly in attacking cells marked for destruction by the immune system.

D

daughter strand   A newly synthesized DNA or chromosome strand. (p. 182)

deamination   Removal of an amino group ($-NH_2$) from a molecule.

decaploid   An organism with ten sets of chromosomes. (p. 261)

deficiency   *See* **deletion.**

degeneracy   The feature of the genetic code in which an amino acid corresponds to more than one codon; also called redundancy. (p. 442)

degenerate code   *See* **degeneracy.**

degree of dominance   The extent to which the phenotype of a heterozygous genotype resembles one of the homozygous genotypes. (p. 656)

degrees of freedom   An integer that determines the significance level of a particular statistical test. In the goodness-of-fit type of chi-square test in which the expected numbers are not based on any quantities estimated from the data themselves, the number of degrees of freedom is one less than the number of classes of data. (p. 112)

deletion   Loss of a segment of the genetic material from a chromosome; also called deficiency. (p. 281)

deletion mapping   The use of overlapping deletions to locate a gene on a chromosome or a genetic map. (p. 336)

deme   *See* **local population.**

denaturation   Loss of the normal three-dimensional shape of a macromolecule without breaking covalent bonds, usually accompanied by loss of its biological activity; conversion of DNA from the double-stranded into the single-stranded form; unfolding of a polypeptide chain. (p. 199)

denaturation mapping   An electron-microscopic technique for localizing regions of a double-stranded DNA molecule by noting the positions at which the individual strands separate in the early stages of denaturation.

deoxyribonuclease   An enzyme that breaks sugar-phosphate bonds in DNA, forming either fragments or the component nucleotides; abbreviated DNase. (p. 225)

deoxyribonucleic acid   *See* **DNA.**

deoxyribose   The five-carbon sugar present in DNA. (p. 174)

depurination   Removal of purine bases from DNA. (p. 570)

derepression   Activation of a gene or set of genes by inactivation of a repressor.

deuteranomaly   *See* **color blindness.**

deuteranopia   *See* **color blindness.**

deviation   In statistics, a difference from an expected value. (p. 672)

diakinesis   The substage of meiotic prophase I, immediately preceding metaphase I, in which the bivalents attain maximum shortening and condensation. (pp. 92, 93)

dicentric chromosome   A chromosome with two centromeres. (p. 260)

dideoxyribose   A deoxyribose sugar lacking the 3' hydroxyl group; when incorporated into a polynucleotide chain, it blocks further chain elongation. (p. 211)

dideoxy sequencing method   Procedure for DNA sequencing in which a template strand is replicated from a particular primer sequence and terminated by the incorporation of a nucleotide that contains dideoxyribose instead of deoxyribose; the resulting fragments are separated by size via electrophoresis. (p. 211)

differentiation   The complex changes of progressive diversification in cellular structure and function that take place in the development of an organism; for a particular line of cells, this results in a continual restriction in the types of transcription and protein synthesis of which each cell is capable.

diffuse centromere   A centromere that is dispersed throughout the chromosome, as indicated by the dispersed binding of spindle fibers, rather than being concentrated at a single point. (p. 246)

dihybrid   Heterozygous at each of two loci; progeny of a cross between true-breeding or homozygous strains that differ genetically at two loci. (p. 42)

dimer   A protein formed by the association of two polypeptide subunits.

diploid   A cell or organism with two complete sets of homologous chromosomes. (p. 82)

diplotene   The substage of meiotic prophase I, immediately following pachytene and preceding diakinesis, in which pairs of sister chromatids that make up a bivalent (tetrad) begin to separate from each other and chiasmata become visible. (pp. 92, 93)

direct repeat   Copies of an identical or very similar DNA or RNA base sequence in the same molecule and in the same orientation. (p. 243)

discontinuous variation   Variation in which the phenotypic differences for a trait fall into two or more discrete classes.

disjunction   Separation of homologous chromosomes to opposite poles of a division spindle during anaphase of a mitotic or meiotic nuclear division.

displacement loop   Loop formed when part of a DNA strand is dislodged from a duplex molecule because of the partner strand's pairing with another molecule. (p. 590)

distribution   In quantitative genetics, the mathematical relation that gives the proportion of members in a population that have each possible phenotype. (p. 671)

disulfide bond   Two sulfur atoms covalently linked; found in proteins when the sulfur atoms in two different cysteines are joined.

dizygotic twins   Twins that result from the fertilization of separate ova and that are genetically related as siblings; also called fraternal twins. (p. 682)

D loop   *See* **displacement loop.**

DNA   Deoxyribonucleic acid, the macromolecule, usually composed of two polynucleotide chains in a double helix, that is the carrier of the genetic information in all cells and many viruses. (p. 2)

DNA cloning   *See* **cloned gene.**

DNA gyrase   One of a class of enzymes called topoisomerases, which function during DNA replication to relax positive supercoiling of the DNA molecule and which introduce negative supercoiling into nonsupercoiled molecules early in the life cycle of many phages. (p. 225)

DNA ligase   An enzyme that catalyzes formation of a covalent bond between adjacent 5'-P and 3'-OH termini in a broken polynucleotide strand of double-stranded DNA. (p. 196)

DNA looping   A mechanism by which enhancers that are distant from the immediate proximity of a promoter can still regulate transcription; the enhancer and promoter, both bound with suitable protein factors, come into indirect physical contact by means of the looping out of the DNA between them. The physical interaction stimulates transcription. (p. 494)

DNA methylase   An enzyme that adds methyl groups ($-CH_3$) to certain bases, particularly cytosine. (p. 487)

DNA polymerase   Any enzyme that catalyzes the synthesis of DNA from deoxynucleoside 5'-triphosphates, using a template strand. (p. 191)

DNA polymerase I (Pol I)   In *E. coli* DNA replication, the enzyme that fills in gaps in the lagging strand and replaces the RNA primers with DNA. (p. 192)

DNA polymerase III (Pol III) The major DNA replication enzyme in *E. coli*. (p.192)

DNA repair Any of several different processes for restoration of the correct base sequence of a DNA molecule into which incorrect bases have been incorporated or whose bases have been chemically modified. (p. 577)

DNA replication The copying of a DNA molecule. (p. 10)

DNase *See* **deoxyribonuclease.**

DNA typing Electrophoretic identification of individual persons by the use of DNA probes for highly polymorphic regions of the genome, such that the genome of virtually every person exhibits a unique pattern of bands; sometimes called DNA fingerprinting. (p. 640)

DNA uracyl glycosylase An enzyme that removes uracil bases when they occur in double-stranded DNA. (p. 565)

domain A folded region of a polypeptide chain that is spatially somewhat isolated from other folded regions. (p. 430)

dominance Condition in which a heterozygote expresses a trait in the same manner as the homozygote for one of the alleles. The allele or the corresponding phenotypic trait expressed in the heterozygote is said to be dominant. (p. 35)

dominance variance The part of the genotypic variance that results from the dominance effects of alleles affecting the trait.

dosage compensation A mechanism regulating X-linked genes such that their activities are equal in males and females; in mammals, random inactivation of one X chromosome in females results in equal amounts of the products of X-linked genes in males and females. (p. 275)

double-stranded DNA A DNA molecule consisting of two antiparallel strands that are complementary in nucleotide sequence. (p. 10)

double-Y syndrome The clinical features of the karyotype 47,XYY. (p. 278)

Down syndrome The clinical features of the karyotype 47,+21 (trisomy 21). (p. 273)

drug-resistance plasmid A plasmid that contains genes whose products inactivate certain antibiotics.

duplex DNA A double-stranded DNA molecule. (p. 10)

duplication A chromosome aberration in which a chromosome segment is present more than once in the haploid genome; if the two segments are adjacent, the duplication is a tandem duplication. (p. 282)

E

ectopic expression Gene expression that occurs in the wrong tissues or the wrong cell types. (p. 67)

editing function The activity of DNA polymerases that removes incorrectly incorporated nucleotides; also called the proofreading function. (p. 193)

effector A molecule that brings about a regulatory change in a cell, as by induction.

electrophoresis A technique used to separate molecules on the basis of their different rates of movement in response to an applied electric field, typically through a gel. (pp. 205, 629)

electroporation Introduction of DNA fragments into a cell by means of an electric field. (p. 363)

elongation Addition of amino acids to a growing polypeptide chain or of nucleotides to a growing nucleic acid chain. (p. 433)

embryo An organism in the early stages of development (the second through seventh weeks in human beings).

embryoid A small mass of dividing cells formed from haploid cells in anthers that can give rise to a mature haploid plant. (p. 266)

embryonic induction Developmental process in which the fate of a group of cells is determined by interactions with nearby cells in the embryo. (p. 517)

embryonic stem cells Cells in the blastocyst that give rise to the body of the embryo. (p. 378)

endonuclease An enzyme that breaks internal phosphodiester bonds in a single- or double-stranded nucleic acid molecule; usually specific for either DNA or RNA. (p. 192)

endoreduplication Doubling of the chromosome complement because of chromosome replication and centromere division without nuclear or cytoplasmic division. (p. 263)

endosperm Nutritive tissues formed adjacent to the embryo in most flowering plants; in most diploid plants, the endosperm is triploid.

endosymbiont theory Theory that mitochondria and chloroplasts were originally free-living organisms that invaded ancestral eukaryotes, first perhaps as parasites, later becoming symbionts. (p. 613)

enhancer A base sequence in eukaryotes and eukaryotic viruses that increases the rate of transcription of nearby genes; the defining characteristics are that it need not be adjacent to the transcribed gene and that the enhancing activity is independent of orientation with respect to the gene. (p. 494)

enhancer trap Mutagenesis strategy in which a reporter gene with a weak promoter is introduced at many random locations in the genome; tissue-specific expression of the reporter gene identifies an insertion near a tissue-specific enhancer. (p. 499)

environmental variance The part of the phenotypic variance that is attributable to differences in environment. (p. 677)

enzyme A protein that catalyzes a specific biochemical reaction and is not itself altered in the process. (p.12)

episome   A DNA element that can exist in the cell either as an autonomously replicating entity or become incorporated into the genome. (p. 314)

epistasis   A term referring to an interaction between non-allelic genes in their effects on a trait. Generally, *epistasis* means any type of interaction in which the genotype at one locus affects the phenotypic expression of the genotype at another locus. In a more restricted sense, *epistasis* refers to a situation in which the genotype at one locus determines the phenotype in such a way as to mask the genotype present at a second locus. (p. 20)

equational division   Term applied to the second meiotic division because the haploid chromosome complement is retained throughout. (p.94)

equilibrium   *See* **genetic equilibrium.**

equilibrium centrifugation in a density gradient   A method for separating macromolecules of differing density by high-speed centrifugation in a solution of approximately the same density as the molecules to be separated; the centrifugation produces a gradient of density inside the centrifuge tube, and each macromolecule migrates to a position in the gradient that matches its own density. (p. 183)

equilibrium density-gradient centrifugation   *See* **equilibrium centrifugation in a density gradient.**

erythroblastosis fetalis   Hemolytic disease of the newborn; blood cell destruction that occurs when anti-Rh^+ antibodies in a mother cross the placenta and attack Rh^+ cells in a fetus.

E site   *See* **exit site.**

estrogen   A female sex hormone; of great interest in the study of cellular regulation in eukaryotes because the number of types of target cells is large. (p. 493)

ethidium bromide   A fluorescent molecule that binds to DNA and changes its density; used to purify super-coiled DNA molecules and to localize DNA in gel electrophoresis.

euchromatin   A region of a chromosome that has normal staining properties and undergoes the normal cycle of condensation; relatively uncoiled in the interphase nucleus (compared with condensed chromosomes), it apparently contains most of the genes. (pp. 145, 241)

Eukarya   One of the major kingdoms of living organisms, in which the cells have a true nucleus and divide by mitosis or meiosis. (p. 22)

eukaryote   A cell with a true nucleus (DNA enclosed in a membranous envelope) in which cell division takes place by mitosis or meiosis; an organism composed of eukaryotic cells. (p. 23)

euploid   A cell or an organism having a chromosome number that is an exact multiple of the haploid number. (p. 269)

evolution   Cumulative change in the genetic characteristics of a species through time, resulting in greater adaptation. (pp. 22, 649)

excision   Removal of a DNA fragment from a molecule. (p. 345)

excisionase   An enzyme that is needed for prophage excision; works together with an integrase. (p. 345)

excision repair   Type of DNA repair in which segments of a DNA strand that are chemically damaged are removed enzymatically and then resynthesized, using the other strand as a template. (p. 581)

exit site   The tRNA-binding site on the ribosome that binds each uncharged tRNA just prior to its release. (p. 431)

exon   The sequences in a gene that are retained in the messenger RNA after the introns are removed from the primary transcript. (p. 425)

exon shuffle   The theory that new genes can evolve by the assembly of separate exons from preexisting genes, each coding for a discrete functional domain in the new protein. (p. 430)

exonuclease   An enzyme that removes a terminal nucleotide in a polynucleotide chain by cleavage of the terminal phosphodiester bond; nucleotides are removed successively, one by one; usually specific for either DNA or RNA and for either single-stranded or double-stranded nucleic acids. A 5'-to-3' exonuclease cleaves successive nucleotides from the 5' end of the molecule; a 3'-to-5' exonuclease cleaves successive nucleotides from the 3' end. (pp. 192,196)

expressivity   The degree of phenotypic expression of a gene.

extranuclear inheritance   Inheritance mediated by self-replicating cellular factors located outside the nucleus—for example, in mitochondria or chloroplasts. (p. 600)

F

fate   In development, the final product of differentiation of a cell or of a group of cells. (p. 514)

fate map   A diagram of the insect blastoderm identifying the regions from which particular adult structures derive. (p. 530)

favism   Destruction of red blood cells (anemia) triggered by eating raw fava beans or by coming into contact with certain other environmental agents; associated with a deficiency of the enzyme glucose-6-phosphate dehydrogenase. (p. 20)

feedback inhibition   Inhibition of an enzyme by the product of the enzyme or, in a metabolic pathway, by a product of the pathway.

fertility   The ability to mate and produce offspring. (p. 655)

fertility factor *See* **F plasmid.**

fertility restorer   A suppressor of cytoplasmic male sterility. (p. 612)

F factor *See* **F plasmid**

F₁ generation   The first generation of descent from a given mating. (p. 33)

F₂ generation   The second generation of descent, produced by intercrossing or self-fertilizing F₁ organisms. (p. 36)

first-division segregation   Separation of a pair of alleles into different nuclei in the first meiotic division; happens when there is no crossing-over between the gene and the centromere in a particular cell. (p. 155)

first meiotic division   The meiotic division that reduces the chromosome number; sometimes called the reduction division. (p. 89)

fitness   A measure of the average ability of organisms with a given genotype to survive and reproduce.

5'-P group   The end of a DNA or RNA strand that terminates in a free phosphate group not connected to a sugar farther along. (p. 175)

five prime end (5' end)   The end of a polynucleotide chain that terminates with a 5' carbon unattached to another nucleotide. (p. 175)

fixed allele   An allele whose allele frequency equals 1.0. (p. 629)

flagella   The whip-like projections from the surface of bacterial cells that serve as the organs of locomotion. (p. 705)

fluctuation test   A statistical test used to determine whether bacterial mutations occur at random or are produced in response to selective agents.

folded chromosome   The form of DNA in a bacterial cell in which the circular DNA is folded to have a compact structure; contains protein. (p. 225)

folding domain   A short region of a polypeptide chain within which interactions between amino acids result in a three-dimensional conformation that is attained relatively independently of the folding of the rest of the molecule. (p. 430)

forward mutation   A change from a wildtype allele to a mutant allele. (pp. 582, 652)

founder effect   Random genetic drift that results when a group of founders of a population are not genetically representative of the population from which they were derived.

F plasmid   A bacterial plasmid—often called the F factor, fertility factor, or sex plasmid—that is capable of transferring itself from a host (F⁺) cell to a cell not carrying an F factor (F⁻ cell); when an F factor is integrated into the bacterial chromosome (in an Hfr cell), the chromosome becomes transferrable to an F⁻ cell during conjugation. (p. 314)

F' plasmid   An F plasmid that contains genes obtained from the bacterial chromosome in addition to plasmid genes; formed by aberrant excision of an integrated F factor, taking along adjacent bacterial DNA. (p. 324)

fragile-X chromosome   A type of X chromosome containing a site toward the end of the long arm that tends to break in cultured cells that are starved for DNA precursors; causes fragile-X syndrome. (p. 279)

fragile-X syndrome   A common form of inherited mental retardation associated with a type of X chromosome prone to breakage (the fragile-X chromosome). (p. 279)

frameshift mutation   A mutational event caused by the insertion or deletion of one or more nucleotide pairs in a gene, resulting in a shift in the reading frame of all codons following the mutational site. (p. 440)

fraternal twins   Twins that result from the fertilization of separate ova and are genetically related as siblings; also called dizygotic twins. (p. 682)

free radical   A highly reactive molecule produced when ionizing radiation interacts with water; free radicals are potent oxidizing agents. (p. 573)

frequency of recombination   The proportion of gametes carrying combinations of alleles that are not present in either parental chromosome. (p. 125)

G

gain-of-function mutation   Mutation in which a gene is overexpressed or inappropriately expressed. (p. 526)

galactosidase   *See* β-**galactosidase.**

gamete   A mature reproductive cell, such as sperm or egg in animals.

gametophyte   In plants, the haploid part of the life cycle that produces the gametes by mitosis. (p. 89)

gap gene   Any of a group of genes that control the development of contiguous segments or parasegments in *Drosophila* such that mutations result in gaps in the pattern of segmentation. (pp. 533, 538)

gastrula   Stage in early animal development marked by extensive cell migration. (p. 514)

G-E association   *See* **genotype-environment association**

G-E interaction   *See* **genotype-environment interaction**

gel electrophoresis   *See* **electrophoresis.**

gene   The hereditary unit containing genetic information transcribed into an RNA molecule that is processed and either functions directly or is translated into a polypeptide chain; a gene can mutate to various forms called alleles. (pp. 2, 40)

gene amplification   A process in which certain genes undergo differential replication either within the chromosome or extrachromosomally, increasing the number of copies of the gene. (p. 481)

gene cloning   *See* **cloned gene.**

gene conversion   The phenomenon in which the products of a meiotic division in an *Aa* heterozygous genotype are in some ratio other than the expected 1*A* : 1*a*—for example, 3*A* : 1*a*, 1*A* : 3*a*, 5*A* : 3*a*, or 3*A* : 5*a*. (p. 586)

gene dosage   Number of gene copies. (p. 481)

gene expression   The multistep process by which a gene is regulated and its product synthesized. (p. 412)

gene flow   Exchange of genes among populations resulting from either dispersal of gametes or migration of individuals; also called migration. (p. 650)

gene library   A large collection of cloning vectors containing a complete (or nearly complete) set of fragments of the genome of an organism. (p. 372)

gene pool   The totality of genetic information in a population of organisms. (p. 628)

gene product   A term used for the polypeptide chain translated from an mRNA molecule transcribed from a gene; if the RNA is not translated (for example, ribosomal RNA), the RNA molecule is the gene product. (p. 412)

generalized transduction   *See* **transducing phage.**

general transcription factor   A protein molecule needed to bind with a promoter before transcription can proceed; transcription factors are necessary, but not sufficient, for transcription, and they are shared among many different promoters. (p. 494)

gene regulation   Processes by which gene expression is controlled in response to external or internal signals. (p. 460)

gene targeting   Disruption or mutation of a designated gene by homologous recombination. (p. 379)

gene therapy   Deliberate alteration of the human genome for alleviation of disease. (p. 387)

genetic code   The set of 64 triplets of bases (codons) corresponding to the twenty amino acids in proteins and the signals for initiation and termination of polypeptide synthesis. (p. 438)

genetic differentiation   Accumulation of differences in allele frequency between isolated or partially isolated populations.

genetic divergence   *See* **genetic differentiation.**

genetic engineering   The linking of two DNA molecules by *in vitro* manipulations for the purpose of generating a novel organism with desired characteristics. (p. 360)

genetic equilibrium   In population genetics, a situation in which the allele frequencies remain constant from one generation to the next. (p. 652)

genetic map   *See* **linkage map.**

genetic marker   Any pair of alleles whose inheritance can be traced through a mating or through a pedigree. (p. 129)

genetics   The study of biological heredity. (p. 2)

genome   The total complement of genes contained in a cell or virus; commonly used to refer to all genes present in one complete haploid set of chromosomes in eukaryotes. (p. 222)

genome equivalent   The number of clones of a given size that contain the same aggregate amount of DNA as the haploid genome size of an organism. (p. 389)

genotype   The genetic constitution of an organism or virus, typically with respect to one or a few genes of interest, as distinguished from its appearance, or phenotype. (p. 40)

genotype-environment association   The condition in which genotypes and environments are not in random combinations. (p. 680)

genotype-environment interaction   The condition in which genetic and environmental effects on a trait are not additive. (p. 678)

genotype frequency   The proportion of members of a population that are of a prescribed genotype. (p. 628)

genotypic variance   The part of the phenotypic variance that is attributable to differences in genotype. (p. 676)

germ cell   A cell that gives rise to reproductive cells. (p. 82)

germinal mutation   A mutation in a cell from which gametes are derived, as distinguished from a somatic mutation. (p. 556)

germ line   Cell lineage consisting of germ cells.

germ-line mutation   *See* **germinal mutation.**

glucose-6-phosphate dehydrogenase (G6PD)   An enzyme used in one biochemical pathway for the metabolism of glucose. In human beings, a defective G6PD enzyme (G6PD deficiency) is associated with a form of anemia in which the red blood cells are prone to burst. (p. 20)

glyphosate   A widely used herbicide. (p. 386)

goodness of fit   The extent to which observed numbers agree with the numbers expected on the basis of some specified genetic hypothesis. (p. 109)

G6PD   *See* **glucose-6-phosphate dehydrogenase.**

G6PD deficiency   *See* **glucose-6-phosphate dehydrogenase.**

G_1 period   *See* **cell cycle.**

G_2 period   *See* **cell cycle.**

gray   A unit of absorbed radiation equal to 1 joule of radiation energy absorbed per kilogram of tissue, abbreviated Gy; equal to 100 rad.

guanine (G)   A nitrogenous purine base found in DNA and RNA. (p. 9)

guide RNA   The RNA template present in telomerase. (p. 248)

gynandromorph   A sexual mosaic; an individual organism that exhibits both male and female sexual differentiation.

H

Haldane's mapping function   A mapping function based on the assumption of no chromosome interference. (p. 145)

haploid   A cell or organism of a species containing the set of chromosomes normally found in gametes. (pp. 83, 262)

haplotype   The allelic form of each of a set of linked genes present in a single chromosome, particularly genes of the major histocompatibility complex.

Hardy-Weinberg principle   The genotype frequencies expected with random mating. (p. 634)

H chain   *See* **heavy chain.**

heavy (H) chain   One of the large polypeptide chains in an antibody molecule. (p. 483)

helicase   A protein that separates the strands of double-stranded DNA (198)

helix   *See* α **helix.**

helix-loop-helix   A DNA-binding motif found in many regulatory proteins.

helper T cell   A type of white blood cell involved in activating other cells in the immune system.

hemizygous gene   A gene present in only one dose, such as the genes on the X chromosome in XY males.

hemoglobin   The oxygen-carrying protein in red blood cells. (p. 15)

hemophilia A   One of two X-linked forms of hemophilia; patients are deficient in blood-clotting factor VIII. (p. 101)

heritability   A measure of the degree to which a phenotypic trait can be modified by selection. *See also* **broad-sense heritability** and **narrow-sense heritability.**

heterochromatin   Chromatin that remains condensed and heavily stained during interphase; commonly present adjacent to the centromere and in the telomeres of chromosomes. Some chromosomes are composed primarily of heterochromatin. (pp. 145, 240)

heterochronic mutation   A mutation in which an otherwise normal gene is expressed at the wrong time. (p. 525)

heteroduplex region   Region of a double-stranded nucleic acid molecule in which the two strands have different hereditary origins; produced either as an intermediate in recombination or by the *in vitro* annealing of single-stranded complementary molecules. (p. 587)

heterogametic   Refers to the production of dissimilar gametes with respect to the sex chromosomes; in most animals, the male is the heterogametic sex, but in birds, moths, butterflies, and some reptiles, it is the female. *See also* **homogametic.** (p. 97)

heterogeneous nuclear RNA (hnRNA)   The collection of primary RNA transcripts and incompletely processed products found in the nucleus of a eukaryotic cell.

heterokaryon   A cell or individual having nuclei from genetically different sources, the result of cell fusion not accompanied by nuclear fusion.

heteroplasmy   The presence of two or more genetically different types of the same organelle in a single organism. (p. 602)

heterosis   The superiority of hybrids over either inbred parent with respect to one or more traits; also called hybrid vigor. (p. 687)

heterozygote superiority   The condition in which a heterozygous genotype has greater fitness than either of the homozygotes. (p. 657)

heterozygous   Carrying dissimilar alleles of one or more genes; not homozygous. (p. 40)

hexaploid   A cell or organism with six complete sets of chromosomes. (p. 261)

H4 histone   *See* **histone.**

Hfr   An *E. coli* cell in which an F plasmid is integrated into the chromosome, making possible the transfer of part or all of the chromosome to an F⁻ cell. (p. 317)

high-copy-number plasmid   A plasmid for which there are usually considerably more than 2 (and often more than 20) copies per cell.

highly repetitive sequence   A DNA sequence present in thousands of copies in a genome. (p.239)

highly significant   *See* **statistically significant.**

histocompatability   Acceptance by a recipient of transplanted tissue from a donor.

histocompatibility antigens   Tissue antigens that determine transplant compatibility or incompatibility.

histocompatibility genes   Genes coding for histocompatibility antigens.

histone   Any of the small basic proteins bound to DNA in chromatin; the five major histones are designated H1, H2A, H2B, H3, and H4. Each nucleosome core particle contains two molecules each of H2A, H2B, H3, and H4. The H1 histone forms connecting links between nucleosome core particles. (p. 229)

Holliday junction resolving enzyme   An enzyme that catalyzes the breakage and rejoining of two DNA strands in a Holliday junction to generate two independent duplex molecules. (p. 590)

Holliday model   A molecular model of genetic recombination in which the participating duplexes contain heteroduplex regions of the same length. (p. 587)

Holliday structure   A cross-shaped configuration of two DNA duplexes formed as an intermediate in recombination. (p. 588)

holocentric chromosome　A chromosome with a diffuse centromere. *See* **diffuse centromere.** (p. 246)

holoenzyme　*See* **RNA polymerase holoenzyme.**

homeobox　A DNA sequence motif found in the coding region of many regulatory genes; the amino acid sequence corresponding to the homeobox has a helix-loop-helix structure. (p. 543)

homeotic gene　Any of a group of genes that determine the fundamental patterns of development. *See also* **homeotic mutation.** (p. 541)

homeotic mutation　A developmental mutation that results in the replacement of one body structure by another body structure. (p.542)

homeotic selector gene　*See* **homeotic gene.**

homogametic　Producing only one kind of gamete with respect to the sex chromosomes. *See also* **heterogametic.** (p. 97)

homologous　In reference to DNA, having the same or nearly the same nucleotide sequence. (p. 88)

homologous chromosomes　Chromosomes that pair in meiosis and have the same genetic loci and structure; also called homologs. (p. 88)

homologous recombination　Genetic exchange between identical or nearly identical DNA sequences.

homothallism　The capacity of cells in certain fungi to undergo a conversion in mating type to make possible mating between cells produced by the same parental organism. (p. 482)

homozygous　Having the same allele of a gene in homologous chromosomes. (p. 40)

H1 histone　*See* **histone.**

Hoogstein base pairing　An unusual type of base pairing in which four guanine bases form a hydrogen-bonded quartet. (p. 250)

hormone　A small molecule in higher eukaryotes, synthesized in specialized tissue, that regulates the activity of other specialized cells. In animals, hormones are transported from their source to a target tissue in the blood. (p. 493)

hot spot　A site in a DNA molecule at which the mutation rate is much higher than the rate for most other sites. (p. 339, 565)

housekeeping gene　A gene that is expressed at the same level in virtually all cells and whose product participates in basic metabolic processes. (p. 488)

H substance　The carbohydrate precursor of the A and B red-blood-cell antigens.

H3 histone　*See* **histone.**

H2A histone　*See* **histone.**

H2B histone　*See* **histone.**

human genome project　A worldwide project to map genetically and sequence the human genome. (p. 392)

Huntington disease　Dominantly inherited degeneration of the neuromuscular system, with onset in middle age. (p. 52)

hybrid　An organism produced by the mating of genetically unlike parents; also, a duplex nucleic acid molecule produced of strands derived from different sources. (p. 32)

hybrid vigor　*See* **heterosis.**

hydrogen bond　A weak noncovalent linkage between two negatively charged atoms in which a hydrogen atom is shared.

hydrophobic interaction　A noncovalent interaction between nonpolar molecules or nonpolar groups, causing the molecules or groups to cluster when water is present.

hypermorph　A mutation in which the gene product is produced in greater abundance than wildtype. (p. 66)

hypomorph　A mutation in which the gene product is produced in less abundance than wildtype. (p. 66)

I

identical twins　*See* **monozygotic twins.**

identity by descent　The condition in which two alleles are both replicas of a single ancestral allele in the recent past. (p. 647)

imaginal disk　Structures present in the body of insect larvae from which the adult structures develop during pupation. (p. 540)

immune response　Tendency to develop resistance to a disease-causing agent after an initial infection.

immunity　A general term for resistance of an organism to specific substances, particularly agents of disease. (pp. 344, 483)

immunoglobulin　One of several classes of antibody proteins. (p. 483)

immunoglobulin class　The category in which an immunoglobulin is placed on the basis of its chemical characteristics, including the identity of its heavy chain. (p. 483)

imprinting　A process of DNA modification in gametogenesis that affects gene expression in the zygote; a probable mechanism is the methylation of certain bases in the DNA. (p. 280)

inborn error of metabolism　A genetically determined biochemical disorder, usually in the form of an enzyme defect that produces a metabolic block. (p. 12)

inbreeding　Mating between relatives. (pp. 634, 645)

inbreeding coefficient　A measure of the genetic effects of inbreeding in terms of the proportionate reduction in heterozygosity in an inbred organism compared with the heterozygosity expected with random mating. (p. 646)

inbreeding depression   The deterioration in a population's fitness or performance that accompanies inbreeding. (p. 687)

incompatible   Refers to blood or tissue whose transfusion or transplantation results in rejection.

incomplete dominance   A situation in which the heterozygous genotype has a phenotype that falls somewhere in the range between the two homozygous genotypes. (p. 67)

incomplete penetrance   Condition in which a mutant phenotype is not expressed in all organisms with the mutant genotype. (p. 71)

independent assortment   Random distribution of unlinked genes into gametes, as with genes in different (nonhomologous) chromosomes or genes that are so far apart on a single chromosome that the recombination frequency between them is 1/2. (p. 45)

individual selection   Selection based on each organism's own phenotype. (p. 684)

induced mutation   A mutation formed under the influence of a chemical mutagen or radiation. (p. 557)

inducer   A small molecule that inactivates a repressor, usually by binding to it and thereby altering the ability of the repressor to bind to an operator. (p. 461)

inducible   A gene that is expressed, or an enzyme that is synthesized, only in the presence of an inducer molecule. *See also* **constitutive.** (pp. 461, 463)

inducible transcription   *See* **inducible.**

induction   Activation of an inducible gene; prophage induction is the derepression of a prophage that initiates a lytic cycle of phage development. (p. 461)

initiation   The beginning of protein synthesis. (pp. 187, 431)

inosine (I)   One of a number of unusual bases found in transfer RNA.

insertional inactivation   Inactivation of a gene by interruption of its coding sequence; used in genetic engineering as a means of detecting insertion of a foreign DNA sequence into the coding region of a gene.

insertion sequence   A DNA sequence capable of transposition in a prokaryotic genome; such sequences usually code for their own transposase. (p. 347)

in situ **hybridization**   Renaturation of a radioactive probe nucleic acid to the DNA or RNA in a cell; used to localize particular DNA molecules to chromosomes or parts of chromosomes and to identify the time and tissue distribution of RNA transcripts. (p. 234)

integrase   An enzyme that catalyzes the site-specific exchange when a prophage is inserted into or excised from a bacterial chromosome; in the excision process, an accessory protein, excisionase, also is needed. (p. 343)

integration   The process by which one DNA molecule is inserted intact into another replicable DNA molecule, as in prophage integration and the integration of plasmid or tumor viral DNA into a chromosome. (p. 314)

intercalation   Insertion of a planar molecule between the stacked bases in duplex DNA. (p. 571)

interference   The tendency for crossing-over to inhibit the formation of another crossover nearby.

intergenic complementation   Complementation between mutations in different genes. *See also* **complementation test.**

intergenic suppression   Suppression of a mutant phenotype because of a mutation in a different gene; often refers specifically to a tRNA molecule able to recognize a stop codon or two different sense codons. (p. 585)

interphase   The interval between nuclear divisions in the cell cycle, extending from the end of telophase of one division to the beginning of prophase of the next division. (p. 83)

interrupted-mating technique   In an Hfr × F$^-$ cross, a technique by which donor and recipient cells are broken apart at specific times, allowing only a particular length of DNA to be transferred. (p. 319)

intervening sequence   *See* **intron.**

intron   A noncoding DNA sequence in a gene that is transcribed but is then excised from the primary transcript in forming a mature mRNA molecule; found primarily in eukaryotic cells. *See also* **exon.** (p. 425)

inversion   A structural aberration in a chromosome in which the order of several genes is reversed from the normal order. A pericentric inversion includes the centromere within the inverted region, and a paracentric inversion does not include the centromere. (p. 286)

inversion loop   Loop structure formed by synapsis of homologous genes in a pair of chromosomes, one of which contains an inversion. (p. 287)

inverted repeat   Either of a pair of base sequences present in the same molecule that are identical or nearly identical but are oriented in opposite directions; often found at the ends of transposable elements. (p. 243)

in vitro **experiment**   An experiment carried out with components isolated from cells.

in vivo **experiment**   An experiment performed with intact cells.

ionizing radiation   Electromagnetic or particulate radiation that produces ion pairs when dissipating its energy in matter. (p. 572)

IS element   *See* **insertion sequence.**

isochromosome   A chromosome with two identical arms containing homologous genes.

isotopes   The forms of a chemical element that have the same number of electrons and the same number of protons but differ in the number of neutrons in the atomic nucleus; unstable isotopes undergo transitions to a more stable state and, in so doing, emit radioactivity.

J

J (joining) region Any of multiple DNA sequences that code for alternative amino acid sequences of part of the variable region of an antibody molecule; the J regions of heavy and light chains are different. (p. 485)

just-so story Made-up stories, not supported by hard evidence, about the presumed adaptive significance of a trait; named after the 1902 book *Just So Stories* by Rudyard Kipling. (p. 24)

K

kappa particle An intracellular parasite in *Paramecium* that releases a substance capable of killing sensitive cells. (p. 614)

karyotype The chromosome complement of a cell or organism; often represented by an arrangement of metaphase chromosomes according to their lengths and the positions of their centromeres. (p. 269)

kilobase pair (kb) Unit of length of a duplex DNA molecule; equal to 1000 base pairs. (p. 222)

kindred A group of relatives.

kinetochore The cellular structure, formed in association with the centromere, to which the spindle fibers become attached in cell division. (p. 86)

Klinefelter syndrome The clinical features of human males with the karyotype 47,XXY. (p. 279)

Kosambi's mapping function A mapping function based on the assumption that the coincidence between crossovers is proportional to the distance between them. (p. 145)

kuru A neurological disease caused by a prion protein. (p. 724)

L

***lac* operon** The set of genes required to metabolize lactose in bacteria. (p. 466)

Lac repressor A protein coded by the *lacI* gene that, in the absence of lactose, binds with the operator and prevents transcription. (p. 464)

lactose Milk sugar; each molecule consists of a joined glucose and galactose. (p. 462)

lactose permease An enzyme responsible for transport of lactose from the environment into bacteria. (p. 462)

lagging strand The DNA strand whose complement is synthesized in short fragments that are ultimately joined together. (p. 195)

lariat structure Structure of an intron immediately after excision in which the 5' end loops back and forms a 5'-2' linkage with another nucleotide. (p. 426)

L chain *See* **light chain.**

leader *See* **leader polypeptide, leader sequence.**

leader polypeptide A short polypeptide encoded in the leader sequence of some operons coding for enzymes in amino acid biosynthesis; translation of the leader polypeptide participates in regulation of the operon through attenuation. (pp. 424, 473)

leader sequence The region of an mRNA molecule from the 5' end to the beginning of the coding sequence, sometimes containing regulatory sequences; in prokaryotic mRNA, it contains the ribosomal binding site.

leading strand The DNA strand whose complement is synthesized as a continuous unit. (p. 195)

leptotene The initial substage of meiotic prophase I during which the chromosomes become visible in the light microscope as unpaired thread-like structures. (p. 92)

lethal mutation A mutation that results in the death of an affected organism.

liability Risk, particularly toward a threshold type of quantitative trait. (p. 690)

library *See* **gene library.**

ligand The molecule that binds to a specific receptor. (p. 528)

light (L) chain One of the small polypeptide chains in an antibody molecule. (p. 483)

lineage The ancestral history of a cell in development or of a species in evolution. (p. 519)

lineage diagram A diagram of cell lineages and their developmental fates. (p. 519)

linkage The tendency of genes located in the same chromosome to be associated in inheritance more frequently than expected from their independent assortment in meiosis. (p. 124)

linkage equilibrium The condition in which the alleles of different genes in a population are present in gametes in proportion to the product of the frequencies of the alleles.

linkage group The set of genes present together in a chromosome. (p.132)

linkage map A diagram of the order of genes in a chromosome in which the distance between adjacent genes is proportional to the rate of recombination between them; also called a genetic map. (pp. 124, 127)

linker In genetic engineering, synthetic DNA fragments that contain restriction-enzyme cleavage sites that are used to join two DNA molecules. (p. 230)

local population A group of organisms of the same species occupying an area within which most individual members find their mates; synonomous terms are *deme* and *Mendelian population.* (p. 628)

localized centromere The type of centromere organization found in most eukaryotic cells in which the centromere is located in one small region of the chromosome. (p. 246)

locomotor activity   Movement of organisms from one place to another under their own power. (p. 714)

locus   The site or position of a particular gene on a chromosome. (p. 132)

loss-of-function mutation   A mutation that eliminates gene function; also called a null mutation. (p. 526)

lost allele   An allele no longer present in a population; its frequency is 0. (p. 629)

lysis   Breakage of a cell caused by rupture of its cell membrane and cell wall. (p. 330)

lysogen   Clone of bacterial cells that have acquired a prophage. (p. 341)

lysogenic cycle   In temperate bacteriophage, the phenomenon in which the DNA of an infecting phage becomes part of the genetic material of the cell. (p. 329)

lysozyme   One of a class of enzymes that dissolves the cell wall of bacteria; found in chicken egg white, human tears, and coded by many phages.

lytic cycle   The life cycle of a phage, in which progeny phage are produced and the host bacterial cell is lysed. (p. 329)

M

macrophage   One of a class of white blood cells that processes antigens for presentation as a necessary step in stimulation of the immune response.

major groove   In B-form DNA, the larger of two continuous indentations running along the outside of the double helix. (p. 179)

major histocompatibility complex (MHC)   The group of closely linked genes coding for antigens that play a major role in tissue incompatibility and that function in regulation and other aspects of the immune response.

map-based cloning   A strategy of gene cloning based on the position of a gene in the genetic map; also called positional cloning. (p. 374)

mapping function   The mathematical relation between the genetic map distance across an interval and the observed percentage of recombination in the interval. (p. 144)

map unit   A unit of distance in a linkage map that corresponds to a recombination frequency of 1 percent. Technically, the map distance across an interval in map units equals one-half the average number of crossovers in the interval expressed as a percentage. Map units are sometimes denoted centimorgans (cM). (p. 128)

masked mRNA   Messenger RNA that cannot be translated until specific regulatory substances are available; present in eukaryotic cells, particularly eggs; storage mRNA. (p. 503)

maternal effect   A phenomenon in which the genotype of a mother affects the phenotype of the offspring through substances present in the cytoplasm of the egg. (p. 603)

maternal effect gene   A gene that influences early development through its expression in the mother and the presence of the gene product in the oocyte. (p. 532)

maternal inheritance   Extranuclear inheritance of a trait through cytoplasmic factors or organelles contributed by the female gamete. (p. 601)

maternal sex ratio   Aberrant sex ratio in certain *Drosophila* species caused by cytoplasmic transmission of a parasite that is lethal to male embryos. (p. 616)

mating system   The norms by which members of a population choose their mates; important systems of mating include random mating, assortative mating, and inbreeding. (p. 632)

mating-type interconversion   Phenomenon in homothallic yeast in which cells switch mating type as a result of the transposition of genetic information from an unexpressed cassette into the active mating-type locus. (p. 481)

Maxam-Gilbert method   A technique for determining the nucleotide sequence of DNA by means of strand cleavage at the positions of particular nucleotides.

MCS   Multiple cloning site. *See* **polylinker.**

MCP   *See* **methyl-accepting chemotaxis protein.**

mean   The arithmetic average. (p. 672)

megabase pair   Unit of length of a duplex nucleic acid molecule; equal to 1 million base pairs. (p. 222)

meiocyte   A germ cell that undergoes meiosis to yield gametes in animals or spores in plants. (p.88)

meiosis   The process of nuclear division in gametogenesis or sporogenesis in which one replication of the chromosomes is followed by two successive divisions of the nucleus to produce four haploid nuclei. (p. 87)

melting curve   A graph of the amount of denatured DNA present in a solution (measured by UV light adsorption) as a function of increasing temperature. (p. 199)

melting temperature   The midpoint of the narrow temperature range at which the strands of duplex DNA denature. (p. 199)

Mendelian genetics   The mechanism of inheritance in which the statistical relations between the distribution of traits in successive generations result from (1) particulate hereditary determinants (genes), (2) random union of gametes, and (3) segregation of unchanged hereditary determinants in the reproductive cells. (p. 32)

Mendelian population   *See* **local population.**

meristem   The mitotically active growing point of plant tissue. (p. 543)

meristic trait   A trait in which the phenotype is determined by counting, such as number of ears on a stalk of corn or number of eggs laid by a hen. (p. 671)

merodiploid   A bacterial cell carrying two copies of a region of the genome and therefore diploid for the genes in this region. (p. 324)

messenger RNA (mRNA)   An RNA molecule transcribed from a DNA sequence and translated into the amino acid sequence of a polypeptide. In eukaryotes, the primary transcript undergoes elaborate processing to become the mRNA. (pp. 12, 423)

metabolic pathway   A set of chemical reactions that take place in a definite order to convert a particular starting molecule into one or more specific products.

metacentric chromosome   A chromosome with its centromere about in the middle so that the arms are equal or almost equal in length. (p. 260)

metaphase   In mitosis, meiosis I, or meiosis II, the stage of nuclear division in which the centromeres of the condensed chromosomes are arranged in a plane between the two poles of the spindle. (pp. 86, 93)

metaphase plate   Imaginary plane, equidistant from the spindle poles in a metaphase cell, on which the centromeres of the chromosomes are aligned by the spindle fibers. (p. 86)

methyl-accepting chemotaxis protein   Alternative designation of chemoreceptors in bacterial chemotaxis; abbreviated MCP. (p. 712)

methylation   The modification of a protein or a DNA or RNA base by the addition of a methyl group ($-CH_3$). (p. 487)

MHC   *See* **major histocompatibility complex.**

micrococcal nuclease   A nuclease that is isolated from a bacterium and that cleaves double-stranded DNA without regard to sequence. (p. 230)

middle repetitive DNA sequence   A DNA sequence present tens to hundreds of times per genome. (p. 239)

migration   Movement of organisms among subpopulations; also, the movement of molecules in electrophoresis. (pp. 205, 650)

minimal medium   A growth medium consisting of simple inorganic salts, a carbohydrate, vitamins, organic bases, essential amino acids, and other essential compounds; its composition is precisely known. Minimal medium contrasts with complex medium or broth, which is an extract of biological material (vegetables, milk, meat) that contains a large number of compounds and the precise composition of which is unknown. (p. 312)

minor groove   In B-form DNA, the smaller of two continuous indentations running along the outside of the double helix. (p. 179)

mismatch   An arrangement in which two nucleotides opposite one another in double-stranded DNA are unable to form hydrogen bonds. (p. 578)

mismatch repair   Removal of one nucleotide from a pair that cannot properly hydrogen-bond, followed by replacement with a nucleotide that can hydrogen-bond. (p. 578)

missense mutation   An alteration in a coding sequence of DNA that results in an amino acid replacement in the polypeptide. (pp. 444, 558)

mitosis   The process of nuclear division in which the replicated chromosomes divide and the daughter nuclei have the same chromosome number and genetic composition as the parent nucleus. (p. 83)

mitotic spindle   The set of fibers that arches through the cell during mitosis and to which the chromosomes are attached. *See also* **spindle.** (p. 86)

monocistronic mRNA   A mRNA molecule that codes for a single polypeptide chain.

monoclonal antibody   Antibody directed against a single antigen and produced by a single clone of B cells or a single cell line of hybridoma cells.

monohybrid   A genotype that is heterozygous for one pair of alleles; the offspring of a cross between genotypes that are homozygous for different alleles of a gene. (p. 33)

monomorphic gene   A gene for which the most common allele is virtually fixed.

monoploid   The basic chromosome set that is reduplicated to form the genomes of the species in a polyploid series; the smallest haploid chromosome number in a polyploid series. (p. 261)

monosomic   Condition of an otherwise diploid organism in which one member of a pair of chromosomes is missing. (p. 269)

monozygotic twins   Twins developed from a single fertilized egg that splits into two embryos at an early division; also called identical twins. (p. 682)

morphogen   Substance that induces differentiation. (p. 515)

mosaic   An organism composed of two or more genetically different types of cells. (p. 276)

M period   *See* **cell cycle.**

mRNA   *See* **messenger RNA.**

multifactorial trait   A trait determined by the combined action of many factors, typically some genetic and some environmental. (p. 670)

multiple alleles   The presence in a population of more than two alleles of a gene. (p. 60)

multiple cloning site   *See* **polylinker.**

multiplication rule   The principle that the probability that all of a set of independent events are realized simultaneously equals the product of the probabilities of the separate events. (p. 50)

multivalent   An association of more than two homologous chromosomes resulting from synapsis during meiosis in a polysomic or polyploid individual.

mutagen   An agent that is capable of increasing the rate of mutation. (p. 556)

mutagenesis   The process by which a gene undergoes a heritable alteration; also, the treatment of a cell or organism with a known mutagen.

mutant   Any heritable biological entity that differs from wildtype, such as a mutant DNA molecule, mutant allele, mutant gene, mutant chromosome, mutant cell, mutant organism, or mutant heritable phenotype; also, a cell or organism in which a mutant allele is expressed.

mutant screen   A type of genetic experiment in which the geneticist seeks to isolate multiple new mutations that affect a particular trait. (p. 55)

mutation   A heritable alteration in a gene or chromosome; also, the process by which such an alteration happens. Used incorrectly, but with increasing frequency, as a synonym for mutant, even in some excellent textbooks. (pp. 15, 556, 650)

mutation pressure   The generally very weak tendency of mutation to change allele frequency.

mutation rate   The probability of a new mutation in a particular gene, either per gamete or per generation. (p. 563)

N

narrow-sense heritability   The fraction of the phenotypic variance revealed as resemblance between parents and offspring; technically, the ratio of the additive genetic variance to the total phenotypic variance. (p. 684)

natural selection   The process of evolutionary adaptation in which the genotypes genetically best suited to survive and reproduce in a particular environment give rise to a disproportionate share of the offspring and so gradually increase the overall ability of the population to survive and reproduce in that environment. (pp. 24, 650)

nature versus nurture   Archaic term for genetics versus environment; usually a false dichotomy because most traits are influenced by both genetic and environmental factors.

negative assortative mating   *See* **assortative mating**

negative chromatid interference   In meiosis, when two or more crossovers occur in a chromosome, a tendency for the nonsister chromatids that participate in one event to also participate in another event. (p. 141)

negative regulation   Regulation of gene expression in which mRNA is not synthesized until a repressor is removed from the DNA of the gene.

negative supercoiling   Supercoiling of DNA that counteracts underwinding of the DNA. (p. 225)

neomorph   A mutation in which the mutant allele expresses either a novel gene product or a mutant gene product that has a novel phenotypic effect. (p. 66)

neutral allele   An allele that has a negligible effect on the ability of the organism to survive and reproduce.

nick   A single-strand break in a DNA molecule. (p. 196)

nicked circle   A circular DNA molecule containing one or more single-strand breaks.

nitrous acid HNO_2, a chemical mutagen. (p. 569)

nondisjunction   Failure of chromosomes to separate (disjoin) and move to opposite poles of the division spindle; the result is loss or gain of a chromosome. (p. 103)

nonhistone chromosomal proteins   A large class of proteins, not of the histone class, found in isolated chromosomes.

nonparental ditype   An ascus containing two pairs of recombinant spores. (p. 150)

nonpermissive conditions   Environmental conditions that result in expression of the phenotype of a conditional mutation.

nonselective medium   A growth medium that allows growth of wildtype and of one or more mutant genotypes. (p. 312)

nonsense mutation   A mutation that changes a codon specifying an amino acid into a stop codon, resulting in premature polypeptide chain termination; also called a chain-termination mutation. (pp. 444, 558)

nonsense suppression   Suppression of the phenotype of a polypeptide chain-termination mutation mediated by a mutant tRNA that inserts an amino acid at the site of the termination codon. (pp. 447, 585)

nontaster   Individual with the inherited inability to taste the substance phenylthiocarbamide (PTC); phenotype of the homozygous recessive.

normal distribution   A symmetrical bell-shaped distribution characterized by the mean and the variance; in a normal distribution, approximately 68 percent of the observations are within 1 standard deviation from the mean, and approximately 95 percent are within 2 standard deviations. (p. 673)

nuclease   An enzyme that breaks phosphodiester bonds in nucleic acid molecules. (p. 192)

nucleic acid   A polymer composed of repeating units of phosphate-linked five-carbon sugars to which nitrogenous bases are attached. *See also* **DNA** and **RNA.** (p. 175)

nucleic acid hybridization   The formation of duplex nucleic acid from complementary single strands. (p. 202)

nucleoid   A DNA mass, not bounded by a membrane, within the cytoplasm of a prokaryotic cell, chloroplast, or mitochondrion; often refers to the major DNA unit in a bacterium. (pp. 225, 308)

nucleolar organizer region   A chromosome region containing the genes for ribosomal RNA; abbreviated NOR.

nucleolus (*pl.* nucleoli)   Nuclear organelle in which ribosomal RNA is made and ribosomes are partially synthesized; usually associated with the nucleolar organizer region. A nucleus may contain several nucleoli. (p. 86)

nucleoside   A purine or pyrimidine base covalently linked to a sugar. (p. 174)

nucleosome   The basic repeating subunit of chromatin, consisting of a core particle composed of two molecules each of four different histones around which a length of DNA containing about 145 nucleotide pairs is wound, joined to an adjacent core particle by about 55 nucleotide pairs of linker DNA associated with a fifth type of histone. (p. 230)

nucleotide   A nucleoside phosphate. (pp. 9, 174)

nucleotide analog   A molecule that is structurally similar to a normal nucleotide and that is incorporated into DNA. (p. 568)

nucleus   The organelle, bounded by a membranous envelope, that contains the chromosomes in a eukaryotic cell.

nullisomic   A cell or individual that contains no copies of a particular chromosome.

nutritional mutation   A mutation in a metabolic pathway that creates a need for a substance to be present in the growth medium or that eliminates the ability to utilize a substance present in the growth medium. (p. 311)

O

ochre codon   Jargon for the UAA stop codon; an ochre mutation is a UAA codon formed from a sense codon. (p. 447)

octoploid   A cell or organism with eight complete sets of chromosomes. (p. 261)

Okazaki fragment   Any of the short strands of DNA produced during discontinuous replication of the lagging strand; also called a precursor fragment. (p. 194)

oligonucleotide   A short, single-stranded nucleic acid, usually synthesized for use in DNA sequencing , as a primer in the polymerase chain reaction, as a probe, or for oligonucleotide site-directed mutagenesis.

oligonucleotide primer   A short, single-stranded nucleic acid synthesized for use in DNA sequencing or as a primer in the polymerase chain reaction. (p. 207)

oligonucleotide site-directed mutagenesis   Replacement of a small region in a gene with an oligonucleotide that contains a specified mutation. (p. 375)

oncogene   A gene that can initiate tumor formation. (p. 298)

open reading frame   In the coding strand of DNA or in mRNA, a region containing a series of codons uninterrupted by stop codons and therefore capable of coding for a polypeptide chain; abbreviated ORF. (pp. 400, 424)

operator   A regulatory region in DNA that interacts with a specific repressor protein in controlling the transcription of adjacent structural genes. (p. 465)

operon   A collection of adjacent structural genes regulated by an operator and a repressor. (p. 600)

ORF   *See* **open reading frame.**

organelle   A membrane-bounded cytoplasmic structure that has a specialized function, such as a chloroplast or mitochondrion. (p. 600)

origin of replication   A DNA base sequence at which replication of a molecule is initiated. (p. 187)

overdominance   A condition in which the fitness of a heterozygote is greater than the fitness of both homozygotes. (p. 657)

overlapping genes   Genes that share part of their coding sequences. (p. 450)

ovule   The structure in seed plants that contains the embryo sac (female gametophyte) and that develops into a seed after fertilization of the egg.

P

pachytene   The middle substage of meiotic prophase I in which the homologous chromosomes are closely synapsed. (p. 92)

pair-rule gene   Any of a group of genes active early in *Drosophila* development that specifies the fates of alternating segments or parasegments. Mutations in pair-rule genes result in loss of even-numbered or odd-numbered segments or parasegments. (pp. 533, 538)

palindrome   In nucleic acids, a segment of DNA in which the sequence of bases on complementary strands reads the same from a central point of symmetry—for example 5'-GAATTC-3'; the sites of recognition and cleavage by restriction endonucleases are frequently palindromic. (pp. 204, 361)

paracentric inversion   An inversion that does not include the centromere. (p. 287)

parasegment   Developmental unit in *Drosophila* consisting of the posterior part of one segment and the anterior part of the next segment in line. (p. 532)

parental combination   Alleles present in an offspring chromosome in the same combination as that found in one of the parental chromosomes. (p. 124)

parental ditype   An ascus containing two pairs of nonrecombinant spores. (p. 150)

parental strand   In DNA replication, the strand that served as the template in a newly formed duplex. (p. 182)

partial digestion   Condition in which a restriction enzyme cleaves some, but not all, of the restriction sites present in a DNA molecule. (p. 391)

partial diploid   A cell in which a segment of the genome is duplicated, usually in a plasmid. (p. 324)

partial dominance   A condition in which the phenotype of the heterozygote is intermediate between the corresponding homozygotes but resembles one more closely than the other.

Pascal's triangle   Triangular configuration of integers in which the *n*th row gives the binomial coefficients in the expansion of $(x + y)^{n-1}$. The first and last numbers in each row equal 1, and the others equal the sum of the adjacent numbers in the row immediately above. (p. 107)

paternal inheritance   Extranuclear inheritance of a trait through cytoplasmic factors or organelles contributed by the male gamete. (p. 601)

path in a pedigree   A closed loop in a pedigree, with only one change of direction, that passes through a common ancestor of the parents of an individual. (p. 646)

pathogen   An organism that causes disease. (p. 348)

pattern formation   The creation of a spatially ordered and differentiated embryo from a seemingly homogeneous egg cell. (p. 513)

PCR   *See* **polymerase chain reaction.**

pedigree   A diagram representing the familial relationships among relatives. (p. 51)

penetrance   The proportion of organisms having a particular genotype that actually express the corresponding phenotype. If the phenotype is always expressed, penetrance is complete; otherwise, it is incomplete. (p. 72)

peptide bond   A covalent bond between the amino group ($-NH_2$) of one amino acid and the carboxyl group ($-COOH$) of another. (p. 412)

peptidyl site   The tRNA-binding site on the ribosome to which the tRNA bearing the nascent polypeptide becomes bound immediately after formation of the peptide bond. (p. 431)

peptidyl transferase   The enzymatic activity of ribosomes responsible for forming a peptide bond. (p. 433)

percent G + C   The proportion of base pairs in duplex DNA that consist of GC pairs. (p. 175)

pericentric inversion   An inversion that includes the centromere. (p. 287)

permease   A membrane-associated protein that allows a specific small molecule to enter the cell.

permissive condition   An environmental condition in which the phenotype of a conditional mutation is not expressed; contrasts with nonpermissive or restrictive condition. (p. 556)

petite mutant   A mutation in yeast resulting in slow growth and small colonies. (p. 610)

PEV   *See* **position effect variegation.**

PFGE   *See* **pulse-field gel electrophoresis.**

P_1 generation   The parents used in a cross, or the original parents in a series of generations; also called the P generation if there is no chance of confusion with the grandparents or more remote ancestors. (p. 35)

P group   *See* **5'-P group.**

phage   *See* **bacteriophage.**

phage-attachment site   The base sequence in a bacterial chromosome at which bacteriophage DNA can integrate to form a prophage. (p. 341)

phage repressor   Regulatory protein that prevents transcription of genes in a prophage.

phenotype   The observable properties of a cell or an organism, which result from the interaction of the genotype and the environment. (p. 41)

phenotypic variance   The total variance in a phenotypic trait in a population.

phenylketonuria   A hereditary human condition resulting from inability to convert phenylalanine into tyrosine; causes severe mental retardation unless treated in infancy and childhood by a low-phenylalanine diet; abbreviated PKU. (p. 730)

phenylthiocarbamide   *See* **PTC tasting.**

Philadelphia chromosome   Abnormal human chromosome 22, resulting from reciprocal translocation and often associated with a type of leukemia. (p. 298)

phosphodiester bond   In nucleic acids, the covalent bond formed between the 5'-phosphate group (5'-P) of one nucleotide and the 3'-hydroxyl group (3'-OH) of the next nucleotide in line; these bonds form the backbone of a nucleic acid molecule. (p. 175)

photoreactivation   The enzymatic splitting of pyrimidine dimers produced in DNA by ultraviolet light; requires visible light and the photoreactivation enzyme. (p. 580)

phylogenetic tree   A diagram showing the genealogical relationships among a set of genes or species. (p. 619)

physical map   A diagram showing the relative positions of physical landmarks in a DNA molecule; common landmarks include the positions of restriction sites and particular DNA sequences. (p. 392)

plaque   A clear area in an otherwise turbid layer of bacteria growing on a solid medium, caused by the infection and killing of the cells by a phage; because each plaque results from the growth of one phage, plaque counting is a way of counting viable phage particles. The term is also used for animal viruses that cause clear areas in layers of animal cells grown in culture. (p. 331)

plasmid   An extrachromosomal genetic element that replicates independently of the host chromosome; it may exist in one or many copies per cell and may segregate in cell division to daughter cells in either a controlled or a random fashion. Some plasmids, such as the F factor, may become integrated into the host chromosome. (p. 314)

pleiotropic effect   Any phenotypic effect that is a secondary manifestation of a mutant gene. (p. 17)

pleiotropy   The condition in which a single mutant gene affects two or more distinct and seemingly unrelated traits. (p. 17)

point mutation   A mutation caused by the substitution, deletion, or addition of a single nucleotide pair.

polarity   The 5'-to-3' orientation of a strand of nucleic acid. (p. 175)

pole cell   Any of a group of cells, set off at the posterior end of the *Drosophila* embryo, from which the germ cells are derived. (p. 529)

poly-A tail   The sequence of adenines added to the 3' end of many eukaryotic mRNA molecules in processing. (p. 425)

polycistronic mRNA   An mRNA molecule from which two or more polypeptides are translated; found primarily in prokaryotes. (pp. 436, 449)

polygenic inheritance   The determination of a trait by alleles of two or more genes.

polylinker   A short DNA sequence that is present in a vector and that contains a number of unique restriction sites suitable for gene cloning. (p. 372)

polymer   A regular, covalently bonded arrangement of basic subunits or monomers into a large molecule, such as a polynucleotide or polypeptide chain.

polymerase   An enzyme that catalyzes the covalent joining of nucleotides—for example, DNA polymerase and RNA polymerase. (p. 191)

polymerase α The major DNA replication enzyme in eukaryotic cells. (p. 192)

polymerase chain reaction   Repeated cycles of DNA denaturation, renaturation with primer oligonucleotide sequences, and replication, resulting in exponential growth in the number of copies of the DNA sequence located between the primers. (pp. 207, 369)

polymerase γ The enzyme responsible for replication of mitochondrial DNA. (p. 192)

polymorphic gene   A gene for which there is more than one relatively common allele in a population. (p. 630)

polymorphism   The presence in a population of two or more relatively common forms of a gene, chromosome, or genetically determined trait.

polynucleotide chain   A polymer of covalently linked nucleotides. (p. 175)

polypeptide   *See* **polypeptide chain.**

polypeptide chain   A polymer of amino acids linked together by peptide bonds. (pp. 15, 412)

polyploidy   The condition of a cell or organism with more than two complete sets of chromosomes. (p. 261)

polyprotein   A protein molecule that can be cleaved to form two or more finished protein molecules.

polyribosome   *See* **polysome.**

polysome   A complex of two or more ribosomes associated with an mRNA molecule and actively engaged in polypeptide synthesis; a polyribosome. (p. 450)

polysomy   The condition of a diploid cell or organism that has three or more copies of a particular chromosome. (p. 267)

polytene chromosome   A giant chromosome consisting of many identical strands laterally apposed and in register, exhibiting a characteristic pattern of transverse banding. (p. 234)

P1 bacteriophage   Temperate virus that infects *E. coli*; in cells lysogenic for P1, the phage DNA is not integrated into the chromosome and replicates as a plasmid.

population   A group of organisms of the same species. (p. 628)

population genetics   Application of Mendel's laws and other principles of genetics to entire populations of organisms. (p. 628)

population structure   *See* **population substructure.**

population substructure   Organization of a population into smaller breeding groups between which migration is restricted. Also called population subdivision. (p. 642)

positional cloning   A strategy of gene cloning based on the position of a gene in the genetic map; also called map-based cloning. (p. 374)

positional information   Developmental signals transmitted to a cell by virtue of its position in the embryo. (p. 515)

position effect   A change in the expression of a gene depending on its position within the genome. (p. 296)

position-effect variegation   Mosaic phenotype (variegation) due to variation in the level of expression of a gene in different cell lineages owing to its position in the genome. (p. 296)

positive assortative mating   See **assortative mating.**

positive chromatid interference   In meiosis, when two or more crossovers occur in a chromosome, a tendency for the nonsister chromatids that participate in one event not to participate in another event. (p. 141)

positive regulation   Mechanism of gene regulation in which an element must be bound to DNA in an active form to allow transcription. Positive regulation contrasts with negative regulation, in which a regulatory element must be removed from DNA. (p. 461)

postmeiotic segregation   Segregation of genetically different products in a mitotic division following meiosis, as in the formation in *Neurospora* of a pair of ascospores that have different genotypes. (p. 150)

postreplication repair   DNA repair that takes place in nonreplicating DNA or after the replication fork is some distance beyond a damaged region. (p. 582)

posttranslocation state   The state of the ribosome immediately following the movement of the small subunit one codon farther along the mRNA; the aminoacyl (A) site is unoccupied. (p. 434)

precursor fragment   *See* **Okazaki fragment.**

prediction equation   In quantitative genetics, an equation used to predict the improvement in mean performance of a population brought about by artifical selection; it always includes heritability as one component.

pretranslocation state   The state of the ribosome immediately prior to the movement of the small subunit one codon farther along the mRNA; the exit (E) site is unoccupied. (p. 434)

Pribnow box   A base sequence in prokaryotic promoters to which RNA polymerase binds in an early step of initiating transcription.

primary transcript   An RNA copy of a gene; in eukaryotes, the transcript must be processed to form a translatable mRNA molecule. (p. 424)

primase   The enzyme responsible for synthesizing the RNA primer needed for initiating DNA synthesis. (p. 196)

primer   In nucleic acids, a short RNA or single-stranded DNA segment that functions as a growing point in polymerization. (p. 191)

primer oligonucleotide   A single-stranded DNA molecule, typically from 18 to 22 nucleotides in length, that can hybridize with a longer DNA strand and serve as a primer for replication. In the polymerase chain reaction, two primers anneal to opposite strands of a DNA duplex with their 3'-OH ends facing. *See also* **polymerase chain reaction.** (p. 207)

primosome   The enzyme complex that forms the RNA primer for DNA replication in eukaryotic cells. (p. 196)

prion protein   Protein infectious agent; a protein that can transmit disease. (p. 723)

probability   A mathematical expression of the degree of confidence that certain events will or will not be realized. (p. 106)

probe   A radioactive DNA or RNA molecule used in DNA-RNA or DNA-DNA hybridization assays. (pp. 206, 372)

processing   A series of chemical reactions in which primary RNA transcripts are converted into mature mRNA, rRNA, or tRNA molecules or in which polypeptide chains are converted into finished proteins. (p. 412)

programmed cell death   Cell death that happens as part of the normal developmental process. *See also* **apoptosis.** (p. 524)

prokaryote   An organism that lacks a nucleus; prokaryotic cells divide by fission. (p. 23)

promoter   A DNA sequence at which RNA polymerase binds and initiates transcription. (pp. 419, 465)

promoter mutation   A mutation that increases or decreases the ability of a promoter to initiate transcription. (p. 422)

promoter recognition   The first step in transcription. (p. 418)

proofreading function   *See* **editing function.**

prophage   The form of phage DNA in a lysogenic bacterium; the phage DNA is repressed and usually integrated into the bacterial chromosome, but some prophages are in plasmid form. (p. 341)

prophage induction   Activation of a prophage to undergo the lytic cycle. (p. 345)

prophase   The initial stage of mitosis or meiosis, beginning after interphase and terminating with the alignment of the chromosomes at metaphase; often absent or abbreviated between meiosis I and meiosis II. (p. 86, 92)

protanomaly   *See* **color blindness.**

protanopia   *See* **color blindness.**

protein   A molecule composed of one or more polypeptide chains. (pp. 15, 412)

protein subunit   Any of the polypeptide chains in a protein molecule made up of multiple polypeptide chains. (p. 414)

prototroph   Microbial strain capable of growth in a defined minimal medium that ideally contains only a carbon source and inorganic compounds. The wildtype genotype is usually regarded as a prototroph. (p. 312)

pseudoautosomal region   In mammals, a small region of the X and Y chromosome containing homologous genes. (p. 273)

pseudogene   A DNA sequence that is not expressed because of one or more mutations but that has a functional counterpart in the same organism; pseudogenes are regarded as mutated forms of ancient gene duplications.

P site   *See* **peptidyl site.**

PTC tasting   Genetic polymorphism in the ability to taste phenylthiocarbamide. (p. 729)

P transposable element   A *Drosophila* transposable element used for the induction of mutations, germ-line transformation, and other types of genetic engineering.

pulse-field gel electrophoresis   A type of electrophoresis in which the electric field is manipulated to make possible the separation of large DNA fragments. (p. 228)

Punnett square   A cross-multiplication square used for determining the expected genetic outcome of a mating. (p. 43)

purine   An organic base found in nucleic acids; the predominant purines are adenine and guanine. (p. 174)

pyrimidine   An organic base found in nucleic acids; the predominant pyrimidines are cytosine, uracil (in RNA only), and thymine (in DNA only). (p. 174)

pyrimidine dimer   Two adjacent pyrimidine bases, typically a pair of thymines, in the same polynucleotide strand, between which chemical bonds have formed; the most common lesion formed in DNA by exposure to ultraviolet light. (p. 571)

Q

QTL   *See* **quantitative trait locus.**

quantitative trait   A trait—typically measured on a continuous scale, such as height or weight—that results from the combined action of several or many genes in conjunction with environmental factors. (p. 670)

quantitative trait locus   A locus segregating for alleles that have different, measurable effects on the expression of a quantitative trait. (p. 692)

R

race   A genetically or geographically distinct subgroup of a species.

rad   Radiation absorbed dose; an amount of ionizing radiation resulting in the dissipation of 100 ergs of energy in 1 gram of matter; equal to 0.01 gray.

random genetic drift   Fluctuation in allele frequency from generation to generation resulting from restricted population size. (p. 650, 658)

random mating   System of mating in which mating pairs are formed independently of genotype and phenotype. (p. 632)

random spore analysis   In fungi, the genetic analysis of spores collected at random rather than from individual tetrads. (p. 154)

reading frame   The phase in which successive triplets of nucleotides in mRNA form codons; depending on the reading frame, a particular nucleotide in an mRNA could be in the first, second, or third position of a codon. The reading frame actually used is defined by the AUG codon that is selected for chain initiation. (p. 440)

reannealing   Reassociation of dissociated single strands of DNA to form a duplex molecule. (p. 200)

recessive   Refers to an allele, or the corresponding phenotypic trait, expressed only in homozygotes. (p. 35)

reciprocal cross   A cross in which the sexes of the parents are the reverse of another cross. (p. 35)

reciprocal translocation   Interchange of parts between nonhomologous chromosomes. (p. 288)

recombinant   A chromosome that results from crossing-over and that carries a combination of alleles differing from that of either chromosome participating in the crossover; the cell or organism containing a recombinant chromosome. (p. 124)

recombinant DNA technology   Procedures for creating DNA molecules composed of one or more segments from other DNA molecules. (p. 360)

recombination   Exchange of parts between DNA molecules or chromosomes; recombination in eukaryotes usually entails a reciprocal exchange of parts, but in prokaryotes it is often nonreciprocal. (p. 124)

recombination repair   Repair of damaged DNA by exchange of good for bad segments between two damaged molecules. (p. 586)

recruitment   The process in which a transcriptional activator protein interacts with one or more components of the transcription complex and attracts it to the promoter. (p. 495)

red-green color blindness   *See* **color blindness.**

reductional division   Term applied to the first meiotic division because the chromosome number (counted as the number of centromeres) is reduced from diploid to haploid. (p. 89)

redundancy   The feature of the genetic code in which an amino acid corresponds to more than one codon; also called degeneracy. (p. 443)

regulatory gene   A gene with the primary function of controlling the expression of one or more other genes.

rejection   An immune response against transfused blood or transplanted tissue.

relative fitness   The fitness of a genotype expressed as a proportion of the fitness of another genotype. (p. 693)

relaxed DNA   A DNA circle whose supercoiling has been removed either by introduction of a single-strand break or by the activity of a topoisomerase. (p. 224)

rem   The quantity of any kind of ionizing radiation that has the same biological effect as one rad of high-energy gamma rays; rem stands for roentgen equivalent man.

renaturation   Restoration of the normal three-dimensional structure of a macromolecule; in reference to nucleic acids, the term means the formation of a double-stranded molecule by complementary base pairing between two single-stranded molecules. (p. 200)

repair   *See* **DNA repair.**

repair synthesis   The enzymatic filling of a gap in a DNA molecule at the site of excision of a damaged DNA segment. (p. 581)

repellent   Any substance that elicits chemotaxis of organisms in a direction away from itself. (p. 704)

repetitive sequence   A DNA sequence present more than once per haploid genome. (p. 235)

replica plating   Procedure in which a particular spatial pattern of colonies on an agar surface is reproduced on a series of agar surfaces by stamping them with a template that contains an image of the pattern; the template is often produced by pressing a piece of sterile velvet upon the original surface, which transfers cells from each colony to the cloth. (p. 561)

replication   *See* **DNA replication;** θ **replication.**

replication fork   In a replicating DNA molecule, the region in which nucleotides are added to growing strands. (p. 187)

replication origin The base sequence at which DNA synthesis begins. (p. 187)

replication slippage The process in which the number of copies of a small tandem repeat can increase or decrease during replication. (p. 559)

replicon A DNA molecule that has a replication origin.

reporter gene A gene whose expression can be monitored. (p. 500)

repression A regulatory process in which a gene is temporarily rendered unable to be expressed. (p. 461)

repressor A protein that binds specifically to a regulatory sequence adjacent to a gene and blocks transcription of the gene. (pp. 344, 461, 464)

repulsion *See trans* **configuration.**

restriction endonuclease A nuclease that recognizes a short nucleotide sequence (restriction site) in a DNA molecule and cleaves the molecule at that site; also called a restriction enzyme. (pp. 147, 202, 360)

restriction enzyme *See* **restriction endonuclease.**

restriction fragment A segment of duplex DNA produced by cleavage of a larger molecule by a restriction enzyme. (p. 205)

restriction fragment length polymorphism (RFLP) Genetic variation in a population associated with the size of restriction fragments that contain sequences homologous to a particular probe DNA; the polymorphism results from the positions of restriction sites flanking the probe, and each variant is essentially a different allele. (p. 147)

restriction map A diagram of a DNA molecule showing the positions of cleavage by one or more restriction endonucleases. (p. 205)

restriction site The base sequence at which a particular restriction endonuclease makes a cut. (pp. 204, 360)

restrictive condition A growth condition in which the phenotype of a conditional mutation is expressed. (p. 556)

retinoblastoma An inherited cancer caused by a mutation in the tumor-suppressor gene located in chromosome band *13q14.* Inheritance of one copy of the mutation results in multiple malignancies in retinal cells of the eyes in which the mutation becomes homozygous—for example, through a new mutation or mitotic recombination. (p. 298)

retrovirus One of a class of RNA animal viruses that cause the synthesis of DNA complementary to their RNA genomes on infection. (p. 378)

reverse genetics Procedure in which mutations are deliberately produced in cloned genes and introduced back into cells or the germ line of an organism. (p. 377)

reverse mutation A mutation that undoes the effect of a preceding mutation. (pp. 583, 652)

reverse transcriptase An enzyme that makes complementary DNA from a single-stranded RNA template. (p. 369)

reverse transcriptase PCR Amplification of a duplex DNA molecule originally produced by reverse trascriptase using an RNA template. (p. 370)

reversion Restoration of a mutant phenotype to the wild-type phenotype by a second mutation. (p. 583)

RFLP *See* **restriction fragment length polymorphism.**

R group *See* **side chain.**

Rh Rhesus blood-group system in human beings; maternal-fetal incompatibility of this system may result in hemolytic disease of the newborn.

Rh-negative Refers to the phenotype in which the red blood cells lack the D antigen of the Rh blood-group system.

rhodopsin One of a class of proteins that function as visual pigments. (p. 283)

Rh-positive Refers to the phenotype in which the red blood cells possess the D antigen of the Rh blood-group system.

ribonuclease Any enzyme that cleaves phosphodiester bonds in RNA; abbreviated RNase.

ribonucleic acid *See* **RNA.**

ribonucleoprotein particle Nuclear particle containing a short RNA molecule and several proteins; involved in intron excision and splicing and other aspects of RNA processing.

ribose The five-carbon sugar in RNA. (p. 195)

ribosomal RNA (rRNA) RNA molecules that are components of the ribosomal subunits; in eukaryotes, there are four rRNA molecules—5S, 5.8S, 18S, and 28S; in prokaryotes, there are three—5S, 16S, and 23S. (p. 12)

ribosome The cellular organelle on which the codons of mRNA are translated into amino acids in protein synthesis. Ribosomes consist of two subunits, each composed of RNA and proteins. In prokaryotes, the subunits are 30S and 50S particles; in eukaryotes, they are 40S and 60S particles. (p. 13)

ribosome-binding site The base sequence in a prokaryotic mRNA molecule to which a ribosome can bind to initiate protein synthesis; also called the Shine-Dalgarno sequence. (p. 432)

ribozyme An RNA molecule able to catalyze one or more biochemical reactions. (p. 427)

ring chromosome A chromosome whose ends are joined; one that lacks telomeres. (p. 260)

RNA Ribonucleic acid, a nucleic acid in which the sugar constituent is ribose; typically, RNA is single-stranded and contains the four bases adenine, cytosine, guanine, and uracil. (p. 12)

RNA editing   Process in which certain nucleotides in RNA are chemically changed to other nucleotides after transcription. (p. 605)

RNA polymerase   An enzyme that makes RNA by copying the base sequence of a DNA strand. (pp. 196, 418)

RNA polymerase holoenzyme   A large protein complex in eukaryotic nuclei consisting of Pol II in combination with at least 9 other subunits; Pol II itself includes 12 subunits (p. 495)

RNA processing   The conversion of a primary transcript into an mRNA, rRNA, or tRNA molecule; includes splicing, cleavage, modification of termini, and (in tRNA) modification of internal bases. (p. 424)

RNase   Any enzyme that cleaves RNA.

RNA splicing   Excision of introns and joining of exons. (p. 425)

Robertsonian translocation   A chromosomal aberration in which the long arms of two acrocentric chromosomes become joined to a common centromere. (p. 293)

roentgen   A unit of ionizing radiation, defined as the amount of radiation resulting in 2.083×10^9 ion pairs per cm^3 of dry air at 0°C and 1 atm pressure; abbreviated R; 1 R produces 1 electrostatic unit of charge per cm^3 of dry air at 0°C and 1 atm pressure.

rolling-circle replication   A mode of replication in which a circular parent molecule produces a linear branch of newly formed DNA. (p. 189)

Rous sarcoma virus   A type of retrovirus that infects the chicken.

R plasmid   A bacterial plasmid that carries drug-resistance genes; commonly used in genetic engineering. (p. 348)

rRNA   *See* **ribosomal RNA.**

RT-PCR   *See* **reverse transcriptase PCR.**

S

S   *See* **cell cycle.**

salvage pathway   A minor source of deoxythymidine triphosphate (dTTP) for DNA synthesis that uses the enzyme thymidine kinase (TK).

satellite DNA   Eukaryotic DNA that forms a minor band at a different density from that of most of the cellular DNA in equilibrium density gradient centrifugation; consists of short sequences repeated many times in the genome (highly repetitive DNA) or of mitochondrial or chloroplast DNA. (p. 235)

scaffold   A protein-containing material in chromosomes, believed to be responsible in part for the compaction of chromatin. (p. 233)

second-division segregation   Segregation of a pair of alleles into different nuclei in the second meiotic division, the result of crossing-over between the gene and the centromere of the pair of homologous chromosomes. (p. 156)

second meiotic division   The meiotic division in which the centromeres split and the chromosome number is not reduced; also called the equational division. (p. 89)

segment   Any of a series of repeating morphological units in a body plan. (p. 532)

segmentation gene   Any of a group of genes that determines the spatial pattern of segments and parasegments in *Drosophila* development. (p. 532)

segment-polarity gene   Any of a group of genes that determines the spatial pattern of development within the segments of *Drosophila* larvae. (pp. 533, 538)

segregation   Separation of the members of a pair of alleles into different gametes in meiosis. (p. 40)

segregational petites   Slow-growing yeast colonies deficient in respiration because of mutation in a chromosomal gene; inheritance is Mendelian. (p. 610)

segregation mutation   Developmental mutation in which sister cells express the same fate when they should express different fates. (p. 523)

selected marker   A genetic mutation that allows growth in selective medium. (p. 319)

selection   In evolution, intrinsic differences in the ability of genotypes to survive and reproduce; in plant and animal breeding, the choosing of organisms with certain phenotypes to be parents of the next generation; in mutation studies, a procedure designed in such a way that only a desired type of cell can survive, as in selection for resistance to an antibiotic.

selection coefficient   The amount by which relative fitness is reduced or increased. (p. 653)

selection differential   In artificial selection, the difference between the mean of the selected organisms and the mean of the population from which they were chosen.

selection limit   The condition in which a population no longer responds to artificial selection for a trait. (p. 686)

selection-mutation balance   Equilibrium determined by the opposing effects of mutation tending to increase the frequency of a deleterious allele and selection tending to decrease its frequency.

selection pressure   The tendency of natural or artificial selection to change allele frequency.

selectively neutral mutation   A mutation that has no (or negligible) effects on fitness. (p. 650)

selective medium   A medium that allows growth only of cells with particular genotypes. (p. 312)

self-fertilization   The union of male and female gametes produced by the same organism.

selfish DNA   DNA sequences that do not contribute to the fitness of an organism but are maintained in the genome through their ability to replicate and transpose. (p. 246)

semiconservative replication The usual mode of DNA replication, in which each strand of a duplex molecule serves as a template for the synthesis of a new complementary strand, and the daughter molecules are composed of one old (parental) and one newly synthesized strand. (p. 183)

semisterility A condition in which a significant proportion of the gametophytes produced by a plant or of the zygotes produced by an animal are inviable, as in the case of a translocation heterozygote. (p. 288)

sense strand Originally defined as the template strand in duplex DNA from which the mRNA is transcribed. Some later authors used the term in the opposite sense to mean the nontemplate DNA strand that has the same nucleotide sequence as the mRNA. Because both usages persist, the term is ambiguous and is best avoided. Use *template strand* or *transcribed strand* instead.

sequence-tagged site (STS) A DNA sequence, present once per haploid genome, that can be amplified by the use of suitable oligonucleotide primers in the polymerase chain reaction in order to identify clones that contain the sequence. (p. 395)

sex chromosome A chromosome, such as the human X or Y, that has a role in the determination of sex. (p. 96)

sex-influenced trait A trait whose expression depends on the sex of the individual.

sex-limited trait A trait expressed in one sex and not in the other.

sex-linked Refers to a trait determined by a gene on a sex chromosome, usually the X. (p. 97)

Shine-Dalgarno sequence *See* **ribosome-binding site.**

shuttle vector A vector capable of replication in two or more organisms—for example, yeast and *E. coli.* (p. 392)

sib *See* **sibling.**

sibling A brother or sister, each having the same parents. (p. 51)

sibship A group of brothers and sisters. (p. 49)

sickle-cell anemia A severe anemia in human beings inherited as an autosomal recessive and caused by an amino acid replacement in the β-globin chain; heterozygotes tend to be more resistant to falciparum malaria than are normal homozygotes. (pp. 17, 558)

side chain In protein structure, the chemical group attached to the α-carbon atom of an amino acid by which the chemical properties of each amino acid are determined; also called R group. (p. 412)

sievert A unit of radiation exposure of biological tissue equal to the dose in gray multiplied by a quality factor that measures the relative biological effectiveness of the type of radiation; abbreviated Sv. Roughly speaking, 1 sievert is the amount of radiation equivalent in its biological effects to one gray of gamma rays; 1 Sv equals 100 rem.

sigma (σ) subunit The subunit of RNA polymerase needed for promoter recognition.

signal transduction The process by which a regulatory signal ultimately determines cell fate by activation of a series of regulatory proteins.

significant *See* **statistically significant.**

silent mutation A mutation that has no phenotypic effect. (pp. 444, 557)

simple tandem repeat polymorphism (STRP) A DNA polymorphism in a population in which the alleles differ in the number of copies of a short, tandemly repeated nucleotide sequence. (p. 150)

single-active-X principle In mammals, the genetic inactivation of all X chromosomes except one in each cell lineage, except in the very early embryo. (p. 275)

single-copy sequence A DNA sequence present only once in each haploid genome.

single-strand binding protein A protein able to bind single-stranded DNA. (p. 198)

single-stranded DNA A DNA molecule that consists of a single polynucleotide chain. (p. 10)

sister chromatids Chromatids produced by replication of a single chromosome. (p. 86)

site-specific exchange *See* **site-specific recombination.**

site-specific recombination Recombination that is catalyzed by a particular enzyme and that always takes place at the same site between the same DNA sequences. (p. 330)

small ribonucleoprotein particles Small nuclear particles that contain short RNA molecules and several proteins. They are involved in intron excision and splicing and other aspects of RNA processing.

somatic cell Any cell of a multicellular organism other than the gametes and the germ cells from which gametes develop. (p. 82)

somatic-cell genetics Study of somatic cells in culture.

somatic mosaic An organism that contains two genetically different types of cells. (p. 160)

somatic mutation A mutation arising in a somatic cell. (pp. 487, 556)

SOS repair An inducible, error-prone system for repair of DNA damage in *E. coli.* (p. 582)

Southern blot A nucleic acid hybridization method in which, after electrophoretic separation, denatured DNA is transferred from a gel to a membrane filter and then exposed to radioactive DNA or RNA under conditions of renaturation; the radioactive regions locate the homologous DNA fragments on the filter. (p. 206)

spacer sequence   A noncoding base sequence between the coding segments of polycistronic mRNA or between genes in DNA.

specialized transduction   *See* **transducing phage.**

species   Genetically, a group of actually or potentially inbreeding organisms that is reproductively isolated from other such groups.

S period   *See* **cell cycle.**

spindle   A structure composed of fibrous proteins on which chromosomes align during metaphase and move during anaphase. (p. 86)

splice acceptor   The 5' end of an exon. (p. 526)

splice donor   The 3' end of an exon. (p. 526)

spliceosome   An RNA-protein particle in the nucleus in which introns are removed from RNA transcripts. (p. 426)

spontaneous mutation   A mutation that happens in the absence of any known mutagenic agent. (p. 557)

spore   A unicellular reproductive entity that becomes detached from the parent and can develop into a new organism upon germination; in plants, spores are the haploid products of meiosis. (p. 89)

sporophyte   The diploid, spore-forming generation in plants, which alternates with the haploid, gamete-producing generation (the gametophyte). (p. 89)

standard deviation   The square root of the variance. (p. 673)

start codon   An mRNA codon, usually AUG, at which polypeptide synthesis begins. (p. 443)

statistically significant   Said of the result of an experiment or study that has only a small probability of happening by chance on the assumption that some hypothesis is true. Conventionally, if results as bad or worse would be expected less than 5 percent of the time, the result is said to be statistically significant; if less than 1 percent of the time, the result is called statistically highly significant; both outcomes cast the hypothesis into serious doubt. (p. 111)

sticky end   A single-stranded end of a DNA fragment produced by certain restriction enzymes capable of reannealing with a complementary sequence in another such strand. (pp. 204, 361)

stop codon   One of three mRNA codons—UAG, UAA, and UGA—at which polypeptide synthesis stops. (p. 436)

structural gene   A gene that encodes the amino acid sequence of a polypeptide chain. (p. 415)

STRP   *See* **simple tandem repeat polymorphism.**

STS   *See* **sequence-tagged site.**

submetacentric chromosome   A chromosome whose centromere divides it into arms of unequal length. (p. 261)

subpopulation   Any of the breeding groups within a larger population between which migration is restricted. (p. 628)

subspecies   A relatively isolated population or group of populations distinguishable from other populations in the same species by allele frequencies or chromosomal arrangements and sometimes exhibiting incipient reproductive isolation.

substrate   A substance acted on by an enzyme.

subunit   A component of an ordered aggregate of macromolecules—for example, a single polypeptide chain in a protein containing several chains. (p. 414)

supercoiled DNA   *See* **supercoiling.**

supercoiling   Coiling of double-stranded DNA in which strain caused by overwinding or underwinding of the duplex makes the double helix itself twist; a supercoiled circle is also called a twisted circle or a superhelix. (pp. 224, 225)

suppressive petites   A class of cytoplasmically inherited, small-colony mutants in yeast that are not corrected by the addition of normal mitochondria.

suppressor mutation   A mutation that partially or completely restores the function impaired by another mutation at a different site in the same gene (intragenic suppression) or in a different gene (intergenic suppression). (p. 583)

suppressor tRNA   Usually, a tRNA molecule capable of translating a stop codon by inserting an amino acid in its place, but a few suppressor tRNAs replace one amino acid with another. (p. 447)

synapsis   The pairing of homologous chromosomes or chromosome regions in zygotene of the first meiotic prophase. (p. 92)

synaptonemal complex   A complex protein structure that forms between synapsed homologous chromosomes in the pachytene substage of the first meiotic prophase.

syncytial blastoderm   Stage in early *Drosophila* development formed by successive nuclear divisions without division of the cytoplasm. (p. 529)

syndrome   A group of symptoms that appear together with sufficient regularity to warrant designation by a special name; also, a disorder, disease, or anomaly.

synteny   The presence of two different genes on the same chromosome.

synteny group   A group of genes present in a single chromosome in two or more species. (p. 398)

T

tandem duplication   A pair of identical or closely related DNA sequences that are adjacent and in the same orientation. (p. 283)

taster   One who has the inherited ability to taste the substance phenylthiocarbamide (PTC); the phenotype of the homozygous dominant and heterozygote.

TATA binding protein (TBP)   A protein that binds to the TATA motif in the promoter region of a gene. (p. 495)

TATA box   The base sequence 5'-TATA-3' in the DNA of a promoter. (p. 420)

tautomeric shift   A reversible change in the location of a hydrogen atom in a molecule, altering the molecule from one isomeric form to another. In nucleic acids, the shift is typically between a keto group (keto form) and a hydroxyl group (enol form).

TBP   TATA-box binding protein. *See* **TATA binding protein.**

TBP-associated factor (TAF)   Any protein found in close association with TATA binding protein. (p. 495)

T cells   A class of white blood cells instrumental in various aspects of the immune reponse. (p. 487)

T **DNA**   Transposable element found in *Agrobacterium tumefaciens,* which produces crown gall tumors in a wide variety of dicotyledonous plants. (p. 381)

telomerase   An enzyme that adds specific nucleotides to the tips of the chromosomes to form the telomeres. (p. 248)

telomere   The tip of a chromosome, containing a DNA sequence required for stability of the chromosome end. (p. 247)

telophase   The final stage of mitotic or meiotic nuclear division. (p. 87, 94)

temperate phage   A phage capable of both a lysogenic and a lytic cycle. (p. 329)

temperature-sensitive mutation   A conditional mutation that causes a phenotypic change only at certain temperatures. (p. 556)

template strand   A nucleic acid strand whose base sequence is copied in a polymerization reaction to produce either a complementary DNA or RNA strand. (pp. 10, 191, 421)

terminal redundancy   In bacteriophage T4, a short duplication found at opposite ends of the linear molecule; the duplicated region differs from one phage to the next. (p. 335)

termination   Final stage in polypeptide synthesis when the ribosome encounters a stop codon and the finished polypeptide is released from the ribosome. (p. 436)

testcross   A cross between a heterozygote and a recessive homozygote, resulting in progeny in which each phenotypic class represents a different genotype. (p. 41)

testis-determining factor (TDF)   Genetic element on the mammalian Y chromosome that determines maleness. (p. 273)

tetrad   The four chromatids that make up a pair of homologous chromosomes in meiotic prophase I and metaphase I; also, the four haploid products of a single meiosis. (p. 93)

tetrad analysis   A method for the analysis of linkage and recombination using the four haploid products of single meiotic divisions. (p. 150)

tetraploid   A cell or organism with four complete sets of chromosomes; in an autotetraploid, the chromosome sets are homologous; in an allotetraploid, the chromosome sets consist of a complete diploid complement from each of two distinct ancestral species. (p. 261)

tetratype   An ascus containing spores of four different genotypes—one each of the four genotypes possible with two alleles of each of two genes. (p. 150)

thermophile   An organism that normally lives at an unusually high temperature. (p. 208)

θ replication   Bidirectional replication of a circular DNA molecule, starting from a single origin of replication. (p. 187)

30-nm fiber   The level of compaction of eukaryotic chromatin resulting from coiling of the extended, nucleosome-bound DNA fiber. (p. 233)

three-point cross   Cross in which three genes are segregating; used to obtain unambiguous evidence of gene order. (p. 141)

3'-OH group   The end of a DNA or RNA strand that terminates in a sugar and so has a free hydroxyl group on the number-3' carbon. (p. 175)

threshold trait   A trait with a continuously distributed liability or risk; organisms with a liability greater than a critical value (the threshold) exhibit the phenotype of interest, such as a disorder. (p. 671)

thymine (T)   A nitrogenous pyrimidine base found in DNA. (p. 9)

thymine dimer   *See* **pyrimidine dimer.**

time of entry   In an Hfr × F⁻ bacterial mating, the earliest time that a particular gene in the Hfr parent is transferred to the F⁻ recipient. (p. 320)

Ti **plasmid**   A plasmid present in *Agrobacterium tumefaciens* and used in genetic engineering in plants. (p. 381)

topoisomerase   An enzyme that introduces or eliminates either underwinding or overwinding of double-stranded DNA. It acts by introducing a single-strand break, changing the relative positions of the strands, and sealing the break. (p. 187)

topoisomerase I   A nick-closing type of topoisomerase enzyme in which one supercoil is removed from DNA per reaction cycle. (p. 225)

topoisomerase II   A topoisomerase enzyme able to remove both positive and negative supercoils from DNA. These enzymes can also cleave duplex DNA and pass another duplex DNA molecule through the gap. (p. 225)

total variance   Summation of all sources of genetic and environmental variation. (p. 678)

totipotent cell   A cell capable of differentiation into a complete organism; the zygote is totipotent.

trait   Any aspect of the appearance, behavior, development, biochemistry, or other feature of an organism. (p. 2)

trans configuration   The arrangement in linked inheritance in which a genotype heterozygous for two mutant sites has received one of the mutant sites from each parent—that is, $a_1+/+a_2$. (p. 126)

transcript   An RNA strand that is produced from, and is complementary in base sequence to, a DNA template strand. (p. 13)

transcription   The process by which the information contained in a template strand of DNA is copied into a single-stranded RNA molecule of complementary base sequence. (pp. 13, 412)

transcriptional activator protein   Positive control element that stimulates transcription by binding with particular sites in DNA. (pp. 462, 491)

transcription complex   An aggregate of RNA polymerase (consisting of its own subunits) along with other polypeptide subunits that makes transcription possible.

transducing phage   A phage type capable of producing particles that contain bacterial DNA (transducing particles). A specialized transducing phage produces particles that carry only specific regions of chromosomal DNA; a generalized transducing phage produces particles that may carry any region of the genome. (p. 325)

transduction   The carrying of genetic information from one bacterium to another by a phage. (p. 325)

transfer RNA (tRNA)   A small RNA molecule that translates a codon into an amino acid in protein synthesis; it has a three-base sequence, called the anticodon, complementary to a specific codon in mRNA, and a site to which a specific amino acid is bound. (p. 13)

transformation   Change in the genotype of a cell or organism resulting from exposure of the cell or organism to DNA isolated from a different genotype; also, the conversion of an animal cell, whose growth is limited in culture, into a tumor-like cell whose pattern of growth is different from that of a normal cell. (p. 3, 363)

transformation mutation   A mutation in which affected cells undergo a developmental fate that is characteristic of other cells. (p. 522)

transgenic   Refers to animals or plants in which novel DNA has been incorporated into the germ line. (p. 378)

trans heterozygote   _See_ **_trans_ configuration.**

transition mutation   A mutation resulting from the substitution of one purine for another purine or that of one pyrimidine for another pyrimidine. (p. 557)

translation   The process by which the amino acid sequence of a polypeptide is synthesized on a ribosome according to the nucleotide sequence of an mRNA molecule. (pp. 14, 412, 430)

translocation   Interchange of parts between nonhomologous chromosomes; also, the movement of mRNA with respect to a ribosome during protein synthesis. _See also_ **reciprocal translocation.** (pp. 288, 434)

transmembrane receptor   A receptor protein containing amino acid sequences that span the cell membrane. (p. 514)

transposable element   A DNA sequence capable of moving (transposing) from one location to another in a genome. (pp. 242, 243)

transposase   Protein necessary for transposition. (p. 246, 347)

transposition   The movement of a transposable element. (pp. 243, 346)

transposon   A transposable element that contains bacterial genes—for example, for antibiotic resistance; also loosely used as a synonym for transposable element. (p. 347)

transposon tagging   Insertion of a transposable element that contains a genetic marker into a gene of interest. (p. 349)

transversion mutation   A mutation resulting from the substitution of a purine for a pyrimidine or that of a pyrimidine for a purine. (p. 557)

trinucleotide repeat   A tandemly repeated sequence of three nucleotides; genetic instability in trinucleotide repeats is the cause of a number of human hereditary diseases. (p. 279)

triplet code   A code in which each codon consists of three bases. (p. 439)

triplication   The presence of three copies of a DNA sequence that is ordinarily present only once.

triploid   A cell or organism with three complete sets of chromosomes. (p. 263)

trisomic   A diploid organism with an extra copy of one of the chromosomes. (p. 267)

trisomy-X syndrome   The clinical features of the karyotype 47,XXX. (p. 278)

trivalent   Structure formed by three homologous chromosomes in meiosis I in a triploid or trisomic chromosome when each homolog is paired along part of its length with first one and then the other of the homologs. (p. 267)

tRNA   _See_ **transfer RNA.**

true-breeding   Refers to a strain, breed, or variety of organism that yields progeny like itself; homozygous. (p. 32)

truncation point   In artificial selection, the value of the phenotype that determines which organisms will be retained for breeding and which will be culled. (p. 684)

tumor-suppressor gene   A gene whose absence predisposes to malignancy; also called an anti-oncogene. (p. 298)

Turner syndrome   The clinical features of human females with the karyotype 45,X. (p. 279)

twin spot   A pair of adjacent, genetically different regions of tissue, each derived from one daughter cell of a mitosis in which mitotic recombination took place. (p. 159)

U

ultracentrifuge   A centrifuge capable of speeds of rotation sufficiently high to separate molecules of different density in a suitable density gradient. (p. 183)

uncharged tRNA   A tRNA molecule that lacks an amino acid. (p. 434)

uncovering   The expression of a recessive allele present in a region of a structurally normal chromosome in which the homologous chromosome has a deletion. (p. 281)

underwound DNA   A DNA molecule whose strands are untwisted somewhat; hence, some of its bases are unpaired. (p. 187)

unequal crossing-over   Crossing-over between nonallelic copies of duplicated or other repetitive sequences—for example, in a tandem duplication, between the upstream copy in one chromosome and the downstream copy in the homologous chromosome. (p. 283)

uniparental inheritance   Extranuclear inheritance of a trait through cytoplasmic factors or organelles contributed by only one parent. *See also* **maternal inheritance.** (p. 600)

unique sequence   A DNA sequence that is present in only one copy per haploid genome, in contrast with repetitive sequences. (p. 239)

univalent   Structure formed in meiosis I in a monoploid or a monosomic when a chromosome has no pairing partner. (p. 269)

uracil (U)   A nitrogenous pyrimidine base found in RNA. (p. 12)

V

variable expressivity   Differences in the severity of expression of a particular genotype. (p. 71)

variable number of tandem repeats (VNTR)   Any region of the genome that, among the members of a population, is highly polymorphic in the number of tandem repeats of a short nucleotide sequence; the result is polymorphism in the size of the restriction fragment that contains the repeated sequence. (p. 148)

variable region   The portion of an immunoglobulin molecule that varies greatly in amino acid sequence among antibodies in the same subclass. *See also* **V region.** (p. 484)

variance   A measure of the spread of a statistical distribution; the mean of the squares of the deviations from the mean. (p. 672)

variegation   Mottled or mosaic expression. (p. 296)

vector   A DNA molecule, capable of replication, into which a gene or DNA segment is inserted by recombinant DNA techniques; a cloning vehicle. (p. 362)

viability   The probability of survival to reproductive age. (p. 655)

viral oncogene   A class of genes found in certain viruses that predispose to cancer. Viral oncogenes are the viral counterparts of cellular oncogenes. *See also* **cellular oncogene, oncogene.** (p. 298)

virulent phage   A phage or virus capable only of a lytic cycle; contrasts with temperate phage. (p. 329)

virus   An infectious intracellular parasite able to reproduce only inside living cells. (p. 308)

VNTR   *See* **variable number of tandem repeats.**

Von Hippel-Lindau disease   Hereditary disease marked by tumors in the retina, brain, other parts of the central nervous system, and various organs throughout the body. The disease is caused by an autosomal dominant mutation in chromosome 3.

V region   One of multiple DNA sequences coding for alternative amino acid sequences of part of the variable region of an antibody molecule. *See also* **variable region.** (p. 485)

V-type position effect   A type of position effect in *Drosophila* that is characterized by discontinuous expression of one or more genes during development; usually results from chromosome breakage and rejoining such that euchromatic genes are repositioned in or near centromeric heterochromatin.

W

Watson-Crick pairing   Base pairing in DNA or RNA in which A pairs with T (or U in RNA) and G pairs with C. (p. 10)

wildtype   The most common phenotype or genotype in a natural population; also, a phenotype or genotype arbitrarily designated as a standard for comparison. (p. 19)

wobble   The acceptable pairing of several possible bases in an anticodon with the base present in the third position of a codon. (p. 446)

X

χ^2 *See* **chi-square.** (p. 109)

X chromosome   A chromosome that plays a role in sex determination and that is present in two copies in the homogametic sex and in one copy in the heterogametic sex. (p. 97)

xeroderma pigmentosum   An inherited defect in the repair of ultraviolet-light damage to DNA, associated with extreme sensitivity to sunlight and multiple skin cancers. (p. 572)

X inactivation   *See* **single-active-X principle.**

X-linked inheritance   The pattern of hereditary transmission of genes located in the X chromosome; usually evident from the production of nonidentical classes of progeny from reciprocal crosses. (p. 99)

Y

YAC   *See* **yeast artificial chromosome.**

Y chromosome   The sex chromosome present only in the heterogametic sex; in mammals, the male-determining sex chromosome. (p. 97)

yeast artificial chromosome (YAC)   In yeast, a cloning vector that can accept very large fragments of DNA; a chromosome introduced into yeast derived from such a vector and containing DNA from another organism. (p. 391)

Z

Z-form DNA   One of the unusual three-dimensional structures of duplex DNA in which the helix is left-handed. (p. 181)

zinc finger   A structural motif, found in many DNA-binding proteins, in which finger-like projections entrap a zinc ion. (p. 492)

zygote   The product of the fusion of a female and a male gamete in sexual reproduction; a fertilized egg.

zygotene   The substage of meiotic phrophase I in which homologous chromosomes synapse. (p. 92)

zygotic gene   Any of a group of genes that control early development through their expression in the zygote. (p. 532)

Index

development (continued)
  nematode, 520–522
  pattern formation, 513, 532–534
  precocious and retarded, 525
deviation, 672–673
  standard, 673, 674–675
de Vries, Hugo, 32
dGTP (deoxyguanosine triphosphate), 191, 192
diakinesis, 92, 93
dicentric, 260
dideoxynucleoside analogs, clinical use of, 213
dideoxyribose, 211
dideoxy sequencing method, 211, 212
differential interference contrast microscopy, 521
diffuse centromere, 246
digestion, partial, 392
dihybrid cross, 42
dihybrid ratios, modified F_2, 61–64
dihybrids, 42
dihybrid testcross, 45
diploid organisms, 262
diploids, 82
  partial, 322
  selection in, 654
diplotene, 92, 93
direct repeats, 243
discontinuous replication, 194–199
discrete trait, 68
displacement loop, 590
distribution, 671–675
  normal, 673–675
diversity, adaptation and, 23–24
DNA cloning, 205, 393
DNA (deoxyribonucleic acid), 2–14
  alteration of, 480–488
  bacterial attachment site of, 341–343
  bacteriophage l, 340–345
  in bacteriophage, genetic role of, 6–9
  base composition of, 175–176, 177
  base pairing in, 9–10, 177–179
  base sequences in, 210–213
  B form of, 177
  chemical agents modifying, 569–571
  chemical composition of, 174–177
  coding sequence, 424
  complementary, 369–370, 386, 397
  conjugation and, 306, 314–325
  denaturation of, 199–200, 201, 208
  discovery of, 2–3
  double helix of, 9–10, 177–181, 410
  duplex, 10, 179
  experimental proof of genetic function of, 3–6
  genetic requirements of, 181–182
  integrated, 314
  linker, 230
  methylation, 487–488
  mitochondrial, 618–621
  nomenclature, 175
  nucleotide sequences in, 235–242
  nucleotides of, 174–175

phage attachment site of, 341–343
physical structure of, 177–181
polymorphisms, 148–150, 630–632
primer, 191–192
programmed rearrangements, 481–483
relaxed, 224
renaturation of, 200, 201, 202, 208, 236–239
repair mechanisms, 577–583
replication of. See replication
RNA compared to, 195–196
satellite, 235
selfish, 246
single-stranded, 10, 222
site-specific DNA cleavage, 202–205
supercoiling of, 224–225
synthesis of, 83
template, 10, 191
transcription of, 12–14, 412, 418–424
transduction and, 308, 325–328
transformation and, 312–314
T (transposable), 381–382
Z form of, 181
DNA fragments
  cloning large, 390–392
  isolation and characterization of, 199–207
  joining, 366, 367, 368
  manipulating large, 390
  production of defined, 360–362
  restriction fragment, 205
DNA ligase, 196
DNA looping, 494–495
DNA methylase, 487
DNA polymerase, 182, 191–192
  I (Pol I), 192
  III (Pol III), 192
DNA polymerase slippage, 579
DNA probe, 206, 372, 630–631
DNase, 225
DNA sequencing, 239–242
  automated, 401–405
  large scale, 400–405
DNA synthesis, 191–194
  priming of, 196, 197
DNA typing, 61, 639–645
DNA uracil glycosylase, 565–566, 578
Dobzhansky, Theodosius, 22
domains, folding, 430
dominance
  complications in concept of, 64–72
  degree of, 656
  incomplete, 67–68
dominant trait, 35, 36
dorsal gene, 537
dosage compensation, 274–277
double crossing-over, 129–130, 131, 138, 140, 141, 142, 143–144, 152–153, 154
double helix, 9–10, 177–181, 410
double-strand break model, 590, 592
double-stranded DNA (duplex DNA), 10, 179

double-Y syndrome, 278
Down syndrome, 271, 274, 275, 281, 294, 730
D-raf gene, 537
Drosophila
  alternative promoters in, 500–501
  aneuploid vs. euploid abnormalities in, 269
  chromosome 2 in, 146
  early development in, 510
  genetic distance and physical distance in, 145
  homeotic mutations in, 523
  polytene chromosomes, 296
  transcriptional activation during development of, 495–497
  transformation in, 377, 378
Drosophila melanogaster, 389
  attached-X chromosome of, 134–137
  Bar duplication in, 283
  behavioral mutants in, 714
  chromosome structure, 281–282
  circadian rhythms in, 714–716
  ClB procedure and, 563–565
  courtship behavior, 716–720
  courtship song in, 716–720
  development in, 513, 517, 529–543
  DNA molecules of, 228
  DNA replication in, 189, 190
  eye color in, 19–20
  gene amplification in, 481
  genetic maps of, 131–132
  genome of, 222
  highly repetitive sequences in, 240–241
  maternal sex ratio in, 616
  metacentric autosomes of, 261
  middle-repetitive sequences in, 241–242
  mitotic recombination in, 158–160
  molecular genetics of Drosophila clock, 720–723
  Notch deletions in, 281–282
  Notch gene product in, 527
  polytene chromosomes of, 234, 281, 283
  positional cloning in, 374
  position-effect variegation (PEV) in, 296–297
  recombination in, 160
  regulation of, 480
  sex determination of, 104–106
  transposable elements in, 243
  white gene, 101, 124–127
  X-linked genes of, 97–101, 124–127
  yellow gene, 126–127
Drosophila simulans, courtship song in, 720
Drosophila virilis
  acrocentric autosomes of, 261
  highly repetitive sequences in, 240
drug resistance, 609–610
  antibiotic-resistant mutants, 311, 348–349, 608, 609–610

dTTP (deoxythymidine triphosphate), 191, 568
duplex DNA, 10, 179
duplication, 282–283

E

easter gene, 537
*Eco*RI, 147, 148
ectopic expression, 67
editing function, 193, 194
E (exit) site, 431
electrophoresis, 390, 401–405, 629–630
    gel, 205–206, 211, 212, 390, 401–405
    protein, 629–630, 631
electroporation, 365
elongation, 421, 431, 433–436
elongation factors, 433
embryoid, 266
embryonic development, in animals, 514–519
embryonic induction, 517
embryonic stem cells, 378, 379, 381
endonucleases, 192
    restriction, 147, 148, 202–205
endoreduplication, 264, 265, 266
endosymbiont theory, 613, 619
engrailed gene, 539
enhancers, 494–499
enhancer trap, 499–500
environment, genes and, 20–22
environmental factors, 670
environmental variance, 675, 676, 677–680
    genotype-environment association, 680
    genotype-environment interaction, 678–680, 726–727
enzyme(s), 12
    Holliday junction-resolving, 590
    mismatch repair, 375–376
    nuclease activity of, 192–193, 194
    "one gene-one enzyme" concept, 420
    protease, 4
    restriction, 147, 148, 202–205, 360–362
    RNase, 4, 5
    starch-branching enzyme I (SBEI), 65, 246
    topoisomerase, 187, 198, 199, 225, 226
enzyme polymorphisms, 629–630
epidermal growth factor (EGF), 527
episome, 314
epistasis, 19–20, 61–64
equational division (second meiotic division), 94–95
equilibrium, 652
equilibrium centrifugation in a density gradient, 183
error-prone DNA repair, 582
Escherichia coli
    bacteriophages of, 6–8, 191
    chemotactic response of, 704–706

chromosome of, 225–227
cloning of genes in, 350
colonies, 311
conjugation in, 306, 314, 315
DNA isolated from, 183
DNA ligase, 196
DNA replication in, 186–187, 192–193
genomes of, 222, 394, 395
infection by bacteriophage l, 476–477
lactose operon, 462
nucleotide sequences in, 614
oligonucleotide site-directed mutagenesis for, 376
physical map of, 392
selection in, 653
sexual process in, 322
time-of-entry mapping of, 322–324
transduction of, 325–328
tRNA suppressors, 449
tryptophan operon, 471–476
wobble rules for, 447
estrogen, 493–494
ethics of cloning, 380
ethyl methane sulfonate, mutagenicity of, 570
euchromatin, 145–146, 241, 296
Eukarya, 22–23
eukaryotes, 23
    chromosomes of, 222, 228–234, 246–252
    crossing-over in, 308
    cytoplasmic transmission in, 614–616
    DNA replication in, 189
    enhancers in, 494, 495
    gene regulation in, 460, 479–480
    genetic apparatus of, 222–224
    genetic organization of, 480, 494, 495
    genomes of, 235–242
    mRNA of, 424
    recombinant DNA technology and, 386
    replication in, 189
    RNA processing in, 424
    size of complex genomes, 388–389
    transcriptional activation in, 495–497
    transcriptional regulation in, 488–501
    translation units in, 450–451
    transposable elements in, 347
euploid, 269
"Eve," mitochondrial, 618–621
even-skipped gene, 538
event, 48
evolution, 22–24, 649–650
evolutionary complexity, genome size and, 222–224
excisionase, 345
excision repair, 581–582
execution mutation, 524
exons, 425
exon-shuffle, 430
exonucleases, 192, 194
expressivity, 71
extinction, mass, 24

extranuclear inheritance, 599–625
    cytoplasmic transmission of symbionts, 614–616
    evolutionary origin of organelles, 613–614
    maternal effect in snail-shell coiling, 617–618
    mitochondrial "Eve," 618–621
    organelle heredity, 604–613
    recognition of, 600–604
extroversion, 731

F

familial fatal insomnia (FFI), 723
fate map, 530, 531, 535
fates, cell, 514
favism, 20
fertility, 655
fertility factor (F factor), 314–317
fertility restorers, 612
fetal alcohol syndrome, 513, 730
fetal hemoglobin, 499
F factor, 314–317
F' (F prime) plasmid, 322–323
F_1 generation, 33–34, 36
F_2 generation, 36, 61–64
first-division segregation, 155, 157
first meiotic division, 89–94
Fischer, Emil, 12
Fisher, Ronald Aylmer, 112, 114, 713
fitness
    components of, 654–655
    relative, 653
5'-P (5' phosphate) of nucleic acids, 175
fixed allele, 629
flagella, 705–706
flagellar components, 707, 708
floral organs, combinatorial determination of, 545–548
flower development in *Arabidopsis*, 545
folded chromosome, 225–227
folding domains, 430
forward mutations, 583, 652
four-o'clock plants, leaf variegation in, 606–609
F pilus, 306
fragile-site mental retardation-1 (FMRI), 280
fragile-X chromosomes, 279
fragile-X syndrome, 279–280, 518, 730
frameshift mutations, 440, 559
Franklin, Rosalind, 180
fraternal twins, 682–683
free radicals, 573
frequency of cotransduction, 326, 328
frequency of recombination, 125, 128, 129, 130, 140, 144, 145
frq (frequency) gene, 723
Fugu fubripes, 480

G

gain-of-function mutation, 526, 542
galactose, 706
  metabolism in yeast, 488–490
b-galactosidase, 462, 463, 468
Galton, Francis, 674
gamete, 37, 82, 83, 89, 629
gametogenesis, 518
gametophytes, 89
gap genes, 533, 536, 538
Garrod, Archibald, 12
gastrula, 514
gastrulation-defective gene, 537
Gates, R.R., 727
Gauss, Carl Friedrich, 674
Gautier, Marthe, 271
gel electrophoresis, 205–206, 211, 212, 390, 401–405
  pulsed-field (PFGE), 228, 390
GenBank, 401
gene(s), 2, 40, 81–121
  anterior, 534
  arrangement in bacteriophages, 340, 341
  candidate, 732–733
  chimeric, 285
  coordinate, 532–537
  defining, 161
  denominator, 105
  environment and, 20–22
  gap, 533, 536, 538
  homeotic, 539, 540–543
  housekeeping, 488
  incomplete penetrance, 71–72
  maternal-effect, 532
  meanings of, 339–340
  number affecting quantitative traits, 681–682
  numerator, 105
  overlapping, 449–450
  pair-rule, 533, 538
  parasegment, 532
  polymorphic, 630
  polypeptides and, 415–418
  posterior, 536
  proteins and, 12–15
  recombination within, 160
  reporter, 500
  segmentation, 532–534
  segment-polarity, 533–534, 538–540
  terminal, 536
  traits and, 16–22
  variable expressivity, 71
  X-linked, 97–103, 124–127, 639
  zygotic, 532
gene amplification, 480–481
gene cloning, 360. *See also* genetic engineering
gene conversion, 586–587
gene dosage, 480–481
gene expression, 411–457
  complex translation units, 450–451
  defined, 412

genetic code, 438–449
  overall process of, 451–452
  overlapping genes, 449–450
  position effects on, 296–297
  relations between genes and polypeptides, 415–418
  RNA processing, 412, 424–430, 460
  transcription, 12–14, 412, 418–424
  translation, 412, 430–438
gene pool, 628
gene product, 412
generalized transducing phage, 325
general nonchemotaxis, 706–707
general transcription factors, 494–495
gene regulation, 459–509
  in bacteriophage l, 476–479
  control points for, 460
  coordinate, 436, 461
  defined, 460
  eukaryotic, 460, 479–480
  general principle of, 504
  negative, 461
  positive, 461, 468–471
  in prokaryotes, 480
  transcriptional, 460–462, 465–467, 488–501
  of tryptophan operon, 471–476
gene splicing. *See* splicing
gene targeting, 379–380, 381
gene therapy, 382, 387–388
genetic analysis, 54–61
  complementation test, 54–60
  multiple alleles, 60–61
genetic code, 438–449
genetic disorders, 388
genetic distance, physical distance and, 145–146
genetic drift, random, 650, 658–661
genetic engineering, 205, 349, 359–388
  applications of, 382–388
  cloning strategies, 365–374
  in plants, 380–382
  protein production and, 386–387
  research, uses in, 386
  restriction enzymes and vectors, 360–365
  reverse genetics, 377–382, 386
  site-directed mutagenesis, 374–376
  viruses and, 387–388
genetic factors, 670
genetic mapping, 124, 127–141
  crossing-over and, 128, 129–130, 132–141
  in human pedigrees, 146–150
  mapping functions, 144–145
  of quantitative-trait loci, 694, 695
  tetrad analysis and, 150–158
  from three-point testcross, 141–146
  time-of-entry, 319–324
  transformation and, 312–314
genetic markers, 129–130, 131, 144, 145, 693
  DNA polymorphisms as, 148–150

genetic prediction
  binomial distribution and, 106–109
  chi-square method and, 109–114
genetic recombination. *See* recombination
genetics
  bacteriophage, 328–340
  defined, 2
  evolution and, 22, 649–650
  history of, 2, 32
  Mendelian, 32
  population. *See* population genetics
  reverse, 377–382, 386
  terminology of, 40–41
genome equivalents, 389–390
genomes, 222
  complex, analysis of, 388–400
  eukaryotic, 235–242
  human, 223–224, 395–397
  zygote, 518
genome size, 388–390
  analysis by renaturation of, 238–239
  evolutionary complexity and, 222–224
genotype, 40
  schizophrenia and, 730
genotype-environment association (G-E association), 680
genotype-environment interaction (G-E interaction), 678–680, 726–727
genotype frequency, 628–632
genotypic variance, 675, 676–677, 678
germ cells, 82
germ-line mutations, 556
germ-line transformation, 377–380
Gerstmann-Straussler disease (GSD), 723
G_1 (growth-1) period, 83
G_2 (growth-2) period, 83
Giemsa, 269
glucose-6-phosphate dehydrogenase (G6PD), 20
glutamic acid, chemical structure of, 413
glutamine, chemical structure of, 413
glycine, chemical structure of, 413
glyphosate, 386
goodness of fit, 109
  chi-square as measure of, 109–114
Gramineae, 398–400
grass family, genome evolution in, 398–400
Greider, Carol W., 249
growth hormone, genetic engineering with, 382–383
G6PD deficiency, 20
guanine, 9, 174, 175, 177, 180
guanosine triphosphate (GTP), 418
guide RNA, 248
Guthrie, Woody, 53

H

H1, 229, 230
H2A, 229, 230
H2B, 229, 230
H3, 229, 230
H4, 229, 230
hairy gene, 538, 539
Haldane's mapping function, 145
haploid, 83, 88
haploid chromosomes, 262
Hardy, Godfrey H., 634, 636
Hardy-Weinberg principle, 634–639, 646
heavy (H) chain, 483, 485
helicase proteins, 198, 199
helix-turn-helix, 492
hemoglobin, 15
    fetal, 499
    sickle-cell, 17–18, 657
    subunits of, 414–415
hemophilia A, 101, 102
hepatitis B, 388
herbicides, 386
hereditary determinants, Mendel's genetic hypothesis of, 36–39
hereditary nonpolyposis colon cancer, 401
heredity, chromosomes and, 96–106
heritability
    broad-sense, 682, 683, 685–686, 731–732, 734
    cultural, 734
    narrow-sense, 684–686, 689–690, 691, 692, 726
    of threshold traits, 690–692
Hershey, Alfred, 6–9, 329
Herskowitz, Ira, 484
heterochromatin, 145–146, 234, 240–241, 296–297
heterochronic mutations, 525–526
heteroduplex region, 587, 588
heterogametic, 97
heteroplasmy, 602–603
heterosis, 687
heterozygotes, frequency of, 637–638
heterozygote superiority, 656–658
heterozygous, 40–41
hexaploidy, 261, 262
Hfr (high frequency of recombination) cells, 317–319
Hicks, James B., 484
high-F disease, 499
highly repetitive sequences, 239, 240–241
highly significant, 111
hip, dislocation of, 691
histidine, chemical structure of, 413
histones, 229–230
HIV (human immunodeficiency virus), 208, 388
Holliday, Robin, 138
Holliday junction-resolving enzyme, 590

Holliday model of recombination, 138, 139, 587–590
Holliday structure, 588–589
holocentric chromosomes, 246
homeobox, 543
homeotic genes, 539, 540–543
homeotic mutations, 523
homogametic, 97
homologous chromosomes, 87, 88
    chromatid exchange in, 93
homothallism, 482
homozygous, 40–41
Hoogstein base pairing, 250–251
hormonal regulation, 493–494
hormones, 493
hot spots of mutation, 339, 565–566
housekeeping genes, 488
human behavior, genetics of, 727–734
human beings
    chromosomes of, 269–281
    genome of, 223–224, 395–397
    pedigree analysis in, 51–54, 146–150
    sex determination in, 106–107
    trisomy in, 274
human blood groups. *See* blood groups
human genetic map, 693
human genome, 223–224, 395–397
human genome project, 392
human immunodeficiency virus (HIV), 208, 388
hunchback, 536, 538, 539
Huntington disease, 52–53, 388
hybridization
    nucleic acid, 200–202
    in situ, 234–235, 240
hybrids, 32–33, 263–266, 383
    traits present in progeny of, 35–36
hybrid vigor, 687
hydrocephalus, 691
hypospadias, 691

I

identical twins, 682–683
identity by descent, allelic, 646–647
imaginal disks, 540
immune system, 487
immunity, 344–345
immunoglobulin, 483
immunoglobulin G (IgG), 483–484, 485
imprinting, 280, 518
inborn error of metabolism, 12, 729, 730
inbreeding, 54, 634, 645–649
    effects of, 649
inbreeding coefficient, 646
    calculation from pedigrees, 647–649
inbreeding depression, 687
incomplete dominance, 67–68
incomplete penetrance, 71–72
independent assortment, principle of, 42–45, 46, 94, 96, 124
independent events, 49–51
individual selection, 684

induced mutations, 557, 566–577
inducers, 461
inducible, 461
inducible transcription, 463–464
industrial wastes, genetic engineering and, 385
influenza virus, 388
inheritance
    cultural, 734
    cytoplasmic. *See* extranuclear inheritance
    extranuclear. *See* extranuclear inheritance
    maternal, 601, 603–604
    Mendelian, 48–51, 600, 628
    paternal, 601
    quantitative, 670–675
    sex limited, 101
    X-linked, 97–103
initiation, 187, 421, 431–433
    RNA primer and, 195–196
initiation factors, 432
insecticides, 386, 563
insertion sequences, 347
in situ hybridization, 234–235, 240
instar, 529
insulin, 385
integrase, 343
integrated, 314
intercalation, 571
interference
    chromatid, 141
    chromosome, 143–144
intergenic suppression, 583, 585
interphase, 83, 85
interrupted-mating technique, 319, 320
intervening sequences, 425
intervening sequences. *See* introns
intragenic suppression, 583–585
introns, 369, 400–401, 425–430
    evolutionary origin, 430
inversion, 286–288
inversion loop, 287, 288, 289
inverted repeats, 243
in vitro experiments, 191
ionizing radiation, mutagenicity of, 572–575
irreversible mutation, 650–651
IS (insertion sequence) elements, 347
isoleucine, chemical structure of, 413

J

Jacob, François, 439, 465, 467, 476
Jeffreys, Alec J., 641
J (joining) region, 485–487
Jukes study, 727–728
just-so stories, evolutionary, 24

K

Kallikak study, 727
kappa particles, 614
karyotype, 269, 270, 272, 273

molecular continuity of life, 22–23
Monod, Jacques, 465, 467, 476
monohybrid crosses, 32–41
monohybrids, 33
monoploidy, 261, 262, 266–267
monosomic, 269
Morgan, Lilian V., 137
Morgan, Thomas Hunt, 97–99, 101, 128
morphogens, 515
mosaicism, 276
motor components, 707
mRNA. *See* messenger RNA (mRNA)
Muller, Hermann J., 67, 248, 566–567, 574
multifactorial traits, 670. *See also* quantitative traits
multiple alleles, 60–61
    Hardy-Weinberg principle and, 638–639
multiple cloning site (MCS), 372
multiple nonchemotaxis, 706, 707, 709
multiplication rule, probability and, 50–51
Murry, Jeffrey C., 693
Mus, 517
mutagenesis, 574
    enhancer-trap, 499–500
    misalignment, 571
    oligonucleotide site-directed, 374–376
    transposable-element, 559–561
mutagens, 556
    base-analog, 567–569
    detection of, 585–586
mutants
    antibiotic-resistant, 311, 348–349, 608, 609–610
    bacterial, 311–312, 348–349
    bacteriophage, 330–331
    carbon-source, 312
    constitutive, 464, 489
    lac⁻, 462–463
    nutritional, 311–312, 322
    petite, 610–612
    respiration-defective mitochondrial, 610–612
    uninducible, 489
mutant screen, 55
mutation(s), 15–16, 23–24, 182, 512, 556–577, 650–652
    affecting cell lineages, 522
    affecting chemotaxis, 706–709
    affecting floral development in *Arabidopsis*, 545, 546
    allelism of, 55–56
    base-substitution, 557–558
    complementation and, 160–161
    conditional, 333, 556
    execution, 524
    forward, 583, 652
    frameshift, 440, 559
    gain-of-function, 526, 527
    germ-line, 556
    heterochronic, 525–526

homeotic, 523
hot spots of, 339, 565–566
induced, 557, 566–577
insertions and deletions, 558–559
irreversible, 650–651
lineage, types of, 522–526
loss-of-function, 526, 527
missense, 444, 558
molecular basis of, 557–561
nonadaptive nature of, 561–563
nonmotile, 707
nonsense, 444, 558
promoter, 422–423
properties of, 556–557
reverse, 583–586, 652
segregation, 523–524
selectively neutral, 650
sickle-cell hemoglobin and, 657
silent, 444, 557
somatic, 556
spontaneous, 557, 561–566
suppressor, 583–586
temperature-sensitive, 556
transformation, 522–523
transition, 557
transversion, 557
mutation rates, 563–565
mutually exclusive events, 48–49
Mycoplasma genitalium, 416–418
myoclonic epilepsy associated with ragged-red muscle fibers (MERRF), 601–602

N

N̆geli, Carl, 32, 82
nanos, 536, 538
narrow-sense heritability, 684–686, 726
    estimation of, 689–690
    of liability, 691, 692
natural selection, 24, 650, 652–658
    in diploids, 654
    in *Escherichia coli*, 653
    selection-mutation balance and, 655–656
nature versus nurture, 730–731
negative assortative mating, 633
negative chromatid interference, 141
negatively supercoiled, 225
negative regulation, 461
negative regulator, 490
nematodes, 246, 515–516. *See also* *Caenorhabditis elegans*
    development in, 520–522
    genome of, 222
neuroblastoma, 481
Neurospora, 150, 157, 420, 586, 614, 723
Neurospora crassa, 154–155, 156, 415
neuroticism, 731, 733–734
neutral petites, 612
Nicholas II, Tsar, 602–603
nick, 225
Nicotiana longiflora, 684

nitrogen mustard, mutagenicity of, 570
nitrous acid, mutagenicity of, 569
N-*myc*, 481
Nomarski microscopy, 521
nonchemotaxis, 706–707
    general, 706–707, 710
    multiple, 706, 707, 709
    specific, 706, 707
noncomplementation, 55
nondisjunction, 103–104, 274
nonmotile mutations, 707
nonparental ditype, 150, 151
nonrecombinant, 588
nonselective medium, 312
nonsense mutations, 444, 558
nonsense suppression, 447–449, 585
nontaster, 713, 729
nopaline, 381
normal distribution, 673–675
Notch deletions, 281–282
Notch gene, 527
nuclease, 192
nucleic acid hybridization, 200–202
nucleic acids, 4
    polynucleotide chains in, 175
    replication of, 182–183
nucleoid, 225–227, 308
nucleoli, 85, 86
nucleoside, 174–175, 213
nucleosomes, 228–230, 231
nucleotide, 9, 174–175
    components of, 174, 175
    repetitive sequences, 235–239
    sequence composition, 239–242
nucleotide analog, 568–569
numerator genes, 105
nurture, nature versus, 730–731
Nÿsslein-Volhard, Christiane, 61, 529, 535
nutritional mutants, 311–312, 322

O

octopine, 381
octoploidy, 261, 262
odd-skipped gene, 538
Oenothera, 292
oil spills, genetic engineering and, 385
Okazaki fragments. *See* precursor fragments
oligonucleotide primer, 207, 208, 209, 370
oligonucleotide site-directed mutagenesis, 374–376
Oliver, C.P., 160
Olson, Maynard V., 393
oncogene, 298, 481
oocytes, 88, 518–519
openness, 731
open reading frame (ORF), 400, 424
operator, 465, 467
operator region, 464–465
operon model, 465–467

operons
    lac, 462–463, 466–471
    trp, 471–476
organelle heredity, 604–613
organelles, 600
    evolutionary origin of, 613–614
    genetic codes of, 605–606
ovalbumin, 369
overdominance. *See* heterozygote superiority
overlapping genes, 449–450
Oxytricha, telomeric DNA of, 251, 252
ozone, 572

P

Pachytene, 92–93
pair-rule genes, 533, 538
palindrome, 204, 631
paracentric inversion, 287, 288
Paramecium aurelia
    autogamy in, 614–615, 616
    killer phenomenon in, 614, 616
parasegment genes, 532
parental combinations, 124
parental ditype, 150, 151
parental strand, 10–11, 182–183
partial digestion, 392
partial diploids, 322
Pascal's triangle, 107
patent ductus arteriosus, 691
paternal inheritance, 601
paternity testing, 641, 644–645
path, 646
pathogens, 348
pattern formation, 513, 532–534
P1 bacteriophage, 390
    vector, 364, 365
pea plant
    flower of, 32, 33
    seeds of, 33–35
Pearson, K., 727
pedigree, 51
    calculation of inbreeding coefficient from, 647–649
    genetic mapping in human, 146–150
    segregation in human, 51–54
P element, 377
penetrance, 71–72
peptide bond, 412
peptidyl transferase, 433–434
percent G + C, 175
pericentric inversion, 287–288, 289
period (per) gene, 716, 717
    molecular genetics of, 720–723
permissive conditions, 556
Perrin, David, 467
personality, genetic factors in, 730, 731–734
petals, 545
Petes, Thomas D., 579
petite mutants, 610–612
PFGE (pulsed-field gel electrophoresis), 228, 390

P_1 generation, 35
phage. *See* bacteriophage(s)
phenotype, 41, 628
    blood-group, 639
    phenotypic change with selection, 684–686
    total variance in, 678
phenylalanine, chemical structure of, 413
phenylketonuria, 61, 730
phenylthiocarbamide (PTC), 713, 729
Philadelphia chromosome, 298
Phlox cuspidata, 636, 646
phosphodiester bonds, 175
phosphorylation, 712
photoreactivation, 580–581
photosynthesis, 606
photosynthetic plastids, 619
phylogenetic tree, 619–621
physical distance, genetic distance and, 145–146
physical mapping, 392–393
    with sequence-tagged sites, 395–397
pin flowers, 633
plants
    development in higher, 543–548
    genetic engineering in, 380–382, 386
    leaf variegation in four-o'clock, 606–609
    male sterility in, 612–613
    random mating in, 636–637
plaque, 330–331
Plasmids, 222, 314–317
    F (fertility) factor, 314–317
    F' (F prime), 322–323
    R, 348
    Ti, 381, 382
plasmid vectors, 364, 365
Plasmodium falciparum, 17–18, 388
plastids, photosynthetic, 619
pleiotropic effects, 17
pleiotropy, 17–18, 695
Pneumococcus. See Streptococcus pneumoniae
pneumonia, 3
polarity, 10
pole cells, 529
pollen, 545
poly-A tail, 425
polycistronic mRNA, 436
polylinker, 372
polymerase a, 192
polymerase l, 192
polymerase chain reaction (PCR), 207–210, 369
polymorphic, 630
polymorphism(s)
    DNA, 148–150, 630–632
    DNA typing and, 639–645
    enzyme, 629–630
    restriction fragment length (RFLPs), 147–150, 631
    serotonin transporter, 733–734

simple tandem repeat (STRP), 150, 372–373, 388, 558–559, 565, 576–577, 631, 692–695
    variable number of tandem repeats (VNTR), 140–144, 148–149, 640–644
polynucleotide chain, 175
polypeptide chain, 413–415, 483, 484–485
    properties of, 412, 414
polypeptides, 15, 412–415
    genes and, 415–418
    leader, 473
polyploidy, 261–266
polysaccharides, 68–69
polysome, 450
polysomy, 267
polytene chromosomes, 234–235, 281, 283, 296
P1 bacteriophage, 390
population, 628
    fixed, 629
    local, 628
population genetics, 627–667, 671
    allele frequencies and genotype frequencies, 628–632, 635–637, 660
    defined, 628
    DNA typing and population substructure, 639–645
    evolution and, 649–650
    inbreeding, 54, 634, 645–649
    mating systems, 632–639
    migration, 650–652
    mutation. *See* mutation(s)
    natural selection, 24, 650, 652–658
    random genetic drift, 650, 658–661
population substructure, 642–645
positional cloning, 372–374, 401
positional information, 515, 516
position effects on gene expression, 296–297
position-effect variegation (PEV), 296–297
positive assortative mating, 633
positive chromatid interference, 141
positive regulation, 461, 468–471
positive regulator, 490
posterior genes, 536
postreplication repair, 582, 583
posttranslocation state, 434
P (peptidyl) site, 431
precursor fragments, 194
    joining of, 196, 198
prenatal diagnosis, restriction mapping and, 388
pretranslocation state, 434
primary transcript, 424
primase, 196
primer, 191–192
    oligonucleotide, 207, 208, 209, 370
primer oligonucleotides, 207, 208, 209, 370
primosome, 196
Primula, 633

prion protein, 723
probability, 48–51
  genetic prediction and, 106–114
probe, 372
probe DNA, 206, 372, 630–631
processing, 412
proflavin, 440
progeny phage, 8
programmed cell death, 524–525
prokaryotes, 23
  chromosomes of, 222
  DNA replication in, 189
  gene regulation in, 480
  genetic organization of, 480
  mRNA of, 424, 449
  RNA processing in, 424
  transcriptional regulation in, 460–462
  translation units in, 450–451
prolactin, 503
proline, chemical structure of, 413
Prolla, Tomas A., 579
promoter mutations, 422–423
promoter recognition, 418–421
promoter region, 465
promoters, 419
  alternative, 500–501
  enhancers and, 497–499
proofreading function, 192–193, 194
prophage, 341
prophage induction, 345
prophase, 85, 86
prophase I, 89, 92–93
prophase II, 94
protanomaly, 284
protanopia, 284, 285–286
protease enzyme, 4
protein electrophoresis, 629–630, 631
protein infectious agent, 723
proteins, 412–415
  amino acids and, 412–415
  genes and, 12–15
  genetic engineering and, 386–387
  helicase, 198, 199
  prion, 723
  repressor, 344, 461–462, 464, 489
  single-strand binding, 198, 199
  synthesis of, 14–15
  TATA-box-binding (TBP), 495
  transcriptional activator, 462, 491–493
protein subunits, 414–415
proto-oncogene, 298
prototrophs, 312
Prozac, 733
pseudoalleles, 161
pseudoautosomal region, 274–275
Pseudomonas fluorescens, 386
PTC (phenylthiocarbamide) tasting, 713, 729
pulsed-field gel electrophoresis (PFGE), 228, 390
Punnett square, 43, 638
purines, 174, 180
pyloric stenosis, 691

pyrimidine dimers, 571–572
pyrimidines, 174, 180
quantitative inheritance, 670–675
quantitative-trait locus (QTL), 692–695
quantitative traits, 68, 669–701
  analysis of, 680–683
  artificial selection and, 683–687
  causes of variation, 675–680
  correlation between relatives, 687–690
  heritability of threshold traits, 690–692
  linkage analysis of quantitative-trait loci, 692–695
  number of genes affecting, 681–682

Q

Quetelet, Adolphe, 674

R

racial designations, in DNA typing, 642
radiation, units of, 573
Ramanis, Zenta, 608
random genetic drift, 650, 658–661
random mating, 632–633
  Hardy-Weinberg principle and, 634–639
  X-linked alleles and, 639
random-spore analysis, 154
reading frame, 440
recessive, 35, 36
reciprocal crosses, 35, 97–100
reciprocal recombinant chromosomes, 160
reciprocal translocation, 288–293
Recombinant DNA technology, 360. See also genetic engineering
recombinant molecules, 588
  detection of, 370–372
  DNA, 362–365
  screening for, 372
recombinants, 124
recombination, 124, 586–592
  frequency of, 125, 128, 129, 130, 140, 144, 145
  within genes, 160
  Hfr cells, 317–319
  Holliday model of, 138, 139, 587–590
  linkage and, 124–127
  lytic cycle and, 330
  mitotic, 158–160
  site-specific, 330
  in temperate bacteriophages, 340–346
  in virulent bacteriophages, 331–336
recombination. See genetic recombination
recombination frequency, 144
recruitment, 495
red-green color blindness, 283–286
reductional division (first meiotic division), 89–94
redundancy, 445–447

regulation. See gene regulation
relative fitness, 653
relatives, correlation between, 687–690
relaxed DNA, 224
renaturation, 200, 201, 202, 208, 236–239
  analysis of genome size and repetitive sequences by, 238–239
  kinetics of, 236–238
repair
  mechanisms of DNA, 577–583
  process, 565–566
repellents, 704–705
repetitive sequences, 235–239
replica plating, 561–563
replication, 10–11, 182–191
  discontinuous, 194–199
  in eukaryotes, 189
  geometry of, 183–191, 194
  multiple origins of, 189
  in nucleic acids, basic rule, 182–183
  other proteins needed for DNA, 196–199
  plasmids and, 315
  rolling-circle, 189–191, 316
  semiconservative, 183–186
  theta (q), 187
replication fork, 187
  fragments in, 194–195
replication origin, 187
replication slippage, 559, 579
reporter gene, 500
repressible, 461
repressor protein, 344, 461–462, 464, 489
repulsion configuration, 126, 127
research, recombinant DNA technology and, 386
respiration-defective mitochondrial mutants, 610–612
restriction endonuclease, 147, 148, 202–205
restriction enzymes, 147, 148, 202–205, 360–362
restriction fragment, 205
restriction fragment length polymorphisms (RFLPs), 147–150, 631
restriction map, 205, 393
restriction site, 204, 360
restrictive conditions, 556
retinoblastoma, 298–299
retroviruses, genetic engineering and, 378, 387
reverse genetics, 377–382, 386
reverse mutations, 583–586, 652
reverse transcriptase, 369–370
reverse transcriptase PCR (RT-PCR), 370
reversion, 583
  for detecting mutagens and carcinogens, 585–586
RFLPs (restriction fragment length polymorphisms), 147–150, 631
R group of amino acids, 412

rhodopsin, 283–284
ribonucleic acid. *See* RNA (ribonucleic acid)
ribose, 195
ribosomal RNA (rRNA), 12–13
ribosome binding site, 432
ribosomes, 13, 430–431, 439
ribozymes, 427
rII gene in bacteriophage T4, 334, 336–340
ring chromosomes, 260
RNA editing, 605
RNA polymerase, 195, 196, 412, 418–419, 421, 422, 423
    subunits of, 414–415
RNA polymerase holoenzyme, 495
RNA primer, 195–196
RNA processing, 412, 424–430, 460
RNA (ribonucleic acid), 4, 12
    DNA compared, 195–196
    guide, 248
    messenger. *See* messenger RNA (mRNA)
    ribosomal, 12–13
    transcription of DNA and, 12–14, 412, 418–424
    transfer. *See* transfer RNA (tRNA)
    translation of, 14–15
RNase enzyme, 4, 5
RNA splicing, 425–426, 427
RNA synthesis, 418–423
Robertsonian translocation, 293–296
rolling-circle replication, 189–191, 316
Romanov, Georgij, 603
Rotman, Raquel, 329
R plasmids, 348
rRNA (ribosomal RNA), 12–13
RT-PCR (reverse transcriptase PCR), 370
rubella virus, 730
Saccharomyces cerevisiae, 482
    centromeric chromatin of, 246–247
    complete sequence of genome, 400–401
    DNA molecules of, 228
    life cycle of, 152
    mating-type interconversion in, 484
    petite mutants of, 610
    topoisomerase II from, 225, 226
    wobble rules for, 447

S

Sager, Ruth, 608
salmon, growth hormone engineered giant, 382–383
Salmonella typhimurium, 386, 586
sampling, random, 658
Sanchez, Carmen, 467
satellite DNA, 235
SBEI (starch-branching enzyme I), 65, 246
scaffold, 233
Schierenberg, E., 521

schizophrenia, 730
Schnieke, Anagelika E., 380
Scottish Finn Dorset ewe ("Dolly"), cloning of, 380
scrapie, 724
second-division segregation, 156, 157
second meiotic division, 89, 94–95
segment, 532
segmentation genes, 532–534
segment-polarity genes, 533–534, 538–540
segregation
    adjacent-1, 289, 291, 295
    adjacent-2, 290, 291, 295
    alternative, 291, 292, 295
    first-division, 155, 157
    Mendelian, 67, 94, 96
    in pedigrees, 51–54
    principle of, 38, 40
    second-division, 156, 157
    of two or more genes, 42–48
segregational petites, 610–612
segregation mutation, 523–524
selected marker, 319
selection
    artificial, 683–687
    correlated response to, 687
    individual, 684
    natural, 24, 650, 652–658
    phenotypic change with, 684–686
selection coefficient, 653
selection limit, 686
selection-mutation balance, 655–656
selectively neutral mutations, 650
selective medium, 312
selective serotonin reuptake inhibitors, 733
self-fertilization, 32, 39, 42, 61, 62, 645–646
selfish DNA, 246
self-splicing, 427
self-sterility, 61
Semeonoff, Robert, 641
semiconservative replication, 183–186
semisterility, 288–293
senile dementia, 730
sensory adaptation, 713–714
sensory perceptions, 713, 729–730
sepals, 545
sequence-tagged sites (STS), 395–397
Sequoia sempervirens, 604
serine, chemical structure of, 413
serine proteases, 537
serotonin, 732–733
serotonin transporter polymorphism, 733–734
sex, chromosomal determination of, 97, 98, 106–107
sex-chromatin body, 276
sex chromosomes, 96–97, 274–277
    abnormalities in, 278–280
sex-determining region Y (SRY), 270, 274–275
Sex-lethal (Sxl) gene, 105

sex limited inheritance, 101
sex-linked inheritance. *See* X-linked inheritance
shuttle vector, 392
siblings (sibs), 51
    covariance and, 690
sibship, 49
sickle-cell anemia, 17–18, 388, 558
sickle-cell hemoglobin, 17–18, 657
significant, statistically, 111
silent mutations, 444, 557
silent substitutions, 557
silkworm, silk fibroin mRNA in, 503
simple tandem repeat polymorphism (STRP), 150, 372–373, 558–559, 565, 576–577, 631, 692–695
single-active-X principle, 275
single-copy sequence, 239, 240
single-strand binding proteins, 198, 199
single-strand break model, asymmetrical, 590, 591
single-stranded DNA, 10, 222
sister chromatids, 85, 86
site-directed mutagenesis, oligonucleotide, 374–376
site-specific DNA cleavage, 202–205
site-specific recombination, 330
sleep cycle, 723
snail-shell coiling, maternal effect in, 617–618
snake gene, 537
somatic cells, chromosomes of, 82
somatic mosaics, 160
somatic mutations, 556
SOS repair, 582
Southern blot, 206–207, 228, 631
specialized transducing bacteriophages, 325, 345–346
specific nonchemotaxis, 706, 707
spermatocytes, 88
spina bifida, 691
splice acceptor, 425
splice donor, 425
spliceosomes, 426
splicing
    alternative, 501–503
    in origin of T-cell receptors, 487
    RNA, 425–426, 427
spongiform encephalopathy, 723
    bovine, 724
spontaneous abortion, chromosome abnormalities and, 280–281
spontaneous mutations, 557, 561–566
spores, 89, 158
    ascospores, 150, 151, 154, 156
sporophyte, 89
SRY (sex-determining region, Y), 270, 274–275
S (synthesis), 83
Stahl, Franklin, 183, 184
stamens, 545
standard deviation, 673, 674–675
Staphylococcus aureus, 230

starch-branching enzyme I (SBEI), 65, 246
statistically significant, 111
sterility
    cytoplasmic male, 612–613
    genetic engineering and, 382, 383–384
    self-sterility, 61
    semisterility, 288–293
Stern, Curt, 134, 158, 159
steroids, 493–494
sticky ends, 204, 361
    identical, 366
stop codons, 436, 437
Strand, Micheline, 579
Strathern, Jeffrey N., 484
Streisinger, George, 332
Streptococcus pneumoniae, 3, 4, 312
streptomycin, 312, 319, 608
streptomycin resistance, 609–610
Strongylocentrotus purpuratus, 517
STRP (simple tandem repeat polymorphism), 150, 372–373, 388, 558–559, 565, 576–577, 631, 692–695
Sturtevant, Alfred H., 127, 131
sublineages, 522
submetacentric, 260, 261
subpopulations, 628
subunits, protein, 414–415
Sulston, John E., 521
supercoiling, 224–225
suppression, nonsense, 447–449, 585
suppressor mutations, 583–586
suppressor tRNA, 585
supreme law of unreason, 674
symbionts, cytoplasmic transmission in, 614–616
symbiosis, 613
synapsis, 92, 286, 288
syncytial blastoderm, 529
synteny group, 398–400
synthetic vaccines, 387

T

tandem duplication, 283
Taq polymerase, 208
taster, 713, 729
TATA box, 420–421
TATA-box-binding protein (TBP), 495
Tatum, Edward L., 322, 415, 420
TBP-associated factors (TAFs), 495
T cell, 487
T-cell receptors, gene splicing in origin of, 487
telomerase, 248, 258
telomeres, structure of, 247–252
telophase, 85, 86, 87
telophase I, 89, 94
telophase II, 95
temperate phage, 329, 340–346
temperature-sensitive mutations, 556
template, 10, 191

template strand, 421, 424
terminal genes, 536
terminal redundancy, 335
termination, 421–422, 431, 436
testcross, 41
    dihybrid, 45
    three-point, 141–146
testis-determining factor (TDF), 274–275
tetrad analysis, 150–158
tetrads, 93, 150–158
    ordered, 154–158
    unordered, 150–154
Tetrahymena, 391, 392
    regulation in, 480
    splicing in, 427
    telomere structure in, 248–252
tetraploidy, 261, 262
tetratype, 150
thermophiles, 208
Thermus aquaticus, 208
theta (q) replication, 187
30-nm fiber, 233
Thomson, J.N., 521
3'-OH, in nucleic acids, 175
three-point cross, 141
three-point testcross, gene mapping from, 141–146
threonine, chemical structure of, 413
threshold, 671
threshold traits, 671
    heritability of, 690–692
thrum, 633
thymidine triphosphate, 568
thymine, 9, 174, 175, 177, 180
thymine dimer, 571
time-of-entry mapping, 319–324
tim (timeless) genes, 721–723
Ti plasmid, 381, 382
tolerance, 70
topoisomerase enzymes, 187, 198, 199, 225, 226
    topoisomerase I, 225
    topoisomerase II, 225, 226
torso gene, 536–537
total variance, 678
traits
    continuous, 671
    defined, 2
    discrete, 68
    distribution of, 671–675
    genes and, 16–22
    meristic, 671, 677
    multifactorial, 670
    quantitative. *See* quantitative traits
    threshold, 671, 690–692
    variation in, 2
trans configuration, 126, 127
transcript, 13
transcription, 12–14, 412, 418–424
transcriptional activator, 462, 491–493
transcriptional enhancers, 494–499

transcriptional regulation
    in eukaryotes, 488–501
    operon model of, 465–467
    in prokaryotes, 460–462
transducing phage, 325, 345–346
    generalized, 325
    specialized, 325, 345–346
transduction, 308, 325–328
transfer RNA (tRNA), 13, 431, 444–445
    charged, 431
    suppressor, 585
    translation and, 14–15
    uncharged, 434, 435
transformation, 3–6, 308, 365, 375
    bacterial, 312–314
    germ-line, in animals, 377–380
transformation mutation, 522–523
transgenic animals, 378
trans-heterozygote, 161, 162
transition mutation, 557
translation, 412, 430–438
translational control, 503–504
translation of RNA, 14–15
translation units, 450–451
translocation, 274, 288–296, 434
    reciprocal, 288–293
    Robertsonian, 293–296
transmembrane receptor, 514
transmission, principles of genetic, 31–79
    dominance, complications in concept of, 64–72
    genetic analysis, 54–61
    Mendelian inheritance and probability, 48–51, 600, 628
    modified dihybrid ratio caused by epistasis, 61–64
    monohybrid crosses, 32–41
    segregation in human pedigrees, 51–54
    segregation of two or more genes, 42–48
transposable–element mutagenesis, 559–561
transposable elements, 242–246, 346–350
    P elements, 377
transposase, 246, 347
transposition, 243, 346–347
transposons, 347, 349–350
transposon tagging, 349, 350, 365–366
transversion mutation, 557
Tribolium, 682
trihybrid cross, 46
trinucleotide repeat, 279–280
triplet code, genetic evidence for, 439–442
triplication, 283
triploidy, 263
trisomy, 267, 268
    in human beings, 274
trisomy-X syndrome, 278
Triturus, 422
trivalent, 267, 268